Lecture Notes in Computer Science 3391

Cheeha Kim (Ed.)

Information Networking

Convergence in Broadband and Mobile Networking

International Conference, ICOIN 2005
Jeju Island, Korea, January 31- February 2, 2005
Proceedings

Springer

Volume Editor

Cheeha Kim
Pohang University of Science and Technology
San 31 Hyoja-Dong, Nam-Gu, Pohang, Gyungbuk 790-784, Korea
E-mail: chkim@postech.ac.kr

Library of Congress Control Number: Applied for

CR Subject Classification (1998): C.2, H.4, H.3, D.2.12, D.4, H.5

ISSN 0302-9743
ISBN 3-540-24467-0 Springer Berlin Heidelberg New York

Springer is a part of Springer Science+Business Media

springeronline.com

Printed in Germany

Typesetting: Camera-ready by author, data conversion by Boller Mediendesign
Printed on acid-free paper SPIN: 11382041 06/3142 5 4 3 2 1 0

Preface

Welcome to ICOIN 2005, the International Conference on Information Networking, held at Ramada Plaza Jeju Hotel, Jeju Island, Korea during January 31–February 2, 2005. ICOIN 2005 followed the success of previous conferences. Since 1986, the conference has provided a technical forum for various issues in information networking. The theme of each conference reflects the historic events in the computer communication industry. (Please refer to www.icoin2005.or.kr for details.) The theme of ICOIN 2004, "Convergence in Broadband and Mobile Networking," was used again for ICOIN 2005 since we believed it was ongoing.

This year we received 427 submissions in total, which came from 22 countries. Upon submission, authors were asked to select one of the categories listed in the Call for Papers. The most popular category chosen was network security, followed by mobile networks and wireless LANs. Other areas with strong showings included QoS and resource management, ad hoc and sensor networks, and wireless multimedia systems. From the outset, we could see where recent research interest lay and could make sure that the theme was still going in the right direction.

The Technical Program Committee members were pleased to work together to present an outstanding program of technical papers. All submissions to ICOIN 2005 underwent a rigorous review process by the TPC members and external reviewers. Each paper was sent to three reviewers and judged based on its originality, significance, contribution, and presentation. The TPC of 37 people and 154 external reviewers was involved in reviewing them. The review process culminated in the meeting at Ewha Womans University, Korea on October 29, 2004, and ended at the meeting at Seoul National University, Korea on November 23, 2004, where the TPC finally accepted 96 papers (an acceptance ratio of 22%) for a three-day technical program.

We thank all the authors who submitted papers to ICOIN 2005. We also thank the external reviewers for their time, effort and timely response. They contributed their expertise a lot to the conference throughout the review process. Their names are provided in the proceedings. We wish to thank the TPC members for the fabulous job they did, especially Profs. Chong-kwon Kim (Seoul National University, Korea) and Sunyoung Han (Konkuk University, Korea) for their devotion to the conference as Vice Chairs of the TPC, and Mr. Ki Yong Park (Konkuk University, Korea) who was in charge of running the system used for submission and review. We extend our sincere thanks to Profs. Pascal Lorenz (Université de Haute Alsace, France) and Nitin Vaidya (UIUC, USA) for their strong support to the TPC.

We wish to express our special thanks to General Chair Prof. Sunshin Ahn (Korea University, Korea) for his advice on all aspects of the conference. We are deeply grateful to all the Organizing Committee members. As the Organizing Committee chair, Prof. Kijoon Chae (Ewha Womans University, Korea) provided wonderful administration support and ensured the smooth operation of the conference.

Thank you also to the attendees for joining us at ICOIN 2005.

November 2004

Cheeha Kim
TPC Chair
ICOIN 2005

Organizing Committee

General Chair	Sunshin An (Korea Univ., Korea)
Organizing Committee Chair	Kijoon Chae (Ewha Womans Univ., Korea)
Vice Chairs	Sung Won Sohn (ETRI, Korea) Hideki Sunahara (NARA Institute of Science and Technology, Japan)
Local Arrangement Co-chairs	Khi Jung Ahn (Cheju National Univ., Korea) Jong Won Choe (Sookmyung Women's Univ., Korea)
Publicity Co-chairs	Mario Marques Freire (Univ. of Beira Interior, Portugal) Chung-Ming Huang (National Cheng Kung Univ., Taiwan) Hyun Kook Kahng (Korea Univ., Korea) Osamu Nakamura (Keio Univ., Japan) Krzysztof Pawlikowski (Univ. of Canterbury, New Zealand)
Publication Co-chairs	Sungchang Lee (Hankuk Aviation Univ., Korea) Hyukjoon Lee (Kwangwoon Univ., Korea)
Registration Chair	Miae Woo (Sejong Univ., Korea)
Financial Chair	Choong Seon Hong (Kyunghee Univ., Korea)
Patron Co-chairs	Sang Chul Shin (NCA, Korea) Yongtae Shin (Soonsil Univ., Korea)
System Administrator	Kiyoung Park (Konkuk Univ., Korea)

한국정보과학회
KOREA INFORMATION SCIENCE SOCIETY

Information Processing Society of Japan

SK Telecom

Program Committee

Chair
Cheeha Kim (Postech, Korea)

Vice Chairs
Sunyoung Han (Konkuk Univ., Korea)
Chong-kwon Kim (Seoul National Univ., Korea)
Pascal Lorenz (De Haute Alsace Univ., France)
Nitin Vaidya (UIUC, USA)

Members
Sanghyun Ahn (Univ. of Seoul, Korea)
William Arbaugh (Univ. of Maryland, USA)
B. Bing (Georgia Institute of Technology, USA)
Raouf Boutaba (Univ. of Waterloo, Canada)
Petre Chemouil (France Telecom R&D, France)
Jun Kyun Choi (ICU, Korea)
Myungwhan Choi (Sogang Univ., Korea)
Il-Yung Chong (Hankuk Univ. of Foreign Studies, Korea)
Michael Devetsikiotis (Carleton Univ., Canada)
Petre Dini (Cisco, USA)
Thierry Ernst (Keio Univ., Japan)
Nelson L.S. Fonseca (State Univ. of Campinas, Brazil)
Mario Marques Freire (Univ. of Beira Interior, Portugal)
H. Guyennet (Univ. of Franche-Comté, France)
A. Jamalipour (Univ. of Sydney, Australia)
Yeong Min Jang (Kookmin Univ., Korea)
Song Chong (KAIST, Korea)
Raj Kettimuthu (Argonne National Lab, USA)
D. Khotimsky (Lucent Bell Labs, USA)
Hwa-sung Kim (Kwangwoon Univ., Korea)
Young-bae Koh (Aju Univ., Korea)
Meejeong Lee (Ewha Womans Univ., Korea)
WonJun Lee (Korea Univ., Korea)
Sanghoon Lee (Yonsei Univ., Korea)
Kyungshik Lim (Kyungpook National Univ., Korea)
G. Omidyar (Computer Sciences Corp, Oman)
J.J. Rodrigues (Univ. of Beira Interior, Portugal)
Jaechul Ryu (Chungnam National Univ., Korea)
Kouichi Sakurai (Kyuchu Univ., Japan)
Winston Seah (Institute for Infocomm Research, Singapore)
Young-Joo Suh (Postech, Korea)
Sung-Ming Yen (National Central Univ.,Taiwan ROC)

Table of Contents

TCP and Congestion Control

Wireless Ad Hoc Network Routing

Network Measurement

Routing

Power Control in Wireless Networks

QoS I

QoS II

Wireless Ad Hoc Networks

Network Design

Peer to Peer Networks

QoS III

Security II

Wireless Networks II

Applications and Services

Security III

Wireless Networks III

Numerical Analysis of IEEE 802.11 Broadcast Scheme in Multihop Wireless Ad Hoc Networks*

Jong-Mu Choi[1], Jungmin So[2], and Young-Bae Ko[1]

[1] School of Information and Computer Engineering
Ajou University, Republic of Korea
{jmc,jso1}@uiuc.edu
[2] Coordinated Science Lab. and Dept. of Computer Science Engineering
University of Illinois at Urbana-Champaign, USA
youngko@ajou.ac.kr

Abstract. In this paper, we study the performance of IEEE 802.11 broadcast scheme in multihop wireless networks using an analytical model. Previous works have evaluated the performance of IEEE 802.11 protocol assuming unicast communication, but there has not been an analysis considering broadcast communication. Analyzing performance of broadcast communication is important because multicast communication is gaining attention in wireless networks with numerous potential applications. Broadcast in IEEE 802.11 does not use virtual carrier sensing and thus only relies on physical carrier sensing to reduce collision. For this study, we define a successful broadcast transmission to be the case when all of the sender's neighbors receive the broadcast frame correctly, and calculate the achievable throughput.

1 Introduction

The IEEE 802.11 standard [3] is widely deployed and used in wireless systems today. Its de facto medium access control (MAC) protocol, called Distributed Coordination Function (DCF) allows multiple nodes to share the wireless medium without any central coordinator. Although IEEE 802.11 DCF was designed for a wireless LAN, it is also used in multihop wireless networks because of its distributed nature.

The major goal of a MAC protocol is to have only a single node in a broadcast domain transmit at a given time. If two nodes that are nearby each other transmit frames at the same time, the frames collide and the channel bandwidth is wasted. To achieve this goal, IEEE 802.11 DCF uses a technology called Carrier Sensing Multiple Access with Collision Avoidance (CSMA/CA). In CSMA/CA, whenever a node has a data frame to transmit, it listens on the channel for a duration of time. This duration of time is called *slot time*. If the channel is sensed

* This work is supported by grant no. R05-2003-000-1607-02004 and M07-2003-000-20095-0 from Korea Science & Engineering Foundation, and University IT research center project (ITRC).

C. Kim (Ed.): ICOIN 2005, LNCS 3391, pp. 1–10, 2005.

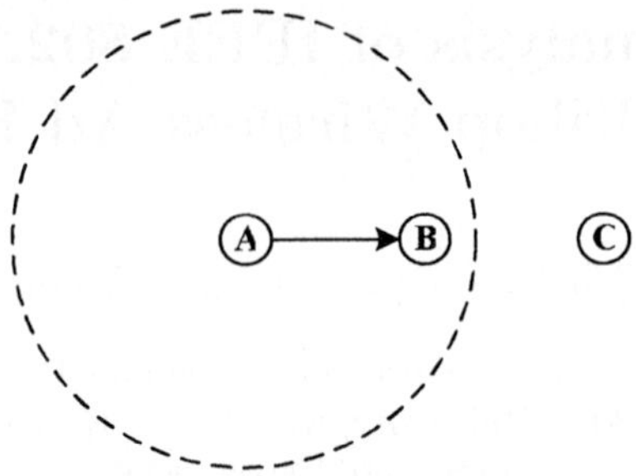

Fig. 1. Illustration of the hidden terminal problem in a multihop wireless network

to be idle during a predefined slot time, the node transmits the data frame. If the channel is busy, then the node defers its transmission and waits for a random delay (called *random backoff interval*) before retrying. Sensing the channel to determine if it is busy or idle is called *physical carrier sensing.*

When a node receives a frame, the node is able to decode the frame correctly only if the received signal power is higher than a threshold called *receiver sensitivity*, which is also called *receive threshold.* Also, when a node senses the channel to see whether it is busy or not, it determines the channel to be busy if the sensed power is greater than the *carrier sense threshold.* The carrier sense threshold is a tunable parameter.

Let us assume for now that the carrier sense threshold is equal to the receive threshold. It means while node A is transmitting a data frame, all nodes in A's transmission range senses the channel to be busy, and the nodes outside of A's transmission ranges senses the channel to be idle. Under this assumption, the scenario in Fig. 1 illustrates a situation where physical carrier sensing cannot prevent collision.

Suppose node A starts transmitting a frame to node B. Since node C is outside of A's transmission range, it senses the channel as idle. So node C can start transmitting its frame, which collides at node B. In this scenario, node C is said to be *hidden* from node A, and C is a *hidden terminal* from A's view [4]. When node A transmits a data frame to B, there is a period of time in which if node C starts transmitting, it will collide with node A's transmission. This period is called a *vulnerable period*[1].[3]

To prevent collisions caused by hidden terminals, a mechanism called *virtual carrier sensing* is used in addition to CSMA/CA. When a node S wants to transmit a frame, it first transmits a Request-To-Send (RTS) frame which is much smaller in size than a data frame. On receiving RTS, the receiver replies with Clear-To-Send (CTS) frame also very small in size. Any node other then the sender and the receiver that receives RTS or CTS defers its transmission while S transmits the data frame. So the RTS/CTS exchange has the effect

[3] The vulnerable period for IEEE 802.11 broadcast scheme is calculated in section II.

of reserving the space around the sender and receiver for the duration of the data transmission. This is called *Virtual Carrier Sensing*, because it provides information on the channel but not by physically sensing the channel.

In IEEE 802.11 DCF, the virtual carrier sensing mechanism is only used in unicast transmissions, and it is an optional feature that can be turned on or off. For a broadcast transmission, only physical carrier sensing is used. Virtual carrier sensing is not directly applicable to broadcast transmissions because CTS messages sent by multiple receivers will result in a collision.

The performance of IEEE 802.11 DCF for unicast transmissions has been studied using mathematical analysis. Cali et al. [5] calculates the throughput of IEEE 802.11 DCF when the basic CSMA/CA scheme is used without RTS/CTS mechanism. Bianchi [6] calculates the throughput of DCF with and without RTS/CTS mechanism, and also a combination of the two. These two studies are for wireless LANs, and so they do not consider hidden terminals. Wu et al. [1] studies the throughput the CSMA protocol in multihop wireless networks, considering hidden terminals.

All of these studies are for unicast communication, and do not consider broadcast communications. We are interested in the performance analysis of broadcast scheme in IEEE 802.11 DCF, operated in a multihop network. Also, we are interested in reliable broadcast, where a broadcast transmission is considered successful only if all of the sender's neighbors receive the broadcast message correctly. Reliable broadcast can be used for numerous applications, such as code distribution, database replication, and a basis for supporting distributed protocols.

The rest of paper is organized as follows: In section 2, we present the analysis of the IEEE 802.11 broadcast scheme. To our knowledge, this is the first analytical study of IEEE 802.11 broadcast scheme in multihop wireless networks. In section 3, we present numerical results from our analysis. Finally, we conclude in Section 4.

2 Numerical Analysis Model

Before analyzing performance of IEEE 802.11 broadcast scheme, we examine the hidden node problem in a broadcast scenario. As you see in the Fig. 2(a), nodes in the receiving region of node T but not in the receiving region of node S, may cause hidden terminal problem. We call this area as a *potential hidden node area*. For unicast communications, the size of the *potential hidden node area* can be calculated using the distance between the sender and receiver. However, in case of broadcast communication (see Fig. 2(b)), the *potential hidden node area* needs to include the receiving range of all the neighbors of the senders. So it is difficult to exactly compute the size of this area. Moreover, as explained earlier, varying the carrier sensing area also change the form of this area. The worst case, where the size of the *potential hidden node area* is maximized, is when there are infinite number of node at the edge of the sender's transmission range. Let R denote the transmission range of a node. As you see in the Fig. 2, maximum size of *potential*

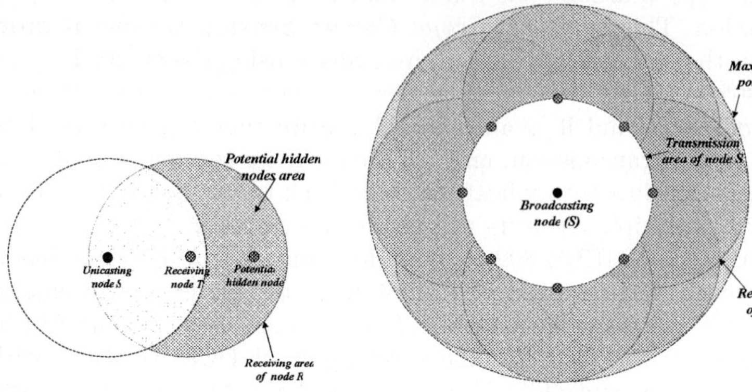

(a) Unicast (b) Broadcast

Fig. 2. Potential hidden node area

hidden node area can be $\pi(2R)^2 - \pi R^2 = 3\pi R^2$. Thus, in case of broadcast, the potential hidden node area can be dramatically larger than that of unicast.

We use the similar approximate approaches used in [1] to achieve the average throughput for multihop wireless networks.

To make our numerical model tractable, we assume followings for the multi-hop wireless network model.

1. All nodes in the network are two-dimensionally Poisson distributed with density λ, i.e., the probability $p(i, A)$ of finding i nodes in an area of size A is given by
$$p(i, A) = \frac{(\lambda A)^i e^{-\lambda A}}{i!}$$
2. All nodes have the same transmission and receiving range, which is denoted as R. N is the average number of neighbor nodes within a circular region of radius R. Therefore, we have $N = \lambda \pi R^2$
3. A node transmits a frame only at the beginning of each slot time. The size of a slot time, τ , is the duration including transmit-to-receive turn-around time, carrier sensing delay and processing time.
4. The transmission time or the frame length is the same for all nodes.
5. When a node is transmitting, it cannot receive simultaneously.
6. A node is ready to transmit with probability p. Let p' denote probability that a node transmits in a time slot. If p' is independent at any time slot, it can be defined to be
$$p' = p \cdot \mathrm{Prob}\{\text{Channel is sensed idle in a slot}\} \approx p \cdot P_I$$
where P_I is the limiting probability that the channel is sensed to be idle.
7. The carrier sensing range is assumed to vary between the range $[R, 2R]$.

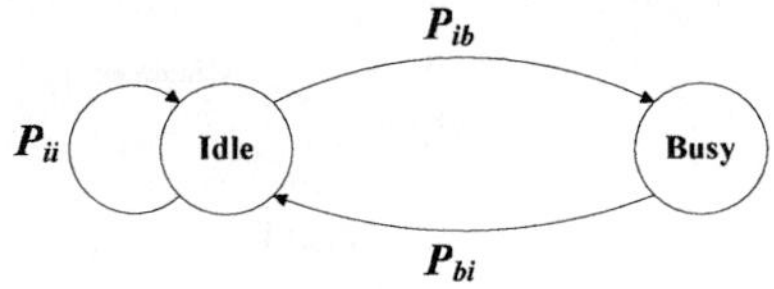

Fig. 3. Markov chain model for the channel

With above assumptions, the channel process can be modeled as a two-state Markov chain shown in Fig. 3. The description of the states of this Markov chain is the following:

Idle is the state when the channel around node x is sensed idle, and its duration T_{idle}, is τ.

Busy is the state when a successful DATA transfer is done. The channel is in effect busy for the duration of the DATA transfer, thus the busy time, T_{busy}, is equal to the data transmission time δ_{data}. ($T_{busy} = \delta_{data}$)

In IEEE 802.11 scheme, all nodes should not transmit immediately after the channel becomes idle. Instead, nodes should stay idle for at least one slot time. Thus the transition probability P_{bi} is 1.

The transition probability P_{ii} is that probability of the neighbor nodes transmits is given by,

$$P_{ii}=\sum_{i=0}^{\infty}(1-p')^i\frac{(\lambda\pi R^2)^i}{i!}e^{-\lambda\pi R^2}=\sum_{i=0}^{\infty}\frac{((1-p')\lambda\pi R^2)^i}{i!}e^{-\lambda\pi R^2(1-p')}e^{-p'N}=e^{-p'N}$$

Let, Φ_i and Φ_b denote the steady-state probabilities of state idle and busy, respectively. From Fig. 3, we have

$$\Phi_i = \Phi_i P_{ii} + \Phi_b P_{bi} = \Phi_i P_{ii} + \Phi_b$$

Since $\Phi_b = 1 - \Phi_i$, we have

$$\Phi_i = \frac{1}{2 - P_{ii}} = \frac{1}{2 - e^{-p'N}}$$

Now the limiting probability P_I can be obtained by

$$P_I = \frac{T_{idle}\Phi_i}{T_{busy}(1-\Phi_i) + T_{idle}\Phi_i} = \frac{\tau}{\delta_{data}(1 - e^{-p'N}) + \tau}$$

According to the relationship between p' and p, p' can be

$$p' = \frac{\tau p}{(\delta_{data})(1 - e^{-p'N}) + \tau}$$

To obtain the throughput, we need to calculate the probability of a successful transmission. The transmission state of a node x can also be modeled by a three-state Markov chain, as shown in Fig. 4. In the figure, *wait* is the state when the

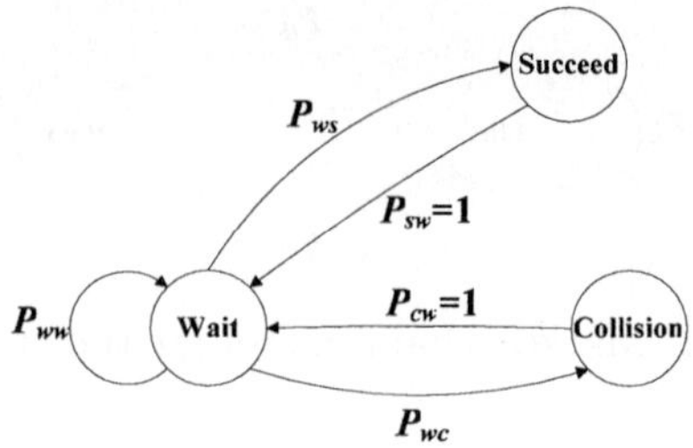

Fig. 4. Markov chain model for the transmission states of node

node in deferring its transmission, *succeed* is the state when the node successfully transmits DATA frame to all of neighbor nodes, and *collision* is the state when a node collides with other nodes.

At the beginning of each time slot, node x leaves the *wait* state with probability p'. Thus the transition probability P_{ww} is given by

$$P_{ww} = 1 - p'$$

and, the duration of a node in *wait* state T_{wait} is τ. The durations of *success* and *collision* states are equal to the frame transmission time, thus T_{succ} and T_{coll} are $\delta_{data} + \tau$. After *success* or *collision* state, node x always enter the *wait* state, thus P_{sw} and P_{cw} are 1.

Let Φ_w, Φ_s, and Φ_c denote the steady-state probabilities of state wait, success, and collision, respectively. From the above Markov chain we have

$$\Phi_w = \Phi_w P_{ww} + \Phi_s P_{sw} + \Phi_c P_{cw} = \Phi_w P_{ww} + 1 - \Phi_w \tag{1}$$

Hence, we have:

$$\Phi_w = \frac{1}{2 - P_{ww}} = \frac{1}{1 + p'}$$

Based on the above condition, transition probability P_{ws} can be

$$P_{ws} = P_1 P_2 P_3 \tag{2}$$

where

$$P_1 = \text{Prob}\{\text{node } x \text{ transmits in a slot}\}$$

$$P_2 = \text{Prob}\{\text{All of node } x\text{'s neighbor nodes do not transmit in the same slot}\}$$

$$P_3 = \text{Prob}\{\text{Nodes in potential hidden nodes area do not transmit for } 2\delta_{data} + \tau\}$$

The reason for the last term is that the *vulnerable period* for an data frame is only $2\delta_{data} + \tau$. As you see in the Fig. 5, this is because the collide from node in *potential hidden node* happen during the period that begin δ_{data} before sending node x begins its transmission and ends one slot after x completes its transmission.

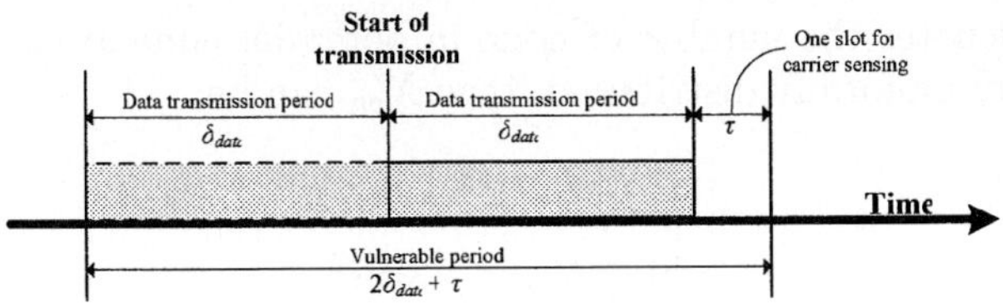

Fig. 5. The vulnerable period for IEEE 802.11 broadcast scheme

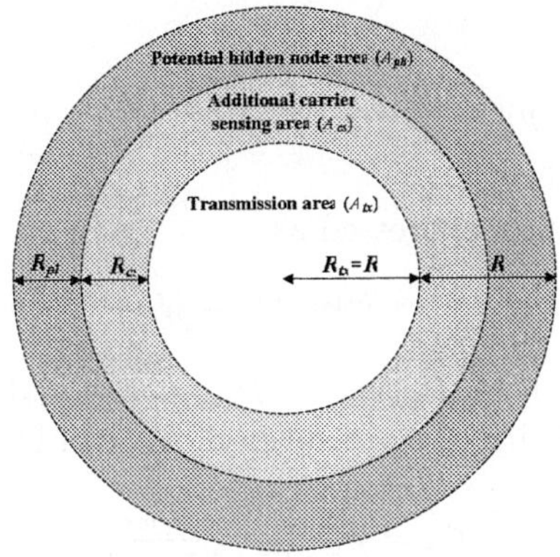

Fig. 6. Illustration of transmission area, additional carrier sensing area, and potential hidden nodes area

Obviously, $P_1 = p'$, while P_2 can be obtained by

$$P_2 = \sum_{i=0}^{\infty}(1-p)^i \frac{(\lambda \pi R^2)^i}{i!} e^{-\lambda \pi R^2} = e^{-p'\lambda \pi R^2} = e^{-p'N}$$

To calculate P_2, we first approximate the number of node in the *potential hidden node area*. Let A_{tx}, A_{cs}, and A_{ph} denote the transmission area, *additional carrier sensing area*, and *potential hidden node area*, respectively. As you see in the Fig. 6, *additional carrier sensing area* is the physical carrier sensing area that is outer of transmission area. We assume that the physical carrier sensing area is larger than transmission range and smaller than *potential hidden node area*. Thus, we have

$$0 \leq A_{cs} \leq 3\pi R^2$$

And, the *potential hidden node area* can be

$$A_{ph} = 2\pi(2R)^2 - A_{tx} - A_{cs} = 2\pi(2R)^2 - \pi R^2 - A_{cs} = 3\pi R^2 - A_{cs}$$

Hence,

$$0 \leq A_{ph} \leq 3\pi R^2$$

Let N_{ph} denotes the number of node in potential node area. As we assume that, nodes are uniformly distributed, thus N_{ph} can be

$$N_{ph} = \lambda A_{ph}$$

$$0 \leq N_{ph} \leq \lambda 3\pi R^2$$

$$0 \leq N_{ph} \leq \lambda 3N \tag{3}$$

With Eq. 3, P_3 is given by

$$P_3 = \{\sum_{i=0}^{\infty}(1-p)^i \frac{(N_{ph})^i}{i!} e^{-N_{ph}}\}^{(2\delta_{data}+\tau)} = e^{-p' N_{ph}(2\delta_{data}+\tau)}$$

Therefore, Eq. 2 can be expressed as

$$P_{ws} = p' e^{-p' N} e^{-p' 3N_{ph}(2\delta_{data}+\tau)} = p' e^{-p'(N+3N_{ph}(2\delta_{data}+\tau))}$$

From the Fig. 4, we have $P_{ws} = 1 - P_{ww} - P_{ws}$ and $P_{cw} = P_{sw} = 1$ Hence, the steady-state probability of state succeed, Φ_s, can be expressed as

$$\Phi_s = \Phi_w P_{ws} = \frac{P_{ws}}{1+p'}$$

According to the definition [2], the throughput equals the fraction of time in which the channel is engaged in successful transmission of user data. Therefore, the throughput Th is equal to the limiting probability that the channel is in state in success.

$$Th = \frac{\Phi_s \delta_{data}}{\Phi_s T_{succ} + \Phi_c T_{coll} + \Phi_w T_{wait}} = \frac{\Phi_s \delta_{data}}{\Phi_s T_{succ} + (1 - \Phi_s - \Phi_w) T_{coll} + \Phi_w T_{wait}} \tag{4}$$

$$= \frac{P_{ws}\delta_{data}}{p' T_{coll} + T_{wait}} = \frac{(p' e^{-p'(N+3N_{ph}(2\delta_{data}+\tau))})\delta_{data}}{\tau + p'(\delta_{data}+\tau)}$$

3 Numerical Results

In this section, we show numerical results based on the models introduced in the previous section. To see the effect of data frame length on throughput performance, we show results relatively large data frames and relatively small data frames. For the long data frame case, we use 100τ as the frame size. For the small frame case, we use 10τ.

We first study the performance of the IEEE 802.11 broadcast scheme by varying the average number of neighboring node (N) and transmission attempt probability (p'). In this scenario, we fix the *potential hidden node area* (R_{ph}) as $3\pi R^2$, which is the worst case.

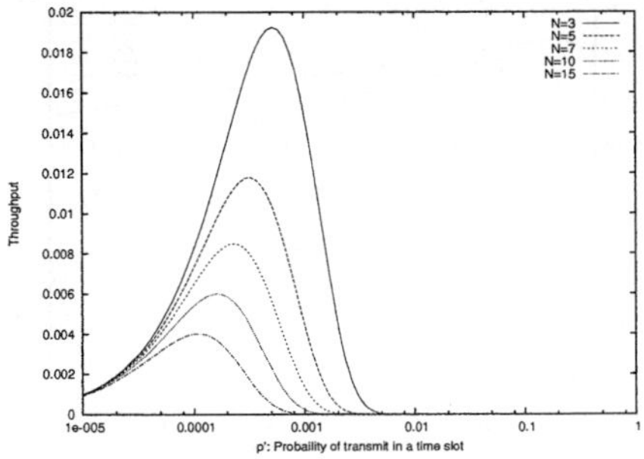

(a) Long data frame: $\delta_{data} = 100\tau$

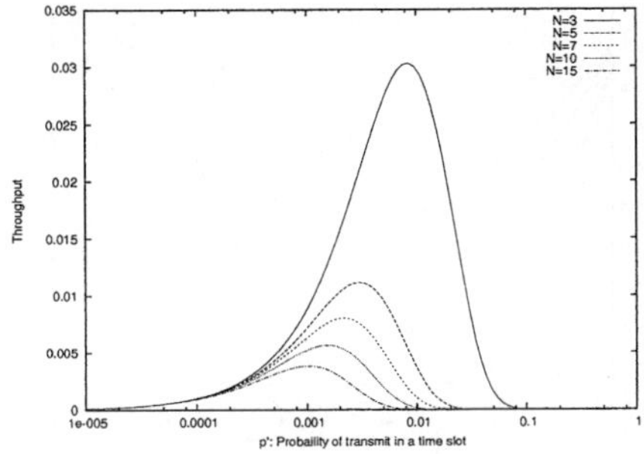

(b) Short data frame: $\delta_{data} = 10\tau$

Fig. 7. The throughput of IEEE 802.11 broadcast scheme varying N and p' (R_{ph}=100%=$3\pi R^2$)

Fig 7 shows the throughput results for the IEEE 802.11 broadcast scheme with different frame sizes. As the average number of neighboring node increases, both case of the IEEE 802.11 broadcast scheme shows very poor throughput performance. The main reason is that the probability of collisions becomes higher as the number of node becomes larger. And as N is increased, p' achieving optimum throughput decreases. This means that, as the number of competing nodes within a region increases, IEEE 802.11 scheme becomes more ineffective. When data frame length is long, the throughput of IEEE 802.11 broadcast scheme is very low. This is the fact that the vulnerability period ($\delta_{data} + \tau$) in equation for P_3 becomes twice the length of the data frame.

Next, we investigate the throughput performance of both case when the *additional carrier sensing area* varies. This is important because IEEE 802.11 broadcast scheme only relies on physical carrier sensing. In this scenario, we fix the probability of transmission in a time slot (p') as 0.001. To see the effect of varying the *additional carrier sensing area* (A_{cs}), we vary the *potential hidden node area* (A_{ph}) since this value is inversely proportional to A_{cs}.

Fig. 8 shows, the throughput versus the percentage of A_{ph} for the IEEE 802.11 broadcast scheme for 3,5,7,10,15 average neighboring nodes. In this result, when the percentage of A_{ph} is 0, i.e., A_{cs} is $3\pi R^2$, throughput performance have maximum value. This means that, by achieving maximum value of A_{cs}, the IEEE 802.11 broadcast scheme minimizes the possibility of hidden node problem. So it is beneficial to set the carrier sensing range large for broadcast communication. However, for unicast communication, a large carrier sensing range leads to reduced spatial reuse, so minimizing hidden node effect and increasing spatial reuse becomes a tradeoff which must be studied further. As the percentage of *potential hidden node area* and number of nodes increase, we observe that throughput of both case decreases more deeper. This again means that IEEE 802.11 broadcast scheme becomes more ineffective as the number of competing nodes within a region increases.

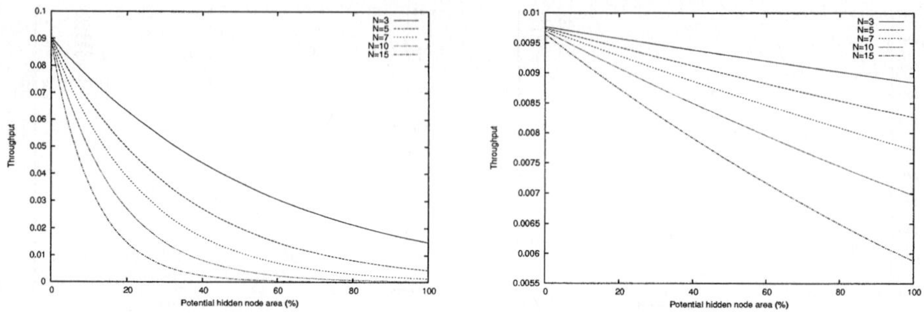

(a) Long data frame: $\delta_{data} = 100\tau$ (b) Short data frame: $\delta_{data} = 10\tau$

Fig. 8. The throughput of IEEE 802.11 broadcast scheme varying N and R_{ph} (p'=0.001)

Our results reveal that hidden terminals degrade the performance of IEEE 801.11 broadcast scheme beyond the basic effect of having larger *potential hidden node area.*

4 Conclusion

Broadcast is an efficient paradigm for transmitting a data from sender to group of receivers. In this paper, we present a performance of the IEEE 802.11 broadcast scheme. To derive the throughput, we have used a simple model based on Markov chains. The result shows that overall performance of IEEE 802.11 broadcast scheme degrades rather rapidly when the number of competing nodes allowed within a region increase.

References

1. L. Wu and P. K. Varshney, *Performance analysis of CSMA and BTMA protocols in multihop netoworks(1). Single channel case*, Elseiver Information Sciences, Vol. 120, pp. 159-177, 1999.
2. R. Rom and M. Sidi, *Multiple Access Protocols: Performance and Analysis*, Springer-Verlag, 1989.
3. IEEE 802.11 Working Group, *Wireless LAN Medium Access Control (MAC0 and Physical Layer (PHY) specifications*, 1999.
4. F. A. Tobagi and L. Kleinrock, *Packet Switching in Radio Channels: Part II - the Hidden Terminal Problem in Carrier Sense Multiple-access Modes and the Busy-tone Solution*, IEEE Trans. on Communications, 1975.
5. F. Cali, M. Conti and E. Gregori, *Dynamic Tuning of the IEEE 802.11 Protocol to Achieve a Theoretical Throughput Limit*, IEEE/ACM Transactions on Networking (TON), 2000.
6. G. Bianchi, *Performance Analysis of the IEEE 802.11 Distributed Coordination Function*, IEEE Journal on Selected Areas in Communications (JSAC), 2000

Design and Performance Evaluation of an Optimal Collision Avoidance Mechanism over Congested and Noisy Channels for IEEE 802.11 DCF Access Method

Dr-Jiunn Deng* and Hsu-Chun Yen

Department of Electrical Engineering, National Taiwan University,
No. 1, Sec. 4, Roosevelt Rd., Taipei, Taiwan, R.O.C.
yen@cc.ee.ntu.edu.tw

Abstract. For the IEEE 802.11 protocol, the basic access method in its medium access control (MAC) layer protocol is the distributed coordination function (DCF). However, this strategy incurs a high collision probability and channel utilization is degraded in bursty arrival or congested scenarios. Besides, when a frame is collided on a wired network, the sender should slow down, but when one is lost on a wireless network, the sender should try harder. Extending the backoff time just makes matters worse because it brings bandwidth wastage. In this paper, we identify the relationship between backoff parameters and channel BER and put forth a pragmatic problem-solving solution. In addition to theoretical analysis, simulations are conducted to evaluate the performance scheme. As it turns out, our design indeed provides a remarkable improvement in a heavy load and error-prone WLANs environment.

1 Introduction

Flexibility and mobility have made wireless local area networks (WLANs) a rapidly emerging field of activity in computer networking, attracting significant interests in the communities of academia and industry [1,2,3,5,6,7,8,9,10,11,13]. In the meantime, the IEEE standard for WLANs, IEEE 802.11 [14], has gained global acceptance and popularity in wireless computer networking markets and has also been anticipated to continue being the preferred standard for supporting WLANs applications. According to the actual version of the standard, the backoff parameters of its collision avoidance mechanism are hard-wired in the physical layer, and are far from the optimal setting in some network configuration conditions especially in congested or noisy scenario. To begin with, this strategy might allocate initial size of CW, only to find out later that it is not enough when the load increased. The size of CW must be reallocated with a larger size, but each increase of the CW parameter value is obtained paying the

* D. J. Deng is with the Department of Information Management, the Overseas Chinese Institute of Technology, Taiwan, R.O.C.

C. Kim (Ed.): ICOIN 2005, LNCS 3391, pp. 11–20, 2005.

cost of a collision (bandwidth wastage). Furthermore, after a successful transmission, the size of CW is set again to the minimum value without maintaining any knowledge of the current channel status.

Besides, the performance of CSMA/CA access method will be severely degraded not only in congested scenarios but also when the bit error rate (BER) increases in the wireless channel. One principal problem also comes from the backoff algorithm. In CSMA/CA access method, immediate positive acknowledgement informs the sender of successful reception of each data frame. This is accomplished by the receiver initiating the transmission of an acknowledgement frame after a small time interval, SIFS, immediately following the reception of the data frame. In case an acknowledgement is not received, as we mention above, the sender will presume that the data frame is lost due to collision, not by frame loss. Consequently, when a timer goes off, it exponentially increases backoff parameter value and retransmits the data frame less vigorously. The idea behind this approach is to alleviate the probability of collision. Unfortunately, wireless transmission links are noisy and highly unreliable. The proper approach to dealing with lost frames is to send them again, and as quickly as possible. Extending the backoff time just makes matters worse because it brings bandwidth wastage.

Although in the past there were adequate discussions on issues about DCF and the performance thereof, there were few papers on the relationship between backoff parameters and channel bit error rate (BER). In fact, as disclosed by research conducted in the past [7,10,13], the performance of DCF will be significantly affected by channel BER in the wireless channels. That is, the proper choice of the CW parameter values has substantial influence on the network performance. In this paper, we attempt to identify the relationship between backoff parameters and channel BER and put forth a pragmatic problem-solving solution. The proposed distributed adaptive contention window mechanism not only dynamically expands and contracts the contention window size according to the current network contention level and channel BER, but also provably optimal in that it achieves optimal channel utilization for IEEE 802.11 DCF access method. The proposed scheme is performed at each station in a distributed manner, and it can be implemented in the present IEEE 802.11 standard with relatively minor modifications. In addition to theoretical analysis, simulations are conducted to evaluate the performance scheme. The performance of our design is examined in detail. As it turns out, our design indeed provides a remarkable improvement in a heavy load and error-prone WLANs environment especially when the bit error rate is severely degraded.

The remainder of this paper is organized as follows. In Section 2, we describe the proposed scheme in detail. Simulation and experimental results are reported in Section 3. Section 4 concludes this paper.

2 Dynamic Optimization for DCF Access Method

An analytical model has been proposed in [2] and [6], which analyze the performance of DCF without BER and frame loss in the WLANs. Here we extend the analytical model for the above purpose. Our scheme is also based on the results of the capacity analysis model of the IEEE 802.11 protocol originally proposed in [1] and [7] as well as the concept introduced in [3] and [5].

In order to exploit the early and meaningful information about the actual congestion status of a channel, we start by defining the utilization factor, α,of a contention window to be the number of transmission attempts , $slot_{busy}$, observed in the latest contention window divided by the size (number of slots) of the current contention window, and it is worth noting that $slot_{busy}$ includes collisions, lost frames, and successful transmission. In practice, the value of α has to be updated in every backoff interval to reflect the actual state of the channel. Assume that there are stations working in asymptotic conditions in the system. This means that the transmission queue of each station is assumed to be always nonempty. The stations transmit frames whose sizes are i.i.d. sampled from a geometric distribution with parameter q, and the size of a frame is an integer multiple of the slot size, t_{slot}. Let t_{frame} , $t_{virtual}$ and $t_{success}$ denote the average frame transmission time, the average temporal distance between two consecutive successful transmission, and the average time required for a successful transmission, respectively. Hence, the protocol capacity, ρ , is $t_{frame}/t_{virtual}$. Also, from the geometric backoff assumption, all the processes which define the occupancy pattern of the channel are regenerative with respect to the sequence of time instants corresponding to the completion of a successful transmission. Hence, the average time required for a successful transmission, $t_{success}$, is bounded above by $t_{frame} + ACK + DIFS + SIFS + 2 \cdot \tau$, where τ denotes the maximum propagation delay. Since an idle period is made up of a number of consecutive slots in which the transmission medium remains idle due to the backoff and the collisions and frame loss might occur between two consecutive successful transmissions, we have

$$t_{virtual} = E\left[\sum_{i=1}^{N} idel\_p_i + coll_i + lost_i + \tau + DIFS\right] \\ + E\left[idel\_p_{N_{collision}+N_{lost}+1}\right] + E\left[t_{success}\right], \qquad (1)$$

where $idel\_p_i$, $coll_i$ and i-th are the lengths of the idle period, frame loss and collision in a virtual time, respectively, and $N_{collision}$ and N_{lost} is the number of collisions and number of lost frame in a virtual time, respectively.

The assumption that the backoff interval is sampled from a geometric distribution with parameter p implies that the future behavior of a station does not depend on the past. Hence, the above equation can be rewritten as

$$t_{virtual} = E\left[N_{collision}\right] \cdot (E[coll] + \tau + DIFS) + E[N_{lost}] \cdot (E[lost] + \tau + DIFS) \\ + E[idle\_p] \cdot (E[N_{collision}] + E[N_{lost}] + 1) + E[t_{success}] \qquad (2)$$

Closed expressions for $E[idle\_p]$, $E[lost]$ and $E[coll]$ have been derived in the literature with $E[N_{collison}]$ and $E[N_{lost}]$:

$$E[idel\_p] = \frac{(1-p)^M}{1-(1-p)^M} \cdot t_{slot} \qquad (3)$$

$$E[N_c] = \frac{1-(1-p)^M}{M \cdot p \cdot (1-p)^{M-1}} - 1 \tag{4}$$

$$E[N_{lost}] = M(1-(1-p)^{M-1} \cdot (1-BER)^{\frac{t_{slot}}{1-q}}) \tag{5}$$

$$E[coll] = E[lost] = \frac{t_{slot}}{1-[(1-p)^M + M \cdot p \cdot (1-p)^{M-1}]} \cdot \\ [\sum_{h=1}^{\infty}(h \cdot ((1-pq^h)^M - (1-pq^{h-1})^M)) - \frac{M \cdot p(1-p)^{M-1}}{1-q})] \tag{6}$$

Hence, $t_{virtual}$ is a function of the system's parameters, the number of active stations (M), the parameter p which defines the geometric-distribution used in the backoff algorithm, and the parameter q that characterizes the frame-size geometric distribution. As mentioned earlier, each station transmits a frame with probability p. This yields:

$$p_{error} = 1-(1-p)^{M-1} \cdot (1-BER)^{\frac{t_{slot}}{1-q}} \tag{7}$$

where p_{error} is the probability that a transmitted frame encounters a collision or is received in error. Using the Markov chain we can obtain an explicit expression for the probability p as a function of probability p_{error}:

$$p = \frac{2(1-p_{error})}{(1-2p_{error})(W+1) + W \cdot p_{error}(1-(2p_{error})^m)} \tag{8}$$

where W is the minimum contention window, and m is the maximum number of backoff stages, i.e., $CW = W \cdot 2^m$. From equation (7), we obtain:

$$M = 1 + \frac{\log(\frac{1-p_{error}}{(1-BER)^{t_{slot}/(1-q)}})}{\log(1-p)} \tag{9}$$

Substituting p, as express by equation (8), into equation (9), we obtain:

$$M = 1 + \frac{\log(\frac{1-p_{error}}{(1-BER)^{t_{slot}/(1-q)}})}{\log(1+\frac{2 \cdot (1-2 \cdot p_{error})}{(2+W) \cdot p_{error} + W \cdot p_{error}(2 \cdot p_{error})^m - (1+W)})} \tag{10}$$

Recall that the probability p_{error} is defined as the probability that a frame transmitted by the considered station fails. Since in each busy slot an eventual frame transmission would have failed, the probability p_{error} can be obtained by counting the number of experienced collision , frame loss, as well as the number of observed busy slot, and dividing this sum by the total number of observed slots on which the measurement is taken, i.e., $p_{error} \approx \alpha$.

In order to maximize the utilization of every slot in a contention window, we still need to engineer the tight upper bound of α to help us complete this scheme. We start with defining p_{opt} to be the value of p parameter that minimizes $t_{virtual}$. Since p_{opt} is closely approximated by the p value that guarantees a balance between the collision and frame loss and the idle periods in a virtual transmission time. Suppose there are M_{tr} stations making a transmission attempt in a slot. Then, we have

$$M \cdot p_{opt} = \sum_{i=1}^{M} i \cdot p\{M_{tr} = i\} \geq 1 - p\{M_{tr} = 0\} = \alpha \tag{11}$$

As a consequence, $M\dot{p}_{opt}$ is a tight upper bound of α in a system operating with the optimal channel utilization level. Substituting M, as express by equation (10), we obtain

$$p_{opt} \geq \frac{\alpha}{M} = \frac{\alpha}{1 + \frac{\log(\frac{1-\alpha}{(1-BER)^{t_{slot}/(1-q)}})}{\log(1 + \frac{2\cdot(1-2\cdot\alpha}{(2+W)\cdot\alpha + W\cdot\alpha\cdot(2\cdot\alpha)^m - (1+W)})}} \tag{12}$$

More precisely, the capacity of 802.11 DCF protocol can be improved to achieve the theoretical throughput limit corresponding to the ongoing network environment, channel BER, and traffic configuration by dynamically adjusting its contention window whose average size is identified by the optimal p value, p_{opt}, that is, when the average size of contention window is $2/p_{opt} - 1$.

A natural strategy for expansion and contraction is to allocate a new contention window size at the end of each transmission time. However, such a common heuristic would conduct the size of contention window to fluctuate rapidly between expansion and contraction. To avoid this undesirable behavior, each station runs the algorithm to estimate the optimal contention window size, and use the following formula to update its contention window:

$$New\_CW = \chi \cdot Current\_CW + (1 - \chi) \cdot Estimate\_Optimal\_CW, \tag{13}$$

where $\chi \in [0, 1]$ is a smoothing factor. Finally, instead of using the backoff time generation function defined in the IEEE 802.11 standard, we refine the backoff time generation function as $\lfloor ranf() \cdot 2^{\lceil \log(New\_CW) \rceil} \rfloor \cdot t_{slot}$ to complete our scheme.

3 Simulations and Performance Evaluation

In this section, we evaluate the performance of the proposed scheme.

3.1 Simulation Environment

Our simulation model is built using the Simscript tool [4]. Performance is measured in terms of the throughput, the average access delay, the dropping probability, the offer-load, among others. The default values used in the simulation are listed in Table I. The values for the simulation parameters are chosen carefully in order to closely reflect the realistic scenarios as well as to make the simulation feasible and reasonable.

Table 1. Default attribute values used in the simulation.

Attribute	Value	Meaning and Explanation
Channel rate	11 Mb/s	Data rate for the wireless channel
Slot_ Time	20 μs	Time needed for each time slot
SIFS	10 μs	Time needed for each short interframe space
DIFS	50 μs	Time needed for each DCF interframe space
MAC header	272 bits	Header length of MAC layer header
PHY header	192 bits	Header length of physical layer header
RTS	160 bits + PHY header	Frame length of each request-to-send frame
CTS	112 bits + PHY header	Frame length of each clear-to-send frame
ACK	112 bits + PHY header	Frame length of each Acknowledgement
Time out	300 μs	ACK/CTS frame time out
χ	0.5	Smoothing factor
r_c	32 kb/s	Voice source data rate
δ	32 ms	Tolerable jitter for voice source
π	5 ms	Time needed for handoff
d	50 ms	Maximum packet delay for video source
buffer	1 frame	Size of buffer for frames
W	16 slots	Minimum contention window size
m	6	Maximum backoff stages

3.2 Simulation Results

In what follows, the performances of the proposed scheme and the conventional IEEE 802.11 DCF protocol are compared based on simulations. Figures 1, 2, and 3 show the effect of channel BER by plotting throughput (average bandwidth utilization), frame drop probability, and average frame delay for two representative network sizes (M=5 and 50). As illustrated in Figure 1, the performance of conventional DCF access method was severely degraded when the channel BER increased, but the performance of proposed scheme was satisfactory all the time until the channel BER approximates to 10^{-5}. In fact, we believe that it is almost impossible to increase the probability of success of transmitting a frame excepting frames fragmentation or FEC (Forward Error Control) in an extremely noisy wireless environment.

Figure 2 presents the frame drop probability as a function of the channel BER. We can see that although there is not much difference in the values of the performance measures when BER is low, however, the proposed scheme provides better performance than the conventional DCF access method when the channel BER increased. For voice and video traffic, the allowable loss probability is about 10^{-2} and 10^{-3}[12], respectively. With this criterion, the proposed scheme can tolerate a BER of 10^{-6}. This simulation result reveals that the proposed scheme is appropriate for transmitting high priority real-time traffic such as voice and video traffic in real-time applications. However, the frame drop probability shows a sharp rise as the BER higher than 10^{-4} for both schemes due to the increased number of error transmissions. Conversely, number of stations only marginally affects frame drop probability for proposed scheme.

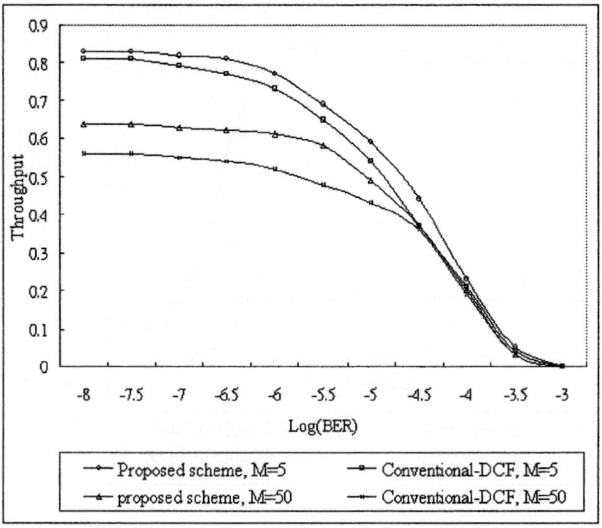

Fig. 1. Throughput against channel BER, for M=5, and 50 respectively.

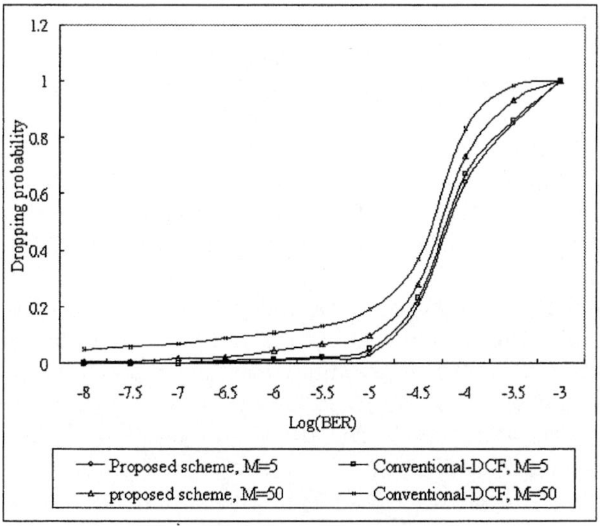

Fig. 2. Dropping probability against channel BER, for M=5, and 50 respectively.

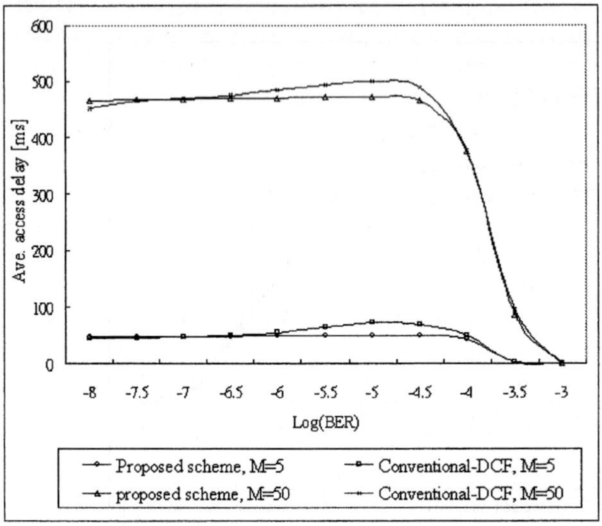

Fig. 3. Average access delay against channel BER, variance=363.65 (proposed scheme, M=5), 561.25 (conventional-DCF, M=5), 29410.25 (proposed scheme, M=50), and 30325.49 (conventional IEEE 802.11, M=50)

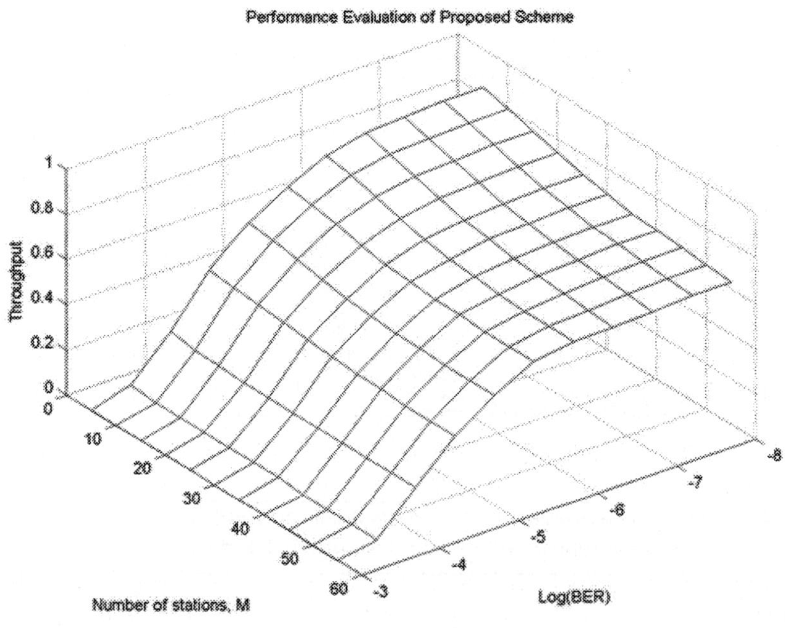

Fig. 4. Throughput versus channel BER and number of stations.

As mentioned earlier, in the noisy and highly unreliable wireless environments, the proper approach to dealing with lost frames is to send them again, and as quickly as possible. Extending the backoff time just makes matters worse because it brings bandwidth wastage. In other words, in an environment full of noise and susceptible to interference, the size of CW should not increase with the number of times packets are sent all over again. Hence, we might have an intuition that CW should be inversely proportional to BER. As a matter of fact, as discovered by our research, the initial size of CW should be marginally positively correlated with BER. In Figure 3, we prove that our hypothesis, a relatively large size of CW is recommended in a high-BER environment, is true. As shown in the figure, with the conventional DCF access method, the average access delay increases with channel BER, whereas the increase is moderate if our method is adopted. The increasing frame access delay could result from either retransmitted frames in conventional DCF access method or relatively large size of CW in proposed scheme. However, please note that it will reach a maximum value and then decreases gradually and finally drops to 0 as the simulation outcome obtained in [7]. The reason is that when channel BER is high enough to result in a significant increase in the drop probability, for example, higher than 10^{-4}, frame delay starts decreasing since the long delays of dropped frames do not contribute to the average frame delay. Thus, the low frame access delay values at high BER concern only a small number of successfully received frames due to high drop probability and, therefore, have a very small significance.

Figure 4 shows the value of the channel BER and the number of stations versus the throughput for the proposed scheme. As the analytical results, it illustrated that the throughput of our scheme is not affected by the channel BER between 10^{-8} to 10^{-5}, but decreases greatly when the channel BER grows to 10^{-4}. Besides, the throughput of our scheme changes little when the number of stations, M, changes. This indicates that, although the proposed scheme has proven it satisfactory superiority in most of the cases, it provides a remarkable improvement over congested and noisy wireless environments.

4 Conclusions

The backoff parameters in IEEE 802.11 DCF access method are far from the optimal setting in heavy-load and error-prone WLANs environment. First, this strategy incurs a high collision probability and channel utilization is degraded in bursty arrival or congested scenarios. Besides, in the noisy and highly unreliable wireless environment, an unacknowledged frame could result from not only collision but also frame loss. When the sender is unable to discriminate the cause of the frame loss, the performance of DCF access method is significantly affected by the channel BER. In this paper, we attempt to identify the relationship between backoff parameters and channel BER and put forth a pragmatic problem-solving solution. The proposed distributed adaptive contention window mechanism not only dynamically expands and contracts the contention window size according to the current network contention level and channel BER, but also provably optimal in that it achieves optimal channel utilization for IEEE 802.11 DCF access method. The proposed scheme is performed at each station in a distributed manner, and it can be implemented in the present IEEE 802.11 standard with relatively minor modifications. Through extensive simulations, we have demonstrated a satisfactory performance of our proposed scheme in a quantitative way. It shows that the proposed scheme has proven it satisfactory superiority in most of the cases. Notable is the remarkable improvement in congested and noisy wireless environments, even with fairly numerous stations.

References

1. L. Alcuri, G. Bianchi, and I. Tinnirello, "Occupancy Estimation in the IEEE 802.11 Distributed Coordination Function," Proc. of ICS2002, 2003, Hualien, Taiwan.
2. G. Bianchi, "Performance Analysis of the IEEE 802.11 Distributed Coordination Function," IEEE Journal on Selected Area of Communications, vol. 18, no. 3, pp. 535-547, Mar. 2000.
3. L. Bononi, M. Conti, and E. Gregori, "Runtime Optimization of IEEE 802.11 Wireless LANs Performance, " IEEE Transactions on Parallel and Distributed Systems, vol. 15, no. 1, pp. 66-80, Jan. 2004.
4. CACI Products Company, Simscript II.5, California 92037, Sep. 1997, http://www.caciasl.com/.
5. F. Cali, M. Conti, and E. Gregori, "IEEE 802.11 Protocol: Design and Performance Evaluation of an Adaptive Backoff Mechanism," IEEE Journal on Selected Area of Communications, vol. 18, no. 9, pp. 1774-1786, Sep. 2000.
6. F. Cali, M. Conti, and E. Gregori, "Dynamic Tuning of the IEEE 802.11 Protocol to Achieve a Theoretical Throughput Limit," IEEE/ACM Transactions on Networking, vol. 8, no. 6, Dec. 2000, pp. 785-799.
7. P. Chatzimisios, A. C. Boucouvalas, and V. Vitsas, "Influence of channel BER on IEEE 802.11 DCF, " Electronics letters, vol. 39, issue 23, pp. 1687-1689, 2003.
8. B. P. Crow, I. Widjaja, J. G. Kim, and P. T. Sakai, "IEEE 802.11 Wireless Local Area Networks," IEEE Commun. Mag., vol. 35, no. 9, pp. 116-126, Sep. 1997.
9. D. J. Deng and R. S. Chang, "A Priority Scheme for IEEE 802.11 DCF Access Method," IEICE Trans. Commun., vol. E82-B, no. 1, pp. 96-102, January 1999.
10. F. Eshghi and A. K. Elhakeem, "Performance Analysis of Ad Hoc Wireless LANs for Real-Time Traffic, " IEEE Journal on Selected Area of Communications, vol. 21, no. 2, pp. 204-215, Feb. 2003.
11. R.O. LaMaire et al., "Wireless LANs and Mobile Networking: Standards and Future Directions," IEEE Commun. Mag., vol. 34, no. 8, pp. 86-94, Aug. 1996.
12. D. Raychaudhuri and N. D. Wilson, "ATM-Based Transport Architecture for Multiservices Wireless Personal Communication Network," IEEE Journal on Selected Areas of Communications,. vol. 12, no. 8, pp. 1401-1414, Oct. 1994.
13. Z. Tang, Z. Yang, J. He, and Y. Liu, "Impact of Bit Errors on the Performance of DCF for Wireless LAN, " Proc. of International Conference on Communications, Circuits and Systems, and West Sino Expositions, vol. 1, pp. 529-533, 2002.
14. Wireless Medium Access Control and Physical Layer WG, IEEE Draft Standard P802.11, "Wireless LAN," IEEE Stds. Dept, D3, Jan. 1996.

On the Load-Balanced Demand Points Assignment Problem in Large-Scale Wireless LANs

Chor Ping Low and Can Fang

School of Electrical and Electronic Engineering
Nanyang Technological University
Singapore 639798
icplow@ntu.edu.sg

Abstract. One of the main issues to be addressed in the design of large-scale wireless LANs is that of assigning demand points to access points (APs) in such a way that each demand point is assigned to one AP and the aggregate traffic demand (which is referred to as load in this paper) of all demand points assigned to any AP does not overload that AP. In this paper, we consider the problem of assigning demand points to APs with the objective of minimizing the maximum load among the set of APs, which qualitatively represents congestion at some hot spots in the network service area. We refer to this problem as the *Load-Balanced Demand Points Assignment Problem (LBDPAP)*. We formulated this problem as an integer linear program (ILP) and show that the problem is NP-hard. We propose an efficient $\frac{4}{3}$-approximation algorithm for the problem.

Keywords: wireless LANs, demand points assignment, load balancing, NP-hard, approximation algorithm.

1 Introduction

A wireless local area network or WLAN is typically comprised of mobile computers with network adapters and access points (APs). A WLAN must be designed so that all of the target space has radio coverage. It must also be designed so that its capacity is adequate to carry the expected load. A typical approach to manage the complexity of the design of WLANs is the Divide-and-Conquer approach [1][2]. Using this approach, the network design problem is decomposed into a number of subproblems which are easier to manage and solve and is comprised of the following steps:

1. Estimation of the demand area map: the WLAN designers should draw the map of service area by investigating the physical space with walls or barriers. The service area map will be divided into smaller demand points where signal is measured from APs and the number of users or traffic demand is estimated.

C. Kim (Ed.): ICOIN 2005, LNCS 3391, pp. 21–30, 2005.

2. Selection of candidate locations for APs: As the physical location of APs may be restricted to particular areas because of the connections to the wired LAN, the power supply, and the installation and administration costs, WLAN designers have to carefully select candidate locations for placement of APs.
3. Signal measurement at the demand point in the service area: In order to provide the maximum coverage and throughput, signal measured or estimated at each demand point should be greater than the threshold with which the minimum rate is guaranteed. For example, in IEEE 802.11b, the automatic rate fallback (ARF) function will provide several kinds of rates such as 1/2/5.5/11 Mbps according to the distance between APs and mobile computers.
4. AP placement: Given the service areas and the candidate locations of APs, a set of APs will be chosen from the list of candidate AP locations to meet the users' traffic demands. This process is referred to as AP placement.
5. Channel Assignment: After AP locations have been determined, channels are assigned to APs in a way that the interference between APs is minimized.
6. Assignment of Demand Points to APs: After the locations of the APs are fixed and channels assigned, the next task is to assign demand points to APs in such a way that each demand point is assigned to one AP and the the aggregate traffic demand of all demand points assigned to any AP does not overload that AP, i.e. is within the capacity of the AP.

Most of the earlier works on the design of WLANs focussed on steps 4 & 5 of the design process [2,3,4,6,7,8,9]. In this paper, we address another problem of the network design process, which is that of assigning demand points to APs (step 6 of the design process). In conventional WLAN design process, this problem is usually ignored and the demand points were automatically assigned to the APs that have the least path loss between them. Due to the non-uniformity of the WLAN users' distribution, some APs may close to many users and are hence heavily loaded (congested) while some other APs may not have any demand point assigned to them. Hence there is a need for an efficient algorithm to assign demand points to APs with the objective of balancing the load among the APs to maximize the overall throughput of the network.

The rest of this paper is organized as follows. In section 2, we present the problem assumptions & formulation and prove that the problem addressed is intractable. Some observations about the characteristics of the problem is describes in section 3. An efficient approximation algorithm is proposed in section 4 and simulation results to evaluate performance of our proposed algorithm are described in Section 5. Section 6 concludes this paper.

2 Load-Balanced Demand Points Assignment Problem

2.1 Problem Assumptions and Formulation

We address the issue of demand points assignment with the objective of minimizing the maximum load of all APs. We call this problem the *Load-Balanced*

Demand Points Assignment Problem (LBDPAP). We adopt the following assumptions and notations in the problem formulation:

- The sites of the demand points and APs are known
- The set of demand points is denoted by $D = \{d_1, d_2 \dots d_n\}$.
- The set of APs is denoted by $A = \{a_1, a_2 \dots a_m\}$.
- The number of demand points is more than the number of APs, i.e. $n > m$.
- The $m \times n$ signal matrix, $S = \{s_{ij}\}$, where s_{ij} represents the Signal-to-Noise Ratio (SNR) value at demand point d_i from access point a_j, is given.
- The traffic demand for each demand point d_i is known and denoted by t_i.
- Each demand point should be assigned to exactly one AP.
- The *load* of an AP is defined to be equal to the sum of traffic demand from all demand points that are assigned to the AP.

Next, we note that a demand point d_i can only be assigned to an AP, say a_j, if the *SNR* s_{ij} from d_i to a_j is greater than a certain threshold. For each demand point d_i, let N_i denote the set of APs onto which it may be assigned. We refer this relationship between APs and demand points as the *assignment constraint.*

LBDPAP may be defined as follows: Let D be the set of traffic demand points to be assigned and let A be the set of APs available. Let $l(a_j)$ denote the load of AP a_j, where $a_j \in A$. The LBDPAP is that of finding an assignment $\Phi : D \rightarrow A$ that assigns each member of D to one of the members of A without violating the assignment constraint and l_{max} is minimized, where:

$$l_{max} = max\{l(a_j)\}, a_j \in A, \quad l(a_j) = \sum_{\Phi(d_i)=a_j} t_i$$

We say that a demand point d_i is *assigned* to access point a_j if $\Phi(d_i) = a_j$.

2.2 Formulation as a Mathematical Program

Prior to the problem formulation, the following variables are defined.

- x_{ij}: a binary variable, 1 if demand point d_i is assigned to access point a_j, otherwise 0
- α : the maximum load that may be assigned to an access point

The Integer Linear Programming (ILP) formulation of the Load-Balanced Demand Points Assignment Problem is as follows:

Objective function:

$$Minimize \quad \alpha \tag{1}$$

Subject to

$$\sum_{j \in N_i} x_{ij} = 1, \qquad \forall i \in D \tag{2}$$

$$\sum_{i \in D} t_i \cdot x_{ij} \leq \alpha, \qquad \forall j \in A \tag{3}$$

The objective (1) is to minimize the maximum load assigned to each AP. Constraint (2) states that each demand point should be assigned to one (and only one) AP. Constraint (3) impose the condition that the total traffic demand of all demand points assigned to a particular AP should not exceed the maximum load permitted.

2.3 The Intractability of LBDPAP

The LBDPAP is related to the following machine scheduling problem.

Problem 1: Minimum Makespan Scheduling Problem on Identical Machines (MMSPIM)
We are given m machines and n jobs with respective processing times $p_1, p_2, \ldots,$ $p_n \in Z^+$. The processing times are the same no matter on which machine a job is run and pre-emption is not allowed. Find an assignment of jobs to m identical machines such that the *makespan* (which is the latest completion time among all machines) is minimized.

Lemma 1. *LBDPAP is NP-hard.*

Proof. Consider a special case of LBDPAP whereby each demand point can be assigned to any APs (i.e. no assignment constraints). It is easy to see that this special case of LBDPAP is identical to Problem 1 (Minimum Makespan Scheduling Problem on Identical Machines) and LBDPAP is thus a generalization of the former problem. Since the Minimum Makespan Scheduling Problem on Identical Machines is known to be NP-hard[5], LBDPAP is also NP-hard.

3 Some Observations About LBDPAP

In this section, we highlight some observations about LBDPAP.

Observation 1
We first observe that that LBDPAP is also related to another machine scheduling problem, namely the Minimum Makespan Scheduling Problem on Unrelated Machines (MMSPUM), which is defined as follows:

Problem 2: Minimum Makespan Scheduling Problem on Unrelated Machines (MMSPUM)
We are given a set J of n jobs and a set M of m machines. The processing time for a job $j \in J$ on machine $i \in M$ is $p_{ij} \in Z^+$ and pre-emption is not allowed. Find an assignment of jobs in J to the machines in M such that the *makespan* is minimized.

In particular, we note that each instance of LBDPAP can be transformed into an instance of the Minimum Makespan Scheduling Problem on Unrelated Machines

(MMSPUM) whereby the demand points and the access points of LBDPAP correspond to the jobs and machines of MMSPUM, respectively. For each demand point $d_i \in D$, let $p_{ij} = t_i \ \forall a_j \in N_i$ and let $p_{ij} = \infty \ \forall a_j \notin N_i$. Then it is easy to see that an optimal solution (least possible maximum load among APs) for LBDPAP corresponds to a schedule for MMSPUM with minimum makespan and vice versa. MMSPUM is also known to be NP-hard[5] and Lenstra et al.[10] gave a 2-approximation algorithm for the problem. This performance bound was further improved to $2 - \frac{1}{m}$ by Shchepin et al.[11] and this is currently the best-known approximation ratio that can be achieved in polynomial time.

Observation 2
We next observe that each instance of LBDPAP can be represented using a bipartite graph as follows. Let $G = (D \cup A, E)$ denote a bipartite graph where E corresponds to a set of edges connecting the vertices in D to the vertices in A. An edge is said to exists between a pair of vertices (d_i, a_j) where $d_i \in D$ and $a_j \in A$ if $a_j \in N_i$. Let $q =\mid E \mid$ and let M be a maximum matching for G.

Lemma 2. *The maximum number of access points that may be used in any assignment of demand points in D to access point points in A is equal to $\mid M \mid$.*

Proof. Let M be a maximum matching for the bipartite graph G. Let D_M and A_M denote the set of matched vertices corresponding to demand points and access points, respectively. We claim that each demand point vertex $d \in D - D_M$ can only be adjacent to one of the matched AP vertex $d \in A_M$, i.e. demand point d can only be assigned to one of the APs in A_M. Suppose otherwise and assume that there exists a demand point vertex $d^* \in D - D_M$ that is adjacent to an access point vertex $a^* \in A - A_M$. In this case, the size of the matching can be increased by one (since both D^* and A^* are unmatched vertices), thus contradicting the fact that M is a maximum matching (refer to Figure 1 for an illustration). Hence the maximum number of APs that may be used in any assignment is equal to $\mid M \mid$.

Lemma 3. *The exists an optimal assignment that uses exactly $\mid M \mid$ access points.*

Proof. Ommitted due to page limit constraint. Intuitively, an optimal assignment will attempt to utilize the maximum number APs possible to distribute the traffic load among the APs.

4 An Approximation Algorithm

The algorithm proposed in [10] and [11] relies on solving a linear programming relaxation of the problem and uses the information obtained from the solution to allocate jobs to machines. In this section, we present a new algorithm, called the

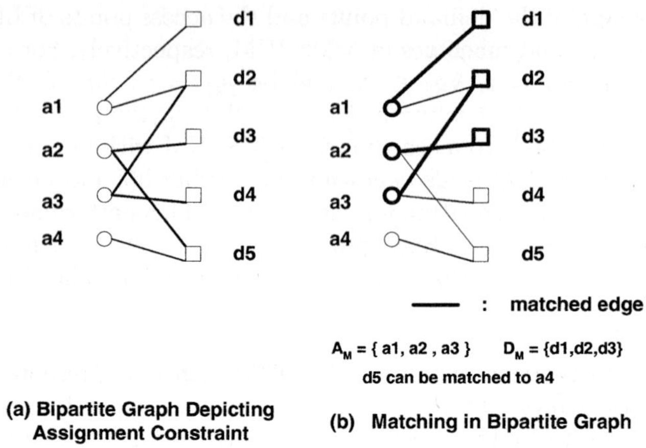

Fig. 1. Maximum number of APs used in an assignment equals $\mid M \mid$

Load-Balanced Demand Point Assignment Algorithm (LBDPAA), for LBDPAP that achieves a lower performance ratio than that of [11] without solving a linear program. In addition, our algorithm runs in $O(n[n+m+q])$ and is combinatorial, hence is more efficient than the algorithm proposed in [11].

4.1 The Load-Balanced Demand Point Assignment Algorithm (LBDPAA)

Our proposed algorithm adopts the approach of utilizing the maximum number of APs possible in the assignment of demand points to APs in order to distribute the traffic load among as many APs as possible. Based on Lemma 2, we know that the maximum number of APs that may be used in any assignment is equal to $\mid M \mid$, where M is the size of a maximum matching in the corresponding bipartite graph G. Hence our algorithm will attempt to find a maximum matching M in the graph G. We first sort the list of demand points in non-increasing order of traffic demand. Let the resultant list be denoted by $D = \{d_1, d_2, \ldots, d_n\}$, where $t_1 \geq t_2 \geq \ldots \geq t_n$. Starting with the first demand point d_1 in the sorted list, we will attempt to match (or assign) d_1 to an AP with zero load in the corresponding bipartite graph G. Next, the algorithm will proceed to match d_2 with another AP with zero load in G. The algorithm will iterate in this manner where in each iteration, we will attempt to find an augmenting path P that connects a given demand point d_i to some unmatched AP (with zero load) in G. If such a path is found , we will augment the edges in P which results in the assignment of d_i to some AP in P and the reassignment of the other demand points to APs in P. In addition the size of the resultant matching will be increased by one. If there does not exists any augmenting path that begins with d_i in G, then we will assign d_i to a least-loaded AP in N_i. The algorithm terminates when all demand points have been assigned. The pseudocode of the algorithm is as follows:

Load-Balanced Demand Points Assignment Algorithm

1.let D = list of demand points sorted in non-increasing order of traffic demand
2.**for** $j = 1$ to m do
  set $l(a_j) = 0$;
 endfor
 set i = 1;
3.**while** $D \neq \emptyset$ do
3.1.find augmenting path P with d_i as one of its end vertex;
  if P exists then
3.2. augment the edges in P;
3.3. **for** each matched edge (d_x, a_v) in P do
    assign demand point d_x to AP a_v
    let d_y be a demand point in P which was assigned to a_v prior to the augmentation of P
    $l(a_v) = l(a_v) + t_x - t_y$
   endfor
3.4.**else**
   let a_j be an AP with the least load in N_i;
   assign d_i to a_j;
   $l(a_j) = l(a_j) + t_i$;
  endif
  $D = D - \{d_i\}$;
  $i = i + 1$;
 endwhile

Lemma 4. *The time complexity of the proposed algorithm is* $O(n[n + m + q])$.

Proof. The sorted list in step 1 can be done in $O(nlogn)$. The initialization of the load of APs in step 2 can be done in $O(m)$. The while loop in step 3 will iterates n times. In step 3.1, each augmenting path can be found in $O(n + m + q)$ using breadth-first search. The augmentation of the edges in step 3.2 can be done in $O(n+m+q)$; the reassignment of demand points to AP and computation of the new load in step 3.3 can be done in $O(n + m)$. In step 3.4, the assignment of a demand point to a least loaded AP can be done in $O(m)$ and the computation of the resultant load can be done in $O(1)$. Hence step 3 can be completed in $O(n[n+m+q])$. Thus, the overall complexity of the algorithm is $O(n[n+m+q])$.

4.2 Performance Ratio

Without loss of generality, we assume that the list of traffic demands $\{t_1, t_2, \ldots, t_n\}$ are all distinct. We will prove that our proposed algorithm is able to achieve a performance ratio of $\frac{4}{3}$ for LBDPAP.

Lemma 5. *The Load-Balanced Demand Point Assignment Algorithm (LBDPAA) is a $\frac{4}{3}$-approximation algorithm.*

Proof. We will construct a proof by contradiction. Suppose that $\frac{4}{3}$ is not a valid bound on the performance ratio. Let OPT denote the maximum load of an optimal assignment. Let I be an instance with the smallest number of demand points such that an assignment which is obtained using LBDPAA has a maximum load $> \frac{4}{3}OPT$. Let d_i be a demand point whose assignment to some access point, say a^*, results in the overall maximum load of the assignment, i.e. $l(a^*) = \max l(a) \; \forall a \in A - \{a^*\}$ and suppose that $i \neq n$. Consider an instance I' which is equal to I without demand point d_n. Then I' is a smaller instance of I for which LBDPAA computes an assignment with maximum load $> \frac{4}{3}OPT$. But this contradicts the choice of I. Hence we can assume that $i = n$.

Let Φ be an assignment obtained using LBDPAA. Let U be an optimal assignment which uses the same number of APs as Φ (i.e. $\mid M \mid$) and has the most number of (demand point, access point) assignments in common with Φ . We first claim that $t_n \leq \frac{OPT}{3}$. Suppose otherwise and assume that $t_n > \frac{OPT}{3}$. Since $t_i \geq t_n \forall i < n$, each AP can be assigned with at most 2 demand points using U. We normalize assignments U and Φ as follows:

- if an AP a has two demand points assigned to it, place the demand point with the higher traffic demand as the first demand point to be assigned to a
- sort the APs so that the first demand points assigned are in decending order of traffic demand.

Next we will compare the assignment obtained by Φ and U as follows. We begin by comparing the assignment of the first demand point using Φ versus using U, in descending order of traffic demand. Following that, we will compare the assignment of second demand point using Φ versus using U in decending order of traffic demand. Let w_j denote the AP that is placed in the jth position of the ordered list of APs. Since Φ and U are not identical, there must exists a (demand point, AP) assignment of Φ that is not in U. Let (d_x, w_p) be the first (demand point, AP) assignment of Φ that is not in U. We consider the following two subcases:

Subcase (i): d_x is the first demand point to be assigned to w_p using Φ. In this case, let the first demand point that is assigned to w_p using U be denoted by d_y. We note that $t_x > t_y$; otherwise d_y would have been considered for assignment before d_x using LBDPAA and be placed as the first demand point to be assigned to w_p. Next we note that the traffic demand of each demand point that is assigned as the first demand point to APs that are ordered after w_p using assignment U must be less than that of d_x. Hence d_x must be assigned as a second demand point to some AP, say w_q, using U. Let the first demand point to be assigned (using U) to w_q be denoted by d_a. Let the second demand point assigned to w_p using U be denoted by d_b. We note that the following must hold: $t_a > t_x$ and $t_y > t_b$. Since $t_x > t_y$, $t_a + t_x > t_y + t_b$. By swapping the assignment of d_x and d_y in U, we have $t_a + t_y \leq t_a + t_x \leq OPT$ and $t_x + t_b \leq t_a + t_x \leq OPT$.

Hence we have another optimal assignment with one additional (demand point, AP) assignment in common with Φ, which is a contradiction.

Subcase (ii): d_x is the second demand point to be assigned to w_p using Φ. In this case, d_x must also be assigned as a second demand point to some AP, say w_q, using U. Let the first demand point to be assigned to w_q be denoted by d_a. Let the first and second demand points assigned to w_p using U be denoted by d_b and d_y, respectively. We note that d_b is also the first demand point to be assigned to w_p using Φ. Since d_x is assigned to w_p (instead of w_q) using Φ, we must have $t_b + t_x \leq t_a + t_x$, which in turn imply that $t_b \leq t_a$. Next we note that since d_x is the first demand point which differs in assignment between Φ and U, $t_x \geq t_y$ (otherwise d_y will be consider before d_y in the assignment Φ in the comparisons of Φ versus U). By swapping the assignment of d_x and d_y in U, we have $t_a + t_y \leq t_a + t_x \leq OPT$ and $t_x + t_b \leq t_x + t_a \leq OPT$. Hence we have another optimal assignment with one additional (demand point, AP) assignment in common with Φ, which is a contradiction.

Since $t_n \leq \frac{OPT}{3}$ and d_n is the demand point whose assignment result in the resultant maximum load L exceeding OPT, we have $L \leq OPT + t_n \leq \frac{4}{3}OPT$.

5 Simulation Results

We study the performance of LBDPAA by comparing its solutions with that of the optimal solutions which are obtained by solving the ILP formulated program using CPLEX. In our empirical studies, we consider the scenario of a wireless local area network (WLAN) whereby the sites for the access points and demand points are generated randomly on a grid plane of 200×200, where each grid represents 10×10 square metre. The traffic demand from each demand point is randomly selected from the range of 100 Kbps to 500 Kbps. We assume that a connection can be established between a AP and a demand point if the distance between them is no larger than 550m (this is typically the case in WLAN).

Figure 2 shows the performance of LBDPAA by varying the number of demand points (which ranges from 20 to 100) while the fixing the number of APs at 20. For each data point in Figure 5, 50 runs are taken and the average calculated. We observe that the solutions obtained using LBDPAA differs from the optimal values by no more than 25%. In particular, we observe that the performance ratio of LBDPAA ranges between 1.17 and 1.24 in all instances generated. This in turn implies that the average case performance of our proposed algorithm is much better than the performance ratio derived.

6 Conclusion

In this paper, we consider the problem of assigning demand points to access points in a WLAN with the objective of distributing the traffic load among the access points so as to ensure that no access point is overloaded. We refer

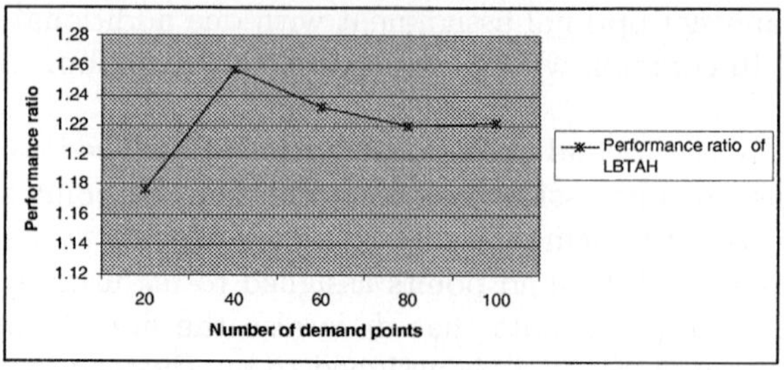

Fig. 2. Performance ratio of LBDPAA

to this problem as the Load-Balanced Demand Points Assignment Problem. We proposed an approximation algorithm for the problem and prove that our proposed algorithm is able to guarantee a performance ratio of $\frac{4}{3}$. Empirical studies have shown that our proposed algorithm is able to perform much better on the average as compared to the performance ratio derived. Hence one direction for future research is to investigate the possibility of tightening the performance bound of our proposed algorithm and to analyse its average-case performance.

References

1. A.Hills, *Large-scale wireless LAN design*, IEEE Communications Magzine, vol. 39, Nov 2001, pp. 98-107.
2. Youngseok Lee and Kyoungae Kim and Yanghee Choi, *Optimization of AP placement and channel assignment in wireless LANs*, Proceedings. LCN 2002. 27th Annual IEEE Conference on Local Computer Networks, Nov 2002, pp. 831-836.
3. L. Nagy and L. Farkas, *Indoor base station location optimisation using genetic algorithms*, IEEE PIMRC 2000, vol. 2, 2000, pp. 843-846.
4. M. Kamenetsky and M. Unbehaun, *Coverage planning for outdoor Wireless LAN system*, International Zurich Seminar on Broadband Communications, 2002.
5. M.R. Garey and D.S. Johnson, *Computers and Intractability: A Guide to the Theory of NP-Completeness*, W.H. Freeman company, New York, 1979.
6. C. Y. Chang, C. T. Chang and P.C. Huang, *Dynamic channel assignment and reassignment for exploiting channel reuse opportunities in ad hoc wireless networks*, The 8th International Conference on Communication Systems(ICCS 2002), 2002.
7. K. L. Yeung and T. P. Yum, *Fixed channel assignment optimisation for cellular mobile networks*, IEICE TRANS. Communication, vol. E83-B, Aug 2000.
8. M.V.S. Shashanka, A. Pati and A. M. Shende, *A characterisation of optimal channel assignments for wireless networks modelled as cellular and square grids*, International Parallel and Distributed Processing Symposium (IPDPS'03), Apr 2003.
9. S. Ramanathan, *A unified framework and algorithm for channel assignment in wireless networks*, Wireless Networks, vol. 5, Mar 1999, pp. 81-94.
10. J.K. Lenstra, D.B. Shmoys and E.Tardos, *Approximation algorithms for scheduling unrelated parallel machines*, Math. Programming, vo. 46, 1990, pp. 259-271.
11. E.V. Shchepin and N.V. Vakhania, *Task distributions on multiprocessor systems*, Lecture Notes in Computer Science, vol. 1872, 2000, pp. 112-125.
12. J. E. Hopcroft and R. M. Karp, *An $n^{2.5}$ algorithm for maximum matching in bipartite graphs*, SIAM Journal on Computing, 1973, pp. 135-158.

Adaptive Window Mechanism for the IEEE 802.11 MAC in Wireless Ad Hoc Networks

Min-Seok Kim, Dong-Hee Kwon, and Young-Joo Suh

Department of Computer Science and Engineering
Pohang University of Science and Technology (POSTECH)
Pohang, Korea
{d0lls, ddal, yjsuh}@postech.ac.kr

Abstract. The IEEE 802.11 MAC protocol adopts the distributed coordination function (DCF) with a binary exponential backoff as a medium access control mechanism. According to previous research results and our simulation study, the performance of IEEE 802.11 is highly dependent upon the number of contending stations and the initial value of contention window (CW_{min}). In this paper, we propose an adaptive contention window mechanism that dynamically selects the optimal backoff window by estimating the current number of contending stations in wireless ad hoc networks. In the proposed scheme, when there are small number of contending stations, a smaller CW_{min} value is selected, while a larger CW_{min} value is selected when there are large number of contending stations. We study the performance of the proposed mechanism by simulation and we got an improved performance over the IEEE 802.11 MAC protocol which uses a fixed CW_{min} value.

1 Introduction

In wireless ad hoc networks, stations communicate with each other without the aid of pre-existing infrastructures. Wireless ad hoc networks can be applied where pre-deployment of a network infrastructure is difficult or impossible. The design of MAC protocols for ad hoc networks has received much attention recently. One of the most popular MAC protocols for ad hoc networks is the IEEE 802.11 MAC [1,2] and it defines the distributed coordinated function (DCF) as a fundamental access mechanism to support asynchronous data transfer on a best effort basis in ad hoc networks. In the DCF mode, before a station starts transmission, it must sense whether the wireless medium is idle for a time period of the Distributed InterFrame Spacing (DIFS) [1]. If the channel appears to be idle for a DIFS, the station generates a random backoff time, and waits until the backoff time reaches 0. The reason for this is to prevent the case that many stations waiting for the medium to be idle can transmit frames at the same time, which may result in high probability of collisions. Thus, the random backoff deferrals of stations before their transmissions can greatly reduce the probability of collisions.

The DCF mode of IEEE 802.11 provides two different handshaking procedures for data transmissions. In the IEEE 802.11 MAC, a sending station must

C. Kim (Ed.): ICOIN 2005, LNCS 3391, pp. 31–40, 2005.

wait for an ACK frame from the receiving station because the sending station can not correctly detect a collision which happens at the receiving station, and it cannot listen to the wireless medium while it is transmitting due to the difference between the transmitted and received signal power strengthes. Thus, the basic handshaking procedure of the IEEE 802.11 MAC for a data transmission follows a DATA-ACK sequence. An optional handshaking procedure requires that the sender and the receiver exchange short RTS (Request-To-Send) and CTS (Clear-To-Send) control frames prior to the basic handshaking procedure. Exchanging of RTS and CTS frames provides a virtual carrier sensing mechanism to prevent the *hidden terminal problem* where a collision can occur at the receiver by the transmission of the sender against the transmissions of other stations which are "out of range" from the sender [3]. For this reason, any stations hearing either a RTS or CTS frame update their Network Allocation Vector (NAV) from the duration field in the RTS or CTS frame. All stations that hear a RTS or CTS frame defer their transmissions by the amount of NAV time.

The DCF adopts a binary slotted exponential backoff mechanism. If there are multiple stations attempting to transmit, the medium may be sensed busy and such stations perform an exponential backoff deferral. Each station waits for a random number of slot times distributed uniformly between [0, CW], where CW is the contention window size and its initial value is CW_{min}. Every time after an unsuccessful transmission, the CW value is doubled until it reaches CW_{max}. After a successful transmission or when the number of retransmission reaches the retry limit and thus the corresponding frame is dropped, the CW value is reset to CW_{min}. Note that a random backoff deferral should be done prior to the transmission of another frame after a successful transmission.

Recently, many researches on wireless LANs investigated the performance of IEEE 802.11 and they showed that the performance of the IEEE 802.11 is heavily dependent on the CW backoff parameters [4-9]. These works also showed that the performance is greatly dependent on the number of stations. In [9], Natkaniec et al. showed the dependency between MAC performance and CW parameters according to the number of contending stations. They also showed that the performance enhancement can be achieved by the selection of the optimal CW_{min} parameter, which depends on the number of contending stations. Bianchi et al. [4] showed that the CSMA/CA protocol suffers from several performance drawbacks and the throughput performance is strongly dependent on both the number of active stations and the total load offered to the network. In the work, the performance can be substantially enhanced if the exponential backoff mechanism is substituted by an adaptive contention adjustment mechanism, depending on the number of contending stations. In [6], Cali et al. derived the theoretical capacity limit of the p-persistent IEEE 802.11 MAC protocol, and they showed that the IEEE 802.11 standard operates very far from the theoretical throughput limit depending on the network configuration. In the work, they achieved throughput improvement toward the theoretical limit by adjusting the contention window size and by estimating the number of contending stations. However, these works considered only infrastructure WLAN environments and focused on the basic handshake (Data+Ack) of the DCF. Furthermore, they assumed an ideal channel condition that there are no hidden terminals. Thus, the research results are not directly applicable to ad hoc networks.

In this paper, we propose an adaptive contention window mechanism that dynamically selects the optimal backoff window using an estimation of the number of contending stations in wireless ad hoc networks using RTS/CTS access mode. In the proposed scheme, when there are small number of contending stations, a smaller CW_{min} value is selected, while a larger CW_{min} value is selected when there are larger number of contending stations. We show the performance of the proposed mechanism by simulation and we compare it with that of the IEEE 802.11 MAC protocol which uses a fixed CW_{min} value.

2 Proposed Mechanism

In this section, we propose a mechanism that dynamically calculates the optimal contention window based on an estimation of the number of contending stations in wireless ad-hoc networks where all stations use the DCF with RTS/CTS scheme.

Before we describe the proposed mechanism, we show by an experiment the effect of the number of contending stations and the contention window size on the throughput performance. Fig.1 shows our *NS-2* simulation result for the throughput performance by changing the number of contending stations in fully connected topologies. The network parameters used in this study are summarized in Table 1 of Section 3. We assume that there are 50 stations in the network and we measure the throughput when the numbers of source stations are 5, 10, and

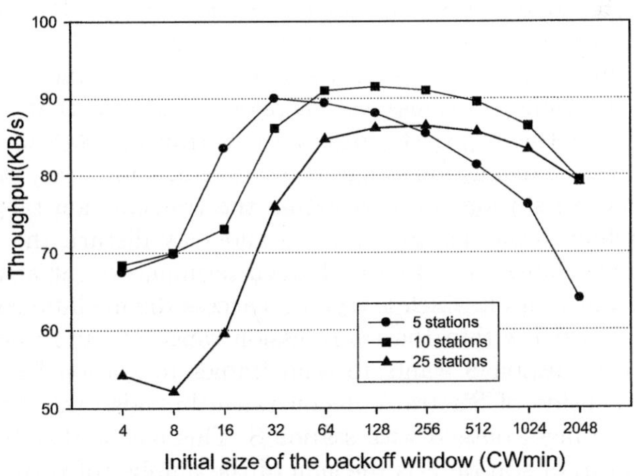

Fig. 1. Throughput versus initial contention window for the RTS/CTS mode

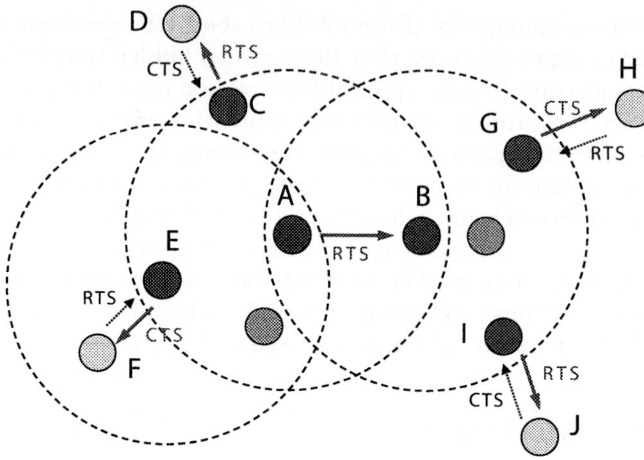

Fig. 2. Estimating the number of contending stations

25. Each source station randomly selects its destination station. Regardless of the number of source stations the network load is fixed to be 100 Kbytes/second. As shown in Fig.1, the throughput performance is highly dependent on both the number of contending stations and CW_{min}. According to Fig.1, the maximum throughput is achieved when CW_{min} is 32 for 5 contending stations, 128 for 10 contending stations, and 256 for 25 contending stations. Therefore, if we can estimate the number of contending stations, then we can select the optimum CW_{min} value which gives the maximum MAC throughput.

First, we describe the way to estimate the number of contending stations. In the RTS/CTS mode, the source station S transmits an RTS frame to the destination station D before transmitting data frames and station D replies with a CTS frame. A station that hears a control frame that is not destined to itself defers its frame transmission. Therefore, station S that has frames to send contends with other active stations (stations transmit RTS or CTS frames) located within the transmission range of station S. Moreover, station S also contends with active stations located within the transmission range of station D. They are hidden terminals to station S and may disturb the transmission of station S. Since collisions caused by hidden terminals occur at the receiving station, the sending station which is trying to access the medium competes with active stations located within the transmission range of the receiving station. As a result, when station S wants to send frames to station D, not only the stations located inside of S's transmission range but also the stations within D's transmission range contend with station S. This means that the number of contending stations of station S is the sum of the number of stations that send control frames within the transmission range of station S and the number of stations that send control frames within the transmission range of station D.

Fig.2 shows contending stations of station A when it has data to send to station B. In the figure, stations A, C, F, H, and I are senders and B, D, E, G, and J are respective receivers. Each of the sender or receiver contends for the medium in the contention period (during the backoff deferral period). Thus, the number of stations contending with station A is four (C, E, G, and I).

Now, we describe how the sender estimates the number of contending stations. The source station S can estimate the number of contending stations by overhearing RTS/CTS control frames or listening to beacon messages from its neighbors. When station S overhears RTS or CTS control frames from other stations, it can notice that those stations that have sent the control frames are active stations. Thus, by overhearing control frames, station S can determine the number of active stations within its transmission range. But, if there are collisions among control frames, the estimation can be inaccurate. More accurate estimation is possible by beacons. For this, we define two additional fields in the beacon frame. One field is one-bit active station flag. If this bit is set, it means that the station sending the beacon is a sender station or receiver station. In Fig.2, station A can know the number of active stations in its transmission range (i.e., 2) by hearing RTS by station C and CTS by station E, or by the beacons sent by stations C and E whose active station flag is set. The other field in the beacon frame is the active station count field. It is required by the sending station S to know the number of contending stations within the transmission range of the destination station D. Each station counts the number of active stations in its transmission range, and records it in the active station count field in its own beacon frame. Station S checks the active station count field in the beacon frame that is transmitted from its destination station D. Thus, the sending station S can estimate the number of contending stations in its transmission range by checking the active station flag in beacons from its neighbors and the number of contending stations within the transmission range of the destination station D by checking the active station count field in the beacon from station D.

In Fig. 2, station B estimates the number of contending stations within its transmission range (i.e., 2) either by hearing RTS by station I and CTS by station G, or by the beacons, in which the active station flag is set, sent by stations I and G. Thus, station B sends its beacon in which active station count field set to 2. Upon listening the beacon from station B, station A sums the value of the active station count field in the beacon from stations B (i.e., 2) and the number of contending nodes in its transmission range (i.e., 2). Now station A knows that the total number of contending stations is 4, and then it calculates the optimum contention window size. According to the related research results [4,5,7], the optimum minimal contention window is obtained by $W = n * \sqrt{2T}$, where n is the number of contending stations and T is the total frame transmission time. The analysis uses a discrete Markov chain that considers a fixed number of contending stations in the 802.11 DCF mode. If a collision occurs, it will occur within a transmission period of RTS and CTS frame, not data frames. Thus, for the RTS/CTS mode, T becomes the transmission time of control frames [4,5,10].

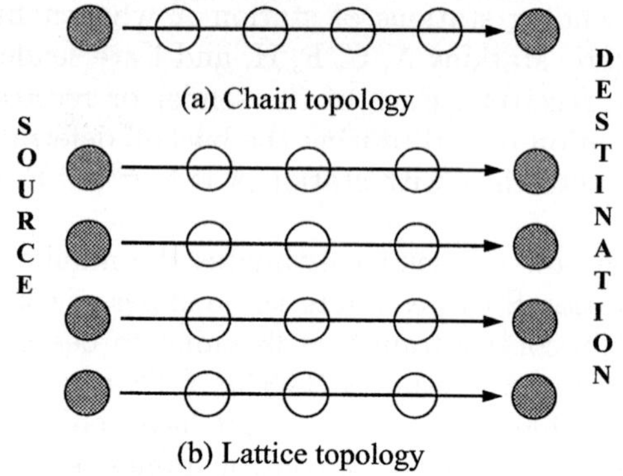

Fig. 3. Network topologies

Table 1. Simulation Parameters

Parameter	Value
SIFS	10 us
DIFS	50 us
Data rate	1 Mbps
Propagation delay	2 us
Packet payload	8184 bits
CW_{min}	32
CW_{max}	1024

So we have $T = RTS + SIFS + \delta + CTS + SIFS + \delta$, where δ is the packet propagation delay.

3 Performance Evaluation

In this section, we evaluate the performance the proposed adaptive mechanism and compare it with the current window mechanism of the IEEE 802.11 MAC. We used the *NS-2* simulator and assumed that data frame size is 1024 bytes and the bandwidth is 1Mbps. The parameters used in our simulation study are summarized in Table 1. For simplicity, the channel is assumed to be error-free. For the IEEE 802.11 MAC, the values of CW_{min} and CW_{max} are set to 32 and 1024, respectively (i.e., m is 5 in $CW_{max} = 2^m CW_{min}$). But, according to our simulation study, we found that the proposed mechanism shows the best performance when m is 3. So, we used this value for the proposed mechanism.

Our simulation study is performed with three different topologies: chain, lattice, and random. The chain topology corresponds to a sparse contending environment where packets travel along the chain of intermediate nodes toward the destination. The lattice topology emulates a dense contending environment,

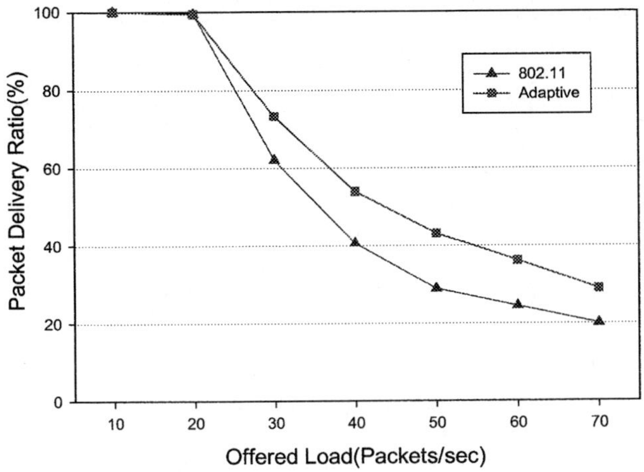

Fig. 4. Packet delivery ratio in the chain topology

and we assume that there are four source/destination pairs. In the chain and lattice topologies shown in Fig.3, we assume that the distance between two neighbor stations is 200 meters. In the random topology, we assume that 100 stations are randomly located in the area of 1000x1000 square meters. Among them, 50 source stations are randomly selected and each source station randomly selects its destination station from its neighbors. The total simulation time is 300 seconds.

Fig.4 shows the packet delivery ratio of the IEEE 802.11 MAC and the proposed contention window mechanism for the chain topology, as a function of offered load. The packet delivery ratio is the number of data packets received by the destinations divided by the number of data packets originated from the sources. As shown in Fig.4, the packet delivery ratio drastically decreases as the offered load increases. Compared to the IEEE 802.11 MAC, the proposed mechanism shows better packet delivery performance. It is due to the fact that there are very few contending stations in the chain topology, and thus the proposed mechanism selects smaller CW_{min} value (ranging between 15 and 30) adaptively, while the IEEE 802.11 MAC uses the fixed larger CW_{min} value (32), which increases idle time, and thus degrades the throughput performance.

Fig.5 shows the packet delivery ratio of the IEEE 802.11 MAC and the proposed mechanism for the lattice topology. As shown in the figure, the proposed adaptive mechanism also shows better performance. In the lattice topology, the number of contending stations is increased compared to the chain topology, and thus larger CW_{min} value is required to reduce the number of collisions. While the proposed mechanism adaptively increases the CW_{min} value (ranging between 40 and 100), the 802.11 MAC keeps the fixed CW_{min} value (32), which increases the number of collisions, and thus the packet delivery ratio (throughput) is lowered.

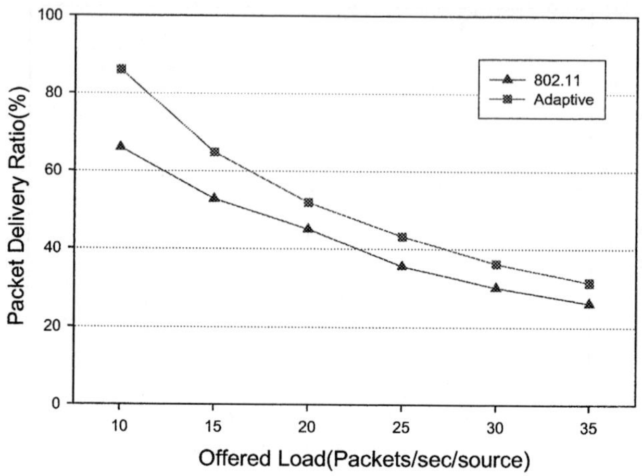

Fig. 5. Packet delivery ratio in the lattice topology

Fig.6 shows the packet delivery ratio for the random topology. As shown in the figure, the proposed mechanism also shows better performance. As discussed above, the proposed scheme adaptively selects the optimum CW_{min} value according to the number of contending stations, which decreases the number of collisions and improves the performance. Fig.7 shows the number of collisions of control frames per source station in the random topology during the total simulation time. As shown in Fig. 7, the IEEE 802.11 MAC shows much more (about 6 times) collisions than the proposed mechanism. Fig.8 shows the delivery delay distribution of the packets (data frames) sent by all source stations when each source station transmits 6 packets per second. The delay is defined to be the elapsed time from the time a packet is transmitted by the MAC layer of the sender to the time the packet is delivered to that of the receiver. About 90% of packets are delivered within 20 *ms* in the proposed mechanism, while about 60% of transmitted packets are delivered within 20 *ms* in the IEEE 802.11 MAC.

4 Conclusion

The performance of the IEEE 802.11 DCF protocol using the RTS/CTS handshake is strongly dependent on the number of active stations in wireless ad hoc networks. In this paper, we proposed an adaptive contention window mechanism that dynamically adapts the optimal backoff window using an estimation of the number of contending stations. In the proposed mechanism, a smaller CW_{min} value is selected when there are small number of contending stations, while a larger CW_{min} value is selected when there are large number of contending stations. We studied the performance of the proposed mechanism by simulation and we got an improved performance over the IEEE 802.11 MAC.

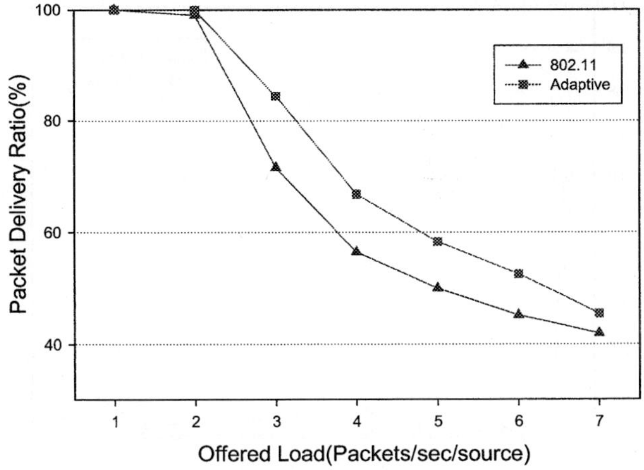

Fig. 6. Packet delivery ratio in the random topology

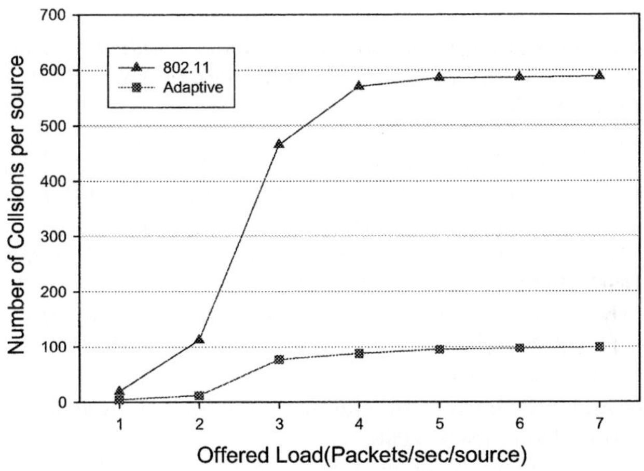

Fig. 7. The number of collisions in the random topology (per source station)

References

1. IEEE 802.11, *Part 11:Wireless LAN Medium Access Control (MAC) and Physical Layer (PHY) Specifications*, Standard, IEEE, Jun. 1997.
2. B.P. Cro and J.G. Kim, "*IEEE 802.11 Wireless Local Area Networks*," IEEE Communications magazine, Dec. 1999.
3. P. Karn, "*MACA - A new Channel Access Method for Packet Ratio*," Proc. ARRL/CRRL Amateur Radio 9th Computer Networking Conference, pp. 134-140, April 1990.

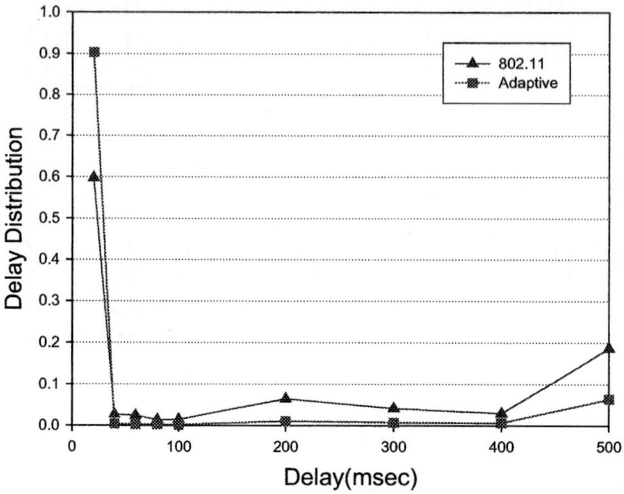

Fig. 8. Delay distribution in the random topology

4. G. Bianchi, L. Fratta, and M. Oliveri, "*Performance Evaluation and Enhancement of the CSMA/CA MAC Protocol for 802.11 Wireless LANs*," Proc. IEEE PIMRC, pp. 392-396, Oct. 1996.
5. Y.C. Tay and K.C. Chua, "*A Capacity Analysis for the IEEE 802.11 MAC Protocol*," ACM/Baltzer Wireless Networks, vol. 7, no. 2, pp. 159-171, Mar. 2001.
6. F. Cali, M. Conti, and E. Gregori, "*IEEE 802.11 Wireless LAN: Capacity Analysis and Protocol Enhancement*," Proc. INFOCOM, Mar. 1998.
7. G. Bianchi, "*Performance Analysis of the IEEE 802.11 Distributed Coordination Function*," IEEE JSAC, vol. 18, Mar. 2000.
8. F. Cali, M. Conti, and E. Gregori, "*Dynamic Tuning of the IEEE 802.11 Protocol to Achieve a Theoretical Throughput Limit*," IEEE/ACM Transactions on Networking, vol. 8, no. 6, Dec. 2000.
9. M. Natkaniec and A.R. Pach, "*An analysis of the backoff mechanism used in IEEE 802.11 networks*," Proc. ISCC 2000.
10. H. Wu, S. Cheng, Y. Peng, K. Long, J. Ma, "*IEEE 802.11 Distributed Coordination Function(DCF) : Analysis and Enhancement*," Proc. ICC, pp. 605-609, 2002.

Experiments on the Energy Saving and Performance Effects of IEEE 802.11 Power Saving Mode (PSM)

Do Han Kwon[1], Sung Soo Kim[1], Chang Yun Park[1], and Chung Il Jung[2]

[1] Dept. of Computer Science and Engineering, Chung-Ang University, Korea
{dohan71, sungsoo}@wm.cau.ac.kr, cypark@cau.ac.kr
Phone: +82–2–816–8757, FAX: +82–2–820–5301

[2] Dept. of Computer Science, Yeojoo Institute of Technology, Korea
cijung@yeojoo.ac.kr

Abstract. This paper presents experimental results for the actual effects of the IEEE 802.11 power saving mode (PSM) on communication performance and energy saving. First, we have measured the throughput and response time of a station working in PSM with various applications and compared them to those of the active mode. Energy consumptions have also been compared by analyzing trace data. Second, the effects of a PSM station to the neighbor stations have been investigated by measuring their performance changes. The experiments show that the amount of effects varies depending on application types due to traffic burstness and congestion control of underlying transport protocols. It may happen that a PSM station running a streaming application, somewhat abnormally, does harm to the performance of the neighbor station. Finally, a primitive solution against this abnormality is presented and experimented. This study could provide a good basis for further studies on utilizing PSM.

1 Introduction

In the last few years the cost of devices confirming to the IEEE 802.11 standard dropped down significantly and these devices are commonly used for the wireless LAN. The IEEE 802.11 was designed for the mobile device such as notebooks and PDAs, which usually rely on limited powers source such as batteries. Therefore, IEEE 802.11 standard should comply with low power communications.

IEEE 802.11 provides power saving mode (PSM) for using power efficiently. In the PSM, the network interface goes into a sleep state and thus its power consumption is significantly decreased. PSM itself is well-known but its utilization is relatively little addressed. Especially, there exists no study, to our best knowledge, about when a station adopts PSM. Two conditions generally accepted are (1) when a station falls into a low power state and (2) when a station has little traffic. However, if there exists a trade-off between power and traffic, for example, if a communication task should be done under a limited power resource, it

C. Kim (Ed.): ICOIN 2005, LNCS 3391, pp. 41–51, 2005.

is a difficult question whether PSM is a good choice. To address this issue, the performance and energy characteristics of PSM should be first investigated.

It is generally believed that PSM is good for energy but harm for performance. However, if a station set in PSM involves in receiving frames very actively, its positive effect on energy saving might be questionable due to control overhead for polling. On the other hand, an application requires low throughput and is not so sensitive to delay (for example, paging), the negative effect on performance of PSM might be negligible.

The other question explored in the study is how much effect a station working on PSM has on the performances of other stations in the same basic service set. General expectation is that what a station goes into PSM is good for other stations with respect to performance because there will be more chances to access the shared link. This will be true as long as a PSM station is in the sleep state, but if the station wakes up and is busy for handling buffered frames other stations may get less share of the link.

This paper concerns the above two questions in a practical approach. Experiments have been performed with real applications in a real 802.11b infrastructure network; throughputs and delays were measured and energy consumptions were calculated based on traces by a sniffer.

With FTP application, a PSM station shows a worse throughput than a active mode station and the decrease is deeper than expected. It can be interpreted that inactivity during sleep periods triggers congestion control of TCP, which also limits the transmission rate at the transport level. With Ping application, the response time can be increased up to one cycle time, i.e., one active period plus one sleep period (200 *msec* in a typical setup). Interestedly, there is a relatively small throughput drop with a MPEG streaming application. It is understood that burstness of MPEG data and no congestion control at the transport (i.e., UDP) caused the difference. A given file transfer task is performed in PSM and active mode, respectively, and their energy consumptions are compared each other. PSM gives better energy saving even though traffic is heavy.

Regarding the effects to other stations, if a PSM station performs FTP, others get a positive effect as expected; in other words, their throughput increase. However, if a PSM station does a streaming application, other stations cannot improve their performances but suffer some loss of their share. The sources of this abnormality are explored and some primitive solutions are presented.

The rest of paper is organized as follows. Section 2 describes the overview of the IEEE 802.11 PSM and related works. Section 3 shows the experiment results on the performance and energy effects of PSM to the station itself. In Section 4 PSM's effect to the neighbor station is experimented and a simple buffering/queueing method at AP is presented to cut off the negative effect. Finally, we conclude the paper with future researches.

2 Overview of the IEEE 802.11 PSM and Related Works

IEEE 802.11 was designed for mobile devices which cannot be constrained by a power cord and usually rely on an internal battery. In the IEEE 802.11, a station can be in one of two power modes: the active mode and power saving (PS) mode[1]. In the active mode a station can receive frames at any time and in power saving mode a station is mainly low-power state and receives frames through buffering at the access point (AP).

A PSM station is synchronized to wake up periodically to listen to Beacons, typically at every other 100 *msec*. If a frame destined for a PSM station arrives at the access point, the frame must be buffered. A PSM station must wakes up and enters the active mode to listen for Beacon frames which includes the traffic indication map (TIM). The TIM indicates the existence of buffered frames by setting bits of corresponding stations. To retrieve buffered frames, stations use PS-Poll frames. Fig. 1 illustrates the process.

Each PS-Poll frame is used to retrieve one buffered frame. That frame must be positively acknowledged before it is removed from the buffer. If multiple frames are buffered for a station, then the More Data bit is set to 1. Stations can then issue additional PS-Poll requests to the AP until the More Data bit is set to 0.

After transmitting the PS-Poll, a stations must remain awake until either the polling transaction has concluded or the bit announcing buffered frame exists is no longer set in the TIM. Once all the traffic buffered for a station is delivered or discarded, the station can resume sleeping.

Many studies have concerned how to lower the energy consumption of mobile devices and have focused on developing low-power communication techniques[2][3][4][5][6][7]. Tseng and et al. proposed three protocols, which could save power with neighbor discovery time[2]. Hsu and et al. presented a new data rate selection protocol and an efficient transmission scheduling protocol for a single-hop mobile ad-hoc network with some PSM stations[3]. Khacharoen proposed a PSM mechanism for ad-hoc network[4]. They introduced a new Beacon interval structure and technique to reduce energy consumption.

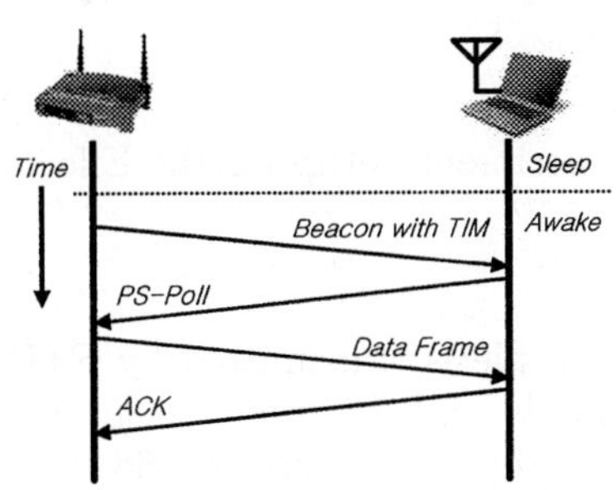

Fig. 1. PSM Operation

Recently, Zheng and et al. presented an analytic characterization of PSM, which investigated energy consumption of PSM in independent (i.e., ad-hoc mode) networks as a function of the traffic load and buffer size in an analytic way[8]. It also includes some results on delay and loss rate of PSM. Our study is different from it in that we have measured real performances in an infrastructure network and also considered the effects of the application types.

R. Krashinsky investigated that PSM's effect to performance (RTT) by simulation and proposed a *Bounded Slowdown* (*BSD*) protocol, which dynamically adapts PSM sleep interval[9].

In summary, most studies on PSM have focused on extension of PSM mechanism such as setting a sleep interval dynamically and with finer granularity. The results are based on analytic or simulation models. Our study addresses real effects of PSM as it is, and the results are based on measurements and traces which reflect all affecting factors in the environment in an integrated way.

3 Performance and Energy Saving of PSM

3.1 Performance Effects of PSM

The experiment setup consists of two server, an access point and two mobile stations (Fig. 2). The access point and two stations are associated with 11 Mbps data rate and a sniffer node is also applied for capturing the wireless traffic. STA2 is doing Ping throughout experiments for traffic generation. STA1 is the target of experiment and its sleep interval in PSM is set to 100 *msec*. We use three representative protocols which generate characteristic traffic for the experiments. The results are shown as follows.

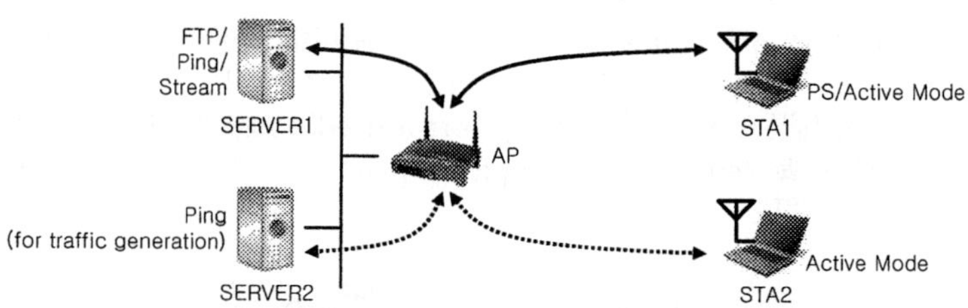

Fig. 2. Experiment Setup for the Effects of PSM

FTP. In the experiment, a file is transmitted by FTP from the SERVER1 to the STA1 in the active mode and PS mode, respectively. Table 1 shows their performance and the throughput decreases to 68 percents when a stations is working on the PSM.

As expected, the throughput while a node is in the PS mode is lower than on the active mode. But a decrease of 68 percents is beyond expectations. The

Table 1. Throughput of FTP (file size : 100 MB)

PS mode	active mode	(PS – active) / active
123 KB/s	384 KB/s	−68 %

worse result is explained that the TCP's congestion control is operated, which is caused by delay at the AP, and total transmission rate is decreased. That is, TCP throttles down data rate because it guesses the frames are lost due to congestion.

Ping. In the experiment, we have measured response times while changing Beacon interval. Table 2 shows the results including the average and the maximum. The results are calculated from 100 response times of Ping request.

Table 2. Response Time of Ping at Various Beacon Intervals

Beacon Interval	PS mode		active mode
	Average (ms)	Max (ms)	Average (ms)
50 ms	31	94	10
100 ms	59	188	10
200 ms	135	407	10

In the experiment setup, the delay from the server to the AP is small, almost constant, and thus negligible. As Beacon interval becomes longer, the average response time grows. The maximum response time is approximately twice Beacon interval. When a Ping request arrives at the access point shortly after it sent a Beacon frame with no pending frames, the destination station immediately sleeps and later receives the Ping request at the next awake period. That is the worst case. Assuming the response is immediate, the delay is two Beacon interval in our experiment setup.

MPEG Stream. We have experimented with a streaming application which uses UDP as a transport protocol. The performance of a station in PSM drops as shown in Table 3. However, compared to the case of FTP, this results are not so bad. The results are explained by two reasons. The one is that streaming data do not go through any congestion control, and therefore the SERVER1 sends data at the same rate as the active mode case. The other is the burstness of

Table 3. Throughput of MPEG Stream (file size : 80 MB)

PS mode	active mode	(PS – active) / active
65 KB/s	73 KB/s	−11 %

streaming data caused by compression. If no data arrives for a sleeping period, it does no harm to the throughput. If a burst of data arrives, they are buffered and delivered for the next awake period. We will explain this later in more detail with Fig. 5.

3.2 Energy Saving Effects at a PSM Station

It is generally believed that PSM is more energy-efficient in light traffics. Our experiment is targeted on a heavy traffic case; a long file is transferred from a server to a station, which generates a lot of frames continuously for a long period time. The energy consumptions in the active mode and PSM are estimated, respectively, and they are compared.

To figure out the amount of energy consumed for a file transfer, we should first know how much power a NIC consumes in each state. Table 4 shows power consumption of NIC that we used (Cisco AIR-PCM350) at 11 Mbps[10].

Table 4. Power Consumption at the NIC operation

transmit state	receiving state	idle state	sleep state
1875 mW	1300 mW	1080 mW	45 mW

The next step is to analyze how long the NIC works (or stays) in each state while the file is transferred. We captured the trace of frame activities for the file transfer by using a sniffer, and calculated how many frames are transmitted/received and how long the NIC sleeps as shown in Table 5. The result is somewhat pessimistically approximated for the case of PSM because the NIC may go to the sleep state in the middle of the awake period if no more data is notified.

With the above ratio and duration, the total energy consumed for the file transfer could be simply calculated as a sum of energies consumed at each state. Table 5 shows the results for each of the normal (i.e, active) mode and PSM. They might not be precisely correct because other power factors, for example, processing power at CPU, are not considered. However, we believe that they are acceptable to address the question whether working in PSM is more energy-efficient.

From the results we could conclude that PSM gives energy saving for a given communication task. Although it takes much longer time, the NIC is mostly in the sleep state where much less power is consumed.

4 Performance Effects of PSM to the Neighbor Stations

4.1 Performance Effects of a Station Working in PSM to the Neighbor Station

In the section 4, we have experimented with the effects of a PSM station to other stations working in the active mode in the same basic service set. The experiment

Table 5. Power Consumed for A File Transfer (file size : 10 MB)

	transmitted data	received data	idle state (ms)	sleep state (ms)	power consumption (J)
PS mode	266	7819	850	31720	21.23
active mode	533	8684	3780	0	24.67

setup is similar to Fig. 2 but since our concern in this section is the effects to the neighbor stations the performances at STA2 are measured and compared. The throughput of FTP is used as a performance measure. It is illustrated in Fig. 3.

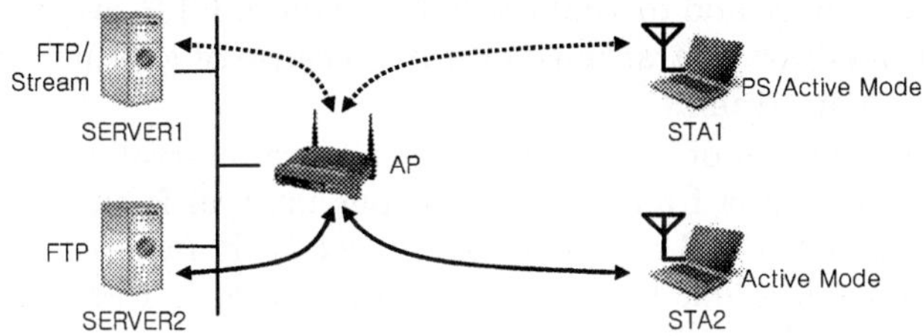

Fig. 3. Experiment Setup for Measuring the Effects of PSM to A Neighbor Station

FTP. When STA1 performs FTP in PSM, the throughput of STA2 is increased as expected (Table 6). What a station is working in PSM is good to the neighbor stations with respect of performance because there will be more chances to access the shared link while the PSM station is sleeping.

Table 6. Performance Difference at A Neighbor Station While Another Station Is Doing FTP in PSM

PS mode	active mode	(PS – active) / active
383 KB/s	261 KB/s	+46 %

MPEG Stream. In experiments with a streaming application, we have had contrary results. Table 7 shows that the performance of a neighbor station is worse when a station works in the PS mode than when it works in the active mode. This is not a result we expected.

We explored how a streaming application generates traffic. The graph of STA1 in Fig. 4 shows a trace of a streaming traffic processed in the experiment setup, where PSM is not applied. The throughput of a neighbor station is also showed as STA2. MPEG stream shapes bursty transmission where no traffic exists for some periods. The throughput of STA2 drops for a moment while

Table 7. Performance Difference at A Neighbor Station While Another Station Is Doing MPEG Streaming in PSM

PS mode	active mode	(PS – active) / active
271 KB/s	305 KB/s	−11 %

STA1 receives MPEG stream, but it is recovered while STA1 does not use the shared link.

Fig. 5 shows the traces when STA1 works in PS mode and STA2 in active mode. In this case, MPEG stream traffic shapes small and steady transmission. Since the burst packets sent to STA1 are buffered, the AP performs continuous operations to announce and to send the buffered data. FTP, eventually TCP, has less chances to increase its transmission rate because of continuous competition with buffered stream traffics.

Counting the number of frames processed in each interval, we have found that the average of total data frames processed per interval is smaller with a PSM station than without a PSM station(20.5 vs. 24.2). It means that the effective bandwidth of a wireless link decreases if a station works in PSM. It is understood that extra overhead for polling makes the difference.

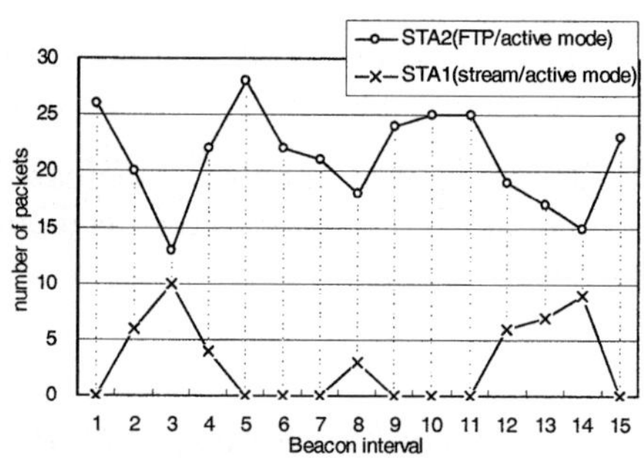

Fig. 4. A Trace of Traffic Processing in the Link without PSM

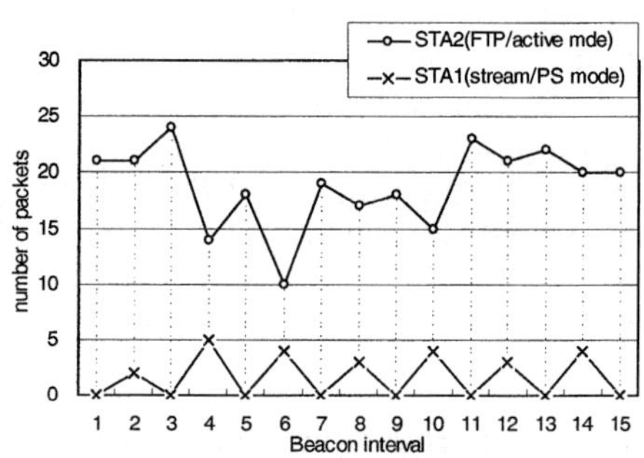

Fig. 5. A Trace of Traffic Processing in the Link with PSM

4.2 Primitive Solutions Against the Negative Effect of PSM

We believe that adopting PSM at a station is a decision for the sake of the station but it should not have any negative effects on the neighbor stations. However, the experiments showed that abnormality might happen depending on traffic behaviors. In this section, we would propose two simple approaches that can prevent or reduce the negative effects of PSM and show some experiment results.

Since we could not modify a commercial AP module for experiments, we have used the *hostap* Linux driver[11] that supports a so called Host AP mode. It takes care of IEEE 802.11 management functions in the host computer and acts as an AP.

First, we have changed the size of the PS buffer, in which the AP stores frames destined to PSM station, from 32 frames to 8 frames per station. The results are shown in Table 8. If the PS buffer size is small, some of bulk data sent to a PSM station will be discarded and the overhead for retrieving buffered frames in an interval can be reduced. Hence, this approach can improve the performance of the neighbor station. However, this solution may give unfair penalty to a PSM station, especially when the traffic in the link is not heavy.

Table 8. Reducing the PSM Buffer at AP and Its Effect

	active mode	PS mode	
buffer size	32 frames	32 frames	8 frames
throughput	446 KB/s	391.2 KB/s	407.2 KB/s
(PS – active) / active		−12%	−8.6 %

The other approach is that AP insert some extra delay between receiving a PS-Poll request and transmitting the buffered frame. This delay is not always applied but only if the number of buffered data is above the threshold which may indicate that handling a burst of buffered frames will happen. To do this, we have modified the part of socket buffer queueing in the *hostap* driver. The basic queueing algorithm is FIFO on the IEEE 802.11 PSM and we have modified FIFO to insert 50 *msec*, the half of Beacon interval, delay. Table 9 shows the results of this approach. Compared to FIFO, the negative effect is reduced to less than the half.

Table 9. Inserting Delay in Queueing A Polled Frame

	active mode	PS mode	
		FIFO	modified FIFO
throughput	446 KB/s	391.2 KB/s	426 KB/s
(PS – active) / active		−12%	−4.5 %

Both approaches described above are not a complete solution but just show some possibility. They also have some fairness problems when traffic in a link is not heavy. A better solution would be applying weighted fair queueing (WFQ) for the PSM buffer. It can separate traffics to a PSM station from traffics to the neighbor stations, and thus the effects of PSM can be isolated.

5 Conclusion and Future Works

IEEE 802.11 PSM is the most well-known technique in wireless LAN to save energy consumption for communication but its utilization is relatively little addressed. We have concerned how much power consumption is saved when the device works in PSM. Secondly, we have investigated the effects of performance to the PSM station itself and to the neighbor station.

Our experiments have showed that PSM is effective in saving energy even in a heavy and continuous traffic situation. The throughput of FTP and the response time of Ping becomes worse in PSM than in the normal active mode, as expected. However, it is interesting that a MPEG streaming application, which generate burst traffics without congestion control, suffers little performance decrease in PSM. We have also found that if a PSM station receives frames actively, its polling overhead reduces the effective bandwidth of the link and has negative effects to the performance of the neighbor stations.

For further studies, more experiments are required to figure out performance characteristics of PSM; changing the sleep interval and experimenting with Web application will be performed next. For a better solution against the negative effects of PSM, implementing WFQ mechanism at AP will also be tested. Finally, we believe that it would be helpful to subdivide the semantics of PS-Polling; if "poll all frames" is defined in addition to the current "poll one frame", it will improve the case a PSM station is involved in a busy receiving session.

Acknowledgement This Research was supported in part by the Chung-Ang University Research Grants in 2002 (#20020018).

References

1. IEEE standard for Wireless LAN-Medium Access Control and Physical Layer Specification, 802.11, November 1997.
2. Y. Tseng, C. Hsu and T. Hsieh, "Power-Saving Protocols for IEEE 802.11-Based Multi-Hop Ad Hoc Networks", *Proceeding of IEEE Infocom '02*, pp.200–209, 2002.
3. C. Hsu, J. Sheu and Y. Tseng, "Minimize Waiting Time and Conserve Energy by Scheduling Transmissions in IEEE 802.11-based Ad Hoc Networks", *Int'l Conf. on Telecommunications*, 2003.
4. T. Khacharoen and A. Phonphoem, "A Power Saving Mechanism in Ad Hoc Network with Quality of Service Support", *Proceeding of ICT 2003*, pp.119–124, 2003.
5. J. Chang and L. Tassiulas, "Energy Conserving Routing in Wireless Ad-hoc Networks", *Proceddings of IEEE Infocom '00*, pp.22–31, 2000.
6. B. Chen, K. Jamieson, H. Balakrishnan and R. Morris, "Span: An Energy-Efficient Coordination Algorithm for Topology Maintenance in Ad Hoc Wireless Network", *ACM Wireless Networks Journal*, Volume 8, Number 5, September 2002.
7. Y. Xu, J. Heidemann and D. Estrin. "Geography-informed Energy Conservation for Ad Hoc Routing", *Proceeding of MobiCom '01*, pp.70–84, 2001.
8. R. Zheng, J. Hou and L. Sha, "Performance Analysis of the IEEE 802.11 Power Saving Mode", *Proceeding of CNDS '04*, 2004.

9. R. Krashisky and H. Balakrishnan, "Minimizing Energy for Wireless Web Access with Bounded Slowdown", *Proceeding of MobiCom '02*, pp.119–130, 2002.
10. E. Shih, P. Bahl and M. Sinclair, "Wake on Wireless: An Event Driven Energy Saving Strategy for Battery Operated Devices", *Proceeding of MobiCom '02*, pp.160–171, 2002.
11. J. Malinen, "Host AP driver for Intersil Prism2/2.5/3 and WPA Supplicant", *http://hostap.epitest.fi*

A High-Performance Network Monitoring Platform for Intrusion Detection

Yang Wu and Xiao-Chun Yun

Computer Network and Information Security Technique Research Center,
Harbin Institute of Technology, Harbin 150001,China
{yangwu, yxc}@pact518.hit.edu.cn

Abstract. This paper presents and implements a high-performance network monitoring platform (HPNMP) for high bandwidth network intrusion detection system (NIDS). The traffic load on a single machine is heavily reduced in an operation mode of parallel cluster. An efficient user-level messaging mechanism is implemented and a multi-rule packet filter is built at user layer. The results of experiments indicate that HPNMP is capable of reducing the using rate of CPU while improving the efficiency of data collection in NIDS so as to save much more system resources for complex data analysis in NIDS. ...

1 Introduction

Network intrusion detection systems (NIDS) are becoming a research hotspot in the fields of network security. Effective intrusion detection requires significant computational resources: widely deployed systems such as snort [1] need to match packet headers and payloads against tens of header rules and often many hundreds of attack signatures. This task is much more expensive than the typical header processing performed by firewalls. So performing effective intrusion detection in high-speed network requires further improvement on performance of data collection and data analysis in NIDS.

This paper focuses on the process of data collection in NIDS and presents a scalable high-performance network-monitoring platform (HPNMP) for NIDS. In HPNMP, multiple node machines operate in parallel, fed by a suitable traffic splitter element to meet the requirement of different network bandwidth. An efficient user-level messaging mechanism (ULMM) and a user-level packet filter (ULPF) are presented and implemented in order to improve packet capture performance and packet processing efficiency on a single machine.

2 Related Work

With the increasing network bandwidth in recent years, it is becoming very important to study and solve related problems about improving data processing performance of high-speed network intrusion detection systems.

C. Kim (Ed.): ICOIN 2005, LNCS 3391, pp. 52–61, 2005.

The traditional endpoint packet capture systems generally use Libpcap (Library of Packet Capture) [2] which is based on in-kernel TCP/IP protocol stack, but the slow network fabrics and the presence of the OS in the critical path (e.g. the system call, in-kernel protocol stack, interrupt handling and data copies) are the main bottlenecks on every packet sending and receiving. Therefore, inefficient Libpcap cannot adapt to the environment of heavy traffic network.

Libpacket [3] reduces system overhead of context switch by saving certain numbers of packets in the allocated kernel buffer and reading multiple packets in a single system call. In essence, the layered structure of user-kernel in Libpacket doesn't remove the kernel from the critical path of data transfer. The main performance bottlenecks are still in existence.

To eliminate main performance bottlenecks during communication completely, zero-copy protocol architectures for cluster systems were presented, including U-Net/MM [4], VIA [5] and VMMC [6]. These architectures adopt flat structure of user-hardware, fully bypassing in-kernel protocol stack in OS and allowing applications direct access to the network interface. The Virtual Interface Architecture (VIA) is connected oriented: each VI instance (VI) is specially connected to another VI and thus can only send to and receive from its connected VI. The U-Net/MM and VMMC architectures integrate a partial virtual address translation look-aside buffer into the network interface (NI) and allow network buffer pages to be pinned and unpinned dynamically, coordinate its operation with the operating system's virtual memory subsystem in case of a TLB miss. This definitely increases the implementation complexity of network card firmware and causes the great overhead because of frequent pinning/unpinning buffer and requesting pages. In addition, VMMC commonly requires customized high-speed interconnection network and NI hardware. Whereas, the NIDS passively monitor the TCP/IP network. So the above communication architectures are not suitable to high-speed network intrusion detection systems.

The traditional packet filter such as BPF [7] commonly is implemented in OS kernel. When the received packets are not interesting to the user application, BPF will drop these packets so as to save system overhead for copying them from kernel to application. BPF matches packets with multiple filter rules one by one in checking packets, thus BPF's processing efficiency is low when the number of rules is many.

3 Architecture of HPNMP

At high-speed network monitoring spot, data processing performance of a single machine may reach threshold so as not to meet the need of real-time intrusion analysis. We make use of load balance technique to build a scalable parallel network monitoring architecture, which adopts computing mode of SPMD to extend or shrink by network traffic. The model of scalable network monitoring platform is shown in Fig. 1, which consists of three parts: IDS load balancer which adopts a load balance algorithm mainly based on connection round robin; packet transfer module which uses user-level messaging mechanism; packet filter module which is a user-level multi-rule packet filter.

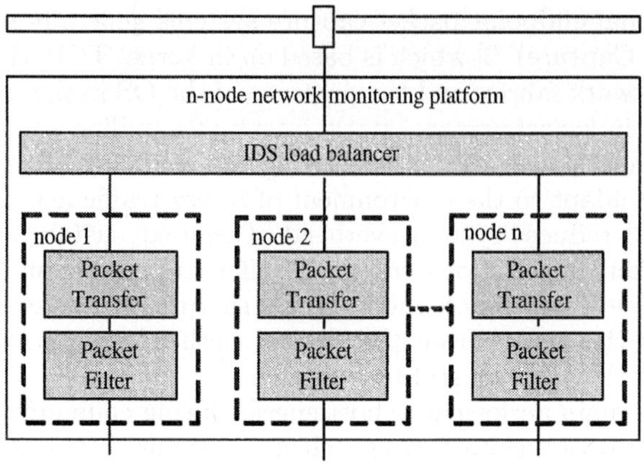

Fig. 1. Scalable network monitoring platform model

3.1 Load Balance Algorithm Based on Connection Round Robin

Since load balancer commonly needs processing a large number of network packets, it adopts a simple efficient load-balancing algorithm for data splitting. The majority of network packets are based on TCP protocol in current heavy traffic backbone network. For network traffic of TCP connection, an ideal load balance algorithm should meet following requirements: 1) data is almost evenly split into every node to ensure approximate load balance among node machines; 2) bidirectional data of any TCP connection is split into single node to ensure data independence among node machines.

```
Given:  N  is the number of node machines;
        m  is one-node machine number allocated recently;
        A  is one-node machine number obtained currently;
Initialize m=1;
For every packet p of TCP protocol {
  If p is the first packet of a connection (SYN packet)
      the obtained entry address A=m mod N+1;
      split p into node machine A;
      record four-tuple of this new connection and A in HASH table;
      m=A;
  Else
      look up HASH table to find entry address A;
      split p into node machine A;
      if p is the last packet of a connection or connection is overtime
        remove this connection record from HASH table;
}
```

Fig. 2. Network traffic load balance algorithm based on connection round robin

For the TCP protocol, a four-tuple with the form of <*source IP, destination IP, source port, destination port*> uniquely defines a connection. The connection round robin scheduling algorithm is described in Fig. 2.

For other protocol type (e.g. ICMP, UDP), the entry address may be computed by simple and direct hashing of two-tuple <*source IP, destination IP*>. The formula is as follows:

$$destination\,node\,number = (source\,IP \oplus destination\,IP)\,mod\,N \quad (1)$$

3.2 Efficient User-Level Messaging Mechanism-ULMM

For improving packet-processing performance of one-node machine and reducing cost of hardware resource, a zero-copy based user-level messaging mechanism (ULMM) for intrusion detection is presented. In ULMM, the OS kernel is removed from the critical path of data communication, thus messages can be transferred directly to and from user-space applications by the network interface without any intermediate steps.

ULMM eliminates the system overhead of dynamic allocating/releasing buffer and pinning/unpinning buffer by allocating a continuous static user-space buffer and pinning corresponding physical memory pages. Caching the whole virtual-to-physical address table removes the overhead from address translation operation of the operating system's virtual memory subsystem in case of partial virtual-to-physical address table miss in U-Net/MM.

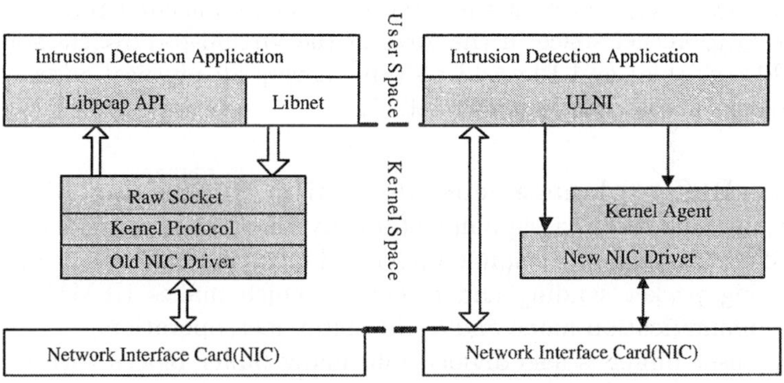

(a) Message Transfer Model in Traditional NIDS (b) Message Transfer Model in ULMM

Fig. 3. Comparison of two message transfer models

Message transfer model of ULMM is compared with that in traditional NIDS, which is shown in Fig. 3. Fig. 3 (a) shows that Libpcap includes four gray modules: Libpcap API, Raw Socket, Kernel Protocol and Old NIC Driver. Fig. 3 (b) clearly shows the architecture of ULMM in three gray modules: Kernel Agent (K-Agent), New NIC Driver, and User-Level Network Interface (ULNI). The block arrows describe the data flow, while the line arrows describe the control flow. The module of ULNI provides the API library for the application. Kernel Agent and New NIC Driver do all the real work to copy the data from NIC memory to user process's memory. Specifically, Kernel Agent obtains the physical addresses of the application's memory range and then creates the buffer ring. This buffer ring holds all the packets copied from NIC waiting to be filtered by the packet filter in Sect. 3.3. And the New NIC Driver asks Kernel Agent for the physical address table of this buffer ring which is used by asynchronous DMA of NIC and initiates DMA to transfer packets between the application's buffer and the on-chip memory of NIC. In comparison with the traditional message transfer model in traditional NIDS, ULMM greatly improves the packet capture performance, which makes packet capture more practical in high-speed network.

Translation Mechanism for Virtual Address in User Space. ULMM directly transfers packets between the application's buffer and the on-chip memory of NIC by asynchronous DMA. Since the DMA facility only accesses the physical memory address, whereas application uses virtual address, one main difficulty in designing ULMM is translation between virtual address of user buffer and physical address accessed by DMA.

In ULMM, The user application statically allocates a continuous user-space memory as message buffer and coordinates with K-Agent to inform it about the starting virtual address and size of user buffer through API library provided by ULNI. Linux kernel currently uses a three level page table. K-Agent completes translation from virtual address to physical address and pins physical pages by using related kernel functions. The translated physical addresses are cached in kernel space in the form of the virtual-to-physical address table (PhyAddressTable) and PhyAddressTable covers all physical addresses of user buffer blocks accessed by network interface.

Message Buffer Management Mechanism Supporting Multi-thread. ULMM saves packets in a big buffer statically allocated in user space. The whole user buffer is divided into sending buffer and receiving buffer that are separately used during packet sending and receiving, which makes ULMM support full duplex communication mode and avoid mutex lock operation.

Every user buffer is also divided into many buffer blocks with size of 2KB, each of which is for saving a network packet. For supporting application's multi-thread access to message buffer without data copies on SMP machine, K-Agent allocates four buffer rings in kernel space to manage the user buffer: sending busy ring (SendQ), receiving busy ring (RecvQ), sending free ring (SFreeQ) and receiving free ring (RFreeQ), each of which includes descriptor items for every buffer block. Each item structure consists of two fields $<index, size>$: 1. *Index* corresponds to block number; 2. *Size* corresponds to size of packet in data block. ULMM maps four kernel rings into user space by calling mmap function provided by memory mapping mechanism in Linux, so as to make user process and kernel module share buffer management rings.

Efficient Packet Sending and Receiving Process. Message transfer process in ULMM is shown in Fig. 4. Packet receiving process is $1 \rightarrow 2 \rightarrow 3 \rightarrow 4 \rightarrow 5 \rightarrow 6$, detailed description of which is as follows: When a new packet arrives, New NIC Driver gets item of a free data block from the head of RFreeQ ring and acquires pinned physical address of this free data block from physical address table (PhyAddressTable) according to block number (*index*) of the item, and then initializes asynchronous DMA to transfer packets. When DMA transfer is finished, an interrupt is generated. New NIC Driver puts the item of the data block just filled with packet at the tail of RecvQ ring in interrupt handler. When the application needs to process packets, application gets the item of a data block from the head of RecvQ ring and reads the packet in this buffer block accordingly. After application processes the packet, it puts item of the just used data block at the tail of RFreeQ ring. Reading or writing buffer ring is in blocking mode. In like manner, packet sending process is $(1) \rightarrow (2) \rightarrow (3) \rightarrow (4) \rightarrow (5) \rightarrow (6)$ in turn.

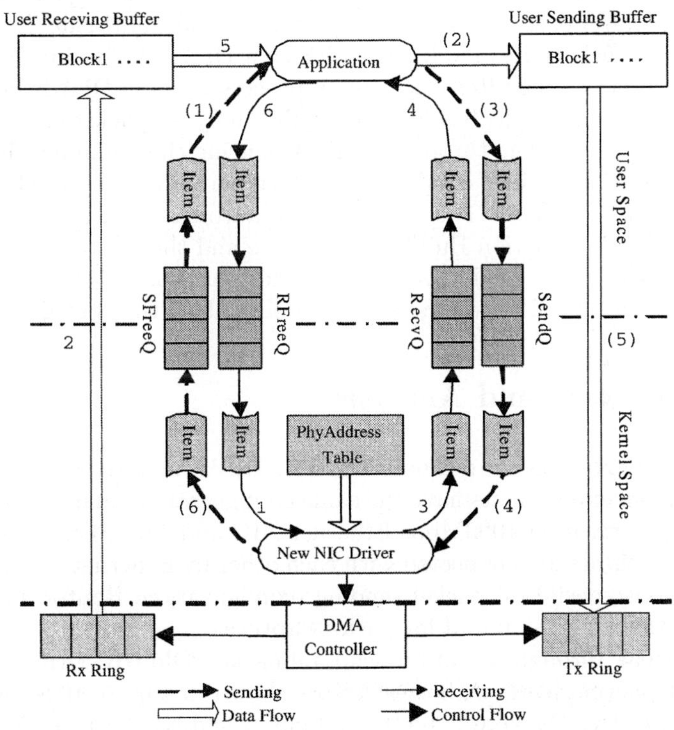

Fig. 4. Packet sending and receiving process in ULMM

3.3 Multi-rule Packet Filter Mechanism at the User Layer

The packet filter is a subsystem to reduce the volume of data to be analyzed by the security analysis module by removing non-interesting network packets, and at the same time protects the NIDS itself from hostile attacks such as DOS/DDOS. Based on ULMM, a multi-rule user-level packet filter (ULPF) is built. Different from the traditional packet filter such as BPF, ULPF has to be implemented at user layer to work with the zero-copy ULMM in Sect. 3.2.

ULPF uses a rule description language like that in [8] to define fields to be checked in the packet headers and the action followed once the packet satisfies a precondition. A precondition is composed of equations and logical operators. The filtering rules are the form of "packet | precondition → action". For example, a rule is defined as follows: packet (p) | (p.e_type = ETHER_IP) && (p.protocol != IP_TCP) && (p.protocol !=IP_UDP) && (p.protocol != IP_ICMP) && (p.protocol != IP_IGMP) → drop, means that Ethernet packet will be dropped if its protocol type of IP layer is unknown.

In practice, the packet filter often uses a large number of filter rules. To satisfy the requirements of both the performance and multi-rule filter, we build the multi-rule packet filter model in ULPF as a DFA (Deterministic Finite Automata). At first, all the filter rules are preprocessed and a DFA is built from all the equations, and then the header fields of the analyzed packet are scanned from the left to the right and go through the DFA. During the scanning, the unrelated fields are skipped as an idea of adaptive pattern matching presented in [9], which speeds up the packet filter. The Algorithm for automaton construction of ULPF is shown in Fig. 5. Function Build() is recursive and the entire automata can be established by invoking Build(root), where the root is associated with an empty matching set and a full candidate set containing all of the specified rules.

4 Experiments and Analysis

The following experiments evaluate ULMM, ULPF and HPNMP. The testing machines in the experiments have the same configuration: dual 1GHz PIII CPU, 2G main memory, Intel 1000Mbps Ethernet NIC and 18G SCSI hard disk.

1. Two machines are connected with each other by Ethernet in the same local domain. One is special for packet sending (configuration: Router Tester GbE/4 1000Base); the other is special for packet capture.

We test peak throughputs of Libpcap, Libpacket, Old NIC Driver and ULMM on different packet size and the results are shown in Fig. 6. It is for analyzing performance of ULMM to test the throughput of Old NIC Driver. Peak throughput of ULMM increases with packet size and reaches its threshold at the point of about 1500B because of zero-copy technique that eliminates the memory copy between the kernel and the user applications, which shows that ULMM is a high-performance packet capture library. Peak bandwidth of Libpcap does not vary too much with packet size and reaches peak value of 196.38Mbps at the point of about 512B, this is the result of traditional kernel protocol used by Libpcap. For

```
void Build (struct node v) {/*v is a node in automaton, extra information are attached
                to each node: p is the field offset to be inspected, m is the set of
                already matched rules and c is the set of candidate rules*/
  If (v.c is empty)
    return;           /*if no candidate rule, terminate the procedure*/
  v.p=select(v.c);    /*select the field offset to inspect in the node v */
  buildbranch (v.p);/*create all the possible branches of node v, each branch
                       has a edge to it from v, with corresponding value*/
  for (each rule r in v.c){
    if (r has test relevant to v.p) {
      if (test for equality) {
        if (r can be matched after this test)
          add r into matched rule set of the branch with corresponding value;
        else
          add r into candidate rule set of the branch with corresponding value;
      }
      if (test for inequality) {
        if (r can be matched after this test)
          add r into matched rule set of the branch with corresponding value;
        else
          add r into candidate rule set of all the branches except the branch
          with corresponding value;
      }
    }
    else
      add r into candidate rule set of all branches;
  }
  for (each branch v')
    Build (v');     /*recursively call Build for v'*/
}
```

Fig. 5. Algorithm for automaton construction of ULPF

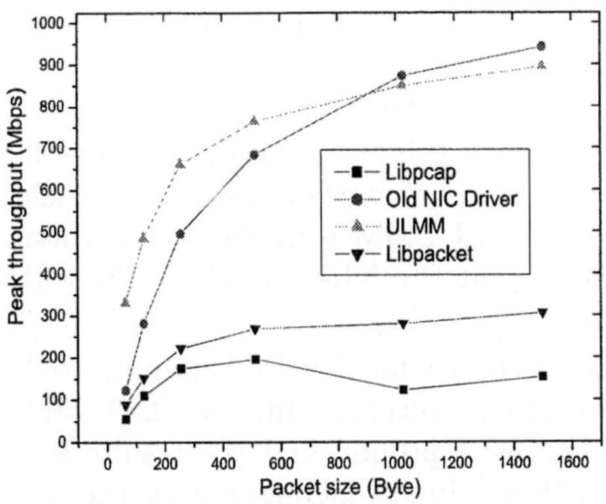

Fig. 6. Peak throughput of Libpcap, Libpacket, Old NIC Driver and ULMM with different packet size

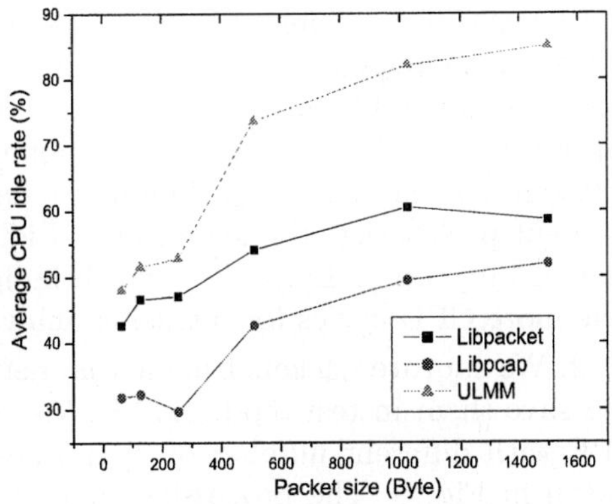

Fig. 7. Average CPU idle rates of LibpcapLibpacket and ULMM with different packet size at peak throughput

each different packet size, the throughput of ULMM is much greater than that of Libpcap or Libpacket. Fig. 6 also indicates that there is a crossover between the curves of ULMM and Old NIC Driver, this is because Old NIC Driver uses limited kernel buffer allocated/released dynamically to cache packets, ULMM employs a big static buffer in user space for saving packets so as to eliminate the overhead of frequent allocating/releasing buffer, whereas the overhead of allocating/releasing kernel buffer for Old NIC Driver declines with the gradual increment of the packet size.

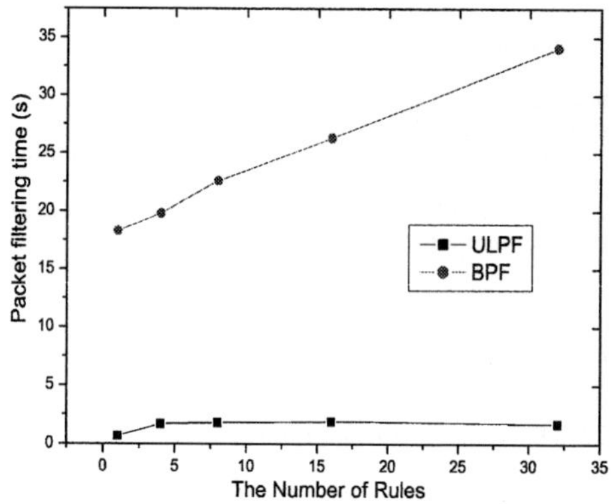

Fig. 8. Time of packet filtering with different number of filter rules

Fig. 7 shows the results of the average CPU idle rates of Libpcap, Libpacket and ULMM with different packet size at peak throughput. The CPU idle rate of ULMM increases with packet size because the system overheads of interrupt handling and DMA initialization declines continually. The CPU idle rate of Libpcap reaches the lowest value at the point of 256B, this is the result of tradeoff between the overhead of hard interrupt handling and that of data copy. For each different packet size, the average CPU idle rate of ULMM is much greater than that of Libpcap or Libpacket, which indicates that the NIDS with ULMM will save more CPU cycles for intrusion analysis.

2. We capture packets from actual network environment with Tcpdump tool and save them in test.tcpdump file. We test packet-filtering time of ULPF and BPF with different number of filter rules on test.tcpdump and the results are shown in Fig. 8. The processing time of BPF gradually increases with the increment of the number of filter rules, because BPF checks rules one by one for every packet. Processing time of ULPF is nearly invariant with the number of rules. For different number of filter rules, processing efficiency of ULPF is much greater than that of BPF.

3. To validate the scalability of HPNMP, we test the relation between the number of node machines and the connecting network bandwidth. The results

are shown in Table 1. The traffic load of each node machine is nearly even in every group of experiment.

Table 1. The results of experiments on HPNMP

The number of node machines	2	6	8	16
Network bandwidth (Mbps)	612	1716	2192	4656

5 Conclusion and Future Work

To meet the need of real intrusion analysis in high traffic network, this paper designs and implements a scalable high-performance network-monitoring platform (HPNMP) for intrusion detection. The application in actual environment indicates that HPNMP is very practical. Future work includes how to improve the efficiency of data analysis algorithm in NIDS.

References

1. Roesch M.: Snort: Lightweight Intrusion Detection for Network. In Proceedings of the 13th Systems Administration Conference, Seattle, Washington, USA (1999) 265-273
2. Libpcap. http://www.tcpdump.org/release/libpcap-0.7.2.tar.gz, 2002
3. Yang Wu, Fang Binxing, et al.: Research and Improvement on the Packet Capture Mechanism in Linux for High-Speed Network. Journal of Harbin Institute of Technology (New Series), 11 (2004) 56-64
4. Matt W., Anindya B., Thorsten V. E.: Incorporating Memory Management into User-Level Network Interfaces. In Proceedings of Hot Interconnects Symposium, Stanford (1997) 618-628
5. Eicken V., Vogels W.: Evolution of the Virtual Interface Architecture. IEEE Computer, Vol. 31. 11 (1998) 61-68
6. Cezary D., Liviu I., Edward W., et al.: Software Support for Virtual Memory-Mapped Communication. In Proceedings of the 10th International Parallel Processing Symposium (IPPS '96), Honolulu (1996) 372-381
7. Steven M., Jacobson V.: The BSD Packet Filter: A New Architecture for User-Level Packet Capture. In Proceedings of The Winter USENIX Conference, San Diego (1993) 259-269
8. Vankamamidi R.: ASL: A Specification Language for Intrusion Detection and Network Monitoring. M.S. Thesis, Department of Computer Science, Iowa State University. 1998
9. Sekar R. C., Ramesh R., Ramakrishnan I. V.: Adaptive Pattern Matching. SIAM Journal on Computing, Vol. 24. 6 (1995) 1207-1234

Experience with Engineering a Network Forensics System

Ahmad Almulhem and Issa Traore

ISOT Research Lab
University of Victoria, Canada
{almulhem, itraore}@ece.uvic.ca

Abstract. *Network Forensics* is an important extension to the model of network security where emphasis is traditionally put on prevention and to a lesser extent on detection. It focuses on the *capture*, *recording*, and *analysis* of network packets and events for investigative purposes. It is a young field for which very limited resources are available. In this paper, we briefly survey the state of the art in network forensics and report our experience with building and testing a network forensics system.

1 Introduction

Most organizations fight computer attacks using a mixture of various technologies such as firewalls and intrusion detection systems [1]. Conceptually, those technologies address security from three perspectives; namely *prevention*, *detection*, and *reaction*. We, however, believe that a very important piece is missing from this model. Specifically, current technologies lack any *investigative* features. In the event of attacks, it is extremely hard to tie the ends and come up with a thorough analysis of how the attack happened and what the steps were. Serious attackers are skillful at covering their tracks. Firewall logs and intrusion detection alerts are unlikely to be adequate for a serious investigation. We believe the solution is in the realm of *Network Forensics* [2]; a dedicated investigation technology that allows for the capture, recording and analysis of network packets and events for investigative purposes. It is the network equivalent of a video camera in a local convenience store.

In this paper, we report our experience with designing, implementing and deploying a network forensics system. First, we review the topic of network forensics in section 2. In section 3, we review some related work. In section 4, a network forensics system will be proposed. In section 5, we will discuss our implementation of the proposed architecture and some interesting results. Finally, we conclude and discuss our future work in section 6.

2 Network Forensics

In 1997, security expert Marcus Ranum coined the term *network forensics* [2]. He also introduced a network forensic system called *Network Flight Recorder* [3].

C. Kim (Ed.): ICOIN 2005, LNCS 3391, pp. 62–71, 2005.

Marcus, however, did not provide a definition for the new term. Therefore, we adopt the following one from [4]:

> *Network forensics* is the *capture*, *recording*, and *analysis* of network packets and events for investigative purposes.

When designing such a system, there are several challenges which include:

1. *Data Capture*:
   (a) Where should the data be captured?
   (b) How much data should be captured?
   (c) How do we insure the integrity of the collected data?
2. *Detection Efficiency*: The system should *detect* attacks efficiently in order to trigger the forensics process. Therefore, it should accommodate for different detection approaches.
3. *Data Analysis*: After collecting the data, the system has to *correlate* them in order to reconstruct an attacker's actions.
4. *Attacker Profiling*: The system has to maintain information about the attacker himself. For instance, it can identify the attacker's operating system through passive OS fingerprinting.
5. *Privacy*: Depending on the application domain, privacy issues can be a major concern.
6. *Data as Legal Evidences*: For the collected data to qualify as evidences in a court of law, they have to be correctly collected and preserved in order to pass *admissibility* tests [5, 6].

3 Related Work

Unlike the traditional *computer forensics* field, network forensics emerged in response to network hacking activities [7]. Typically, it is conducted by experienced system administrators rather than by law enforcement agencies [8].

The current practice in investigating such incidents is generally a manual brute-force approach. Typically, an investigation proceeds by examining various types of logs which are located in a number of places. For instance, a unix network is usually equipped with a dedicated auditing facility, such as *Syslogd*. Also, applications like web servers and network devices like routers, maintain their own logs. Various tools and homemade scripts are typically used to process these logs.

Brute force investigation, however, is a time consuming and error-prone process. It can also be challenging because the mentioned logs are usually scattered everywhere over the network. Also, these logs are not meant for thorough investigation. They may lack enough details or contrarily have lots of unrelated details. They also come in different incompatible formats and levels of abstractions.

On the high-end, there are commercial tools known as *network forensic analysis tools* which can be used for investigations in addition to varieties of tasks like network performance analysis [3, 9]. Generally, these tools are combined hardware/software solutions which continuously record network traffic. They also provide convenient GUI front-ends to analyse the recorded data.

The main problem with these commercial tools is dealing with encrypted traffic. Currently, the general approach is to install modified (trojaned) encrypted services like *ssh*. So if an attacker uses these services, his sessions can be decrypted. This, however, can be defeated, if the attacker installs his own encrypted service.

4 A Network Forensics System

In this section, we propose an architecture of a network forensics system that records data at the host-level and network-level. It also manages to circumvent encryption if an attacker chooses to use it. At first, we will provide an overview of the system, then discuss its main components in more details. Implementation and results will be postponed to the next section.

4.1 System Overview

In a typical network with multiple hosts, the proposed system consists of three main modules which are arranged as shown in Fig. 1.

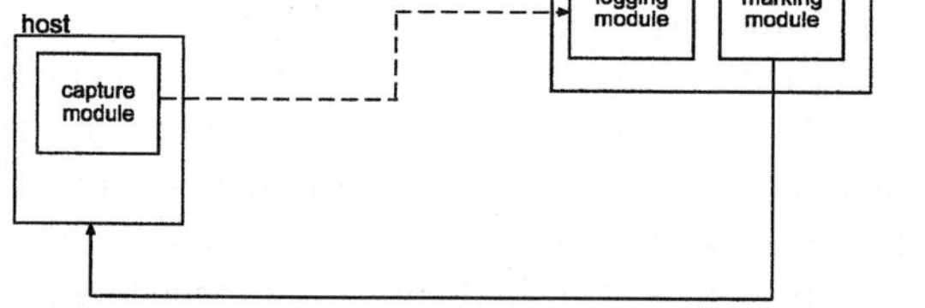

Fig. 1. The overall architecture of the system

The modules are

1. a *marking module*; a network-based module for identifying and marking suspicious packets as they enter the network,
2. *capture modules*; host-based modules which are installed in all the hosts in order to gather marked packets and post them to a logging facility, and

3. a *logging module*; a network-based logging facility for archiving data.

Together, these modules form a kind of closed circuit. An incoming packet first passes through the marking module which marks "suspicious" packets. Subsequently, when a host receives a marked packet, it posts the packet to the logging module. Each module will now be explained in further details.

4.2 Marking Module

This module is the entry point to our system. It is in charge of deciding whether a passing-by packet should be considered friendly or malicious. Then, it marks the packet accordingly. In nutshell, this module relies on a group of sensors to maintain a list of suspicious IP addresses. Then, it marks a passing-by packet if it's source IP address is in the list.

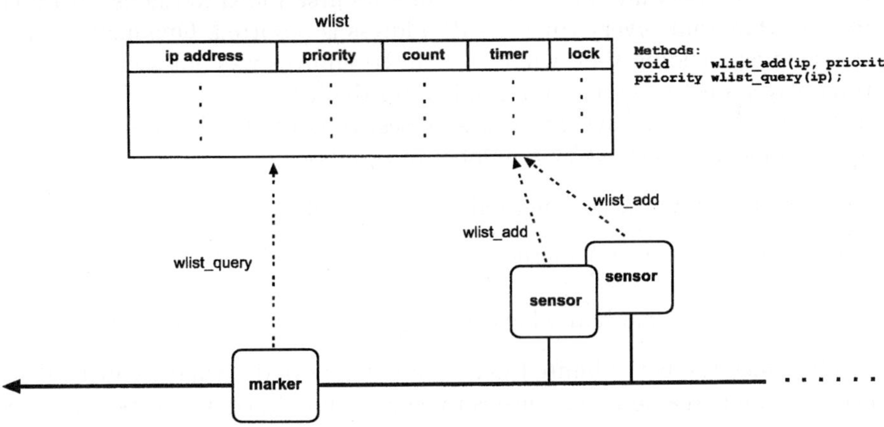

Fig. 2. The marking module

Figure 2 depicts the architecture of this module, which consists of the following three components:

1. *Sensors*: One or more sensor(s) to report suspicious IP addresses to a watch list (*wlist*). It is important to note that a sensor is not limited to a network-based IDS. It can be any process that can report suspicious IP addresses. This is essential to increase the system's detection efficiency.
2. *A Watch List (wlist)*: A list of suspicious IP addresses. We will explain it in more details shortly.
3. *A Marker*: A process to mark packets. Before sending a packet to its way, it queries the watch list to check whether the packet's source IP address is in the list. It accordingly modifies the *type of service field (TOS)* in the IP header.

The *watch list (wlist)* is basically a data structure which maintains a list of the current system's offenders. One may think of it as a cache memory of suspicious IP addresses. Each row corresponds to a unique IP address that has been reported by at least one of the sensors. For every suspicious IP address, the list also maintains the following information:

1. *priority*: A measure which indicates the *current* offence level of the corresponding IP address. Three levels are defined; namely HIGH, MEDIUM and LOW. A sensor must be able to classify an attack into one of these three levels. When different priorities are reported for a given IP address, the list only keeps the highest.
2. *count*: A counter which is incremented every time the corresponding IP address is reported.
3. *timer*: A count-down timer which is automatically decremented every second. If it reaches zero, the corresponding row is removed from the list. This field is set to a certain value when an IP address is first added to the list. It is also reset to that value every time the IP address is reported. One may think of this field as a sliding time window. If an IP address was not seen for a long time (say 1 week), we may remove it from the list.
4. *lock*: This field is to synchronize accesses. It is needed because the list is asynchronically accessed by a number of processes.

To interact with *wlist*, two methods are provided:

1. *wlist_add(ip, priority)*: A method to add an attacker's IP address to the list.

2. *wlist_query(ip)*: A method which returns the priority of a given IP address.

Finally, since the list is limited in size, one may wonder what happens if the list becomes full and a newcomer needs to be accommodated. Obviously, we need to decide which row should be replaced. Specifically, we should replace the *least important* row. A row with a low *timer* value indicates that the corresponding IP address was not seen for a long time. On the other hand, a high *count* value suggests that the corresponding IP address is suspicious. Thus, finding the least important row is a function of the three fields; namely *priority*, *count* and *timer*. Formally, let the priority, count, and timer be p_i, c_i and t_i respectively for a given row i. Then, the least important row (l) is

$$l = f(p_i, c_i, t_i)$$

The exact definition of this function is implementation specific. We will show an example definition when we discuss our implementation.

4.3 Capture Module

The second major component in our architecture is a collection of lightweight capture modules, which reside silently in the hosts waiting for marked packets. They, then, arrange to reliably transport them to the logging module for safe archival. This transportation is necessary because we cannot store the data locally. Once a system has been compromised, it cannot be trusted.

Installing capture modules in hosts is essential for two reasons. First, there is no guarantee that a suspicious packet will actually compromise or damage a host. In fact, the packet may be directed to a nonexistent service or even a nonexistent host. Installing capture modules in the hosts insures logging only relevant packets.

The second and more important reason is the fact that attackers increasingly use *encryption* to hide their activities. As a result, sniffing their traffic or trying to break-it is either useless or impractical. We may choose to install trojaned encrypted services; say *ssh*. However, careful attackers usually avoid these services and use their own encrypted channels. Therefore, only at the host, we can circumvent encryption and fully record an attacker's actions. This can be done by intercepting certain system calls [10].

4.4 Logging Module

The logging module is our system's repository where attack data are being stored. Ideally, one would turn to this module for reliable answers and documentation about any attack.

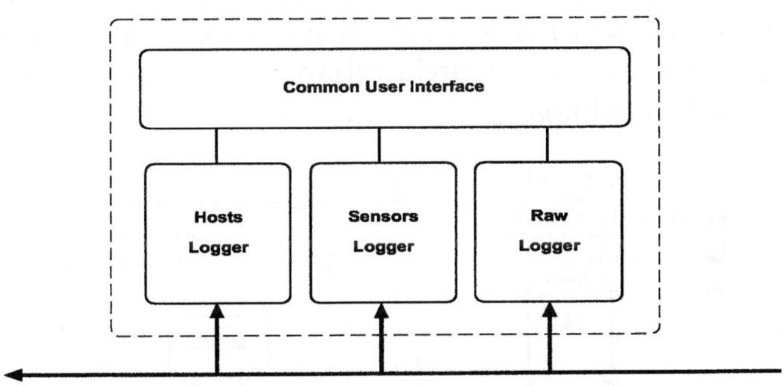

Fig. 3. The logging module

Figure 3 shows the architecture of a network-based logging module. We propose to use the following three loggers:

1. Hosts Logger: This logger is responsible for storing data sent by the capture modules. It is expected to log detailed data pertaining to real attacks. Therefore, storage requirements should be low.
2. Sensors Logger: This logger stores the sensors' alerts. Although, a typical alert is only a one-line text message, it provides a quick diagnosis about an attack. This logger is also expected to require low storage requirement.

3. Raw Logger: This is an *optional* logger which provides a last resort solution when other loggers fail. It archives raw packets straight off the line. In busy networks, however, this logger is expected to require an excessive amount of storage.

The last part in this module's architecture is a layer that should provide a common user interface to access these loggers.

5 Implementation and Results

5.1 Implementation

To test our approach, we built a prototype of the proposed architecture using two PCs; a host and a bridge configured as shown in Fig. 1. The host is a PC with a 400MHz Pentium II processor, 132MB RAM and 6GB hard drive. To allow break-in, we installed a relatively old Linux distribution; namely RedHat 7.1 Also, we enabled a vulnerable service; namely FTP (wu-ftpd 2.6.1-16). We also installed a *capture* module called *sebek* [10] from the *Honeynet* project [11]. It is a kernel-based data capture tool which circumvent encryption by intercepting the *read* system call.

The bridge is a PC with a 1.7GHz Celeron processor, 512MB RAM, 40GB hard drive and 2 NICs. We installed a custom Linux operating system and a collection of tools and homemade programs which reflect the proposed architecture. Figure 4 shows the internal architecture of this bridge. It hosts both the *marking* and *logging* modules.

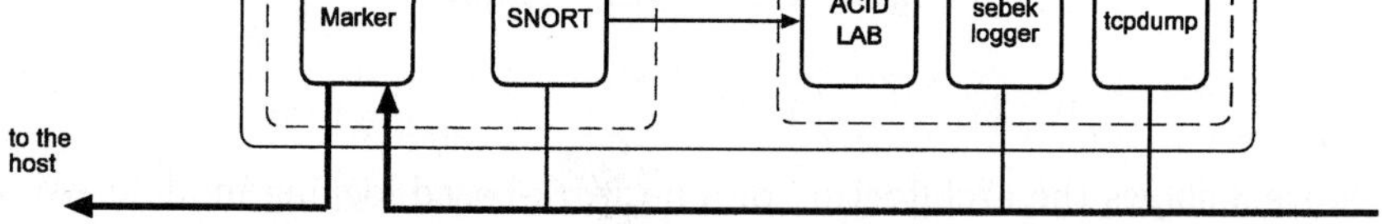

Fig. 4. The Bridge Internal

The marking module follows the architecture described earlier. Only one sensor was used; namely SNORT [12]. Both the watch list *(wlist)* and the marker were implemented in C language. When *wlist* becomes full, we used the following function: $l = min(\{t_i \mid t_i$ is the *timer* value of the ith row in *wlist* $\})$ where *min* is the minimum function.

The logging module also follows the proposed architecture. It consists of three loggers and MySQL [13] as a backbone database. The first logger records packets captured by the host. Since *sebek* was used to capture packets there, we used its corresponding server-side tools. The second logger is for the sensor; i.e. SNORT. We used SNORT's *barnyard* tool to log alerts in MYSQL and *ACID Lab* [14] for analysis. Finally, we chose *tcpdump* [15] as a raw logger just in case we miss something.

5.2 Results

The prototype was connected to the Internet for 12 days from March 17th until March 28th of 2004. Its IP address was not advertised. It was, however, given to members of our research lab who were interested in participating in the experiment. During the experiment, the host was compromised three times using a known FTP exploit.

General Statistics: Once the prototype was connected to the Internet, the host started receiving traffic. Table 1 lists the number of received packets grouped by protocol type.

Table 1. Number of friendly and strange packets directed to the host

	friendly packets	strange packets
tcp	70130	133216
udp	8928	9581
icmp	5150	6986
total	84208	149783
	36%	64%

Table 2. Storage requirement for each logger

	count	size
SNORT	3482 alerts	$111KB$
sebek	336132 packets	$38MB$
tcpdump	734500 packets	$69MB$

The first column lists the number of *friendly* packets; i.e. packets generated by participating members of our research lab. The second column lists the number of *strange* packets; i.e. packets coming from uninvited strangers. Overall, 64% of the traffic was not friendly. The table also shows that TCP is more frequent than other protocols. For the strangers' traffic, *TCP* packets are about 10 times (20 times) more than *UDP* (*ICMP*).

Table 2 sorts the storage requirement for the three used loggers in ascending order. SNORT requires the least amount, while tcpdump requires the most. Although, sebek is a powerful tool in honeypot settings, it actually did not fit our need. It captures far more data than we need. In the future, we are planning on developing our own capture module.

A Detailed Attack: We now discuss an attack by some stranger who successfully compromised the host and installed a rootkit. Overall, he generated about 1100 packets and caused a 158 *SNORT* alerts: 2 high priority, 154 medium priority and 2 low priority. Using the collected data, we were able to reconstruct his attack. Figure 5 shows a time-line diagram of his attack's steps.

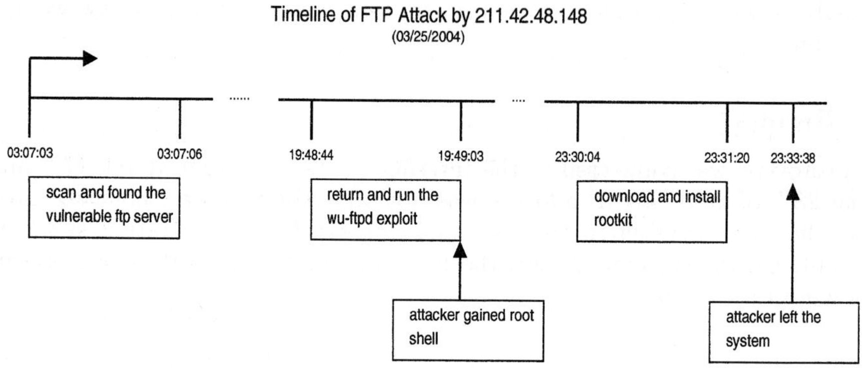

Fig. 5. Time Analysis of one of the attacks on our ftp server

At first, he scanned and found the vulnerable ftp server. After about 16 hours, he returned back with an effective exploit. He run the exploit and immediately gained a root shell. He then left the connection open for about 4 hours. When returned, he typed a number of commands and then exited. The following is a recreation of those commands.

```
[23:28:52] w
[23:29:54] wget
[23:30:04] wget 65.113.119.148/l1tere/l1tere.tgz
[23:30:19] ls
[23:30:24] tar xzvf l1tere.tgz
[23:31:20] ./setup
```

Those commands discloses the attacker's steps to downloading and installing a rootkit. Further analysis of this rootkit revealed the following main impacts:

1. creates directories and files under */lib/security/www/*.
2. removes other known rootkits.
3. replace some binaries with trojaned ones; many to mention!
4. installed a sniffer and a fake *SSHD* backdoor.
5. disable the anonymous vulnerable ftp server.
6. send an email to l1tere@yahoo.com with detailed information about the host.
7. cleans up and delete downloaded files.

Assessing the Results: Assessing the results is informal at this stage. We, however, can safely argue that we were able to detect and reconstruct all the compromises of the host. The proof pertains to using *sebek* at the host which was setup not to be accessed remotely. In particular, *sebek* can capture keystrokes. Therefore, seeing any keystrokes means a compromise. Also, *SNORT* (our sensor) is aware of the relatively old vulnerable *ftp* service. This gave us another indication of an ongoing attack.

6 Concluding Remarks

A network forensics system can prove to be a valuable investigative tool to cope with computer attacks. In this paper, we explored the topic of network forensics and proposed an architecture of network forensics system. We then discussed our implementation and obtained results. The proposed system manages to collect attack data at hosts and network. It is also capable of circumventing encryption if used by a hacker.

In the future, we plan to extend our system architecture with a fourth module named it expert module. The *expert module*, to be implemented as an expert system, will analyze the logged data, assess and reconstruct key steps of attacks. There are several facts that can be used to systematically characterize ongoing attacks and thereby may serve to construct the knowledge base of such expert system. For instance, the fact that some keystrokes are detected while only remote access is possible not only shows that a target has been compromised, but can also be used to partially reconstruct the attack.

References

[1] Richardson, R.: 2003 csi/fbi computer crime and security survey (2003)
[2] Ranum, M.: Network forensics: Network traffic monitoring. NFR Inc. (1997)
[3] Ranum, M., et al.: Implementing a generalized tool for network monitoring. Proceedings of the Eleventh Systems Administration Conference (LISA '97) (1997)
[4] searchSecurity.com: Definitions. (searchsecurity.techtarget.com)
[5] Sommer, P.: Intrusion detection systems as evidence. Computer Net. **31** (1999)
[6] Brezinski, D., Killalea, T.: Guidelines for evidence collection and archiving. BCP 55, RFC 3227 (2002)
[7] Fennelly, C.: Analysis: The forensics of internet security. SunWorld (2000)
[8] Berghel, H.: The discipline of internet forensics. Comm. of the ACM (2003)
[9] King, N., Weiss, E.: Analyze this! Information Security Magazine (2002)
[10] Balas, E.: Know Your Enemy: Sebek. Honeynet Project. (2003)
[11] Spitzner, L.: Honeynetproject. (www.honeynet.org)
[12] Roesch, M., Green, C.: Snort Users Manual. (2003)
[13] MySQL. (www.mysql.com)
[14] Danyliw, R.: Analysis console for intrusion databases. (acidlab.sourceforge.net)
[15] tcpdump/libpcap. (www.tcpdump.org)

An Alert Reasoning Method for Intrusion Detection System Using Attribute Oriented Induction

Jungtae Kim[1], Gunhee Lee[1], Jung-taek Seo[2], Eung-ki Park[2], Choon-sik Park[2], and Dong-kyoo Kim[1]

[1] Graduate School of Information Communication, Ajou University, Suwon, Korea
{coolpeace, icezzoco, dkkim}@ajou.ac.kr
[2] National Security Research Institute, Hwaam-dong, Yuseong-gu, Daejeon, Korea
{seojt, ekpark, csp}@etri.re.kr

Abstract. The intrusion detection system (IDS) is used as one of the solutions against the Internet attack. However the IDS reports extremely many alerts as compared with the number of the real attack. Thus the operator suffers from burden tasks that analyze floods of alerts and identify the root cause of them. The attribute oriented induction (AOI) is a kind of clustering method. By generalizing the attributes of raw alerts, it creates several clusters that include a set of alerts having similar or the same cause. However, if the attributes are excessively abstracted, the administrator does not identify the root cause of the alert. In this paper, we describe about the over generalization problem because of the unbalanced generalization hierarchy. We also discuss the solution of the problem and propose an algorithm to solve the problem.

1 Introduction

Recently various attacks using system vulnerabilities are considered as a serious threat to the network. To control those threats properly, most network administrators employ several information security systems such as intrusion detection system (IDS), firewall, VPN, and network scanner. These systems monitor the network status. In the abnormal situation, those systems generate an alert. Periodically or continuously the administrator analyzes those alerts, and he/she looks for the cause of them. According to the cause, he/she responds against the threat in order to remove it. However, it is difficult and burden work since there are extremely many alerts on a small networks.

In the severe situation, identifying the root cause is one of the important tasks to prevent the system securely. If the administrator knows the correct cause of the problem, he/she responds against it timely. Otherwise, he/she tries several solutions with some misunderstood causes to prohibit the abnormal behavior. During performing these wrong trials, the network and the system are exposed to the threat. However, there is no effective way to identify the unique cause of the attack. Moreover, the IDS reports extremely many alerts as compared with the number of the real attack [1], [2].

C. Kim (Ed.): ICOIN 2005, LNCS 3391, pp. 72–79, 2005.

To solve these problems, there are some researches on the efficient alerts handling method such as aggregation and correlation [3]. The attribute oriented induction (AOI) method is a kind of the aggregation method [4]. The clustering is an aggregation method. It makes several clusters that contains similar or the same alerts. Since it uses all the alert attributes as the criteria, it properly identifies and removes the most predominant root causes. However, it has over-generalization problem that confuses the administrator by generalizing the attributes value excessively. Although there is an attempt to solve it by K. Julisch, it still has the problem because of the unbalanced generalization hierarchy [5], [6].

In this paper, we discuss about the solution that reduces or removes over-generalization, and then propose an algorithm based on the AOI. It reduces the over generalization problem of the existing AOI algorithm effectively. Therefore the administrator identifies the root causes of the alert more easily.

This paper is organized as follows. We describe the AOI and its over generalization problem in section 2. In section 3, we discuss some possible solutions. In section 4, we describe proposed algorithm in detail. This is followed by the experimental results of the proposed method in section 5. Section 6 concludes.

2 Analysis on the Previous Works

Attribute Oriented Induction is operated on relational database tables and repeatedly replaces attribute values by more generalized values. The more generalized values are taken from user defined generalization hierarchy [4]. By the generalization, previous distinct alerts become identical, and then it can be merged into single one. In this way, huge relational tables can be condensed into short and highly comprehensible summary tables.

However the classic AOI algorithm has over-generalization problem that is important detail can be lost. K. Julisch modified classic AOI algorithm to prevent this problem. It abandons the generalization threshold d_i. Instead, it searches alerts $a \in T$ that have a count bigger than min_size (i.e. $a[count] > min\_size$) where $min\_size \in N$ is a user defined constant.

Fig. 1 represents an example generalization hierarchy and alert table having the attributes *Src-IP* and *Dest-IP*. Table 1 shows the comparision between clustering results of both algorithms, the classic AOI algorithm and K. Julischs algorithm using the example alerts in Fig. 1. For the classic AOI, the threshold d_i is set to 2, and for the K. Julisch's algorithm, the min_size is set to 10. While the *Src-IP* of first record is represented with $ANY\text{-}IP$ in the result of the classic AOI, it is represented with $ip3$ in the result of the K. Julisch's algorithm. The latter is more specific and informative.

Though K. Julisch's algorithm improves over-generalization of classic AOI, it does not properly handle the problem owing to the configuration of generalization hierarchy. For example, for the node $MyCompanyIPs$ in the Fig. 1-(b), the depths of sub trees are not equal. The depth of its left sub tree DMZ[3] is

[3] In this paper, when we refer a tree, we will use the name of its root node.

Src-IP	Dst-IP	Count
ipA1	ip1	1
...	...	...
ipJ1	ip1	1
ipK1	ip4	1
...	...	...
ipZ1	ip4	1
ip3	ipA1	1
...	...	...
ip3	ipZ1	1

a) Alert table

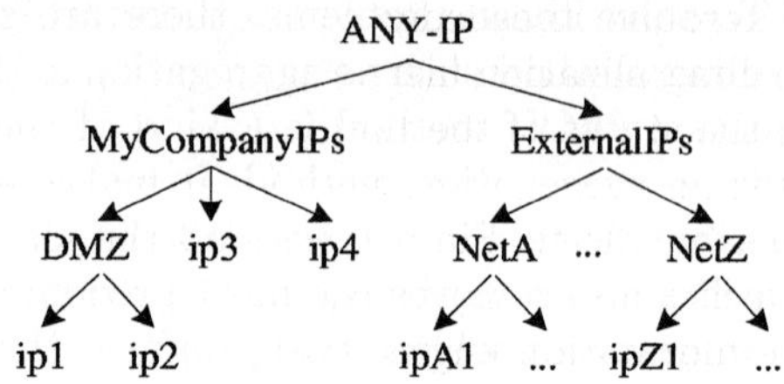

b) Generalization hierarchy for IP address

Fig. 1. A schematic of existing system

Table 1. Clustering result from classic AOI

Algorithm	Src-IP	Dest-IP	Count
Classic AOI algorithm	ANY-IP	ExternalIPs	26
	ExternalIPs	ANY-IP	26
K. Julisch's algorithm	ip3	ExternalIPs	26
	ExternalIPs	ANY-IP	26

2 and another two (*ip*3 and *ip*4) are 1. These nodes are denoted by *unbalanced node.* The *unbalanced generalization hierarchy* is a tree that has more than one *unbalanced node.*

In the result of K. Julisch's algorithm, the *Dest-IP* of the second record is represented with *ANY-IP.* However, it is more reasonable and meaningful that *Dest-ip* is represented with *MyCompanyIPs.* If *Dest-IP* is abstracted to *ANY-IP*, system administrator can't identify whether target of attack is home network or external network. It is another over-generalization problem, caused by the *unbalanced node, MyCompanyIPs.*

3 Considerations for the Solution

To solve the problem, we considers constructing well-formed generalization hierarchy and modifying the algorithm. The former is to construct the balanced generalization hierarchy and the latter is to modify the generalization process of the algorithm.

In the balanced hierarchy, all sub trees of a node in the hierarchy, except the root, should have the same depth. If the hierarchy is balanced, the nodes at the same level are abstracted simultaneously at a generalization step. Therefore, we can prevent over-generalization problem caused by the *unbalancednode.*

However, for more meaningful generalization hierarchy, the system administrator constructs his/her own generalization hierarchy according to his/her background knowledge about the related application domain. For example, for the generalization hierarchy of the IP address, the administrator should deliberate the status of the target network such as network topologies, variation of the network traffic, and providing services. For the continuous information such as time, he/she should decide the interesting interval (such as afternoon, Tuesday, a week, and so on) based on the significant alert and situation. As the most of the information engineering has no clear way to do this, so does the construction of informative generalization hierarchy. Moreover, it is inappropriate that the administrator changes the physical environment such as network topology to construct balanced generalization hierarchy.

If the administrator does not build a balanced hierarchy because of the physical environment, virtual dummy node can be employed. The virtual node is inserted into the *unbalanced hierarchy* between the *unbalanced node* and its sub tree in order to construct balanced generalization hierarchy. However, the use of the virtual nodes causes unnecessary overheads. As the number of nodes in a hierarchy is increased, the number of database update is also increased. When the amount of data is small enough, this could not be a serious problem. However, it could be a big problem where data size is numerous. In our experiments, if the data size is up to 50,000, the computing time was increased by 4% for each new virtual node.

In this paper, we modify the algorithm that works effectively as if the algorithm performs the generalization on the balanced hierarchy. The modified algorithm is able to efficiently prevent over-generalization without extra overheads. We explain the algorithm in the next section.

4 Proposed Algorithm

Fig. 2 shows the proposed algorithm to prevent over-generalization. To handle the *unbalanced generalization hierarchy*, we added a new attribute *hold count* in generalization hierarchy. This value means the remaining count of the generalization until all the lower level nodes of the current node are generalized to the current node. If the levels of sub trees are the same or the number of sub tree is less than one, the *hold count* is initialized to zero. Otherwise, it is initialized to maximum value among the differences between the level of current node and the level of leaf nodes belonged to its sub tree. For example, in the Fig. 1-(b), since the level of the sub-trees for the $ExternalIPs$ are equal, the *hold count* is set to 0. For the $MyCompanyIPs$, on the other hand, levels of its sub-tree are not equal, thus the *hold count* is set to 2, which is the difference between level of $MyCompanyIPs$ and $ip1$.

The algorithm starts with the alert logs L and repeatedly generalize the alerts in L. Generalizing alerts is done by choosing an attribute A_i of the alert and replacing the attribute values of all alerts in by their parents in generalization hierarchy of H_i. At this point, the *hold count* of a node k in H_i which is matched with attribute value $a[A_i]$ is must be zero. This process continues until an alert has been found to which at least min_size of the original alerts can be generalize.

```
Input: An alert clustering problem (L, min size, H_1, ..., H_n)
Output: A heuristic solution for (L, min size, H_1, ..., H_n)
1: T := Store log L in table T
2: H_ik[V] := Value of node k in generalization hierarchy H_i
3: H_ik[HC] := hold count of node k in generalization hierarchy H_i
4: for all alerts a in T do a[count] := 1                 // initialize counts
5: while all a ∈ T : a[count] < min size do {
6:    Use heuristics to select an attribute A_i, i ∈ {1..... n}
7:    for all alerts a in T do
8:       for all attribute A_i of alert a do
9:          if a[A_i] = H_ik [V] and H_ik[HC] = 0          // if hold count is zero
10:            a[A_i] := father of a[A_i] in H_i          // generalize attribute A_i
11:               if H_ik[HC] > 0
12:            H_ik[HC] := H_ik[HC] - 1                   // decrease hold count
13:   while identical alerts a, a' exist do               // merge identical alerts
14:      Set a[count] := a[count] + a'[count] and delete a' from T
   }
```

Fig. 2. Modified alert clustering algorithm

The algorithm considers the *hold count* during the generalization step and it is decreased by one where it is bigger than zero at each generalization step (line $6 \sim 12$ in the Fig. 2). When a node is to be generalized in the hierarchy, it first needs to check the *hold count* of the node. If the *hold count* of the node is not zero, then it implies that there are several records that should be generalized to the node. Therefore, it waits until no such nodes left. In other words, it should wait until the generalization level of all sub trees is to the current node.

Table 2. Clustering result from the proposed algorithm

Src-IP	Dest-IP	Count
ip3	ExternalIPs	26
ExternalIPs	MyCompanyIPs	26

For example, in the Fig. 1-(b), *ip*1 and *ip*2 is generalized to DMZ and *ip*3 and *ip*4 is generalized to $MyCompanyIPs$ at the first generalizations. At the second generalization, there exist several records that are not generalized to $MyCompanyIPs$ in the alert table though they belong to the sub tree of $MyCompanyIPs$. Thus, the records having $MyCompanyIPs$ are waiting until

DMZ is generalized to $MyCompanyIPs$. Table 2 shows the result of the clustering using the proposed algorithm with the data in Fig. 1. In the result, the *Dest-IP* of the second record is represented with the $MyCompanyIPs$.

5 Experimental Results

For the experiment, we constructed simulation network logically as shown in the Fig. 3-(a). $FireWall$ has two IP addresses that are $ip1$ and $ip2$. The $ip3$ represents an application server such as HTTP, FTP and SMTP. The $ip4$ is a network intrusion detection system. We regard other IP addresses $(ip5, ip6, \ldots)$ as the $Internet$.

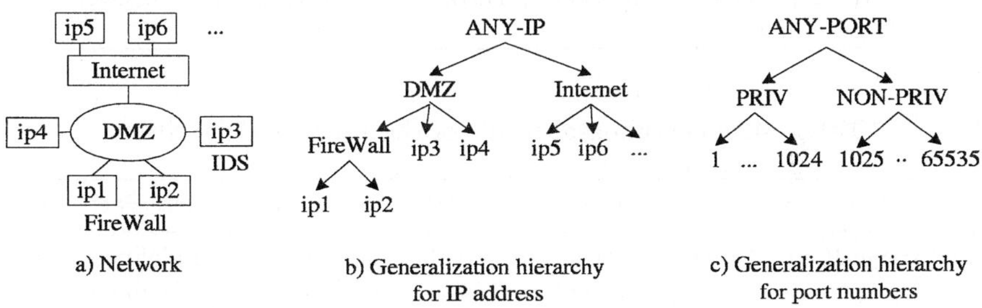

Fig. 3. Network and generalization hierarchies of the simulation

Snort was used as a network sensor [7]. The set of alerts generated by the sensor are logged into mySQL database. For a day, we randomly generated attacks against the simulation network using attack generation tool, *Snot* [8]. We clustered 15,670 alerts that are gathered for a day using Snort.

In the experiment, we used five attributes of *DMZ* alerts for the clustering: alert type, source IP, destination IP, source port, and destination port. Based on those attributes, the program created an initial set of the alerts from the raw alert generated by the sensors. For the attribute generalization, we used the generalization hierarchy as shown in Fig. 3-(b) for IP addresses and in Fig. 3-(c) for ports. Since $FireWall$ has two IP addresses such as $ip1$ and $ip2$, $ip1$ and $ip2$ should be abstracted to $FireWall$. $FireWall$, $ip3$, and $ip4$ are could be abstracted to DMZ. For more informative alert cluster, we did not generalize the value of alert type. Since alert type is critical information, if it is abstracted, it is difficult to identify what kind of attack is generated.

Table 3 shows the result of the clustering using the proposed algorithm and Table 4 shows the result of the K. Julisch's algorithm. Each row in both tables represents a cluster. Both results have the same number, 10, of the cluster. At the attack generation, we assume that most of the attacks are come from external networks. Therefore, most clusters have the same abstracted value, *Internet*, of the attribute *Src-IP* except for the third cluster.

Table 3. Experimental result of the proposed algorithm

Alert Type	Src-IP	Dest-IP	Src-Port	Dest-Port	Count
ICMP Large ICMP Packet	Internet	FireWall	undefined	undefined	845
	Internet	DMZ	undefined	undefined	862
	ANY-IP	DMZ	ANY-PORT	ANY-PORT	429
Bare Byte Unicode Encoding	Internet	FireWall	NON-PRIV	80	2937
	Internet	DMZ	NON-PRIV	80	3007
	Internet	Internet	NON-PRIV	80	982
Apache Whitespace	Internet	FireWall	NON-PRIV	80	725
	Internet	DMZ	NON-PRIV	80	707
Unknown Datagram decoding problem	Internet	DMZ	undefined	undefined	421
BAD-TRAFFIC data in TCP SYN packet	Internet	DMZ	ANY-PORT	ANY-PORT	413

Table 4. Experimental result of the proposed algorithm

Alert Type	Src-IP	Dest-IP	Src-Port	Dest-Port	Count
ICMP Large ICMP Packet	Internet	FireWall	undefined	undefined	845
	Internet	DMZ	undefined	undefined	862
	ANY-IP	ANY-IP	ANY-PORT	ANY-PORT	429
Bare Byte Unicode Encoding	Internet	FireWall	NON-PRIV	80	2937
	Internet	DMZ	NON-PRIV	80	3007
	Internet	Internet	NON-PRIV	80	982
Apache Whitespace	Internet	FireWall	NON-PRIV	80	725
	Internet	DMZ	NON-PRIV	80	707
Unknown Datagram decoding problem	Internet	ANY-IP	undefined	undefined	621
BAD-TRAFFIC data in TCP SYN packet	Internet	ANY-IP	ANY-PORT	ANY-PORT	413

The first 8 records excepting 3rd one are the same in the both Table 3 and Table 4 since the over-generalization occurs when an attribute value is abstracted higher than *unbalanced node* such as *DMZ*. However, these cluster's level of *Dest-IP* is lower than *DMZ* in the generalization hierarchy. Thus they are not affected by the unbalance of generalization hierarchy.

While *Dest-IP* attribute has been generalized to *ANY-IP* for the 3rd, 9th and 10th clusters in the existing algorithm, the generalization of those clusters are stopped at the *DMZ* in the proposed algorithm. It means that existing algorithm still has over-generalization problem. Thus proposed algorithm is more effective to prevent this problem than the existing algorithm. For example, when the system operator analyzes the 3rd cluster of the existing algorithm, he/she can't know whether the target of *Large ICMP Packet* is home network or extra network. On the other hand, from the result of the proposed algorithm, we can easily know that the target of the attack is the home network.

The number of alerts in the 9th cluster by the proposed algorithm is much smaller than one of the K. Julisch's algorithm. It means that the cluster does not contain unnecessary data as compared with the existing algorithm. In the existing algorithm, the 9th cluster includes many unrelated alerts since it contains all alerts from the *Internet* and the *DMZ*.

6 Conclusions

We proposed an algorithm that handles over-generalization problem of AOI algorithm owing to *unbalanced generalization hierarchy*. It addresses the problem of over-generalization in the alert clustering techniques of intrusion detection system (IDS) by applying the existing AOI. From the experimental results, the proposed algorithm prevents the problem of over-generalization more efficiently than existing algorithm. And it generate more informative cluster than previous algorithm. Therefore, system administrator identifies the root causes of the alert more easily and more effectively. We currently only considers the generalization hierarchy in the form of a tree. In the future work, the algorithm should employ the graph-based expression of the generalization hierarchy.

References

1. Valdes, A., Skinner, K.: Probabilistic Alert Correlation, Proceedings of Recent Advances in Intrusion Detection, LNCS 2212 (2001) 54-68
2. Axelsson, S.: The Base-Rate Fallacy and the Difficulty of Intrusion Detection, ACM Transactions on Information and System Security, Vol. 3, No. 3 (2000) 186-205
3. Debar, H., Wespi, A.: Aggregation and Correlation of Intrusion-Detection Alerts, Proceedings of Recent Advances in Intrusion Detection, LNCS 2212 (2001) 85-103
4. Han, J., Cai, Y.: Data-Driven Discovery of Quantitative Rules in Relational Databases. IEEE Transactions on Knowledge and Data Engineering, Vol. 5, No. 1 (1993) 29-40
5. Julisch, K.: Clustering intrusion detection alarms to support root cause analysis, ACM Transactions on Information and System Security, Vol. 6, No. 4 (2002) 443-471
6. Julisch, K.: Mining Intrusion Detection Alarms for Actionable Knowledge, Proceedings of the eighth ACM SIGKDD international conference on Knowledge discovery and data mining (2002) 366-375
7. Snort user manual, http://www.snort.org/docs/snort_manual/
8. Snot program, http://www.solenshoes.net/sniph/index.html

SAPA: Software Agents for Prevention and Auditing of Security Faults in Networked Systems

Rui Costa Cardoso and Mário Marques Freire

Department of Informatics, University of Beira Interior,
Rua Marquês d'Ávila e Bolama,
P-6201-001 Covilhã, Portugal
{rcardoso, mario}@di.ubi.pt

Abstract. This paper describes the design and implementation of a multi-agent system to detect and audit host security vulnerabilities. The system uses agent platforms allocated through the network to scan and interact with each host. The information collected by each agent is then used to build a common knowledge base that together with data retrieved from vulnerabilities information sources is used to improve the overall security. This approach reduces the total time to scan the network and the processing time overhead associated. The amount of traffic involved is also reduced. It allows the dissemination of updated knowledge about the network security status and reduces the communication with the network administrator. This solution provides an autonomous and proactive distributed system. It acts as a vulnerability assessment tool to make security notifications only if needed.

1 Introduction

In order to guarantee a global security solution in a given enterprise network it is necessary to take into account several issues such as: security mechanisms for exchange and access to remote information, mechanisms for protection of networked systems and administrative domains, detection of new vulnerabilities and exposures, monitoring and periodic audit of the implemented security mechanisms, and disaster recovery plans. This paper is focused on the problem of detection of security vulnerabilities in a proactive way, using software agents. Network security risks are rising continuously [1,2]. As networks become more interconnected, the number of entry points increases and therefore exposes each network to threats. The widespread of Internet access allows the dissemination of new vulnerabilities and hackers know-how. The Networks and applications are becoming more complex and difficult to manage and there is not a significantly increase in the number o human resources allocated. Also software development lifecycle is shorter resulting in flawed or poorly tested releases. Hackers have better tools, which require less technical skills and allow large scale attacks. The time between the identification of new vulnerabilities and the exploit attempt

C. Kim (Ed.): ICOIN 2005, LNCS 3391, pp. 80–88, 2005.

has been substantially reduced, giving less time to administrators to patch the vulnerabilities. Moreover, hackers often access to that information before the vendors are able to correct the vulnerabilities and it is difficult for network administrator to keep update with the patches. On the other hand, detection of security faults (holes) in hosts can anticipate the occurrence of service failures and compromises.

There are several approaches to the problem of vulnerability assessment. NIST (National Institute of Standards and Technology) [3] gives a good overview of the subject. Basically, there are open source approaches based on software packages like Nessus [4] and SARA [5] and commercial approaches by SecurityMetrics [6], Qualys [7] and ICSA [8]. The approach of security companies is mainly concentrated on the development of automated security programs capable to analyze the attacks within a single system such as Nessus [4], Nmap [9], SAINT [10], SARA [5] and SNORT [11]. All of these software products use a standalone approach and never share knowledge except when downloading updates from the central server. They examine system and network configurations for vulnerabilities and exploits that unauthorized users could use. Usually they act from a specific host in the network scanning all the others. Although most of them are programs and scripts run periodically by network administrators, its use lead to a rise in consuming time in which they are installed and also in the bandwidth availability, other negative remark is that this procedures may eventually lead to overhead in the performance of systems which could cause instability and crash in the scanned systems. Because there is not any exchange of data between applications there is not any guaranty that all share the same knowledge. On the other hand, some research work has been reported using software agents for network vulnerability scanning [12,13]. It uses a mobile agent approach and therefore can, if implemented correctly reduce the overall communication traffic in the network. Because mobile agents are not always in contact with its origin host, they have a reduced interaction when doing some task. They also allow the user to create specialized services by tailoring the agent to a specific task. The drawback of this approach is the security issues that arise when using mobile agents in an insecure network; this could lead to eventual content alteration. Is the agent trustable? And its content is authenticated? An interesting approach to the security problem in FIPA Agents platforms was made by Min Zhang et al. [14], which could be used to solve some of the mobile agents problems previously refereed. This paper addresses the problem of using a secure FIPA Agents platform for vulnerability analysis.

In this paper we present the design and implementation of an agent-based system, built using JADE [15], in which agents main task is detecting vulnerabilities and exposures [16]. Each agent can exchange knowledge with others in order to determine if certain suspicious situations are actually part of an attack, this procedure allow them to warn each other about possible threats. For external source of vulnerabilities [17] used to keep update the agent system we consider to use the ICAT Metabase [18], a search index of vulnerabilities in computerized systems. The ICAT binds the users the diverse public databases of vulnerabilities as well as patch sites, thus allowing us to find and to repair the existing vulnerabilities in a given system [19]. ICAT is not properly a database of vulnerabilities, but an index pointing to some reports of vulnerabilities as well as the information about patches currently available.

The remainder of this paper is organized as follows: Section 2 describes SAPA Agents. Section 3 discusses the distributed architecture. Section 4 is devoted to FIPA-ACL interactions. The Knowledge building is addressed in section 5. Main conclusions are presented in Section 6.

2 SAPA Agents

As more and more computers are connected to the network the security risk increases accordingly [20]. Although vulnerability assessment tools [21,22] and intrusion detection systems [23] are good solutions, they lack several features that we consider important, according [24]. Network Administrators needed to be permanently updated, and in control of everything that appended in their network, but it is widely known that this is not feasible in real time. An administrator has many routine and time-consuming tasks that could be delegated with advantages. Our solution is based in the delegation and cooperation; by delegating tasks to specific applications capable of autonomous behavior we can enhance the overall performance of the security in a network. In figure 1 we present a sketch of the platform. Our approach to the problem used JADE, a Java-based agent development framework [15,25], to evaluated the feasibility of the system. Jade is a FIPA-Compliant Agent Platform [26] in which we can deploy Java agents. The platform includes an AMS (Agent Management System), a DF (Directory Facilitator) and a sniffer RMA (Remote Management Agent). The features available includes: FIPA interaction protocols, automatic registration of agents with MAS, FIPA-compliant naming service and FIPA-compliant IIOP (Internet Inter-Orb Protocol) to connect to different APs (Agent Platforms). A GUI is also available to manage several agents and APs. SAPA Agents, acts as wrappers of VA (Vulnerability Assessment tools). JADE provide the infrastructure where the agents coexist and interact.

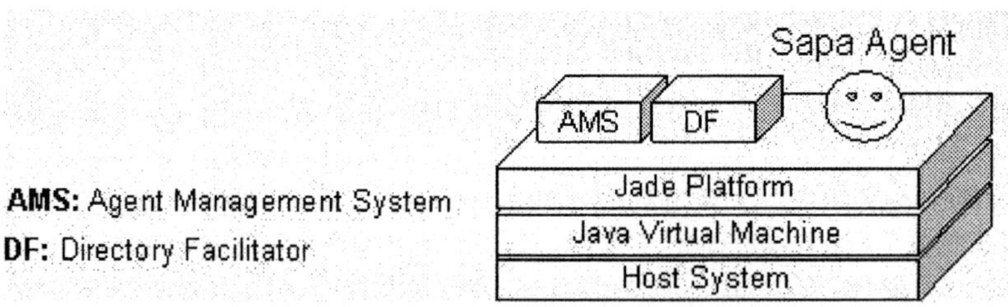

Fig. 1. SAPA Agent in a JADE Platform

2.1 Architecture

The SAPA agent has three main modules: DENOFAS module, FIPA-ACL Module, and Knowledge Module. The DENOFAS module [16] is used by SAPA to interact with the network. His main task is gathering valuable data to be processed. The FIPA-ACL [27] module used to exchange knowledge with other agents, and the Knowledge module used to maintain updated information about which Agents are active. In figure 2 we present the SAPA modules

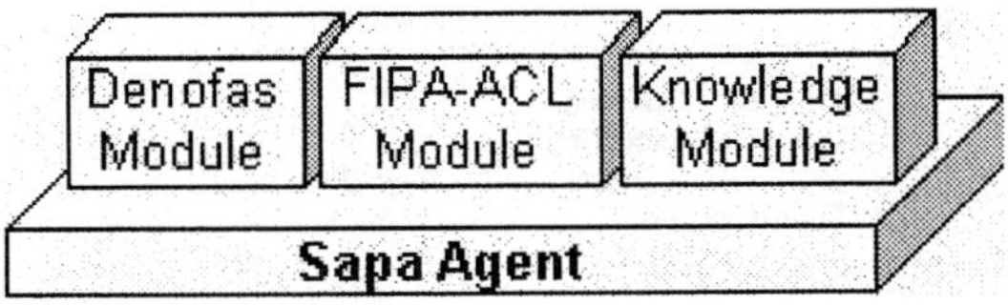

Fig. 2. SAPA Components

2.2 Denofas Module

The DENOFAS module acts has a vulnerability assessment tool and is used to scan the ports and services (not only in a specific host, but also the specific network allocated to him). All the data gathered is used by in the knowledge building. Each SAPA agent can by its initiative access the external vulnerability database [18] to extract new vulnerability data. Figure 3 represents the interactive process between SAPA agent and a Host.

2.3 FIPA-ACL Module

This module is used to interact with other agents. It allows the exchange of FIPA-ACL messages [27,29], between agents. The types of messages send are requests to fulfill several actions. The requests can be: Ask to acquire the knowledge of a specific SAPA Agent, disseminate warnings of possible attack based in new vulnerability posted in the database, inform about new updates in ICAT database, propagate vulnerability detection information, among others. Figure 4 represents the full FIPA-ACL compliant per formatives used in the exchange of FIPA Compliant ACL messages.

2.4 Knowledge Module

A brief overview of the Knowledge module follows. The data used to build the knowledge is supplied by the Denofas and by direct interaction with others agents through the FIPA-ACL module. The knowledge base allows the agent to act autonomously or in cooperation with others in the persecution of is tasks.

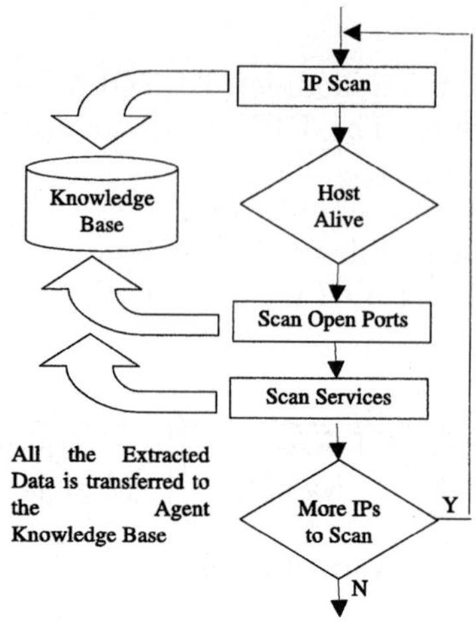

Fig. 3. DENOFAS Vulnerability Assessment procedures

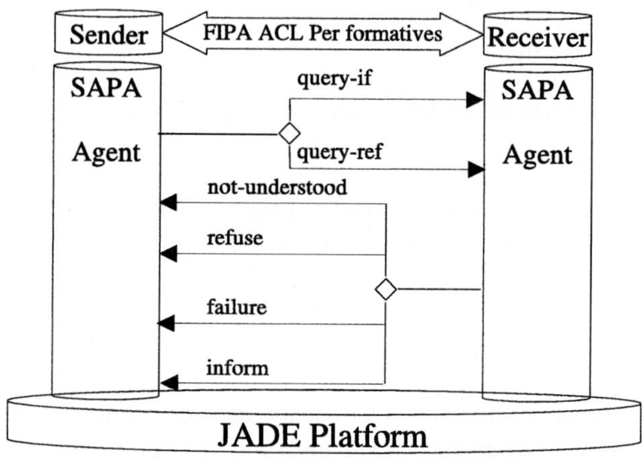

Fig. 4. FIPA-ACL Performatives

3 Distributed Architecture

In Figure 5 we present the overall interconnection solution based in distributed SAPA Agents (Software Agents for Prevention and Auditoring of Security Faults in Networked Systems). The SAPA Agent, receive solicitations to perform a specific task. The tasks currently supported are: host scan, network scan and host/network monitoring, in which the agent perform a detailed scan of the network, the open ports and the services active in those ports. After it collects all the requested data, it will use data from previous scans and data stored from several external sources to build is knowledge base. These sources are the ICAT [18] database of known vulnerabilities and exposures, the PortsDB [28] that tell us which services are associated with specific ports. After the inference process is complete and the agent has gathered sufficient knowledge about the situation, it will create an output result. Finally, presenting the output to the requester ends the process. The following Figure presents a summarized view of the Multi-agent System based in SAPA Agent interactions.

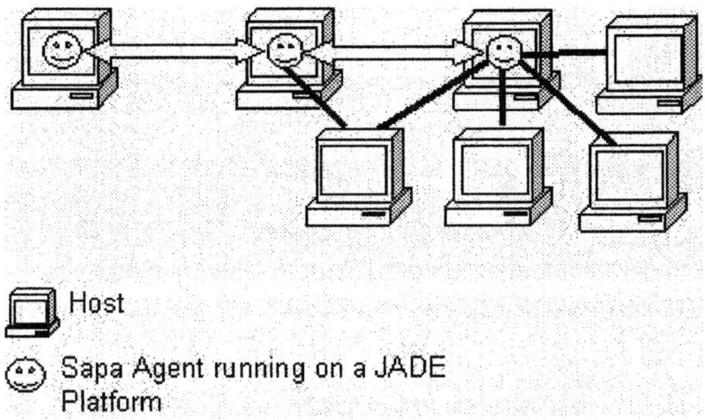

Fig. 5. SAPA MAS Multi-Agent System

The system is deployed in JADE [15] platforms based in hosts dispersed in the network. Each SAPA agent then can be given a task to monitor a specific host or group of hosts. The agents can also share information among them using FIPA-ACL messages [27]. The SAPA Agent main goal is to help network/host administrators to improve the security of computer systems networks. As already it was seen, the security problem arises, not only at the host level, where the agent is used to improve the security of a particular host, but also at network level. Being able to be used by network administrators, they could improve the overall security assessment of the network administrators. SAPA agents present a detailed assessment of the weak points in a network; making possible future security solutions to implemented. In this section we presented the architecture of our system.

4 Agent Interactions

In our multi-agent system we used FIPA-ACL messages to exchange information between Agents. Each Agent could: notify others that a specific host was listening in a suspect port, that ICAT data has been recently updated and also share is part of knowledge about the networks. Following we present an example of message exchange in our MAS using FIPA ACL (Agent Communication Language). The following messages represent a request/answer communication:

FIPA ACL Request:

```
(request :sender(agent-identifier:agentsapa1)
:receiver(agent-identifier :agentsapa2)
:content (action (agent-identifier:agentsapa2) open(data.txt))
:language fipa-sl)
```

The sender agent agentsapa1 send a request to the receiver agent agentsapa2, to update his data, using FIPA Semantic Language

FIPA ACL Failure:

```
(:performative failure :sender(agent-identifier:agentsapa2)
:receiver (set (agent-identifier :agentsapa1))
:content ((action (agent-identifier :agentsapa2) (open data.txt)
 (error-message No such file:data.txt))
:language fipa-sl)
```

The receiver agent agentsapa2 send a failure response to the sender agentsapa1 using FIPA-SL

5 Knowledge Building

The SAPA Knowledge Building process, integrate several data sources: online resources, information from others SAPA agents and the scanned data. ICAT [19] data used by the agent is automatic updated from the web. The agent uses the complete ICAT [18] database (exporticat.htm) and the information about all vendors, products, and version numbers contained within ICAT (vpv.htm). All this data is processed internally and is used as input to the Knowledge Base. In the knowledge building the agent also use the PortsDB [28] to specify what service is using what port. All this information is processed internally by the knowledge base. When the agent starts another scan it will use the data available to perform is task. Figure 6 illustrate the components of the knowledge building of the SAPA Agent.

6 Conclusions and Future Work

With this approach we achieve several goals: Reduce the response time, Propagate information more effectively. Free network administration from time consuming and routine tasks. Reduce substantially the burden of processing power in the overall by distributing it by several AP (Agent Platforms). The exploit of the same vulnerability in different systems became more difficult, because the agents help to detect it on time.

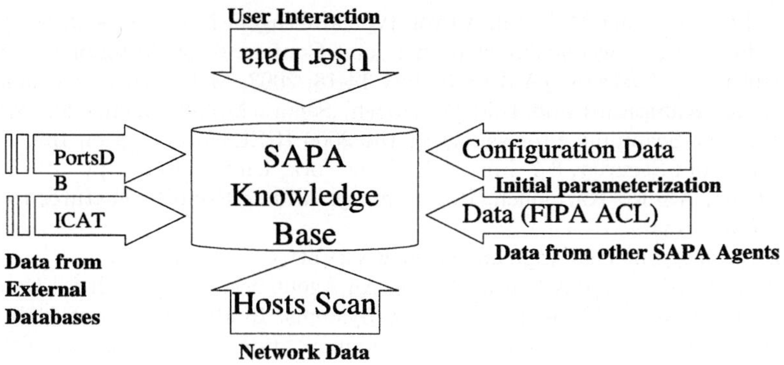

Fig. 6. SAPA Knowledge Building

7 Acknowledgements

The authors would like to thank the Jade Developer Community for useful discussions and precious suggestions. Part of this work has been supported by the Group of Networks and Multimedia of the Institute of Telecommunications - Covilh Lab, Portugal, within SAPA (Software Agents for Prevention and Auditing of Security Faults in Networked Systems) Project and by the EU IST EuroNGI (Design and Engineering of the Next Generation Internet) Network of Excellence, Sixth Framework Program, Priority 2, Information Society Technologies.

References

1. A. Householder, K. Houle, and C. Dougherty, Computer Attack Trends Challenge Internet Security, IEEE Computer, Security and Privacy - Supplement, April 2002, pp. 5-7.
2. CERT, http://www.cert.org. Accessed: 05/30/2004.
3. NIST: National Institute of Standards and Technology http://www.nist.org. Accessed: 05/30/2004.
4. Nessus, http://www.nessus.org. Accessed: 05/30/2004.
5. SARA: The Security Auditors Research Assistant http://www-arc.com/sara/. Accessed: 05/30/2004.
6. Securitymetrics. Integrated Vulnerability Assessment, Intrusion Detection and Prevention. Technical White Paper, Securitymetrics, 2003.
7. Qualys. On-Demand Security Audits and Vulnerability Management: A Proactive Approach to Network Security. Technical White Paper, Qualys. 2003.
8. Bace, R. , An Introduction to Intrusion Detection & Assessment. Technical White Paper, ICSA, 1999.
9. Nmap http://www.nmap.org. Accessed: 05/30/2004.
10. Saint http://www.saintcorporation.com. Accessed: 05/30/2004.
11. Snort: Open source network intrusion detection system http://snort.org. Accessed: 05/30/2004.

12. Taraka Pedireddy, Jos M. Vidal, A Prototype Multiagent Network Security System, Proceedings of the Second International Joint Conference on Autonomous Agents and Multiagent Systems AAMAS03, July 14-18, 2003, Melbourne, Australia.
13. Jeffrey W. Humphries and Udo W. Pooch ,Secure Mobile Agents for Network Vulnerability Scanning, Proceedings of the 2000 IEEE Workshop on Information Assurance and Security, 6-7 June 2000, New York, United States, pp. 19-25.
14. Min Zhang, Ahmed Karmouch, Roger Impey, Adding Security Features to FIPA Agent Platforms.
15. JADE (Java Agent DEvelopment Framework) http://jade.tilab.com.
16. Rui Costa Cardoso and Mário M. Freire. An Agent-based Approach for Detection of Security Vulnerabilities in Networked Systems. In Proceedings of 11th International Conference on Software, Telecommunications and Computer Networks (SoftCom2003), Split, Dubrovnik (Croatia), Venice, Ancona (Italy), October, 7-10, 2003, pp. 49-53.
17. CVE: Common Vulnerabilities and Exposures http://www.cve.mitre.org. Accessed: 05/30/2004.
18. ICAT: Internet Categorization of Attacks Toolkit http://icat.nist.gov. Accessed: 05/30/2004.
19. Peter Mell, Understanding the World of your Enemy with I-CAT (Internet-Categorization of Attacks Toolkit), in 22nd National Information System Security Conference, October 1999.
20. J. P. Anderson, Computer Security Threat Monitoring and Surveillance, James P. Anderson, Co. FortWashington, PA, 1980.
21. Robert A. Martin, Managing Vulnerabilities in Networked Systems, in IEEE Computer, November 2001 (Vol. 34, No. 11), pp. 32-38.
22. R. A. Kemmerer and G. Vigna, Intrusion Detection: A Brief History and Overview, IEEE Computer, Security and Privacy - Supplement, April 2002, pp. 27-29.
23. C. Manikopoulos and S. Papavassiliou, Network Intrusion and Fault Detection: A Statistical Anomaly Approach, IEEE Communications Magazine, Vol. 40, No. 10, pp. 76-82.
24. B. Kim, J.Jang, and T. M. Chung, Design of Network Security Control Systems for Cooperative Intrusion Detection, in Information Networking, I.Chong (Ed.), Heidelberg, Springer Verlag, LNCS 2344, 2002, pp. 389-398.
25. F. Bellifemine at al., JADE - A FIPA-compliant agent framework, Proceedings of PAAM99, London, April 1999, pg.97-108.
26. FIPA http://www.fipa.org. Accessed: 05/30/2004.
27. FIPA ACL Message Structure Specification, http://www.fipa.org/specs/fipa00061/. Accessed: 05/30/2004.
28. PortsDB (Ports Database) http://www.portsdb.org. Accessed: 05/30/2004.
29. Y. Labrou, T. Finin, and Y. Peng. Agent communication languages: The current landscape, IEEE Intelligent Systems, March/April 1999, pp. 45-52.

CIPS: Coordinated Intrusion Prevention System*

Hai Jin, Zhiling Yang, Jianhua Sun, Xuping Tu, and Zongfen Han

Cluster and Grid Computing lab
Huazhong University of Science and Technology, Wuhan, 430074, China
hjin@hust.edu.cn

Abstract. In this paper, we present the design and implementation of *Coordinated Intrusion Prevention System* (CIPS), which includes *Parallel Firewall* (PFW), *Flow Detection* (FD) and *Multiple Intrusion Detection System* (MIDS) to against large-scale or coordinated intrusions. The PFW consists of several firewalls working in parallel mainly by means of packet filtering, state inspection, and SYN proxy. The FD and MIDS detect and analyze the flow at the same time. The former one uses artificial neural network to analyze network traffic and detect flow anomaly. The latter one adopts traditional techniques such as protocol flow analysis and content-based virus detection to detect and prevent conventional intrusions and virus. Taking load balancing into account, CIPS also has *Flow Scheduler* (FS) for dispatching packets to each parallel component evenly. In addition, there is a *Console & Manager* (CM) aiming to reduce redundant alerts and to provide a feedback mechanism by alert clustering and to recognize the potential correlation rules among coordinated intrusion through mining large amounts of alerts.

1 Introduction

Coordinated intrusion implies one or multiple intruders take various means according to the policies and steps anticipated in advance and coordinate intrusion to one or multiple aspects of the same target by seeking leaks, communicating and uniting intrusions to bring some breakage. Coordinated intrusions are difficult to detect and defend against effectively for their great flexibility and complex attack methods. Distributed Denial of Service (DDoS) attack [1] is such an example. In this circumstance, the traditional approaches of building a protective shield such as a firewall around the protected targets or configuring IDS at the backbone of network or on personal computer are not sufficient to provide perfect security. New or coordinated intrusion detection techniques, methods and architectures are needed.

In this paper, we present our design and implementation of *Coordinated Intrusion Prevention System* (CIPS) aiming to defend against tremendous or coordinated intrusions. It is configured at the entry of a region to monitor and

* This paper is supported by Wuhan Hi-Tech project under grant 20031003027 and Key Nature Science Foundation of Hubei Province under grant 2001ABA001

C. Kim (Ed.): ICOIN 2005, LNCS 3391, pp. 89–98, 2005.

control all the network flows. CIPS includes *Parallel Firewall* (PFW), *Flow Detection* (FD) and *Multiple Intrusion Detection System* (MIDS), which provides a large-scale and expansible architecture and a coordinated response mechanism.

The PFW is the real channel that network flow passes through, and it parallelizes multiple loose-coupled firewalls by using many techniques such as stateful inspection and SYN proxy, which provides a middle detection granularity compared to MIDS and FD to detect and prevent DDoS in the session level. The FD detects large-scale intrusions by means of detecting the whole flow anomaly in coarse detection granularity in the traffic level. The MIDS composed of multiple IDSs providing a fine detection granularity mainly uses conventional intrusion detection techniques such as protocol flow analysis and pattern match to detect general intrusions and viruses in the packet level.

In order to achieve high performance and scalability of CIPS, three types of *Flow Scheduler* (FS) components are deployed in CIPS. One is front-end FS for PFW, which dispatches incoming flow to PFW evenly. Another is back-end FS for PFW, which forces the response flow to be received by the firewall that handles the corresponding request flow. And the other FS is for MIDS, which dispatches mirrored flow to IDS in balance.

CIPS also has a *Console & Manager* (CM) to control and manage the whole system, especially to reduce redundant alerts from the different detectors and to provide a coordinated response mechanism by alert clustering and to implement intelligent prevention by alert correlation by mining the potential correlation rules among coordinated intrusion.

The rest of the paper is organized as follows. Section 2 refers to related works. Section 3 presents the architecture of CIPS, describes its components and working flows and their mechanisms. Section 4 presents evaluation to the performance and section 5 gives the conclusion.

2 Related Works

It is quite necessary to carry out a thorough research and bring forward a feasible solution for coordinated intrusions since they have already become the major threat to the security of the Internet. But there is few published work that directly address this problem.

CIDS (*Collaborative Intrusion Detection System*) [2] employs Snort, a network level IDS, *Libsafe*, an application level IDS, a new kernel level IDS called *Sysmon* and a manager based framework for aggregating the alarms from the different detectors to provide a combined alarm for an intrusion.

CARDS (*Coordinated Attack Response & Detection System*) [3] is a prototype with the signature-based detection model, which represents coordinated attacks as generic event patterns over the structures of the typical information that can be found on target systems. The system contains three types of components (signature managers, monitors, and directory services), which are distributed at various places in the network.

GrIDS (*Graph Based Intrusion Detection System for Large Networks*) [4] is designed to analyze network activity on TCP/IP networks. GrIDS models a network as a hierarchy of departments and hosts where hosts consist of a data source and a software controller and departments are collections of hosts and software console and a graph engine. SHOMAR [5] provides a system of distributed intrusion detection clusters that are independent of each other yet collaborate to form a collective IDS.

SHOMAR assumes a logical hierarchy across an autonomous system. It can also facilitate communications between intrusion detection services across autonomous system.

EMERALD (*Event Monitoring Enabling Responses to Anomalous Live Disturbances*) [6][7] is an environment for anomaly and misuse detection. It addresses intrusion detection issues associated with large, loosely coupled enterprise networks. EMERALD uses a hierarchical approach to provide three levels of analysis by a three-tiered system of monitors: service monitors, domain monitors, and enterprise monitors.

3 CIPS Framework and Mechanisms

Figure 1 illustrates the CIPS framework with PFW, FD, MIDS, FS, and CM. PFW works in parallel by employing techniques such as packet filtering, state inspection and SYN proxy. The FD and MIDS detect and analyze the mirrored flow from the shared-media hub. For the FD, it detects intrusions through monitoring and auditing the network traffic in real time. For the MIDS, it detects general intrusions and viruses using multiple IDSs. The FS takes the characteristics of various intrusions into account and provides an algorithm to dispatch network traffic for parallelism and load balancing. The CM manages the whole system, providing an intelligent and coordinated response mechanism by alert clustering and correlation. CM plays a key role in CIPS system. It must have the capability of real-time processing all kinds of alert messages, providing coordinated response mechanism, and giving mined rules to the PFW.

Traditional security system that mainly adopts single detection mechanism such as IDS or FW appears deficient and weakly facing more and more complicated and coordinated intrusions. To address this problem, CIPS integrates PFW, MIDS and FD that work in different levels with various detection granularities to provide powerful detection function. By means of the integration of three components, coordinated intrusion can hardly occur since PFW and FD can detect and prevent DDoS attacks and MIDS can detect conventional intrusion and besides there is a coordinated response mechanism among them through CM.

The function of alert clustering is significant to reduce some redundant alerts owing to the parallel detection components and it is also helpful to implement coordinated response.

Although PFW is the only prevention component in the system, but due to its parallelism and expandability, it can prevent large-scale attacks and avoid coordinated intrusions occurring since coordinated intrusions are always in large scale. On the other hand, the CM can make some decision and give some hints for PFW through alert clustering to take some measures correspondingly both for PFW itself and FD or MIDS.

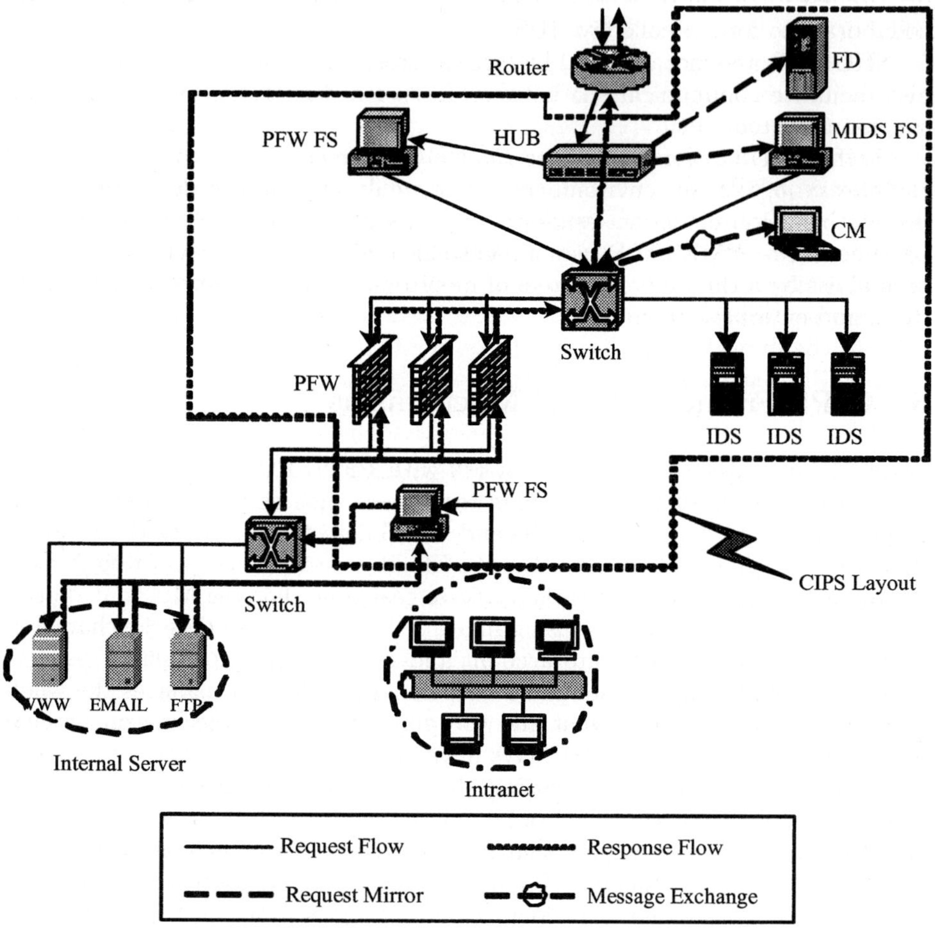

Fig. 1. CIPS Architecture

3.1 PFW (Parallel Firewalls)

In PFW, we integrate many techniques such as packet filter, state inspection, and DDoS prevention.

State inspection delivers the ability to create virtual session information for tracking connectionless protocols. We design the programmable state inspect engine, which allows PFW to add support for new and custom applications. For TCP protocol, combined with SYN proxy, system constructs finite automation of TCP protocol state conversion in advance. Under each condition, only the TCP packet satisfying the state conversion can be passed through, otherwise, PFW discards the packet and records it for later statistics and analysis.

For UDP and ICMP protocol, they do not contain any connection information (such as sequence number). However, UDP contains port pairs, and ICMP has type and code information. All of these data can be analyzed in order to build "virtual connections" in the state table. For a short period of time, UDP packets that have matching UDP information in state table will be allowed through the firewall. Same situation exists for ICMP. For application protocols, some complex protocols (such as FTP and audio/video) utilize multiple network connections simultaneously. Corresponding protocol modules monitor connections for these protocols, and instruct the firewall to allow related connections. With such assistance, the firewall does not need to perform additional checks against the security policy, and all connections required for the protocol are handled without the firewall inadvertently rejecting them.

In CIPS, some mechanisms are adopted to detect and prevent classical DDoS attacks. For SYN flood attack, PFW proxies the requests and not passes them on to the victim server unless the 3-way handshake is complete. To protect SYN flood efficiently, an adaptive approach is proposed, which uses the state-based approach (TCP half-connection hash table) under the low rate of attacks and stateless approach (SYN Cookie) under the high rate of attacks. For TCP flood attack, PFW filters TCP flooding packets such as ACK attack, RST attack and hybrid attack based on TCP state conversion diagram, meanwhile PFW records the information of attack packets accordingly. Through analysis for discarded packet periodically, PFW can detect the type of TCP attack based on threshold policy. For UDP flood attack, an attacker hooks up one system's UDP service with another system's UDP echo service by spoofing, and PFW can detect UDP flood indirectly by analysis the ICMP protocol packet which has the type of port unreachable. For ICMP/Smurf flood attack, PFW filters ICMP packets using state inspection based on ICMP protocol analysis and records the information of attack packets. Through statistics and analysis for discarded packet periodically, PFW can detect ICMP flood.

3.2 MIDS (Multiple Intrusion Detection System)

MIDS are several conventional intrusion detection systems working together to detect, filter and block conventional attack, virus mails and malicious content. Each IDS element adopts network-based architecture such as snort [8] to analyze and inspect all the network traffic by protocol flow analysis, content-based virus detection and content filtering using pattern matching.

Protocol flow analysis gives detection engine essential knowledge of a particular application protocol. Since protocol flow analysis is performed at high level and is usually only concerned with a few important aspects of a particular protocol flow such as a client request type, it provides special data for further analysis so as to facilitate the whole process. The essence of protocol flow analysis is pattern matching. The premise of protocol flow analysis is that the data from servers is secure and reliable. It is easy to see that it is applicable for the services with small request data and large response data such as Web service.

3.3 FD (Flow Detector)

The network traffic will appear an obvious anomaly if DDoS attack happens in large-scale and high-speed network, so anomaly detection on network traffic can be a viable method to network security. In CIPS, we propose anomaly detection model based on neural networks. Compared with MIDS and PFW, FD works in a coarse detection granularity to find intrusions mainly by monitoring and auditing the whole network traffic in real time.

A limited amount of researches has been conducted of using neural networks to detecting computer intrusions. Artificial neural networks have been proposed as alternatives to the statistical analysis component of anomaly detection systems [9][10]. The establishment of network flow model based on artificial neural network is to train large-scale training data with BP algorithm, in which the learning process of network is the alternation of spread and reverse spread. In order to improve the precision of the FD model, we should take advantage of the important attributes that is most related with connection such as related protocol, source IP, source port and the persisting time of connection.

3.4 FS (Flow Scheduler)

The objective of FS is to dispatch data packet to each parallel component evenly. It includes front-end FS for PFW, back-end FS for PFW and FS for MIDS. FS is different from traditional server schedule such as LVS [11]. There are two important issues that we must consider. (1) Since the detection mechanism based on protocol state is introduced in PFW, how to make sure that request flow and response flow belong to the same connection flows through the same FW, thus FW does not lose the information of protocol connection state. (2) The FS must schedule data packet to the current available FW under the circumstances of dynamic join or depart and unpredictable failure of FW. To improve the performance and throughout of FS, FS adopts Direct Routing [12] technique which is completely implemented in Linux kernel.

To avoid the delay induced by searching information table, we propose an efficient SourceIP Hashing Scheduling Algorithm (SHSA), which is based on the theory of multiplicative hash [13]. We use HashKey = (source_ip * 2654435761UL) mod FW_Count, where FW_Count is the total number of available FW.

We denote n as the total number of current "live" FW, and FWk (0k¡n) as the kth FW. Figure 2 (a) presents the case about the failure of FW. Two scenarios are considered: (1) Failure Scheduling: Calculate the hashkey = (source_ip * 2654435761UL) mod (n-1), supposed to be i. If corresponding FWi is failure, FS selects a FW (supposed to be j) with lightest load from the PFW and registers the scheduling info with the characteristics of TTL (Time-To-Live) in global Failure Hash Table, otherwise, sends packet to FWj directly via Direct Routing. (2) Recovery Scheduling: When the FWi is "live" now, FS lookups the Failure

Hash Table to examine if the incoming packet should be scheduled based on existed scheduling info, if found, supposed to be k, FS sends packet to FWk via Direct Routing; if not, calculate the hashkey = (source_ip * 2654435761UL) mod n, supposed to be m, FS sends packet to FWm directly via Direct Routing.

Fig. 2 (b) presents the case about the dynamic join of FW. Two scenarios are considered: (1) Join Scheduling: FS still adopts mod n scheduling algorithm, Calculate the hashkey = (source_ip * 2654435761UL) mod n, supposed to be i, send packet to FWi and register the scheduling info with the characteristics of TTL (Time-To-Live) in global Transition Hash Table. (2) Transition Scheduling: When the time elapsed exceeds TTL, then FS begins to adopt mod (n+1) scheduling algorithm. For the incoming packet, FS firstly lookups the Transition Hash Table to examine whether the incoming packet can be scheduled based on existed scheduling info, if found, supposed to be k, FS sends packet to FWk via Direct Routing, if not, calculate the hashkey = (source_ip * 2654435761UL) mod (n+1), supposed to be m, FS sends packet to FWm directly via Direct Routing. When the Transition Hash Table is NULL, the transition phase finishes and FS schedules consequent incoming packet based on mod (n+1) algorithm.

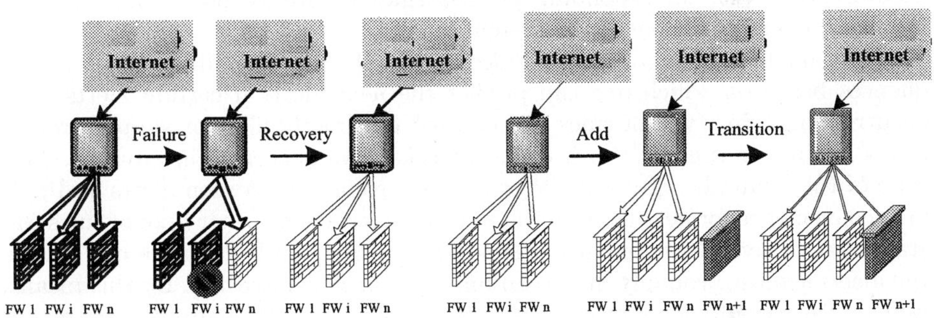

Fig. 2. ROC Curves on Detection Rates and False Alarm Rates.

3.5 CM (Console & Manager)

CM is to manage the whole process of the system including the detectors' register and logout, collect and illustrate state information of all the modules. In addition, there are two novelties which are worth emphasizing compared with others, one is the function of automatic clustering on gathered alerts coming from detect modules to decide whether the whole alerts should be given by means of similarity evaluation and layered clustering, and the other is the alerts correlation function which will improve the intelligence of system.

Current intrusion detection systems generally produce large volumes of alerts, including false alerts. In situations where there are intensive intrusions, not only will actual alerts be mixed with false alerts, but also the amount of alerts will become unmanageable and consequently the whole system will be too stressed to work efficiently [14]. This same problem appears serious in CIPS since there are multiple partial detectors working in parallelism. We address this problem by means of alert clustering which employs similarity evaluation and layered clustering to reduce redundant alerts as many as possible compared to individual detector without a substantial degradation in performance. In addition, alert clustering is crucial to implement coordinated response among multiple modules by means of analysis the essence of intrusion. For every alert coming from detectors, the console first compares it to the other alerts coming in close time and evaluates their similarities according to their alerts attributes and accordingly decides whether to give a whole alert.

The alert correlation model is based on the observation that in a series of attacks, the partial attacks are usually not isolated, but related to different stages of the larger attacks, with the early ones preparing for the later ones. The approach of CIPS is to employing association rules analysis and sequence patterns mining [15] techniques to find the correlations between large amounts of alerts, which can help users to understand why alerts are triggered and how they should be handled in the future. Based on this knowledge, the users can accordingly devise corresponding process rules for future use to improve the intelligence and efficiency of the system.

We want to find two types of rules: one is the maximal frequent alerts sequences based on which one can predict the next alerts if certain alerts have occurred, and this type of rules can be used to directly filter or simply process alerts in time or prevent the essential attacks from occurring in advance. This type of rules can also discover the repeated correlation between alerts and provide convenience for dealing with the repeated alerts. The other type of rules can state that if several alerts emerge simultaneous frequently then we can induce the alerts is most probably due to an unknown or new intrusion. By this means, we can discover unknown or coordinated intrusions and take appropriate means to deal with them or prevent more damage occurring if one of them occurs.

These patterns are of direct relevance to alert handling. Knowing the rules of alerts, it is easy to predict the next alert and filter the related alerts out in the future. Similarly, if an alert systematically entails other redundant alerts, then one can reduce the overall alert load by fusing these alerts into a single, semantically richer alert. We can detect or destroy future intrusions by advanced consciousness of them in a more reliable and timely manner which can also discover new or unknown intrusion patterns.

4 Performance Evaluation

To simulate the real environment, we divide the 100M Ethernet LAN into three sections, one is to simulate the external Internet from which the attacker appears, another is the internal protected LAN where there are web servers, email servers and FTP servers and CIPS is configured at the entry of it, and the other is the section in which CIPS works and the PFW, MIDS and FD are configured.

Table 1. Maximal parallel connections evaluation of the CIPS

Description of Scenario	Without Protection	With traditional FW on the Gateway	With 2-PFW CIPS	With 3-PFW CIPS
Maximal connections per second	10,000	50,000	70,000	80,000 -90,000

Table 2. Livability of normal user evaluation of CIPS

Connections per Second	Number of Users Attempt to Connect	With 2-PFW CIPS	With 3-PFW CIPS
100,000	400	54%	74%
125,000	400	45%	63%

Security evaluation includes the detection of conventional intrusion and coordinated intrusion. For conventional intrusion, we evaluate the efficiency by detecting port-scan, denial of service, ICMP flood, web attack, and FTP attack.

For coordinated intrusion, we consider several scenarios: 1) Pure DDoS attack. 2) Firstly implement the DDoS attack to consume the resource and weaken the detection ability of the target and then use conventional mean to intrude. 3) Take mixed DDoS attacks combined with SYN flood, TCP flood, ICMP flood and UDP flood, greatly consume the system resource and make the system denial of legal services.

The capacity of maximal parallel connections illuminates the security performance since coordinated intrusion mainly floods the target with tremendous packets. We assume the size of every connection is average, thus the number of connections can reflect the capacity. The contrast of maximal parallel connections per second is showed in Table 1.

Performance evaluation involves evaluation of throughput, response time, delay ratio and livability of normal user. For each case, we suppose several different cases: 1) Without any protection. 2) Without CIPS, but with one conventional firewall on the host. 3) Only with CIPS at the entry of the protected LAN. Throughput evaluation is done by comparing the average download speed of the FTP servers configured in the internal LAN under the different conditions illustrated above. The result shows the average throughput is decreased 5

5 Conclusions

In this paper, we present the framework and the related key techniques of CIPS to against large-scale or coordinated intrusions. In CIPS, PFW, FD and MIDS work in different levels with various detection granularities to provide powerful detection function. PFW consists of several special firewalls working in parallel by means of packet filtering, stateful inspection and SYN proxy. FD captures and analyzes network traffic to detect flow anomaly. MIDS uses traditional detection techniques to detect and prevent conventional attacks and virus.

References

1. J. Mirkovic, J. Martin, and P. Reiher, "A taxonomy of DDoS attacks and DDoS defense mechanisms", *Technical Report*, University of California at Los Angeles, 2001.
2. Y. S. Wu, B. R. Foo, Y. G. Mei, and S. Bagchi, "Collaborative Intrusion Detection System (CIDS): A Framework for Accurate and efficient IDS", *Proceedings of the 19th Annual Computer Security Applications Conference*, December 8-12, 2003.
3. J. Yang, P. Ning, X. S. Wang, and S. Jajodia, "CARDS: A distributed system for detecting coordinated attacks", *Proceedings of IFIP TC11 Sixteenth Annual Working Conference on Information Security*, 2000.
4. S. S. Chen, S. Cheung, R. Crawford, M. Dilger, J. Frank, J. Hoagland, K. Levitt, C. Wee, R. Yip, and D. Zerkle "Grids-a graph based intrusion detection system for large networks", *Technical report*, 1996.
5. J. Undercoffer, F. Perich, and C. Nicholas, "SHOMAR: an open architecture for distributed intrusion detection services", *Technical Report*, CSEE UMBC, Sept., 2002.
6. P. A. Porras and P. G. Neumann, "EMERALD: event monitoring enabling responses to anomalous live disturbances", *Proceedings of the Nineteenth National Computer Security Conference*, Baltimore, Maryland, October 22-25, 1997.
7. P. G. Neumann and P. A. Porras, "Experience with emerald to date", *Proceedings of First USENIX Workshop on Intrusion Detection and Network Monitoring*, April, 1999.
8. M. Roesch, "Snort - lightweight intrusion detection for networks", *Proceedings of the 1999 USENIX LISA Conference*, November, 1999.
9. H. Debar, M. Becke, and D. Siboni, "A Neural Network Component for an Intrusion Detection System", *Proceeding of the IEEE Symp. on Research in Security and Privacy*, 1992.
10. H. Debar and B. Dorizzi, "An Application of a Recurrent Network to an Intrusion Detection System", *Proceeding of the Int'l Joint Conference on Neural Networks*, 1992.
11. W. Zhang, "Linux Virtual Server for Scalable Network Services", Ottawa Linux Symposium, 2000.
12. S. Alstrup, J. Holm, K. Lichtenberg, and M. Thorup, "Direct routing on trees", *Proceedings of the Ninth Annual ACM-SIAM Symposium on Discrete Algorithms (SODA 98)*, pp.342-349, 1998.
13. C. Lever and S.-N. Alliance, "Linux kernel hash table behavior: analysis and improvements", *Proceedings of 4th Annual Linux Showcase*, 2000.
14. P. Ning and D. B. Xu, "Learning attack strategies from intrusion alerts", *Proceeding of the 10th ACM CCS 2003*, October, 2003.
15. H. Mannila, H. Toivonen, and AI Verkamo, "Discovery of frequent episodes in event sequences", *Data Mining and Knowledge Discovery*, 1(3):259-289, 1997.

A Two-Phase TCP Congestion Control for Reducing Bias over Heterogeneous Networks

Jongmin Lee[1], Hojung Cha[1], and Rhan Ha[2]

[1] Dept. of Computer Science, Yonsei University,
134 Shinchon-dong, Seodaemun-gu, Seoul 120-749, Korea
{jmlee, hjcha}@cs.yonsei.ac.kr

[2] Dept. of Computer Engineering, Hongik University,
72-1 Sangsoo-dong, Mapo-gu, Seoul 121-791, Korea
rhanha@cs.hongik.ac.kr

Abstract. This paper presents a sender side TCP congestion control scheme that reduces biases in wired as well as wireless networks. TCP has a problem utilizing the full bandwidth in high speed networks with a long delay. Moreover, competing flows with different roundtrip times share the bandwidth unfairly; a flow with long RTT experiences a throughput penalty. The throughput penalty is severe in wireless networks since TCP treats packet losses caused by link error as an indication of network congestions that trigger transfer rate reductions. The proposed scheme controls the network congestion in two phases - a fair convergence phase and a congestion avoidance phase - both of which are based on the application's transfer data patterns. The transfer rate is then adjusted adaptively by considering the current transfer rate and the estimated bandwidth in order to reduce bias and throughputs. The scheme has been implemented in the Linux platform and experimented with various TCP variants in real environments. The experimental results show that the mechanism reduces biases, and the network bandwidth is shared fairly among the proposed and the traditional TCP flows.

1 Introduction

TCP has widely been adopted as a data transfer protocol in wired and wireless networks. The current network infrastructure does not provide much information to a TCP source. Information such as the network bandwidth or explicit congestion feedback may not be provided. TCP uses an Additive Increase Multiplicative Decrease (AIMD) congestion control scheme to determine the network capacities [1]. TCP increases the transfer rate continuously until it detects any implicit feedback signals, which are timeouts and duplicate acknowledgements that indicate the network has reached its capacity. If TCP detects the signals then it halves the transfer rate and increases it again slowly to prevent successive packet losses. This window based congestion control scheme poses some problems such as a slow increase rate and a bias against flows with long roundtrip times.

C. Kim (Ed.): ICOIN 2005, LNCS 3391, pp. 99–108, 2005.

TCP increases the transfer rate slowly to prevent massive packet losses [2]. The slow increasing rate, however, causes a big problem - especially in high speed networks with long delays - because it takes long time to accomplish full network utilizations. A bias against flows with long roundtrip times is that the throughput of a flow with a long roundtrip time decreases significantly [3]. In the window based congestion control scheme, the window of a flow with shorter roundtrip times grows faster than that of a flow with longer roundtrip times. If two flows with different roundtrip times compete in the same bottleneck link, then the flow with longer roundtrip times experiences severe throughput penalties. Another performance penalty imposed in wireless networks is that frequent packet losses caused by link errors reduce the TCP transfer rate.

High Speed TCP (HSTCP) [4] and Scalable TCP (STCP) [5] are proposed to solve the network underutilization problem of TCP in high speed networks. Both protocols adjust their transfer rate adaptively based on the current congestion window size, so that the larger the window is, the faster it grows. The network utilization in high speed networks can be improved; however, it still poses a bias against flows with long roundtrip times. Constant Rate Window Increase (CRWI) [6], which increases the congestion window in proportion to the square of roundtrip times in each roundtrip time, can reduce a bias against flows with long roundtrip times. CRWI, however, does not scale well in heterogeneous networks because of the bandwidth variations of networks provided. Binary Increase Congestion (BIC) [7] control is proposed to solve the network underutilization problem and a bias against flows of long roundtrip times. BIC increases the congestion window additively or logarithmically based on the current window size and the designated window size. The mechanism improves the network utilization but still has bias. TCP Westwood+ (TCPW+) [8], which is the TCP Westwood (TCPW) [9] with enhanced bandwidth estimations, controls the congestion window based on the estimated network bandwidth at a TCP source. Both TCPW and TCPW+ do not scale high in high speed networks with large delays because they are based on the traditional AIMD congestion control scheme.

This paper proposes a new TCP congestion control scheme called TCP-TP (TCP with Two-Phase congestion control). TCP-TP enhances bandwidth utilizations in high speed networks as well as a bias against flows with long roundtrip times. Moreover, TCP-TP improves transmission performance in wireless networks by increasing transfer rate quickly when packet losses are occurred. It also works well with the traditional TCP flows as well as the TCP-TP flows. TCP-TP has been implemented in real systems and is experimented among other TCP variants such as BIC or TCPW+ in order to validate its performance characteristics. The paper is organized as follows: Section 2 describes the design of TCP-TP, Section 3 describes the details of the implementations, experiments, and results of the experiment, and Section 4 concludes the paper.

2 Two-Phase Congestion Control

TCP-TP controls the network congestions in two phases. One phase is called a fair convergence phase and the other is a congestion avoidance phase. But both are based on the application's transfer data patterns that consist of a bulk data transfer mode and a non-bulk data transfer mode. In a non-bulk data transfer mode, TCP-TP adopts the AIMD scheme to offer fairness, whereas in a bulk data transfer mode, TCP-TP adjusts transfer rate adaptively considering both the current transfer rate and the estimated bandwidth to reduce the possibility of a bias. This section describes how TCP-TP estimates the network bandwidth and a detailed description of how two-phase congestion control works. Table 1 below shows the definitions of the parameters used in this paper.

Table 1. Parameter definitions

Parameters	Definitions
ACK	Acknowledgement message
BDP	Bandwidth Delay Product
RTT	Packet roundtrip time (sec)
BaseRTT	Minimum RTT during the session (sec)
IT_{ack}	Inter-arrival time between two successive immediate ACKs (sec)
S_{pkt}	Size of the packet ($Bytes$)
S_{rcv}	Bytes the receiver received during IT_{ack} ($Bytes$)
W_{cn}	Congestion window ($Bytes$)
B_m	Measured bandwidth at the sender ($Bytes/sec$)
B_s	Smoothed bandwidth ($Bytes/sec$)
N_p	Number of packets contained in one packet bunch
α	Constant for B_m
β	Number of bytes to transmit in one RTT ($Bytes$)
γ	Constant for β
δ	Maximum interval of packets sent to the network in bulk data transfer mode ($microseconds$)

2.1 Bandwidth Measurement

TCP-TP measures network bandwidth at the sender and increases the transfer rate considering the measured bandwidth B_m and the current congestion window W_{cn} in order to prevent bandwidth underutilization, which is the one of the problems with the traditional TCP. The sender can measure the network bandwidth using Equation 1.

$$B_m = \frac{S_{rcv}}{IT_{ack}}. \tag{1}$$

B_m is calculated by the inter-arrival time between successive immediate ACKs IT_{ack} and the bytes the receiver received during the time S_{rcv}. The standard TCP uses delayed ACKs, and the delay is up to 500 milliseconds [10]. The receiver delays ACKs to reduce the number of packets sent on to the network. The delayed ACKs, however make B_m differ from the real network bandwidth so that the sender should consider the delayed ACKs to calculate B_m correctly. TCP-TP transmits several packets together to induce an immediate ACK based on the fact that the receiver transmits the immediate ACK when it receives two full sized segments successively [11]. This scheme is similar to the packet bunch scheme [12], and at least four packets must consist of one packet bunch to induce at least two successive immediate ACKs. Figure 1 shows the process of transmitting packet bunches and the procedure of calculating B_m.

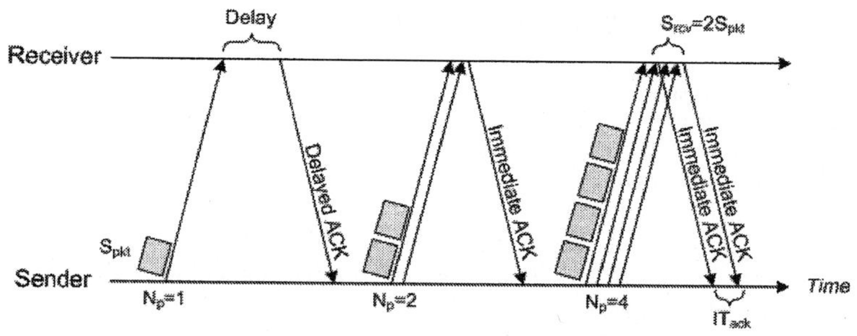

Fig. 1. Bandwidth measurement based on ACK inter-arrival times

The measured bandwidth B_m varies according to the network status changes, such as the appearance of competing traffics, the increase of system overhead, or the variety of loss rates, and so on. Therefore, B_m should be stabilized. Equation 2 explains the procedure to obtain the smoothed bandwidth (B_s is calculated to stabilize B_m).

$$B_{s(i)} = \alpha B_{s(i-1)} + (1 - \alpha) B_m, \quad where \quad 0 \leq \alpha \leq 1. \tag{2}$$

The current smoothed bandwidth $B_{s(i)}$ reflects the previously smoothed bandwidth $B_{s(i-1)}$ and the current measured bandwidth B_m. Even if B_m fluctuates largely due to the unstable network, B_s remains stabilized because B_s reflects the previous smoothed bandwidth. When the network bandwidth changes, B_s becomes B_m as B_s reflects B_m.

2.2 Determining the Transfer Rate

Bandwidth Delay Product (BDP), which is the minimum size of the send window to maximize the efficiency of transmission, can be calculated by multiplying the network bandwidth and roundtrip times RTT [13]. The sender can calculate the optimal size of the congestion window W_{cn} based on BDP. When the sender calculates W_{cn}, it uses the smoothed bandwidth B_s instead of the measured bandwidth B_m since B_m fluctuates largely when the network is unstable. The standard TCP sender measures RTT when it receives ACK. The measured RTT includes the buffering delays of routers involved. The sender uses the minimum RTT during the session. That is, BaseRTT, instead of the measured RTT, is used in calculating W_{cn} to reduce the buffering overhead at intermediate routers. Equation 3 shows the calculation of W_{cn}.

$$W_{cn} = B_s \times BaseRTT. \tag{3}$$

Increasing the congestion window W_{cn} to BDP at once may raise the network overhead especially when BDP is large. TCP-TP increases W_{cn} gradually whenever it receives ACK to reduce the network overhead. Assume that TCP doubles W_{cn} each RTT in a slow start phase, the transfer rate increases $\frac{W_{cn}}{RTT}$ $bytes/RTT$ or $\frac{W_{cn}}{RTT^2}$ $bytes/sec$ [3]. The increasing rate of throughput is inversely proportional to RTT^2, hence if the competing flows share the same bottleneck link, the flow with longer RTT experiences severe throughput penalties. To reduce the bias against the flow with long RTT, TCP-TP increases the congestion window in proportion to RTT^2 whenever the sender receives ACK. In the traditional TCP, the larger the window is, the faster it grows. If a new flow competes with other flows then the new flow is hard to take bandwidth from other flows. To provide throughput fairness to a new flow, TCP-TP increases the congestion window in proportion to the difference between BDP and W_{cn}.

$$\beta = \gamma RTT^2 (B_s \times BaseRTT - W_{cn}). \tag{4}$$

Equation 4 calculates β, the number of bytes to transmit in one RTT. The increasing rate of throughput is $\gamma(B_s \times BaseRTT - W_{cn})$ $bytes/sec$. The throughput increases regardless of RTT; accordingly the proposed mechanism can reduce the bias against flows with long RTT. Moreover, the fast convergence to the full network bandwidth is possible even if the network has large BDP, since β considers BDP. The longer the difference between BDP and W_{cn} is, the larger the throughput increasing rate is. The throughput unfairness of a new flow can be improved because the new flow easily takes a bandwidth from other competing flows.

$$\frac{W_{cn}}{RTT^2} = \gamma(B_s \times BaseRTT - W_{cn}), \quad where \quad W_{cn} = \frac{B_s \times RTT}{2}. \tag{5}$$

The larger γ is, the faster the transfer rate of TCP-TP increases. The throughput increasing rate should consider a fair sharing of the network bandwidth with the traditional TCP. The throughput increasing rate of the traditional TCP in the slow start phase is $\frac{W_{cn}}{RTT^2}$, and the throughput increasing rate of TCP-TP is $\gamma(B_s \times RTT - W_{cn})$. Let these two throughput increasing rates equal one another when W_{cn} is a half of BDP in order to provide friendliness between the traditional TCP and TCP-TP as Equation 5. The extract of Equation 5 is that

γ equals $\frac{1}{RTT^2}$. The average RTT of the Internet is considered 200 milliseconds [14], and it makes γ a constant.

TCP-TP transmits packets whenever the sender receives ACK. β is the additional number of bytes to transmit in one RTT. The number of ACK in one RTT is roughly $\frac{W_{cn}}{2S_{pkt}}$ with delayed ACK and $\frac{W_{cn}}{S_{pkt}}$ without delayed ACKs. Therefore, the number of packets to transmit to the receiver when the sender receives ACK is $\frac{2S_{pkt}(\beta+W_{cn})}{W_{cn}}$ with delayed ACK and $\frac{S_{pkt}(\beta+W_{cn})}{W_{cn}}$ without delayed ACK.

2.3 Congestion Control

TCP-TP transmits several packets in a bunch. However, it is possible that the data, which the sender transmits to the receiver, is less than the number of packets contained in one packet bunch N_p. In this case, the receiver delays ACK and the sender can not measure the network bandwidth correctly. To solve this problem TCP-TP divides the applications into two modes according to their current transfer behaviors by monitoring the intervals of packets sent to the network [6]. If the number of packets sent to the network with δ intervals is equal to or more than N_p, TCP-TP initiates a fair convergence phase or begins a congestion avoidance phase. Figure 2 illustrates the mechanism.

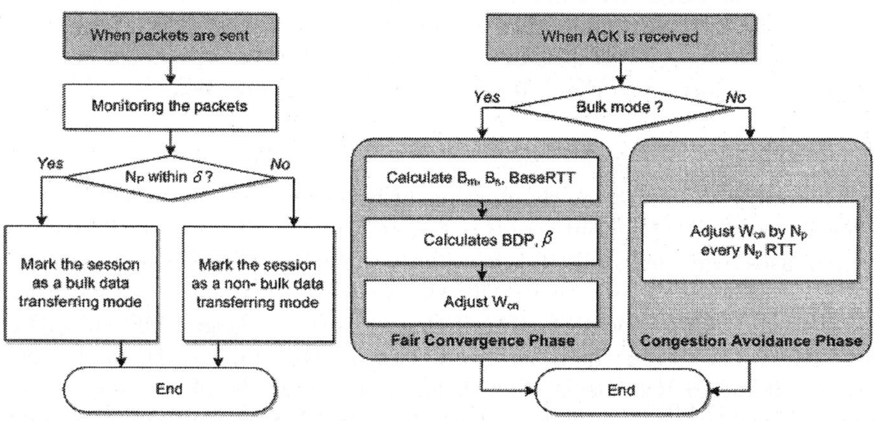

Fig. 2. Two-Phase congestion control algorithm

In a fair convergence phase, TCP-TP measures the network bandwidth, calculates the number of packets to transmit to the receiver. In a congestion avoidance phase, TCP-TP works the same as the traditional TCP except that TCP-TP increases and decreases W_{cn} by N_p number of packets at once, whereas the traditional TCP does it by one packet to detect the transfer mode changes. TCP-TP adjusts W_{cn} every N_p RTT for the purpose of making the transfer rate of TCP-TP in a congestion avoidance phase equal to that of the traditional TCP. Consequently, TCP-TP and the traditional TCP become friendly.

3 Experiments

TCP-TP has been implemented by modifying the Linux kernel 2.6.7. The experimental systems used to verify TCP-TP consist of the server and the client. In addition, the network emulator called NIST Net, is used to emulate various packet roundtrip times. The server and client are connected directly to NIST Net using 100Base-T Ethernet and Wireless LAN networks. The well known protocol analysis programs netperf, tcpdump and tcptrace are used to gather and analyze the experimental data. Table 2 shows the parameters and values used in the experiments. We set α to 0.8, which is the same value in calculating a smoothed RTT in the traditional TCP.

Table 2. Experimental parameters

Parameters	Definitions
S_{pkt}	1448 Bytes
N_p	4
α	0.8
γ	0.000025
δ	100 microseconds
Network Bandwidth	11Mbps (Wireless), 100 Mbps (Wired)
RTT	50, 250, 500, 1000 milliseconds (emulated)

A total of 10 groups of experiments are conducted to evaluate the performance of TCP-TP. In addition, the performance of TCP-Reno, BIC, and TCPW+ are also experimented for comparison purposes. Figure 3 shows the changes of the transfer rate in TCP-Reno and TCP-TP by varying RTTs. The transfer rate of TCP-Reno decreases greatly as RTT gets longer. This is because the transfer rate of TCP-Reno increases inversely proportional to the square of RTT. However, the transfer rate of TCP-TP decreases slightly as RTT gets longer since TCP-TP adjusts the transfer rate regardless of RTT. The small transfer rate of TCP-TP at the beginning of the session is due to the fact that TCP-TP takes some time to measure the network bandwidth.

Throughput comparisons of TCP-TP with other TCP variants in wired networks as well as wireless networks are shown in Figure 4. When RTT is small, the throughputs of other TCP variants are similar in both wired and wireless networks. As shown in Figure 4(a), the throughput of TCP-Reno significantly decreases when RTT is longer in wired networks. For that reason, TCP-Reno has a bias against flows with long RTT. TCP-TP adjusts the transfer rate regardless of RTT, but the throughput of TCP-TP also decreases because RTT is longer. The throughput decreases in this instance is because TCP-TP requires some time to measure the network bandwidth at the beginning of the session. The transmission performance of TCP-TP in wireless network is also higher than

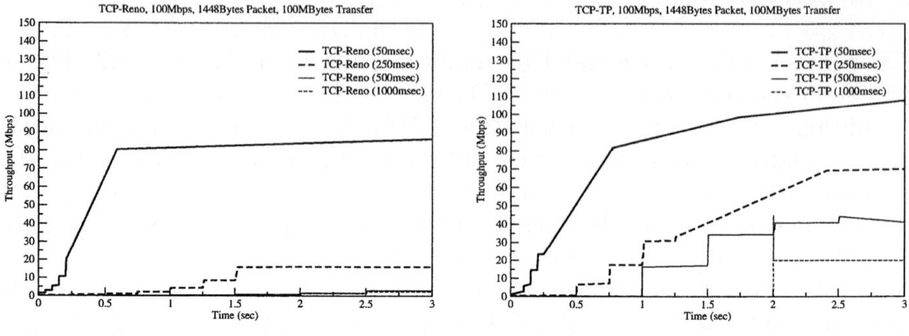

(a) TCP-Reno (b) TCP-TP

Fig. 3. Throughput trace of TCP-Reno and TCP-TP with different RTTs

the other TCP variants as shown in Figure 4(b). However, the throughput is fluctuated highly due to the unstable network condition. The wireless network has a high loss rate which causes a transfer rate reduction, so the congestion window rarely increases. TCP-TP increases transfer rate faster than the others when the congestion window is small. When packets are lost, TCP-TP increases the transfer rate fast; hence improves the transmission performance compared to the others.

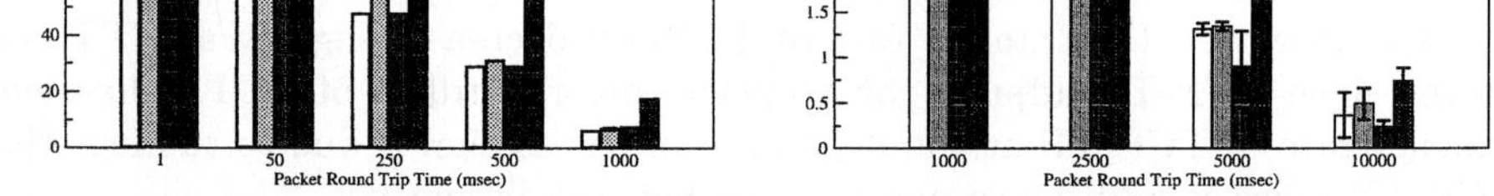

(a) Wired Networks (b) Wireless Networks

Fig. 4. Throughput comparison with TCP variants

An additional 10 groups of experiments were also conducted to evaluate the fairness among TCP-TP flows, and the friendliness between TCP-TP and the traditional TCP. The fairness among TCP-TP flows implies that competing TCP-TP flows sharing the same link have equal opportunities to transfer data.

The friendliness between TCP-TP and the traditional TCP means that TCP-TP can coexist with the traditional TCP without decreasing the throughput of the traditional TCP. Both fairness and friendliness are important properties of a TCP protocol and must be provided [9].

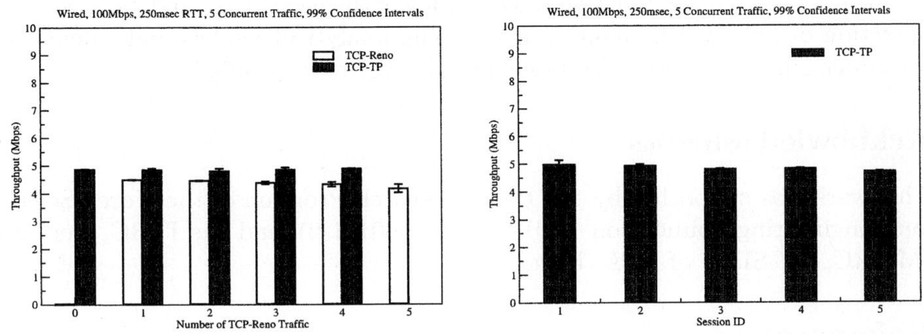

(a) Concurrent TCP-Reno and TCP-TP (b) Concurrent TCP-TP

Fig. 5. Throughput comparison of five simultaneous flows

Figure 5 shows the throughput comparisons when five simultaneous flows share the same links. Figure 5(a) shows that the TCP-Reno flows achieve a similar throughput regardless of the numbers of TCP-LP flows. TCP-TP is friendly with TCP-Reno because TCP-TP does not decrease the throughput of TCP-Reno. When five TCP-TP flows share the same link, they also achieve a similar throughput as shown in Figure 5(b). TCP-TP is, therefore, considered to provide the fairness among TCP-TP flows.

4 Conclusions

This paper proposed a two-phase congestion control (TCP-TP) scheme where the sender measures the network bandwidth and controls the transfer rate adaptively according to the application's transfer mode, which is either a bulk data transfer mode or a non-bulk data transfer mode. TCP-TP controls the network congestion in two phases, a fair convergence phase and a congestion avoidance phase, and switches between phases according to the application's transfer mode. TCP-TP in a fair convergence mode controls the transfer rate in regards to BDP and the congestion window in order to reduce a bias against flows with long RTT and improve network utilizations. TCP-TP in a congestion avoidance phase controls the transfer rate similar to the traditional TCP in order to share bandwidth fairly. TCP-TP also improves the transmission performance in the wireless networks due to the increasing transfer rate quickly in the case when packet losses caused by link errors reduce the congestion window.

TCP-TP has been implemented and experimented in real environments. As the experimental result shows, TCP-TP reduces bias against flows with long RTT and improves throughputs compared to other TCP variants. The experimental results also show that TCP-TP share the network bandwidth fairly among TCP-TP flows as well as TCP-Reno flows.

Future research will improve the accuracy of the measured bandwidth and the detection of bandwidth changes. Developing analytical models and calculating the processing overhead of TCP-TP are also yet to be studied.

Acknowledgements

This work was supported by the Basic Research Program of the Korea Science and Engineering Foundation (R01-2002-000-00141-0), and the ITRC programs (MMRC, HY-SDR) of IITA, Korea.

References

1. Van Jocobson: Congestion Avoidance and Control, Proceedings of ACM SIGCOMM, (1998) 314–329
2. Kevin Lai, Mary Baker: Measuring Bandwidth, Proceedings of IEEE INFOCOM'99, New York (1999) 235–245
3. Sally Floyd, Van Jocobson: On Traffic Phase Effects in Packet-Switched Gateways, Journal of Internetworking: Practice and Experience, **3(3)** (1992) 115–156
4. Sally Floyd: High Speed TCP for Large Congestion Windows, IETF Internet Draft, draft-floyd-tcp-highspeed-02.txt (2002)
5. Tom Kelly: Scalable TCP: Improving Performance in High Speed Wide Area Networks, ACM SIGCOMM Computer Communication Review, **33(2)** (2003) 83–91
6. Sally Floyd: Connections with Multiple Congested Gateways in Packet-Switched Networks Part 2: Two-way Traffic, ACM Computer Communication Review, **21(5)** (1991) 30–47
7. Lisong Xu, Khaled Harfoush, and Injong Rhee: Binary Increase Congestion Control (BIC) for Fast, Long Distance Networks, Proceedings of IEEE INFOCOM'04, Hong Kong (2004)
8. Luigi A. Grieco, Saverio Mascolo: Performance Evaluation and Comparison of Westwood+, New Reno, and Vegas TCP Congestion Control, ACM SIGCOMM Computer Communica-tion Review, **34(2)** (2004) 25–38
9. Saverio Mascolo, Claudio Casetti, Mario Gerla, M. Y. Sanadidi, and Ren Wang: TCP West-wood: Bandwidth Estimation for Enhanced Transport over Wireless Links, Mobile Comput-ing and Networking (2001) 287–297
10. M. Allman, V. Paxon, and W. Stevens: TCP Congestion Control, RFC2581, IETF, (1999)
11. R. Braden: Requirements for Internet Hosts - Communication Layers, RFC1122, IETF (1989)
12. V. Paxon: End-to-End Internet Packet Dynamics, IEEE/ACM Transactions on Networking, **7(3)** (1999) 277–292
13. Brian L. Tierney: TCP Tuning Guide for Distributed Applications on Wide Area Networks, Usenix ;login: journal, **26(1)** (2001) 33–39
14. Network Service & Consulting Cooperation: Internet Traffic Report, http://www.internettrafficreport.com

A New Congestion Control Mechanism of TCP with Inline Network Measurement

Tomohito Iguchi, Go Hasegawa, and Masayuki Murata

Graduate School of Information Science and Technology, Osaka University
1-5 Yamadaoka, Suita, Osaka 565-0871, Japan
{t-iguti, hasegawa, murata}@ist.osaka-u.ac.jp

Abstract. In this paper, we propose a novel congestion control mechanism of TCP, by using an inline network measurement technique. By using information of available bandwidth of a network path between sender and receiver hosts, we construct quite a different congestion control mechanism from the traditional TCP Reno and its variants, based on logistic and Lotka-Volterra models from biophysics. The proposed mechanism is intensively investigated through analysis and simulation evaluations, and we show the effectiveness of the proposed mechanism in terms of scalability with the network bandwidth, convergence time, fairness among connections, and stability.

1 Introduction

Transmission Control Protocol (TCP) is the de facto standard transport layer protocol of the Internet. It was first designed in the 1970s, and the first Request for Comments (RFC) on TCP was released in 1981 [1]. Since the Internet has undergone such developmental changes as link bandwidth and number of nodes, TCP has also been frequently modified and enhanced according to such changes in the network.

One of the most important functions of TCP is its congestion control mechanism [2]. Its main purpose is to avoid and resolve network congestion, and to distribute network bandwidth equally among competing connections. TCP employs a window-based congestion control mechanism that adjusts data transmission speed by changing the window size. TCP's window updating mechanism is based on an Additive Increase Multiplicative Decrease (AIMD) policy: a TCP sender continues increasing window size additively until it detects a packet loss(es) and decreases it multiplicatively when a packet loss occurs. In [3], the authors argue that an AIMD policy is suitable for efficient and fair bandwidth usage in a distributed environment.

However, there are many problems in the congestion control mechanism of the current version of TCP (TCP Reno), which have emerged with increases of heterogeneity and the complexity of the Internet ([4-6] for some instances). The main reason is the fixed AIMD parameter values in increasing/decreasing window size, whereas they should be changed according to the network environment. For example, many previous papers [7-9] described that the throughput of TCP connections decreases when it traverses wireless links, since TCP cannot distinguish a congestion-oriented packet loss and a wireless-oriented (link loss and/or handoff) packet loss. In this case, the AIMD parameters, especially the decreasing parameters, must be changed dynamically according to the origins of the packet loss.

C. Kim (Ed.): ICOIN 2005, LNCS 3391, pp. 109–121, 2005.

Another problem is the low throughput of TCP connections in high-speed and long-delay networks. In [10], the authors argued that a TCP Reno connection cannot fully utilize the link bandwidth of such networks, since the increasing parameter (1 packet per a Round Trip Time (RTT)) is too small and the decreasing parameter, which halves the window size when a packet loss occurs, is too large for networks with a large bandwidth-delay product.

Although there are many solutions against the above problems [8-13], almost all inherit the basic mechanism of the congestion control mechanism of TCP: the AIMD mechanism triggered by the detection of packet losses in the network. Most previous papers focused on changing the AIMD parameters according to the network environment. Since those methods may employ ad hoc modifications for a certain network situation, their performance is not clear when applied to other network environments.

TCP's performance is incomplete because the TCP sender does not have an effective mechanism to recognize the available bandwidth of the network path between sender and receiver hosts. In a sense, a traditional TCP Reno can be considered a tool that measures available bandwidth because of its ability to adjust the congestion window size to achieve a transmission rate appropriate to the available bandwidth. However, it is ineffective since it only increases window size until a packet loss occurs. In other words, it induces packet losses to obtain information about the available bandwidth(-delay product) of the network. All modified versions of TCP using AIMD policy contain this essential problem.

If a TCP sender recognizes an available bandwidth quickly and adequately, we can create a further better mechanism for congestion control in TCP. Many measurement tools have been proposed in the literature [14-16] to measure the available bandwidth of network paths. However, we cannot directly employ those existing methods into TCP mechanisms since they utilize a lot of test probe packets; they also require a long time to obtain one measurement result. Fortunately, we have a method called Inline measurement TCP (ImTCP) that does not include these problems [17, 18]. It does not inject extra traffic into the network, and it estimates the available bandwidth from data/ACK packets transmitted by an active TCP connection in an inline fashion. Furthermore, the ImTCP sender can obtain the information of available bandwidth every 1–4 RTT that follows well the traffic fluctuation of the underlying IP network. Therefore, we can make a novel congestion control mechanism of TCP by using an inline network measurement mechanism.

In this paper, we propose a new congestion control mechanism of TCP that utilizes available bandwidth information obtained from inline measurement techniques. The proposed mechanism does not use ad hoc algorithms such as TCP Vegas [19], instead employs algorithms which have a mathematical background, by which we are able to mathematically discuss and guarantee its behavior even though it poses a simplification of the target system. More importantly, it becomes possible to give a reasonable background on our selection of control parameters within TCP, instead of conducting intensive computer simulation and/or choosing parameters in an ad-hoc fashion. We borrowed the algorithm from biophysics; a logistic equation and a Lotka-Volterra competition model [20] that describe changes in the population of species are applied to the window updating mechanism of our TCP. This application can be done by considering the number of a single species as a window size of a TCP connection, a carrying capacity as a bottleneck link bandwidth, and interspecific competition among species as a bandwidth share among competing TCP connections. We present in detail how to apply the logistic equation and the Lotka-Volterra competition model to the congestion control algorithm of TCP as well as analytic investigations of the proposed algorithm. Then, we can utilize the existing discussions and results on various characteristics of the model, including stability, fairness, and robustness. Giving those characteristics to TCP is our main purpose of the current study. We also present some preliminary simulation results to evaluate the proposed mechanism and show that, compared with traditional TCP Reno and other TCP variants it utilizes network bandwidth effectively, quickly, and fairly.

The rest of this paper is organized as follows. In Section 2, we import two mathematical models from biophysics: the logistic equation and the Lotka-Volterra competition model. The transition of those models to the data transmission rate control algorithm in computer networks is presented. Then we propose a new congestion control mechanism with inline network measurement and discuss its characteristics in Section 3. In Section 4, we show some simulation results to evaluate the performance of the proposed mechanism. We finally conclude this paper and offer future work in Section 5.

2 Mathematical Models Applied to a Congestion Control Mechanism

In this section, we briefly summarize the mathematical models from biophysics utilized by our proposed mechanism to control the congestion window size of a TCP connection.

2.1 Logistic Equation

The logistic equation is a formula that approximates the evolution of the population of a species over time. Generally, the increasing rate of a species population becomes larger as the species population becomes larger. However, since there are various restrictions about living environments, the environmental capacity, which is the maximum of the population of the species, exists. The logistic equation approximates such changes in the species population:

$$\frac{dN}{dt} = \varepsilon \left(1 - \frac{N}{K}\right) N$$

where t is time, N is the number of species, K is the carrying capacity of the environment, and ε is the intrinsic growth rate of the species. Fig. 1 shows changes of the species population (N) as a function of time where $K = 100$ and ε changes to 0.6, 1.8, 2.4, and 3.0. Looking at lines with $\varepsilon = 0.6$ and 1.8, we can observe the following characteristics of the logistic equation; when N is much smaller than K, the increasing speed of N becomes larger as N increases. On the other hand, when N becomes close to K, the increasing rate decreases and N converges to K. As ε increases from 0.6 to 1.8, the convergence time becomes small at an expense of some overshoot. When ε is 2.4 or 3.0, however, N does not converge to K and remains unstable. This is a well-known characteristic of the logistic equation, where ε should be less than 2.0 to successfully converge N to K.

We consider that the increasing trend of N in the logistic equation can be applied to the control of the data transmission speed of TCP. That is, by considering N as the transmission rate of a TCP sender and K as the physical bandwidth of the bottleneck link, rapid and stable link utilization can be realized. However, the logistic equation describes the population of one species, whereas there are two or more TCP connections in the practical network. In the next subsection, we introduce an extended model that describes the changes of the population of two species with interaction between themselves.

2.2 Lotka-Volterra Competition Model

The Lotka-Volterra competition model is a famous model for the population growth of two species including interspecific competition between them. In the model, a logistic equation is modified to include the effects of interspecific competition as well as intraspecific competition. The Lotka-Volterra model of interspecific competition is comprised of the following equations for the population of species 1 and 2, respectively:

$$\frac{dN_1}{dt} = \varepsilon_1 \left(1 - \frac{N_1 + \gamma_{12} \cdot N_2}{K_1}\right) N_1 \tag{1}$$

$$\frac{dN_2}{dt} = \varepsilon_2 \left(1 - \frac{N_2 + \gamma_{21} \cdot N_1}{K_2}\right) N_2 \tag{2}$$

where N_i, K_i, and ε_i are the population, the environmental capacity, and the intrinsic growth rate of the species i, respectively. γ_{ij} is the ratio of the competition coefficient of species i on species j.

In this model, the population of species 1 and 2 does not always converge to some value larger than 0, and in some cases one of them sometimes dies out. It depends on the value of γ_{12} and γ_{21}. It is a well-known characteristic that when the following conditions are satisfied, the two species survive in the environment:

$$\gamma_{12} < \frac{K_1}{K_2}, \ \gamma_{21} < \frac{K_2}{K_1} \tag{3}$$

Assuming that the two species have the same characteristics, they have the same values of K, ε, and γ, Equations (1) and (2) can be written as follows:

$$\frac{dN_1}{dt} = \varepsilon \left(1 - \frac{N_1 + \gamma \cdot N_2}{K}\right) N_1 \tag{4}$$

$$\frac{dN_2}{dt} = \varepsilon \left(1 - \frac{N_2 + \gamma \cdot N_1}{K}\right) N_2 \tag{5}$$

Note that the conditions in Equation (3) become $\gamma < 1$. Fig. 2 shows the change in the population of the two species by using Equations (4) and (5), where species 2 join the environment in 10 time units after species 1. We can observe from this figure that the population of the two species converges quickly at the same value, which is considered an ideal behavior for the control of the transmission rate in computer networks.

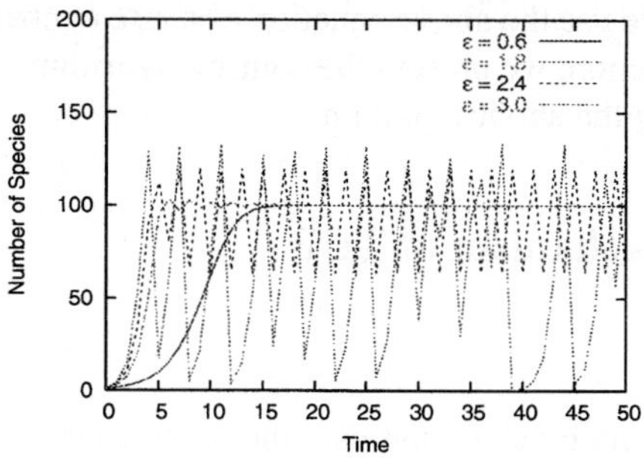

Fig. 1. Logistic equation ($K = 100$)

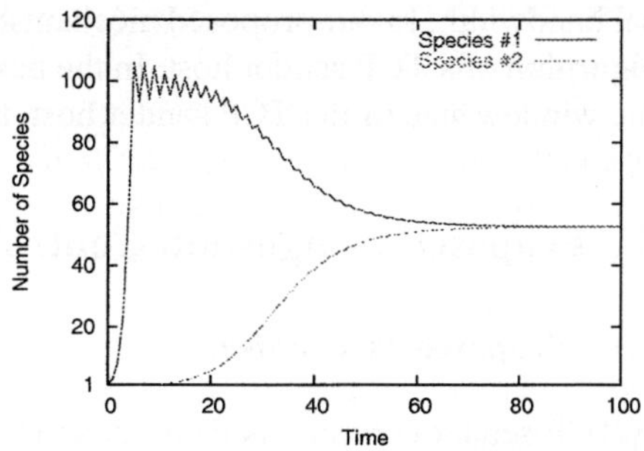

Fig. 2. Changes in species population with Lotka-Volterra competition model ($\varepsilon = 1.95$, $\gamma = 0.9$, $K = 100$)

2.3 Application to Transmission Rate Control Algorithm

In a practical network there are usually more than two TCP connections sharing a network bandwidth. We can easily extend Equations (4) and (5) for n species as follows:

$$\frac{dN_i}{dt} = \varepsilon \left(1 - \frac{N_i + \gamma \cdot \sum_{j=1,i\neq j}^{n} N_j}{K}\right) N_i \tag{6}$$

When we consider Equation (6) as the control algorithm for the data transmission rate for TCP connection i (N_i), it is necessary for connection i to know the data transmission rates of all other connections that share the same bottleneck link. This assumption is quite unrealistic in the current Internet. However, when we obtain the available bandwidth for connection i with the inline measurement mechanism [17], we can approximate the sum of the data transmission rates of all of other connections as follows:

$$\sum_{j=1,i\neq j}^{n} N_j = K - A_i$$

Thus, Equation (6) becomes as follows;

$$\frac{dN_i}{dt} = \varepsilon \left(1 - \frac{N_i + \gamma \cdot (K - A_i)}{K}\right) N_i \tag{7}$$

where N_i, and A_i are the data transmission rate and the available bandwidth for connection i. K is the physical bandwidth of the bottleneck link, where we assume that all connections share the same bottleneck link. Our proposed mechanism assumes that we can obtain A_i and K by using the inline network measurement. The current version of ImTCP [17, 18] can measure A_i with high accuracy in various conditions of the network. Therefore, we consider that the proposed mechanism can set A_i by ImTCP. On the other hand, because a physical bandwidth measurement algorithm is now under consideration, we directly set K to the correct value. However, the change of the physical bandwidth of the network path is smaller than that of the available bandwidth, so we can expect that the measurement error is also smaller. Hence, we consider that the effect of the measurement error of the physical bandwidth (K) on the performance of the proposed mechanism is negligible when we use the measurement results of the physical bandwidth. In our proposed mechanism, we use the above equation as a rate control algorithm of a TCP sender host. In the next section, we present the control algorithm of the window size of the TCP sender host, using the above equation.

3 Proposed Congestion Control Mechanism of TCP

3.1 Proposed Mechanism

A TCP sender controls its data transmission rate by changing its window size when it receives an ACK packet. Here we convert Equation (7) to obtain an increasing algorithm of the window size in TCP. The window size of connection i, w_i, is calculated from N_i, the transmission rate, by the following simple equation:

$$w_i = N_i \cdot base\_rtt_i$$

where $base\_rtt_i$ is the minimum value of the RTTs of connection i. Then Equation (7) can be rewritten as follows:

$$\frac{dw_i}{dt} = \varepsilon \left(1 - \frac{w_i + \gamma \cdot base\_rtt_i \cdot (K - A_i)}{K \cdot base\_rtt_i}\right) w_i$$

We next change the equation in RTT.

$$\frac{dw_i}{drtt} = \varepsilon \left(1 - \frac{w_i + \gamma \cdot base\_rtt_i \cdot (K - A_i)}{K \cdot base\_rtt_i}\right) w_i \qquad (8)$$

Finally, we derive the amount of the increase in window size when an ACK packet is received at the TCP sender by considering that w_i ACK packets are received in one RTT:

$$\Delta w_i = \varepsilon \left(1 - \frac{w_i + \gamma \cdot base\_rtt_i \cdot (K - A_i)}{K \cdot base\_rtt_i}\right)$$

This is the fundamental equation in increasing window size in our proposed mechanism. Since this equation requires the measurements of the available bandwidth and physical bandwidth of a network path, we use the same algorithm as TCP Reno for window updating algorithm until the measurement results are obtained through inline network measurements. In cases of packet loss(es), window size is decreased in identical way to TCP Reno. When a timeout occurs, sender TCP discards all measurement results, window size is reset to 1, and the slow-start phase begins as a TCP Reno sender does.

3.2 Characteristics of the Proposed Mechanism

Here we briefly summarize the characteristics of the proposed mechanism:

- Scalability with network bandwidth
  When we consider one TCP connection in the network, the window size $w(t)$ is represented as the following formula from the Equation (8):

$$w(t) = \frac{w_0 \cdot K \cdot base\_rtt_i}{w_0 + (K \cdot base\_rtt_i - w_0) \cdot e^{-\varepsilon t}}$$

where w_0 is an initial value of window size. Here we assume $w_0 = (1-b) \cdot K \cdot base\_rtt_i$ and $w_t = c \cdot K \cdot base\_rtt_i$, and calculate the time T it takes to increase the window size from w_0 to w_t. Then,

$$T = \frac{1}{\varepsilon} \cdot log\left(\frac{c}{1-c} \cdot \frac{b}{1-b}\right)$$

Note that T is independent on K, the physical bandwidth of the bottleneck link, meaning that our proposed mechanism can increase its window size in the same time regardless of the network bandwidth.

- Convergence time
  As shown in Figures 1 and 2, the transmission rate size quickly converges to a certain value. Note that $\varepsilon \leq 2$ is required for stable convergence.
- Stability
  In the original Lotka-Volterra competition model, $\gamma < 1$ is required for the survival of the two species in the environment when the environmental capacity of the two species are the same. This characteristic can also be satisfied in our proposed mechanism. However, in a practical network, all connections do not always have the same physical link capacity, especially when the access link bandwidth is relatively small. We comment on this issue in Section 4.
- Lossless behavior
  When n TCP connections exist in the network, the sum of the converged window sizes of all connections becomes:

$$\sum_{i=1}^{n} w_i = \frac{n}{1+(n-1)\cdot\gamma} \cdot K \cdot base\_rtt_i$$

  This means that the sum of the window size increases when n increases. However, it is limited by $\frac{K \cdot base\_rtt_i}{\gamma}$, obtained by calculating $\lim_{n\to\infty} \sum_{i=1}^{n} w_i$ from the above equation. That is, when the buffer size of the bottleneck router is enough large, no packet loss occurs. Note that the traditional TCP Reno cannot avoid periodic packet losses due to its window control algorithm.
- Fairness among connections
  From Equation (7), it is obvious that the converged window sizes of all TCP connections become identical when they have the same value as the physical bandwidth K. However, a problem may emerge when K is different among connections, which is discussed in Section 4.

4 Simulation Results

In this section, we present some simulation results to evaluate the performance of the congestion control mechanism proposed in Section 3. We used ns-2 [21] for the simulation experiments. The traditional TCP Reno, HighSpeed TCP (HSTCP) [10] and Scalable TCP [12] were chosen for performance comparison. In our proposed mechanism, the available bandwidth information was obtained through an inline measurement mechanism. Note that we directly give the physical bandwidth information to the TCP sender since there is currently no effective mechanism to measure the physical bandwidth in an inline fashion. We set $\varepsilon = 1.95$ and $\gamma = 0.9$ for the proposed mechanism. For HSTCP, we use the parameters described in [10].

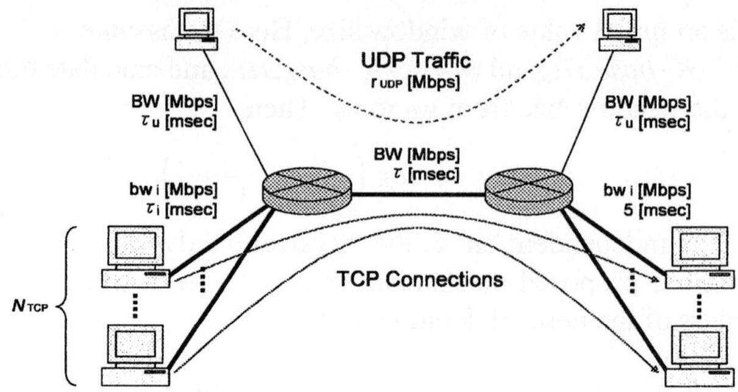

Fig. 3. Network topology in simulation experiments

The network model used in the simulation is depicted in Fig. 3. It consists of sender/receiver hosts, two routers, and links between the hosts and routers. $N_{\rm tcp}$ TCP connections are established between TCP sender i and TCP receiver i. For creating the background traffic, we inject UDP packets at the rate of $r_{\rm udp}$ into the network. That is, $N_{\rm tcp}$ TCP connections and an UDP flow share the bottleneck link between the two routers. The bandwidths of the bottleneck link and the access link for the UDP sender/receiver are all set to BW, and the propagation delays are τ and $\tau_{\rm u}$, respectively. The bandwidth and the propagation delay of the access link for TCP sender i are bw_i and τ_i, respectively. We deployed a Taildrop discipline at the router buffer, and the buffer size is set to twice the bandwidth-delay product of the bottleneck link between the two routers.

We first confirm the fundamental behavior of the proposed mechanism with one TCP connection ($N_{\rm tcp} = 1$). Fig. 4 shows the change in the window size of TCP Reno, HSTCP, Scalable TCP and the proposed mechanism, where we set $bw_1 = 100$ Mbps, $\tau_1 = 5$ msec, $BW = 100$ Mbps, $\tau = 40$ msec, $\tau_{\rm u} = 5$ msec and $r_{\rm udp} = 50$ Mbps. This result shows that TCP Reno, HSTCP and Scalable TCP connections experience periodic packet losses due to buffer overflow, since they continue increasing the window size until packet loss occurs. On the other hand, the window size of the proposed mechanism converges to an ideal value quickly and no packet loss occurs. Furthermore, the increasing speed is much larger than that of HSTCP and Scalable TCP, meaning that the proposed mechanism effectively utilizes the link bandwidth.

We next investigate the scalability with link bandwidth of the proposed mechanism by checking the convergence time, which is defined as the time it takes for the TCP connection to utilize 99% of the link bandwidth. We set $N_{\rm tcp} = 1$, $\tau_1 = 5$ msec, $\tau = 40$ msec and $\tau_{\rm u} = 5$ msec. Fig. 5 shows the change of the convergence time when we change BW from 10 Mbps to 1 Gbps, where $r_{\rm udp}$ is set to $(0.2 \cdot BW)$ Mbps and bw_1 is set equal to BW. In the figure, the average values and the 95% confidence intervals for 10 simulations experiments are shown. From this figure, we can see that the TCP Reno connection requires quite a large time to fully utilize the link bandwidth since the increasing speed of the window size is fixed at a small value regardless of the link bandwidth. HSTCP dramatically reduces the convergence time, but the larger the link bandwidth becomes, the larger convergence time requires to fill the bottleneck link bandwidth. This means

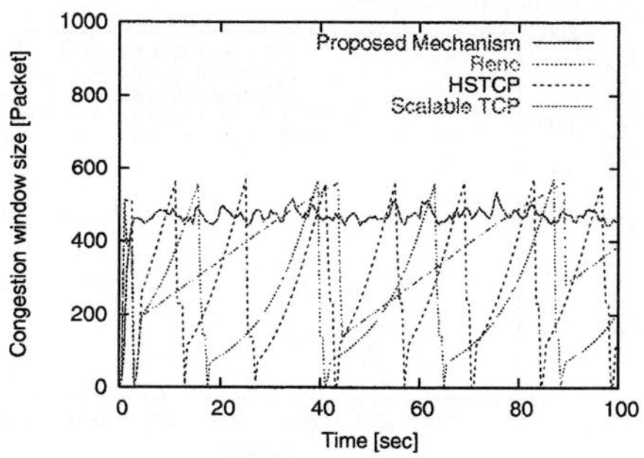

Fig. 4. Change of window size in one connection case

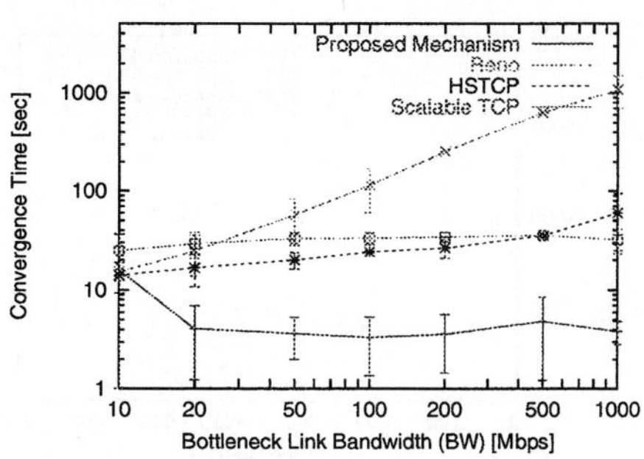

Fig. 5. Convergence time as a function of bottleneck link bandwidth

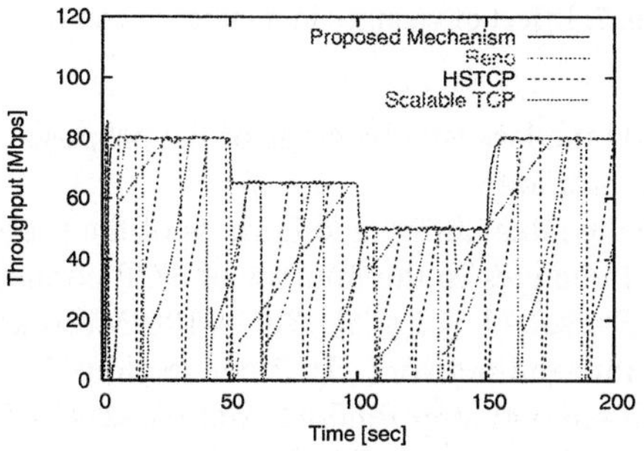

Fig. 6. Adaptability to change in available bandwidth

that HSTCP is fundamentally unable to resolve the scalability problem of TCP Reno. In the case of Scalable TCP, the convergence time remains constant regardless of the link bandwidth, which was confirmed in [12]. However, it is quite larger than that of the proposed mechanism. The proposed mechanism, however, keeps the smallest and the almost constant convergence time regardless of the link bandwidth, which shows good scalability with the network bandwidth as described in Subsection 3.2. The confidence interval of the proposed mechanism is large because the measurement results have some errors.

Adaptability to changes in the available bandwidth is also an important characteristic of the transport layer protocol. To confirm, we set N_{tcp} = 1, bw_1 = 100 Mbps, τ_1 = 5 msec, BW = 100 Mbps, τ = 40 msec, and τ_u = 5 msec, and change r_{udp} so that the available bandwidth of the bottleneck link is 80 Mbps from 0 to 50 sec, 65 Mbps from 50 to 100 sec, 50 Mbps from 100 to 150 sec, and 80 Mbps from 150 to 200 sec. Fig. 6 presents the change of the throughput of a TCP connection in TCP Reno, HSTCP, Scalable TCP and the proposed mechanism. The results obviously show the effectiveness of the proposed mechanism, which gives good adaptability to the changes of the available bandwidth. Furthermore, no packet loss occurs even when the available bandwidth suddenly decreases. On the other hand, TCP Reno, HSTCP and Scalable TCP connections experience many packet losses during simulation time, and the link utilization is much

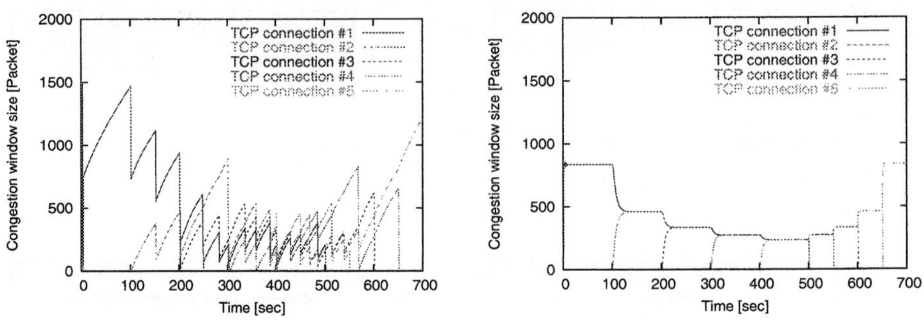

(a) TCP Reno (b) Proposed mechanism

Fig. 7. Effect of changes in number of connections

lower than 100%. This is the largest advantage of the proposed mechanism which uses an inline measurement technique.

We also investigate the adaptability and fairness of the proposed mechanism by investigating the effect of changes in the number of TCP connections. We set $N_{tcp} = 5$, $bw_i = 100$ Mbps, $\tau_i = 5$ msec ($1 \leq i \leq 5$), $BW = 100$ Mbps and $\tau = 40$ msec. We do not inject UDP traffic into the network. TCP connections 1–5 join the network at 0, 100, 200, 300, and 400 sec and stop sending data packets at 500, 550, 600, 650, and 700 sec, respectively. Fig. 7 shows change of window size for the five TCP connections as a function of simulation time in TCP Reno (Fig. 7(a)) and the proposed mechanism (Fig. 7(b)). This figure shows that TCP Reno cannot maintain the fairness among connections at all, mainly because it takes long time for all connections to have the fair window sizes. Furthermore, TCP Reno connections suffer from cyclic packet losses. On the other hand, the proposed mechanism converges the window size very quickly and no packet loss occurs when a new connection joins the network. Furthermore, when the TCP connection leaves the network, the proposed mechanism quickly fill the unused bandwidth.

Finally we investigate the effect of the heterogeneity of the access network such as the difference of the access link bandwidth. We set $N_{tcp} = 2$, $bw_1 = 10$ Mbps, $bw_2 =$ 20 Mbps, $\tau_1 = \tau_2 = 5$ msec, $\tau = 40$ msec, and we change BW from 5 Mbps to 30 Mbps. We do not inject UDP traffic into the network. Fig. 8 shows the change in the throughput of the two TCP connections in TCP Reno and the proposed mechanism, as a function of BW. We observe from the figure that TCP Reno equally shares the bottleneck link bandwidth regardless of the value of BW. On the other hand, the proposed mechanism shows an interesting characteristic. When $BW < bw_1$, the two TCP connections equally share bottleneck link bandwidth. When $bw_1 < BW < bw_2$, however, the bottleneck link bandwidth is distributed proportionally to the ratio of bw_1 and bw_2. This property can be explained from the equation utilized by the proposed mechanism. By using Equation (7), the converged transmission rate for connection i, denoted by $\hat{N}_i$, which have different physical link bandwidth (K_i), can be calculated as follows (a detailed calculation is omitted due to space limitations):

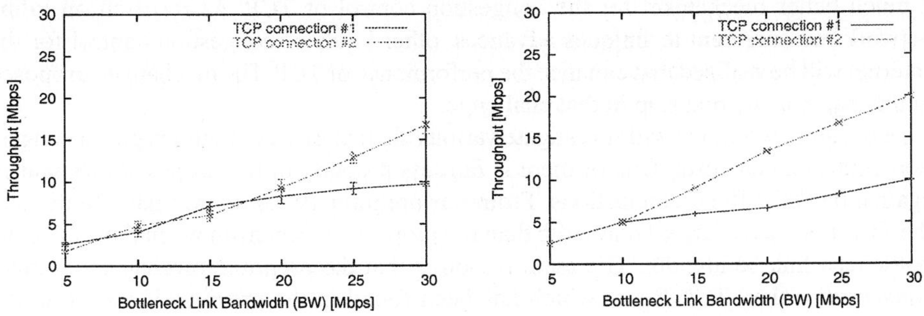

(a) TCP Reno (b) Proposed mechanism

Fig. 8. Effect of different access link bandwidths

$$\hat{N}_i = \frac{K_i}{\sum_{i=1}^{n} K_i} \cdot BW \quad (9)$$

It is under the condition of $\gamma < 1$. That is, the bottleneck link bandwidth is shared proportionally to the physical bandwidth of each TCP connection. Since a physical bandwidth of the network path is defined as a bandwidth of the tightest link between TCP hosts (a sender and a receiver), the simulation results in Fig. 8 that matches Equation (9).

We consider that this characteristic is ideal for a congestion control strategy on the Internet; in the history of the Internet, the ratio of the bandwidth of access network to that of backbone network has been changing over time [22]. Therefore, compared with access networks, the resources amount of backbone network are sometimes scarce and sometimes plentiful. We believe that when backbone resources are small, they should be shared equally between users regardless of their access link bandwidth. When they are sufficient, on the other hand, they should be shared according to the access link bandwidth. The characteristic of the proposed mechanism found in Fig. 8 and Equation (9) realizes such a resource sharing strategy.

5 Conclusion and Future Work

In this paper, we proposed a new congestion control mechanism of TCP which utilized an inline network measurement technique. The proposed mechanism is based on the logistic equation and the Lotka-Volterra competition model that represents population changes of species. We applied the two models to the transmission rate control in the computer network and constructed a new algorithm to change the window size of the TCP connections. Through analysis and simulation evaluations, we confirmed the effectiveness of the proposed mechanism for scalability, convergence speed, fairness, stability, and so on.

We consider that by obtaining the important information for congestion control, for example, the available and physical bandwidth of the network path, we can create a much better mechanism for the congestion control of TCP. As research on inline network measurement techniques advances, other kinds of congestion control for the Internet will be realized that enhance the performance of TCP. The mechanism proposed in this paper is the first step in that challenge.

For future work, we will investigate various characteristics of our proposed congestion control mechanism. One of them is fairness property of the proposed mechanism against the TCP Reno connections. From our preliminary results, we have found that the TCP Reno uses larger bandwidth than the proposed mechanism when they share the bottleneck link bandwidth. The main reason is that the proposed mechanism is more *conservative* than TCP Reno, which has been found in the previous literature in the fairness between TCP Reno and TCP Vegas [23, 24]. Now we consider improving the proposed mechanism to solve this problem, based on the ideas that we regulate the parameters of the proposed mechanism or we switch the behavior of the proposed mechanism according to the existence of TCP Reno connections in the network. Another research plan is to compare the performance of the proposed mechanism with the other kinds of congestion control algorithm such as FAST TCP [13], which has similar characteristics in terms of lossless behavior with window size stability. Additionally, we will investigate fairness among connections with different RTTs, the effect of measurement errors of available/physical bandwidth, and so on.

Acknowledgement

This work was partly supported by the Ministry of Education, Science and Culture, Grant-in-Aid for (A)(16680003), 2004.

References

1. J. B. Postel, "Transmission control protocol," *Request for Comments 793*, Sept. 1981.
2. W. R. Stevens, *TCP/IP Illustrated, Volume 1: The Protocols*. Reading, Massachusetts: Addison-Wesley, 1994.
3. D.-M. Chiu and R. Jain, "Analysis of the increase and decrease algorithms for congestion avoidance in computer networks," *Journal of Computer Networks and ISDN Systems*, pp. 1–14, June 1989.
4. S. Shenker, L. Zhang, and D. D. Clark, "Some observations on the dynamics of a congestion control algorithm," *ACM Computer Communication Review*, vol. 20, pp. 30–39, Oct. 1990.
5. J. C. Hoe, "Improving the start-up behavior of a congestion control scheme of TCP," *ACM SIGCOMM Computer Communication Review*, vol. 26, pp. 270–280, Oct. 1996.
6. L. Guo and I. Matta, "The War Between Mice and Elephants," *Technical Report BU-CS-2001-005*, May 2001.
7. Z. Fu, P. Zerfos, H. Luo, S. Lu, L. Zhang, and M. Gerla, "The impact of multihop wireless channel on TCP throughput and loss," in *Proceedings of IEEE INFOCOM 2003*, Apr. 2003.
8. K. Xu, Y. Tian, and N. Ansari, "TCP-Jersey for wireless IP communications," *IEEE Journal on Selected Areas in Communications*, vol. 22, pp. 747–756, May 2004.
9. E. H.-K. Wu and M.-Z. Chen, "JTCP: Jitter-based TCP for heterogeneous wireless networks," *IEEE Journal on Selected Areas in Communications*, vol. 22, pp. 757–766, May 2004.

10. S. Floyd, "HighSpeed TCP for large congestion windows," *RFC 3649*, Dec. 2003.
11. M. Gerla, M. Y. Sanadidi, R.Wang, A. Zanella, C. Casetti, and S. Mascolo, "TCP Westwood: Congestion window control using bandwidth estimation," in *Proceedings of IEEE GLOBECOM '01*, Nov. 2001.
12. T. Kelly, "Scalable TCP: Improving performance in highspeed wide area networks," in *proceedings of PFLDnet '03: workshop for the purposes of discussion*, Feb. 2003.
13. C. Jin, D. X. Wei, and S. H. Low, "FAST TCP: motivation, architecture, algorithms, performance," in *Proceedings of IEEE INFOCOM 2004*, Mar. 2004.
14. B. Melander, M. Bjorkman, and P. Gunningberg, "A new end-to-end probing and analysis method for estimating bandwidth bottlenecks," in *Proceedings of IEEE GLOBECOM 2000*, Nov. 2000.
15. M. Jain and C. Dovrolis, "End-to-end available bandwidth: Measurement methodology, dynamics, and relation with TCP throughput," in *Proceedings of ACM SIGCOMM 2002*, Aug. 2002.
16. V. Ribeiro, R. Riedi, R. Baraniuk, J. Navratil, and L. Cottrell, "pathChirp: Efficient available bandwidth estimation for network paths," in *Proceedings of NLANR PAM2003*, Apr. 2003.
17. M. L. T. Cao, "A study on inline network measurement mechanism for service overlay networks," Master's thesis, Graduate School of Information Science, Osaka University, Feb. 2004.
18. M. L. T. Cao, G. Hasegawa, and M. Murata, "Available bandwidth measurement via TCP connection," to be presented at IFIP/IEEE MMNS 2004, Oct. 2004.
19. L. S. Brakmo, S. W.O'Malley, and L. L. Peterson, "TCP Vegas: New techniques for congestion detection and avoidance," in *Proceedings of ACM SIGCOMM'94*, pp. 24–35, Oct. 1994.
20. J. D. Murray, *Mathematical Biology I: An Introduction.* Springer Verlag Published, 2002.
21. The VINT Project, "UCB/LBNL/VINT network simulator - ns (version 2)." available from http://www.isi.edu/nsnam/ns/.
22. J. Crowcroft, S. Hand, R. Mortier, T. Roscoe, and A. Warfield, "QoS's downfall: At the bottom, or not at all!," in *Proceeding of ACM SIGCOMM 2003 Workshop on Revisiting IP QoS (RIPQOS)*, Aug. 2003.
23. J. Mo, R. J. La, V. Anantharam, and J. Walrand, "Analysis and comparison of TCP reno and vegas," in *Proceedings of IEEE INFOCOM'99*, Mar. 1999.
24. G. Hasegawa, K. Kurata, and M. Murata, "Analysis and improvement of fairness between TCP Reno and Vegas for deployment of TCP Vegas to the Internet," in *Proceedings of IEEE ICNP 2000*, Nov. 2000.

V-TCP: A Novel TCP Enhancement Technique for Wireless Mobile Environments

Dhinaharan Nagamalai[1], Dong-Ho Kang[2], Ki-Young Moon[2], and Jae-Kwang Lee[1]

[1] Department of Computer Engineering, Hannam University, 306 791, Daejeon, South Korea
dhinaharan2000@yahoo.com
jklee@netwk.hannam.ac.kr

[2] E-Government Security Research Team, Information Security Division, ETRI, 161, Gajeong-dong, Yuseong-gu,Daejeon, South Korea
{dhkang, kymoon}@etri.re.kr

Abstract. Transmission Control Protocol (TCP) is a reliable transport protocol tuned to perform well in habitual networks made up of links with low bit-error rates. TCP was originally designed for wired networks, where packet loss is assumed to be due to congestion. In wireless links, packet losses are due to high error rates and the disconnections induced are due to mobility. TCP responds to these packet losses in the same way as wired links. It reduces the window size before packet retransmission, initiates congestion control avoidance mechanism and resets its transmission timer. This adjustment results in unnecessary reduction of the bandwidth utilization causing significant degraded end-to-end performance. A number of approaches have been proposed to improve the efficiency of TCP in an unreliable wireless network. But researches only focus on scenarios where TCP sender is a fixed host. In this paper we propose a novel protocol called V-TCP (versatile TCP), an approach that mitigates the degrading effect of host mobility on TCP performance. In addition to scenarios where TCP sender is fixed host, we also analyze the scenario where TCP sender is a mobile host. V-TCP modifies the congestion control mechanism of TCP by simply using the network layer feedback in terms of disconnection and connection signals, thereby enhancing the throughput in wireless mobile environments. Several experiments were performed using NS-2 simulator and the results were compared to the performance of V-TCP with Freeze-TCP [1], TCP Reno and with 3-dupacks [2]. Performance results show an improvement of up to 50% over TCP Reno in WLAN environments and up to 150% in WWAN environments in both directions of data transfer.

1 Introduction

The research community has been working very hard to find the solutions to the poor performance of TCP over the wireless networks. Such work can be classified into 4 main approaches: a) Some researchers have focused on the problem

C. Kim (Ed.): ICOIN 2005, LNCS 3391, pp. 122–131, 2005.

at the data link layer level (LL) by hiding the deficiencies of the wireless channel from the TCP. b) Others believe in splitting the TCP: one for the wired domain and another for the wireless domain. c) A third group of researchers believe in modifying the TCP to improve its behavior in the wireless domain. d) A final group is those who believe in the creation of new transport protocols tuned for the wireless networks. While several approaches have been proposed for mitigating the effect of channel conditions,[3] [4] [5] [6] [7] of late approaches that tackle mobility induced disconnections have also been proposed [1] [2] [10]. But all these approaches only concentrate on scenarios where TCP sender is a fixed host (FH) and they do not perform good when the sender is a mobile host (MH). Moreover some of these approaches require modifications to the TCP at FH [5] as well as some are susceptible to scalability issues. [4] In this paper we propose a novel protocol V-TCP (versatile TCP) that mitigates the degrading affect of mobility on TCP performance. V-TCP is designed to improve the TCP performance in mobile environments in duplex ways, i.e. from MH to FH as well as FH to MH. This protocol requires only modification to TCP at MH and is based on network providing feedback about the mobility status in terms of connection and disconnection event signals. V-TCP uses the connection and disconnection signals to freeze/continue ongoing data transfer and changes the action taken at RTO event, thereby leading to enhanced throughput. Through simulation we have shown that V-TCP performs better than freeze-TCP, TCP Reno and 3-dupack. The results show that V-TCP achieves an improvement of up to 50environments in both directions of data transfer. This paper is organized as follows. In section 2 we analyze the motivation and the related approaches. In section 3 V-TCP's mechanism is introduced. In section 4 experiments using the simulator are presented. Finally in section 5 we compare our approach with other approaches that were proposed earlier. We wrap up our contributions in section 6.

2 Design Alternatives and Related Work

There have been many wireless enhancements proposed over the last decade. Many approaches that attempt to reduce the detrimental effects of mobility on TCP performance have been proposed [1] [2] [10]. Our main focus is the approach that requires modifications only in MH [1] [2]. Our approach V-TCP falls in this category. Let's now examine the weakness of the existing approaches.

The Freeze-TCP Approach [1]. It requires an indication of the looming disconnection by the network layer at the MH. The disadvantage of this approach is that how fast a prediction about the disconnections by the network layer is needed to be available to the TCP at the MH. If it is available faster than RTT of the connections, this may lead to degraded performance by the freeze-TCP. Moreover RTT values differ depending on the connections and this adds to the difficulty in accurate predictions.

3-Dupacks Approach [2]. It requires information about the ongoing mobility to the TCP layer at MH by the network layer. The drawback in this approach is that, the TCP (sender) at FH reduces the congestion window (cwnd) and slow start threshold (ssthresh) parameters when it enters fast recovery phase, thus resulting in degraded throughput.

TCP Reno. It retains all the enhancements of TCP Tahoe and also incorporates a new algorithm, the fast recovery algorithm. Fast recovery is based on the fact that a dupack indicates that a segment has left the network.One performance problem of the Reno TCP is that, if multiple packets are lost from one window of data during the Fast Retransmit and Fast Recovery, the TCP source has to wait for the RTO to expire.

3 V-TCP Mechanism

The main idea of designing V-TCP is to improve the performance of TCP in wireless mobile environments in the presence of temporary disconnections caused by mobility. Unlike previous research approaches, V-TCP not only improves the performance when the TCP sender is a FH, but also when TCP sender is a MH. The only change required in the V-TCP mechanism is to modify the network stack at the MH and also it requires feedback regarding the status of the connectivity. V-TCP makes reasonable assumptions which are similar to the network layer in wireless mobile environments like mobile IP [11]. V-TCP assumes that a network layer sends connection_event signal to TCP, when MH gets connected to the network and a disconnection_event signal, when the MH gets disconnected from the network. V-TCP utilizes these signals for freeze/continue data transfer and changes the actions taking place at RTO (Retransmission Time Out_event), leading to enhanced TCP throughput. The mechanism of V-TCP is explained as follows.

3.1 Data Transfer from MH to FH

We consider 3 event signals here for our analysis.

Disconnection_Event Signal. V-TCP's behavior under different disconnection scenarios is explained in the Fig 1. *Case 1 Sending window open.* V-TCP cancels the retransmission timer and does not wait for Ack for the packets that were sent before disconnection. *Case 2 Sending window closed.* V-TCP waits for ack, and does not cancel the retransmission timer, but waits for the RTO to occur.

Connection_Event Signal. V-TCP assumes that a network layer sends connection_event signal to TCP when MH gets connected to the network. *Case1 Sending window open.* V-TCP sets the retransmission timer after the data is

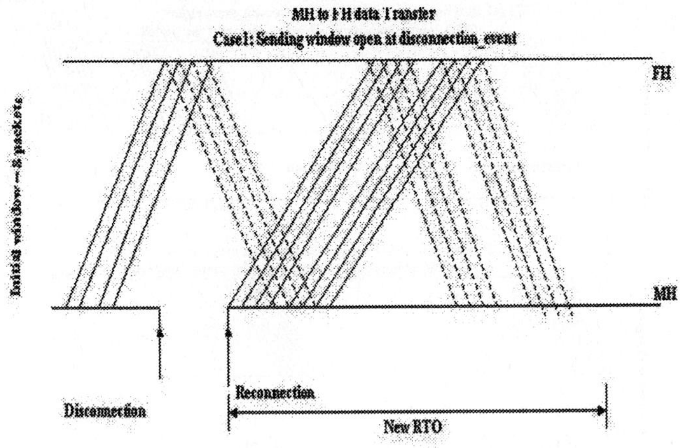

Fig. 1. V-TCP's behavior under different disconnection scenarios

sent. Because all the Acks are cumulative, any Ack for the data that had been newly sent also acknowledges the data sent before disconnection. *Case2 Sending window closed (RTO occurred).* V-TCP retransmits if the sending window is closed and RTO occurred.*Case3 Sending window closed (RTO not occurred).* V-TCP waits for an RTO event to occur.

Retransmission Time Out_Event Signal. V-TCP utilizes these signals for freeze/continue data transfer and changes the action taken place at RTO (Retransmission Time Out_event), leading to enhanced TCP throughput. V-TCP first checks, if there had been any disconnection in the network. If the disconnection had taken place (case2 as seen in Fig 2) then, V-TCP sets the ssthresh=cwnd at the time of disconnection, instead of reducing the ssthresh (behavior of TCP) and also sets cwnd=1. But in the case where a connection had occurred (case 3 as seen in Fig 2), V-TCP retransmits the lost packets without any modification to ssthresh and cwnd parameters. Thus V-TCP promptly salvages the cwnd value prior to disconnections, thus reducing under utilization of the available link capacity.

3.2 Data Transfer from FH to MH

V-TCP will delay the Ack for the last two bytes by "x' milliseconds (nearly 800 milliseconds)[8].

Disconnection_Event Signal. Once disconnected the network connectivity status is updated.

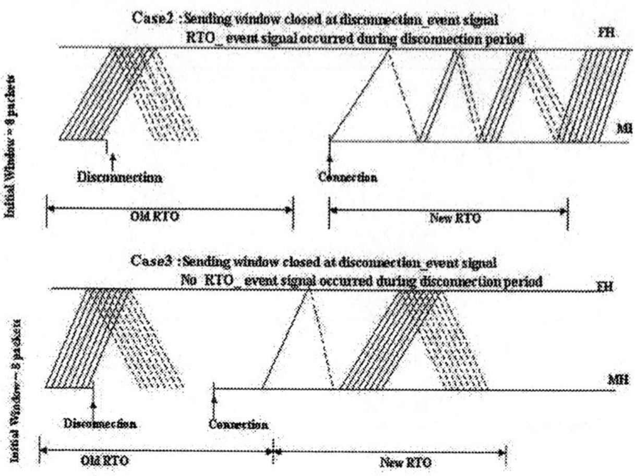

Fig. 2. V-TCP's behavior under different disconnection scenarios

Connection_Event Signal. TCP acknowledges the first bytes with ZWA (zero window advertisement) and second bytes with a FWA (full window advertisement). TCP at FH will process these acks as they have a higher sequence number than the previous Acks received [8]. The ZWA will cause the TCP sender at FH to freeze its retransmission timer, without reducing the cwnd. Thus TCP at FH is prevented from entering into congestion control mechanism when packets are lost and when the MH is disconnected.

4 Simulations

We have performed several experiments using simulation. V-TCP, freeze-TCP and 3-dupack were implemented in the network simulator ns-2 [9]. In the ns-2 simulator TCP Reno is already implemented. The only modification required is mobile IP [11] for providing mobility information to the TCP agent. The mobility of the MH is maintained by a variable, Network status, in the TCP agent whose values changes from connection to disconnection or vice versa, determined by a disconnection timer handler.

4.1 Network Topology

The network topology is shown in Fig.3. An FTP application simulated a large data transfer, with a packet size 1000 of bytes and the throughput of TCP connections was measured. Values ranging from 50ms to 5s were chosen as disconnection duration. Larger values occur in WWAN and smaller values for WLAN. The disconnection frequency was chosen as 10seconds, indicating a high mobility. The RTT was chosen as 8ms for WLAN 800ms for WWAN. The link capacity

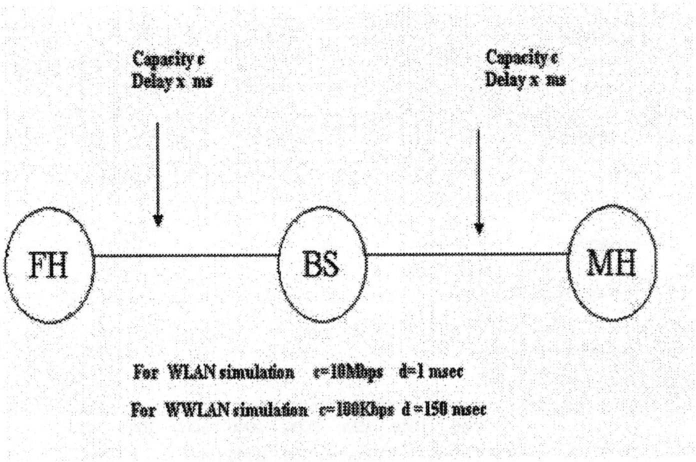

Fig. 3. Network Topology

(c) is 10Mbps for 8ms RTT (WLAN) and 100 kbps for 800ms RTT (WWAN). The capacity of both links, i.e. FH to BS and BS to MH are maintained equal to avoid any packet loss due to buffer overflow in the routers. The simulations were carried out for 100sec for WLAN environment and 1000sec for WWAN environment.

5 Performance Evaluation

The performance of V-TCP is explained for both directions and the results are compared with TCP Reno, Freeze-TCP and 3-dupack.

5.1 Data Transfer from MH to FH

As seen from Fig 4 and 5 V-TCP shows an increased throughput when compared to TCP Reno. Since there are no approaches in freeze-TCP and 3-dupacks for MH being TCP sender, therefore only TCP Reno is compared V-TCP in this case. We point the factors that lead to the increase throughput of V-TCP.

No Idle Period. When a RTO event occurs due to disconnection TCP Reno retreats exponentially. Upon MH reconnection, TCP Reno waits for the retransmission timer (RTX) to expire. Thereby the TCP Reno has to be idle until the RTX expires. As the disconnection period increases, the number of RTO events also increases. This result in exponentially increasing RTX values, thereby increasing the idle period for TCP Reno before it tries for retransmission. But in the case of V-TCP it does not have this idle period and thereby increases the performance.

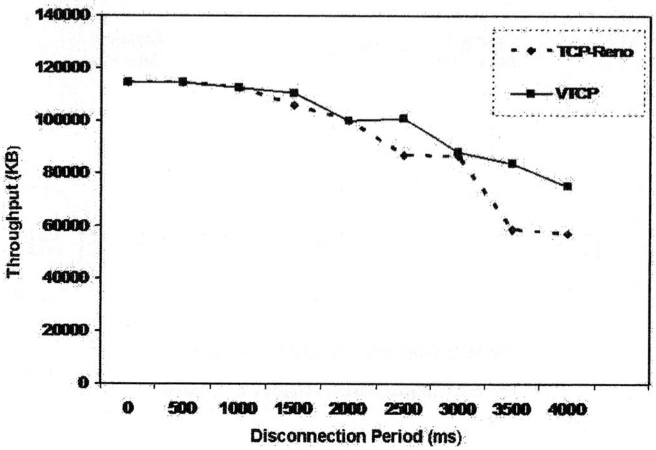

Fig. 4. Data transfer from MH to FH RTT ~8ms

RTO Event. At each event of RTO, TCP Reno reduces the ssthresh to half. If a RTO occurs when a MH is disconnected it is undesirable. But in the case of V-TCP, ssthresh is not reduced, instead it sets ssthresh equal to cwnd value reached at the time of disconnection. This results in V-TCP attaining a full window capacity faster than TCP Reno. There is a significant increase in throughput for V-TCP over TCP Reno for large RTT connections. This is because connections with large RTT have analogous large values of RTX,thereby increasing the idle period for TCP Reno. Performance results show an improvement of up to 50%

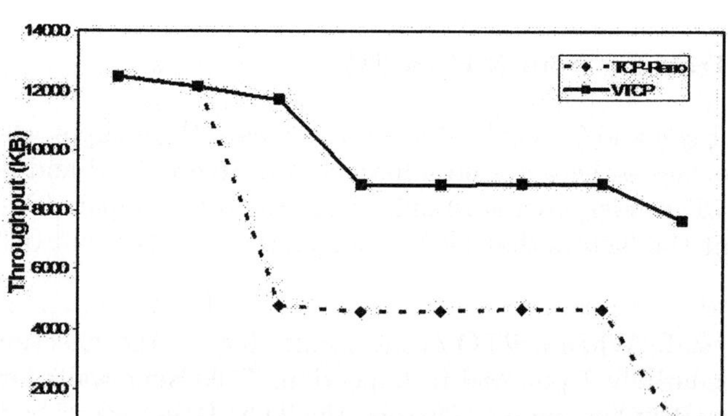

Fig. 5. Data transfer from MH to FH RTT ~800ms

improvement over TCP Reno for short RTT connections and up to 150% in the case of long RTT connections, with long periods of disconnection.

5.2 Data Transfer from FH to MH

We now compare the performance for FH to MH of V-TCP with TCP Reno, 3-dupacks and freeze-TCP. As seen in Fig 5 and 6, there is significant improvement in performance of V-TCP over TCP Reno and 3-dupacks.

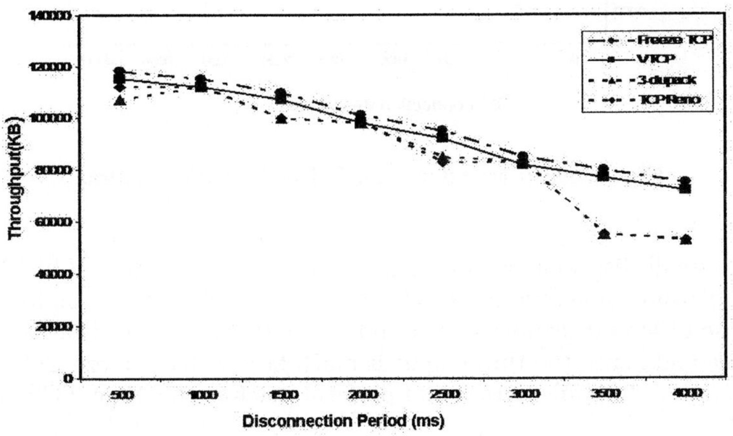

Fig. 6. Data transfer from FH to MH RTT 8ms

WLAN Environment. For long disconnection intervals we see that V-TCP and freeze-TCP showed an improvement of 50 %. But in WLAN environments, the idle period after reconnection is the prime factor for degraded throughput rather than reduction of cwnd. Both V-TCP and freeze-TCP perform better than TCP Reno by reducing the idle period.

WWAN Environment. In the case of small disconnections periods the performance of V-TCP and freeze-TCP are almost very similar. For disconnections up to 1000ms, both V-TCP and freeze-TCP showed up to 150% improvement over TCP Reno.But for longer disconnection period V-TCP showed only 65% improvement whereas freeze-TCP showed 150% improvement over TCP Reno. However freeze-TCP depends on predicting impending disconnection and its throughput was observed to be sensitive to variations in the prediction period. Fig 7 shows the V-TCP output for various values of 'x'. In WWAN environments, the main features that degrade the performance are idle period and reduction of

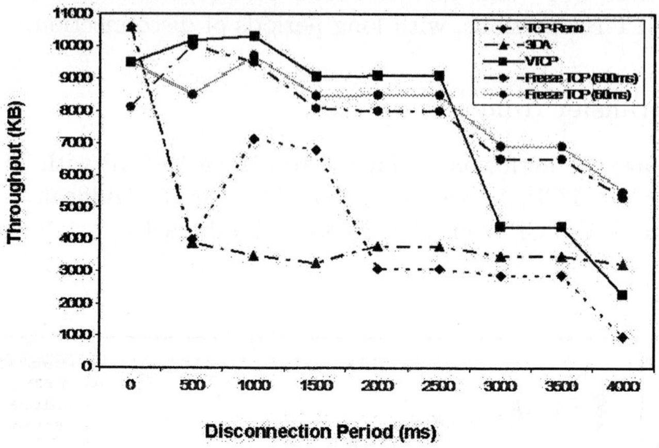

Fig. 7. Data transfer from FH to MH RTT 800ms

cwnd. For small disconnections periods where no RTO occurs, V-TCP doesn't change the cwnd value and hence achieves the same throughput as freeze-TCP. But in case of long disconnection period, V-TCP can prevent the reduction in cwnd value and hence the throughput is much better than Freeze TCP. V-TCP is also able to reduce idle period and perform much better than TCP Reno.

6 Conclusion

We have illustrated the V-TCP mechanism to alleviate the degrading effect of host mobility on TCP performance. It requires modifications only to TCP at MH and is optimized for data transfer from MH to FH as well as FH to MH. V-TCP uses feedback from the network layers at MH in terms of disconnection and connection signals, to swiftly regain the full window after the MH gets reconnected. Several simulated experiments were performed and the results of V-TCP were compared with 3-dupack, TCP Reno and freeze-TCP. V-TCP significantly performs better than TCP Reno in both directions of data transfer. Performance results show an improvement of up to 50% over TCP Reno in WLAN environments and up to 150% in WWAN environments in both directions of data transfer. As mentioned earlier 3-dupack and freeze-TCP approaches do not deal with data transfer from MH to FH and hence we compare them only for FH to MH. We wrap up by proposing a new approach called V-TCP, that performs better than TCP Reno and 3-dupack and generally analogous to that of freeze-TCP. Thus this new approach alleviates the degrading effect of host mobility in TCP.

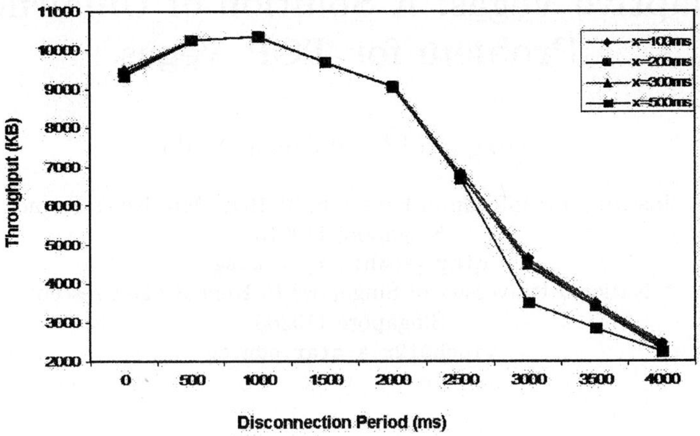

Fig. 8. Data transfer from FH to MH for various x values

References

1. Goff,T., Moronski,J., Phattak,D.S., Gupta,V.: Freeze-TCP: A true end-to-end TCP enhancement mechanism for mobile environments. Infocom, Israel (2000)
2. Ramon Caceres, Liviu Iftode.: Improving the performance of reliable transport protocol in mobile computing environments.ACM Computer Communication review, vol 13. (1995)
3. Mascolo,S., Claudio Caseetti.: TCP Westwood: Bandwidth Estimation for enhanced Transport over wireless links. ACM SIGMOBILE 7/01 Rome Italy (2001)
4. Balakrishnan,H., Padmanbham,V.N, Katz,R.: Improving Reliable Transport and Handoff Performance in Cellular Wireless Net. Wireless Networks, vol.1. (1995)
5. Sinha,P., Venkataraman,N., Sivakumar,R., Bharghavan,V.: WTCP: A reliable transport protocol for WWANs. ACM MOBICOM 99, Seattle, Washington (1999)
6. Ajay Bakre, Badrinath,B.R.: I-TCP:Indirect TCP for Mobile Hosts. Tech Rep, Reuters university (1995)
7. Balakrishnan,H., Padmanabhan,V.N, Seshan,S., Katz,R.H.: A Comparison of Mechanisms for Improving TCP Performance over Wireless Links.IEEE/ACM Transactions on Networking (1997)
8. Braden,R.:RFC 1122 for Internet Hosts-Communication Layers.(1989)
9. The network Simluator ns-2.1b8a, http://www.isi.edu/nsnam/ns
10. Brown,K., Singh,S.: M-TCP: TCP for Mobile Cellular Networks. ACM Computer Communications Review, vol27, (1997)
11. Perkins,C.:RFC 2002. IP Mobility Support (1996)
12. Montenegro,G., Dawkins,S.: Wireless Networking for the MNCRS. Internet drafts, (1998)

Adaptive Vegas: A Solution of Unfairness Problem for TCP Vegas

Qing Gao[1,2] and Qinghe Yin[1]

[1] Institute for Infocomm Research, 21 Heng Mui Keng Terrace
Singapore 119613
qing_gao@hicorp.com.sg
[2] National University of Singapore,10 Kent Ridge Crescent
Singapore 119260
yinqh@i2r.a-star.edu.sg

Abstract. We study the unfairness problem for TCP Vegas. There are three sources to cause the unfairness for Vegas: $\alpha < \beta$, over-estimation of base RTT, and multiple congested gateways. To solve the unfairness caused by multiple congested gateways, we propose a new version of Vegas–Adaptive Vegas(A-Vegas) which assigns the value of parameters adaptively according to the number of congested gateways. Simulation shows that A-Vegas not only can solve the unfairness caused by multiple congested gateways, it also reduces the unfairness caused by over-estimation. We also introduce a new fairness index, RNBS (normal bandwidth sharing ratio), which indicates the ratio of the amount of bandwidth grabbed by a connection to the uniformly distributed bandwidth.

1 Introduction

TCP Vegas was proposed by Brakmo, O'Malley and Peterson as a replacement of TCP Reno in 1994([3]). Recently, several researches pointed out that TCP Vegas exhibits unfairness problems (see [1], [2], [5],[7], [8] and [11]). The unfairness can be caused by three reasons. The first two of them are: $\alpha < \beta$ and over-estimation of the base RTT. Hasegawa et al. in [7] proposed an enhanced Vegas by setting $\alpha = \beta$ and showed that, in both homogeneous and heterogeneous cases, TCP Vegas with $\alpha = \beta$ can achieve a better fairness. In [2], the authors pointed out that taking $\alpha = \beta$ can not cancel the unfairness caused by over-estimation. The authors of [5] noticed another type of unfairness: the unfairness for flows with multiple congested gateways.

Notice that the third type unfairness is different from the other two types. The unfairness problems caused by $\alpha < \beta$ and overestimation of base RTT are "pure" unfairness. By "pure" we mean that they do not depend on the standard of fairness. In other words, they are unfair according to any common used fairness definition. But the unfairness caused by multiple congested gateways is according to max-min fairness. As it is well known, Vegas is proportionally fair (see, e.g., [10]). Although a flow with multiple congested gateways can only get a very small amount of bandwidth comparing with a flow with a single congested gateway,

C. Kim (Ed.): ICOIN 2005, LNCS 3391, pp. 132–141, 2005.

we would call it "fair" according to the proportional fairness. In our point of view, the proportional fairness is not a good criterion in this situation.

The motivation of this paper is to find a solution for the unfairness problem for TCP Vegas. We start with analyzing the unfairness from the three sources by throughput rate of Vegas flows in steady state. We notice that the reason of unfairness for flows with multiple congested gateways is that Vegas uses the same parameters α and β for all TCP connections. This unfairness problem can be relieved if we allow a Vegas source to choose parameters (α and β) according to the number of congested gateways. Hence we suggest to set the values of the parameters α, β as increasing functions of the number of congested gateways. However, it is impossible to detect the number of congested gateways in an end-to-end manner. We propose a way to estimate the number of congested gateways with the aid of extra information from the gateways along the path. We let all the congested gateways mark[3] an incoming packets randomly with the same marking probability and the source estimates the number of congested gateways according to the marking information. We find that by applying dynamically adaptive parameter α ($= \beta$), it not only can solve the unfairness caused by multiple congested gateways, it also improves the fairness problem caused by the overestimation of the propagation delay. Evidently, it also eliminates the unfairness caused by $\alpha < \beta$.

In section 2, we first briefly recall the TCP Vegas's congestion control algorithm and then analyze the unfairness from three different sources in a unified way: to analyze the throughput rate of Vegas flows in steady state. Section 3 presents our resolution for the unfairness problem caused by multiple congested gateways: Adaptive Vegas (A-Vegas). In section 4 we evaluate the performance of A-Vegas by simulation results. We conclude this paper and point out the limitation of A-Vegas in section 5.

2 Unfairness Problem in TCP Vegas

2.1 TCP Vegas's Congestion Control

TCP Vegas defines a variable "*Diff*" as the product of the minimum round trip time and the rate difference between the actual rate and the non-congestion rate

$$Diff := \left(\frac{W}{baseRTT} - \frac{PktsTrans}{rtt}\right) \times baseRTT \qquad (1)$$

where W is the current window size–the number of packets currently in transit, *PktsTrans* is the number of packets transmitted during the last Round Trip Time(RTT), *rtt* is the average RTT of the segments acknowledged during the last RTT and *baseRTT* is the RTT when there is no congestion for the connection.

[3] Here packet marking is realized as [5], the gateway marks the packet by setting the CE bit in the IP packet and the TCP sender can read that bit through ECN bit in the TCP packet header.

In practice, Vegas uses the minimum RTT the TCP connection ever experienced so far as the $baseRTT$.

In the steady state where $W = PktsTrans$, (1) can be simplified as:

$$Diff = \left(\frac{W}{baseRTT} - \frac{W}{RTT}\right) \times baseRTT. \quad (2)$$

By (2), "*Diff*" can be explained as the number of extra packets buffered on the gateways among all packets in flight (window size). The number of packets buffered in the connection path will decide whether the TCP source should increase or decrease the congestion window size to adjust the throughput according to the available bandwidth for that connection.

When *Diff* is less than α, Vegas will increase the congestion window size by 1 in next RTT; when *Diff* is greater than β, Vegas will decrease the window size by 1 in next RTT; otherwise, when *Diff* is between α and β, Vegas keeps the congestion window unchanged. Putting all together,

$$W = \begin{cases} W + 1, & \text{if } Diff < \alpha, \\ W, & \text{if } \alpha \le Diff \le \beta, \\ W - 1, & \text{if } Diff > \beta. \end{cases}$$

Roughly speaking, α, β represent respectively the lower and upper bound of the bocklogged packets from a TCP connection without changing its window size.

2.2 Analysis of the Unfairness Problem in TCP Vegas

As stated in Section 1, there exist three sources causing the unfairness problem for TCP Vegas: $\alpha < \beta$, over estimation of base RTT, and multiple congested gateways. In the following we give an analysis of the unfairness problem of TCP Vegas by analyzing the throughput rate of a Vegas flow in the steady state.

Use D_q to denote the queuing delay of a TCP connection. Then $D_q = RTT - baseRTT$. Thus (2) can be rewritten as

$$Diff = \frac{D_q \cdot W}{RTT}. \quad (3)$$

When Vegas reaches its steady state, the window size, queuing delay and round trip time will keep unchanged. Then *Diff* will be fixed at a value in the interval $[\alpha, \beta]$. Then by (3) we can get a formula for the throughput rate:

$$R_{\text{throughput}} = \frac{W}{RTT} = \frac{Diff}{D_q} \quad (4)$$

(4) gives us the following consequence:

Consequence. *The throughput rate of a TCP Vegas flow is proportional to the value of Diff and is inversely proportional to its queuing delay. The propagation delay does not affect the throughput.*

We analyze the three sources of unfairness using the above consequence.

1. $\alpha < \beta$. If we take $\alpha < \beta$, when Vegas approaches to the steady state, it is possible that different flows have different value of *Diff*. If two flows have the same queuing delay then we have

$$\frac{R_{\text{throughput}}(\text{flow}_1)}{R_{\text{throughput}}(\text{flow}_2)} = \frac{\mathit{Diff}(\text{flow}_1)}{\mathit{Diff}(\text{flow}_2)} \tag{5}$$

The ratio $\frac{\mathit{Diff}(\text{flow}_1)}{\mathit{Diff}(\text{flow}_2)}$ can be fixed at any value from α/β to β/α. If $\mathit{Diff}(\text{flow}_1) = \alpha$ and $\mathit{Diff}(\text{flow}_2) = \beta$, then the throughput of flow_2 is β/α times of that of flow_1.

In addition, in the implementation of Vegas, it is not really to compare *Diff* with α and β, but use the nearest integer of *Diff*, $\lfloor \mathit{Diff}+0.5 \rfloor$ to replace the value of *Diff*. Then the ratio $\frac{\mathit{Diff}(\text{flow}_1)}{\mathit{Diff}(\text{flow}_2)}$ can reach to $\frac{\beta+0.5}{\alpha-0.5}$. In particular, when $\alpha = 1$ and $\beta = 3$, the throughput of one flow can be 7 times of the other. Nevertheless, the larger the value of the ratio β/α, the heavier the unfairness from this source.

To analyze the other two sources we take $\alpha = \beta$.

2. Overestimation of base RTT. Because Vegas taking the minimum round time RTT_{min} as the $baseRTT$ and the existing of the queuing delay, it is almost impossible for a Vegas source to measure the $baseRTT$ correctly, especially for a latterly joined flow. The measured *Diff* value is given by

$$\mathit{Diff}_{\text{measured}} = \frac{(RTT - RTT_{min}) \cdot W}{RTT} \tag{6}$$

Because of the over-estimation, the real value of *Diff* is greater than the measured value. We have $\frac{\mathit{Diff}}{\mathit{Diff}_{\text{measured}}} = \frac{RTT-baseRTT}{RTT-RTT_{min}}$. If RTT_{min} is very close to RTT, this ratio can be very large. For a flow without over-estimation of the $baseRTT$, the throughput rate is given by $\frac{\alpha}{D_q}$ in the steady state. But for a flow with over-estimation we have $\mathit{Diff}_{\text{measured}} = \alpha$ in the steady state, the throughput is given by

$$R_{\text{throughput}}(\text{flow}_{\text{over\_est}}) = \frac{\mathit{Diff}}{D_q} = \frac{\alpha}{D_q} \cdot \frac{\mathit{Diff}}{\alpha} = \frac{\alpha}{D_q} \cdot \frac{RTT - baseRTT}{RTT - RTT_{min}} \tag{7}$$

From (7) we see that a flow with over-estimation can have very large throughput comparing with a flow without over-estimation.

3. Multi-congestion. Now we consider the case that a flow which has n ($>$ 1) congested gateways shares a congested gateway with another flow which has only one congested gateway. To emphasize the unfairness caused by multiple congested gateways, we neglect the effect of overestimation. Let $D_q(\text{flow}_1)$ and $D_q(\text{flow}_2)$ be the queuing delays of the two flows respectively. Then, if the two flows use the same value of α, then by (4) we get

$$\frac{R_{\text{throughput}}(\text{flow}_1)}{R_{\text{throughput}}(\text{flow}_2)} = \frac{\alpha/D_q(\text{flow}_1)}{\alpha/D_q(\text{flow}_2)} = \frac{D_q(\text{flow}_2)}{D_q(\text{flow}_1)} \tag{8}$$

If the first flow encounters the same queuing delay in each of the n congested gateways, then its throughput is only $1/n$ of that of the second flow.

Now it is clear that the fact a flow with multiple congested gateways having a very small throughput rate is because that it has a large queuing delay and it uses the same value of α as other flows.

It seems that the problem can be solved by setting the value of α proportional to the number of congested gateways. However, the congestion level of different gateways may not be the same: while some of them are congested severely the others may have very mild congestion. Because of this, setting the value of α proportional to the number of congested gateways may benefit the flow with multiple congested gateway more than necessary and causes unfairness in opposite direction.

The three sources of unfairness may interact with each other. The unfairness caused by one source might be cancelled by other sources. It is also possible they all work together and make the situation much worse.

To evaluate the fairness of bandwidth sharing in one link, Jain et al in [9] defined a fairness index $F = \frac{[\sum_{i=1}^{n} X_i]^2}{n \sum_{i=1}^{n} X_i^2}$. But F is not efficient to reflect the fairness for an individual. For example, when n is large, if we have $X_1 = 0$ and $X_2 = \cdots = X_n = 1$ then $F = 1 - \frac{1}{n}$ which is very close to 1. But it is absolutely unfair for X_1.

We introduce a metric, normal bandwidth sharing ratio (RNBS), to quantify the fairness for an individual flow. Let

$$\mathrm{RNBS}(\mathrm{flow}_i) = \frac{\text{average throughput of flow}_i}{\text{bandwidth}/N} \tag{9}$$

where N is the total number of flows of the gateway in consideration. That the RNBS of certain flow is close to 1 means that it obtains a fair share, while greater than 1 means that the flow is too greedy and less than 1 means that it is too conservative. If the average value of RNBS for all flows is less than 1, it means that the bandwidth is under utilized. The normal bandwidth sharing ratio can sharply describe how fair it is to an individual.

3 Adaptive Vegas – A Way to Solve the Multi-congestion Unfairness Problem

As stated in last section, the unfairness in multi-congestion case is due to the fixed boundaries (α and β) for different TCP connections. Then one possible solution is to have dynamic boundaries, which make α and β change according to the number of congested gateways, i.e. set α and β as increasing functions of n, the number of congested gateways. The functions $\alpha(n)$ and $\beta(n)$ will be based on the following principles:

1. $\alpha(n) = \beta(n)$;
2. $\alpha(n)$ is increasing with respect to n;
3. for small n, $\alpha(n)$ will be close to $\alpha_0 n$; for large n, $\alpha(n)$ will be increasing much slower than $\alpha_0 n$. Here $\alpha_0 > 0$ is a constant.

To take $\alpha(n) = \beta(n)$ is to avoid the unfairness caused by $\alpha < \beta$. The reason of the second principle is evident. As pointed out in last section, because the congestion level in different gateways are different, if we take $\alpha(n)$ proportional to n, it may benefit the flow with multiple congested gateways more than necessary and causes unfairness in opposite direction. This is the reason behind the last principle.

Based on the above principles, we choose

$$\alpha(n) = \alpha_0 n^\gamma, \tag{10}$$

for $n \geq 1$, where $0 < \gamma < 1$ is a constant.

Now the key problem is converted to how to measure n for each TCP connection. It seems that it is impossible to solve it in an end-to-end way.

We propose a method to estimate the number of congested gateways with the aid of extra information from gateways along the path. We let each congested gateway (the buffer of which is non-empty) to mark an incoming packet independently with a uniform probability p. If there are n congested gateways along the path, then the end-to-end probability for a packet being marked from the source to the destination is

$$p_{\text{end-to-end}} = 1 - (1-p)^n \tag{11}$$

From (11), we can get n as

$$n = \frac{\ln(1 - p_{\text{end-to-end}})}{\ln(1-p)} \tag{12}$$

Since the end-to-end marking probability $p_{\text{end-to-end}}$ can be estimated from the TCP source, it is easy for the source to estimate n according to (12) and then the source can adjust window size according to $\alpha(n)$. Since n is an estimated value, it is possible that $n < 1$ or even $n = 0$. We need to amend the definition of $\alpha(n)$ for $0 \leq n < 1$. One simple way is to extend the definition of $\alpha(n) = \alpha_0 n^\gamma$ to include $0 \leq n < 1$, but this would cause other problem. When a flow has only one congested gateway, and the window size and/or the value of p are/is small, it is hardly to get a marked acknowledge in one round trip time. It is also not reasonable to let $\alpha = 0$ and reduce window according to it, which may cause under-utilization of the link. We set $\alpha(0) = \frac{1}{2}\alpha(1)$ and a linear function for $0 \leq n \leq 1$ to make $\alpha(n)$ continuous at $n = 1$, i.e.

$$\alpha(n) = \begin{cases} \alpha_0 n^\gamma & \text{if } n \geq 1, \\ \frac{\alpha_0(1+n)}{2} & \text{if } 0 \leq n \leq 1. \end{cases} \tag{13}$$

In each round trip time, a TCP Vegas source counts the total number of received acknowledges, N, and the number of marked acknowledges N_m. Then it estimate the number of congested gateways by

$$\hat{n} = \frac{\ln(N - N_m) - \ln N}{\ln(1-p)} \tag{14}$$

Next it sets the value of α as $\alpha(\hat{n})$. And lastly it adjusts the window size according to $Diff$ greater than or less than α. We name this version of Vegas *Adaptive Vegas* (A-Vegas).

4 Performance of A-Vegas

In this section we evaluate the performance of A-Vegas by simulations. We design the simulation scenario as shown in Figure 1 (take $n = 10$): We have 10 gateways from G_1 to G_{10} which are connected side by side. The propagation delay between any two directly connected nodes is 10ms; the bandwidth between any two directly connected gateways is 2Mbps; the bandwidth between a host and its connected gateway is 10 Mbps. In this topology, the long TCP connection 0 shares the bandwidth with all short TCP connections on gateway-gateway links. Connection 0 starts from host H_0 and ends at host H_{11}. Connection 1 to 9 are short TCP connections that start from host H_i and end at its right neighbor host H_{i+1}, for $1 \le i \le 9$. For example, short connection 1 starts from host H_1 and ends at host H_2. To reflect different congestion conditions at different congested gateways, we set the number of flows in each connection differently. Without loss of generality, we let the long connection 0 be made up of 2 flows, the short connection 1 of 18 flows, short connection 2 of 16 flows, short connection 3 of 14 flows and so on until short connection 9 of 2 flows. Each flow starts a FTP traffic independently. The simulations run in NS2[10].

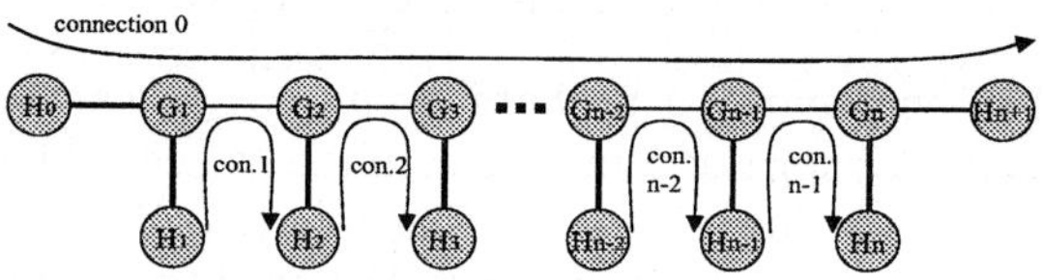

Fig. 1. Simulation Topology

Our simulation makes comparisons among three cases of TCP clients:
Case 1: TCP Vegas with $\alpha = 1$, β=3;
Case 2: TCP Vegas with $\alpha = \beta = 2$; and
Case 3: TCP A-Vegas.
For A-Vegas, we take $\alpha_0 = 2$, $\gamma = 0.3$, i.e.

$$\alpha(n) = \begin{cases} 2n^{0.3} & \text{for } n \ge 1 \\ 1 + n & \text{for } n \le 1 \end{cases}$$

and $p = 0.15$. We run two groups of simulations. First we let all TCP flows start at the same time. In this way, the effect of over-estimation of *baseRTT* can be reduced to the lowest level, since there is no congestion at all at the beginning. Then we let different flow start at a different time.

Table 1. Comparison of Throughputs and RNBS for Vegas and A-Vegas (without over-estimation)

Connection	Flow	Throughput (bytes/s)			RNBS		
		Case 1	Case 2	Case 3	Case 1	Case 2	Case 3
C0	0-1	1745	2970	12215	0.1396	0.2376	0.9772
	0-2	1745	2965	12920	0.1396	0.2372	1.0336
	Average	**1745**	**2967**	**12567**	**0.1396**	**0.2374**	**1.0054**
C1	1-1	18285	13600	12835	1.4628	1.0880	1.0268
	1-2	18285	13600	12660	1.4628	1.0880	1.0128
	1-3	18280	13600	12610	1.4624	1.0880	1.0088
	1-4	18265	13600	12615	1.4612	1.0880	1.0092
	1-5	18265	13600	12850	1.4612	1.0880	1.0280
	1-6	18265	13600	12195	1.4612	1.0880	0.9756
	1-7	18265	13600	12160	1.4612	1.0880	0.9728
	1-8	18265	13600	12375	1.4612	1.0880	0.9900
	1-9	18265	13600	12420	1.4612	1.0880	0.9936
	1-10	9140	13550	12415	0.7312	1.0840	0.9932
	1-11	9140	13550	12230	0.7312	1.0840	0.9784
	1-12	9140	13550	12525	0.7312	1.0840	1.0020
	1-13	9140	13540	12365	0.7312	1.0832	0.9892
	1-14	9140	13535	12860	0.7312	1.0828	1.0288
	1-15	9140	13535	12265	0.7312	1.0828	0.9812
	1-16	9140	13535	12380	0.7312	1.0828	0.9904
	1-17	9140	13535	12275	0.7312	1.0828	0.9820
	1-18	9140	13535	12985	0.7312	1.0828	1.0388
	Average	**13705**	**13570**	**12501**	**1.0964**	**1.0856**	**1.0001**

Note: C0: the long connection 0; C1: the short connection 1.
Case 1: Vegas, $\alpha = 1$, $\beta = 3$; Case 2: Vegas, $\alpha = \beta = 2$;
Case 3: A-Vegas.

Table 1 presents the throughput and RNBS for each flow from the long connection C0 and short connection C1 when all the TCP start simultaneously. The simulation duration is 200 second. As expected, Case 1 (Vegas with $\alpha = 1$, $\beta = 3$) has the worst fairness, because it has two sources of unfairness: $\alpha < \beta$ and multiple congestions. The long connection flows can only reach 14% of their fair share (RNBS=0.1396) where the short connection flows can grab 109% on average. The differences among short connection flows are very large: while some flows only obtain about 73%, some others grab more than 146%, 2 times of the former. This is caused by the $\alpha < \beta$ unfairness source. Case 2 (Vegas with $\alpha = \beta = 2$) shows that taking $\alpha = \beta$ has no help for the unfairness caused by multi-congestions, but it improves the fairness significantly among the short connection flows. This time the smallest among the short connection flows is about 108.2% while the largest is about 108.8%, almost the same as the former. We can see that Case 3, the A-Vegas, gives the best fairness results. Among all flows, the largest throughput which is 12985 is only 1.07 times of the smallest one, which is 12160, very close to the average bandwidth which is 12500 byte/s.

To investigate the improvement of A-Vegas to the fairness problem caused by over-estimation, we let each flow starts at a different time: flow 1 of C1 starts at 0 second, then each second after, a new flow starts until the last one, the flow 2 of C9 has started. This duration is 90 seconds. We let the 2 flows of the long connection C0 start at 40 second and 60 second. This duration of this group of simulations is 300 seconds.

Table 2. Comparison of Throughputs and RNBS for Vegas and A-Vegas (with over-estimation)

		Throughput (bytes/s)			RNBS		
Connection	Flow	Case 1	Case 2	Case 3	Case 1	Case 2	Case 3
C0	0-1	3845	3575	6405	0.3076	0.2860	0.5124
	0-2	8925	5630	13160	0.7140	0.4504	1.0528
	Average	**6385**	**4602**	**9782**	**0.5108**	**0.3682**	**0.7826**
C1	1-1	5640	4500	7205	0.4512	0.3600	0.5764
	1-2	7535	4500	7255	0.6028	0.3600	0.5804
	1-3	7520	6750	7360	0.6016	0.5400	0.5888
	1-4	7540	6750	7315	0.6032	0.5400	0.5852
	1-5	7520	6750	7990	0.6016	0.5400	0.6392
	1-6	7520	6750	7870	0.6016	0.5400	0.6296
	1-7	9400	6750	8825	0.7520	0.5400	0.7060
	1-8	9425	6750	8945	0.7540	0.5400	0.7156
	1-9	9400	9000	9100	0.7520	0.7200	0.7280
	1-10	11310	9000	10200	0.9048	0.7200	0.8160
	1-11	11310	9000	9915	0.9048	0.7200	0.7932
	1-121	15080	11250	10030	1.2064	0.9000	0.8024
	1-13	15080	11250	13940	1.2064	0.9000	1.1152
	1-14	16955	18000	17165	1.3564	1.4400	1.3732
	1-15	24440	18020	16475	1.9552	1.4416	1.3180
	1-16	18850	20250	21340	1.5080	1.6200	1.7072
	1-17	18850	33750	27330	1.5080	2.7000	2.1864
	1-18	33840	51775	32275	2.7072	4.142	2.5820
	Average	**13178**	**13377**	**12807**	**1.0542**	**1.0702**	**1.0246**

From table 2 we see that the over-estimation of baseRTT can cause very severe unfairness. In short connection C1, the throughput rate for the lastly joined flow, which is 33840 bytes/s, is 6 times of that for firstly joined flow (5640 bytes/s) when $\alpha = 1$ and $\beta = 3$. This ratio reaches to 11.51 times when $\alpha = \beta = 2$! In this case $\alpha \neq \beta$ relieve the effect of over-estimation. But it is different for short connection flows. When $\alpha \neq \beta$ the throughput rate of latterly joined flow is 2.32 times of that for earlier joined flow. But when $\alpha = \beta$ this ratio becomes 1.57. This time $\alpha \neq \beta$ enhances the unfairness.

In this group of simulations, the ratios of average throughput rate for flows in C0 to that for flows in C1 are 0.48 when $\alpha = 1$ and $\beta = 3$, 0.34 when $\alpha = \beta = 2$, and 0.77 for A-Vegas. This is a significant improvement although the result is not that good comparing with the no over-estimation case. Besides of this, we note that for A-Vegas the ratio of largest throughput rate to smallest throughput rate for short connection C1 is $32275/7205 \approx 4.48$, and that for the long connection C0 is $13160/6405 \approx 2.05$. Although the unfairness is still very heavy, comparing with Vegas, we can see that the unfairness due to over-estimation is significantly reduced.

5 Conclusion

In this paper, we have studied the unfairness problems of TCP Vegas. To solve the unfairness caused by multiple congested gateways, we have proposed Adaptive Vegas (A-Vegas). A-Vegas assigns the value of parameters dynamically according the estimated number of congested gateways. A-Vegas not only improves the fairness for TCP connection with multiple congested gateways, it also helps to reduce the unfairness caused by over-estimation of base RTT. But the unfairness caused by over-estimation is still very large. However, it seems that it is impossible to eliminate the over-estimation. One possible resolution is to introduce some random perturbation to the minimal RTT from time to time.

References

1. T. Bonald,"Comparison of TCP Reno and TCP Vegas via fluid approximation", Technical Report, INRIA, 1998.
2. C. Boutremans and J. Le Boudec, "A NOTE ON THE FAIRNESS OF TCP VEGAS", In Proceedings of International Zurich Seminar on Broadband Communications (Feb. 2000) pages 163-170.
3. L. Brakmo, S. O'Malley and L. Peterson, "TCP Vegas: New techniques for congestion detection and avoidance", In Proceedings of the SIGCOMM '94 Symposium (Aug. 1994) pages 24-35.
4. L. Brakmo and L. Peterson. "TCP Vegas: End to End Congestion Avoidance on a Global Internet", IEEE Journal on Selected Areas in Communication, Vol 13, No. 8 (October 1995) pages 1465-1480.
5. H. Cunqing and T. Yum, "The Fairness of TCP Vegas in Networks with Multiple Congestion Gateways", High Speed Networks and Multimedia Communications 5th IEEE International Conference on , 2002.
6. S. Floyd, "TCP and Explicit Congestion Notification", ACM Computer Communication Review, V. 24 N. 5, October 1994, p. 10-23.
7. G. Hasegawa, M. Murata and H. Miyahara, "Fairness and stability of congestion control mechanisms of TCP", Telecommunication Systems Journal, pp.167-184, November 2000.
8. U. Hengartner, J. Bolliger and T. Gross, "TCP Vegas Revisited", in Proceedings of Inforcom'2000.
9. R. Jain, A. Durresi, G. Babic, "Throughput Fairness Index: An Explanation", ATM Forum/99-0045, February 1999.
10. D.D. Luong and J. Bíró, "On the Propotional Fainess of Vegas", in Proceedings of Globecom'01, 2001.
11. J. MO, R. La, V. Anantharam and J. Walrand, "Analysis and Comparison of TCP Reno and Vegas", in Proceedings of Globecom'99, 1999.
12. http://www.isi.edu/nsnam/ns

RED Based Congestion Control Mechanism for Internet Traffic at Routers*

Asfand-E-Yar, Irfan Awan, and Mike E. Woodward

Performance Modelling and Engineering Research Group
Department of Computing, School of Informatics, University of Bradford
BD7 1DP, Bradford, West Yorkshire, England, UK
{A. Yar, I.Awan, M.E.Woodward}@bradford.ac.uk

Abstract. This paper begins with a brief literature review of various approaches to congestion avoidance and control of Internet traffic. It focuses mainly on one of Active Queue Management (AQM) schemes known as Random Early Detection (RED) mechanism for congestion control at routers. Towards the end of paper, an approximate analytical performance model has been proposed based on standard RED mechanism as an effective congestion control technique. Methodology adopted is based on Principle of Maximum Entropy (ME) to model RED mechanism. To model the bursty input traffic, Generalized Exponential (GE) distribution has been used. Closed form expressions for the state and blocking probabilities have also been presented. Numerical examples have been presented. By comparing results obtained from ME (Analytical Model) and Simulation in QNAP-2 [22], it validates the credibility of ME solution.

1 Introduction

Large Content is hosted all over the world and increasing demand from users of Internet to access that large content is contributing in form of congestion to the Internet traffic. Thus, in order to ensure the efficient content access to all the users with minimum loss rate at gateway during busy hours [4], speedy connectivity, and negligible packet loss with high bulk throughput [8] is becoming of a high consideration by Internet Engineering Task Force (IETF) as the Internet Protocol Performance Metrics (IPPM) [8]. Various approaches have been adopted to solve the Quality of Service (QoS) issues in Internet traffic particularly the problem of congestion. Figure 1, provides a generic analogy of a complex Internet setup for ease of understanding.

These approaches are classified into two main blocks namely: Congestion Avoidance and Congestion Control Techniques. *Congestion Avoidance* is a preventive technique [1,3,5,7], which comes into play before network is congested by overloading. *Congestion Control* [3, 9] comes into play after the congestion

* Supported by the Engineering and Physical Sciences Research Council (EPSRC), UK, under grant GR/S01658/01.

C. Kim (Ed.): ICOIN 2005, LNCS 3391, pp. 142–151, 2005.

at a network has occurred and the network is overloaded. Due to unresponsive and non TCP-compatible flows, the danger of congestion collapse still exists [9]. Therefore router mechanisms should support end-to-end congestion control and stimulate the use of it [10].

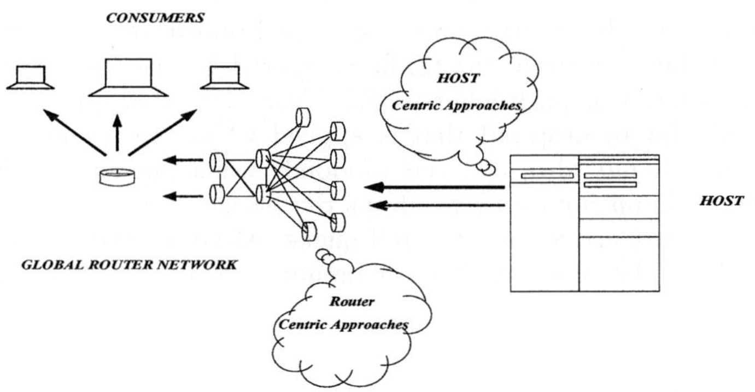

Fig. 1. Internet Traffic Congestion and Approaches to Congestion avoidance and control.

1.1 Approaches at Host

In 1988, Jacobson pioneered the concepts of TCP congestion avoidance and control [5]. TCP was later augmented with fast retransmission and fast recovery algorithms in 1990 to avoid inefficiency caused by retransmission timeouts (RTOs) [6,7]. These basic TCP principles were designed based on the assumption that the gateway drops at most one packet per flow when the sender increases the congestion window by one packet per round trip time. Sally Floyd has summarized some recent developments in TCP congestion control in [11]. Although the TCP Congestion Control algorithms and TCP variants[1] are necessary, but they are not enough to control the Internet congestion in all the circumstances due to the limit of how much control can be exercised from the edge of the network (i.e. from host). There are some mechanisms developed to work at the routers (known as Router-centric approaches) inside the network in order to complement congestion avoidance and control solutions at host.

1.2 Approaches at Routers

Basically there are two congestion avoidance mechanisms at routers, namely *Scheduling algorithm,* which regulates the allocation of bandwidth among flows and determines which packet to send next and *Queue management algorithm,* which manages the length of the queue by dropping packets when necessary and is deployed in routers these days.

[1] Tahoe, Reno, Modified New – Reno and SACK.

1.3 Drop-Tail: Traditional Congestion Control Technique

Drop Tail sets a maximum length for each queue at the router and accepts packet until the maximum queue length is reached. Once the maximum queue length is achieved, the algorithm drops packets until queue length is again below the maximum set value. Drop tail mechanism has some serious limitations to its use like low end-to-end delay, jitter and global synchronization problem. Although there is TCP flow control in the traffic generated by the sources, the packets arrive to the routers as packet bursts [12]. There are other queue management algorithms similar to drop-tail that is applied when the queue becomes full. "*Random drop on full*" drops packets randomly when the queue length reaches its maximum. "*Drop front on full*" drops the packet at the front of the queue when a new packet arrives into the full queue. Although these two algorithms solve the lockout phenomenon, they still do not solve all problems caused by full queue.

1.4 Active Queue Management (AQM)

An approach where packets are dropped before queue becomes full is called Active Queue Management (AQM) and it provides a solution to overcome demerits of the tail drop scheme. Active Queue Management maintains a small size steady state queue, thus results in reduced packet loss, decreased end-to-end delay, and the avoidance of lock out behavior thus using the network resources more efficiently. Also, by keeping the average queue size small results in the efficient use of bandwidth and thus increases the link utilization by avoiding global synchronization. Furthermore, due to availability of extra queue space packet bursts will be absorbed as well. Finally, the bias of routers against flows that use small bandwidth due to monopolize flows will be prevented, which will result in prevention of lockout behavior. The AQM techniques to maintain acceptable level of service (in particular RED) are strongly recommended in RFC 2309 [10].

2 Random Early Detection (RED) Mechanism

Sally Floyd and Van Jacobson in 1993 [2] presented Random Early Detection (RED's) design objectives are to minimize packet loss and queuing delay, maintain high link utilization, and remove biases against bursty sources. Random Early Detection, RED, is used in routers to notify the network about congestion [2]. It is designed also to avoid global synchronization, which occurs when all sources detect congestion and reduce their sending rates at the same time, resulting in a fluctuation of link utilization. RED achieves these goals by introducing an enhanced control mechanism involving randomized packet dropping and queue length averaging. In this way RED drops packets in proportion to the input rates of the connections. This enables RED to maintain equal rate allocation and remove biases against bursty connections. By using probabilistic packet dropping RED also eliminates global synchronization. Although research on RED applications and its variants[2] seem to be very elaborate and promising still further investigations are necessary. Variants of RED try to solve some of the problems of RED. One of the main weaknesses pointed out in literature is the variation of the average queue size of RED (variants) with the level of congestion and parameter settings. Some researchers [14,15] even claim that there is no advantage in using RED over Drop Tail as in their opinion throughput may be very poor for RED if the parameters are not tuned properly. RED is widely implemented in routers today [16]; however, doubts have arisen about the degree to which it can improve the network performance [13].

Hence, from the literature survey it is evident that a little is known about the analytical analysis of RED performance evaluation while dealing with traffic scenarios where traffic consists of correlated inter-arrival time and is bursty in nature. Therefore, it has formed the motivation for our research. Thus, we propose an analytical solution to model RED mechanisms for implementing AQM scheme for Internet congestion control. The model is based on principle of maximum entropy (ME) [17-19] and Generalized Exponential (GE)- type queuing systems [8] to capture burstiness of input traffic.

3 Proposed Model

Our research focuses on performance modeling of a basic analytical model based on theoretical model of RED [2,10]. We use quantitative analysis techniques, and specifically focus on bursty traffic scenarios, for which performance results are difficult to capture. In this context, a finite capacity GE/GE/1/$\{N_i, ...N_R\}$ queue (i = 1,2,3...R) with N_i to be intermediate threshold value, $N_R$second threshold value and total buffer capacity, First-Come-First-Serve (FCFS) scheduling discipline and censored arrival process for single class jobs [21] is analyzed and closed form analytical expressions for the state probabilities and blocking probabilities have been presented. Figure 2, shows one such RED mechanism for AQM.

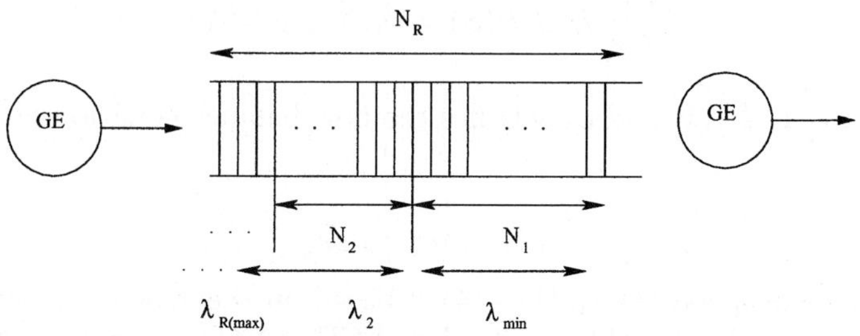

Fig. 2. RED Mechanism for AQM.

[2] Random Early Marking (REM), BLUE, Stabilized-RED (SRED), Flow-RED (FRED), Adaptive RED (ARED), Dynamic-RED (DRED), Adaptive Virtual Queue (AVQ) and RED with ECN

3.1 Analysis of Proposed Model

Notation: Let: 'S' be the state of the queue ;'**Q**' the set of all feasible states of S; 'λ' be the arrival rate, if number of jobs in the queue are less than equal to N_1 and if Number of jobs in the queue exceeds the Threshold, N_2be the arrival rate; 'μ_i' be the service rate; 'π_i' be the blocking probability that an arrival finds the queue full; P(S) is the stationary state probability.

The entropy maximization is based on knowledge about prior information which in our case are auxiliary functions and mean value constraints. For each state S, SεQ the following auxiliary functions are define as: $n_i(S)$ = the number of jobs present in state S, $s_i(S) = 1$ if $n_i(S) > 0$ & 0, otherwise; $f_i(S) = 1$, if $n_i(S) = N_i$, &0, otherwise. Where i = 1, 2...R. Suppose the Mean value constraints about the state probability P(S) are known to exist:

(i) Normalisation

$$\sum_{S\in\mathbf{Q}} P(S) = 1, \tag{1}$$

(ii) Utilisation

$$\sum_{S\in\mathbf{Q}} s_i(S)P(S) = U_i, 0 < U_i < 1, \tag{2}$$

(iii) Mean queue length

$$\sum_{S\in\mathbf{Q}} n_i(S)P(S) = L_i,\ U_i < L_i < N_i, \tag{3}$$

(iv) Full buffer state probability

$$\sum_{S\in\mathbf{Q}} f_i(S)P(S) = \phi_i, 0 < \phi_i < 1, \tag{4}$$

Where i = 1, 2...R and ϕ_i satisfies the flow balance equation namely:

$$\lambda_i(1 - \pi_i) = \mu_i U_i. \tag{5}$$

The choice of mean values (1) - (4) is based on the type of constraints used for the ME analysis of stable single class FCFS G/G/1/N queue [20]. If additional constraints are used, it is no longer feasible to capture a closed-form ME solution at the building block level, with clearly, adverse implications on the efficiency of an iterative queue-by-queue decomposition algorithm for general QNMs. Conversely, if one or more constraints from the set (1) - (4) are missing, it is expected that the accuracy of the ME solution will be generally reduced.

Provided the mean value constraints for utilization, mean queue length and full buffer state about the state probability P(S) are known, it is implied that after some manipulation, the ME state probability distribution for the proposed model can be given by:

$$P(S_o) = \frac{1}{Z} \tag{6}$$

$$P(k) = \frac{1}{Z}\sum_{i=1}^{R} x_i g_i \left(\prod_{j=1}^{R-1} \sum_{m(i)=0}^{n_j-1-\sum_{j=1}^{i} m(j)} x_i^{m(i)} C_{m(i)}^{n_j-1-\sum_{j=1}^{i} m(j)} x_R^{k-1-\sum_{j=1}^{R-1} n_j} \right) \tag{7}$$

For: k = 1, 2... $N_R - 1$, $n_j = \min(k, N_i)$ and m(i)= m(1), m(2) ...

$$P(N_R) = \frac{1}{Z}\sum_{i=1}^{R} x_i g_i y_i^{f_i(k)} \left(\prod_{j=1}^{R-1} \sum_{m(i)=0}^{n_j-1-\sum_{j=1}^{i} m(j)} x_i^{m(i)} C_{m(i)}^{n_j-1-\sum_{j=1}^{i} m(j)} x_R^{k-1-\sum_{j=1}^{R-1} n_j} \right) \tag{8}$$

Where x_i, g_i and y_i are lagrangian co-efficient for above-mentioned constraints. Then, 'Z', which is the normalizing constant, can be expressed as:

$$Z = 1+\sum_{k=1}^{N_R} \left(\sum_{i=1}^{R} x_i g_i y_i^{f_i(k)} \right) \left(\prod_{j=1}^{R-1} \sum_{m(i)=0}^{n_j-1-\sum_{j=1}^{i} m(j)} x_i^{m(i)} C_{m(i)}^{n_j-1-\sum_{j=1}^{i} m(j)} x_R^{k-1-\sum_{j=1}^{R-1} n_j} \right) \tag{9}$$

By using the flow balance equation,$\lambda_i(1-\pi_i) = \mu_i U_i$, the blocking probabilities of censored GE/GE/1/ $\{N_i, ...N_R\}$ queue will be as follows:

$$\pi_i = \sum_{k=0}^{N_i} \delta_i(k)(1-\sigma_i)^{N_i-k} P(k) \tag{10}$$

where:

$$\delta_i(k) = \begin{cases} \frac{r}{r(1-\sigma)+\sigma}, & k=0 \\ 1, & ow \end{cases} \tag{11}$$

Where $\sigma_i = {}^{2}/_{(1+C_{ai}^2)}$, and$r_i = {}^{2}/_{(1+C_{si}^2)}$, where C_{ai}^2 and C_{si}^2 are the squared coefficients of variation for the inter-arrival and service times, respectively. Similarly, assuming the lagrangian coefficients $x_i$and g_i are invariant to the buffer capacity of size 'N_R', it can be established that [20]:

$$x_i = \frac{L_i - \rho_i}{L} \tag{12}$$

$$g_i = \frac{\rho_i(1-X)}{(1-\rho)x_i} \tag{13}$$

Where: $X = \prod_{i=1}^{R} x_i, \rho = \prod_{i=1}^{R} \rho_i, \rho_i = {}^{\lambda_i}/_{\mu_i} and L_i, \{i = 1, 2...R\}$ is the asymptotic mean queue length of a GE/GE/1 queue. Note that, statistically$L_i$can be determined by using an established relation involving the normalizing constant Z and lagrangian coefficientsx_i, namely $L_i = ({}^{x_i}/_Z)\,{}^{\partial Z}/_{\partial x_i}$ [17]. Finally, the Lagrange coefficient y_i, can be derived by substituting the value of state probabilities, P (k), k = 0, 1 ... N, and blocking probabilitiesπ_i, into the flow balance condition, (5),[23].

4 Numerical Results

4.1 Analytical Model Results

For the proposed model, the analytical solution results obtained for effects of threshold on mean queue length (Figure 3), utilizations (Figure 4), delay (Figure 5), throughput (Figure 6) and on probability distribution (Figure 7) can be seen. From the results it is clear that as we increase the threshold value, the mean queue length increases, which in turn increase the utilization of the system resulting in high throughput. Similarly, with increase in threshold a linear reduction is achieved for the number of jobs in the system with a certain probability, which is RED standard. Also, it is evident from results that the mean queue length can be maintained by setting the threshold value in order to prevent congestion.

4.2 Validation of Results: QNAP Simulation Results Vs Analytical Model

The Credibility of ME solution involving GE-Type queue against simulation based on QNAP-2 is at 95 % confidence [22]. For the proposed model, the ME solution results obtained for effects of threshold on utilizations (Figure 8), mean queue length (Figure 9), and on probability distribution (Figure 10) can be seen very comparable to those of results obtained from simulation. From the results it is clear that as soon as the mean queue length approaches the threshold value, the probability for the number of jobs present in the queue rapidly decrease. Also, the mean queue length can be maintained by setting the threshold value in order to prevent congestion.

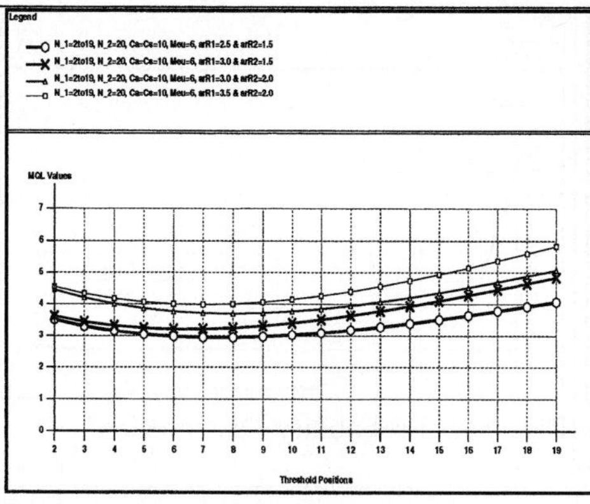

Fig. 3. Effects of Threshold on Mean Queue Length

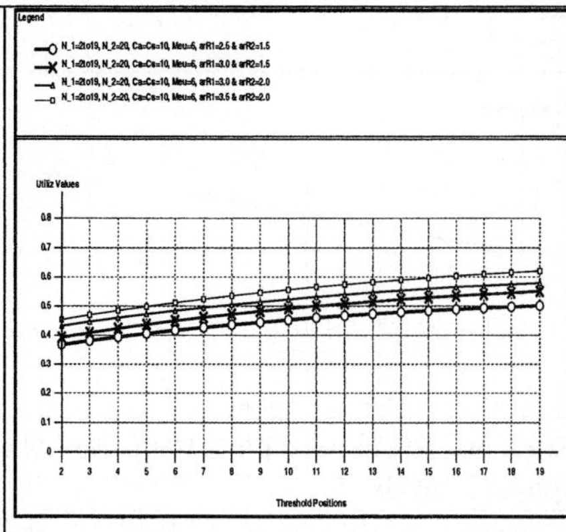

Fig. 4. Effect of Threshold on Utilization

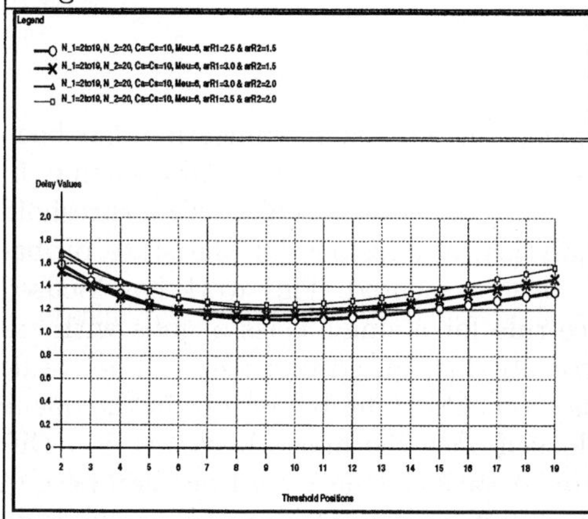

Fig. 5. Effects of Threshold on Delay

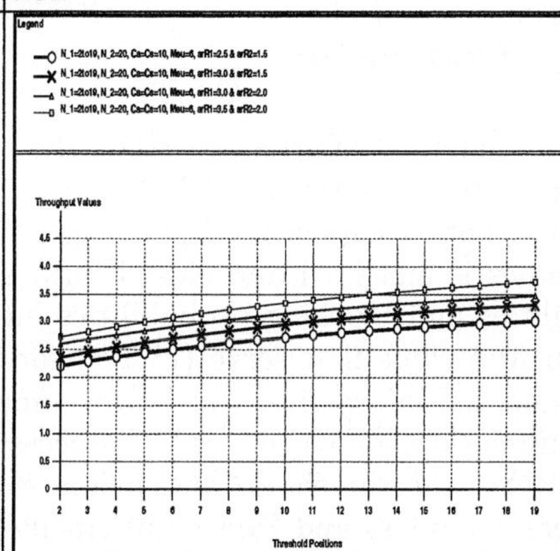

Fig. 6. Effect of Threshold on Throughput

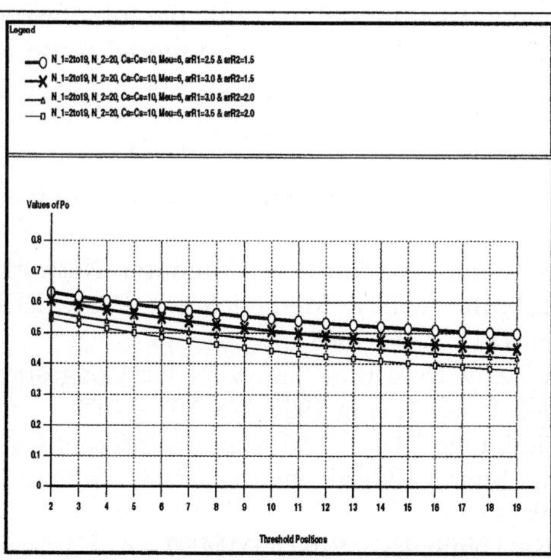

Fig. 7. Effect of Threshold on Probability

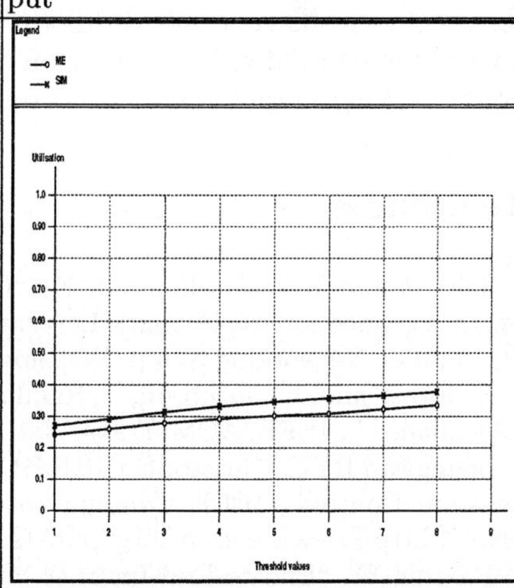

Fig. 8. Effect of threshold on utilization

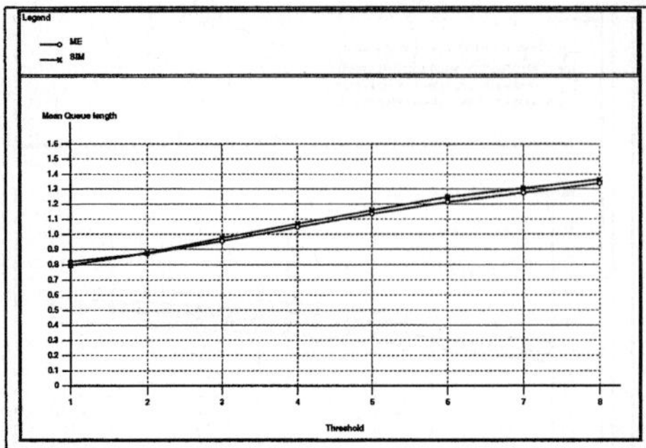

Fig. 9. Effect of Threshold on Mean Queue Length

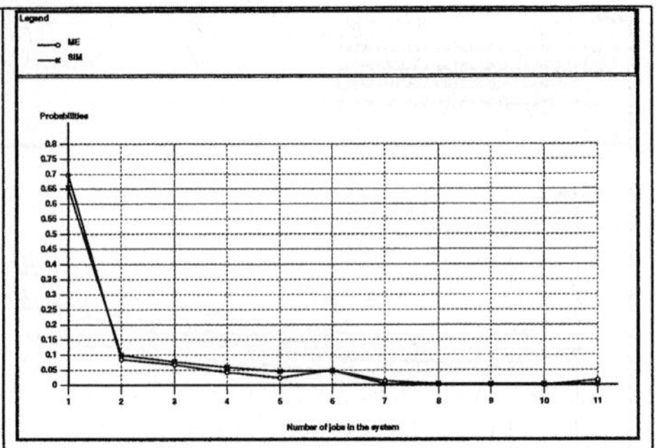

Fig. 10. Probability Distribution

5 Conclusion

An analytical solution, based on the principle of entropy maximization, has been presented to model the RED mechanism for implementing the AQM scheme. In this context capacity GE/GE/1/ $\{N_i, ...N_R\}$ queue with $\{N_i, ...N_R\}$ threshold values is analyzed and closed form analytical expressions for the state probabilities and blocking probabilities have been presented. This work has focused on first come first serve (FCFS) service rule for a single class of jobs with exponential interarrival time and geometrically distributed bulks sizes. The traffic source slows down the arrival process as soon as the number of jobs in the queue reaches the thresholds and jobs are blocked once the queue becomes full. Different job loss and QoS requirements under various load conditions can be met by adjusting the threshold value. Typical numerical examples were included to demonstrate the credibility of ME solution against simulation results. Future work includes the generalization of the analytical model for multiple job classes. It will then be followed by extension of the generalized analytical model to input scenarios where input traffic is correlated and bursty in nature.

References

1. Jain R and Ramakrishnan, K.K, "Congestion avoidance in Computer Networks with a connectionless Network Layer: Concepts, Goals & Methodology", Proc. IEEE Computer Networking Symp. Washington D C, April 1988, pp 134-143
2. S. Floyd and V. Jacobson, " Random Early Detection Gateways for Congestion Avoidance ", IEEE/ACM Trans. Net., vol. 1, no. 4, Aug 1993, pp. 397 – 413
3. Seungwan RYU, Christopher RUMP, Chunming QIAO, "Advances in Internet Congestion Control", IEEE Communications Survey, Third Quarter 2003, Volume 5, No.1, http://www.comsoc.org/pubs/surveys
4. Paxson, V., "End-to-End Internet Packet Dynamics", SIGCOMM'97
5. Jacobson, V., "Congestion Avoidance and Control", In Proceedings of SIGICOMM, Volume 1101, pages 34-45, Stanford, CA, USA, Mar 27-28 1988.

6. Zhang, L., Shenker, S., Clark, D., "Observations on the Dynamics of a congestion control algorithm: The effects of Two-Way Traffic", SIGCOMM'91.
7. Stevens, W.R., "TCP Slow Start, Congestion Avoidance, Fast Retransmit, and Fast Recovery Algorithms", RFC 2001.
8. Internet Engineering Task Force, "Internet Protocol Performance Metrics", http://www.advanced.org/IPPM/
9. J. Nagle, Congestion Control in IP/TCP Internetworks, January 1984-1989, RFC 896.
10. S. Floyd, V. Jacobson, B. Barden, D. Clark, et al, "Recommendations on Queue Management and Congestion Avoidance in Internet", IETF- RFC 2309, April 1998.
11. S. Floyd, "A report on some recent developments in TCP congestion control", IEEE Communication Magazine, April 2001.
12. W. Willinger, M. S. Taggu, R. Sherman, and D. Wilson, "Self-similarity through high-variability: statistical analysis of Ethernet LAN traffic at the source level", IEEE/ACM Transactions on Networking, 5(1): 71-86,1997. http://citeseer.nj.nec.com/willinger97selfsimilarity.html
13. M. May, J. Bolot, C. Diot, and B. Lyles, "Reasons Not to Deploy RED", Proceedings of 7th International Workshop on Quality of Service IWQoS'99, 1999. pp.260-262.
14. Mikkel Christiansen, Kevin Jeffay, David Ott, F. Donelson Smith, "Tuning RED for Web Traffic", IEEE/ACM Transactions on Networking, Volume 9, Number 3, (June 2001), pages 249-264
15. M. May, T. Bonald, and J. Bolot, "Analytical Evaluation of RED Performance", In proceedings of IEEE INFOCOM, Tel Aviv, Israel, March 2000.
16. Cisco Systems, "Congestion Avoidance Overview", and web page: http//www.cisco.com
17. E. T. Jaynes, "Information Theory and Statistical Mechanics", Phys, Rev 106, (1957), pp. 620-630.
18. E. T. Jaynes, "Information Theory and Statistical Mechanics", II Phys, Rev 108, (1957), pp. 171-190.
19. D. D. Kouvatsos, "Entropy Maximization and Queuing Network Models", Annals of Operation Research 48, (1994), pp. 63-126.
20. D. D. Kouvatsos, "Maximum Entropy and G/G/1/N Queue", Acta Informatica, Vol. 23, (1986), pp. 545-565.
21. D. D. Kouvatsos, Spiros G. Denazis, "Blocking and Multiple Job Classes", Performance Evaluation17, (1993), pp. 189-205.
22. M.Veran and D. Potier; QNAP – 2: A Portable Environment for Queuing Network Modeling Holland, (1985) Asfand E Yar; "Stochastic Analysis of Internet Traffic Congestion Control using RED Mechanism with QoS Constraints",Technical Report, (2004)

Selective Route Discovery Routing Algorithm for Mobile Ad-Hoc Networks

Tae-Eun Kim[1], Won-Tae Kim[2], and Yong-Jin Park[1]

[1] Division of Electrical and Computer Engineering, Hanyang University,
Haengdang-dong Sungdong-gu, Seoul 133-791, Korea
Tel. +82-2-2290-0355
{tekim, park}@hyuee.hanyang.ac.kr
http://nclab.hanyang.ac.kr

[2] Rostic Technologies, Inc., B207, HIT Building, Hanyang University,
Haengdang-dong Sungdong-gu, Seoul 133-791, Korea
Tel. +82-2-2290-0519
wtkim@rostic.com
http://www.rostic.com

Abstract. In mobile ad-hoc networks the real traffic of a node is commonly concentrated in a small number of particular nodes. This characteristic has not been considered in the design of the existing routing algorithms. Therefore, it is difficult to guarantee performance in a simulation environment with realistic accuracy. To resolve this problem we propose a new routing algorithm called the Selective Route Discovery (SRD) algorithm. In this algorithm, each node selects frequently accessed nodes and periodically sends additional RREQ messages. Therefore, it can quickly adapt to the changes in network topology according to the movement of the nodes. This paper shows that the SRD algorithm has a shorter packet delivery time than the AODV algorithm when the simulation condition is improved so that the traffic concentration for each destination node varies.

1 Introduction

The routing algorithms used in wired networks are not well-suited for mobile ad-hoc networks(MANETs) since the accuracy of the route information decreases and the overhead produced by the periodic route update messages increases. Therefore, the design of an efficient routing algorithm for MANETs have been actively researched in the last few years. These can be classified into two categories: table-driven, and on-demand according to the maintenance method of the route information. The table-driven algorithms are induced by traditional wired networks and periodically refresh the routing table to get new route information. The corresponding protocols are Destination-Sequenced Distance-Vector (DSDV) [1], Wireless Routing Protocol (WRP) [2], and Clusterhead Gateway Switch Routing (CGSR) [3]. However, the on-demand algorithms maintain only the routing paths that have changed and are needed to send the data packets currently in the network. The corresponding protocols are the Ad-hoc On-Demand

C. Kim (Ed.): ICOIN 2005, LNCS 3391, pp. 152–159, 2005.

Distance Vector (AODV) [4], Dynamic Source Routing (DSR) [5], and the Temporally Ordered Routing Algorithm (TORA) [6]. In recent research, many performance comparisons have shown that on-demand algorithms are generally more suitable for MANETs than table-driven algorithm since the on-demand algorithms have less traffic overhead than table-driven algorithms [7,8,9,10,11]. Hybrid algorithms combining table-driven with on-demand algorithms have been proposed such as Zone Routing Protocol (ZRP) [12] and Core Extraction Distributed Ad-hoc Routing (CEDAR) [13].

This paper proposes a new routing algorithm called a Selective Route Discovery (SRD) routing algorithm. It maintains the features of on-demand algorithm and improves it by using periodical RREQ messages. This algorithm is very efficient for real traffic patterns and can be used together with any other on-demand routing protocol. We show that the SRD algorithm has the smaller packet delivery latency than DSDV and AODV.

2 Selective Route Discovery Routing Protocol

Through analyzing the patterns of network traffic, it is generally known that most traffic of a node is concentrated on a few particular nodes [14]. Figure 1 shows an example of traffic concentration where the size of the arrow is directly proportional to the amount of traffic between different nodes. However, existing routing protocols are designed without considering this phenomenon. In most simulation environments, it is assumed that the probability of accessing a node is as uniform as possible. To address this deficiency, we have proposed a new routing algorithm called the Selective Route Discovery (SRD) routing algorithm.

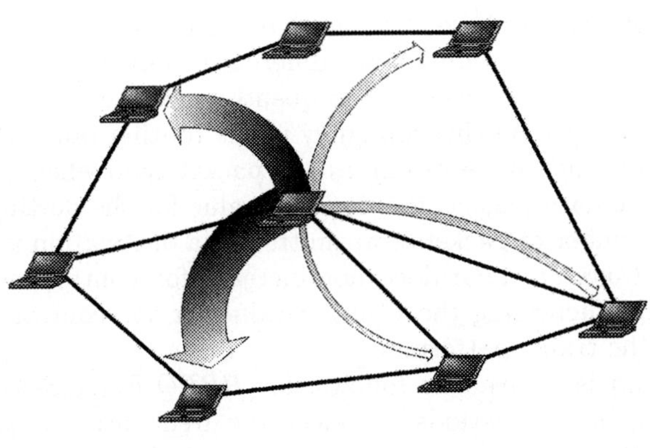

Fig. 1. The flow of network traffic (NOTE: The size of the arrow represents the amount of traffic)

2.1 Design of Selective Route Discovery Routing Algorithm

Since the traffic of a node is concentrated at a few particular nodes, the performance can be improved to a level greater than the on-demand algorithm by periodically updating the routing paths for these nodes receiving the greatest concentration. Therefore, our SRD routing algorithm uses additional RREQ messages for frequently accessed nodes to continuously maintain the routing paths in the routing table.

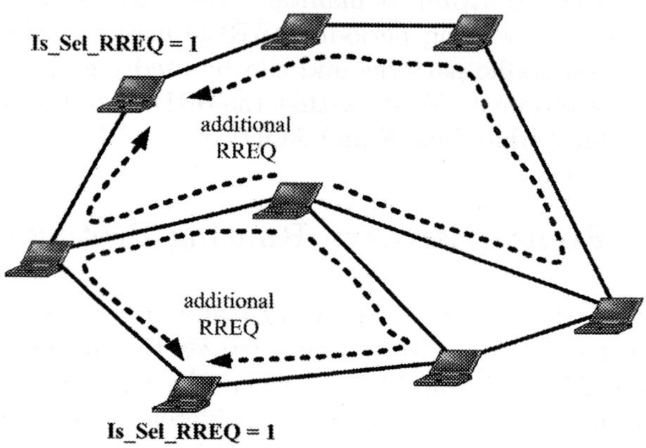

Fig. 2. Additional RREQ messages for frequently used nodes

SRD Algorithm. SRD is used in combination with the AODV routing protocol since this algorithm is not a fundamental routing protocol but a supplementary algorithm to improve the base routing algorithm. For SRD, the basic process is the same as AODV. The important difference from AODV is that the additional RREQ messages are used for frequently accessed nodes. There are two additional fields in each destination entry in the routing table: the *Counter* and the *Is_Sel_RREQ*. Each node examines the packet type when transmitting the packet. If it is the data packet, the *Counter* value for the destination increases. Therefore, the number of packets transmitted to a destination is easily grasped. Note that the *Counter* value does not increase for control messages such as RREQ or RREP. Increasing the *Counter* value for the control packet disturbs the analysis of the traffic pattern.

The algorithm is derived as follows: let *RREQ_Entry_Selection_Time* and *Sel_RREQ_Time* be the periods to select the frequently accessed nodes and to send the additional RREQ messages, respectively. Each host searches the *Counter* values in its routing table entries and determines the *RREQ_Entry_Number* nodes with the largest values. The *Is_Sel_RREQ* values of these nodes are set to one. Each node sends additional RREQ messages to the destination nodes every *Sel_RREQ_Time* if the *Is_Sel_RREQ*s corresponding to those nodes equals one. This process is shown in Fig. 2. Then, the Counter values in the routing entries are initialized to zero in order to rapidly adapt to the changes in the network. These procedures are consecutively repeated every *RREQ_Entry_Selection_Time*. By doing so, the packet is sent faster and the data communication is more stable for a network topology that changes rapidly.

SRD Parameters. It is important to determine the appropriate parameter values for network environments. In the proposed algorithm, there are three important parameters, the *RREQ_Entry_Number*, *Sel_RREQ_Time*, and *RREQ_Entry_Selection_Time*. If the *RREQ_Entry_Number* is large, additional RREQ messages become excessive causing too much overhead. On the other hand, if the *RREQ_Entry_Number* is small, the SRD algorithm is very similar to the on-demand algorithm. Therefore, the enhanced performance compared to the on-demand algorithm is poor. To resolve this problem, the *RREQ_Entry_Number* should be determined by considering the tradeoff between performance and overhead. A similar tradeoff is applied to *Sel_RREQ_Time* and *RREQ_Entry_Selection_Time*, respectively. If *Sel_RREQ_Time* is short, the load on the network increases because of the number of RREQ messages. In the opposite case, it is difficult to transmit the data packet quickly. For *RREQ_Entry_Selection_Time*, if it is long, it is difficult to cope with the changes in the network environment quickly. Therefore, the performance of our algorithm fully depends on the parameters, so suitable parameters must be selected carefully.

2.2 Features of the Selective Route Discovery Routing Protocol

The SRD routing algorithm maintains the basic operations of AODV and partially combines the table-driven algorithm that periodically updates the route. However, the SRD routing algorithm can be applied to other on-demand routing algorithms as well as AODV. The advantages of SRD are node-independence and protocol-compatibility. Node-independence is guaranteed since each node can independently select *Sel_RREQ_Time* and *RREQ_Entry_Selection_Time* according to the environment of the node. Nodes using the SRD routing protocol can coexist with other on-demand routing protocols since they do not send new types of control messages and it is only the internal algorithm that is used in each node. Therefore, it is possible to communicate between nodes using the SRD routing algorithm and others if both nodes use on-demand as the base routing protocol. This allows protocol-compatibility.

3 Performance Analysis

3.1 Simulation Environments

For the simulations, the ns2 simulator and the CMU Monarch extension for wireless mobility modeling were used. The network environment for the ns2 simulator is given in Table 1. Especially, the ratio accessed to each destination was differently generated to consider a real traffic pattern and is represented in Table 2. In

Table 1. Simulation Setting

MAC Layer	IEEE 802.11 b
Simulation Area	500m x 500m
Simulation Time	300 seconds
Mobile Nodes	20 nodes
Node Mobility Speed	0 50 m/s
Node Moving Pattern	Random Way Point Model
Traffic Source Type	UDP(CBR)
Packet Size	512 bytes
Number of Connection	100
ACTIVE_ROUTE_TIMEOUT(AODV)	10 seconds
RREQ_Entry_Number	5
RREQ_Entry_Selection_Time	3
Sel_RREQ_Time	15,30,40,50 seconds

Table 2. Traffic Pattern Type by Destination Concentration Rate

(a)Traffic Pattern 1

Destination concentration	
Top 1 node	41%
Top 3 nodes	89%
Top 5 nodes	98%

(b)Traffic Pattern 2

Destination concentration	
Top 1 node	25%
Top 3 nodes	68%
Top 5 nodes	90%

order to show the performances according to the different concentration ratios, the simulation was done using Traffic pattern 1 and 2, respectively. The SRD routing algorithm is combined with AODV as a base on-demand routing protocol and we call it the SRD-AODV routing protocol in this paper. To evaluate the SRD-AODV, it is compared with the AODV and DSDV routing protocols.

3.2 Simulation Results

When the SRD routing algorithm is applied, the average packet delivery time and packet overhead are shown in this simulation. Figure 3 shows the average packet delivery times for various routing protocols according to traffic patterns. In Fig. 3 (a) and (b), DSDV shows the shorter average packet delivery time than AODV and it is consistent with existing simulation results [7,8,9]. The SRD outperforms the DSDV and AODV except that *Sel_RREQ_Time* is 15sec since RREQ messages increased excessively.

Figure 4 indicates the ratio of routing overhead according to the mobility speed of the node where (a) and (b) correspond to traffic patterns 1 and 2, respectively. Note that the routing overhead of DSDV is nearly fixed regardless of the mobility speed. However those of AODV and SRD-AODV show slight variation according to mobility speed. In addition, SRD-AODV has higher routing overhead than AODV since the RREQ messages increase by the process of selective route discovery.

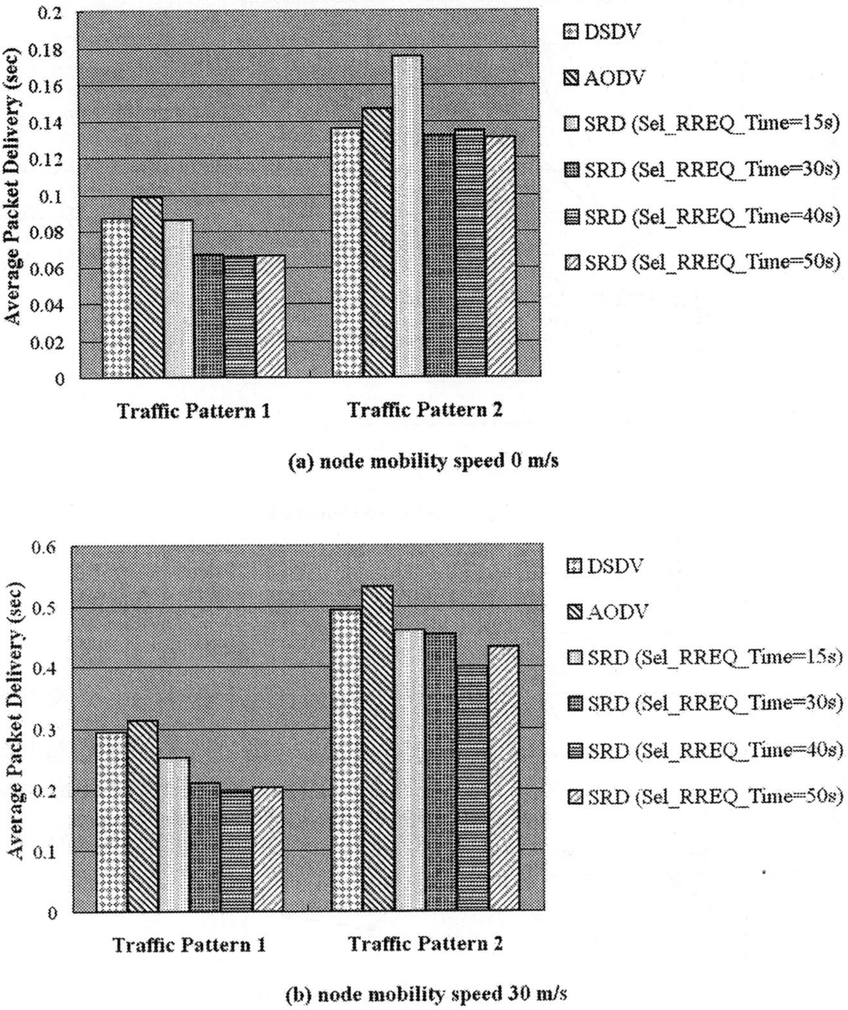

Fig. 3. Average Packet Delivery

4 Conclusion

SRD periodically searches the routing paths to the destinations which are frequently accessed. Each node can configure the appropriate SRD parameters by its traffic pattern. Therefore, it provides a quick response to changes in the network and minimizes the waste of network resources. From the simulations, it is shown that SRD has a shorter packet delivery time than AODV or DSDV. Although it has a larger routing overhead than AODV, this is not problematic due to the small gap between the overheads.

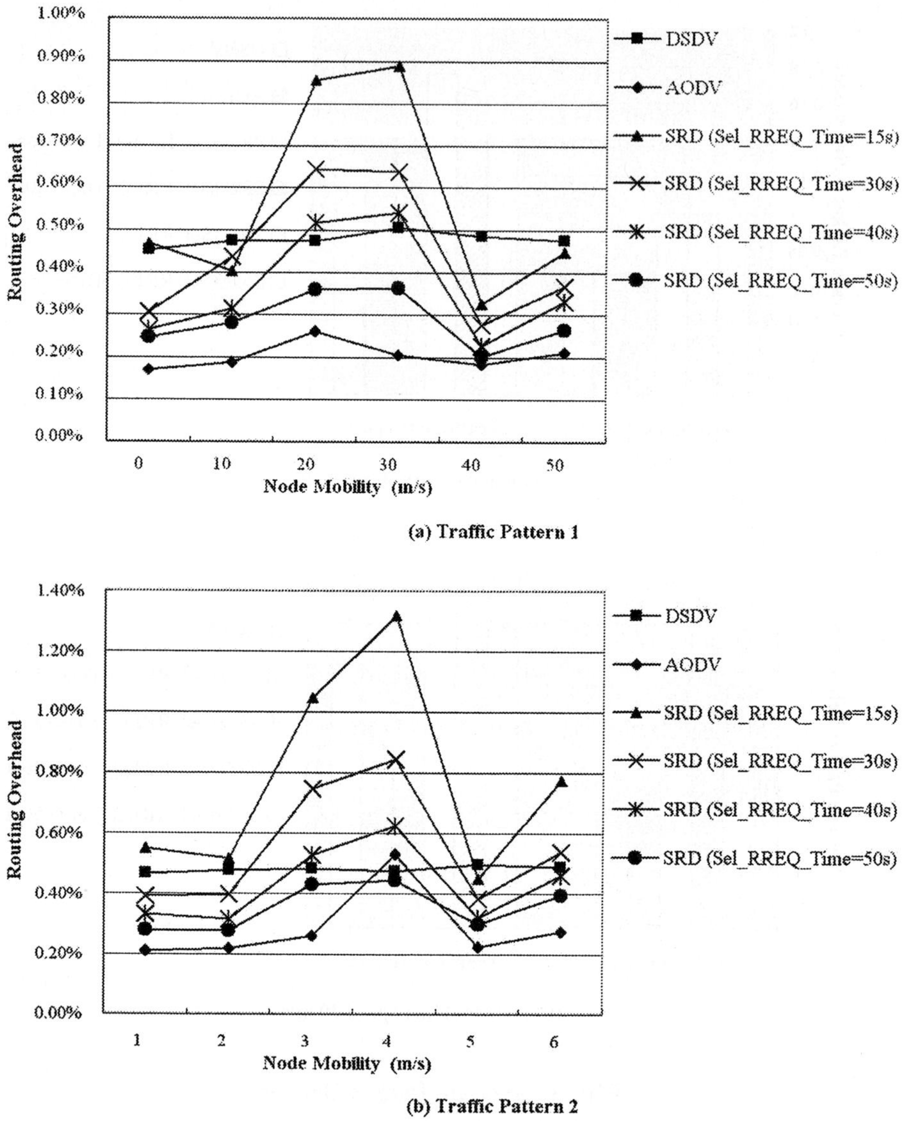

Fig. 4. Routing Overhead

In future studies, it is necessary to analyze the relationship between the SRD parameters (e.g. *RREQ_Entry_Selection_Time*, *Sel_RREQ_Time*, and *RREQ_Entry_Number*) and the network traffic patterns. It should be demonstrated that SRD can be efficiently applied to other on-demand routing protocols. Furthermore, it would be interesting if each node was able to carry out self-analysis and self-configuration.

References

1. C. E. Perkins, P. Bhagwat, "Highly dynamic Destination-Sequenced Distance-Vector routing (DSDV) for Mobile Computers," SIGCOMM Symposium on Communications Architectures and Protocols, (London, UK), Sept. 1994.
2. S. Murthy and K.K. Garcia-Luna-Aceves, "An Efficient Routing Protocol for Wireless Networks," ACM Mobile Networks and App. J., Special Issue on Routing in Mobile Communication Networks, Oct. 1996.
3. C.-C. Chiang, "Routing in Clustered Multihop, Mobile Wireless Networks with Fading Channel," Proc. IEEE SICON '97, Apr. 1997.
4. C. E. Perkins and E. M. Royer, "Ad-hoc On-Demand Distance Vector Routing," Proc. 2nd IEEE Wksp. Mobile Comp. Sys. and Apps., Feb. 1999.
5. D. B. Johnson and D. A. Maltz, "Dynamic Source Routing in Ad-Hoc Wireless Networks," Mobile Computing, T. Imielinski and H. Korth, Eds., Kluwer, 1996.
6. V. D. Park and M. S. Corson, "A Highly Adaptive Distributed Routing Algorithm for Mobile Wireless Networks," Proc. INFOCOM '97, Apr. 1997.
7. E. M. Royer and C-K Toh, "A Review of Current Routing Protocols for Ad Hoc Mobile Wireless Networks," IEEE Personal Communications Magazine, pp. 46-55, April 1999.
8. I. Gerasimov and R. Simon, "Performance Analysis for Ad Hoc QoS Routing Protocols," IEEE MobiWac'02.
9. J. Broch et al., "A Performance Comparison of Multi-hop Wireless Ad hoc Network Routing Protocols," ACM Mobicom '98, Oct. 1998.
10. . S. Lee, Mario Gerla, and C.K. Toh, "A Simulation Study of Table-Driven and On-demand Routing Protocols for Mobile Ad hoc Networks," IEEE Network Magazine, Aug. 1999.
11. P. Johansson et al., "Scenario-based Performance Analysis of Routing Protocols for Mobile Ad-hoc Networks," ACM Mobicom '99, Aug. 1999.
12. Z. J. Haas and M. R. Pearlman "The Zone Routing Protocol (ZRP) for ad hoc networks," IETF Internet Draft, http://www.ietf.org/internet-drafts/draft-ietf-manet-zone-zrp-00.txt, 1997.
13. R. Sivakumar et al., "CEDAR: Core Extraction Distributed Ad hoc Routing," IEEE Journal on Selected Areas in Communication, Special Issue on Ad hoc Networks, Vol 17, No. 8, 1999
14. Barabasi, Albert-Laszlo, "LINKED", Penguin, June 2003.

LSRP: A Lightweight Secure Routing Protocol with Low Cost for Ad-Hoc Networks*

Bok-Nyong Park, Jihoon Myung, and Wonjun Lee**

Dept. of Computer Science and Engineering
Korea University, Seoul, Republic of Korea
wlee@korea.ac.kr

Abstract. Ad-hoc networks consist of only mobile nodes and have no support infrastructure. Due to the limited resources and frequent changes in topologies, ad-hoc network should consider these features for the provision of security. We present a lightweight secure routing protocol (LSRP) applicable for mobile ad-hoc networks. Since the LSRP uses an identity-based signcryption scheme, it can eliminate public or private key exchange, and can give savings in computation cost and communication overhead. LSRP is more computationally efficient than other RSA-based protocols because our protocol is based on the properties of pairings on elliptic curves. Empirical studies are conducted using NS-2 to evaluate the effectiveness of LSRP. The simulation results show that the LSRP is more efficient in terms of cost and overhead.

1 Introduction

Ad-hoc network is a collection of mobile nodes with wireless interface dynamically forming a network without the use of any preexisting network infrastructure [9]. To provide the mobility of nodes, ad-hoc network should have efficient routing protocol for mobile ad-hoc network environment. Although the research on routing and communication topology in ad-hoc networks has progressed a great deal, there are still many open problems in ad-hoc networks. Moreover, the security of ad-hoc network is more vulnerable than that of wireless networks using fixed infrastructure so that the security services in the ad-hoc network faces a set of challenges [9]. Ad-hoc network security research often focuses on secure routing protocols. However, such routing protocols neglect the inheritance features of ad-hoc network such as limited resources, computational ability, and so on. In this paper, our aim is to evaluate and optimize the effect in ad-hoc networks when we apply the security schemes to be integrated with the ad hoc routing protocols. To improve the efficiency of communication and computation, the proposed protocol uses the identity-based signcryption scheme [2] based on pairings (the Weil and Tate pairings) over elliptic curves [5], which is named as

* This work was supported by KOSEF Grant (No. R01-2002-000-00141-0), Korea Research Foundation Grant (KRF-2003-041-D00509), SKT, and ETRI.
** Corresponding Author.

C. Kim (Ed.): ICOIN 2005, LNCS 3391, pp. 160–169, 2005.

LSRP. The LSRP does neither need to authenticate a public key nor maintain a public key directory, rather it simultaneously fulfills both functions of encryption and signature [12] using identity-based signcryption. Also, the LSRP can guarantee the efficiency of computation cost and communication overhead. Moreover, it can reduce the load of computation and reply faster because it operates over elliptic curves [6]. This paper is organized as follows. Section 2 will give a brief overview of widely-known secure routings in ad-hoc networks. Section 3 explains the proposed secure routing protocol using identity-based signcryption scheme and Section 4 shows the simulation results and the efficiency analysis. Finally Section 5 concludes the paper.

2 Related Work

Using its wireless interfaces, an ad-hoc network is much easily attacked than wired networks. In an ad-hoc network, it is difficult to identify the reliable node and to protect data delivered through the multi hops. The studies of secure routing in ad-hoc networks have been carried out by ARAN [1], Ariadne [11], SRP [10], and so on [8]. ARAN protocol [1] consists of a preliminary certification process, a mandatory end-to-end authentication stage, and an optional second stage that provides secure shortest paths. Fundamentally, it requires the use of a trusted certificate server because each node has to request a certificate signed by a trusted certificate server before entering the ad-hoc network. This protocol uses the digital signature and certificate based on public key cryptography to guarantee authentication, message integrity, non-repudiation, and other security goals. However, it has a serious problem of high overhead to sign and verify the message. Ariadne protocol [11] is an on-demand secure ad-hoc routing protocol based on DSR that withstands node compromise and relies on only highly efficient symmetric cryptography like hash functions. It provides point-to-point authentication of a routing message using Message Authentication Code and a shared key between the two parties. However, it has a problem such that it must have all information of discovery routing paths. SRP [10] provides correct routing information. The requirement of SRP is that any of two nodes have a security association. Thus, it does not require that any of the intermediate nodes perform cryptography operators. However, the most serious problem in SRP is that it cannot provide any authentication process for the intermediate nodes between the source node and the destination node.

3 LSRP: The Lightweight Secure Routing Protocol with Low Cost

The inheritance features of ad-hoc networks pose opportunities for attack ranging from passive eavesdropping to active impersonation, message replay, and message distortion. To cope with these attacks, we propose to employ features of network security in routing protocols.

Table 1. Notations used for LSRP

Symbol	Definition	Symbol	Definition
ID_X	Identification of node X	P	Generator
Sig_X	Digital signature of node X	P_{pub}	System master secret key · P
H	One-way hash function	k, r, Z	Security papameters
σ	Authentication information	$\hat{e}$ (P, Q)	Bilinear map based on the Weil Pairing

3.1 Overview of the Protocol

The LSRP is an extensive protocol of AODV. Thus, LSRP retains most of the AODV mechanisms, such as route discovery, reverse path setup, forward path setup, route maintenance, and so on. The protocol is abstracted as the process of two mechanisms: route discovery and route maintenance. The table 1 describes the notations used throughout this paper and Fig. 1 represents the algorithm of our protocol in a diagram. Our protocol is based on the following assumptions:

- The proposed protocol is satisfied with the managed-open environment [1].
- The proposed protocol is based on the identity-based signcryption scheme which makes use of Weil parings on elliptic curves [2]. At the time of network formation of ad-hoc network, nodes need to form a system master key and a system master secret key.
- All links between the nodes are bi-directional.
- We do not differentiate compromised nodes from attackers in the security point of view, where the power of the attackers are limited.

3.2 Route Discovery Process

The route discovery process is initiated whenever a source node needs to communicates with another node which does not have routing information in its routing table. The route discovery process is abstracted as the exchange of two messages: *route request* and *route reply.*

A sender achieves the route discovery to establish a path to its destination.

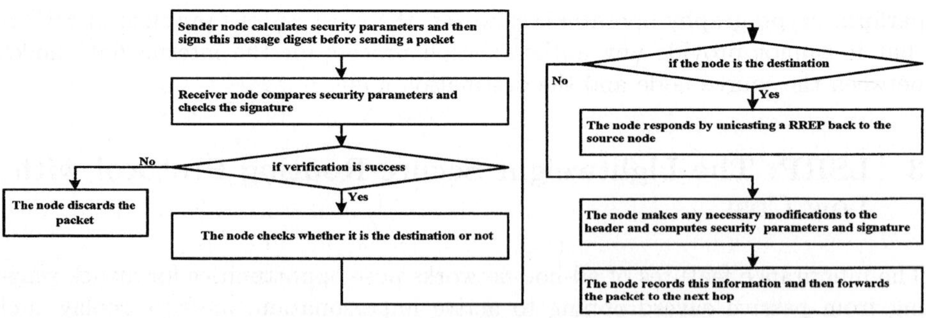

Fig. 1. Flow diagram for LSRP.

Source → Intermediate: $< RREQ \parallel ID_S \parallel \sigma \parallel Sig_S\{H(M)\} >$

A Source node begins route instantiation to a destination node by broadcasting its RREQ packet with a message for authentication. The functions of RREQ fields in this protocol are the same as those of RREQ fields of the general AODV [4]. Using ID, the source node computes public key, $PK_S = H(ID_S)$ and private key $SK_S = S^* PK_S$ where S^* is a system master secret key. It chooses a random number, x, and it computes $k = \hat{e}\ (P, P_{pub})^x$ for the authentication of the origin and both $r = H(k \parallel PK_S \parallel RREQ)$ and $Z = xP_{pub} - rSK_S \in G_1$ for the authentication of nodes. It sends routing request messages, $\sigma = (r, Z)$, and the created values, all of which are signed where M is defined as follows: $M = (RREQ \parallel ID_S \parallel \sigma)$.

Intermediate → Destination: $< RREQ \parallel ID_S \parallel ID_X \parallel \sigma \parallel \sigma_X \parallel Sig_S\{H(M)\} >$

An intermediate node $X_i (1 \leq i \leq n)$ computes $\bar{k} = \hat{e}(P, Z)\hat{e}(P_{pub}, PK_S)^r$ for the authentication of the node which sends the message, and it checks $r = H(\bar{k} \parallel PK_S \parallel RREQ)$ for the validity of σ after verifying the sign. When this procedure is finished successfully, the intermediate node can trust the received message and then it computes σ_X. Finally, it broadcasts the message to the next nodes. When the destination node receives this message, it checks the destination address. If the destination address is the same as its address, it verifies the signature, σ and σ_X. If the verification process is successful, it is ready to reply a message. The destination node sends a RREP message to a source node.

Destination → Intermediate: $< RREP \parallel ID_D \parallel \sigma \parallel \sigma_D \parallel Sig_D\{H(M')\} >$

The destination node unicasts a RREP packet with a message for authentication back along the reverse path to the source node. To confirm the destination, σ_D is added in RREP packet. The computation method of σ_D in the route reply follows the similar way in RREQ. It computes $k = \hat{e}(P, P_{pub})^x$, and then it computes $r = H(k \parallel PK_D \parallel RREP)$ and $Z = xP_{pub} - rSK_D$ using the result of $k = \hat{e}(P, P_{pub})^x$. The M' of signature is defined as follows: $M' = (RREP \parallel ID_D \parallel \sigma \parallel \sigma_D)$.

Intermediate → Source: $< RREP \parallel ID_D \parallel ID_X \parallel \sigma \parallel \sigma_D \parallel \sigma_X \parallel Sig_D\{H(M')\} >$

Neighbor nodes that receive the RREP message forward the packet back to the predecessor from which they received the original RREQ message. The messages for authentication are signed by the sender. When an intermediate node receives the message, it judges the message delivered by the legal node via verifying the digital signature in the message. Also, it checks the message integrity using a hash value. If the digital signature and σ_D are valid, it can trust the message. It signs the result of the σ_{D-X} and then it unicasts a RREP message to the node from which it received the original RREQ message. This can avoid attacks where malicious nodes instantiate routes by impersonation and replay of their message.

When the source node receives the RREP packet with a message for authentication, it verifies that a correct hash value is returned by the destination node as well as the destination node's signature. If the verification of the digital signature and σ value is successful, security can be established on the route.

3.3 Route Maintenance Process

If the path within source and destination is broken due to link failure because either the destination or some intermediate node moves during an active session, the node X, which detects the link failure of node R, sends the RERR message to source node.

Intermediate → Source: $< RERR \parallel ID_X \parallel ID_R \parallel \sigma \parallel \sigma_D \parallel Sig_X\{H(M'')\} >$

Upon receiving a notification of the link failure of node R, intermediate nodes subsequently relay that message to their active neighbors. The nodes which receive the RREP message update their routing table. This process continues until all active nodes are notified. The M'' of signature is defined as follows: $M'' = (RERR \parallel ID_X \parallel ID_R \parallel \sigma \parallel \sigma_X)$.

4 Simulation Experiments and Performance Analysis

The goal of this section is to evaluate the effects of integration of the security scheme into ad-hoc network routing protocol without degradation of performance and to analyze the efficiency of the proposed protocol.

4.1 Simulation Environment and Parameters

The performance of the LSRP is evaluated by simulation using NS-2 simulator [7]. The AODV protocol simulation is available as part of the simulator. The RREQ packets are treated as broadcast packets in the MAC. RREP and data packets are all unicast packets with a specified neighbor as the MAC destination. RERR packets are treated as broadcast packets. Table 2 shows a summary of simulation parameters.

The radio model uses characteristics similar to a commercial radio interface, the 914MHz Lucent's WaveLAN [3] DSSS radio interface. WaveLAN is modeled as shared-media radio with a nominal bit rate of 2 Mb/s and nominal radio range of 250 meters. In our experiments, 25 nodes move around in a rectangular area of 900m×800m according to a mobility model i.e., the random waypoint model. For the work related to energy-aware routing, we assume long-lived sessions. The session sources are CBR and generate UDP packets with each packet being 512 bytes long in 900 second simulated time. The source-destination pairs are spread randomly over the network. Each node starts its journey from a random location to a random destination with randomly chosen speed. Once the destination is reached, another random destination is targeted after a pause. We vary the pause time which affects the relative speeds of the mobiles. Also, we vary the data rate. The number of data sources is maintained at 10. The traffic and mobility models are the same as [8].

Table 2. Parameters used for all simulations.

Parameter	Value	
	Varying mobility	Varying offered load
Transmitter range	250m	
Bandwidth	2 Mb / s	
Simulation time	900 s	
Environment size	900m × 800m	
Traffic type	CBR (Constant Bit Rate)	
Packet (data) rate	4 packets / s	4, 12, 20, 40, 100, 200, 400, 800 kb / s
Packet size	512 byte	
Pause time	10, 50, 100, 200, 400, 800 s	200 s

4.2 Simulation Results

We start the simulations in order to compare the original AODV routing protocol without any security requirements with the AODV routing protocol with routing authentication extension. Our simulation codes set up the wireless simulation components. The first set of experiments uses 10 sources with a moderate packet rate and varying pause time. For the 25 node experiments, we used 10 traffic sources and a packet rate of 4 packets/s. We varied the pause time until the simulating time that means high pause time is low mobility and small pause time is high mobility. The next set of experiment uses 10 traffic sources and a pause time of 200 s with varying date rate in order to see the throughput in 900 second simulation time. We have done this study to illustrate that our scheme works for many security issues in the routing protocol, without causing any substantial degradation in the network performance. Three key performance metrics are evaluated in our experiments:

- Packet delivery faction: The ratio of the data packets delivered to the destinations to those generated by the CBR sources; also, a related metric, received throughput (in kilobits per second) at the destination has been evaluated

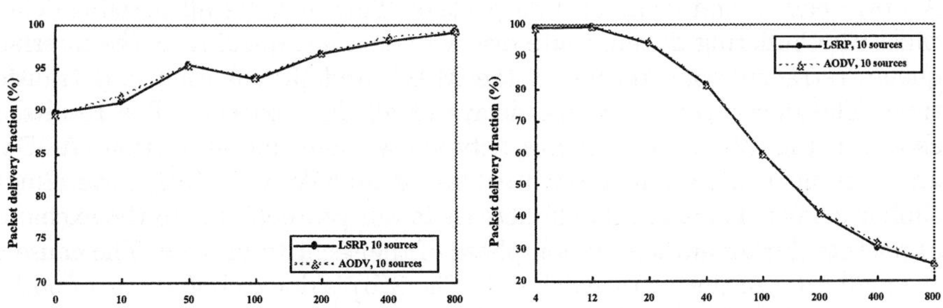

(a) Packet delivery fraction vs Pause time

(b) Packet delivery fraction vs Data rate

Fig. 2. Packet delivery fraction for the 25-node model with various pause time and data rate.

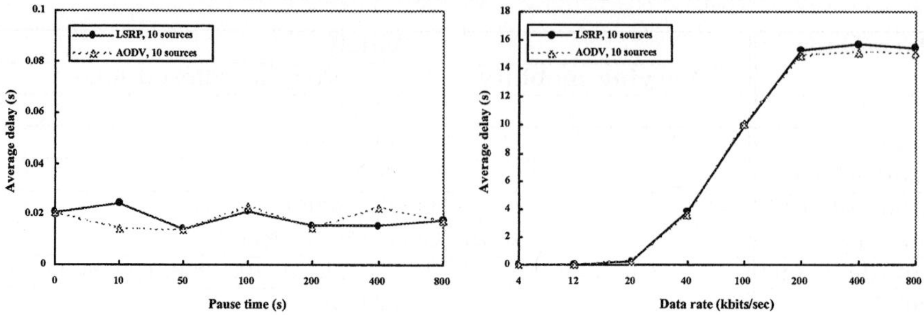

(a) Average delay vs Pause time (b) Average delay vs Data rate

Fig. 3. Average data packet delays for the 25-node model with various pause time and data rate.

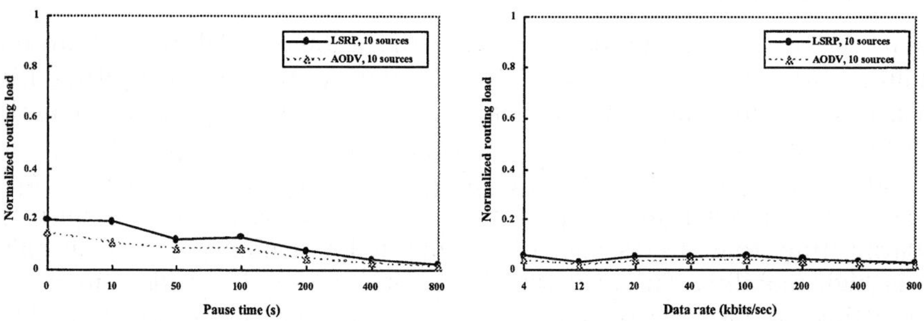

(a) Normalized routing load vs Pause time (b) Normalized routing load vs Data rate

Fig. 4. Normalized routing loads for the 25-node model with various pause time and data rate.

in some cases. From the results shown in Fig. 2, our LSRP protocol could work well in experiment because the effect of throughput of the network is fairly small (around 2-10%). The packet delivery fractions for AODV and LSRP are very similar in two cases. As data rates increase, the data packets reaching to the destination decreases (Fig. 2(b)).

- Average end-to-end delay of data packets: This includes all possible delays caused by buffering during route discovery latency, queuing at the interface queue, retransmission delays at the MAC, and propagation and transfer time. The delay is the average delays of all data packets. The results as shown in Fig. 3(a) are fairly low between without authentication (AODV) and with authentication (LSRP) extension. AODV and LSRP have almost similar delays. There is a small increase in our protocol due to the exchange of packets during authentication phase of the security process. The cause of the high ratio of the average delays in Fig. 3(b) is due to buffer size of nodes. In this simulation, we did not limit the buffer size.
- Normalized routing load: The number of routing packets transmitted per data packet delivered at the destination. Each hop-wise transmission of a routing packet is counted as one transmission. We increased the traffic rates

and varied the mobility in each simulation. The number of routing packets increases when our scheme is incorporated. The increase in routing load is higher at lower pause time (Fig. 4(a)). This is because at lower pause time routes need to be found more frequently. The Fig. 4(b) shows that the routing load is very low because it does not require any additional route discoveries. The normalized routing load of AODV and LSRP is fairly stable with an increasing number of sources. A relatively stable normalized routing load is a desirable property for scalability of the protocols, since this indicates that the actual routing load increase linearly with the number of sources.

4.3 Protocol Efficiency and Safety

Ad-hoc network has some features such as resource-constrained in bandwidth, computational ability, low energy, and frequent changes in topologies. Therefore, secure routing protocols in ad-hoc network should not waste the time and resources for the computation and authentication of active nodes. To satisfy these requirements, we propose a lightweight secure routing protocol using identity-based signcryption scheme based on pairings on elliptic curves [2]. The proposed protocol, LSRP, has some obvious advantages. LSRP can reduce the amount of storage and computation because identity-based cryptography needs no public certificate exchange and no verification of signature. In addition, the identity-based signcryption scheme used in the proposed protocol combines both the functions of digital signature and encryption, and therefore it is more computationally efficient than other schemes with sign-then-encrypt approach [12]. Also, the elliptic curve cryptography used in this protocol has smaller parameters including faster computations and smaller keys.

ARAN [1] uses the RSA public key cryptosystem for authenticating a node, while Ariadne [11] uses only the symmetric cryptography. Although identity-based cryptography is a public key cryptography like RSA algorithm, the performance of it is much better than that of RSA in aspects of the computational speed and cost. Generally, an elliptic curve whose order is a 160bit prime offers approximately the same level of security as RSA with 1024bit [6]. Thus LSRP can use smaller parameters than with old discrete logarithm systems but with equivalent levels of security. The overhead for each protocol is as follows:

$$Overhead = \sum_{i=1}^{x} = ((n \mid p \mid +n \mid q \mid +n \mid H \mid +n(packet \times 8)) \times ComputationCost)_i \quad (1)$$

where *Overhead* is value of communication overhead and computation cost, n is the number of execution, i is i^{th} node, $\mid p \mid$ is encryption, $\mid q \mid$ is signature, H is hash function, and packet is RREQ (44byte) or RREP (36byte). The computation cost is 2.17, 4.5, and 5.17, respectably [12].

We exclude the Ariadne protocol in the comparison of computation because it is based on the different encryption scheme. Fig. 5 compares the communication overhead and computation cost of the previous protocols with that of LSRP. In

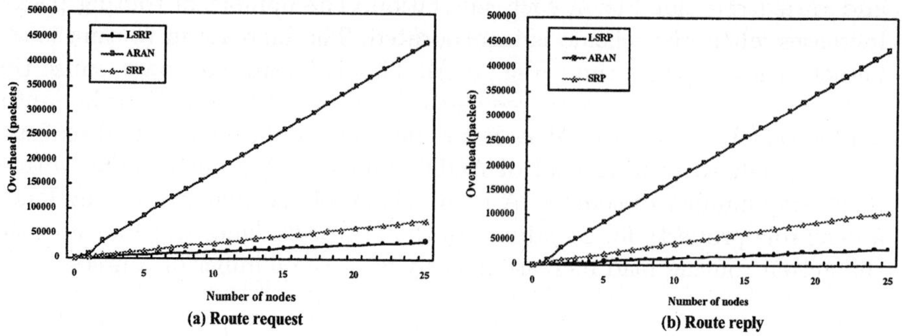

Fig. 5. Overhead for communication overhead and computation cost

Fig. 5, the communication overhead of ARAN is higher than our protocol and SRP due to public key certificate and many sign and verification message. The graphs show that the overhead of the LSRP in terms of routing load is very low because the computation cost of signcryption is very low compared to the other schemes.

Table 3 summarizes all the comparisons that we have carried out in this paper, in terms of savings in communication overhead and computation cost.

Security is important and necessary to protect messages during their transmission, and it guarantees that message transmissions are authentic. It means protecting the information and the resources from both the outside network and the inside network [9]. To authenticate the messages, we use the identity-based signcryption scheme. The LSRP can authenticate all of the nodes on routes with σ generated by parameters based on identity based signcryption scheme while ARAN and SRP cannot authenticate nodes on routes. The safety of LSRP results in Bilinear Diffie-Hellman Assumption. We assume two cyclic groups, G_1 and G_2 which have a large prime order, q. P, an element of G_1, is selected randomly. aP, bP, cP are defined when a, b, c is selected randomly. We assume that it is difficult to compute $\hat{e}(P, P_{pub})^{abc}$. The safety of identity-based scheme and signcryption scheme is verified by [2] [5]. Moreover, we used elliptic curves

Table 3. Comparison of LSRP and other protocols.

Scheme	LSRP	ARAN	SRP
Key distribution	Public key (Id-based signcryption)	Public key (RSA)	Public key (ECC)
Intermediate node authentication	Yes	No	No
Certification	Not need	Need	Not need
Communication Overhead	Low	High	Low
Computation cost	Low	High	High
Total Overhead	Low	High	Middle

in this paper since the elliptic curve discrete logarithm problem appears to be significantly harder than the discrete logarithm system.

5 Conclusions and Future Work

In this paper, we have focused on the efficiency of computation and communication for a secure routing protocol, LSRP, in the ad-hoc network environment. The proposed protocol can protect modification, impersonation, fabrication and other attacks using an identity-based signcryption without involving any significant overheads. Furthermore, this protocol has an advantage that it does not need to exchange the certificate public keys. Also, the LSRP reduce the network resources and communication overheads than conventional secure routing protocols. However, our protocol is operated in managed-open environment so that it can just guarantee the security in small local areas. We will study a secure protocol to guarantee robustness in wide area, which not only protects external attacks but also detects serious attacks from the compromised nodes and selfishness nodes.

References

1. B. Dahill, B. N. Levine, E. Royer, C. Shields, "ARAN: A secure Routing Protocol for Ad Hoc Networks", UMass Tech Report 02-21, 2002.
2. B. Libert, J-J. Quisquater, "New identity based signcryption schemes from parings", full version, available at *http://eprint.iacr.org/2003/023/*.
3. B. Tuch, "Development of WaveLAN, and ISM Band Wireless LAN," AT&T Tech. J., vol. 72, no. 4, July/Aug 1993. pp. 27-33.
4. C. Perkins and E. Royer, "Ad-Hoc On-Demand Distance Vector Routing", in Proceedings of 2nd IEEE Workshop on Mobile Computing Systems and Applications, February 1999.
5. D. Boneh, M. Franklin, "Identity Based Encryption From the Weil Pairing", Advances in Cryptology-Crypto'01, LNCS 2193, Springer, 2001.
6. J. Lopez and R. Dahab, "Performance of Elliptic Curve Cryptosystems", Technical report IC-00-08, 2000., *http://www.dcc.unicamp.br/ic-main/publications-e.html.*
7. K. Fall and K. Varadhan, Eds., "ns Notes and Documentation", 2003; available from *http://www.isi.edu/nsnam/ns/*.
8. L. Venkatraman and D. P. Agrawal, "Strategies for enhancing routing security in protocols for mobile ad hoc networks," J. Parallel Distrib. Comput. 63, February 2003.
9. M. Ilyas, The Handbook of Ad-Hoc Wireless Networks, CRC PRESS, 2002.
10. P. Papadimitratos, Z. Haas, "Secure Routing for Mobile Ad Hoc Networks", in proceedings of CNDS 2002, San Antonio, TX, January 27-31, 2002.
11. Y. C. Hu, A. Perrig, D. B. Johnson, "Ariadne: A secure On-Demand Routing Protocol for Ad Hoc Networks", in proceedings of MOBICOM 2002.
12. Y. Zheng, "Digital Signcryption or How to Achieve Cost (Signature & Encryption) $<<$ Cost (Signature) + Cost (Encryption)", Advances in Cryptology-Crypto'97, LNCS 1294, Springer, pp. 165-179, 1997.

Cost-Effective Lifetime Prediction Based Routing Protocol for MANET

Huda Md. Nurul[1], M. Julius Hossain[2], Shigeki Yamada[3], Eiji Kamioka[3], and Ok-Sam Chae[2]

[1] The Graduate University for Advanced Studies
2-1-2 Hitotsubashi, Chiyoda-ku, Tokyo 101-8430, Japan
huda@grad.nii.ac.jp
[2] Department of Computer Engineering, Kyung Hee University
1 Seochun-ri, Kiheung-eup, Yongin-si, Kyunggi-do, Korea, 449-701
[3] National Institute of Informatics, 2-1-2 Hitotsubashi, Chiyoda-ku, Tokyo 101-8430, Japan

Abstract. Almost every node of a Mobile Ad-hoc Network (MANET) has to perform the function of a router. The lifetime of participating nodes affects the stability of the network. Recent MANET routing protocols are greedy on network lifetime because of battery power limitations. Although, these algorithms help to maintain the stability of the network, they are not as much cost effective as traditional existing routing algorithms. Our proposed method considers both the routing cost and network lifetime issues in route selection, which is a good compromise between these two conflicting interests. The simulation results show that the proposed scheme selects a path with less cost than a path in lifetime prediction based routing algorithms and results more stable network than cost-effective routing algorithms do.

1 Introduction

In mobile Ad hoc Networks it is assumed that all the devices acting as intermediate nodes in a routing path would forward data for other network nodes. An energy efficient routing protocol ensures that a packet from a source node to a destination nodes gets routed along the most energy efficient path possible [1]. Selection of the least power cost route may possess a harmful impact on the network connectivity when the selected path contains some node with small remaining energy. The energy of the poor node is likely to be used up soon and it would die. This may result disconnection of the network.

We encounter two conflicting goals: on one hand, in order to optimize cost, a least cost routing path is desirable, while on the other hand, use of a least cost route means that nodes with higher path degree might die soon since they are likely to be used in most cases. We define path degree of a node as the count of paths between any two nodes through that node. The cost can be considered as energy, hop count, delay, link quality as well as other factors. Another metrics used in lifetime predictive routing is the lifetime of nodes, which is a function of

C. Kim (Ed.): ICOIN 2005, LNCS 3391, pp. 170–177, 2005.

the remaining battery energy. As in [2], lifetime of a node is predicted based on the residual battery capacity and the rate of energy discharge. These techniques maximize the network lifetime by finding routing solutions that minimize the variance of the remaining energies of the nodes in the network.

The routing protocol proposed in this paper is a reactive routing protocol like DSR [3] There are two objectives in our scheme; one is to minimize the cost of routing and the other one is to maximize the network lifetime. To achieve these goals, we find a tradeoff between the cost and the lifetime of each of the possible paths. The proposed technique results to a more stable network than the power-aware routing algorithms and offers a much less routing cost than those of the existing lifetime predictive routing protocols.

The remainder of this paper is organized as follows: The next section contains review of some recent related research work along with problems of routing in mobile ad hoc networks. Section 3 describes the rationale and details of the proposed Cost-effective Lifetime Prediction based Routing (CLPR) technique. Section 4 elaborates on the simulation environment, the implementation and experimental results comparing CLPR with Lifetime Prediction based Routing and Power-Aware Routing. Finally section 5 concludes the paper.

2 Related Works

Routing in mobile ad hoc networks has been the subject of intense research efforts over the past few years; these efforts have resulted in numerous proposals for routing protocols. Among the two types of routing protocols, proactive (table driven) routing protocols [4], [5] are similar to and come as a natural extension of those of wired networks. Each node contains the latest information of the routes to any node in the network. Any change in the topology is updated and propagated through all nodes in the network. Reactive (on-demand) routing protocols do not maintain or constantly update their route tables with the latest route topology. Examples of reactive routing protocols are the dynamic source Routing (DSR) [3], ad hoc on-demand distance vector routing (AODV) [6] and temporally ordered routing algorithm (TORA)[7].

Some researchers concentrate on energy efficient broadcast/multicast algorithms [8], [9], [10]. One major approach for energy conservation is to route a communication session along the path which requires the lowest total energy consumption. This optimization problem is referred to as Minimum-Energy Routing [11]. While the minimum-energy unicast routing problem can be solved in polynomial time by shortest-path algorithms, it remains open whether the minimum-energy broadcast routing problem can be solved in polynomial time, despite the NP-hardness of its general graph version. Recently three greedy heuristics were proposed in [12]: MST (minimum spanning tree), SPT (shortest-path tree), and BIP (broadcasting incremental power). They have been evaluated through simulations in [10].

It has recently been recognized that medium access control (MAC) schemes can significantly increase the energy efficiency of mobile batteries [13]. If a mobile device T transmits data to another mobile device R, neighboring mobiles do not listen to the data from mobile T since listening causes unnecessary power consumption. Another energy efficient MAC scheme has been proposed in [14].

In paper [15] the authors have used new power-aware matrix for determining routes in wireless ad-hoc networks. The main disadvantage of power aware routing techniques is that they always select the least power cost routes. Since the same node may be selected repeatedly in this scheme, there is a large possibility of selecting a node that has a very little lifetime; hence it would die early. So the network would get disconnected and the network lifetime will be adversely affected.

A lifetime prediction based routing technique is proposed in [2] which is an on demand source routing protocol that uses battery lifetime prediction. The objective of this routing protocol is to extend the service life of MANET with dynamic topology. This protocol favors the path whose lifetime is maximum. The authors calculated the lifetime of a route with the following equation.

$$Max(T_{\pi}(t)) = Min(T_i(t)).....\{i \epsilon \pi\} \tag{1}$$

Where, $T_{\pi}(t)$: lifetime of path π

$T_i(t)$: predicted lifetime of node i in path π

In this algorithm, the lifetime of a path is predicted by the minimum lifetime of all nodes along the path. In this way the minimum lifetimes of all the paths from the source to the destination are calculated. The path that has maximum value of calculated minimum lifetimes is selected. The main objective of LPR is to minimize the variance in the remaining energies of all the nodes and thereby prolong the network lifetime.

Although, LPR increases the stability of the network, this technique has totally overlooked the cost of routing. As a result in most of the cases it may select a path with a higher cost than the minimum. As we have stated in the introduction section, there are many cost factors. If the cost factor is energy, this will result in more energy loss. If the cost factor is hop count, this selection scheme may cause more traffic in the network, more delay, and more security threat to data.

3 Cost-Effective Lifetime Prediction Based Routing (CLPR)

To achieve a tradeoff between the routing cost and network stability, we propose a new routing technique that combines the best features of cost efficient routing and lifetime prediction based routing. We want to minimize the routing cost as well as to maximize the network lifetime. In most of the cases when we try to achieve both of these two goals they become conflicting. For example, improvement of routing cost (i.e., less cost) degrades the stability (i.e., lifetime) of the network and improvement of the network stability degrades the routing cost. In such situations, to achieve the best performance, we need to find a tradeoff between the two contradictory parameters.

Cost is a general term and is a function of many other parameters like hop count, transmission power etc. In minimum hop count routing the path having minimum number of hops from the source to the destination is selected. To achieve this goal we have to compromise network lifetime because use of minimum hop count paths may use nodes with less lifetime frequently and some nodes may die soon causing decrease of network lifetime. Again, in some other algorithms, like lifetime prediction base routing (LPR), the objective is to maximize the network lifetime. To achieve this goal sometimes it needs to use paths with larger hop count than the minimum.

The lifetime of an ad hoc network is reflected by the lifetime of nodes. When a node in the network dies the network suffers from some loss of connectivity. A disconnected network is useless in ad hoc environment because of the infrastructure nature of the network. Hence we should keep alive as much nodes as possible. Besides, loss of a node implies loss of many paths that run through that node. The effect on the network of losing a node depends on the number of connections through that node. Absence of a high degree node would compel to use longer routing paths and in the worst case, disconnection of part of the network from the rest.

In power-aware routing algorithms, the selected path of transmission is the most cost-effective (here, the cost factor is power) whereas LPR algorithms select a path with maximum lifetime and hence results in more stability of the network. But in these two different types of goal oriented protocols, because of ignoring other objectives, power-aware routing algorithms suffer from poor network stability and lifetime prediction based routing algorithms suffer from high routing cost. As we use a tradeoff between the routing cost and the lifetime of the network, our proposed CLPR technique results to a more stable network than the power-aware routing algorithms and also needs less routing cost than the LPR protocols.

In our network model we consider a mobile ad hoc network N=(V,E,C) consisting of a set of nodes $V = \{v_1, ..., v_n\}$ that represent mobile devices, a set $E \subseteq VxV$ of edges $\{(v_i, v_j), 1 \leq i, j \leq n\}$ that connect all the nodes, and a weight function $C : E \rightarrow R$ (Rational number) for each edge (v_i, v_j) that indicates the transmission cost of a data packet between node v_i to v_j. Each node has a unique identification number, but it is not a priori known which nodes are currently in the network, nor is edge set E or weight function ω known. A node can not control the direction in which it sends data, and thus data are broadcast to all nodes inside its transmission range. Nodes can move and the edge cost between any two nodes can change over time. Also the lifetime of any node can change over time. However, for the ease of presentation, we assume a static network during the route discovery phase.

Let us assume that the maximum possible lifetime of any node is L and the maximum possible transfer cost between any two nodes is C. We define a scaling factor ξ as the ratio of the two parameters.

$$\xi = \frac{L}{C} \qquad (2)$$

Let there be n paths ($\pi_1, \pi_2, ...\pi_n$) from source to destination. The lifetime of a path is bounded by the lifetime of all the nodes along the path. When a node dies along a path we can say that the path does not exist any longer. So we can consider the life-time of a path is the same as the minimum lifetime among all the nodes along the path. The lifetime τ_i of a path π_i can be defined as:

$$\tau_i = Min(T_j(t)).......\{j\varepsilon i\} \tag{3}$$

$T_j(t)$: predicted lifetime of node j in path π_i

The cost of a path is the sum of all the costs calculated between two consecutive nodes along the path from source to the destination. Cost of a path i can be defined as:

$$\varsigma_i = \sum_{j=1}^{\pi_{i_m}-1} C_{\pi_{i_{j,j+1}}}(t) \tag{4}$$

where π_{i_m} is number of nodes in path π_i and

$C_{\pi_{i_{j,j+1}}}$ is the cost between node j and j+1 of the path π_i

Our path selecting parameter β is represented by

$$\beta_i = \frac{\tau_i}{\xi\varsigma_i} \tag{5}$$

CELP selects a path, which has the largest β i.e. $\max(\beta_i)$. If more than one path having highest β is found, any one of them can be selected. Thus, the proposed method is inclined to select a path having higher lifetime τ and lower cost ς.

Figure 1 shows an instance of an ad hoc network represented by a graph. Nodes are labeled with their lifetime values and the edges are labeled with the cost between its two adjacent nodes. In this instance there are six paths from

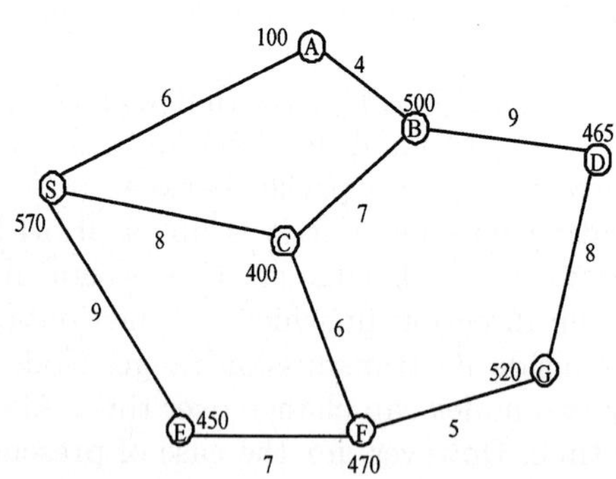

Fig. 1. An instance of the MANET.

source (S) to destination (D). They are S→A→B→D, S→A→B→C→F→G→D, S→E→F→C→B→D, S→E→F→G →D, S→C→F→G→D, and S→C→B→D.

If we calculate the total cost along each path, we get the cost 19 for the path S→A→B→D, 36 for the path S→A→B→C→F→G→D, 40 for the path

S→E→F→C→B→D, 29 for the path S→E→F→G→D, 27 for the path S→C→F →G→D and 24 for the path S→C→B→D. Similarly we calculate lifetimes of each paths and get the lifetime 100 for the path S→A→B→D, 100 for the path S→A→B→C→F→G→D, 400 for the path S→E→F→C→B→D, 450 for the path S→E→F→G→D, 400 for the path S→C→F→G→D and 400 for the path S→C→B→D.

If we select the path with minimum cost among them, as done in cost-effective routing, we get the path S→A→B→D having cost 19 and lifetime 100. While in LPR, the route S→E→F→G→D is chosen having lifetime 450 and cost 29. The minimum cost routing is greedy for cost minimization and LPR is greedy for highest lifetime. Hence minimum cost routing suffers from poor lifetime of the path and LPR suffers from high routing cost.

For our CLPR algorithm let us assume maximum possible cost (C) between any two nodes is 15 and maximum possible lifetime (L) of any node is 600. So the scaling factor ξ becomes 40. Hence, using CLPR algorithm the selecting parameter β for the paths S→A→B→D, S→A→B→C→F→G→D, S→E→F→C→B→D, S→E→F→G→D, S→C→F→G→D, and S→C→B→D are 0.1316, 0.069, 0.25, 0.3879, 0.3704 and 0.4166 respectively. The path S→C→B→D has the highest β value. So the selected path is S→C→B→D having cost 24 and lifetime 400.

We find that CLPR is better than LPR in cost perspective and also better than cost-effective routing in stability perspective. Although CLPR may select a path with cost little higher than a path with least cost and a path having little less lifetime than a path having highest lifetime, to achieve the balance between the two contradictory goals, this is acceptable considering both the stability and the cost-effectiveness of the route.

4 Simulation

We have had experiments on the performance of the proposed CLPR and have com-pared it with that of the LPR and Power aware routing. In this section we describe the simulation environment, experimental results and comparison of the three related protocols.

4.1 Simulation Setup

In our discrete event driven simulation we used 25 nodes. The lifetime of a node is varied between 1 and 600 while the transmission cost to neighboring nodes is varied between 1 and 15. Every node has fixed transmission power resulting in a 40 m transmission range. The sources and sinks were spread uniformly over the simulation area; the size of the area varies between simulations from 100 X 100 to 200 X 200. Random connections were established between nodes within the transmission range. The simulation was run 200 times. Nodes followed random waypoint mobility model. Each packet relayed or transmitted has a cost factor and this cost is considered as the cost at the transmitter node.

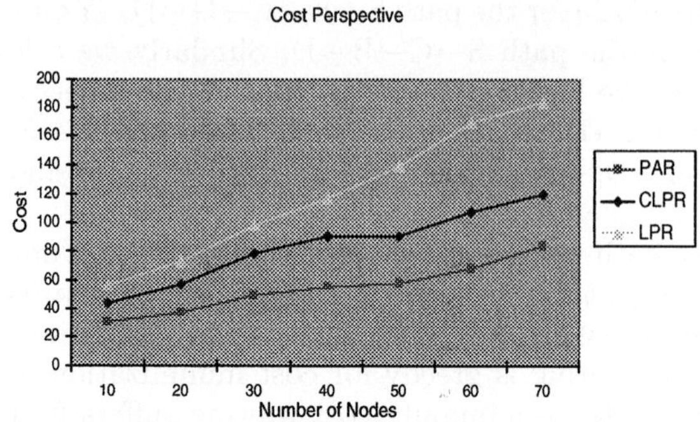

Fig. 2. Comparison of Cost among three related protocols.

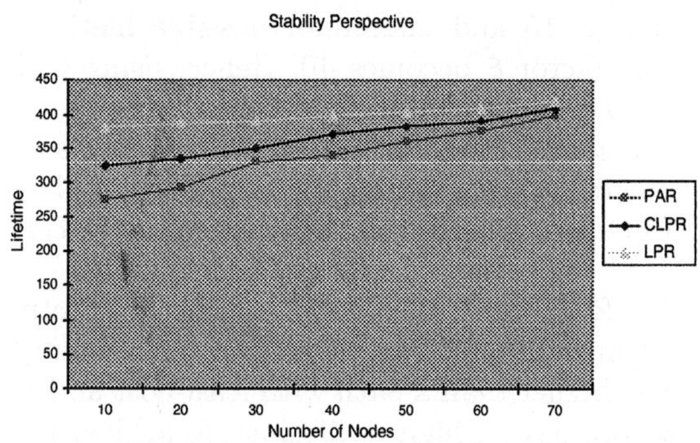

Fig. 3. Comparison of Lifetime/stability among three related protocols.

The results of our simulation have been projected in figure 2 and figure 3. From the figures shown above, we see that as the number of nodes increase the routing cost increases. Power aware routing (PAR) needs minimum cost but its network stability is poor (i.e. minimum). On the other hand, lifetime prediction based routing (LPR) has maximum network lifetime or stability but it suffers from highest routing cost. The Cost-effective Lifetime Prediction based routing algorithm does not suffer extremely from either routing cost or network stability. It maintains a balance between the two and offers cost-effective routing maintaining maximum network stability. The network lifetime is defined as the time taken for a fixed percentage of the nodes to die due to energy resource exhaustion. In our simulation, we considered network lifetime until 65 percent of total nodes die. Some of the nodes, alive at this point are also rendered unreachable since many of the nodes have exhausted their energy and hence they cannot reach other nodes consistently.

5 Conclusion

A Cost-effective Lifetime Prediction based Routing protocol for mobile ad hoc networks that optimizes the network stability and routing cost has been proposed in this paper. Simulation results show that the proposed CLPR can increase the lifetime up to about 20 percent than that of power-aware routing and can cut routing cost up to 33 percent than that of lifetime prediction based routing. Thus the proposed method cuts the cost short while it still tries to maintain maximum lifetime of the network. The lifetime and cost of such a network are two contradictory functions and improvement of one factor has a negative effect on the other. But if any one of these parameters is ignored totally, the network will suffer from poor efficiency. Our proposed method makes a tradeoff between the two and ensures a balanced use of both of them so that maximum utilization is achieved.

References

1. Anderegg L., and Eidenbenz S.: Ad hoc-VCG: A Truthful and Cost-Efficient Routing Protocol for Mobile Ad hoc Networks with Selfish Agents, Proceedings of MobiCom (2003) 245–259
2. Maleki M., Dantu K., and Pedram M.: Lifetime Prediction Routing in Mobile Ad-Hoc Netowrks. Proceedings of IEEE WCNC vol.2 (2003) 1185–1190
3. David B. Johnson: The Dynamic Source Routing for Mobile Ad Hoc Wireless Networks. http://www.ietf.org/internet-drafts/draft-ietf-manet-dsr-09.txt, IETF Internet Draft, Apr. (2003)
4. C. Perkins and P. Bhagwat: Highly Dynamic Destination-Sequenced Distance-Vector Routing (DSDV) for Mobile Computers. Proc of ACM SIGCOMM (1994) 234–244
5. Murthy and J.J. Garcia-Luna-Aceves: An Efficient Routing Protocol for Wireless Networks. MONET vol. 1, (1996) 183–197
6. Charles E. Perkins, Elizabeth M. Belding-Royer, and Samir Das: Ad Hoc On Demand Distance Vector (AODV) Routing. IETF Internet draft November (2001)
7. V.Park and S.Corson: Temporally-Ordered Routing Algorithm (TORA). IETF Internet Draft, July (2001)
8. J. H. Chang and L. Tassiulas: Energy Conserving Routing in Wireless Ad Hoc Networks. Proc. of INFOCOM vol. 1(2000) 22–31
9. A. Michail and A. Ephremides: Energy Efficient Routing for Connection-Oriented Traffic in Ad Hoc Wireless Networks. Proc of the PIMRC vol.2 (2000) 762–766
10. J. E. Wieselthier, G. D. Nguyen and A. Ephremides: Energy-Efficient Broadcast and Multicast Trees in Wireless Networks. MONET vol. 7, (2002) 481–492
11. J. WAN, G. C Alinescu LI and O. Frieder: Minimum-Energy Broadcasting in Static Ad Hoc Wireless Networks. Wireless Networks vol. 8, (2002) 607–617
12. C. Oliviera, J. Kim, and T. Suda: An Adaptive Bandwidth Reservation Scheme for High Speed Multimedia Wireless Networks. IEEE J. Selected Areas in Comm vol. 16 (1998) 858–874
13. Woesner. H, Evert. J, Schlager. M, and Wolisz A: Power-saving mechanisms in emerging standards for Wireless LANs: the MAC level perspective. IEEE Personal. Communication, vol. 5 (1998) 40–48
14. Jin, K. and Cho D: A MAC algorithm for energy-limited ad-hoc networks. IEEE Vehicular Technology Conference, vol. 1, Boston (2000) 219–222
15. Suresh Singh, Mike Woo and C.S. Raghavendra: Power-Aware routing in Mobile Ad-hoc Networks. Proceedings of MOBICOM (1998) 181–190

Design and Simulation Result of a Weighted Load Aware Routing (WLAR) Protocol in Mobile Ad Hoc Network

Dae-In Choi[1], Jin-Woo Jung[1], Keum Youn Kwon[1], Doug Montgomery[2], and Hyun-Kook Kahng[1]

[1] Rm342 Bio-Technology Bldg. 1 Anam-Dong-5-Ka Songbuk-Ku Seoul, Korea 136-701
nbear@korea.ac.kr

[2] National Institute of Standards and Technology, 100 Bureau Drive, Stop 8920, Gaithersburg, MD 20899, USA

Abstract. An ad hoc network has notable features such as frequent mobility, bandwidth limitation, and power constraints. Especially, due to the low bandwidth, there is a strong possibility to cause congestion. If there is congestion, however, power depletion and queuing delay will be serious problems in mobile nodes. In this paper, we propose a Weighted Load Aware Routing (WLAR) routing protocol, which shows excellent performance in an ad hoc network. WLAR Protocol, an extension of AODV, is to distribute the traffics among ad hoc nodes through load balancing mechanism. A new term of total traffic load is defined as a route selection metric, which is the product of average queue size and number of sharing nodes. Using NS2 simulator and real implementation, we show the performance, comparing to AODV.

1 Introduction

The hardware specific such as long-time use, mass storage space, and high speed process of potable computers like a laptop computer and a PDA causes high interest and demand of mobile communication. To satisfy those things, studying mobile networking is actively progressing. To satisfy those things, research on mobile networking is actively progressing. Among them, Mobile Ad Hoc Network (MANET) can make mobile nodes communicate through multi-hop communication. Since this network can communicate without a base station and a fixed cable network, the network can be configured dynamically and can be used in the case of wartime, emergencies and conference. However, this ad hoc network has many problems; high interferences, transmission errors, frequency band sharing, limited power source, and route changes caused by the change of topology and limited power source.

* This research was supported by the MIC(Ministry of Information and Communication), Korea, under the ITRC(Information Technology Research Center) support program supervised by the IITA(Institute of Information Technology Assessment)

C. Kim (Ed.): ICOIN 2005, LNCS 3391, pp. 178–187, 2005.

In ad hoc networks, there are lots of proposals to set the route from mobile node to destination node. They can be classified as proactive protocols and reactive protocols. Proactive protocols maintain the route table to each destination by periodic route information exchanges, whereas reactive protocols retrieve the route by the request of mobile node whenever needed. Both proactive protocols and reactive protocols choose a route based on the metric, the smallest number of hops to the destination. However it may not be the most significant route when there is congestion or bottleneck in the network. It may cause the packet drop rate, packet end-to-end delay, or routing overhead to be increased. This phenomenon surely shows up when traffics are concentrated on a special node, or a gateway through which mobile nodes from ad hoc network can connect to Internet.

For load balancing, there are various proposed mechanisms [Section 2.2], considering traffic load as a route selection. However these mechanisms reflect neither burst traffic nor transient congestion. To work out this problem, we propose Weighted Load Aware Routing (WLAR) Protocol, which selects the route based on the information from the neighbor nodes which are on the route to the destination. In WLAR, a new term of traffic load of node is defined as the product of average queue size of the interface at the node and the number of sharing nodes which are declared to influence the transmission of their neighbors. Average number of queue in the interface is measured and calculated by Exponentially Weighted Moving Average (EWMA) formaular. In section 2, we present related works. In section 3, we describe WLAR Protocol in detail. And, in the following sections, we describe the method of performance comparison and conclusion.

2 Related Works

WLAR Protocol proposed in this paper enhances the AODV (On-Demand Protocol) in ad hoc network environment with a load balancing mechanism. Here, we describe AODV first, and then Load-Balanced Ad hoc Routing (LBAR), Dynamic Load-Aware Routing (DLAR) and Load-Sensitive Routing for Mobile Ad Hoc Networks (LSR) as Load Balance Routing Protocols.

2.1 AODV

AODV [7] relies on routing table entries to propagate an RREP back to the source and to route data packets to the destination. AODV uses sequence numbers maintained at each destination to determine freshness of routinf information and to prevent routing loops. An important feature of AODV is the maintenance of timer-based states in each node, regarding utilization of individual routing table entries.

2.2 Load Balance Routing Protocols

It is easy to implement routing protocols in ad hoc networks which are based on number of hops. It also easily adapts to the change of topology. However, it cannot reflect queuing delay and contention delay in route selections at intermediate nodes. Depending on sepical environments, a route with smallest number of hops may have the longer end-to-end delay comparing to other routes. To solve this problem, there are several methods proposed.

Load-Balanced Ad hoc Routing (LBAR) protocol [4] extends DSR. LABR defines a new metric for routing known as the degree of nodal activity to represent the load on a mobile node. In LBAR routing information on all paths from source to destination are forwarded through setup messages to the destination. Setup messages include nodal activity information of all nodes on the traversed path. After collecting information on all possible paths, the destination then makes a selection of the path with the best-cost value and sends an acknowledgement to the source node.

Dynamic Load-Aware Routing (DLAR) protocol [5] considers intermediate node routing loads as the primary route selection metric. When a route is required but no information to the destination is known, the source floods the RREQ packet. When nodes other than the destination receive a non-duplicate RREQ, they build a route entry for the <source, destination> pair and record the previous hop to that entry. Nodes then attach their load information (the number of packets buffered in their interface) and broadcast the RREQ packet. After receiving the first RREQ packet, the destination waits for an appropriate amount of time to learn all possible routes.

Load-Sensitive Routing (LSR) protocol [6] is based on the DSR. At each node, network load is defined as the sum of the number of queuing packets at mobile host and its neighboring hosts. If destination node receives RREQ messages, it responds to the source node through reverse path including RREQ message by RREP packet. Since destination node doe not wait for all possible routes, the source node can quickly obtain the route information and it quickly responds to calls for connections.

3 Weighted-Load Aware Routing (WLAR) Algorithm

WLAR is to extend the AODV and to distribute the traffics among ad hoc nodes through load balancing mechanism. WLAR adopts basic AODV procedure and packet format.

In WLAR, each node has to measure its average number of packets queued in its interface, and then check wether it is a sharing node to its neighbor or not. If it is a sharing node itself, it has to let its neighbors know it. After each node gets its own average packet queue size and the number of its sharing nodes, it has to calculate its own total traffice load. Now when a source node initiates a route discovery procedure by flooding RREQ messages, each node receiving an RREQ will rebroadcast it based on its own total traffic load so that the flooded RREQ's which traverse the heavily loaded routes are dropped on the way or at the destination node. Destination node will select the best route and replies RREP.

3.1 Definitions

Average number of packets queued in interface is calculated by Exponentially Weighted Moving Average (EWMA). The reason to use average number of packets queued in interface is to avoid the influence of transient congestion of router.

Traffic load at a node is defined as the product of its average packet queue size and the number of sharing nodes.

Sharing node is defined as nodes whose average queue size is greater than or equal to some predetermined threshold value. Sharing node is expected to give some transmission influence to its neighbors. If its average queue size is not greater than a threshold value, it is assumed that its effect is negligible.

Total traffic load in node is defined as its own traffic load plus the product of its own traffic load and the number of sharing nodes

$$TLk = Lk + WL * Lk * SNk \quad ifLk > 0$$

$$TLk = 0 \quad ifLk = 0$$

Where, TLk is a total traffic load at the node k, Lk average queue size at node k, SNk number of the sharing node of the node k, and WL the load weight constant.

WL indicates how much the total traffic load is influenced by the number of sharing node. Here, 0.4 is recommended as the value ofWL.

Path load is defined as sum of total traffic loads of the nodes which include source node and all intermediate nodes on the route, except the destination node.

3.2 Route Selection Procedure

When a source node initiates a route discovery procedure by flooding RREQ messages, each node that receives the RREQ looks in its routing table to see if it has a fresh route to the destination. Even though it does, it does not unicast a route reply (RREP) message back to the source. It adds its traffice load value to the value of the traffic path load field in RREQ, holds it for some delay, then rebroadcasts the RREQ.

Delaying RREQ is motivated by the fact that each node accepts only an earlier RREQ and discards other duplicate RREQ's. With delaying RREQ technique, RREQ's with non-optimal route are more likely to be discarded than the

RREQ messages from nodes with the optimal value. The time of holding RREQ and delaying the RREQ rebroadcast at an intermediate node is proportional to its total traffic load. After delaying, the RREQ to be rebroadcasted is put in priority queue. Consequently RREQ traversing nodes with low traffic loads will arrive at the destination early.

Each node keeps track of its local connectivity and its information, i.e., which neighbors are sharing nodes. This is performed either by using periodic exchange of so-called HELLO messages.

4 Simulation Results and Implementation

Here we show the simulation results, design architecture and its implementation. To validate the implementation, we measure the performance.

4.1 Simulation Results

The Simulation environments and methods are described in [13]. In [13] we showed the end-to-end delay and packet delivery ratio. Here we show the TCP traffic in Figure 1. WLAR, compared with AODV, generally show superior performance. Especially, it shows better performance in the network where nodes do not move much and the network traffic is heavy.. Also, the number of effective TCP conncections for WLAR is more than AODV. Under the light traffic load, as network congestion is not generated, there is similar performance.

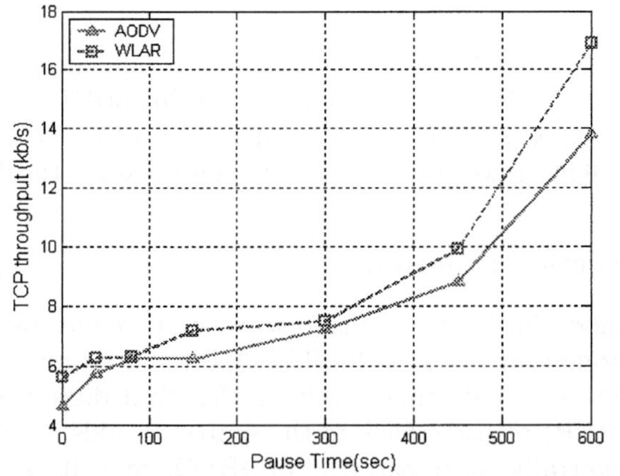

Fig. 1. Throughput for TCP Reno connections

4.2 Implementation of WLAR

The WLAR has been implemented by modifying the source code of AODV-UU, which is developed at Uppsala University. WLAR is developed for laptop computers on Linux kernel version 2.4.20-8, the kernel version provided by the Red Hat Linux version 9.0, and for PDA on Linux kernel version 2.4.29, kernel version 0.7.0. The Linux operation system is chosen due to its availability and familiarity. For Wi-Fi equipments, laptop computers are equipped with Lucent Orinoco (Wave LAN) Silver, and iPAQ H5450 of HP with built-in WI-Fi

4.3 Architecture of WLAR

Figure 2 and 3 illustrate the logical strucute of he WLAR implementation, shades are modified parts from AODV-UU.

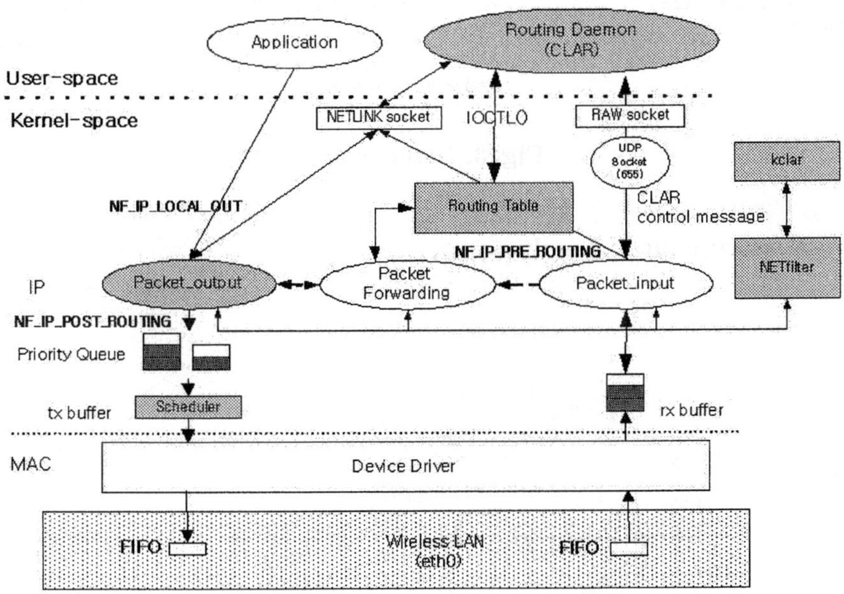

Fig. 2. WLAR Routing Aritecture

The kernel routing table is maintained by the routing daemon according to the WLAR algorithm. WLAR typically maintains two different routing tables: kernel routing table and routing table cache. The routing table cache is used in user domain to optimize the route discovery overhead. Each entry in this route cache has an expiration timer, which needs to be reset when the corresponding route is used. The entry should be deleted when the timer for that entry expires. k_add_rte() and k_del_rte() functions add or delete routes to the kernel routing table using the ioctl() interface. k_chg_rte() function updates the kernel routing table entry using ioctl() interface.

Figure 3 shows filve hooks defined in the Netfilter architecture. A datagram may enter to the IP layer either from the upper layers or through a network interface.

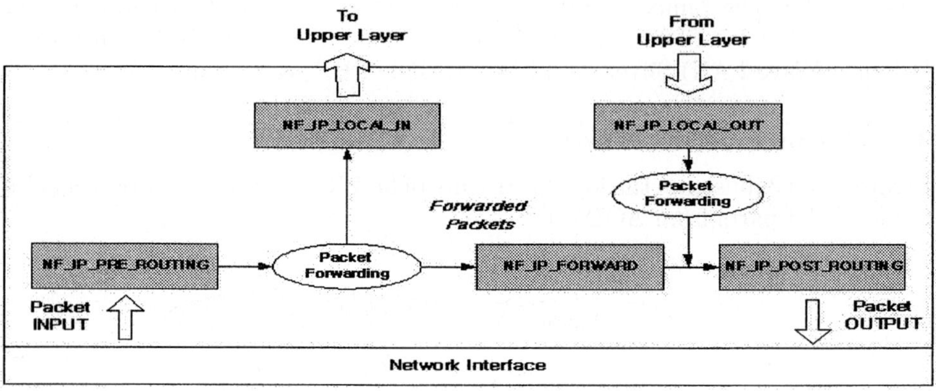

Fig. 3. Netfilter hooks

4.4 Experimemtal Results

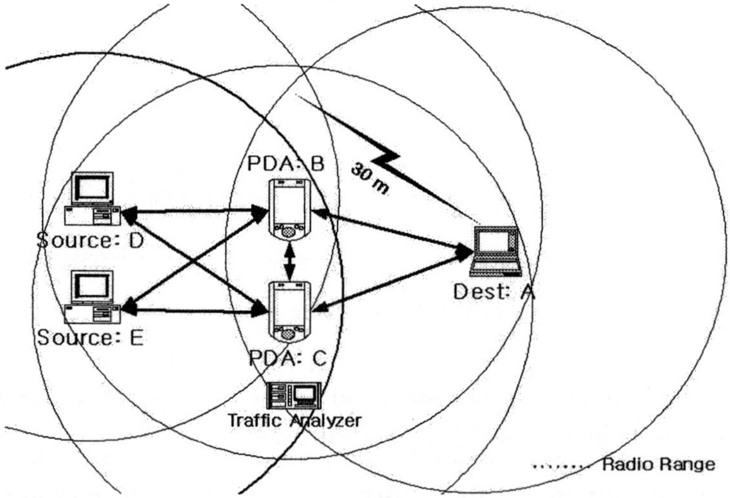

Fig. 4. Ad hoc wireless testbed for experimental evaluation

Figure 4 shows an ad hoc wireless testbed used in experiments. Test network topology and test scenarios are as follows:

- Node B and C are one-hop neighbors
- Node D and E are in the radio range of B and C
- The distance between B and A is equal to the distance between C and A. Also, the distance between D and B (or E and B) is equal to the distance between D and C (or E and C).
- Node D and E are not direct reachable from node A
- Scenario is that Node A makes a CBR communication with node D and then makes another CBR communication with Node E.
- Data rate is 50 packets per second and data size is 20 bytes.
- We assume that Node A, B, C, D, and E do not change the topology during the test.

WLAR reduces the end-to-end delay up to one-tenth. WLAR can improve the throughput of TCP than other implementations by 65% since the reduction of end-to-end delay reduces the round trip delay of data packets. The WLAR improves the Packet since it reduces the congested areas by distributing two CBR traffic to different routes.

Version Metrics	NIST-AODV	AODV-UU	WLAR
Average End-to-End Delay	0.2176	0.1514	0.0221
Throughput (Kbytes/sec)	40	48	79
Packet delivery Ratio	0.8152	0.8542	0.920304

Table 1. Experimental Results

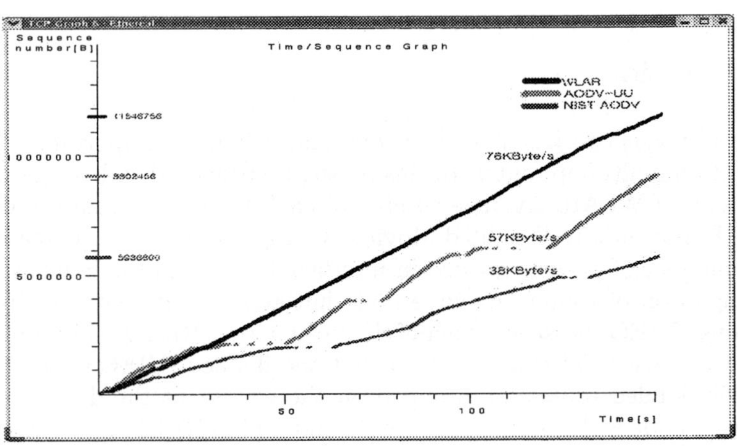

Fig. 5. 50 packets per second TCP performance

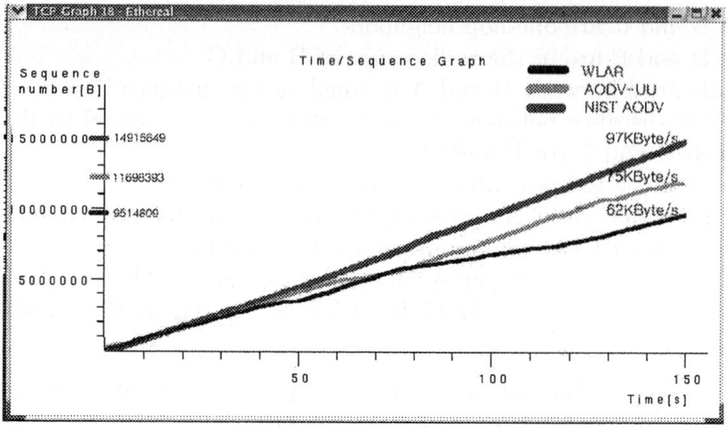

Fig. 6. 16 packets per second TCP performance

Figure 5 and 6 show throughput performance for each implementation, where throughput is calculated to be the number of bytes delivered to the destination node during 150 seconds. And Figure 5 show the TCP throughput over ad hoc network with offered load: For the offered load, the data rate is equal to 16 packets per second and data size is 480 bytes (G.711 voice traffic). Figure 6 date rate is 50 packets per scond and data size is 20 bytes.

WLAR improves the TCP performance by reducing the occurrence of congestion. It is reduces the RTT of the TCP traffic and results in improving the throughput. In Figure 5 and 6, while the slope of WLAR is similar to that of AODV-UU WLAR has lower delay than that of AODV-UU since all traffic in AODV-UU are delivered on one route.

5 Conclusions

Since an ad hoc network is multi-hop communication, it is important to generate and maintain multi-hop routes. In this paper, we proposed and verified excellent performance of WLAR. Average number of packets queued in interface is calculated by Exponentially Weighted Moving Average (EWMA). The reason to use average number of packets queued in interface is to avoid the influence of transient congestion of router. a source node initiates a route discovery procedure by flooding RREQ messages, each node receiving an RREQ will rebroadcast it based on its own total traffic load so that the flooded RREQ's which traverse the heavily loaded routes are dropped on the way or at the destination node. Destination node will select the best route and replies RREP. Accordingly, it can choose the better route to reflect not only its state but also near-by node state. To measure the performance, seen from the simulation, WLAR algorithm shows excellent performance compared with existing AODV in ad hoc network. It also avoids the influence of burst traffic and transient congestion. Route selection of this method provides the near-by nodes of gateway with load balance through one gateway when mobile node connects Internet [3]. Accordingly, among near-by nodes, it produces the reduction of power consumption and end-to-end delay time, advancement of general throughput and improvement of delivery ratio for specific node.

References

1. C.E Perkins, "ad hoc networking," Addison Wesley, 2001
2. U. Jonsson, F. Alriksson, T. Larsson, P. Johansson, and G. Maguire, Jr., "MIP-MANET - Mobile IP for Mobile Ad hoc networks", in Proceedings of IEEE/ACM Workshop on Mobile and Ad Hoc Networking and Computing, Boston, MA USA, August 1999
3. Y. Sun, E. M. Belding-Royer, and C. E. Perkins, "Internet connectivity for ad hoc mobile networks", International Journal of Wireless Information Networks special issue on Mobile Ad Hoc Networks (MANETs): Standards, Research, Applications, 2002
4. H. Hassanein and A. Zhou, "Routing with Load Balancing in Wireless Ad Hoc Networks," in Proc. ACM MSWiM, Rome, Italy, July 2001
5. S. J. Lee and M. Gerla, "Dynamic Load-Aware Routing in Ad hoc Networks," the Proceeding of ICC 2001, Helsinki, Finaland, June.
6. K. Wu and J. Harms, "Load-Sensitive Routing for Mobile Ad Hoc Networks", in Proc. IEEE ICCCN'01, Scottsdale, AZ Oct. 2001.
7. C. E. Perkins, E. M. Royer and S. R. Das, "Ad hoc On-Demand Distance Vector (AODV) Routing", July 2003, IETF rfc3561
8. S. R. Das, C. E. Perkins, E. M. Royer, and M. K. Marina, "Performance comparison of two on-demand routing protocols for ad hoc networks", IEEE personal Communications Magazine, special issue on Mobile Ad Hoc Networks, vol. 8, no. 1, p. 16-29, February 2001.
9. R. Wakikawa, C. E. Perkins, A. Nilsson, and A. J. Tuominen, "Global connectivity for IPv6 Mobile Ad Hoc Networks", Internet Engineering Track Force, Internet Draft (Work in Progress), October 2003.
10. Floyd, S., Jacobson, V., "Random early detection gateways for congestion avoidance," Networking, IEEE/ACM Transactions on Volume 1, Aug. 1993.
11. The VINT Project, "The network simulator – ns-2," Available at http://www.isi.edu/nsnam/ns
12. K. Fall and K. Varadhan, "The ns Manual (formerly ns Note and Documentation), April 2002, http://www.isi.edu/nsnam/ns/ns-documentation.html
13. Jin-Woo Jung, Daein Choi, Keumyoun Kwon, Ilyoung Chong, Kyungshik Lim, Hyun-Kook Kahng: "A Correlated Load Aware Routing Protocol in Mobile Ad Hoc Networks". ECUMN 2004: 227-236, October 25-27, 2004

Modeling the Behavior of TCP in Web Traffic

Hyoung-Kee Choi[1] and John A. Copeland[2]

[1] School of Information and Communication, Sungkyunkwan University,
Suwon, Korea 440-746
hkchoi@ece.skku.ac.kr

[2] School of Electrical and Computer Engineering, Georgia Institute of Technology,
Atlanta, Georgia 30022
copleand@ece.gatech.edu

Abstract. The growing importance of Web traffic on the Internet makes it important that we have an accurate understanding of this traffic source in order to plan and provision. In this paper we present an empirical model of TCP connections used in delivering Web objects. Our TCP model takes advantage of a unique behavior of TCP that it alternates between inactive and active periods of transmitting data segments in bursts. Based upon the bursts in a TCP connection, we characterize TCP by defining the period between the starts of adjacent bursts and measuring the number of data segments transmitted and the time spent in this period. From the characterization we develop a TCP model that attempts to capture the major aspects of the real TCP connection.

1 Introduction

The exponential growth rate of the World Wide Web (Web) has led to a dramatic increase in Internet traffic as well as a significant degradation in user-perceived latency while accessing the Web. This has spawned significant research aimed at improving Web performance [8]. Fundamental to improve Web performance is a solid understanding of Web traffic characteristics. Systems designed without proper understanding of traffic patterns can result in unpredictable outputs and biased performance in certain situations. Consequently, it is important that we have deep and broad understanding of this source of traffic.

The performance of the Web heavily depends on the performance of TCP and HTTP. Hence, it is important that we study the behaviors of TCP and HTTP layers for complete understanding of Web traffic. We have presented a study of Web traffic in the HTTP layer in our early study [2]. This early study covered the behavior of Web traffic in the boundary of the Web pages: e.g., the number of objects and connections in a page, individual object size, and so on. In this study, we characterize and model Web traffic in the TCP layer. The characterization in the TCP layer analyses the behaviors of Web traffic in connections used in delivering objects.

Numerous studies have developed analytical models of TCP behavior in an accurate manner [5],[6]. These analytical models of TCP have two weaknesses

C. Kim (Ed.): ICOIN 2005, LNCS 3391, pp. 188–197, 2005.

to be used in our study. First, most analytical TCP models have concentrated on the steady-state throughput of TCP. A TCP model in our study necessitates the transient behavior because TCP under HTTP rarely reaches the steady state because of a small object size [2],[8]. Secondly, some assumptions made in the past studies are a distance from being realistic: for example, delayed ACK is suppressed; segments experience constant RTT; and the receiver has a large buffer space.

The question we try to address in this study is how to model a complete TCP connection used in delivering Web o bjects. In particular, how can we incorporate a transient behavior of a TCP connection into our Web traffic model? This question is a challenging problem because a system behavior in the transient state is hard to predict.

We have observed that TCP alternates between inactive and active periods of transmitting data segments in bursts. Based upon the bursts in a TCP connection, we model TCP by: (1) defining the period between the starts of adjacent bursts and (2) measuring the number of data segments transmitted and the time spent in this period. We have also observed that inter-arrival times of segments inside bursts are a lot smaller than those across bursts. The inter-arrival times of segments across bursts are comparable to a round trip time (RTT) of a connection. From this observation, we have developed a scheme to estimate the RTT of a connection and then to distinguish different bursts in a connection using the estimated RTT.

The remainder of this paper is organized as follows. Section 3 describes how we distinguish the different bursts in a connection. Section 4 characterizes TCP connections, and Section 5 describes the generation of traffic based on the model. We conclude, in Section 6, with a discussion of remaining work.

2 Related Work

Mathis et al.[6] and Padhye et al.[5] derived analytical models of TCP Reno congestion avoidance in the steady state. Mathis observed an envelope of the window-size evolution traverses a perfectly periodic saw tooth under certain circumstances. They considered the area surrounded by this envelope was the throughput and derived a closed-form equation of the throughput with respect to RTT and the probability of segment drops. Padhye also used the envelope but his model was more general as some assumptions made by Mathis were relaxed. They validated their models by showing the closeness in the throughput between the model and a real connection.

Not many studies have attempted to model the transient behavior of TCP. Caldwell et al.[7] extended Padhye's work to derive new models for two aspects that can dominate TCP latency: the connection establishment three-way handshake and slow start. Hiedemann et al.[4] modeled the performance of HTTP over several transport protocols. To evaluate HTTP performance over TCP, they modeled slow-start behavior of TCP and proposed the number of segments in burst for different acknowledgment policies.

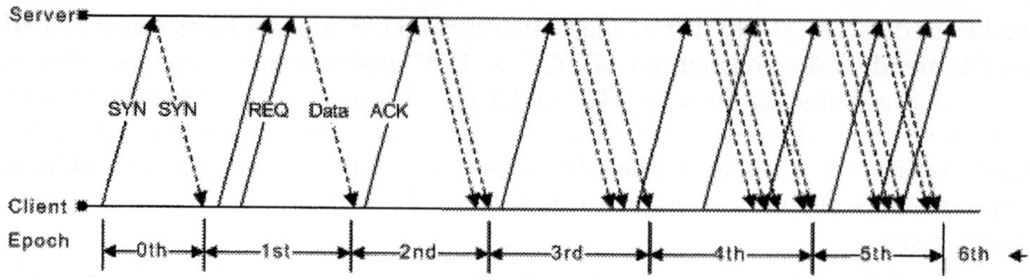

Fig. 1. A typical transaction of a TCP connection transferring a Web object

3 Epoch

TCP alternates between inactive and active periods of transmitting data segments in bursts. After a TCP sender transmits an entire window-size worth of data segments in a burst, it pauses until the first ACK in a burst returns. That is because window size is still too small to completely fill the pipe. Based upon the bursts in a TCP connection, we characterize TCP by: (1) defining the period between the starts of adjacent bursts and (defined as epoch) (2) measuring the number of data segments transmitted and the time spent in this period (defined as the number of data segments in an epoch and epoch time, respectively). The number of segments and the epoch time form the primary parameters in our study of TCP characterization and modeling.

A typical transaction of a TCP connection retrieving a Web object is shown in Fig.1. The TCP connection in Fig.1 contains six epochs. In Fig.1, the epoch time in the second epoch is the period between the first and the third data segments. The number of data segments in the second epoch (epoch number two) is two.

3.1 Epoch Delimitation

At epochs the number of data segments doubles in size in slow-start and increases by one segment in congestion avoidance. Thus the relationship in the slow start stage is exponential relation between the epoch number and the number of segments in an epoch. Hence, one might conclude that a new epoch is initiated after transmitting the fixed number of data segments in the current epoch. However, this exponential relation is not always guaranteed, resulting in an obscurity of epoch boundaries.

The delayed ACK scheme [3], variable RTTs, slow servers and segment drops are the factors contributing to epoch variability. In reality, RTTs vary and the variance is quite significant. Unlike Fig.1, time lines between subsequent data segments and ACKs are not parallel any more. Slow servers make TCP wait for data segments to arrive in clients, increasing the periods between subsequent segments. When a segment drop is detected, TCP enters a special stage to repair the drop. This stage lasts a number of times longer that an RTT in a connection.

If data segments in a burst were delayed long enough to be comparable to RTT because of aforementioned factors, the delayed data segments would arrive at a client as if they were the start of a new burst.

All of these cases may create the circumstances where the epoch boundaries are hard to distinguish. We are given nothing but a series of TCP segments with timestamps passing in two directions. Nevertheless, from this limited information, we should be able to tell which segments initiate new epochs.

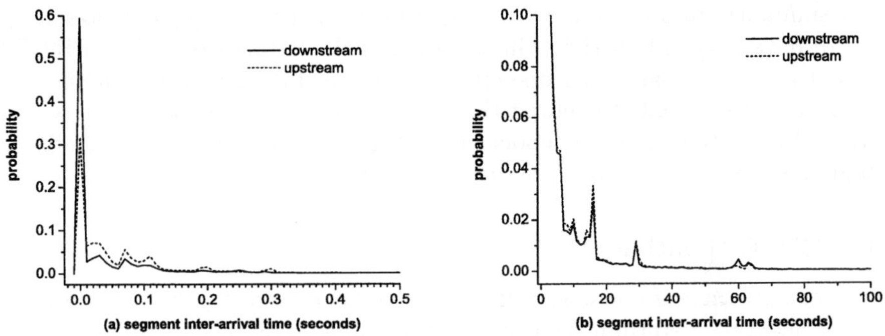

Fig. 2. Two probability distributions of inter-arrival time of segments (a) between 0 and 500ms and (b) above 500ms

3.2 Segment Inter-arrival Time

To achieve accurate epoch delimitation, we take advantage of the inter-arrival time of segments. Fig.2 shows the distribution of inter-arrival times of segments in two directions measured from the trace: upstream (a client to a server) and downstream (a server to a client). The inter-arrival time of segments can be divided into three regions: up to 20ms, between 20ms and 500ms, and above 500ms. We will consider these three regions in turn.

The peak in each distribution between 0 and 20ms corresponds to the segment arrival in bursts. The quantile difference in Fig.2.a - 0.32 and 0.59, respectively - is caused by the TCP acknowledgement strategy. While data segments are transmitted immediately after they are available to TCP, ACK segments can be delayed no more than 200ms because of the delayed ACK. Hence, the inter-arrival time of ACK segments in the upstream tends to be longer than that in the downstream.

Relatively small mounds between 20ms and 500ms are mainly attributed to a connection waiting for an ACK segment after sending a burst of data segments. We contend that this waiting period should be comparable to an RTT in a connection. That is because this waiting period lasts until the ACK of the first segment in the burst returns to the server and, by definition, this period is an RTT.

Segments in a connection experiencing a drop or segments in keep-alive connections can also cause inter-arrival time larger than 500ms. When a segment drop is detected a client abstains from transmitting segments for a retransmission timeout. This timeout period lasts at least one second to a few seconds. As a result, the inter-arrival time increases proportionally. Fig.2.b shows the distribution of inter-arrival time between 0.5 to 100 seconds. In this figure, relatively small mounds are shown near 15, 30, and 60 seconds. These inter-arrival times were caused by connections forced to close by a server after the expiration of a keep-alive timer.

In summary, in an epoch, a relatively long inactive period is followed by a burst of data segments arriving in short intervals. It takes roughly an RTT for the next epoch to start after one epoch ends. We take advantage of the inter-arrival time of segments to delimit the epochs in a connection: that is, a segment is treated as initiating a new epoch when the segment is transmitted after the period close to the RTT from the previous segment.

3.3 RTT Estimation

A new epoch starts if two consecutive data segments are separated by an RTT. For the accurate determination of epoch boundaries a good estimation of the RTT is essential. However, the RTT is neither known a priori nor a constant value throughout the connection. For an unknown RTT we use an estimation technique to assess the RTT in a connection. As explained later in this Section , we select conservative parameter values in the RTT estimation. For a conservative RTT estimation, we introduce the use of a value that we call the threshold of the epoch. The threshold is used instead of the estimated RTT to identify epochs to compensate for possible error in the RTT estimation. As a result we use seventy five percent of the RTT estimation for the threshold.

We can measure a number of RTT samples in a connection. A measured RTT value does not necessarily equal to the previous value because some components of RTT may vary depending upon the condition of a connection. We accommodate varying RTTs by using an adaptive adjustment scheme. In essence, this scheme monitors the condition of a connection and deduces reasonable values for RTT. As the condition of a connection changes, this scheme revises its RTT value.

We illustrate the measurements of RTT samples and the RTT updating instances in our scheme with Fig.1. The SYN interval is the first measurement of the RTT in a connection. This measurement is marked as"0th" (zeroth) in Fig.1. The first epoch is initiated when a client sends a request. After the zeroth epoch until the first data segment arrives at the client it takes another RTT (first epoch in Fig.1), and we measure the second RTT sample. The RTT estimation is updated with the second RTT sample. The second data segment arrives at the client after the threshold, and, hence, this segment initiated the second epoch. At this time, the third RTT sample is measured and the RTT estimate is updated accordingly. According to Fig.1, as the connection evolves, we have four more RTT samples.

We do not update the new RTT estimate when the RTT sample is 2.5 times larger than the current RTT estimate. This long period is caused by the segment drop but not by a dramatic change in the condition of a connection. A segment drop mostly causes an abnormally long inter-arrival time in TCP. Therefore, we do not want to update the RTT estimate. However, by definition, the segment drop eventually initiates a new epoch.

Because the RTT in a connection may vary over time we developed a scheme, called "exponential averaging technique", to adapt the RTT estimate promptly to the new condition. Our adaptive adjustment scheme is similar to the TCP timeout and retransmission [3]. The RTT estimate and the threshold can be derived from the following simple equations:

$$diff = \widetilde{RTT} - RTT\left[i\right]. \tag{1}$$

$$RTT\left[i+1\right] = RTT\left[i\right] + \delta \times diff, \text{if} \widetilde{RTT} \leq 2.5 \times RTT\left[i\right]. \tag{2}$$

$$RTT\left[i+1\right] = RTT\left[i\right], \text{if} \widetilde{RTT} > 2.5 \times RTT\left[i\right]. \tag{3}$$

$$\text{threshold} = \beta \times RTT\left[i+1\right]. \tag{4}$$

where δ is a fraction between 0 and 1 that controls how quickly the new sample affects the RTT estimate. β is a factor that controls how much the estimate affects the epoch threshold. We use 0.75 for the value of β and $1/2^3$ for δ. A variable, called *diff* in Equation (1), keeps track of the difference between the current RTT sample and the previous RTT estimate.

We evaluated the performance of the selected δ and β values. We implemented the differential averaging technique with different β values of 0.65, 0.75, and 0.9. Our conclusion was that the changes to the β value within 0.1 from 0.7 would not degrade the performance of the proposed technique significantly. We have selected the δ value of 1/8 that minimizes the difference in the distributions between the epoch time and the downstream inter-arrival time.

4 TCP Characterization

Having defined an epoch in a connection, we collected statistics of the two primary parameters from a trace. We would like to measure the traffic close to a client because we are modeling the traffic sent by, and to, the client. At the same time, we want a large cross-section of traffic. We meet these objectives by recording a trace of traffic on the backbone network of the Georgia Tech campus. We use primarily a trace that was collected on the backbone network of the Georgia Tech campus from 11 a.m. to 12 p.m. on Wednesday, March 15 2000.

For better understanding of the statistical properties we introduce a "modulated-joint distribution (MJD)". Let X denote one of the primary parameters and Y denote the epoch number. The statistics of X changes at different Y because X is a non-stationary random process with respect to the epoch number. We represent the statistics of X in a single plot as a joint distribution of X and Y. In this representation we modulate the probability of $p(x, y)$ to $p(x|y)$. One benefit is a compact representation of random process X in a single plot. We can also represent detailed statistics of X at large Ys.

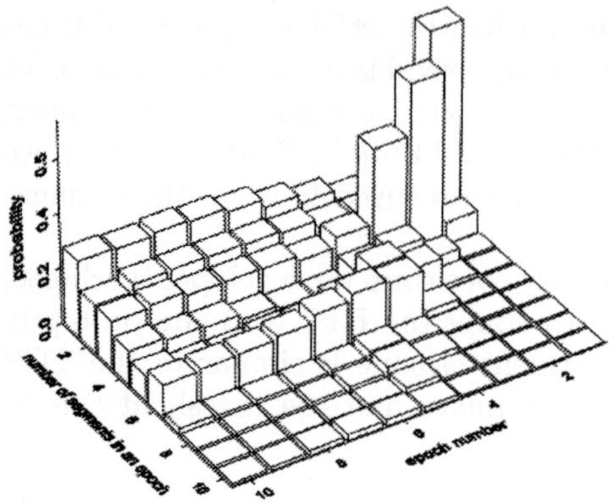

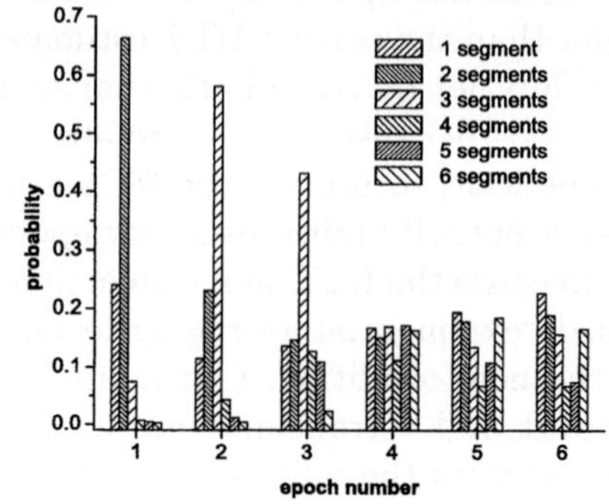

Fig. 3. MJD of epoch number and the number of segments in an epoch

Fig. 4. The number of data segments up to the six epochs

4.1 The Number of Segments in an Epoch

Fig.3 shows a MJD of the epoch number and number of segments in an epoch. The figure shows up to epoch number 10 and 10 segments in an epoch. Changes in the distribution are small for epoch numbers greater than six.

Fig.4 shows the probabilities of the first six segments per epoch. At the first epoch more than 60 percent of the epochs serve the two segments. This measurement attracts our attention because the number of segments in the first epoch is equal to the value of the initial congestion window (ICW). From this measurement we can access different TCP implementations on the ICW. Two segments are most common in the first epoch. A newer version of Linux and the beta version of Solaris 8 set their ICWs to four segments.

The two data segments in the first epoch result in either one ACK acknowledging the two segments or two ACKs acknowledging the individual data segment. In either case the first epoch is closed, and the window size would be increased to three or four. A server turns to the second epoch and can transmit three or four data segments in the second epoch. Our result shows that three segments in the second epoch are most common. After transmitting three segments in a row a server waits for an ACK in the second epoch. A client is supposed to acknowledge immediately when receiving two-full-MSS-size segments even though the delayed ACK feature is set to active. The ACK for the first two segments is returned immediately when the second epoch finishes. This ACK increases the congestion window size to four. At the third epoch in Fig.4, three segments are also most common, although the current window size is four. At the beginning of the third epoch, the server still has one unacknowledged data segment transmitted in the second epoch. Hence, the server can only send three data segments.

At the fourth epoch the probability is more or less uniformly distributed. We expected by inductive reasoning that the probabilities would be large at four, five, and six segments and that the rest of the probabilities would be small. The difference from our expectation is because of the large number of one, two, and three segments in the fourth epoch. When alerted to a segment drop, TCP decreases the congestion window size either to one or to half of the size before the drop. One and two segments occurring in the fourth epoch are partially the result of the window size being reset to one. This observation suggests one, two, and three segments are possible in epoch number four and greater.

At the epoch number greater than six, the probabilities on one, two, and six segments become prominent. The statistical properties of the number of data segments at these higher epochs are quite similar to that at epoch number four.

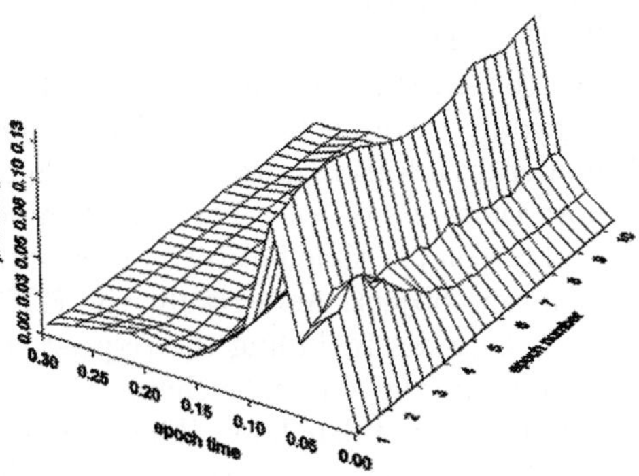

Fig. 5. MJD of epoch number and the epoch time.

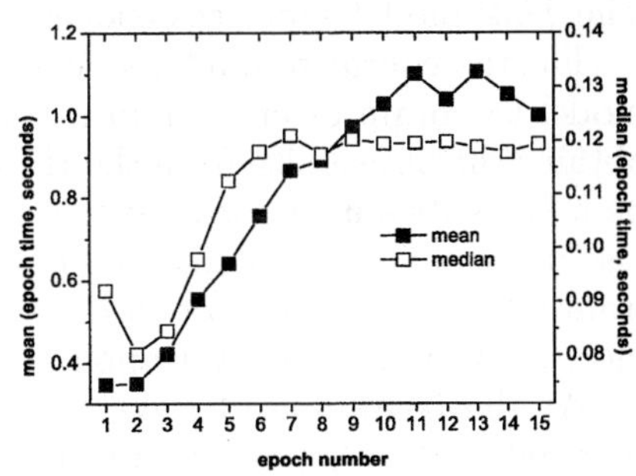

Fig. 6. Means and medians of epoch time at different epochs.

4.2 Epoch Time and Epoch Ratio

Fig.5 shows the modulated-joint distribution of epoch numbers up to 10 and the epoch time smaller than 300ms. The variation in the epoch time is small in the epoch numbers greater than six. Fig.6 shows means and medians of the epoch time along with the epoch number. The average epoch time increases with the epoch number until ten and is saturated to 1.1 seconds. A mean is affected by outliers that is, in our case, the relatively large keep-alive timer. The median decreases between the first and second epochs, and then increases up to the sixth epoch. The large median at the first epoch is because of the long processing time at the server. A client sends a request for an object at the first epoch. The server parses this request and reads the corresponding object to download the object. The first epoch includes this processing time and tends to be longer than neighbor epochs.

We introduce another parameter, called an epoch ratio. An epoch ratio is a ratio of a SYN interval to an epoch time in a connection. This parameter helps us to understand how epoch times in a connection varies with respect to a SYN interval. We measured SYN intervals from the time taken by the three-way handshake. We did not collect any statistics of the parameters from connections where either SYN segment was retransmitted. The distribution of the epoch time shown in Fig.5 is transformed to that of the epoch ratio by dividing a unique SYN interval in a connection by the epoch time in the same connection.

5 Traffic Generation in TCP Model

We constructed a TCP model based upon the TCP characterization. The number of segments in an epoch, the epoch ratio and the additional parameter of the SYN interval are used primarily to generate the synthetic traffic in the model. The TCP model simulates epochs in a TCP connection.

For the complete model of Web traffic, the TCP model relies on the HTTP model to obtain essential information regarding a connection. The TCP model obtains the object size from the HTTP model. In addition, a real connection can be a keep-alive connection, delivering more than one object. The HTTP model also determines a keep-alive connection and the elapse time between objects: i.e., from the end of an object to the start of a new object. As a result, the HTTP model determines the four application-level statistics including the request size.

At the beginning of a connection a client in the model exchanges a SYN segment with a server, mimicking the three-way handshaking. The three-way handshaking lasts for a period drawn from the SYN interval distribution. After the three-way handshaking a client enters at the first epoch by sending a request segment. At this point, the HTTP model informs of the request size and the object size. The model calculates the total number of segments in a connection by dividing the object size by a MSS of 1,460 bytes.

The number of segments in the first epoch is determined from the distribution of number of segments in an epoch given epoch number one. For the time the first epoch lasts, we use the multiplication of epoch ratio drawn from the distribution given epoch number one and the SYN interval for this connection. The generations of the two parameters in the successive epochs follow the same manner as the first epoch, but are assumed to be independent across epochs.

In an epoch the model generates a burst of segments followed by an idle period until the next epoch starts. At the end of each epoch the model checks if a server has transmitted enough segments for the given object size. The model succeeds to the next epoch as long as the cumulative number of transmitted segments is less than the total number of segments. At the last epoch, when the model finishes downloading the current object, the HTTP model informs the TCP model of the decision on whether the connection should remain open for the delivery of the next object or should be closed.

The TCP model developed in this study was combined with the HTTP model developed in [2]. This complete Web traffic model can substitute for an actual client and can be used in simulations as a Web traffic generator. The complete Web traffic model simulates an ON/OFF source where the ON state represents the activity of a Web page and the OFF state represents a silent time after all objects in a Web page are retrieved. In general, the HTTP model determines the number of connections in a Web page. The TCP model characterizes behaviors of the model in the connections.

6 Conclusion

In this paper we have presented an empirical model of a TCP connection used to deliver Web objects. For a complete Web traffic model, it is essential to characterize and model a TCP connection from the beginning to the end. However, an analytical model of the entire TCP connection is difficult because of its transient behavior in the slow-start stage and the many factors that go into determining TCP behavior.

We characterized a TCP connection including its transient state using the two primary parameters: the number of data segments in an epoch and the epoch time. Based upon these parameters and their statistics we built a TCP model that attempted to capture the major aspects of the real TCP connection.

A number of avenues for future work remain. First our model can be enhanced to design a better adaptive scheme to respond quickly to possible fluctuation on the RTT in a connection. Our accurate epoch delimitation was possible because the RTT variation was moderate. The performance of the proposed adaptive scheme might be degraded on applying to TCP connections over wireless or dial-up channels where the connections may experience more severe RTT variation. Second we can add more values on our model by extending the model to cover other types of applications where TCP operates at a bulk transfer mode. File transfer protocol (FTP), simple mail transfer protocol (SMTP) and peer to peer (P2P) are those applications.

References

1. Clark, D.: Window and acknowledgement strategy in TCP. RFC 813, July 1982.
2. Choi, H. K., Limb, J. O.: A behavioral module of Web traffic, In *Proceedings of the IEEE ICNP '99*, October 1999.
3. Postel, J.: Transmission control protocol, RFC 793, 1981.
4. Heidemann, J., Obraczia, K.: Modeling the performance of HTTP over several transport protocols, *IEEE/ACM Transaction on Networking*, August 1997.
5. Padhye, J., Firoui, V., Towsley, D.: Modeling TCP throughput: A simple model and its empirical vallidation, In *Proceedings of ACM SIGCOMM '98*, August 1998.
6. Mathis, M., Semske, J., Ott, T.: The macroscopic behavior of the TCP congestion avoidance altorithm, *ACM Computer Communication Reivew*, July 1997.
7. Caldwell, N., Savage, S., Anderson, T.: Modeling TCP latency, In *Proceedings of IEEE INFOCOM*, March 2000.
8. Padmanabhan V. N., Mogul, J. C.: Improving HTTP latency, In *Proceedings of 2nd International WWW Conference*, October 1994.

Using Passive Measuring to Calibrate Active Measuring Latency

Zhiping Cai, Wentao Zhao, Jianping Yin, and Xianghui Liu

Department of Computer Science and Technology,
National University of Defense Technology, Changsha, 410073 China
{xiaocai, bruce_zhao}@163.net
jpyin@nudt.edu.cn
liuxh@tom.com

Abstract. The network performance obtained from the active probe packets is not equal to the performance experienced by users. We can obtain more exact result by using the characteristics of packets gained by passive measuring to calibrate the result of active measuring. The method of combining passive and active approaches has some advantages such as protocol-independent, negligible extra traffic, convenience and being able to estimate individual user performance. Considering the number of user data packets arriving between probe packets and the latency alteration of neighborhood probe packets, we propose the Pcoam (Passive Calibration of Active Measurement) method. It could reflect the actual network status more exactly, especially in the case of network congestion and packet loss, which has been validated by simulation.

1 Introduction

Latency is clearly a key performance parameter and utilization indicator for any modern IP network[1,2,3]. In general, conventional schemes to measure the network delay are classified into two types, active and passive measurements. Active measurement measures the performance of a network by sending probe packets and monitoring them. In passive measurement, the probe device accessing the network records statistics about the network characteristics of data packets. Unfortunately, both types have drawbacks especially when they are applied to delay measurement[4].

Although active measurement is easy to implement, the result of active measuring depends on the network traffic load, measuring time, measuring interval, sampling methods and even the size of measuring packets. To measure the perflow generated by individual users or applications, active measurement must insert many active probe packets. This method would make the network congestion to be more serious. On the other hand, passive measurement can get more exact measure result. But the passive measurement requires all measuring devices to be synchronized. It does not have good scalability when it is adapted to a large-scale network.

C. Kim (Ed.): ICOIN 2005, LNCS 3391, pp. 198–206, 2005.

Using short active probe packets and sending them at certain intervals, like ping, could get the approximate performance of network. But the performance obtained from the probe packets is not equal to the performance experienced by users [5].

Masaki Aida *et al.* have proposed a new measuring technique, called CoMPACT Monitor[4,5,6]. Their scheme requires both active and passive monitoring using easy-to-measure methods. It is based on change-of-measure framework and is an active measurement transformed by using passively monitored data. Their scheme can estimate not only the mixed QoS/performance experienced by users but also the actual QoS/performance for individual users, organizations, and applications. In addition, their scheme is scalable and lightweight.

The CoMPACT scheme supposes that the interval of sending probe packets can be very short and thus ensures that the interval of receiving those probe packets is also short. The error of estimator would increase as the interval of the probe packets arriving increases, especially when the network is in congestion or the loss ratio is high.

We can use the characteristics of user packets gained by passive measuring to calibrate the active measuring for more exact measuring result. We consider not only the number of user data packets, but also the relationship of the adjacent probe packets' delay. We propose the Pcoam method (Passive Calibration of Active Measurement) and our method could reflect the actual network status more exactly in the case of network congestion and packet loss.

The paper proceeds as follows. Section 2 describes the active measuring method and the CoMPACT monitor. We analyze the adjacent packet latency alteration and propose the Pcoam method (Passive Calibration of Active Measurement) in section 3. In section 4, we show the validity of the Pcoam method by simulation. Finally, conclusions deriving from the effort are presented and future work is discussed in Section 5.

2 Active Measurement and CoMPACT Monitor

The network performance obtained from the active probe packets is not equal to the performance experienced by users. The simple active measurement cannot estimate the delay experienced by users. Fortunately, we could obtain more exact evaluation by using the characteristics of packets gained by passive measurement to calibrate the result of active measuring.

2.1 Active Measurement

Let X be the measurement objective, measurement period is $[0, T]$. Let $V(t)$ be the delay, the indicator function is defined: $\phi(t,a) = \begin{cases} 1, V(t) > a \\ 0, V(t) \le a \end{cases}$. So the distribution function of delay is as follows:

$$\Pr(X > a) = \frac{\int_0^T \phi(t,a)dt}{T}. \tag{1}$$

Suppose there are n user data packets and $m$probe packets arriving in the measuring period. Let A_i be the delay of packet i, the indicator function ϕ is as: $\phi(i,a) = \begin{cases} 1, A_i < a \\ 0, A_i \leq a \end{cases}$. Then the delay distribution function obtained by user packets as follows:

$$\Pr(X > a) = \frac{1}{n}\sum_{t=1}^{n}\phi(i,a) \tag{2}$$

The mean delay obtained by user packets is as:

$$M_u(X) = \frac{1}{n}\sum_{i=1}^{n}A_i \tag{3}$$

The delay distribution function obtained by active probe packets is as:

$$\Pr(X > a) = \frac{1}{m}\sum_{i=1}^{m}\phi(i,a) \tag{4}$$

The mean delay obtained by probe packets is as:

$$M_{pr}(X) = \frac{1}{m}\sum_{i=1}^{m}A_i \tag{5}$$

2.2 CoMPACT Monitor

It is difficult to measure user packets delay directly in that not only should the time clocks of the monitoring devices be synchronized, but also the identification process is hard as the packet volume is huge in a large-scale network.

Although we could not measure user packets delay directly, we can assume the change of delay is little at any measuring time t if the interval of sending probe packets Δt is enough short. Then

$$\forall s, s' \in [t, t+\Delta t) \Longrightarrow V(s) \cong V(s') \tag{6}$$

We can obtain the number of user packets between the neighborhood probe packets through the simplified passive monitoring device. The simplified passive monitoring device only monitors the arrival of the probe packets and counts the number of the user data packets.

Supposing the number of user data packets is ρ_i between active probe packet i and active probe packet $i-1$. As the CoMPACT monitor[4,5,6] , which is proposed by Masaki Aida *et al.*, the delay distribution function is:

$$\Pr(X > a) = \sum_{i=1}^{m}\phi(i,a)\frac{\rho_i}{n} \tag{7}$$

The mean delay is:

$$M_c(X) = \frac{1}{n}\sum_{i=1}^{m} A_i \rho_i \tag{8}$$

The CoMPACT monitor can use the number of user packets monitored during the short neighbourhood around the arrival of active probe packets to replace the number of user packets monitored between active probe packets [4]. This method does not need care about the interval of active probe packets, but it makes the passive monitoring device more complex.

3 Pcoam Method

3.1 Pcoam Method

Even though the interval of sending probe packets could be very short, it could not be ensured that the interval of receiving or monitoring those probe packets is short enough. The error of estimator would increase as the interval of the probe packets arriving increases, especially when the network is in congestion or the loss ratio is high.

When the network is busy or in congestion, we can assume that the change of link delay is continuous in the period of $[t, t+\Delta t)$, where the interval Δt is short enough compared to the time variance of $V(t)$.

Furthermore, we can assume the delay of packets in the period $[t, t+\Delta t)$ is a linear relationship with the time. Then

$$\forall s \in [t, t+\Delta t) \Longrightarrow \frac{V(s)-V(t)}{s-t} \cong \frac{V(t+\Delta t)-V(t)}{\Delta t} \tag{9}$$

Then the weight of active probe packets delay would not merely be the number of user data packets between the former probe packet and this packet, for the difference between the delays of the adjacent probe packets should be taken into account necessarily. We propose the passive calibration of active measurement–Pcoam method. It could help to do more exact evaluation of network performance even with slightly complex computations comparing to the CoMPACT monitor[5]. The indicator function of the Pcoam method is as follows:

$$\phi'(i,a) = \begin{cases} 1 & : \; A_i > a, A_{i-1} > a \\ \frac{A_i - a}{A_i - A_{i-1}} & : \; A_i > a, A_{i-1} \le a \\ \frac{A_{i-1} - a}{A_{i-1} - A_i} & : \; A_i \le a, A_{i-1} > a \\ 0 & : \; A_i \le a, A_{i-1} \le a \end{cases} \tag{10}$$

Then, the delay distribution is as follows:

$$\Pr(X > a) = \sum_{i=1}^{m} \phi'(i,a)\frac{\rho_i}{n} \tag{11}$$

The mean delay is: $M_{pc}(X) = \frac{1}{n}\sum_{i=1}^{m} \frac{(A_i + A_{i-1})}{2}\rho_i$, where $A_0 = 0$.

3.2 Implementation

Because the supposal (9) should not be applied to the network being idle, we can set up a delay threshold based on experience to decide whether the network is busy or not. As the delay of both adjacent probe packets is beyond the threshold, we can suppose that each router is busy when this probe packets and data packets traverse each hop. In this case we can presume the network is busy. Otherwise we presume the network is not busy.

We could get the better measurement result to combine two methods to deal with different cases. With establishing the threshold value D, when the adjacent packets delay is higher than D, we should adopt Pcoam method, whereas the CoMPACT method should be adopted. Therefore the indicator function is as follows:

$$\phi''(i,a) = \begin{cases} \phi'(i,a) & : \; A_i > D, A_{i-1} > D \\ \phi(i,a) & : \; otherwise \end{cases} \tag{12}$$

The delay distribution function is:

$$\Pr(X > a) = \sum_{t=1}^{m} \phi''(i,a) \frac{\rho_i}{n} \tag{13}$$

To combine the CoMPACT method with the Pcoam method, it is important to determine an appropriate threshold. The threshold depends on network topology, network congestion condition and the interval of sending active probe packets.

4 Simulation

We use the *ns2* [7] network simulator to demonstrate our schemes. We measure the queueing delay at the bottleneck router, which does not include the service time for the packets themselves. We use a few of ON-OFF sources to generate and simulate the network traffic.

Let us consider a network model with multiple bottlenecks as shown in Fig. 4 The model has 20 pairs of source/destination hosts. Twenty sources are categorized four different types described in Table 1. Each application traffic type is assigned to five hosts. As the transport protocol for the user packets, both TCP and UDP are evaluated.

Link capacity between hosts and edge routers is 1.5 Mbps, that between the edge routers and core routers is 8 Mbps, and that between the core routers is 10 Mbps. The queue discipline of this links is FCFS. Each host on the left is a source and transfers data to the corresponding host on the right.

We conducted a 2400-s simulation using ns2 and measured the distributions of delay between ingress and egress routers for both active probe packets and user packets. Simultaneously, we calculated the delay distribution by CoMPACT Monitor and Pcoam Method.

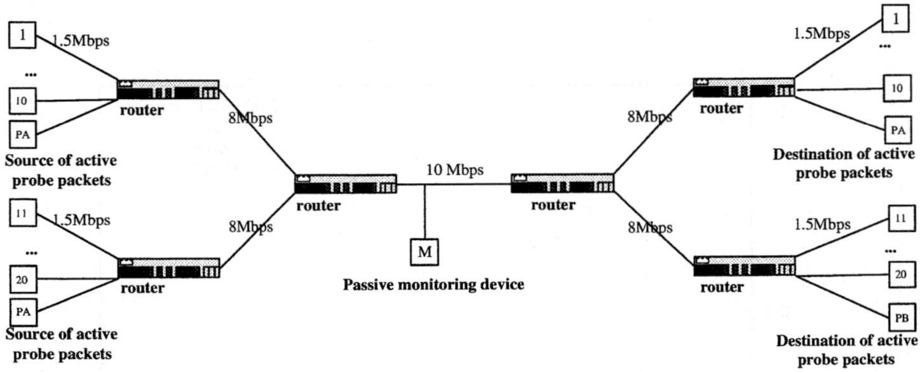

Fig. 1. Simulative multiple hop network model

The active probe packets are fixed at 64 bytes long. To evaluate the active measuring, there are two pairs of hosts for sending and receiving the active probe packets, *PA*'s and *PB*'s in Fig. 4. They are connected in the same manner as the user hosts. Active probe packets sending by host *PA* are generated every 2 second. Active probe packets sending by host *PB* are inserted into the network according to a Poisson process and the mean interval of active probe packets inserted into the network is 2s. The extra traffic caused by the active probe packets is only about 0.0032% of the link capacity of 8 Mbps and 0.0052% of the link capacity of 10 Mbps, so the influence on user traffic is negligible.

Table 1. Simulative multiple hop network node traffic configure

Node	Protocol	Packet length	ON period	OFF period	ON/OFF distribution	Shape	Rate
#1–#5	TCP	1.5KB	10s	5s	Exponential	–	1Mbps
#6–#10	UDP	1.5KB	5s	5s	Pareto	1.5	1.5Mbps
#11–#15	TCP	1.5KB	5s	10s	Exponential	–	1Mbps
#16–#20	UDP	1.5KB	2s	8s	Pareto	1.5	1.5Mbps

We chose the connection #6 and connection #11 to do the analysis. Fig. 2 and Fig. 3 show the queueing delay distribution of connection #6 and connection #11 of the active probe packets, data packets, and the delay distribution obtained by CoMPACT Monitoring and Pcoam method. We found that Pcoam method could acquire better evaluation than the CoMPACT method in the case of network congestion.

Table 2 lists the mean delay of the four types connection obtained by four methods. We found the implementation of CoMPACT Monitor and Pcoam Method could estimate the queueing delay according to user traffic characteris-

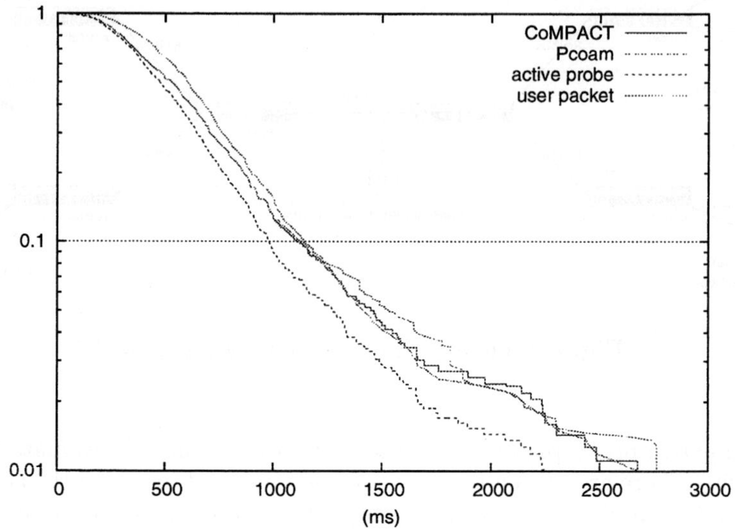

Fig. 2. Delay distribution for connection #6

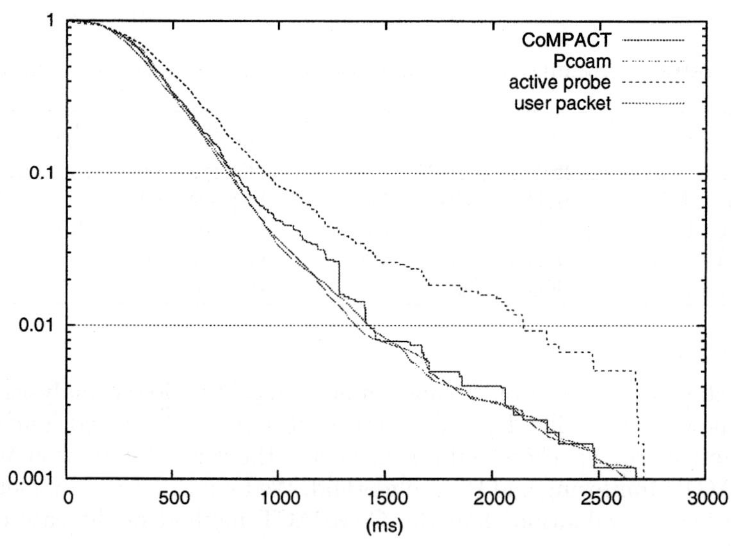

Fig. 3. Delay distribution for connection #11

tics. The Pcoam method is more exact than CoMPACT method, especially in the case of network congestion.

Table 2. Mean Delay

Node	Mean delay (ms)				Number of user data packets
	user packets	active probe	CoMPACT Monitor	Pcoam Monitor	
#1	412.621	563.464	425.081	421.817	100741
#6	697.176	563.464	622.457	626.341	156301
#11	445.468	544.013	466.591	461.027	80761
#16	705.745	544.013	671.574	679.249	58499

5 Conclusions

The passive calibration of active measuring latency method improves the CoMPACT method with considering the number of user data packets arriving between probe packets and the latency alteration of neighborhood probe packets. This effective method is able to overcome the difficulties both in active and passive schemes as CoMPACT method could. It has some advantages such as protocol-independent, negligible extra traffic, convenience and being able to estimate individual user performance. This method could reflect the actual network status more exactly, especially in the case of network congestion and packet loss. This method is useful for IP network and Virtual private network.

We used simulation to validate the proposed method. As a result, our scheme and the CoMPACT method have been shown could give a good estimation of the performance experienced by the user. Our scheme could get better result especially when the network is busy or in congestion. So it would be better to combine two methods to get more accurate measurement results. We would like to improve our method to deduce more exact performance by using less data in the future.

Acknowledgements

The research reported here has been supported by the National Natural Science Foundation of China(No.60373023). And the authors would like to thank Dr. Masaki Aida for his help and encouragement.

References

1. Almes, G., Kalidindi, S., Zekauskas, M.: A one-way delay metric for ippm. Technical Report RFC2679, IETF (1999)

2. Paxon, V., Almes, G., Mahdavi, J., Mathis, M.: Framework for ip performance metric. Technical Report RFC 2330, IETF (1998)
3. Breibart, Y., Chan, C.Y., Carofalakis, M., Rastogi, R., Silberschatz, A.: Efficiently monitoring bandwidth and latency in ip networks. In: IEEE INFOCOM 2000. (2000)
4. Aida, M., Miyoshi, N., Ishibashi, K.: A scalable and lightweight qos monitoring technique combining passive and active approaches. In: IEEE INFOCOM 2003. (2003)
5. Aida, M., Ishibashi, K., Kanazawa, T.: Compact-monitor: Change-of-measure based passive/active monitoring — weighted active sampling scheme to infer qos—. In: IEEE SAINT 2002 Workshop. (2002)
6. Ishibashi, K., Aida, M., Kuribayashi, S.: Estimating packet loss-rate by using delay information and combined with change-of-measure framework. In: IEEE GLOBECOM 2003. (2003)
7. ns 2: (The network simulator - ns-2) http://www.isi.edu/nsnam/ns.
8. Ishibashi, K., Kanazawa, T., Aida, M.: Active/passive combination type performance measurement method using change-of-measure framework. In: IEEE GLOBECOM 2002. (2002)
9. Lindh, T.: A new approach to performance monitoring in ip networks —— combining active and passive methods. In: Passive and Active Measurements 2002. (2002)
10. Duffield, N.G., Lund, C., Thorup, M.: Properties and prediction of flow statistics from sampled packet streams. In: ACM SIGCOMM Internet Measurement Workshop 2002. (2002)
11. Liu, X., Yin, J., Tang, L.: Analysis of efficient monitoring method for the network flow. Journal of Software (2003) 300–304
12. Willinger, W., Paxson, V., Taqqu, M.S.: Self-Similarity and Heavy Tails: Structural Modeling of Network Traffic. Birkhauser Verlag (1998)
13. Pasztor, A., Veitch, D.: On the scope of end-to-end probing methods. IEEE Communications Letters (2002)
14. H.Patel, S.: Performance inference engine (pie) —— deducing more performance using less data. In: Passive and Active Measurements 2000. (2000)
15. Paxon, V.: End-to-end internet packet dynamics. IEEE/ACM Trans. Networking (1999) 277–292
16. Liu, X., Yin, J., Cai, Z.: The analysis of algorithm for efficient network flow monitoring. In: 2004 IEEE International Workshop on IPOM. (2004)

Topological Discrepancies Among Internet Measurements Using Different Sampling Methodologies

Shi Zhou[1] and Raúl J. Mondragón[2]

[1] University College London, Adastral Park Campus,
Ross Building, Ipswich, IP5 3RE, United Kingdom
s.zhou@adastral.ucl.ac.uk

[2] Queen Mary, University of London,
Mile End Road, London, E1 4NS, United Kingdom
r.j.mondragon@elec.qmul.ac.uk

Abstract. Studies on Internet topology are based on measurement data. There are three types of Internet topology measurements using different sampling methodologies. This paper compares all of the three measurements together by examining their topological properties, in particular the recently introduced structural metric of rich-club connectivity. Numerical results show that, although having similar degree distribution, the topological discrepancies among the three Internet measurements are significant. This work provides a "graph-centred" analysis on the limitations of each sampling methodologies that are responsible for the measurement deficiencies.

1 Introduction

Topology is the connectivity graph of a network, upon which the network's physical and engineering properties are based. Effective engineering of the Internet is predicated on a detailed understanding of issues such as the large-scale structure of its underlying physical topology, the manner in which it evolves over time, and the way in which its constituent components contribute to its overall function [1]. Unfortunately, developing a deep understanding of these issues has proven to be a challenging task, since it in turn involves solving difficult problems such as mapping the actual topology, characterizing it, and developing models that capture its emergent behavior. First of all, a complete and accurate measurement is vital because studies on the Internet topology are based on measurement data [2], [3], [4], [5], [6].

Practical measurements on the Internet topology became available only recently. There are three types of data sources of Internet topology at the Autonomous Systems (AS) level, namely the BGP AS graph [7], [8], [9], the Extended BGP AS graph [10], [11], [12] and the Traceroute AS graph [13], [14], [15], [16], which are collected using different methodologies of inferring AS connectivity information. There have been studies comparing between the BGP

C. Kim (Ed.): ICOIN 2005, LNCS 3391, pp. 207–214, 2005.

AS graph and the Extended BGP AS graph [11], [12], and between the BGP AS graph and the Traceroute AS graph [16]. This paper compares all of the three measurements together by examining topological properties, including degree distribution, shortest path length, triangle coefficient and in particular, the recently reported rich-club connectivity [17], [18], [19]. Although the three measurements have similar degree distribution, numerical results show that they exhibit significant topological discrepancies. This work provides a "graph-centred" analysis on limitations of each sampling methodologies that are responsible for the measurement deficiencies.

2 Internet Topology Measurements

The vertices (nodes) of the Internet connectivity topology are:

- Hosts that are the computers of users;
- Servers that are computers or programs providing a network service, which also can be hosts;
- Routers that arrange traffic across the Internet;
- Domains or Autonomous Systems (AS) that are subnetworks in the Internet.

The global structure of the Internet is not determined by the hosts, but by the routers (the router-level) and by domains (the AS-level). This paper is focused on the topology of AS-level Internet (AS graph). After all, the Internet traffic is routed using Border Gateway Protocol (BGP) among ASes. The followings are the three types of measurements of AS-level Internet topology collected using different sampling methodologies.

2.1 BGP AS Graph

The Internet passive measurement [7], [8], [9] produces BGP AS graphs, which are constructed from Internet inter-domain BGP routing tables, which contain the information of links from an AS to its immediate neighbors. The widely used BGP data are available from the Passive Measurement Project at National Laboratory for Applied Network Research [7] and the Route Views Project at University of Oregon [8]. Both projects connect to a number of operational routers within the Internet for the purpose of collecting BGP routing tables. Though the BGP AS path does not reflect how traffic actually travels in network on the IP-level, it is due to show the forward AS-level path followed by the IP packets. BGP data have widespread public availability and BGP tables have the advantage that they are relatively easy to parse, process and comprehend. However due to the way BGP advertise its paths, the BGP data in fact "hide" some peerings.

2.2 Extended BGP AS Graph

The Topology Project at University of Michigan [10] provided the extended version [11], [12] of BGP AS graph by using additional data sources, such as the

Internet Routing Registry (IRR) data and the Looking Glass (LG) data. The IRR maintains individual ISP's (Internet Service Provider) routing information in several public repositories to coordinate global routing policy. The LG sites are maintained by individual ISPs to help troubleshoot Internet-wide routing problems. Extended BGP AS graphs typically have 40% more links (and about 2% more nodes) than the original BGP AS graphs.

Most studies on the AS-level Internet topology were based on the above two BGP-derived graphs, such as power-law degree distribution [20], hierarchical structure [21], error and attack tolerance [22], [23] and degree-degree correlations [24], [25].

2.3 Traceroute AS Graph

Another source of Internet connectivity data are the so-called Traceroute AS graphs, which are produced by the active measurement methodology using traceroute probing data. From 1998, the Cooperative Association for Internet Data Analysis (CAIDA) [13] began its Macroscopic Topology Project to collect and analyze Internet-wide topology at a representatively large scale. In the course of this project CAIDA has created several innovative measurement, analysis and visualization tools. The primary topology measurement tool is skitter [14], [15], [16], which implements the Internet Control Message Protocol (ICMP) collect the forward path from the monitor to a given destination and capture the addresses of intermediate routers in the path. The skitter runs on more than 20 monitors around the globe and actively collects forward IP path to over half a million destinations. Traceroute AS graph extracts interconnect information of ASes from the massive traceroute data (on the router level) collected by Skitter. Traceroute AS graphs capture about 30% more links than BGP AS graphs due to the visibility of peering at exchange points in traceroute paths, which rarely appear in BGP tables.

Table 1. Network Properties of the three AS graphs

AS graphs	Traceroute	Extended BGP	BGP
Number of nodes, N	11122	11461	11174
Number of links, L	30054	32730	23409
Characteristic path length, l^*	3.13	3.56	3.62
Average triangle coefficient, $\langle k_t \rangle$	12.7	23.4	5.3

3 Numerical Comparison and Graph-Centred Analysis

In this paper we study a BGP AS graph and an Extended BGP AS graph measured in May 2001 [10], and a Traceroute AS graph collected in April 2002 [26]. The three AS graphs have similar numbers of nodes whereas the Extended BGP

AS graph and the Traceroute AS graph have significantly more links than the BGP AS graph (see Table 1).

3.1 Degree Distribution

Degree k of a node is the number of links (or immediate neighbors) that the node has. While degree is a local property, the probability distribution of the degree gives important information of the global properties of a network. Users of Internet measurements often did not pay enough attention on topological discrepancies between different Internet measurements due to the fact that all three AS graphs exhibit similar power-law degree distributions, $P(k) \propto k^{-2.22}$ [20]. However degree distribution characterize only one topological aspect of the network's extremely complex structure.

3.2 Rich-Club Connectivity

The Internet has a power-law degree distribution, which means the majority of nodes have only a few links (low-degree nodes), whereas a small number of nodes have large numbers of links (high-degree nodes). The Internet also exhibits a so-called disassortative mixing behavior [27], where high-degree nodes tend to connect with low-degree ones. However this property does not imply how high-degree nodes are interconnected to each other (see Fig. 1).

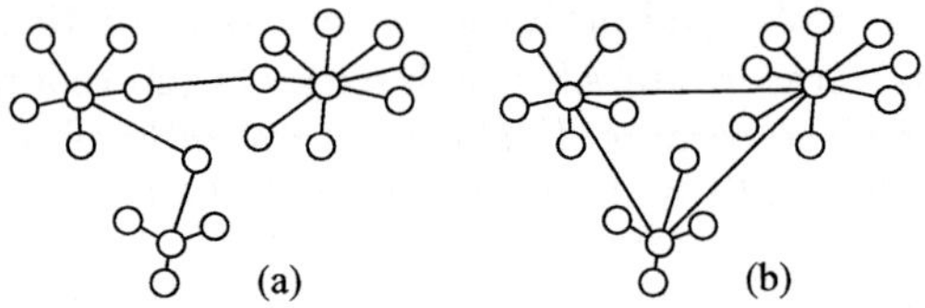

Fig. 1. Two disassortative power-law networks. (a) High-degree nodes are loosely interconnected. (b) High-degree nodes are tightly interconnected.

Recently Zhou and Mondragon [17], [18], [19] introduced the concept of *rich-club phenomenon* to characterize the Internet hierarchical structure, where high-degree nodes (rich nodes) are tightly interconnected with each other and form a rich-club. The average hop-distance between club members is very small (1 to 2 hops). The club membership is defined as "the r richest guys", where r is node rank sorted by decreasing order of node degree. A quantitative assessment of rich-club property is obtained by measuring the *rich-club connectivity*, ϕ, defined as the fraction of allowable links[3] that actually exist among the club members. The rich-club connectivity indicates how well club members "know" each other. A rich-club connectivity of 1 means that all the members have a direct link to any other member, i.e. they form a fully connected subgraph.

[3] The number of allowable links in a r-node subgraph is $r(r-1)/2$.

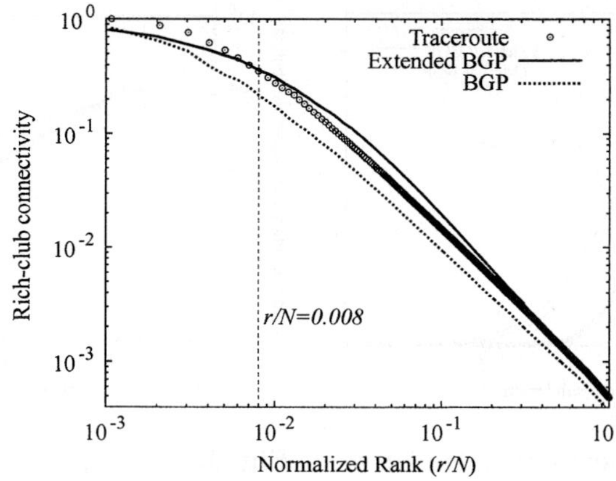

Fig. 2. Rich-club connectivity ϕ as a function of normalized rank r/N. The top 0.8% best-connected nodes are marked with the vertical hash line.

Fig. 2 shows rich-club connectivity as a function of node rank r normalized by the total number of nodes N. Fig. 2 shows that rich-club connectivity of the BGP AS graph is significantly lower than the other two measurements. It is expected because the BGP AS graph contains much less links. What is interesting is that the top 0.8% richest nodes of the Traceroute AS graph are more tightly interconnected than those in the Extended BGP AS graph although the Traceroute AS graph contains less links than the Extended BGP AS graph.

The reason that many of the interconnections among rich nodes only appear in the Traceroute AS graph is because they are actually "peer-peer" peerings among large ISPs (Tier-1), who are not willing to provide the information of their peerings since this information is economically critical. Therefore these peerings are not transitive, i.e. not advertised to BGP peers except customers at best, hence will not be seen in the BGP tables. Whereas traceroutes rely on much more source-destination pairs than BGP data, so they sample the core of the Internet better.

Fig. 2 shows that the Extended BGP AS graph contains more links connecting among less connected nodes (see $10^{-}2 < r/N < 10^{-}1$) than the other two measurements. This is because the peerings added by IRR or LG data in the Extended BGP AS graph are mainly peerings among small ISPs that are not in the core of the network. These peerings are not present in the BGP AS graph because small ISPs peer among each other so that except by getting the BGP table from these ASes themselves, these "peer-peer" peerings can not be inferred. Whereas the BGP tables collected by the passive measurements are mainly from large ISPs in the core of the Internet. The reason that many links among less connected nodes are not exhibited in the Traceroute AS graph is because traceroutes sample "best routes" actually used by the traffic, so that some peerings in the Extended BGP AS graph simply cannot be seen by active measurements since they'll only be used when other links fail.

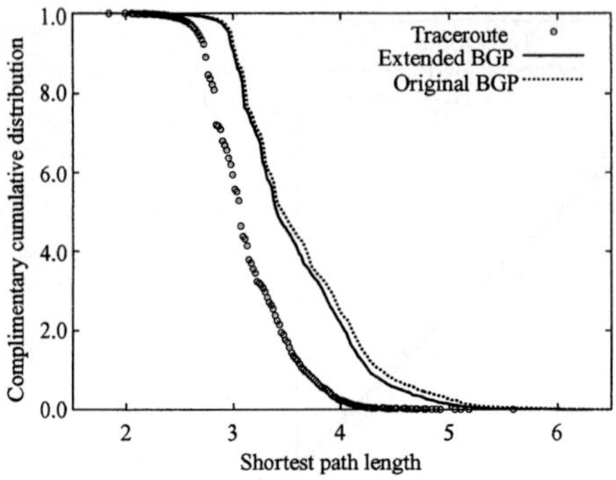

Fig. 3. Complimentary cumulative distribution of shortest-path length.

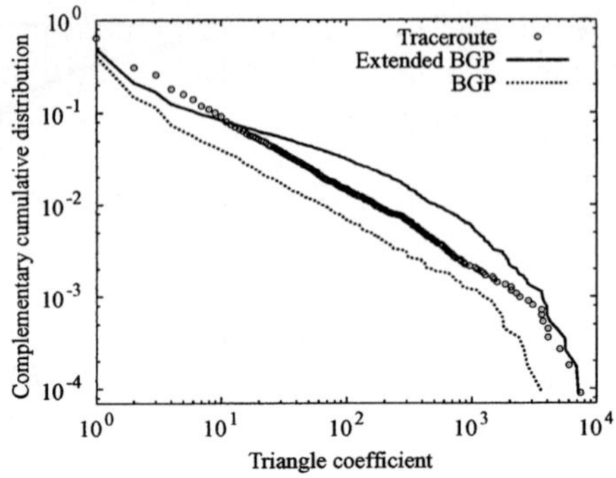

Fig. 4. Complimentary cumulative distribution of triangle coefficient.

3.3 Shortest-Path Length (Routing Efficiency)

The shortest-path length, l, of a node is defined as the average of shortest hop-length between the node and all other nodes. Fig. 3 shows that the BGP AS graph and the Extended BGP AS graph (with 40% more links) have nearly the same complimentary cumulative distribution shortest-path length, which is notably displaced to the right of the Traceroute AS graph. The characteristic path length, l^*, of a network is the average of shortest hop-length between all pairs of nodes [28]. As shown in Table 1, the characteristic path length of the two BGP-derived graphs are 0.5 hop longer than that of the Traceroute AS graph.

This difference can be explained by the above analysis on rich-club connectivity. While the rich-club is a "super" traffic hub of the network, the disassortative mixing ensures that peripheral nodes are always near the hub. These two structural properties together contribute to the network routing efficiency. The Traceroute AS graph measures more "peer-peer" peerings among large ASes and contains a more tightly interconnected rich-club than the two BGP-derived graphs. Therefore the Traceroute AS graph exhibits a higher degree of network routing efficiency. However the extra links captured by the Extended BGP AS graph are connections among small ISPs, which provide few shortcuts for routing and therefore have very limited contribute to the network routing efficiency. In fact the distribution of shortest-path length and the characteristic path length of the Extended BGP AS graph are fairly close to those of the BGP AS graph, which does not use the IRR or LG data.

3.4 Triangles Coefficient (Network Redundancy)

The quantity of short cycles [29] (e.g. triangle and quadrangle) is relevant because the amount of alternative reachable routes in a network increases with the density of short cycles. The triangle coefficient, k_t, of a node is defined as the number of triangles (or the number of inter-neighbor links) that the node has [18]. Fig. 4

shows the complimentary cumulative distribution of triangle coefficient of the three AS graphs. Fig. 4 and Table 1 show that the average triangle coefficient of the Extended BGP AS graph is larger than that of the other two graphs, because the IRR data and the LG data used by the Extended BGP AS graph contains a large number of "peer-peer" peerings among small ISPs, which construct large amount of triangles in the network.

4 Discussion and Conclusion

This paper compares all of the three measurements of AS-level Internet topology. Each of them was collected by using different connectivity-inferring methodologies. Although the three AS graphs exhibit similar degree distributions, they contain significantly different topological properties. The principal disparities are characterized by the recently reported structural metric of rich-club connectivity. Comparing with the Traceroute AS graph, the two BGP-derived graphs lack of links among highly connected nodes, whereas the Extended BGP AS graph contains more links among less connected nodes than the other two graphs. This paper provides a "graph-centred" analysis on the limitations of the very way each type of measurement samples the AS-level Internet.

Numerical results presented in this paper suggest that none of the three available AS graph measurements is complete. Since they do not overlap with each other, it is not sensible to judge which measurement is more complete or accurate. But it is clear that their structural differences are non-trivial because they are relevant to network behaviors, such as routing efficiency (shortest-path length) and routing flexibility (density of short cycles). Studies based on these measurements should investigate the sensitivity of the findings to the deficiencies and inaccuracies of the measurement data. Also there is a need of improving the techniques of measuring the Internet.

Acknowledgments

This work is funded by the U.K. Engineering and Physical Sciences Research Council (EPSRC) under grant no. GR-R30136-01.

References

1. Floyd, S., Kohler, E.: Internet research needs better models. ACM SIGCOMM Computer Communications Reviews **33** (2003) 29–34
2. Albert, R., Barabási, A.L.: Statistical mechanics of complex networks. Rev. Mod. Phys. **74** (2002) 47–97
3. Barabási, A.L.: Linked: The New Science of Networks. Perseus Publishing (2002)
4. Bornholdt, S., Schuster, H.G.: Handbook of Graphs and Networks - From the Genome to the Internet. Wiley-VCH, Weinheim Germany (2002)
5. Dorogovtsev, S.N., Mendes, J.F.F.: Evolution of Networks - From Biological Nets to the Internet and WWW. Oxford University Press (2003)

6. Pastor-Satorras, R., Vespignani, A.: Evolution and Structure of the Internet - A Statistical Physics Approach. Cambridge University Press (2004)
7. National Laboratory for Applied Network Research, (http://moat.nlanr.net/)
8. Route Views Project, University of Oregon, Eugene. (http://www.routeviews.org/)
9. Routing Information Service, RIPE (http://www.ripe.net/), Network Coordination Center.
10. Topology Project (http://topology.eecs.umich.edu/), University of Michigan, Ann Arbor.
11. Chen, Q., Chang, H., Govindan, R., Jamin, S., Shenker, S.J., Willinger, W.: The origin of power laws in Internet topologies (revisited). In: Proc. of IEEE INFOCOM 2002. (2002) 608–617
12. Chang, H., Govindan, R., Jamin, S., Shenker, S., Willinger, W.: Towards capturing representative AS-level Internet topologies. Computer Networks Journal **44** (2004) 737–755 Elsevier Publisher.
13. Cooperative Association For Internet Data Analysis, (http://www.caida.org/)
14. Broido, A., kc Claffy: Internet topology: connectivity of IP graphs. In: SPIE International symposium on Convergence of IT and Communication 2001. (2001)
15. Huffaker, B., Plummer, D., Moore, D., kc Claffy: Topology discovery by active probing. In: Proc. of the 2002 Symposium on Applications and the Internet. (2002)
16. Hyun, Y., Broido, A., claffy, k.: Traceroute and BGP AS path incongruities. (http://www.caida.org/outreach/papers/2003/ASP/)
17. Zhou, S., Mondragón, R.J.: The rich-club phenomenon in the Internet topology. IEEE Comm. Lett. **8** (2004) 180–182
18. Zhou, S., Mondragón, R.J.: Redundancy and robustness of the AS-level Internet topology and its models. IEE Elec. Lett. **40** (2004) 151–152
19. Zhou, S., Mondragón, R.J.: Accurately modelling the Internet topology. to appear in Physical Review E (2004) (preprint: arXiv.cs.NI/0402011)
20. Faloutsos, M., Faloutsos, P., Faloutsos, C.: On power-law relationships of the Internet topology. Comput. Commun. Rev. **29** (1999) 251–262
21. Subramanian, L., Agarwal, S., Rexford, J., Katz, R.H.: Characterizing the Internet hierarchy from multiple vantage points. In: Proc. of IEEE INFOCOM 2002. (2002) 618–627
22. Albert, R., Jeong, H., Barabási, A.L.: Error and attack tolerance of complex networks. Nature **406** (2000) 378–381
23. Park, S.T., Khrabrov, A., Pennock, D.M., Lawrence, S., Giles, C.L., Ungar, L.H.: Static and dynamic analysis of the Internet's susceptibility to faults and attacks. In: Proc. of IEEE INFOCOM 2003. Volume 3. (2003) 2144–2154
24. Pastor-Satorras, R., Vázquez, A., Vespignani, A.: Dynamical and correlation properties of the Internet. Phys. Rev. Lett. **87** (2001)
25. Vázquez, A., Pastor-Satorras, R., Vespignani, A.: Large-scale topological and dynamical properties of Internet. Phys. Rev. E **65** (2002)
26. The Data Kit #0204 was collected as part of CAIDA's Skitter initiative. Support for Skitter is provided by DARPA, NSF, and CAIDA membership.
27. Newman, M.E.J.: Mixing patterns in networks. Phys. Rev. E **67** (2003)
28. Watts, J.: Small Worlds: The Dynamics of Networks between Order and Randomness. Princeton Univeristy Press, New Jersey, USA (1999)
29. Bianconi, G., Capocci, A.: Number of loops of size h in growing scale-free networks. Phys. Rev. Lett. **90** (2003)

Time and Space Correlation in BGP Messages

Kensuke Fukuda[1], Toshio Hirotsu[2], Osamu Akashi[1], and Toshiharu Sugawara[1]

[1] Nippon Telegraph and Telephone Corp., Tokyo 180-8585, Japan
[2] Toyohashi University of Technology, Toyohashi 441-8580, Japan
{fukuda, hirotsu, akashi, sugawara}@t.ecl.net

Abstract. To quantify the statistical dynamics of the BGP, we analyze the temporal and spatial correlation of macroscopic BGP message flows obtained by passive measurement. We show that the time series for the number of announcement and withdrawal messages has little correlation in time, unlike the statistical behavior of traffic volumes. This indicates that there is little possibility of a cascading failure, in which a failure causes following failures, and that the occurrence of burst of BGP messages has a Poisson nature. We also point out that there is space correlation with the delay between the flows for the different measurement points. Namely, even from macroscopic and passive measurement, we show that the propagation delay of routing information from one measurement point to another point can be statistically estimated.

1 Introduction

The Border Gateway Protocol (BGP) 4 [1] is the de facto inter-domain routing protocol in the current Internet. It is responsible for routing between organizations called autonomous systems (ASes). Because of the complexity of the structure of the AS network, and also the financial/contractual agreements among ASes, inter-domain routing has difficulties in managing to provide users with efficient and correct routing information. In BGP, each AS only exchanges routing information with neighboring (or peer) ASes, when the network topology or routing policy changes. Each BGP router maintains routing entries in its routing table, according to BGP update messages, consisting of announcements and withdrawals. An announcement message is generated when a BGP router in an AS discovers a new path for an existing route or for a previously unavailable destination. Each announcement message consists of the prefix of the destination network and the list of AS numbers along the router to the destination. A withdrawal message is an advertisement that a certain destination network is no longer reachable.

Quantifying the dynamics—especially the stability—of BGP behavior in the real Internet is an important research issue, because BGP dominates the stability and performance of the current Internet. Historically, the Internet backbone has been widely believed to be robust against individual failures in links or routers. However, recent analyses of BGP have shown that it is not so stable because of algorithmic, implementation, or operational problems. Ref. [2] reported convergence delay on the order of minutes in calculating of routing from fault-injection

C. Kim (Ed.): ICOIN 2005, LNCS 3391, pp. 215–222, 2005.

and active measurement. Moreover, recently, to quantify the BGP dynamics, an architecture for broadcasting BGP messages as beacon has been proposed [3] to characterize the microscopic behavior of BGP.

In this paper, we focus on the macroscopic behavior of BGP rather than its microscopic event-driven behavior from passively measured BGP traces, to quantify the effects of failures in live data. From the viewpoint of temporal (i.e., possibility of a cascading failure) and spatial (i.e., propagation delay for failure information among measurement points) correlation of BGP messages, we analyze the time series for the number of BGP announcement / withdrawal messages obtained from four different measurement points in the U.S., U.K., and Japan.

2 Data Trace

We collected BGP messages with time stamps at four monitoring points peering with eBGP routers at ISPs for 3 days in June 2003: two Tier-1 ISPs in the U.S. (US1 and US2), an AS which multihomed to two Tier-1 ISPs in Japan (JP), and a Tier-1 ISP in the U.K. (UK). A BGP update message contains multiple announcement / withdrawal messages, so for every measurement point, we constructed a time series for the number of prefix-based announcement messages and that for the number of withdrawal messages in one-minute bins. We omitted the data corresponding to the first 20 min., because the first part of the data included a large amount of routing information exchanged to initialize the router at the measurement point. Also note that all the clocks at the measurement points were synchronized by NTP.

Figure 1 is an example of a time series for the number of announcement / withdrawal messages at four measurement points. The figure suggests that the fluctuation in the number of messages is bursty; most fluctuations were small, however, there was a few large spikes over 2000 messages in one min. In addition, it is obvious that the times when bursts occurred in the four sub figures were not always synchronized. For example, for $t \approx 1500$, there are bursts of announcement messages indicating huge changes in route in (b). However, we cannot find the same kind of behavior in (a) and (d), indicating that the announcement messages were not generated by ASes near US1 and UK. This is likely due to the effect of multihomed ASes, having multiple links to other ASes. In this situation, another path is primarily selected as the best path, so the primary route does not change even for a link failure along the alternative path. Update messages are absorbed at such ASes.

3 Results

3.1 Time Correlation

We analyzed the time correlation, or long-range dependence (LRD), for BGP message flows. In other words, we investigated how many BGP messages we

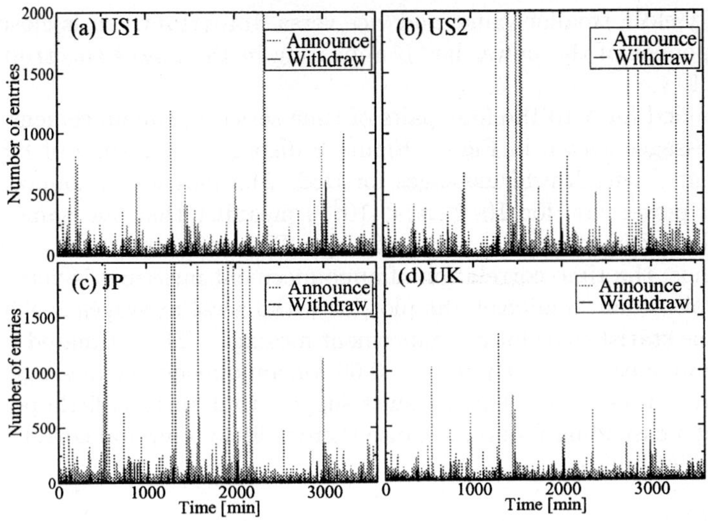

Fig. 1. Number of BGP messages in a one-min. bin: (a) US1, (b) US2, (c) JP, and (d) UK.

could statistically observe in the following time steps, if we observed a burst of BGP messages at time t. We expect to observe a stronger time correlation when a failure causes following failures, though the correlation is weaker for a single (and isolated) failure.

In quantifying time correlation, we used Detrended Fluctuation Analysis (DFA) [4]. DFA, which is based on the root-mean-square (rms) method of random walk, is a well-known method for analyzing complex time series in physiology, the stockmarket, and the Internet. An advantage of using DFA rather than traditional methods like the power spectrum analysis or the R/S analysis is that the DFA can prevent pseudo time correlations from being detected.

A detailed description of DFA is given in Ref [4]. Here, we will only briefly explain the method. We first integrate the time series, then divide this into "boxes" of length n. In each box, we calculate the least-squares polynomial fit of order p to the integrated signal (we used $p = 1$ in this study). Finally, in each box, we calculate the rms deviations of the integrated signal from the polynomial fit. We repeat the above procedure for different size boxes. For a given time series, we find the power law relationship $F(n) \sim n^{\alpha}$ between the average magnitude of rms deviations $F(n)$ and box size n. The value of exponent α is the parameter that quantifies the statistical properties of the time series: $\alpha = 0.5$ corresponds to white noise, meaning that bursts in the time series are not correlated with each other. For $0.5 < \alpha \leq 1.0$, a burst in the time series is positively correlated in time, i.e., if one observes a burst, there is a high probability of observing a similar size of burst in the following time steps. On the other hand, $0 < \alpha < 0.5$ means that the time series has negative correlation, indicating that a larger value will yield a smaller value, and vice versa. Importantly, α is closely related to the exponent of the power law β appearing in the power spectrum analysis: $\beta = 2\alpha - 1$.

We applied DFA to the four pairs of time series of announcement and withdrawal messages shown in Fig. 1. Figure 2 displays the results of DFA for announcement / withdrawal messages for US2. The plot for announcement messages is a power law for $10 < n < 10^{3.5}$, meaning that the same statistical property lasted from 10 min. to over 2 days. The estimated value of the slope α is 0.58. Thus, the time correlation of announcement messages is very weak, and close to Poissonian. Similarly, the plot for withdrawal messages is characterized by the same statistics as for announcement messages. The estimated mean value of α for four measurement points is 0.60 for announcement messages and 0.59 for withdrawal messages. These results suggest that there is little possibility of generating a cascading failure, because there is little temporal correlation in the appearance of bursts.

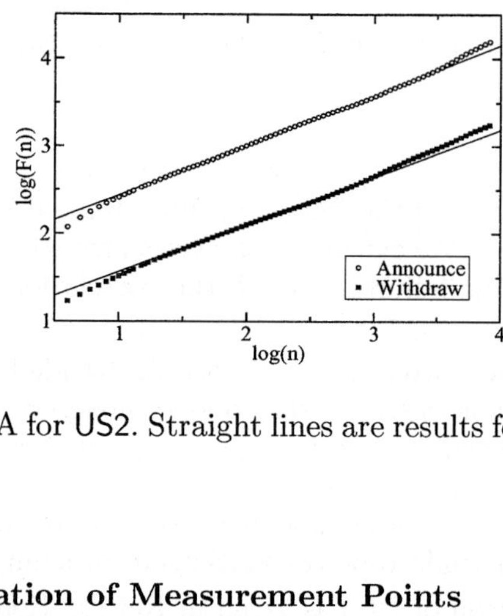

Fig. 2. Results of DFA for US2. Straight lines are results for least squares fitting.

3.2 Space Correlation of Measurement Points

Eyeball verification indicates that the pattern of spikes in Fig. 1 is not always synchronized, despite the simultaneous observation. Of course, all the time series do not need to be synchronized completely, because BGP messages that are not related to other ASes are absorbed at routers along the path. However, if there is a failure affecting most of the ASes, we expect to observe some temporal and spatial correlation between ASes. We compared the temporal patterns of flows for BGP messages for four measurement points, to quantify the level of the spatial difference between measurement points.

The correlation coefficient was used for this purpose. For two time series $F(t)$ and $G(t)$, the generalized correlation coefficient $C(F, G)$ is defined as

$$C(F(t), G(t+\tau)) = \frac{\sum\{(F(t_i)-E[F(t)])(G(t_i+\tau)-E[G(t)])\}}{\sqrt{V[F(t)]}\sqrt{V[G(t)]}}, \tag{1}$$

where $E[F(t)]$ and $V[F(t)]$ are the mean and variance of time series $F(t)$, respectively; τ represents the delay of the time series; and Eq. (1) for $\tau = 0$ is known as the correlation coefficient. $C(F, G)$ characterizes the degree of similarity between two time series as follows. $C = 0.0$ indicates that there are not correlated. $0.0 < C \leq 1.0$ corresponds to a positive correlation, showing that both time series statistically resemble each other. By definition, $C = 1.0$ for $F(t_i) = G(t_i)$. Moreover, $-1.0 \leq C < 0$ indicates an anti-correlation, i.e., $G(t_i)$ takes a smaller value for larger $F(t_i)$, and vice versa.

Table 1 lists the values of the correlation coefficient for $\tau = 0$ between different measurement points: (a) withdrawal and (b) announcement messages. For withdrawal messages, the values of the coefficient are larger than 0.2, except for the values related to JP, meaning that the patterns of withdrawal message statistically resemble each other for the three measurement points. However, the values for JP are smaller than 0.1, so the temporal pattern of the spikes is different from the others. Conversely, for announcement messages, the values of the correlation coefficient are close to 0 for three measurement points, although US1 and UK are still correlated ($C \approx 0.3$). This suggests that the message flows in two different measurement points are characterized by different patterns, despite the simultaneous observation.

Table 1. Values of correlation coefficient for two time series of (a) withdrawal and (b) announcement messages.

(a)	US1	US2	JP	UK
US1	1.00	0.29	0.05	0.58
US2	0.29	1.00	0.03	0.23
JP	0.05	0.03	1.00	0.06
UK	0.58	0.23	0.06	1.00

(b)	US1	US2	JP	UK
US1	1.00	0.03	0.01	0.31
US2	0.03	1.00	0.01	0.05
JP	0.01	0.01	1.00	0.06
UK	0.31	0.05	0.06	1.00

Next, we focus on the correlation coefficient with delay τ. If two time series statistically resembled each other with delay τ, we would observe the peak value of the correlation coefficient at τ. Figure 3 displays the values of correlation coefficients with the delay τ for withdrawal messages. Figure 3 (a) indicates that the messages obtained from US1 were synchronized to those of US2 and UK at $\tau = 0$. However, the peak shifted to $\tau = 3$ for JP, indicating that there was a three-min. delay for propagating the burstiness of messages from US1 to JP. Interestingly, for US2, all the peaks deviate from $\tau = 0$. The BGP messages relayed from US2 took one min. to US1, two min. to UK and four min. to JP. Therefore, the result means that US2 was the closest to the source of burstiness of BGP messages among the four measurement points. On the other hand, we cannot confirm any peak points in Fig. 3 (c), showing that there were no positive delays from JP to other measurement points. Figure 3 (d) also indicates that messages arrived at UK before JP. We also confirmed that the propagation delay is observable in the time series of announcement messages, although the value of the coefficient is smaller than that for the withdrawal. We can conclude that for both withdrawal and announcement messages, there are minute-order propagation delays between measurement points.

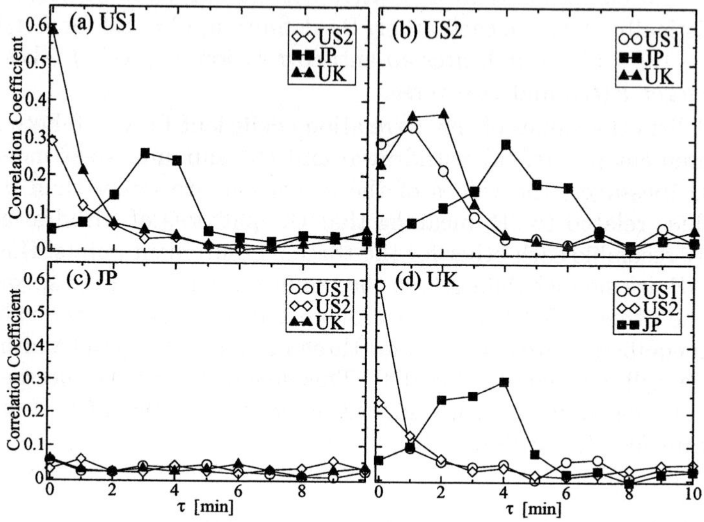

Fig. 3. Values of correlation coefficients with delay τ for withdrawal messages.

Figure 4 (a) shows the inferred propagation delays τ. The arrows with numbers show the delay in minutes estimated from the correlation coefficients of withdrawal messages between two locations. Here, τ was defined by the time step where the coefficient has the maximum value. The dotted lines indicate that the value of the correlation coefficient is less than 0.1 for $0 \leq \tau \leq 50$, i.e., non-correlation from one point to another. The map shows that US2 was the closest to the sources of the burst in the observed periods. Moreover, few bursts of withdrawal messages originated from JP.

To investigate the relationship between the propagation delay and the topological map of ASes, we checked the AS path length between monitoring points, which is obtained from the announcement messages. US1, US2, and UK were connected to each other by one AS, i.e., the length is 2. Also, the length from US2 to JP is 4, and the lengths from US1 and UK to JP are 5. The result indicates that the delay relates to the distance between measurement points in terms of the number of AS hops.

Note that the map of the delay propagation is a snap-shot, which means that the direction of the propagation delay is likely not always stable, because it depends on the location of the failure. To clarify its dependence, we estimated the propagation delay for every 1-hour withdrawal time series split from the original time series. Figure 4 (b) displays the dynamics of the propagation delay from US1 to JP and from US1 to UK, both corresponding to 2.5 days long. We

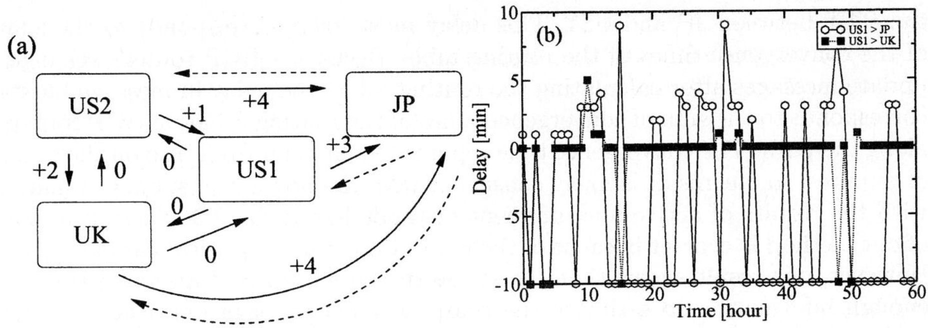

Fig. 4. (a) Delay propagation map and (b) dyanmics of delay propagation.

treated the delay τ as the time step where the value of the correlation coefficient was maximum and also larger than 0.4. Here, $\tau = -10$ indicates that the time series for two measurement points were spatially non-correlated. For US1 to JP, most plots are larger than 0 and there is no 0 value, meaning that the routing information from US1 to JP always had an observable delay. We also found some temporally non-correlated periods. This is probably because there were few routing events propagating from US1 to JP. On the other hand, for US1 to UK, most of the values were close to 0 and few were negative. Thus, the routing information from US1 to UK arrived without a delay of more than 1 min., meaning that both routing table were synchronized each other without delay.

4 Discussion

For the time correlation, our DFA results revealed that an observed burst triggered by a certain point has little dependence on the following time steps. In other words, the appearance of a burst has a Poissonian nature for the observed time scale. Concerning the value of α, it has been reported that the time series for traffic volume in WAN exhibits self-similarity, i.e., strong positive correlation ($\alpha \approx 0.8 - 1.0$) [5]. Generally, the probability density of an individual failure occurring is characterized by an exponential distribution, which is memory-less. Thus, small values of α for BGP messages directly reflect the result of random failures of links or routers. Consequently, we can conclude that there is little possibility of generating a cascading failure in our observations.

For the space correlation, larger correlation coefficients for withdrawal messages indicate that link failure information tends to propagates from one AS to the other ASes. Conversely, the announcement message bursts have a locality within an AS, characterized by a smaller value of a coefficient. This locality of the burst is due to the effect of a multihomed AS as explained in Section II. We also observed a propagation delay of over one min. between message flows at different measurement points. The delay we observed is 100 times larger than the RTT between JP and US1. This delay most likely corresponds to the sum of the convergence times of the routing table. Because a BGP router exchanges update messages after calculating the routing table, the delay in message bursts corresponds to the sum of convergence times of the routing tables in BGP routers along the path. The convergence time reported in [2] in an AS is also on the order of minutes, so our result is likely consistent with the previous results. Compared with the results of active measurement methods like Refs. [2,3], the estimation of our method is coarser because of the restriction of using passive measurement. However, our result shows that the spikes in the withdrawal messages provide enough information to estimate the coarse-grained propagation delay without probe messages. Moreover, though the sources of a beacon message are fixed, our method is simple and places no restrictions on the location of the source of a failure. Thus, it can be a complementary method to the active method. Finally, we discuss a possible improvement of the estimation of the delay. We checked the time series in 10-s bins instead of 1-min. bins. However, it is difficult to identify the peak position for τ unlike Fig. 3, suggesting that there is an appropriate bin size for estimating the propagation delay. Further study of this method is needed to obtain the propagation delay more accurately.

5 Conclusion

To investigate the macroscopic behavior of BGP, we analyzed the time series for the number of BGP messages through statistical tests. We found that there was little temporal correlation between bursts of messages, indicating that the probability of observing a cascading failure is very small. We also pointed out that the time series from two measurement points had little temporal correlation. However, we estimated of the propagation delays in bursts of messages between measurement points from the macroscopic view. The order of delays was consistent with the order of routing table convergence times. Finally, we revealed the dynamics of the temporal and spatial dependence of the BGP messages between the measurement points.

References

1. Rekhter, Y., Li, T.: Border gateway protocol 4. RFC 1771. (1995)
2. Labovitz, C., Ahuja, A., Bose, A., Jahanian, F.: Delayed Internet routing convergence. Proc. ACM SIGCOMM 2000. (2000) 175-187
3. Mao, Z., Bush, R., Griffin, T., Roughan, M.: BGP beacons. Proc. ACM SIGCOMM Measurement Workshop 2003. (2003) 1–14
4. Peng, C-K., Havlin, S., Stanley, H. E., Goldberger, A. L.: Quantification of scaling exponents and crossover phenomena in nonstationary heartbeat time series. Chaos **5** (1995) 82–87
5. Park, K., Willinger, W. (eds.): Self-similar network traffic and performance evaluation. John Wiley & Sons, New York (2000)

A Framework to Enhance Packet Delivery in Delay Bounded Overlay Multicast

Ki-Il Kim[1], Dong-Kyun Kim[2], and Sang-Ha Kim[1]

[1] Department of Computer Science, Chungnam National University,
220 Gung-dong, Yuseong-gu, Daejeon, 305-764, Korea
kikim@cclab.cnu.ac.kr
shkim@cclab.cnu.ac.kr**

[2] KISTI, 52 Eoeun-Dong, Yuseong-gu, Daejeon, 305-806, Korea
mirr@kreonet2.net

Abstract. Overlay multicast has been proposed as an alternative scheme to provide one-to-many or many-to-many data delivery on Internet. However, since data delivery is entirely dependent on replications on each group member, if one member cannot receive a data packet, none of its children can receive that packet. Furthermore, the higher the member's level is, the more nodes cannot also receive data packet. In this paper, we give a detailed framework to enhance packet delivery ratio in overlay multicast. Unlike previous efforts based on duplicated forwarding, our scheme builds another type of overlay data delivery tree (DDT), which is adaptively reconstructed based on the number of group member's measured packet delivery ratio while guaranteeing end-to-end delay bound. Through practical simulation results, we analyzed packet delivery ratio, control overhead, and end-to-end delay.

1 Introduction

Due to several complex deployment issues [1] such as state maintenance on intermediate routers, overlay multicast schemes [2] have been proposed as the alternative solution for one-to-many and many-to-many data delivery on Internet. They shift multicast forwarding functionality from a router to the host, an each group member. So far, various works have been proposed to efficiently construct overlay DDT. Since they generally consider the shortest hop distance as major metric for parent selection, other metrics have not been relatively emphasized. In this paper, we focus on packet delivery ratio in overlay multicast. In general overlay multicast, packet delivery ratio is dependent on robustness of participants on data delivery tree as well as amount of influence caused when a group member cannot correctly receive data packet. Unlike native IP multicast where routers are components of DDT, overlay multicast scheme constructs DDT based on hosts which are inherently more susceptible to failures than the routers. On the other hand, a node not receiving data packet has no influence

** Sang-Ha Kim is corresponding author

C. Kim (Ed.): ICOIN 2005, LNCS 3391, pp. 223–232, 2005.

on other group members' packet delivery ratio in IP multicast. However, since overlay DDT is organized with group members, if one member node cannot receive data packet, none of its children receive multicast data. Furthermore, the higher the node's level is, the more nodes cannot receive data packet.

To achieve high data delivery ratio, several schemes [3-8] have been recently proposed. For example, S. Banerjee et al. [3] proposed Probabilistic Resilient Multicast (PRM) which can guarantee arbitrarily high data delivery ratios and low latency bounds in proactive or reactive manners. In a proactive manner, a randomized forwarding randomly sends data to other group members with a low probability. On the other hand, triggered NAKs are introduced to handle data loss due to link errors and network congestion. Similarly, L. Xie et al. [4] presented random jump (RJ) concept. With RJ, a node receives data not only from its parent, but also from some other nodes in the tree.

As mentioned above, the previous schemes attempt to provide higher data delivery mainly by using duplicated forwarding in order to recover lost packets. By utilizing multiple deliveries, they can achieve higher packet delivery ratio. However, this duplicated forwarding not only wastes network resources, but also significantly increases each group member's maintenance overhead.

In this paper, we give a framework to enhance packet delivery ratio in delay bounded overlay multicast. We call it Reducing delivery failure influence on Overlay Multicast (ROM). The objective of ROM is to enhance packet delivery ratio by reducing the influence of delivery failure on overlay DDT, instead of using fast recovery or retransmission for lost packets. To achieve this objective, compared to the previous schemes, ROM initially constructs another type of overlay DDT with measured packet delivery ratio as major metric in a centralized manner.

The remainders of this paper are organized as follows. The section 2 describes a framework of ROM and detailed join and leave operations. In section 3, we demonstrate the performance of ROM through simulation results. Finally, in section 4, we make concluding remarks.

2 Reducing Delivery Failure Influence on Overlay Multicast

In ROM, overlay DDT is constructed based on measured packet delivery ratio instead of shortest hops. Overlay DDT is built as source-based tree. So, each group source manages entire group membership in a centralized manner. Each source constructs overlay DDT in a form of min heap, where the value in each node is not smaller than in its children. The key value in each group member is set to node's measured packet delivery ratio. Based on the min heap structure, if a node hopes to join group, it is handled by the similar algorithm to insert a node into a min heap. Similarly, algorithm to delete one node from a min heap is applied when a group member is willing to leave a group. But, major difference from general min heap structure is that the degree on each group member is not strictly bounded as 2.

2.1 Local Data Structure

Each group member in the network should maintain the following variables and data structures:

- Parent and children (if any): its parent and its children on overlay DDT.
- Sequence number: packet sequence number in every *TIME_WINDOW*.
- *npd(normalized packet delivery ratio)*: how many data packets and *Keep_Alive* messages are dropped. Since *Keep_Alive* messages are periodically generated to check reachability from parent_relay_node, it indicates route stability between group members. Each member should keep track of *npd* in every period of *TIME_WINDOW* and update this variable.

$$E_{packets} = (L_s - F_s) + w * L_k + C$$

$$L_{packets} = (L_D + w * L_k)$$

$$npd_{cur} = \frac{L_{packets}}{E_{packets}} \quad (1)$$

$$npd = \alpha * npd_{cur} + (1 - \alpha) * npd \quad (2)$$

L_s	Sequence number of late received packet in *TIME_WINDOW*
F_s	Sequence number of first received packet in *TIME_WINDOW*
w	When a *Keep_Alive* message is not correctly received, each node should re-designate a new parent_relay_node. During this rejoin procedure, a lot of data packets are lost. So, loss of one *Keep_Alive* message should be handled differently from loss of one data packet.
C	The number of *Keep_Alive* messages in *TIME_WINDOW*. This value is given by *TIME_WINDOW* / Time period between two consecutive *Keep_Alive* messages
L_k	Sum of lost *Keep_Alive* messages
L_D	Sum of lost data packets
α	Smoothing factor lying between 0 and 1

Initially, *npd* is set to 0. By taking the number of *Keep_Alive* messages into account, unstable group members not receiving *Keep_Alive* message from parent_relay_node are differentiated with other group members. Furthermore, when a group member is recovered from out of order, this value is set to *previous_npd * 10* (weight).

- Group-id: stands for sequence of packet delivery.
- Delay: stands for end-to-end delay from source to itself.
- A node's impact factor for failure: stands for the influence caused by node's failure. This value is affected by the total number of children as well as a node's failure probability.

2.2 Protocol Description

In this section, we describe group join and leave operation. In ROM, it is initially assumed that the multicast group address, source address and port number are communicated or announced to all members through online or offline methods such as URL, a directory service or an e-mail message.

2.3 Group Join

When a group member wants to join a multicast group, it first contacts source by sending JOIN_REQUEST packet encoding delay bound. When a source receives JOIN_REQUEST packet, it searches already constructed overlay DDT and a source sends back JOIN_REPLY packet to requesting node. The parent_relay_node for requesting group member is determined as follows.

Since there is no measured group member's *npd*, requesting group member may be located with only delay bound. This algorithm is described as below Algorithm 1. When potential parent_relay_node is determined, this information is back to requesting group member. With this parent_relay_node, a requesting group member can designate parent_relay_node.

We use $d_h(i)$ to denote the minimum delay incurred from a source to a group member i over overlay DDT where M_h represents group members at level h.

Algorithm 1 : DelayBound

1. Set $d_i(k) = \infty$, for $1 <= h <= log_t(v{+}1)$; set $d_0(\text{source}) = 0$. Let $h = log_t(v{+}1)$;
2. while the optimal path is not found for group member $i$at the tree level$h$, update

$$d_{h}'(i) = \min(d_h(i), \min_{j \in M_h} (d_h(j) + d(i,j))).$$

If $d_h(i) <= DelayBound$, then the minimum hop has been found with h edges. Otherwise, $h = h$ - 1.

Above steps begin under situation that a node is connected as leaf node's child, and end if optimal path is found or all the group members are spanned. At the first step, attempts to find location on DDT with end-to-end delay bound are accomplished by comparing delay through direct connection to source with delay via other group members at the same tree level. If the path is not found, comparison moves to next level on the tree. Despite all spanning, if parent_relay_node is not found, a source designates requesting group member as one of child_relay_nodes.

2.4 Overlay DDT Maintenance

When a node initially becomes a group member, overlay DDT is newly reconfigured with only end-to-end delay bound since there is no measured node failure probability as described above. To reflect a node's measured *npd*, each group member periodically contacts with a source. On the other hand, a source reconstructs overlay DDT according to definition of min heap. If we assumes that group member with lower *npd* is considered as more reliable, constructing overlay DDT to enhance packet delivery ratio is defined as following, where P_i indicates probability not receiving data packet on the path to the group member i, and D_i defines delay from source to member i. In problem definition, M is a list of group members.

$$\min \sum_{i=M}^{v} P_i(t)$$

$$s.t. \quad for \quad \forall_i \in M, D_i \leq D_0$$

In order to construct overlay DDT with delay bound, we propose a heuristic centralized algorithm. This algorithm includes the following steps.

Algorithm 2: MinHeap_DelayBound

1. Constructs a min heap in the form of complete binary tree.
2. For each group member, if the end-to-end delay from source to a group member is larger than *DelayBound*, this group member is removed from the min heap. When an intermediate group member is removed, the current min heap is adjusted by the same algorithm applied for group leave procedure.
3. The group members violating end-to-end delay begin to find adequate location on the min heap by Algorithm 1 until end-to-end delay is bounded within a specific threshold. In spite of complete spanning the whole min heap, if no group member is designated, a group member requests connection to the group member that can minimize end-to-end delay.

2.5 Data Forwarding and Establishing Redundant Paths

Once overlay DDT is constructed, data packets are forwarded along established overlay DDT. In ROM, each group member has its own impact factor. Impact factor is defined by the amount of influence caused by a group member's delivery failure. This impact factor increases as the total number of children increases and each node's *npd* has larger value. Larger *npd* indicates node's higher instability. So, we define each node's impact factor of failure (*NIF*) as follows.

$$NIF = total\ number\ of\ children * npd$$

If this *NIF* is larger than specified threshold, this fact indicates high node's delivery failure probability or forwarding responsibility for a lot of group members. Thus, protection for delivery failure on this important node by establishing

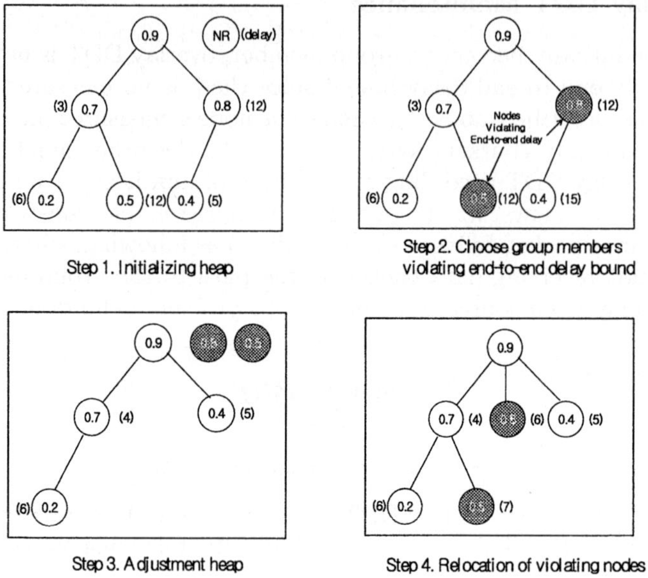

Fig. 1. Example of overlay DDT maintenance

multiple parents is required. Otherwise, since the influence of this member's failure is minor, an attempt to fastly redesignate a new parent_relay_node is applied instead of duplicated forwarding in order not to waste resources. When we try to designate multiple parents, it is very critical and necessary problem to prevent looping as mentioned earlier. In ROM, a sender initiates a random group-id and piggybacks this value into data packet. When group members receive this packet, they generate its own group-id by computing a value that is larger and not consecutive. Therefore, there are gaps between the group-id of a sender and a receiver. Each group member repeatedly propagates its own group-id to its children. As a result, overlay DDT constructs another min heap with different node value, group-id. If a group member with large *NIF* wishes to find multiple parents, it sends MULTIPARENT_REQUEST packet that includes parent group-id. When a source receives this request, it returns MULTIPARENT_REPLY with other group members located at the upper level than the level of current parent. To find the closest parent, a source returns the group member making the smallest difference between requesting group-id and itself.

3 Performance Evaluation

In this section, we analyze the performance of the proposed ROM using simulation. The used simulation tool is NS-2 [9]. The simulation is performed toward two directions; performance enhancement gained by adapting 1) min heap structure with *npd*, 2) allowing redundant paths according to *NIF*. The performance is evaluated in terms of packet delivery ratio, control overhead, and end-to-end delay. The simulation results are compared with NICE and PRM over NICE[2] because they also have hierarchical cluster property. The brief explanations for NICE protocol are as follows. The NICE protocol arranges the set of members into a hierarchical control topology. As new members join and existing members leave the group, the basic operation of the protocol is to create and maintain the hierarchy. The hierarchy implicitly defines the multicast overlay data paths and is crucial for scalability of this protocol to large groups. The members at the bottom of the hierarchy maintain (soft) state about a constant number of other members, while the members at the top maintain such state for about $O(logN)$ other members.

We run our simulations using Transit-Stub graph model topologies obtained using the GT-ITM topology generator [10]. We use topologies of 1,000 nodes and multicast groups of 128 members. Members are attached to random routers and links are assigned a random delay of 1 - 4ms.

During a warm-up time of 300 seconds, a group member is randomly selected and becomes out of order for a certain amount of time period, which follows exponential distribution (with mean equal to 5 sec.). Through this warm-up time, a group member has its own node failure probability. To model arbitrary node failure, our simulation scenarios follow the steps shown below. Basic idea of following :

1. Randomly chooses failure frequency, for example, 1 node per second or 2 nodes per second.
2. Generates one random variable, $0 <= r <= 1$.
3. Gathers group members whose node failure probability is larger than r.
4. Sorts them in decreasing order and removes group members, which go through failure in the previous three periods.
5. Chooses nodes as many as failure frequency in remaining group member and makes those nodes fail.

In our simulation, we evaluate three different ROM versions, ROM0, ROM15, and ROM20 according to *NIF* value. ROM0 means that if the *NIF* value is larger than 0, multiple paths are established. That is, all group members have multiple parents because *NIF* is always larger than 0. With similar meaning, ROM15 establishes multiple parents when a group member's *NIF* is larger than 15. Because *npd* is ranged from 0 to 1, multiple paths are established on one group member when at least more than 15 group members are children in ROM15. To evaluate PRM, randomized forwarding is employed.

Fig. 2 shows successful packet delivery ratio as a function of failure frequency which is defined by the number of failed nodes per second. In general, packet delivery ratio decreases as failure frequency increases. Since ROM0 always establishes multiple paths, it guarantees almost 98.7% packet delivery ratio. The 1.3% packet drop happens when both parent group member and protection group member simultaneously fail. Since ROM does not support recovery for lost packets, the perfect reliable data delivery is not achievable in this simulation. On the other hand, since NICE does not consider reliability, it shows the worst packet delivery ratio among the comparative protocols. PRM shows similar or little lower packet delivery ratio than ROM15. PRM randomly chooses and then utilizes additional edges for reliable data delivery. Therefore, considerable effect from group members at upper level remains. One thing worthwhile mentioning is that ROM makes a little difference even though failure frequency increases. It is largely because the more group member's failures occur, the more accurate failure probability of the node is obtained. Thus, the enhanced min heap structure can be created. Similarly, the location of a group member with low failure probability is adaptively and dynamically determined to lower level of the tree. Consequently, influence of failure is minimized. Fig. 3 shows end-to-end delay with varying failure frequency. Since NICE uses delay as major metric when selecting parent group members, it has the shortest end-to-end delay. ROM shows shorter end-to-end delay than PRM. While PRM uses a randomly chosen parent group member, ROM selects the closest member among the group members located at higher level on OMT, with topology awareness property as well as group-id. Accordingly, in case of PRM, data dissemination along the path established from protected parent and group member takes longer end-to-end delay than ROM. However, both PRM and ROM take end-to-end delay into account in constructing overlay DDT, and a little difference is not a critical problem.

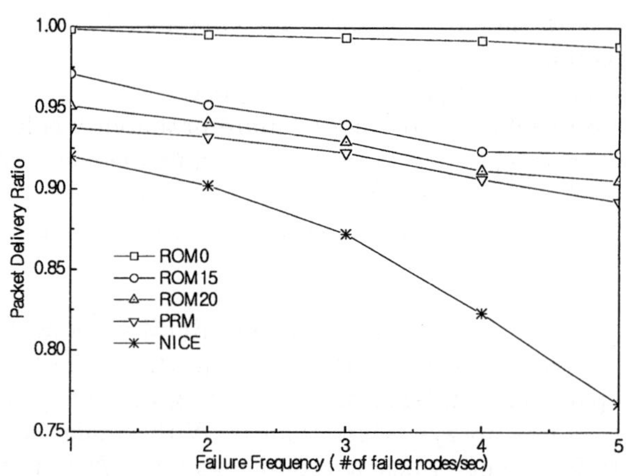

Fig. 2. Successful packet delivery ratio vs. failure frequency

Control overhead is illustrated in Fig. 4. In this paper, control overhead is defined as number of required control packets for delivering one data packet. Control overhead consists of two parts. One is control packet to repair broken links, and the other is control packet to maintain overlay DDT. In term of the former, the more control packets are incurred in NICE. When a group member fails, since ROM and PRM have already established redundant routes, these

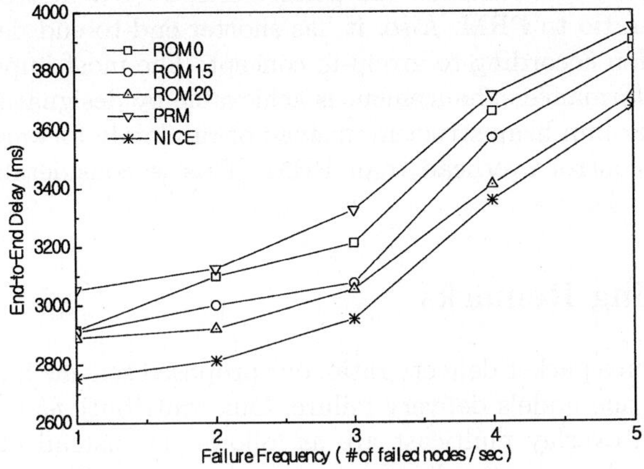

Fig. 3. End-to-end delay vs. failure frequency

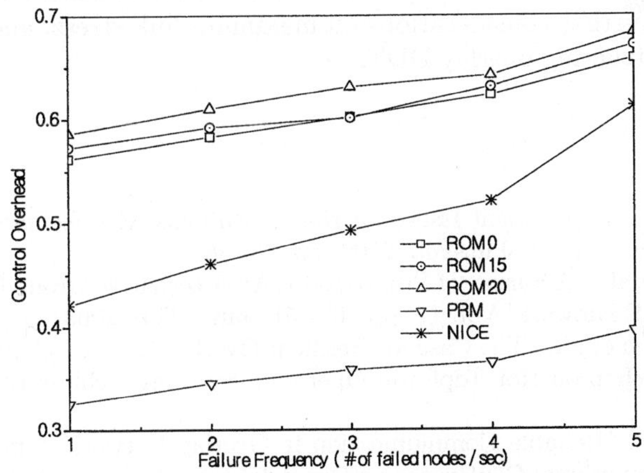

Fig. 4. Control overhead vs. failure frequency

control packets takes a little portion in total control overhead. On the contrary, there are many control packets to maintain overlay DDT in ROM. ROM should periodically exchange group member's measured *npd*. In addition, a source sends control packets to each group member according to recomputed min heap structure. This recomputed min heap structure should be addressed to group members so that each group member attempts to request establish new connection. Due to this control overhead, ROM has the largest control overhead. On the contrary, PRM provides redundant routes without periodical message exchanges. With above two factors, PRM incurs less control overhead than NICE and ROM.

As a result, we can conclude the performance of ROM. ROM has similar packet delivery ratio to PRM. Also, it has shorter end-to-end delay than PRM due to designation according to group-id concept. The most important result is that similar performance enhancement is achievable by designating a few multiple parents under min heap structure instead of randomly forwarding. However, it incurs more control overhead than PRM. This is considerable deployment issues.

4 Concluding Remarks

In order to enhance packet delivery ratio, our proposed scheme aims to minimize the influence of one node's delivery failure. Our contributions to higher packet delivery ratio in overlay multicast are as follows. 1) Instead of overlay DDT construction based on shortest hops, new type overlay DDT is constructed with measured packet delivery ratio in a form of min heap, 2) the looping problem caused by additional edges is prevented by the concept of logical height, 3) Throughout the simulation results, we investigate the higher packet delivery ratio while periodical message exchanges in-creases control overhead. Related to this work, it needs further considerations on maximum link stress, and transmission cost while constructing overlay DDT.

References

1. C. Diot et al., "Deployment Issues for the IP Multicast Service and Architecture," IEEE Network, Vol. 14, Jan.-Feb. 2000, pp. 78 -88.
2. A. El-Sayed et al., "A Survey of Proposals for An Alternative Group Communication Service," IEEE Network, Vol. 17, pp. 46 - 51, Jan. - Feb. 2003.
3. D. G. Andersen et al., "The Case for Resilient Overlay Networks," in Proceedings of the 8th Workshop on Hot Topics in Operating Systems, Schloss Elmau, Germany, May 2001.
4. Y. Amir et al., "Reliable Communication in Overlay Networks," in Proceedings of the IEEE International Conference on Dependable Systems and Networks (DSN03), San Francisco, Jun. 2003.
5. S. Banerjee et al., "Resilient Multicast using Overlays," in Proceedings of ACM SIGMETRICS, San Diego, CA, Jun. 2003, pp. 102 - 113.
6. S. Birrer et al., "Resilient Overlay Multicast from the Ground Up," Tech. Report NWU-CS-03-22, Department of Computer Science, Northwestern University, 2003.
7. L. Xie et al., "An Approach to Reliability Enhancement of Overlay Multicast Trees," in Proceedings of PostGraduate Networking Conference, Jun. 2003.
8. G. I. Kwon et al., "ROMA: Reliable Overlay Multicast with Loosely Coupled TCP Con-nections," in Proceedings of IEEE INFOCOM, Hong Kong, Mar. 2004.
9. Network Simulation, http://www.isi.edu/nsnam/ns/
10. E. W. Zegura et al., "How to Model an Internetwork," In Proceedings of IEEE INFOCOM, Vol. 2, Mar. 1996, pp. 594-602.

A Rerouting Scheme with Dynamic Control of Restoration Scope for Survivable MPLS Network*

Daniel Won-Kyu Hong[1] and Choong Seon Hong[2]

[1] Operations Support System Lab., KT
62-1 Hwaam-Dong Yuseong-Gu, Daejeon 305-718 KOREA
wkhong@kt.co.kr

[2] School of Electronics and Information, Kyung Hee Univerity
1 Seocheon Giheung Yongin, Gyeonggi 449-701 KOREA
cshong@khu.ac.kr

Abstract. This paper proposes a rerouting scheme that can be applied to the restoration of working Label Switched Paths (LSPs) and pre-provisioned backup LSPs, which consists of two subsequent algorithms. The first algorithm for the dynamic deterimination of the restoration scope (RS)increases the restoration speed by minimizing the complexity of the network topology and by maximizing the reusability of the existing working LSP. The second newly proposed concept of RS extension minimizes the probability of restoration failure by dynamically widening the restoration scope until the RS is equal to the whole network topology. Through simulation, we evaluate the performance of our restoration scheme and the existing protection schemes in terms of the restoration speed, packet loss, network resource utilization, and resource reusability of the existing working LSP.

1 Introduction

The rerouting method for traffic engineering in IP networks became the driving force behind MPLS. The ability to protect traffic against failure or congestion in an LSP can be important in mission-critical MPLS networks [6,7,9]. Restoration is necessary for two different reasons: one is fast restoration and the other is optimized restoration. Fast restoration minimizes service disruptions for the flows affected by an outage using a backup path [1,9]. Optimized restoration serves to alternatively optimized traffic flow in line with a changed network topology [2,9]. However, fast restoration cannot provide fast change of traffic flows any more when the backup and working paths go down simultaneously and cannot increase network resource utilization. In this paper, we propose a LSP rerouting scheme that dynamically aligns the restoration scope in MPLS network. Because this scheme dynamically adjusts the restoration scope depending on the fault and

* This work was supported by University ITRC Project of MIC. Dr. C.S. Hong is the corresponding author.

C. Kim (Ed.): ICOIN 2005, LNCS 3391, pp. 233–243, 2005.

congested location and the overall network status, we may provide the most reasonable bypassing path for avoiding congestion or for restoring fault. This paper proposes a rerouting algorithm to determine the restoration scope taking into account the bandwidth, delay, and hop count. This algorithm can be a scalable one because it expands the restoration scope when it fails to calculate the reasonable bypassing path within the found restoration scope. Our model can expand the restoration scope until the overall network is a restoration scope. Our restoration scheme focuses on the maximization of network resource utilization better than the restoration speed. However, it does not prominently degrade the performance of restoration speed compared with the existing backup path driven approaches [4,5], which establishes the working and backup paths simultaneously for fast restoration and lacks resource utilization and backup path protection schemes. Our model may provide a moderated restoration speed compared with the existing backup path approaches [4,5], maximize network resource utilization and protect the backup path without requiring much longer restoration time. We define a rerouting model that dynamically establishes a transient backup path, taking into account the current network status and topology when the node or link goes down and automatically releases it when the node or links goes up. We propose a algorithm to determine the restoration scope, an algorithm to find reasonable bypassing paths, and demonstrate a procedure that restores the faulty working path without prominent performance degradation while maximizing network resource utilization better than the existing backup path driven restoration methods [4,5]. Through simulation, the performance of the proposed restoration scheme is measured and compared with the existing schemes in terms of packet loss, restoration speed, resource utilization and the reusability of the existing working LSP.

2 The Provision Process of the Alternative LSPs

To meet the requirements of the maximization of network resource utilization and the minimization of the restoration speed, we propose a rerouting model that dynamically provides an alternative path bypassing the fault location with the concept of dynamic RS arrangement. The process of dynamic RS arrangement is composed of two subsequent steps: (1) the generation of the Intermediate Weighted Network Graph (IWNG) from the Network Graph (NG) taking into account the fault location and the traffic metrics, such as requested bandwidth (B_{REQ}), traffic class (T), and hop count (H); and (2) the determination of intermediate ingress node for confine the restoration scope.

Fig. 1 shows the overall process to create an alternative LSP. The detail procedure for the creation of an alternative LSP avoiding the fault location is as follows:

`Preconditions:`

- A network graph, $NG(N, L, B_{AVA}, R)$, where N is a set of node, L is a set of link, B_{AVA} is avariable bandwidth of link, and R is the rerouting option.

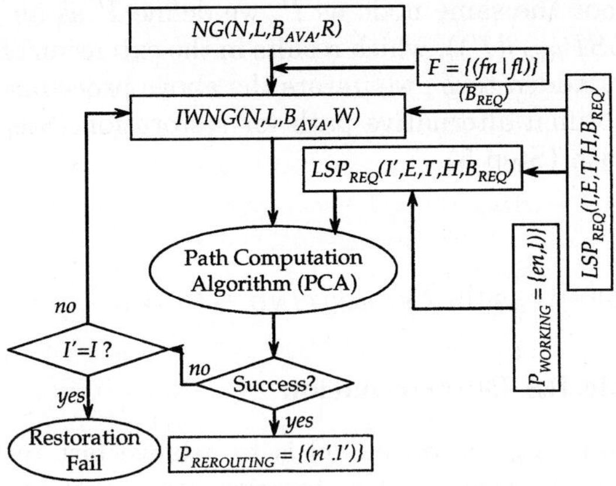

Fig. 1. The overall process to create an alternative LSP

- A list of fault locations, $F = (f_n|f_l)$, which is a set of abnormal nodes (n) or abnormal links (l)
- An LSP request, $LSP_{REQ}(I, E, T, H, B_{REQ})$, where I is the ingress LSR, E is the egress LSR, T is the traffic class (gold, silver, and bronze), H is the hop count that can hopefully be the end-to-end delay if the delay between each link is constant, and B_{REQ} corresponds to the bandwidth requirements.
- A working LSP, $P_{WORKING} = (n, l)$, where n is node and l is link traversing the working LSP

`Procedure:`

[Step 1] First, we create an Intermediate Weighted Network Graph (IWNG) with such information as F, $LSP_{REQ}(I, E, T, H, B_{REQ})$, and $P_{WORKING}$. We will propose the algorithm for creation of IWNG in next section. The weight (W) of IWNG is assigned by the IWNG creation algorithm.

[Step 2] We determine intermediate ingress node (I') to generate an alternative LSP, that is to say, we confine the restoration scope that will be the most reasonable boundary to create an alternative path for the restoration of the occurred faults. In other words, we create $LSP_{REQ}(I', E, T, H, B_{REQ})$. The I' should be one of the nodes in $P_{WORKING}$.

[Step 3] We find the most optimal alternative path from $LSP_{REQ}(I')$ to $LSP_{REQ}(E)$ with the routing constraints such as T, H, and B_{REQ}.

[Step 4] If the path computation algorithm generates an optimal path between $LSP_{REQ}(I')$ and $LSP_{REQ}(E)$, we select it as an alternative path $(P_{REROUTING})$ for the restoration of the fault and stop the procedure.

[Step 5] However, if there is no any reasonable alternative path between $LSP_{REQ}(I')$ and $LSP_{REQ}(E)$, we compare the original ingress node (I) with I'. If I is the same node as I', we stop the procedure because there can be no more wide restoration scope.

[Step 6] If I is not the same node as I', we define I' as an implicit abnormal node ($F \leftarrow LSP_{REQ}(I')$), which results in the extension of restoration scope. As I' is newly added to F, we iterate the above procedure from Step 1 until we find an optimal alternative path for restoration (Step 4) or there is no alternative path (Step 5).

`Output:`

An optimal alternative path, $P_{REROUTING} = (n', l')$.

2.1 A Dynamic RS Determination

The purpose of our restoration model is to achieve fast restoration and high resource utilization. However, fast restoration comes into conflict with high resource utilization. Therefore, this paper proposes a leverage algorithm that reconciles these two factors. In our restoration algorithm, we narrow down the restoration scope as soon as possible for the rapid path computation. However, the wide restoration scope is better than the narrow restoration scope in terms of resource utilization.

To determine the reasonable restoration scope based on the fault location, this paper proposes an algorithm for dynamic RS arrangement. This algorithm generates the WNG for the working LSP provision and the IWNG for the alternative LSP provision.

`Preconditions or Definitions:`

- A network graph, $NG(N, L)$, which is composed of nodes, $n \in N$, and links, $l \in L$.
- A node, $n(w, d, v)$, where w is a weight, d is an accumulated delay, and v is a visiting flag.
- A link, $l(w, r, d)$, where w is a weight, r represents reachability (*yes* or *no*), and d is a delay.
- n_{active} represents the active node allocating the proper weights to all of its neighbor nodes and links connected to it.
- $n_{passive}$ represents the passive node order which is assigned by an active node, that is to say, $n_{passive}$ is a neighbor node of n_{active}.

`Algorithm:`

Refer to Fig. 2.

`Output:`

An Intermediate weighted network graph($IWNG(N, L, B_{AVA}, W)$), where N is a set of node, E is a set of link, B_{AVA} is the avaiable bandwidth of the link, and W is the assigned weight on a link.

```
For each n ∈ N do
    n(w,v,d) ← (∞,no,0);
For each l ∈ L do
        if (l(r)==no and (l(r)==yes and l(B_AVA) < B_REQ))) trims l from NG ;
        if (l(r)==yes and l(B_AVA) < B_REQ)) l(w) ← (∞);
Define the ingress node (LSP(I)) as the initial active node (n_active);
Define the egress node (LSP(D)) as the destination node (n_dest);
n_active(w,d) ← (0,0);
Procedure Allignment(n_active, n_dest)
        if(n_active == n_dest) return;
        n_active(v) ← yes;
        n_passive ← ladj;
        for each l connected to n_active do
                if( l(w) > n_active(w)+1)
                        l(w) ← (n_active(w)+1);
                if( n_passive(v) == no)
                        l(d) ← l(d)+n_active(d);
                if(n_passive(v) == yes) continue;
                if(n_passive(w) > l(w)) {
                        n_passive(w,d) ← (l(w),l(d));
                        Allignment(n_passive, n_dest);
                }
        end of for
end of procedure
```

Fig. 2. An algorithm for determiantion of restoration scope

Starting at the ingress node, our algorithm visits all the neighbors, then visits all the neighbors of these neighbors, and so on, until there are no neighbors left to visit. Our algorithm is very simple but there are some rules for assigning the appropriate weight of each node and link with the following steps:

[Step 1] We initialize the nodes of the network using infinity (∞), zero (0) and *no* of visit flag to weight, delay and visiting flag, respectively.

[Step 2] We trim the unfavorable links from NG. If the reachability of link is no or the reachability is yes but its available bandwidth is less than the requested bandwidth, we trim the link from NG. If the link is favorable, we assign zero to the weight of the link.

[Step 3] We define the ingress node ($LSP(I)$ representing the ingress node terminating the LSP) as a first active node (n_{active}) and define the egress node ($LSP(D)$ representing the destination node terminatiing the LSP) as a destination node (n_{dest}).

[Step 4] Initially, we assign zero to the weight and delay of $n_{active}(n_{active}(w,d) \leftarrow (0,0);)$. From now on, we traverse NG until there are no nodes to visit, calling the procedure of Alignment(n_{active}, n_{dest}).

[Step 5] We assign yes to the visiting flag of n_{active} ($n_{active}(v) \leftarrow yes;$) in order to avoid the duplicated traverse of NG.

[Step 6] We determine the weight and the accumulated delay of each links ($l(w,d)$) connected to the $n_{active}$with the following rule:

$$l(w) = \begin{cases} n_{active}(w) + 1, & if(l,w) > n_{active} + 1 \ or \ l(w) == 0 \\ l(w), & if(l(w) \leq n_{active} + 1 \ and \ l(w) \neq \infty \end{cases} \tag{1}$$

$$l(d) = \begin{cases} n_{active}(d) + l(d), & if(n_{passive}(v) == no) \\ l(d), & if(n_{passive} == yes) \end{cases} \tag{2}$$

[Step 7] If the visiting flag of $n_{passive}$ is yes, then we traverse another link that is not traversed. If $n_{passive}$ is no, we assign the weight and accumulated delay of $n_{passive}$ with the following rule:

$$n_{passive}(w) = \begin{cases} l(w),\ if(n_{passive}(w) == \infty \ or \ n_{passive}(w) > l(w) \ or \ n_{passive}(v) == no); \\ n_{passive}(w),\ \ if(n_{passive}(w) \leq n_{passive}(v) == no) \end{cases} \tag{3}$$

$$n_{passive}(d) = \begin{cases} l(d),\ if((n_{passive}(d) == 0 \ or \ n_{passive}(v) == no \ and \ n_{passive}(d) > l(d); \\ n_{passive}(d),\ \ if(n_{passive}(d) \leq l(d)); \end{cases} \tag{4}$$

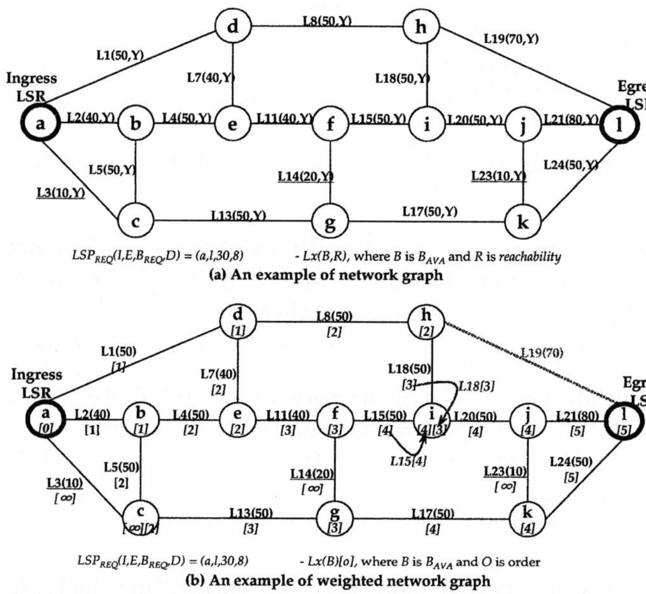

Fig. 3. An example of network graph (NG) and weighted network graph (WNG)

Let's assume that the working path, $LSP_{WORKING}$, traverses *(a-L1-d-L8-h-L19-l)* and there is a fault at *L19*. Fig. 3 (a) shows an NG, where link *L9* is faulty, $LSP(I, E, B_{REQ}, D)$ is *(a,l,30,8)*, B_{AVAs} of *L3*, *L14* and *L23* are less than B_{REQ}. We trim the unfavorable links of *L3*, *L14*, and *L23* and traverse NG from the ingress node of *l* until all visiting flags of nodes can be *yes* in accordance with the rules described in Step 6 and Step 7. As a result of NG traversing, we can generate WNG as shown in Fig. 3 (b).

After generating the WNG, we determine the reasonable RS. At first, we determine I', which can be the node connected to the abnormal link along the reverse traffic flow. In the case of fault at *L19* as shown in Fig. 3 (a), the first candidate I' shall be h because nodes of *h* and *l* are connected to abnormal link,

L19. But h is in the reverse traffic flow of the working LSP. So, we select h as I'. Once the I' is selected, we define the restoration scope as the set of nodes and links whose weights are greater than $I'(w)$ and the I' itself as following rule:

$$RS = (n(w), l(w)) \cup (I'), \quad where\ w \supset I'(w) \tag{5}$$

For example, the first RS to restore the link fault (*L19*) can be *(h, L18, i, L20, j, L21, l)* according to the rule for the determination of RS as shown in Fig. 3 (a). Thus, we find an alternative path avoiding the abnormal link of *L19* between h and l. The alternative path can be *(h-L18-i-L20-j-L21-l)*. The algorithm for selecting an alternative path will be given in the next section.

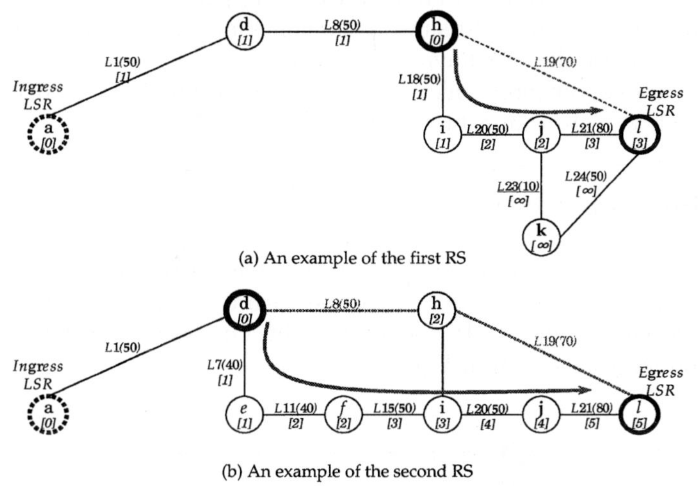

Fig. 4. An example of the intermediate weighted network graph (IWNG)

If we failed to find an alternative path avoiding the abnormal link of *L19* within the first RS for some reason or another, we need to rearrange the RS, which is the concept of the extension of the restoration scope. Rules for extension of the restoration scope are as follows:

- We suppose that there is an abnormality at I'.
- We propagate the abnormality of I' to its connected links.
- We select one link among the links connected to I', which is a part of the working LSP.

Having selected the link for the extension of RS, we determine the restoration scope with the same rule of equation (5)applied to the determination of the first RS. As a result of determination of the second RS, the RS can be *(d,L7,e,L11,f,L15,i,L20,j,L21,l)* as shown in Fig. 3 (b). Our algorithm can extend RS until I' is equal to the original ingress node. As we extend the restoration scope, we enhance the resource utilization and reduce the restoration speed. From this perspective, the RS is gradually widened.

2.2 An Alternative Path Computation

Once we determine RS, we should find an alternative path avoiding the abnormal link. In this section, we describe the algorithm to find an optimal alternative path within RS. This algorithm is very simple because it utilizes the weight in WNG. We traverse WNG from the egress node until we reach the intermediate ingress node in accordance with the following rules.

1. Select the link having the least weight among the links connected to a node.
2. If there are two or more links whose weights are equal, we select the link having the largest residual bandwidth ($B_{RESIDUAL} = B_{AVA} - B_{REQ}$).
3. If there are two or more links whose weights and the residual bandwidth are equal, we select the link having the least delay.
4. If there are two or more links whose weights, residual bandwidth, and delay are equal, we select an arbitrary link.

3 Performance Issues

In order to simulate the proposed restoration scheme and compare the restoration performance of our restoration scheme with two existing backup protection schemes, global backup [4,11] and reverse backup [5,11], we used the simple network topology as shown in Fig. 5.

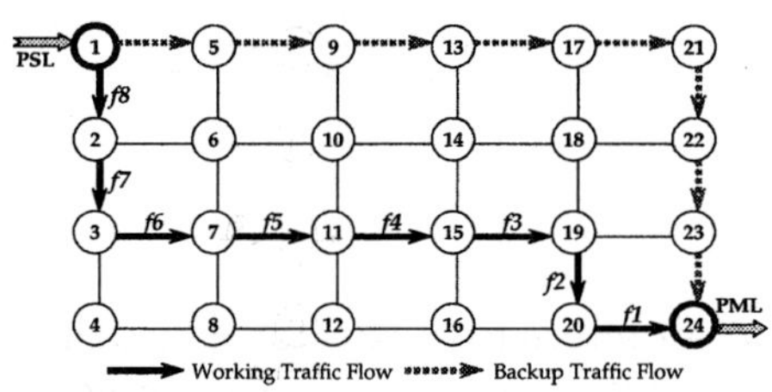

Fig. 5. A network topology for simulation

In this topology, there are 24 nodes and 38 links. With $LSP_{REQ}(1, 24, 24ms, 10Mbps)$, we create a working LSP traversing *(a-c-l-m-p)* and create a global backup path traversing *(1-2-3-7-11-15-19-20-24)* and a reverse backup path traversing *(1-5-9-13-17-21-22-23-24)*. Each node is connected with a duplex link with a 50Mbps bandwidth, 5ms delay and a CBQ queue. We use one pair of real-time traffic that is inserted into node *a* corresponding to the PSL and escapes through node *p* corresponding PML. We use a real-time traffic model for setting up LSPs from a working LSP, a global backup LSP and a reverse backup LSP with specific bandwidth requirements as the QoS traffic. For this, the traffic generator [10] generates 256-byte packets at 10 Mbps and at a constant bit rate. We define the 8 fault locations *f1, f2, f3, f4, f5, f6, f7* and *f8* along the working LSP to measure the restoration performance according to the different fault location. The measured performances are shown in Fig. 6.

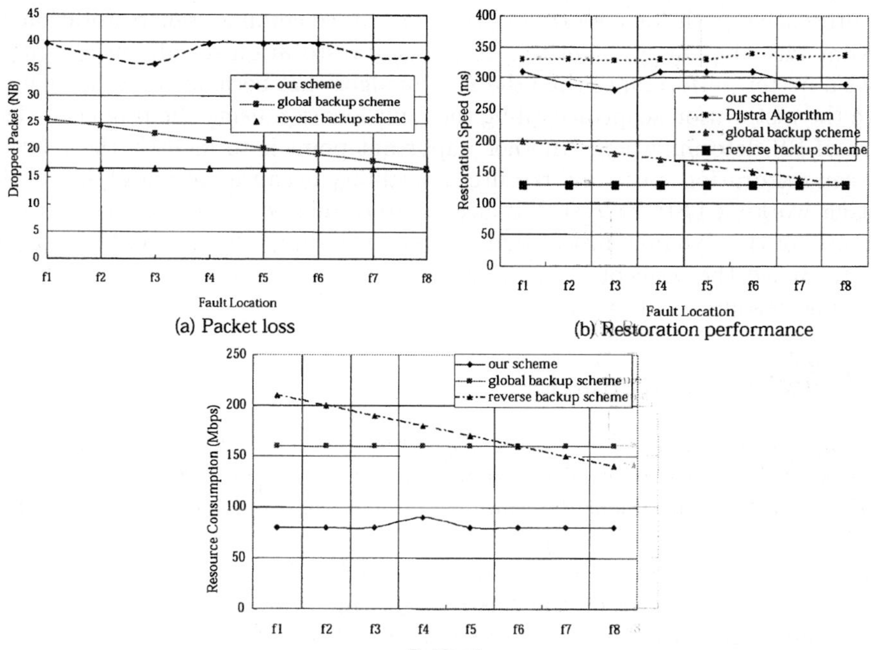

(a) Packet loss

(b) Restoration performance

(c) Resource consumption

Fig. 6. Restoration performance comparison

There are various performance evaluation criterions in relation to MPLS path restoration, such as recovery time, full restoration time, setup vulnerability, backup capacity, additive latency, reordering, state overhead, packet loss, and coverage [8]. Because our algorithm focuses on the maximization of the network resource utilization and the moderation of the restoration speed, we evaluate packet loss, resource utilization and restoration performance in the proposed restoration scheme. Fig. 6 (a) shows the packet loss depending on the four different fault locations along the working LSP. From the perspective of packet loss, our scheme shows the worst performance comparing with the two protection schemes of the global backup scheme and the reverse backup scheme, which is a natural result because our scheme dynamically restores the fault. Fig. 6 (b) shows the resource utilization performance at the eight different fault locations. In order to measure the resource utilization, we define three metrics, label, bandwidth and buffer used, for the working LSP and the two backup paths. Our scheme shows the best performance compared to other schemes in terms of resource utilization because our scheme finds the alternative path avoiding the fault location, taking into account the global network status. Fig. 6 (c) shows the restoration performance. If we compare our scheme with other protection schemes, it is a natural result that our scheme shows the worst performance compared with the protection schemes because our scheme dynamically finds the alternative path for restoration. However, if we compare our restoration algorithm with the Dijstra algorithm in terms of restoration speed, our algorithm shows a better performance than the *Dijstra* algorithm because our scheme minimize the restoration scope according the fault location using the proposed algorithm of RS determination. Another important point is to identify the relation between restoration speed and resource reusability because the reusability of the existing working LSP for restoration is entirely related to the restoration performance of the dynamic restoration scheme, including our scheme. In addition, we introduced the concept of RS extension in order to enhance the restoration speed and resource utilization.

4 Concluding Remarks

This paper has proposed a rerouting algorithm to enhance restoration speed and resource utilization compared to existing rerouting schemes. The algorithm for the dynamic RS adjustment determines the most reasonable candidate nodes or links to be applied to the restoration and assigns the appropriate weights to the candidate nodes and links. Our algorithm showed a higher performance than the rerouting approach based on the Dijstra algorithm in terms of restoration speed and resource utilization. On the other hand, we can enhance the restoration speed with the concept of the dynamic adjustment of the restoration scope that minimizes the complexity of network topology which will be used for restoration. Also, by the extension of the RS, we reduced the restoration failure probability. Our restoration scheme showed that the fault location directly affected the restoration speed and the resource reusability of the working LSP. On the basis of the performance evaluation results, we concluded that our scheme can be applicable for the protection of bronze-class working LSPs and all kinds of backup LSPs, such as the global backup LSP and the reverse working LSP.

References

1. E. Rosen, A. Viswanathan, and R. Callon, "Multiprotocol Label Switching Architecture," IETF RFC3031, 2001.
2. D. Awduche, J. J. Malcolm, J. Agogbua, M. O'Dell and J. McNabus, "Requirements for Traffic Engineering over MPLS," IETF RFC2702, 1999.
3. D. Awduche et al., "Requirements for Traffic Engineering Over MPLS," IETF REC 2702, 1999.
4. C. Huang, V. Shrma, S. Makam, Ken Owens, "A Path Protection/Restoration Mechanism for MPLS Networks," Internet Draft, draft-chang-mpls-path-protection-02.txt, 2000.

5. D. Haskin, R. Krishnan, "A Method for Setting an Alternative Label Switched Paths to Handle Fast Reroute," Internet Draft, draft- haskin-mpls-fast-reroute-05.txt, 2000.
6. V. Sharma, B.M. Crane, K. Owens, C. Huang, F.Hellstrand, B. Cain, S. Civanlar and A. Chiu, "Framework for MPLS-based Recovery," Internet Draft, draft-ietf-mpls-recovery-frmwrk-03.txt, 2001.
7. C. Huang, V. Sharma, K. Owens, and S. Makam, "Building Reliable MPLS Networks Using a Path Protection Mechanism," IEEE Communications Magazine, pp156-162, March 2002.
8. G. Ahn and W. Chun, "MPLS Restoration Using Least-Cost Based Dynamic Backup Path," ICN2001, Springer LNCS 2094, pp319-328, 2001.
9. E. Harrison, A. Farrel, and B. Miller, "Protection and Restoration in MPLS Networks," Data Connection (http://www.dataconnection.com), October 2001.
10. Helsinki University of Technology, "QoS Routing Simulator (QRS) Version2.0," Available at http:://www.tct.hut.fi/ pgzhang/QRS/index.html.
11. G. Ahn and J. Jang, "An Efficient Rerouting Scheme for MPLS-based Recovery and Its Performance Evaluations," Telecommuincation Systems, Vol. 3, pp.481-495, 2002.

QoS-Aware and Group Density-Aware Multicast Routing Protocol

Hak-Hu Lee, Seong-Chung Baek, Dong-Hyun Chae, Kyu-Ho Han, and Sun-Shin An

Department of Electronics Engineering, Korea University, 1,5-Ka, Anam-dong Sungbuk-ku, Seoul, 136-701, Korea
{hhlee, scbaek, hsunhwa, garget, sunshin}@dsys.korea.ac.kr

Abstract. Traditionally, there is an assumption that the QoS-aware multicast routing protocol acquires QoS information via the QoS-aware unicast routing protocol. Mostly, the research focuses on about managing group dynamics and failure recovery on multicast and there is no interest in the QoS-aware unicast routing protocol for multicast. In fact, researches of the QoS-aware multicast based on QoS-aware unicast are uncommon. Hence, we focus on the QoS-aware unicast routing algorithm and apply it to multicast. In this paper, we use a novel algorithm, called QoS Restricted and Distributed Generic Shortest Path Algorithm (QRDGSPA). As it is a very simple and measured method without complex computation Hence, it can make a shortest path or QoS-aware multiple paths. For this purpose, we use a specific routing algorithm, Multicast Candidate Selection Method (MCSM), which can choose the optimal candidate path from a receive node to a candidate node. Based on integration of QRDGSPA and MCSM, we propose QoS-aware and group Density-aware Multicast Routing Protocol (QDMRP).

1 Introduction

In recent years, global Internet and data traffic including voice, video, multimedia data, etc has been exponentially expanding. Although single-direction unicast data was predominant in the past, the importance of multicast, such as teleconferencing, tele-education, and video-conferencing has recently surged.

Currently, multicast routing protocols such as CBT, PIM, and DVMRP [2] compute the shortest path based hop count without regarding QoS over a single path. On the other hand, QoS-aware multicast routing protocols such as YAM (spanning join) [3], QoSMIC [4], and QMRP [5] consider various constraints on multiple path. Before considering QoS-aware multicasting, we will review methods to obtain and manage in networks. If we can get QoS information by means of routing protocols, it is easy to treat QoS-aware multicast. Therefore, we need to study routing protocols above all. The QoS-aware unicast routing protocol is essential to treat QoS-aware multicast.

There are two basic approaches of IP multicast routing protocols according to the expected distribution of multicast group members throughout the network [2]. One is called dense-mode which is based on the assumptions that the

C. Kim (Ed.): ICOIN 2005, LNCS 3391, pp. 244–253, 2005.

multicast group members are densely distributed throughout the network and the bandwidth is plentiful. Examples of this approach are DVMRP, PIM-DM, and MOSPF [6]. The other is called sparse-mode which assumes that the multicast group members are sparsely distributed throughout the network and that bandwidth is not necessarily widely available. Some examples of this approach are PIM-SM and CBT.

A multicast group can have multiple sources and the distribution of the packets can be done in two ways. First, each source can create its own distribution tree, called a source-based tree, with itself as the root. Second, all sources can distribute their packets using the same tree called a shared tree. Source-based trees have better end-to-end performance and distribute the traffic of each group across the network. However, this approach leads to large routing tables. Examples of it are DVMRP, PIM-DM, and MOSPF.

On the other hand, shared trees concentrate a traffic group within a few links in the network. This concentration is bad in the case where all sources are active simultaneously, but can be beneficial when sources take turns transmitting. PIM-SM and CBT are examples of shares trees. The two approaches have complementary behavior and are both useful depending on the situation. According to the above description, group density and tree type are strongly interrelated. If group density is high, source-based shortest path tree is useful. If group density is sparse mode, shared tree is more efficient than source-based tree.

In this paper, we introduce a special routing algorithm, Quality Restricted and Distributed Generic Shortest Path Algorithm (QRDGSPA) [1], and use several of candidate selection methods in multicast. Using these, we propose QoS-aware and group Density-aware Multicast Routing Protocol (QDMRP). If we know the QoS of the new joining path and group density, joining any multicast group is very simple and easy. Hence, it is possible to save network resources and improve network utilization. Finally, we can achieve QoS-aware, Density-aware, efficiency, and robust multicasting. We trust that our proposed protocol is a useful and powerful algorithm.

The remainder of the paper is organized as follows. Section 2 describes our proposed scheme, QoS-aware and group Density-aware Multicast Routing Protocol (QDMRP) architecture in detail. Section 3 illustrates the complexity of QDMRP with analytical methods. Section 4 shows simulation results of our proposed protocol. Finally, Section 5 provides our conclusions and future research direction.

2 The Proposed Scheme

Researchers who have studied QoS-aware multicast routing protocol simply assume that QoS-aware multicast routing protocols get QoS information in the network via unicast routing protocol. Hence, they are not interested in the unicast routing protocol. However, we strongly focus on QoS-aware unicast routing algorithm and seek to apply it to multicast. For this purpose, we use a specific routing algorithm that can process the QoS-aware metric and support the optimized path. According to a novel routing algorithm, when a node joins to a multicast group, if it knows the multicast group density, it joins quickly resulting in low control overhead. When the node joins the candidate nodes, if it finds the best QoS path, it is possible to optimize the management path and save network resources, enabling efficient network. Therefore, we propose a novel multicast routing protocol applying the method of multicast candidate selection.

2.1 QoS Restricted and Distributed Generic Shortest Path Algorithm

Each node sends routing information involving node and QoS information to neighboring nodes via interfaced link. The neighboring node saves this information to its routing table, accumulates its QoS value. We call such accumulated QoS as Total_QoS, which inserts itself to the node information, it is Nodelist, and sends the routing information to its neighboring nodes. As a consequence, all nodes know the path and QoS information of the other nodes.

Each node checks the node information to prevent "Counting to infinity". If each node is found in the Nodelist which is received from neighboring nodes, there must be a loop. In this case, the received information is unnessary therefore discarded. Intermediate nodes, they can receive several routing information that has the same source node in node information and has the same sequence number including different Total_QoS in QoS information. In that case, the intermediate node should decide whether later routing information has superior Total_QoS or not. If it is better Total_QoS, the intermediate node saves this information to its routing table, adds its routing information to the received routing information and sends the updated routing information to its neighboring nodes. Otherwise, the node can discard the received routing information to reduce control overhead, resulting in saved processing power and resources. However, this is depends on Policy.

Using the routing table, the forwarding table is created. Each node can easily take the optimized route in this table when the node computes a route for new data traffic. If there are multiple paths that satisfy the required QoS, the node can choose the shortest path or multiple paths, depending on the policy. According to circumstances, some paths that have minimum hops and satisfy the required QoS are selected, not necessarily the shortest paths. This is also depends on Policy.

Even if the shortest path is used exclusively, if failure occurs on that path, rerouting to other multiple paths is possible. Therefore, fault recovery can be fast. Hence, if data traffic load is shared by multiple paths, we can efficiently manage the network and speed up the data transmission. Actually, each node is a source node, sending routing information to its neighboring nodes. Neighboring nodes are both intermediate and destination nodes. Neighboring nodes save the routing information to their routing tables and send the updated routing information to their neighboring nodes. Consequently, all nodes in the network know the other QoS metrics value and they can be optimal routed without complex computation. It is a measured method where the nodes simply accumulate QoS value and insert node information.

QRDGSPA can be hierarchically extended and is applied to various types of networks as autonomous systems, domains, and areas. Hence, it is useful for QoS-aware multicast. The network consists of a source node that generates the routing information, intermediate node that handles and relays the routing information, and destination node that processes the routing information. As mentioned above, each node can be a source node, intermediate node, and destination node concurrently. Source node generates the originated routing information and sends it to its neighboring nodes. Intermediate nodes check Total_QoS and Nodelist in the received routing information. If it is useful, intermediate nodes save that information to its routing table, updates the routing information and sends it to its neighbor nodes. Otherwise, the received routing information is discarded. All nodes measure their available resources. When they generate originated routing information or update the received routing information, they apply these available resources. If the received routing information is useful, the node sends the updated routing information to its neighboring nodes except the transmitting node.

2.2 QoS-Aware and Group Density-Aware Multicast Routing Protocol

In existing QoS-aware multicast routing protocols, YAM has a large overhead because it broadcasts request-messages to other nodes to search for candidate nodes. QoSMIC needs an extra manager for tree search and has difficulty in deploying different Autonomous Systems. Hence, QMRP may increase overhead in the case of multiple path. However, QoS-aware and Density-aware Multicast Routing Protocol (QDMRP) does not need an additional manager, which satisfies various kinds of QoS constraints, making possible efficient multicast by using density-aware search and QRDGSPA.

QDMRP has two searching procedures, similar to a QoSMIC, local search and tree search. However, QDMRP is based on QRDGSPA as a routing protocol using MCSM and does not need an extra manager. The receive node first finds the candidate node within a limited scope by using time-to-live (TTL). If the receive node finds the candidate node which is satisfies the required QoS for the given multicast group, a new candidate path is created and the receive node joins the candidate node (Fig. 1-(a)). If local search failes, the receive node try to do a tree search (Fig. 1-(b)). The receive node requests to join the source node. According to MCSM operation, the receive node recognizes candidate node and joins the multicast group. When an existing member leaves the multicast group, it sends a PRUNE message to the multicast tree to leave the branch. As shown in Fig. 1, the detailed description of the QDMRP operation is as follow, in case of local-search.

1. If new member joins, the receive node broadcasts a JOIN-REQUEST message with specified time-to-live (TTL) occurs.

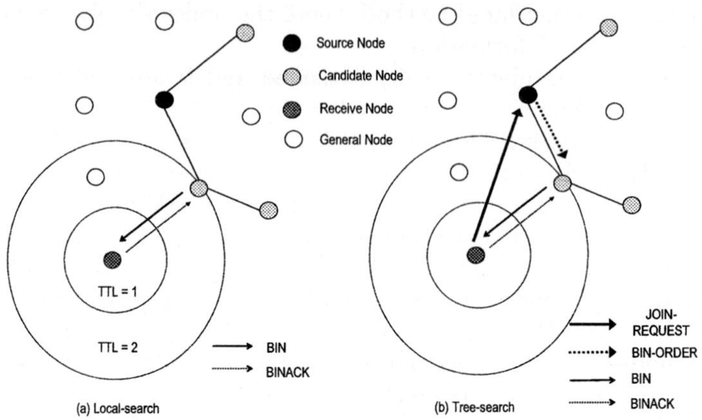

Fig. 1. Overview of QDMRP

2. If there are candidate nodes that are within TTL, they unicast BID message to the receive node. BID messages contain Total_QoS and Nodelist information.
3. If the receive node receives a BID message, the receive node chooses the candidate node that has the best Total_QoS and sends an ACK message back to the candidate node.
4. If the receive node does not receive the BID message within the specific time, the receive node tries to a tree search.

QDMRP generally uses tree-search. After a JOIN-REQUEST message transmission, this message traverses to the source node. Before arriving at the source node, if there is a candidate node, the candidate node can process this JOIN-REQUEST message. Although a candidate has not received BID-ORDER message from the source node, the candidate node can send a BID message to the receive node. We name this process QDMRP2. The detailed description of the QDMRP operation is as follows, in the case of tree- search.

1. The receive node sends JOIN-REQUEST message to the source node (or to the core node).
2. Using MCSM, the source node chooses candidate node.
3. The selected candidate node unicasts BID message to the receive node. BID message contains Total_QoS and Nodelist information. In some cases, BID message can pass through another candidate node. At this time, the candidate node that receives BID message interrupts the BID message and sends its own BID message to the receive node. This is very useful and reasonable technique to supply better Total_QoS.
4. After receiving BID message, the receive node sends ACK message back to the candidate node.

3 The Analysis of QDMRP

To further illustrate the efficiency of our proposed protocol, we compare QDMRP with several existing multicast routing protocols, which are YAM and QoSMIC, in term of control overhead complexity. Table 1 [7] shows the results of control overhead complexity.

Table 1. Complexity comparison of QoS-aware multicast routing protocols

Routing Protocol	Control message overhead
YAM	$\sum_{i=1}^{t'} \omega(\omega-1)^{i-1} + (c+1) \cdot H_{op_c}, t' >> t$
QoSMIC-Distr.	$\omega \cdot (\omega-1)^{t-1} + H_{op_s} + \lvert T \rvert + (c+1) \cdot H_{op_c}$
QoSMIC-Centr.	$\omega \cdot (\omega-1)^{t-1} + H_{op_s} + c \cdot H_{op_{avg}} + (c+1) \cdot H_{op_c}$
QDMRP	$\omega \cdot (\omega-1)^{t-1} \cdot p + (H_{op_s} + H_{op_b} + 2 \cdot H_{op_c}) \cdot (1-p)$

The abbreviations used in Table 2 are listed below:
t = The maximum TTL value of the local-search (default t = 2),
ω = The average degree of a node,
c = The number of candidate nodes,
$|T|$ = The average size of a multicast tree,
$H_{op_{avg}}$ = The average hop count between Manager and candidate nodes,
H_{op_c} = The average hop count from a new node to the candidate node,
H_{op_s} = The average hop count from a new node to the source node,
H_{op_b} = The average hop count from the source node to the candidate node,
p = The success probability of local-search.

YAM only uses local-search to search for a candidate node; therefore, the complexity is very large. In the case of local-search, the complexity of the above protocols except YAM is the same. However, in the case of tree-search, the complexities are different. Of course, the complexity of QDMRP is less than the others. Since QoSMIC needs a manager for tree-search, there are many control messages. However, QDMRP does not need a manager and so QDMRP saves the control messages between the manager and candidate nodes. In QDMRP, hops mean the number of messages that is join message which the receiving node and the new joining node, sends to the source node. Hence, 2 * H_{op_c} represents BID message and BIDACK message between the selected candidate node and the receiving node.

4 Performance Evaluation

In this section, we evaluate QDMRP and compare it with other multicast protocols: YAM and QoSMIC. We conduct a performance evaluation of our proposed protocol (QDMRP) through simulations using Network Simulator 2 (ns-2) that has been developed by VINT group [10]. We also use BRITE as topology generator developed at Boston University [9] that supplies various kinds of network topology. We choose flat random Waxman's probability model [8]. To ensure fairness, we use a unicast routing algorithm applied QRDGSPA while we simulate with other QoS-aware multicast protocols.

In this paper, we evaluate the performance of QoS-aware multicast protocols by three metrics: control message overhead, average join latency and average path length. Hence, we examine this evaluation in sparse mode and dense mode.

4.1 Network Topology

We evaluate the performance of our proposed protocol using network topology applied artificial graphs. It is a number of flat-random Waxman graphs and random topology using Waxman's probability model for interconnection among nodes over network topology, which is given by:

$$P(u,v) = \alpha \cdot e^{\frac{-d}{(\beta L)}}$$

where $0 < \alpha,\ \beta \leq 1$, d is the Euclidean distance from node u and node v, and L is the maximum distance between any two nodes. Table 2 shows the parameters of flat random Waxman topology. The network topology consists of nodes between 50 and 300 nodes and hundreds of links. We simulate performance evaluation according to the changing number of nodes.

Table 2. Parameters of flat random Waxman topology

Parameter	Meaning	Values
HS	Size of one side of the plane	≥ 1
N	Number of nodes	$1 \leq$ N $\leq$ HS * HS
α	Waxman-specific exponent	$0 < \alpha \leq 1, \alpha \in \mathbf{R}$
β	Waxman-specific exponent	$0 < \beta \leq 1, \beta \in \mathbf{R}$
Node Placement	Placement of nodes in the plane	random or heavy-tailed

4.2 Simulation Results

QDMRP Family We consider three types of QDMRP for evaluation. First, we evaluate QDMRP that uses Centralized Candidate Selection. Second, we evaluate QDMRP2 which is an improved version of QDMRP and is described in the previous section. Finally, we evaluate the distributed version, QDMRP-Distr., which uses Distributed Candidate Selection. Hence, we evaluate the QDMRP family on sparse mode. Performance results are presented in Fig. 2. QDMPR-Distr. is worst and QDMRP2 is best. They show similar results in message overhead. However, Average Join is very different.

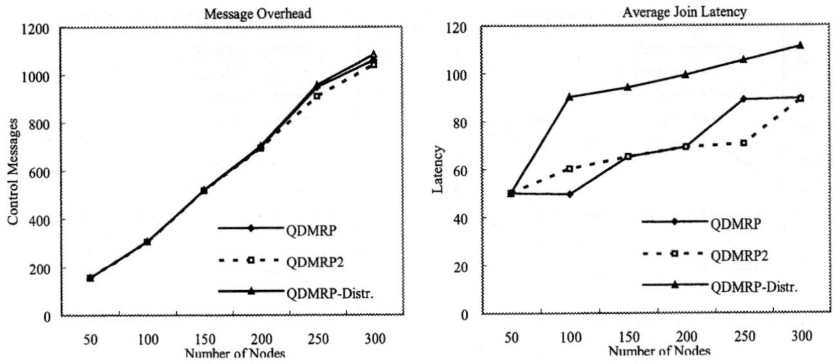

Fig. 2. Message Overhead and Average Join Latency of QDMRP family

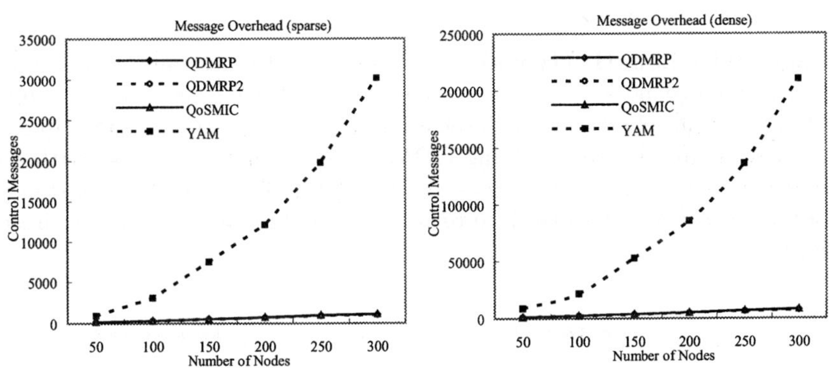

Fig. 3. Message Overhead of sparse and dense mode

Message Overhead We compare message overhead of QDMRP with other protocols, which are QoSMIC-centr. and YAM. For sparse mode, we determine that 10% of total nodes join their multicast group in sparse mode, and 80% of total nodes joined in dense mode. In Fig. 3, the control message of YAM increases dramatically. Therefore, we additionally examine QDMRP, QDMRP2 and QoSMIC, except YAM. Performance results of QDMRP2 are better than the others in sparse and dense mode. The difference in performance increases with increasing node number. In both cases, sparse and dense mode, the results show a similar pattern.

Average Join Latency Join latency indicates the period between the time when the receive node sends JOIN-REQUEST and the time when the source node or the candidate node receives BIDACK message. As shown in Fig. 4, the average join latency of QoSMIC is worst in both sparse and dense modes. Consequently, QDMRP and QDMRP2 show better performance than YAM or QoSMIC.

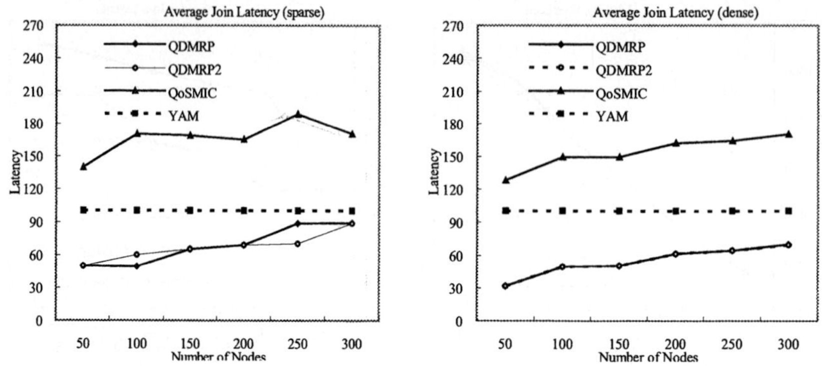

Fig. 4. Average Join Latency of sparse and dense mode

Average Path Length (Hops) Average hops is the average path length between the receive node and the source node. QDMRP uses the strength of QRDGSPA effectively during path selection. QRDGSPA supports multiple-paths that satisfy required QoS, enabling QDMRP to select the shortest path among those multiple paths. For this reason, the QDMRP family has shorter paths than other protocols. Performance results on average path length are presented in Fig. 5.

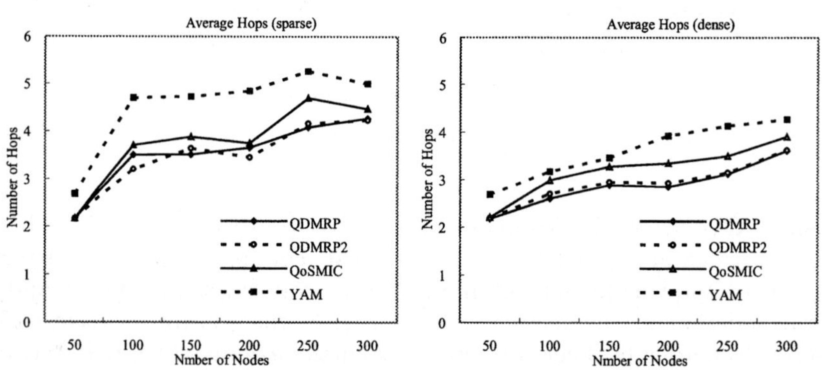

Fig. 5. Average Hops of sparse and dense mode

5 Conclusion

In this paper, we use QRDGSPA as a routing algorithm because each node easily recognizes the optimized QoS-aware path from itself to each other node and the path can be either single or multiple. QRDGSPA is a quantifiable measured method without complex computation. Furthermore, it can improve performance and can process exactly some data that is assumed in other QoS-aware multicast protocols. We introduce MCSM (Multicast Candidate Selection Method), the first is a centralized method and the second is a distributed method. The centralized method is efficient and takes full advantage of QRDGSPA. Finally, we propose a QDMRP applied QRDGSPA considering QoS and multicast group density.

We present the QDMRP family. The first, QDMRP-Distr. is a distributed version, and the second is QDMRP2, an applied centralized version. It is classified by MCSM. In the evaluation result, QDMRP2 shows the best performance among them. QDMRP saves network resources and improves network utilization by decreasing control message overhead, shortening join latency and reducing the path length. When a new node joins any multicast group, as mentioned above, the three performance metrics result in performance improvement.

Since QDMRP uses the routing information of QRDGSPA, a manager node that uses QoSMIC protocol is unnecessary. When a new joining node, the receive node, requests to join the source node during tree-search, the source node recognizes each Total_QoS from on-tree node to the receive node. Hence, the source node selects a specific on-tree node as candidate node. As a result, QDMRP achieves QoS-aware, Density-aware, efficiency, and robust multicasting.

After this research, we would like to study inter-domain multicast and additional types of network like Ad-Hoc wireless network, MPLS, and so on.

References

1. S. An: A New Generic Shortest Path Algorithm. Technical Report, UC, Canada, Oct. 1998
2. A. Striegel and G. Manimaran: Survey of QoS Multicasting Issues. IEEE Communications Magazine, June 2002
3. K C. and J. C.: Building shared trees using a one-to-many joining mechanism. ACM Computer Communication. Review, pp. 5-11, Jan. 1997
4. S. Y. and M. F.: QoS-Aware Multicast Routing for the Internet: The Design and Evaluation of QoSMIC. IEEE/ACM Transactions on networking, vol. 10, no. 1, February
5. S. Chen, K. Nahrstedt, and Y. Shavitt: A QoS-Aware multicast routing protocol. in Proc. IEEE INFOCOM, vol. 3, pp. 1594-1603, Mar. 2000
6. John Moy: Multicast routing extensions for OSPF. ACM Communication, vol. 37, no. 8, pp.61-66, 1994
7. C. Tseng and C. Chen: The Performance of QoS-aware IP Multicast Routing Protocols. IEEE, 2001
8. Kenneth L. Calvert, Matthew B. Doar, and Ellen W. Zegura: Modeling Internet Topology. IEEE Communications Magazine, June 1997
9. A. Medina, A. Lakhina, I. Matta, and J. Byers: BRITE: Universal Topology Generation from a User's Perspectiv. Apr. 2001
10. University of Berkeley: NS2 network simulator, http://www.isi.edu/nsnam/ns2

A Minimum Cost Multicast Routing Algorithm with the Consideration of Dynamic User Membership

Frank Yeong-Sung Lin, Hsu-Chen Cheng, and Jung-Yao Yeh

Department of Information Management, National Taiwan University
50, Lane 144, Keelung Rd., Sec.4, Taipei, Taiwan, R.O.C.
hccheng@ieee.org

Abstract. In this paper, we attempt to solve the problem of constructing a minimum cost multicast tree with the consideration of dynamic user membership. Unlike the other minimum cost multicast tree algorithms, this problem consists of one multicast group of fixed members and each destination member is dynamic and has a probability of being active as which was gathered by observation over some period of time. With the omission of node join/leave handling, this model is suitable for prediction and planning purpose than for online maintenance of multicast trees. We formally model this problem as an optimization problem and apply the Lagrangean relaxation method and the subgradient method to solve the problem. Computational experiments are performed on regular networks and random networks. According to the experiment results, the Lagrangean based heuristic can achieve up to 37.69% improvement compared to the simple heuristic.

1 Introduction

The power of Internet comes from its openness that interconnects computers around the world as long as they follow the protocols. After about one decade of continuous development, this global network has somehow revolutionized the way people communicate and the way businesses are done. However the application involving online audio and video require higher quality of transmission and may consume much more bandwidth over its transmission path, therefore it's worthwhile that we pay more attention to the problems that were aroused by such applications.

A very common scenario is that a source may try to send data to a specific group of destinations, for example a server of video streaming service sending its video stream to all of its service subscribers. Such traffic group communication is called multicast, as opposed to unicast and broadcast. The multicast traffic over IP often follows the route of a spanning tree over the existing network topology, called a multicast spanning tree, taking advantages of sharing common links over paths destined for different receivers. The efficiency of multicast is achieved at the cost of losing the service flexibility of unicast, because in unicast each destination can individually negotiate the service contract with the source. ¿From

C. Kim (Ed.): ICOIN 2005, LNCS 3391, pp. 254–263, 2005.

the viewpoint of network planning, each link in the network can be assigned with a cost, and the problem of constructing a multicast spanning tree with its cost minimized is called Steiner tree problem, which is known to be NP-complete. Reference [1] and [2] surveyed the heuristics of Steiner tree algorithms.

¿From the multicast protocols surveyed in [3], we can see that most complexity of these protocols comes from dealing with the changing of group members, that is, the joining and leaving of nodes. The motivation of this paper would be creating a mechanism for finding and evaluating the cost-efficiency of a multicast tree with a given network and fixed set of group members. Also the group members are dynamic in that they might shut-off for a while, and turn on later. Such probability may be acquired by observation of user behavior over a certain a period of time.

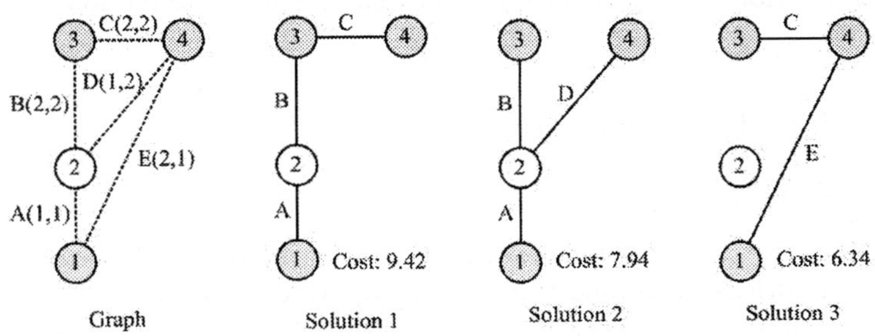

Fig. 1. Example network

Consider the network in Figure 1 with node 1 as the source and nodes 3 and 4 as the destinations which have active probabilities 0.7 and 0.8 respectively. The connection setup costs and transmission costs of the links are labeled in the parentheses beside the links. Figure 1 shows three possible solutions to construct the multicast tree. Consider the solution 1, because nodes 3 and 4 have active probabilities 0.7 and 0.8, the probability that links A and B have no traffic are 0.06. The probability that link C has traffic is 0.8. Consequently, the total cost of solution 1 is 9.42. The cost of solution 2 and 3 are 7.94 and 6.34 respectively. The details of the result are shown in Table 1.

In this paper, however, we do not deal with the complexity node joining and leaving in our heuristic, instead, the activity for a node in summarized as a probability. Therefore, the model proposed here tends to be of analytical and planning use. Still, the problem of multicasting has strong connection with the Steiner tree problem, which is a NP-complete problem, the approach of Lagrangean relaxation is taken to achieve accurate approximation with significantly reduced computation time.

The rest of this paper is organized as follows. In Section 2, we formally define the problem being studied, as well as a mathematical formulation of min-cost

Table 1. Total cost of example network

Cost	Link A	Link B	Link C	Link D	Link E	Total
Solution 1	1+1×(1-(0.3×0.2))	2+2×(1-(0.3×0.2))	2+2×(1-0.2)	x	x	9.42
Solution 2	1+1×(1-(0.3×0.2))	2+2×(1-0.3)	x	1+2×(1-0.2)	x	7.94
Solution 3	x	x	2+2×(1-0.3)	x	2+1×(1-(0.3×0.2))	6.34

optimization is proposed. Section 3 applies Lagrangean relaxation as a solution approach to the problem. Section 4, illustrates the computational experiments. Finally, in Section 5 we present our conclusions and the direction of future research.

2 Problem Formulation

2.1 Problem Description

In this paper we consider, for a network service provider, the problem of constructing a multicast spanning tree that sends traffic to receivers (destinations), while at the same time, the total cost resulted by the multicast tree is minimized. The network is modeled as a graph where the switches are depicted as nodes and the links are depicted as arcs. A user group is an application requesting transmission in this network, which has one source and one or more destinations. Given the network topology and bandwidth requirement of every destination, we want to determine the routing assignment (a tree for multicasting or a path for unicasting) of the user group.

By formulating the problem as a mathematical programming problem, we intend to solve the issue optimally by obtaining a network that will enable us to achieve our goal, i.e. one that ensures the network operator will spend the minimum cost on constructing and servicing the multicast tree. The notations used to model the problem are listed in Table 2.

Each destination $d \in D$ has a given probability Q_d that indicated the fraction of time that the destination is active, and thus the traffic is to be routed to that node. Such probability may be acquired by observation of user behavior over a certain a period of time. The cost associated with a link consists of two parts: 1) fixed cost of connection setup and 2) transmission cost proportional to link utilization. At the determination of the multicast tree, utilizations for all links may be computed, which are used to estimate the total cost.

2.2 Mathematical Formulation

According to the problem description in pervious section, the min-cost problem is formulated as a combinatorial optimization problem in which the objective

Table 2. Description of Notations

Notation	Description
	Given Parameters
Notation	Description
D	The set of all destinations of multicast group
r	The source of multicast group
N	The set of all nodes in the network
L	The set of all links in the network
I_i	The set of all incoming links to node i
q_d	The probability that the destination d is active
a_l	The transmission cost associated with link l
b_l	The connection maintenance cost associated with link l
P_d	The set of all elementary paths from r to $d \in D$
δ_{pl}	The indicator function which is 1 if link is on path p
	Decision Variables
Notation	Descriptions
y_l	1 if link l is included in the multicast tree and 0 otherwise
x_p	1 if path p is included in the multicast tree and 0 otherwise
g_l	The fraction of time that the link l is active on the multicast tree
f_{dl}	1 if link l is used by destination $d \in D$ and 0 otherwise

is to minimize the total cost associated with the multicast tree, including the accumulated transmission costs (pay per time unit) and setup cost (pay per connection) on each link used.

Objective function (IP):

$$Z_{IP} = \min \textstyle\sum_{l \in L} (b_l y_l)(a_l g_l) \quad . \tag{1}$$

subject to:

$$g_l \geq 1 - \textstyle\prod_{d \in D} (1 - q_d f_{dl}) \qquad \forall l \in L \ . \tag{2}$$

$$\textstyle\sum_{l \in I_i} y_l \leq 1 \qquad \forall i \in N - \{r\} \ . \tag{3}$$

$$\textstyle\sum_{l \in I_r} y_l = 0 \ . \tag{4}$$

$$\textstyle\sum_{p \in P_d} \delta_{pl} x_p \leq f_{dl} \qquad \forall l \in L, \forall d \in D \ . \tag{5}$$

$$\textstyle\sum_{p \in P_d} x_p = 1 \qquad \forall d \in D \ . \tag{6}$$

$$f_{dl} \leq y_l \qquad \forall l \in L, \forall d \in D \ . \tag{7}$$

$$f_{dl} = 0 \text{ or } 1 \qquad \forall l \in L, \forall d \in D \ . \tag{8}$$

$$y_l = 0 \text{ or } 1 \qquad \forall l \in L \ . \tag{9}$$

$$x_p = 0 \text{ or } 1 \qquad \forall p \in P_d, \forall d \in D \ . \tag{10}$$

$$0 \leq g_l \leq 1 - \textstyle\prod_{d \in D} (1 - q_d) \qquad \forall l \in L \ . \tag{11}$$

The objective function of (1) is to minimize the construction cost and total transmission cost of servicing the maximum bandwidth requirement destination through a specific link for the multicast group.

Constraint (2) is referred to as the utilization constraint, which defines the link utilization as a function of q_d and f_{dl}. Since the objective function is strictly an increasing function with g_1 and (1) is a minimization problem, each g_1 will equal the aggregate flow in an optimal solution. Constraints (3) and (4) are both tree constraints. Constraint (3) requires that the number of selected incoming links y_1 to node is less than 1 , while constraint (4) requires that there are no selected incoming links y_1 to the node that is the root of multicast group. Constraint (5) and (6) require that only one path is selected for each multicast source-destination pair. Constraint (7) requires that if link l is not included in the multicast tree, then it won't be used by any destination.

Furthermore, here is an example of many possible extensions that could be made to this problem but not discussed in this paper. Say the dependency among destinations, e.g., the members of the group can be further divided into subgroups such that the group members within each subgroup behave identically. The link utilization can be modeled as follows:

$$g_l = 1 - \prod_{m \in G} \left(1 - q_m \left(1 - \prod_{i \in M_m} (1 - f_{il})\right)\right) \quad . \tag{exp}$$

Where G is the set of subgroups

As you may notice that the structure of this formula resembles the constraint for link utilization of constraint (1), with its f_{dl} replaced with $\left(1 - \prod_{i \in M_m} (1 - f_{il})\right)$.

3 Solution Approach

3.1 Lagrangean Relaxation

Lagrangean methods were used in both the scheduling and the general integer programming problems at first. However, it has become one of the best tools for optimization problems such as integer programming, linear programming combinatorial optimization, and non-linear programming [4] [5].

By using the Lagrangean Relaxation method, we can transform the primal problem (IP) into the following Lagrangean Relaxation problem (LR) where Constraints (2), (5) and (7) are relaxed.

Optimization problem (LR):

$$\begin{aligned} Z_d(\alpha, \beta, \theta) = \min & \textstyle\sum_{l \in L} (b_l a_l + a_l g_l) \\ & + \textstyle\sum_{l \in L} \alpha_l \left(\sum_{d \in D} \log(1 - q_d \cdot f_{dl}) - \log(1 - g_l)\right) \\ & + \textstyle\sum_{l \in L} \sum_{d \in D} \beta_{dl} \left(\sum_{p \in P_d} \delta_{pl} \cdot x_p - f_{dl}\right) \\ & + \textstyle\sum_{l \in L} \sum_{d \in D} \theta_{dl} (f_{dl} - y_l) \quad . \end{aligned} \tag{12}$$

subject to: (3) (4) (6) (8) (9) (10) and (11).

Where $\alpha_l, \beta_{dl}, \theta_{dl}$ are Lagrangean multipliers and $\beta_{dl}, \theta_{dl} \geq 0$. To solve (12), we can decompose (12) into the following four independent and easily solvable optimization subproblems.

Subproblem 1: (related to decision variable x_p)

$$Z_{SUB1}(\beta) = \min \sum_{d \in D} \sum_{p \in P} \left(\sum_{l \in L} \beta_{dl} \cdot \delta_{pl} \right) \cdot x_p \ . \tag{13}$$

subject to: (6) (10).

Subproblem 1 can be further decomposed into $|D|$ independent shortest path problems with nonnegative arc weights β_{dl}. Each shortest path problem can be easily solved by Dijkstra's algorithm.

Subproblem 2: (related to decision variable y_l)

$$Z_{SUB2}(\theta) = \min \sum_{l \in L} \left(b_i - \sum_{d \in D} \theta_{dl} \right) \cdot y_l \ . \tag{14}$$

subject to: (3) (4) (9).

The algorithm to solve Subproblem 2 is:

Step 1 Compute the number of negative coefficients $\left(b_l - \sum_{d \in D} \theta_{dl} \right)$ for all links.

Step 2 Sort the links in ascending order according to the coefficient.

Step 3 According to the order and complying with constraints (3) and (4), if the coefficient is less than zero, assigns the corresponding negative coefficient of y_1 to 1 and 0 otherwise.

Subproblem 3: (related to decision variable g_l)

$$Z_{SUB3}(\alpha) = \min \sum_{l \in L} \left(a_l g_l - \alpha_l \cdot \log(1 - g_l) \right) \ . \tag{15}$$

subject to: (11).

This subproblem of minimization can be solved by substituting with its lower and upper bound because the minimum of this function appears at endpoints.

Subproblem 4: (related to decision variable f_{dl})

$$Z_{SUB4}(\alpha) = \min \sum_{l \in L} \sum_{d \in D} \left(\alpha_l \log(1 - q_d \cdot f_{dl}) + (\theta_{dl} - \beta_{dl}) f_{dl} \right) \ . \tag{16}$$

subject to: (8).

This subproblem of minimization can be solved by simply substitute f_{dl} with 0 and 1 and keep the one that yields the minimum.

According to the weak Lagrangean duality theorem [6], for any $\beta_{dl}, \theta_{dl} \geq 0$, $Z_D(\alpha_l, \beta_{dl}, \theta_{dl})$ is a lower bound on Z_{IP}. The following dual problem (D) is then constructed to calculate the tightest lower bound.

Dual Problem (D):

$$Z_D = \max Z_D(\alpha_l, \beta_{dl}, \theta_{dl}) \tag{17}$$

subject to:

$$\beta_{dl}, \theta_{dl} \geq 0$$

3.2 Getting Primal Feasible Solutions

During solving the dual problem, a simple algorithm is needed to provide an adequate initial upper bound of the primal problem Z_{IP}. Dijkstra algorithm is used to generate a minimum cost spanning tree over the given network, using the connection setup cost b_l as the arc weight of link l. The result yielded thereby is feasible and expected to give solution of better quality than a random guess. We also use the result of this simple heuristic to compare with Lagrangean relaxation based result in section 4 to prove our improvement.

To calculate the primal feasible solution of the minimum cost tree, the solutions to the Lagrangean Relaxation problems are considered. By solving the dual problem optimally we get a set of decision variables that may be appropriate for being the inputs of getting primal heuristics. However that solution might not be feasible and thus takes some more modifications. The set of g_l obtained by solving (15) may not be a valid solution to problem (IP) because the utilization constraint is relaxed. However, the utilization constraint may be a valid solution for some links. Also, the set of f_{dl} obtained by solving (16) may not be a valid solution because of the path and link constraints are relaxed and the union of y_l may not be a tree.

Here we propose a heuristics to obtain a primal feasible solution. While solving the Lagrangean relaxation dual problem, we may get some multipliers related to each links. According to the information, we can make our routing more efficient. Two of our getting primal heuristics are created by taking the LR multiplier β_{dl} as the source of arc weight in Dijkstra algorithm. We describe the Lagrangean based heuristic below.

[Lagrangean multiplier based heuristic]

Step 1 Calculate $\sum_{d \in D} \beta_{dl}$ as link l's arc weight.
Step 2 Use the arc weight obtained from step 1 and run the Dijkstra algorithm.

4 Computational Experiments

In this section, computational experiments on the Lagrangean relaxation based heuristic and simple primal heuristics are reported. The heuristics are tested on two kinds of networks- regular networks and random networks. Regular networks are characterized by low clustering and high network diameter, and random networks are characterized by low clustering and low diameter.

Two regular networks shown in Figure 2 are tested in our experiment. The first one is a grid network that contains 25 nodes and 40 links, and the second is a cellular network containing 19 nodes and 42 links. Random networks tested in this paper are generated randomly, each having 25 nodes. The candidate links between all node pairs are given a probability following the uniform distribution. In the experiments, we link the node pair with a probability smaller than 2%. If the generated network is not a connected network, we generate a new network.

For each testing network, several distinct cases, which have different predetermined parameters such as the number of nodes, are considered. The traffic

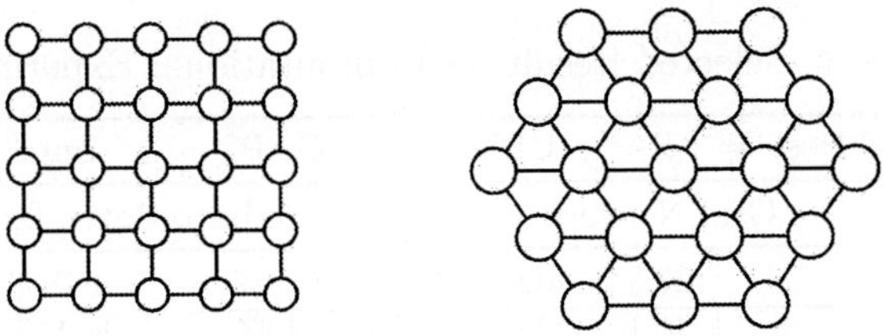

(a) Grid Network (b) Cellular Network

Fig. 2. Regular Networks

demand for multicast group is drawn from a random variable. The link connection maintenance costs and transmission cost are randomly generated between 1 and 5. The active probability of each destination is randomly generated between 0.1 and 1. The parameters used for all cases are listed in Table 3. The cost of the multicast tree is decided by multiplying the link transmission cost and the bandwidth requirement of the multicast group plus link maintenance costs. We conducted 200 experiments for each kind of network. For each experiment, the result was determined by the group destinations and link costs generated randomly. Table 4 summaries the selected results of the computational experiments.

Table 3. Parameters for Lagrangean Relaxation

Number of Iterations	1,000
Initial Multipliers	0
Improvement Counter	15
Delta Factor	2
Optimal Condition	Gap < 0.001

For each testing network, the maximum improvement ratio between the simple heuristic and the Lagrangean based heuristic is 20.17%, 20.77%, and 37.69%, respectively. In general, the Lagrangean based heuristic performs well compared to the simple heuristic. There are two main reasons of which the Lagrangean based heuristic works better than the simple algorithm. First, the Lagrangean based heuristic makes use of the related Lagrangean multipliers which include the potential cost for routing on each link in the topology. Second, the Lagrangean based heuristic is iteration-based and is guaranteed to improve the solution quality iteration by iteration. Therefore, in a more complicated testing environment, the improvement ratio is higher. To summarize, by relaxing constraints in the primal problem and optimally solving dual problem, the set of LR multipliers revealed iteration by iteration became unique sources for improving our solutions in getting primal heuristics.

Table 4. Selected Results of Computational Experiments

CASE	Dest. #	SA[a]	UB[b]	LB	GAP[c]	Imp.[d]
	Grid Network				Max Imp. Ratio: 20.17 %	
A	5	27.34	26.05	25.99	0.22%	4.73%
B	5	40.10	32.01	31.54	1.49%	20.17%
C	10	66.40	54.78	53.83	1.76%	17.51%
D	10	72.24	66.75	63.83	4.57%	7.60%
E	15	53.42	48.01	47.04	2.06%	10.13%
F	15	103.34	98.02	92.56	5.90%	5.15%
G	20	164.34	145.43	144.61	0.57%	11.50%
H	20	156.61	132.82	113.68	16.84%	15.19%
	Cellular Network				Max Imp. Ratio: 20.77 %	
A	5	16.62	16.62	16.62	0.00%	0.00%
B	5	32.16	25.48	23.34	9.17%	20.77%
C	10	56.37	46.98	45.30	3.71%	16.66%
D	10	48.88	40.87	38.33	6.63%	16.39%
E	15	58.12	47.31	38.09	24.20%	18.60%
F	15	85.76	82.30	74.45	10.54%	4.03%
G	20	124.35	118.83	111.34	6.73%	4.44%
H	20	143.33	128.98	124.43	3.66%	10.01%
	Random Networks				Max Imp. Ratio: 37.69 %	
A	5	7.75	7.75	7.75	0.00%	0.00%
B	5	14.32	12.94	12.48	3.69%	9.64%
C	10	53.83	42.47	39.89	6.47%	21.10%
D	10	59.16	36.86	33.04	11.56%	37.69%
E	15	60.38	57.82	53.37	8.34%	4.24%
F	15	76.42	68.88	62.34	10.49%	9.87%
G	20	93.46	74.83	73.08	2.39%	19.93%
H	20	103.46	92.64	83.78	10.58%	10.46%

[a] SA: The result of the simple heuristic

[b] UB and LB: Upper and lower bounds of the Lagrangean based modified heuristic

[c] GAP: The error gap of the Lagrangean relaxation

[d] Imp.: The improvement ratio of the Lagrangean based heuristic

To claim optimality, we also depict the percentile of gap in Table 4. The results show that most of the cases have a gap of less than 20%. We also found that the simple heuristic performs well in many cases, such as the case A of Cellular network and case A of random network.

The contribution of this research would be quite academic, with the innovative idea of constructing a multicast tree that adapts to the activity of end users in a minimization problem, making the model itself aware of the phenomenon of dynamic user join and leave without all the fuss of dealing with it in our heuristic. For this reason, this model is ideal for network planning purpose. Still the computational results show that the structure of the problem is suitable for the methodology of Lagrangean relaxation. However this model is still in a simple form and interested researchers may come up with quite a few extensions to this simple model with ease.

5 Conclusions

In this paper, we attempt to solve the problem of min-cost multicast routing with the consideration of dynamic user membership. Our achievement of this paper can be expressed in terms of mathematical formulation and experiment performance. In terms of formulation, we propose a precise mathematical expression to model this problem well. In terms of performance, the proposed Lagrangean relaxation and subgradient based algorithms outperform the primal heuristics.

Some additional topics to this problem might be 1) Multiple groups of users and the behaviors of the members within one group are identical or somewhat correlated. 2) Multiple trees may be constructed over the network at the same time, with different data-rate demands. 3) Quality-of-service constraints may be added such as: link capacity, hop count and delay constraints. 4) Different getting primal feasible heuristics can be invented to produce solutions with better optimality.

References

1. Winter, P.: Steiner Problem in Networks: A Survey. Networks (1987) 129-167
2. Hwang, F.K.: Steiner Tree Problems. Networks (1992) 55-89
3. Alvarez-Hamelin, J.I., Fraigniaud, P., and D. Alberto, Survey of multicast trees construction, Algotel 01, Saint Jean de Luz, France (2001)
4. Fisher, M.L.: The Lagrangian Relaxation Method for Solving Integer Programming Problems. Management Science, Vol. 27 (1981) 1-18
5. Ahuja, R.K., Magnanti, T.L., Orlin, J.B.: Network Flows: Theory, Algorithms, and Applications. Prentice Hall (1993)

Optimal Multi-sink Positioning and Energy-Efficient Routing in Wireless Sensor Networks*

Haeyong Kim, Yongho Seok, Nakjung Choi,
Yanghee Choi, and Taekyoung Kwon

School of Computer Science and Engineering
Seoul National University
San 56-1, Shillim-dong, Gwanak-gu, Seoul, Korea
{hykim, yhseok, fomula, yhchoi, tk}@mmlab.snu.ac.kr

Abstract. In wireless sensor networks, the sensors collect data and deliver it to a sink node. Most of the existing proposals deal with the traffic flow problem to deliver data to the sink node in an energy-efficient manner. In this paper, we extend this problem into a multi-sink case. To maximize network lifetime and to ensure fairness, we propose (i) how to position multiple sink nodes in a sensor network and (ii) how to route traffic flow from all of the sensors to these multiple sink nodes. Both of the problems are formulated by the linear programming model to find optimal locations of the multiple sink nodes and the optimal traffic flow rate of routing paths in wireless sensor networks. The improved lifetime and fairness of our scheme are compared with those of the multi-sink aware minimum depth tree scheme.

1 Introduction

In a wireless sensor network, each sensor node has a small sensing coverage and a small communication range because increasing the sensing coverage and communication range would consume more battery power. Since each sensor node has a small communication range, each sensor node relays the sensed events to a sink node [1]. As the number of sink nodes is increased, the path length from sensor node to sink node is decreased and the lifetime of the sensor nodes is increased. However, the number of sink node is constrained financially because the cost of the sink node is more expensive than the sensor node.

There has been much research done on traffic engineering in wired networks. The main purpose of traffic engineering is to increases the link utilization by equally balancing the traffic over the network. It is also possible to apply the traffic engineering in a wireless sensor network. However, the purpose should be to increase the network lifetime instead of improving the link utilization,

* This work was supported in part by the Brain Korea 21 project of Ministry of Education and in part by the National Research Laboratory project of Ministry of Science and Technology, 2004, Korea.

C. Kim (Ed.): ICOIN 2005, LNCS 3391, pp. 264–274, 2005.

because the critical resource of wireless sensor networks is the battery power of the sensor node instead of the link bandwidth although the sink nodes do not have an energy constraint since they are connected to a wired network.

In the past, most researches focused on the energy conservation [2][3] or data aggregation [5][6]. Recently, several studies [7][8][9] handle locating the multiple sinks in large-scale wireless sensor networks and optimizing the placement of integration points in multi-hop wireless networks, but any of papers did not consider traffic engineering.

In this paper, a formulation is proposed to improve the lifetime and the fairness of wireless sensor network with multiple sink nodes. We assume that the wireless sensor network shows the cluster architecture [2][3]. In this architecture, the cluster header has data forwarding functionality. The proposed formulations solve two problems, the location of sink nodes and the traffic flow rates of routing paths in the wireless sensor network. We formulate this problem into two types of LP (linear programs). In the first LP formulation, it is assumed that the location and the number of the sink nodes are fixed. The solution of the first LP formulation shows the optimal traffic flow rates of routing paths in the wireless sensor network. In the second LP formulation, instead of assuming the pre-fixed location and number of the sink nodes, the constraint of the maximum number of sink nodes is only assumed. In this case, the solution of the LP formulation finds both the optimal location of sink nodes and the optimal traffic flow rates of routing paths in wireless sensor network. Our main focus is the second LP formulation. It is shown that the proposed formulation increases the lifetime and the fairness of a wireless sensor network by using CPLEX [4] program that is a type of ILP (Integer Linear Programming) solver.

The organization of the paper is as follows: In Section 2, related work is shown such as the sink node location problem in multi-hop wireless network and wireless sensor network. The assumed wireless sensor network model and the proposed LP formulations are presented in Section 3 and in Section 4. In Section 5, the performance of the proposed LP formulation is evaluated by using the CPLEX tool. The conclusion of the paper is in Section 6.

2 Problem Definition

The wireless sensor network model used in this paper refers to [10]. Fig. 1 illustrates the layered sensor model. There are *Application Sensor Nodes* (ASNs) that collect data at the bottom layer. An ASN is a very low cost sensor, and there is a cluster of ASNs that belong to a clusterhead, or *Aggregation and Forwarding Node* (AFN). AFNs are logically located at the higher layer than the lower layer consisting of ASNs only. The AFN aggregates data from a group of ASNs and reduces redundancy of data. AFNs also relay data to sink nodes (or base stations).

Most previous schemes related to the layered sensor networks assume that the location of a sink node is fixed and seek to find the optimal routing paths and traffic flow rates. To the best of our knowledge, there is no work on finding

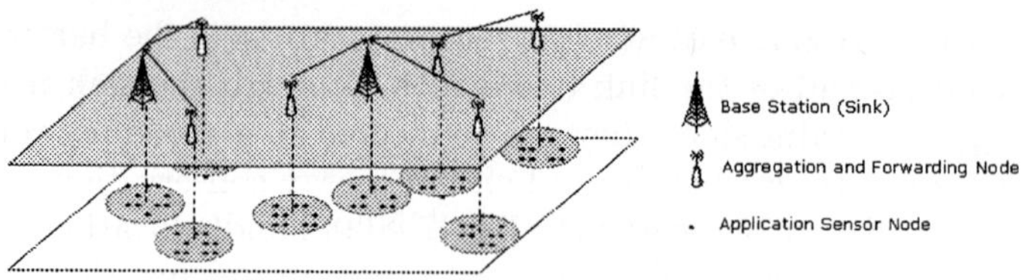

Fig. 1. The layered sensor network model

optimal locations of multiple sink nodes. The optimal locations of sink nodes, also routing paths and traffic flow rate through each path, should be determined to maximize network lifetime and to ensure fairness from the input of the locations of AFNs and the number of sinks to be deployed.

We suppose following scenario. Several areas to be investigated are determined. Each area needs an AFN and forms an cluster at least. The optimal AFNs as the sink, routing paths and traffic flow rates are computed before sensor nodes are deployed. AFNs (battery powered) are pre-configured to computed paths and flow rates. AFNs, AFNs as sink nodes (Base Station, wire-connected powered) and ASNs are deployed. Finally, The network starts to collect data. We assume that the number of sink nodes is limited by other environment (e.g., commercial cost).

The entire network is abstracted by two kinds of nodes hereafter. The sensor node(AFN) aggregates data and delivers it to sink nodes(Base Station) and the sink node receives data from sensor nodes. The data volume is also used instead of the data rate and time is disregarded in modeling the traffic flow, since the lifetime of the sensor node is proportional to the traffic volume (number of packets) transmitted from a sensor node if the power consumption of a sensor node in idle mode is negligible.[1]

3 The Proposed Formulation

The proposed formulations in this paper considers network lifetime and fairness. That is, under a constrained sensor node energy, first, to maximize the minimum among data volume generated by each sensor node for ensuring fairness, and then to maximize total data volume produced by nodes for maximizing the network lifetime. The MAX-MIN scheme is known to give good fairness. The given specific network lifetime can be satisfied by limiting sending data volume per unit time with regard to initial energy of sensor node.

[1] In wireless system, the power consumption of interface in idle mode is lower than both the transmission power and the reception power. In addition, the wireless sensor nodes use the power saving mechanism by switching between active state and sleep state. The data transmission/ reception is performed only in active period and we assume the ratio of active period is sufficiently small.

If the location of sink nodes is fixed, routing paths of each sensor node and the data volume through each path can be obtained by Formulation 1.

`Formulation 1.`

Maximize
First, Vol_{min}
Second, Vol_{total}

Subject To

$$\sum_{j \in N \cup S, j \neq i} X_{ij} - \sum_{j \in N, j \neq i} X_{ji} = \Delta_i \quad (\forall i \in N) \tag{1}$$

$$\Delta_i \geq 0 \quad (\forall i \in N) \tag{2}$$

$$X_{i,j} = 0 \quad (\forall i \in S) \tag{3}$$

$$Vol_{min} \leq \Delta_i \quad (\forall i \in N) \tag{4}$$

$$Vol_{total} = \sum_{i: i \in N} \Delta_i \tag{5}$$

$$\sum_{j \in N, j \neq i} (P^t_{ij} \cdot X_{ij}) + \sum_{j \in N, j \neq i} (P^r \cdot X_{ji}) \leq E_{init} \quad (\forall i \in N) \tag{6}$$

Variables
Δ_i : Data volume that node i produce
Vol_{min} : The minimum of Δ_i
Vol_{total} : Thetotal sum of Δ_i
X_{ij} : Data volume transmitted from node i to node j

Constants
N : Set of sensor nodes
S : Set of sink nodes
P^t_{ij} : Transmission power per unit data volume from node i to node j
P^r : Recieve power per unit data volume
E_{init} : Initial energy of a seonsor node

Each line means,

(1) Define Δ_i which is data volume produced by node(i).
(2) A sensor node should transmit data more than 0 bit.
(3) A sink node should not transmit any data to other sensor nodes.
(4) The Vol_{min} is the minimum among data volume produced by each sensor node.
(5) The Vol_{total} is total data volume produced by sensor nodes.
(6) A sensor node consumes power when send or receive data. The consumed power cannot be larger than initial energy. Idle power is assumed to be negligible.

The Formulation 1 is an LP (Linear Programming) problem because the location of sink nodes is fixed, so it can be solved in polynomial time by using LP solver. Formulation 1 guarantees the network performance of two type.

Fairness - When the Vol_{min} is maximized, all sensor nodes are guaranteed to generate the data volume of at least Vol_{min} and to communicate with sink nodes. Consequently, the each sensor node can produce the data volume of Vol_{min} regardless of the data volume produced by other sensor nodes.

Lifetime - When the idle power of sensor node is negligible, the lifetime of sensor network is dependent on the data volume of transmission and reception. Therefore, if the Vol_{total} is maximized, the lifetime of sensor network is also maximized. Here, the network lifetime means the duration that the batteries of all sensor node have been depleted.

Formulation 1 is modified to apply to cases in which the location of sink nodes is not fixed to get Formulation 2.

`Formulation 2. (for CPLEX)`

Maximize

$$C \cdot Vol_{min} + Vol_{total} \quad (C \cdot Vol_{min} \gg Vol_{total})$$

Subject To

$$\sum_{j \in N, j \neq i,k} X_{ij}^k - \sum_{j \in N, j \neq i} X_{ji}^k = \Delta_i^k \quad (\forall i, \forall k \in N, i \neq k) \tag{7}$$

$$\sum_{j \in N, j \neq i} X_{ij}^i = \Delta_i^i \quad (\forall i \in N) \tag{8}$$

$$\sum_{k \in N, k \neq i} \Delta_i^k \geq -C \cdot S_i \quad (\forall i \in N) \tag{9}$$

$$Vol_{min} \leq \Delta_i^i + C \cdot S_i \quad (\forall i \in N) \tag{10}$$

$$Vol_{total} = \sum_{i \in N} \Delta_i^i \tag{11}$$

$$\sum_{i: i \in N} S_i \leq N_{sink} \tag{12}$$

$$\sum_{j \in N, j \neq i} \left(P_{ij}^t \cdot \sum_{k \in N, k \neq j} X_{ij}^k + P^r \cdot \sum_{k \in N, k \neq i} X_{ji}^k \right) \leq E_{init} + C \cdot S_i \quad (\forall i \in N) \tag{13}$$

$$X_{ij}^k \leq C \cdot (1 - S_i) \quad (\forall i, \forall j, \forall k \in N, i \neq j, j \neq k,) \tag{14}$$

Bounds

$0 \leq \Delta_i^i \leq C \quad (\forall i \in N)$

$-C \leq \Delta_i^k \leq 0 \quad (\forall i, \forall k \in N, i \neq k)$

$0 \leq X_{ij}^k \leq C \quad (\forall i, \forall j, \forall k \in N, i \neq j, j \neq k,$ node j is in the transmission range of node $i)$

Binaries

$S_i \quad (\forall i \in N)$

Variables

Vol_{min} : The minimum data volume generated by each sensor node

Vol_{total} : The total data volume generated by sensor nodes

X_{ij}^k : Data volume transmitted from node i to node j,
node k is the source of data
Δ_i^k : Data volume that node i produce, node k is the source of data
S_i : If node i is a sinknode, $S_i = 1$. else $S_i = 0$

Constants
N : Set of all nodes in network
C : Infinite constant
N_{sink} : The maximum number of sink nodes
P_{ij}^t : Transmission power from node i to node $j$per unit data volume (bit)
P^r : Recieve power per unit data volume (bit)
E_{init} : Initial energy of a sensor node

A binary variable S_i is added to distinguish which node is selected as a sink node or not because the sink nodes is not decided yet. Variable k in X_{ij}^k, Δ_i^k means source node of data. It is just used to reduce useless equations to solve a problem more quickly and to distinguish the data source for debugging. It will also be used for advanced formulation (e.g., limiting hop-count of transmitted data) as future work. Each line means,

(7) Δ_i^k is data volume relayed by node i when the source of data is k.
(8) Δ_i^i is data volume produced by sensor node i.
(9) Only sink nodes can receive data of other nodes. If node i is a sensor node, it should relay received data to other nodes. (see also *2nd line of Bounds*)
(10) Vol_{min} is the minimum among data volume produced by each sensor node.
(11) Vol_{total} is total data volume produced by sensor nodes.
(12) N_{sink} is the maximum number of sink nodes.
(13) A sensor node consumes power when send or receive data. The consumed power cannot be larger than initial energy. Idle power is assumed to be negligible. The energy of sink node is not limited.
(14) Sink nodes do not send data to other nodes. (see also *3rd line of Bounds*)

The Formulation 2 is an M-ILP (Mixed-Integer Linear Program) problem because of integer variable S_i which is used to select sensor nodes for the role of sink nodes. Since M-ILP problems are NP-hard, it will take a long time to solve a problem if the wireless sensor network is huge. However, if the wireless sensor network consists of about 30 sensor nodes, we can get a solution quickly for that. Moreover, 30 sensor nodes (AFNs) will cover very large area in cluster architecture (layered sensor networks). In this paper, Formulation 2 was applied to a sample wireless sensor network and the result is shown in Section 5. Proposing an approximate algorithm that finds a solution in polynomial time will be left to future works.

4 Performance Evaluation

In this chapter, the result of a simulation by Formulation 2 is analyzed. Fig. 2 is a sample network that is used for simulation. 20 nodes are deployed randomly

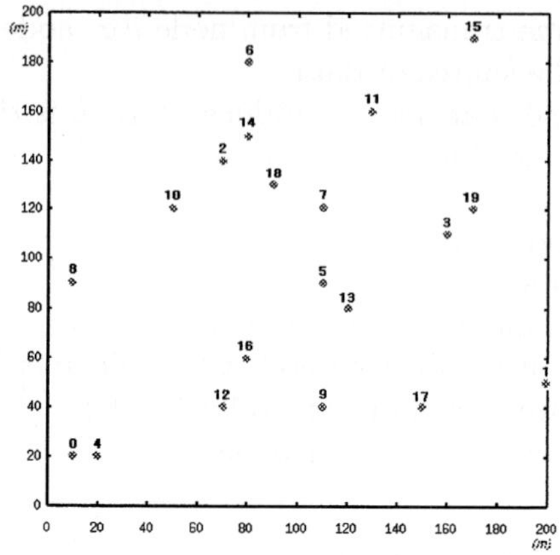

Fig. 2. A sample sensor network

Table 1. The node's number selected as sink node in fully-connected sample network

The number of sink nodes	Formulation 2	m-MDT
1	5	5
2	11, 16	9, 14
3	0, 11, 17	8, 11, 17
4	4, 10, 11, 17	1, 8, 11, 12
5	1, 4, 10, 11, 13	1, 4, 5, 10, 11

in 200 x 200(m). The parameters of Formulation 2 are |N| = 20 (the number of nodes), $C = 7000$, $N_{sink} = 1, 2, 3, 4$ or 5, $P^t_{ij} = 0.5 + 0.00013 * dist(i, j)^4$(nJ/bit) [11], $P^r = 0.5$(nJ/bit), $E_{init} = 100$(nJ). For the comparison with the proposed formulation, m-MDT (multi-sink aware Minimum Depth Tree) is used. The link cost is energy that is consumed when unit data is transmitted through the link. It has been known that MDT can route packets with minimal energy consumption.[12]

The simulation result in the sample network are compared in Table 1 and Table 2. We simulate two cases, one in fully-connected network and the other in which the transmission range of a sensor node is 60m. Although the node's number to be selected as sink node is the same as No.5 when there is only one sink node, the result shows that there is a difference in selected sensor nodes as sink nodes between Formulation 2 and m-MDT in case that the number of sink nodes is 2,3,4 or 5.

Fig. 3 and Fig. 4 show that Vol_{min} by Formulation 2 is much bigger than that by m-MDT though both select the same sensor node as sink node when there is only one sink node in the network. This is because m-MDT allows a sensor node to select only one path which is the shortest path (low energy consumption) to communicate with the sink node. In the m-MDT algorithm,

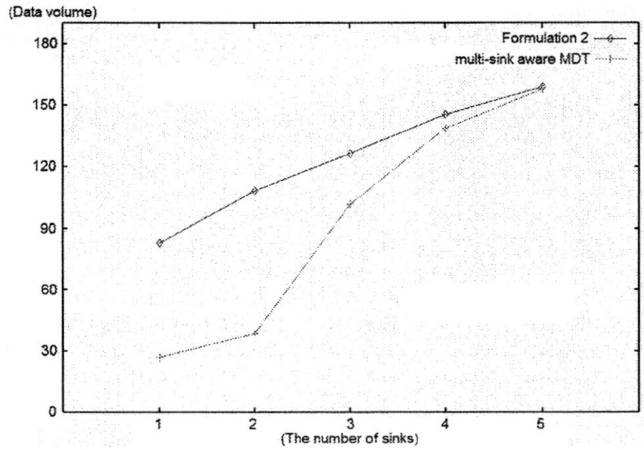

Fig. 3. The comparison of Vol_{min} in fully-connected network

sensor nodes around sink node always consume more energy than nodes far from the sink node. Therefore, sensor nodes far from the sink node cannot send data to the sink node even if they have a lot of energy. On the other hand, Formulation 2 considers not only the shortest path but also other available paths when a sensor node communicates with the sink node, so the maximum of Vol_{min} can be found by Formulation 2. Fig. 5 and Fig. 6 are detailed figures to show that Formulation 2 allows many paths unlike m-MDT, which allows only the shortest path of the tree architecture.

The proposed formulation in this paper significantly increases Vol_{min} by using many paths for communications and by admitting determined data volume to each link. In case there is only one sink node in a fully-connected network, the Vol_{min} by proposed formulation, 82.97, is about 3 times bigger than the Vol_{min} by m-MDT, 28.56. The reason that the location of the sink node selected by each algorithm is different is also caused by whether multi-path is available or not. Moreover, most of the nodes transmit almost the same data volume in Fig. 5 and Fig. 6. This means that the MAX-MIN scheme works well for fairness in this scenario, and network fairness is increased to compare with m-MDT. Since

Table 2. The node's number selected as sink node in sample network, the transmission range of a node is $60m$

The number of sink nodes	Formulation 2	m-MDT
1	5	5
2	7, 12	6, 9
3	4, 11, 17	0, 13, 14
4	4, 10, 11, 17	4, 10, 11, 17
5	1, 4, 10, 11, 13	1, 4, 5, 10, 11

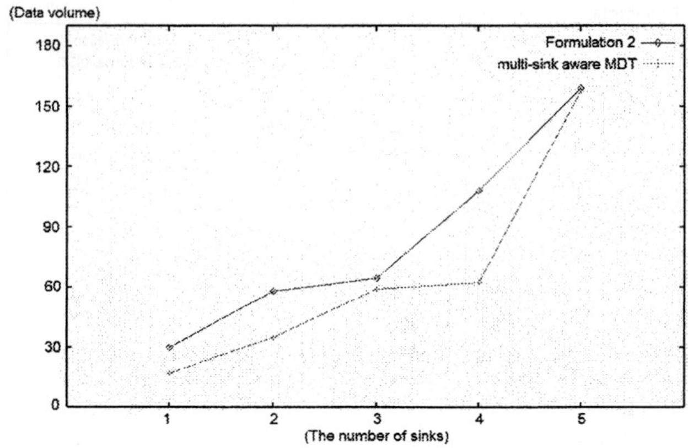

Fig. 4. The comparison of Vol_{min} when the transmission range of sensor node is $60m$

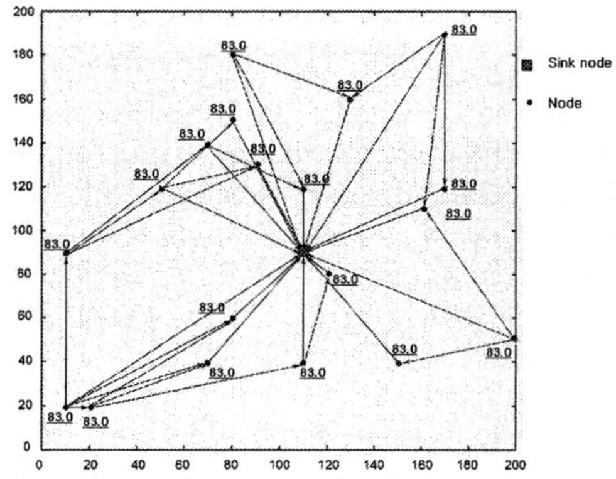

Fig. 5. Routing paths and data volume by Formulation 2 in fully-connected network, the number is data volume produced by each sensor node

m-MDT organizes the network into a tree architecture, there is a wide difference of the data volume produced by each node.

5 Conclusion

In this paper, the formulation to find the optimal locations of the multiple sink nodes and to find the optimal traffic flow rate is proposed. Maximizing network lifetime and ensuring fairness are the main objectives of this linear programming formulation. The proposed scheme is compared with m-MDT (multi-sink aware Minimum Depth Tree), and the results show that the proposed scheme improves network lifetime and fairness significantly. The proposed formulation allows sensor nodes to communicate with the one or more sink nodes through multiple

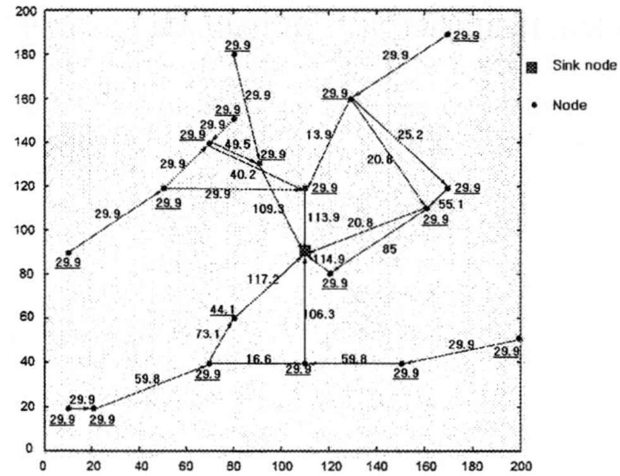

Fig. 6. Routing paths and data volume by Formulation 2 when the transmission range of node is $60m$, <u>$xx.x$</u> is data volume produced by each sensor node, $xx.x$ is data volume of each path

paths. The numerical results reveal that the number of the sink nodes is vital in the performance evaluation, so that the trade-off between the performance improvements and the deployment cost of the sink nodes should be taken into account carefully.

References

1. C. Intanagonwiwat, R. Govindan and D. Estrin, "Directed diffusion: A Scalable and robust communication paradigm for sensor networks," ACM MOBICOM 2000.
2. S. Bandyopadhyay and E. Coyle, "An Energy Efficient Hierarchical Clustering Algorithm for Wireless Sensor Networks," IEEE INFOCOM 2003.
3. B. Chen, K. Jamieson, H. Balakrishnan, R. Morris, "Span: An Energy-Efficient Coordination Algorithm for Topology Maintenance in Ad Hoc Wireless Networks" Proc. of the 6th ACM MOBICOM , Rome, Italy, July 2001.
4. ILOG CPLEX, http://www.cplex.com
5. Bhaskar Krishanamachari, Deborah Estrin and Stephen Wicker, "The Impact of Data Aggregation in Wireless Sensor Networks," In International Workshop of Distributed Event Based Systems (DEBS), Vienna, Austria, July 2002.
6. S. Chatterjea and P. Havinga, "A Dynamic Data Aggregation Scheme for Wireless Sensor Networks," ProRISC 2003, Veldhoven, Netherlands, November 2003.
7. E. I. Oyman and C. Ersoy, "Multiple Sink Network Design Problem in Large Scale Wireless Sensor Networks," Proceedings of the International Conference on Communications (ICC), Paris, France, June 2004.
8. L. Qiu, R. Chandra, K. Jain, and M. Mahdian, "Optimizing the Placement of Integration Points in Multi-hop Wireless Networks," Proceedings of the International Conference on Network Protocols (ICNP), Berlin, Germany, October 2004.
9. M. Younis, M. Bangad and K. Akkaya, "Base-Station Repositioning For Optimized Performance of Sensor Networks," Proceedings of the Vehicular Technology Conference (VTC), Orlando, Florida, October 2003.

10. Thomas Hou, Yi Shi, Hanif Sherali, “On Rate Allocation in Wireless Sensor Networks with Network Lifetime Requirement,” MobiHoc2004.
11. W.Heinzelman, “Application-specific Protocol Architectures for Wireless Networks,” PH.D. thesis, MIT, 2000.
12. Patrick Y.H. Cheung, and Nicholas F. Maxemchuk, “lpha Tree in Sensor Network,” IEEE Technical Report, Aug 2003.

An Efficient Genetic Algorithm for the Power-Based QoS Many-to-One Routing Problem for Wireless Sensor Networks*

Pi-Rong Sheu, Chia-Hung Chien, Chin-Pin Hu, and Yu-Ting Li

Department of Electrical Engineering
National Yunlin University of Science & Technology
Touliu, Yunlin 640, Taiwan, R.O.C.
sheupr@yuntech.edu.tw

Abstract. Since the operations of sensors in a wireless sensor network mainly rely on battery power, power consumption becomes an important issue. In this paper, we will consider the problem of searching for multiple paths between multiple source sensors and the sink such that any sensor in a path from a source sensor to the sink does not run out of its power during the transmission of packets. The problem has been proved to be NP-complete. Based on the principle of genetic algorithms, in this paper, we will design an efficient heuristic algorithm for it. Computer simulations verify that the suboptimal solutions generated by our genetic algorithm are very close to the optimal ones.

1 Introduction

A wireless sensor network (WSNET) is formed by a large number of tiny sensing devices (or *called sensors*)[5, 6]. A sensor in a WSNET can generate as well as forward data, which are gathered from every sensor's vicinity and will be delivered to the remote base station (or called *the sink*). WSNETs are useful in a broad range of environmental sensing applications such as vehicle tracking, seismic data, and so on.

The research of WSNETs has attracted a lot of attentions recently. In particular, since WSNETs are characterized by their limited battery-supplied power, extensive research efforts have been devoted to the design of power-aware routing protocols [5], such as LEACH, Directed Diffusion, SPIN, and MLDA & MLDR [6]. Most of the existing power-aware routing protocols are designed primarily to maximize the lifetimes of WSNETs. To achieve the goal, the key mechanism adopted by them is either to evenly distribute packet-relaying loads to each sensor to prevent the battery power of any sensor from being overused or to minimize the total power consumption of the entire WSNET. Little attention is paid to the issues related to the power-based QoS requirement of a route, i.e., to

* This work was supported by the National Science Council of the Republic of China under Grant # NSC 93-2213-E-224-023

C. Kim (Ed.): ICOIN 2005, LNCS 3391, pp. 275–282, 2005.

provide guaranteed battery power for the transmission of packets along a path from a source sensor to the sink such that any sensor in the path does not run out of its power during the transmission of packets. Only recently, a power-based QoS routing algorithm is proposed for the unicast routing problem in a mobile ad hoc network [7].

In WSNETs, the basic operation is the periodic gathering and transmission of sensed data to the sink for further processing. To be more specific, during a period of time (called *a round*), the sink first broadcasts a queue for its interested data, then the sensors which possess the appropriate data (called *the source sensors*) deliver their data to the sink. Obviously, it is possible that a lot of source sensors want to communicate with the sink simultaneously. In this paper, we will consider the problem that given multiple source sensors and the sink in a WSNET, find a path between each source sensor and the sink satisfying the power requirements, namely, guaranteeing that the transmission of data packets during a round can be supported by the residual power capacity of each sensor along the path. The problem has been named as *the power-based QoS many-to-one routing (PQMOR) problem*. The PQMOR problem is first defined and discussed in [4] and has been proved to be NP-complete [4]. Therefore, for the time being, the best way to deal with the PQMOR problem is to develop a heuristic algorithm and evaluate its performance through computer simulations.

It has been justified that the genetic algorithms (GAs) are a promising approach to handle the NP-complete problems in communication networks, such as the multicast routing problem [3]. Recently, many researchers have attempted to adopt GAs to solve various problems existing in wireless networks [1]. As a result, it is worthy to develop efficient GA to yield the better solutions for the PQMOR problem. In this paper, we will use the principle of GAs to design an efficient heuristic algorithm for the PQMOR problem. Computer simulations verify that our GA performs well and its solutions are very close to the optimal ones.

The rest of the paper is organized as follows. In Section 2, the formal definition of the PQMOR problem is given. In Section 3, an efficient GA for the PQMOR problem is proposed. In Section 4, the performance of our GA is evaluated through computer simulations and compared to the optimal solutions. Lastly, Section 5 concludes the whole research.

2 The Definition and Complexity of the PQMOR Problem

2.1 Traffic Model and Data Aggregation

In the following, we will define *a round* as a period of time in which the sink first broadcasts a queue for its interested data, then the source sensors, which own the related data, deliver their data packets to the sink. We assume that during a round, each source sensor v_{s_i} generates $\gamma(v_{s_i})$ data packets to be transmitted to the sink.

Data aggregation has been recognized as a useful routing paradigm in WSNETs. The main idea is to combine data packets from different sensors to eliminate redundant messages and to reduce the number of transmissions such that the total power of network is saved. One of the simplistic data aggregations is that an intermediate sensor always aggregates multiple incoming packets into a single outgoing packet. However, as discussed in [6], data aggregation is not applicable in all sensing environments. In the following, for simplicity, we will deal with our PQMOR problem without data aggregation.

2.2 Problem Formulation

In the following, we assume that the WSNET's topology would not change, i.e. no sensor gets move, that each sensor is able to estimate its current residual power capacity, and that only the transmitting power is considered(i.e., the reception power is ignored). We represent a WSNET by a weighted graph $G = (V, E)$, where V denotes the set of sensors and the sink, and E denotes the set of communication links connecting the sensors and/or the sink. For V , we define *a residual power capacity function* $\alpha : V \to R^+$. For source sensor set V_s, we define *a packet number function* $\gamma : V_s \to R^+$. For E , we define *a transmission power consumption function* $\beta : E \to R^+$. The value $\alpha(v_i)$ represents the current residual power capacity of sensor v_i. The value $\gamma(v_{s_i})$ represents the number of packets in v_{s_i} to be transmitted to the sink. The value $\beta(i,j)$ associated with link $(i,j) \in E$ represents the transmission power that one packet will consume on that link.

Based on these notations and definitions, we can now formally describe the PQMOR problem in our paper: given a weighted graph $G=(V,E)$, k source sensors v_{s_i}, the sink v_{sink} , α , γ , β, find a set R of k feasible paths r_i connecting each source sensor v_{s_i} to the sink v_{sink} such that for each sensor $v_i \in V$, $\alpha(v_i) \geq \sum_{(i,j)\in E} \beta(i,j) \times \mu(i,j)$, where $\mu(i,j)$ = the number of packets passing link (i,j). In other words, a set of feasible paths must guarantee that the transmission of packets along each path r_i from its source sensor v_{s_i} to the sink v_{sink} does not run out of power of any sensor in r_i until the completion of the round.

As an illustration of the above definitions and notations, let us consider the example shown in Fig. 1. In Fig. 1, let v_{s_1}, v_{s_2}, and v_{s_3} be the source sensors and v_{sink} be the sink, respectively. The number within a sensor represents the current residual power capacity of the sensor, the number next to a node represents the number of packets to be transmitted to the sink, and the number next to a link represents the transmission power consumption of the link. It is not difficult to observe that there is no feasible solution in Fig. 1 for the 3 paths to be established. In fact, at most two feasible paths can be discovered (as shown by the two bold-faced lines which connect v_{s_1} to v_{sink} and connect v_{s_2} to v_{sink}). On the other hand, if the current residual power capacity of node v_4 is increased from 10 to 15, then there exists a set of 3 feasible paths for the current PQMOR problem (the new path connecting v_{s_3} to v_{sink} is shown by the dotted line).

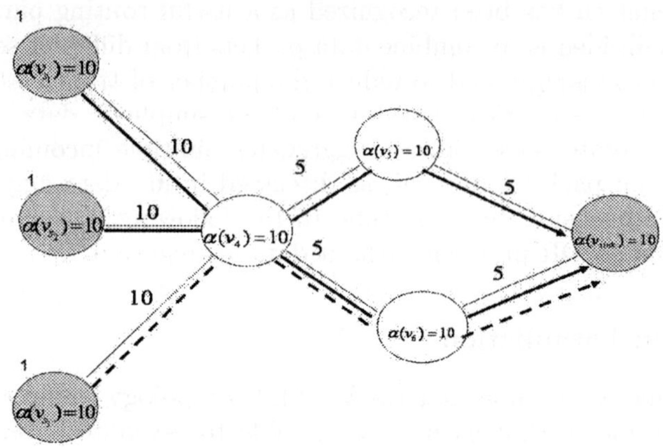

Fig. 1. An example to illustrate the PQMOR problem

3 An Efficient GA for the PQMOR Problem

The PQMOR problem has been shown to be NP-complete (reduced from the known NP-complete problem: the partition problem)[4]. When a problem is proved to be NP-complete, the follow-up quest will be to search for various heuristic algorithms and evaluate them by computer simulations. In this section, we will use the principle of GAs to design an efficient heuristic algorithm for the PQMOR problem.

3.1 Representation of Chromosomes

Given an instance of the PQMOR problem, where the given graph is $G = (V, E)$, $n_v = |V|$, and the k *pairs of source sensors and the sink* $= \{ (v_{s1}, v_{sink}), (v_{s2}, v_{sink}),\ldots, (v_{s_k}, v_{sink}) \}$, without loss of generality, the nodes in G is numbered from v_1 to v_{n_v}, v_{n_v} denotes the sink, and the k pairs of source sensors and sink are numbered as (v_i, v_{n_v}), $1 \leq i \leq k$.

A possible solution (i.e., a subgraph consisting of a set of possible paths) is represented by a chromosome c, which is a string of genes with length n_v. A gene is a bit and the ith gene g_i is corresponding to node v_i in the given graph G. Gene g_i being 1 (0) means its corresponding vertex v_i being (being not) included in the constructed subgraph. Because any possible subgraph must include all the source sensors and the sink, the first k genes and the last gene should always be set to 1. Therefore, at some places in our GA, we will be only concerned with the genes between the $(k+1)$th and the $(n_v - 1)$th in a chromosome. The subgraph induced by a chromosome c is the graph G_c only consisting of the node set V_c whose corresponding genes are set to 1, and the edges E_c whose both endpoints belong to V_c.

For a certain pair of source sensor and the sink in the subgraph induced by a chromosome, there may exist zero or more than one path between them. Therefore, we need a method to decode a chromosome into a set of feasible paths. Our decoding method is to establish a shortest path P_i between each source sensor v_{s_i} and the sink v_{sink} from the subgraph G_c by Dijkstra's algorithm [2].

3.2 Fitness Function

Obviously, under such a representation, the subgraph G_c induced by a chromosome c, which is generated at random and may experience the crossover and mutation operations, may include no or few feasible paths between the given source sensors and the sink. As a result, to measure the quality of a chromosome, we need to define a fitness function such that when a subgraph has more feasible paths, the corresponding chromosome will be assigned a higher fitness value. In our GA, the fitness value of a chromosome is related to be the number of feasible paths between the given source sensors and the sink in the subgraph induced by the chromosome. Given a chromosome c_i, the fitness function of c_i is computed by the algorithm DAHA (Dijkstra's algorithm based Heuristic Algorithm) shown in Fig. 2.

3.3 The Detailed Procedure of Our GA

In the following, we will explain each step in our GA

Step 1. Initialization of chromosomes: Step 1 generates n_p different chromosomes at random, which form the first generation of chromosomes. Note that only the genes between the $(k+1)$th and the $(n_v$ -1)th in a chromosome are randomly set to "0" or "1".

Step 2. Evaluation of chromosomes: At step 2, we compute the fitness value of each chromosome using the DAHA.

Step 3. Termination criteria: We adopt the common terminal rule, the maximum generation number, to terminate the evolution of our GA.

Step 4. Duplication: At this step, we make one copy S_n' of the current generation of chromosomes. This copy will be used at step 8.

Step 5. Selection: Our selection is implemented by using a roulette wheel selection scheme, where the probability of any chromosome c_i to be selected from the population is defined as $\frac{f(c_i)}{\sum_{j=1}^{n_p} f(c_i)}$, where $f(c_i)$ is the fitness function.

Step 6. Performing crossover operation on the selected chromosomes: For each chromosome, a random number between 0 and 1 is generated. If the random number is less than the given crossover rate r_c, the chromosome will be marked to indicate that crossover will be executed with it.Our GA adopts one-cut-point crossover [3] in which two selected chromosomes exchange their genes according to the cut point.

Step 7. Performing mutation operation on the selected genes: A random number between 0 and 1 is generated for each gene. If the random number

The DAHA :

Input: (1) a WSNET $G = (V, E)$, a *residual power capacity function* α, a packet number function γ, a *transmission power consumption function* β, *k source sensor* v_{s_i} and the sink v_{sink}
(2) a chromosome c_i

Output: the fitness value $f(c_i)$ of c_i

01. from G, construct the subgraph $G_{c_i} = (V_{c_i}, E_{c_i})$ corresponding the chromosome c_i;
02. $A = \{(v_1, v_{n_v}), (v_2, v_{n_v}), \cdots, (v_k, v_{n_v})\}$; $R = \Phi$;
03. while $A \neq \Phi$
04. for each $(v_i, v_{n_v}) \in A$
05. find a path r_i with the minimum total transmission power consumption for (v_i, v_{n_v}) in G_{c_i} by using Dijkstra's algorithm;
06. if r_i is a feasible route then $\{ A = A - (v_i, v_{n_v})$; $R = R \cup r_i$; for each $(j_1, j_2) \in r_i$ do $\alpha(v_{j_1}) = \alpha(v_{j_1}) - \beta(j_1, j_2)\mu(j_1, j_2)$ end $\}$
07. end of for each $(v_i, v_{n_v}) \in A$ loop
08. for each $(i, j) \in E$ do if $\alpha(v_i) < \beta(i, j)$ or $\alpha(v_j) < \beta(i, j)$ then $E_{c_i} = E_{c_i} - (i, j)$;
09. end of for each $(v_i, v_j) \in E$ do loop
10 $G_{c_i} = V_{c_i} \cup E_{c_i}$;
11. for each $(v_i, v_{n_v}) \in A$
12. if v_i is not connected to v_{n_v} in G_{c_i} (checked by Dijkstra's algorithm) then $A = A - (v_i, v_{n_v})$;
13. end of for each $(v_i, v_{n_v}) \in A$ loop
14. end of while $A \neq \Phi$ loop
15. output $f(c_i) = \dfrac{|R|}{k}$;

Fig. 2. DAHA: an algorithm to computer the fitness value of chromosome c_i

is less than the given mutation rate r_m, then the corresponding gene will do a mutation, i.e., the value of the gene will be inversed.

Step 8. Reproduction: Our reproduction operation will generate the next generation of chromosomes by picking up the top 50% chromosomes with higher fitness from the current generation S_n' of chromosomes and the set S_n'' of chromosomes, which is obtained from S_n' through the selection, crossover, and mutation operations.

4 Computer Simulations

In this section, by means of computer simulations, we will examine the efficiency of our GA and compare the suboptimal solutions generated by our GA with those obtained by the DAHA and the optimal solutions.

Our simulation environments are set as follows: the WSNET consists of n sensors located in a $100 \times 100m^2$ area randomly. The number of links is set to $m \times n$. The power capacity of a sensor is set to $q \times p_{max}$, where p_{max} denotes the maximum value among the transmission power consumptions of all the links connected to the sensor. The number of packets generated by a source sensor is between 1 and 10. The transmission power consumption of a link is assigned to a value between 10 and 40 randomly. The number of source sensors is set to k .

4.1 Determining the Proper Values for Different Parameters in Our GA

Like any other GAs, in order to yield the best performance, the four main parameters in our GA: the size n_p of population, the crossover rate r_c, the mutation rate r_m, and the maximum number of generations MAX_{ng}, must be properly selected. In our simulation environments to determine these parameters, $n = 50, m = 10, q = 5$ and $k = 30$. According to the simulation results [4], it can be found that the more proper values for theses parameters are as follows: $n_p = 70, r_c = 0.5, r_m = 0.5$, and $MAX_{ng} = 50$.

4.2 Simulation Results

In this subsection, we will compare our GA with the DAHA and the optimal solution in two different simulation environments. As the PQMOR problem is NP-complete, its optimal solution is hard to find. Therefore, we will use the upper bound instead of the optimal solution. Note that the upper bound indeed is the desired number k of feasible paths.

In our first simulation, where $n = 60, m = 10$, and $k = 20$, we will observe how the power capacity of a sensor impacts on the average number of feasible paths discovered by our GA. Fig. 3 shows the simulation results. From Fig. 3(a), it can be found that the average number of feasible paths discovered by our GA is approaching to the upper bound k after the power capacity of sensor achieves $6 \times P_{max}$.

In our second simulation, where $n = 60, q = 5$, and $k = 20$, we will observe how the total number of links impacts on the performance of our GA. From Fig. 4(a), we can see that the average number of feasible paths discovered by our GA is approaching to the upper bound k after the total number of links achieves $11 \times n$.

From Fig. 3(a) and Fig. 4(a), it can been observed that our GA outperforms the DAHA. From Fig. 3(b) and Fig. 4(b), we find that although the execution time of our GA is larger than that of the DAHA, it is within an acceptable region. In summary, these simulation results indicate that our GA is an efficient heuristic algorithm for the PQMOR problem.

5 Conclusions

Based on the principle of GAs, in this paper, we have designed an efficient heuristic algorithm for the PQMOR problem, which has been proved to be NP-complete. Computer simulations verify that the suboptimal solutions obtained by our GA are better than those obtained by the DAHA and very close to the optimal ones.

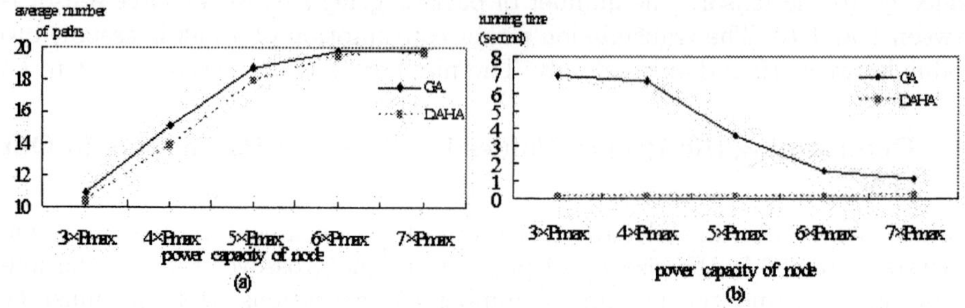

Fig. 3. The influence of power capacity of a sensor on our GA's performance

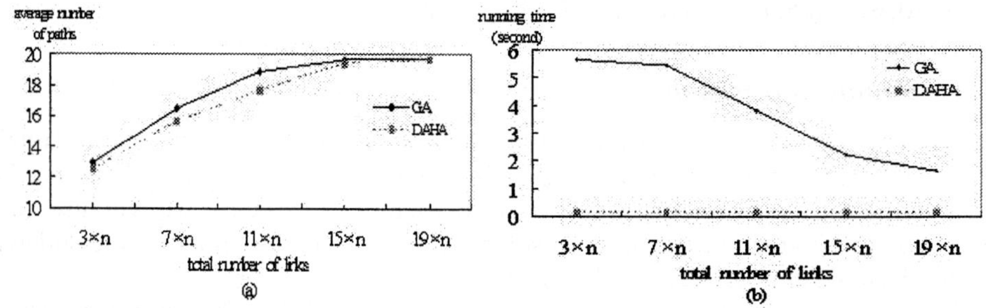

Fig. 4. The influence of total number of links on our GA's performance

References

1. Banerjee, N., Das, S. K.: Fast Determination of QoS-Based Multicast Routes in Wireless Networks Using Genetic Algorithm. ICC 2001, Vol. 8 (2001) 2588-2592
2. Dijkstra, E. W.: A Note on Two Problems in Connection with Graphs. Numerische Mathematik, Vol. 1 (1959) 269-271
3. Gen, M., Cheng, R.: Genetic Algorithms & Engineering Design. Wiley-Interscience (1997)
4. Hu, C. P.: An Efficient QoS Power-Aware Reverse Multicast Routing Protocol in Wireless Sensor Networks. Master's Thesis, National Yunlin University of Science and Technology, Yunlin, Taiwan, R.O.C (2004)
5. Jiang, Q., Manivannan, D.: Routing Protocols for Sensor Networks. IEEE Consumer Communications and Networking Conference (2004) 93-98
6. Kalpakis, K., Dasgupta, K., Namjoshi, P.: Maximum Lifetime Data Gathering and Aggregation in Wireless Sensor Networks. In the Proceedings of the 2002 IEEE International Conference on Networking, Atlanta, Georgia (2002) 685-696
7. Tragoudas S., Dimitrova, S.: Routing with Energy Considerations in Mobile Ad-Hoc Networks. IEEE Wireless communications and Networking Conference, Vol. 3 (2000) 1258-1261

Advanced MAC Protocol with Energy-Efficiency for Wireless Sensor Networks

Jae-Hyun Kim[1], Ho-Nyeon Kim[1], Seog-Gyu Kim[1], Seung-Jun Choi[2], and Jai-Yong Lee[1]

[1] Department of Electrical and Electronics Engineering, Yonsei University, 134 Shinchon-dong Seodaemun-gu, Seoul, 120-749, Korea
{jaykim, major06, sgkion}@nasla.yonsei.ac.kr
jyl@yonsei.ac.kr

[2] Department of Computer Engineering, Hankuk Aviation University, 200-1, Hwajeon-dong, Deogyang-gu, Goyang-city, Gyeonggi-do, 412-791, Korea

Abstract. This paper proposes E^2-MAC, a contention-based energy-efficient Medium Access Control (MAC) Protocol for wireless sensor networks. Energy efficiency is primary goal in wireless sensor networks. Existing MAC protocols for sensor networks attempt to solve energy consumption problem caused by idle listening using an active/sleep duty cycle. Since there are various traffic conditions, however, they may not always provide improvements in energy consumption. We propose a MAC protocol algorithm that stores data in a buffer and transmits data when the buffer exceeds a threshold value so that energy efficiency is always guaranteed for any network traffic conditions. Analytical results show that our proposed algorithm has significant improvements in energy consumption compared to the existing MAC protocols for sensor networks.

1 Introduction

In general, the main design goal of typical MAC protocols is to provide high throughput and QoS (Quality of Service). On the other hand, the primary design goal of wireless sensor MAC protocol is to minimize power consumption of sensor nodes to maximize the lifespan of the network. Other important design considerations include such factors as scalability and self-organization capability [4][5].

Energy is not efficiently consumed in the MAC layer of wireless sensor network for the following reasons. Firstly, from idle listening where a node must keep its radio on at all times since it does not know when it will receive messages from its neighbors. Secondly, from collisions where interferences occur between nodes when they transmit packets at the same time. Packets are corrupted as a result. Thirdly, from control packet overhead where the control packets are transmitted and received between nodes but they do not contain application data. These packets are considered as over-head. Finally, from overhearing where a node may receive packets that are not destined for it since the wireless link is a shared medium. It could then as well have turned off its radio. Among these reasons,

C. Kim (Ed.): ICOIN 2005, LNCS 3391, pp. 283–292, 2005.

idle listening most adversely affect the energy efficiency. Simulation results from Stemm and Katz [2] and Digital Wireless LAN Module (IEEE802.11) specification [3] show that the energy consumption ratio between idle-mode, receive-mode and send-mode is 1:1.05:1.4 and 1:2:2.5, respectively. In recent researches, duty cycle is used in sensor nodes to reduce unnecessary power consumption caused by idle listening.

IEEE 802.11 MAC protocol uses contention-based CSMA and has power-saving characteristics. However, all nodes need to be located in the same network and thus communication is limited to one hop. Thus, it is not adequate for sensor network which requires multi-hop communications [1]. One other protocol uses separate wake-up signal and communication signal. The wake-up signal is only used for waking up nodes and thus no additional data processing is necessary. Although, the protocol is highly energy efficient, additional devices need to be implemented on nodes and a separate frequency need to be allocated for the signal [6].

So far S-MAC (Sensor-Medium Access Control) [4] and T-MAC (Timeout-Medium Access Control) [5] protocols is proposed to improve energy efficiency in wireless sensor network. S-MAC is a contention-based MAC protocol that uses a single frequency. Time is divided into frames that are consist of active and sleep periods. During the sleep period, radio is turned off and data is not transmitted. They are stored instead. During the active period, the stored data is communicated with neighboring node. However, S-MAC uses fixed duty cycles. This causes energy waste if there is too little data to send and not all packets are transmitted if there is too much data to send. To improve energy efficiency, T-MAC uses a timer to switch to sleep mode after a certain period of time when it detects that there is no data to send or receive.

2 E^2-MAC Protocol

In this paper, we propose E^2-MAC (Energy Efficiency-MAC) protocol that specifies a threshold value to buffers in sensor nodes. Data is transferred if the buffer value exceeds the specified threshold value. Otherwise, the timer is set to a smaller value then in T-MAC. Since the protocol switches to sleep mode when the node does not transmit RTS (Request-to-Send) control packet or data during the timer, it improves energy efficiency compared with existing MAC protocols. Moreover, where as the efficiency of existing protocols degrades in some network traffic conditions, proposed protocol uses threshold value and optimizes energy efficiency in various traffic conditions.

Fig. 1 shows the operation algorithm of the E^2-MAC protocol. q_l is the accumulated data in each frame, q_t is the total buffer size of the sensor node, α is the threshold weight where $0< \alpha \leq 0.5$, and q_{th} is the buffer threshold value and expressed as below.

$$q_{th} = \alpha \times q_t \tag{1}$$

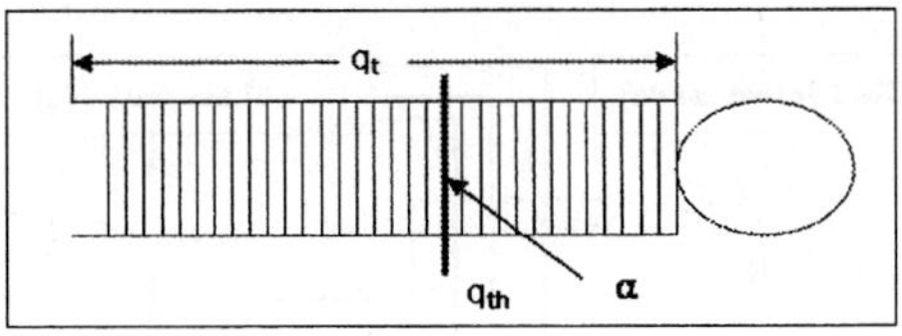

Fig. 1. Proposal system using a threshold of buffer

2.1 Operation of E^2-MAC Protocol and Determining Time Out

In E^2-MAC protocol, sensor nodes check the value of the accumulate data in their buffers and transfer data only if the condition $q_l \geq q_{th}$ is met.

In E^2-MAC protocol, B-RTS (Booking RTS) is used to reserve multi-hop data transfer within the same frame. This significantly improves the data transmission delay along with the energy efficiency.

To reduce unnecessary idle listening time after data transmission, T-MAC uses a timer defined in (2) to switch to sleep mode and conserve energy when there is no data to be received.

$$T_o > C + R + T \tag{2}$$

C is the contention interval, R is the RTS packet length and T is the very short time interval between RTS and CTS (Clear-to-Send) that is identical to SIFS (Short Inter Frame Space) in 802.11 MAC. The time it takes for E^2-MAC to detect RTS packet from neighboring nodes is determined as in (3).

$$T_E > R + (n-1)C \qquad (n > 2) \tag{3}$$

R is the RTS packet length, C is the CTS packet length, n is the number of hops. For multi-hop data transmission, $T_E = R + (n-1)C$ is applied and in case of 1st and 2nd hop, $T_E = R$ is applied.

In E^2-MAC, Timer operates after contention period. Since shorter than timer of T-MAC protocol, therefore, E^2-MAC can reduce a waste of energy according to idle listening.

3 Performance Analysis

We have analyzed mathematically and show results for energy efficiency and transmission delay which are key factors in designing sensor network protocols.

3.1 Active Time

In wireless sensor network, energy is consumed during the active period. So, we com-pare active time between S-MAC, T-MAC and E^2-MAC protocol to com-pare energy efficiency of these protocols. Equal number of frames is transmitted

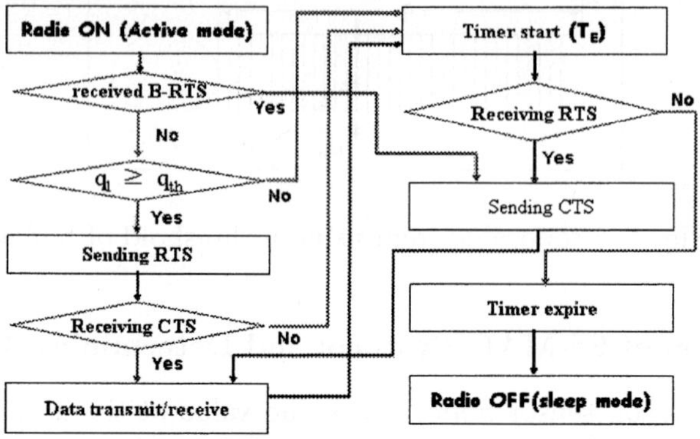

Fig. 2. Proposal algorithm

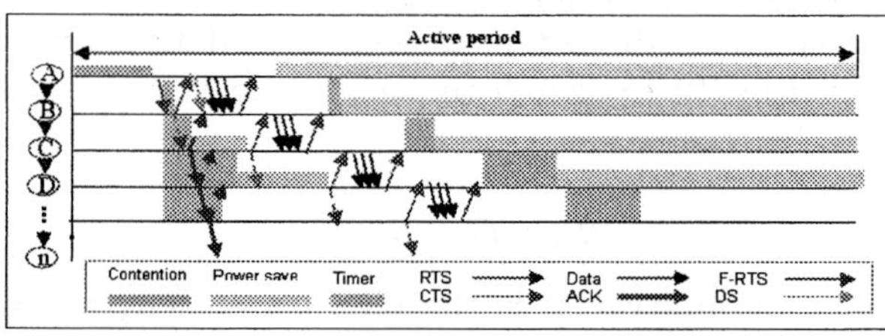

Fig. 3. Modified control packet and timer

Table 1. Parameter for Analysis

Parameter	Description
T_t	Total radio ON period in active mode
t_p	Contention period time
t_c	Time of control packet transmitting and receiving
t_r	Data packet receiving time
t_T	Data transmission time
t_R	Acknowledgement packet transmitting and receiving time
T_A	Active time of each time
T_o	Timeout interval in T-MAC
T_E	Timeout interval in E^2-MAC
E_s	Energy consumption for packet transmission
E_r	Energy consumption for packet reception
E_d	Energy consumption for idle listening
p	Probability in T-MAC : no data in active time
q	Probability in T-MAC : exist data in active time
ρ	Probability in E^2-MAC : $0 \leq q_l < q_{th}$ and no RTS received
τ	Probability in E^2-MAC : $q_l \geq q_{th}$

and represents the number of frames transmitted. First, the total active time in S-MAC protocol is

$$T_{total} = \sum_i T_{A_i} = \sum_i (T_{off_i} - T_{on_i}) \tag{4}$$

T-MAC protocol is classified into two states. The probability that there is no data to be transmitted is p

$$T_{A_p} = p \times T_o \tag{5}$$

and the probability that there is data to be transmitted is q

$$T_{A_q} = q \times (T_l - T_{on}) + T_o \tag{6}$$

The total active time in T-MAC protocol is therefore (where, $p + q = 1$)

$$T_{total} = \sum_i (T_{A_{p_i}} + T_{A_{q_i}}) = p\sum_i T_{o_i} + q\sum_i [(T_{l_i} - T_{on_i}) + T_{o_i}] \tag{7}$$

E^2-MAC protocol is classified into three states. If $0 \leq q_l < q_{th}$ and the probability that RTS packet is not received is ρ, then

$$T_{A_\rho} = \rho \times T_E \tag{8}$$

If the probability that $q_l \geq q_{th}$ is τ, then

$$T_{A_\tau} = \tau \times [(T_l - T_{on}) + T_E] \tag{9}$$

If RTS packet is received and data are received, Also, the probability that data are transmitted to a neighbor node or not is $1 - \rho - \tau$, then

$$T_{A_{(1-\rho-\tau)}} = (1 - \rho - \tau) \times [(T_l - T_{on}) + T_E] \tag{10}$$

The total active time in E^2-MAC protocol is

$$\begin{aligned} T_{total} = {} & \rho\sum_i T_E + \tau\sum_i [(T_{l_i} - T_{on_i}) + T_{E_i}] \\ & + (1 - \rho - \tau)\sum_i [(T_{l_i} - T_{on_i}) + T_{E_i}] \end{aligned} \tag{11}$$

Fig. 4(a) shows the accumulated active times of each protocol when equal numbers of frames are transmitted. The longest active time is required in S-MAC protocol to transfer since the active time is set to a fixed value by the duty cycle. E^2-MAC protocol can complete transfer in a smaller active time, about 17% less than in T-MAC.

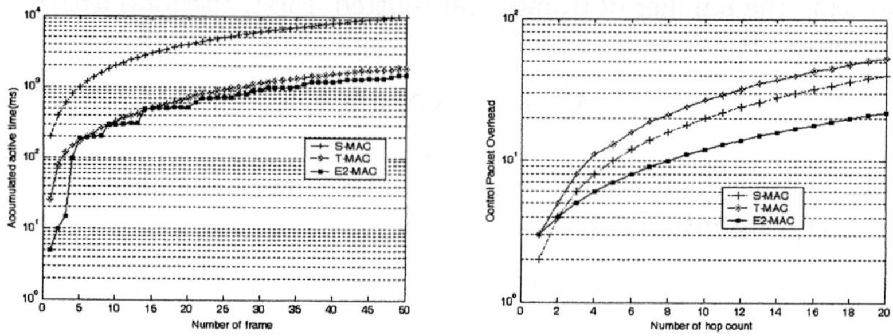

(a) Active time (b) # of Control packet overhead

Fig. 4. Avtive time and Number of CPO in S-MAC, T-MAC, E^2-MAC

3.2 Data Transmission Delay in 1-Frame

In this section, we analyze data transmission delay of each protocol. For S-MAC, the maximum number of hops that can be transmitted in one frame is [4]. The number of hops which can be transmitted is determined the following equation.

$$[N] = \frac{activetime}{[contention\ period + n(t_{RTS} + t_{CTS} + t_{Data} + t_{ACK})]} \tag{12}$$

(where, n is the data transmission count between nodes.)

Therefore, the data transmission delay is

$$\begin{aligned} t_{D_T} &= t_{D_1} + \sum_{i=2}^{n} t_{D_i} = t_p + 2t_c + t_T + t_R \\ &+ [(n-1)(2t_c + t_T + t_R)] = t_p + 2nt_c + n(t_T + t_R) \end{aligned} \tag{13}$$

In T-MAC protocol, the maximum number of hops that can be transmitted in one frame is three [5]. Therefore, the data transmission delay is

$$\begin{aligned} t_{D_T} &= \sum_{ijk}^{n} [(\frac{n-1}{3} T_{sleep} + t_{p_i} + 3t_{c_i} + t_{T_i} + t_{R_i} + T_{O_i}) + (\frac{n-2}{3} T_{sleep} \\ &+ 2t_{c_j} + t_{T_j} + t_{R_j} + T_{O_j}) + (\frac{n-3}{3} T_{sleep} + 3t_{c_k} + t_{T_k} + t_{R_k} + T_{O_k})] \end{aligned} \tag{14}$$

(where, i=1,4,7,10,$\cdots j$=2,5,8,11,$\cdots$ k=3,6,9,12,$\cdots$)

Since, E^2-MAC protocol, as illustrated in Fig. 3, uses modified control packet, it can transmit n-hops in one frame duration. Therefore, the data transmission delay is

$$t_{D_{E^2}} = t_{D_1} + \sum_{i=2}^{n} t_{D_i} = t_p + (n+2)t_c + n(t_T + t_R) + (\frac{n^2 - n + 2}{2})T_E \quad (15)$$

3.3 Control Packet Overhead

In terms of a waste of energy, control packet overhead is one of main factors in wire-less sensor networks. The E^2-MAC protocol supports advanced algorithm and modified control packets to minimize unnecessary traffic. Since minimizing unnecessary control packets, the proposed scheme can reduce a waste of energy and data transmission delay compared to the existing protocols.

In S-MAC, the maximum number of hops that can be transmitted in active period of one frame is n.

That is, the maximum number of hops that can be transmitted is in proportion to the sized of transmitting data. From eq.(13), two control packets occur in order to transmit data at each hop. Therefore, CPO (Control Packet Overhead) for n-hops transmission are

$$CPO = \sum_{i=1}^{n} 2i = n(n+1) \quad (16)$$

In T-MAC, the maximum number of hops that can be transmitted in one frame is three. For T-MAC, control packets differently occur according to the number of hops transmitted for each frame. Control packets for n-hops transmission are

$$CPO = \sum_{ijk}^{n} 3t_{c_i} + 2t_{c_j} + 3t_{c_k} \quad (17)$$

(where, i=1,4,7,10,$\cdots j$=2,5,8,11,$\cdots$ k=3,6,9,12,$\cdots$)

In E^2-MAC, multi-hop communication is reserved by using Booking-RTS and except for first hop, only one control packet is used for all other hops. Therefore, control packet overhead can be reduced in E^2-MAC. Control packets for n-hops transmission are

$$CPO = \sum_{i=1}^{n} (i+2) = \frac{n^2 + 5n}{2} \quad (18)$$

Fig. 4(b). shows control packet overhead occurred according to the number of hops for one frame. In Fig. 4(b), average control packet is T-MAC > S-MAC > E^2-MAC. We can see that reduction of control packet results in decrease of energy consumption, so that the proposed E^2-MAC protocol has the most significant benefit with energy-efficiency.

3.4 Energy Consumption

To analyze energy consumption we have referred to EYES nodes [5] for data transmission and reception. In S-MAC protocol, data can be transmitted until n-hops for fixed active period. The energy consumption for nth-hop transmission is

$$E_S = t_p E_d + n[t_c(E_s + E_r) + t_T E_s + t_R E_r] + E_d[T_{A_p} - n(2t_c + t_T + t_R)] \quad (19)$$

The Energy consumption for n-hops is

$$E_S(n) = t_p E_d + \sum_{i=1}^{n} i[t_{c_i}(E_s + E_r) + t_{T_i} E_s + t_{R_i} E_r]$$
$$+ E_d[T_A - i(2t_c + t_T + t_R)] = t_p E_d + \frac{n(n+1)}{2}[t_c(E_s + E_r) + t_T E_s + t_{R_i} E_r]$$
$$+ E_d[T_A - \frac{n(n+1)}{2}(2t_c + t_T + t_R)] \quad (20)$$

In T-MAC protocol, data can be transmitted until 3-hops for fixed active period. Energy consumption for n-hops transmission is

$$E_T(n) = \sum_{ijk}^{n}[[t_{p_i} E_d + 3t_{c_i}(E_s + E_r) + t_{T_i} E_s + t_{R_i} E_r + (\frac{i+2}{3})T_{o_i} E_d]$$
$$+ [t_{p_i} E_d + 2t_{c_j}(E_s + E_r) + t_{T_j} E_s + t_{R_j} E_r + (\frac{j+1}{3})T_{o_j} E_d]$$
$$+ [t_{p_i} E_d + 2t_{c_k}(E_s + E_r) + t_{T_k} E_s + t_{R_k} E_r + (\frac{k}{3})T_{o_k} E_d]] \quad (21)$$

(where, i=1,4,7,10,$\cdots$ j=2,5,8,11,$\cdots$ k=3,6,9,12,$\cdots$)

In E^2-MAC protocol, data can be transmitted until n-hops for fixed active period. In case of 1-hop transmission at each frame, since the data is transmitted only when the buffer is equal or greater than the threshold value, three cases of probability exist. If and the probability that the node does not receive RTS packet from neighboring nodes is ρ , then

$$E_\rho = \rho(t_\rho E_d + T_E E_d) \quad (22)$$

If the probability that $q_l \geq q_{th}$ is τ is, then the energy consumption is

$$E_\tau = \tau[t_p E_d + t_c(E_s + E_r) + t_T E_s + t_R E_r + T_E E_d] \quad (23)$$

And if $0 \leq q_l < q_{th}$, the probability of the rest case is $1-\rho-\tau$, then

$$E_{\sigma} = (1-\rho-\tau)[t_p E_d + t_c(E_s + E_r) + t_r E_s + t_R E_r + T_E E_d] \quad (24)$$

Therefore, energy consumed for 1-hop transmission in E^2-MAC protocol is

$$\begin{aligned} E_{E^2} &= (t_p + T_E)E_d + [(1-\rho)(E_s + E_r)(t_c + t_R)] \\ &+ (1-\rho-\tau)(t_r E_r + t_R E_s) + \tau(t_T E_s + t_R E_r) \end{aligned} \quad (25)$$

The energy consumed for n-hops transmission in E^2-MAC protocol is

$$\begin{aligned} E_{E^2(n)} &= \sum_{i=1}^{n}(E_{\rho_i} + E_{\sigma_i} + E_{\tau_i}) = t_p E_d + (\frac{n^2-n+2}{2})T_E E_d + n[(1-\rho) \\ &(E_s + E_r)(t_c + t_R)] + n(1-\rho-\tau)(t_r E_r + t_R E_s) + n\tau(t_T E_s + t_R E_r) \end{aligned} \quad (26)$$

Fig. 5(a),(b) shows energy consumption in various traffic conditions. E^2-MAC protocol yields best efficiency in any traffic volume. Furthermore, Fig. 5(c),(d) shows that the life-span of the sensor nodes in E^2-MAC protocol is the longest among the protocols.

4 Conclusion and Future Work

In this paper, we have proposed an algorithm that uses threshold value in buffers of sensor nodes to improve energy efficiency in sensor networks which is the key factor in designing MAC protocols for sensor networks. Analytical results show that the proposed E^2-MAC protocol significantly improves energy efficiency compared with existing MAC protocols. Also, data transmission delay is reduced by using enhanced control packet and timer. We are currently in the process of various simulations with NS-2 for analyzing other results among S-MAC, T-MAC and our scheme. We are confident that those results will prove the proposed E^2-MAC is practical and very efficient for wireless sensor networks. More research on the optimal threshold value in E^2-MAC is needed. In the future, we are planning to apply the proposed algorithm onto practical environment as integrating our algorithm and simple sensor nodes.

Acknowledgments

This research was supported by the MIC(Ministry of Information and Communication), Korea, under the ITRC(Information Technology Research Center) support program supervised by the IITA(Institute of Information Technology Assessment)

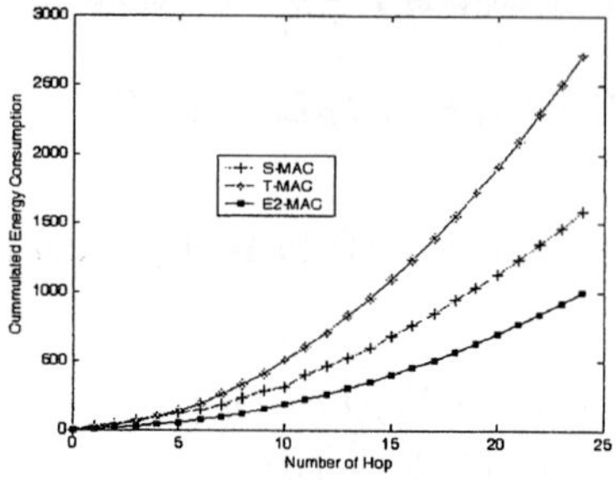

(a) low traffic

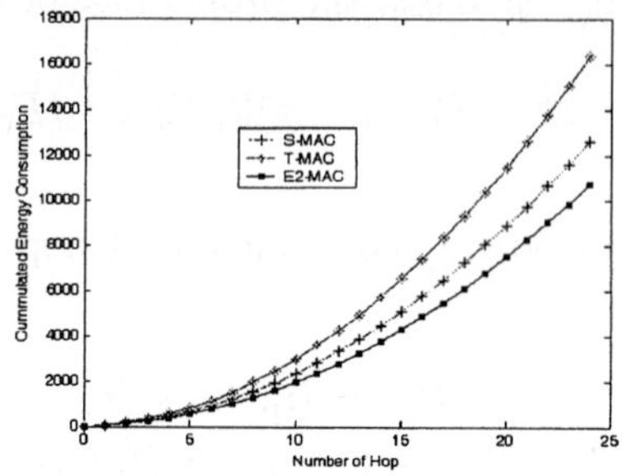

(b) high traffic

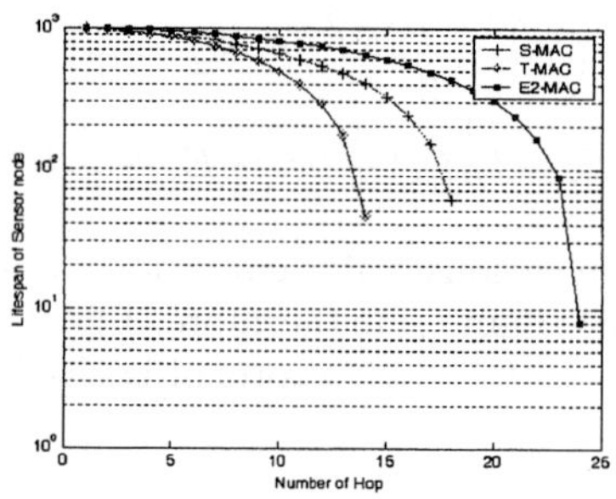

(c) low traffic

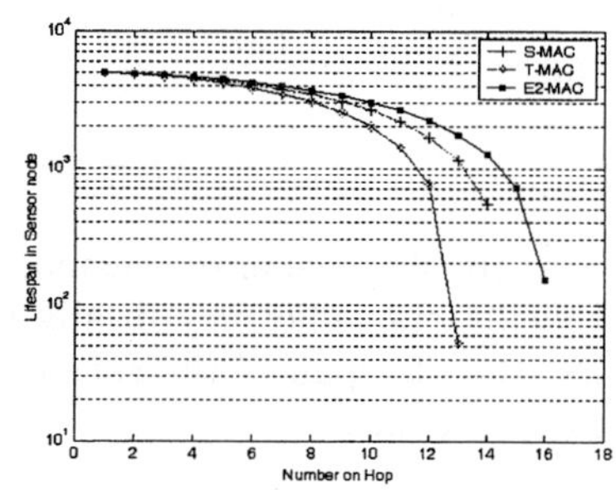

(d) high traffic

Fig. 5. Energy consumption(Top) and Lifespan(Bottom) in various traffic condition

References

1. LAN MAN Standards Committee of the IEEE computer Society.: IEEE Std 802.11-1999, wireless LAN Medium Access Control(MAC) and Physical layer(PHY) specification. IEEE 1999
2. Mark Stemm and Randy H Katz.: Measuring and reducing energy consumption of network interfaces in hand-held devices. IEICE Transactions on communicationsm, Vol. E80-B. no.8. pp.1125-1131, Aug 1997
3. Oliver Kasten.: Energy Consumption, http://www.Inf.ethz.ch/~kasten/research/bathtub/energy_comsumption.html. Eidgenössische Technische Hochschule Zürich
4. Wei Ye, John Heidemann, Devorah Estrin.: Medium Access Control with Coordinated Adaptive Sleeping for wireless Sensor networks. USC/ISI Technical Report ISI-TR-567, January 2003
5. Tijs van Dam, Koen Langendoen.: An adaptive Energy-Efficient MAC protocol for wireless Sensor Networks. Sensys'03, November 2003
6. C. Guo, L. Zhong, and J. Rabaey.: Low-Power Distributed MAC for Ad Hoc Sensor Radio Networks. Proc. Internet Performance Symp. (Globecorn '01), Nov. 2001

The Energy-Efficient Algorithm for a Sensor Network

Saurabh Mehta, Sung-Min Oh, and Jae-Hyun Kim

School of Electrical and Computer Engineering,
AJOU University, San 5 Wonchon-Dong, Youngtong-Gu, Suwon 442-749, Korea
{saurabh, smallb01 and jkim}@ajou.ac.kr

Abstract. We considered a network that consists of massively deployed tiny energy constrained sensors with one sink node to communicate with the outer world. The key challenge in design of wireless sensor networks is maximizing its lifetime. We explored this challenge and proposed our algorithm to increase a network lifetime. A node lifetime and numerical analysis show the comparison of the proposed algorithm with existing algorithm to increase a network lifetime. The proposed algorithm increases a critical node lifetime by 38% and hence a network lifetime.

1 Introduction

The rapid development in small, low-power, low-cost microelectronic and micro-electromechanical (MEMs) sensor technology along with the advances in wireless technology have enabled wireless sensor networks to be deployed in large quantities to form wireless sensor networks for a wide uses. There are multiple scenarios in which such networks find uses, such as environmental monitoring/controlling and interactive toys, etc.

In [1] authors describe about characteristics and challenges of WSN. Due to energy constrain, energy efficiency is a critical consideration for designing the sensor networks and its routing protocols. In [2] authors describe upper bound on the lifetime of sensor networks, while in [3] the lifetime of a cluster based sensor that provides periodic data is studied. In a large sensor network all the nodes send their data to sink node for the further processing as shown in figure 1, due to this fact the nodes near to sink node consumed their energy more rapidly compared to other sensor nodes. In [4] authors describe the problem of developing an energy efficient operation of a randomly deployed multi-hop sensor network by extending the lifetime of the critical nodes and as a result the overall network's operation lifetime, were considered and analyzed but they didn't propose any solution for the same. In this paper we are extending our work further from [4] and proposing algorithm to increase the critical node lifetime and hence a network lifetime. In [5, 6, 7, 8] authors suggested power aware multiple paths algorithms to distribute the relay load equally among all the nodes.

In single path algorithm there is only one path available from source to sink and normally it is a minimum hop path. Due to single path there is always very

C. Kim (Ed.): ICOIN 2005, LNCS 3391, pp. 293–302, 2005.

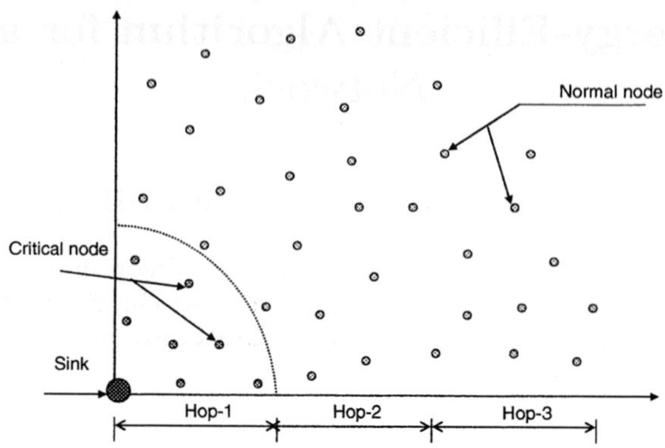

Fig. 1. Sensor networks

heavy traffic on the route and also its lifetime is short but we can overcome these disadvantages by implementing the proposed algorithm which distributes the load among the nodes. In multiple paths every node is connected to the number of paths and it is not practical to make them work in just one mode, such as sense or relay mode. Further we evaluated the efficiency of each algorithm from mathematical and numerical analysis.
Based on our study, all proposed solutions for the senor networks lifetime are categorized as follows [2, 3, 5, 6, 7, 8, 9, 11, 12, 15].

1. Using an energy efficient and multiple path routing algorithm.
2. Using higher battery capacity relay node or cluster based method.
3. Using the different working modes for a node.

This paper is organized as follows. In section 2, we introduced our sensor networks model. Section 3 described our proposed algorithm. In section 4 we described a node lifetime analysis, followed by numerical analysis in section 5. Finally conclusion and future work are in section 6.

2 Sensor Network Model

All nodes in a sensor network are static, same in size, battery capacity, etc. Every node has a static ID (Not IP) and does the relying and sensing. A node in hop 2 will always find the critical path in hop 1. A network consists of randomly but uniformly deployed nodes. We make our further assumption from [4] as follows.

1. E_o = The energy of a node.
2. E_s = The energy needed to sense one bit. It depends on the power dissipation of the internal circuitry. It is denoted by ε_{si}. Where i represents i^{th} node (s_i).
3. $E_{b,rx}$ =The energy needed to receive a bit. It is denoted by ε_{rxi}.

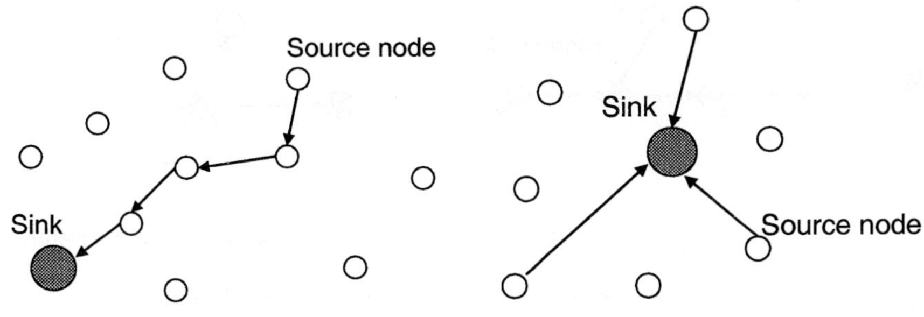

(a) When sink is away from source (b) When sink is near to source

Fig. 2. Sink node at different positions

4. $E_{b,tx}$ = The energy required to transmit a bit. It is given by the $E_{b,tx} = \varepsilon_{tx} + \varepsilon_{rf}(d/d_0)^n$, where ε_{rf} is the energy consumed to transmit a bit to the reference distance d_0 and n is the path loss index.
5. $E_{b,process}$ = The energy consumed per bit for data processing, such as aggregation and special functions required to relay data. Let us denote by γ the data aggregation ratio. The energy per bit for aggregation is a function of γ that is given by $E_{b,process} = \varepsilon_p + \varepsilon_a f(\gamma)$, where $f(\gamma) = 0$ if $\gamma = 1$.
6. $\lambda_{org,i}$ = The number of packets generated per unit time by s_i. It is indicated by λ_{s_i}.
7. $\lambda_{re,i}$ = The number of packets relayed per unit time by s_i. It is indicated by λ_{ri}.
8. L = Length of a data packet.

Figure 1 shows a sensor network model's first quadrant, here we considered only one collector node which is placed at the center of a network. Now based on the above definitions and assumptions, the power dissipation of node s_i is given by

$$P_i = \varepsilon_{si}\lambda_{si}L + \varepsilon_{rxi}\lambda_{ri}L + [\lambda_{si} + \lambda_{ri}]\varepsilon_p L + [\lambda_{si} + \lambda_{ri}]\gamma\varepsilon_{tx}L\,. \tag{1}$$

For simplicity we considered $\gamma = 1$ and $d = d_0$. Still we can simplify above terms by assuming that $\varepsilon_s = \varepsilon_p = \varepsilon/2$ and $\varepsilon_{rx} = \varepsilon_{tx} = \varepsilon$ [4]. From all above assumption we can rewrite (1) in the following way

$$P_i = [2\varepsilon + \varepsilon_{rf}]\lambda_{si}L + [2.5\varepsilon + \varepsilon_{rf}]\lambda_{ri}L\,. \tag{2}$$

And let $E(t_i)$ be i^{th} node life time that we have

$$E(t_i) = E_0/P_i\,. \tag{3}$$

From (2) we can observe that power consumed by a node is divided into two terms. First term is used only for sensing and transmitting its own data and second term used for the relaying purpose. Figure 2 (a) shows the above condition.

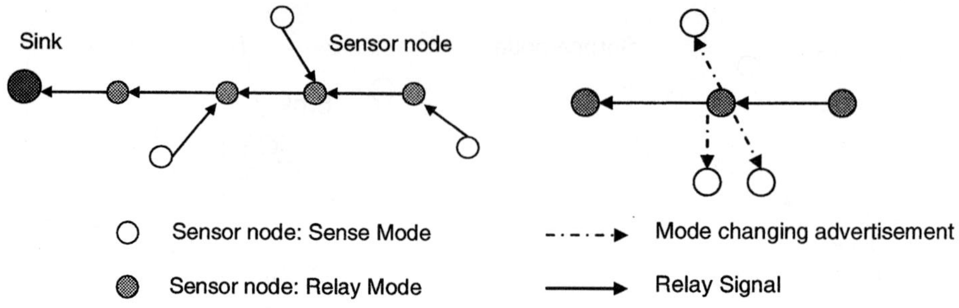

(a) A sensor node working in two modes (b) A sensor node change its state

Fig. 3. A node's transition

From (2) we can conclude that 65% of its energy gets used only for the relaying data that is actually waste for a node [12]. If we can cut or make the second term as low as possible, we can increase the node lifetime.
We can make second term equal to zero only if a node doesn't need to relay any external data packet. This can be possible only in one case when a sink node is in range of all sensors as shown in figure 2 (b). Here we proposed the algorithm which helps nodes to create an energy efficient routing backbone based on energy consumption metric.

3 The Proposed Energy-Efficient Algorithm

Here we considered a large area network for a monitoring application which contains the number of sensor nodes with one sink. All nodes have to transfer their data to sink that means a common destination address for all the nodes. As we mentioned above to maximize networks lifetime, load distribution of a relay packet is very important. In [6,7,8,11] authors follow multiple route paths algorithm. The basic idea of the proposed algorithm and detailed procedure can easily be understood from the following steps.

1. All nodes are working under two mode, relay and sense and keep changing their modes as per the algorithm's set condition. Figure 3 gives clear idea about these two modes and their transition.
2. At first, sink node will broadcast an advertisement for hop count information and after some delay all nodes will know its hop position.
3. After the hop information, sink node will run 3 color's algorithm [14] to choose random nodes which are one hop apart from each other for carrying relay loads.
4. Let us denote all randomly chosen nodes as Cluster Node (CN). All CN nodes advertise about their status and thus inform its cluster members to send all their data.

5. All CN nodes will establish connection with just one node which is only one hop apart and has lower hop count than CN and it will be also consider as CN.
6. Whenever any sensor node wants to send data, it will send to its reachable CN and from CN to lower hop CN and thus to sink.
7. All CN will operate in relay mode and the rest of the nodes are in sense mode.
8. CN will operate in relay mode until the remaining energy reach to its set energy threshold. Then it will broadcast an advertisement for a mode change. This advertisement contents its ID, upper and lower hop CN ID and its energy level.
9. CN will wait for some random delay. During this delay time some response will come from the near by node and CN will change its mode from relay to sense furthermore it will choose new CN for the relay.
10. Now old CN will work in sense mode till predetermined energy threshold. If any advertisement comes for relay and if that advertisement satisfied all decision condition then it will again enter into the relay mode otherwise remain in a same mode.

3.1 Proposed Algorithm Pseudo Code

Here we specified the algorithm in pseudo code for a single path energy-efficient scheme.

```
Begin
    If (mode==sense) sense_mode ();
    else relay_mode ();
End
    sense_mode ();
Begin
    If (senser_pkt_int==arrived)
       ++ sense;
       energy;
       create_data_tx ();
    Begin
       If (energy>adv_energy)
       change_mode ();
       else cont_sense_mode ();
    End
End
    relay_mode ();
Begin
    If (relay_pkt_int==arrived)
       ++ relay;
       energy;
       next_hop_tx ();
```

```
    Begin
        If (enrgy<limit)
        energy_adv_tx ();
        else cont_relay_mode ();
    End
End
```

where,

1. *create_data_tx*() = Procedure for creating data frame and transmitting to near by CN.
2. *change_mode*() = Procedure for changing the operation mode.
3. *cont_sensing_mode*() = Procedure for continuing the operation in sensing mode.
4. *next_hop_tx*() = Procedure for transmitting packet to next hop.
5. *energy_adv_tx*() = Procedure for transmitting an advertisement for changing the mode and change the mode on receiving acknowledgement.

4 A Node Lifetime Analysis

As we proposed in the algorithm all nodes will work in two modes and their energy consumption will change according to their operating mode. Energy consumed in relay mode is given by

$$P_{ri} = [2.5\varepsilon + \varepsilon_{rf}]\lambda_{ri}L\,. \tag{4}$$

Energy consumed in sense mode is given by

$$P_{si} = [2\varepsilon + \varepsilon_{rf}]\lambda_{si}L\,, \tag{5}$$

but in our proposed algorithm sensor node also need to transmit and receive some overhead signals. We added 0.5ε in the above term. So modified equation is as follows.

$$P_{si} = [2.5\varepsilon + \varepsilon_{rf}]\lambda_{si}L\,. \tag{6}$$

In multiple paths a node has to consider some overhead energy consumption and is given by

$$P_i = [\lambda_{si} + \lambda_{ri}][2.5\varepsilon + \varepsilon_{rf}]L\,. \tag{7}$$

In the proposed algorithm nodes change its state according to energy threshold set, so to find life time of a node we need to find average power consumption at node during all modes and it is given by

$$P_i = \sum_{i}^{n/2}\left\{\sum_{H_{i+1}}^{H_t}\lambda_{rhti}(2.5\varepsilon + \varepsilon_{rf})L + (2.5\varepsilon + \varepsilon_{rf})\lambda_{si}L\right\}/n, \tag{8}$$

where n is threshold interval.

To know networks lifetime we need to calculate average lifetime of critical node. Here critical node means a node which connects sink with other sensor nodes. To calculate average lifetime of a node we need to calculate maximum traffic rate arriving at critical node in multiple paths as well as single path case.

4.1 Multiple Paths Analysis

Maximum traffic that can arrive at critical node is given by

$$Relay\,packets + Own\,generated\,packets = \left[\sum_{i}^{n_p} \lambda_{ri} f(e) + \sum_{i}^{n_{cp}} \lambda_{ri} + \lambda_{si}\right] . \quad (9)$$

Where, n_p is the number of path connected to critical node and n_{cp} means the number of critical paths connected to critical node. If we consider that multiple paths algorithm is energy aware, n_p is depends on a function of energy and its value varies from 1 to 0. Here critical path means, a path don't have any other routing path except one connected to critical node. Now from (1), (7) and (9), power consumed at multiple paths node is given by

$$P_i = \left[\sum_{i}^{n_p} \lambda_{ri} f(e) + \sum_{i}^{n_{cp}} \lambda_{ri} + \lambda_{si}\right] \times [2.5\varepsilon + \varepsilon_{rf}]L\,, \quad (10)$$

from (4) and (10) node lifetime is given by

$$E(t_i) = E_0 / \left\{ \sum_{i}^{n_p} \lambda_{ri} f(e) + \sum_{i}^{n_{cp}} \lambda_{ri} + \lambda_{si} \right\} (2.5\varepsilon + \varepsilon_{rf})L. \quad (11)$$

4.2 Single Path Analysis

Maximum traffic that can arrive at node when it is in relay mode is given by

$$\lambda_{ri} = \sum_{H_{i+1}}^{H_t} \lambda_{rhti} \,. \quad (12)$$

Where, H_t means total number of hop count and H_i is individual node hop count. Maximum traffic at node when it is in sense mode is given by λ_{si}. If we set energy threshold to E_0/n, average P_i is given from (8).From (4) and (8) node lifetime is given by

$$E(t_i) = E_o n / \left[\sum_{i}^{n/2} \left\{ \sum_{H_{i+1}}^{H_t} \lambda_{rhti}(2.5\varepsilon + \varepsilon_{rf})L + (2.5\varepsilon + \varepsilon_{rf})\lambda_{si} L \right\} \right] , \quad (13)$$

from (11) and (13) we can compare a critical node lifetime for single and multiple path algorithms.

5 Numerical Analysis

All nodes have 6 joule battery capacity which can support 9000 packets of 32 byte long. In sensor networks relay rate is always higher than the packet generating

Table 1. Parameter's Value

Parameters	Assumed value
E_0	6J
ε	$50n$J
λ_s	$5pkt/hr$
L	$32byte$
ϵ_{rf}	2.5μJ

rate. From [4,12,15] we assumed some parameter's values and summarized them in table 1.

Figure 4 shows the average number of packet processed by a node in the proposed algorithm. From figure 4 we can observe that the number of packet processed by a critical node changes according to the operating mode. Normally packet arrival rate is very high compared to packet generating rate. We divided the total energy into the number of threshold level, as we proposed in the algorithm node will change its state on every threshold level. As we can see it from figure 4, the number of packet arrival rate will also change accordingly. When a node is in a relay mode it has to process on a larger number of packets than in sense mode. This is the key factor of the proposed algorithm. Because of two operating modes average arrival rate of packets is low compared to energy aware multiple paths algorithm.

Figure 5 and 6 show some important numerical results which are based on a node lifetime analysis and assumptions. Figure 5 shows the critical node lifetime. From figure 5 we can observe that the proposed algorithm increases the critical node lifetime and hence a networks lifetime. From figure 5 we can observe that the arrival rate of packets in multiple paths algorithm is depends on a function of energy. As the energy level decreases the number of packet process by a node decreases but the average arrival rate of packet is higher than a single path algorithm. This is the main difference between the two algorithms. Figure 6 shows the energy consumed by 1 packet to reach destination from source. The proposed algorithm always creates a minimum hop path from source to

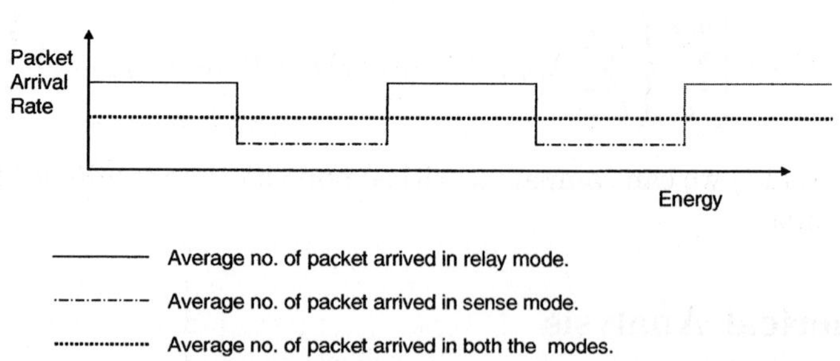

Fig. 4. Average number of packets process by a node

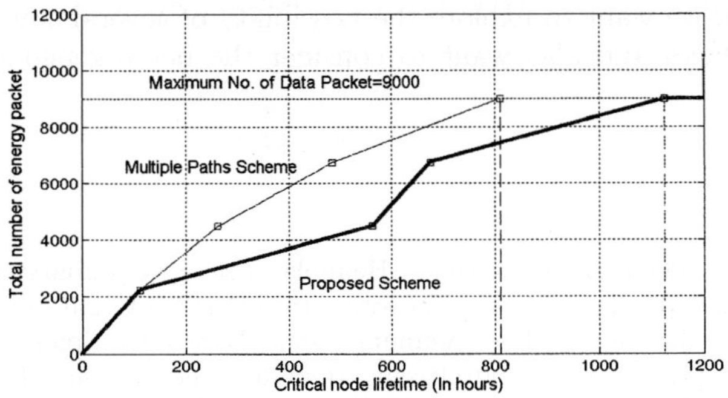

Fig. 5. Critical node lifetime

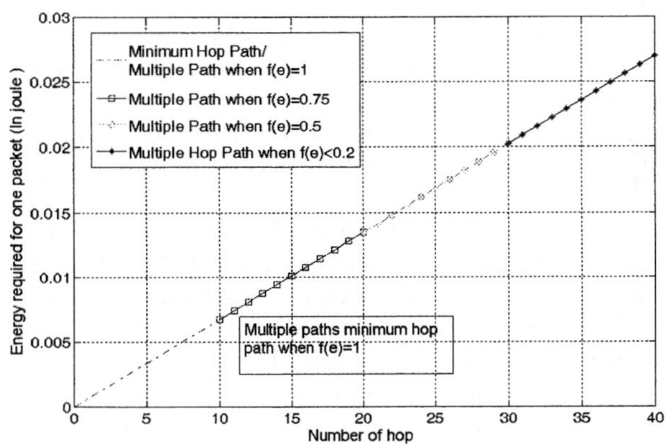

Fig. 6. Hop path for a node

destination. But in multiple paths algorithm it is a function of energy.For real time application the number of hop required to transmit the data from source to destination is very important because it involves the delay factor in transmission.

6 Conclusion and Future Work

In this paper we proposed the algorithm to increase a network lifetime and we compared it with multiple energy aware paths algorithm. The proposed algorithm increases a network lifetime by fairly distributing the relay load among the nodes with the help of two different operating modes. So our approach is suitable for a large number of sensor network. Numerical result shows the good improvement in a critical node lifetime. The proposed algorithm increase the lifetime by around 38% which looks quite promising and the proposed algorithm generate a minimum hop path which is very important result for the real time data applications.

In future work we want to explore the possibility of several sink nodes located in different places and also want to consider the heterogeneous nodes in the networks.

References

1. Z. Shelby, C. Pomalaza-Raez and J. Haapola.: Energy Optimaization in Multihop Wireless Embedded and sensor Networks. 15th IEEE International symposium on Personal, Indoor, and Mobile communications, Barcelona, Spain, September 5-8
2. M. Bhardwaj and A.P. Chandrakasan : Bounding the Lifetime of sensor Networks Via Optimal Role Assignments.IEEE INFOCOM 2002 (2002), Vol. 3, 1587-1596
3. E.J. Duarte-Melo and M. Liu.: Analysis of Energy Consumption and Lifetime of Hetrogenous Wireless Sensor Networks. IEEE GLOBCOM 2002 (2002), Vol. 1, November, 21-25
4. J. Zhu and S. Papavassilios.: On the Energy-Efficient Organization and the Life Time of Multi hope Sensor Networks. IEEE Communication letters (2003), Vol. 7, No. 11, November
5. S. Coleri, M. Ergen and T.J. Koo.: Life Time Analysis of a Sensor Network with Hybrid Automata Modelling. WSNA 2002 (2002), ACM publication, Atlanta, September
6. R. Shah and J.M. Rabaey.: Energy Aware Routing for Low Energy Ad hoc Sensor Networks. IEEE WCNC'02 (2002), Orlando, Vol. 1, March, 350-355
7. S.C. Huang and R.H. Jan.: Energy-Aware, Load Balanced Routing Schemes for Sensor Networks. In Proc. Tenth international Conference on Parallel and Distributed System (ICPAD'04) (2004), July, 419-425
8. E. Gelenbe and R. Lent.: A power-Aware Routing. International Symposium on Performance Evalution of Computer and Telecommunication Systems (SPECTS'03)(2003), Montreal, Canada, July
9. H. Saito and H. Minami.: Performance Issues and Network Design for Sensor Networks. IEICE Trans.Communication (2004), Vol. E87-B, No. 2, February
10. A.F. Raquel, B. Nath and A.A.F. Loureiro.: A Probabilistic Approach to predict the Energy Consumption in Wireless Sensor Networks. Comunicacao Sem Fio e Computacao Movel (2002), Sao Paulo, Brazil, October
11. J.F. Chamberland and V.V. Veeravalli.: Decentralized Detection in Sensor Notworks. IEEE Trans. Signal Processing (2003), Vol. 51, February, 407-416
12. H. Xiaoyan, G. Mario and W. Hanbiao.: Load balanced, Energy-aware communications for Mars sensor networks. In Proc. IEEE Aerospace Conference (2002), Vol. 3, 1109-1115
13. B. Bhardwaj, G. Timothy and C.P. Anantha.: Upper Bounds on the lifetime of sensor networks. IEEE ICC'01 (2001), Vol. 1, June, 1633 - 1639
14. B. Deb, S. Bhatnagar and B. Nath.: A Topology Discovery Algorithm for Senssor Network with Application to Network Management. In IEEE CAS workshop (2002), September
15. K. Akkaya and M. Younis.: Energy and QoS Aware Routing in Wireless Sensor Networks. In Proc. of the IEEE MWN'03 (2003), Providence, Rhode Island, May

Utility Based Service Differentiation in Wireless Packet Network*

Jaesung Choi[1] and Myungwhan Choi[2]

[1] Samsung Electronics, Seoul, Korea
js98.choi@samsung.com
[2] Dept. of Computer Science and Eng., Sogang Univ., Seoul 121-742, Korea
mchoi@ccs.sogang.ac.kr

Abstract. A new wireless packet scheduling scheme is proposed to support differential services to different users over the forward wireless links whose transmission speed changes due to the ever-changing channel conditions. The scheme is intended to maximize the overall system satisfaction level using utility function. Internet applications of elastic, delay adaptive real-time, and rate adaptive real-time types are considered. Obtaining the optimal control scheme is a complicated problem, requiring detailed information on the channel statistics. Even if it were known, the computations would be significantly hampered by the curse of dimensionality. So, a heuristic method, which can run online, is proposed and its performance is evaluated.

1 Introduction

A new wireless packet scheduling scheme is proposed to support differential services to different users over the forward wireless links whose transmission speed changes due to the ever-changing channel conditions.

With the need for multimedia services to be supported over the same wireless network infrastructure, the scheduler (or resource allocator) should determine the application that would derive the maximum benefit from a certain quantity of resource, and then make a scheduling decision based on this information. The idea of utility functions from micro-economics captures this notion of importance of the quantity of resource allocated to the application very well.

There are several resource allocation and scheduling schemes maximizing aggregate utility in the literature [1,2,3,4,5,6]. Schemes proposed in [1,2,3,4] were investigated in wireline domain. However they cannot be directly carried over to wireless system because of unique characteristics of the wireless channel such as channel-condition-dependent performance.

By using adaptation technique such as controlling coding rate or spreading factor, different wireless spectral efficiency can be attained depending on the channel condition [7]. All the major wireless standards have included procedures

* This work was supported by the ministry of information and communications, Korea, under the information technology research center (ITRC) support program.

C. Kim (Ed.): ICOIN 2005, LNCS 3391, pp. 303–312, 2005.

to exploit this: adaptive modulation and coding schemes are implemented in the 3G TDMA standard, and variable spreading and coding are implemented in the 3G CDMA standards. In general, a user is served with better quality and/or at a higher bit rate when the channel condition is better. However, not many literatures have reflected this feature in the scheduling schemes.

The schemes in [5,6] were investigated in wireless network having a characteristic of channel condition dependent capacity. Proportional fairness scheduling (PFS) in [5] was proposed by the QUALCOMM and is being used for the CDMA2000 1xEV-DO service. It maximizes the average aggregate utility when the utility function is logarithmic [1]. The authors in [6] focused on the delay performance of schedulers. They proposed a scheme to maximize the time averaged utility, where utility is a decreasing function of the delay incurred when serving a request. In both works, however, it is assume that the utility function is concave.

In [8], the author classified the Internet multi-media applications into four classes (elastic, hard real-time, delay adaptive real-time, and rate adaptive real-time) and modeled their utility functions. Among them the utility functions of delay adaptive real-time and rate adaptive real-time types are partially convex and the schemes proposed in [5,6] cannot be used for these types of applications.

In this paper, Internet applications of elastic, delay adaptive real-time, and rate adaptive real-time types are considered. Obtaining the optimal solution of the problem is a complicated problem, requiring detailed information on the channel statistics. Even if it were known, the computations would be significantly hampered by the curse of dimensionality. Instead, a heuristic method, which can run online, is proposed in this work.

An extension to support multiple subscription levels, for example, for premium and light services, is also proposed. It enables to assign different subscription levels to different users, that is, it can allocate more bandwidth to premium users than light users.

2 Wireless Network Model

In this paper, networks having a characteristic of channel condition dependent capacity are considered. Namely, users with better channel conditions are served at higher data rates via adaptation technique. Specifically, networks providing services such as high data rate (HDR) are considered. HDR is specified as the service to provide for the high speed forward Internet access in the CDMA2000 1xEV-DO (IS-856) [5]. It uses the modulation and coding scheme adaptive to the varying channel conditions. The proposed scheme, however, may be applied to other networks having a characteristic of channel condition dependent capacity with minor modifications.

A time-slotted forward-link system is considered in this paper. The slot duration is fixed and the access point (AP) (or base station) can transmit data to only one access terminal (AT) (or mobile terminal) during each time slot. Differently from the conventional time division multiplexing (TDM) system, the data rate of the forward link during a time slot is not fixed. Instead, it changes as the channel state to the corresponding AT changes.

The data rate used in a time slot is determined as follows. The AP transmits pilot burst of the fixed transmit power level every time slot. Each AT i measures the signal to noise ratio (SNR) of the pilot burst and determines the data rate $r_i(k)$ to be used during time slot k. As the SNR of the pilot signal may change as the AT moves around, the data rate $r_i(k)$ may change continuously. The AT transmits the information about that data rate to the AP via the backward link. Then, the AP selects an appropriate AT for transmission based on the data rate information. For the HDR service, the SNR bound, the transmission frame length, and the number of time slots needed to transmit the frame for a given data rate are shown in [9]. For example, the AT whose data rate is 38.4 kbps will receive a packet of 1024 bits over 16 consecutive time slots and the AT whose data rate is 2457.6 kbps will receive four packets of 1024 bits each over only one time slot.

3 Proposed Scheme

3.1 Introduction to Utility

The service satisfaction level differs for different applications for a given delay and/or throughput. In [8], the author classified the Internet applications into four classes and modeled their utility functions as shown in Fig. 1. Utility function is the increasing function of the bandwidth allocated to the corresponding applications. With this formalism, the goal of scheduler design is to maximize the sum of the utilities.

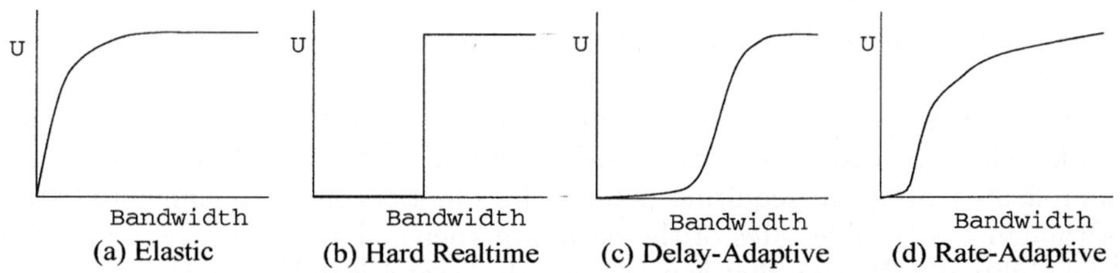

Fig. 1. Utilities as a function of bandwidth

3.2 Scheduling Algorithm

Among the above four classes, elastic, delay adaptive real-time, and rate adaptive real-time applications are considered in this paper. Notice that hard real-time applications need fixed predetermined bandwidths. The utility functions of these three types may be modeled by the unified model as shown in Fig. 2. In the figure, $U_a(R)$ is the utility function of an application a as a function of achieved throughput R, and $R_{a,F}$ is a point of inflection for application a. The utility function is convex in the range $[0, R_{a,F}]$ and concave in the range $[R_{a,F}, \infty]$. The utility functions of elastic type applications are for the case of $R_{a,F} = 0$, where there is no convex region.

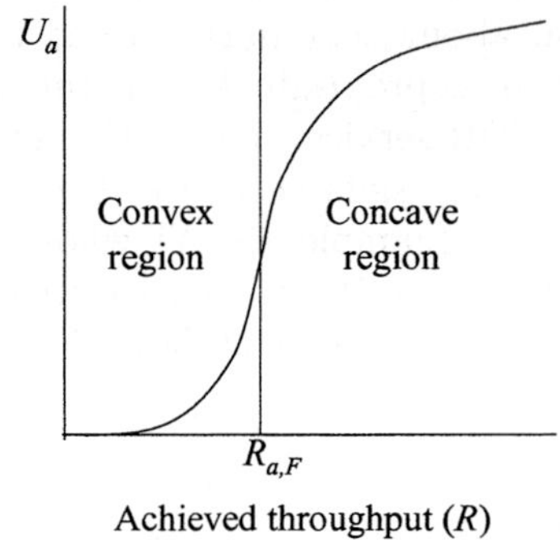

Fig. 2. Unified model of utility functions

In this work, the objective of the scheduling scheme used for the downstream traffic is to maximize the overall system utility. That is, the objective is represented by (1).

$$\text{maximize} \sum_{i \in B(k)} U_{a(i)}(R_i) \tag{1}$$

where $B(k)$ is the set of backlogged sessions at time slot k and $a(i)$ is the application of session i.

Obtaining the optimal solution of the above problem for the wireless networks where a wireless link is shared by many users as in HDR system is a complicated problem, requiring detailed information on the channel statistics. Although the feasible rates of the users are known slot by slot, the underlying probability distribution that is producing these rates is unknown. Even if it were known, feasible rates might be correlated, so that the computations would be significantly hampered by the curse of dimensionality. To avoid these obstacles, a heuristic method is proposed in this paper.

The session i's achieved throughput, $R_i(k)$, at time slot k is computed by (2): the moving average with average filter time constant k_c.

$$R_i(k) = \begin{cases} \left(1 - \frac{1}{k_c}\right) \times R_i(k-1) + \frac{1}{k_c} \times r_i(k-1) & \text{, if session } i \text{ is served at } k-1 \\ \left(1 - \frac{1}{k_c}\right) \times R_i(k-1) & \text{, otherwise} \end{cases} \tag{2}$$

Define $R_i^+(k)$ and $R_i^-(k)$ as follows:

$$\begin{aligned} R_i^+(k) &= \left(1 - \tfrac{1}{k_c}\right) \cdot R_i(k) + \tfrac{1}{k_c} \cdot r_i(k) \\ R_i^-(k) &= \left(1 - \tfrac{1}{k_c}\right) \cdot R_i(k) \end{aligned} \tag{3}$$

That is, if session i is served during time slot k, $R_i(k+1) = R_i^+(k)$, and if not, $R_i(k+1) = R_i^-(k)$. At decision epoch of time slot k, the proposed algorithm selects the session which will maximize the overall system utility at time slot $k+1$. That is, the objective function is

$$\text{maximize} \sum_{i \in B(k)} U_{a(i)}(R_i(k+1)) \tag{4}$$

Assume that a session j is selected to be served during the time slot k. By applying (3) to (4), we can get

$$\sum_{i \in B(k)} U_{a(i)}(R_i(k+1)) = \sum_{i \in B(k)} U_{a(i)}(R_i^-(k)) + U_{a(j)}(R_j^+(k)) - U_{a(j)}(R_j^-(k)) \tag{5}$$

To maximize the above sum, j should be

$$j = \arg \max_{i \in B(k)} \{U_{a(i)}(R_i^+(k)) - U_{a(i)}(R_i^-(k))\} \tag{6}$$

Therefore, the selection rule of this algorithm is to find j using (6).

This selection rule may work well for elastic class of applications. However, it doesn't work for delay adaptive real-time class and rate adaptive real-time class of applications, because of the convex region. In the convex region, the increasing rate of utility around R=0 is almost zero, and therefore $U_a(R^+) - U_a(R^-)$ is also almost zero around R=0. This means that the delay and rate adaptive real-time classes of applications are hardly selected for transmission before the elastic class of applications are satisfied fully. To solve this problem, the potential utility curve is proposed to be used in this paper. It is a straight line starting from the origin with the slope P_a, which is obtained as follows:

$$P_a = \max_R \left\{ \frac{U_a(R)}{R} \right\} \tag{7}$$

As shown in Fig. 3, P_a is the minimum slope satisfying $U_a(R) \le P_a \times R$. That is, the actual utility curve is below the potential utility curve $P_a \times R$. In this figure, $(R_{a,P}, U_a(R_{a,P}))$ is the intersection point of the actual utility curve and the potential utility curve.

P_a is the maximum average utility increment per unit achieved throughput, which is obtained when the achieved throughput is $R_{a,P}$. Therefore if the achieved throughput goes over $R_{a,P}$, P_a may be regarded as the increasing rate of utility in the region $[0, R_{a,P}]$. This conception is used to serve delay and rate adaptive real-time classes of applications in the proposed algorithm. That is, for the delay and rate adaptive real-time classes of applications,

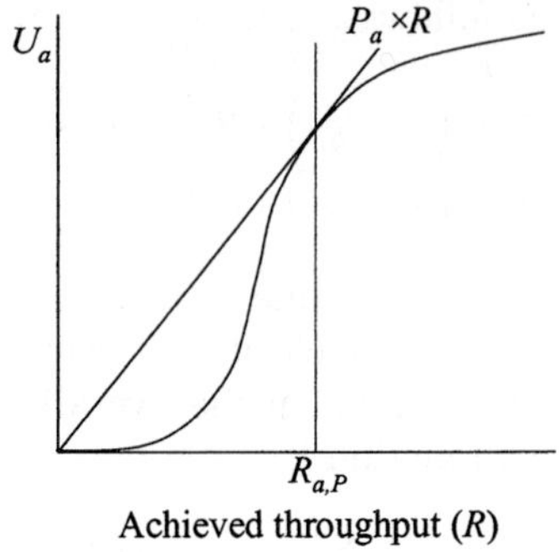

Fig. 3. Potential utility curve ($P_a \times R$)

if $R_i(k) \geq R_{a(i),P}$
use $U_{a(i)}(R_i(k))$ for the utility function
else
use $P_{a(i)} \times R_i(k)$ for the utility function

3.3 Extension to Support Multiple Subscription Levels

In the above section, the service satisfaction level was defined for Internet applications and a heuristic algorithm to maximize the sum of utilities for the applications of elastic, delay adaptive real-time, and rate adaptive real-time types was proposed. With this scheme, all sessions serving the same kind of application are treated impartially. That means, if any two sessions serving the same kind of application experience exactly the same channel conditions, the achieved throughputs of them are identical.

A question to ask is whether this equalitarianism is always desirable? The answer is, it may not be. For example, present wired broadband data network service providers have classified service levels of subscription into premium (or gold) and light (or silver), where the premium service (provided to customers paying higher subscription fee) supports more bandwidth than the other. The same thing can happen to wireless data network.

To support multiple levels of subscription, it will be a simple and efficient way to define different utility functions for different subscription levels. One way to do this is to multiply utility by some weight assigned to each subscription level. With this approach, however, the saturation point $R_{a,H}$ which represents the minimum throughput beyond that the utility of the application a is saturated doesn't change. In some case, the higher level user may want bigger $R_{a,H}$ to be assigned. To support this requirement, it will be a simple way to multiply some expansion factor to the achieved throughput.

Let $s(i)$ be the subscription level of session i, $w_{a,s}$ be the weight of subscription level s of application a, and $e_{a,s}$ be the expansion factor of subscription level s of application a. Then, the objective function and the selection rule can be rewritten as (8) and (9), respectively.

$$\text{maximize} \sum_{i \in B(k)} w_{a(i),s(i)} \cdot U_{a(i)}(e_{a(i),s(i)} \cdot R_i(k+1)) \tag{8}$$

$$j = \arg\max_{i \in B(k)} \{w_{a(i),s(i)} \cdot \{U_{a(i)}(e_{a(i),s(i)} \cdot R_i^{+}(k)) - U_{a(i)}(e_{a(i),s(i)} \cdot R_i^{-}(k))\}\} \tag{9}$$

4 Simulation

4.1 Simulation Model

In this section, the performance of the proposed algorithm is presented. In the simulation, it is assumed that the traffic source of each session is persistent and every time slot is completely filled by the traffic source. To observe the throughput performance experienced by users when the channel conditions change, we conducted the simulation under the time-varying channel conditions.

We considered two kinds of applications a and b. Application a is an elastic type and b is an adaptive real-time type (delay adaptive or rate adaptive). For each type of applications, we assumed that there are two subscription levels, i.e. premium and light. Figure 4 shows the utility functions used in the simulation. In Fig. 4, $w_{a,light} = 1$, $w_{a,premium} = 2$, $w_{b,light} = 1$, $w_{b,premium} = 2$, $e_{a,light} = 1$, $e_{a,premium} = 0.5$, $e_{b,light} = 1$, $e_{a,premium} = 1$.

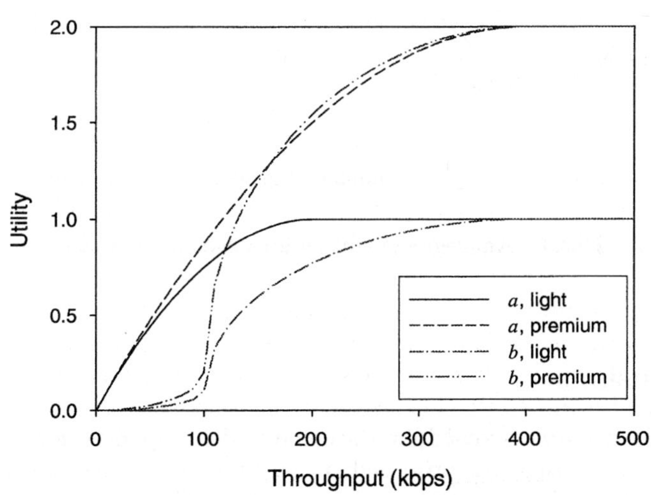

Fig. 4. Utility function

4.2 Results

The experiments are classified into three cases: case a) when all sessions are serving application a, case b) when all sessions are serving application b, and case c) when half of the sessions are serving application a and the others are serving application b. In each case, half of the sessions of each application are premium subscriber and the others are light subscriber. Figure 5 shows the aggregate utility for cases a, b, and c as a function of the number of users. The PFS scheme was implemented and simulated for comparison. In all cases, the achieved aggregate utility of the proposed scheme is greater than that of PFS. Note that the PFS is identical to the proposed scheme using logarithmic utility function [1]. Figures 6 and 7 show the average achieved throughput per session for cases a and b. Figures 6 and 7 demonstrate that the proposed scheme can provide differentiated services among different subscription levels, premium and light.

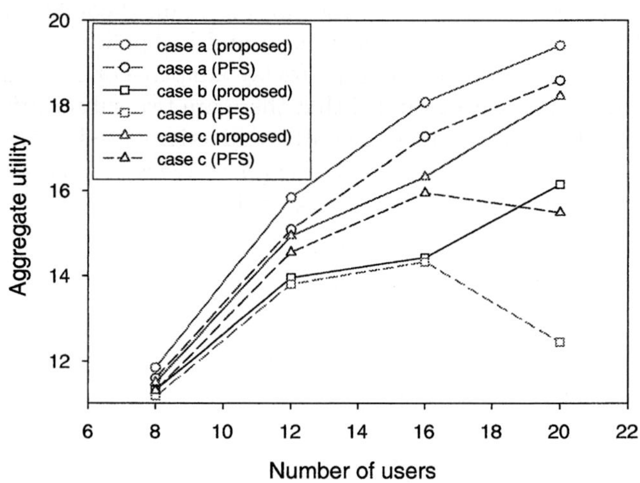

Fig. 5. Aggregate utilities for cases a, b, and c

5 Conclusions

In this paper, a wireless scheduling scheme which supports service differentiation to different users are proposed. It is intended to maximize the overall system satisfaction level in terms of aggregate utility for the applications of elastic, delay adaptive real-time, and rate adaptive real-time types. Obtaining the optimal solution is a complicated problem. So, a heuristic method, which

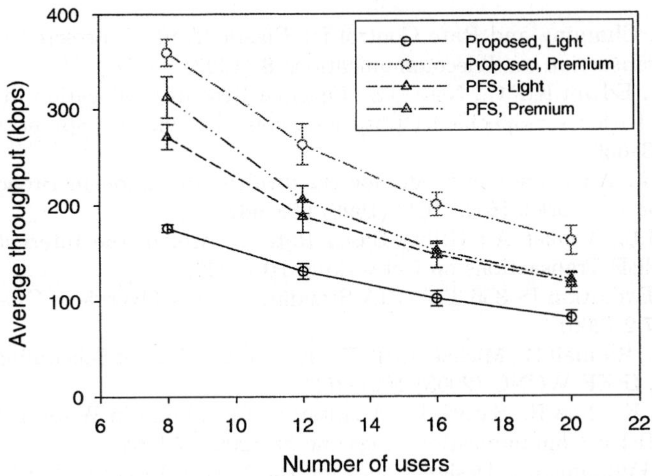

Fig. 6. Average achieved throughput per session for case a)

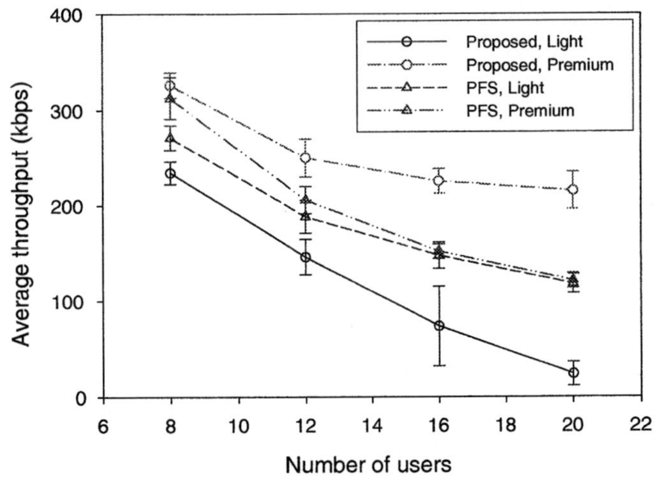

Fig. 7. Average achieved throughput per session for case b)

can run online, was proposed. Study on how much the performance of the proposed scheme differs from the optimal solution is reserved for future work. It is also important to determine the appropriate utility functions. It should reflect the user requirements. The system operator may intend to maximize the profit properly determining the utility function. Therefore, tradeoff between these two requirements will be needed in determining the utility functions.

References

1. Frank K.: Charging and Rate Control for Elastic Traffic (corrected version), European Transaction on Telecommunications **8** (1997) 33–37
2. Suresh K., Edwin K.P.C., Ness B.S.: Optimal Resource Allocation in Multi-Class Networks with User-Specified Utility Functions, Elsevier Computer Networks **38** (2002) 613–630
3. Hongbin J.: An Economic Model for Bandwidth Allocation in Broadband Communication Networks, IEEE ICC (1996) 658–662
4. Richard J.L., Venkat A.: Utility-Based Rate Control in the Internet for Elastic Traffic, IEEE Transactions on Networking **10** (2002)
5. 1xEV:1x Evolution IS-856 TIA/EIA Standard Airlink Overview, Qualcomm Inc., Revision 7.2 (2001)
6. Peijuan L., Randall B., Michael L.H.: Delay-Sensitive Packet Scheduling in Wireless Networks, IEEE WCNC (2003) 1627–1632
7. Sanjiv N., Krishna B., Sarath K.: Adaptation Techniques in Wireless Packet Data Services, IEEE Communication Magazine **38** (2000) 54–64
8. Scott S.: Fundamental Design Issues for the Future Internet, IEEE Journal on Selected Areas in Communications **13** (1995) 1176-1188
9. Paul B., Peter B., Matthew G., Roberto P., Nagabhushana S., Andrew V.: CDMA/HDR: A Bandwidth-Efficient High-Speed Wireless Data Service for Nomadic Users, IEEE Communications Magazine **38** (2000) 70–77

ComBAQ: Provisioning Loss Differentiated Services for Hybrid Traffic in Routers*

Suogang Li[1], Jianping Wu[2], and Ke Xu[3]

Department of Computer Science, Tsinghua University,
Beijing, 100084 P. R. China
[1]lsg@csnet1.cs.tsinghua.edu.cn
[2]jianping@cernet.edu.cn
[3]xuke@mail.tsinghua.edu.cn

Abstract. The hybrid types of traffic containing traditional data and multimedia streams are delivered in networks. Besides the Best Effort service, Internet has to supply other different services to various types of traffic with multiple priorities. DiffServ architecture is considered to be the promising Internet framework meeting QoS requirements. One of crucial elements in DiffServ is its network node scheme. In this paper, we propose a novel scheme which combines dynamic buffer management with active queue management to assist routers in provisioning loss differentiated services. In order to support several levels of different traffic classes, we extend a dynamic threshold buffer management method in the case of multiple loss priorities. We detail the implementation of the scheme on routers. Simulation results confirmed that our scheme is able to adapt to load changing conditions and offer different guarantees for hybrid traffic flows in terms of their loss priorities.

1 Introduction

Nowadays there is an increasing demand for streaming multimedia applications over the Internet. The traffic in networks is mingled with traditional data and multimedia streams. So Internet is needed to supply different services with multiple priorities, in addition to the BE (Best Effort). IETF specified DiffServ (Differentiated Services) [1] as a current trend in the Internet community which supports QoS, since it is a scalable QoS architecture not burdened with the complex task of reservation states creation and maintenance. The IETF DiffServ working group has specified the Assured Forwarding (AF) per hop behavior (PHB) [2], which is intended to provide different levels of forwarding assurances for IP packets in routers.

Using appropriate buffer allocation and AQM (Active Queue Management) methods we can achieve the AF PHB and support various levels of loss assurances. A buffer-sharing method with Dynamic Threshold improves the router

* This work is supported by the National Natural Science Foundation of China under Grant No. 60303006 and No. 60203025 and the National Grand Fundamental Research 973 Program of China under Grant No. 2003CB314801.

C. Kim (Ed.): ICOIN 2005, LNCS 3391, pp. 313–320, 2005.

throughput and the AQM method detects incipient congestion early to avoid packet loss, e.g. RED (Random Early Detection) [3]. In the literature, SPRED [4] proposed by Hou et al., generalizes buffer management algorithms of Drop-Tail, Pushout and RED. This mechanism provides different services for real-time flows and traditional TCP flows. However, this scheme has not capability of offering more different service levels for real-time flows or TCP flows, respectively. Aweya et al. propose an AQM algorithm using dynamic buffer thresholds in a shared-memory architecture [5]. It targets to ensure the fair sharing of buffer under changing traffic conditions. But it is not suitable for the real-time traffic requiring high loss-priority service, for its packet drop routines treat them just like the TCP traffic. Moreover, this scheme and the above one use RED that has some disadvantages detailed in Section 2.2.

To remedy these deficiencies, we propose a scheme combining dynamic buffer allocation with active queue management to support differentiated services in routers, called *comBAQ* (*com*bining *B*uffer Management and *A*ctive *Q*ueue Management). Our scheme can provide loss-differentiated services for the real-time multimedia streams and congestion control for traditional TCP flows as well. For scalability, our scheme conforms to the DiffServ principle of no per-flow state maintained in core routers. According to IETF's advices that the network does not re-order packets belonging to the same flow [6], our scheme stores and services all packets in the same queue. Furthermore, the simple implementation of this scheme and its adaptability to dynamic network is considered an important virtue by us.

This paper is organized as follows. In Section 2, we simply explain why we select this or that kind of buffer management and AQM methods in our scheme. In Section 3, we adapt a dynamic threshold to the multiple priorities case, and then we define in detail packet service type and illustrate the implementation of comBAQ. Section 4 demonstrates the performance of our scheme in supporting differentiated services through simulation experiment results. Section 5 concludes this paper.

2 Background

2.1 Buffer Management

So far, many buffer management methods have been proposed that can be broadly categorized into three classes as static methods, Pushout methods, and dynamic methods. Firstly, conventional schemes allocate buffer space to each queue through static buffer thresholds, which adapt poorly to dynamic network. Secondly, the so-called Pushout mechanism allows an incoming packet to enter the buffer by discarding some other packets in the buffer. We integrate it with the methods described later to guarantee that high priority packets can seize buffer preemptively. Moreover, dynamic buffer methods are heuristic and adaptive. Since our scheme intends to support differentiated loss services, we will describe a dynamic threshold method scheme adapted for the cases of multiple loss priorities in Section 3.

2.2 Active Queue Management

Most of AQM algorithms are devised only for the data flows supporting retransmit, such as TCP traffic. The traditional technique for managing router queue is the DropTail mechanism which is unable to offer service differentiation for our targets. RED is one of the most famous AQM schemes. Its randomization in packet dropping avoids the global synchronization effect of all connections and maintains high throughout for TCP traffic in the routers. Some key problems associated with RED are: (*i*) RED fails to prevent buffer over flow as sources increase [8]; (*ii*) Probability calculation triggered by packet arrival is not applicable for high-speed link; (*iii*) Some parameters of RED are dissatisfactory, including that the performance is extremely sensitive to parameter setting.

Another adaptive AQM algorithm called WSAP (Weighted Simple Adaptive Proportional) [9] has been proposed by authors in our lab. WSAP algorithm is shown in Fig. 1. It has following three advantages: (*i*) dropping probability is calculated periodically for the given interval; (*ii*) it has better scalability with less parameters to be set for each queue priority; (*iii*) the loss weight configuration is easier. We choose WSAP to manage the queues of the packets that can be retransmitted in comBAQ.

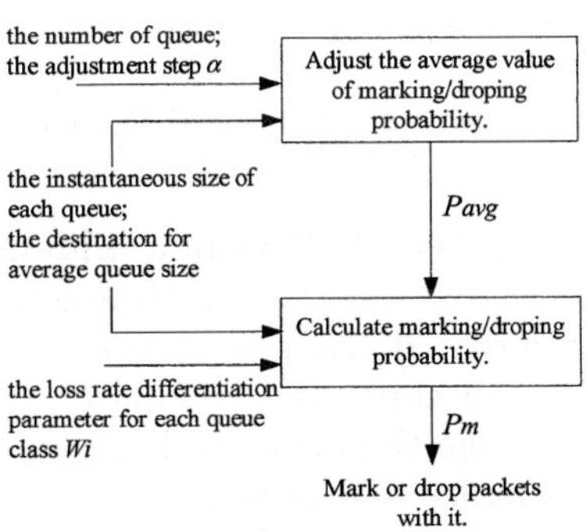

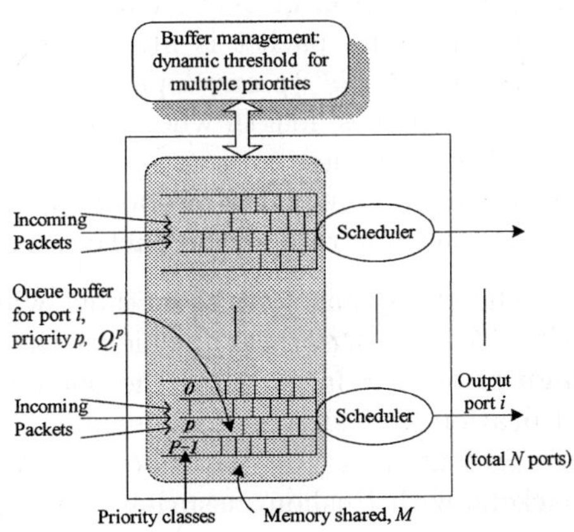

Fig. 1. The WSAP algorithm routine **Fig. 2.** Shared buffer architecture model

3 Service Framework of ComBAQ

3.1 Adaptation to Multipriority

Fan et al. propose a buffer management method with dynamic threshold [7]. Considering the case of various priorities, we extend the method to handle traffic with a number of loss priority classes. Fig. 2 shows the shared buffer architecture model with queueing at output port, using the buffer management with dynamic threshold for multipriority classes. In this figure, overall buffer size is M, shared by N ports. We assume there are P loss priority classes marked as 0 through $P-1$, where class 0 packets belong to the most loss-sensitive class, i.e., with the highest priority and class $P-1$ packets belong to the most loss-tolerant, i.e., with the lowest priority. The output queues with such loss priority classes may be contained in any output port. The queue length with priority p at port i is $Q_i^p(t)$, and thus overall queue length at time t is $Q(t) = \sum_{i=1}^{N}\sum_{p=0}^{P-1} Q_i^p(t)$, where $Q(t) \le M$, obviously. Q_0 denotes the targeted amount of buffer to be used. We set the threshold $Th^p(t)$ for the queues with priority class p at time t to take the place of the common threshold for all port queues in the approach in [7]. When a packet with priority class p destined to port i at time t, the threshold can be computed as the following algorithm:

$$Th_{new}^p(t) = \begin{cases} \max_{1\le i\le N}\{\sum_{q\ge p} Q_i^q(t)\}, & Q(t) < Q_0; \\ \max(Th_{old}^p - c, Th_{min}), & Q(t) \ge Q_0. \end{cases} \quad (3.1)$$

where c is arrival packet size in bytes and Th_{min} represents the minimum threshold. If α denotes the rate of available buffer, then $Q_0 = \alpha M$. Equation 3.1 implies that:

1. when the overall queue length, $Q(t)$, is less than the utilizable buffer, Q_0, the arriving packet is always accepted into its respective queue, and the threshold Th^p (p enveloped in the packet) is updated to the largest sum of those queues' length whose priority classes are equal to or greater than p, in all port queues.
2. when $Q(t)$ is equal to or above Q_0, the packet is dropped and Th^p is decreased at the same rate as the packet arrival rate.

Therefore, this control method, which we call DTMP (Dynamic Threshold with Multiple Priorities), achieves the objective that ensures the threshold for high priority is larger than the one for low priority. It has advantages of good utilization and simple implementation, and detail simulation results about it are presented in [10]. We choose DTMP to manage the queue of high priority packets, with Pushout assuring them to access buffer preemptively.

3.2 Service Types and Priorities

In our node scheme, we define two types of services to support differentiated services, namely, the More Guaranteed (MG) service and the Less Guaranteed (LG) service, which is similar to SPRED. Packets under MG service are those requiring reliable delivery, such as some packet critical to multimedia application; packets under LG service may be able to be retransmitted in traditional data flows.

Suppose there are N and M priority classes in MG and LG, labeled as MG[i] ($0 < i \le N$) and LG[j] ($0 < j \le M$), respectively. The loss rates of each class service are Loss(MG[i]) and Loss(LG[j]). The smaller i or j, the higher loss priority is, i.e., the lower loss rate is. The node scheme need achieve followings:

- $Loss(MG[i]) \leq Loss(LG[j]),\ \forall i, j, 0 < i \leq N, 0 < j \leq M$;
- $Loss(MG[i]) < Loss(MG[j])$, if $i < j, 0 < i \leq N, 0 < j \leq M$;
- $Loss(LG[i]) < Loss(LG[j])$, if $i < j, 0 < i \leq N, 0 < j \leq M$.

Every value of MG[i] or LG[j] maps to a DSCP value in the "type of service (TOS) octet" of the IPv4 packet header or in "Flow Label" field of the IPv6 packet header. We designate corresponding DSCP value to their priorities, as detailed in Table 1.

Table 1. The corresponding DSCP value of AF PHB to each comBAQ service priority

Loss priorities	Class 1	Class 2	Class 3
More Guaranteed	001010 (AF11)	001100 (AF12)	001110 (AF13)
Less Guaranteed	010010 (AF21)	010100 (AF22)	010110 (AF23)

Table 2. The decision to accept an arriving packet P with size c, priority i or not

P is	buffer state		decision
MG	$B(MG[i]) < Th_{MG}[i]$	$B_{FREE} \geq c$	accept P;
		otherwise	pushout LG packets in buffer to accept P;
	$B(MG[i]) \geq Th_{MG}[i]$		drop P;
LG	$B_{LG} > (1 - T_{MG})B_{ALL}$, or $B_{FREE} < c$		drop P;
	otherwise		WSAP decides to accept P with probability;

3.3 The Scheme Implementation in Routers

We first give comBAQ algorithm flow as follows. Then, we specify packet data structure and organization in buffer.

Decision Algorithm We define B_{ALL} as the size of overall buffer in router, $B(MG[i])$ as the size of occupied buffer by MG packets with priority i, $B(MG)$ as the size of buffer used by all MG packets and similarly $B(LG)$ as buffer size by LG packets, all in unit of bytes. And then $B_{FREE} = B_{ALL} - B(MG) - B(LG)$ represents the size of unused buffer in bytes. Define T_{MG} to denote the minimum threshold of buffer for MG service packets and the entire buffer is available for MG service packets except the TCP target buffer of WSAP. Table 2 shows how comBAQ scheme decides on whether to accept an incoming packet or not based on current buffer state in router. In this table $Th_{MG}[i]$ is the threshold for MG[i] service packets, computed by Equation 3.1 in DTMP algorithm.

Queue Structure in Buffer In our scheme implementation, we maintain two doubly linked lists: the linked list of all packets in the buffer, L_{ALL} and the linked list of LG service packets, L_{LG}. The latter is embedded in the former, that is to say L_{LG} is part of L_{ALL}. Fig. 3 shows the linked list structure for packets in the buffer at a node.

Discussions The reason why L_{ALL} is a doubly linked is that it is easy to locate the LG packet to be discarded in L_{ALL}. If a singly linked list is used for L_{ALL}, we would spend more time and traverse the list more than once. In addition, there are two policies on which routers can depend to pick packets to drop when congestion occurs: wine (drop new packets and keep old) and milk (drop old packets and keep new). For real-time multimedia a new packet is more important than the old one. In consideration of dropping old packets when necessary, a doubly linked list for L_{ALL} is preferred.

Adding and deleting packets of various sizes cause much memory fragmentation in actual routers, that is the cost of employing doubly linked lists to implement this scheme. A concept can be developed to reduce the cost of memory management, which involves pre-allocating a pool of memory and reserving it until it is actually needed. The memory pre-allocation will be called at system startup.

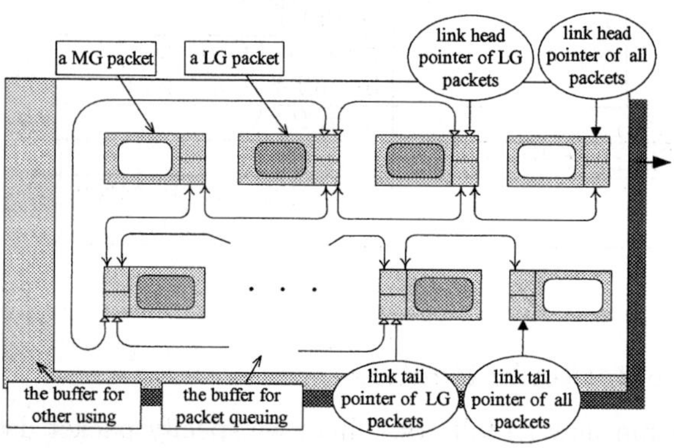

Fig. 3. Linked list data structure

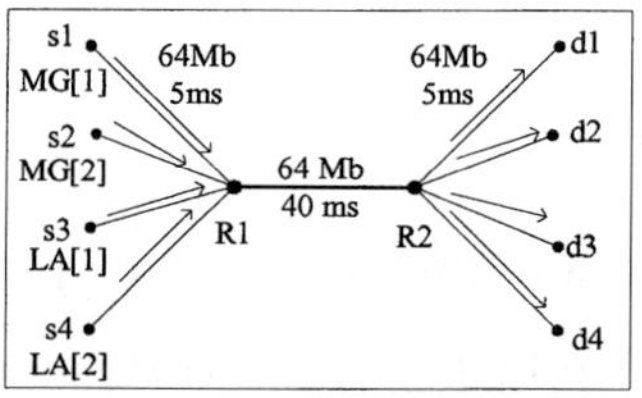

Fig. 4. Network topology

4 Simulation Experiments

4.1 Scenario Setting

In this section we implement our comBAQ scheme on the network simulator, NS-2. The simulation topology is shown in Fig. 4 with a single bottleneck link that has a bandwidth capacity of 64 Mbps. There are four source nodes and four destination nodes. Each traffic from different source has different priorities, for example, Source 1 sends MG[1] service packets. When congestion occurs, we directly drop arriving TCP packets instead of marking them. In comBAQ scheme, B_{ALL} is 40 KB. The target of average queue size for WSAP algorithm is 20 KB, and loss rate differentiation parameters are 1.0 for LG[1] and 4.0 for LG[2]. For DTMP algorithm, $T_{MG} = B_{ALL}*0.1 = 4$ KB. So the maximum buffer space occupied by MG packets is 20 KB.

4.2 Experiment Results

Firstly, 350 applications of burst flows upon UDP are set up in source 1 and source 2, while 40 FTP applications upon TCP are set up in source 3 and source 4 respectively. UDP applications are activated at time 0 s and TCP ones are activated at time 100 s. They are all stopped at time 200 s. As illustrated in Fig. 5, the MG[1] queue is longer than that of MG[2] queue due to the differentiation functionality provided by DTMP algorithm. In addition, we observe some LG packets pushed out at time 101 s.

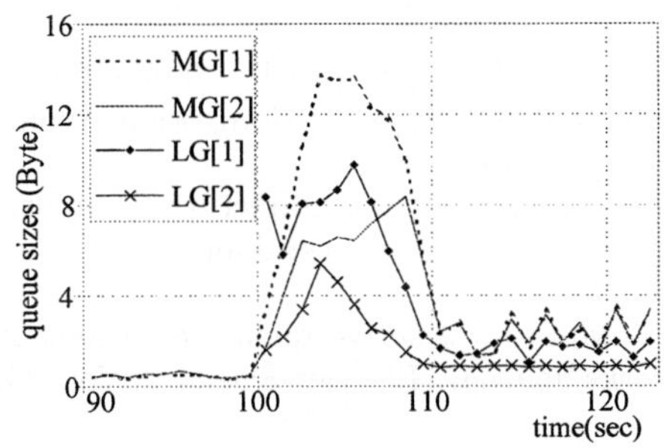

Fig. 5. Queue sizes of 4 flows

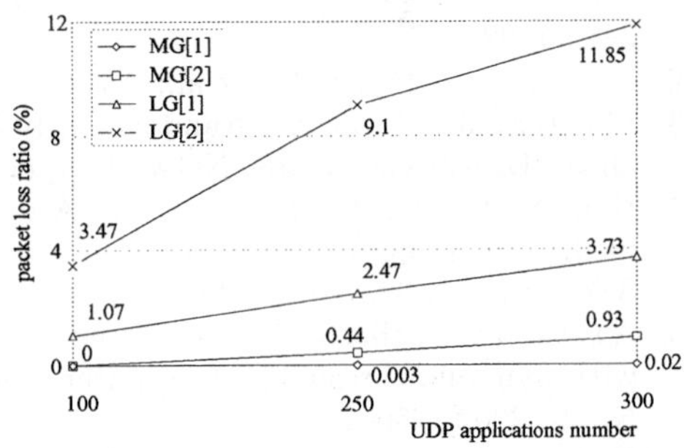

Fig. 6. The loss ratio for 4 flows

Secondly, we set up and immediately activate 100, 250, 300 UDP applications at 0 s for MG[1] and MG[2] respectively, and set up 40 TCP applications for each LG service class at 0 s and activate them at 100 s. Fig. 6 shows the packet loss ratio for four flows. It is easy to observe that MG[1] packet loss ratio is almost zero and MG[2] packets discarded are no more than a few. As MG packet applications increase, some MG packets are discarded due to buffer's insufficiency to accept so many incoming packets. The loss ratio of MG[1] packets here is still lower than that of MG[2]. While LG packet loss ratio, especially that of LG[2], is much higher than MG packet loss ratio.

5 Conclusion and Future Work

In this paper, we propose a novel scheme named as comBAQ to assist routers in providing loss-differentiated services for hybrid traffic containing traditional

TCP flows and real-time multimedia streams. To achieve this objective, we design DTMP for buffer management and choose WSAP as active queue management. We detail implementation of the scheme and give the algorithm routine. The simulation experiment results demonstrated that comBAQ can provide service differentiation according to packet loss priorities. In addition, our scheme is easy to implement in routers due to its simple configuration of parameters.

FUTURE WORK. Our scheme focuses only on the packet-loss aspect of differentiated QoS, while other aspects of QoS such as delay are also important for the scheme. The simulation results indicate that the end-to-end delay of TCP flows increases resulting from many packet losses, for the bandwidth would be mainly occupied by the increasing UDP applications. With limited buffer resource, the throughput improvement of high priority traffic is achieved inevitably at the sacrifice of packet losses. We would give an analysis of the trade-off in our future study.

References

1. Blake, S., Black, D., et al.: An Architecture for Differentiated Services. RFC 2475(1998)
2. Heinanen, J., Baker, F., et al.: Assured forwarding PHB group. RFC 2597 (1999)
3. Floyd, S., Jacobson, V.: Random Early Detection Gateways for Congestion Avoidance. IEEE/ACM Trans. Networking (1993) 397–413
4. Hou, Y. T., Wu, D., Li, B., et al.: A differentiated services architecture for multimedia streaming in next generation Internet. Computer Networks Journal (Elsevier Science), Vol. 32, No. 2 (2000) 185–209
5. Aweya, J., Ouellette, M., Montuno, D. Y.: Multi-level active queue management with dynamic thresholds. Computer Networks Journal (Elsevier Science), Vol. 25, No. 8 (2002) 756–771
6. Nichols, K., Jacobson, V., et al.: A Two-bit Differentiated Services Architecture for the Internet. RFC 2638 (1999)
7. Fan, R., Ishii, A., et al.: An Optimal Buffer Management Scheme with Dynamic Threshold. Proc. IEEE Globecom (1999) 631–637
8. Athuraliya, S., Li, V. H., Low, S. H., Yin, Q.: REM: Active Queue Management. IEEE Network, Vol. 15, No. 3 (2001) 48–53
9. Zhang, M., Wu, J., Lin, C., Xu, K.: WSAP: Provide Loss Rate Differentiation with Active Queue Management. Proc. IEEE ICCT, Vol. 1 (2003) 385–391
10. Li, S., Xu, K., Wu, J.: Buffer Management Algorithm with Dynamic Thresholds for Multiple Priorities. Technical report, THU CS Technical Report (2002)

Multiresolution Traffic Prediction: Combine RLS Algorithm with Wavelet Transform

Yanqiang Luan

School of Electrical and Information Engineering,
University of Sydney,
Sydney, NSW 2006, Australia
yqluan@ee.usyd.edu.au

Abstract. Numerous research in the literature has convincingly demonstrated the widespread existence of self-similarity in network traffic. Self-similar traffic has infinite variance and long range dependence (LRD) which makes conventional traffic prediction method inappropriate. In this paper, we proposed a traffic prediction method by combining RLS (recursive least square) adaptive filtering with wavelet transform. Wavelet has many advantages when used in traffic analysis. Fundamentally, this is due to the non-trivial fact that the analyzing wavelet family itself possesses a scale invariant feature. It is also proved that wavelet coefficients are largely decorrelated and only has short range dependence (SRD). In this paper, We investigate the computation characteristics of discrete wavelet transform (DWT) and shows that the *à trous* algorithm is more favorable in time series prediction. The proposed method is applied to real network traffic. Experiment results show that more accurate traffic prediction can be achieved by the proposed method.

1 Introduction

Recent measurements and simulation studies have revealed that wide area network traffic has complex multifractal characteristics on short timescales, and is self-similar on long timescales [1]. The widespread existence of self-similarity is also demonstrated [2], [3], [4], [5]. These measurement works collectively revealed that self-similar and long-range dependence phenomena widely exist in network traffic. One of important properties of self-similarity which may have great impact on traffic forecasting and manipulation is long range dependence (LRD). The autocorrelation function $r(k)$ of a self-similar process decay hyperbolically rather than exponentially fast, implying a non-summable autocorrelation function $\sum_k r(k) = \infty$. As a result of this LRD, conventional prediction algorithm is not valid for self-similar network traffic. Some models which are capable of describing traffic's multiscale characteristics have been proposed and used for prediction. These tools include linear and nonlinear methods, FARIMA (fractional autoregressive integrated moving average) models [6], neural network approach [7], fuzzy logic approach [8] and methods based on α-stable models [9], etc.

In this paper, we propose a traffic prediction method based on the idea of multiresolution analysis. The key feature of multiresolution analysis is to decompose

C. Kim (Ed.): ICOIN 2005, LNCS 3391, pp. 321–331, 2005.

whole function space into subspaces, then the decomposed signal pieces express the original signal on different time scale. The mathematical implementation of this multiresolution idea is wavelet transform, which take the input of a continuous function $f(t)$ or a discrete time series $\mathbf{X}(n)$ with proper initial process [18] and gives out a set of coefficients. These coefficients provide information about the original signal in frequency domain as well as in time domain. Intuitively, this multiresolution analysis is capable of capturing more characteristics of a self-similar traffic. Abry *et al.* proved that the wavelet coefficients of a self-similar traffic is largely decorrelated and do not exhibit LRD any more [12]. Therefore, predict algorithms for short range dependence (SRD) time series can be used for these coefficients. Several approach for coefficients prediction has been studied, which include methods based on neural network [7], [10], [19], Kalman filtering [20], or an ARIMA (autoregressive integrated moving average) model [21]. In this paper, we used another coefficient prediction method based on recursive least square (RLS) algorithm.

The rest of this paper is organized as follows: Section II focus on the idea of multiresolution analysis and wavelet transform and the reason to choose *à trous* algorithm to perform wavelet transform; Section III presents the proposed prediction framework by combining RLS algorithm with wavelet transform; Section IV gives the experiment result and analysis; Section V concludes the paper.

2 Wavelet Transform: From the Traditional to À Trous

Generally speaking, wavelet transform provides a way of analyzing a signal both in time domain and frequency domain. The discrete wavelet transform (DWT) gives out a set of coefficients $\{d_j(k), c_J(k),\ j \in [1, J]\}$ to describe the signal $f(t)$ at different time scale:

$$d_j(k) = \langle f(t), \psi_{jk}(t) \rangle\ ,\ k \in Z \tag{1}$$

$$c_J(k) = \langle f(t), \varphi_{Jk}(t) \rangle\ \ ,\ k \in Z \tag{2}$$

where $\langle *\,, *\,\rangle$ denotes inner product, function $\psi_{jk}(t)$ is so-called wavelet, which is usually orthogonal and constructed from a reference pattern $\psi(t)$ called mother-wavelet by a time-shift operation and a dilation operation:

$$\psi_{jk}(t) = 2^{-j/2}\psi(2^{-j}t - k) \tag{3}$$

and $\varphi_{Jk}(t)$ is the scaling function for corresponding wavelet, the integer j represents scales and k is time index. Therefore, a signal $f(t)$ can be expand as

$$f(t) = \sum_k c_J(k)\varphi_{Jk}(t) + \sum_{j=1}^{J}\sum_k d_j(k)\psi_{jk}(t) \tag{4}$$

(1), (2) and (4) are actually the definition of discrete wavelet transform and inverse discrete wavelet transform (IDWT). The wavelet coefficients $d_j(k)$ and the scaling function coefficients $c_J(k)$ conveys information about the behavior of the function $f(t)$ concentrating on effects of scale around 2^j and near time $k \times 2^j$.

From a point view of signal spaces and multiresolution analysis, the scaling function $\varphi_k(t) = \varphi(t-k)$ construct a subspace $\mathcal{V}$ of $L^2(\mathbf{R})$ as

$$\mathcal{V} = \overline{Span_k\{\varphi_k(t)\}} \tag{5}$$

where the over-bar denotes closure and *Span* denotes function span, which means defining a space as the set of all functions that can be expressed by linear combination of basis function. Here the scaling function is the basis function of subspace $\mathcal{V}$. Multiresolution analysis require scaling functions at each scale and their corresponding subspace have relationship as

$$\mathcal{V}_j \subset \mathcal{V}_{j-1} \ for \ all \ j \in Z \tag{6}$$

with

$$\mathcal{V}_\infty = \{0\}, \ \mathcal{V}_{-\infty} = L^2 \tag{7}$$

then wavelet $\psi_{jk}(t)$ is basis function of another type of subspaces $\mathcal{W}_j$, which is the orthogonal complement between $\mathcal{V}_j$ and $\mathcal{V}_{j-1}$:

$$\mathcal{V}_{j-1} = \mathcal{V}_j \bigoplus \mathcal{W}_j \tag{8}$$

then in general this gives:

$$L^2 = \mathcal{V}_J \bigoplus \mathcal{W}_J \bigoplus \mathcal{W}_{J-1} \bigoplus \cdots \tag{9}$$

This idea of multiresolution analysis is actually the heart of Mallat's fast wavelet transform algorithm, or so-called fast pyramid algorithm. In this algorithm, the wavelet coefficients and scaling function coefficients are calculated by bank filtering followed by down-sampling:

$$d_j(k) = \sum_m h_1(m-2k)c_{j-1}(m) \tag{10}$$

$$c_j(k) = \sum_m h_0(m-2k)c_{j-1}(m) \tag{11}$$

where $h_0(n)$ is the impluse response of a low-pass filter related to scaling function and $h_1(n)$ is the impluse response of a high-pass filter related to wavelet. Because of the down-sampling or so-called decimation, the coefficients becomes about half length as scale j goes one level higher, thus form a pyramid shaped coefficients set. This traditional wavelet transform involving bank filtering and decimation is efficient at a wide range of signal processing such as de-noising and compressing, but have problems in time series prediction.

One of these problems is the lack of stability at the end boundary of coefficients. When the available traffic data series become longer so that the DWT coefficients get extended by one, we can find that last a few coefficients become different from those obtained from last time. The longer the filter length or the higher the scale, the longer this non-stability appears at the end boundary:

$$L_j = \begin{cases} fix\,((lf-1)/2) & j=1 \\ fix\,((L_{j-1}+lf-1)/2) & 1<j\leq J \end{cases} \tag{12}$$

where L_j denotes the length of non-stable coefficients at scale j, lf denotes the filter length and fix is a operator that round the number to nearest integer towards zero. This problem is vital when we perform prediction for DWT coefficients, in which last a few taps of data are used to make predict. The appearance of this non-stability result from the lack of shift invariance of traditional wavelet transform. Two methods can be used to tackle this problem, one is multi-step prediction so that to avoid using unstable coefficients to make prediction, another is adopt a redundant or nondecimated version of wavelt transform, which will essentially eliminate this non-stability. Here we choose the later one in our prediction framework.

The *à trous* algorithm is implementation of aforementioned nondecimated wavelet transform [16]. The *à trous* wavelet transform can be simply describe as follows. First, perform successive convolutions with the discrete low-pass filter h:

$$c_{j+1}(k) = \sum_{l=-\infty}^{+\infty} h(l)c_j(k+2^j l) \tag{13}$$

where the finest scale is the original series: $c_0(t) = f(t)$. The low-pass filter we choose is a B_3 spline, defined as $(\frac{1}{16}, \frac{1}{4}, \frac{3}{8}, \frac{1}{4}, \frac{1}{16})$. To deal with the end boundary of data, we shift the low-pass filter so that only known data are involved in the convolutions. The resulting phase shift of $c_j(k)$ can be inherently compensated by the following step. Secondly, the wavelet coefficients are calculated by

$$d_j(k) = c_{j-1}(k) - c_j(k) \tag{14}$$

Thus the coefficients set is $\{d_j(k), c_J(k)\,,\, 1 < j \leq J\}$, the original signal is expanded as

$$f(t) = c_J(t) + \sum_{j=1}^{J} d_j(t) \tag{15}$$

thus we readily obtain a series of coefficient sets with equal length and the reconstruction is a simply additive procedure. In (15) we change of the index k of the coefficients $\{d_j(k)\,,\, c_J(k)\}$ to t because they just refer to the same time point. This is indeed one of the advantages of *à trous* algorithm, the information of the signal at a given time point can be located in coefficients definitely and uniquely. Another advantage is that the coefficients are calculated only from data obtained previously in time. This advantage actually prevent non-stability phenomena at the end boundary of coefficients.

The scaling function $c_J(k)$ is a smoothed version of original signal and wavelet coefficients $d_j(k)$ contains high frequency component of the original signal at each scale. Thus the signal are decompose according to different time scale and the frequency behaviors of the original signal are separated. In practice we can arbitrarily choose a max decomposed level J so that the coefficients are smooth enough for prediction algorithm.

3 Coefficient Prediction Algorithm

The framework of the proposed traffic prediction algorithm combining RLS algorithm with wavelet transform consists three steps. Firstly, the original traffic data $X(t)$ is decomposed with *à trous* wavelet transform by using (13) and (14). Thanks to the shift invariance property of *à trous* algorithm, the coefficient index k strictly correspond to the time index t of the original signal, and the coefficients of any segment of a signal are strictly equal to the coefficients in the same segment. This property is very favorable because we do not have to recompute the whole length of coefficient sets when new traffic data are regularly obtained. Secondly, these coefficients at each scale are predicted by RLS algorithm. Finally, all predict values are summed to make prediction for the original traffic data.

Recursive least square adaptive filter algorithm is one of most widely used linear predict methods. As the DWT coefficients only exhibit short range dependence (SRD) and has finite variance, RLS algorithm is valid for make prediction for these coefficients. Given the filter input vector as the most recent known values of DWT coefficients, the output of the filter provides the predicted valued of future coefficient. The adaptive mechanism is achieved in every instant of time when new traffic data become available [15]. The computations in each iteration n ($n = 1, 2, ...$) are as following:

1. Compute the updated gain vector $\mathbf{k}(n)$ by

$$\mathbf{k}(n) = \frac{\lambda^{-1}\mathbf{P}(n-1)\mathbf{u}(n)}{1 + \lambda^{-1}\mathbf{u}^H(n)\mathbf{P}(n-1)\mathbf{u}(n)} \tag{16}$$

   with

$$\mathbf{u}(n) = [d_j(k)\ d_j(k-1)\ ...\ d_j(k-p)]^T \tag{17}$$

   where $\mathbf{u}$ is the input vector and $\mathbf{P}$ represents inverse correlation matrix computed at the last iteration, λ is a forgetting factor, which will be introduced later in this section.

2. Compute the predicted value of next coefficient by

$$\widehat{d}_j(k+1) = \widehat{\mathbf{w}}^T(n-1)\mathbf{u}(n) \tag{18}$$

   where $\widehat{\mathbf{w}}$ is weight vector.

3. Compute the prediction error $e(n)$ when new data become available by

$$e(n) = d_j(k+1) - \widehat{d}_j(k+1) \tag{19}$$

   where d_j is the real value and $\widehat{d}_j$ is predicted value.

4. Update $\widehat{\mathbf{w}}$ and $\mathbf{P}$ by

$$\widehat{\mathbf{w}}(\mathbf{n}) = \widehat{\mathbf{w}}(n-1) + \mathbf{k}(n)e(n) \tag{20}$$

$$\mathbf{P}(n) = \lambda^{-1}\mathbf{P}(n-1) - \lambda^{-1}\mathbf{k}(n)\mathbf{u}^T(n)\mathbf{P}(n-1) \tag{21}$$

thus RLS algorithm adaptively decreases the prediction error in sense of minimizing the sum of the square of the difference between predicted value and real value. The algorithm is initialized by setting

$$\widehat{\mathbf{w}}(0) = 0 \tag{22}$$

$$\mathbf{P}(0) = \delta^{-1}\mathbf{I} \tag{23}$$

where $\mathbf{I}$ denotes the identity matrix.

Compared to the LMS (Least Mean Square) family adaptive filter, RLS filter has a number of advantages such as faster convergence speed and smaller mean-square error. There are several adjustable parameters for RLS filter:

Forgetting Factor λ: Let $\varepsilon(n)$ denote the sum of all prediction error up to step n, in RLS algorithm, $\varepsilon(n)$ has following:

$$\varepsilon(n) = \sum_{i=1}^{n} \lambda^{n-i}|e(i)|^2 \tag{24}$$

Therefore λ play a role of exponentially diminishing the total error result from the long distant past. Parameter λ should be typically set between 0.95 and 1, when λ is 1, it means the adaptive filter has infinite memory; λ is usually set to less than 1 for better performance when the process is non-stationary.

The regularization parameter δ: This parameter is set to small positive constant for high SNR or large positive constant for low SNR.

The order of RLS filter M: This parameter is closely related to the order p of AR(p) (autoregressive) model of the series. It decide how many most recent data are used for computing the predicted value. Parameter M is required to set to large value for process with slow decay autocorrelation function.

4 Simulation Result and Analysis

In this section we use the method describe above to make prediction on real network traffic. The network traffic used in our analysis was collected by WAND research group at the University Of Waikato Computer Science Department. It is the LAN traffic trace at University of Auckland on campus level. The traffic trace was colleted on June 11, 2000 on a 100Mbps Ethernet link. IP headers in the traffic trace are GPS synchronized and have a time accuracy of 1 μs. Total length of the traffic using for our simulation is 1000 seconds. More information

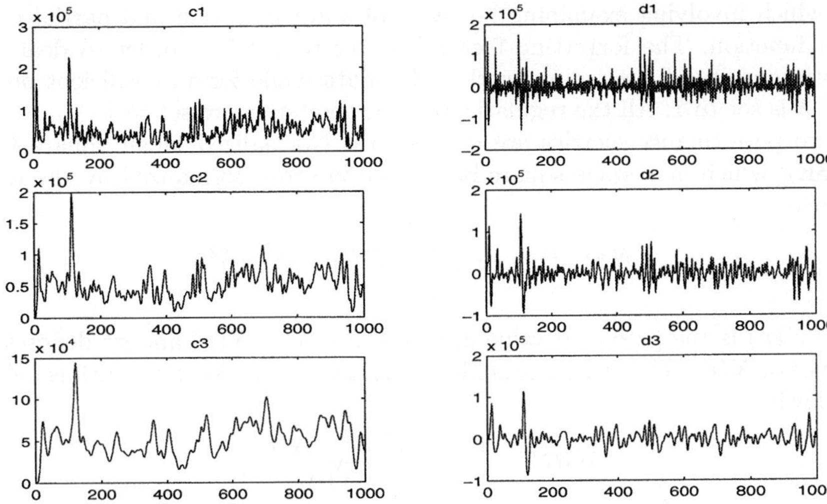

Fig. 1. Illustration of multiresolution analysis for traffic data: smoothed traffic c_j and details d_j at three consecutive decompose level.

on the traffic trace and the measurement infrastructure can be found on their webpage [17].

In simulation, we use the proposed algorithm to make one-step prediction on network traffic load in term of bit rate, the unit is bit per second. Based on the recent traffic bit rate, our algorithm gives the predicted value for next second. Longer time prediction can be achieved by aggregating the traffic data into larger time interval. We perform our prediction method with three consecutive wavelet decompose level to show the effect of wavelet transform. A direct RLS prediction without DWT decomposition is also applied to this network traffic for comparing purpose.

Table 1. RLS Parameter Settings

	$d_1(k)$	$c_1(k)$	$d_2(k)$	$c_2(k)$	$d_3(k)$	$c_3(k)$	$X(t)$
λ	1	1	1	1	1	1	0.995
δ	1	1	1	1	1	1	1
M	5	5	10	10	10	10	20

Fig. 1 shows all coefficient series obtained by three level wavelet decomposition for this network traffic. It is clear from this figure that the scaling function coefficients c_j become more and more smooth as decompose level goes higher. Then each coefficient series are predicted individually by RLS algortithm. Parameter settings for RLS algorithm are shown in Table 1. The RLS filter order M for each data series are decided by preliminary model identification [15],

[22], which involving examining behavior of autocorrelation and partial correlation function. The forgetting factor λ is set to 0.995 in order to deal with non-stationarity of the real network traffic data while for all coefficient predictions, λ is set to 1. All the regularization parameter δs are set to 1.

Two performance metrics are used, one is Normalized Mean Square Error (NMSE), which is average square of prediction error normalized by the traffic variance:

$$NMSE = \frac{1}{\sigma^2}\frac{1}{N}\sum_{n=1}^{N}(X(n) - \hat{X}(n))^2 \tag{25}$$

where $\hat{X}(n)$ is the predicted value of true traffic load $X(n)$ and σ^2 denotes the variance of $X(n)$. The other one is Mean Relative Error (MRE), which is defined as following:

$$MRE = \frac{1}{N}\sum_{n=1}^{N}\left|\frac{X(n) - \hat{X}(n)}{X(n)}\right| \tag{26}$$

In computing MRE, it is sensible to set a threshold so that only $X(n)$ larger than this certain value are taken into count, because too small $X(n)$ value as the denominate will lead to undesired large relative error, which is no meaning for evaluate the performance. Here we choose the data series mean as this threshold. In all RLS prediction procedures, the first 100 samples of total data are used for training the adaptive filter. Therefore, these data are not counted in for evaluating prediction error.

The result is shown in the Table 2, Table 3 and Fig. 2. In Table 3, we see that the proposed method achieves more accurate prediction than directly applying RLS to traffic data in terms of both NMSE and MRE. In fact, we see that only one level wavelet transform can make RLS performance improved by about 0.10. Moreover, the prediction become even more accurate when the level goes higher although the improvement is limited. Therefore, it is proved that combining with wavelet transform make RLS algorithm more proper for traffic prediction.

Examining the individual RLS prediction performance at each scale, which is shown in Table 2, we find that significant improvement can be achieved when the data series become smooth. But the performance for $d_1(k)$ is even worse than performance of a direct RLS prediction for original traffic. Therefore, better performance of the proposed method is achieved only by coefficients prediction on level 2 and 3. The reason is that high frequency component of the original traffic is extracted by wavelet decompose and concentrate in $d_1(k)$. Therefore $d_1(k)$ is even more bursty than the original traffic data. This high bursty behavior is hardly capture by RLS algorithm. Also, primary model identification

Table 2. RLS Prediction Performance At Each Scale

	$d_1(k)$	$c_1(k)$	$d_2(k)$	$c_2(k)$	$d_3(k)$	$c_3(k)$
NMSE	0.8657	0.0126	0.020	0.0002	0.003	0.000004
MRE	1.8390	0.0356	0.4132	0.0040	0.0614	0.000495

Table 3. Performance of proposed method: RLS combining with different level of wavelet decomposition, where Level 0 means direct RLS method without wavelet transform.

	Level 0	Level 1	Level 2	Level 3
NMSE	0.7274	0.6291	0.6272	0.6271
MRE	0.2709	0.2693	0.2668	0.2667

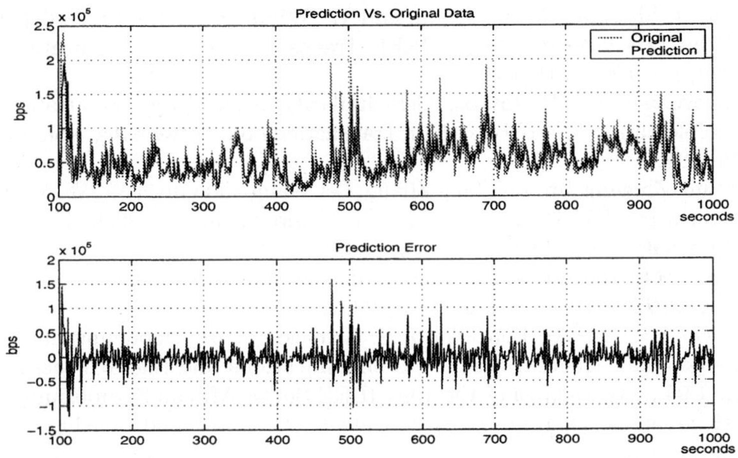

Fig. 2. Performance of Proposed Method: RLS Combined with Level 3 Wavelet Transform

result shows that $d_1(k)$ can be describe as an ARMA model with large variance, suggesting the same conclusion.

5 Conclusion

In this paper, we introduced the self-similarity in network traffic and idea the multiresolution analysis. A traffic prediction method based on the idea of multiresolution was proposed. The mathematical implement of multiresolution analysis is wavelet transform and its computation characteristics is investigated and we shown that the *à trous* algorithm is more favorable for time series prediction than the traditional discrete wavelet transform. The proposed method is applied to real network traffic. Experiment result shows that more accurate traffic prediction can be achieved by the proposed method.

References

1. A. Feldmann, A. Gilbert, W. Willinger, and T. Kurtz, "The changing nature of network traffic: Scaling phenomena", Computer Communication Review, vol. 28, no. 2, pp. 5-29, 1998.
2. W. E. Leland, M. S. Taqqu, W. Willinger, and D. V. Wilson, "On the self-similar nature of ethernet traffic (extended version)", IEEE/ACM Transactions on Networking, vol. 2, no. 1, pp. 1-15, 1994.
3. V. Paxson and S. Floyd, "Wide area traffic: the failure of poisson modeling", IEEE/ACM Transactions on Networking, vol. 3, no. 3, pp. 226-244, 1995.
4. J. Beran, R. Sherman, M. S. Taqqu, and W. Willinger, "Long-range dependence in variable-bit-rate video traffic", IEEE Transactions on Communications, vol. 43, no. 2/3/4, pp. 1566-1579, 1995.
5. M. E. Crovella and A. Bestavros, "Self-similarity in world wide web traffic: Evidence and possible causes", IEEE/ACM Transactions on Networking, vol. 5, no. 6, pp. 835-846, 1997.
6. Yantai Shu, Zhigang Jin, Lianfang Zhang, Lei Wang and Oliver W. W. Yang, "Traffic prediction using FARIMA models", in Communications, 2000. ICC 2000. 2000 IEEE International Conference, vol.3, pp1325-1329
7. S. Soltani, "On the use of the wavelet decomposition for time series prediction", Neurocomputing, vol. 48, pp. 267-277, 2002.
8. M. F. Scheffer, J. J. P. Benekem J. S. Kunicki, "Fuzzy modeling and prediction of network traffic fluctuations", Communications and Signal Processing, 1994. COMSIG-94., Proceedings of the 1994 IEEE South African Symposium , pp41-45,
9. F. C. Harmantzis, D. Hatzinakos, I. Katzela, "Shaping and policing of fractal α-stable broadband traffic", Electrical and Computer Engineering, 2001. Canadian Conference on , Volume: 1 , 13-16 May 2001 pp.697 - 702, vol.1
10. Amir B. Geva, "ScaleNet-multiscale neural-network architecture for time series prediction", IEEE Transaction on Neural Networks, vol.9, No.5, September 1998.
11. G. Mao, "Finite timescale range of interest for self-similar traffic measurements, modelling and performance analysis," in IEEE International Conference on Networks, Sydney, 2003, pp. 7-12.
12. P. Abry, P. Flandrin, M. S. Taqqu, and D. Veitch, "Wavelets for the analysis, estimation, and synthesis of scaling data", in Self-Similar Network Traffic and Performance Evaluation, K. Park and W. Willinger, Eds., pp. 39-88. John Wiley and Sons, Inc., 2000.
13. Yanqian Fan, "On the approximate decorrelation property of the discrete wavelet transform for fractionally differenced processes", IEEE Transactions On Information Theory, Vol.49, NO.2, pp516-521.
14. G. Strang and T. Nguyen, Wavelets and Filter Banks, Wellesley-Cambridge Press, 1996.
15. S. Haykin, Adaptive Filter Theory, the fourth edition, Prentice Hall, 2002.
16. Mark J. Shensa, "The discrete wavelet transform: wedding the à trous and mallat algorithms," IEEE Transcation On Signal Processing , Vol.40, No.10, October 1992
17. Website: http://wand.cs.waikato.ac.nz/wand/wits/index.html
18. P. Abry, P. Flandrin, "On the initilization of the discrete wavelet transfrom algorithm", Singal Processing Leters 1:15-30, 1994.
19. G. Zheng, J. L. Starck, J. Campbell and F.Murtagh, " The wavelet transform for filtering financial data streams," Journal of Computational Intelligence in Finance, 1(3): 18-35, 1999.

20. L. Hong G. Cheng, C.K. Chui, "A filter-bank-based Kalman filtering technique for wavelet estimation and decomposition of random signals," IEEE Transactions on Circuits and Systems II: Analog and Digital Signal Processing, Vol 45, Issue 2, pp 237-241.
21. K. Papagiannaki, N. Taft, Z. Zhang, C. Diot, "Long-term forecasting of internet backnoe traffic: observations and initial models," INFOCOM 2003. Twenty-Second Annual Joint Conference of the IEEE Computer and Communications Societies.
22. George E. P. Box and Gwilym M. Jenkins, Time Series Analysis: Forecasting and control, San Francisco: Holden-Day, 1976

Proportional Fairness Mechanisms for the AF Service in a Diffserv Network

Sangdok Mo and Kwangsue Chung

School of Electronics Engineering, Kwangwoon University, Korea
sdmo@adams.kw.ac.kr, kchung@daisy.kw.ac.kr

Abstract. Previous works for the AF (Assured Forwarding) service in a Differentiated Service (Diffserv) network have no sufficient consideration on the proportional fairness of bandwidth share based on RTTs, the target rates, and the impact of UDP against TCP. In order to solve these problems, we propose the Fair Differentiated Service Architecture (FDSA), Target rate and RTT Aware Three Color Marking (TRA3CM), and Target Rate Based Dropping (TRBD) mechanisms. These mechanisms provide three color marking and proportional fair bandwidth share among flows by considering RTT, target rate, and UDP flows simultaneously. In the results of comparing the performance among existing and proposed mechanisms, the proposed mechanisms are able to mitigate the RTT and UDP effect better than the existing ones. In addition, the proposed mechanisms are shown to provide the fair bandwidth share proportional to various target rates.

1 Introduction

Recently, multimedia applications such as VoIP, VoD, and Visual Conference are increasing their traffics in the Internet. The requirements of these services are different from best effort services. The Differentiated Service (Diffserv) architecture has become the preferred method to meet the requirements of these services. An end-to-end differentiated service in a Diffserv network is obtained by the concatenation of per-domain services and Service Level Agreements (SLAs) between adjoining domains along the source-to-destination traffic path. Per domain services are realized by traffic conditioning at the edge and differentiated forwarding mechanisms at the core of the network. Two forwarding mechanisms generally used in a Diffserv network are the Expedited Forwarding (EF) and Assured Forwarding (AF) Per Hop Behaviors (PHBs). The basis of the AF service is differentiated dropping of packets during congestion at a router. The differentiated dropping is achieved via Multiple-RED Active Queue Management (AQM) technique. The RFC of the IETF for AF service specifies four classes and three levels of drop precedences per class. Three drop precedences can be represented by Green, Yellow, and Red in order of lower drop precedence. In this basic Diffserv architecture, it is difficult to provide real differentiated services to end users, because of different RTTs and target rates among flows and TCP/UDP interaction, etc. These issues need to be resolved for real differentiated services.

C. Kim (Ed.): ICOIN 2005, LNCS 3391, pp. 332–341, 2005.

In this paper, we propose the Fair Differentiated Service Architecture (FDSA), Target rate and RTT Aware 3 Color Marker (TRA3CM), and Target Rate Based Dropper (TRBD). These mechanisms provide three color marking and proportional fair bandwidth share among flows by considering RTT, target rate, and TCP/UDP interaction at the same time. In the results of simulations for comparing existing mechanisms with the proposed mechanisms, we show that the proposed mechanisms are able to mitigate the impacts of RTT and TCP/UDP interaction better than existing ones. The proposed mechanisms are shown to provide fair bandwidth share proportional to various target rates as well.

The rest of this paper is organized as follows. Background and related work are examined in the next section. Section 3 describes the proposed mechanisms and the analysis of it, the discussion and evaluation of the proposed solutions are addressed in Sect. 4. Section 5 contains concluding remarks and points to areas of future work.

2 Background and Related Work

There have been a number of simulation studies that focused on a RED-based Diffserv scheme. Clark and Fang in [1], showed that sources with different target rates can approximately achieve their targets using RIO (RED with IN/OUT) even with different RTTs, whereas simple RED routers cannot do. But if two flows have the same target rate and different RTTs, short RTT flows consume most of the excess bandwidth. One of our goals is to distribute the excess bandwidth among all flows such that short RTT flows do not steal all the extra bandwidth. Ibanez and Nichols [2], via simulation studies, confirmed that RTT and TCP/UDP interaction were key factors in the throughput of flows that obtain an Assured Service using a RIO-like scheme. Seddigh, Nandy, and Pieda showed that target rates and TCP/UDP interaction were also critical for the distribution of excess bandwidth in an over-provisioned network [3]. Fang, Seddigh and Nandy proposed the Time Sliding Window Three Color Marker (TSW3CM) [4], which we refer to as the standard conditioner.

Nandy et al extended the TSW marker to design RTT and target rate aware traffic conditioners [5]. These conditioners are RTT Aware Traffic Conditioner (RATC), Target rate Aware Traffic Conditioner Two Drop precedences (TATC2D), and TATC Three Drop precedences (TATC3D). The basic ideas of these conditioners are to adjust the packet drop rate in relation to the RTT and target rate. But their model does not consider RTT, target rate, and the impact of UDP flows simultaneously, and some of their conditioners provide only two color marking. Feroz et al proposed a TCP-Friendly marker [6]. As TCP applications over Diffserv are influenced by a bursty packet loss behavior, they used TCP characteristics to design their marker. Their conditioner protects small-window flows from packet losses by marking such traffic as IN. Habib et al designed and evaluated a conditioner based on RTT as well as the Retransmission Time-Out (RTO) [7]. However, their conditioner cannot alleviate the impact of TCP/UDP interaction. Su and Atiquzzaman proposed a new TSW based three-color marker (ItswTCM), which provide proportional fair share of excess bandwidth among aggregates in a Diffserv network [8]. ItswTCM lacks of consideration for RTT and the impact of TCP/UDP interaction, etc.

2.1 Time Sliding Window Three Color Marker

To take advantage of three color marking, the TSW based TSW3CM has been proposed. For TSW3CM, whenever a packet arrives, the marker calculates an estimated arrival rate. If the estimated arrival rate is less than the Committed Target Rate (CTR), arriving packets are marked as Green; otherwise, they are marked as Green, Yellow or Red according to the calculated probabilities. The TSW3CM algorithm is shown in Fig. 1. If there are various RTTs or target rates in a Diffserv network, this algorithm cannot provide fair bandwidth share for end users, because it cannot mitigate the impacts of different RTTs and target rates among flows.

R_{avg} = Estimated average traffic rate
R_C = Committed Target Rate
R_P = Peak Target Rate

If ($R_{avg} \leq R_C$):
    the packet is marked as Green;
else if ($R_{avg} \leq R_P$) AND ($R_{avg} > R_C$):
    $P_G = R_C/R_{avg}$
    $P_Y = (R_{avg} - R_C)/R_{avg}$
    the packet is marked as Green with P_G probability;
    the packet is marked as Yellow with P_Y probability;
else:
    $P_G = R_C/R_{avg}$
    $P_Y = (R_P - R_C)/R_{avg}$
    $P_R = (R_{avg} - R_P)/R_{avg}$
    the packet is marked as Green with P_G probability;
    the packet is marked as Yellow with P_Y probability;
    the packet is marked as Red with P_R probability;

Fig. 1. Marking Algorithm for the TSW3CM Marker

2.2 Intelligent Traffic Conditioner

The RATC (RTT Aware Traffic Conditioner), one of Intelligent Traffic Conditioners (ITC), avoids unfair bandwidth share between the TCP flows with short and long RTTs through marking packets with the high drop priority inversely proportional to the square of their RTTs [5]. This is based upon the steady state TCP behavior modeled in [9]. Equation (1) shows that, in this model, bandwidth is inversely proportional to the RTT (MSS is the maximum segment size and p is the packet loss probability):

$$BW \propto \frac{MSS}{RTT\sqrt{p}} \tag{1}$$

Considering two flows with achieved rates R_1 and R_2, for fair bandwidth share of two flows, R_1 must be equal to R_2 as (2). If the packet sizes for two flows are the same, then the (2) becomes as (3).

$$R_1 = R_2 \tag{2}$$

$$RTT_1\sqrt{p_1} = RTT_2\sqrt{p_2} \tag{3}$$

If two flows have different RTTs, then:

$$\frac{p_2}{p_1} = \left(\frac{RTT_1}{RTT_2}\right)^2 \tag{4}$$

Therefore, in order to achieve the same rate for two flows, the ratio of their packet drop probabilities should have an inverse squared relationship to the RTTs. If assuming that the ratio of out-of-profile packet marking is directly proportional to the ratio of packet drop probabilities at the core, out-of-profile marking schemes for two flows at the edge become as follows:

$$q = \frac{R_C}{R_{avg}} \tag{5}$$

$$p_{out,1} = q \text{ and } p_{out,2} = \left(\frac{RTT_1}{RTT_2}\right)^2 q \tag{6}$$

where R_C is a target rate, R_{avg} is an estimated average rate, $p_{out,1}$ and $p_{out,2}$ are the OUT marking probabilities of flow *1* and *2* respectively. Based on (6), the generalized marking scheme for out-of-profile packets would be:

$$p_{out,i} = \left(\frac{RTT_{min}}{RTT_i}\right)^2 \tag{7}$$

where RTT_i is the RTT for the flow i and RTT_{min} is the minimum RTT of all the flows in the network. The marking probabilities in TATC (Target rate Aware Traffic Conditioner), which is another of intelligent traffic conditioners, are derived from the equations similar to the ones of RATC, but use target rates instead of RTTs.

Using these approaches, intelligent traffic conditioners lighten the effect of RTT and different target rates. However, intelligent traffic conditioners cannot mitigate all of them simultaneously and also lack of consideration for TCP/UDP interaction.

3 Proportional Fair Bandwidth Sharing Mechanism

In this section, we discuss the design concept of Fair Differentiated Service Architecture (FDSA), Target rate and RTT Aware Three Color Marker (TRA3CM), and Target Rate Based Dropper (TRBD).

3.1 Design of Fair Differentiated Service Architecture

Most of existing works on fairness issues in a Diffserv network, have focused on marking methods at an edge node. However, it is difficult to provide the fair share of excess bandwidth by simply applying marking methods to an edge node. In order to effectively provide proportional fair share of bandwidth, an edge node and core nodes need to interact by deliberately considering RTT, target rate, TCP/UDP interaction, and so on. In this paper, we propose the Fair Differentiated Service Architecture (FDSA), which can provide proportional fair share of excess bandwidth through cooperation between edge nodes and core nodes. Figure 2 shows the FDSA.

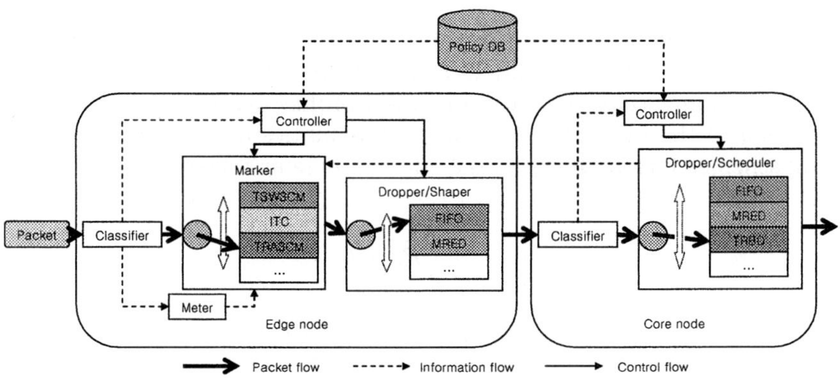

Fig. 2. Fair Differentiated Service Architecture

The FDSA operates as follows:

- The packet injected into an edge node is classified with SLA and the information related with the packet is sent to the controller and the meter.
- The meter estimates RTT and arrival rate of the flow for the packet and gives the information to the marker.
- The controller refers to a policy DB, selects a marker for the packet, and configures the marker with the information needed to process the packet.
- In order to mark the packet, the marker uses the information from the controller, meter, and core nodes.
- At the core node, the dropper selected by the controller processes the injected packet and sends the marker the information needed to process a packet. The information consists of drop probabilities for Green, Yellow, and Red packets.

3.2 Target Rate and RTT Aware Three Color Marker

The TRA3CM takes a similar approach with the RATC. However, it can mitigate the impact of RTT, and concurrently provide fair bandwidth share proportional to target rates among TCP flows. In addition, the TRA3CM marks a packet with three colors and adjusts marking probabilities between Yellow and Red for alleviating the effect of RTT. The TRA3CM marks a packet with marking probabilities, P_G, P_Y, and P_R for each color. Our goal is to obtain P_G, P_Y, and P_R. The packet drop probability P_2 is derived from (1) as follows:

$$P_2 = \left(\frac{RTT_1}{RTT_2}\right)^2 \left(\frac{R_{C1}}{R_{C2}}\right)^2 P_1 = a^2 P_1 \tag{8}$$

where P_1 and P_2 are packet drop probabilities for flow *1* and *2*, R_{C1} and R_{C2} are committed target rates for each flow. The sum of marking probabilities for each color is the same as (9) and the g, y, and r, the packet drop probabilities for each color from a core node, can be represented as (10).

$$P_G + P_Y + P_R = 1 \tag{9}$$

$$g = g_1 = g_2,\ y = y_1 = y_2,\ r = r_1 = r_2 \tag{10}$$

The marking probabilities for flow *2*, P_{G2}, P_{Y2}, and P_{R2}, are derived from (8), (9), (10), and the marking probabilities of P_G, P_Y, and P_R in Fig. 1.

$$P_{G2} = P_{G1} = \frac{R_{C2}}{R_2} \tag{11}$$

$$\begin{aligned} P_{Y2} &= a^2 P_{Y1} + \frac{1-a^2}{1-y}(1 - P_{G1}) \\ &= a^2 \frac{R_{P2} - R_{C2}}{R_2} + \frac{1-a^2}{1-y}\left(\frac{R_2 - R_{C2}}{R_2}\right) \end{aligned} \tag{12}$$

$$\begin{aligned} P_{R2} &= a^2 P_{R1} - \frac{y(1-a^2)}{1-y}(1 - P_{G1}) \\ &= a^2 \frac{R_2 - R_{P2}}{R_2} - \frac{y(1-a^2)}{1-y}\left(\frac{R_2 - R_{C2}}{R_2}\right) \end{aligned} \tag{13}$$

R_2 is an estimated average arrival rate of flow *2*. With the marking probabilities of P_{G2}, P_{Y2}, and P_{R2} in (11), (12), and (13), the TRA3CM marks a packet in the marking algorithm described in Fig. 1. However, if y equals to 1, the TRA3CM marks a packet to Green with P_{G2} and Red with (1-P_{G2}). And if R_2 is less than R_{P2}, it calculates P_{Y2} and P_{R2} with R_{P2} equal to R_{C2}.

3.3 Target Rate Based Dropping (TRBD) Mechanism

At a core node, the TRBD provides proportional fair bandwidth share among TCP and UDP flows. TCP packets are processed by MRED and UDP packets by the TRBD. The TRBD drops UDP packets with drop probability P_d, which is derived from packet history for each color. The drop probability P_d can be obtained as follows:

If the numbers of total and Green packets of flow i in the history are N_{ti} and N_{gi} respectively, the ratio of the injected rate to the target rate of flow i, α_i, becomes as follows:

$$\alpha_i = \frac{N_{ti}}{N_{gi}} \tag{14}$$

If the α_i is greater than the α_a, which represents the ratio of the average throughput to the target rate at a core node, it indicates that packets of flow i are over-injected as more as the α_d of (15). Therefore, if packets of the flow i are dropped with the drop probability P_d of (16), the proportional fair share of TCP and UDP flows can be achieved.

$$\alpha_d = \alpha_i - \alpha_a \tag{15}$$

$$P_d = \frac{\alpha_d}{\alpha_i} = 1 - \frac{\alpha_a}{\alpha_i} \tag{16}$$

4 Simulations and Evaluation

In order to evaluate the performance of our mechanisms, a number of experiments have been performed on the basis of the ns (Network Simulator) of LBNL (Lawrence Berkley National Laboratory) [10]. The network topology shown in Fig. 3 is used to show the performance of the TRA3CM-TRBD mechanism. Table 1 shows the RED parameters for the MRED in edge and core nodes. To evaluate simulation results, we define the generalized fairness index of (17). In (17), x_i is the throughput of flow i, and R_{Ci} is the target rate of flow i. Greater gfi represents better proportional fairness of bandwidth share.

Table 1. RED Parameters for MRED

	Green(IN)	Yellow(OUT)	Red
Min_{th}	40 pkts	25 pkts	10 pkts
Max_{th}	55 pkts	40 pkts	25 pkts
Max_p	0.02 pkts	0.05 pkts	0.1 pkts
w_q	0.002	0.002	0.002

$$\text{Generalized fairness index(gfi)} = \frac{\left(\sum_{i=1}^{n} \frac{x_i}{R_{Ci}}\right)^2}{n \sum_{i=1}^{n} \left(\frac{x_i}{R_{Ci}}\right)^2} \tag{17}$$

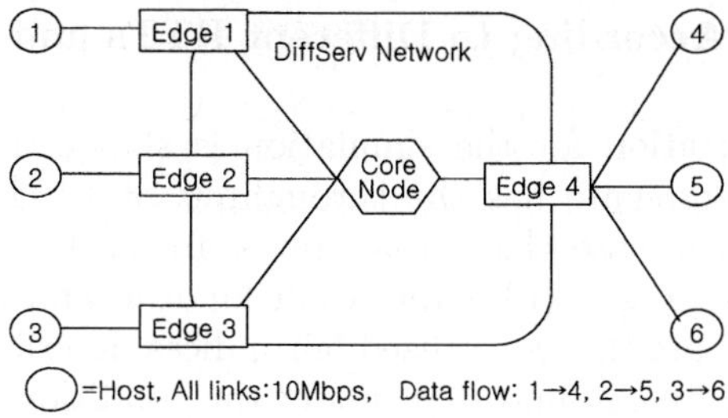

Fig. 3. Topology of the Simulation Network

4.1 Comparison According to Different Target Rates

The configuration of network for this experiment is shown in the Table 2. Each microflow of a UDP aggregate has sending rate of 1.5Mbps. In this configuration, two simulations are performed, one is the case of increasing the target rate of a TCP aggregate flow and another is the case of increasing the target rate of a UDP aggregate flow. In the results of Fig. 4, the line closer to the Ideal has a better proportional fairness. Therefore, we can find that the TRA3CM-TRBD provides proportional fair bandwidth share better than other mechanisms.

Table 2. Network Configuration for the Simulation on the TCP and UDP Flows with Different Target Rates

Marker(Dropper)	flows	RTT(ms)	CTR(Mbps)	PTR(Mbps)	microflows
RATC/TATC (MRED)	UDP(9Mbps)	20	1(1∼10)	x	6
	TCP	20	1∼10(1)	x	6
TSW3CM(MRED)	UDP(9Mbps)	20	1(1∼10)	CTR+1	6
TRA3CM(TRBD)	TCP	20	1∼10(1)	CTR+1	6

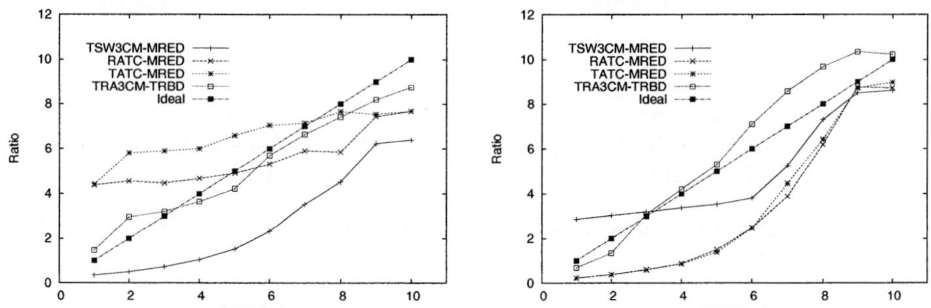

(a) TCP(x Mbps)/UDP(1 Mbps)   (b) UDP(x Mbps)/TCP(1 Mbps)

Fig. 4. Throughput Ratio of the Increasing CTR Flow to the Fixed CTR Flow

4.2 Comparison According to Different RTTs and Target Rates

The network configuration for the simulation is shown in the Table 3. Each microflow of UDP aggregate has the throughput of 1.5Mbps. The Fig. 5 (a) shows the result of the case that target rates are 2, 4, and 1Mbps for UDP, TCP1 and TCP2. The Fig. 5 (b) is the result for 1, 4, and 2Mbps for each. The vertical axes of Fig. 5 are the generalized fair indices described in (17). Based on experimental results, we can find that, under the condition with different RTTs and target rates and TCP/UDP interaction, the TRA3CM and TRBD are able to provide proportional fair bandwidth share better than other mechanisms.

Table 3. Network Configuration for Simulation on the TCP and UDP Flows with Different RTTs and Target Rates

Marker(Dropper)	flows	RTT(ms)	CTR(Mbps)	PTR(Mbps)	microflows
RATC/TATC (MRED)	UDP(9Mbps)	20	2(1)	x	6
	TCP1	20	4(4)	x	6
	TCP2	20∼200	1(2)	x	6
TSW3CM(MRED)/ TRA3CM(TRBD)	UDP(9Mbps)	20	2(1)	CTR+1	6
	TCP1	20	4(4)	CTR+1	6
	TCP2	20∼200	1(2)	CTR+1	6

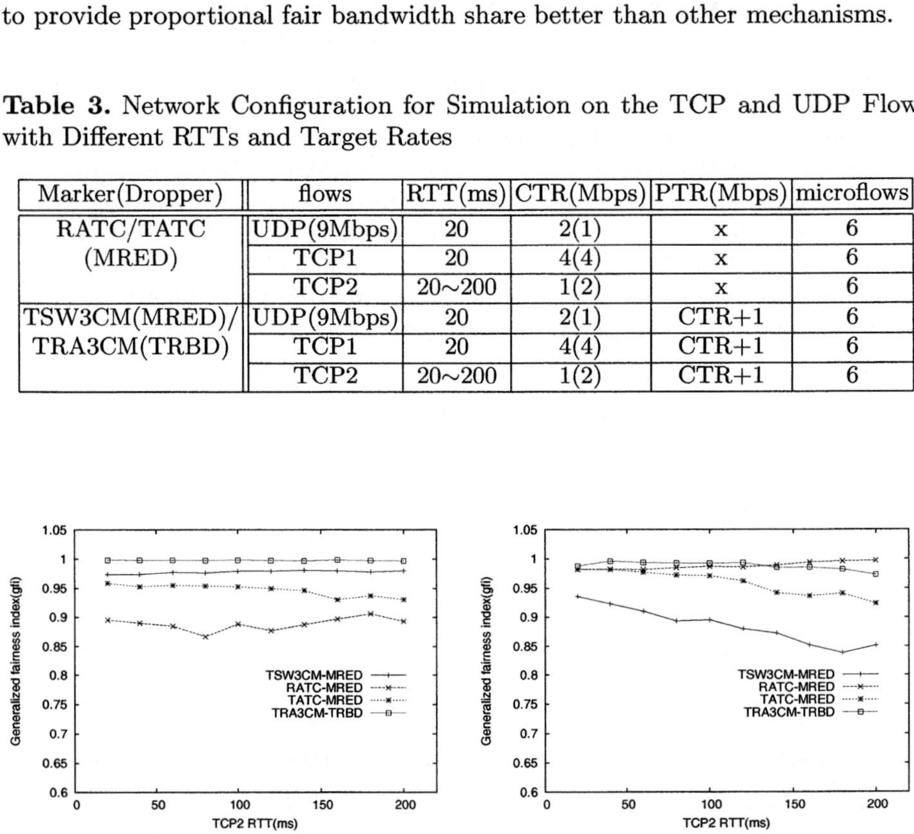

(a) UDP(2Mbps), TCP1(1 Mbps), TCP2(2 Mbps)

(b) UDP(1 Mbps), TCP1(4 Mbps), TCP2(2 Mbps)

Fig. 5. Generalized Fairness Indices on the Flows with Different RTTs and Target Rates

The results show that the TRA3CM-TRBD mechanism consistently outperforms other mechanisms in the proportional fairness. This is because the TRA3CM and TRBD mechanisms coordinate for solving the various problems against the proportional fairness.

5 Conclusion and Further Work

It is difficult to provide proportional fairness with only marking mechanisms of an edge node, because unresponsive flows such as UDP do not adjust the sending rate according to marking probabilities and drop probabilities. Therefore, in order to achieve the proportional fairness, a marker at an edge node and a dropper at a core node need to interact with each other. In this paper, we have designed the FDSA, which is composed of the proposed TRA3CM and TRBD mechanisms. The TRA3CM at an edge node and the TRBD at a core node cooperate to achieve the proportional fair bandwidth sharing. The simulation results show that the TRA3CM and TRBD mechanisms are able to effectively provide the proportional fairness for aggregate flows.

Further work involves enhancing the proposed mechanisms to fit a high speed network and to consider relationship between aggregate flows and microflows.

Acknowledgement

This research has been conducted by the Research Grant of Kwangwoon University in 2004. This research was supported by the MIC(Ministry of Information and Communication), Korea, under the ITRC(Information Technology Research Center) support program supervised by the IITA(Institute of Information Technology Assessment).

References

1. D. D. Clark and W. Fang: Explicit Allocation of Best Effort Packet Delivery Service. IEEE/ACM Transactions on Networking, vol. 1. (1998)
2. J. Ibanez and K.Nichols: Preliminary Simulation Evaluation of an Assured Service. Internet Draft, draft-ibanez-diffserv-assured-eval-00.txt (1998)
3. N. Seddigh, B. Nandy, and P. Pieda: Bandwidth Assurance Issues for TCP Flows in a Differentiated Services Network. Globecom. (1999)
4. W. Fang, N. Seddigh, and B. Nandy: A Time Sliding Window Three Color Marker. Internet RFC 2859. (2000)
5. B. Nandy, N. Seddigh, P. Pieda, and J. Ethridge: Intelligent Traffic Conditioners for Assured Forwarding based Differentiated Services Networks. IFIP High Performance Networking. (2000)
6. A. Feroz, S. Kalyanaraman, and A. Rao: A TCP-Friendly Traffic Marker for IP Differentiated Services. Proc. of the IEEE/IFIP IWQoS. (2000)
7. A. Habib, B. Bhargava, and S. Fahmy: A Round Trip Time and Time-out Aware Traffic Conditioner for Differentiated Services Networks. IEEE ICC. (2002)
8. H. Su and M. Atiquzzaman: ItswTCM: a New Aggregate Marker to Improve Fairness in DiffServ. Computer Communications, vol. 26. (2003)
9. M. Mathis, J. Semke, J. Mahdavi, and T. Ott: The Macroscopic Behavior of the TCP Congestion Avoidance Algorithm. ACM SIGCOMM Computer Communication Review, vol. 27. (1997)
10. UCB LBNL VINT: Network Simulator ns (Version 2). http://www-mash.cs.berkeley.edu/ns/

RWA on Scheduled Lightpath Demands in WDM Optical Transport Networks with Time Disjoint Paths*

Hyun Gi Ahn, Tae-Jin Lee, Min Young Chung, and Hyunseung Choo

Lambda Networking Center
School of Information and Communication Engineering
Sungkyunkwan University
440-746, Suwon, Korea +82-31-290-7145
{puppybit, tjlee, mychung, choo}@ece.skku.ac.kr

Abstract. In optical networks, traffic demands often demonstrate periodic nature for which time-overlapping property can be utilized in routing and wavelength assignment (RWA). A RWA problem for scheduled lightpath demands (SLDs) has been solved by combinatorial optimal solution (COS) and graph coloring, or heuristic sequential RWA (sRWA). Such methods are very complex and incurs large computational overhead. In this paper, we propose an efficient RWA algorithm to utilize the time disjoint property as well as space disjoint property through fast grouping of SLDs. The computer simulation shows that our proposed algorithm indeed achieves up to 54% faster computation with similar number of wavelengths than the existing heuristic sRWA algorithm.

1 Introduction

Optical virtual private networks (OVPNs) are the key service networks provided by an optical transport network (OTN) [1]. In OVPNs, connection requests offered by clients can be classified into three different types: static, scheduled, and dynamic. A set of static lightpath demands is provided by OVPN clients in order to satisfy their minimal connectivity and capacity requirements. When connection requests are dynamically established and released in time, such traffic demands are called dynamic lightpath demands. Scheduled lightpath demands may be required to increase the capacity of a network at specific times and/or on certain links. For example, suppose that periodical backups of database are required between the headquarter and production centers during office hours or between data centers during nights. Then, the lightpath demands for the backups of database are called scheduled lightpath demands (SLDs).

In real OTNs, we believe that most of demands will be considered as static or scheduled for the time being. The reason is that the traffic load in a transport network is fairly predictable because of its periodic nature [2][7]. Fig. 1 gives an

* This work was supported in parts by Brain Korea 21 and the Ministry of Information and Communication, Korea. Dr. Lee is the corresponding author.

C. Kim (Ed.): ICOIN 2005, LNCS 3391, pp. 342–351, 2005.

indication of this phenomenon. The figure shows the traffic on the New York-Washington link of the Abilene backbone network during a typical week. A similar periodic pattern was observed on all the other links of the network in the same period (This trend becomes greater during working hours). The figure shows clear evidence of the connection between the traffic intensity and the human usage pattern.

Routing and wavelength assignment (RWA) finds an appropriate route for a traffic demand and assigns a wavelength to the route, and the problem is one of the most important issues in wavelength division multiplexing (WDM) optical networks [3]. Since RWA has a great impact on performance and cost of optical networks, various approaches have been proposed [3][4][5][6]. The typical objectives of RWA research are 1) to minimize the required number of wavelengths under static connection requests, 2) to minimize the blocking probability under given number of wavelengths and dynamic connection requests, or 3) to minimize overall the network cost, e.g., wavelength converters. In the conventional RWA research, traffic demands has been assumed to be either static or dynamic. Noting the nature of the scheduled lightpath demands, we can utilize more efficient RWA for SLDs.

An SLD can be represented by 4 tuple (s, t, μ, ω), where s and t are source and destination nodes of a demand, μ and ω are setup and teardown times of a demand. The SLDs for which setup and teardown times are known in advance can take the advantage of the time scheduling property. That is, unless two lightpaths overlap in time, they can be assigned the same wavelength since the paths are disjoint in time. So, in this paper, we propose an efficient RWA algorithm in which SLDs with non-overlapping service times are grouped in order to enhance the performance of RWA. Our algorithm is shown to achieve up to 52% performance improvement compared to the conventional RWA algorithms.

This paper is organized as follows. First, we discuss related works on RWA in Section 2. We propose a RWA algorithm based on time disjoint path (TDP) in Section 3. Performance evaluation of the proposed algorithm is presented in Section 4. Finally, we conclude in Section 5.

2 Related Works

BGAforEDP is a simple and heuristic RWA algorithm [4][5]. It is a simple edge disjoint paths scheme based on the shortest path algorithm [5]. Let $G_B = (V_B, E_B)$ be the graph of a physical network, where V_B and E_B are the set of vertices and the set of edges, respectively. And let τ be a demand set, $\tau = \{(s_1, t_1), \ldots, (s_k, t_k)\}$. BGAforEDP operates with G_B, τ, and d, where d is $max(diam(G_B), \sqrt{|E_B|})$ [9]. The parameter d is used to limit the number of hops for the assigned paths. First, the BGAforEDP algorithm randomly selects a demand $\tau_i = (s_i, t_i)$ from the demand set τ and finds the shortest path P_i for this request. If the path length of P_i is less than the bound d, add (τ_i, P_i) to the allocated path set P, and τ_i to the set of routed demands $\alpha(G_B, \tau)$. And then it deletes the edges on P_i from G_B and removes τ_i from τ. If the path length of

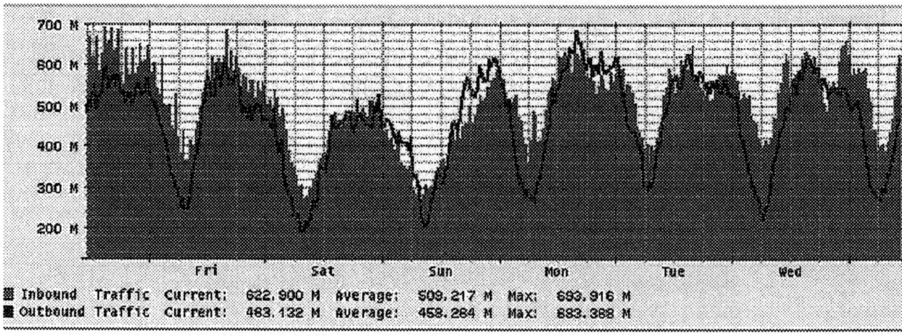

Fig. 1. Traffic on the New York-Washington link of the Abilene backbone network from April 2, 2003 to April 10, 2003 [7].

P_i is greater than *d*, the demand τ_i is not assigned the path. This is repeated until the paths are not assigned to the remaining demands in τ. The set $\alpha(G_B, \tau)$ then contains the demands that are assigned the same wavelength. Next, it removes $\alpha(G_B, \tau)$ from τ, and obtain the set of unassigned lightpaths τ'. Then BGAforEDP performs RWA on the original G_B and τ' to obtain the set of assigned lightpaths with another wavelength. This is repeated until τ' becomes empty. The total number of assigned wavelengths is the result of this algorithm.

The BGAforEDP algorithm is suitable for static demands, but not appropriate for SLDs, since SLDs have setup and teardown times in addition to source and destination nodes [3]. In [7], the combinatorial optimal solution (COS) for SLDs has been proposed. COS and graph coloring [6] approach has great complexity and requires much time cost. Especially, as the number of demands increases, the cost increases exponentially. For example, if there are 30 demands and each demand has 3 possible shortest paths, the necessary number of operations is 3^{30}. The branch and bound (B&B) search algorithm is considered to reduce the amount of calculation [10][11]. And Kuri et al. [7] proposed the meta-heuristic tabu search algorithm, since B&B still incurs a lot of computational cost [12][8]. The performance of the tabu search algorithm is somewhat low and requires high complexity. They also proposed sRWA [7] based on the first fit (FF) wavelength assignment algorithm [4]. They, however, did not provide a mechanism to take the property of time overlapping into consideration. Thus, in this paper, we propose a new heuristic algorithm for SLDs, which has very little time cost and complexity while achieving commensurate RWA performance with the others.

3 Proposed Time Disjoint Path RWA Algorithm

In static demands, one wavelength can not be assigned to two or more lightpaths with overlapping links on their routes. If, however, the service times of two SLDs do not overlap, two SLDs can use the same wavelength on the overlapping links. In Fig. 2, there is an overlapping link between the shortest lightpaths of demands

1 and 2. In the case of static demands, they can not be assigned the same wavelength, but in the case of SLDs, they can be assigned the same wavelength due to the time disjoint SLDs. We take the property into consideration.

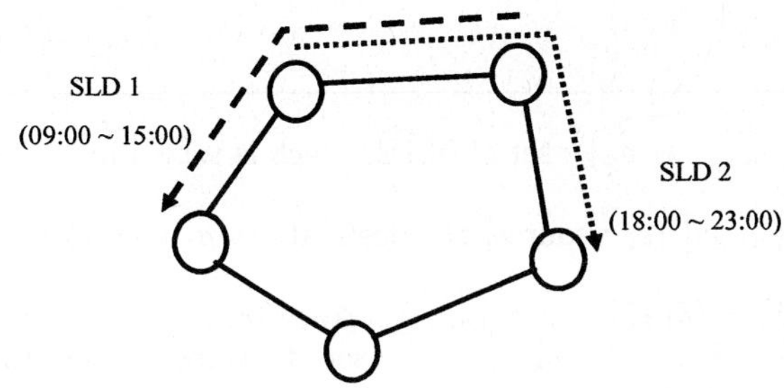

Fig. 2. Time disjoint paths for SLD1 and SLD2.

/* $G(V, E)$: network, λ : wavelength number, $\varDelta$: set of SLDs, $\varDelta_T$: sets of time-disjoint SLDs
$P(G, \varDelta)$: set of the assigned shortest paths of demands, $\alpha(G, \varDelta)$: set of assigned SLDs */

Input : G, $\varDelta$
Ouput : λ
01: **Algorithm TDP-RWA**$(G, \varDelta)$
02: TDP-Selector$(G, \varDelta)$
03: $d = max(diam(G), \sqrt{|E|})$
04: $\lambda = 0$
05: While $(\varDelta \neq \phi)$
06: $\lambda = \lambda + 1$
07: RWAforTDP$(G, \varDelta_T, d)$
08: Assign λ to all paths in $P(G, \varDelta)$
09: $\varDelta = \varDelta - \alpha(G, \varDelta)$

Fig. 3. Proposed TDP-RWA algorithm.

Our algorithm, TDP-RWA, consists of two phases, grouping of time disjoint SLDs (TDP-Selector) and RWA (RWAforTDP). Fig. 3 represents the pseudo code of our proposed TDP-RWA algorithm. Let $G(V, E)$ denote a network with set of nodes V and set of links E. In the algorithm, λ, $\varDelta$, $\varDelta_T$, $P(G, \varDelta)$ and $\alpha(G, \varDelta)$ denote wavelength number, the set of demands, set of grouped demands, set of assigned shortest paths of demands, and set of assigned SLDs, respectively. First, TDP-Selector groups SLDs according to setup and teardown times of demands, and it returns grouped demands $\varDelta_T$ (line 2). We utilize $d =$

max(diam(G), $\sqrt{|E|}$*)* to limit unnecessarily long paths [9]. The steps from line 5 to line 9 are iterated until Δ becomes empty. The RWAforTDP function finds appropriate paths for SLDs and returns $\alpha(G, \Delta)$ and $P(G, \Delta)$ (line 7). Then all the paths of $P(G, \Delta)$ are assigned a wavelength (line 8) and the assigned SLDs are removed from Δ (line 9).

Input : $G(V, E)$
$\Delta=\{\delta_1, \delta_2, \delta_3, \ldots, \delta_n\}$: set of SLDs, which is sorted in an increasing order of ω_i of δ_i
$\delta_i=[s_i, t_i, \mu_i, \omega_i]$ (s_i : source, t_i : destination, μ_i : setup time, ω_i : teardown time)
Output : $\Delta_T=\{\Delta_{T1}=\{\delta_{1,1}, \ldots, \delta_{1,|\Delta_{T1}|}\}, \Delta_{T2}=\{\delta_{2,1}, \ldots, \delta_{2,|\Delta_{T2}|}\}, \ldots, \Delta_{Tk}=\{\delta_{k,1}, \ldots, \delta_{k,|\Delta_{Tk}|}\}\}$: set of grouped sets of time-disjoint SLDs
01: **TDP-Selector**(G, Δ)
02: j = 1
03: $\Delta_{Tj} = \phi$
04: While $(\Delta \neq \phi)$
05: i = 1
06: $\delta_x = i_{th}$ element in Δ, $\delta_z = i_{th}$ element in Δ
07: $\Delta_{Tj} = \Delta_{Tj} \cup \{\delta_x\}$
08: While ($\delta_x \neq$ last element in Δ)
09: i = i + 1
10: $\delta_x = i_{th}$ element in Δ
11: If $(\mu_x \geq \omega_z)$
12: $\Delta_{Tj} = \Delta_{Tj} \cup \{\delta_x\}$
13: $\delta_z = \delta_x$
14: $\Delta = \Delta$ - Δ_{Tj}
15: j = j + 1

Fig. 4. Grouping algorithm (TDP-Selector).

In the grouping phase (TDP-Selector), we make sets of non-time-overlapping SLDs as illustrated in Fig. 4. First, we find a maximal set of time disjoint SLDs. If there are some SLDs left after this first grouping, we group the time disjoint SLDs among the remaining SLDs into another set. This procedure continues until all SLDs are grouped into sets of time disjoint SLDs. The detailed operation of the algorithm is as follows. The demands of the set Δ are sorted in an increasing order of teardown time ω_i. At first, group index j is set to 1 (line 2) and j_{th} set of time disjoint SLDs, Δ_{Tj}, is set to ϕ (line 3). Line 4 $\sim$ line 15 are iteratively performed until Δ becomes empty. In line 5, current SLD index i is initialized to 1. Then, δ_z and δ_x are set as the i_{th} element in the sorted Δ (line 6). And δ_x becomes the first element of Δ_{Tj}. After $(i+1)_{th}$ element of Δ is set to δ_x, teardown time ω_z of δ_z is compared with setup time μ_x of δ_x (line 9 $\sim$ 11). If $\mu_x \geq \omega_z$, then δ_x becomes an element of group Δ_{Tj}, since this indicates that δ_x and δ_z are not time-overlapping (line 12). And the reference δ_z becomes δ_x (line 13). Then it continues comparison with the next target δ_x while δ_x is not

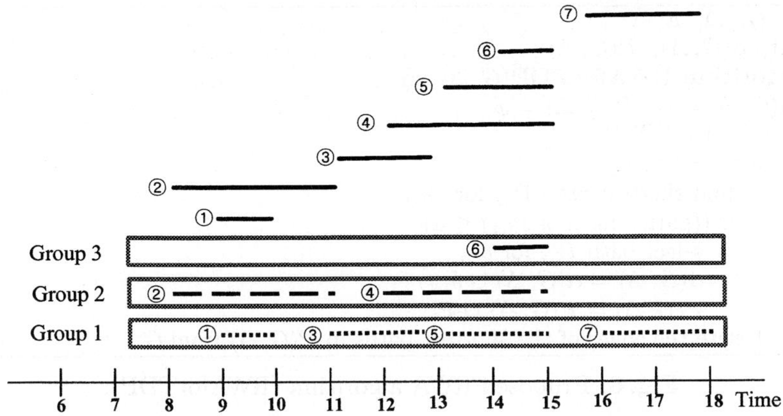

	SLD	μ	ω
①	δ_1=(3, 6, 09:00, 10:00)	09:00	10:00
②	δ_2=(4, 2, 08:00, 11:00)	08:00	11:00
③	δ_3=(2, 5, 11:00, 13:00)	11:00	13:00
④	δ_4=(1, 7, 12:00, 15:00)	12:00	15:00
⑤	δ_5=(3, 5, 13:00, 15:00)	13:00	15:00
⑥	δ_6=(1, 5, 14:00, 15:00)	14:00	15:00
⑦	δ_7=(4, 7, 16:00, 18:00)	16:00	18:00

Fig. 5. An example of TDP-Selector for seven SLDs.

the last demand in Δ. After that, the SLDs in the set Δ_{Tj} are removed from Δ (line 14). If there are still SLDs in Δ, j is increased and the algorithm continues from line 4. This algorithm can group as many as possible time-disjoint SLDs.

Fig. 5 shows an example of the TDP-Selector algorithm. SLDs are sorted in an increasing order of teardown time. At first, δ_1 becomes the first element of group Δ_{T1}. And teardown time of δ_1 is compared to the setup time of δ_2. As the setup time of δ_2 is earlier than the teardown time of δ_1, SLDs 1 and 2 are time-overlapping. So the teardown time of δ_1 is now compared to the setup time of δ_3. As the setup time of δ_3 is later than or equal to the teardown time of δ_1, SLD δ_3 is assigned to group Δ_{T1}, and the new basis for comparison is now set to δ_3. Repeating comparisons, SLDs δ_1, δ_3, δ_5 and δ_7 are allocated to group Δ_{T1}, SLDs δ_2 and δ_4 are allocated to group Δ_{T2}, and SLD δ_6 is allocated to group Δ_{T3}.

In the RWA phase (RWAforTDP), paths and wavelengths for SLDs are allocated. Fig. 6 is the pseudo codes of RWAforTDP. At first, $\alpha(G, \Delta)$ and $P(G, \Delta)$ are initialized. Line 5 $\sim$ line 10 are iterated as many as the number of groups times the number of elements (SLDs) in each group. After finding a shortest path of an SLD j in group i (line 5), this shortest path is compared to d. If the length of the shortest path is not longer than d, $\alpha(G,\ \Delta)$ includes the SLD and

Input : $G,\ \Delta,\ d$
Output : $\alpha(G,\Delta),\ P(G,\Delta)$
01: **Algorithm RWAforTDP**(G, Δ_T, d)
02: $\alpha(G,\Delta) = \phi,\ P(G,\Delta) = \phi$
03: for i = 1 to $|\Delta_T|$
04: for j = 1 to $|\Delta_{Ti}|$
05: find shortest path $P_{i,j}$ for $\delta_{i,j}$
06: if ((path length of $P_{i,j}$) $\leq d$)
07: select path $P_{i,j}$ for $\delta_{i,j}$
08: $\alpha(G,\Delta) = \alpha(G,\Delta) \cup \delta_{i,j}$
09: $P(G,\Delta) = P(G,\Delta) \cup P_{i,j}$
10: Delete the edges of the shortest paths in $P(G,\Delta)$ from G

Fig. 6. Proposed RWA algorithm (RWAforTDP).

$P(G,\ \Delta)$ includes the edges passed by the shortest path. Then all the edges of the shortest paths in $P(G,\ \Delta)$ are removed from G in line 10.

Fig. 7 shows an example of the procedure of the overall TDP-RWA algorithm. It first generates three groups of time disjoint SLDs. Then, it performs RWA for group 1, i.e., δ_1, δ_3, δ_5 and δ_7, with the 1st wavelength. Note that since they are not overlapping in time, all of them can be assigned the same wavelength. The edges of the assigned paths are removed from the graph. Thus, in the 2nd group, only δ_8 is assigned the wavelength since δ_2 and δ_4 cannot find paths (Fig. 7(b), (c)). Similarly, in the 3rd group, δ_6 is not able to be assigned a path (Fig. 7(c)). Since there are still SLDs waiting for RWA, another new wavelength is considered. At this point, original graph is recovered. Then δ_2 and δ_4 (group 2) and δ_6 (group 3) are assigned the paths and the 2nd wavelength (Fig. 7(d), (e)).

4 Performance Evaluation

We evaluate and compare the performance of TDP-RWA with that of BGA-forEDP [5], sRWA [7], and COS [7] in terms of the number of wavelengths and running time. Network topologies used for performance evaluation are randomly generated networks. We generate a random network by specifying the number of nodes ($|V|$) in a graph G and the probability of an edge between any two nodes p_e. The demands are generated according to the probability of a demand between any two nodes p_l. On any source-destination pair, we assume that there can be multiple demands N_c. We denote $T_{service}$ as the average service time of demands and service time is assumed to be uniformly distributed among 0 and 24 hours. The smaller this value, the smaller the probability of time-overlapping among SLDs becomes. We conduct simulations 1000 times for each simulation condition and obtain the average number of wavelengths assigned.

Fig. 8(a) shows the number of wavelengths as p_l increases in random networks with 20 nodes when N_c is 3 or 5, p_e is 0.4 and $T_{service}$ is 4 hours. Since the number of demands increases as N_c and p_l increase, the number of wavelengths increases in general. In case of BGAforEDP the amount of increase in the number

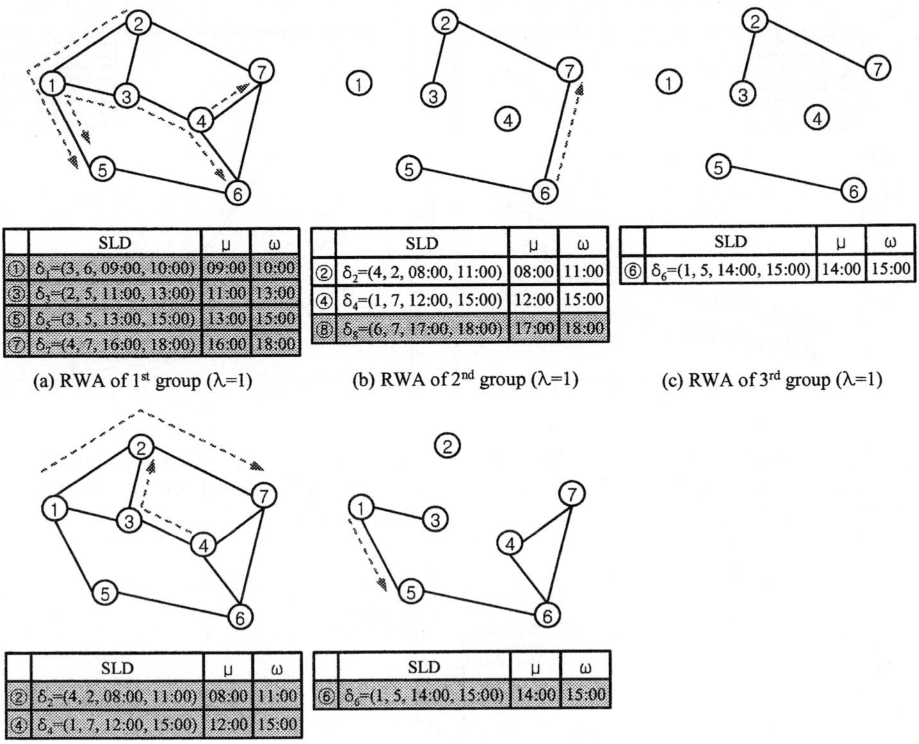

(a) RWA of 1st group (λ=1) (b) RWA of 2nd group (λ=1) (c) RWA of 3rd group (λ=1)

(d) RWA of 2nd group (λ=2) (e) RWA of 3rd group (λ=2)

Fig. 7. An example of the proposed TDP-RWA algorithm.

of wavelengths increases more as p_l increases than TDP-RWA does. Especially, for $N_c = 5$, this phenomenon becomes much greater. There is little difference between TDP-RWA and sRWA. TDP-RWA is shown to reduce the number of wavelengths up to about 25% than BGAforEDP. Our TDP-RWA uses slightly more wavelengths than optimal COS.

Fig. 8(b) shows the number of wavelengths as p_e increases in random networks with 18 nodes when N_c is 3 or 5, p_l is 0.3 and $T_{service}$ is 4 hours. Since increasing the number of edges makes the network more connected and thus generates more candidate paths, the number of wavelengths is shown to decrease. The performance of BGAforEDP is very low comparing to other algorithms and the performance of our proposed TDP-RWA is similar sRWA (up to 2.1% difference). COS is shown to reduce the number of wavelengths up to about 9.1% $\sim$ 9.3% than TDP-RWA or sRWA.

In Fig. 8(c), the number of wavelengths in random networks ($|V| = 20$) is presented when N_c is 3 or 5, p_l is 0.3 and p_e is 0.3. Varying $T_{service}$ affects the property of time overlapping for SLDs. Since the probability of time overlapping

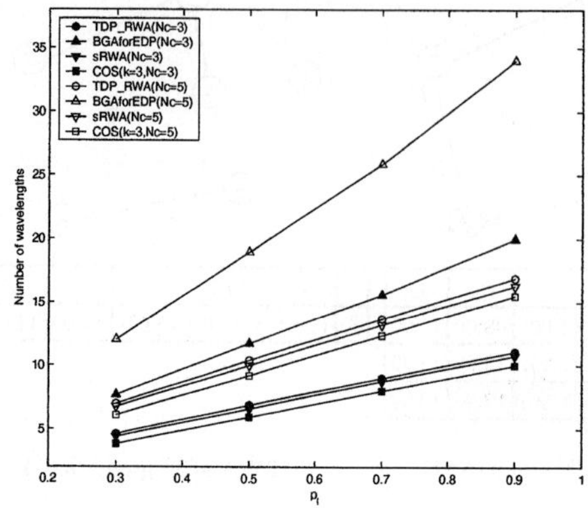

(a) Number of wavelengths as p_l increases (p_e=0.4, N_c=3 or 5, $T_{service}$=4).

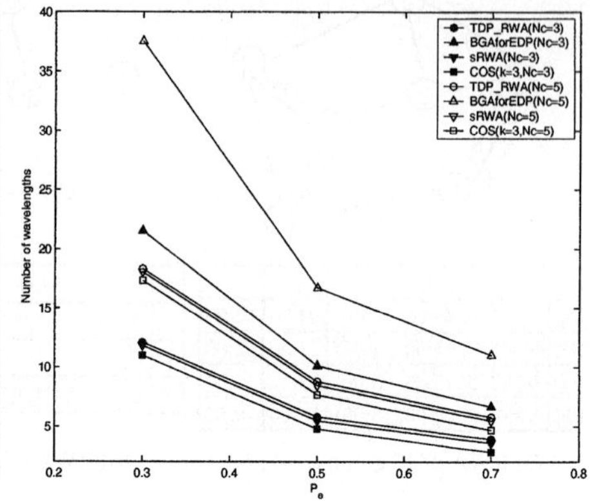

(b) Number of wavelengths as p_e increases (N_c=3 or 5, p_l=0.3, $T_{service}$=4).

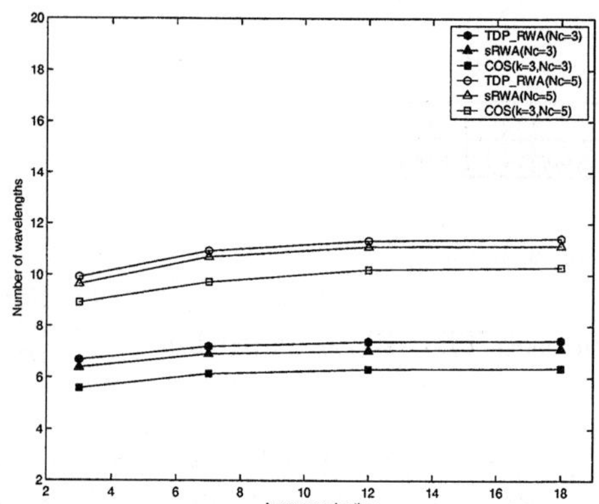

(c) Number of wavelengths as $T_{service}$ increases (N_c=3 or 5, p_l=0.3, p_e=0.3).

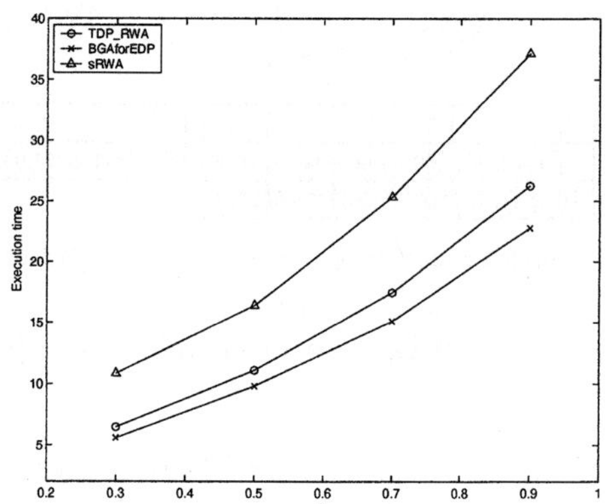

(d) Average execution time as p_l increases (N_c=3, p_e=0.4, $T_{service}$=4).

Fig. 8. Performance evaluation of TDP-RWA.

is low in case of smaller $T_{service}$, the number of wavelengths is shown to decrease as $T_{service}$ becomes smaller. Our TDP-RWA is shown to utilize almost the same wavelengths as sRWA (up to 2.98% difference).

In Fig. 8(d), we illustrate average execution time as p_l increases in random networks ($|V| = 20$) when N_c is 3, p_e is 0.4 and $T_{service}$ is 4 hours. The average execution time of COS is more than 2000 sec, which can not be shown in the figure. But the average execution time of other heuristic algorithms is 5 sec $\sim$ 37 sec. BGAforEDP is the fastest algorithm since it does not utilize any time overlapping property. Our TDP-RWA is faster than sRWA up to 54%. Because of fast grouping, our proposed TDP-RWA has similar execution time with BGAforEDP. The reason why sRWA is slow is that it requires edge comparison for already assigned paths when a demand is assigned a path and a wavelength.

5 Conclusion

In this paper, we have proposed the TDP-RWA algorithm to solve the RWA problem efficiently for SLDs. The optimal COS has very high complexity and requires large time cost, and sRWA based on the FF algorithm requires additional execution time cost and huge memory overhead due to edge comparison. The proposed TDP-RWA is shown to be a fast algorithm to utilize time disjoint paths through fast grouping as well as conventional space disjoint paths. The simulation results for random networks show that our TDP-RWA has the faster execution time than sRWA without additional memory overhead while its performance is commensurate with sRWA.

References

1. Architecture of Optical Transport Network, ITU-T Recommendation, G. 872. (2001)
2. Advanced Networking for Research and Education, Online, Available: http://abilene.internet2.edu/
3. Ramaswami, A., Sivarajan, K.: Routing and Wavelength Assignment in All-Optical Network, IEEE/ACM Transactions on Networking, Vol. 3. No. 5. (1995) 489–500
4. Zang, H., Jue, J.P., Mukherjee, B.: A Review of Routing and Wavelength Assignment Approaches for Wavelength-routed Optical WDM Networks, Optical Networks Magazine, Vol. 1. No. 1. (2000) 47–60
5. Manohar, P., Manjunath, D., Shevgaonkar, R.K.: Routing and Wavelength Assignment in Optical Networks from Edge Disjoint Path Algorithms, IEEE Communications Letter, Vol. 5. (2002) 211–213
6. Kirovski, D., Potkonjak, M.: Efficient Coloring of a Large Spectrum of Graphs, Proc. 35th Conf. Design Automation, (1998) 427–432
7. Kuri, J., Puech, N., Gagnaire, M., Dotaro, E., Douville, R.: Routing and Wavelength Assignment of Scheduled Lightpath Demands, IEEE Journal on Selected Areas in Communications, Vol. 21. No. 8. (2003) 1231–1240
8. Kuri, J., Puech, N., Gagnaire, M., Dotaro, E.: Routing Foreseeable Lightpath Demands Using a Tabu Search Meta-heuristic, Proc. GLOBECOM 2002, Taipei, Taiwan, (2002) 2803–2807
9. Kleinberg, J.: Approximation Algorithms for Disjoint Paths Problems, Ph. D. dissertation, MIT, (1996)
10. Clausen, J.: Branch and Bound Algorithm-Principles and Examples, Online, Available : http://www.imm.dtu.dk/ jha/
11. Clausen, J., Perregaard, M.: On the Best Search Strategy in Parallel Branch-and-Bound-Best-First-Search vs. Lazy Depth-First-Search, Annals of Operations Research, No. 90. (1999) 1–17
12. Glover, F., Laguna, M.: Tabu Search, MA:Kluwer-Academic, (1997)

Performance Implications of Nodal Degree for Optical Burst Switching Mesh Networks Using Signaling Protocols with One-Way Reservation Schemes

Joel J.P.C. Rodrigues[1], Mário M. Freire[1], and Pascal Lorenz[2]

[1] Department of Informatics, University of Beira Interior
Rua Marquês d'Ávila e Bolama, 6201-001 Covilhã, Portugal
{joel, mario}@di.ubi.pt
[2] IUT, University of Haute Alsace
rue du Grillenbreit, 68008 Colmar, France
lorenz@ieee.org

Abstract. This paper investigates the role of nodal degree (meshing degree) in optical burst switching (OBS) mesh networks using signaling protocols with one-way reservation schemes. The analysis is focused on the following topologies: rings, degree-three chordal rings, degree-four chordal rings, degree-five chordal rings, mesh-torus, NSFNET, ARPANET and the European Optical Network. It is shown that when the nodal degree increases from 2 to around 3, the largest gain is observed for degree-three chordal rings (slightly less than three orders of magnitude) and the smallest gain is observed for the ARPANET (less than one order of magnitude). On the other hand, when the nodal degree increases from 2 to around 4, the largest gain is observed for degree-four chordal rings (with a gain between four and five orders of magnitude) and the smallest gain is observed for the European Optical Network (with a gain less than one order of magnitude). Since burst loss probability is a key issue in OBS networks, these results clearly show the importance of the way links are connected in this kind of networks.

1 Introduction

Optical burst switching (OBS) [1]-[4] has been proposed to overcome the technical limitations of optical packet switching, namely the lack of optical random access memory and to the problems with synchronization. OBS is a technical compromise between wavelength routing and optical packet switching, since it does not require optical buffering or packet-level processing and is more efficient than circuit switching if the traffic volume does not require a full wavelength channel. In OBS networks, IP (Internet Protocol) packets are assembled into very large size packets called data bursts. These bursts are transmitted after a burst header packet, with a delay of some offset time. Each burst header packet contains routing and scheduling information and is processed at the electronic

C. Kim (Ed.): ICOIN 2005, LNCS 3391, pp. 352–361, 2005.

level, before the arrival of the corresponding data burst. Several signaling protocols have been proposed for optical burst switching networks. In this paper, we concentrate on just-in-time (JIT) [3], JumpStart [4]-[6], JIT^+ [7], just-enough-time (JET) [1], and Horizon [2] signaling protocols.

A major concern in OBS networks is the contention and burst loss. The two main sources of burst loss are related with the contention on the outgoing data burst channels and on the outgoing control channel. In this paper, we consider bufferless networks and we concentrate on the loss of data bursts in OBS networks.

The reminder of this paper is organized as follows. In section 2, we present an overview of signaling protocols with one-way reservation schemes. In section 3, we describe the model of the OBS network under study, and in section 4 we discuss performance implications of the nodal degree for OBS networks with mesh topologies. Main conclusions are presented in section 5.

2 Signaling Protocols with One-Way Reservation Schemes

In OBS networks, the burst offset is the interval of time, at the source node, between the transmission of the first bit of the setup message and the transmission of the first bit of the data burst. According to the length of the burst offset, signaling protocols may be classified into three classes: no reservation, one-way reservation and two-way reservation. In the first class, the burst is sent immediately after the setup message and the offset is only the transmission time of the setup message. This first class is practical only when the switch configuration time and the switch processing time of a setup message are very short. The Tell And Go (TAG) protocol [8] belongs to this class. In signaling protocols with one-way reservation, a burst is sent shortly after the setup message, and the source node does not wait for the acknowledgement sent by the destination node. Therefore, the size of the offset is between transmission time of the setup message and the round-trip delay of the setup message. Different optical burst switching mechanisms may choose different offset values in this range. JIT, JIT+, JumpStart, JET and Horizon are examples of signaling protocols using one-way reservation schemes. The offset in two-way reservation class is the time required to receive an acknowledgement from the destination. The major drawback of this class is the long offset time, which causes the long data delay. Examples of signaling protocols using this class include the Tell And Wait (TAW) protocol [8] and the scheme proposed in [9]. Due to the impairments of no reservation and two-way reservation classes, we concentrate the study in one-way reservation schemes. Therefore, the remaining of this session provides an overview of signaling protocols with one-way wavelength reservation schemes for optical burst switching networks. One-way reservation schemes may be classified, regarding the way in which output wavelengths are reserved for bursts, as immediate and delayed reservation. JIT and JIT^+ are examples of immediate wavelength reservation, while JET and Horizon are examples of delayed reservation schemes. The JumpStart signaling protocol may be implemented using either immediate or delayed reservation.

The JIT signaling protocol considers that an output wavelength is reserved for a burst immediately after the arrival of the corresponding setup message. If a wavelength cannot be reserved immediately, then the setup message is rejected and the corresponding burst is dropped. JIT^+ is a modified version of the immediate reservation scheme of JIT. Under JIT^+, an output wavelength is reserved for a burst if (1) the arrival time of the burst is later than the time horizon of the wavelength and (2) the wavelength has at most one other reservation. According to the authors, this signaling protocol does not perform any void filling. Comparing JIT+ with JET and Horizon, last ones permit an unlimited number of delayed reservations per wavelength, whereas JIT+ limits the number of such operations to at most one per wavelength. On the other hand, JIT+ maintains all the advantages of JIT in terms of simplicity of hardware implementation.

Delayed reservation, exemplified by JET and Horizon signaling protocols, considers that an output wavelength is reserved for a burst just before the arrival of the first bit of the burst. If, upon arrival of the setup message, it is determined that no wavelength can be reserved at the suitable time, then the setup message is rejected and the corresponding burst is dropped. In this kind of reservation scheme, when a burst is accepted in an OBS node, the output wavelength is reserved for an amount of time equal to the length of the burst plus T_{OXC}, being T_{OXC} the amount of time needed to configure the switch fabric of the OXC in order to set up a connection from an input port to an output port.

The Horizon considers that an output wavelength is reserved for a burst only if the arrival time of the burst is later than the *time horizon* of the wavelength. If, upon arrival of the *setup* message, it is determined that the arrival time of the burst is earlier than the smallest time horizon of any wavelength, then the *setup* message is rejected and the corresponding burst is dropped.

On the other hand, JET signaling protocol is the most known delayed wavelength reservation scheme *with void filling*, which uses information to predict the start and the end of the burst. In this protocol, an output wavelength is reserved for a burst if the arrival time of the burst (1) is later than the *time horizon* of the wavelength, or (2) coincides with a void on the wavelength, and the end of the burst (plus the OXC configuration time T_{OXC}) occurs before the end of the void. If, upon arrival of the setup message, it is determined that none of these conditions are satisfied for any wavelength, then the *setup message* is rejected and the corresponding burst dropped.

3 Network Model

We consider OBS networks with the following mesh topologies: chordal rings with nodal degrees between 3 and 5, mesh-torus with 16 and 20 nodes, the NSFNET with 14-node and 21 links [10], the NSFNET with 16 nodes and 25 links [11], the ARPANET with 20 nodes and 32 links [10], [12], and the European Optical Network (EON) with 19 nodes and 37 links [13]. For comparison purposes bi-directional ring topologies are also considered. These topologies have the following nodal degree: ring: 2.0; degree-three chordal ring: 3.0; degree-four chordal ring: 4.0; degree-five chordal ring: 5.0; mesh-torus: 4.0; NSFNET with 14-node and 21 links: 3.0; the NSFNET with 16 nodes and 25 links: 3.125; the ARPANET with 20 nodes and 32 links: 3.2; and the EON: 3.895.

Chordal rings are a well-known family of regular degree three topologies proposed by Arden and Lee in early eighties for interconnection of multi-computer systems [14]. A chordal ring is basically a bi-directional ring network, in which each node has an additional bi-directional link, called a chord. The number of nodes in a chordal ring is assumed to be even, and nodes are indexed as 0,1,2,...,N-1 around the N-node ring. It is also assumed that each odd-numbered node i (i=1,3,...,N-1) is connected to a node $(i+w) mod\ N$, where w is the chord length, which is assumed to be positive odd. For a given number of nodes there is an optimal chord length that leads to the smallest network diameter. The network diameter is the largest among all of the shortest path lengths between all pairs of nodes, being the length of a path determined by the number of hops. In each node of a chordal ring, we have a link to the previous node, a link to the next node and a chord. Here, we assumed that the links to the previous and to the next nodes are replaced by chords. Thus, each node has three chords, instead of one. Let $w1$, $w2$, and $w3$ be the corresponding chord lengths, and N the number of nodes. We represented a general degree three topology by D3T($w1$, $w2$, $w3$). We assumed that each odd-numbered node i (i=1, 3, ..., N-1) is connected to the nodes $(i+w1) mod\ N$, $(i+w2) mod\ N$, and $(i+w3) mod\ N$, where the chord lengths, $w1$, $w2$, and $w3$ are assumed to be positive odd, with $w1 \leq N-1, w2 \leq N-1$, and $w3 \leq N-1$, and $w_i \neq w_j, \forall i \neq j$ and $1 \leq i,j \leq 3$. In this notation, a chordal ring with chord length w is simply represented by D3T(1,N-1,$w3$).

Now, we introduce a general topology for a given nodal degree. We assume that instead of a topology with nodal degree of 3, we have a topology with a nodal degree of n, where n is a positive integer, and instead of having 3 chords we have n chords. We also assume that each odd-numbered node i (i=1,3,...,N-1) is connected to the nodes $(i+w) mod\ N$, $(i+w2) mod\ N$, ..., $(i+wn) mod\ N$, where the chord lengths, $w1$, $w2$, ..., wn are assumed to be positive odd, with $w1 \leq N-1, w2 \leq N-1, ..., wn \leq N-1$, and $w_i \neq w_j, \forall i \neq j$ and $1 \leq i,j \leq n$. Now, we introduce a new notation: a general degree n topology is represented by DnT($w1$, $w2$,...,wn). In this new notation, a chordal ring family with chord length w is represented by D3T(1,N-1,w). In this new notation, a chordal ring family with a chord length of $w3$ is represented by D3T(1,N-1,$w3$) and a bi-directional ring is represented by D2T(1,N-1).

We assume that each node of the OBS network supports F+1 wavelength channels per unidirectional link. One wavelength is used for signaling (carries setup messages) and the other F wavelengths carry data bursts. Each OBS node consists of two main components [7]: i) a signaling engine, which implements the OBS signaling protocol and related forwarding and control functions; and ii) an optical cross-connect (OXC), which performs the switching of bursts from input to output. It is assumed that each OXC consists of non-blocking space-division switch fabric, with full conversion capability, but without optical buffers. It is assumed that each OBS node requires [12]: i) an amount of time, T_{OXC}, to configure the switch fabric of the OXC in order to set up a connection from an input port to an output port, and requires ii) an amount of time, $T_{setup}(X)$ to process the setup message for the signaling protocol X, where X can be JIT, JET, horizon, and JumpStart. It is also considered the offset value of a burst under reservation scheme X, $T_{offset}(X)$, which depends, among other factors, on the signaling protocol, the number of nodes the burst has already traversed, and if the offset value is used for service differentiation. In this study, it is assumed that [7]: $T_{OXC} = 10ms$, $T_{setup}(JIT) = 12.5\mu s$, $T_{setup}(JIT+) = 12.5\mu s$, $T_{setup}(JumpSart) = 12.5\mu s$, $T_{setup}(JET) = 50\mu s$, $T_{setup}(Horizon) = 25\mu s$, the mean burst size, $1/\mu$, was set to $50ms$, and the burst arrival rate λ, is such that $\lambda/\mu = 32$.

4 Performance Assessment

In this section, we investigate the influence of nodal degree on the performance of OBS mesh networks for JIT, JIT$^+$, JumpStart, JET, and Horizon signaling protocols. Details about the simulator used to produce simulation results can be found in [15].

In chordal ring topologies, different chord lengths can lead to different network diameters, and, therefore, to a different number of hops. One interesting result that we found is concerned with the diameters of the D3T($w1$,$w2$,$w3$) families, for which $w2$=($w1$+2)mod N or $w2$=($w1$-2)mod N. Each family of this kind, i.e. D3T($w1$,($w1$+2)mod N, $w3$) or D3T($w1$, ($w1$-2)mod N, $w3$), with $1 \leq w1 \leq 19$ and $w1 \neq w2 \neq w3$, has a diameter which is a shifted version (with respect to $w3$) of the diameter of the chordal ring family (D3T(1, N-1, $w3$)). For this reason, we concentrate the analysis on hordal ring networks, i. e., D3T(1, 19, $w3$).

In order to quantify the benefits due to the increase of nodal degree, we introduce the nodal degree gain, $G_{(n-1),n}(i,j)$, defined as:

$$G_{n-1,n}(i,j) = \frac{P_i(n-1)}{P_j(n)} \tag{1}$$

where $P_i(n-1)$ is the burst blocking probability in the i-th hop of a degree (n-1) topology and $P_j(n)$ is the burst blocking probability in the j-th hop of a degree n topology, for the same network conditions (same number of data wavelengths per link, same number of nodes, etc), and for the same signaling protocol.

Figures 1, 2, 3, 4, and 5 show, respectively for JIT, JET, Horizon, JIT$^+$, and JumpStart, the nodal degree gain, in the last hop of each topology, due to the increase of the nodal degree from 2 (D2T(1,15)) to: 3 (D3T(1, 15, 5)), 3.125 (NSFNET), 4 (D4T(1,15,5,13) and mesh-torus), and 5 (D5T(1,15,7,3,9)). Concerning chordal rings, we have chosen among several topologies with smallest diameter the ones that led to the best network performance. As may be seen in those figures, the considered topologies may be sorted from the best performance for the worst performance as: D4T(1,15,5,13, D5T(1,15,7,3,9), D3T(1, 15, w3), mesh-torus, and NSFNET.

We observed that the performance of the NSFNET is very close to the performance of degree-three chordal rings with chord length of w3=3 or w3=7 (figure not shown due to space limitations). This results reveals the importance of the way links are connected in the network, since chordal rings and NSFNET have similar nodal degrees and therefore a similar number of network links. Also interesting is the fact that degree-three chordal rings with w3=5 have better performance than mesh-torus networks, which have a nodal degree of 4, i. e., more 25% of network links. Results presented in these figures (1 to 5) 2 were obtained for the JIT, JIT$^+$, JumpStart, JET, and Horizon protocols, and, as may be seen, their performance is very close, except for D5T in which a variation within one order of magnitude is observed.

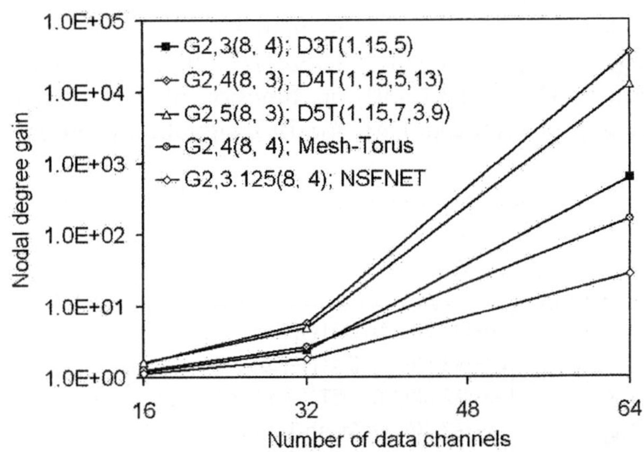

Fig. 1. Nodal degree gain due to the increase of the nodal degree from 2 (D2T(1,15)) to: 3 (D3T(1, 15, w3)), 3.125 (NSFNET), 4 (D4T(1,15,5,13) and mesh-torus), and 5 (D5T(1,15,7,3,9)), as function of the number of data channels, in the last hop of each topology, for JIT signaling protocol; N=16.

Since the burst blocking probability is a major issue in OBS networks, clearly ring topologies are the worst choice for these network due to very high blocking probabilities and, surprisingly, degree-three chordal rings with smallest diameter have a very good performance with burst blocking probabilities ranging from 10^{-3}–10^{-5}, depending on the number of hops. For 16 nodes and 64 data channels per link, the nodal degree gain due to the increase of nodal degree from 2 (rings) to 3 (chordal ring with smallest diameter) is about three orders of magnitude in the last hop. This nodal degree gain increases to between 4 and 5 orders of

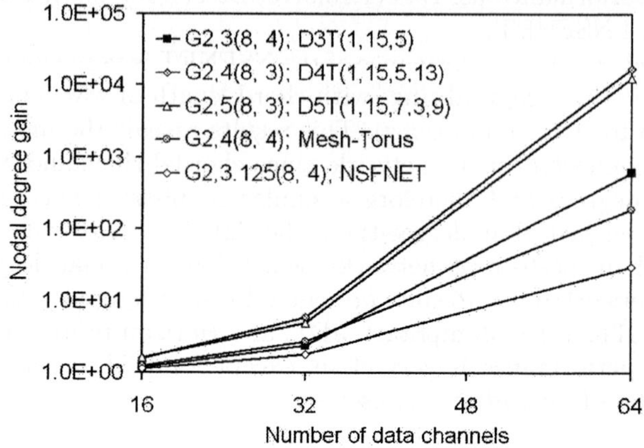

Fig. 2. Nodal degree gain due to the increase of the nodal degree from 2 (D2T(1,15)) to: 3 (D3T(1, 15, w3)), 3.125 (NSFNET), 4 (D4T(1,15,5,13) and mesh-torus), and 5 (D5T(1,15,7,3,9)), as function of the number of data channels, in the last hop of each topology, for JET signaling protocol; N=16.

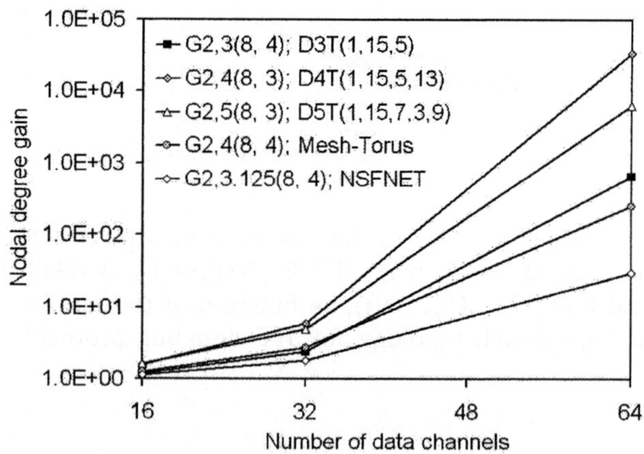

Fig. 3. Nodal degree gain due to the increase of the nodal degree from 2 (D2T(1,15)) to: 3 (D3T(1, 15, w3)), 3.125 (NSFNET), 4 (D4T(1,15,5,13) and mesh-torus), and 5 (D5T(1,15,7,3,9)), as function of the number of data channels, in the last hop of each topology, for Horizon signaling protocol; N=16.

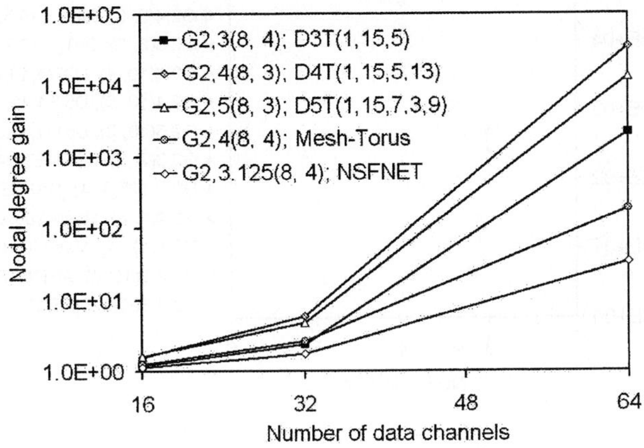

Fig. 4. Nodal degree gain due to the increase of the nodal degree from 2 (D2T(1,15)) to: 3 (D3T(1, 15, w3)), 3.125 (NSFNET), 4 (D4T(1,15,5,13) and mesh-torus), and 5 (D5T(1,15,7,3,9)), as function of the number of data channels, in the last hop of each topology, for JIT$^+$ signaling protocol; N=16.

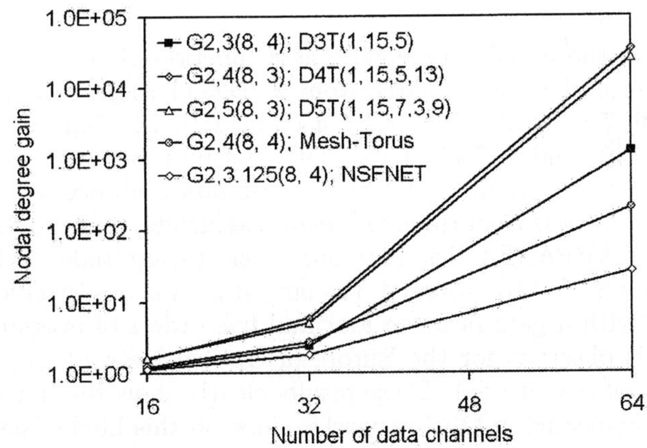

Fig. 5. Nodal degree gain due to the increase of the nodal degree from 2 (D2T(1,15)) to: 3 (D3T(1, 15, w3)), 3.125 (NSFNET), 4 (D4T(1,15,5,13) and mesh-torus), and 5 (D5T(1,15,7,3,9)), as function of the number of data channels, in the last hop of each topology, for JumpStart signaling protocol; N=16.

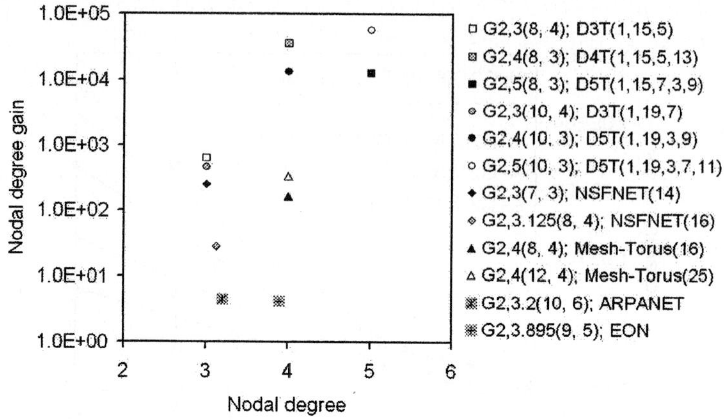

Fig. 6. Nodal degree gain in the last hop of each topology, as a function of the nodal degree, due to the increase of the nodal degree from 2 (D2T(1,15)) to: 3 (D3T(1, 15, 5) and D3T(1, 19, 7)), 3.125 (NSFNET), 4 (D4T(1,15,5,13) and mesh-torus with 16 and 25 nodes), and 5 (D5T(1,15,7,3,9)), for JIT signaling protocol; F=64.

magnitude if the nodal degree increases from 2 (rings) to 4 (D4T(1,15,5,13)), and increases to around 4 orders of magnitude if the nodal degree increases from 2 (rings) to 5 (D5T(1,15,7,3, 9)).

Fig. 6 shows the nodal degree gain, as a function of the nodal degree, due to the increase of the nodal degree from 2 (D2T(1,15)) to: 3 (D3T(1, 15, 5) and D3T(1, 19, 7)), 3.125 (NSFNET), 4 (D4T(1,15,5,13) and mesh-torus with 16 and 25 nodes), and 5 (D5T(1,15,7,3,9)). As may be seen, when the nodal degree increases from 2 to around 3, the largest gain is observed for degree-three chordal rings (a bit less than three orders of magnitude) and the smallest gain is observed for the ARPANET (less than one order of magnitude). When the nodal degree increases from 2 to around 4, the largest gain is observed for degree-four chordal rings (with a gain between four and five orders of magnitude) and the smallest gain is observed for the European Optical Network (with a gain less than one order of magnitude). These results clearly show the importance of the way links are connected in OBS networks, since, in this kind of networks, burst loss probability is a key issue.

5 Conclusions

In this paper, we discussed performance implications of the nodal degree for OBS mesh networks with the following topologies: rings, chordal rings, mesh-torus, NSFNET, ARPANET and the EON. It was shown that when the nodal degree increases from 2 to around 3, a larger gain of slightly less than three orders of magnitude is observed for degree-three chordal rings and a smaller gain less than one order of magnitude is observed for the ARPANET. When the nodal degree increases from 2 to around 4, a larger gain between four and five orders of magnitude is observed for degree-four chordal rings and a smaller gain less than one order of magnitude is observed for the European Optical Network.

References

1. Qiao, C., Yoo, M.: Optical burst switching (OBS)-A New Paradigm for an Optical Internet. Journal of High Speed Networks, Vol. **8**, No. 1 (1999) 69-84.
2. Turner, J.S.: Terabit Burst Switching. Journal of High Speed Networks, Vol. **8**, No. 1 (1999) 3-16.
3. Wei, J.Y., McFarland, R.I.: Just-in-time signaling for WDM optical burst switching net-works. In Journal of Lightwave Technology, Vol. **18**, No. 12 (2000) 2019-2037.
4. Baldine, I., Rouskas, G.N., Perros, H.G., Stevenson, D.: JumpStart: A just-in-time signaling architecture for WDM burst-switched networks. In IEEE Communications Magazine, Vol. **40**, No. 2 (2002) 82-89.
5. Zaim, A.H., Baldine, I., Cassada, M., Rouskas, G.N., Perros, H.G., Stevenson, D.: The JumpStart Just-In-Time Signaling Protocol: A Formal Description Using EFSM. In Optical Engineering, Vol. **42**, No. 2, February (2003) 568-585.
6. Baldine, I., Rouskas, G.N., Perros, H.G., Stevenson, D.: Signaling Support for Multicast and QoS within the JumpStart WDM Burst Switching Architecture. In Optical Networks, Vol. **4**, No. 6, November/December (2003)
7. Teng, J., Rouskas, G. N.: A Detailed Analysis and Performance Comparison of Wavelength Reservation Schemes for Optical Burst Switched Networks, Photonic Network Communications (to appear).
8. Widjaja, I. Performance Analysis of Burst Admission Control Protocols. IEE Proceeding of Communications, Vol. 142, pp. 7-14, February 1995.
9. Duser M.; and Bayvel P. Analysis of a Dynamically Wavelength-Routed Optical Burst Switched Network Architecture. J. Lightwave Technol., Vol. 20, No. 4, (2002), 574-585.
10. Sridharan, M., Salapaka, M. V., and, Somani, A. K.: A Practical Approach to Operating Survivable WDM Networks. IEEE Journal on Selected Areas in Communications, Vol. **20**, No. 1, (2002) 34-46.
11. Ramesh, S., Rouskas, G. N., and Perros, H. G.: Computing Blocking Probabilities in Multi-class Wavelength-Routing Networks With Multicast Calls. IEEE Journal on Selected Areas in Communications, Vol. **20**, No. 1, (2002) 89-96.
12. Nayak, T. K., and Sivarajan, K. N.: A New Approach to Dimensioning Optical Networks. IEEE Journal on Selected Areas in Communications, Vol. **20**, No. 1, (2002) 134-148.
13. O'Mahony, M. J.: Results from the COST 239 Project: Ultra-high Capacity Optical Trans-mission Networks. Proc. 22nd European Conf. on Optical Communication (ECOC), Oslo, Norway, Vol. 2, (1996) 2.11-2.18.
14. Arden, B.W., Lee, H.: Analysis of Chordal Ring Networks. IEEE Transactions on Computers, Vol. **C-30**, No. 4 (1981) 291-295.
15. Rodrigues, J.J.P.C., Garcia, N.M., Freire, M.M. and Lorenz, P.: Object-Oriented Modeling and Simulation of Optical Burst Switching Networks, accepted for IEEE Global Telecommunications Conference (GLOBECOM'2004), Dallas, Nov. 29-Dec. 3, 2004.

Offset-Time Compensation Algorithm – QoS Provisioning for the Control Channel of the Optical Burst Switching Network

In-Yong Hwang[1], Jeong-Hee Ryou[2], and Hong-Shik Park[1]

[1] Optical Internet Research Center (OIRC), Information and Communications University, 119, Munjiro, Yuseong-gu, Daejeon, 305-732, Korea
{iyhwang, hspark}@icu.ac.kr

[2] Land Information Center, Korea Land Corporation 217, Jeongja-Dong, Bundang-Gu, Sungnam City, Kyunggi-Do, 463-755, Korea
viva@iklc.co.kr

Abstract. Optical Burst Switching (OBS) has intrinsically time-sensitive nature with the separation of the header and the payload. Thus, minute care for offset-time is definitely required, which has not been a highlighted issue in an existing OBS research. In this paper, we focus on the relatively early arrival of the data burst due to the excessive queueing delay of the burst control packet (BCP) under a heavily loaded network. We propose a new scheduling algorithm for the OBS control channel, the Offset-Time Compensation (OTC) algorithm to solve the early arrival problem incurring the data burst drop. In the OTC algorithm, the offset-time is determined by the existing static offset-time scheme and the proposed dynamic offset-time reflecting the offered load of the OBS network. scheme varying with the network condition. By scheduling the BCP, the OTC can reduce the data burst loss rate due to early arrival. With the dynamic offset-time scheme, the data burst loss rate decreases significantly and maintains regular rate regardless of the offered load. We also expend the OTC, QoS-Aware (QA) OTC to provides controllable QoS differentiation in terms of data burst loss rate due to early arrival. The service objective of higher class is satisfied even in a heavily loaded situation.

1 Introduction

Optical burst switching (OBS) is an attractive technology for increasing network utilization in wavelength paths because of the potential of the fine-granularity optical switching. An original feature of the OBS is the physical separation of the optical data transport and the electronic control of the switch about data burst, which can facilitate the electronic processing of Burst Control Packet (BCP) at OBS core nodes and provides end-to-end transparent optical paths for transporting the data burst [1]. A data burst may enter into the optical switching fabric before its control packet has been fully processed due to excessive processing delay of the BCPs. This event is called 'early arrival' which can result

C. Kim (Ed.): ICOIN 2005, LNCS 3391, pp. 362–369, 2005.

in the data burst loss in the OBS core nodes. We propose a new scheduling algorithm for the OBS control channel, the Offset Time Compensation (OTC) algorithm, which schedules the BCP to compensate the offset time difference. Thus, it makes the BCP arrive at the switching fabric of a node at a scheduled time. We introduce two schemes for determining the initial offset time value: 1) the static offset time scheme, and 2) the dynamic offset time scheme. In Section 2, we describe the offset-time issues, while in Section 3, we propose the OTC algorithm to reduce the early arrival rate and the QoS-Aware (QA) OTC algorithm to provide differentiate service. Section 4 presents our simulation results of the OTC and QA-OTC algorithm related to the data burst loss rate due to early arrival. And finally, in Section 5, we offer our conclusions.

2 Offset-Time Issues

2.1 Early Arrival Problem in OBS Network

The offset time indicates the difference between the arrival time of the BCP and that of the data burst [1, 2]. It is necessary because the BCP incurs a processing delay at each switch while the data burst does not. In a conventional manner, the offset time T is set to be at least Hd, where H is the number of hops between the source and the destination, and d is the defined (expected) processing time incurred by the BCP packet at a core node which is determined as the same value at each core node [1, 2]. When congestion occurs in the control channel, the elapsed time in the switching module of each core node increases. If a data burst enters into the optical switching matrix before its BCP has been processed, the data burst is simply dropped because the optical switch is not configured for the early arrived data burst. This is referred to as the so called 'early arrival' phenomenon [10]. Remarkably, the difference between the defined offset-time and the actual (experienced) offset-time results in inevitable data burst loss.

2.2 Related Works

There have been some related studies on the offset-time in the OBS network. First, offset time-based QoS techniques [4, 5] use the fact that bursts with larger offset times are blocked less, and that the high priority class achieves significant blocking reduction, while the low priority class experiences increased blocking. It actually relies on the fact that requests for reservation further into future are most likely to succeed since the number of bursts already in the schedule is smaller on average. However, as bursts proceed through a network, their residual offset time decreases. The further a burst goes, the more likely it is to be blocked, leading to many bursts consuming many network resources only to be blocked not far from their destinations. There are two scheduling algorithms used to reduce the blocking of bursts with long route lengths by finding a way to give them priority over short route bursts in the wavelength scheduling algorithm.

A merit-based scheduling algorithm [6], which ranks an arriving burst against those which have already been scheduled for transmission, preempts the one which will cause the least impact in terms of lost resources in favor of the new arrival. Even though it uses the offset time information, it only concentrates on data channel scheduling with no consideration of control channel scheduling. Thus, it can not solve the early arrival problem. In the BSCOT algorithm [7], a BCP with a short residual offset-time is served prior to the BCPs with long residual offset-time, so it can reduce the data burst loss rate at each node and the total loss rate over the entire network. Both algorithms assume that the offset-time can vary along with the offered load to the OBS network; however, they are only focused on the residual offset time proportional to the remaining hops, and do not consider the time difference between the estimated offset time and actual offset time. As the experienced offset-time is larger than the defined offset-time, it is difficult to avoid the early arrival problem due to congestion in the control channel.

3 Offset-Time Compensation (OTC) Scheduling Algorithm

We introduce our proposed the offset-time decision scheme and the OTC algorithm to avoid the early arrival problem and provide differentiated QoS in the OBS networks.

3.1 Offset-Time Decision

Static offset-time In the static offset-time scheme the defined offset-time is fixed, similar to the existing approach. Because the network situation dose not reflected on the application of the algorithm. If the network load increases, the congestion occurs in the control channel, which naturally results in increasing data burst loss rate due to the early arrival problem [1, 2].

Dynamic offset-time The offset-time varies according to the network condition. The BCP is transmitted backward from the destination node to the source node. The destination node continues to monitor the BCP arrival and maintain the information of defined end-to-end delay of the BCP D_d . When the BCP arrives at the destination node, experienced end-to-end delay D_e is obtained by using the defined offset-time field T_{D_offset} of the BCP: $D_e = D_d - T_{D\_offset}$. D_d is updated by average D_e for a certain time. This updated D_d is transmitted to a source node by the backward BCP, then the source node uses this updated D_d as the defined offset-time T_{D_offset}.

3.2 OTC Algorithm

The OTC is a priority scheduler for the OBS control channel in which the priority of a packet increase proportionally with its compensation time and waiting time. We define two offset-time fields of the BCP for the OTC.

- T_{D_offset} : defined offset-time
- T_{E_offset} : experienced offset-time

The BCP carries the information about experienced offset-time and defined offset-time. Initially, these fields have the same value at a source node, and are updated at each core node via the route from the source to the destination.

$$T_{D\_offset,t'} = T_{D\_offset,t} - \Delta \tag{1}$$

$$T_{E\_offset,t'} = T_{E\_offset,t} - (t_{out} - t_{in}) \tag{2}$$

The defined offset-time field decreases as the defined per-hop processing delay of the BCP Δ, at each node. However, the experienced offset-time field decrease actually suffers a processing delay of the BCP, which is the difference between the departure time of a BCP t_{out} at a node, and the arrival time of the BCP t_{in} at a node.

The priority of a packet in queue i at time t is determined by compensation time $t_{i,t}$, which is the offset-time difference between the defined and the experienced, and waiting time $d_{i,t}$.

$$t_{i,t} = T_{D\_offset,t} - T_{E\_offset,t} \tag{3}$$

$$p_{i,t} = t_{i,t} + d_{i,t} \tag{4}$$

The OTC scheduler consists of various numbers of queues. When a BCP arrives at switch fabric of an OBS core node, it can be time stamped with its arrival time and then entered into any one of the multiple queues. When a scheduler has to dequeue a BCP, it computes its priority $p_{i,t}$, of each first packet of each queue using (4), and the BCP with the highest priority is served first. In terms of scalability and performance, the OTC requires at most N-1 comparisons for each packet transmission, which is a minor overhead. An important requirement is that packets have to be time-stamped upon arrival so that delays can be measured.

3.3 QA OTC

We propose a new scheme with controllable QoS differentiation on delay variance and data burst loss rate due to the early arrival problem by using scheduling in the control channel. The basic principles are the same as the OTC scheduling algorithm which is that instant priority of packets at the head of queue is determined by the difference between the defined offset-time and the experienced offset-time. There are some modifications for supporting the QoS differentiation. Fist of all, The QA-OTC scheduler maintain n queues Q_1, Q_2,... ,and Q_n, with Q_i being used to store the BCPs of Class-i data burst. The priority $p_{i,t}$, of a packet in queue i at time t is determined by compensation time $t_{i,t}$, which is the difference between the defined offset-time and the experienced offset-time, and the waiting time $d_{i,t}$. The weight w_i determines the rate with which the priority of the packets of a certain class increases.

$$p_{i,t} = t_{i,t} + d_{i,t} \cdot w_i \tag{5}$$

4 Performance Evaluation

We present some simulation results to evaluate our algorithm comparing existing FIFO for the OBS control channel. The performance metrics we use for this comparison is the data burst loss rate due to early arrival. We simulate the algorithm in a 5*5 lattice topology. In this topology, user traffic is generated at each node which traverses the various hops through the shortest path. To generate self-similar traffic which is considered a real OBS control channel, all traffic sources have a pareto-distributed ON/OFF process with a mean t_{on} (0.2ms) of the ON period, a mean t_{off} (0.2ms) of the OFF period, and 1.2 for shape parameter α reporting the measured Hurst parameter of 0.9. All source agents generate packets with a fixed size of 64 bytes [3, 4, and 5].

4.1 Data Burst Loss Rate due to EA with OTC

We present the data burst loss rate due to EA performance for the 8, and 16 hops as a function of load in Fig. 1, and 2, respectively. Fig. 1 (a), and 2 (a) show the data burst loss rate due to EA with the static offset-time scheme. Fig. 1 (b), and 2 (b) show the data burst loss rate due to EA when we use the dynamic offset-time scheme. The results of data burst loss due to EA are the accumulated effects of every node in the route of packet. As the load increases, the data burst loss rate increases. These results show that the OTC scheduler has better performance than the FIFO scheduler. With the static offset -time scheme, the defined offset-time has fixed value. So, the criterion of the EA is also fixed regardless of the load. When network load increases, the data burst loss rate due to the EA increases accordingly. With the dynamic offset-time scheme, the defined offset-time value varies and the criterion of the EA also varies according to the network load.

4.2 Data Burst Loss Rate due to EA with QA-OTC

We present data burst loss rate due to EA performance of the QA-OTC for 4 and 16 hops as a function of load in Fig. 3 and 4, respectively. Fig. 1 (a) and 2 (a) show the data burst loss rate due to EA with the static offset-time scheme. Fig. 3 (b) and 4 (b) show the data burst loss rate due to EA when we use the dynamic offset-time scheme. The data burst loss rate due to EA in the case of 16 hops is generally larger than in the case of 4 hops. The results show that the OTC provides the differential service in terms of the data burst loss rate due to EA. More specifically, the class1 and class 2 have a lower data loss rate than the classless case (or average); while class 3 and class 4 have a higher loss rate

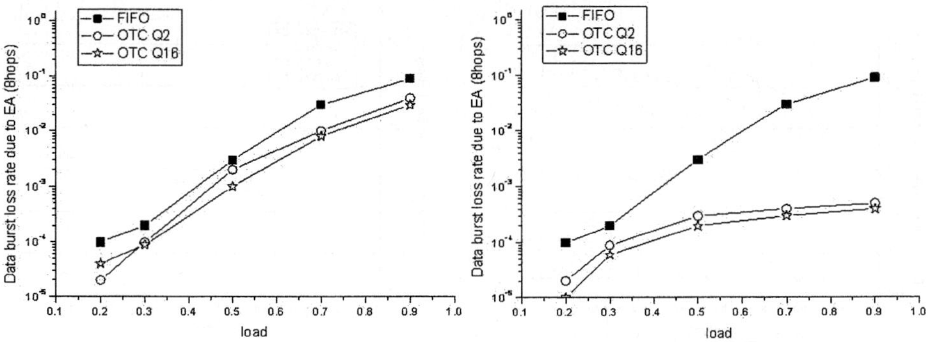

(a) Static offset time scheme (b) Dynamic offset time scheme

Fig. 1. Data burst loss rate due to EA for FIFO vs. OTC in 8 hops.

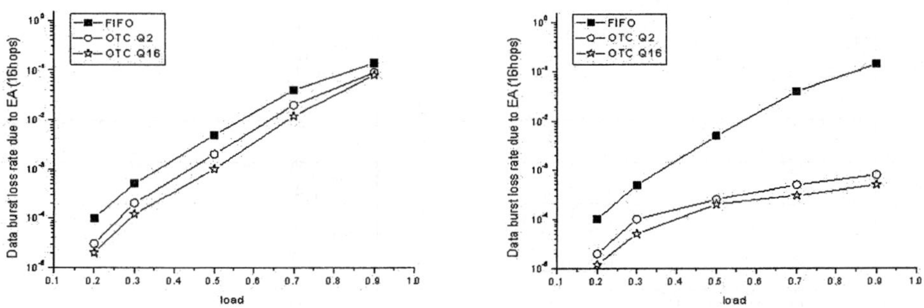

(a) Static offset time scheme (b) Dynamic offset time scheme

Fig. 2. Data burst loss rate due to EA for FIFO vs. OTC in 16 hops.

because of their low priority. The results show that the performance satisfies the service objective in terms of the data burst loss rate due to EA, 10^{-2} in the case of class 1, and 10^{-1} in the case of class 2, respectively. Data burst loss rate due to the EA of a higher priority packet remarkably decreases in both the static offset-time scheme and the dynamic offset-time scheme. In the case of the static offset-time scheme, as the load increases, the data burst loss rate due to EA increases in all hops of the network. We know that class 1 and class 2 have a lower loss rate than the criterion of the service objective in terms of data burst loss rate due to EA. In the case of the dynamic offset-time scheme, the performance of the QA-OTC is much better than the static offset-time scheme from two points of view. First, the data burst loss rate due to EA is generally small in the every class with regardless of network condition and also provide the differential rate of this performance according to the class of traffic. And second, the degree of differential performance is larger than in a heavily loaded network condition.

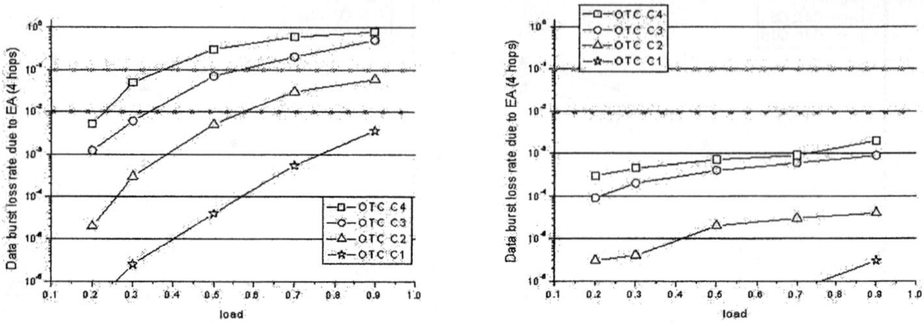

(a) Static offset time scheme (b) Dynamic offset time scheme

Fig. 3. Data burst loss rate due to EA for FIFO vs. OTC in 4 hops.

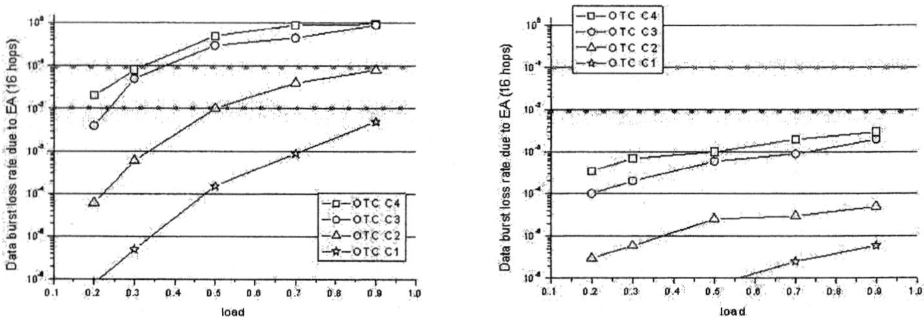

(a) Static offset time scheme (b) Dynamic offset time scheme

Fig. 4. Data burst loss rate due to EA for FIFO vs. OTC in 16 hops.

5 Conclusion

We proposed an Offset-time Compensation (OTC) scheduling algorithm for an OBS control channel. It is mainly focused on the offset-time differential problem of an OBS network. To reduce the data burst loss due to early arrival, we attempted to make the BCP arrive at the defined offset-time by compensating the excessive time. An extended OTC algorithm, QA-OTC, can support differentiated service by prioritizing BCPs with multiple queues. In our simulation results, the OTC shows better performance than the traditional FIFO for OBS control channel in terms of the data loss rate due to early arrival, In particular, in the case of using the dynamic offset-time scheme for QA-OTC, the data burst loss rate is almost under 10^{-3}, regardless of the offered load.

Acknowledgement

This work was supported in part by the Korea Science and Engineering Foundation (KOSEF) through OIRC project.

References

1. C. Qiao, M. Yoo: Optical Burst switching (OBS) - a new paradigm for an optical Internet, Journal of High Speed Networks, Vol. 8, no 1, (1999) 68-84
2. Jonathan S. Turner: Terabit Burst Switching, Journal of High Speed Networks, Vol. 8, No.1, January (1999) 3-16
3. Y. Xing, M. Vanderhoute, C.C. Cankaya: Control architecture in optical burst-switched WDM networks, IEEE Journal on Selected Areas in Communications, Vol.18, No.10, October (2000) 1838-1851
4. M.Yoo and C. Qiao: A new optical burst switching protocol for supporting quality of service, SPIE'98 Conf. All Optical Communication Syst.: Architecture, Control, Network Issues, vol. 3531, Boston, Nov. (1998) 396-405
5. M. Yoo, C. Qiao, and S. Dixit: QoS performance of optical burst switching in ip-over-wdm networks, IEEE Journal on Selected Areas in Communication, October (2000) 2062-2071
6. J. White, R. Tucker, K. Long: Merit-base Scheduling Algorithm for Optical Burst Switching, COIN 2002, July, (2002)
7. J. Kim, H. Yun, J. Choi, M. Kang: A Novel Buffer Scheduling Algorithm for Burst Control Packet in Optical Burst Switching WDM Networks, APOC, Oct., (2002)
8. A. Ge, F. Callegati, L.S Tamil: On optical burst switching and self-similar traffic," IEEE Communications Letters, vol. 4, No. 3, March (2000) 98-100
9. F. Xue, S. J. B. Yoo: Self-similar traffic shaping at the edge router in optical packet switched network, ICC 2002, (2002)

A Mapping Algorithm for Quality Guaranteed Network Design Based on DiffServ over MPLS Model over UMTS Packet Network*

Youngsoo Pi[1], Miyoun Yoon[1], and Yongtae Shin[2]

Dept. of Computing, Graduate School, Soongsil University,
Sangdo5-Dong,Dongjak-Gu, Seoul, Republic of Korea
[1]{coolps, myyoon}@cherry.ssu.ac.kr
[2]shin@comp.ssu.ac.kr

Abstract. Previous researches are restricted to scalability or complexity of implementation and are occurred both a high transmission delay and a delay variation resulted from resource reservation over UMTS packet network. Moreover, these techniques are limited to guarantee quality of service over UMTS packet network because these don't define mapping relation between service classes based on traffic characteristic to maintain consistent QoS characteristic between network elements in accordance with use different protocols and service domains. This paper proposes service architecture of a boundary node for design quality guaranteed network that be able to offer high-speed wireless multimedia service and a mapping algorithms based on traffic characteristic, also evaluates to verify reasonableness of mapping algorithms between service classes and proves excellence of quality guaranteed network.

1 Introduction

As converging current heterogeneous networks into an all IP network, UMTS packet network [1] has been requested to guarantee satisfied QoS for various traffics [2]. Many researches have been studied for the issues [3, 4, 5, 6, 7]. However, current UMTS packet network only provides best-effort service for IP based packet service, so it can not guarantee various data quality requested from users due to some problems such as network congestion and packet loss. Thus, UMTS packet network should support IP QoS for IP based packet service oriented for all IP network. For this, several researches [3, 4, 5, 6] are proposed. They propose IntServ or DiffServ model over UMTS packet network. They have several problems such as implementation difficulty, constraints of scalability, increase of resource reservation overhead. Furthermore, they have a critical problem that relationship between its own class of UMTS and class for internet QoS model is not clear.

We apply DiffServ over MPLS model for UMTS packet network and propose simple and fast one-pass mapping algorithm with clear definition among the service classes for the model. In section 2, we introduce related works [3, 4, 5, 6]

* This Work Was Supported By The Soongsil University Research Fund

C. Kim (Ed.): ICOIN 2005, LNCS 3391, pp. 370–379, 2005.

and analyze and compare with between them. In section 3, we suggest service architecture of a boundary node for design quality guaranteed network that be able to offer high-speed wireless multimedia service. In addition, we propose our mapping algorithm on DiffServ over MPLS model over UMTS packet network. Finally, we simulate performance of IP QoS model for showing validity of applying DiffServ over MPLS model. Besides, we simulate the mapping algorithm compared with other mapping mechanisms in section 4, and conclude the paper and suggest further works in section 5.

2 Related Works

3GPP [2, 9] proposed hierarchical UMTS QoS framework. It has recommended applying internet QoS technology defined by IETF [10] for its network. Furthermore, UMTS packet network also has been recommended to be based on DiffServ model for QoS of UMTS packet network. [4] proposed a mechanism to apply IntServ model for GPRS core network. The study is focused on solving scalability problem of IntServ model. It performs resource reservation not flow by flow but aggregated traffic per MS (Mobile Station). It totally results for reducing control overhead and a number of updating times. [5] proposed a system structure applying DiffServ and RSVP at the same time for providing strict QoS, but it has no definition of mapping between service classes over UMTS packet network. [3, 6] defined relation between DiffServ class and UMTS class, and proposed router structure providing PHB (Per Hop Behavior) of DiffServ. However, they did not propose detailed basis requirements for mapping two classes. Because [3, 6] are based on DiffServ model, they are not able to update SLA aggregation and exchange resource status, so network congestion is incurred frequently and they can not guarantee bandwidth resource by flow. Although [4, 5] which are based on RSVP solves above problems, they have high implementation complexity and low scalability. Furthermore, [3, 6, 7] did not define QoS requirement factors to map UMTS QoS and IP QoS, so they can not keep up consistent QoS characteristics. Besides, they lack mapping scalability and flexibility as they map two classes as one-to-one relation. Thus, we propose simple and fast one-pass mapping function on improved IP QoS model to solve these problems for UMTS packet network. Table 1 shows exist models' comparisons and constraints.

3 Mapping Algorithm for Quality Guaranteed Network

The SGSN, that is edge node, of our QoS model has a mapping table for mapping UMTS and DiffServ Classes and also performs mapping function between DiffServ classes and MPLS labels. For applying Diffserv over MPLS QoS model for UMTS packet network, we design network model to support the related functions and then, we propose mapping algorithms for that.

Table 1. Comparison of existed related works

Factors	[12]	[13]	[14]	[15]
IP QoS model	DiffServ	IntServ /RSVP	RSVP /DiffServ	DiffServ
Complexity	low	high	high	low
Scalability	high	low	moderate	high
Transmission Delay	X	O	O	X
Bandwidth	X	O	O	X
Mapping Function	O	X	X	O

3.1 Network Model

We design DiffServ over MPLS model for satisfying a variety of service requests and providing high speed wireless multimedia services. This model takes some advantages of DiffServ and MPLS mode each other. DiffServ model is able to explicitly defined differentiated service using DSCP over network layer. However, the model does not provide signaling protocol for resource reservation over network. MPLS can support the functions over between 2 and 3 layer. It also provides high speed packet forwarding and traffic engineering. So, the two technologies can be used at the same time because they operate on different layer each other.

Thus as we apply DiffServ over MPLS model on UMTS packet network, we can support packet classification function of DiffServ and signaling protocol of MPLS. For apply the model for UMTS packet network, we design service structure of edge node, which is SGSN, over this network. Fig. 1 shows proposed service structure of SGSN.

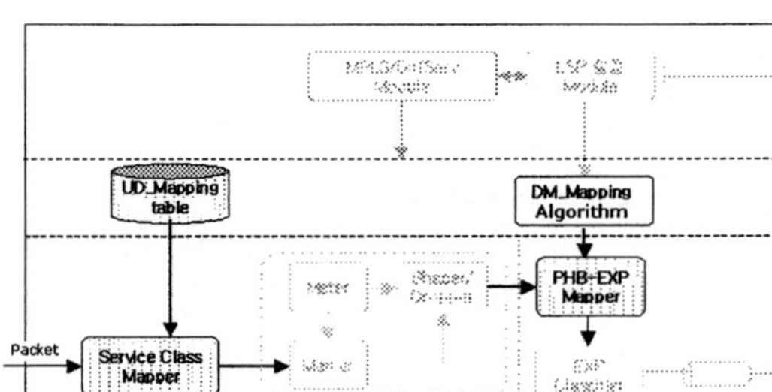

Fig. 1. Service Structure of Edge Node

For maintaining data quality consistency among different service domains, SCM(Service Class Mapper) performs mapping between corresponding DiffServ classes and UMTS classes depending on predefined UD(UMTS DiffServ)_Mapping table. That is, SCM performs packet classification based on DiffServ model.

Then, PHB-EXP Mapper provides mapping between DiffServ class and MPLS class through DM(Diffserv MPLS)_mapping algorithm. The service framework of SGSN is able to support consistent QoS characteristics and high speed forwarding wireless multimedia service.

3.2 Mapping algorithms for Different QoS Service Models

As a function of defined Service Class Mapper, we propose a simple mapping algorithm. We define EF, AF$xy(1 \le x \le 4, 1 \le y \le 3)$ and DF Class that are commonly used[11] at DiffServ model. For mapping 4 classes – CC(Conversational Class), SC(Streaming Class), IC(Interactive Class) and BC(Background Class) into 12 classes of DiffServ model, we aggregate the 12 DiffServ Classes to 5 Classes relying on our defined service range at first. The service range is shown like Definition 1 and 2.

Definition 1 We define AFxy=(Mxy, Exy, Dxy) *s.t.* $M_{DF} < M_{xy} < M_{EF}$, $E_{EF} < E_{xy} < E_{DF}$, $D_{EF} < D_{xy} < D_{DF}$ where M_{EF} and M_{DF} are allowed maximum bit rate of EF and DF class, respectively. In addition, E_{EF} and E_{DF}, D_{EF} and D_{DF} are maximum bit error rate and maximum transfer rate of EF and DF, respectively.

Definition 2 For AFxy=(Mxy, Exy, Dxy), we suppose $0 < \alpha_i < \alpha_s < 1$, $0 < \beta_s < \beta_i < 1$, $0 < \gamma_s < \gamma_i < 1$ are experimental rate values by ISP.

(i) EF=(M_{EF}, E_{EF}, D_{EF}) s.t the highest service quality determined by ISP.
(ii)Gold=$\{AF_{xy} \mid 3 < x \le 4, 1 \le y \le 3, x, y \in N\}$ if ($\alpha_s M_{EF} \le M_{xy} < M_{EF}$ and $D_{EF} < D_{xy} \le \beta_s D_{EF}$) or $E_{EF} < E_{xy} \le \gamma_s E_{EF}$
(iii)Silver=$\{AF_{xy} \mid 2 \le x \le 3, 1 \le y \le 3, x, y \in N\}$ if ($\alpha_i M_{EF} \le M_{xy} < \alpha_s M_{EF}$ and $\beta_s D_{EF} < D_{xy} \le \beta_i D_{EF}$) or $\gamma_s E_{EF} < E_{xy} \le \gamma_i E_{EF}$
(iv)Bronze= $\{AF_{xy} \mid 1 \le x < 2, 1 \le y \le 3, x, y \in N\}$ if ($M_{DF} \le M_{xy} < \alpha_i M_{EF}$ or $\beta_i D_{EF} < D_{xy} < D_{DF}$) and $\gamma_i E_{EF} < E_{xy} < \gamma_i E_{DF}$
(v) DF=(M_{DF}, E_{DF}, D_{DF}) s.t. the lowest service quality determined by ISP.
When x=4 and y=3, the AF class is the highest class of AFxy, and when x=1 and y=1, the class is the lowest class of AFxy. We first examine DiffServ service class depending on definition 2, and then we perform our proposed algorithm 1 at the edge node. By definition 1 and 2, we identify IP resource requirements(M, E, D) of the input traffic based on UMTS service class specification. Then we can get DSCP code. Finally, we determine corresponding MPLS label(MPLS EXP field) as a result of truncating the DSCP code and applying not bit-operation to the DSCP like Fig 2. Our proposed algorithm just needs only one mapping table and one algorithm on DiffServ over MPLS QoS model even though we need two mapping procedures- UMTS to DiffServ and DiffServ to MPLS. After that, the edge node performs pure MPLS operation as defined service class requirements.

Fig. 2 and 3 show proposed mapping algorithms from UMTS class to MPLS label. The SGSN(edge node) checks service requirement of incoming UMTS traffic and stamps appropriate DiffServ service class. Based on UMTS class specification, we define that CC traffic should be EF Class and SC and IC should be Gold class and Silver class depending on definition 2 respectively. In addition, BC should be Bronze and DF class. We are able to map the two classes with consistent QoS condition by our proposed algorithm and definition 1and 2.

```
Algorithm 1 : UMTS_DiffServ Mapping Algorithm

Input P=[ m e d];
      switch(P) {
            case ' m≥ M_EF && e≤ E_EF && d≤ D_EF' :
                  P ∈ Conversational Class;
                  mapping (P, EF);
                  break;
            case ' M_s≤m< M_EF && E_EF<e≤ E_s && D_EF<d≤ D_s' :
                  P ∈ Streaming Class;
                  mapping (P, Gold);
                  break;
            case ' M_i≤m< M_s && E_s<e≤ E_i && D_s<d≤ D_i' :
                  P ∈ Interaction Class;
                  mapping (P, Silver);
                  break;
            case ' M_DF< m< M_i && E_i<e< E_DF && D_i<d< D_DF' :
                  P ∈ Background Class;
                  mapping (P, Bronze);
                  break;
            case ' m≤ M_DF && e≥ E_DF && d≥ D_DF' :
                  P ∈ Background Class;
                  mapping (P, DF);
                  break;

            default:              /* P != Service Class of UMTS */
                  mapping fail;
      }
```

Fig. 2. Mapping UMTS Class into DiffServ Class

In Fig. 1, P is input data request, and Ms and Mi are bit rate through $\alpha_s \mathrm{M}_{EF}$and $\alpha_i \mathrm{M}_{EF,}$respectively. E_s and E_i are also bit error rate through $\alpha_s \mathrm{E}_{EF}$ and $\alpha_i \mathrm{E}_{EF,}$and D_s and D_i are transmission delay through $\alpha_s \mathrm{D}_{EF}$ and $\alpha_i \mathrm{D}_{EF}$. Besides, Gold is service set for AF_{xy} with range $\{3 < x \leq 4, 1 \leq y \leq 3\}$, and Silver and Bronze are also service sets for AF_{xy} with range $\{2 \leq x \leq 3, 1 \leq y \leq 3\}$ and $\{1 \leq x < 2, 1 \leq y \leq 3\}$. In addition, mapping() returns DSCP code of DiffServ.

Like Fig. 3, for forwarding input traffic marked by DiffServ model over MPLS network, we propose to map DiffServ class to MPLS label with no mapping table and only simple operation using the marked DSCP code. As you see the Fig 4, we do not have to maintain any mapping table to convert DiffServ class into MPLS label. We only perform not bit-operation for 3 bits string of the DSCP code from a left bit. Then the SGSN establishes LSP depending on marked label. Here, we assume E-LSP [5, 6, 10] with EXP field of MPLS shim header. Because UMTS packet network has 4 different classes and our DiffServ class is consist of 6 sets,

```
Algorithm 2 : DiffServ_MPLS Mapping Algorithm
Input P;
code = check(bit_string);
code = ~code;
      switch(code) {
            case 010 :
                  P ∈ EF Service class;
                  mapping (P, committed);
                  break;
            case 011 :
                  P ∈ AF_4y service class;    /* 1≤y≤3 */
                  mapping (P, premium);
                  break;
            case 100 :
                  P ∈ AF_3y service class     /* 1≤y≤3 */
                  mapping (P, Business_H);
                  break;
            case 101 :
                  P ∈ AF_2y service class     /* 1≤y≤3 */
                  mapping (P, Business_L);
                  break;
            case 110 :
                  P ∈ AF_1y service class     /* 1≤y≤3 */
                  mapping (P, standard_H);
                  break;
            case 111 :
                  P ∈ DF service class        /* 1≤y≤3 */
                  mapping (P, standard_L);
                  break;
            default :
                  mapping fail;
      }
```

Fig. 3. Mapping DiffServ Class into MPLS Label

the EXP-field is enough to support them as using 3 bits which can provide 8 different service classes. The output label for the mapping function is identified by definition of our DiffServ sets. Thus, each LSP should be established relying on the service requirement defined by the DiffServ sets. The core router can identify the service class of input traffic as the MPLS label. Our algorithm helps to keep consistent service quality from UMTS class to MPLS label via DiffServ class as performing our one-pass mapping algorithm. Furthermore, the proposed algorithm just only requests one mapping table, and simple and fast operation for mapping three different QoS models.

In Fig. 3, for traffic P, bit_string means marked DSCP code, and check() returns a head of three bits of bit_string – code. In addition, mapping() performs translating DSCP code to MPLS label for packet forwarding. Lastly, six sets – committed, premium, business_H, business_L, standard_H and standard_L – are based on QoS policy relying on defined DiffServ service for MPLS label. They have EXP field value, 010, 011, 100, 101, 110 and 111, respectively.

4 Performance Analysis

We simulate our one-pass algorithm on the DiffServ over MPLS model over UMTS packet network using OPNET[8]. We analyze transport delay and receipt rate from SGSN to GGSN for CC, and receipt rate for SC traffic of UMTS packet network, For IC and BS traffic, we measure response time of FTP and E-mail Server, respectively. We show that the input traffic has satisfied data quality as mapping by proposed one-pass algorithm over experimental topology like Fig. 4.

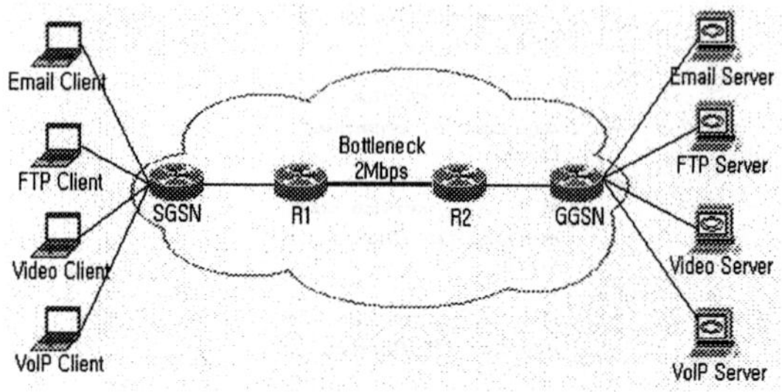

Fig. 4. Experimental Topology

Table 2. Mapping Definitions and Experimental Value

mechanism	CC Class	SC Class	IC Class	BC Class
Mapping 1	AF_{3y}	AF_{2y}	AF_{1y}	DF
Mapping 2	AF_{4y}	AF_{3y}	AF_{2y}	AF_{1y}
Proposal	EF	AF_{4y}	AF_{3y},AF_{2y}	AF_{1y},DF
Input Traffic	Real-Time	Real-Time	Non Real-Time	Non Real-Time
WFQ	0.45	0.35	0.15	0.5

At experimental topology, all link capacity is 10Mbps except for the bottle neck point – 2Mbps, and VoIP and Video client, FTP and Email client transmits 1Mbyte and 2Mbyte at the same time respectively. We apply WFQ(Weighted Fair Queuing) for the experimental topology like Table 2.

4.1 Analysis of mapping algorithm

CC Class is for voice traffic, and transmission delay and receipt rate are important characteristics. So we measure transmit delay and receipt rate from VoIP Client to VoIP Server for 390 seconds. And we simulate receipt rate and response time for SC class, IC and BC Class from each client to the server respectively.

Fig. 5 and 6show transmission delay and receipt rate of CC Class, and Fig. 7 describes receipt rate for SC class. In mapping 1, it shows high transmission delay and low receipt rate compared with other mechanisms. Our proposal shows the lowest transmission delay $8ms$ of them. Also, our receipt rate is higher than the mapping2's one like Fig 6. Therefore, mapping to EF shows better performance for CC class. In Fig 6 shows results for mapping SC class. The average receipt rate shows 584Kbps, 2878Kbps and 3500Kbps in mapping 1, mapping 2 and

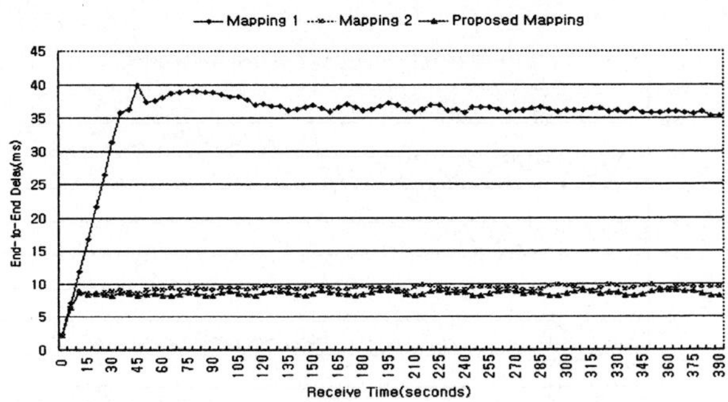

Fig. 5. Transmission Delay of VoIP Traffic(CC Class)

proposed mapping respectively. Our proposal shows the best performance of them. It is better to map CC and SC class into separate DiffServ class EF and AF_{4y}.

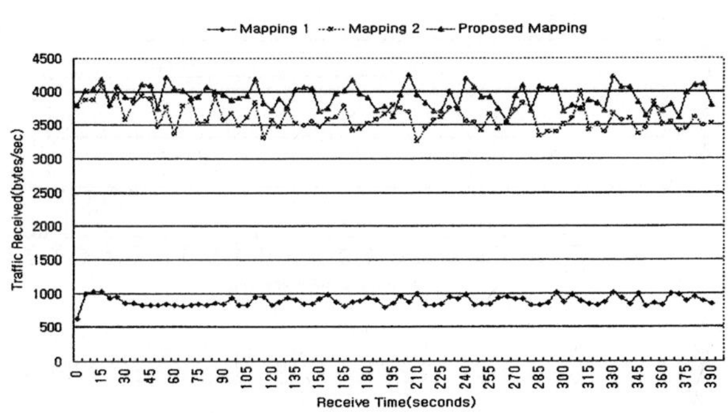

Fig. 6. Receipt Rate of VoIP Traffic(CC Class)

Fig. 8 and 9 show response times for E-mail and FTP traffic respectively. As increasing simulation attempts, all of them converges almost same response time. Thus, mapping into each DiffServ class by our proposal is appropriate.

For mapping DiffServ class into MPLS label, we compare our algorithm with [12] in time complexity. [12] has one mapping table, so it needs table search time. When using binary search algorithm, it costs $O(\log_2 d)$, and costs $O(d)$ when using sequential search mechanism. Here, d is a number of DiffServ service class. However our algorithms just costs O(1) because it has no mapping table, only once not bit-operation. Also, table size of the SGSN needs d+m, when each d and m is a number of diffServ classes and mpls label. However, we need not any table to convert diffServ class to MPLS label. Thus, our algorithms are fast and simple.

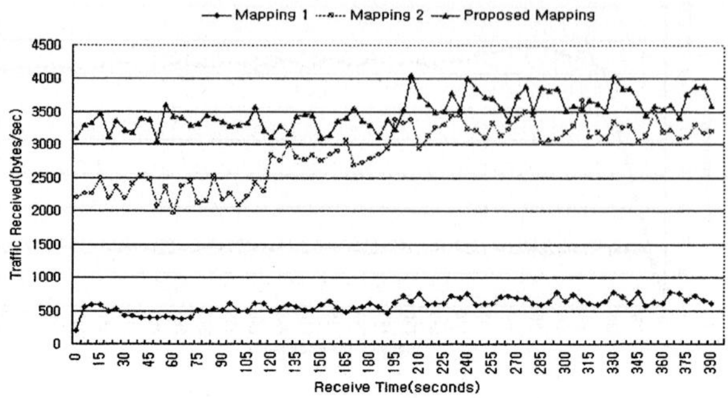

Fig. 7. Receipt Rate of Video Traffic(SC Class)

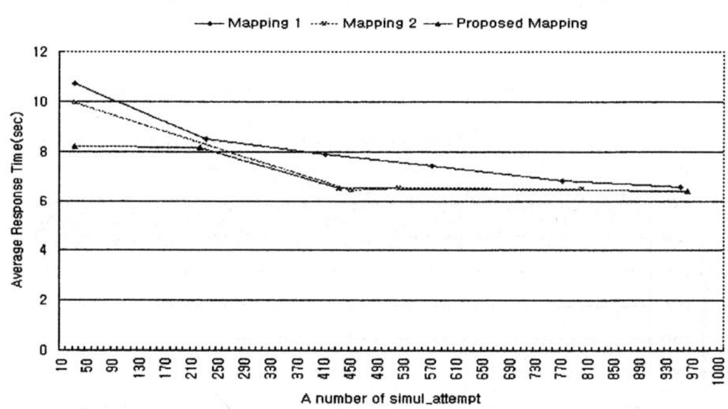

Fig. 8. Response Time of FTP Traffic(IC Class)

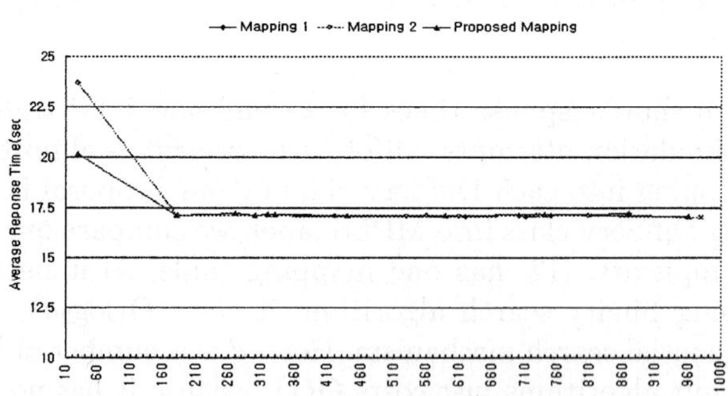

Fig. 9. Response Time of e-mail Traffic(BC Class)

5 Conclusion and Further Works

In this paper, we propose mapping algorithms which make traffic to be transmitted with keeping various QoS requested by each user over UMTS packet network based on DiffServ over MPLS. Through performance analysis from the viewpoint of that whether satisfied QoS is kept, we show our algorithms are able to guarantee data QoS with lasting QoS depending on a specification of each user. In addition, the proposed algorithms are performed fast through one mapping procedure and once not bit-operation without necessity of twice mappings. Therefore, we look forward to becoming a basic research to provide high speed wireless multimedia service over UMTS packet network. We need a further study for a packet scheduling algorithm considering UMTS packet.

References

1. 3GPP.: General UMTS Architecture, 3G TS23.101 V5.0.1, (2004)
2. 3GPP.: Quality of Service(QoS) Concept and Architecture, 3G TS23.107 V6.0.0, Release, (2003)
3. F. Agharebparast and V.C.M. Leung.: QoS support in the UMTS/GPRS backbone network using DiffServ, in Proc. IEEE Globecom'02, Teipei, ROC, (2002)
4. G. Priggouris, S. Hadjiefthymiades, L. Merakos: Supporting IP QoS in the General Packet Radio Service, IEEE Network (2000) 8–17
5. M. Puuskari: Quality of Service Framework in GPRS and Evolution towards UMTS, in Proc. 3rd Euro. Pres. Mobile Communication Conference, Paris, France, (1999)
6. Rajeev Koodli: Supporting Packet Data QoS in Next Generation Cellular Networks, IEEE Communications Magazine (2001) 180–188
7. S.Maniatis, et. al: DiffServ-based Traffic Handling Mechanisms for the UMTS Core Network, In Proc. of IST Mobile and Wireless Telecommunications Summit 2002, Thessaloniki, Greece, (2002)
8. Optimized Network Engineering Tool-OPNET IT Guru Academic Edition 9.1 URL : http://www.opnet.com (2004)
9. 3GPP: End-to-End Quality of Service(QoS) Concept and Architecture, 3G TS23.207 V 6.1.1 Release, (2004)
10. http://www.ietf.org
11. K. Nichols, S. Blake, F. Baker and D. Black: Definition of the Differentiated Services Field(DS Field) in the IPv4 and IPv6 Headers, RFC 2474, (1998)
12. E. Horlait, N. Rouhana: Differentiated Services and Integrated Service use of MPLS, Fifth IEEE Symposium on Computers and Communications (ISCC 2000), France, (2000)
13. B. Davie, J.C.R. Bennett, K. Benson, J.Y. Le Boudec, W. Courtney, S. Davari, V. Firoiu, D. Stiliadis: An Expedited Forwarding PHB(Per-Hop Behavior), RFC 3246, (2002)
14. BJ. Heinanen,F. Baker,W. Weiss, J. Wroclawski: Assured Forwarding PHB Group, RFC 2597, (1999)
15. L. Andersson, P. Doolan, N. Feldman, A. Fredette and B. Thomas: LDP Specification, RFC 3036, (2001)

A Route Optimization Scheme by Using Regional Information in Mobile Networks

Hee-Dong Park[1], Jun-Woo Kim[2], Kang-Won Lee[2], You-Ze Cho[2], Do-Hyeon Kim[3], Bong-kwan Cho[4], and Kyu-Hyung Choi[4]

[1] Department of Computer Engineering, Pohang College, Pohang, 791-711, Korea
hdpark@pohang.ac.kr

[2] School of Electrical Engineering & Computer Science, Kyungpook National University, Taegu, 702-701, Korea
juitem@eecs.knu.ac.kr
kw0314@palgong.knu.ac.kr
yzcho@ee.knu.ac.kr

[3] Faculty of Telecommunication & Computer Engineering, Cheju National University, Jeju-do, 690-756, Korea
kimdh@cheju.ac.kr

[4] Korea Railroad Research Institute
{bkcho, khchoi}@krri.re.kr

Abstract. NEMO basic solution supports network mobility by using a bi-directional tunnel between a mobile router and its home agent. Yet, the multiple levels of bi-directional tunnels in nested mobile networks lead to significant routing overhead (so-called pinball routing). Non-optimal routing increases bandwidth consumption and transmission delays. The current paper proposes Regional Information-based Route Optimization (RIRO) in which mobile routers maintain a Nested Router List (NRL) to obtain next hop information. RIRO supports nested mobility without the nested tunnel overheads. This is accomplished by using a new routing header, called RIRO Routing Header (RIRO-RH). RIRO has the minimum packet overhead that remained constant, irrespective of how deep the mobile network was nested, in comparison with two earlier proposed schemes - Reverse Routing Header (RRH) and Bi-directional tunnel between HA and Top-level mobile router (BHT).

1 Introduction

There have been significant technological advancements in the areas of portable and mobile devices. And the rapid growth of wireless networks and Internet services drives the need for IP mobility. Traditional work in this topic is to provide continuous Internet access to mobile hosts only. Host mobility support is handled by Mobile IP and specified by the IETF Mobile IP working group [1, 2].

Despite this, there is currently no means to provide continuous Internet access to nodes located in a mobile network. In this case, a mobile router (MR) changes its point of attachment, but there is a number of nodes behind the MR.

C. Kim (Ed.): ICOIN 2005, LNCS 3391, pp. 380–389, 2005.

The ultimate objective of a network mobility solution is to allow all nodes in the mobile network to be reachable via their permanent IP addresses, as well as maintain ongoing sessions when the MR changes its point of attachment within the Internet. The IETF working group on network mobility (NEMO) is currently standardizing a basic support for moving networks. The basic NEMO protocol suggests a bi-directional tunnel between a MR and its home agent (HA) [3]. A unique characteristic of network mobility is nested mobility. Network mobility support should allow a mobile node or a mobile network to visit another mobile network (this is the example of a PAN in a train). Such scenarios may lead to multilevel aggregations of mobile networks. Although the NEMO basic solution with the MR-HA tunnel supports nested mobility, it suffers from sub-optimal multi-angular routing (so-called pinball routing) in the topology, and severe header overheads and transmission delays as the packets from the correspondent node (CN) are repeatedly encapsulated by all the HAs of the MRs in the nested mobility hierarchy. Therefore, in nested mobility, the support of route optimization is very important to allow packets between a CN and a mobile network node (MNN) to be routed along the optimal path. In order to solve the sub-optimal routing and header overheads introduced by basic network mobility support, many solutions are presented in the NEMO working group. Two good examples for comparing the performance with our proposal are a Reverse Routing Header (RRH) and Bi-directional tunnel between HA and Top-level mobile router (BHT) [4, 5]. Although the above two solutions provide route optimization for nested mobility, they still suffer from an increasing overhead as the depth of the nested mobile network increases.

To support network mobility and route optimization for nested mobility, we propose the Regional Information-based Route Optimization (RIRO) scheme, which has the advantage of a small and constant size of packet overhead, irrespective of how deep the mobile network is nested. To support RIRO, mobile routers need to maintain a Nested Router List (NRL) and be able to read RIRO-Routing Headers (RIRO-RHs). NRL is used to record regional information and to obtain next hop information, and the new routing header, RIRO-RH, is used to avoid nested tunnel overheads.

The remainder of this paper is organized as follows. First, related work is surveyed in section II, then section III presents the proposed scheme RIRO and section IV provides a performance analysis. Finally, the conclusion is given in section V.

2 Related Work

The NEMO WG has already proposed various route optimization schemes. For example, the RRH, NPI, and ARO schemes use an extended type-2 routing header in common, while the BHT and HMIP-based schemes have similar characteristics from the viewpoint of making multiple encapsulations within nested mobile networks [5-7]. Among the above solutions, this section gives a brief overview of the RRH and BHT schemes, because they have good features for comparing the performance with RIRO.

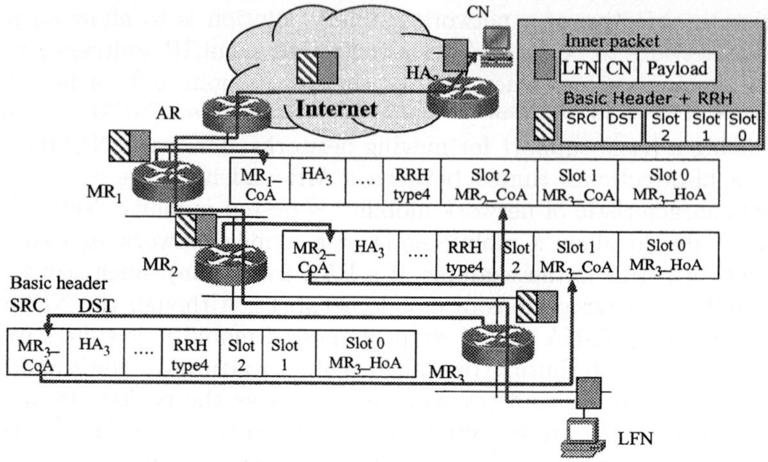

Fig. 1. Packet transmission in RRH scheme.

2.1 RRH

Fig. 1 shows the process of how the routing information is delivered to an HA using an RRH (type 4 routing header). The first mobile router on the path, MR3, adds a reverse routing header with N=3 pre-allocated slots, where N represents the nested depth levels. MR3 puts its home address, MR3_HoA, in slot 0. The outer packet's source and destination fields are then inserted with MR3's care of address and MR3's HA address, respectively. Each of above the mobile routers, MR2 and MR1, overwrites the source field with its own care of address after putting the old source address in the free slot of the RRH. Therefore, when HA3 receives the packet, it can obtain optimal path information to MR3 from the packet's source field and RRH. The HA inserts every intermediate MR's CoA into an extended type-2 routing header based on the path information in the RRH of the packets sent from the MR. In contrast with an RRH, the destination address field is exchanged with the next-hop MR's CoA.

2.2 BHT

The BHT supports route optimization based on bi-directional tunneling between an HA and the top-level mobile router (TLMR). The TLMR has the path information to MRs and creates a tunnel to each MR based on registration messages from the MRs. As shown in Fig. 2, the data generated by an LFN or MR3 is encapsulated by MR3 to be sent to HA3. Then, MR3 does one more encapsulation to directly communicate with the TLMR (equal to MR1 in Fig. 2). Plus, the packet is re-encapsulated by each intermediate MR on the path. When this packet arrives at the TLMR, all the encapsulation headers having TLMR's CoA as the destination address are decapsulated, then the packet is re-encapsulated to send to HA3. When HA3 transmits a packet to MR3, HA3 encapsulates the

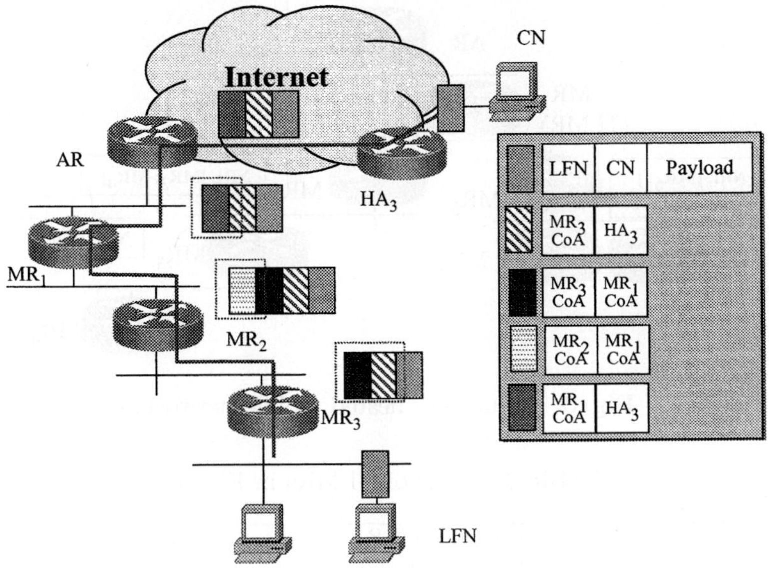

Fig. 2. Packet transmission in BHT scheme.

data twice to send to TLMR as well as MR3. When the TLMR receives this packet, it decapsulates the packet and finds a path to MR3. The TLMR then makes multiple encapsulation headers, based on one header to one intermediate MR, in order to send the packet to MR3 via all intermediate MRs.

As outlined above, even though an RRH and BHT both support route optimization for nested mobility, they still suffer from an increasing overhead as the depth of the nested mobile network increases. The more MRs that are nested, the more slots that need to be made in the RRH, and the more nested tunnels that need to be made within the nested mobile networks in the BHT, which can lead to a severe degradation of transmission efficiency.

3 RIRO

The proposed scheme, RIRO, supports route optimization for network mobility based on the use of Nested Router Lists and RIRO-RHs.

3.1 Nested Router List (NRL)

To support RIRO, each MR, as well as the TLMR, maintains a list of the CoAs of all the MRs located below it along with the nested MRs' tree. As shown in Fig. 3 and Table 1, the next-hop information and a sub MR address list is cached together in the NRL. Therefore, when a MR receives a packet, it can determine the next-hop MR by retrieving the NRL. For example, if MR1 receives a packet

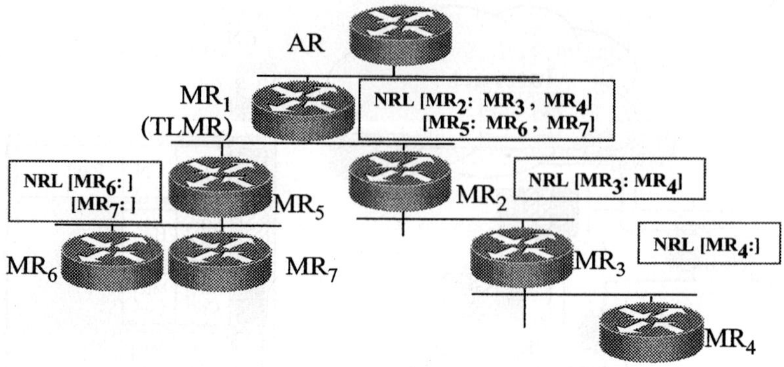

Fig. 3. Example of nested mobile network.

Table 1. NRLs of all MRs in Fig. 3.

Items / NRL location	Attached MR address (Nest hop)	Sub MR address list (Destination)
$TLMR$	MR_2_CoA	MR_3_CoA, MR_4_CoA
	MR_5_CoA	MR_6_CoA, MR_7_CoA
MR_2	MR_3_CoA	MR_4_CoA
MR_3	MR_4_CoA	...
...	...	...

destined for MR4, it forwards the packet to MR2. And MR2 forwards the packet to MR3, and finally MR3 forwards the packet to MR4. All MRs located within the RIRO domain send extended router advertisement (RA) messages, including the TLMR's CoA. If an MR does not receive an extended RA message, it acts as a TLMR and advertises its address as the TLMR address. When a new MR moves into an RIRO domain and receives the extended RA message, its address will be registered in the NRLs of all the MRs located from the parent MR to TLMR along with the nested MR's tree. The NRL is not updated by a general routing protocol message, but rather by an RIRO-RH with a binding update flag 'B'. It is also possible to use special NRL update message.

3.2 Packet Transmission with RIRO-RH

The new routing header, an RIRO-RH, is used to encapsulate packets avoiding nested tunnel overheads. It prevents each MR from making nested tunnels. As shown in Fig. 4, the type value of an RIRO-RH is 7 or 8. Only the default MR's address is put in each header. The packets destined for an HA and an MR are encapsulated with an RIRO-RH7 and with an RIRO-RH8, respectively. While RIRO-RH7 includes the source MR field, RIRO-RH8 includes the destination MR field.

An RIRO-RH with a 'B' flag set is used to update the NRLs of the MRs located on the destination path. Based on a 'prefix length' field, it is possible to register using the prefix of the address as well as the host.

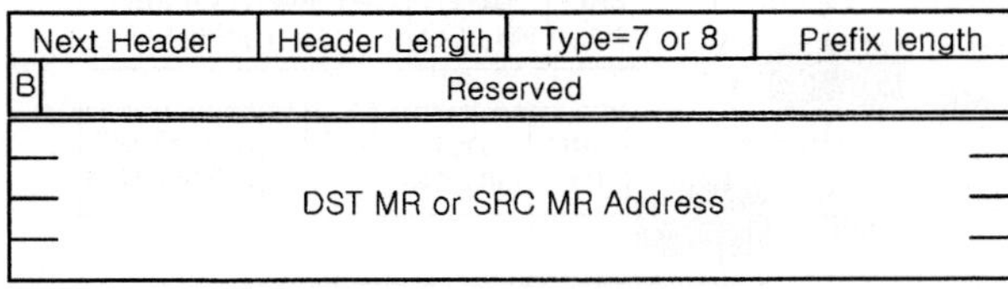

Fig. 4. Structure of RIRO-RH.

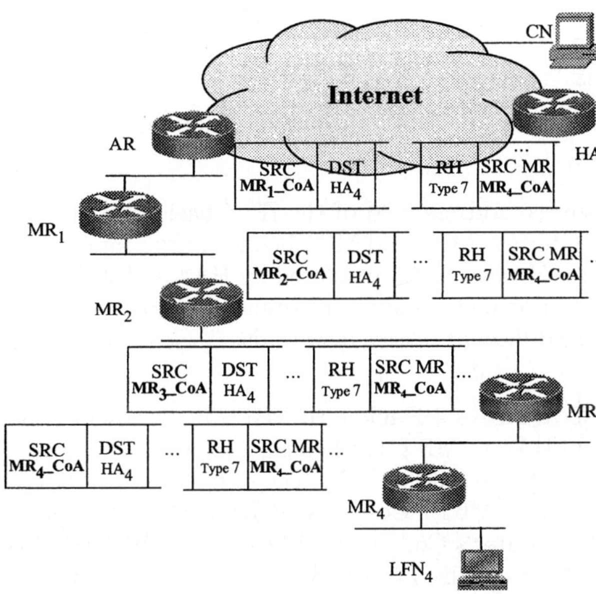

Fig. 5. Packet transmission from MR to HA.

1) When MR Transmits a Packet to HA Fig. 5 shows the process of packet transmission from LFN4 to the CN using an RIRO-RH7. MR4 encapsulates the data generated by LFN4 using an IPv6 basic header and RIRO-RH7 to send to HA4. The source and destination address fields are the MR4's CoA and HA4's address, respectively. Plus, the default source mobile router's CoA, MR4's CoA, is put in the SRC MR field of the RIRO-RH7. After receiving the packet, MR3

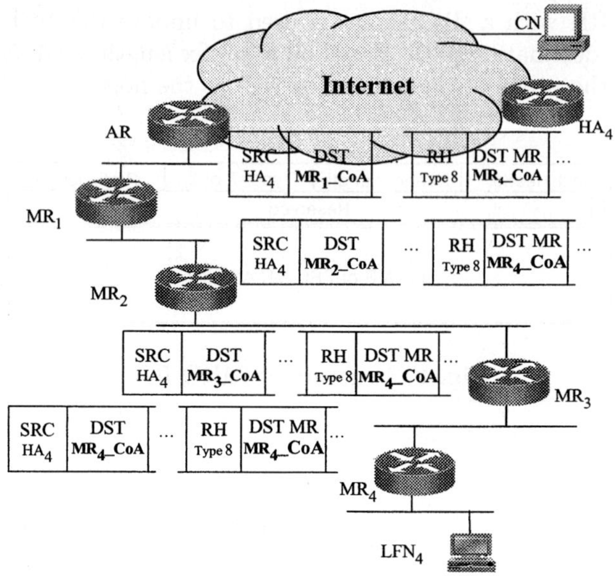

Fig. 6. Packet transmission from HA to MR.

overwrites the source address field of the IPv6 basic header using its own CoA to avoid ingress filtering. MR2 and MR1 behave in the same way as MR3. When the packet reaches HA4, it is decapsulated by HA4 and delivered to the CN. As such, the MRs do not make additional tunnels for packets with an RIRO-RH, thereby avoiding additional packet overheads.

2) When HA Transmits a Packet to MR Fig. 6 shows the process of packet transmission from HA4 to MR4. HA4 encapsulates the data using an IPv6 basic header and RIRO-RH8 to send to MR4. The source and destination address fields are HA4's address and MR1's CoA, respectively. Plus, MR4's CoA, the default destination mobile router's CoA, is put in the DST MR field of the RIRO-RH8. In the figure, MR1 acts as the TLMR. When MR1 receives the packet, it retrieves the NRL to check whether the address of the DST MR field is on the list. By the result, MR1 can determine whether or not to deliver the packet.

4 Performance Analysis

This section analyzes the performance of RIRO by estimating its overhead in comparison with RRH and BHT. The network model used for the estimation is shown in Fig. 7. It is supposed that the depth of the nested mobile network is N, and that the LFNs and CNs do not have any mobility function. Table 2 shows the overhead for each scheme in communication between an LFN and an CN. In the table, the word 'overhead' means the number of bytes of additional headers

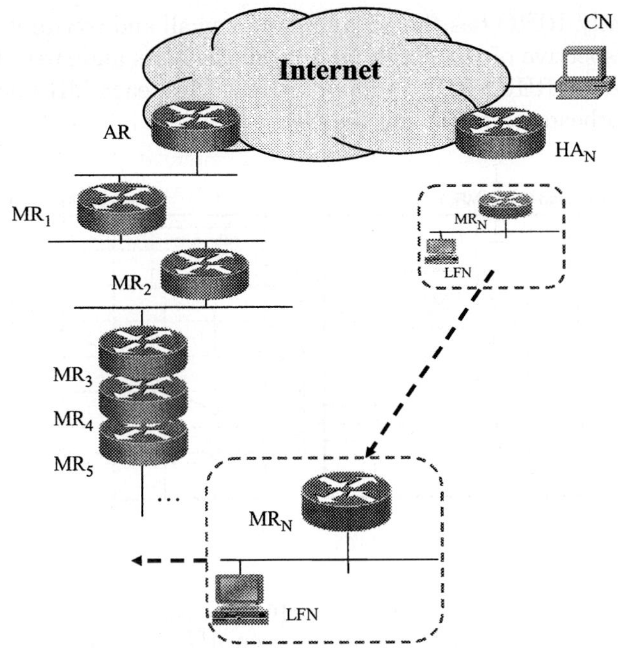

Fig. 7. Mobile network with depth N.

added to the packet generated by an LFN or CN for the purpose of routing. The capital 'N', as mentioned earlier, represents the total depth of the nested mobile network, and 'n' represents the depth of the MR a packet is passing through. To make a fair comparison, only the headers that are indispensable for routing are considered for each scheme. Therefore, the MIPv6 Home Address option is not considered in Table 2. This means that the HoA field inserted into slot 0 of an RRH is also not taken into consideration. When considering the IPv6 basic header of the packet generated by an LFN or CN, 40 bytes should be added to each overhead.

First, RIRO has a constant overhead of 64 bytes, irrespective of the depth N and transmission sections. The 64 bytes consist of an IPv6 basic header with 40 bytes, an RIRO-RH with 8 bytes, and DST MR (or SRC MR) with 16 bytes. Meanwhile, in the case of the RRH scheme, a basic header with 40 bytes, an RRH with 8 bytes, and N-1 slots with 16(N-1) bytes, add up to 16N+32 bytes, whereas the BHT scheme consists of 80 bytes in the Internet or 40(N-n+2) bytes in a nested mobile network, because an IPv6 basic header of 40 bytes is added or subtracted every time a packet pass through an MR.

Fig. 8 illustrates the overhead variations as the depth N increases from 2 to 10. As a result, the RIRO scheme exhibits the minimum overhead that remained constant, while the overhead for the BHT and RRH schemes increases most in the Internet and in the nested mobile networks, respectively.

Consequently, RIRO has the advantage of a small and constant size of packet overhead, irrespective of the depth N. Yet, all the MRs and HAs should be able to read and use an RIRO-RH to support RIRO. Plus, each MR may have a little processing overhead for retrieving the NRL.

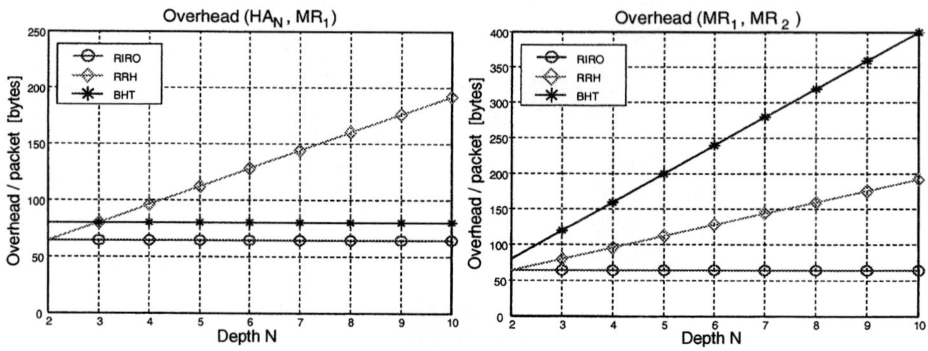

(a) (b)

Fig. 8. Packet overhead according to depth N. (a) Packet overhead between HA_N and MR_1. (b) Packet overhead between MR_1 and MR_2

Table 2. Comparison of packet overhead.

Schemes	RIRO	RRH	BHT
$HA_N \leftrightarrow MR_1$	64	$16N + 32$	80
$MR_N \leftrightarrow MR_{N-1}$	64	$16N + 32$	$40(N - n + 2)$

5 Conclusions

The current paper proposed RIRO for network mobility and route optimization. To support route optimization, each MR in RIRO domain maintains a NRL, and encapsulates packets with RIRO-RHs. In NRL, the CoAs of underlying MRs' are recorded. Therefore, the MR can obtain next hop information by retrieving the NRL. With RIRO-RH, RIRO can avoid nested tunneling. Estimation results demonstrated that RIRO had the minimum packet overhead that remained constant, irrespective of how deep the mobile network was nested, in comparison with two earlier proposed schemes - RRH and BHT.

Acknowledgement

This work is supported in part by the KOrea Science and Engineering Foundation (KOSEF) under contract R01-2003-000-10155-0 and supported by ITRC (Information Technology Research Center) Project.

References

1. C. Perkins, "IP Mobility Support for IPv4," RFC3344, Aug. 2002.
2. D. Johnson *et al.*, "Mobility Support in IPv6," $Internet\ Draft$, $< draft-ietf-mobileip-ipv6-24.txt >$, Jun. 2003.
3. V. Devarapalli *et al.*, "Nemo Basic Support Protocol," $Internet\ Draft$, $< draft-ietf-nemo-basic-support-01.txt >$, Sep. 2003.
4. P. Thubert *et al.*, "IPv6 Reverse Routing Header and its application to Mobile Networks," $Internet\ Draft$,$< draft-thubert-nemo-reverse-routing-header-03.txt >$, Oct. 2003.
5. Hyunsik Kang *et al.*, "Route Optimization for Mobile Network by Using Bi-directional Between Home Agent and Top Level Mobile Router," $Internet\ Draft$, $< draft-hkang-nemo-ro-tlmr-00.txt >$, Jun. 2003.
6. Jongkeun Na *et al.*, "Secure Nested Tunnel Optimization using Nested Path Information," $Internet\ Draft$, $< draft-na-nemo-nested-path-info-00.txt >$, Sep. 2003.
7. H. Ohnishi *et al.*, "HMIP based Route optimization method in a mobile network," $Internet\ Draft$, $< draft-onishi-nemo-ro-hmip-00.txt >$, Oct. 2003.

An Efficient Broadcast Scheme for Wireless Data Schedule Under a New Data Affinity Model

Derchian Tsaih[1], Guang-Ming Wu[2], Chin-Bin Wang[1], and Yun-Ting Ho[3]

[1] Department of Electronic Commerce Management,
Nan Hua University, Chiayi, Taiwan
{dtsaih,cbwang}@mail.nhu.edu.tw

[2] Department of Information Management, Nan Hua University, Chiayi, Taiwan
gmwu@mail.nhu.edu.tw

[3] Department of Information Management, National Chi-Nan University, Taiwan
s1213019@ncnu.edu.tw

Abstract. Data schedule in broadcasting is playing a more importance role due to increasing demand for large client popularity and vast amount of information. For system with multipoint queries, data records which queried by same query are broadcasted contiguously to reduce the average access time. Several techniques have been used in clustering data by defining the affinity between them. The data affinity function defined was mainly aiming at minimizing the linear Query Distance. However, our work showing that in order to minimize the average access time, the objective function shall be in quadratic form. We propose a *Minimum Gap* algorithm(MG) which merge relevant segments base on this new affinity function. Through extensive experiments, the results show not only the query's access time can be reduced by using this new affinity function, by using a dummy segment to speed our algorithm, the scheme we proposed have significant saving on both time complexity and memory space complexity.

1 Introduction

With the growing advances of wireless technology, a mobile client can access data from wireless network any time any where through portable devices. Differ from conventional wire line network, the communication between the server and clients are asymmetrical in nature. Due to the limited bandwidth and small battery power of each mobile client, the research issue on broadcasting is gaining more interest from wireless environment. In order to speed the client's data access, the efficient data allocation in broadcast channel must be performed to lower the client access time from broadcast sequence.

Several techniques which used in linear placement problem can be modified and used in our broadcast schedule problem, [7] [8] use the spectral technique in partitioning and clustering, [9] builds data allocation by mining the user's moving pattern, [1] uses the bottom-up clustering algorithm and merge segments by linear affinity function.

C. Kim (Ed.): ICOIN 2005, LNCS 3391, pp. 390–400, 2005.

For accessing multiple data records in broadcast sequence, [2] proposed a measure named *Query Distance* (*QD*) and broadcast schedule method called *Query Expansion Method* (*QEM*) to generate a broadcast sequence targeting the minimum *Query Distance*. Many research issues on this problem are based on reducing this Query Distance, including the latter work of their *Gray Code Method* (*GCM*)[4] which using the gray code scheme to minimize the *QD*, a modified *QEM* algorithm from Lee[5] and a genetic algorithm used to arrange broadcast sequence with multiple channel is from [6]. Although this *Query Distance* serves as a very good measurement in evaluating the average access time, however, develop a schedule algorithm based on it will ignore the quadratic effect of distance between each co-accessed data.

Since the goal of the schedule problem is to minimize the average access time and from next section we know that average access time is inversely proportional to sum of square distance between each nearest co-accessed data record pair, we propose a new measure named as *Quadratic Query Distance*(*QQD*) which represent the closeness of data set accessed by same query. The results will show that by using this quadratic objective function we will get a more accurate result than the previous linear objective function.

2 Preliminaries

Given a set of N data D=$\{d_1$,d_2 ,d_3 .....,$d_N\}$ and a set of K queries Q =$\{q_1$,q_2,q_3, ...$q_K\}$. each query q_i accesses a set of data records called *Query Data Set*, represented by $QDS(q_i)$, where $QDS(q_i) \subset D$ and D=$\bigcup_{1<i<K} q_i$. Assuming client's query can only start at beginning of each broadcasted data and all data record size is equal. We denote a broadcast sequence on broadcast channel by $\sigma =< d_i$, d_j ...., $d_k >$ and the goal of wireless scheduling problem is to find the optimum broadcast schedule σ_{opt} which minimum the average query access time.

Considering the single query case such that query q_1 accessing d_1 and d_3 from the broadcasted data set, that is $QDS(q_1)$=$\{d_1, d_3\}$. Under the schedule σ_1=$< d_1, d_2, d_3, d_4, d_5, d_6, d_7 >$ as showing from Fig 1a, the access time will be 3 if this query start at beginning of d_1 and the access time will be 7 if this query start at beginning of d_2 since this client has to wait for the next broadcast cycle to access d_1. Depend on the instance this query start, the access time for query start at each beginning data record is then <3,7,6,7,6,5,4> with average as 38/7. If the broadcast sequence is changed to $\sigma_2 =< d_2, d_1, d_3, d_4, d_5, d_6, d_7 >$ as showing from Fig 1b, the access time for query start at each beginning data record is then <3,2,7,7,6,5,4> with average reduced to 34/7.

Let N denote the number of total data records and $\mid q \mid$ denote the number of data records which are queried by query q , the average access time for query q under a broadcast schedule σ is

$$\mathrm{AT}^{\mathrm{avg}}(q, \sigma) = N - \sum_{j=1}^{|q|} \frac{\delta_j(\delta_j + 1)}{2N} \tag{1}$$

where δ_j is number of data records which are not queried between j th queried data record and $j+1$ th queried data record from σ. Note that, this formula of average access time is the discrete version of [2].

In a multiple query system with a broadcast schedule σ, let the average access time for query q_k is denoted as $\mathrm{AT}^{\mathrm{avg}}$ (q_k,σ) and reference frequency of q_k is denoted as freq(q_k). the total average query access time which denoted as $\mathrm{TAT}^{\mathrm{avg}}$ (σ) is

$$\mathrm{TAT}^{\mathrm{avg}}(\sigma) = \sum_k \mathrm{freq}(q_k)\mathrm{AT}^{\mathrm{avg}}(q_k,\sigma) / \sum_i \mathrm{freq}(q_i) \tag{2}$$

Query Distance (QD) is the minimal access time for query starting at all possible positions. If N data records is broadcasted under schedule σ, the QD of a query q which access $|q|$ data records is defined as

$$QD(q,\sigma) = N - \max(\delta_1, \delta_2, ..., \delta_{|q|}) \tag{3}$$

Consider the two queries case where $QDS(q_1)=\{d_2,\ d_4\}$ with freq(q_1)= 10 and $QDS(q_2)=\{d_2,\ d_7\}$ with freq(q_2)= 9. If two broadcast schedule $\sigma_1 = < d_1, d_2, d_3, d_4, d_5, d_6, d_7 >$ and $\sigma_2 = < d_2, d_1, d_3, d_4, d_5, d_6, d_7 >$ were to be considered for the average query distance(Equ 3), we have $QD(\sigma_1)$ = 57/19 and $QD(\sigma_2)$=58/19, which showing the σ_1 shall be the better schedule. However, since $\mathrm{AT}^{\mathrm{avg}}(q_1,\sigma_1)$= $\mathrm{AT}^{\mathrm{avg}}$ (q_2,σ_1)= 7-(1*2+4*5)/14, $\mathrm{AT}^{\mathrm{avg}}$ (q_1,σ_2)= 7-(2*3+3*4)/14 and $\mathrm{AT}^{\mathrm{avg}}$ (q_2,σ_2)= 7-(5*6)/14, the average access time under broadcast schedule σ_1 is $\mathrm{TAT}^{\mathrm{avg}}(\sigma_1)$=38/7 and the average access time under broadcast schedule σ_2 is $\mathrm{TAT}^{\mathrm{avg}}(\sigma_2)$= 10*(40/7)/19+9*(34/7)/19= (37.1)/7, which show that the schedule σ_2 is actually the better one with less average access time.

This inconsistency between the linear objective function and average query access time function is due to linear objective function under estimate the cost of smaller query distance and tend to choose the schedule with larger query distance.

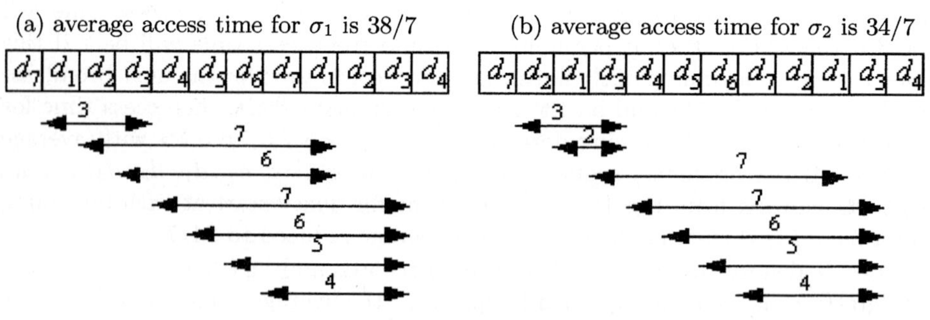

Fig. 1. Two different broadcast schedules

3 Affinity Function

In this section we explore the data affinity model which can be used in latter section to cluster data records. The data affinity represents degree of association between data pair. In previous works, the data affinity between two set of data records is depend on the access frequency and number of access data records from queries which access both sets. This affinity function have difficulty in measuring appropriate cost of merging if number of access records from each query is large. The data affinity we proposed is based on minimizing the worse case average access time and will not influenced by the diversity of client's queries.

3.1 Quadratic Query Distance

Definition 1. *Given a query q which access a set of data records $QDS(q)$ from D, the Quadratic Query Distance (QQD) of q from schedule σ is defined as:*

$$QQD(q,\sigma) = \sum_{j=1}^{|q|} \delta_j^2 \tag{4}$$

Lemma 1. *Given a query q and two schedules σ_1 and σ_2*

$$\text{if } QQD(q,\sigma_1) \geq QQD(q,\sigma_2) \text{ then } \mathrm{AT}^{\mathrm{avg}}(q,\sigma_1) \leq \mathrm{AT}^{\mathrm{avg}}(q,\sigma_2)$$

Proof. Replace $\sum_{j=1}^{|q|} \delta_j$ with $N - |q|$ in (1), we can easily see that average access time is monotonously decreased when quadratic query distance increase.

Definition 2. *Suppose a set of data records D which accessed by a set of queries Q is $\{d_1, d_2, d_3, .. d_N\}$, $\delta_{k,j}$ is the number of j's consecutive data records which is not queried by q_k and $|q_k|$ is number of data records which are queried by q_k. The Total Quadratic Query Distance(TQQD) of Q under schedule σ is defined as:*

$$TQQD(Q,\sigma) = \sum_k \mathrm{freq}(q_k) \sum_{j=1}^{|q_k|} \delta_{k,j}^2$$

Lemma 2. *Given a set of query Q and two schedules σ_1 and σ_2,*

$$\text{if } TQQD(Q,\sigma_1) \geq TQQD(Q,\sigma_2) \text{ then } \mathrm{TAT}^{\mathrm{avg}}(Q,\sigma_1) \leq \mathrm{TAT}^{\mathrm{avg}}(Q,\sigma_2)$$

Proof. : from Lemma 1.

For notation convenience, let $m_k = N - |q_k|$ denote the number of data records which are not queried by q_k and since $\sum_j \delta_{k,j} = N - |q_k| \ \forall k$, the extreme cases of theoretical maximum and minimum of total quadratic query distance are

$$TQQD_{max} = \sum_k \text{freq}(q_k) m_k^2 \tag{5}$$

$$TQQD_{min} = \sum_k \text{freq}(q_k) m_k^2 / |q_k|. \tag{6}$$

Theoretical maximum can be reach if there exist a broadcast schedule such that all the data record queried by same query can be broadcasted consecutively, that is, for every query q_k there is only one interval j such that $\delta_{k,j} = m_k$ and $\delta_{k.l} = 0$ for $l \neq j$. Theoretical minimum can be reach if there exist a broadcast schedule such that all data records which been queried by same query can be broadcasted equally spread, that is for every query q_k , $\delta_{k,j} = m_k \ / |q_k|$, $\forall j$.

If data are broadcasted without any schedule optimization and the data been queried by each query q_k, are spread randomly, by strong law of large number, the number of consecutive data records which are not queried by q_k is geometrical distributed with parameter $|q_k|/(m_k + |q_k|)$. Under this randomly broadcast schedule $TQQD$ will become $TQQD_{random} = \sum_k \text{freq}(q_k)(2m_k^2/|q_k| + m_k)$ and without schedule optimization this result will deteriorate fast when $|q_k|$ increase.

In a multiple queries system, the broadcast schedule can be determined by cascading data record pair and the data pair chosen at each step is base on the value of $TQQD$ increased. However, since there are $N - 1$ steps for choosing best data pair and the increased size between N^2 data pairs could change after each step, the complexity will be $O(N^3K)$ which is not reasonable when N is large. By defining the affinity between two data records which are queried by same query we can develop a heuristic algorithm to find the optimum broadcast schedule.

3.2 The Data Affinity and Segment Affinity

The data affinity between each data records pair represent the value of $TQQD$ increased when two data records are broadcasted consecutively and defined as following:

Definition 3. *:The data affinity between any two data records d_i and d_j is defined as*

$$\text{aff}(d_i, d_j) = \begin{cases} \sum_k \frac{m_k^2}{|q_k|} \text{QryHas}(q_k, d_i)\ \text{QryHas}(q_k, d_j) \quad \text{for } i \neq j \\ \qquad\qquad\qquad\qquad\quad \text{freq}(q_k) \\ 0 \qquad\qquad\qquad\qquad\qquad\qquad \text{otherwise} \end{cases} \tag{7}$$

$$\text{where } \text{QryHas}(q_k, d_i) = \begin{cases} 1 & if \ \ d_i \in q_k \\ 0 & \text{otherwise} \end{cases}$$

The data affinity between any data pair is composed of the affinity from all queries. The affinity from each query is the average cost for increasing $TQQD$ from $TQQD_{min}$ to $TQQD_{max}$ at $|q_k|-1$ attempts. In a multiple queries system, all costs need to be weighted in order to maximize $TQQD$. If two data records requested by same query is not worth to broadcast them consecutively, however, minimize the broadcast distance between them can still raise $TQQD$ to some extent.

Since the cost of broadcasting queried data pair with distance δ_k is same as the cost of broadcasting queried data pair consecutively with $m_k - \delta_k$ unrequested data, the average cost to broadcast each queried data pair with distance δ_k will be $(m_k - \delta_k)^2/|q_k|$. In order to put the broadcast distance into consideration, data are treated as segment and cascading data pair will merge two segments into a new segment. If only data affinity is considered, the cost is based on the decision of broadcasting two queried data records consecutively or not. For segment affinity, the cost will depends on the distance of queried data pair. By doing so, the affinity between them can be more accurately measured.

With groups of data records which are already ordered as segment, the affinity between them can be defined as following:

Definition 4. *For segment affinity between the right side of S_i and left side of S_j is defined as*

$$\text{SegAff}(S_i, S_j) = \begin{cases} \sum_k \frac{(m_k - (R(S_i,k)+L(S_j,k))^2}{|q_k|} \text{QryHas}(q_k, d_i) \text{ for } i \neq j \\ \qquad \text{QryHas}(q_k, d_j) \quad \text{freq}(q_k) \\ 0 \qquad\qquad\qquad\qquad\qquad\qquad \text{otherwise} \end{cases} \tag{8}$$

where $\text{SegHas}(S_i, q_k)$ is one if there exist at least one data record in segment S_i which is queried by q_k and zero otherwise, $R(S_i, k)$ is number of consecutive data records counted from the rightmost of S_i which are not queried by q_k and $L(S_j, k)$ is number of consecutive data records counted from the leftmost of S_j which are not queried by q_k. Note that, if there are no data records which are queried by q_k. then $R(S_i, k)$ and $L(S_j, k)$ are both equal to segment length of S_i.

The segment affinity between any segment-side pair is compose of the affinity of each query q_k and this affinity is only considered on the condition when two data segment both have the data record queried by q_k. Since there are two side for each segment (left and right), there will be four different merging order for merging two segments. Among all the segments, the affinity values between each pair of segment-side may be different.

As with traditional affinity function, segment affinity is sum of the data affinity in two segments and each data affinity is then multiplied by a distance factor. The segment affinity function we proposed depends only on the affinity between each adjacent side of segments. For each query q_k, we use the number of consecutive data which is not queried by q_k to measure the affinity between two segment-sides.

4 Minimum Gap Algorithm

Starting with each segment being initialized with only single data record. Each merging process cascade two segments from the segment-sides pair which the affinity value between them is maximum.

For simplicity we always merge segment with larger index into segment with smaller index. During the merge, we place the segment with smaller index on the left of new formed segment and the segment with larger index on the right of new formed segment. Let S_l^{-1} denote the inverse segment of S_l and the new formed segment by merging the right side of S_i and left side of S_j is denoted as $S_i \oplus S_j$. For $i \leq j$, the four possible new S_i by merging S_i and S_j will be formed as Case i: $S_i \oplus S_j$ Case ii: $S_i^{-1} \oplus S_j$ Case iii: $S_i \oplus S_j^{-1}$ Case iv: $S_i^{-1} \oplus S_j^{-1}$.

In order to explain the complexity in our algorithm the affinity value are saved in matrixes. From Chung [1], the four different segment affinity between all segment pairs are first determined and used to construct two auxiliary affinity matrixes which are segment affinity matrix(A) and inverse segment affinity matrix(I) as following:

$$A = [A_{lm}] = \begin{cases} \text{SegAff}(S_l, S_m) & \text{for } l < m \qquad \text{case i} \\ \text{SegAff}(S_m^{-1}, S_l^{-1}) & \text{for } m \geq l \qquad \text{case iv} \end{cases}$$

$$I = [I_{lm}] = \begin{cases} \text{SegAff}(S_l, S_m^{-1}) & \text{for } l < m \qquad \text{case ii} \\ \text{SegAff}(S_m^{-1}, S_l) & \text{for } m \geq l \qquad \text{case iii} \end{cases}$$

That is for $i < j$, The segment affinity between right side of S_i and left side of S_j is saved in A_{ij}, The segment affinity between left side of S_i and right side of S_j is saved in A_{ji}. The segment affinity between right side of S_i and right side of S_j is saved in I_{ji}. The segment affinity between left side of S_i and left side of S_j is saved in I_{ij}. At each step of merging, A and I are scanned for largest affinity value and its corresponding segment-side pair is cascaded to form a new segment.

Depend on the entry of this value found, merge can be made between two particular segment-side pair. If segment reverse is need before merge, the $L(S_i, k)$, $R(S_i, k)$ shall exchange and its broadcast sequence shall reverse. After merging S_j into S_i, The segment parameters of S_i need to updated with following:

$$L(S_i, k) = L(S_i, k) + (1 - \text{SegHas}(S_i, q_k))L(S_j, q_k)$$

$$R(S_i, k) = R(S_j, k) + (1 - \text{SegHas}(S_j, q_k))R(S_i, q_k)$$

$$\text{SegHas}(S_i, q_k) = \text{SegHas}(S_i, q_k)|\text{SegHas}(S_j, q_k)$$

Initially all segment contain only single data record, set $S_i = (d_i)\ \forall i$. If $d_i \in q_k$ then set SegHas(S_i, q_k) to one and set $R(S_i, q_k)$, $L(S_i, q_k)$ to zero. If $d_i \notin q_k$ then set SegHas(S_i, q_k) to zero and set $R(S_i, q_k)$, $L(S_i, q_k)$ to one. Second, build A and I base on the segment-side affinity between each pair. At last, in each merging step keep merging segment-sides which have maximum segment affinity between them and the affinity value from A and I which related to this new segment shall be updated after each merge.

The complexity of constructing the affinity matrix A and I is $O(N^2K)$. At each merging step, complexity of each recomputing segment affinity is $O(NK)$ and finding maximum data segment affinity is $O(N^2)$. The total complexity is then $O(\max(N^3, N^2K))$. This value can be further reduced by reducing the search space for maximum affinity value and partially constructing the affinity matrixes.

5 Reduce the Search Space and Partially Construct the Affinity Matrix

Definition 5. *Let S_{dummy} denote as a dummy data segment with only single data record which is queried by all queries. The dummy affinity of segment S_i is defined as segment affinity between the right side of S_i and left side of S_{dummys}*

$$\text{Dummy}(S_i) = \text{SegAff}(S_i, S_{dummy})$$

Note that, in the formulas above, the affinity between the left side of S_i and right side of S_{dummy} will be the dummy affinity of a segment which is inverse of S_i, that is $\text{SegAff}(S_{dummy}, S_i) = \text{SegAff}(S_i^{-1}, S_{dummy}) = \text{Dummy}(S_i^{-1})$

Proposition 1. *For any data segment S_i, S_j, S_k, S_l, following statement is hold:*

$$\textit{if } \text{SegAff}(S_i,S_j) \geq \text{Dummy}(S_k) \textit{ then } \text{SegAff}(S_i,S_j) \geq \text{SegAff}(S_k,S_l)$$

Proof. Since the segment affinity in each query is decreasing with number of consecutive data records which are not queried, the segment affinity between any segment-side and S_{dummy} is no less than segment affinity between this segment-side and others.

Before processing the merging process we build a dummy list *DumList*. It is built with each item containing *Seg*, which is the link to segment, *SegId* which is the segment id, *indication* which indicate this segment is either in normal or inverse sequence, *dummy* which is dummy affinity of this segment-side and *twin* which is a link to its twin item where this twin item has a link to the same segment but in reverse sequence. This list is sorted in descending order of *dummy* and the complexity of constructing *DumList* is $O(\max(NK, N\log(N)))$.

Each *dummy* value in *DumList* is the segment-side affinity between the segment it link and the dummy segment. If *Seg* link to S_i with normal sequence, the dummy value is the segment affinity between right side of S_i and dummy segment, which is $\text{Dummy}(S_i)$. However, if *Seg* link to S_i with inverse sequence(which is S_i^{-1}), the *dummy* value will be $\text{Dummy}(S_i^{-1})$, which is same as the segment affinity between the left side of S_i and dummy segment. Each *dummy* value is the upper bound of affinity value between the segment-side in this item and other segment-sides from other items(from proposition 1).

Determining the maximum affinity among all affinity value from A and I is comparable to determine the largest affinity value among all the value which are the affinity between the right side of each segment indicated in *DumList*. Since the segment in *DumList* could be either in normal or inverse sequence, the four possible pairs from any two items are (S_i , S_j), (S_i^{-1} , S_j), (S_i , S_j^{-1}) and (S_i^{-1}, S_j^{-1}) for $i < j$. The merge are made between the right side of first segment and left side of second segment. By inverting the second segment and merged into first segment, the corresponding new formed segments are $S_i \oplus S_j^{-1}$, $S_i^{-1} \oplus S_j^{-1}$, $S_i \oplus S_j$ and $S_i^{-1} \oplus S_j$.

Our search strategy is starting from finding two segment-sides in items which located at beginning of *DumList*. Whenever a larger affinity between two segment-sides is found, we keep saving it to w_{min}, which is the minimum pruning affinity. The search space can be vastly reduced due to segment-sides needed to be considered are from items which their dummy value larger than w_{min}.

At each merging step, the segment-side pair which has the largest affinity value between them are merged to formed a new segment. After the merge, the *DumList* and affinity matrixes will be updated before next search. However, the affinity value determined and updated in affinity matrixes can also limited to segment-sides from items in *DumList* where their dummy value are larger than w_{min}, the number of affinity updated is then also reduced.

Due to the affinity value between any two segment-sides can be computed only when needed. The affinity matrixes need not to be fully constructed at beginning and most of value in affinity matrixes need not to be determined at all. By doing so the whole schedule problem can be solved very efficiently and since the affinity matrixes can be constructed in sparse form, the memory space is also vastly reduced.

6 Performance Evaluation

In this section, we evaluate the performance of the proposed algorithms through experiments in comparison with Chung[1]. The simulation models are running under 1000 data records and 100 queries. Each query access 20 data records and data records queried by each query are randomly selected among all data records. The access frequency of each query are investigated under zipf distribution with different skewness parameters. From Fig. 2, we find the performance of our *MG* algorithm is better than Chung's method in all skewness considered. From Fig. 3, the results show that system time can be reduced drastically with our algorithm. From Fig. 4, the results show a great of reduction in memory space especially when number of broadcasted data records is large.

7 Conclusions

In this paper we proposed a clustering algorithm for merging the wireless broadcast data for multipoint queries. Previous work on segment affinity need to evaluate every data affinity between data of different segment. However, our work on data affinity is developed based on minimizing the average query access time

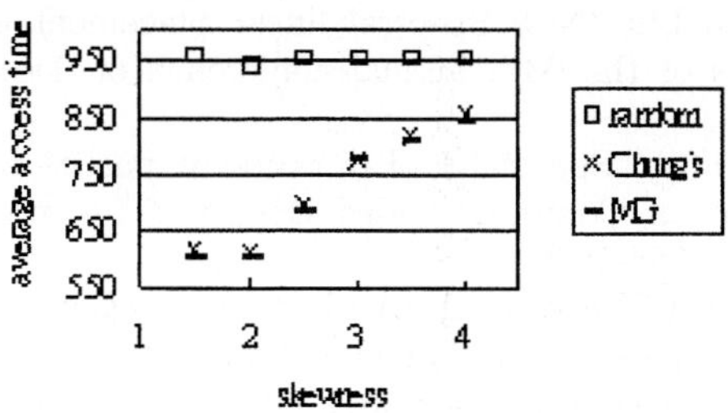

Fig. 2. average access time according to different skewness of query's access frequency

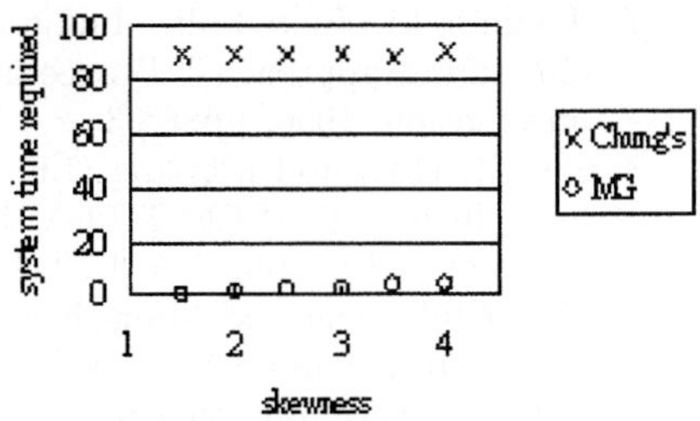

Fig. 3. the system time required from different algorithm according to different skewness

Number of data records	100	200	400	600	800	1000
Fully construct A and I	2E4	8E4	3.2E5	7.2E5	1.2E6	2E6
MG	91	98	153	378	465	496

Fig. 4. Memory space required according to different total data records

and our segment affinity model is developed from the modification of this data affinity model. Without considering every data affinity between data of different segment, the model we proposed is more efficient than previous work. Through experiments, the results show that our model using this segment affinity function perform better than previous algorithm. By using a dummy linked list to reduce the search space and reduce the number of affinity computed by partially constructing the affinity matrixes, the results shown in our experiments have proven a vast reduction in both time complexity and space complexity.

References

1. Y.D. Chung, S. Bang and M. Kim "An efficient broadcast data clustering method for multipoint queries in wireless information systems," The journal of Systems and Software 2002, pp173-181.
2. Y.D. Chung, M. Kim "Effective data Placement for wireless Broadcast," Distributed and Parallel databases, 9,2001, pp133-150.
3. S.sahni and T.Gonzalez. "P-complete Approximation Problem," Journal of the ACM,23 pp555-565,1976.
4. Y.D. Chung, M. Kim "A wireless data clustering method for multipoint queries," Decision support System 30 (4), pp469-482.
5. GuanLing Lee, Meng-Shin Yeh Shou-Chin Lo and Arbee L.P. Chen "A strategy for efficient access of multiple data items in mobile environments," Proceeding of the thid international conference on Mobile Data Management, 2002.
6. Jiun-long Huqng and Ming-Syan chen "Dependent data broadcasting for unorder queries in a multiple channel mobile environment," IEEE transactions on Knowledge and Data Engineering, 2004, pp1143-1156.

7. Jianmin Li, John Lillis, Lung-Tien Liu "New Spectral linear placement and clus-tering approach," Proceedings of the 33rd annual conference on Design automation, 1996, pp88-93.
8. Pak K. chan and Martine D.F. Schlag "Spectral K-way ratio-cut partitioning and clustering" IEEE Transactions on Computer-Aided Design of Integrated Circuits and System," Sep 1994, pp1088-1096.
9. Wen-Chih Peng and Ming-Syan Chen "Developing data allocation schemes by incremental mining of user moving patterns n a mobile computing system," IEEE Transactions on Knowledge and Data Engineering. Jan/Feb 2003, pp70-85.

S-RO: Simple Route Optimization Scheme with NEMO Transparency

Hanlim Kim[1], Geunhyung Kim[2], and Cheeha Kim[3]

[1] Convergence Laboratory, Korea Telecom (KT)
[2] Technology Network Laboratory, Korea Telecom (KT),
463-1 Jeonmin-dong, Yusung-gu, Daejeon, 305-811, Korea,
{nangel, geunkim}@kt.co.kr
[3] Department of Computer Science and Engineering,
Pohang University of Science and Technology (POSTECH),
San 31 HyoJa-Dong, Nam-Gu, Pohang, 790-784, Korea
chkim@postech.ac.kr

Abstract. Network mobility (NEMO) basic support protocol maintains the connectivity when mobile router (MR) changes its point of attachment to the Internet by establishing a bidirectional tunnel between MR and Home Agent (HA). However, it results in pin-ball routing and multiple encapsulations especially in the nested NEMO. In order to solve these problems, we propose a simple route optimization scheme for NEMO and nested NEMO. In the proposed scheme, a correspondent node (CN) maintains network prefix binding information of intermediate MRs to obtain optimal path to the mobile network node (MNN). For a CN to receive network prefix bindings, each intermediate MR updates its network prefix binding to the inner source of the encapsulated packet, when it receives an encapsulated packet from its HA. The proposed scheme uses the routing header type 0 (RH0) and the routing header type 2 extension (RH2 extension) to deliver packets to the MNN through the optimal path and to remove multiple encapsulations. In addition, we extend the Router Alert Option (RAO) so that the MR knows the original source address of the packet, when ingress filtering is applied.

1 Introduction

To maintain the continuity of the sessions of MNNs within mobile networks, the NEMO working group proposes NEMO basic support protocol [1]. It extends Mobile IPv6 (MIPv6), since network mobility is not supported properly by MIPv6. However, there are two main problems in NEMO basic support [2,3]. One is the pin-ball routing that results in the use of unnecessary network resources and the other is multiple encapsulation which may cause delayed packet delivery. Also, there are three sub-problems. First, multiple encapsulations may divide the original packet if the packet size is limited to link-MTU. Second, on the assumption that HA manages several MRs, the HA of the MR will be overloaded by MNN's packets when many MNNs exist. Third, if the MR is closer to

C. Kim (Ed.): ICOIN 2005, LNCS 3391, pp. 401–411, 2005.

the root MR, the HA of the MR will be overloaded much more. Route optimization for network mobility solves two main problems and three sub-problems are solved incidentally.

In this paper, we introduce simple route optimization based on four main ideas. First, the network mobility support transparency for MNNs in NEMO basic support is maintained so that the movement and location management of a mobile network should be managed only by its MR. The (MR_prefix, MR_CoA) network prefix binding in NEMO basic support is used for those managements. Second, we extend NEMO basic support so that a MR can update the network prefix binding to CNs like the host binding in Mobile IPv6. Third, a MR as well as a mobile host updates binding information to the inner source of the packet, when the encapsulated packet came from its HA. Especially, the MR updates network prefix binding to the inner source of the encapsulated packet when the inner packet is for its MNNs. In case of a nested mobile network, the NP_BUs of nested MRs are performed from the first MR, the MR of the network that contains the MNN, to the common upper MR or the root MR in turn. Finally, a MR would support avoiding ingress filtering of upper MR by encapsulating the packet from its network nodes. Multiple encapsulations may occur when mobile networks are nested. To avoid multiple encapsulations in nested mobile network, Gu et al. [4] extended RAO [5]. In this RAO extension, the first MR adds the RAO in encapsulating the packet from its network nodes, then intermediate MRs do not encapsulate the packet and only change the outer header's source address with their CoAs. In this case, the MR, which receives the RAO laden packet from its HA, has to know the original source address of the packet for binding update. However, Gu et al. did not mention how to obtain the original source address of the packet in detail. For the MR to obtain the original source address of the packet without additional information, we extended the functionality of the MR. This additional functionality entails the MR updating the network prefix binding to the source address of the inner-inner packet when the MR receives the RAO laden packet from its HA.

In addition, to avoid both unnecessary pin-ball routing and multiple encapsulations, we use two routing header types. The node directly uses the routing header type (RH0) [6] to send packet to the fixed host in the mobile network and the routing header type 2 extension (RH2 extension) [7] to send packets to the mobile host in the mobile network, respectively. the RH0 and RH2 extension are similar to loose source routing.

The remainder of this paper is organized as follows. We first discuss related work in section 2. Our proposed route optimization scheme for NEMO and nested NEMO is explained in Section 3. An evaluation of our proposed scheme is presented in Section 4. Finally, we conclude in section 5.

2 Related Work

Currently route optimization schemes in NEMO have been discussed at the NEMO working group in the IETF and at various conferences. Also, network mobility support terminology [8] and the taxonomy of route optimization models in NEMO context [2] help to understand the current problem space of route optimization. In this section, we will review protocols [1,4,7,10] that are similar to our scheme.

The NEMO basic support [1] provides a basic routing scheme to support network mobility. Since Mobile IPv6 does not consider network mobility, packets can not be forwarded to nodes in the MR's network using Mobile IPv6. Hence, NEMO basic support considers a bidirectional tunnel between the MR and the HA of the MR. However, when the mobile networks are nested, it suffers from pin-ball routing and multiple encapsulations.

RRH [7] proposed a new routing header, called RRH and RH2 extension, to record intermediate MR's addresses into the packet header and to avoid packet delivery through all HAs of the intermediate MRs. Therefore, in this scheme, there is only one bidirectional tunnel between the first MR and its HA. Although the authors argue that this scheme performs route optimization between a CN and a MR, there is no description of it in detail. Moreover, their scheme suffers from no secured binding update mechanism using RRH and it has to modify the router advertisement (RA)message to count the number of intermediate MRs.

The Optimized Route Cache (ORC) scheme of Wakikawa et. al [10] has two main features. One is a proxy router that intercepts packets whose source addresses belong to target network prefixes using IGP protocols in the Autonomous System (AS), encapsulates them, and tunnels them to the MR. This feature reduces the number of binding updates efficiently when the mobile network moves. The other is that the upper routers in the nested mobile network should know the network prefix information of lower routers and the upper routers update its lower routers' network prefix information to its ORC router. To do that, it modifies the RA message by adding a mobility flag to inform the MR that the upper network is a mobile network and the information about its ORC routers may be delivered to the upper MRs. However, since modified RA with a mobility flag will not pass fixed router in the mobile network, the fixed router in the mobile network also should be modified.

Gu's proposal [4] supports CN-to-MR route optimization. This performs route optimization and minimal encapsulation between the node outside a mobile network and the node in a nested mobile network. However, it suffers from the lack of a secured binding update mechanism and requires RA modification like RRH. Additionally, it suffers from non-optimal routing through the root MR in the intra-mobile network communication, which means communication between nodes in different nested mobile networks behind the root MR.

In the next section, we describe our proposed route optimization scheme for network mobility, which is backward compatible with both Mobile IPv6 and NEMO basic support and solves the intra-mobile network communication problem.

3 Proposed Route Optimization Scheme

3.1 General Operations

Similar to NEMO basic support, a MR is considered as a mobile host in the visited network and performs the role of router for its network nodes in our proposed scheme. Therefore, a MR does not exchange routing information with the routers in the attached network unlike the ORC scheme. Also, to provide network mobility support transparency, a MR does not handle any information about the movement and location management of other mobile networks. So, in the proposed scheme a MR does not need hierarchical information of the nested mobile network unlike others [4,7,10]. We use MR_Prefix-MR_CoA bindings like NEMO basic support, for a CN to obtain optimal path to the MNNs in nested mobile network.

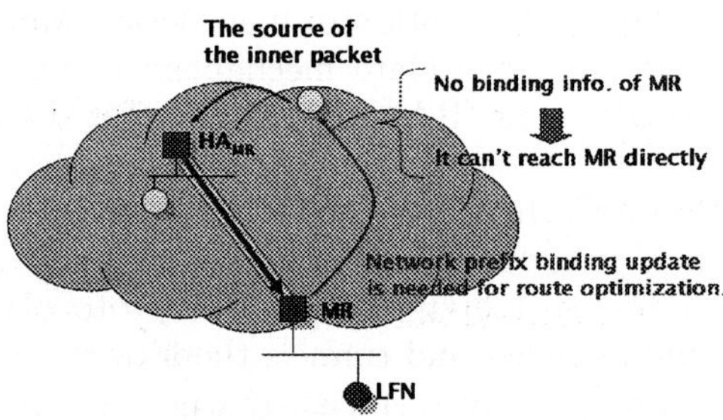

Fig. 1. Binding Update of Mobile Router.

Moreover, as shown in Fig. 1, a mobile host and a MR update binding information to the inner source of the packet which may be HA or CN, when they receive encapsulated packet. Especially, the MR updates network prefix binding to the inner source of the packet when the inner packet is for its MNNs. Based on the binding information of MRs updated by the above mechanism and following extension, we can achieve route optimization.

Type 2 Routing Header Extension (RH2 extension) [7] : In our scheme, we extend the destination option header with multiple slots to be filled with the CoAs of intermediate MRs prior to the CoA and HoA of MNN, while Mobile IPv6 defines one slot as a destination option header.

Router Alert Option (RAO) [5] Extension : This option is used to prevent multiple encapsulations in a nested mobile network, when a MR encapsulates a packet from an ingress interface to avoid ingress filtering. It enables the MR not to encapsulate the packet at every ingress interface in a nested mobile network and to replace the source address of an outgoing packet with its CoA.

NEMO basic support [1] Extension : In our scheme, we extend the MR to update its network prefix bindings to CN, while the MR updates its network prefix bindings to HA in NEMO basic support.

Mobile IPv6 [6] Extension : In our scheme, the MR updates its network prefix binding to the inner source address of the packet, similar to host binding in Mobile IPv6, when a MR receives an encapsulated packet. Additionally, we need to extend the CN's capability of decapsulation due to ingress filtering.

3.2 Packet Delivery

In this section, we describe the routing mechanism that is backward-compatible with Mobile IPv6. In our scheme, when a CN has the binding update of a mobile node, the CN creates packets with a RH2 extension [7] that supports multi-hop loose source routing. In addition, when the CN does not have the binding update of a destination node and has the network prefix binding updates (NP_BUs) of intermediate MRs, the CN creates packets with a RH0 [6]. Otherwise, the CN creates packets without a RH0 or RH2 extension.

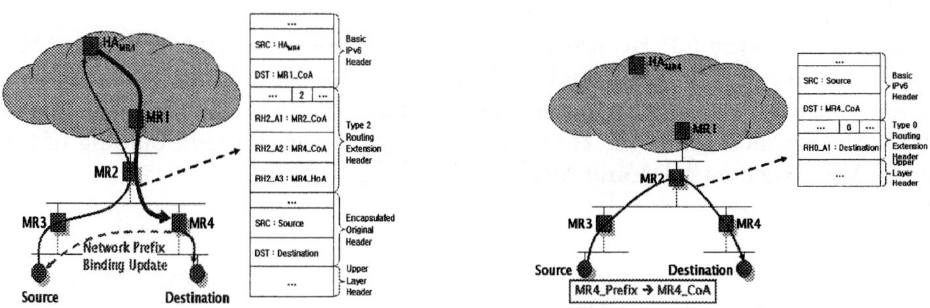

(a) Packet delivery through the HA and NP_BU update between LFNs.

(b) Packet delivery through the optimal route between LFNs.

Fig. 2. Packet delivery in intra communication.

Fig. 2 shows the packet delivery between LFNs when the source does not have NP_BUs of MR4 in the beginning and the ingress filtering is not applied to MRs. In Fig. 2(a), the NP_BU of MR4 is delivered to the source when MR4 receives the packet of the source through the HA of MR4. After the source receives the NP_BU of MR4, it sends packets with RH0 to the destination. The packet with RH0 will be delivered directly to the destination without encapsulation as shown in Fig. 2(b). Based on this scenario, when the destination is a mobile node, we can ensure that the communication between MNNs can be adapted equally if the source sends packets with RH2 extension.

3.3 Binding Update Scenarios in a Nested NEMO

In our scheme, a CN obtains NP_BUs of intermediate MRs from the first MR to the root MR in turn during the communication with nested mobile network.

We will show how the HA of a MR obtains its upper MR's NP_BUs and how it acts when a CN communicates with a MNN whose mobile network is nested.

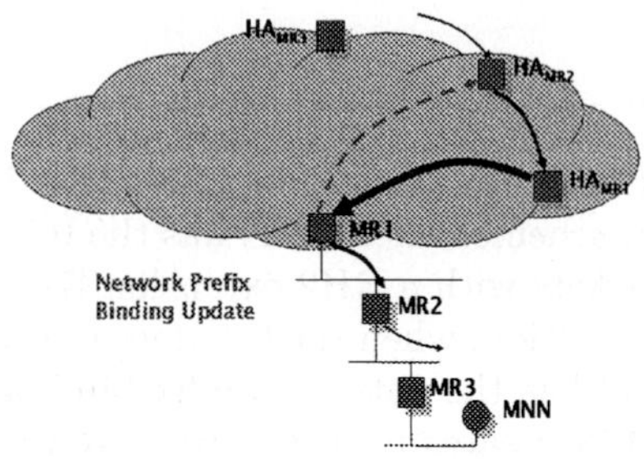

Fig. 3. The NP_BU scenario for the HA_{MR2}.

Fig. 3 shows the NP_BU scenario in the case where the MR1, the upper MR of MR2, performs NP_BU of MR1 to the HA_{MR2}, the HA of MR2, after it receive an encapsulated packet whose inner source address is that of the HA_{MR2}. Fig. 4 shows the scenarios that the HA_{MR3} obtains the network prefix binding of the upper MRs, such as MR1 and MR2.

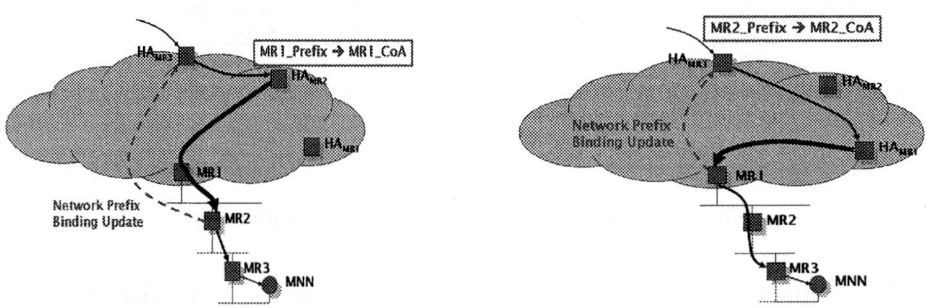

(a) The NP_BU of MR2 to the HA_{MR3}

(b) The NP_BU of MR1 to the HA_{MR3}

Fig. 4. NP_BU scenario for the HA_{MR3}.

Fig. 3 and 4 show the scenarios that the binding cache in the HA of a MR is updated by upper MRs, when a node sends packets to the MR's network through its HA and the HA has no NP_BU of upper MRs. When the packet for a MNN arrives at the HA_{MR3}, the HA encapsulates the packet using RH2 extension because the MR is a kind of a mobile host in the HA-to-MR tunneling. After the NP_BUs in Fig. 4(a) and Fig. 4(b), the HA encapsulates the packet for MNN using RH2 extension with the NP_BUs of MR1 and MR2 and sends it to the MNN directly.

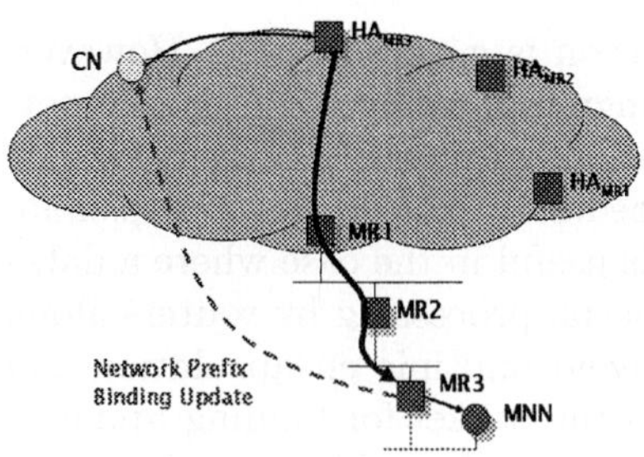

(a) NP_BU of MRs to a CN (step 1)

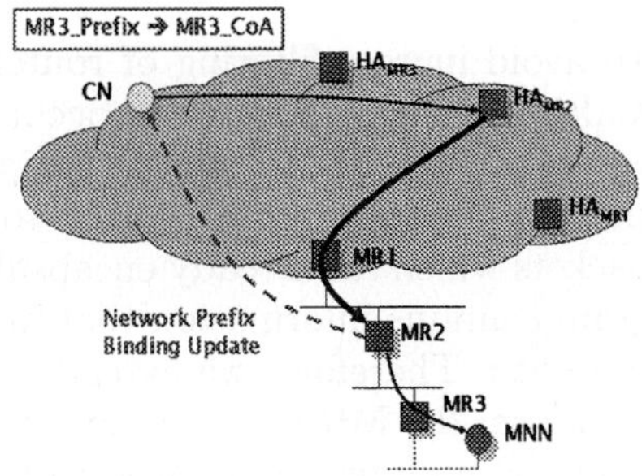

(b) NP_BU of MRs to a CN (step 2)

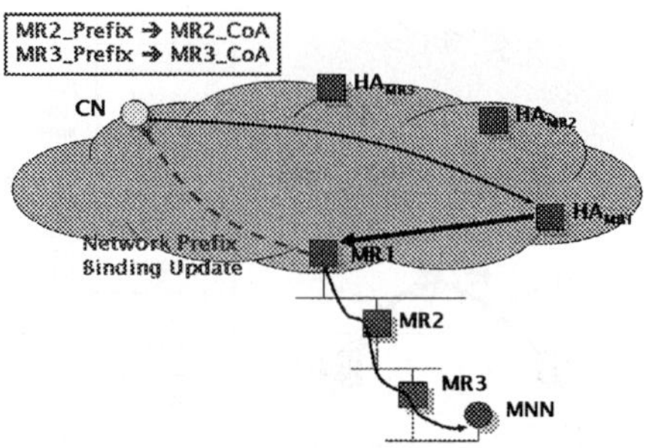

(c) NP_BU of MRs to a CN (step 3)

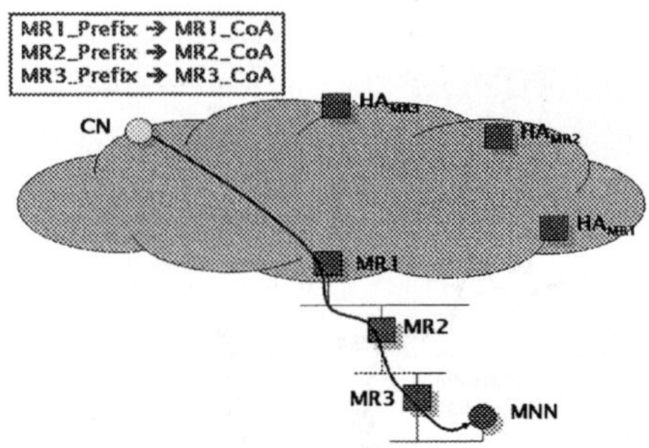

(d) Packet delivery from a CN

Fig. 5. NP_BU scenario for a CN.

Fig. 5 shows how a CN obtains the NP_BUs of intermediate MRs. This scenario assumes that the HA of a MR knows the NP_BUs of upper MRs. However, it is not difficult to adapt Fig. 3 and 4 to Fig. 5, when the HA of a MR does not have the NP_BUs.

We have illustrated the binding update scenarios to optimize route between a CN or HA and a MNN in Fig. 3, 4, and 5 and the binding update scenario between LFNs in Fig. 2. The route optimization between MNNs using the binding update scenario explained above can be obtained although we did not show the full scenario of route optimization.

As a result, the proposed scheme supports enough binding information for route optimization, when a MR updates the network prefix binding to the node, which wants to send packets to the MR's network and does not know how to reach the mobile network directly. With these NP_BUs and the proposed routing mechanism we can obtain route optimization in mobile network environment.

3.4 Consideration of Ingress Filtering

To avoid ingress filtering of routers, encapsulation is generally used. However, multiple encapsulations will occur because of ingress filtering when mobile networks are nested. To avoid multiple encapsulations in nested mobile networks, the hop-by-hop option is used so that intermediate MRs do not encapsulate packets which are already encapsulated. RAO is useful in the case where a datagram contains information that may require special processing by routers along the path. Therefore, we extend this RAO to avoid multiple encapsulation and to inform the MR the original source address of the packet for binding updates. The MR updates the network prefix binding to the source address of the inner-inner packet when it receives the packet whose inner packet has RAO through its HA.

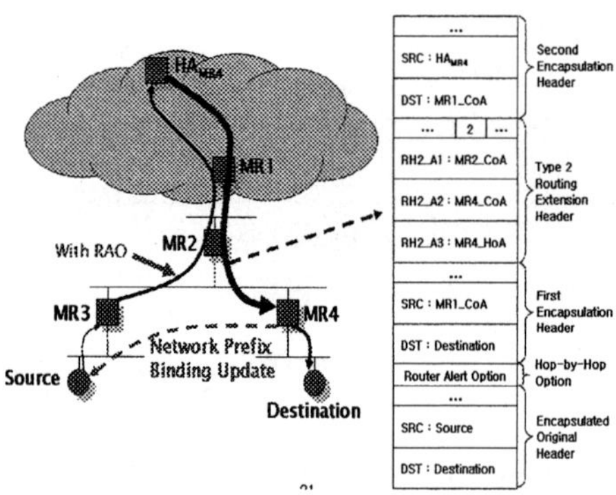

(a) Packet delivery through HA and NP_BU between LFNs (with ingress filtering)

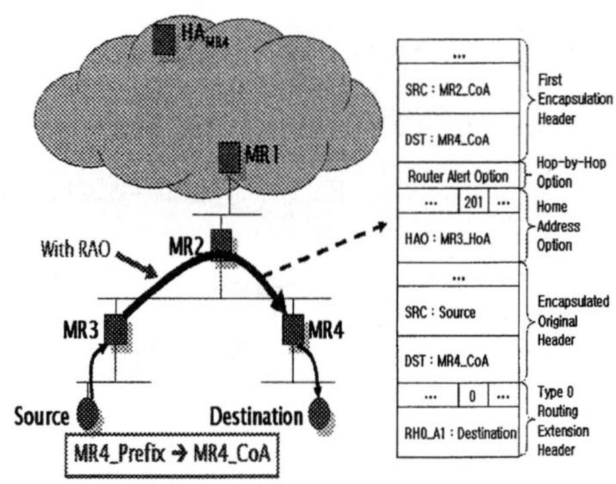

(b) Packet delivery through optimal route between LFNs (with ingress filtering)

Fig. 6. Packet delivery and NP_BU considering ingress filtering.

Fig. 6(a) shows the packet delivery from the source to the destination through the HA_{MR4} and the NP_BU scenario when ingress filtering is considered and MR4 receives the packet from its HA. Fig. 6(b) shows the route optimization between LFNs after MR4 sends NP_BU to the source. In the MR3's encapsulation processing, MR3 does not add a home address option (HAO) because it operates as a mobile router and the binding of MR3 does not exist in MR4. Throughout these scenarios, we conclude that the optimal route between any nodes can be obtained by extending these scenarios in our scheme even though ingress filtering is considered.

4 Evaluation

In Table 1, we compare the proposed scheme with other schemes that are mentioned in section 2.

Table 1. The comparison with other schemes.

	NEMO	RRH	ORC	Gu's	Proposed scheme
tunneling	HA↔ MR	HA↔ MR	ORC R↔ MR	HA←MR	CN←MR
NEMO routing	pin-ball	triangular	semi-optimal	optimal	optimal
nested NEMO routing	pin-ball	triangular	triangular or semi-optimal	optimal	optimal
intra-mobile network routing	pin-ball	routing through HA↔HA	optimal	routing through root MR	optimal
encap. degree	nested level	one or two	one or two	zero/one	zero/one
additional function with NEMO basic	-	RRH, RH2 ext. RA mod.	ORC R, RA mod.	NCO, RAO, RH2 ext., RA mod.	RAO, RA mod.
additional info. with NEMO basic	-	nested level of MRs in MRs	lower level MR's NP in MRs	upper level MR's CoA in MRs	None

From a tunneling point of view, RRH scheme requires a bidirectional tunneling between the first MR and its HA. ORC scheme requires both the tunnel from the first MR to the discovered ORC router and the one from the discovered ORC router to the root MR. In Gu's proposal and the proposed scheme, the unidirectional tunnel from the first MR to the CN is required when ingress filtering is applied to MRs and a tunnel is not required otherwise.

In NEMO and nested NEMO routing, ORC scheme supports semi-optimal routing if an ORC router is in the AS that a CN belongs to or the triangular routing if not. Gu's proposal and proposed scheme provide the optimal routing between a CN and a MNN.

In the case of an intra-mobile network routing between MNNs which share root MR, RRH scheme performs routing through both HAs of each first MR. Also, Gu's proposal performs routing through root MR. ORC scheme and the proposed scheme provide optimal routing.

Considering the encapsulation degree, only Gu's proposal and the proposed scheme provide minimal encapsulation (zero if ingress filtering is not applied to MRs, one if ingress filtering is applied). Additionally, all three schemes except the proposed scheme modify RA messages of Mobile IPv6 and enable a MR to manage the information related to the nested mobile network hierarchy.

In RRH and Gu's proposal, the topology change information is delivered by RA message. When the intermediate mobile network moves to another network with its sub-mobile networks, the movement acknowledgements of the sub-mobile networks are achieved by the propagated RA messages. Therefore the calculation of slot size in RRH and the creation of binding update information in Gu's proposal can be delayed if the nested level is large. It can be problematic if the mobile network moves fast. In proposed scheme and NEMO basic support, because of NEMO transparency, the delivery of topology change information is not needed and the binding update is performed correctly and immediately.

5 Conclusion

In this paper, we have discussed the problems of the NEMO basic support in a nested mobile network, multiple encapsulations and pin-ball routing. Although schemes to solve these problems have been proposed, they did not provide optimal solutions or they require the modifications of NEMO basic support. Also, most approaches did not consider the intra-mobile network communication that can be possible with the popularization of a PAN.

Therefore, we proposed a new route optimization scheme as the optimal solution for the multiple encapsulations and pin-ball routing problem in a nested mobile network. On these main ideas, we show the proposed scheme privides optimal routing in NEMO, nested NEMO, and intra-mobile network communication using minor extensions.

In the future work, we will study the scheme to reduce NP_BUs by aggregation of NP-BU or the deferment or omission of NP_BUs according to the applications.

References

1. V. Devarapalli, R. Wakikawa, A. Petrescu, and P. Thubert, "Network Mobility (NEMO) Basic Support Protocol," IETF draft, December 2003, Work in Progress.
2. P. Thubert, M. Molteni, C.-W. Ng, and H. Ohnishi, "Taxonomy of Route Optimization models in the Nemo Context," IETF draft, October 2004, Work in Progress.
3. F. Zhao, S. F. Wu, and S. Jung, "MEMO Route Optimization Problem Statement, Requirements and Evaluation Considerations," IETF draft, October 2004, Work in Progress.
4. Z. Gu, D. Yang, and C. Kim, "Mobile Ipv6 Extensions to support Nested Mobile Networks," IEEE Advanced Information Networking and Application, March 2004.
5. C. Partridge and A. Jackson, "IPv6 Router Alert Option," RFC 2711, October 1999.
6. D. B. Johnson, C. E. Perkins, and J. Arkko, "Mobility Support in IPv6," IETF draft, June 2003, Work in Progress.

7. P. Thubert and M. Molteni, "IPv6 Reverse Routing Header and its application to Mobile Networks," IETF draft, October 2003, Work in Progress.
8. T. Ernst and H.-Y. Lach, "Network Mobility Support Terminology," IETF draft, February 2004, Work in Progress.
9. T. Ernst, "Network Mobility Support Requirements," IETF draft, October 2002, Work in Progress.
10. R. Wakikawa, S. Koshiba, K. Uehara, and J. Murai, "ORC: Optimized Route Cache Management Protocol for Network Mobility," IEEE Information & Communication technologies, February 2003.
11. T. Ernst, "Network Mobility Support in IPv6," PhD thesis, University of Joseph Fourier, France, October 2001.

Decreasing Mobile IPv6 Signaling with XCAST

Thierry Ernst

Keio University, Japan*
ernst@sfc.wide.ad.jp

Abstract. Mobile IPv6 is the IETF proposition to support host mobility in the Internet. It provides routing optimization for packets sent to the mobile node at the expense of signaling messages that are periodically sent to the correspondent of the mobile node. We propose using XCAST, a multicast technique best designed for small groups, as a means to decrease this signaling. XCAST records the addresses of all the destinations in the packet itself. We evaluate the performance of our proposition against Mobile IPv6 route optimization. We obtain good results even when a very few routers, well located in the network, are able to process the extension.

1 Introduction

Fixed nodes are permanently attached to the same subnetwork and are identified by a permanent IP address which determines the subnetwork where they are attached to. Unlike fixed nodes, mobile nodes (MNs) change their point of attachment in the Internet topology. They are moving from subnetwork to subnetwork and are reachable at different locations in the Internet topology. As a result from this, MNs must change their addresses. However, the IP address has a dual semantic at the network layer (used both as a locator and as an identifier) and is also used at upper layers as a node identifier. There is a therefore a trade-off between retaining the IP address which fails routing and changing the address which breaks upper layer connections [1, 2]. Mobile IPv6 (MIP6) [3] proposes *two-tier addressing* as the solution to this conflicting dual semantic and use of IP addresses. *Two-tier addressing* associates a MN with two addresses, one is permanent and used as a location invariant identifier, and the other one is temporary and used for routing. An address translation mechanism offers migration transparency to upper layers and insures backward compatibility with transport protocols. Connections are not disrupted as a result of mobility. Interestingly, Mobile IPv6 allows routing optimization (direct routing) for packets sent to the MN. This saves bandwidth and preserves delays, at the expense of signaling messages to update the current location of the MN.

* The ideas and simulations presented in this paper were originally performed under the supervision of Hong-Yon Lach (Motorola Labs Paris, France) and Claude Castelluccia (INRIA Rhône-Alpes Grenoble, France) and co-financed by Motorola Labs Paris, France and the French government under a PhD program. The paper was later written at Keio University. The author is also grateful to Christophe Janneteau for his careful review on an earlier version of the present paper.

C. Kim (Ed.): ICOIN 2005, LNCS 3391, pp. 412–421, 2005.

Table 1. Traditional Multicast vs Explicit Multicast

Metric	Traditional Multicast	Explicit Multicast
Group Members	Unknown to the source	Known to the source
Identifier	Group identified by single address	No group identifier
Membership Management	Protocol needed	No group membership management
Distribution Tree	Multicast routing protocol responsive for distribution tree establishment	No need to build a tree, use standard unicast routing table
Multicast Address	Multicast address discovery	No multicast address discovery
Packet Overhead	No overhead	List of group members self contained in the transmitted data packet
Packet Processing	Check forwarding interface for each branch of the tree	Check forwarding interface for each destination
Signaling	Large amount of signaling	No signaling
Memory	State in each multicast router	No state in routers
Scalability	Large number of group members	Large number of small groups

Multicast aims at minimizing bandwidth consumption when there are multiple destinations for a given packet. Hence, the aim of multicast is to avoid duplicate information flowing over the same link. The packet is only duplicated when all destinations are not reachable via the same next hop. In the traditional multicast model, as defined in [4, 5], a *multicast address* is assigned to a collection of nodes that form a *multicast group*. A *multicast routing protocol* constructs a *multicast delivery tree*. We commonly distinguish two kinds of multicast delivery tree, the *Shortest Path Tree* (SPT), and the *Shared Tree*, or *Core-Based Tree* (CBT). The SPT is a *minimum spanning tree* rooted at the source. Each source in the group has its own SPT. The CBT is a single delivery tree built per multicast group, and is shared by all senders in this group. This tree is rooted at a single *core* router. Packets are thereafter sent to the multicast address and forwaded along the delivery tree.

Explicit Multicast (Xcast) [6, 7, 8, 9] (also known as *Small Group Multicast* or *List-Based Multicast*), is an orthogonal and recent multicast technique, designed to complement traditional multicast. The basic idea is to carry the list of recipients of a packet in the packet itself. Intermediate routers must read the list of destinations to check if they have distinct next hops. Compared to traditional multicast, Xcast is very simple. There is no multicast routing protocol, no multicast address, and no group membership protocol. Both techniques are indeed complementary to one another since a "one size fits all" protocol seems unable to meet the requirements of all applications. As shown on tab.1, Xcast seems more appropriate for a large number of multicast groups with a small number of members, whereas *traditional multicast* is more appropriate for a large number of group members. Applications of Xcast include *narrowcast-like* (*few-to-few*) applications (e.g. IP telephony, collaborative applications, etc), whereas traditional multicast is targeted to *broadcast-like* (*one-to-many*) applications (e.g. TV and radio programs, weather forecast, etc).

In this paper, we propose Xcast as a means to carry Mobile IPv6 signaling while maintaining routing optimization. Mobile IPv6 is outlined in section 2 whereas the Xcast extensions brought to Mobile IPv6 are described in section 3. The performance of this solution is evaluated in section 4.

2 Mobile IPv6 and Its Shortcomings

Mobile IPv6 associates a MN with two addresses. The Home Address MN_{HoA} is permanent and obtained on a link in the home network (home link). In addition, the MN obtains a new temporary Care-of Address MN_{CoA} on each subsequent visited link. This terminology is illustrated on fig.1. The MN may own several Care-of Addresses at anytime, one of which is selected as the primary MN_{CoA}. The binding between MN_{HoA} and the primary MN_{CoA} is registered with the Home Agent (HA), a special node on the home link. This registration is performed by means of a *Binding Update* (BU) message containing both addresses. Once it receives a BU, the HA adds or update an entry in its *Binding Cache*. The Home Address is used as the key for searching the *Binding Cache*. As a result of this registration, the HA is able to intercept all packets intended for the MN and to encapsulate them to the current Care-of Address MN_{CoA}.

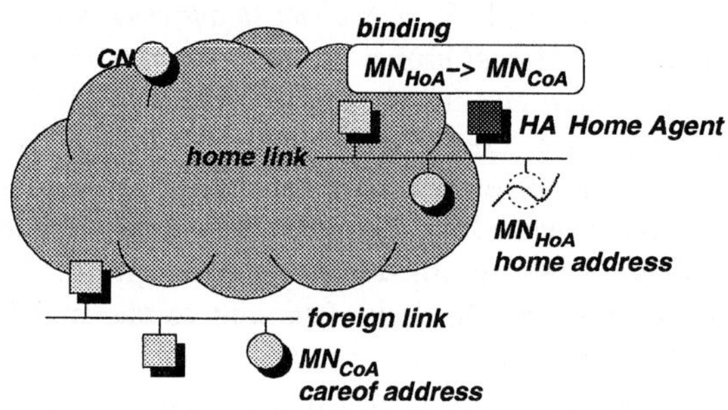

Fig. 1. Mobile IP Terminology

A correspondent node (CN) willing to communicate with a MN first calls the *DNS* which returns the MN's Home Address. Packets are then routed to the home link where they are intercepted and encapsulated by the HA to MN_{CoA}. The packet is decapsulated by the MN and the CN is inserted in the *Binding List*. At this point, the MN may also send a BU containing its primary MN_{CoA} to some or all CNs recorded in its *Binding List* to avoid triangle routing via the HA (fig.2.a). The CN authenticates the packet and adds an entry in the *Binding Cache* like the HA. Forthcoming packets are then directly sent to the MN_{CoA} using an *IPv6 Routing Extension Header*.

BUs are sent alone in separate packets containing no payload, but according to [3], BUs may be piggybacked in payload datagrams. However, this has so far not been specificied. Typically, the MN sends 5 consecutive BUs with a 1-second interval just after forming a new primary Care-of Address. After this, the MN keeps sending BUs at a lower rate, typically every 10 seconds, in order to refresh *Binding Cache* entries.

Routing optimization from the CN to the MN is one of the most interesting Mobile IPv6 features, particularly when MNs are not located in their native administrative domain. In this situation, simulations results in [10] have shown that the mean distance, expressed in number of hops, when routing optimization is used between the CN and the MN, is half the distance via the HA when routing optimization is not used. Not only the *transmission cost* (total number of bytes transmitted over the network) increases, but also the mean delay. Simulations have also outlined that the distance varies much more on the triangle route than on the direct route, thus further exhibiting the need for routing optimization. When the transmission cost is compared against the *mobility cost* (total number of bytes consumed by Mobile IPv6 signaling), the simulations show that the aggregated cost remains less important over the direct route with routing optimization than over the triangle route with no routing optimization, even for a very low data rate between the CN and the MN.

However, since BUs are sent periodically, we have also observed that the MN sends short packet bursts, separated by silent periods on the order of several seconds. This periodic burst is propagated to the entire network and consumes a significant amount of the available bandwidth in situations where the MN has a large number of CNs. This is particularly true on the wireless link between the MN and its Access Router when the MN has formed a new Care-of Address (5 consecutive BUs are sent to each CN). This problem is referred to as a *binding update explosion* in [10] or a *binding update storm.*

We conclude that while there is no doubt that Mobile IPv6 provides for optimal routing advantageouly, the transmission cost has to be balanced against the mobility cost which increases linearly with the number of CNs. Solutions to minimize this mobility cost must thus be looked into.

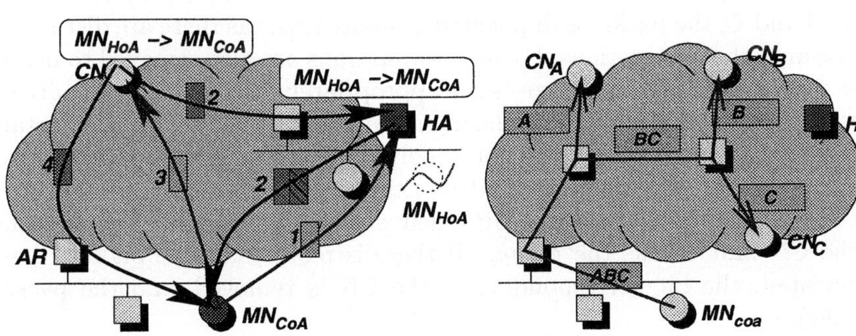

(a) With Unicast BUs (b) With XCAST BUs

Fig. 2. Routing Optimization with Mobile IPv6

3 XCAST Delivery of Binding Updates

In this section, we propose extending Mobile IPv6 with Xcast to deliver BUs as a means to minimize signaling. The addresses of several CNs are recorded in the BU, instead of sending individual BUs. This only requires minor extensions to Mobile IPv6. For doing so, we define a new *Xcast Header*, implemented as an *Hop-by-Hop Extension Header*. As such, it should be processed by all routers that understand this option. Its length varies according to the number of CNs, number that may only be limited for performance considerations. Each destination recorded in the header corresponds to a rank in a *bitmap* field. When set, a bit indicates that the corresponding destination still remains undelivered. This field must be updated by each duplicating router. Its purpose is to prevent loops and to avoid the processing of destinations to which a duplicate packet was already transmitted. A *remove flag* specifies if destinations not reachable from the interface where a BU is going to be forwarded must be removed. The MN must be able to fill the *Xcast Header* and is provided a basic decision mechanism to decide whether BUs should be sent individually or by means of Xcast. In the latter case, the *Xcast Header* is filled with the corresponding CNs as found in the *Binding List*. The *header length* and *bitmap* fields are filled appropriately. CNs may also be split into separate Xcast BUs.

The *Xcast Header* must be processed by a number of routers and the CNs, as illustrated on fig.2.b. A single BU is sent to *A, B and C.* On receiving a packet with an *Xcast Header*, a Xcast-enabled router checks if there is still destinations to which the packet remains undelivered. If all bits in the *bitmap* are unset, no duplication is required anymore. For each bit set, the Xcast-enabled router reads the corresponding address and interrogates its routing table for ascertaining the next hop towards this address. As the next hop to *A* diverges from the next hop toward *B* and *C*, the packet is duplicated as many times as there are distinct next hops towards the destinations. Useless destinations are also removed to decrease packet overhead, and the *bitmap* is set appropriately. A copy is transmitted on the interface toward *A* and one on the interface toward *B* and *C.* In circumstances where Xcast-enabled routers are not widely deployed, a CN may receive a BU with a number of CNs remaining in the header. In such a case, processing the *Xcast Header* at the CN ensures BUs' delivery to all CNs. On receiving such a BU, the CN first checks the *bitmap*. If there is more than one bit set, the BU is duplicated, the bit corresponding to the CN is removed, and the packet is forwarded.

4 Performance Evaluation

In this section, we compare the Xcast delivery of BU against the standard unicast delivery. Simulations were conducted using *NS-2* which has been extended on purpose [?]. A 1050-routers hierarchical topology was generated with *GT-ITM* This topology comprises 10 backbones and a total of 100 sites. Border routers (BRs in figures) are routers connecting sites to a backbone and transit routers

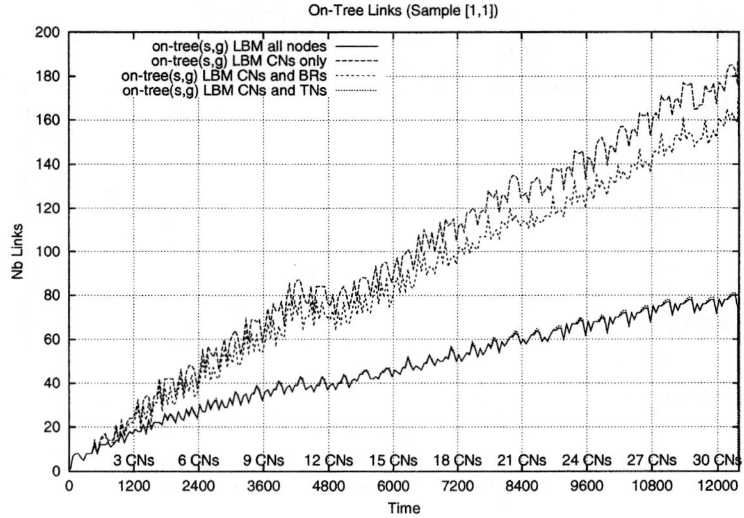

Fig. 3. On-Tree Links

(TNs in figures) are routers in a backbone. CNs are randomly attached to a site router. Given the size of sites, CNs are located 1 to 3 hops away from a border router.

The MN is displaced for 50 seconds in a specific site and is communicating with a number of x randomly selected CNs varying from 1 to 32. The MN doesn't move between subnetworks in a given site. In order to obtain uniform results, independently from the location of the MN and the selected CNs, the simulation is performed 8 times for each set of CNs, positioning the MN in a different randomly selected site. We show all the 8 positions on figures to emphasize that the location of the MN and CNs has no significant impact on the results. This explains why curves are always irregular between any two numbers of CNs. This indeed corresponds to a scenario where the MN is performing wide-area mobility (i.e. topologically distant displacements between sites or access networks), i.e. changing site every 50 seconds and increasing the number of CNs every 8×50 seconds.

The same seed numbers are always retained for selecting CNs and generating displacements of the MN. The performance of Xcast is evaluated under four situations, when all routers and CNs are Xcast-enabled; when only CNs are Xcast-enabled; when both CNs and BRs are Xcast-enabled; and when both CNs TNs are Xcast-enabled. The *remove flag* is always set.

More details about the NS-2 extensions and the simulation results can be found in [10]. In fig. 3 to 7, Xcast is referred to as *LBM* (*List Based Multicast*) whereas Mobile IPv6 is referred to as *unicast*.

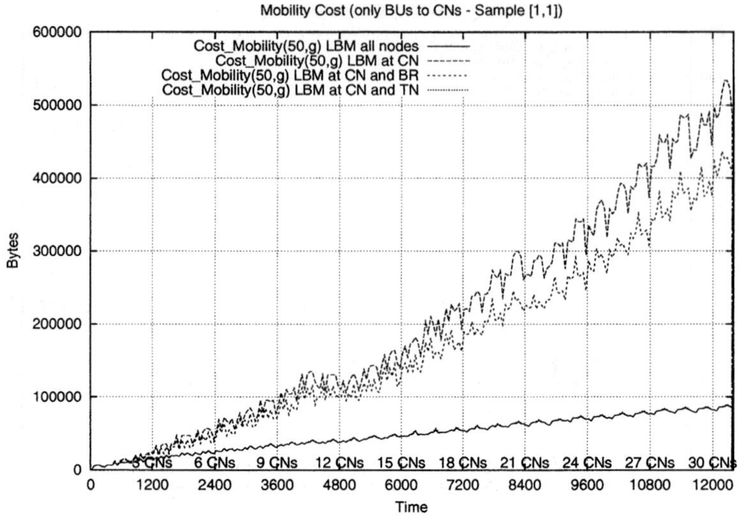

Fig. 4. Mobility Cost: BUs to CNs only

4.1 Which Routers Should Be XCAST-enabled

As shown on fig.3, the number of *on-tree links* (total number of links consumed by periodic BUs transmitted over the entire network from a position s of the MN to the group g of x CNs during interval of time t=50) when all routers are Xcast-enabled is equivalent to the situation where only transit routers and CNs are Xcast-enabled (lower two curves). Similarly, this number is nearly equal when only CNs are Xcast-enabled and when both border routers and CNs are Xcast-enabled (upper two curves). This is due to the uniform distribution of CNs in the topology (the probability that two CNs are in the same site is low). We also see that the slope of the upper curves is higher than the lower curves. The former two increase linearly whereas the latter two look like log(nb CNs).

The *mobility cost* (total number of bytes consumed by periodic BUs transmitted over the entire network from a position s of the MN to the group g of x CNs during interval of time t=50) as shown on fig.4, is not proportional to the number of *on-tree links* since the packet length increases with the number of CNs. Xcast performs badly when the feature is not well deployed in the network since BUs are bounced from one Xcast speaker to another.

We note that the slope of all curves is not constant and slows down between 9 and 12 CNs, as for the *on-tree links*, particularly the upper two curves. On the *on-tree links* curves, the slope is higher before 9 CNs, and lower after 12 CNs; on the *mobility cost* curves, the slope is lower before 9 CNs and higher after 12 CNs. This seems to indicate first an optimal value when a BU is most likely to be duplicated, and second an optimal value when the list of CNs is most likely to be split into two equivalent number of CNs. As the BU progresses in the network, the combination of these two events is most unlikely to happen.

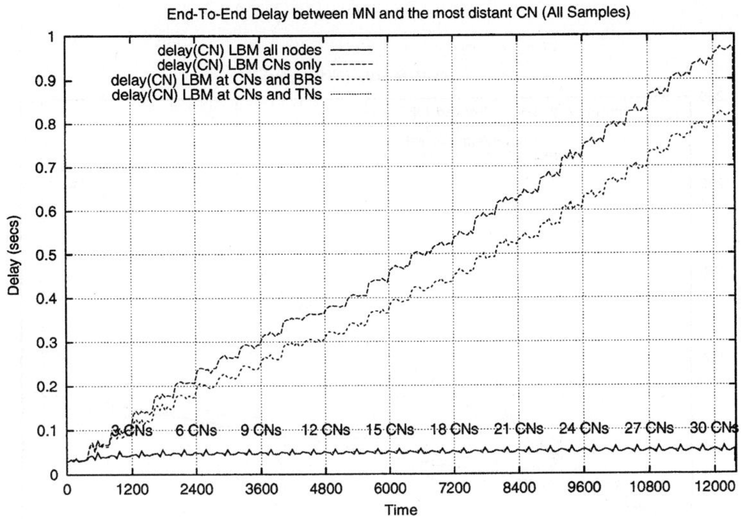

Fig. 5. End-To-End Delay

The *maximum end-to-end delay* between the MN and the most distant CN is shown on fig.5. This delay has an upper limit when all routers or at least TNs are Xcast-enabled (lower two curves). It obviously tends to increase rapidly to a very large *end-to-end delay* when only CNs are Xcast-enabled (upper curve). In this case, the packet traverses all CNs before reaching its ultimate destination.

4.2 XCAST Vs Unicast

When we compare the number of *on-tree links* for the Xcast delivery of BUs against standard Mobile IPv6 (fig.6), we see that Xcast-enabled at both transit routers and CNs (lower curve) is more appropriate than unicast from a number of CNs turning around 5 (upper curve). In terms of *mobility cost*, fig.7, shows that Xcast only enabled at CNs (upper curve) compares very poorly against standard Mobile IPv6 (second curve from top). However, the benefit of Xcast over unicast is obvious when there exists a minimum number of Xcast-enabled routers well located in the network, i.e. at CNs and TNs (third curve from top).

5 Conclusion

Routing optimization between correspondent nodes and mobile nodes is a necessary feature to reduce the network load and the transmission delays. However, routing optimization using Mobile IPv6 is made at the expense of periodic BUs that must be sent individually and periodically to every CNs. When these messages are sent to an increasing number of CNs, a periodic signaling burst is propagated in the network. Links close to the node that emits these BUs are most

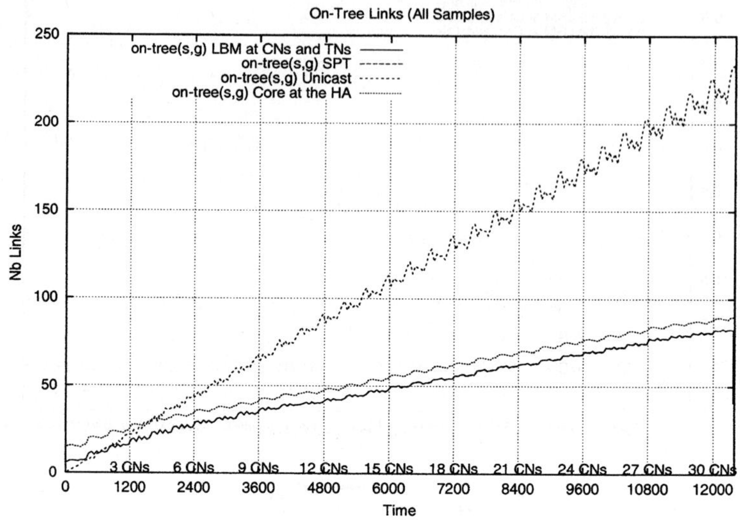

Fig. 6. XCAST vs Unicast: On-Tree Links

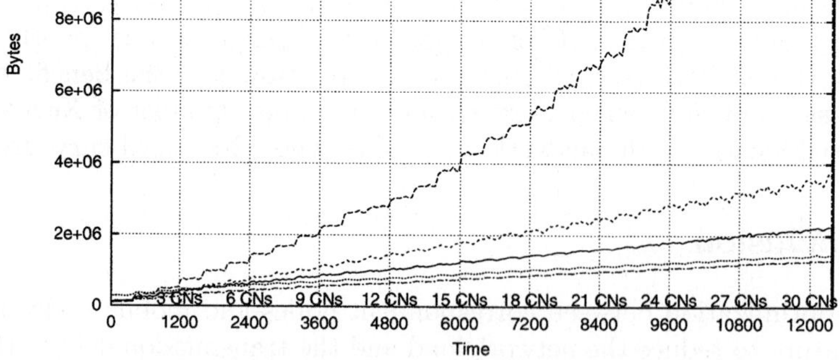

Fig. 7. XCAST vs Unicast: Mobility Cost

likely to suffer from this, particularly over the air where a significant amount of the available bandwidth is consumed. It is therefore proposed to use Xcast as a means to minimize Mobile IPv6 signaling. The comparison of the unicast delivery of BUs against Xcast shows that Xcast performs rather well provided that a number of routers are Xcast-enabled. On the other hand, Xcast is clearly inefficient when only CNs are able to process the header. We note that the *Xcast Header* requires more processing of the packet at intermediate routers, thus it is not recommended to process the header at each router. The tradeoff is certainly, as highlighted by the performance results, to process this header only at routers well located in the network, probably at transit points like in the backbone. In order to ensure the delivery of BUs to all CNs, all CNs listed in the *Xcast Header* must also be Xcast-enabled. Of course, Xcast is inevitably restricted to a limited number of CNs since the more CNs in the packet, the larger the packet length. To overcome this, CNs may be split into separate groups. As a side note, we observe that our proposal faces a number of security issues that need to be considered in future work. Also, a possible area of application could be network mobility (NEMO).

References

[1] Bhagwat, P., Perkins, C., Tripathi, S.: Network Layer Mobility: an Architecture and Survey. IEEE Personal Communications **3** (1996) 54–64

[2] Ioannidis, J., Duchamp, D., Maguire Jr., G.Q.: IP-based Protocols for Mobile Internetworking. In: Proc. ACM SIGCOMM, Department of Computer Science, Columbia University (1991) 233–45

[3] Johnson, D.B., Perkins, C., Arkko, J.: Mobility Support in IPv6. Request For Comments 3775, IETF (2004)

[4] Deering, S.: Multicast Routing in Datagram Internetwork. PhD thesis, Stanford University, US (1991)

[5] Deering, S., Cheriton, D.: Multicast Routing in Datagram Internetworks and Extended LANs. ACM Transactions on Computer Systems (1990) 85–111

[6] Boivie, R., Feldman, N., Metz, C.: Small Group Multicast: A New Solution for Multicasting on the Internet. IEEE Internet Computing **4** (2000) 75–79

[7] Braun, T.: Multicast for Small Conferences. Technical Report IAM-00-008, University of Berne, Switzerland (2000) http://www.iam.unibe.ch/ rvs/publications.

[8] Boivie, R., Feldman, N., Imai, Y., Livens, W., Ooms, D., Paridaens, O.: Explicit Multicast Xcast Basic Specification. Internet Draft draft-ooms-xcast-basic-spec-02.txt (2001) Work in progress.

[9] Imai, Y.: XCAST Related Researches, Web Page (As of October 2004) http://wiki.xcast.jp/cgi-bin/xcast-wiki.pl.

[10] Ernst, T.: Network Mobility Support in IPv6. PhD thesis, Université Joseph Fourier (2001) http://www.inria.fr/rrrt/tu-0714.html.

Downconversion of Multiple Bandpass Signals Based on Complex Bandpass Sampling for SDR Systems

Junghwa Bae and Jinwoo Park

Department of Electronics Engineering, Korea University
5-1, Anam-Dong, Sungbuk-Gu, Seoul 136-701, Korea
{iruntop, jwpark}@korea.ac.kr

Abstract. Bandpass sampling is a useful digital method for direct downconversion without analog mixer. Therefore, this sampling method is essential for the software-defined radio (SDR) system to be required minimization of RF front-end. Also, the bandpass sampling system can be extended to accommodation of multiple standards for the future wireless communication systems. This paper proposes a novel downconversion scheme for multiple standards based on a complex bandpass sampling method. It includes generalized formulae derivation for the available sampling range, the signal's intermediate frequency (IF) and the minimum sampling frequency for the multiple RF signals. And we verify from simulation results that the proposed system is more flexible and suitable than a real bandpass sampling system.

1 Introduction

The reconfigurable and flexible system, or the SDR system, has been beginning to get into the spotlight to next generation multimedia communication system for 4G. The SDR system is defined that a radio transceiver can be reconfigured via software, by replacing a analog circuit for a digital circuit such as DSP chip and FPGA. So, different air interfaces or standards can co-exist on a common hardware platform. Therefore, the SDR can dynamically change protocols and update communications systems over the air as a service provider allows [1]. In other words, the role of digital signal processing to replace the functionalities that have been implemented with analog component such as mixers and filters becomes very important. Accordingly, the requirement for such system is to minimize the components of the RF front-end, that is, to place the analog-to-digital converter (ADC) as near the antenna as possible as shown in Fig. 1.

To design the SDR system, we can employ a bandpass sampling technique. This is a method that a RF bandpass signal can be directly downconverted to a baseband or a low IF signal without analog mixers. That is to say, this sampling uses the intentional aliasing of the information bandwidth of the signal. Accordingly, the sampling frequency required is no longer based on the frequency of the RF carrier, but rather on the information bandwidth. Fig. 2 shows the spectrum of signal downconverted by only such sampling technique without mixers.

C. Kim (Ed.): ICOIN 2005, LNCS 3391, pp. 422–431, 2005.

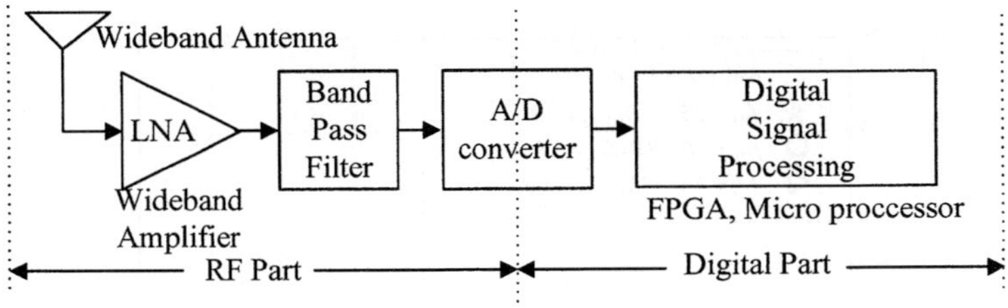

Fig. 1. The basic block diagram of the software-defined radio system

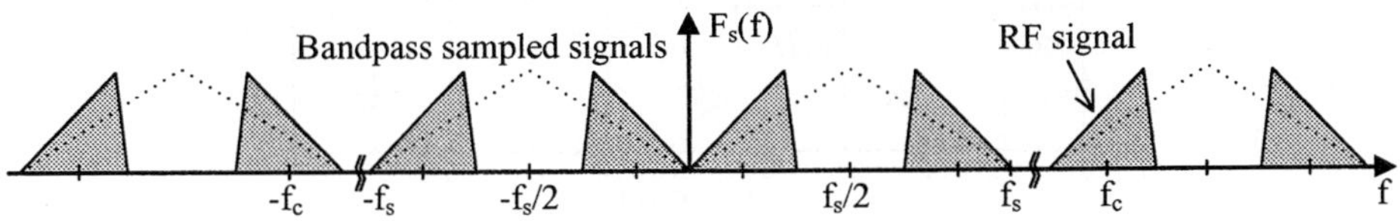

Fig. 2. The signal spectrum sampled by the bandpass sampling technique

Therefore, to minimize the RF front-end, this technique can be an alternative choice for such SDR system. Generally, a bandpass sampling, called real or first-order bandpass sampling, has been constrained due to aliasing by the negative frequency part of self-signal [2]. This can cause some ambiguity in the choice of the suitable sampling frequency. In contrast, downconversion by a complex bandpass sampling is not known to have the aliasing problem because the negative parts of the bandpass signal are rejected due to the characteristics of Hilbert transform [3]. Hence, this sampling technique can provide larger sampling range and normal spectral placement, not inverse spectral, in the lowest spectral band [4]. In the result, more flexible SDR system can be designed through employment of complex bandpass sampling.

In order to fully exploit the benefits of the SDR system, these sampling techniques can be used for simultaneous downconversion of multiple communication standards to be processed in a transceiver. In [5], the downconversion of multiple signals by using real bandpass sampling are introduced. However, in spite of the potential advantages of the complex bandpass sampling, solid investigation for direct downconversion using such sampling theory has not been reported in any literature so far. In this paper, thus a novel scheme which downconvert multiple signals based on the complex bandpass sampling technique is proposed, and several formulas are developed for an available sampling range, a minimum sampling frequency and IF of multiple signals. Moreover, through comparing with the real bandpass sampling, we verify the proposed system is more flexible and has lower sampling rate.

The structure of this paper is as follows. In Section 2, the downconversion system for multiple signals using real bandpass sampling is reviewed. Section 3 describes the novel proposed system using the complex bandpass sampling and

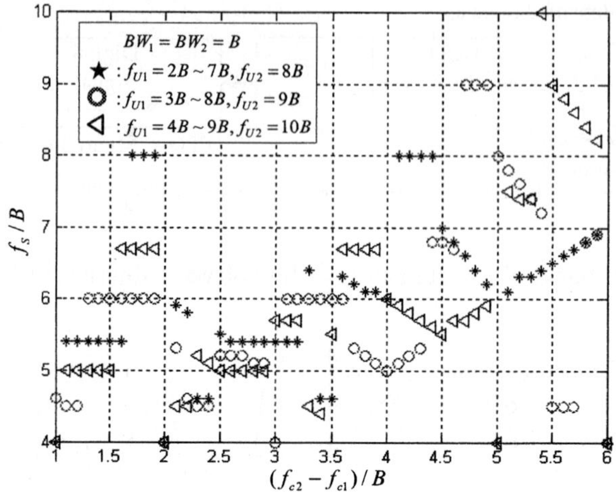

Fig. 3. The minimum sampling frequency required by changing the difference of center frequency of two signals when using the real bandpass sampling

the formulas related to the proposed system are derived. And through simulation results we demonstrate the usage of the proposed method in Section 4. Finally, Section 5 concludes the paper.

2 Downconversion of Multiple Signals Using Real Bandpass Sampling

One of the most important factors in a digital radio system is the choice of a suitable sampling frequency. A requirement of the sampling frequency should be translated all distinct RF signals with different wireless standard, into digital IF signals in the resultant sampled bandwidth without aliasing and mutual overlapping. N multiple bandpass signals $f_i(t)(i = 1, 2, ..., N)$ can be defined as follows. Let f_{ci} represent the center frequency, f_{Ui}, f_{Li}, f_{IFi} , upper bound, lower bound and IF in the sampled bandwidth, and BW_i is the bandwidth of the bandpass signal $f_i\,(t)$, respectively. And it assumes that $f_{ci} \leq f_{c(i+1)}$, $1 \leq i \leq N-1$.

When we simultaneously downconvert two signals $f_1(t)$ and$f_2(t)$ using the real bandpass sampling, the available sampling frequency must satisfy with the three following conditions [2].

$$(2f_{U1}/n_1) \ \leq f_s \leq \ \{2f_{L1}/\,(n_1-1)\,\} \tag{1}$$

$$(2f_{U2}/n_2) \ \leq f_s \leq \ \{2f_{L2}/\,(n_2-1)\,\} \tag{2}$$

$$|\ f_{IF1} - f_{IF2}| \geq \ (BW_{1,2}/2) \tag{3}$$

where $BW_{1,2} = BW_1 + BW_2$, $1 \leq n_1 \leq \lfloor f_{U1}/BW_1 \rfloor$, and $1 \leq n_2 \leq \lfloor f_{U2}/BW_2 \rfloor$, respectively. Here, $\lfloor \cdot \rfloor$defines the floor function.

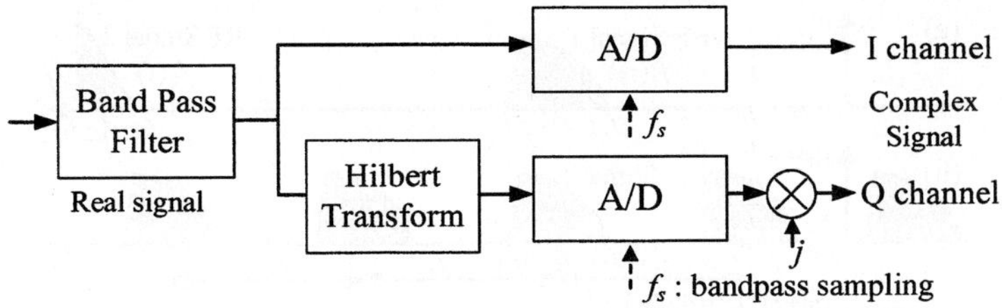

Fig. 4. The block diagram for the complex bandpass sampling

It is also noted that the minimum sampling frequency in the available sampling range is required owing to computational complexity of the system. Fig. 3 displays the result of an example that shows the smallest sampling frequency in the range satisfied with equation (1), (2) and (3). With the assumption that two signals have the same bandwidth of $1B$, and the position of $f_1(t)$ can be varied from $2B$ to $9B$ while $f_2(t)$ is fixed. It is found that the minimum sampling frequency is in irregular distribution. This results from that it should be avoided the aliasing occurred due to the negative frequency part of self-signal as well as overlap of the positive and the negative frequency parts of another signal. Therefore, it is difficult to derive a generalized formula to express the available sampling frequency.

3 Downconversion of Multiple Signals Using Complex Bandpass Sampling

The complex bandpass sampling technique is illustrated in Fig. 4. This sampling is simply to sample an analytic signal obtained by a Hilbert transformer. So, the signal in the negative frequency is rejected, and then the required sampling frequency is equal to the signal bandwidth.

3.1 Sampling Range for Two Signals

Firstly, let's consider two bandpass filtered signals of $N = 2$. In order to derive a formula as a function with respect to sampling frequency, we used a scheme as follows. The upper limit for the available sampling range can be determined by the condition that the lower bound f_{L2_m} of the mth left shift of the RF signal $f_2(t)$ is larger than the upper bound f_{U1} of the RF signal $f_1(t)$. Similarly, the lower limit is determined by that the upper bound f_{U2_m+1} of the $(m+1)^{th}$ left shift of $f_2(t)$ is smaller than the lower bound f_{L1} of $f_1(t)$. These expressions are illustrated in Fig. 5, where presents two complex bandpass sampled signals after Hilbert transform. These can be represented as

$$\{f_{c2} - (BW_2/2) - mf_s\} - \{f_{c1} + (BW_1/2)\} \geq 0 \quad (4)$$

$$\{f_{c2} + (BW_2/2) - (m+1)f_s\} - \{f_{c1} - (BW_1/2)\} \leq 0 \quad (5)$$

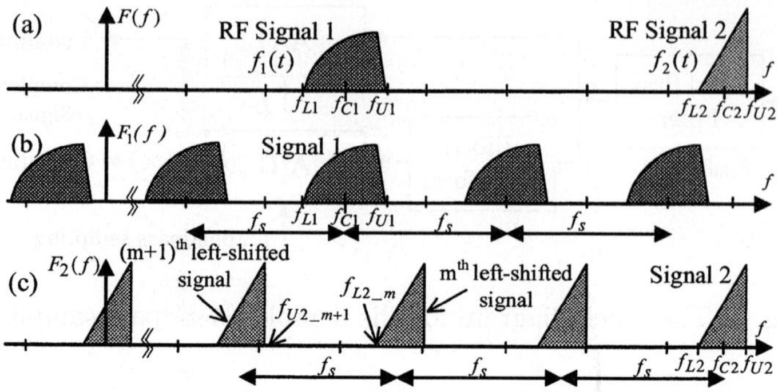

Fig. 5. The signal's spectrum by using complex bandpass sampling after Hilbert transform, (a) Two RF distinct signals, (b) The signal $f_1(t)$ sampled by the complex bandpass sampling , (c) The signal $f_2(t)$ sampled by the complex bandpass sampling

Combining these two equations, the resultant sampling range is rewritten as

$$\frac{(f_{c2}-f_{c1})+(BW_{1,2}/2)}{m+1} \leq f_s \leq \frac{(f_{c2}-f_{c1})-(BW_{1,2}/2)}{m} \tag{6}$$

where m is an integer given by

$$0 \leq m \leq \left\lfloor \frac{(f_{U2}-f_{U1})-BW_2}{BW_{1,2}} \right\rfloor = \left\lfloor \frac{f_{L2}-f_{U1}}{BW_{1,2}} \right\rfloor \tag{7}$$

We can recognize from this derived formula that the available sampling frequency is a function of both the sum of the bandwidth and the difference of the center frequency of two signals. Fig. 6 is the graph obtained by the proposed method. This figure displays the available sampling areas obtained by three kinds of the sum of two signals' bandwidth, such as $2B, 2.5B$ and $3B$. And the axes are normalized by the bandwidth B of any one of two signals. Note that the area which m denotes, not white area, presents the available sampling zone, while the white-colored area means the unavailable sampling region due to overlap of two signals each other. Here, the parameter m, which does not mean a frequency shift coefficient n in the real bandpass sampling [3], presents how many times both signals can be placed without mutual overlapping in the interval $f_{L2}-f_{U1}$. Thus, in the case that m is 0, it should be treated as one combined signal since an interval of two signals is very close. Hence, according to (6), the condition in this case becomes $f_s \geq (f_{U2}-f_{L1})$. Moreover, the maximum value of m is dependent on the distance between two signals. Hence, as the distance is longer, the m becomes larger but the signals are not sampled by lower sampling frequency. Although the m has small value, lower sampling frequency can be selected if the distance is short. In the case that two signals are fixed, however, the larger m the lower the sampling frequency we can obtain.

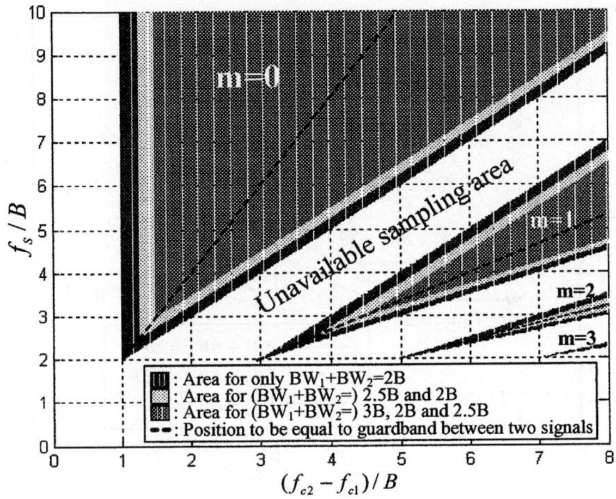

Fig. 6. The relationship between f_s and $f_{c2} - f_{c1}$ when using complex bandpass sampling

Next, we can consider an interval or guard band between two signals. In the case of complex bandpass sampling, the guard band becomes a maximum when the intervals between two signals are the same. Accordingly, the sampling frequencies to be equal to guard bands should be chosen to put the center frequency of $f_1(t)$ in the center of the mth and $(m+1)^{th}$ left shift of $f_2(t)$, as the following expression.

$$(f_{c2} - mf_s) - [f_{c2} - (m+1)\,f_s] = 2f_{c1} \tag{8}$$

This equation means

$$f_s = 2\,(f_{c2} - f_{c1})\;/\;(2m+1) \tag{9}$$

Hence, using these sampling frequencies, the adjacent channel interference between signals can be minimized owing to maximizing the distance between signals, and the channel selection filter can be also designed easily. These sampling frequencies are depicted as the dashed lines in Fig. 6.

The minimum sampling frequency is obtained by using combination of (6) and (7). The resultant equation is written as

$$f_{s\_\min} = \frac{(f_{c2} - f_{c1}) + (BW_{1,2}/2)}{\lfloor \{(f_{c2} - f_{c1}) - (BW_{1,2}/2)\}\;/\;BW_{1,2} \rfloor + 1} \tag{10}$$

Fig. 7 shows the plot depicted by above formula. Unlike real bandpass sampling, note that the lowest bound on this sampling frequency can be expressed as one formula with respect to the sampling frequency, and its value is the sum of the bandwidth of two signals at specific positions. In addition, though we compare results about two signals of Fig. 3 and 7, we can recognize that the system

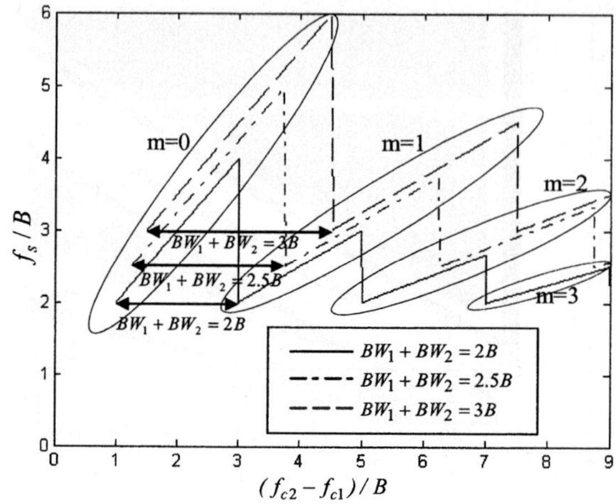

Fig. 7. The minimum sampling frequency required for two signals

using the complex bandpass sampling has lower sampling frequency and wider sampling range.

3.2 Sampling Range for Multiple Signals

Using the formula (6), we can extend to N multiple signals to get a generalized formula. Firstly, the available sampling range for two signals $f_i(t)$ and $f_j(t)$ is represented as

$$\frac{(f_{cj}-f_{ci})+(BW_{i,j}/2)}{m_{i,j}+1} \le f_{s\,ij} \le \frac{(f_{cj}-f_{ci})-(BW_{i,j}/2)}{m_{i,j}} \tag{11}$$

where $0 \le m_{i,j} \le \lfloor (f_{Lj}-f_{Ui})/BW_{i,j} \rfloor$ and $f_{ci} < f_{cj}$. Based on this constraint, each available sampling range for all possible combinations of two signals taken from N signals should be obtained, and then the common ranges out of these sampling ranges are found. Accordingly, the range $f_{s,all}$ must be satisfied with all constrains as follows.

$$f_{s,all} = f_{s\_1} \cap f_{s\_2} \cap f_{s\_3} \cap \cdots \cap f_{s\_N-1} \tag{12}$$

$$where \begin{cases} f_{s\_1} = f_{s\ 1,2} \cap f_{s\ 1,3} \cap \cdots \cap f_{s\ 1,N}, \\ f_{s\_2} = f_{s\ 2,3} \cap f_{s\,2,4} \cap \cdots \cap f_{s\ 2,N}, \\ \vdots \qquad \vdots \\ f_{s\_N-1} = f_{s\ (N-1),(N)}. \end{cases} \tag{13}$$

Consequently, above conditions are the same result that N signals in the complex sampled bandwidth $[-f_s/2, f_s/2]$ satisfy the following two equations.

$$|f_{IF\,b} - f_{IF\,a}| \ge (BW_{a,b})/2, \ a = 1, 2, \cdots, N-1, \ b = a+1 \tag{14}$$

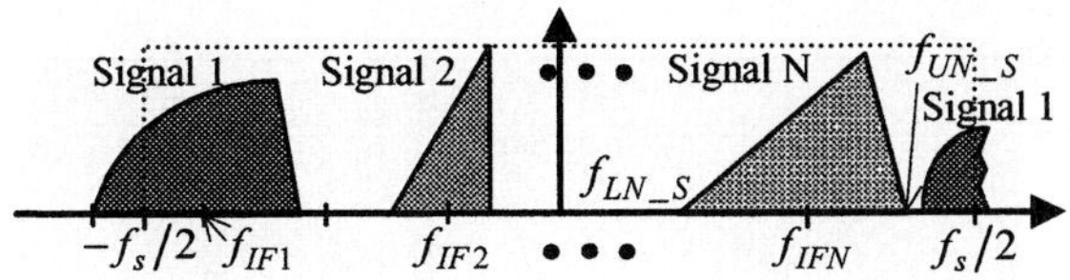

Fig. 8. N multiple signals downconverted by the complex bandpass method

$$|\,(rem(f_{c1}, f_s) - \ (rem(f_{cN}, f_s)\,| \ \geq (BW_{1,N})\,/2 \tag{15}$$

where $rem\ (f_{c1}, f_s)$ is the remainder after division of f_{c1} by f_s, f_{IFN} denotes IF of $f_N(t)$, and $f_{IF\,i} < f_{IF\,(i+1)}$. These signals are then placed as shown in Fig. 8.

3.3 The Signal's IF by Complex Bandpass Sampling

The position of the signal changed by sampling, namely the signal's IF of the complex bandpass sampled signal, is placed at the positive or negative frequency part. The position can be decided according to $\lfloor f_{ci}/\,(f_s/2)\rfloor$, and this value is required to be even number for placing positive frequency region and odd number for placing negative frequency region. Consequently, the IF in the complex sampled bandwidth is given by

$$\lfloor f_{ci}/\,(f_s/2)\rfloor \ is \ \begin{cases} even\ :\ f_{IF\,i} = rem\ (f_{ci}, (f_s/2)) \\ odd\ :\ f_{IF\,i} = -\,[(f_s/2) - rem\ \{f_{ci}, (f_s/2)\}] \end{cases} \tag{16}$$

Furthermore, if the value within $\lfloor\cdot\rfloor$ of (16), or $f_{ci}/\,(f_s/2)$, is even integer number, not fraction, the signal can be directly placed at baseband. So this system does not require an additional digital mixer for baseband downconversion. In the case of odd integer number, the signal is placed at $f_s/2$.

4 Simulation Results

In this section, we compare two bandpass sampling method when two wireless communication standards are downconverted. Let's consider two bandpass signals $f_1(t)$ with $f_{c1} = 1.9B$ and $BW_1 = B$, and $f_2(t)$ with $f_{c2} = 7B$ and $BW_2 = 2B$. It assume that these parameters are normalized to the bandwidth B of $f_1(t)$. The variation of the position of two signals by using the proposed complex bandpass sampling method is depicted in Fig. 9. The circle and triangle mark denote the center frequency of $f_1(t)$ and $f_2(t)$, respectively. And, the shaded areas show the available sampling regions. Here, the lowest point in the shaded regions denotes $3.3B(Hz)$ as the required minimum sampling frequency. The line '-' presents the bound of the sampled bandwidth denoting $-f_s/2(Hz)$ in the negative frequency and $f_s/2(Hz)$ in the positive frequency.

In the other hand, the result of the case to be applied to real bandpass sampling is presented in Fig. 10. Unlike the complex bandpass sampling, the

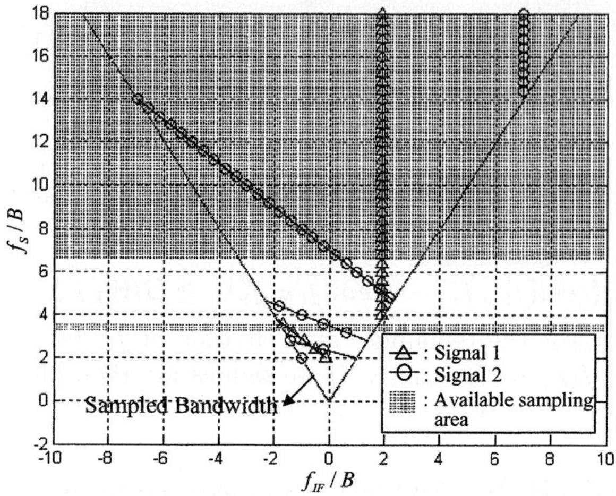

Fig. 9. The available sampling range and IF of two signals by using the complex bandpass sampling

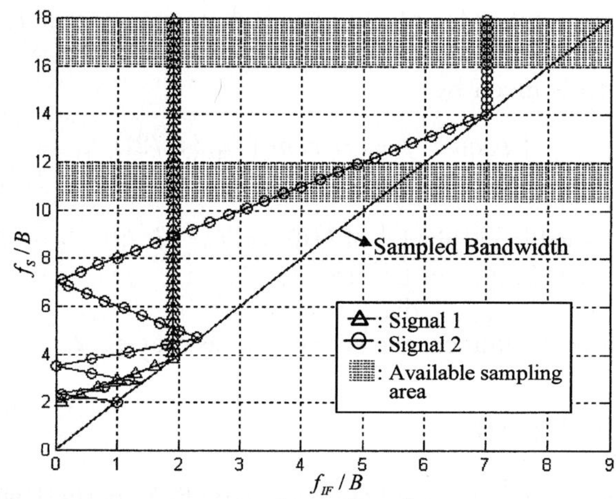

Fig. 10. The available sampling range and IF of two signals by using the real bandpass sampling

range of the sampled bandwidth is from $0(Hz)$ to $f_s/2(Hz)$. And according to the selected sampling frequency, the signal's spectrum can be inversed in the sampled bandwidth since the signal of the negative frequency part by real bandpass sampling cannot be removed. In Fig. 10, the sampled signal is presented as the form of backslashes '\' for inverse spectral placement and the form of slashes '/' for normal spectral placement. Also, The form of '|' denotes the signal sampled by Nyquist sampling rate. In this case, the lowest point in the shaded regions, namely, the minimum sampling frequency is $10.4B(Hz)$. This frequency is approximately three times as large as that of the complex method. Therefore, Two results certainly shows that the complex sampling method is more suitable for the system that can accommodate various communication environment.

5 Conclusions

When expanding digital signal processing of mobile terminals toward the antenna while making more wideband to be able to cope with different communication standards, the designer is faced with critical requirements to implement a SDR system, namely, flexibility and reconfigurability of RF components. By employing a bandpass sampling technique, however, this problem can be considerably solved since analog mixers is no longer required. Therefore, The system that many communication standards can be accommodated is obtained.

In this paper, applying such bandpass sampling, a novel downconversion method based on complex bandpass sampling technique for multiple RF signals has been presented. And, we have derived formulas for the ranges of the available sampling frequency, the minimum sampling rate and the changed position of signal. From the derived formulas, it is noted that the relation between the bandwidths and the differences of the center frequency of signals is an important factor. In addition, we verify from numerical analysis and simulation results that this proposed method has even lower sampling frequency than real bandpass sampling. This is critical in the design of a SDR system, as the sampling rate directly determines the computational requirements. Therefore, the SDR system using the complex bandpass sampling provides us many merits in various aspects.

Acknowledgement

This work was supported by University IT Research Center Project in Korea University.

References

1. Hentschel, T., Henker, M., Fettweis, G.: The Digital Front-End of Software Radio Terminals, IEEE Personal Communications, **6** (1999) 40–46
2. Vaughan, R.G., Scott, N.L., White, D.R.: The Theory of Bandpass Sampling, IEEE Transactions on Signal Processing, **39** (1991) 1973–1983
3. Valkama, M., Pirskanen, J., Renfors, M.: Signal Processing Challenges for Applying Software Radio Principles in Future Wireless Terminals: An Overview, International Journal of Communication Systems, **15** (2002) 741–769
4. Liu, J., Zhou, X., Peng, Y.: Spectrum Arrangement and Other Topics in First-Order Bandpass Sampling Theory, IEEE Transactions on Signal Processing, **49** (2001) 1260–1263
5. Akos, D.M., Stockmaster, M., Tsui, J.B.Y., Caschera, J.: Direct Bandpass Sampling of Multiple Distinct RF Signals, IEEE Transactions on Communications, **47** (1999) 983–988

An Enhanced Traffic Marker for DiffServ Networks

Li-Fong Lin[1], Ning-You Yan[1], Chung-Ju Chang[1], and Ray-Guang Cheng[2]

[1] Dept. of Communication Eng., National Chiao-Tung University,
Hsinchu 300, TAIWAN, ROC
{kawai.cm85g, mozi.cm90g}@nctu.edu.tw, cjchang@cc.nctu.edu.tw
[2] Dept. of Electronic Eng., National Taiwan University of Science and Technology,
43, Sec. 4, Keelong Rd., Taipei 106, TAIWAN, ROC
crg@et.ntust.edu.tw

Abstract. In this paper, we propose an enhanced traffic marker (ETM) to amend the inappropriate marking in DiffServ networks. The ETM is based on the Two-Rate-Three-Color-Marker (TRTCM) scheme and introduces the feature of aggressive promotion and fair share marking. Simulation results show that the ETM fairly allocates bandwidth among micro-flows and achieves a higher throughput than TRTCM does.

1 Introduction

There are increasing demands for the supporting of quality-of-service (QoS) over Internet. Internet Engineering Task Force (IETF) proposed two models, named the Integrated Services (IntServ) [1] and the Differentiated Services (DiffServ) [2], respectively, to support Internet QoS. The IntServ model reserves the network resource before using it. It ensures the end-to-end QoS for each application (i.e. micro-flow) but has the scalability problem [3]. The DiffServ model focuses on the QoS of aggregate micro-flows in order to reduce the complexity. The micro-flows that require a similar QoS level would be assigned to the same class. In a DiffServ network, an edge router is responsible for classifying the traffic into different micro-flows, conditioning the micro-flows in the same class, and processing the packet according to the QoS requirement. The conditioning and processing functions are handled by a model named a traffic conditioner. The traffic conditioner consists of a meter, a marker and a shaper (or a dropper) [3]. The traffic meter determines the conforming level according to the measured traffic flow and its QoS profile. The traffic marker assigns a notation to the incoming packet based on the determined conforming level and the additional information (e.g. the existing packet notation) to meet the QoS profile. Then, the traffic shaper (or dropper) would shape (or discard) the packet according to the packet notation and the network status.

Several marking schemes such as Single-Rate-Three-Color-Marker (SRTCM) [4], Two-Rate-Three-Color-Marker (TRTCM) [5] and Time-Sliding-Window-Three-Color-Marker (TSWTCM) [6] were proposed in RFC to implement the

C. Kim (Ed.): ICOIN 2005, LNCS 3391, pp. 432–442, 2005.

traffic conditioner. In TRTCM, the packet notation assigned by the traffic marker is defined as three colors, denoted as green, yellow, and red, which are corresponding to different conforming level of the packet. The green packet stands for the best conforming level and has the lowest dropping precedence (or the least shaping delay); the red packet represents the worst confirming level and has the highest dropping precedence (or the largest shaping delay). The TRTCM polices the arrived packets of aggregate traffic according to the metered conforming level and the existing packet notation, and ensure the individual output traffic rate of green packets as well as the aggregate output rate of green and yellow packets to conform to the traffic QoS profile. A natural demotion capability that re-marks a packet with higher conforming level color to be the one with the lower conforming level color is inevitable. However, a packet that is demoted due to congestion may not have the chance to restore its conforming level and the bandwidth fair share among micro-flows is uncertain after the packet demotion take places. Also, the traffic rate of green packets might be impaired due to excessive incoming yellow packets in TRTCM. A random early demotion and promotion (REDP) techniques [7] was proposed to overcome the unfair-marking problem. It allows packet promotion in addition to the demotion nature of the RED-In/Out (RIO) [8] marking mechanism and introduces fair marking during packet demotion and promotion by appropriately allocating the demotion/promotion probabilities among packets. In order to enhance the throughput of the aggregate traffic flow and provide the better fairness among micro-flows for TRTCM, a TC_PFG marking scheme [9] was proposed. However, in TC_PFG, only yellow-packet promotion is allowed and this limits its application. Moreover, TC_PFG has the problem of unjust-promotion that a demoted high-priority packet is not guaranteed to be promoted first when the excess network resource is available. In this paper, we propose an enhanced traffic marker (ETM) to deal with the problems of TC_PFG and improve the performance on traffic throughput and fairness.

2 Enhanced Traffic Marker

The ETM is based on TRTCM scheme. It adopts the concept of RED [7] and provides functions of *promotion*, *fairness-guarantee*, and *green-packet protection.* The promotion function remarks the low-conforming packets into high-conforming ones when there are excessive resources of the network, and this would improve the throughput of the aggregate flow. Based on the natural demotion capability and the proposed promotion function, the fairness-guarantee function further improves the fair share among the micro-flows by appropriately determining reasonable demotion/promotion probabilities for the green and yellow packets of the micro-flows. The green-packet protection function allows the token number in the bucket to be in deficit for incoming green packets to protect them from being affected by excessive incoming yellow packets.

TRTCM is composed of two token buckets denoted as T_P and T_C. The Peak Information Rate (PIR), the Peak Burst Size (PBS), the Committed Information

Rate (CIR), and the Committed Burst Size (CBS) are four parameters to be configured. The size and the token generation rate of T_P (T_C) are set to be PBS and PIR (CBS and CIR), respectively. Initially, both the token buckets T_P and T_C are set to be full. An incoming packet is marked as green if both T_P and T_C are not empty. A packet is marked as yellow if T_P is not empty and T_C is empty. If T_P is empty, the incoming packet is marked as red. After marking, the number of tokens consumed from T_P and T_C is depend on the size of the packet.

The functional block diagram of the proposed ETM is illustrated in Fig. 1. We adopt the same architecture and parameters used in TRTCM, but modify its marking algorithm. The ETM also consists of two token buckets, denoted as T_P and T_C, respectively, and a marking algorithm processor, the *fair traffic marker with aggressive promotion* (FTM_AP). The size and token generation rate of T_P (T_C) are also set to be PBS and PIR (CBS and CIR), respectively. The FTM_AP works with a record unit and a promotion/demotion probability generator. The *record unit* stores the flow-id, the existing color, and the arrival time for incoming packets every Δt. The number of green, yellow, and red packets of micro-flow j, denoted by $g(j)$, $y(j)$, and $r(j)$, respectively, are then recorded. The *promotion/demotion probability generator* uses the statistics to estimate the distribution of incoming packets and determines the promotion/demotion probability for each micro-flow based on the available tokens.

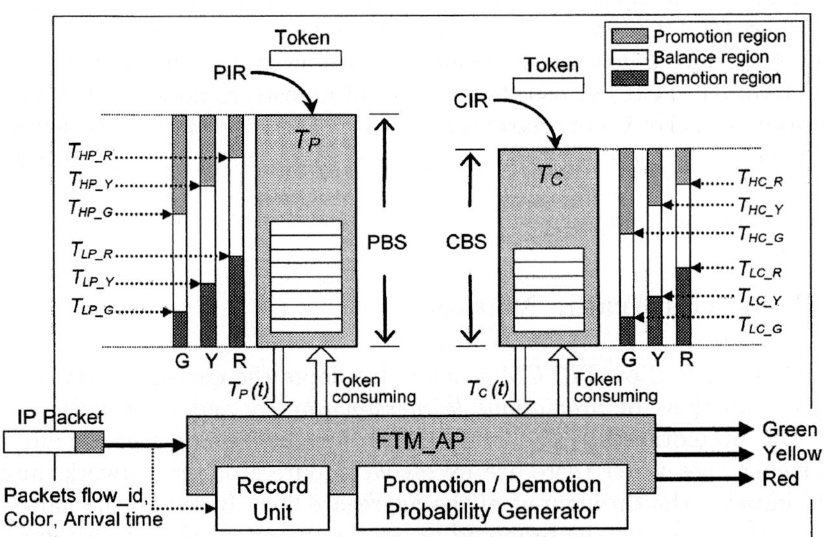

Fig. 1. ETM scheme

FTM_AP supports the promotion of yellow and red packets to enhance the throughput as well as achieve better fairness. FTM_AP further uses the original color to reduce the unjust promotion. For each incoming packet, currently unused

(CU) bits in the DS field [2] are used to store the information of its original color and current color. The original color is assigned at the source end and the current color could be remarked at any intermediate node. For simplicity, the G, Y, and R are used to denote the green, yellow, and red colors, respectively. Y_G, for example, represents that the current color is yellow and the original color is green.

Twelve thresholds are defined and are further categorized into four groups, denoted by T_{HC}, T_{LC}, T_{HP}, and T_{LP}, respectively. T_{HC} and T_{LC} are used in T_C and they divide T_C into the promotion, balance, and demotion regions. Similarly, T_{HP} and T_{LP} divide T_P into three regions. Each threshold group defines three sub-thresholds for packets with different original color. For example, the thresholds T_{HC_G}, T_{HC_Y}, and T_{HC_R} defined in the group T_{HC} are specified for the original color of green, yellow and red packet, respectively. In order to mitigate the influence of unjust promotion, the constraint $T_{X\_G} \leq T_{X\_Y} \leq T_{X\_R}$, where $X \in \{HP, LP, HC, LC\}$, should be met. The constraint is to assure that a packet with a higher original conforming level color at the source would be demoted with a lower probability.

Assume $T_C(t)$ and $T_P(t)$ denote the number of tokens in T_C and T_P observed at time t, respectively. In our design, FTM_AP demotes an incoming packet from green to yellow when $T_C(t) < T_{LC}$. The demotion probability P_d^G is given by

$$P_d^G = Max_d^G \times \frac{T_{LC\_X} - T_C(t)}{T_{LC\_X}}, \tag{1}$$

where Max_d^G is the maximum demotion ratio defined by the system and $X \in \{G, Y, R\}$ is corresponding to the original color of the incoming packet. It can be found that a large T_{LC} will result in a higher demotion probability. The demotion probability is increased as the decrease of available tokens. We further use P_d^G and the packet number statistics $g(i)$ to estimate the actual demotion probability applied on the incoming green packet. At first, we can obtain the amount of green packets that can pass through ETM without demotion, denoted by g_{pass}, by

$$g_{pass} = \sum_{i=1}^{n} g(i) \times (1 - P_d^G), \tag{2}$$

where $g(i)$ is the amount of green packets of micro-flow i; n is total number of micro-flows. According to the max-min fairness [10], we have to guarantee the sending rate of the "micro-flow that need less (MFNL)" traffic. The remaining bandwidth is then equally shared by the "micro-flow that need more (MFNM)" traffic. It means that we shall not demote any green packet for MFNL micro-flows and then share the remaining bandwidth of g_{pass} to MFNM micro-flows. Assume that there are n_{MFNM} MFNM micro-flows and the remaining bandwidth of g_{pass} is g_{MFNM}. Then we can recursively obtain the demotion probability of the green packet for micro-flow j until the following condition is fulfilled:

$$P_d^G(j) = \begin{cases} 0 & \textit{if } j \textit{ belongs to MFNL traffic,} \\ 1 - \frac{(g_{MFNM}/n_{MFNM})}{g(j)} & \textit{if } j \textit{ belongs to MFNM traffic.} \end{cases} \tag{3}$$

That is, a micro-flow with more green packets (i.e. larger $g(j)$) its green packets will have a higher probability to be demoted in ETM.

Similarly, a yellow packet is demoted to be red when $T_P(t) < T_{LP}$. The demotion probability P_d^Y is given by

$$P_d^Y = Max_d^Y \times \frac{T_{LP\_X} - T_p(t)}{T_{LP\_X}}, \tag{4}$$

where Max_d^Y is the maximum demotion ratio defined by the system and $X \in \{G, Y, R\}$ is corresponding to the original color of the incoming packet. The amount of yellow packets that can pass through ETM without demotion, denoted by y_{pass}, is then given by

$$y_{pass} = \sum_{i=1}^{n} y(i) \times (1 - P_d^Y). \tag{5}$$

And we can recursively obtain the demotion probability of the yellow packet for micro-flow j as

$$P_d^Y(j) = \begin{cases} 0 & \textit{if } j \textit{ belongs to MFNL traffic,} \\ 1 - \frac{(y_{MFNM}/n_{MFNM})}{y(j)} & \textit{if } j \textit{ belongs to MFNM traffic.} \end{cases} \tag{6}$$

We will promote the yellow and red packets when there is available bandwidth (i.e. sufficient token number in T_P and T_C). Based on the concept of max-min fairness, the micro-flow i that consumes the smallest bandwidth among the micro-flows will be promoted first. In ETM, a packet is promoted by FTM_AP from yellow to green when $T_C(t) > T_{HC}$. The promotion probability P_p^Y is given by

$$P_p^Y = Max_p^Y \times \frac{T_C(t) - T_{HC\_X}}{CBS - T_{HC\_X}}, \tag{7}$$

where Max_p^Y is the maximum promotion ratio defined by the system and $X \in \{G, Y, R\}$ is corresponding to the original color of the incoming packet. It can be found that the promotion probability is increased as the increase of the available tokens. The excess bandwidth results from the promotion of yellow packets, denoted as y_{prom}, can be obtained by

$$y_{prom} = \sum_{i=1}^{n} y(i) \times P_p^Y. \tag{8}$$

The excess bandwidth is then equally shared by the micro-flows that do not exceed their requested bandwidth. We can recursively obtain the promotion probability of the yellow packet for micro-flow j, denoted as $P_p^Y(j)$, until the following condition is fulfilled:

$$P_p^Y(j) = \begin{cases} 0 & \textit{if } j \textit{ exceeds its requested } BW, \\ \dfrac{\left(\left(y_{prom} + \sum_{i=1}^{k} g(i)\right) \Big/ k\right) - g(j)}{y(j)} & \textit{otherwise,} \end{cases} \tag{9}$$

where k is the total number of micro-flows that do not exceed their requested bandwidth.

Similarly, a red packet can be promoted to be yellow when $T_P(t) > T_{HP}$. The promotion probability P_p^R is given by

$$P_p^R = Max_p^R \times \frac{T_P(t) - T_{HP\_X}}{CBS - T_{HP\_X}}, \tag{10}$$

where Max_p^R is the maximum promotion ratio defined by the system and $X \in \{G, Y, R\}$ is corresponding to the original color of the incoming packet. The excess bandwidth results from the promotion of red packets, denoted as r_{prom}, is obtained by

$$r_{prom} = \sum_{i=1}^{n} r(i) \times P_p^R. \tag{11}$$

The promotion probability of red packets for micro-flow j, denoted as $P_p^R(j)$, is given by

$$P_p^R(j) = \begin{cases} 0 & \textit{if } j \textit{ exceeds its requested } BW, \\ \dfrac{\left(\left(r_{prom} + \sum_{i=1}^{k} (g(i)+y(i))\right) \Big/ k\right) - (g(j)+y(j))}{g(j)+y(j)} & \textit{otherwise.} \end{cases} \tag{12}$$

In this equation, k is the total number of micro-flows that do not exceed their requested bandwidth and $g(j) + y(j)$ is the bandwidth that has been used by the micro-flow j.

3 Simulation Results

In this section, two simulation scenarios were presented to verify the marking accuracy and fairness of the ETM. The results were then compared with the TRTCM. The network configuration for simulation is demonstrated in Fig. 2. N micro-flows belonging to the same service class originate from the sources and traverse across three DiffServ domains to reach their destinations (i.e. the "sink" node). The link capacity and delay parameter for each link are directly noted in the figure, and the round trip time (RTT) of a connection is assumed to be 36ms.

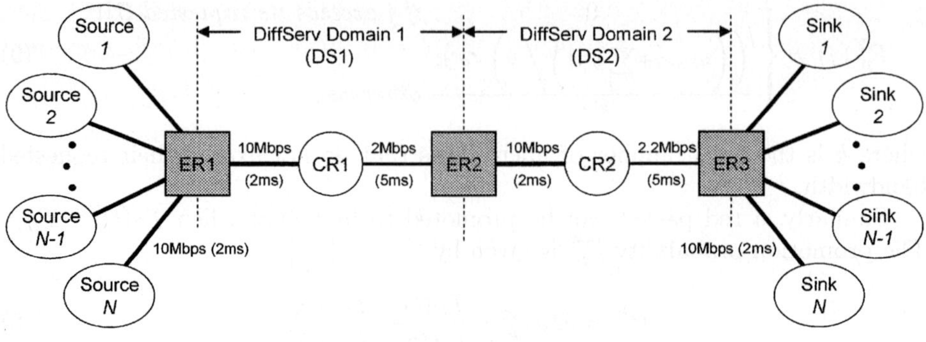

Fig. 2. Simulation topology

3.1 Accuracy of the Marking

The first simulation scenario we take is to verify the marking accuracy of traffic markers. In this scenario, only single traffic source is necessary (i.e. $N = 1$ in Fig. 2.) but diverse traffic parameter conditions of the source would be considered to explore the marker's performance on accuracy. In this paper, a Pareto traffic source with ten different traffic rate combination conditions is employed. The QoS profile specified at the ER1 is *CIR* equals to 5Mbps and *PIR* equals to 10Mbps and no profiles are specified ar ER2 and ER3. That is, the maximum ideal green and yellow rates observed at the output of ER1 are 5Mbps and 5Mbps, respectively, and the ER2 and ER3 are transparent for the traffic. The other system parameters and the simulation results are listed in Table 1 and Table 2, respectively, and the results are observed and measured at the output of ER1.

Table 1. System parameters of scenario 1

	Parameter	Value
TRTCM	CBS	60 packets
	PBS	60 packets
ETM	T_{L_P}	17 packets
	Δt	0.432 ms
	CBS	60 packets
	PBS	60 packets
	$(Max_d^G, Max_d^Y, Max_P^Y, Max_P^R)$	(1, 1, 1, 1)
	$(T_{LC\_G}, T_{LC\_Y}, T_{LC\_R})$	(10, 17, 24) packets
	$(T_{LP\_G}, T_{LP\_Y}, T_{LP\_R})$	(10, 17, 24) packets
	$(T_{HC\_G}, T_{HC\_Y}, T_{HC\_R})$	(36, 43, 50) packets
	$(T_{HP\_G}, T_{HP\_Y}, T_{HP\_R})$	(36, 43, 50) packets

In Table 2, it can be found that the ETM and TRTCM have similar results in traffic conditions 1, 3, and 4. Traffic condition 2 demonstrates the case that a greedy source generates an excess amount of yellow packets than its QoS profile. In this case, the TRTCM does not take action for the excess yellow packets. Therefore, the yellow packets consume most of the tokens in T_P and result in the starvation of green packets. With ETM, it can be found that both the green and yellow packets are conformed to the QoS profile. Traffic conditions 5, 8, 9, and 10 simulate the congestion at source nodes such that the input green rate is smaller than the QoS profile. It's found that the ETM marks more green packets via aggressive promotion to meet the profile. In summary, the proposed ETM meets the traffic profile and achieves a highest throughput than TRTCM does.

Table 2. Simulation results of scenario 1

Scenario 1 Traffic Conditions		Input Rate (Mbps)	QoS Profile (Mbps)	Output rate of TRTCM (Mbps)	Output rate of FTM (Mbps)
	Green	5.0	5.0	4.9145	4.9461
1	Yellow	5.0	5.0	4.8082	4.7701
	Red	2.0	—	2.2773	2.2838
	Green	5.0	5.0	4.5504	5.0033
2	Yellow	8.0	5.0	5.1563	4.8674
	Red	2.0	—	5.2934	5.1293
	Green	8.0	5.0	4.9371	4.9941
3	Yellow	5.0	5.0	4.8152	4.8559
	Red	2.0	—	5.2477	5.1500
	Green	10.0	5.0	4.9605	5.0010
4	Yellow	12.0	5.0	4.9035	4.9313
	Red	2.0	—	14.1359	14.0678
	Green	4.0	5.0	3.7645	4.8717
5	Yellow	8.0	5.0	6.0348	4.9541
	Red	2.0	—	4.2008	4.1742
	Green	3.0	5.0	2.9781	4.7209
6	Yellow	6.0	5.0	5.8557	4.2170
	Red	2.0	—	2.1662	2.0621
	Green	6.0	5.0	4.9578	4.9787
7	Yellow	3.0	5.0	3.8930	4.0682
	Red	2.0	—	2.1492	1.9531
	Green	3.0	5.0	3.0488	4.6262
8	Yellow	2.0	5.0	2.0203	3.2920
	Red	6.0	—	5.9309	3.0818
	Green	3.0	5.0	3.0447	4.6838
9	Yellow	4.0	5.0	4.0328	3.1402
	Red	2.0	—	1.9195	1.1760
	Green	2.0	5.0	1.9969	4.1533
10	Yellow	2.0	5.0	1.9582	1.3479
	Red	2.0	—	2.0449	0.4988

3.2 The Fairness of the Marking

The second simulation scenario is to verify the fair share making capability of traffic markers. Therefore, several micro-flows with different traffic characteristics and parameters are employed. The QoS profile in DiffServ domain 1 (DS1) is then configured as the bottleneck for the incoming green traffic. Thus, some of the green packets would be demoted or discarded. Besides, the Multiple-RED (MRED) [11] scheme is adopted in the core router (CR1 and CR2) to handle the congestion for both TRTCM and ETM. In MRED, packets marked as the lowest conforming level will be dropped first if congestion happens.

In the simulation, 45000 packets (about 20 seconds) were simulated and the size of all packets is 512 bytes. Four UDP and two TCP micro-flows are assumed. The UDP sources are implemented as Constant Bit Rate (CBR) traffic. The TCP sources are adaptive traffic with varied sending rates. The round trip time (RTT) of a TCP connection is assumed to be 36ms. The traffic parameters, the output (green, yellow, red) traffic rate in Mbps, for each source are as follows: UDP1=(1.9, 0.3, 0.3), UDP2=(1.0, 0.9, 0.9), UDP3=(0.7, 0.35, 0.35), UDP4=(0.3, 0.35, 0.35), TCP1=(1.0, 0.5, 0.5) and TCP2=(1.0, 0.5, 0.5). The QoS profile in each edge router is the same and is set as $CIR = 2.0$ Mbps and $PIR = 2.5$ Mbps. The other system parameters used in the simulation are the same with scenario 1 in Table 1.

In order to evaluate fairness among micro-flows, we adopt the *fairness_index* defined by [11]

$$x_i = \frac{achieved\_rate_i}{ideal\_rate_i}, \tag{13}$$

$$fairness\_index = \frac{\left(\sum_i x_i\right)^2}{n \times \sum_i x_i^2}, \tag{14}$$

where $achieved\_rate_i$ and $ideal\_rate_i$ are the practical average throughput and the ideal throughput for micro-flow i, respectively; n is the number of active micro-flows. The *fairness_index* falls into the range between 0 and 1. For the perfect fairness, the *fairness_index* should be equal to 1.

The average throughput for each traffic source obtained during the simulation is shown in Fig. 3. In Fig. 3, the *fairness_index* of TRTCM and ETM are 55.51% and 97.13%, respectively. It's because that TRTCM does not protect the TCP traffic and, thus, a large number of packets are demoted in ER1 and dropped in the core routers. This leads to the re-transmission mechanism of TCP and results in a low throughput for TCP users; therefore, the *fairness_index* is decreased.

4 Concluding Remarks

In this paper, we proposed an enhanced traffic marker (ETM) for DiffServ networks. The primary feature of the proposed ETM is that it can allocate bandwidth for both green and yellow packets according to the QoS profile. It also

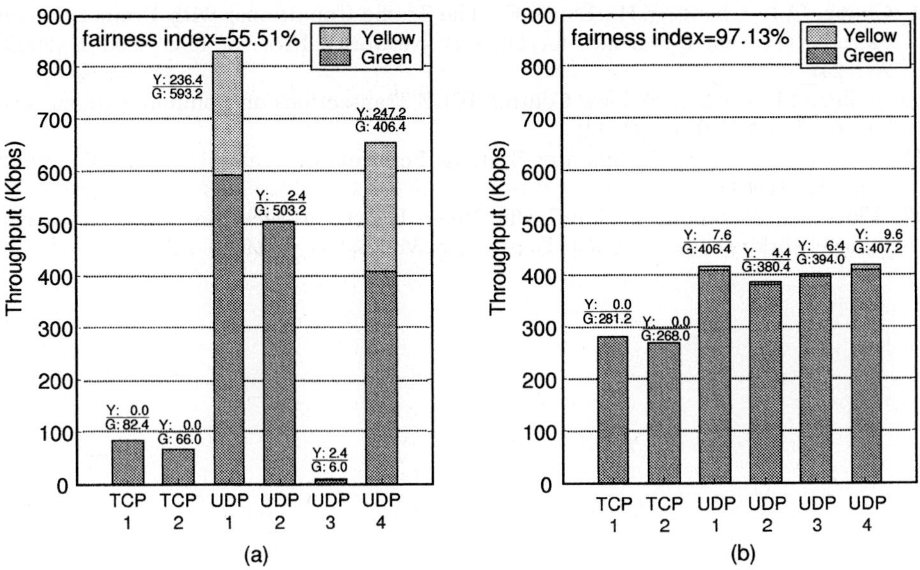

Fig. 3. Throughput distribution of different traffic marker schemes in simulation scenario 2: (a) TRTCM scheme, (b) ETM scheme

promotes yellow and red packets to enhance the throughput when there is available bandwidth. In the paper, the operation of ETM as well as the promotion/demotion probabilities is defined. The performance of the proposed ETM was verified via simulation and the results were compared with TRTCM. Simulation results show the ETM outperform the TRTCM in both congested networks and under-loaded networks.

References

1. Braden, R., Clark, D., Shenker, S.: Integrated Services in the Internet Architecture: An Overview. RFC 1633 (1994)
2. Blake, S., Black, D., Carlson, M., Davies, E., Wang, Z., Weiss, W.: An Architecture of Differentiated Services. RFC 2475 (1998)
3. Xiao, X., Ni, L.: Inernet QoS: The Big Picture. IEEE Magazine on Network (1999) 8–18
4. Heinanen, J., Guerin, R.: A Single Rate Three Color Marker. RFC 2697 (1999)
5. Heinanen, J., Guerin, R.: A Two Rate Three Color Marker. RFC 2698 (1999)
6. Fang, W., Seddigh, N., Nandy, B.: A Time Sliding Window Three Color Marker. RFC 2859 (2000)
7. Wang, F., Mohapatra, P.: A Random Early Demotion and Promotion Marker for Assure Service. IEEE Journal of Selected Areas in Communications, Vol. 18, No. 12 (2000) 2640–2650
8. Clark, D.D., Wenjia Fang: Explicit Allocation of Best-effort Packet Delivery Service. IEEE/ACM Transactions on Networking, Vol. 6, No. 4 (1998) 362–373

9. Chang, C.J., Cheng, Y.H., Lin, L.F.: The Traffic Conditioner with Promotion and Fairness Guarantee Schemes for DiffServ Networks. Proc. of ICC 2003 **1** (2003) 238–242
10. Jaffe, J.M.: Bottleneck Flow Control. IEEE Transactions on Communications, Vol. 29, No. 7 (1981) 954–962 [11]
11. Jain, R.: The Art of Computer Systems Performance Analysis. John Wiley and Sons Inc. (1991)
12. Floyd, S., Jacobson, V.: On Traffic Phase Effects in Packet-switched Gateways. Internetworking: Research and Experience, Vol. 3, No. 3 (1992) 115–156

Adaptive Bandwidth Control Using Fuzzy Inference in Policy-Based Network Management*

Hyung-Jin Lim[1], Ki-jeong Chun[2], and Tai-Myoung Chung[1]

[1] Internet Management Technology Laboratory & Cemi: Center for Emergency Medical Informatic,
Scool of Information and Communication Engineering,
SungKyunKwan University, Korea
{hjlim, tmchung}@imtl.skku.ac.kr
[2] Dept. of Media, Sangmyung University, Korea
chunkj@smu.ac.kr

Abstract. This paper presents the fuzzy logic-based control structure for incoming traffic from an arbitrary node to provide admission control in a policy-based IP network management structure. The proposed control structure uses a scheme for deciding the network resource allocation depending on the requirement of predefined-policies and network states. The proposed scheme enhances policy adapting methods of existing binary methods, and can use resource of network more effectively to provide adaptive admission control, according to the unpredictable network states for predefined QoS policies. Simulation results show that the proposed controller improves the ratio of packet rejection up to 17%, because it performs the soft adaptation based on the network states instead of accept/reject actions in a conventional CAC(Connection Admission Controller).

1 Introduction

TCP/IP provides essential network connection infrastructure in the world of internet where IP network has a property that is non-deterministic and does best efforts for transferring traffics. Under such Internet architecture, however, this does not guarantee QoS for delay-sensitive multimedia applications. Differentiated services [1] proposed a class based service for each packet, as a mechanism to ensure QoS in the IP network. Also, InterServ [2] proposed a resource reservation mechanism through the signaling protocol such as RSVP (Resource Reservation Protocol) to offer the flow based on guaranteed QoS. Generally these mechanisms require a frequent re-configuration to control a classification of the traffic at the network edge. The administrator has a difficulty to configure some static policies about unforeseeable network traffics. The policy-based network management (PBNM) was proposed as a suitable approach for these administration

* This study was supported by a grant of the Korea Health 21 R&D Project, Ministry of Health & Welfare, Republic of Korea(02-PJ3-PG6-EV08-0001)

C. Kim (Ed.): ICOIN 2005, LNCS 3391, pp. 443–451, 2005.

problems. That is, it has required the control mechanism based on predefined policies to guarantee QoS. However, the QoS management requires to monitor the predefined policies being properly achieved in the network.

We designed the adaptive admission control module using a fuzzy inference that accommodates PDP (Policy Decision Point) roles in PBNM. Examples of applying fuzzy theories [7][10] can be found in a engineering field on fuzzy control and on computer communication[4][5][6][9]. Therefore, we also proposed the admission controller that reflects the present resource status through a network monitoring with SNMP (Simple Network Management Protocol)[14], and decides availability of a predefined traffic policy. Therefore, Section 2 explains the control architecture for the fuzzy controller. Section 3 covers a simulation to verify the efficiency of the fuzzy control for the adaptive QoS policy. Section 4 describes analysis of the control algorithms. In the final section, the conclusion of this study is followed.

2 Fuzzy Control Architecture

In the IP network which uses IntServ and DiffServ [3], the edge network nodes must perform admission control through reception of a packet or a signaling protocol, when the control module decides processing availability of a packet or a flow. The control requires negotiation with predefined policy according to resources monitoring of the present node. The controller proposed in this study has a control ability through fuzzy inferencing, and it can resolve the policy conflicts between a network status and predefined policies for arrival traffic in DiffServ, or for a resource request in IntServ. We propose IAC (Intelligent Admission Controller) that offer an adaptation capability according to the network condition. Fig. 1 shows IAC component that is consists of FBE, NCE, NRE, FCAC and PAC. As space is limited, we haven't shown detail modules and fuzzy membership functions for each IAC components, but it is shortly introduced as follows.

The Fuzzy Bandwidth Estimator (FBE) is the module that estimates the required bandwidth that is calculated from parameters, such as packet delay, packet duration and packet rate limit. The parameters express as Rd, Rdu, Rr. The FBE define Ce as a service level of the traffic. The Ce is an output linguistic variable depending on expert knowledge [11] [12] [13] to compose the FBE.

The Network Congestion Estimator (NCE) generates a congestion indicator Ci according to the measured system statistics that use SNMP, such as the queue length, the change rate of the queue length, and packet loss ratio. The congestion indicator is based on buffer-threshold approach [8].

The Network Resource Estimator (NRE) performs the role to calculate traffic throughput which is processing at present node through monitoring using SNMP. The NRE calculates an acceptance possibility by present processing capacity for incoming traffic, and defines the acceptance possibility as Ca.

The Fuzzy Connection Admission Controller (FCAC) calculates acceptance level of a policy to crisp value through input value from the NRE and the NCE. That is, FCAC takes acceptance level through packet loss ratio (pl) as Feedback

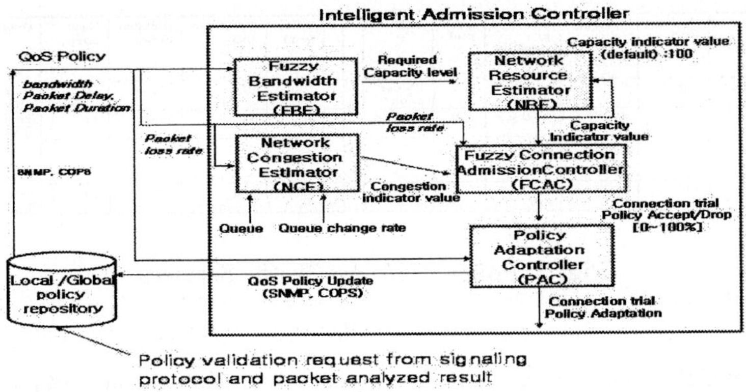

Fig. 1. Intelligent Admission Controller

performance [9] value and Capacity indicator value (Ca). Then, it defines the acceptance level as z. An inferencing process for the FBE and the FCAC can be obtained through Matching, Inferencing, Combination, Defuzzification. Combination defined as μ_{wi}(Ci) = min(μ_s(Rp), μ_L(y), μ_{pl}(Rm), μ_h(Tp)) and μ_{WA}(z) = min(μ_{NE}(Ca), μ_N(Ci), μ_s(pl), respectively. They used CoA (Center of Area) approach for Defuzzification. The area means a membership function size that is the Combination result.

$$C_i = \sum_{i=0}^{n} Area(C_i) \times CoA(C_i) / \sum_{i=0}^{n} Area(C_i) \quad (1)$$

The Policy Adaptation Controller (PAC) evaluates a proper request level according to the current network state. That is, the PAC decides the resources assignment policy for arrival traffic, and operates as PDP in PBN system. Therefore, the PAC must update the policies through SNMP or COPS (Common Open Policy Service) [3] to the local policy repository. The proposed fuzzy controller infers a adaptation availability considering network state about predefined policy, and can have soft control architecture that is adapted by suitable policy level when network situation is inadequate in the required policy.

3 Simulation Experiments

For the control performance provided by the fuzzy controller, the simulation results are observed during the gradual increase in the arrival traffic amount to network edge node, which functions as admission control. For this purpose, the policy for the service level required at individual traffic predefined by the administrator is set as the input value for the node formed in virtual environment. The output value as the result of control adapted by current packet processing

Source IP Address	Source Port	Protocol	Dest IP Address	Dest port	CoS			Delay (ms)	Percentile (%)	Duration (time)	Rate limit (Mbps)	Overflow
12.0.0.3	256	6	*	265	Precedence			100	91	1hour	10	Best Effort
*	*	6	12.0.0.3	300	Preferred			300	99	1day	12	Best Effort
⋮	⋮	⋮	⋮	⋮	⋮			⋮	⋮	⋮	⋮	⋮
12.0.0.3	11	6	11.0.0.0/24	17	Default			500	100	1month	50	Drop

Fig. 2. Policy Table for Traffic Services

capacity, which takes into account the current congestion status and packet drop rate. The algorithm of the fuzzy controller is implemented using C language. The maximum processing capacity of the control node is set at 100 Mbps, with one input and output port.

We assumed that the control node does not reject the resource request from arrival traffic when the congestion degree in the node is high; can adaptively allocate the resource according to the node situation even if it cannot allocate the all request resource at an initial period, since it allocates the remaining resources requested when the network status is improved. Traffic policy as an input variable is predefined by the administrator, and the node status of network can be obtained through monitoring by SNMP. Fig. 2 shows the policy table for traffic service request given as an input variable to be used in simulation.

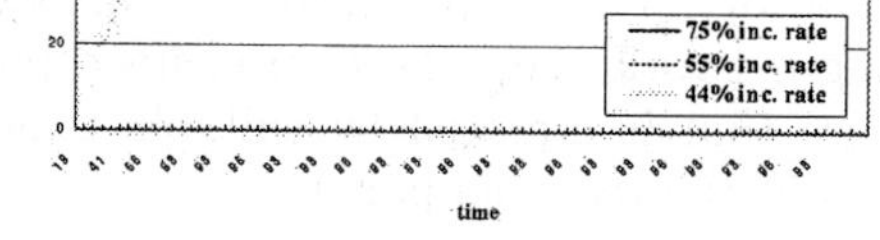

Fig. 3. Traffic Adaptation Degree in Non-Fuzzy Environment

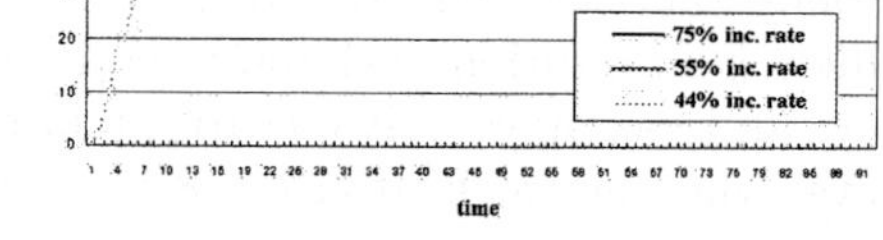

Fig. 4. Traffic Adaptation Degree in Fuzzy Environment

Simulation results show that a request service level is adapted according to node status as well as rate limit on fuzzy controller. However, packet drop rate for the service requirements is high on the non-fuzzy controller. The results are shown when there is an arrival traffic rate approximating to threshold for capacity that a node can process. Accordingly, an experiment scenario is prepared by separating into general traffic and burst traffic.

3.1 Adaptation Ratio of General Traffic Environment

To observe the adaptation degree of fuzzy admission controller according to a change in arrival traffic rate as input variables, the adaptation degree is measured while Rr value is changed. Other variables on policy table are given as fixed values. Fig. 3 and 4 shows the traffic adaptation degree for a service level request when the node has the throughput close to processing capacity that the node can accept. In the non-fuzzy environment, traffic acceptance ratio is found to drop for the traffic near threshold (i.e., 100 Mbps) on the control node. There is a difference in time of arriving at threshold, depending on the arrival traffic increase ratio. However, the same pattern is shown for the traffic acceptance at the throughput status approximating to the threshold. Therefore, the higher the increase ratio of arrival traffic the higher the admission rejection ratio.

There is a difference in the adaptation period for a service level of traffic required at node, depending on the increase ratio of arrival traffic in fuzzy environment. When the increase ratio of arrival traffic is high, it reaches the threshold of node throughput faster. Therefore, the fuzzy adaptation period grows quicker. The interval of adaptation points and thresholds shown in Fig. 4 is found to be affected by the rule of admission controller reflected in fuzzy controller. The adaptation degree formed around 80Mbps is because the rejection or acceptance is determined according to the request level by control algorithm, when the control node has an idle bandwidth about 20 Mbps. Accordingly, it can be adjusted to optimize the control result through the control algorithm, and the algorithm consists of the membership function of fuzzy controller for the required policies.

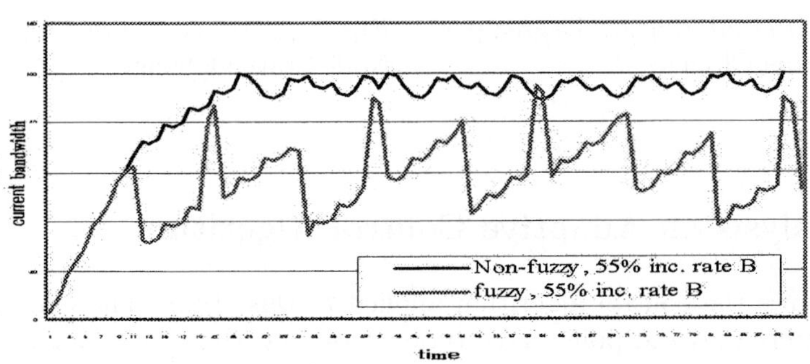

Fig. 5. Burst Adaptation Degree per Control Structure

3.2 Adaptation Ratio of Burst Traffic Environment

Fig. 5 shows the adaptation degree according to control structure for burst traffic in fuzzy/non-fuzzy environment. Fig. 6 and 7 show the adaptation degree that appears during the adjustment of values of the input variables affecting the node status when the service for burst traffic is requested. When there is an increase in congestion degree and packet drop ratio affecting node status in fuzzy controller, the adaptation degree for burst traffic is found to be higher than in non-fuzzy environment. The controller selects and operates the higher adaptation degree when a node is in the congestion status, since an acceptance of continuous traffic resource requests at this status may cause a severe congestion.

No special adaptation degree has been found when other variable values are changed during the simulation. This is because the adaptation degree is not determined according to the only output values (i.e., for the requested service level) from FBE module, but is determined together with the current network status. That is, lots of adaptation degrees are found when the congestion degree or the loss ratio of network is changed. It shows the result that a higher adaptation degree is found in a congestion status.

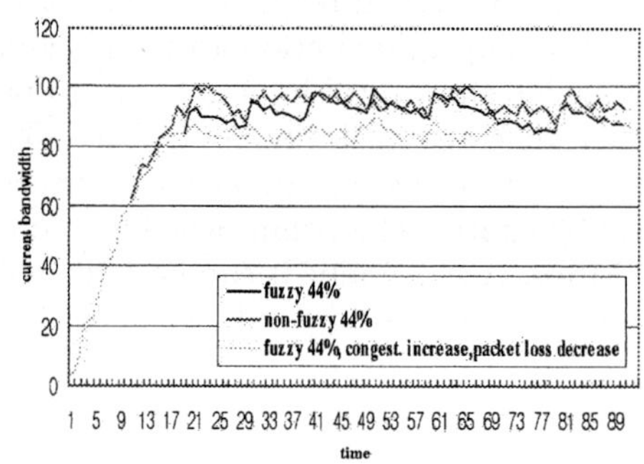

Fig. 6. Burst adaptation Degree per Node Status Change

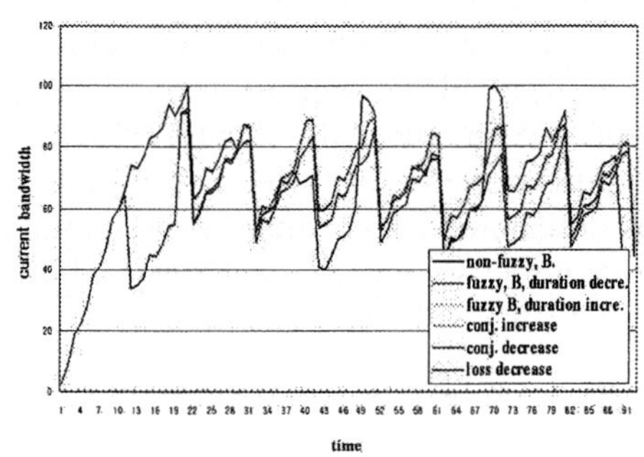

Fig. 7. Burst adaptation Degree per Node Status Change

4 Analysis on Adaptive Control Algorithm

Fig. 8 shows the control algorithm applied to this study. The current request policy determines the policy adaptation based on the result of monitoring on the inference module. It determines whether to execute the policy directly, or whether to execute an adaptive policy through such results.

The resource allocation request for a new traffic policy will be rejected if the current network processing capacity is not sufficient. Therefore, when an adaptive determinant module has to determine whether to apply the result value from the inference module, an adaptation will be determined based on the network resource status and the current policy request. Accordingly, the processing capacity of the current network controlled by the IAC can be descried as follows:

$$\theta = \lambda \times C_a + (1 - \lambda) \times C_{max}, (0 \leq \lambda \leq 1) \tag{2}$$

$$C_{m1} = \sum_{i=1}^{k} R_{ri}(C_m < C_a \leq C_{max}, k \geq 1) \tag{3}$$

$$C_{m2} = \sum_{i=1}^{k} R_{ri} + \sum_{i=k+1}^{n} R_{rj} \times \mu_{wa}(z), (C_a \leq \theta, n \geq 1) \tag{4}$$

The determination of the FCAC may reflect the adaptation degree by definition of membership function designed according to an administration policy. Through the simulation, it has been found that the change ratio of attributes value requested for traffic flow does not have significant influence on the total adaptation degree of fuzzy controller, and the adaptation degree also is determined through correlation between current node status and the traffic service level defined through the FBE. If the control node has an idle bandwidth, the adaptation degree displayed through fuzzy controller shows the same acceptance ratio as in the non-fuzzy environment. In case of a throughput status close to threshold, the fuzzy controller shows that it operates according to the adaptation value. Accordingly, it is possible to enhance the node efficiency and to reduce the loss ratio by a drop in packet flow that may occur in the network congestion status, through an increase in an average acceptance ratio at the node.

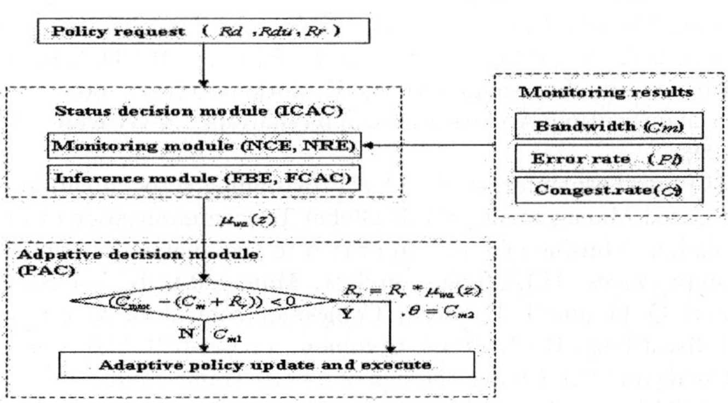

Fig. 8. Adaptive Policy Control Algorithm

5 Conclusion and Future Works

In this study it is proposed that a control structure determines the network resource allocation according to the network status. As a result of executing the simulation with the fuzzy controller designed in this study, it improved the packet rejection ratio by 17% on the average according to traffic patterns compared with non-fuzzy environment. This is because it performs a soft adaptation based on network status by fuzzy controller, rather than accept/reject action in the non-fuzzy environment.

The fuzzy controller shows the same acceptance ratio as in the non-fuzzy environment when an idle bandwidth exists in the node, while it operates according to its adaptation value when it reaches the threshold of the node throughput. Accordingly, it demonstrates that the node processing efficiency can be improved and the loss ratio by packet drop which may occur in a network congestion status can be reduced as the average acceptance ratio is raised.

In a practical network, setting values of the fuzzy membership function set in our proposal should be newly adjusted to fit. In a future study, we expect to expand to a fuzzy control structure that has learning functions according to the network status. Therefore, the control structure will be able to automatically adjust to proper setting values in a practical network.

References

1. S. Blake, D. Black, M. Carlson, E. Davies, Z. Wang, and W. Weiss, 'An Architecture for Differentiated Service.', RFC2475, December 1998.
2. S. Shenker, C. Partridge, and R. Guerin, 'Specification of Guaranteed Quality of Service', RFC 2212, September 1997.
3. . Dinesh C. Verma, 'Policy-Based Networking: Architecture and Algorithms', New Riders, pp139-181, November 2000.
4. Yao-Ching Liu and Christos Douligeris, 'Nested Threshold Cell Discarding with Dedicated Buffers and Fuzzy Scheduling', IEEE GLOBCOM, pp.429-432, 1996.
5. A. Vasilakos and K. Anagnostakis, 'Evolutionary-fuzzy prediction for strategic inter-domain routing: Architecture and mechanism', in WCCI 98, Anchorage, USA, May 1998.
6. Marcial Porto Fernandez, et al., 'QoS Provisioning across a DiffServ Domain using Policy-Based Management', IEEE Global Telecommunications Conference, 2001.
7. L. A. Zadeh, 'Outline of a new approach to the analysis of complex systems and decision processes', IEEE Trans. on Syst., Man, and Cyb., Vol. SMC-3, No.1, 1973.
8. N. Yin, S. Q. Li, and T. E. Stern, 'Congestion control for packet voice by selective packet discarding', IEEE Trans. Commun., pp.674-683, May 1990.
9. R.G. Cheng and C.j. Chang, 'Design of a fuzzy traffic controller for ATM network', IEEE/ACM Trans. Networking, Vol.4, No.3, pp.460-469, June 1996.
10. James C. Bezdek, Sankar K. Pal, 'Fuzzy Models For Pattern Recongnition', IEEE Press, 1991.
11. R. Guerin, H. Ahmadi, and M. Naghshineh, 'Equivalent capacity and its application to bandwidth allocation in high-speed networks', IEE J. Select. Areas Commun., Vol.9. No.7, pp.968-981, September 1991.

12. A.I. Elwalid and D. Mitra, 'Effective bandwidth of general Markovian traffic sources and admission control of high speed network', IEEE/ACM Trans. Networking, Vol.1, No.3, pp.329-343, June 1993.
13. G. Kesidis, J. Walrand, and C.S. Chang, 'Effective bandwidths for multiclass Markov fluids and other ATM source', IEEE/ACM Trans. Networking, Vol.1, No.4, pp.424-428, August 1993.
14. Allan Leinwand, et. al., 'Network Management: A Practical Perspective', ADDISON-WESLEY, 1996.

Link Layer Assisted Multicast-Based Mobile RSVP (LM-MRSVP)

Hongseock Jeon, Myungchul Kim, Kyunghee Lee, Jeonghoon Mo, and Danhyung Lee

School of Engineering, Information and Communications University,
Yuseong P.O. Box 77, 305-600, Daejeon, Korea
{kanjuk95, mckim, leekhe, jhmo, danlee}@icu.ac.kr

Abstract. Even though several RSVP extensions have been proposed to support QoS guarantee in mobile Internet, they still suffer from issues such as excessive advanced reservations or a nonoptimal path generated by Mobile IP. We propose an algorithm called "Link Layer Assisted Multicast based Mobile RSVP (LM-MRSVP)" to resolve the issues. Our implementation shows practicability of our proposal and the simulation study confirms our claims.

1 Introduction

Provision of seamless Quality of Service (QoS) in a mobile environment has been a challenging issue for researchers. When a mobile node (MN) moves from one subnet to another, rerouting of packets is required within a reasonable amount of time to support QoS. Two protocols, Mobile IP [1] and RSVP [2], have been considered to support seamless QoS in the mobile Internet. The first provides an alternative IP address to the MN moving out of one subnet for rerouting to be possible. The second provides a way to reserve network resources to guarantee QoS.

However, the combination of the two caused RSVP invisibility problem. The IP-in-IP encapsulation mechanism of Mobile IP makes RSVP messages invisible. The modification of IP header due to the encapsulation causes RSVP messages invisible to routers. As a consequence, routers cannot reserve the path. To address this problem, RSVP Tunnel [3] was proposed. The underlying idea of RSVP Tunnel is that additional messages, tunnel PATH and tunnel RESV, are sent by tunnel entry and exit points on top of the existing RSVP messages to make a reservation in the tunnel area. The new messages are used to establish path between the entry and exit points and the original RSVP messages are used to establish end-to-end path.

Simple combination of the two protocols is not enough to provide seamless QoS. Seamless QoS requires establishment of a new RSVP session within a reasonable time limit. Otherwise, an MN would experience disruption in service. Mobile RSVP (MRSVP) [4] introduced the concept of advanced resource reservation (ARR). Instead of waiting until an MN moves to a new subnet, it makes

C. Kim (Ed.): ICOIN 2005, LNCS 3391, pp. 452–462, 2005.

advanced reservations at multiple potential locations to save time for the session establishment. However, MRSVP has possibility of wasting network resources by making too many reservations as the number of MN increases. Moreover, it inherits the long triangular route of Mobile IP. The packets destined to the MN should go through the home agent of the node, which incurs additional delays. It may cause the service disruption.

Hierarchical MRSVP (HMRSVP) [5] integrates RSVP with Mobile IP registration. It makes advance resource reservations only when an inter-region handover may possibly occur. Therefore, HMRSVP can reduce the number of ARRs. However, HMRSVP still inherits the inefficient routing path due to the triangle routing.

In this paper, we propose Link Layer Assisted Multicast-based Mobile RSVP (LM-MRSVP) to address the above issues: excessive advance reservation, inefficient routing of Mobile IP, and the invisibility problem. It uses Layer 2 trigger [6] to predict potential movement. With the prediction, it only makes advance resource reservation of the potential cell not to make an excessive reservation. It uses a multicast session when handoff happens. Packets are forwarded more than two subnets simultaneously to avoid service disruptions. We implemented the protocol and perform experiments to support our ideas.

The rest of this paper is organized as follows. Section 2 describes related work. In section 3, we describe the main ideas of our proposal. Sections 4 and 5 present the performance evaluation of LM-MRSVP. We conclude the paper in section 6.

2 Related Work

In this section, we describe existing approaches to resolve issues of RSVP in mobile Internet.

RSVP Tunnel RSVP Tunnel [3] was proposed to address RSVP messages invisibility problem. In RSVP Tunnel, both tunnel entry router (Rentry) and tunnel exit router (Rexit) send newly designed RSVP messages in addition to the existing RSVP messages. The new messages, tunnel PATH and tunnel RESV, are just used for reservation of the IP tunnel area, not end-to-end QoS. With these newly designed messages, RSVP Tunnel mechanism can establish a RSVP session over the IP tunnel as well as end-to-end path. However, RSVP Tunnel still suffers from some problems because it is based on Mobile IP. First, more time and resources are required to establish a RSVP session due to a nonoptimal routing path of Mobile IP. Second, a home agent (HA) always becomes an anchor point of all RSVP sessions for MNs which are registered the same HA. In such a situation, we encounter the bottleneck problem of RSVP sessions at HA. Finally, additional messages of RSVP Tunnel stir up a heavy traffic problem of signaling messages more and more.

Multicast Based RSVP RSVP Mobility Support [7] is a new RSVP model based on IP multicast in order to support mobility. In the approach, all the data and RSVP messages are delivered over the multicast session and the mobility of an MN is modeled as multicast group join and leave operation. Each BS, called Mobile proxy, initiates join operation and reserves a multicast path in place of an MN. RSVP Mobility Support also uses the advanced resource reservations to support seamless QoS guarantee. When an MN launches a reservation, join operation is initiated from all the neighboring BSs as well as the current BS. Therefore, in the hierarchical network structure, RSVP Mobility Support always provides the optimized reservation paths. However, this approach also does not address the excessive advance reservation problem. In paper [8], Huang expanded the above scheme over the hierarchical Mobile IPv6 structure.

MRSVP Talukdar proposed the Mobile RSVP (MRSVP) [4] to support mobility for RSVP. In MRSVP, each MN maintains Mobility Specification (MSPEC), which indicates a set of locations the MN wishes to make ARRs. Then MRSVP reserves the resources in advance according to the MSPEC. MRSVP divides a reservation into two types: active and passive reservation. An active reservation denotes a conventional reservation of RSVP and a passive reservation represents a state of resources only being reserved but not transmitting actual data packets. MRSVP makes an active reservation at the current location of an MN and makes passive reservations at the locations within its MSPEC. Such passive reservations make MRSVP practicable to guarantee seamless QoS for an MN. However, if MSPEC includes multiple locations (MSPEC might indicate neighboring but also further away cells), too many ARRs may be required. It wastes network resources excessively and causes a scalability problem as the number of MN increases.

HMRSVP Tseng proposed Hierarchical MRSVP (HMRSVP) [5], an enhanced mechanism based on MRSVP. HMRSVP employs RSVP Tunnel and Mobile IP regional registration [9]. In HMRSVP, ARRs are only established at the boundary cells of a region, which is possibly represented as a routing domain, when an inter-region handover occurs. Therefore, HMRSVP can reduce the number of ARRs. When an MN moves in a single region, HMRSVP guarantees seamless QoS with Mobile IP regional registration scheme. Mobile IP regional registration decreases the setup time of resource reservation for new path. However, though Mobile IP regional registration can reduce the cost of re-establishing reservation path, it still establishes the inefficient routing path caused by the triangle routing of Mobile IP.

3 Link Layer Assisted Multicast-Based Mobile RSVP

The proposed mechanism, Link Layer Assisted Multicast-based Mobile RSVP (LM-MRSVP), reduces the number of ARRs and avoids the problem incurred by using Mobile IP (that is, using RSVP Tunnel). It can be achieved by predicting a potential cell using Layer 2 trigger and RSVP sessions established independently with Mobile IP by Multicast and RSVP Agent (MRA).

Figure 1 shows steps to establish RSVP session using LM-MRSVP. When the MN moves close to a new BS, it receives a beacon signal from the new BS. By this beacon signal, the MN learns about the new BS and then informs the current BS of the new BS through an *InformNewBS* message. After receiving the *InformNewBS*, the current BS sends a *JoinResvReq* message to the new BS in order to request to build a new multicast RSVP session. Upon receiving the *JointResvReq* message, the new BS sets up a new multicast RSVP session between the new BS and the crossing router.

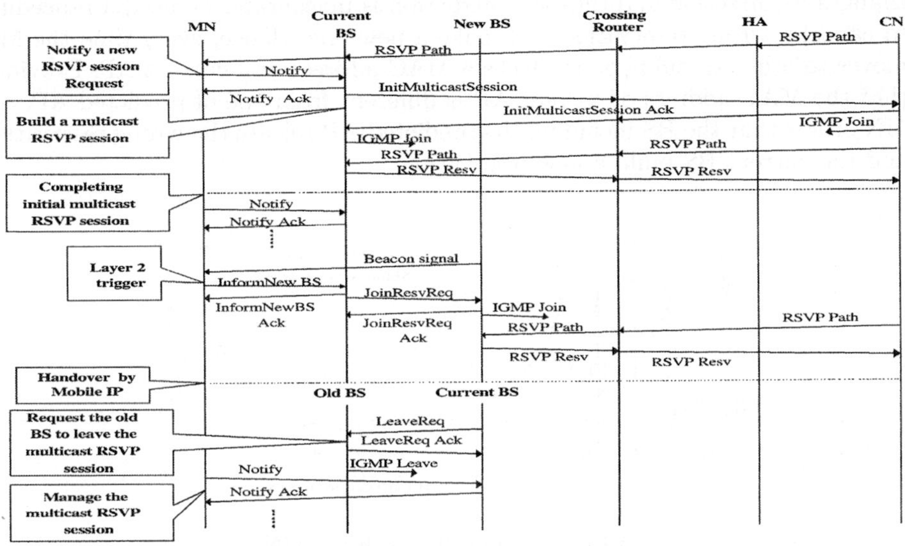

Fig. 1. Procedure of LM-MRSVP

3.1 Movement Prediction Using Layer 2 Trigger

To reduce excessive advance resource reservations, it makes only one ARR to the potential cell. In LM-MRSVP, the Layer 2 trigger is used to find a potential cell. Generally, we can identify a new access point (AP) through the Layer 2 trigger (e.g., mobile trigger). In case of IEEE 802.11b Wireless LAN [10], an AP broadcasts beacon signal messages periodically. The beacon signal message contains a MAC address of the AP. Such information in the beacon signal allows MN to identify a new AP. Also Service Set ID (SSID), name of wireless local area network, in the beacon signal helps an MN to distinguish whether the new AP is attached to other IP subnet or same IP subnet. In case the beacon signal from the new AP contains different SSID from current one, we can consider the new AP is included in other IP subnet, therefore we need to set up a new RSVP session. Otherwise, we donot have to build a new RSVP session even though we find a new AP.

For predicting a new AP, the signal strength value of the beacon frames is used as an important measure. In Figure 2, as an MN closes to a new AP, the signal strength from an old(current) AP becomes weak gradually while the signal strength from the new AP becomes stronger. When the signal strength from the old AP is below *Scan Start Threshold*, the MN starts to find a new AP. If it happens that the signal strength from the new AP is stronger than that of the old AP in some degrees, we can consider the MN will start immediately a layer 2 roaming process to the new AP. In Figure 2, such a point time is when the signal strength of the old AP reaches *Prediction Threshold*. Unfortunately, if the MN changes its direction and thus the prediction is inaccurate, we need a procedure to cancel a wrong reservation and make a new one. Hence, every time the MN moves to other IP subnets, it checks a MAC address of a current AP. Provided that the MAC address of current AP is different from one of predicted AP, the MN orders that the BS including the predicted AP release the wrong reservation and the current BS build a new reservation.

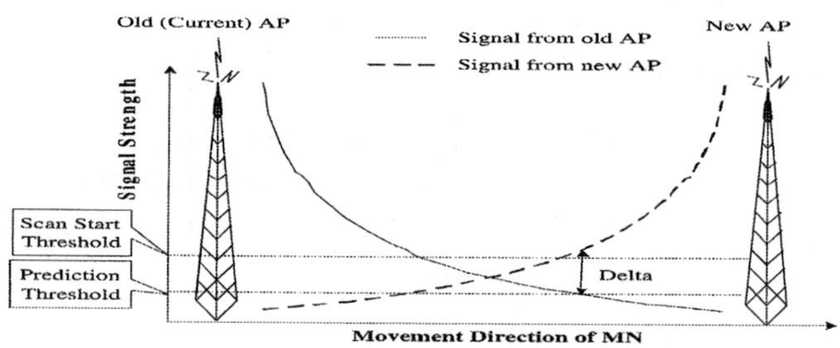

Fig. 2. Signal Strength for MN

3.2 Multicast and RSVP Agent (MRA)

Though LM-MRSVP is supported by Mobile IP to manage host mobility, it delivers all guaranteed traffic for an MN without Mobile IP. It comes from that Multicast and RSVP Agent (MRA) in BS performs all the processes related to IP multicasting and resource reservation in substitute for an MN.

MRA manages information about MNs with a visitor list table, which contains following columns: RSVP session information (MN IP address/port number, CN IP address/port number, reservation style and flow descriptor), multicast group address, and life time. When a BS receives a *Notify* message from an MN in order to build a multicast RSVP session, MRA in the BS allocates a multicast group address and adds on a new entry in the visitor list table with the multicast group address as well as RSVP session information conveyed by the *Notify* message. The lifetime field in the visitor list is refreshed by a periodic *Notify* message. An entry in the visitor list can be deleted with expiration of lifetime or a *LeaveReq* message. In LM-MRSVP, an MN lets the current BS know a MAC address of the new AP via an *InformNewBS* message. Hence BSs must be able to find a new BS's IP address corresponding to the MAC address of the new AP. Thus, all the BSs in LM-MRSVP manage a mapping table which matches neighboring AP's MAC addresses to neighboring BS's IP addresses. In order to build the mapping table, BSs have to exchange some information with each other to discover their neighboring BSs and APs. In paper [11], there was a discussion about a protocol to perform such information exchanges. Each entry in the mapping table is composed of following tuple: Essid, AP MAC address, BS IP address.

In LM-MRSVP, MRA allocates dynamically multicast group addresses for multicast RSVP sessions. Such a multicast group address allocation is an important issue of IP multicast. RFC 2908 [12] proposes an architecture for multicast address allocation.

4 Experimental Results

We built a testbed and implemented LM-MRSVP in C on the Linux. The testbed consists of a router, a CN, a HA, an MN, and two BSs as shown in Figure 3. The router is enabled to handle the RSVP messages over IP multicast session and uses a class based queuing (CBQ) scheme. The HA, MN, and BSs are equipped with Mobile IP modules. In addition, the BSs have an MRA module supporting LM-MRSVP. The wired and wireless communications are based on the IEEE 802.3 and IEEE 802.11b technology, respectively.

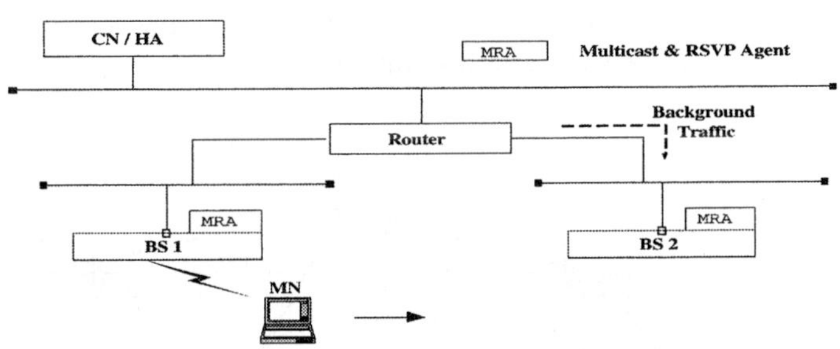

Fig. 3. Testbed Configuration for LM-MRSVP

In our experiment, an MN moves from non-congested subnet to congested one which suffers from massive background traffic. We evaluated a performance of LM-MRSVP in terms of throughput and PSNR [1]. Figure 4(A) shows the throughput for LM-MRSVP and RSVP. The throughput of RSVP is considerably

[1] Peak Signal to Noise Ratio(PSNR): a measure of error used to determine the quality of compressed image and video. When no error, the value of PSNR is 78.13 dB. And the value decreases if damaged. In general, anything below 25 dB is unacceptable to the human visual system.

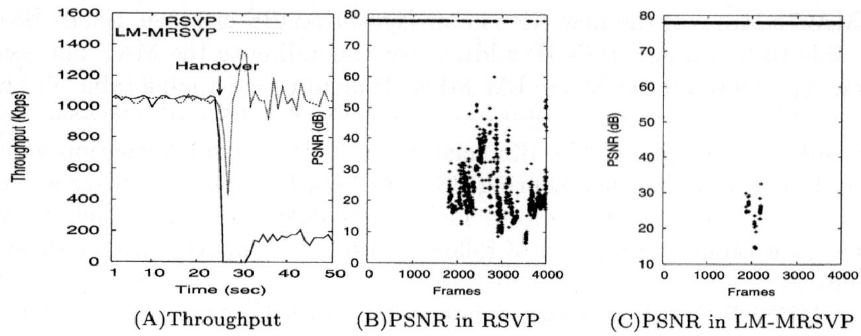

Fig. 4. Performance for LM-MRSVP and RSVP

dropped from 1,000 Kbps to 200 Kbps after handover. It indicates that data destined to an MN are not guaranteed against the network congestion. However, the throughput of LM-MRSVP does not suffer from the network congestion after handover. In LM-MRSVP, the throughput is dropped instantaneously while an MN performs a handover and then is rapidly recovered to the original value. Figure 4(B) and Figure 4(C) show PSNR value of streaming MPEG-1 file[2] with RSVP and LM-MRSVP. Before a handover, an average PSNR value is 78.13 dB in both approaches. However, with only RSVP, video quality is significantly reduced after a handover. After a handover, the average PSNR value in RSVP becomes 35.75 dB and 1390 frames obtain PSNR values below 25dB. On the other hand, the average PSNR value of LM-MRSVP is 75.67 dB after a handover and only 187 frames have PSNR values below 25 dB. Those results show the practicability of LM-MRSVP in mobile Internet.

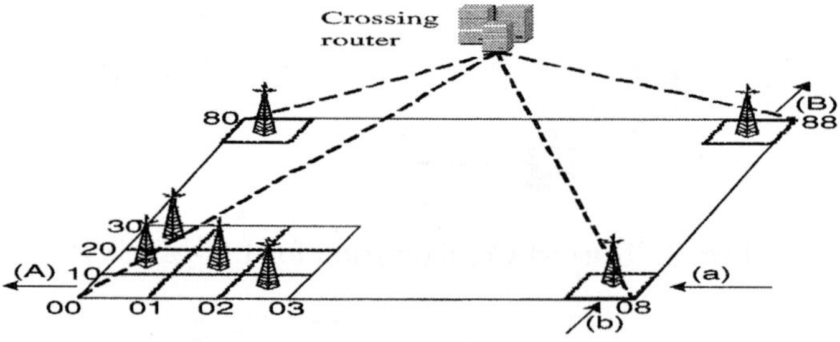

Fig. 5. 8 by 8 Mesh Toplogy

[2] Video Clip Spec.: Frame size - 352 * 240, Average data rate - 1600Kbps, Frame rate - 29.97fps, Frame number - 4034, Video size - 22.4 Mbyte

5 Simulation Results

Based on the simulator of Tseng [5], we made appropriate modifications for LM-MRSVP. We used 8 by 8 topology as shown in Figure 5. Each rectangular shaped cell has one BS and total of 64 cells form one region. All BSs in a region are connected via a crossing router. To simulate inter-region handovers, when an MN moves toward (A) in Figure 5, we assume that inter-region handover occurs and the MN enters the region again with the direction of (a). An MN moves randomly and all handovers are layer 3 handovers. New RSVP requests are generated at the rate of λ and assumed to be Poisson distributed. The RSVP session holding time follows an exponential distribution with mean value of 180 sec. Table 1 shows parameters in the simulation. Note that the number of MN per cell and speed of MN are used with several values.

Table 1. Simulation Parameters

New RSVP request arrival rate (λ)	1/180 RSVP/sec
Mean RSVP session holding time ($1/\mu$)	180 sec
Radius of cell	500 m
Number of MN per cell	varied
Speed of MN (km/h)	varied

Figure 6(A) to (C) show the performances of LM-MRSVP, HMRSVP, and MRSVP as the number of MNs per cell increases, when the speed of MN is 80 km/h. Figure 6(A) describes the reservation blocking rate at which a new RSVP session request of an MN is blocked. The reservation blocking rate of LM-MRSVP is in between the other approaches. This is because, though HMRSVP builds ARRs just when an inter-region handover occurs, MRSVP sets up those

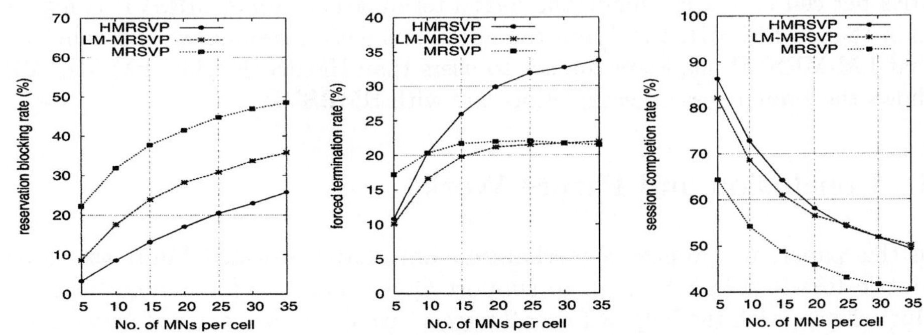

(A)Reservation blocking rate (B)Forced termination rate (C)Session completion rate

Fig. 6. Performance for LM-MRSVP, HMRSVP and MRSVP.

at all neighboring cells in every handover and LM-MRSVP builds an ARR only at the predicted cell unlike MRSVP.

Figure 6(B) shows the forced termination rate. The forced termination rate is the probability that a RSVP session which is not blocked is terminated in force when an MN performs a handover. Typically, the more ARRs provide the lower forced termination rate because the ARR allows an MN to have one more chance for making reservation when handover occurs. Therefore, the forced termination rates of LM-MRSVP and MRSVP, which build ARRs in every handover, are generally lower than that of HMRSVP. When the number of MNs in a cell is 5, LM-MRSVP and HMRSVP have similar forced termination rates, which are about 10 %. However, if the number of MNs in a cell increases to be 35, the forced termination rate of HMRSVP becomes 34 % and the corresponding value of LM-MRSVP becomes only 21 %.

In Figure 6(C), LM-MRSVP shows little lower performance than HMRSVP in terms of the session completion rate, which is a probability that a RSVP session is completed without any reservation blocking and forced termination. However, when the number of MNs in a cell is larger than 25, the session completion rate of LM-MRSVP becomes similar to that of HMRSVP. This is because the session completion rate is a combinational result of the reservation blocking rate and the forced termination rate. Thus, the session completion rate of MRSVP appears the lowest because MRSVP has the considerably higher reservation blocking rate comparing LM-MRSVP and HMRSVP.

Generally, when an MN experiences many handovers, the forced termination rate is a more significant indication than other metrics because dropping ongoing session is more annoying than blocking a new session from the user's point of view. Figure 7 depicts the forced termination rates of LM-MRSVP, HMRSVP and MRSVP, which vary with the speed of an MN. All approaches show the similar forced termination rates in case of the low speed of an MN, 4 km/h. However, as the velocity of the MN reaches 120 km/h, the forced termination rate of HMRSVP becomes higher than that of LM-MRSVP. Such a phenomenon becomes outstanding as the number of MNs per cell increases. LM-MRSVP and MRSVP have the similar forced termination rate. However, if the number of MNs per cell is relative small, the forced termination rate of MRSVP is higher than that of LM-MRSVP. From the above simulation results, we can conclude that LM-MRSVP has lower impact to users than HMRSVP while LM-MRSVP shows the similar session completion rate with HMRSVP.

6 Conclusion and Future Work

In this paper, we propose a mechanism, Link Layer Assisted Multicast-based Mobile RSVP (LM-MRSVP), to support QoS guarantees for mobile Internet. Our scheme, with the help of Layer 2 trigger, avoids excessive use of network resources by making only one advanced resource reservation at the most probable cell. Moreover, all BSs in LM-MRSVP build a RSVP session not along a routing path of Mobile IP but along an IP multicast tree including a CN and BSs as its

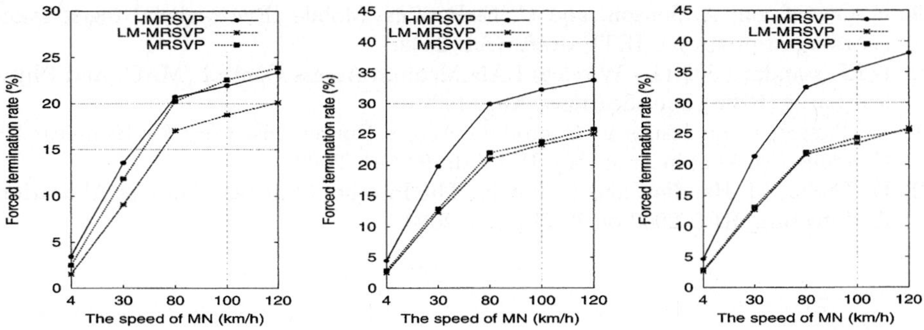

(A)No. of MNs per cell is 10 (B)No. of MNs per cell is 20 (C)No. of MNs per cell is 30

Fig. 7. Forced Termination Rate for LM-MRSVP, HMRSVP and MRSVP.

members. Thus LM-MRSVP offers an optimized RSVP session and overcomes the problems incurred when using Mobile IP. Through implementation and simulation results, we showed the practicability of LM-MRSVP and analyzed the performance of LM-MRSVP comparing HMRSVP and MRSVP. For the future work, we plan to study the advantages of LM-MRSVP in terms of the bottleneck problem in HA and signaling overhead due to RSVP tunneling.

7 Acknowledgement

This paper was supported in part by the Grid Middleware Center in OITRC.

References

1. C. E. Perkins: IP Mobility Support for IPv4, RFC 3220 on IETF, Aug. 2002.
2. R. Braden, L. Zhang, S. Berson, S. Herzog, S. Jamin: Resource ReSerVation Protocol (RSVP), RFC 2205 on IETF, Sep. 1997.
3. A. Terzis, J. Krawczyk, J. Wroclawski and L. Zhang: RSVP Operation Over IP Tunnels, RFC 2746 on IETF, Jan 2000.
4. A. K. Talukdar, B. R. Badrinath, A. Acharya: MRSVP - A resource reservation protocol for an integrated service network with mobile hosts, Wireless Networks v.7, page 5-19, 2001.
5. Chien-Chao Tseng, Gwo-Chuan Lee, Ren-Shiou Liu: HMRSVP - A Hierarchical Mobile RSVP Protocol, IEEE International Conference on Distributed Computing Systems Workshop, p.467-472, 2001.
6. J. Kempf, et al: Supporting Optimized Handover for IP Mobility - Requirements for Underlying Systems (working in progress), IETF draft, June. 2002.
7. Wen-Tsuen Chen, Li-Chi Huang: RSVP Mobility Support - A Signaling Protocol for Integrated Services Internet with Mobile Hosts, IEEE INFOCOM 2000.
8. N. F. Huang and W. E. Chen: RSVP Extensions for Real-Time Services in Hierarchical Mobile IPv6, Mobile Networks & Applications v.8, page 625-634, 2003.

9. E. Gustafsson, A. Jonson, and C. E. Perkins: Mobile IPv4 regional registration (working in progress), IETF draft, Oct. 2002.
10. IEEE standard 802.11 - Wireless LAN Medium Access Control (MAC) and Physical Layer (PHY) Specifications, Aug. 1999.
11. D. Trossen et al: Issues in Candidate Access Router Discovery for Seamless IP Handoffs (working in progress), IETF draft, Oct. 2002.
12. D. Thaler, M. Handley, and D. Estrin: The Internet Multicast Address Allocation Architecture, RFC 2908 on IETF. Sep. 2000.

Comparison of Multipath Algorithms for Load Balancing in a MPLS Network

Kyeongja Lee, Armand Toguyeni, Aurelien Noce, and Ahmed Rahmani

LAGIS, UMR CNRS 8146, Ecole centrale de Lille
BP48 59651 Villeneuve d'ASCQ, France
{Kyeong_Ja.Lee, Armand.Toguyeni, Ahmed.Rahmani}@ec-lille.fr
Aurelien.Noce@centrale-lille.net

Abstract. Traffic Engineering aims to optimize the operational performance of a network. This paper focuses on multipath routing for traffic engineering that routes the demand on multiple paths simultaneously for balancing the load in the network. According to the schematic approach of multipath routing that we propose in this paper, a multipath routing algorithm selects candidate paths using multicriteria simultaneously and then distributes the traffic demand among selected paths. Among multipath routing algorithms, we select four algorithms which fit our expectations in terms of architecture: WDP, MATE, multipath-AIMD and LDM. We focused exclusively on technologies designed for MPLS networks. Each algorithm is compared with respect to complexity and stability of two stages: the computation of multiple paths and the traffic splitting among multiple paths.

1 Introduction

The current Internet service is often referred as best effort. Because best effort service treats all packets equally, when a link is congested, packets are simply lost and any flows could not get a priority. Congestion derives from insufficient network resources and unbalanced traffic distribution. Adding more bandwidth to networks is not the solutions for solving the congestion problems in the long term. To improve quality of service (QoS) of actual network, we focus on Traffic Engineering [1] and more specifically on multipath routing. This study aims to compare different algorithms in order to verify if they are scalable and efficient. This can be done indirectly by the study of their complexity and their stability. In this study, we do not want to compare different algorithms with specific criteria such as packet delay or blocking ratio. To obtain the routing that gives an average QoS for different type of applications, we think that the real criterion consists in minimizing the maximum utilization ratio of each network link. After that, diffserv model is suitable to prioritize the different traffics according to specific criteria such as packet delay, jitter and so on.

This paper is organized as it follows. Section 2 states the study problem. Thus in section 3, we propose a functional generic model that enables to explain the role of the different algorithms in the literature and to compare them. In

C. Kim (Ed.): ICOIN 2005, LNCS 3391, pp. 463–470, 2005.

section 4, we compare different recent algorithms with regard to their complexity and their stability. We also propose a new modified LDM algorithm. In section 5, the paper ends with our conclusions and some perspectives.

2 The Study Context

Actual IP routing in Internet does not use efficiently the network resources within backbone. This poor utilization is primarily caused by two properties. The one is that IP routing is destination based and the other is that decision-making in current routing is based on local optimization. In order to engineer the traffic effectively in IP networks, network administrators must be able to control the complete paths of packets instead of hop-by-hop. This requires some kinds of connections in the connectionless IP networks. MPLS (Multi-Protocol Label Switching) can make it more efficient. So network traffic is distributed more evenly than best effort and the probability of network congestion can be reduced [2]. An MPLS system for traffic engineering in a large ISP (Internet Service Provider) network can be deployed now.

This paper focuses on multipath routing that routes demands on multiple paths simultaneously for balancing the load in the network instead of routing all the traffic on the only one shortest path. Multipath routing algorithm selects candidate paths using one or more criteria defining a metric and then distributes the traffic demand among selected paths.

In literature, there are a lot of proposed algorithms for multipath routing. Each algorithm is declared very efficient by its authors but generally with respect to restricted conditions. This study is a preliminary work to propose a general framework that will allow developing a multi-model approach for a multipath routing depending on the network status. Since all actual propositions depend on specific conditions, the real problem is not to define a unique routing model. Our approach consists in adapting the routing model according to the traffic features. More specifically this study compares some recent algorithms such as MATE [9] or LDM [11] both with regard to their respective model and with regard to the scalability and the stability of each solution.

3 Functional Model Used by Multipath Algorithms

There are basically two main stages in a multipath routing algorithm (Fig. 1): *computation of multiple paths and traffic splitting among these multiple paths.*

The first stage computes the set of candidate paths which is a subset of all the paths between a pair of considered routers. According to the nature of a cost function, different algorithms can be applied to determine these candidate paths. The authors consider various static criteria such as bandwidth, hop count, delay, error ratio, and so on for a cost function. This problem of a cost function definition is typically a multicriteria problem [3].

$$Cost_{static} = f(bandwidth,\ hopcount,\ delay,\ error\ ratio) \quad (1)$$

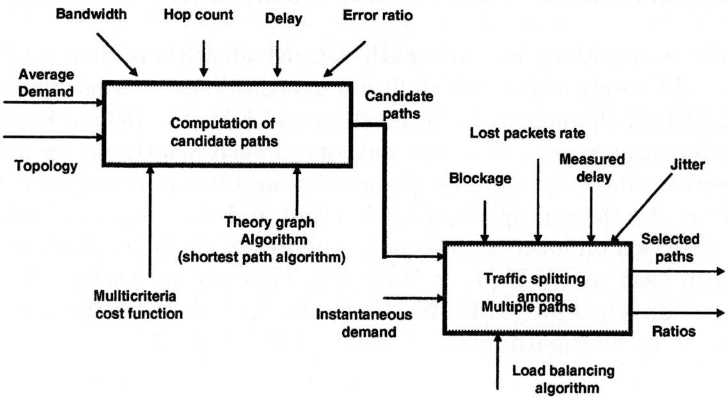

Fig. 1. Functional decomposition of multipath algorithm

The second stage consists in splitting traffic among multiple candidate paths. These paths are qualified of candidate paths because all of them are not necessary to be used at a given time. The utilization ratio of selected paths depends on the evaluation of dynamic criteria such as blockages, the packet loss ratio, the measured delay, the jitter, and so on. This also requires the definition of a cost function based on dynamic criteria what is still a multicriteria problem.

$$Cost_{dynamic} = f'(blockage,\ packet\ lost\ ratio,\ measured\ delay,\ jitter) \quad (2)$$

If multiple criteria must be optimized simultaneously, the complexity of the algorithms usually becomes very high [4]. A lot of heuristic algorithms are proposed to solve this problem. A common method is called sequential filtering, under which a combination of metrics is ordered in some fashion, reflecting the importance of different metrics. First, paths based on the primary metric are computed and then a subset of them is eliminated based on the secondary metric until a subset of good paths [5]. This is a trade-off between performance optimization and computation simplicity. As examples of sequential filtering, there are SWP (Shortest Widest Path) [6] and WSP (Widest Shortest Path) [7] that can give an opposite path selection when they are applied to the same network.

Last example shows that heuristic approaches such as filtering approach cannot guarantee that all QoS requirements can be guarantee to all types of traffic. To improve IP routing, our opinion is to mix both load balancing and diffserv prioritizing approach. Therefore, a good load balancing approach must minimize the maximum utilization ratio of each link in the network. This study focuses on four load balancing models that respect this requirement. Consequently, our problem is to determine if the corresponding routing algorithms are scalable and efficient to be deployed in a large network.

4 Comparison of Algorithm's Complexity and Stability

Among the propositions for multipath routing algorithms that we found, we have selected four algorithms which fit our expectations in terms of architecture: WDP [8], MATE [9], multipath-AIMD [10] and LDM [11]. Behind the respective objective function expressed by the authors of each algorithm, we find a common objective that is to minimize the maximum utilization ratio of each link of the network. All these propositions are recent and they seem scalable. Indeed, they are all based on local or infrequent updates of the network state, contrary to approach such as SWP [6] or WSP [7]. They are all exclusively based on technologies designed for MPLS Networks. Each of these algorithms is studied according to the multipath model introduced in section 3.

4.1 WDP [8]

WDP (Widest Disjoint Paths) is not a full multipath-routing algorithm, but focuses on the selection of good paths. This approach is mainly based on two concepts: path width and path distance. Path width concept is a way to detect bottlenecks in the network and to avoid them if possible. Path distance is original because contrary to most approaches, it is not a hop-count measure but it is indirectly dependent on the utilization ratio of each link defining the path. In this way, WDP is an improvement of SWP. This approach is very promising when considering a practical implementation with numerous ingress-egress router pairs.

WDP algorithm performs candidate paths selection based on the computation of the width of the good disjoint paths with regard to bottleneck links. The width of a path is defined as the residual bandwidth of its bottleneck link. The principle of WDP is to select a restricted number of paths. A path is added to the subset of good paths if its inclusion increases the width of this subset. At the opposite, a path is deleted if this does not reduce the width of the subset of good paths. This is a heavy computation to perform on every path, and the algorithm is very time-consuming: computing a set of n paths will take $O(n^3)$ cycles because the selection procedure proposed in [8] is clearly in $O(n^2)$ and allows selecting one path at each iteration considering all potential paths between the pair of ingress-egress routers.

For the traffic splitting stage, the authors propose EBP (Equalizing Blocking Probability) that is a localized approach. The complexity of this algorithm is $O(n)$ since it consists in updating the blocking probability b_{ri} of each path ri depending on the relative distance of b_{ri} from the current average blocking probability $\bar{b}$.

4.2 MATE [9]

The MATE paper presents a traffic engineering scheme called "MPLS Adaptive Traffic Engineering". This approach uses a constant monitoring of the links using probe packets to evaluate link properties such as packet delay and packet loss.

Using these statistics the MATE algorithm is able to optimize packets repartition among paths to avoid link congestion.

Contrary to other algorithms compared here, there is no selection of candidate paths in MATE: all available paths are considered. Anyway, another step is required before proceeding to the load distribution: the incoming packets are regrouped into a fixed number of bins. The number of bins determines the minimum amount of data that can be shifted. The repartition of the packets among the bins can be done by different approaches, as shown in [9]. These approaches differ in their complexity and the way they handle packet sequencing. The better repartition method is *Using flow hash* because its complexity is $O(n)$ (with n the number of bins) and this method preserves packet sequencing.

The load balancing stage splits the content of the bins among LSPs, by using a technique such as the gradient projection algorithm. The complexity of this algorithm is $O(n^2)$ where n is the number of LSPs between an ingress-egress pair of nodes. Finally, since the two stages are in sequence and if we assume that the numbers of bins and the number of LSPs are comparable, MATE complexity is in $O(n^2)$. The designers of MATE have proved in [9] that MATE's algorithm converges to an optimal routing when specific conditions are verified (see Theorem 2 page 4 in [9]).

4.3 Multipath-AIMD [10]

In the multipath-AIMD (Additive Increase Multiplicative Decrease) paper, the authors present an algorithm based on the notion of primary path. A primary path is a preferred path associated to each source. The data will then be sent mainly on the primary path, but can also use other LSPs when the bandwidth of the primary path is not sufficient.

The selection of paths in multipath-AIMD consists in selecting n LSPs equal to the number of current sources. This can be done by sorting the LSPs using their metric and then extracting the better paths. For a set of n sources, the complexity of the treatment is in $O(n \, ln(n))$. There is no stability issue in this step of the routing procedure.

The traffic splitting uses an Additive Increase/Multiplicative Decrease: starting on an average repartition, the iterative process increases the data to be sent to a path with a constant value if the link is not saturated (additive increase) and divides the amount of data to be sent to a path by a certain value if the path is saturated (multiplicative decrease). This approach is done in a $O(n)$ complexity for n chosen paths. The paper [10] also presents a modified AIMD algorithm that is closer to PPF-optimality. This solution, called multipath-AIMD with PPF correction, is better in the way it comes closer to the expectations of the authors in terms of behavior, but it is also more expensive in resources. Its complexity is in $O(n^2)$ for n chosen paths.

4.4 Original Proposition of LDM [11]

The LDM algorithm, which stands for "Load Distribution over Multipath", has the particularity to use a flow based approach of multipath routing. This particularity is interesting to avoid some problems encountered with lower-level load distribution by preserving packet sequencing.

The algorithm tries to find a minimal set of *good* paths. The set is built on two criteria: the metric hop-count associated to each path must be as low as possible while maintaining links utilization inferior to a certain parameter ρ. This is done in a $O(n^2)$ time in the worst case, the number of iterations growing with the utilization of the network. Here n refers to the number of paths available for the considered ingress-egress pair of nodes. In terms of convergence, the number of iterations has an upper limit defined by a given parameter δ, so the number of iterations is bounded and stability issues avoided.

The traffic splitting is then done using a heuristic to determine a repartition policy for incoming flows. Each path is adjoined a probability of selection using the formula (3). Once probabilities are defined, each incoming flow is directed to its route with the highest probability. The h(l) and d(l) functions refer to the length and the remaining capacity of link l, while C_0 and C_1 are constants computed to make $P(l)$ a probability. The a0 and a1 factors are to be defined by the administrator to fit its needs.

$$P(l) = a_0 \frac{C_0}{h(l)} + a_1 \frac{d(l)}{C_1} \text{ with } a_0 + a_1 = 1 \tag{3}$$

The complexity of the whole procedure is clearly $O(n)$. Here n refers to the number of paths selected at the end of the previous step.

4.5 Our Proposition of a Modified LDM with Two Thresholds

LDM suffers potential stability issues. LDM algorithm for selecting candidate paths converges necessarily to a solution in a finite delay because of the limitation of the number of extra-hops that is admissible to augment the number of selected paths. In the path selection algorithm given in [13], this is expressed by the condition $m > \delta$ that allows stopping the first loop. However the real problem of stability that can have LDM can be caused by oscillations due to candidate path selection. Each time there are changes in path utilization values, the whole algorithm is applied without taking account of previous selections. Let us assume first that $\eta(t) < \rho$ and that the load of the network becomes too important and $U(A_{ij})$ becomes superior to ρ . The computation at $t + \Delta T$ will give more candidate paths than at time t. Consequently the load will be distributed on new paths (with length superior to shortest paths) implying the decrease of the utilization of each path. If the load of shortest paths decreases under $\eta(t)$, the set of computed candidate paths will come back to the situation of the time t and so on.

Our idea is that the path selection algorithm can be improved by using two thresholds. The first threshold ρ_1 allows adding new paths in candidate path set. The second threshold ρ_2 must be lower than the first one. If the minimum of candidate path utilization goes under this threshold, this enables to reset the candidate paths selection by restarting the whole algorithm (Fig. 2).

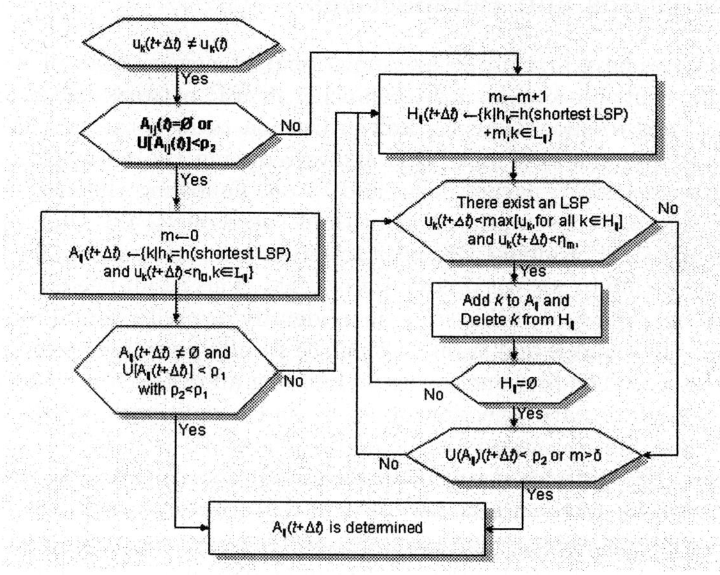

Fig. 2. Modified LDM candidate path selection algorithm

4.6 Comparison Results

The results mentioned above can be summarized in Fig. 3. This work opens new perspectives: by combining different approaches it may be possible to use an hybrid approach, for example by using a WDP selection of paths and a multipath-AIMD optimized PPF approach on a very powerful system, or the simple path selection of multipath-AIMD with the traffic splitting of LDM on old systems.

Algorithm	Candidate paths computation		Traffic splitting	
	Author criteria	complexity	Author criteria	complexity
WDP+erp	Hop count, Residual bandwidth	$O(n^3)$	Blocking probability	$O(n)$
MATE	-	-	Delay and Packet loss	$O(n^2)$
AIMD	-	$O(n \ln(n))$	Binary feedback value of congestion	$O(n^2)$
LDM	Hop count	$O(n^2)$	Link utilization	$O(n)$

Fig. 3. Comparison of the 4 algorithms with regard to the two stages

5 Conclusions and Perspectives

In this study we have first proposed a functional framework that allows comparing different contributions for load balancing in ISP network based on MPLS. This study tries to show that the link utilization ratio is the best criterion to guarantee an average QoS since this allows reducing the network congestions (and then packet loss), packet delay and so on. Following this assessment, we have studied four recent multipath algorithms based on MPLS. Our study shows that the stability of original LDM path selection is not guarantee. We propose a modified algorithm to correct this aspect of LDM. This result needs to be verified by simulations. The goal of this study is to identify formally, actual routing algorithms that are scalable and stable. Using the results given here two approaches can be envisaged. The first one consist in building an hybrid approach that combine the candidate path selection of WDP and a multipath-AIMD optimized PPF approach, and the second is the multipath-AIMD with LDM.

Generally the authors justify a new routing algorithm by the necessity to adapt the routing to the traffic features. In this context, instead of developing a new routing algorithm we propose a multimodel approach. In this study we have shown that all the four propositions can be used in a multimodel approach. The interest of such type of approach is to obtain a real adaptative model. This perspective must be verified by simulation, and notably the capacity to adapt dynamically the multipath algorithm depends on the network functioning point.

References

1. Zheng Wang, Internet QoS: Architectures and Mechanisms for Quality of Service, Morgan Kaufmann Publishers, Lucent Technology (2001)
2. Xipeng Xiao, Providing QoS in the Internet, Ph.D thesis, Michigan state Univ. (2000).
3. Vincent T'kindt, Jean-Charles Billaut, Multicriteria Scheduling: Theory, Models and Algorithms. Springer, 300 pages (2002).
4. Shigang Chen, Routing Support for Providing Guaranteed End-to-End Quality-of-Service, Ph.D. thesis, UIUC, 207 pages (1999).
5. E.Crawley et al. A Framework for QoS-based Routing in the Internet, Internet RFC 2386 (1998).
6. Wang, Z., Crowcroft, J. QoS Routing for Supporting Multimedia Applications. IEEE Journal of Selected Areas in Communications 14 (1996) 1228-1234.
7. Guerin, R., Orda, A., Williams, D. Qos Routing Mechanisms and OSPF Extensions. In Proc. of the Global Internet Miniconference. Phoenix, USA (1997).
8. Srihari Nelakuditi, Zhi-Li Zhang, On Selection of Paths for Multipath Routing, In Proc. IWQoS'01, Karlsruhe, Germany (2001).
9. A. Elwalid, C. Jin, S. Low, and I.Widjaja, MATE: MPLS Adaptive Traffic Engineering, INFOCOM'2001, Alaska (2001).
10. Jianping Wang, Stephen Patek, Haiyong Wang, and Jrg Liebeherr, Traffic Engineering with AIMD in MPLS Networks, LNCS (2002).
11. J. Song, S. Kim, M. Lee, Dynamic Load Distribution in MPLS Networks, LNCS Vol. 2662 (2003) 989-999.

A Buffer-Driven Network-Adaptive Multicast Rate Control Approach for Internet DTV

Fei Li, Xin Wang, and Xiangyang Xue

Department of Computer Science, Fudan University, 200433 Shanghai, China
{021021107, xinw, xyxue}@fudan.edu.cn

Abstract. The current Internet does not offer any quality of service guarantees or support to Internet multimedia applications. There are two requirements of multicast sending rate for Internet DTV: (1) it can adapt well to the change of network congestion; (2) it can meet the requirements of decoding rate. The difficulties of Internet DTV stream multicast are analyzed, then a multicast rate control approach for network DTV stream based on buffer management is given, which can suit for the change of network traffic and satisfy the requirement of DTV decoder by controlling sending rate logically. The test results show the good practical value.[1]

1 Introduction

With the rapid development of Internet, Network DTV is more and more becoming a research hotspot. High quality of Internet DTV program, such as video's continuity and audio's synchronization, needs enough bandwidth guarantees. Currently Internet only offers "best effort" service, which is hardly any quality of service guarantee. As we know, TCP is not well-suited for streaming applications and real-time audio and video because the reliability and ordering semantics it ensures increases end to-end delays and delay variations [1]. For effectively using network bandwidth, multicast technology is adopted increasingly to transmit Internet DTV data. IP Multicast delivers source traffic to multiple receivers without adding any additional burden on the source or the receivers while using the least network bandwidth of any competing technology, so it can reduce the possibility of network congestion and increase data transmission efficiency. But the difficulties of IP multicast, such as expansibility, reliability, validity, feedback implosion and complexity of group managing, can not be neglected. So Internet DTV multicast is facing many technology problems.

In this paper we analyze the difficulties of Internet DTV server multicast rate control and propose an improved TCP-friendly multicast sending rate control approach which is based Network condition and sender buffer occupancy. We call it Buffer-driven Network-adaptive Multicast Rate Control approach (BNMRC). It is adaptive to network TCP friendly available bandwidth according to feedback

[1] This work was supported in part by NSFC-60402007, NSFC-60373020, 863-2002AA103065, 863-2002AA103011-5, Shanghai Municipal R&D Foundation under contracts 035107008, 03DZ15019 and 03DZ14015, MoE R&D Foundation.

C. Kim (Ed.): ICOIN 2005, LNCS 3391, pp. 471–478, 2005.

information by using dynamic single rate to control multicast sending, and give users better video quality by guaranteeing sending buffer not to be overflow.

This paper is organized as follows: section 2 discussed the related works of TCP friendly multicast congestion control theory briefly. Section 3 describes our Internet DTV multicast sending rate control system. The simulation studies including the results are discussed in section 4. Section 5 concludes the paper.

2 TCP-friendly Multicast Congestion Control

In the multimedia applications, burst loss and overtime delay will lead to bad video quality, especially for compressed video files. And these are because of network congestion. So multimedia multicast streams need suitable congestion control strategy. The stability of the Internet to date has in large part been due to the congestion control and avoidance algorithms [2]implemented in its dominant transport protocol, TCP [3][4]. For multicast to be successful, it is crucial that multicast congestion control mechanisms be deployed that can co-exist with TCP in the FIFO queues of the current Internet[5]. For real time application, such as video or audio, sending rate must not change abruptly as it can noticeably reduce the user-perceived quality. In our judgement, equation-based TCP friendly multicast congestion control, which uses a control equation that explicitly gives the maximum acceptable sending rate as a function of the recent loss event rate, is a viable mechanism to provide relatively smooth congestion control for Internet DTV traffic. There has been significant previous researches on equation based congestion control mechanisms[6][7][8], TFMCC (TCP Friendly Multicast Congestion Control) proposal is one of them. In TFMCC protocol [9],which is based on TFRC (TCP-friendly Rate Control protocol) for multicast scheme, a control equation derived from a model of TCP's long-term throughput is used to control the sender's transmission rate according to feedback information from the receiver called CLR (Current Limiting Receiver) who is experiencing the worst network conditions. At the same time, each receiver continuously determines a desired receive rate that is TCP-friendly for the path from the sender to this receiver. If it was selected as CLR, then it reported the rate to the sender in feedback packets. TFMCC is designed to be reasonably fair when competing for bandwidth with TCP flows. A multicast flow is "reasonably fair" if its sending rate is generally within a factor of two of the sending rate of a TCP flow from the sender to the slowest receiver of the multicast group under the same network conditions. In general, TFMCC has a low variation of throughput, which makes it suitable for streaming media where a relatively smooth sending rate is of importance. Since help to decrease network loss, end to end TCP friendly multicast congestion control mechanism is important for Internet real time applications. But in the implementation of TFMCC, the key challenges lie in scalable RTT ($round - trip\ time$) measurements, appropriate feedback suppression, and in ensuring that feedback delays in the control loop do not adversely affect fairness towards competing flows. Since loss event rate and RTT are important parameters in the traffic calculation formula, if they are not accurate the rate control will be too bad. According to the characteristic of Internet DTV transmission system, we didn't directly use TFMCC as rate control mechanism. We use BNMRC to control sending rate, which is based network condition and server sending buffer occupancy. And under this mechanism, the server sending rate change is smoother than TFMCC.

3 Buffer-Driven Network-Adaptive Multicast Rate Control

3.1 System Model

Currently Internet DTV data is transported by MPEG-2 Transport Stream, which is tailored for communicating or storing one or more programs of coded data and other data in environments in which significant errors may occur. DTV TS rate is about 4 $\sim$ 8 Mbps .Since TS is variable bit rate, directly sending data to network instead of using suitable rate control may result in network traffic fluctuating and worsen network condition. This will increase data loss possibility. According to video coding's basic compressing technology, even if one bit loss will influence the quality of correlative frame. This will result in errors of video images and debase severely visual quality. So it is key issue in streaming multimedia applications that adopting suitable sending rate control approach to utilize adequately network bandwidth and sending data to network smoothly to reduce data loss. Further more, the data transmitting of Internet DTV is different from generic VOD as its source can be from real time streaming of secondary planet or disk database of server. Since the DTV server's storage and transmitting buffer is limited, the data stream must transmit to network in time, otherwise it may be lost in the server and this will worsen video quality of user. Therefore two key issues must be taken into account: one is that sending buffer must not be overflow, the other is sending rate must be adaptive to network situation.

Our proposal details are: firstly, a basic sending rate from the timestamps of video data was calculated , then a TCP friendly network bandwidth was detected by using TFMCC, at last the basic sending rate, the practicable network bandwidth and the data occupancy of sending buffer were assembled to realize sending rate control. The system model is illustrated in figure 1. The DTV server sending rate control model is made up of congestion watch module, rate feedback module and multicast sending control module.

3.2 Congestion Watch

Congestion watch is important for getting network bandwidth information. The congestion watch module at server is responsible for receiving feedbacks from clients and assisting receiver-side RTT measurements. At the same time, it also control receiver's feedbacks and get expecting rate from feedback and remit to rate control module. The functions of congestion watch module at client are

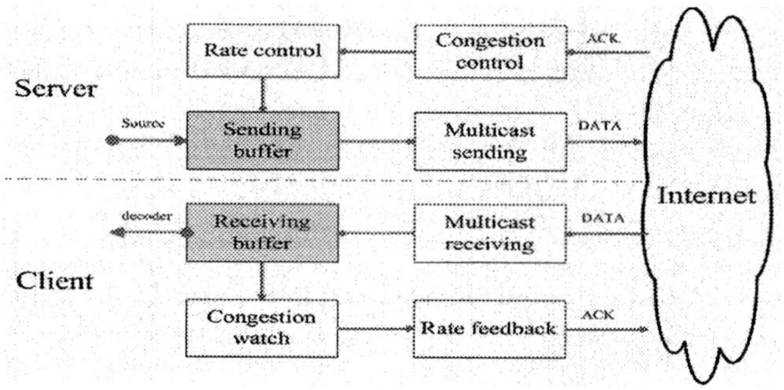

Fig. 1. Buffer-driven Network-adaptive Multicast Rate Control System

measure of RTT, calculate the loss event rate P and expecting rate T_{TCP}. At the same time it also judges whether it has qualification to send feedback. If has, it will send feedback. Otherwise, it will cancel feedback. The adopted rate calculation formula is:

$$T_{TCP} = \frac{s}{RTT \cdot I_{total}}. \tag{1}$$

Where, T_{TCP} is the transmit rate in bits/second, s is the packet size in bytes. RTT is the round-trip time in seconds. P is the loss event rate, between 0.0 and 1.0, of the number of loss events as a fraction of the number of packets transmitted. The parameters P and RTT reflect the network's conditions.

RTT was measured at receiver by a timestamp in a receiver report which is echoed by the sender. When a receiver gets a data packet that carries the receiver's own ID, the receiver updates its RTT estimate, that is: $RTT = t_{now} - ts_r$.where t_{now} is the time the data packet arrives at the receiver and ts_r is the receiver report timestamp echoed in the data packet. If the data packet has not carried the receiver's ID, then it must do two-way measurement and determine its RTT by history measurement information. So RTT measurement is not only important but also complicated.

Obtaining an accurate and stable measurement of the loss event rate is of primary importance for congestion watch. Loss event rate P is performed at receivers based on the detection of lost from the sequence numbers of arriving packets. A loss event was defined as one or more lost packets from the packets received during one RTT. The number of packets between consecutive loss events is called a loss interval. The average loss interval size can be computed as the weighted average of the m most recent loss intervals $l_k \ldots\ldots l_{k-m+1}$: $l_{avg}(k) = \frac{\sum_{i=1}^{m-1} \cdot w_i \cdot l_{k-i}}{\sum_{i=1}^{m-1} \cdot w_i}$, $w = \{5, 5, 5, 5, 4, 3, 2, 1\}$. The weights w are chosen so that very recent loss intervals receive the same high weights. The loss event rate P is defined as the inverse of $l_{avg}(k)$.

Since multicast group have so many members, feedback Suppression is very important for multicast congestion control. For avoidance of feedback implosion and ensure the receiver's report with lowest bandwidth, it is not that every receiver can send feedback packets in every feedback round. Only CLR and the receiver whose calculated rate is lower than the currently sending rate can send feedback in a feedback round. And when a new feedback round begins, outstanding feedbacks for the old round are cancelled.

3.3 Sending Rate Control

In the packet elementary stream of TS, both of the decoding timestamp and presentation timestamp include time information, which can be used to calculate sending rate theoretically. But because frame resetting exist in the during of MPEG-2 decoding, PTS is not ordered in the time. So PTS is not fit for calculate sending rate. Since DTS is ordered, it can be used to calculate sending rate. When TS are received from DVB-S card, or when multimedia data are read from files, the video data can be copied to another buffer. TS packets can be parsed for getting DTS information. Then, on the assumption that the last bit of DTS_i is ended on the P_i position, and the last bit of DTS_{i+1} is ended on the P_{i+1} position, then $R_i = \frac{P_{i+1}-P_i}{DTS_{i+1}-DTSi}$, R_i is the mean sending rate of the TS between P_i and P_{i+1}. It can be used to write received data to sending buffer. To the decoder, receiving data at the rate R_i can assure that the receiving buffer is neither overflow nor underflow. Therefore if only on the demand of video's factor, $R_i$is the best sending rate theoretically to the decoder. But on the factor of network transmission, according to TFMCC , T_{TCP} is the best sending rate . Now the two factors are integrated. We define:

$$t_i = DTS_{i+1} - DTS_i. \tag{2}$$

$$S_i = P_{i+1} - P_i. \tag{3}$$

Suppose during t_i the data has entered into the sending buffer for S_i bytes, the available bandwidth got from feedback is T_{TCP} , and the sending timer's interval is Δt $(t_i > \Delta t)$,then during $t_i$the server's mean sending data R is:

$$R = R_i \times \Delta t + (T_{TCP} - R_i) \times \Delta t \times O(b), \qquad (0 < b < 1) \tag{4}$$

where

$$O(b) = \begin{cases} b & \text{if } T_{TCP} >= R_i \\ 1 - 1.2 \times b & \text{if } T_{TCP} < R_i \end{cases} \tag{5}$$

Here b represents the present data occupancy of the buffer, which is the quotient of the amount of existing data and the capability of sending buffer.$(T_{TCP} - R_i)\times$ Δt shows the largest range of adjustable flux, which may be positive number or negative. $O(b)$ shows the rate feeble adjusting function, which can make sending buffer not to be overflow or underflow. Here it uses linearity function to be adjusting parameters, which can guarantee sending rate increase when network bandwidth is enough and data is buffer is close to full. And when network congestion is occurred and available bandwidth is smaller than the bandwidth needed by TS, the actual traffic rate will be decreased with the buffer occupancy variety . If the buffer occupancy is big, the sending rate decreased range is less.

And it the buffer occupancy is small, the sending rate decreased range is bigger more.

Implementation steps as below: In every t_i period

Step1 Calculate the sending rate R_i;

Step2 When server has not received feedbacks, it sends data as the rate of R_i in the t_i period ;

Step3 When server has received feedbacks, it sends data as the rate of R ;

Step4 Repeat 3 until the end.

4 Experiment

We have tested BNMRC across the public Internet and in the ns network simulator. These results give us confidence that BNMRC is effective for Internet DTV server data transmitting. At the same time, we have tested TFMCC's effect in Internet DTV server data transmitting. The experiment's results is below.

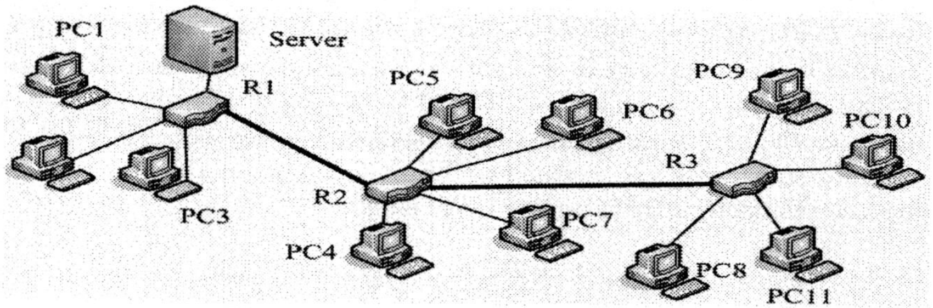

Fig. 2. Network Simulation Topology

Figure 2 shows the network simulation topology. It includes 2 level multicast trees and 3 subnets. Besides a server, all the others are DTV clients and multicast receivers. Three TCP links exist among the different subnets. The link bandwidth between router 1 and router 2 is 10Mbps. The link bandwidth between router2 and router3 is 5Mbps which is the bottleneck. The data source is a segment of DTV program which we record its coding stream's character parameters based DTS information. The other parameters are:$\Delta t = 5ms, B = 2Mbytes$.

Figure 3,4,5 tell us the simulation results. The sending rate curve got from BNMRC and the rate curve from TFMCC were contrasted in figure 3. We can see that the BNMRC's rate was limited in the available bandwidth range which is calculated according to TCP traffic formula. This shows that the rate is TCP friendly. At the same time, its rate changing range is much smaller than the TFMCC's and it shows that the impact of BNMRC to the network is smaller than TFMCC's. Figure 4 contrasts the BNMRC rate and TS rate. We can see that BNMRC rate is changed with the change of network condition. Figure 5 shows the buffer's data occupancy of BNMRC and the buffer's data occupancy of TFMCC respectively. We can see that BNMRC's sending buffer has not been overflow, which kept a good wave range. But TFMCC's buffer occupancy has fluctuated in a bigger range and sometimes it was overflow which maybe results in data loss in the sending buffer.

Table 1. Loss rate of BNMRC and TFMCC

	Loss rate of buffer	Mean loss rate of network
BNMRC	0%	0.025%
TFMCC	0.25%	0.02%

We have also counted the loss rates from BNMRC and TFMCC respectively. The results tell us BNMRC is more suitable for data transmitting than TFMCC. See table 1. In generally, the data loss of data transmitting of Internet DTV server may be occurred in two sections of transmission way. One is because of sending buffer's overflow which can make data loss when data have not entered into network. The other is the data loss in the network. From table 1, we can find that the mean network loss rate from BNMRC is bigger than TFMCC appreciably, but BNMRC can guarantee no loss in the server's sending buffer. TFMCC can not guarantee no data loss in sending buffer since its rate change is so waved. When the network is congested, the sending rate decreased very low and sending buffer may be overflow when new data was coming constantly. Some data was lost in the sending buffer. Integrate the two loss factors, we consider BNMRC is more suitable for Internet DTV server, which can adaptively send packets with the network condition and advance the video quality by decreasing the total loss rate for receivers.

5 Conclusion

Instead of using TFMCC absolutely for congestion control, we use BNMRC to control multicast sending rate in the Internet DTV system. This approach uses the timestamp on TS data to calculate a basic sending rate, then uses TFMCC to detect TCP friendly available network bandwidth, at last it integrates the basic sending rate, the available network bandwidth and sending buffer's data occupancy to choose a network adaptive sending rate. The experiments show that BNMRC has good useful value. Since this approach use CLR's feedback to determine network condition, CLR's choosing strategy is very important. Further work will continue to deeply research CLR's perfect choosing strategy.

References

1. D. Bansal and H. Balakrishnan, TCP-Friendly Congestion Control for Real-time Streaming Applications, May 2000.MIT Technical Report MIT-LCS-TR-806.
2. Jacobson, V. Karels,M., "Congestion Avoidance and Control", In Proc.ACM SIGCOMM Aug 1988.
3. Postel, J. B., "Transmission Control Protocol", Internet Engineering Task Force, September 1981. RFC 793.
4. Stevens, W. R., "TCP/IP Illustrated", Vol. 1. Addison-Wesley, MA, Nov 1994.

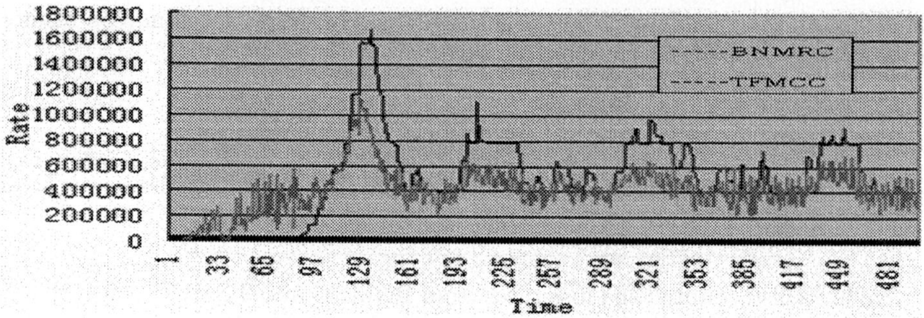

Fig. 3. BNMRC Rate and TFMCC Rate

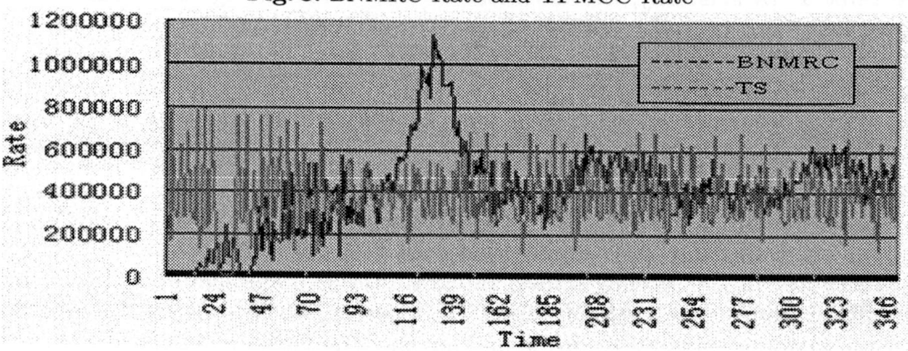

Fig. 4. BNMRC Rate and TS Rate

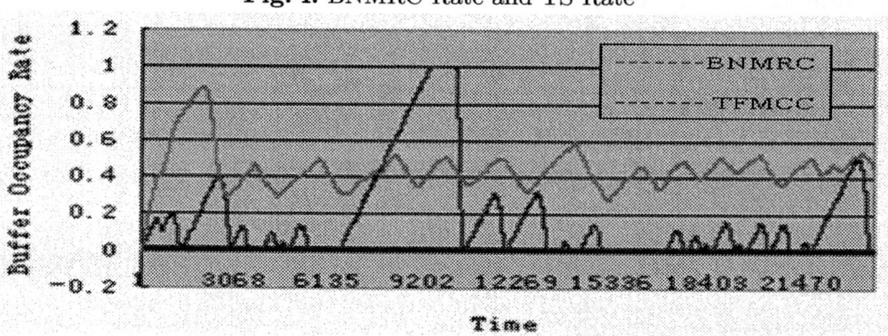

Fig. 5. BNMRC Buffer Occupancy and TFMCC Buffer Occupancy

5. J.Widmer and M.Handley, "Extending Equation-Based Congestion Control to Multicast Applications", Proc ACM SIGCOMM 2001, SanDiego, August 2001
6. S.Floyd and K.Fall, "Promoting the use of end-to-end congestion control in the Internet", IEEE/ACM Trans.Networking, vol.7, pp.458-472, Aug. 1999.
7. S. Floyd, M. Handley, J. Padhye, and J. Widmer.Equation-based congestion control for unicast applications, In Proc. ACM SIGCOMM, pages 43 - 56, Aug. 2000.
8. Todd Montgomery, "A Loss Tolerant Rate Controller for Reliable Multicast", Technical Report:NASA-IVV-97-011, West Virginia University, August 1997.
9. J. Widmer and M. Handley, " TCP-Friendly Multicast Congestion Control (TFMCC):Protocol Specification", draft-ietf-rmt-bb-tfmcc-02.txt, July 2003

On the Hidden Terminal Problem in Multi-rate Ad Hoc Wireless Networks*

Joon Yoo and Chongkwon Kim

School of Electrical Engineering and Computer Science,
Seoul National University, Seoul 151-742, Republic of Korea
{joon, ckim}@popeye.snu.ac.kr

Abstract. Multi-hop ad hoc wireless networks generally use the IEEE 802.11 Distributed Coordination Function (DCF) MAC protocol, which utilizes the request-to-send/clear-to-send (RTS/CTS) mechanism to prevent the hidden terminal problem. It has been pointed out that the RTS/CTS mechanism cannot completely solve the hidden terminal problem in ad hoc networks because the interference range could exceed the basic rate transmission range. In this paper we provide a worst-case analysis of collision probability induced by the hidden terminal problem in ad hoc networks with multi-rate functionality. We show that the interference caused by the nodes in the area that is not covered by the RTS/CTS is bounded by $C'R^{-4}$, where C' is a constant and R is the distance between the two transmitting nodes. The analytic result showed that the interference could shorten the data transmission range up to 30 percent. We then propose a simple multi-rate MAC protocol that could prevent the hidden terminal problem when transmit power control (TPC) is employed.

1 Introduction

Ad hoc wireless networks consist of wireless mobile hosts which form a multi-hop wireless network without the support of established infrastructure or centralized administration. Each mobile host in an ad hoc network functions as a router to establish end-to-end multi-hop connection between any two nodes. Typical application areas include battlefields, emergency search and rescue sites, and data acquisition in remote access.

The hidden terminal problem is a common phenomenon due to the multi-hop nature of ad hoc networks. For example, in Fig. 1, when node A is transmitting data to node B, the hidden terminal problem occurs when node D, which is unaware of the ongoing transmission, attempts to transmit, thus causing collision at node B. The IEEE 802.11 standard [1] distributed coordination function (DCF) medium access control (MAC) protocol employs the request-to-send / clear-to-send (RTS/CTS) option to prevent the hidden terminal problem. Nodes A and B in Fig. 1, can exchange the RTS/CTS frames prior to the data/ack transmissions so that their neighbor nodes defer for the duration of the data/ack transmissions.

* This work was supported by the Brain Korea 21 Project in 2004 and grant No.R01-2001-00360 from the Korea Science and Engineering Foundation.

C. Kim (Ed.): ICOIN 2005, LNCS 3391, pp. 479–488, 2005.

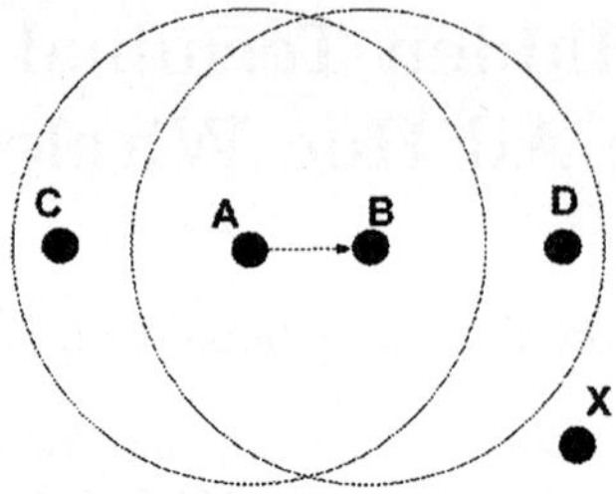

Fig. 1. Node A is transmitting a data frame to B. Node C is a hidden terminal to node A and nodes D and X are hidden terminals to node B

Even when the RTS/CTS handshake completes its role and in turn, all the neighboring nodes of A and B are deferring their transmissions, the hidden terminal problem may not be completely solved. For example, node X in Fig. 1 is beyond the basic rate transmission range of A's RTS and B's CTS, so it can initiate transmission freely assuming it does not sense the carrier busy. K. Xu et al [4] pointed out that node X can also be a hidden terminal; therefore interfere with the ongoing transmission. The area where a node could cause interference is called the *interference range.*

The emerging radio interfaces such as IEEE 802.11a/b/g [1] can provide a multi-rate capability to the ad hoc networks. For instance the popular IEEE 802.11b can dynamically select between 1, 2, 5.5 and 11Mbps according to the channel condition. Higher data rates can be utilized when the signal-to-interference and noise ratio (SINR) value is sufficiently high enough to meet the threshold of the specific modulation scheme. On the other hand, the interference range can grow larger when using multi-rate, since the receiver requires a higher SINR value when it intends to receive at higher data rates. As shown in Fig. 2, the interference range is largest when data rate of 11Mbps in IEEE 802.11b is used. As a result, the hidden terminal problem has become a much serious issue in multi-rate environments.

To assure that a hidden node, for example node X in Fig. 1, does not interfere with the on going transmission, we need to consider the SINR of the receiver especially in multi-rate environments. The SINR value should exceed a certain threshold value depending on the selected data rate for a node to receive a data frame without error.

In this paper we analyze the effect of interference and collision probability due to the hidden terminal problem in multi-rate ad hoc wireless networks. We assume the worst-case where all the nodes in the network are active and the node density is high so that the transmissions can cover the whole network space. We show that the interference caused by the nodes in the area that is not covered by the RTS/CTS is bounded by $C'R^{-4}$, where C' is a constant and R is the distance between the two transmitting nodes. The analytic result showed that the interference could shorten the data transmission range about 30 percent.

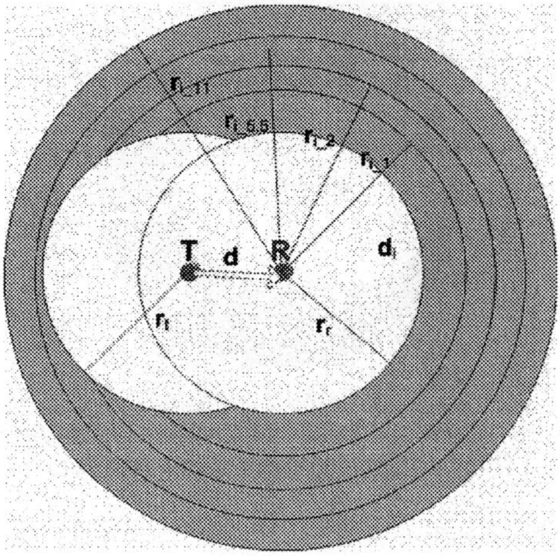

Fig. 2. r_t, and r_r are the transmission range of the RTS and CTS respectively. The interference range is denoted as r_{i_1}, r_{i_2}, $r_{i\_5.5}$ and r_{i_11}, when 1, 2, 5.5 and 11Mbps data rate for IEEE 802.11b is used respectively

We then propose a simple way to prevent this kind of hidden terminal problem in multi-rate when transmit power control (TPC) is employed. To prevent hidden terminal problems, we control the CTS transmit power to cover the interference range depending on the selected data rate.

The rest of the paper is organized as follows. We briefly review the previous work related to the multi-rate aware MAC in Sect. 2. In Sect. 3 we present the analysis of the interference and collision probability due to the hidden terminals and show a numerical example. After proposing a new method that prevents the hidden terminal problem in Sect. 4, we conclude our paper in Sect. 5.

2 Related Work

Many MAC protocols/algorithms have been developed to utilize the multi-rate functionality. The Auto Rate Fallback (ARF) [2] is typically implemented in commercial 802.11 products. ARF chooses to raise or lower its transmission rate according to consecutive transmission successes or failures, respectively. In the Receiver Based Auto Rate (RBAR) [3], the receiver selects an adequate transmission rate according to the channel quality measured from the received request-to-send (RTS) frame.

K. Xu et al [4] pointed out that the RTS/CTS cannot effectively prevent the hidden terminal problem in a single-rate environment. They propose to reduce the data transmission range so that the RTS/CTS can be effective enough to prevent a *single* hidden terminal from interfering. On the other hand, we present results based on *multiple* hidden terminals that can interfere in multi-rate environments.

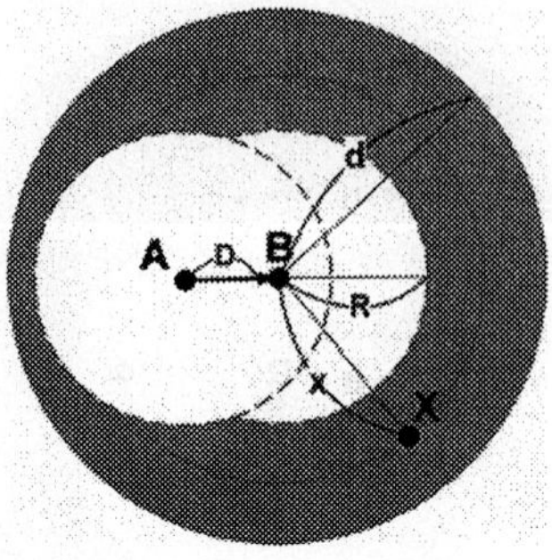

Fig. 3. The shaded area shows where a hidden terminal can be located in. node X is a hidden terminal located x (m) away from receiver node B

3 Analysis of the Hidden Terminal Problem in Multi-rate Ad Hoc

We consider two nodes A and B, and the distance between the two nodes is D as shown in Fig. 3. The multi-rate physical and MAC protocol used here is the IEEE 802.11b [1] and RBAR [3]. We assume that the transmit power is fixed at the maximum level, and basic rate (1Mbps) is used for the RTS/CTS transmission. The inner dashed line in Fig. 3 indicates the transmission range of node A and B when sent at the basic rate. According to [5], the receive power of a signal at the receiver can be modeled as:

$$P_r = P_t G_t G_r \frac{h_t^2 h_r^2}{D^4} . \tag{1}$$

where P_t is the transmit power, G_t, G_r , h_t and h_r are antenna gains and height of antennas of transmitter and receiver respectively. D is the distance between the transmitter and the receiver. To simplify our analysis, we assume that the ad hoc network is homogeneous, and the physical conditions are all equal at each node. Thus,

$$P_r = \frac{C}{D^4}, \text{ where } C = P_t G_t G_r h_t^2 h_r^2 . \tag{2}$$

Note that C is a constant based on our assumptions. The receiver node B determines the transmission rate according to the channel condition measured by the RTS frame as in RBAR [3]. The channel condition can be estimated by the measured SNR. Let's assume that node B decided to use n Mbps data transmission rate since it estimated that the current SNR is above SNR_n, which is the SNR threshold of data rate n Mbps.

$$\frac{C/D^4}{\eta} \geq SINR_n . \tag{3}$$

C is the constant shown in equation (2), and η is the theremal noise. The randomly generated network topology is modeled, where the nodes are uniformly

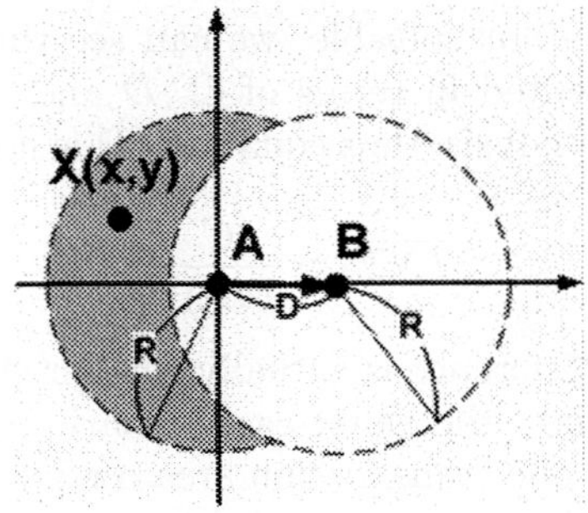

Fig. 4. The shaded area shows the area covered by the RTS but not covered by the CTS

placed on an infinitely large two dimensional area. As shown in Fig. 3, node X is a hidden terminal located in the shaded area. The shaded area is the area not covered by the RTS/CTS. The distance from the receiving node B and the hidden terminal X is x, where x is ranged form R to d. d is the maximum distance that can cause interference, which will be extended to infinity later on. Using equation (2) for receive power, the average interference caused by node X is

$$\overline{INT_x} = \int_R^d Pr\{L = x\} \cdot INT(L = x)dx - INT_{RTS}$$

$$= \int_R^d \frac{2x}{d^2 - R^2} \cdot \frac{C}{x^4} dx - INT_{RTS} = \frac{C}{(dR)^2} - INT_{RTS}. \quad (4)$$

where $\overline{INT_x}$ denotes the average interference that receiver node B can suffer from a hidden terminal X, $\Pr\{L = x\}$ is the probability of node X being located at distance x and $INT(L = x)$ is the interference caused by node X. INT_{RTS} is the average interference caused by the area covered by the RTS but not covered by the CTS, which is the shaded area shown in Fig. 4. The calculation of INT_{RTS} is as follows. The area we are interested in is

$$x^2 + y^2 \leq R^2 \text{ and } (x - D)^2 + y^2 \geq R^2.$$

If we order the above equation in terms of x,

$$-\sqrt{R^2 - y^2} \leq x \leq min(\sqrt{R^2 - y^2}, -\sqrt{R^2 - y^2} + D). \quad (5)$$

Using equations (2) and (5), we obtain the average interference caused by the shaded area shown in Fig. 4.

$$INT_{RTS} = \frac{\int_{-R}^{R} \int_{-\sqrt{r^2-y^2}}^{min(\sqrt{R^2-y^2}, -\sqrt{R^2-y^2}+D)} \frac{C}{(D+x)^2+y^2)^2} dxdy}{\int_{-R}^{R} \int_{-\sqrt{r^2-y^2}}^{min(\sqrt{R^2-y^2}, -\sqrt{R^2-y^2}+D)} 1 \cdot dxdy}. \quad (6)$$

Although this equation is not solvable, we can see that INT_{RTS} is a positive value that can be represented in terms of C, D and R. We omit the effect of INT_{RTS} to simplify our analysis, thus equation (4) is now represented as follows,

$$\overline{INT_x} \approx \frac{C}{(dR)^2}. \tag{7}$$

Note that the average interference is actually smaller since we omitted INT_{RTS} in equation (4). To look into the worst-case scenario, we assume that one node attempts to transmit per one transmission area $\alpha\pi R^2 (\alpha \approx 1.609)$. The constant is obtained by averaging the area covered by the two communicating nodes. The calculations of α and A_1 are presented in the Appendix. Using equation (20) in the Appendix, the maximum number of nodes attempting to transmit is,

$$N = \frac{\pi d^2 - \pi R^2 - (\pi R^2 - 2A_1)}{\alpha\pi R^2} = \frac{(\pi d^2 - 2\pi R^2 + 2A_1)}{\alpha\pi R^2}, \text{ where } \alpha \approx 1.609. \tag{8}$$

So, the worst-case average total interference that the transmitting nodes in the shaded area in Fig. 3 can affect receiver B is

$$\overline{INT_{Tot}} = \overline{INT_x} \cdot N \approx (\frac{C}{(dR)^2})(\frac{\pi d^2 - 2\pi R^2 + 2A_1}{\alpha\pi R^2}). \tag{9}$$

where $\overline{INT_x}$ and N are obtained from equations (7) and (8) respectively. Since we should consider the worst-case interference effect of the hidden nodes in the entire network, d should diverge to infinity. Thus,

$$\overline{INT_{HT}} \leq \frac{C'}{R^4}, \text{ where } d \to \infty \text{ and } C' = \frac{C}{\alpha\pi}. \tag{10}$$

Equation (10) shows the worst-case total interference bound of the hidden nodes in the entire network that are not covered by the RTS/CTS. This shows that the worst-case interference can be bounded by $C'R^{-4}$. Now by inserting equation (10) into equation (2), we obtain

$$\frac{C/D^4}{\overline{INT_{HT}} + \eta} = \frac{CD^{-4}}{C'R^{-4} + \eta} \geq SINR_n. \tag{11}$$

where C is defined in equation (2), D is the distance between the sender and receiver, R is the basic transmission range and η is the noise. This result shows when considering the worst-case, the effect of the hidden node could be much worse than expected.

Next, we give numerical examples to show the effect of the worst-case interference scenario. The physical parameters used in the example are shown in Table 1. We use SNR threshold values of Agere Systems Chipset, 802.11b W-LAN card [6]. Inserting these values into equation (11) yields,

$$D \leq [(1.25 \times 10^{-10}) \times SINR_n]^{-\frac{1}{4}}. \tag{12}$$

Table 1. Physical parameters used in the example

Transmit Power	15.0 dBm
Antenna height	1 m
Antenna gain	1
Constant background noise	-91.0 dBm
11Mbps SNR threshold	15 dB
5.5Mbps SNR threshold	12 dB
2Mbps SNR threshold	9 dB
1Mbps SNR threshold	6 dB

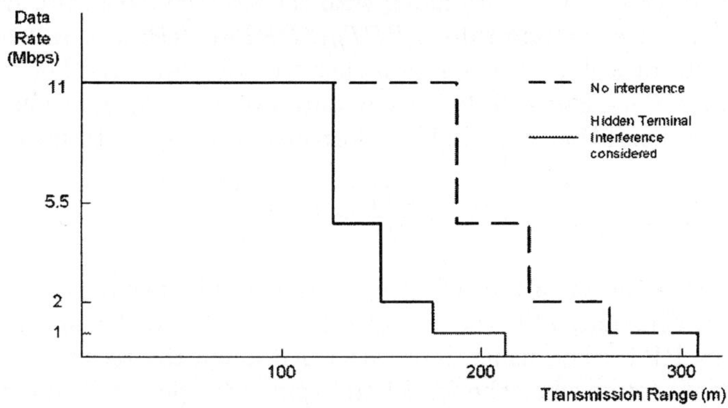

Fig. 5. This figure shows the transmission range for each rate i.e. modulation schemes in 802.11b. The dashed line represents when we assume that there is no interference whereas the solid line shows when the worst-case interference is considered

Fig. 5 shows how the transmission range should be reduced for each transmission rate i.e. modulation schemes in 802.11b. The dashed line shows the transmission range acquired by equation (3) where we assume that the RTS/CTS completely prevents the hidden terminal problem, so that no interference is present. The solid line shows the transmission range acquired by equation (12), where the worst-case interference is considered. We can easily see that the transmission range will considerably decrease due to the interference of hidden terminals not covered by RTS/CTS transmissions. In essence, the maximum transmission range that will not be affected by the interference is in-between the best and worst case.

4 Transmit Power Controlled Multi-rate MAC Protocol

In this section we propose a simple multi-rate MAC protocol that can be used when transmit power control (TPC) is employed. We use the simple intuition that for a node to correctly receive a data frame, it must satisfy two conditions. First, the receive power should exceed a certain receive power threshold (RPT). Second, the SINR should also surpass a certain threshold. We will call the two thresholds as, RPT and SINR threshold respectively. Similar to Sect. 3, the RBAR [3] protocol and the IEEE 802.11b [1] is used for multi-rate physical and MAC, so that the available data rates are 1, 2, 5.5 and 11Mbps.

4.1 Transmit Power Control (TPC)

For a node to correctly receive a data frame, the receive power should go beyond the RPT. The RPT should vary along with the selected data rate. Assume that the RPT value for each data rate is RPT_{R1}, RPT_{R2}, $RPT_{R5.5}$ and RPT_{R11}. As the higher data rate should require a higher receive power, $RPT_{R1} < RPT_{R2} < RPT_{R5.5} < RPT_{R11}$. Say a node is using data rate i, using equation (2) in Sect. 3 the following condition must hold to correctly receive a data frame:

$$P_R = \frac{CP_t}{D^4} \geq RPT_{R_i}. \tag{13}$$

P_R is the receive power of the RTS frame, P_t is the transmit power of the RTS frame, C is a constant, D is the distance between the sender and receiver, and RPT_{Ri} is the RPT when using data rate i. Although the receiver can correctly receive the data frame when the sender transmits with power P_t, it can still properly receive it even when the sender transmits with a lower power P_t' ($P_t' \leq P_t$), as long as it satisfies the above equation. Using this idea, we perform transmit power control (TPC) by adjusting the transmit power of the sender.

$$P_t' = P_t \cdot \frac{RPT_{R_i}}{P_R}. \tag{14}$$

P_t' indicates the transmit power of the sender when using rate i. The receiver should send this information to the sender by adding it to the CTS frame along with the selected data rate used in RBAR.

4.2 Preventing Hidden Terminal Interference

For a node to correctly receive a data frame, the SINR should surpass the SINR threshold ($SINR_{th}$). Similar to the RPT value discussed in Sect. 3.1, the SINR threshold value should change with the selected data rate. For data rates 1, 2, 5.5 and 11Mbps, the $SINR_{th}$ values are $SINR_{th\_1}$, $SINR_{th\_2}$, $SINR_{th\_5.5}$ and $SINR_{th\_11}$, respectively. We assume that the current SINR value of the RTS reception can be estimated as in [3]. We also assume that there is no interference other than noise in the current RTS reception.

$$SINR = \frac{P_R}{\eta}. \tag{15}$$

η is the value of thermal noise. Say a node uses data rate R_i selected by RBAR The node should follow the next constraint to correctly recieve a data frame:

$$\frac{P_{Ri}}{\eta + \frac{CP_t}{d^4}} \geq SINR_{th\_i}. \tag{16}$$

P_{R_i} is the receive power for each data rate i when TPC is employed as shown in section 4.1. The right hand part of the denominator denotes the interference that a node outside the CTS transmission range can cause (See Sect. 2). In other words, equation (16) takes into account the effect of the hidden terminal that is outside the CTS transmission range. Therefore, d is the interference range of this transmission. We control the transmit power of the CTS frame to cover the interference range. So the following should hold:

$$\frac{CP_{t\_CTS}}{d^4} \geq RPT_{R1}. \tag{17}$$

P_{t_CTS} is the controlled transmit power of the CTS frame, and RPT_{R1} is the receive power threshold of the CTS frame. Therefore, combining equations (16) and (17), the controlled CTS transmit power should be

$$P_{t\_CTS} \geq \frac{d^4}{C} RPT_{R1} \geq \frac{P_{t_{RTS}} \cdot RPT_{R1}}{SNR_{th\_i} \cdot P_{R_i} - \eta}. \tag{18}$$

We can avoid the hidden terminal's interference by using the above equation. Although this method assures the CTS frame to cover the interference range, one argument that can come out is that the CTS frame transmitted with a higher power level can also interfere some other data receptions. This argument was also presented by D. Qiao [7]. We use the similar idea that CTS frames are normally shorter than data frames, and it would not be severe as the interference caused by the data frames. We leave the evaluation of this protocol as future work.

5 Conclusion

In this paper, we analyzed the worst-case scenario of collision probability induced by the hidden terminal problem in multi-rate ad hoc networks. We show that the interference caused by the nodes in the area that is not covered by the RTS/CTS is bounded by $C'R^{-4}$, where C' is a constant and R is the distance between the two transmitting nodes. Analytic results showed that the interference could shorten the transmission range about 30 percent even when the basic RTS/CTS handshake mechanism is used. We also propose a simple multi-rate MAC protocol to prevent the hidden terminal problem when transmit power control (TPC) is employed. The proposed protocol should be very effective when using multi-rate data transmission, since the CTS frames would effectively cover the interference range of the receiver.

References

1. IEEE standard for Wireless LAN Medium Access Control (MAC) and Physical Layer (PHY) specifications. ISO/IEC 8802-11: (1999(E))Aug. 1999
2. A. Kamerman, L. Monteba: WaveLAN II: A high-performance wireless LAN for the unlicensed band. Bell Labs Technical Journal. (1997) 118–133
3. G. Holland, N. Vaidya, P. Bahl: A Rate-Adaptive MAC Protocol for Multi-Hop Wireless Networks. In Proc. ACM Mobicom'01 (2001)
4. K. Xu, M. Gerla, and S. Bae: How Effective is the IEEE 802.11 RTS/CTS Handshake in Ad Hoc Networks? In Proc. IEEE Globecom'02 (2002)
5. T. Rappaport: Wireless Communications: Principles and Practice. Prentice Hall, New Jersey (1996)
6. Agere Systems Chipset http://www.agere.com
7. D. Qiao, S. Choi, A. Jain and K. Shin: MiSer: An Optimal Low-Energy Transmission Strategy for IEEE 802.11a/h In Proc. ACM Mobicom'03. (2003)

Appendix: The Average Transmission Area

In this section we show how the average transmission area is obtained. As shown in Fig. 6, D is the distance between the two communicating nodes and R is the transmission range. The overlapping area over the two circles, A_1, can be obtained as follows.

$$A_1 = \frac{R^2(2\theta)}{2} - \frac{R^2\sin(2\theta)}{2}, \text{ where } \theta = \cos^{-1}\frac{D}{2R}. \quad (19)$$

So, the average area can be obtained by averaging the distance between the communicating nodes as follows.

$$A = 2(\pi R^2) - A_1.$$

$$\overline{A} = \int_0^R A dx = \int_0^R \{(\pi R^2) - A_1\} dx = \alpha \pi R^2, \text{ where } \alpha \approx 1.609 \quad (20)$$

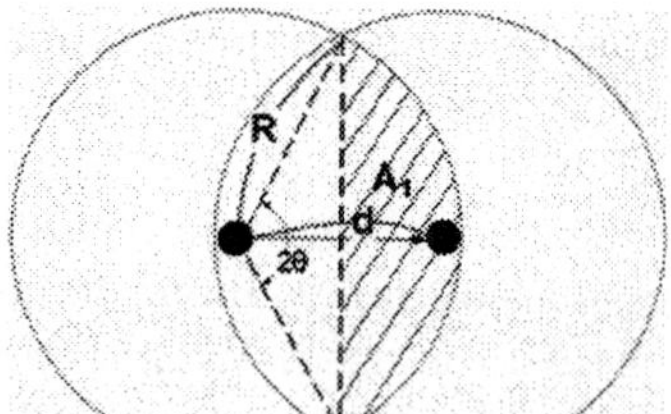

Fig. 6. This figure shows how the average transmission area is obtained

IPv6 Addressing Scheme and Self-configuration for Multi-hops Wireless Ad Hoc Network*

Guillaume Chelius[1], Christophe Jelger[2], Éric Fleury[1], and Thomas Noël[2]

[1] CITI/INSA de Lyon – ARES/INRIA
Villeurbanne 69621 Cedex - France
Guillaume.Chelius@insa-lyon.fr
Eric.Fleury@inria.fr

[2] LSIIT - UMR 7005 CNRS-ULP
Universite Louis Pasteur
Strasbourg - France
{jelger, noel}@dpt-info.u-strasbg.fr

Abstract. Next generation mobile communication systems will comprise both WLAN technologies and ad hoc networks. Ad hoc networks are formed by the spontaneous collaboration of wireless nodes. When communication to the Internet is desired, one or more nodes must act as gateways for the ad hoc network. In this case, global addressing of ad hoc nodes is required. In this paper, we present an IPv6 addressing architecture in order to be able to support "pure" spontaneous IPv6 ad hoc networks but also to allow seamless integration between wireless LANs and ad hoc networks. It implies the possibility to discover a gateway/prefix pair which is used in order to build an IPv6 global address and, when necessary, to maintain a default route towards the Internet.

1 Introduction

Research efforts aiming at merging wireless LAN and ad hoc networking by considering advantages of both WLAN and ad hoc principles have been recently increasing [1, 2, 10, 15]. Hybrid networks, the extension of WLAN/cellular networks using ad hoc connectivity, offer obvious benefits. On one hand, they allow an extension of the WLAN coverage using ad hoc connectivity and on the other hand they provide a global Internet connectivity to ad hoc nodes. The convergence of both ad hoc networks and infrastructures will be possible if the ad hoc architecture is flexible enough. By ad hoc architecture, we denote a set of rules dealing with the addressing and routing schemes that must be set up for the ad hoc network to offer basic services. In this paper we will try to give an appropriate answer to the two following basic questions: What is an ad hoc address ? What element is identified by an ad hoc address ? We also present a protocol which can be used by an ad hoc node to dynamically select a gateway and

* This work was supported by the French Ministry of the Telecommunications and Research. It was part of the RNRT SAFARI project.

C. Kim (Ed.): ICOIN 2005, LNCS 3391, pp. 489–498, 2005.

create an associated IPv6 global address. The originality of our proposal is the introduction of the concept of prefix continuity in an ad hoc network.

The article is organized as follows. In section 2 we describe the need for an IPv6 ad hoc architecture in order to fulfill the ad hoc fundamentals. Section 3 describes a key issue when dealing with Internet connectivity and ad hoc network. Section 4 presents our IPv6 architecture named ana6 [3]. We describe its main concepts that were designed to offer IPv6 ad hoc networking following the MANet philosophy and enabling global Internet connection. Section 5 presents our prefix dissemination that guarantees a prefix continuity inside the ad hoc network in order to ensure that there exists, between a node A and its gateway G, a path of nodes such that each node on this path uses the same prefix P than the node A and its gateway G. Sub-networks (with respect to prefixes) are automatically created and dynamically maintained when multiple gateways are available. Moreover, this concept ensures that each sub-network forms a connected graph of nodes which all use an identical prefix. We conclude this article with section 6.

2 Request for an IPv6 Ad Hoc Architecture

The fundamental service that must be offered by an ad hoc environment is to allow communication between all mobile nodes of the network. One node must be able to reach any other node. Since some nodes may be out of range or since some nodes may not share the same medium, it is necessary to define specific routing mechanisms that allow multi-hop routing. The MANet group of the IETF – *Internet Engineering Task Force* – proposes an architecture in which the basic element is the MANet node: "a MANet node principally consists of a router, which may be physically attached to multiple IP hosts (or IP-addressable devices), which has potentially *multiple* wireless interfaces–each interface using a *different* wireless technology". As stated in [5], a MANet node using wireless technologies A and B (*e.g.* frequency A and frequency B) can communicate with any other node possessing an interface with technology A or B. This means that the unicast routing algorithm must operate over a multi-graph composed of several physical graphs and that an ad hoc node is the union of all its interfaces involved in the ad hoc network. The unicast routing must offer a global connectivity over all the ad hoc interfaces

Given the previous remarks, we argue that it should be possible to address an ad hoc node regardless of the interface it will receive the packet from. The IP address(es) used to identify an ad hoc node should be associated to all its interfaces involved in the ad hoc network. This differs from a classical use of IP where an IP address usually identifies an interface and not a node (a set of interfaces). Gathering several interfaces under a common IP address not only enables routing over a multi-graph or multi-interface topology but also provides interface mobility. It seems important to provide interface mobility (*a.k.a.* vertical hand off) inside an ad hoc node without requiring to set up a costly IP mobility process. Basically, it reinforces the interactions between the layer 2 and the layer

3 and between the end user and the technology. From the user/application point of view, it provides the opportunity to set up and insure network connectivity whatever the available interfaces and to switch smoothly from one interface to another. It also introduces the possibility to optimize several user/host parameters such as the cost of the connection, the QoS and the energy.

3 Internet Connectivity and Related Work

3.1 Internet Connectivity

Another key issue with an ad hoc network is Internet connectivity. It is indeed of the highest importance that such a network can be connected to the Internet in order to offer Internet services (e.g. email and web access) to its users. Furthermore and to be natively reachable from outside the ad hoc network (*i.e.*, without any network address translation mechanism), each node in the ad hoc network must have a global IPv6 address. The presence of a gateway to the Internet therefore implies the diffusion, in the ad hoc network, of an IPv6 prefix which can be used by each node to build its global IPv6 address. Within the Internet, routing to the site owning this prefix is assumed to be in place. Depending on the routing protocol in use within the ad hoc network, ad hoc nodes must also be configured to be able to communicate with nodes in the Internet. Unfortunately, the particular nature of an ad hoc network makes it impossible to use the classical IPv6 mechanisms used in wired networks in order to propagate prefix information, mainly because they have been designed to work on a shared broadcast link.

In this paper we therefore propose a protocol that can be used in a multi-hop ad hoc network, with both proactive and reactive protocols, in order to propagate gateway and prefix information. As in classical IPv6 wired networks, gateways are responsible for prefix announcement. This information propagates in a hop-by-hop manner with each intermediate node being in charge of updating it. Actually, a node only forwards the information sent by a gateway if it decides to use the announced prefix to build its IPv6 global address. This original propagation method naturally leads to a concept that we call *prefix continuity*. Moreover, the protocol allows a node to quickly react to topological changes. It also supports multiple gateways and multiple prefixes, and an extended version also permits to propagate Domain Name Server DNS information.

3.2 Related Work

Wakikawa *et al.* [12] have proposed a method that could also be used with any kind of routing protocols. With their proposal, an ad hoc node broadcasts a request to obtain a prefix with global scope. The gateway replies to the originator of the request with a message containing the prefix. The node receiving this information creates a global address and adds a particular entry in its routing table. They also give technical details in the particular case of the AODV routing protocol. Xi *et al.* [16] also propose a similar mechanism based on a broadcasted request followed by a reply. They also extend this model with periodical broadcasts (containing prefix information) sent by each gateway, and with the possibility for an intermediate node to respond to request messages. The two papers also consider the use of Mobile IPv6 within the ad hoc network. They also shortly introduce the notion of gateway selection, but none of them gives details about how this would be achieved.

While these two papers have proposed some promising ideas, they both have few weaknesses. First and in the case of multiple prefixes, the problem of prefix continuity is not considered. This constraint, detailed in Section 5.2, imposes that nodes which share a common prefix must always for a continuous neighborhood. Second, both proposals do not consider the unpredictable topological changes that occur in an ad hoc network, in the sense that they do not specify how the prefix information is updated (or changed) in time, a crucial consideration with ad hoc networks. Third, when multiple gateways are present and periodical broadcast is considered, each gateway floods the entire ad hoc network with its prefix announcement, leading to unnecessary bandwidth consumption. Our proposal introduces a number of mechanisms which aim to solve all of the above mentioned problems. They are defined in Section 5.

4 Ana6 Architecture

In order to fulfill the requests described in section 2 we need to introduce new notions and some features dedicated to ad hoc networks inside IPv6. The first goal is to enable an IPv6 support for pure autonomous ad hoc networks (where no IPv6 global prefix is available) and also for ad hoc networks connected to the Internet. In consequence, we introduce ad hoc-local addresses whose validity is limited to ad hoc networks and that can be auto-configured without any infrastructure. The second goal is to easier support for multi-interface routing or interface mobility.

4.1 Ad Hoc Local Address

The IPv6 addressing architecture proposes two local unicast addresses and their equivalent multicast scope: link-local and site-local [7]. Unfortunately, the use of IPv6 link-local unicast and multicast addresses is unsuitable to ad hoc networks. A link-Local unicast address refers to a single interface and its validity is limited to the interface link. Thus link local addresses "**should not**" be routed, preventing their use in a flat multi-hop ad hoc environment. One may imagine to use site local addresses in order to solve the problem of addressing ad hoc nodes. However, since an ad hoc network may be included in a larger site or spread over several sites, a specific ad hoc use of site-local addresses appears to be inappropriate. Moreover, site locals addresses *may* probably be deprecated by the IETF [8].

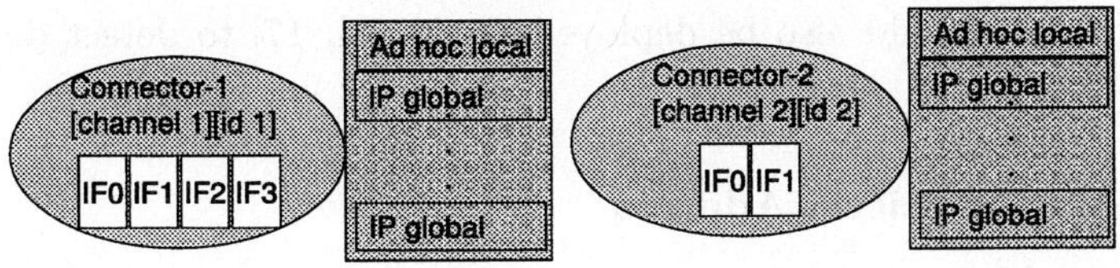

Fig. 1. An ad hoc connector gathers several interfaces. It is associated to pool of addresses (1 ad hoc-local, $k \geq 0$ global). Several connectors may be set up in one ad hoc node. Each connector is defined by the channel number and its Id.

To locally address ad hoc nodes (more precisely ad hoc connectors as we will see in the next section), we introduce a third IPv6 local-use unicast address: "*ad hoc-local addresses*". The validity of an ad hoc-local address is limited to an ad hoc network. We define an ad hoc network as a maximal connected set of ad hoc interfaces. Note that this definition allows no controversy about ad hoc networks boundaries, as opposed to the tricky problem of defining a site frontier which will probably be responsible for the deprecation of site-local addresses. The introduction of ad hoc local addresses provides a basic identification support for ad hoc nodes that can be extended by other configuration mechanisms such as stateless global addresses configuration. Ad hoc-Local addresses have the following format: `fe40::[connector id]/128` where the *connector id* is 64 bit long and will be defined below. The ad hoc local scope is for use in a single ad hoc network and is valid in all ad hoc sub-networks of an ad hoc network.

4.2 Ad Hoc Connector

Now, let us consider an ad hoc network as a multi-graph composed of several physical graphs. As already said, this architecture is made possible by the attribution of one/several common IP address(es) to all interfaces involved in the ad hoc network. To gather several network interfaces in a single addressable entity, we introduce the notion of ad hoc connectors. An ad hoc connector is the basic element of ad hoc networks. It virtualizes several network interfaces into a single addressable object. A host may have several ad hoc connectors and an interface may be bound to several ad hoc connectors. The ad hoc connector is associated to a set of addresses which identify indistinctly all bounded interfaces. This set is composed of an ad hoc-local address and eventually zero, one or more global addresses. The ad hoc-local address ensures connectivity in the ad hoc network and the global ones enable Internet connectivity. Note that each ad hoc interface, *i.e.*, a network interface bound to an ad hoc connector, use and recognize all addresses associated to its connector. In the network, an ad hoc connector is identified by a 64bits value, the ad hoc identifier, and a 16bits channel value. For the ad hoc network to correctly behave, it is desired for ad hoc IDs to be unique. It is the user responsibility to ensure uniqueness of its IDs. One can setup pseudo-unique IDs based on host interface MAC addresses or cryptographic mechanisms such as crypto-unique identifiers [11]. Note that specific protocols could also be deployed [9, 13, 14, 17] to detect duplicate ad hoc addresses.

4.3 Ad Hoc Multicast Address

In order to address multiple ad hoc connectors and to limit the scope of a multicast group within an ad hoc network, we use the subnet multicast scope as defined in [6, 7] to define ad hoc-local multicast addresses. An ad hoc local multicast address has the following format `ff03:0:0:0:[group id]`. Ad-hoc local multicast information should not be forwarded through an interface that is not involved in the ad hoc network, *i.e.*, that is not connected to an ad hoc connector.

As for classical IPv6 multicast scopes, *e.g.*, link or site scope, we set up predefined multicast addresses in addition to the ones given in [6, 7]. We first predefine the "*all ad hoc nodes*" address `ff03::1` that identifies the group of all IPv6 ad hoc nodes within an ad hoc network. We predefine the "*All ad hoc routers*" address `ff03::a` that identifies the group of all IPv6 ad hoc routers (*i.e.*, an ad hoc node which may route packets between ad hoc network(s) and non ad hoc network(s)) within an ad hoc network. Finally, we predefine the "*all ad hoc sub-routers*" address `ff03::b` that will identify the group of all IPv6 ad hoc sub-routers (*i.e.*, an ad hoc node which may route packets between two or more ad hoc sub-networks or channels) within the ad hoc network.

4.4 Channel Multicast Addresses

As said in the previous section, each connector is associated to a 16bits value called the *channel value*. The channel value is used to support logical ad hoc sub-networks inside an ad hoc network. This value indicates which ad hoc sub-network (*a.k.a.* channel) the ad hoc connector is connected to. Once again, we define a channel as a maximal connected set of ad hoc connectors sharing a common channel value. This definition based on the maximal connectivity avoids any ambiguity on sub-network boundaries. By default, a connector has a channel value of 0. It means that the connector does not belong to any ad hoc sub-network. The channel value may change during the ad hoc connector life but it is important to notice that channel mobility does not lead to ad hoc local address changes and thus a node does not need to perform any kind of IP mobility when moving from one channel to another.

Channels are used to limit the diffusion of information and the scope of a multicast group to a subset of ad hoc connectors, *i.e.*, the subset of connected ad hoc connectors sharing the same channel value. We introduce the "*channel-local multicast addresses*" as a subset of the ad hoc-local multicast addresses. Format of a channel-local multicast address is `ff03:0:0:[channel value]:[group id]`. Information broadcasted to a channel-local multicast address is limited to the channel given by the channel value of the multicast address. More precisely, ad hoc nodes must not forward any multicast packet limited to an ad hoc sub-network with channel value X through an interface that is not connected to an ad hoc connector with channel value X. For more informations on how channels can be useful in hybrid architectures or Internet-ad hoc services continuum please refer to [4].

5 Dissemination of Global Prefix Information

In addition to the ANA6 addressing architecture, we have also designed a simple and efficient method in order to disseminate global prefix information in ad hoc networks.

5.1 Forwarding/Propagation of Prefix Information

Our proposal relies on a periodical hop-by-hop exchange of information between each node and its directly connected neighbors. Each gateway is responsible to initiate the sending of gateway and prefix information, which then propagates away from it in a hop-by-hop manner. Depending on the network topology and on the number of gateways, each node may receive multiple gateways and prefixes information. In short, each intermediate node selects the most appropriate information from one of its neighbors (which becomes what we define as its *upstream neighbor*). The node subsequently increases the *distance* field of the selected information and finally propagates the updated information to its neighbors. This field is set to zero by a gateway, as it gives the distance (in hops) between the sender of a GW_INFO message and the gateway that originally sent the message. Also note that a node only forwards the information (*i.e.*, prefix) that it decided to use to create its IPv6 global address. With proactive routing protocols, the node also creates a default routing table entry with its upstream neighbor as next hop. The messages which contain gateway and prefix information are noted GW_INFO messages. The propagation technique itself is illustrated by Fig. 2. The format of such messages is shown in [9].

5.2 Prefix Continuity

An inherent consequence of the propagation technique used to disseminate the GW_INFO messages is what we call *prefix continuity.* Our proposal ensures that any node A that selected a given prefix P has at least one neighbor with prefix P on its path to the selected gateway G. The prefix continuity feature ensures that there exists, between the node A and its gateway G, a path of nodes such that each node on this path uses the same prefix P and gateway G than the node A. While prefix continuity permits to avoid routing problems, it also establishes a topological organization within an ad hoc network, *i.e.*, the network becomes divided in sub-networks, each being formed by a contiguous gathering of nodes using the same prefix. It can be explained as follows. Say, if a node A uses the prefix/gateway pair (P,G) advertised by its upstream neighbor B, B necessarily uses the same pair as advertised by its own upstream neighbor C. Recursively,

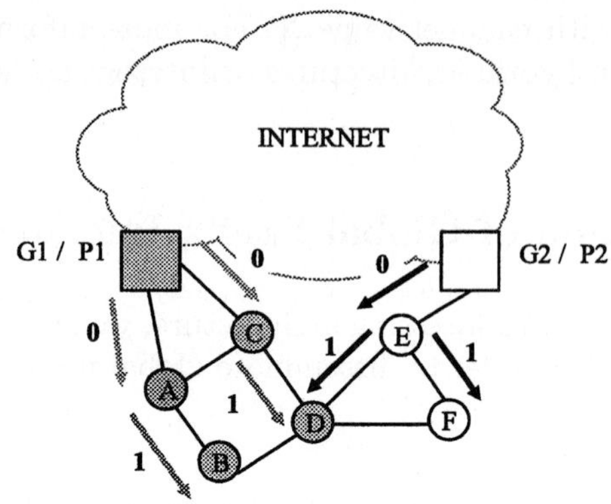

Fig. 2. Hop-by-Hop propagation of GW_INFO messages

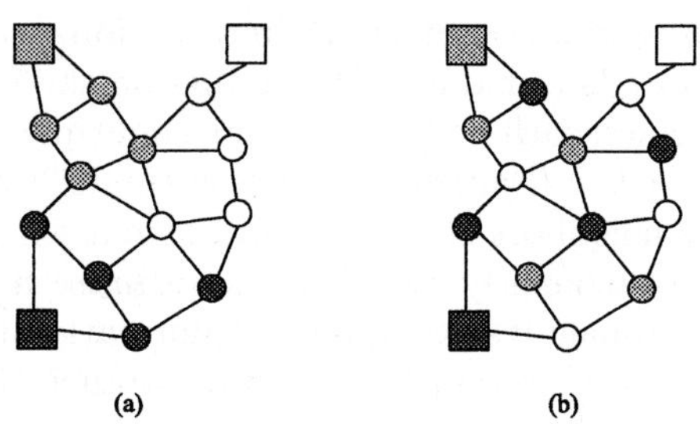

Fig. 3. Ad hoc network with (a) and without (b) prefix continuity

there must exist a path of nodes that use the pair (P,G) between A and G. An example of ad hoc network with and without prefix continuity is shown on Fig. 3.

To maintain this continuity, each node must ensure that it does not become isolated from other nodes which share the same prefix. In contrast to previous proposed work ([12, 16]), our protocol ensures that the prefix continuity requirement is satisfied. Each node is responsible to permanently check its neighborhood to detect the loss of neighbors which share the same prefix. The periodical sending of GW_INFO messages and a neighborhood list maintained by each node easily allows to detect such an event.

A first advantage of prefix continuity is that it permits to avoid some routing problems and overhead. For example and in contrast to other proposals, a node does not need to use an IPv6 routing header in order to specify via which gateway its packets must go through when the destination is outside the ad hoc network. This is because the default route of a node points to its upstream neighbor which necessarily uses the same gateway (recursively the packet will eventually reach the gateway). Without prefix continuity, a node must indeed specify via which gateway its packets must go through in order to avoid ingress filtering. Our proposal is also very robust to network partitioning and it moreover does not require any special mechanism in order to handle such situations. If a network partition occurs and if a node becomes isolated from its current gateway, it will quickly receive GW_INFO messages from a new gateway and will eventually acquire a new global address. Another advantage of prefix continuity is that it establishes a topological organization within an ad hoc network, *i.e.*, the network becomes divided in sub-networks, each being formed by a contiguous gathering of nodes using the same prefix.

5.3 Prefix Selection

We have proposed two different algorithms used by a node to select a prefix/gateway. The first algorithm ensures that a node always selects the closest gateway, whatever prefix it uses. In contrast, the second algorithm ensures that a node keeps its current prefix as long as it has neighbors with the same prefix, whatever distance it is from its current gateway. To do so, each node maintains and updates a list of pairs of the form (SRC_ADDR ; PREFIX) for each GW_INFO packet received. SRC_ADDR is the source address in the IPv6 header of the packet containing the GW_INFO message, and PREFIX is the IPv6 prefix contained in the message. This list allows a node to detect if it does not become isolated from nodes which share the same prefix. We also consider that the global address acquired by an ad hoc node should be used as the Mobile IPv6 care-of address of the node. MIPv6 is used with mobile nodes to maintain connections at the transport layer. Each change of global address in the ad hoc network will therefore trigger the sending of at least one binding update message.

5.4 DNS Extension

As an extension to the proposed protocol, domain name server (DNS) information can also be sent in GW_INFO messages. This allows a node to select a gateway that also permits to reach a DNS server. The GW_INFO extension can easily be extended to include a field which contains the IPv6 global address of a DNS server. Upon reception of such a message, an ad hoc node simply uses the address of the DNS server in order to resolve names and addresses of hosts that are out of the ad hoc network.

6 Conclusion

The main goal of this paper was the applicability of IPv6 for the efficient design of ad hoc network architectures that support both full spontaneous mode, i.e., without any infrastructure and thus requiring an autonomous local ad hoc addressing scheme, and a seamless coexistence with existing WLAN networks that provide a global connection to the Internet. Our proposal, ana6, fulfills all requirements and introduces a number of innovative concepts related to the architecture of ad hoc networks. We have proposed the use of a new multicast scope, namely "*adhoc-local*", in order to limit the diffusion of multicast data within an ad hoc network. This feature is closely related to the use of "*ad hoc connectors*", which are a logical abstraction of physical interfaces. This allows in particular to address nodes within an ad hoc network. Finally, we have proposed a protocol which can be used in order to create sub-networks (with respect to global prefixes) that respect the concept of prefix continuity. Ana6 [3] and the prefix delegation protocol have been implemented for the Linux and FreeBSD systems and are currently deployed and tested in the SAFARI project.

References

[1] R-S. Chang, W-Y. Chen, and Y-F. Wen. Hybrid wireless network protocols. *IEEE TVT*, 52(4):1099–1109, July 2003.

[2] H-C. Chao and C-Y. Huang. Micro-mobility mechanism for smooth handoffs in an integrated ad-hoc and cellular ipv6 network under high-speed movement. *IEEE TVT*, 52(6):1576–1593, November 2003.

[3] G. Chelius and E. Fleury. Ipv6 addressing architecture support for ad hoc. Internet draft, IETF, August 2002.

[4] G. Chelius and E. Fleury. IPv6 addressing architecture support for ad hoc networks. In *IEEE IWWAN 2004*, Oulu, Finland, June 2004. IEEE.

[5] S. Corson and J. Macker. Mobile Ad hoc Networking (MANET): Routing Protocol Performance Issues and Evaluation Considerations. IETF RFC 2501, January 1999.

[6] S. Deering and R. Hinden. Internet protocol, version 6 (ipv6) specification. IETF RFC 2460, IETF, February, 1998.

[7] R. Hinden and S. Deering. Internet protocole version 6 (ipv6) addressing architecture. IETF RFC 3513, IETF, April, 2003.

[8] C. Huitema and B. Carpenter. Deprecating site local addresses. Internet draft, IETF, November, 2003.

[9] C. Jelger and T. Noël. Gateway and address autoconfiguration for IPv6 ad hoc networks. Internet draft, IETF, April 2004.

[10] H. Li and D. Yu. Performance comparison of ad-hoc and cellular based routing algorithms in multihop cellular networks. In *WPMC 2002*, October 2002.

[11] G. Montenegro and C. Castelluccia. Statistically unique and cryptographically verifiable identifiers and addresses. In *NDSS'02*, San Diego, USA, February 2002.

[12] R. Wakikawa, J. Malinen, C. Perkins, A. Nilsson, and A. Tuominen. Internet Connectivity for Mobile Ad hoc Networks. *Wirel. Comm. and Mobile Computing*, 2(5):465–482, August 2002.

[13] K. Weniger. Passive duplicate address detection in mobile ad hoc networks. In *IEEE WCNC 2003*, New Orleans, USA, Mars 2003.

[14] K. Weniger and M. Zitterbart. Ipv6 autoconfiguration in large scale mobile ad-hoc networks. In *European Wireless 2002*, Florence, Italy, February 2002.

[15] C. Wijting and R. Prasad. Evaluation of mobile ad-hoc network techniques in a cellular network. In *IEEE VTC*, pages 1025–1029, 2000.

[16] J. Xi and C. Bettstetter. Wireless Multihop Internet Access: Gateway Discovery, Routing, and Addressing. In *3GWireless*, May 2002. San Francisco, CA, USA.

[17] H. Zhou, L. Ni, and M. Mutka. Prophet address allocation for large scale manets. *Ad Hoc Networks*, 1:423–434, 2003.

[3] http://sourceforge.net/projects/anax

SDSR: A Scalable Data Storage and Retrieval Service for Wireless Ad Hoc Networks

Yingjie Li and Ming-Tsan Liu

Computer Science & Engineering Department
The Ohio State University, Columbus, Ohio, 43210, USA
{yingjie, liu}@cis.ohio-state.edu

Abstract. We present the framework of SDSR, a Scalable, efficient, robust and load-balanced Data Storage/Retrieval service for large scale wireless ad hoc networks. SDSR hashes each data key to normalized geographical coordinates (x,y), which can be seen as a *rendezvous* point for storing/retrieving the data in a unit grid. SDSR achieves scalability, robustness and load-balancing by partitioning the network into hierarchical grids of increasing sizes and replicating the data item into each grid. The storage location in each grid is determined by scaling the normalized coordinates to the corresponding grid size. SDSR retrieves a data item in the same way as it stores the data item. We show that in query dominant, large scale wireless ad hoc networks, SDSR performs better than existing schemes in terms of energy efficiency, query latency, hotspot usage, and resilience to clustering failures. It scales well when the network size and the number of queries increase.

1 Introduction

Wireless ad hoc networks are constructed for sharing information among wireless hosts. The energy constrained wireless devices as well as the frequent network topology change make traditional information sharing schemes such as broadcasting, storing data centrally (CS) or locally (LS) undesirable, and thus necessitate and challenge the design of efficient and robust data dissemination schemes for large scale wireless ad hoc networks.

Previous research on data dissemination schemes such as [1] [2] [3] mainly focused on improving data accessibility/reliability for wireless ad hoc networks with high frequency of network partitions; however, less has been done for achieving data access scalability and efficiency for wireless ad hoc networks with reasonable node density and thus few occurrences of network partitions.

In this paper, we consider the problem of storing/retrieving data in a wireless ad hoc network in a scalable, efficient, and robust manner, where data can be resources of any interests, such as service advertisements [4], files, locations of nodes, presence of an object, etc. We assume a peer to peer network architecture where each node can store/retrieve data to/from the network. Such a network can be constructed either by personal devices such as notebooks or PDAs in a wide area civilian environment, or by military devices in a large battlefield.

C. Kim (Ed.): ICOIN 2005, LNCS 3391, pp. 499–509, 2005.

Contributions of the Paper. We present a distributed and Scalable Data Storage and Retrieval service, SDSR, as a solution to the problem addressed above. SDSR performs similar functions as Geographic Hash Table (GHT) proposed in [5], since both schemes hash each data key to a geographical location, which serves as a *rendezvous* point for storing/retrieving the data item. The difference is that in GHT, the hashed location is one fixed location in the current network, while in SDSR, the hashed location is fixed (x,y) coordinates in a unit grid and can be scaled to get a *rendezvous* point for a network of any size. SDSR achieves scalability, robustness and load-balancing by partitioning the network into hierarchical grids of increasing sizes and replicating the data item into each hierarchical grid. The storage location in each grid is determined by using the normalized coordinates as an offset and scaling it to the corresponding grid size. SDSR utilizes Greedy Perimeter Stateless Routing (GPSR) [6] to store the data item to nodes closest to the storage locations. These nodes serve as storage servers for the data item. SDSR retrieves a data item in a similar hierarchical way as to store the data item. SDSR has the following notable properties:

- **Scalability:** (i) the number of storage servers for a data item increases logarithmically to the increase of network size; (ii) the per node data storage and communication costs increase as a small fraction of the increase of the network size; (iii) the task of storing and serving data items inserted into the network is distributed among nodes in the network.
- **Fault-tolerance:** (i) it handles node joins/fail-stops and node mobility locally with an indexing service; (ii) it is resilient to clustering failures by replicating each data item at nodes distributed in different network regions.
- **Efficient resource utilization:** (i) it constructs a multicast tree to route each data item to its storage locations, which performs better than rooted shortest path and minimum spanning trees in terms of the combination of communication cost, network delay, and data delivery success probability; (ii) it employs a light weight indexing service to maintain the consistent view of data storage while avoiding unnecessary broadcast in a grid.
- **Query locality friendly:** queries in networks usually exhibit locality [7], meaning that nodes are more interested in data generated nearby than data generated far away. To comply with query locality as well as to increase the data accessibility and avoid bottlenecks, SDSR places more storage servers near where most queries are generated and fewer far away.
- **Traffic localization:** the hierarchical structure of storage servers for a particular data item constructed by SDSR guarantees that a query message for the data is propagated locally within a quadrant of the smallest grid containing both the *requester* node and the *source* node.

Our analysis show that in a query dominant large scale wireless ad hoc network, SDSR performs better than existing schemes in terms of energy efficiency, data query latency, hotspot usage, and resilience to clustering failures. It scales well when the network size and the number of queries increase.

The paper is organized as follows: Section 2 presents system model and SDSR problem statement. Section 3 describes SDSR components in detail. We present numerical analysis of SDSR and compare its performance with the performances of existing schemes in Section 4. Section 5 concludes the paper.

2 Preliminaries

System Model. We consider a large scale wireless ad hoc network in a 2-D coordinate plane. Nodes are connected iff they are within unit distance of each other. Nodes and edges are represented by the set V and E, respectively, and the resultant undirected graph by G, where $G = (V, E)$.

Assumptions. We assume a connected network where: 1) with high probability, there are multiple nodes in each unit square area; 2) each node knows its geographic location via certain location service such as GPS [8]; 3) each data item has a unique key; 4) a hash function uniformly maps each key to normalized (x,y) coordinates; 5) energy is a scarce resource for nodes in the network. Since communication consumes most energy of wireless networks [9], minimizing communication cost should be an important criteria in designing energy efficient data dissemination schemes; 6) the longer the Euclidean distance between two nodes, the more the network-level hops required to communicate between them.

Definitions. We use j and k to denote the nodes or locations in the network. Let $dist(j,k)$ denote the Euclidean distance between j and k in G and $e(j,k)$ denote the edge between j and k. *Data delivery success probability,* $P_{j,k}$ is defined as the probability of successfully delivering a data item from j to k. It is computed as $p^{len(j,k)}$, where p is the success probability of delivering the data item between two 1-hop neighbors, and $len(j,k)$ is the length of the path (in hops) traversed by the data from j to k. *Communication cost,* $C_{j,k}$, is defined as $O(m * dist(j,k))$ where m is the number of messages transmitted between j and k. *Network delay* $t_{j,k}$ between j and k is defined as $O(dist(j,k))$.

Problem Statement. The data storage and retrieval problem is to design a distributed, scalable and fault-tolerant scheme that, given a wireless ad hoc network, constructs a hierarchical partitioning of the network such that:

- a *source* node replicates the data at multiple locations in the network in increasing levels of the hierarchy,
- the distance of a level-i storage location relative to its level-i (i>0) grid is always half of the distance of a level-(i+1) storage location relative to its level-(i+1) grid,
- a query for a data item is propagated locally within a quadrant of the smallest grid containing both the *requester* node and the *source* node,
- the request latency experienced by data requester is in proportional to the distance between the *requester* node and the *source* node,
- data is routed from the *source* node to its storage locations with low *total communication cost* and high *data delivery success probability.*

3 SDSR Service

In this section, we introduce each component of SDSR service in detail.

3.1 Hierarchical Partitioning

We construct hierarchical partitioning of the network into squares of increasing sizes similar to the grid partitioning described in GLS [10]. As shown in Fig. 1(a), the smallest square is a level-1 grid and four level-1 grids form a level-2 grid, and so on. The largest grid covering the whole network area is a level-N grid (e.g., level-4 in Fig. 1(a)). Grids do not overlap in the sense that each level-i grid belongs to exactly one level-(i+1) grid. Such a partitioning guarantees that each node belongs to exactly one grid in each level of the hierarchy. All four level-i grids of the same level-(i+1) grid are *neighboring* level-i grids of each other. Since each node knows its location, it knows which grid it belongs to.

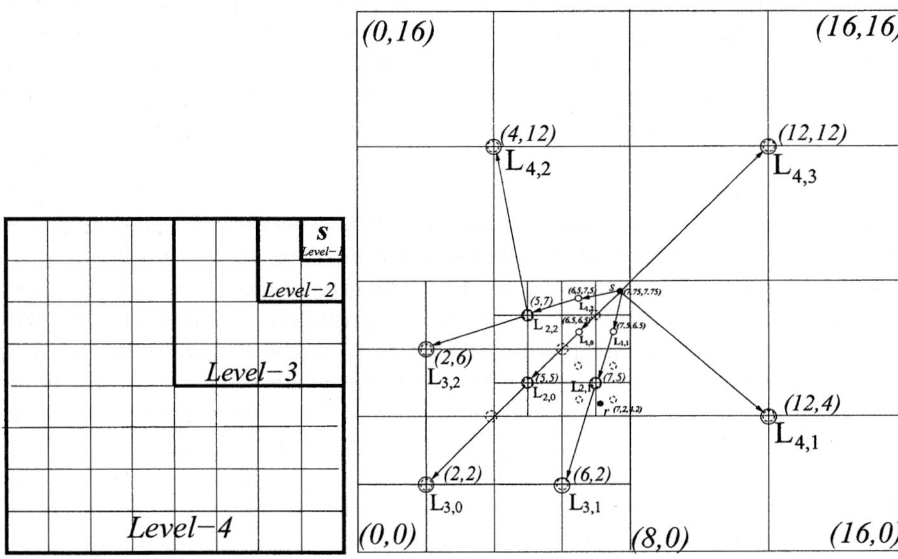

(a) Hierarchical partitioning of the network into four levels

(b) The solid circles are the storage locations selected by *source* node s at (7.75,7.75), the dotted circles are the query locations selected by *requester* node r at (7.2,4.2)

Fig. 1. Example of SDSR hierarchical partitioning (a) and storage location selection, multicast routing tree construction and query location selection (b)

3.2 Data Storage Server Selection

The *source* node s of a data item f selects one storage server for f in each of its three *neighboring* level-i ($0 < \mathrm{i} < \mathrm{N}$) grids as follows: first, s hashes $key(f)$ to get a normalized geographic location (x,y) such that $x \in [0, 1]$ and $y \in [0, 1]$, then, s computes the storage location for f relative to a *neighboring* level-i grid by scaling the normalized geographic location with $2^{(i-1)} * l$, where l is the length of the level-i grid; finally, s stores f, in each of its *neighboring* level-i grids, at the node nearest to the geographic location $(x * 2^{(i-1)} * l,\ y * 2^{(i-1)} * l)$ relative to the lower left corner of the corresponding level-i grid. To guarantee that each computed storage location belongs to exactly one level-1 grid, we specify that the storage locations along the border of two horizontally adjacent grids belong to the left grid , and locations along the border of two vertically adjacent grids belong to the bottom grid. As shown in Fig. 1(b), s with location of (7.75,7.75) hashes $key(f)$ to get $(x, y) \equiv$ (0.5,0.5) and then stores f in its *neighboring* level-1 grids at locations (6.5,7.5), (6.5,6.5) and (7.5,6.5), which are at (0.5,0.5) offsets from the (x,y)-coordinates of the lower left corners of the corresponding grids - (6,7), (6,6), and (7,6) respectively. s repeats the process at each of the *neighboring* level-i grids by scaling the offset by $2^{(i-1)} * l$, i.e., for *neighboring* level-2 grids the offset is (1.0,1.0), for *neighboring* level-3 grid the offset is (2.0,2.0), etc. s stores f at nodes nearest to locations (5,7), (5,5), and (7,5) in each *neighboring* level-2 grids, and at locations (2,6), (2,2) and (6,2) as well as (4,12), (12,4) and (12,12) in *neighboring* level-3, level-4 grids respectively.

3.3 Multicast Storage Tree Routing

The goal of multicast tree routing is to construct a tree, T, with *source* node, s, and the corresponding data storage locations, $L_{1,2}$, $L_{1,0}$, $L_{1,1}$, etc., as the nodes in the tree so that data can be routed from s to the storage locations along the edges of T with optimal *total communication cost* as well as optimal *network delay/data delivery success probability.*

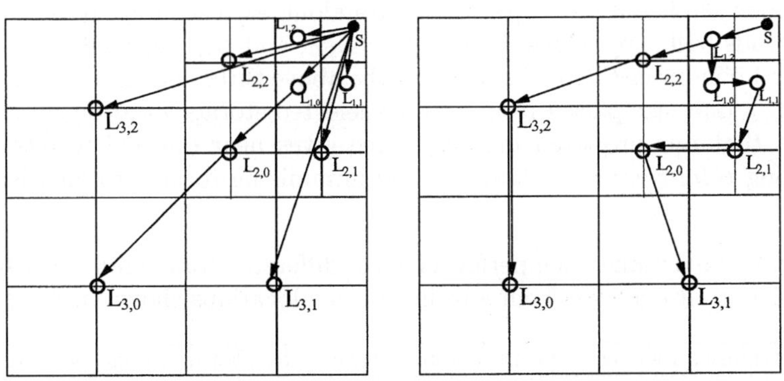

(a) Shortest path tree (b) Minimum spanning tree

Fig. 2. Examples of shortest path tree and minimum spanning tree from root s to the selected storage locations of the bottom left grid of the network shown in Fig. 1(b)

Traditional tree construction schemes such as rooted shortest path tree or rooted minimum spanning tree optimize either the *network delay/ data delivery success probability* or the *total communication cost*, but not both. Specifically, the rooted shortest path tree yields optimized *network delay/data delivery success probability* since it routes data along the shortest path between a *source* node and a storage server; however it suffers from increased *total communication cost* since the *source* has to send m copies of the data items where m is close to the total number of destinations. On the other hand, the minimum spanning tree minimizes the *total communication cost* in storing the data at the destination locations by enabling destinations along the same path to share the same data copy and thus the communication cost; however, it increases the network delay and data loss probability by introducing longer paths from the *source* to the destinations. For example, in Fig. 2(b), one possible path from s to $L_{3,1}$ is s, $L_{1,2}$, $L_{1,0}$, $L_{1,1}$, $L_{2,1}$, $L_{2,0}$, $L_{3,1}$ which is much longer than the shortest path s, $L_{3,1}$ adopted by shortest path tree in Fig. 2(a).

We propose a new multicast tree construction algorithm that aims to compromise between minimizing the *total communication cost* and the *data delivery path length.* The algorithm is as follows: initially, the tree T only contains the *source* node s, the tree grows iteratively, such that in the ith-iteration all level-i storage locations join the tree by forming an edge with a node in the tree which is at minimum Euclidean distance from them, under the constraint that no two nodes at the same level can have an edge between them. This constraint gives us the property that the depth d of a level-i node in the tree is always $\leq i$. As shown in Fig. 1(b), in the first iteration, all level-1 storage locations, $L_{1,2}$, $L_{1,0}$, and $L_{1,1}$, join the tree such that the edges formed are $e(L_{1,2}, s)$, $e(L_{1,0}, s)$, and $e(L_{1,1}, s)$. In the second iteration, all level-2 storage locations, $L_{2,2}$, $L_{2,0}$, $L_{2,1}$, join the tree such that the edges formed are $e(L_{2,2}, L_{1,2})$, $e(L_{2,0}, L_{1,0})$, and $e(L_{2,1}, L_{1,1})$. Similarly, other storage locations in the network join the tree. The *source* node sends out one copy of the data along each separate path. As in Fig. 1(b), s sends one copy of the data to $L_{1,2}, L_{1,0}$, and $L_{1,1}$ separately.

Table 1 compares the performance of the above three storage routing schemes in storing a unit size packet from s to the selected storage locations in Fig. 1(b). It shows that our proposed multicast tree gives near optimal results both in minimizing *total communication cost* and minimizing total path length.

Table 1. Comparison of the performance of different storage routing schemes in routing 1 unit size packet from s to its storage locations shown in Fig.1(b)

Performance metrics	Total communication cost	Total storage path length
Multicast tree	36.37	55.96
Shortest path tree	50.35	50.35
Minimum spanning tree	34.96	80.3

3.4 Data Query and Update

To locate a particular data item f, a *requester* node r first hashes $key(f)$ to get the normalized (x, y) coordinates. Next, it computes the possible storage locations for f in the network using the (x, y) coordinates as an offset.

Query Location Selection. In SDSR, a *requester* node r computes the query locations for f in a similar way as the *source* node computes storage locations for f. Starting from the level-1 hierarchy, r selects one query location from each of its 3 *neighboring* level-i (0<i<N) grids as well as the level-i grid containing r itself, since all 4 level-i grids have the same probability of holding a valid location server for f. For example, in Fig. 1(b), the set of query locations chosen by r located at (7.2,4.2) are (6.5,4.5), (6.5,5.5), (7.5,4.5) and (7.5,5.5) for level-1 hierarchy, (5,5), (7,5), (7,7) and (5,7) for level-2 hierarchy, (2,6), (2,2), (6,2) and (6,6) for level-3 hierarchy, etc.

Data Query. r queries the set of computed query locations hierarchically, starting from its level-1 grids. In each level, instead of querying all the query locations, r only queries the storage location closest to itself. The query terminates when r gets f from a query location or if r does not get f from the highest level query locations. For example, in Fig. 1(b), the query path for f from r is: (7.5,4.5), (7,5), (12,4). The storage server closest to location (7,5) answers the query during second query step. Querying only the closest query location in each level does not work when the *source* node s, the *requester* node r and the closest level-1 query location of f are in the same level-1 grid. To handle this, if r can not locate f in the query location of its own level-1 grid, r broadcasts within the grid before querying the level-2 storage location.

SDSR query scheme is scalable since it not only limits the number of steps needed to satisfy a query but also bounds the geographic region in which the query will propagates to one of the four low level squares of the smallest grid containing both the *source* and the *requester*. It is important to mention that the above query scheme works in a static network with no faults. In section 3.5, we provide additional solutions to handle faults introduced by node mobility, node failure and clustering failure.

Data Update. In our model, each data item can only be updated by its *source* node. After a node updates a data item, it stores the updated item at its storage locations. The set of storage locations selected may change from time to time as the node moves within the network. But this change is local and is bounded by $O(D)$ where D is the diameter of the smallest grid which contains the node's current location and its previous location where the last update occurs.

3.5 Fault Tolerance

Wireless ad hoc networks have frequent topology changes (modeled as faults) due to node joins, node fail-stops (energy exhaustion), and node mobility. The storage service of SDSR, stores a data item f at the node j nearest to the corresponding storage location. But, when a new node k joins the network, it may be closer to the storage location of f than the current storage server j.

Transferring f from the current storage server to the node nearest to its storage location is undesirable in presence of frequent node joins (and moves) as it results in high communication cost especially when f is large. On the other hand, not storing f at the node nearest to the storage location results in broadcasting the query for f. Thus, we need a scheme that avoids high communication cost due to either frequent data transfer or frequent query broadcast.

Indexing Service. The new node k, instead of getting f from the storage server j, serves as an *index* for f and only stores the information that there exists a storage server of f in the grid. Since a query for a data item is routed to the node nearest to the storage location, k, on receiving the query for f, broadcasts the query in the grid as k has the information that f exists in the grid. Storage server j on receiving the query, replies back to k with f, which replies f back to the *query* node. With f stored, k is able to serve the following queries for f directly. Another advantage of *indexing* approach is that if f is not in the grid, then the query is not broadcast by k thus avoiding unnecessary communication cost. The indexing list for a storage location is copied to nodes along the perimeter of the storage location, thus when the closest node to the storage location fails, the second closest node can serve the following queries.

Node Mobility. Node movement has minimal effect on the data storage structure as long as storage servers move in their original level-1 grids. If storage server j of data f moves outside its level-1 grid, j transfers f to the node nearest to the storage location in the original grid, j then can delete f and will not serve as a storage server for f.

Node Fail-stop. A *source* node periodically sends soft state hello messages with low frequency to its storage servers to detect storage server fail-stops and recruit new server accordingly. In order to handle clustering failure, a *requester* node queries multiple storage servers in each hierarchy level and retrieves the data item from the first server which replies with the data.

4 Analytical Results

In this section, we first compare SDSR with existing schemes in the performance of reducing communication cost and hotspot usage. Then, we compare SDSR with GHT and SR-GHT regarding their resilience to clustering failure. Table 2 lists the notations used in computing the hotspot usage and communication cost. Table 3 lists the performance metrics used in the comparison.

4.1 Communication Cost

Local Storage (LS): data is stored in the source node only.
Centralized Storage (CS): data is stored in a centralized server.
GHT: each data key is hashed to one fixed storage location in the network.
Structured Replication GHT (SR-GHT)[5]: to reduce storage cost of GHT, SR-GHT hashes each data key to a *root* location in the network, then divides the whole network uniformly into 4^d subregions (d is the replication hierarchy

depth) and computes $4^d - 1$ mirrors of the *root* location, with one mirror in one subregion. The source node of the data stores the data to the closest mirror. The data itself is not replicated in each subregion.

Table 2. Notations

N	the number of nodes in the network. As in [5], $O(N)$ is the cost of flooding a data item in the network and $O(\sqrt{N})$ is the cost of sending a message between two nodes
d	the replication depth used by SR-GHT and SDSR
M	the number of data items stored in the network
Q	the number of queries generated in the network
R_{sdsr}	the number of replicas per data item in SDSR ($R_{sdsr} = 3^d$)
R_{sr-ght}	the number of mirrors per data item in SR-GHT ($R_{sdsr} = 4^d - 1$)

Table 3. Performance metrics

Total storage msg complexity C_s	Total number of storage msgs generated
Total query msg complexity C_q	Total number of query msgs generated
Total reply msg complexity C_r	Total number of reply msgs generated
Total msg complexity C_{total}	Total number of msgs generated
Hotspot msg complexity H	max number of msgs processed per node

Table 4 lists the performance comparison of the above schemes regarding the hotspot usage and the total communication costs. We observe that: 1) LS incurs the largest *total communication cost* with increased N; 2) CS yields the highest hotspot message counts; 3) SDSR has the lowest hotspot usage since first, the task of storing data is distributed among all the nodes; second, each data item has R_{sdsr} replicas distributed across the network; 4) SR-GHT has the lowest storage cost since it replicates storage locations uniformly in the network. The tradeoff is the highest query cost. Thus SR-GHT doesn't work well in a network where $Q > M$; 5) SDSR introduces the highest storage cost since it replicates data in multilevel hierarchy for achieving better fault tolerance and reduced query latency, yet it yields the lowest query cost as well as reply cost. And the increase of the total cost grows slowly as the network grows. Thus SDSR is preferable to other schemes regarding the *total communication cost* when (1) N is large enough; (2) $Q > M$; (3) network query exhibits locality. For example, Table 5 shows the total communication cost of each scheme when $N = 2500$, $M = 100$, $Q = 1000$ and $d = 4$.

4.2 Resilience to Clustering Failure

Clustering failure occurs when all nodes within the same grid fail. GHT and SR-GHT are vulnerable to clustering failure since they replicate each data item only

Table 4. Performance comparison of hotspot usage and total communication cost

Schemes	Hotspot usage	Total message complexity in the form of $C_s+C_r+C_q$
LS	$2Q$	$0 + Q\sqrt{N} + QN$
CS	$M + 2Q$	$M * \sqrt{N} + Q\sqrt{N} + Q\sqrt{N}$
GHT	$max(M/N, 1) + 2Q$	$M\sqrt{N} + Q\sqrt{N} + Q\sqrt{N}$
SR-GHT	$max(M/N, 1) + 2Q$	$M\sqrt{N}/2^d + Q\sqrt{N} + Q\sqrt{N}2^d$
SDSR	$max(M/N, 1) + 2Q/(3^d + 1)$	$6M\sqrt{N}(1 - 1/2^d) + Q\sqrt{N}/d(1 - 1/2^{d-1})$ $+Q\sqrt{N}(2^d + 2^{d-1} - 2d + 1)/((d - 1)2^{d-1})$

locally around its hashed location. SDSR performs better in handling clustering failure by replicating each data item at nodes that are distributed in different network regions. If each level-1 grid has the same clustering fault probability q, then the probability that all the grids containing data item f fails is q for GHT and SR-GHT, while only q^{3^d+1} for SDSR.

Table 5. Comparison of different schemes regarding the total messages generated

Schemes	LS	CS	GHT	SR-GHT	SDSR
Total messages	255000	105000	105000	850312	77827

5 Conclusion

we present the design and evaluation of SDSR, a data centric, location based data storage/ retrieval service for large scale wireless ad hoc networks. Our analytical results show that in a query dominant, large scale wireless ad hoc network, SDSR performs better than existing schemes in terms of total communication cost, data query latency, hotspot usage, as well as resilience to clustering failures. It complies well with query locality and scales well when the network size and the number of queries increase. Some directions we are currently exploring are 1) extending SDSR to work with node's storage limits and 2) investigating the performance of SDSR on hierarchically clustered wireless ad hoc networks.

References

1. Hara, T.: Effective replica allocation in ad hoc networks for improving data accessibility. Proc. of IEEE INFOCOM (2001) 1568–1576
2. Wang, K., Li, B.: Efficient and guaranteed service coverage in partitionable mobile ad-hoc networks. Proc. of IEEE INFOCOM (2002) 1089–1098

3. Karumanchi, G., Muralidharan, S., Prakash, R.: Information dissemination in partitionable mobile ad hoc networks. Proc. of IEEE Symposium on Reliable Distributed Systems (1999) 4–13
4. Cheng, L.: Service advertisement and discovery in mobile ad hoc networks. Workshop on Ad hoc Communications and Collaboration in Ubiquitous Computing Environments (2002)
5. Shenker, S., Ratnasamy, S., Karp, B., Estrin, D., Govindan, R.: Data-centric storage in sensornets. ACM SIGCOMM Computer Communication Review **33** (2003) 137–142
6. Karp, B., Kung, H.T.: Greedy perimeter stateless routing. Proc. of ACM MOBICOM (2000) 243–254
7. Kempe, D., Kleinberg, J.M., Demers, A.J.: Spatial gossip and resource location protocols. ACM Symposium on Theory of Computing (2001) 163–172
8. : USNO GPS operations. (2001) http://tycho.usno.navy.mil/gps.html.
9. Sailhan, F., Issarny, V.: Cooperative caching in ad hoc networks. Proc. of the 4th International Conference on Mobile Data Management(MDM) (2003) 13–28
10. Li, J., Jannotti, J., DeCouto, D., Karger, D., Morris, R.: A scalable location service for geographic ad hoc routing. Proc. of ACM MOBICOM (2000) 120–130

An Efficient Multicast Data Forwarding Scheme for Mobile Ad Hoc Networks*

Youngmin Kim[1], Sanghyun Ahn[1]**, and Jaehwoon Lee[2]

[1] Department of Computer Science
University of Seoul, Seoul, Korea
{blhole, ahn}@venus.uos.ac.kr
[2] Dongguk University, Seoul, Korea
jaehwoon@dongguk.edu

Abstract. In order for multicast data packets to be transmitted efficiently, the multicast data forwarding is required at each router. Within a wired network, a network interface of a node has a one-to-one connection to one of the incoming interfaces of other nodes. If the packet coming into an incoming interface needs to be forwarded, it is delivered to the corresponding outgoing interface. However, in a mobile ad hoc network (MANET), in most cases each node has only one network interface, so the incoming interface is the same as the outgoing one and the wireless network interface of a node has a one-to-many connection to those of neighboring nodes. This difference may cause problems such as routing loops and packet duplication. Therefore, in this paper, we propose a multicast data forwarding scheme which can be used in the multi-hop wireless ad hoc network without causing either routing loops or packet duplication. In the proposed scheme, a table is defined for the prevention of packet duplication that can happen when the tree-based multicast routing protocol is used. We have implemented our proposed scheme by using the netfilter[1] of the Linux OS.

1 Introduction

The mobile ad-hoc network (MANET) consists of mobile nodes that self-organize to form a network topology without any wired network infrastructure support. Because a MANET can offer an access to network resources to portable devices in any place and at any time, it is attracting much attention and growing quickly. In this paper, we address a multicast data forwarding scheme that sends multicast packets to next hops in a MANET using a multicast routing table generated by a MANET multicast routing protocol.

Multicast routing protocols for MANET can be broadly classified into two categories; the tree-based and the mesh-based approaches. In the tree-based approach, a multicast tree is constructed for the delivery of multicast packets and

* This work was supported by the University IT Research Center.
** Corresponding Author

C. Kim (Ed.): ICOIN 2005, LNCS 3391, pp. 510–519, 2005.

the examples are MAODV [2] and AMRIS [3]. In the mesh-based approach, multiple forwarding paths are constructed among multicast members, and ODMRP [4], CAMP [5] and FGMP [6] belongs to this category.

In a wired network, when a node forwards multicast packets based on a multicast routing tree, the packets are forwarded only to child nodes (i.e., nodes in the downstream of the tree). However, in a multi-hop wireless network (i.e., a wireless ad hoc network) where a network interface of a node doesn't have a one-to-one connection with that of a neighboring node, multicast packets sent by a node can be received by all of its neighbors including its parent. Hence the parent will receive duplicate packets and this will result in routing loops. Therefore we need a multicast data forwarding scheme which can detect and discard duplicate packets so that they can not be forwarded any further. With this mechanism, a node on a multicast tree will forward each multicast packet only once, hence the limited wireless bandwidth can be efficently utilized.

In order to prevent duplicate packets from being forwarded, each packet needs to be assigned with a unique sequence number and a header field for the inclusion of this information is required. With regard to the multicast data forwarding, ODMRP prevents packet duplication by keeping the source IP address and the source-specific packet identifier for each received multicast packet so that the packet can be uniquely identified later on. In this scheme, however, additional information (i.e., the source-specific identifier) need be stored in the IP header. In IPv4, the "IDENTIFICATION" field which is originally defined for identifying fragments belonging to the same IP datagram may be used for the source-specific packet identifier, however this violates the original purpose of the "IDENTIFICATION" field. Also, in IPv6, to provide this functionality, a new extension header must be defined. Using the source-specific identifier for the duplicate packet detection requires each packet to be assigned with a unique source-specific identifier and all forwarding nodes of a packet to store the identifier for the packet, which may not be practical especially in the MANET environment. In this paper, we propose MDF-TM (Multicast Data Forwarding for Tree-based Multicasting in MANET) which doesn't require any additional information in the packet and provides a simple and efficient multicast data forwarding mechanism for any tree-based MANET multicast routing protocols. We have implemented MDF-TM and verified the correctness of MDF-TM.

This paper is organized as follows. Section 2 presents an efficient multicast data forwarding scheme for the MANET. How the proposed multicast data forwarding scheme is implemented is presented in Section 3. We analyze the memory requirement of MDF-TM in Section 4. The paper is concluded in Section 5.

2 Multicast Data Forwarding for Tree-Based Multicasting in MANET

In the wired network, upon receiving a unicast packet destined to another node, a node first looks up its routing table for the next hop to the given destination and obtains the MAC address of the next hop and forwards it via the interface

Fig. 1. A multi-hop network for observing PDRT where the multicast tree is constructed in A-B-C.

to the next hop. In MANET where links are broadcast medium, packets sent by a node can be heard by all of its neighboring nodes. However, since the neighbors with different MAC addresses filter out the packet, unicast packet forwarding in MANET can be carried out without causing packet duplication.

For the multicast data forwarding in the wired network, a node acquires the next hops by referring to its multicast routing table, puts the corresponding multicast MAC address as the L2 destination address and forwards it to the next hops except for that one from which the packet has come. Hence the multicast data forwarding in the wired network does not cause packet duplication. However, in MANET, when a node forwards a packet with a multicast MAC address as the destination through its wireless network interface, all one-hop neighbors hear the packet and those who are on the multicast tree accept the packet. Thus, the parent node of the packet also accepts the packet, which will result in packet duplication and, in turn, routing loops. We name this type of duplication the "packet duplication by reverse transmission" (PDRT). To see PDRT in a real MANET environment, we have configured a test network using SMCRoute (Static Multicast Route) [9] (as shown in Fig. 1) and made all three nodes join the NTE application group using SDR/NTE [10]. In this test environment, we have observed that PDRT happens when node A starts transmission of multicast packets.

In Fig. 2, solid lines indicate links involved in a multicast tree and node A is a source of the multicast group and D, E, and F are members of the group. When A sends a multicast packet, B and C receive it, and B forwards it to D and E that are next hops of the multicast tree. On the other hand, the packet

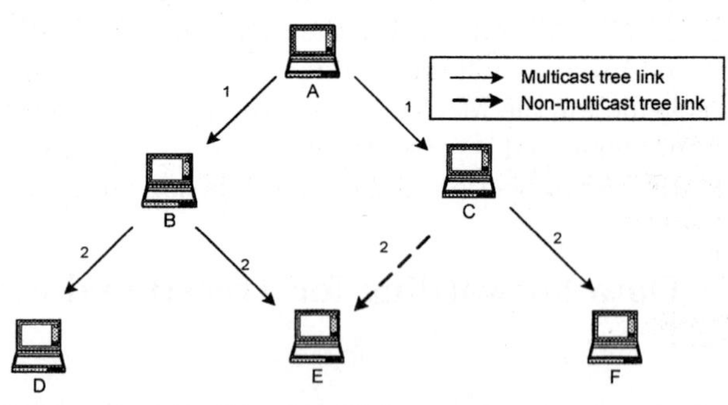

Fig. 2. An example of PDTN.

forwarded by C is received by E and F, where E is not an intended receiver. Since dotted lines do not belong to the multicast tree, node E should not receive multicast packets from node C. We name this type of duplication the "packet duplication by transmission from a non-neighbor" (PDTN).

A transmitting node plays an important role in the multicast data forwarding in the wired network. When a transmitting node forwards a packet to the corresponding network interfaces, the role of a receiving node is simply to accept it. However, in MANET, the scope of the multicast data forwarding can not be limited to a specific receiving node, i.e., a multicast packet is forwarded to all the neighbors within its coverage area. When a transmitting node forwards a multicast packet, the information regarding individual receivers is not included and the transmitting node only needs to check the existence of child nodes and, if there exists at least one, forwards it. So we can say that the functionality of a transmitting node in MANET is relatively simple compared with that in the wired network.

On the other hand, a receiving node in MANET has to prevent both PDRT and PDTN. As described later, the prevention of PDTN makes multicast packets be delivered only along the multicast routing path. That is, the multicast data forwarding is easily accomplished by performing the prevention of PDTN.

Fig. 3 shows the operation of a transmitting and a receiving node involved in the multicast data forwarding in MANET. The multicast data forwarding is carried out by the prevention of PDTN and routing loops are blocked by the prevention of PDRT. Locally generated packets are forwarded only when the number of neighbors in the given multicast tree is one or more, and packets coming from outside are forwarded only when the number of neighbors in the given multicast tree is more than one.

3 Implementation of MDF-TM

MAODV6 used in our implementation is based on the implementation of AODV [7][8] by NIST [11], which is extended to support IPv6 and MAODV. We have implemented MDF-TM on Linux kernel 2.4 [12] according to the design concept in Section 3, and carried out tests with using SDR/NTE as the application. Even though MDF-TM is for MAODV and IPv6 networks, the design concept of MDF-TM can also be applied to IPv4 networks, other OS environments, and other tree-based MANET multicast routing protocols. MDF-TM makes use of packet filtering to prevent data packet duplication. Linux kernel provides netfilter to filter packets, and the framework of MDF-TM is composed of netfilter hooks and callback functions.

3.1 Prevention of Packet Duplication

A possible approach to prevent data duplication in MANET is to store the source L3 address and the source-specific identifier for each incoming multicast packet and use this information for the duplicate packet detection. This scheme can

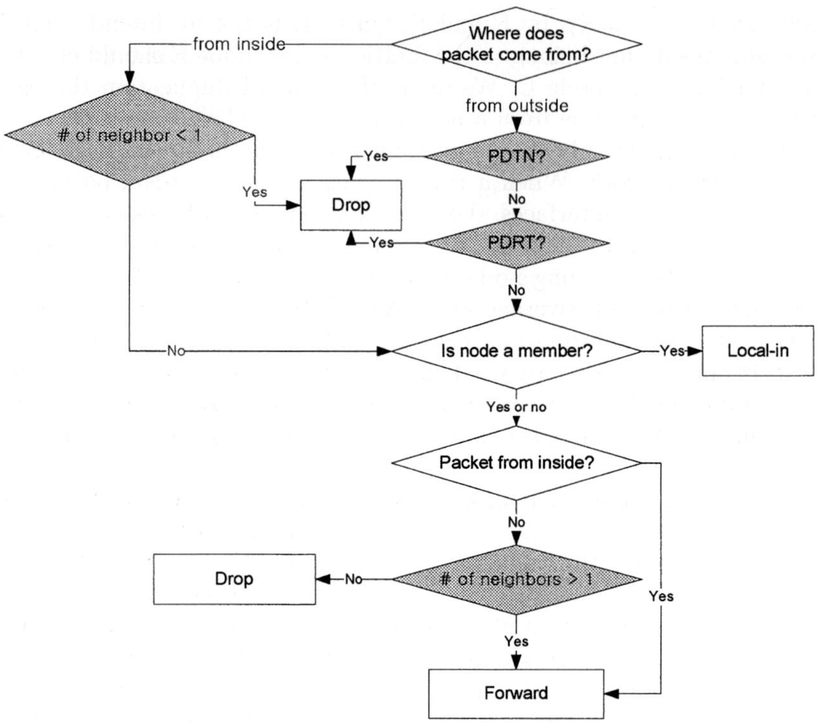

Fig. 3. Operation of the multicast data forwarding in MANET.

prevent PDRT of the tree-based multicast routing protocol, but not PDTN. In Fig. 2, if C, ahead of B, transmits a packet received from A, E receives the packet from C, but drops the packet from B since E has already received the same one from C. Table 1 shows the format of the multicast data forwarding table and that of the table for duplicate packet detection with storing the source L3 address and the source-specific identifier. Table 1 shows fields necessary for the multicast data forwarding but not for table management and statistics. Fields in Table 1. (a) come from the mandatory options of SMCRoute which is a command to create a multicast tree. However, this scheme requires sources to assign unique source-specific IDs to transmitting packets, and receiving nodes to store the source L3 address and the source-specific ID for each received packet. If a node doesn't delete unnecessary entries in a timely manner, the memory resource can be wasted. Furthermore, there is no field to carry the source-specific ID in the IPv6 header, so an user-defined extension header has to be used and this can be some overhead especially in the wireless environment.

In order to prevent PDRT, the information on the parent node of each multicast session needs to be maintained at each tree node so that those packets from child nodes can be filtered out. We have come up with two possible approaches

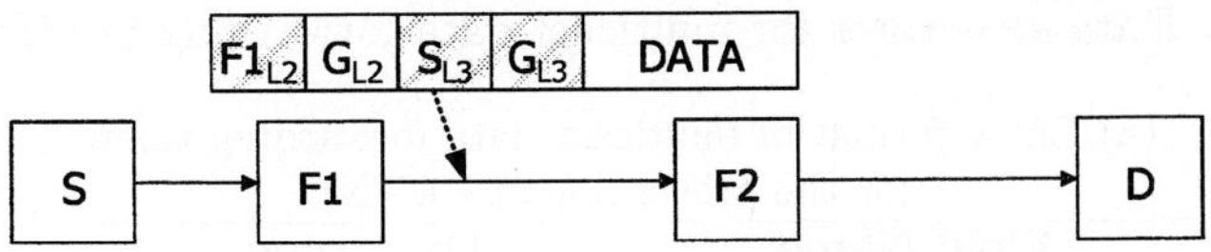

Fig. 4. Prevention of PDRT.

to identify the parent node of a received packet; one is to use the source L2 address and the other the L3 address of the parent node. In order to be able to use the L3 address of the parent node, a means to store the L3 address in the packet header is required. In case of IPv4, there's no such field available and, in case of IPv6, an IPv6 extension header may be used and this additional IPv6 extension header is to be an overhead especially to the wireless environment, so we have decided to use the L2 address of the parent node for this purpose. Thus, in MDF-TM, each node in a multicast tree maintains a table with (Source IPv6 Address, Multicast Group IPv6 Address, Source MAC Address) entries. Fig. 4 illustrates how this table is used; node D is a member of group G, and while node S send a packet to G, it is forwarded from F1 to F2. After F2 receives the packet, it compares (S, G) of the L3 header with the (Source IPv6 Address, Multicast Group IPv6 Address) table and, if there is no match, the node stores $(S_{L3}, G_{L3}, F1_{L2})$ in the table and forwards the packet to D. If there is a match, it compares the source MAC address of the packet with that of the matched entry. Only if they are the same, the packet is accepted. Since the source doesn't have the parent node, it sets the source L2 address of its own entry to null. Hence, when a node encounters null in the source MAC address of the corresponding

Table 1. Data structures for the multicast data forwarding in the scheme using the source specific unique ID.

(a) Entry format of the multicast data forwarding table

Field Name	Data type
Originator Address	L3 (IPv4 or IPv6) Address
Multicast Group Address	L3 (IPv4 or IPv6) Address
Incoming Interface	Integer
Outgoing Interface	Integer

(b) Entry format of the table to prevent data packet duplication

Field Name	Data Type
Originator Address	L3 (IPv4 or IPv6) Address
Unique Identifier 1	Integer
Unique Identifier 2	Integer
...	...
Unique Identifier n	Integer

Table 2. Data structures for multicast data forwarding in MDF-TM.

(a) Entry format of multicast data forwarding table (for the prevention of PDTN)

Field Name	Data type
Multicast Group Address	L3 (IPv4 or IPv6) Address
Neighbor L2 Address 1	L2 (MAC) Address
Neighbor L2 Address 2	L2 (MAC) Address
...	...
Neighbor L2 Address n	L2 (MAC) Address

(b) Entry format of the table to prevent PDRT

Field Name	Data Type
Originator Address	L3 (IPv4 or IPv6) Address
Multicast Group Address	L3 (IPv4 or IPv6) Address
Parent Node of Multicast Flow	L2 (MAC) Address

entry for a received packet, it just drops the packet since the node itself is the source.

For the prevention of PDTN, each node maintains a multicast neighbor table with (Multicast Group L3 Address, Neighbor L2 Address List) entries. This entry has the meaning that a multicast packet is accepted if it comes from the node with the same address as one of the addresses (i.e., neighbors) in the neighbor L2 address list for the multicast group IPv6 address. When MAODV6 adds, deletes and updates entries of the multicast routing table, the multicast neighbor table should be updated so that the routing changes can be reflected.

Table 2 shows the entry format of the multicast data forwarding table and that of the table to prevent data packet duplication by using MDF-TM.

3.2 Correctness Test of MDF-TM

For the operational test of MDF-TM, we have loaded MAODV for IPv6 and MDF-TM on linux machines and verified the correctness of MDF-TM in preventing duplicate multicast packets from being forwarded. SDR is used to create a multicast session and inform other nodes of the existence of the multicast session, and NTE is used for multicast data communication. The testbed used for this experiment consists of three laptops connected with an IEEE 802.11b wireless LAN, as shown in Fig. 1.

Table 3 and 4 show the information for the prevention of PDRT and PDTN at node A. Those multicast groups in the tables are for the session management and NTE communications. Table 3 maintains the L2 address of each neighbor node for each multicast group which is extracted from the path information provided by MAODV6. The entries in table 4 are created when a node receives new multicast data traffic. Throughout the test, we have observed that MDF-TM does not produce any routing loops.

Table 3. Table information for prevention of PDTN (node A)

Entry #	Entry	Description
1	FF0E::0002:7FFE 00:03:47:15:21:7F	Group to inform multicast session L2 address of node B
2	FF0E::0002:CEE5 00:03:47:15:21:7F	Group for NTE session L2 address of node B

Table 4. Table information for prevention of PDRT (node A)

Entry #	Entry	Description
1	3FFE:FFFF:0100:F102::0011 FF0E::0002:7FFE 00:03:47:15:21:7F	L3 address of node B : group leader Group to inform multicast session L2 address of node B
2	3FFE:FFFF:0100:F102::0013 FF0E::0002:CEE5 00:00:00:00:00:00	L3 address of node A Group for NTE session NULL because node A is multicast source
3	3FFE:FFFF:0100:F102::0012 FF0E::0002:CEE5 00:03:47:15:21:7F	L3 address of node C Group for NTE session L2 address of node B
4	3FFE:FFFF:0100:F102::0011 FF0E::0002:CEE5 00:03:47:15:21:7F	L3 address of node B Group for NTE session L2 address of node B

Table 5. Notations for the equations.

G : The number of multicast groups that a node is joining
$\boldsymbol{S} = \{s_1, s_2, ..., s_G\}$: The number of multicast sources for each multicast group
P : The total number of multicast sources of multicast groups that a node is joining
$\boldsymbol{K} = \{k_1, k_2, ..., k_P\}$: The number of unique IDs that a node has for each multicast source
$\boldsymbol{N} = \{n_1, n_2, ..., n_G\}$: The number of neighboring nodes of multicast tree for each multicast group

4 Analysis of Memory Requirement

MDF-TM reduces the amount of memory that is a limited resource in a portable wireless node and, to show this, the comparison between two previously-mentioned schemes is shown in this section through equations. The notations for the equations are appeared in Table 5.

Sum of Eq. (1) and Eq. (2) is the amount of memory used by the approach using source-specific ID. Eq. (1) is the total memory size of the multicast table in Table 1. (a) and Eq. (2) is the total memory size of the table in Table 1. (b) which prevents data packet duplication.

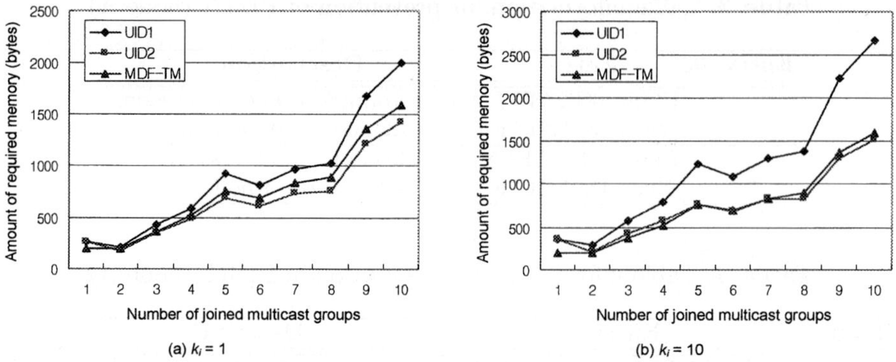

Fig. 5. The amount of required memory.

$$T1a = \sum_{i=1}^{G} s_i \times 2 \times (sizeof(L3Address) + sizeof(Integer)), \quad s_i \in S \quad (1)$$

$$T1b = P \times sizeof(L3Address) + \sum_{i=1}^{P} k_i \times sizeof(Integer), \quad k_i \in K \quad (2)$$

Sum of Eq. (3) and Eq. (4) is that used by MDF-TM in a node. Eq. (3) is the total memory size of the multicast forwarding table in Table 2. (a) and Eq. (4) is the total memory size of the table in Table 2. (b) which prevents PDRT.

$$T2a = G \times sizeof(L3Address) + \sum_{i=1}^{G} n_i \times sizeof(L2Address), \quad n_i \in N \quad (3)$$

$$T2b = \sum_{i=1}^{G} s_i \times 2 \times (sizeof(L3Address) + sizeof(L2Address)), \quad s_i \in S \quad (4)$$

For the simplification of the analysis of the amount of required memory, it is assumed that multicast trees are not changed and all multicast sources transmit packets in CBR. For the analysis, G is incremented by 1 starting from 1 to 10, and s_i and n_i are randomly selected from the set {1, 2, ..., 5}. The amount of memory is analyzed for the maximum and the minimum of P. The maximum of P is obtained when the sources of multicast groups do not overlap, and the minimum of P when the sources of a multicast group which has the largest number of sources among all multicast groups include all other multicast sources. For the approach using the multicast source L3 address and the source-specific ID, the size of an integer is set to 2 bytes and k_i to 1 and 10.

In Fig. 5, UID1 indicates the amount of required memory with the maximum of P and UID2 that with the minimum of P for the approach using the multicast source L3 address and the source-specific ID. The memory requirement of UID2 is almost the same as that of MDF-TM, but since UID1 has more multicast sources due to using the maximum of P, the memory requirement of UID1 is larger than that of MDF-TM. Even when k_i is 1, MDF-TM requires less amount of memory than UID1.

5 Conclusion

In MANET, due to the characteristics of multi-hop wireless transmission, the tree-based multicast data forwarding causes problems like packet duplication and routing loops. In this paper, we have pointed out two problems related to the multicast data forwarding, the packet duplication by reverse transmission (PDRT) and the packet duplication by transmission from a non-neighbor (PDTN). To resolve these problems, we have proposed MDF-TM in which two tables are newly defined and one of the tables is used for the multicast data forwarding itself. For the implementation of MDF-TM, we have modified Linux kernel and verified that MDF-TM with MAODV6 performs correctly in forwarding multicast packets.

References

1. Netfilter, http://www.netfilter.org/.
2. E. Royer, C. Perkins, "Multicast Ad hoc On-Demand Distance Vector (MAODV) Routing," Internet Draft, draft-ietf-manet-maodv-00.txt, July. 2000.
3. C. W. Wu, Y.C. Tay, and C.-K. Toh, "Ad Hoc Multicast Routing Protocol Utilizing Increasing id-numberS (AMRIS) Functional Specification," Internet draft, Nov. 1998.
4. M.Gerla, S.-J. Lee, and W. Su. "On-Demand Multicast Routing Protocol (ODMRP) for Ad Hoc Networks," Internet draft, draft-ietf-manet-odmrp-04.txt, 2002.
5. J. J. Garcia-Luna-Aceves and E.L. Madruga, "The Core-Assisted Mesh Protocol," IEEE JSAC, pp. 138094, Aug. 1999.
6. C.-C. Chiang, M. Gerla, and L. Zhang, "Forwarding Group Multicast Protocol (FGMP) for Multihop, Mobile Wireless Networks," AJ. Cluster Comp, Special Issue on Mobile Computing, vol. 1, no. 2, pp. 187-96, 1998.
7. C. Perkins, E. Royer, and S. Das, "Ad hoc on demand Distance Vector (AODV) routing," RFC 3561, July. 2003.
8. C. Perkins, E. Royer, and S. Das, "Ad hoc On-Demand Distance Vector (AODV) Routing for IP version 6," Internet Draft, draft-ietf-manet-aodv6-01.txt, Nov. 2000.
9. SMCRoute, http://www.cschill.de/smcroute/.
10. UCL Network and Multimedia Research Group, http://www-mice.cs.ucl.ac.uk/multimedia/software/.
11. Kernel AODV, http://w3.antd.nist.gov/wctg/aodv_kernel/.
12. Linux kernel, http://www.kernel.org/.

Design of Heterogeneous Traffic Networks Using Simulated Annealing Algorithms

Miguel Rios, Vladimir Marianov, and Cristian Abaroa

Department of Electrical Engineering, Universidad Catolica de Chile, Casilla 306, Correo 22, Santiago-Chile
{mrios, marianov, cabaroa}@ing.puc.cl

Abstract. We propose a global design procedure for heterogeneous networks , which includes the definition of their topology; routing procedures; link capacity assignment; transfer mode; and traffic policy. We discuss the network model, which minimizes the cost of interconnecting a number of nodes whose locations are known; the traffic model, where we use the concept of Equivalent Bandwidth, and the resolution algorithms, where we compare a Simulated annealing algorithm (SAA) with a commercial solver, with good results. Results show that SAA gets to the optimum solution over 10 times faster than the commercial solver. Experiments also show that, for networks with more than 50 nodes, the SAA still delivers good feasible solutions while the commercial solver is unable to deliver results.

1 Introduction

Current communications systems rely largely on networks that are able to carry heterogeneous traffic. INTERNET itself is a communications network that can transmit data, voice and video. ATM (Asynchronous Transfer Mode) networks are an example of broadband networks that also deal with heterogeneous traffic. The design of heterogeneous broadband networks includes the definition of their topology, routing procedures, capacity assignment, transfer mode, fault tolerance methods and traffic policy. Therefore, models for the design of these networks are very complicated, and involve generally a very large number of integer and continuous variables. Most of the known approaches to this problem deal with these issues as separate problems. There is a considerable body of literature that focuses on the design of broadband networks. Models and solution procedures for network design include the capacity assignment through multiple origin-destination pairs [5], fault tolerant network dimensioning [1], and uncapacitated network design [3]. Several network design problems are presented in textbooks such as [8] and [9]. In [6] a virtual hierarchical network is designed that uses Fractional Brownian Motion (FBM). We use the same distribution in this paper. However, there are no design methods that consider all of the aspects together, being at the same time practical and efficient enough for their use by the industry. We address the global design of a heterogeneous broadband network that carries data, voice and video, including all the aspects except for

C. Kim (Ed.): ICOIN 2005, LNCS 3391, pp. 520–530, 2005.

the fault tolerance, which is left as a subject for future research, and propose and solve models for the design of these networks. Although ATM networks are taken as an example, the proposed methods can be used for the design of any broadband network. In ATM networks, the traffic is divided in cells of a certain length. These cells can become lost or delayed, due to congestion in the network. Thus, in addition to the cost and heterogeneous traffic issues, there is also the Quality of Service (QoS) issue. QoS in these networks refers to the cell loss probability and cell delay. We impose constraints on the QoS, bounding the delay and, consequently, the cell loss and cell delay. Node locations and traffic load for each traffic type are assumed to be known. We first address the network model, minimizing the cost of interconnecting a number of nodes whose locations are known. Assumed known is the amount of traffic of each type between each pair of nodes. In addition, for each information type there must be an acceptable delay. Secondly we address the traffic model, where we use the concept of Equivalent Bandwidth [7]. Although several traffic-modelling techniques can be used, we choose FBM, because it adapts better to the real time network behaviour. Finally we solve the model by using two tools: a) AMPL-CPLEX, a commercial package which finds the optimal solution of the linear model in a time that is exponentially related to the size of the problem, and b) a simulated annealing algorithm (SAA), which does not guarantee an optimal solution, but has a good performance, in terms of finding solutions that are close to the optimum. The main contributions of this paper are: a) a network model that considers most of the aspects of the network design problem starting from the node locations and traffic data; b) we propose an adaptation of the FBM traffic model to the conditions of the network design problem; c) we apply the SAA to solve the model, with good results. We first present and discuss the traffic model. Then, we address the network model. Afterwards, the solution methods are discussed. The next section presents the computational experiments. For this problem, the simulated annealing algorithm finds a solution (which corresponds to the optimum in most cases) around 10 times faster than AMPL-CPLEX. The experiments also show that for networks with more than 50 nodes, AMPL-CPLEX is unable to deliver results due to the enormous processing time, while the SAA still delivers good feasible solutions. Finally, conclusions are drawn and future work is proposed.

2 Traffic Model

An ATM broadband network, allocates its resources according to the amount of traffic of each type, present in the network, which means the resources to be allocated need to be determined in real time. Since the decision has to be taken very fast, the idea of using an equivalent bandwidth (EB) has become attractive [16] [7]. In fact, methods are required that generate timesavings at switches, while maintaining an adequate QoS. The EB is the result of applying statistical theories, and it is defined as the bandwidth ensuring a certain QoS (usually taken as the cell loss probability, α).

2.1 Equivalent Bandwidth and the Fractional Brownian Model

There are four basic models of EB, each one with several variants: Poisson Model [2], Gaussian Model [7], ON/OFF Model [7] and FBM [12]. We use the FBM model because it better represents the network behaviour, and because it allows the characterisation of different traffic types using a small number of parameters.

Several studies have pointed out the existence of a fractal or self-similar behaviour in data traffic [4]. Regarding traffic modelling, the variance of burst length is so large that, in many cases, it tends to infinity. This makes the decay of α, when the buffer size is increased, to be much lower than the one inferred from a Poisson model. Several researchers have attempted to model this behaviour, and the one used in this paper is due to Norros [12]. The model describes the traffic for a connectionless network. The Norros model can be applied to allocate resources in variable bit rate (VBR) and available bit rate (ABR) broadband networks. To model an accumulative system that complies with the self-similarity characteristics, the following equation is used:

$$A_t = mt + \sqrt{am}Z_t \quad t \in (-\infty, \infty) \tag{1}$$

where A_t represents the amount of traffic (number of packets) arrived until time t, m is the mean arrival rate, a is the variance coefficient of the mean per time period (bps per sec). Since t initially has no units, it has to be normalized as $t = T/t_u$ where T is the real time and t_u is the unit. Z_t is a non-dimensional number representing a normalized self-similar Gaussian process (or FBM) with parameter H. The Hurst parameter H, defines the level of self-similarity of the traffic. Its range goes from 0.5 to 1. The fractional Brownian storage with input parameters m, a and H and output capacity $C > m$ is the stochastic process X_t defined as:

$$X_t = \sup_{s \le t}(A_t - A_s - C(t-s)), \quad t \in (-\infty, \infty) \tag{2}$$

Then, the approximate queue length distribution is given by a lower bound [12]:

$$P(X_t > x) \ge \overline{\Phi}\left(\frac{(C-m)^H x^{1-H}}{\kappa(H)\sqrt{am}}\right) \tag{3}$$

where

$$\kappa(H) = H^H \cdot (1-H)^{1-H} \quad \textit{and} \quad \overline{\Phi} = P(Z_1 > y) \tag{4}$$

Function $\overline{\Phi}$ is approximated by a Weinbull distribution, and a lower bound for the probability is obtained. Considering B as the maximum queue length (in bits), the cell loss probability (α) is:

$$\alpha = P(x > B) \ge e^{-\left(\frac{(C-m)^{2H}}{2\kappa(H)^2 am}\right)\cdot B^{2-2H}} \tag{5}$$

From this equation, the equivalent bandwidth can be written as:

$$EB = C = m + \left(\kappa(H) \cdot \sqrt{-2 \cdot \ln(\alpha)}\right)^{\frac{1}{H}} \cdot a^{\frac{1}{2H}} \cdot B^{-\frac{(1-H)}{H}} \cdot m^{\frac{1}{2H}} \tag{6}$$

It can be seen that the equivalent bandwidth is a function of a small number of parameters (m, a, H, B and α), and then it can be used to solve the network model with less complexity.

2.2 Traffic and Network Parameters

Each one of the types of traffic (data, audio and video) is characterized by its flow mean and variance (parameter a of the FBM model), network distribution type, Hurst parameter, α and maximum buffer length B. Most of parameters are related to the EB chosen model. Since in a broadband network resource reservation is used, the parameters are defined according to the EB, allowing a controlled α.

The nodes of the network represent switches that generate, receive or redirect the traffic. The arcs of the network represent the links that connect the switches, and they have a certain capacity or bandwidth. Three discrete capacities (622, 155 and 45 Mbps) were considered. One of the outcomes of the problem is deciding what links have to be built, what is the capacity of each link, and which switches are to be located.

2.3 Delay

The delay considers two components: queuing delay at switches and transmission delay at links. We use a typical optical fibre network, with a transmission delay of 4[microsec/Km]. The queuing delay depends on the traffic model. For the FBM model, each switch has a service queue and independent virtual connections (VC). Then, every connection has its own queue with a Weinbull distribution (equation 9), and a service rate equal to the EB. Defining β and γ as follows:

$$\beta = \left(\frac{(C-m)^{2H}}{2(H^H(1-H)^{1-H})am} \right)^{-1/\gamma} \tag{7}$$

$$\gamma = 2 - 2H \tag{8}$$

$$P(X < x) = \begin{cases} 0 & x < 0 \\ 1 - e^{-\left(\frac{x}{\beta}\right)^{\gamma}} & x \geq 0 \end{cases} \tag{9}$$

Assuming X is the number of elements on this queuing system, equation (9) can be written as follows, considering the maximum capacity of a buffer, B:

$$P(X > B) = e^{-\left(\frac{x}{\beta}\right)^{\gamma}} \quad x \geq 0 \tag{10}$$

Using these equations, it is possible to obtain the expected number of individuals in the queue as:

$$E(X) = \beta \cdot \Gamma \left(1 + \frac{1}{\gamma}\right) \tag{11}$$

The average queue length $\overline{x}$ is obtained by replacing the EB (equation 6) into C of equation 7. This value does not depend of the mean or variance of the traffic type, but only on the maximum buffer size B and the cell loss probability, α. Finally using Little's law, we obtain the queuing delay at the switches, D_S , as:

$$D_s = \frac{\overline{N}}{\lambda} = \frac{\overline{x}}{m} = \frac{(-\ln(\alpha))^{-\frac{1}{2-2\cdot H}} \cdot B \cdot \Gamma\left(1+\frac{1}{2-2\cdot H}\right)}{m} \tag{12}$$

3 Network Model

The model minimizes the costs of establishing the network, given the node locations, traffic loads of each origin - destination pair for the three types of traffic, subject to QoS constraints (node and arc delay). Its formulation follows. First we introduce the notation used. The following terms are given as inputs to the problem:

n : number of nodes in the network
$Node$: Set of all nodes
Arc : Set of all possible undirected arcs among network nodes.
$Link$: Set of all possible directed arcs among nodes.
Cap : Set of arc capacities: 622, 155 or 45 Mbps.
OD : Set of all possible origin-destination (od) pairs.
C_a : Cost of arc, per length unit.
C_s : Fixed costs of node
C_c : Cost of link capacity hardware
x_{ij} : Binary variable that takes value 1 if the arc between nodes i and j exists, and 0 otherwise.
y_{ij}^r : Integer variable. It is the quantity of links of type r in arc ij.
n_i : Binary variable that counts the number of terminal equipments located at node i.
$Eflow_{ij}^{od,k}$: Binary variable equal to 1 if there is flow of traffic type k through arc ij of od pair of type k, 0 otherwise.
$flow_{ij}^{od,k}$: Real variable. Corresponds to the flow through link i, j of pair od and traffic type k.
$traf_{od}^k$: Amount of traffic of type k going from origin to destination node.
d_{ij} : Distance between node i and j.
D_s^k : Queuing delay at switches, for traffic type k
D_T : Transmission delay in the optical fiber, per distance unit
$Vcap_r$: Capacity of link of type r (45, 155 or 622 Mbps)

3.1 Mathematical Formulation

The objective function takes into account the total topology building cost, the total switching equipment cost of nodes, and the total transmission equipment cost.

Objective: Minimize the network building cost:

$$\min \sum_{(i,j)\in Arc} C_a d_{ij} x_{ij} + \sum_{i\in Node} C_s n_i + \sum_{(i,j)\in Link} \sum_{r\in Cap} C_c^r y_{ij}^r \tag{13}$$

Subject to: i) *Connection of nodes.* This restriction tells that if an arc exists, switches must exist at both ends of the arc.

$$n_i + n_j \geq 2x_{ij} \quad \forall\ (i,j) \in Arc \tag{14}$$

ii) *Traffic flow.* This restriction sets the flow passing for a given arc of a given od pair.

$$flow_{ij}^{od,k} = traf_{od}^{k} Eflow_{ij}^{od,k} \quad \forall\ (i,j) \in Arc, (o,d) \in OD, k \in typeT \tag{15}$$

iii) *Arc presence.* If flow exists, in any direction and of any type, then an arc must exist.

$$\sum_{k\ \in\ typeT} \left(flow_{ij}^{od,k} + flow_{ji}^{od,k} \right) \leq \sum_{k\ \in\ typeT} traf_{od}^{k} x_{ij} \ \ \forall\ (i,j) \in Arc, (o,d) \in OD \tag{16}$$

iv) *Flow balance.* There are three cases for the nodes: they can be origin, destination or transfer nodes. The restriction indicates that the traffic is generated, absorbed or transferred at the node, according to its type.

$$\sum_{(i,j)\ \in\ Arc} flow_{ij}^{od,k} - \sum_{(j,l)\ \in\ Arc} flow_{jl}^{od,k} = \begin{cases} traf_{od}^{k} & if\, j = d \\ -traf_{od}^{k} & if\, j = o \\ 0 & otherwise \end{cases};$$

$$\forall\ j \in Node, (o,d) \in OD, k \in typeT \tag{17}$$

v) *Link capacity assignment.* The sum of all flows passing through an arc has to be less or equal than the total capacity of the arc.

$$\sum_{(o,d)\ \in\ Path} \sum_{k\ \in\ typeT} \left(flow_{ij}^{od,k} + flow_{ji}^{od,k} \right) \leq \sum_{r\ \in\ cap} Vcap_r y_{ij}^{r} \ \ \forall\ (i,j) \in Arc \tag{18}$$

vi) *Delay constraint.* The delay is composed of two components: one is related with the number of nodes of the path and the other with the distance.

$$D_s^k \left(\sum_{(i,j)\ \in\ Arc} Eflow_{ij}^{od,k} - 1 \right) + D_T \sum_{(i,j)\ \in\ Arc} d_{ij} Eflow_{ij}^{od,k} \leq Delay^k$$

$$\forall\ (o,d) \in OD, k \in typeT \tag{19}$$

vii) *Unidirectional flow constraint.* This constraint forces the flow of an od pair to go through a given arc in only one direction.

$$Eflow_{ij}^{od,k} + Eflow_{ji}^{od,k} \leq 1 \quad \forall\ (o,d) \in OD, (i,j) \in Arc, k \in typeT \tag{20}$$

The distance was calculated using the Euclidian formula, since node coordinates are known. Traffic was defined by using equation 6.

4 Simulated Annealing Methods

Simulated annealing, a technique introduced by Kirkpatrik [10], is a Monte Carlo approach for minimizing multivariate functions. The term simulated annealing derives from the roughly analogous physical process of heating and then slowly cooling a substance to obtain a strong crystalline structure. In simulation, a minimum of the cost function corresponds to this ground state of the substance. The simulated annealing process lowers the temperature by slow stages until the system "freezes" and no further changes occur. At each temperature the simulation must proceed long enough for the system to reach a steady state or equilibrium. This is known as thermalisation. The time required for thermalisation is the decorrelation time; correlated microstates are eliminated. The sequence of temperatures and the number of iterations applied to thermalise the system at each temperature comprise an annealing schedule. To apply simulated annealing, the system is initialised with a particular configuration. A new configuration is constructed by imposing a random displacement. If the energy of this new state is lower than that of the previous one, the change is accepted unconditionally and the system is updated. If the energy is greater, the new configuration is accepted probabilistically. This is the Metropolis step, the fundamental procedure of simulated annealing [17]. This procedure allows the system to move consistently towards lower energy states, yet still "jump" out of local minima due to the probabilistic acceptance of some upward moves, using the Boltzmann probability distribution. If the temperature is decreased logarithmically, simulated annealing guarantees an optimal solution.

5 Computer Experiments

The proposed model was tested by first trying it with small networks, comparing the results of a commercial package AMPL-CPLEX and a SAA heuristic on MATLAB. Then the heuristic was used on a large 48 node network. The first step validated the heuristic showing its efficiency and speed. All the tests were done in a Digital DEC Alpha 433 cluster.

To test the model with small networks, the TSP library [15] was used. From that library 8 problems were chosen that appear to have random node locations (Att48, Bier127, Ch130, Eil51, KroB100, Rd100 and St70). From these problems the coordinates and Euclidean distances were calculated. In all the problems, a subset of the first 8 node locations was used to define each small network.

The solving time for the proposed model using AMPL-CPLEX was quite large. Up to 8 nodes, the run time was reasonable (hundreds of thousands of seconds). However, for 9 nodes, the run time grew to over 1 million seconds.

In the implementation of the SAA, some adaptations were performed to specialize it to the proposed model. Two variables that are important are the number of heating cycles and the MetropolisR function adaptation constant [14]. In the first case, the method consists in re-heating the system, with the best solution found in the previous processing cycle as the initial network. Using an

8-node subset of the ATT48 problem, the optimum solution of AMPL-CPLEX was compared with the results of running 100 times the SAA with two cycles per run. It was observed that increasing the number of cycles improves the percentage of cases reaching the optimum. This result is very important in large networks. One of the characteristics of the SAA is its ability of jumping out from local optima by using the probability function defined by Metropolis [14]. This constant was defined in such a way that, in some cases, a new higher cost state can be accepted. We tried values of 0.3 and 0.4 observing that the latter improves the results.

5.1 Running Tests

The first observed result of the tests is that SAA solution times are clearly shorter than AMPL-CPLEX, though not always reaching the optimum. In 5 of the 8 networks more than 77% of the SAA runs reached the optimum.The AMPL-CPLEX results were found by using an uppercut restriction between 0.5 and 10% of the optimum. Cases with poor results (Bier127, Krob100 and Rd100) have all a common characteristic, which is that the optimum solution is clearly a star, far from the initial MST solution used as a starting solution. That explains the SAA not providing very good results.

Other cases such as Ch130 and Ch150, show very high percentages of reaching the optimum, since the final solution topology is very similar to the initial MST solution. The main conclusion of all the tests is that the SAA delivers good results when the networks have more uniform traffic distributions and without very hard delay restrictions.Table 1 shows for the Att48 test, and links connected to node n1, the link capacity assignment and flows for each traffic type. All tests used the same traffic matrices for each data type.

Table 1. Tests results for node n1 connections on 8-node subset of problem ATT48.

Link	Number of network cards			Total capacity [Mbps]	Traffic flow [Mbps]		
	622	155	45		Data	Sound	Video
n1 - n3	2	1	0	1399	6.71	526.15	804.93
n1 - n7	2	3	0	1709	5.59	394.61	1207.39
n1 - n8	0	3	0	465	1.68	230.19	201.23

5.2 Full-Scale Problem

To evaluate the model for a larger size network the full ATT48 problem was solved. Here the optimal solution is not known, and the number of runs is limited as each cycle takes around 35 hours.

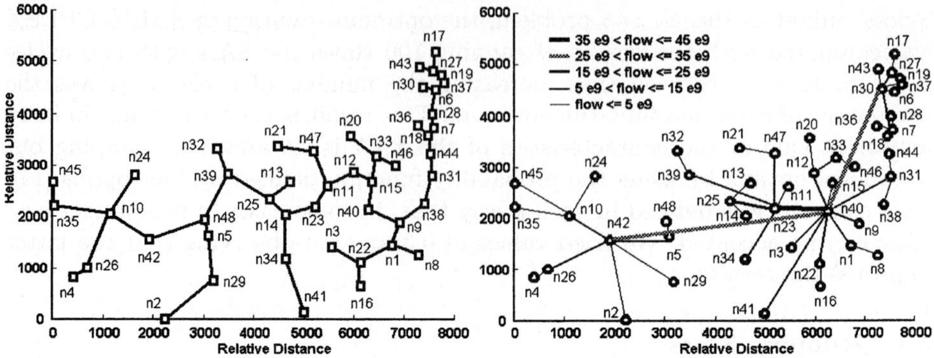

Fig. 1. Initial network and best feasible solution for the ATT48 problem.

Figure 1 shows the best-found solution and the initial MST. While this solution may not be the optimum solution, the SAA heuristic ensures the solution found would be a network that will comply with all the QoS requirements and with building costs within reasonable limits. A larger number of cycles may improve this solution but the processing time necessary to find feasible networks will grow considerably. This means that in the time assigned to a cycle less and less low cost solutions are found. Table 2 shows the capacity assignment and traffic flow obtained in the best solution for node n1.

Table 2. Capacity assignment and resulting flow for node n1 of the ATT48 problem.

Link	Number of network cards			Total capacity [Mbps]	Traffic flow [Mbps]		
	622	155	45		Data	Sound	Video
n1 - n8	5	0	0	3110	9.5	1545.55	1408.62
n1 - n40	23	3	0	14771	19	3025.34	11671.46

6 Conclusions

The model presented considers all the basic elements used in a broadband network design: topology, routing and capacity assignment. It only requires knowledge about node locations and traffic distributions for each information type. The solution algorithm, which uses Simulated Annealing techniques, provides good solutions in reasonable computer times, even for a 48-node example carrying three information types. Three topics were discussed: a) the network model, which minimizes the cost of interconnecting a number of nodes whose locations are known, b) the traffic model, where we used the concept of Equivalent Bandwidth and c) the resolution algorithms, where we used SAA and a commercial solver. The main contributions of this paper are: a) a network model that includes most of the aspects of the network design problem starting from the node locations and traffic data, b) we propose an adaptation of the Fractional Brownian Motion traffic model to the conditions of the network design problem, and c) we apply the simulated annealing algorithm to solve the model. Computational experiments show that SAA gets to the optimum solution over 10 times faster than a commercial AMPL-CPLEX solver. The experiments also show that for networks with more than 50 nodes, AMPL-CPLEX is unable to deliver results due to the enormous processing time, while the SAA still delivers good feasible solutions. In future developments of the model we will try to incorporate fault tolerance capability to the design[1].

References

1. Alevras, D., Grotschel, M., Wessaly, R.: Capacity and Survivability Models for Telecommunication Networks. Preprint SC97-24. Konrad-Zuse-Zentrum fur Informationstechnik. Berlin (1997).
2. Babic, G., Vandelore, B., and Jain, R.: Analysis and Modeling of Traffic in Modern Data Communication Networks. Ohio State University, Department of Computer and Information Science, Technical Report OSU-CISRC-1/98-TR02, Feb 1998.
3. Balakrishnan, A., Magnanti, T., and Wong, T.: A Dual-Ascent Procedure for Large-Scale uncapacitated Network Design. INFORMS Operation Research, 37,(1989) 716-740.
4. Beran, J. et al.: Long Range Dependence in Variable Bit Rate Video Traffic. IEEE Transactions on Communications , 43,(1995) 1566-1579.
5. Bienstock, D. et al.: Minimum Cost Capacity Installations for multicommodity network flows. Mathematical Programming 81 , no. 2-1,(1998) 177-199.
6. Bienstock, D., and Saniee, I.: ATM Network Design: Traffic Models and Optimation-Based Heuristics. DIMACS Report 98-20, Telecom Systems, June 2001.
7. Courcoubetis, C., Fouskas, G., and Weber, R.: On Performance of an Effective Bandwidths Formula. Proc. 14th Int. Teletraffic Cong., 6-10 June 1994 North-Holland Elsevier Science, (1994) 201-212.
8. Ball , M.O. et al. (Eds.): Network Models, Handbook in Operations Research and Management Science, Vol. 7, Amsterdam: Elsevier, (1995).
9. Ball , M.O. et al. (Eds.): Network Routing, Handbook in Operations Research and Management Science, Vol. 8, Amsterdam: Elsevier, (1995).
10. Kirkpatrik, S., Gelatt, C.D., and Vecchi, M.P.: Optimization by Simulated Annealing. Science, 220, 4598,(1983) 671-680.
11. Laarhoven, P. J. M. Van., and Aarts, E. H. L.: Simulated annealing : theory and applications. Kluwer Academic (1987).
12. Norros, I. : On the Use of Fractional Brownian Motion in The Theory of Conneccionless Networks". JSAC, 13-6, August 1995
13. Nurmela, K.J., and Ostergard R.J. :Constructing Covering Designs by Simulated Annealing. Technical report B10, Digital Systems Laboratory, Helsinki University of Technology, (1993).

[1] This work was funded, in part, by grant 1040577 from Fondecyt

14. Press, W. H., et al.: Numerical Recipes in C (2 Ed.). Cambridge University Press (1992).
15. Reinelt, G.:TSPLIB95. Universitat Heidelberg, Institut fur Angewandte Mathematik,Im Neuenheimer Feld 294(1995).
16. De Veciana, G., Kesidis, G., and Walrand, J.: Resourse Management in Wide-Area ATM Networks Using Effective Bandwidths. JSAC, 13-6, August 1995.
17. Metropolis, N., Rosenbluth, A., Rosenbluth, M., Teller, A., and Teller, E.: Simulated Annealing. Journal of Chemical Physics, Vol. 21, (1953) 1087-1092.

Power-Efficient TCAM Partitioning for IP Lookups with Incremental Updates

Yeim-Kuan Chang

Department of Computer Science and Information Engineering
National Cheng Kung University
Tainan, Taiwan R.O.C.
ykchang@mail.ncku.edu.tw

Abstract. Ternary Content-Addressable Memories (TCAMs) provide a fast mechanism for IP lookups and a simple management for route updates. The high power consumption problem can be resolved by providing a TCAM partitioning technique that selectively addresses smaller portions of a TCAM. This paper proposes a 2-level TCAM architecture using prefix comparison rule to partition the routing table into a number of buckets of possibly equal sizes. The 2-level architecture results in a 2-step lookup process. The first step determines which bucket is used to search the input IP. The second step only searches the IP in the determined bucket from the first step. The prefix partitioning algorithm provides an incremental update. Experiments show that the proposed algorithm has lower power consumption than the existing TCAM partitioning algorithms.

1 Introduction

Backbone routers have to forward millions of packets per second at each port. The IP lookups of the routers becomes the most critical operation to reach the capability of forwarding millions of packets per second. In [1], a large variety of routing lookup algorithms were classified and their worst-case complexities of lookup latency, update time, and storage usage were compared. However, these schemes using DRAMs or SRAMs can hardly meet the wire speed requirements needed for current terabit router design since they usually need many memory access cycles to look for a matched routing entry.

TCAMs are fully associative memories that allow a "don't-care" state for each memory cell, in addition to the states of 0 and 1. Each entry of a TCAM consisting of multiple cells is long enough to store a prefix for IP lookups or a rule for packet classification. TCAM is designed in such a way that all the entries are looked up in parallel against the incoming IP address. Thus, a matched entry if it exists can be found in a single TCAM access cycle. If multiple entries match the IP address, the entry with lowest address (i.e., longest prefix) in TCAM is typically returned as the result. Moreover, the update process in TCAMs is in general very simple [5].

C. Kim (Ed.): ICOIN 2005, LNCS 3391, pp. 531–540, 2005.

There are two major disadvantages for TCAMs, the high cost-to-density ratio and power consumption. TCAM designs from IDT and Netlogic have effectively solved the issue of high cost-to-density ratio. The cost of their TCAM designs is very competitive with other hardware alternatives.

The high power consumption comes from the fact that the hardware circuits of all the entries in a TCAM are activated in parallel to perform the matching operations. Therefore, TCAM vendors today provide entry selection mechanisms to reduce power consumption by selecting fixed regions of TCAM called buckets for matching process.

In this paper, we propose a prefix partitioning scheme that performs better than subtree-split and postorder-split in most of the cases. The proposed scheme is based on the prefix comparison mechanism by which we can compare two prefixes of different lengths. Using the prefix comparison mechanism, we can sort the original prefixes and divide them evenly into K groups by selecting $K-1$ pivot prefixes with some small number of duplicated pivot prefixes. We will show by performance evaluation on real routing tables that the proposed partitioning scheme performs better than subtree-split and postorder-split in power consumption reduction.

The rest of the paper begins with the definition of IP lookup problem and related works in section 2. Section 3 illustrates the basic ideas of the proposed partitioning algorithms and the detailed design. The results of performance comparisons using real routing tables available on the Internet are presented in section 4. Finally, a concluding remark is given in the last section.

2 Problem Definition and Related Works

A TCAM consists of a large number of multi-cell entries. Each cell of an entry in a TCAM is a ternary bit which is implemented by a value bit and a netmask bit. A cell is compared with the corresponding bit in the target IP. A cell match is found if the netmask bit is 0 or the value bit is the same as the corresponding bit in the target IP. A TCAM entry matches the target IP if all the cells match the target IP. The TCAM is designed in such a way that when an IP is input, all the TCAM entries are activated to compare against the input IP. Only the match in the TCAM entry with the lowest address is returned as the final result. To ensure the TCAM entry with the lowest address is the longest prefix match, the prefixes in the TCAM must be stored in order of decreasing prefix length. The prefix length ordering slows down the TCAM update process. Shah and Gupta [5] have proposed efficient update schemes for TCAM updates.

In [4], Zane, et. al, took advantage of the entry selection mechanisms and developed two sets of prefix partitioning schemes to reduce the number of prefixes searched in an IP lookup. Their first partitioning scheme called bit selection architecture combines simple glue logic with TCAM. The glue logic uses a simple hashing function to select a set of input bits called hashing bits as an index to the appropriate TCAM bucket. The bit selection schemes are not suitable for real implementations because of the following drawbacks. The first drawback is that it only deals with prefixes of length 16-24. No better scheme to work with the prefixes of length < 16 and > 24. The second drawback is that no prefect hashing function can be obtained in advance and thus the worst-case power consumption is still too high. In order to eliminates the drawbacks of bit selection schemes, the authors proposed another set of schemes, the trie-based partitioning schemes.

Their trie-based partitioning schemes use a two-level TCAM organization. The first-level TCAM is responsible for matching the target IP and obtaining an index of the second-level TCAM bucket. The main idea is to find a subtrie or a set of subtries and put the prefixes in the subtrie(s) in a single TCAM bucket such that all the buckets are balanced. Two different split algorithms were proposed. They are *subtree-split* and *postorder-split* algorithms. Given a parameter b, the subtree-split algorithm can partition the routing table into K buckets, where K $\in [\lceil N/b \rceil, \lceil 2N/b \rceil]$ and each bucket contains $\lceil b/2 \rceil$ to b prefixes. In addition, an index TCAM of size K and an index SRAM are needed. The drawbacks of the subtree-split algorithm are as follows. Usually, K is fixed in advance. There is no simple and efficient method to estimate how big is b for obtaining the expected K. Also, the numbers of prefixes stored in TCAM buckets are not balanced; the worst-case difference between the numbers of prefixes in two TCAM buckets is $\lceil b/2 \rceil$. The worst is that the subtree-split algorithm may fail when K is fixed and TCAM memory pressure to store all the prefixes is high. Another drawback is that the subtree-split algorithm does not provide an incremental update when a new prefix is going to be inserted into a full TCAM bucket. For this situation, only re-partitioning is the solution. The postorder-split algorithm remedies most of the drawbacks of the subtree-split algorithm. It partitions the routing table into K buckets each containing exactly $\lceil N/K \rceil$ prefixes, except possibly the last bucket. However, the postorder-split algorithm comes at the cost of large index TCAM, especially when K is large. The postorder-split algorithm can move the prefixes in an overflowed TCAM bucket to a neighboring bucket. However, it is not clear that both neighboring buckets of an overflowed bucket can be used as the buffering space to hold overflowed prefixes. Also, when both neighboring buckets of an overflowed bucket are full, no incremental update can be achieved.

Based on their performance evaluations in terms of power consumption reduction, subtree-split performs better than postorder-split when the number of buckets grows beyond 64. On the contrary, postorder-split performs better than subtree-split when the number of buckets is less than 64.

In this paper, we use the prefix comparison technique proposed in [6] to sort the prefixes. With the sorted prefixes, we can easily partition the prefixes into K groups by selecting K - 1 pivot prefixes. Thus, all the prefixes can be compared in parallel with the target IP address to determine which group that this target IP belongs to.

Before going on, we make the following assumptions used in the paper. Similar to the assumption used in [4], we assume that an original TCAM can be divided into K buckets. An additional control line is used to select which TCAM bucket is only activated for searching the input IP. Also, we assume that the same prefix can be distributed over many buckets. When a prefix is to be deleted, we assume that we can select a TCAM bucket or all the TCAM buckets from which the prefix is deleted.

3 Proposed Partitioning Algorithms

It is known that if a list of data items can be sorted then they can be easily partitioned into groups such that all the data items in one group are smaller/larger than all items in another group. Next, a list of pivot items that are the largest or smallest items in all the groups can be selected as the first level data items. As a result, the search process for a data is operated in two steps. The first step searches the first level pivot list and determines which group the searched data belongs to. The second step then searches the determined group for the data. Before we can apply the same technique to reduce power consumption in TCAM, we need a comparison mechanism to sort the prefixes in routing tables and partition them into groups. Comparing prefixes is not easy because the lengths of the prefixes may be different. In [6], we have proposed a systematic method to compare prefixes of various lengths. Based on the proposed comparison mechanism for the prefixes of different lengths, the proposed prefix partitioning algorithm can be developed.

In order to devise a new partitioning technique based on the prefix comparison, we also need to consider the enclosure property among the prefixes since shorter prefixes may cover more than one group. The enclosure prefixes must be put into the groups that contain longer prefixes covered by them. Therefore, we need to duplicate the enclosure prefixes and put them in both groups. In summary, the problem of partitioning the prefixes into K groups becomes (1) how to select the pivot prefixes to cut the tree evenly into groups, (2) which enclosure prefixes are duplicated, and (3) which groups to put the duplicated prefixes. Before describing the detailed design of the proposed algorithm, we give the definition of prefix comparisons as follows.

Definition 1 of prefix comparison: The inequality $0 < * < 1$ is used to compare two prefixes in the ternary representation of prefixes.

Instead of sorting the prefixes in a routing table by a fast sorting algorithm such as quick sort, we use the binary trie to complete the sort operation. The binary trie will also be used for determining the enclosure prefixes. Given a routing table, the binary trie is first constructed. Then we perform an inorder traversal in the binary trie and obtain the sorted prefixes. Assume there are N prefixes in the routing table and we want to partition them into K groups. We simply divide the sorted prefixes evenly into K groups. Each group contains $\lceil N/K \rceil$ prefixes. We select the largest prefix in each group as the pivot, except the last group. There are $K - 1$ pivot prefixes, *pivot[i]* for i=0 ... $K - 2$.

Now we describe how to determine which enclosure prefixes should be duplicated and inserted into which group. For a group *i*, we consider the largest and smallest prefixes L_i and S_i. We first traverse the binary trie bottom-up from L_i to the root. If a valid enclosure prefix, P_{enc}, that encloses L_i is found and P_{enc}

$>$ pivot[i] or $P_{enc} \leq$ pivot[i−1], then P_{enc} is duplicated and put into group i. Secondly, similar traversal from S_i to the root can be done. The two traversals for L_i of group i and S_{i+1} of group i+1 can be combined into one by performing the traversal for the longest common ancestor of L_i and S_{i+1}. This is because no valid prefix exists between L_i and S_{i+1} and the longest common ancestor of L_i and S_{i+1} must locate between L_i and S_{i+1}. We call the above procedure for duplicating enclosure prefixes as the *group cut procedure.* Theoretically, each group cut generates at most W extra prefixes. We present the detailed algorithm for the proposed prefix partition as follows.

```
Prefix-partition(K, RoutingTable)
{
    Construct the binary trie (BinaryTrie) from the routing table (RoutingTable)
    Inorder traverse BinaryTrie and store all valid prefixes
        in array Plist in sorted order.
    Select K − 1 prefixes, pivot[i] for i = 0 .. K − 2 such that
        pivot[i] = Plist[i * b − 1], where b= ⌈N/K⌉.
    For (i = 0; i < K − 1; i++) {
        Put prefixes, Plist[i*b] to Plist[i*b + b − 1] in group i.
        For each of the ancestor prefixes (P) of the longest common
            ancestor of Plist[i*b + b − 1] and Plist[i*b + b] do
                If P is not in group i then put P in group i.
                If P is not in group i + 1 then put P in group i + 1.
    }
    Put prefixes Plist[K*b − b] to Plist[size(Plist) − 1] in group K − 1.
}
```

By using $\lceil N/K \rceil$ to be the number of prefixes in each group, we may underestimate the impact of duplications of enclosure prefixes. Therefore, in real implementation, we can use a kind of greedy algorithm by increasing the size of $\lceil N/K \rceil$ to balance the number of prefixes in each TCAM bucket. However, based on our experiments, the greedy algorithm produces no bigger difference. Also in order to make insertion process simple, the pivot prefixes are selected in such a way that they are disjoint with each other. The selected pivot prefixes are restricted not to be that of length 32. This restriction can be fulfilled by selecting a pivot prefix between groups i and i + 1 that is larger than the largest prefix in group i and smaller than the smallest prefix in group i + 1. This restriction for selecting a pivot prefix is for using 32 bits to represent a prefix and will be explained later. Based on the above analysis, we have the following result.

Theorem 1: The size of each group created by the prefix partitioning algorithm is at most $\lceil N/K \rceil + W$, where the size of the routing table is N and maximum length of prefixes in the routing table is W.

The architecture of the proposed prefix partitioning algorithm is illustrated in Figure 1. Each partitioned group of prefixes is in fact treated as a TCAM bucket. The total power consumption for each lookup is equal to the power consumed by the selected TCAM bucket and the first level circuitry. As shown

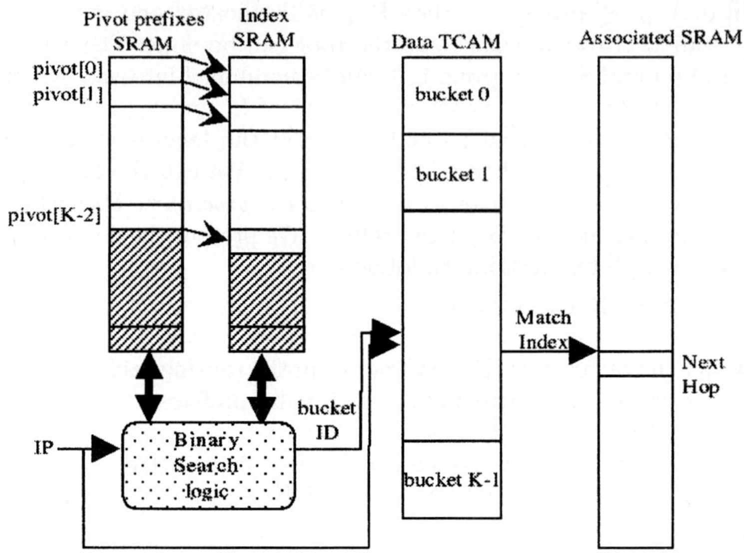

Fig. 1. TCAM architecture for the proposed prefix partitioning algorithm.

in Figure 1, the binary search logic first lookups the pivot array and obtains the TCAM bucket number from the associated index array. Then the TCAM bucket number is input in data TCAM and only that TCAM bucket is looked up.

Route Update. The route update consists of two steps. The first step is to determine if the inserted prefix needs to be duplicated and which groups to insert. The second step is to avoid recomputed the prefix partition process because re-partitioning involves excessive TCAM write operations.

Recall that the prefixes in the pivot array are disjoint and are in an increasing order. When inserting a prefix P, we perform the following steps. First we use the binary search to locate the position at which P is supposed to reside. Assume that pivot[i−1] < P < pivot[i]. We consider two conditions. The first condition is when P is disjoint with both pivot[i−1] and pivot[i] or P is enclosed by either pivot[i−1] or pivot[i], but not both. The second condition is when prefix P encloses one or more pivot prefixes. If the first condition is met, prefix P is inserted into group i since no other group will be affected. If the second condition is met, we need to search for a set of successive pivot prefixes that are enclosed by prefix P. Then we duplicate prefix P and put a copy of P in each TCAM bucket that is enclosed by P. The deletion of an old prefix requires only a delete command issued to all the TCAM buckets to remove the prefix if it exists.

The TCAM partitioning algorithms proposed in [4] assumed that TCAM buckets need to be re-partitioned each time any bucket overflows. Therefore, the frequency of re-partitioning depends on two factors. One is the route addition or deletion rate. The other factor is the prefix occupation pressure in each TCAM bucket. As shown in [4], if we have a low prefix occupation pressure in any

TCAM bucket, the number of times that a re-partitioning process is needed will be very small and thus the updating impact will be negligible. In this section, we shall propose an update scheme that avoids re-partitioning even when the TCAM bucket to be inserted is full.

Assume a prefix P is supposed to be inserted in TCAM bucket i. We first consider the case when TCAM bucket i is full and TCAM bucket i + 1 is not full. Let L_i and M_i be the largest and the second largest prefix in TCAM bucket i. We select another pivot Q_i such that $M_i \leq Q_i < L_i$. Next, we move L_i to group i + 1 and let pivot[i] = Q_i. Similar operations can be performed when TCAM bucket i − 1 is not full. Notice that the smallest bucket (bucket 0) or largest bucket (bucket K) has only one neighboring bucket instead of two. Assume that L_i and S_i are the largest and smallest prefixes in bucket i, respectively. By using S_0 and L_K as additional pivot prefixes, buckets 0 and K can have two neighboring buckets and thus less chance to perform the routing table re-partitioning.

As stated in [4], there are occasional floods of up to a few thousand route additions in one second in the real update router traces. Also, these route additions are very close to each other in a subtrie of the binary trie built from the routing table. These routes are often subsequently withdrawn. These floods often cause a single bucket or even three consecutive buckets to repeatedly overflow. Therefore, we may need a better scheme to avoid repartitioning the routing table even when buckets i − 1, i, and i + 1 are full. The proposed prefix update algorithm is as follows. Assume the new routing prefixes need to be inserted in bucket i. Firstly, we select a new pivot prefix Q_i in bucket i such that is $M_i \leq Q_i < L_i$, where L_i is the largest and M_i is the second largest in bucket i. Secondly we select a non-full bucket j. The prefix L_i is moved to bucket j. Bucket j can be selected to be the one containing the smallest number of prefixes at the time of insertion. Let P_i = pivot[i]. Thirdly, we assign pivot[i] = Q_i and add P_i as an extra pivot between pivot[i] and pivot[i + 1]. Notice that the pivot[j] for j = i + 1 to K − 2 must be shifted to allocate an empty entry to put P_i. The corresponding entry in index SRAM must also be allocated by shift operations and set this index TCAM entry to point to bucket j.

32-bit prefix representation. The two most important operations needed in the IP lookup are matching a prefix with a 32-bit IP address and comparing two prefixes of various lengths. The matching operations have already been taken care by the hardware logics in TCAM. The prefix comparison involves ternary operations where 0, 1, and the don't care bit (denoted as *) are checked. In [6], we proposed a method to represent the prefixes in a binary format using 33 bits in the context of IPv4. The method can be straightforwardly expanded to IPv6. We formally define the binary representation of a prefix as follows.
Definition 2 of (n+1)-bit representation for the prefixes in an n-bit address space: For a prefix of length i, $b_{n-1}b_{n-2}b_{n-i}{}^*$, where b_j=0 or 1 for $n - i > j \geq 0$, its binary representation is $b_{n-1}b_{n-2} \ldots b_{n-i}1a_{n-i-1} \ldots a_0$, where a_j=0 for $j = 0 \ldots n-i-1$.
Since current processors usually have 32-bit wide instructions. The 33-bit prefix representation thus needs two 32-bit storages. A naive implementation of the

comparison operations for the 33-bit representation can lead to inefficiency under 32-bit instructions. Two 32-bit binary comparison operations may be needed in the worst case since only 32-bit arithmetic and logic operations are available in current 32-bit processors. It also means that two 32-bit memory reads are needed if the size of registers is 32 bits.

If there is no prefix of length 32, we can use only 32 bits by ignoring the least significant bit that must always be zero. This is the case for computing the pivot prefixes in the proposed prefix partition algorithm. Note that a pivot is the common ancestor of two prefixes and thus has a length of at most 31. Therefore, when we use 32 bits to represent a pivot prefix of length less than 32, the 33th bit of a pivot prefix must be zero. However, if we use 32 bits to represent a prefix of length 32 (in fact, an IP address) the 33th bit of the prefix is assumed to be one. Therefore, when we compare an IP with a pivot prefix using 32-bit representation, equality means the IP is larger than the pivot prefix. Therefore, the 32-bit representation for a 32-bit ternary prefix is sufficient.

Table 1: general information of the routing tables used in the paper.

	Funet-40k	Oix-80k	Oix-120k	Oix-150k
# of prefixes	41,709	89,088	120,635	159,930
Length distr.	8-30-32	6-32	8-32	8-32
Date	1997-10-30	2003-12-28	2002-12-01	2004-2-1

4 Performance Evaluations

In this section, we present the performance results for the subtree-split and postorder-split algorithms [4] and the proposed T-bit expanded and prefix partitions. Table 1 shows the general information of the routing tables used in the paper. Three routing tables are the snapshots of the routing table traces obtained from University of Oregon Route Views Archive Project [oix]. The fourth table, i.e., the smallest among the four routing tables, is obtained from the web site provided by the authors of the LC trie [3]. For example, Funet-40k routing table contains only prefixes of length 8 to 30 and 31. With these routing tables of various sizes, we can obtain complete understanding on the performance when applying different TCAM architectures. We take K, the total number of TCAM buckets, as the input to the algorithms. The output will be the reduction in *power consumptions* which is defined as *the ratio of the total number of routing table prefixes to the maximum number of prefixes searched in all TCAM buckets for an IP lookup.*

For subtree-split and postorder-split algorithms and the proposed prefix partition, the maximum number of prefixes searched is the sum of the total number of routing entries in the index TCAM or in the pivot engine and the number of prefixes in the largest bucket in the data TCAM. As for the index SRAM and associated SRAM with routing information, its power consumption is small and can be ignored in the power consumption calculation.

The reductions in power consumption for four routing tables are shown in Figure 2. The proposed prefix partition performs consistently better than sub-

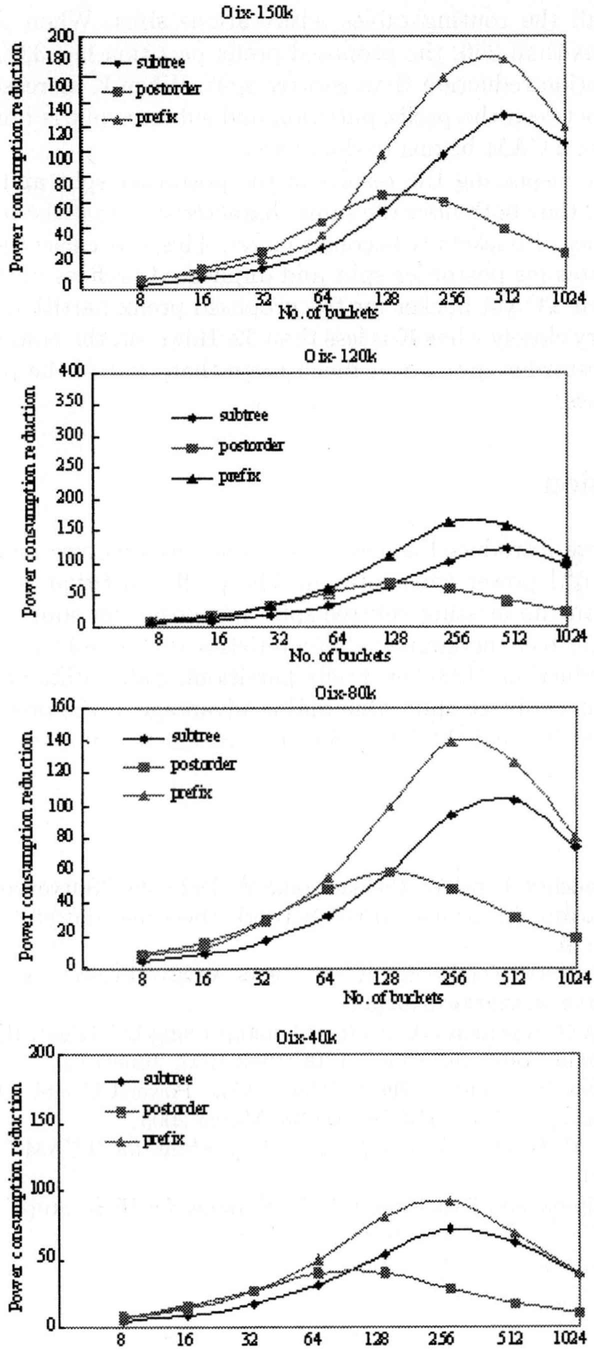

Fig. 2. Power consumption reduction for subtree-split, postorder-split, and prefix-partition algorithms.

tree split for all the routing tables with various sizes. When the number of buckets K is less than 256, the proposed prefix partition has 33% to 91% more power consumption reduction than subtree split. When K increases up to 1024, the difference between the prefix partition and subtree split is minimal because the size of index TCAM begins to dominate.

By carefully inspecting the results of the postorder split and the proposed prefix partition, they both have the same characteristics that the reduction drops when the number of buckets K becomes larger. This is because the size of index TCAM dominates for postorder split and duplicated prefixes of shorter lengths increases in each TCAM bucket for the proposed prefix partition. Their performance stays very closely when K is less than 32. However, the power consumption reduction of postorder split drops much more than that of the prefix partition when K increases.

5 Conclusion

In this paper, we introduced a new partitioning architecture and algorithm to reduce the TCAM power consumption. The prefix partition is shown to perform better than the existing subtree-split and postorder-split schemes. Prefix partition scheme has the similar characteristics to the subtree-split in power consumption reduction. However, prefix partitioning algorithm performs consistently better than subtree-split. One major advantage of the prefix partitioning algorithm is that it is capable of performing incremental updates.

References

1. M. A. Ruiz-Sanchez, Ernst W. Biersack, and W. Dabbous, "Survey and taxonomy of IP address lookup algorithms", IEEE Network Magazine, 15(2):8–23, March/April 2001.
2. D. Meyer, "University of Oregon Route Views Archive Project" at `http://archive.routeviews.org/`.
3. S. Nilsson and G. Karlsson "IP-Address Lookup Using LC-Tries", IEEE Journal on selected Areas in Communications, 17(6):1083-1092, June 1999.
4. F. Zane, G. Narlikar, and A. Basu, "CoolCAMs: Power-Efficient TCAMs for Forwarding Engines," in Proc. INFOCOM 03, March 2003.
5. D. Shah and P. Gupta, "Fast Updating Algorithms for TCAMs", IEEE Micro, pp.36-47, 2001.
6. Y. Chang, "Simple and Fast Binary Prefix Searches for IP Lookups", submitted for publication.

Hardness on IP-subnet Aware Routing in WDM Network

Ju-Yong Lee[1], Eunseuk Oh[2], and Hongsik Choi[2]

[1] Department of Computer Science,
Duksung Women's University, Ssangmoon-dong, Tobong-gu, Seoul, KOREA
jylee@duksung.ac.kr

[2] Department of Computer Science, School of Engineering
Virginia Commonwealth University Richmond, VA 23284-3068, USA
{eoh, hchoi}@vcu.edu

Abstract. In this paper, we study the hardness of the IP-subnet aware routing problem in WDM network. IP-subnet aware routing attempts to reduce routing overhead by grouping a set of addresses under a single subnet. However, routing in WDM network with subnets imposes neighboring IP interfaces along the path to be on the same subnet. This subnet constraint forces the routing path to be longer unless routing permits reconfiguration of subnets. The hardness of this routing problem is first studied in [1]. We investigate further the problem of finding paths accounting for subnets and its hardness. Regardless of subnets, we argue that a shortest IP hop path can be constructed in a polynomial time. We provide efficient routing algorithms to justify our argument.

1 Introduction

Routing in IP over WDM network incorporates traffic and topology information at both IP and optical layers. In turn, the routing protocol enables to leverage IP connectivity and massive WDM bandwidth capacity. In IP over WDM network, IP routers are directly connected to an optical core consisting of cross-connect switches (OXCs), where OXCs are interconnected via high-speed Wave Division Multiplexing (WDM) line system network. These IP routers create a virtual topology at the IP layer. When a routing request arrives, this request is routed along the virtual topology, and thus multiple lightpaths in the optical core. Routing overhead of IP networks can be reduced by grouping a set of addresses under a single subnet [7,8]. On the other hand, in the IP subnet model, two routers can send packets each other when their connected interfaces are on the same subnet. This restriction is known as *subnet constraint.* The subnet constraint impacts IP over WDM routing which aims to permit the flexibility on the IP demands by reconfiguring the optical core dynamically.

Consider a network with four IP routers R_i, $1 \leq i \leq 4$, where routers are connected to an optical core consisting of four optical cross-connect switches (OXCs) O_i, $1 \leq i \leq 4$, respectively. This network is shown in Figure 1. Each interface in the network in Figure 1 is assigned a physical IP address. Since each

C. Kim (Ed.): ICOIN 2005, LNCS 3391, pp. 541–550, 2005.

interface with an address is on a specific subnet, we identify an interface by color associated with its subnet, where each interface is marked by the first letter of its color. Assume that a routing request from R_1 to R_3 arrives when both IP links between R_1 and R_4, and between R_1 and R_2 are at full capacity. Interfaces on R_1 and R_3 are on the same subnet (Red). Thus, if two interfaces on R_1 and R_3 are free, then an optical lightpath $O_1 \rightarrow O_2 \rightarrow O_3$ would connect these two interfaces. In this case, the length of the IP-hop path is one. If two interfaces connected to routes are on the different subnets, then it is possible to create longer network routes. For example, if R_3 is not on the subnet (Red), say on the subnet (Blue), then the shortest possible IP-hop path is $R_1 \xrightarrow{R} R_2 \xrightarrow{Y} R_3$, or $R_1 \xrightarrow{R} R_2 \xrightarrow{B} R_3$. This routing path would be shorter if the protocol permits to reconfigure the subnet on R_3 from B to R. We will discuss the issue related to the subnet reconfiguration in section 4.

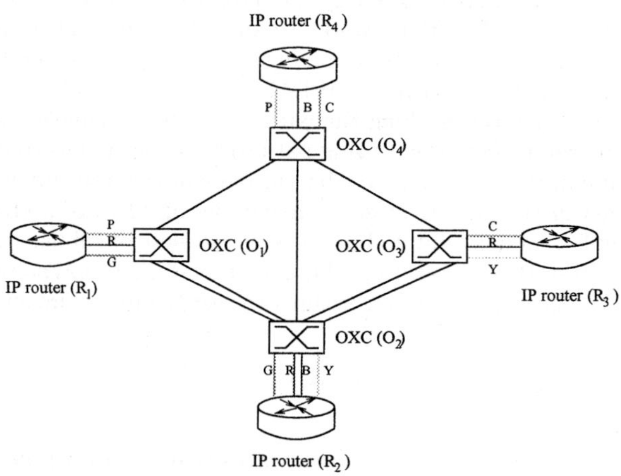

Fig. 1. IP over WDM optical routing with subnets: example network

Routing problems and their numerous variations in the IP context have been extensively studied [6,10]. When the subnet constraint is not considered, a polynomial time algorithm that finds a shortest IP-hop path satisfying the bandwidth requirement was proposed in [4]. Their routing algorithm was developed to handle as many potential future demands as possible. It measures which paths would more interfere with future demands and is designed to avoid those paths. The impact of subnets on IP over WDM routing was first addressed in [1]. They investigated the inherent hardness of the problem of finding the optimal shortest path with IP subnet constraint. In this paper, we argue that the shortest IP hop path accounting for subnets can be constructed in polynomial time. We present shortest path algorithms and analyze their efficiency to route a single demand.

The rest of the paper is organized as follows. In section 2, we outline the network model considered in our problem and give a problem formulation. Also, the reduction presented in [1] is briefly addressed along a counterexample. In section 3, we present an algorithm that finds a shortest path accounting for IP subnet. In section 4, we discuss a generalization that permits some subnet reconfiguration. In section 5, we conclude the paper.

2 Problem Formulation and Motivation

2.1 Network Model

One of our goals is to further investigate the hardness of the routing problem with IP subnet constraint as described in [1]. Thus, we use an identical network model to theirs. We briefly outline the network model under consideration. A WDM mesh network consists of IP routers directly connected to a switched optical core. The assumption on network model is as follows:

- OXCs are capable of wavelength conversion [9].
- All routing requests with a bandwidth requirement are MPLS LSP requests.
- The routing engine has up-to-date knowledge about the available bandwidth.
- MPLS is available for switching on the IP layer.
- RSVP or CR-LDP is available for resource reservation [2].

2.2 Problem Statement

Given a WDM mesh network, a source router s, and a destination router t, a path P between s and t traverses through routers and OXCs. A basic mechanism to find a bandwidth guaranteed path in IP over WDM networks is similar to other bandwidth guaranteed paths. Suppose a router is connected to another router by a lightpath. This connection can be represented as a logical link (or a virtual link) between them. That is, IP over optical network creates a virtual topology consisting of routers and logical links among them. Now, when a routing request with a specific bandwidth requirement is arrived, we eliminate all links that do not satisfy bandwidth requirement, and then apply an algorithm that finds a shortest path on the virtual topology.

In order to find a path on a virtual topology under the subnet constraint, every pair of interfaces connected to neighboring routers along the path must be on the same subnet. Also, they are needed to be free and optically reachable. Since we eliminate all links that violate bandwidth requirement before we attempt to find a path accounting for subnets, we will focus on constructing a shortest IP-hop path that satisfies the subnet constraint on a given WDM mesh network. In this paper, we do not consider the case that a routing request with a bandwidth requirement can be bifurcated on paths with lower bandwidths. For the sake of simplicity, we will regard all paths addressed in the following discussion as bandwidth guaranteed paths without confusion.

Definition For two routers R_i and R_j in a given WDM mesh network, they are said to be *subnet connected* if interfaces connected to them are free and in the same subnet, and are connected by an optical lightpath.

If every two neighboring routers in a path P are subnet connected, we say P is *subnet feasible.* From the definition of subnet connectability, the subnet shortest path ($\mathcal{SSP}$) can be defined as follows.

Subnet Shortest Path Problem ($\mathcal{SSP}$) : Given a WDM mesh network, a source s and destination router t, find a subnet feasible shortest IP-hop path P.

2.3 Motivation

The $\mathcal{SSP}$ problem was claimed to be NP-hard in [1], which is reduced from the well known NP-complete problem, Constrained Shortest Path ($\mathcal{CSP}$) problem [5].

Constrained Shortest Path Problem ($\mathcal{CSP}$) : Given a graph whose edges are associated with nonnegative lengths and weights, and a source node s and t, $\mathcal{CSP}$ problem finds the shortest path between s and t such that the total weight is less than a specified value W.

In Figure 2, we use the same example graphs used in [1] We will not explore the mapping construction from the $\mathcal{CSP}$ problem to the $\mathcal{SSP}$ problem, and the interested reader would refer [1]. In the process of transformation from $\mathcal{SSP}$ to $\mathcal{CSP}$, we observed that the requirement of determining the shortest path in the $\mathcal{CSP}$ problem was not reflexed into the final graph transformation for $\mathcal{SSP}$ problem, which leads to a pitfall of the reduction.

In Figure 2, we use the same example graphs used in [1] in order to extract a counterexample. It has been known that if all lengths or weights are equal, then $\mathcal{CSP}$ can be solved in polynomial time. We point out that the reduction shown in [1] does not require the equal lengths or weights on the $\mathcal{CSP}$ graph. Figure 2(a) shows an example graph for $\mathcal{CSP}$ problem where the values associated with each edge represent the length and the weight, respectively. Figure 2(b) shows a transformed WDM mesh network with subnets for $\mathcal{SSP}$ problem used in [1] where circles represent routers regardless of color. Note that V_{01} and V_{02} represent OXCs.

Consider the graph in Figure 2 (a), where the source node is V_1, the destination node is V_4, and the weight limit W is 3. The optimal solution of $\mathcal{CSP}$ is the path $V_1 \rightarrow V_3 \rightarrow V_4$ of length 2 and weight 3. The path $V_1 \rightarrow V_2 \rightarrow V_3 \rightarrow V_4$ requires longer path, and the path $V_1 \rightarrow V_2 \rightarrow V_4$ does not satisfy the weight constraint. Again, consider the graph in Figure 2 (b) to find a shortest path from V_1 to V_4 such that the ingress color and the egress color on V_{01} and V_{02} are the same. We write the path between two notes V_i and V_j going through V_{01} and V_{02} with the color C as $V_i \stackrel{C}{\rightarrow} V_j$. Then there are two paths from V_1 to V_4 that satisfy the subnet constraint. One is $V_1 \stackrel{S}{\rightarrow} V_2 \stackrel{G}{\rightarrow} V_3 \stackrel{R}{\rightarrow} V_4$ and the other is $V_1 \stackrel{C}{\rightarrow} \bullet \stackrel{M}{\rightarrow} V_3 \stackrel{R}{\rightarrow} V_4$. Both represent subnet feasible shortest paths of IP-hop

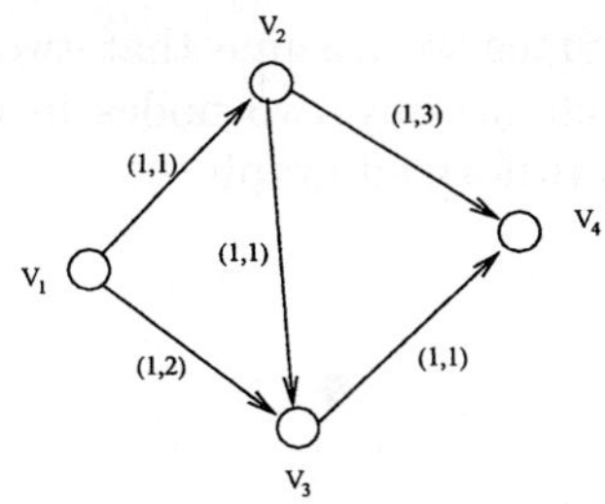

(a) Example graph for $\mathcal{CSP}$ problem

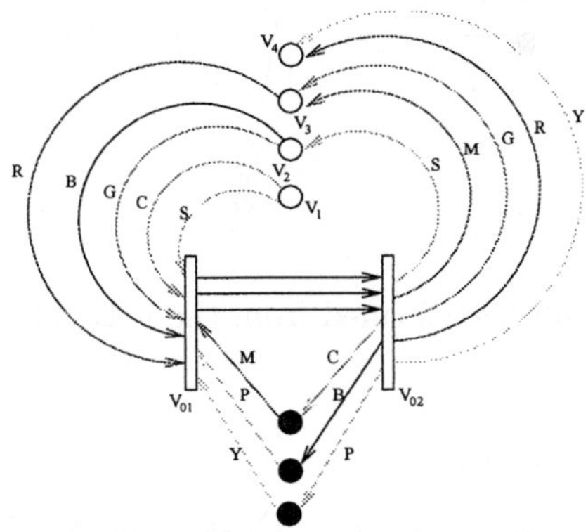

(b) The transformed graph from (a) for $\mathcal{SSP}$ problem

Fig. 2. Example graphs to demonstrate the mapping construction

length 3. Suppose we find the former path as a solution of the $\mathcal{SSP}$ problem. Then the sequence of nodes along the path does not solve the $\mathcal{CSP}$ problem, which is a path $V_1 \to V_2 \to V_3 \to V_4$ of length 3 and weight 3. This observation motivates us to investigate NP-completeness of the original $\mathcal{SSP}$ problem and its generalization.

3 The Subnet Shortest Path Routing Algorithm

First, we describe how to generate a graph on which our algorithm applies. Next, we will formulate a shortest path problem, which is equivalent to $\mathcal{SSP}$ problem, on the generated graph.

Phase 1: Graph Transformation

Let N be a WDM mesh network of IP routers and OXCs. Then a WDM mesh network N can be modeled as a graph, where nodes correspond to routers. There is a link between nodes if and only if two corresponding routers are subnet connected. Let G be such a graph model for the network N. In addition, we label each link of G as a specific subnet (here, a color) assigned to the interfaces on corresponding routers. Since we assume that two interfaces on two subnet connected are free, a link connecting two nodes in G represent a one IP-hop path. Figure 3 shows the transformed graph.

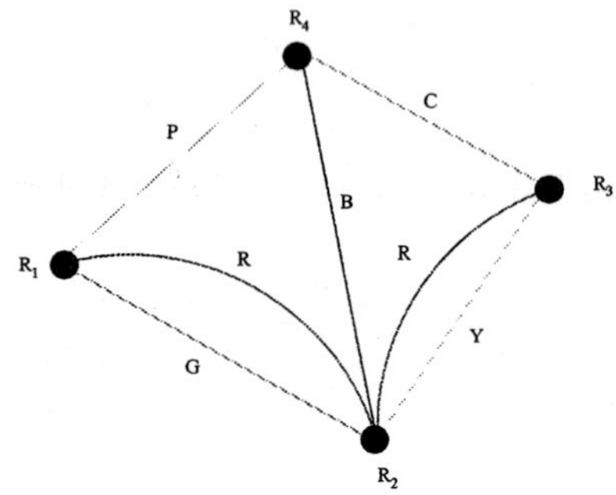

Fig. 3. Graph transformed from the network in Figure 1

Phase 2: Graph Augmentation

In the phase 2, we augment edges to the graph to include the following topology information. Consider a route from R_1 to R_3 in Figure 1. Since two interfaces on R_1 and R_3 are on the same subnet, a one IP-hop path through the optical core exits. However, there are some situations that require more IP-hops on a path. For example, a one IP-hop path through the optical core may not be possible if the interface on R_3 is not free. Also, if interfaces on two routers are on the different subnets, then it requires the subnet conversion. Since the subnet can be converted in IP-routers, the IP-hop is increased by one whenever the subnet conversion occurs. Though interfaces on two routers R_i and R_j are on the same subnet and two routers are optically reachable, we might need to convert subnets (change colors) along the path from R_i to R_j.

Let G be a graph obtained in Phase 1. Consider two neighboring links (i, j) and (j, k) such that i and j are subnet connected and also j and k are subnet connected. If labels of links are the same, say l, then we connect nodes i and k by a link whose label is l. We continue this process until we can not find such subnet connected neighboring links. Let G' be the final graph obtained from G in the phase 2. For each label l in G, consider subgraphs such that all nodes in a subgraph are connected by links labeled as l. In Figure 3, there are six labels, P, G, R, B, C, and, Y. In this example, each label l associates with a unique subgraph $S(l)$. Note that this subgraph may not connected. Phase 2 can be viewed as a process of making each component of a subgraph into a clique. Figure 4 shows a subgraph $S(R)$ in G and its cliques, where $S(R)$ consists of three components. Since any two nodes which are connected by an edge or path in $S(R)$ are eventually connected by an edge in a clique, such two nodes are connected by a link in G'.

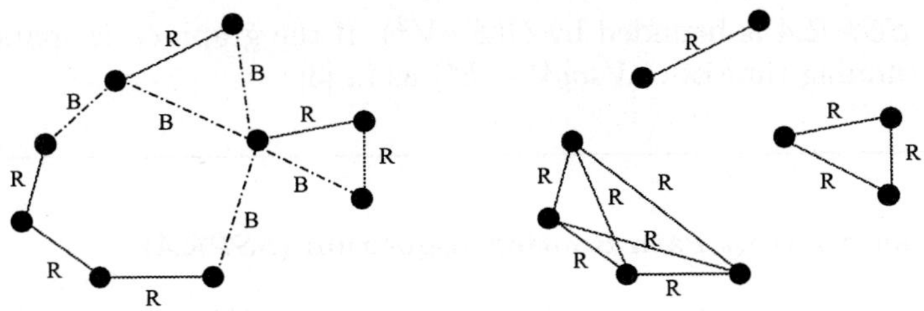

Fig. 4. Subgraphs $S(R)$ and $S(B)$ in G, and cliques of $S(R)$ in G'

Figure 5 shows the final graph obtained in the phase 2. Notice that nodes R_1, R_2, and R_3 form a clique of size 3.

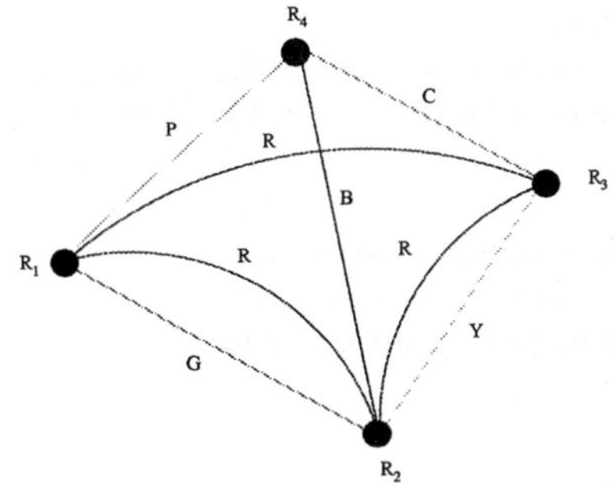

Fig. 5. Final graph after the transformation

Let G' be a graph obtained in the phase 2 as we described. We reformulate $\mathcal{SSP}$ as the shortest path problem on G'.

Subnet Shortest Path Problem ($\mathcal{SSP}$) : Given a graph G', a source s and a destination node t in G', find a shortest path P from s to t.

The subnet shortest path routing algorithm is given below. Suppose the graph G obtained in step 1 consists of V nodes and E links. Since the total number of labels in G is L, $L \leq E$, the minimum number of links in G' is obtained when all labels in G are different. In this case, the number of links in G' is $L = E$, and $G = G'$. On the other hand, the maximum number of links in G' is obtained when any two nodes in G are connected by L different links, which result in L cliques of size V. In this case, the number of links in G' is $L \cdot V(V-1)/2$. Thus, step 2.2 takes $O(L \cdot V^2)$ time. In addition, cliques constructed in step 2.2 for a label l are disjoint. In other words, if we find m disjoint subgraphs $s_1, \ldots, s_m$ in step 2.1, then step 2.2 produces m disjoint cliques of size $k_1, \ldots, k_m$, respectively. Let E' be the number of links in G'. Since $E' = O(L \cdot V^2) = O(E \cdot V^2)$, a shortest path between s to t in G' can be constructed in time $O(V^2 + E') = O(E \cdot V^2)$ by using Dijkstra's shortest path algorithm [3]. Therefore, the total time of the algorithm $\mathcal{SSPRA}$ is bounded by $O(E \cdot V^2)$. If the graph G' is sparse, we can achieve a running time of $O(VlogV + E')$ as in [3].

Subnet Shortest Path Routing Algorithm ($\mathcal{SSPRA}$)

Input: a WDM mesh network N of IP routers and OXCs, a source router s, a destination router t.

Output: a subnet feasible shortest path from s to t in N.

1. transform an original network N into a graph G by mapping routers of N to nodes in G. For any two subnet connected routers in N, connect two corresponding nodes in G by a link, and label the link accordingly.
2. for each label l in G

2.1 find a set of disjoint components $\mathcal{S}(l) = \{s_1, \ldots, s_m\}$ such that all nodes in $\mathcal{S}(l)$ are connected by links labeled as l.

2.2 for each component s_i of size k_i in $\mathcal{S}(l)$, construct a clique of size of k_i.

2.3 assign label l to each new added link.

3. let G' be a graph obtained in step 2.
4. find a shortest path between s to t in G'.
5. return an analogous path from s to t in N.

4 Generalized Subnet Shortest Path Problem

The subnet constraint often forces network routing paths to be longer. In Figure 1, if the IP interface on the router R_3 is not free, here on a subnet (B), and IP links between $R_1 \overset{P}{\rightarrow} R_4$, and between $R_1 \overset{G}{\rightarrow} R_2$ are at full capacity, then the possible shortest path would be $R_1 \overset{R}{\rightarrow} R_2 \overset{B}{\rightarrow} R_4 \overset{C}{\rightarrow} R_3$ unless we reconfigure network interfaces. Suppose we can reconfigure an IP interface on R_4 from B to R, then R_1 and R_4 may be connected by a one-hop IP path. In this case, the possible shortest path is $R_1 \overset{R}{\rightarrow} R_4 \overset{C}{\rightarrow} R_3$. It has been known that there is a trade-off between subnet reconfiguration and the length of the shortest path [1]. How can we balance between the network reconfiguration overhead and finding a shorter path which waste less network resources such as router interfaces and OXC ports? This question can be formulated into a problem as follows:

Generalized Subnet Shortest Path Problem ($\mathcal{GSP}$) : Given a WDM mesh network, a source s and destination router t, and a nonnegative integer K, find a shortest IP-hop path P such that P allows at most K subnet reconfiguration.

In Figure 1, suppose subnets P, C, G, and Y are part of the virtual topology. Let L_f be the total number of subnets which is not in virtual links in a WDM network N where V is the number of routers and E is the number of virtual links. For each subnet which is not in virtual links, we then construct a quasi-clique of size V with a slight modification of Dijkstra's algorithm. For each label l, first, we set a cost 0 to each link labeled with l. Otherwise, we set a cost 1. At this point, each link in G has a cost 0 or 1. Next, we construct a clique of size N, $G''(l)$ such that new added edges have a cost ∞, and then we apply Dijkstra's algorithm on $G''(l)$ where the shortest path length between from s to the other node associates with the minimum number of required subnet reconfiguration. If connecting two nodes requires more than K subnet reconfiguration, then we do not connect such a link, that is, the cost on that link is ∞. This graph forms a clique such that links with a cost $r > K$ are removed. If we can not find a shortest path from s to t on a graph $G''(l)$, then we reconstruct a graph G'' with other labels until we find a shortest path from s to t whose length is at most K. Notice that each link in G'' is associated with a single cost $r \leq K$, which makes $\mathcal{GSP}$ different from $\mathcal{CSP}$.

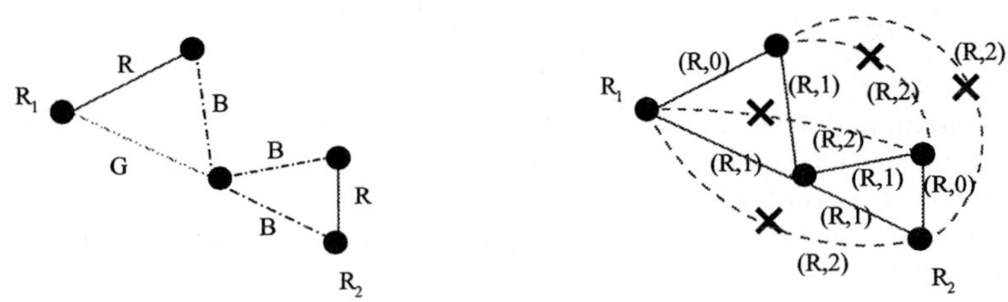

Fig. 6. Subgraphs $S(R)$, $S(B)$, and $S(G)$ in G, and a quasi-clique $G''(R)$

Figure 6 shows subgraphs associated with the labels R, B, and G, and a quasi-clique $G''(R)$ when $K = 1$. In Figure 6, we can not find a path from R_1 to R_2 whose length is at most 1. However, such a path exists on a graph $G''(B)$ or $G''(G)$. Since the total number of links in $G''(l)$ is $O(V^2)$, a shortest path from s to t in $G''(l)$ can be constructed in time $O(V^2)$. Since we may execute the modified Dijkstra's algorithm for each label, the total time takes $O(L_f \cdot V^2) = O(E \cdot V^2)$.

5 Concluding Remarks

The main contribution of the paper is to settle the argument about the hardness of the routing problem with IP subnet constraint. We provided a counterexample of the reduction used in [1], which motivated us to further investigation of its hardness. We developed a polynomial time algorithm that finds a shortest IP-hop path with subnet constraint. Our argument is further reinforced by performance results presented in [1]. Their performance results show that in practice, the shortest path accounting for subnet is obtained in polynomial time. In this paper, regardless of the length of lightpath, we simply assume that two routers are connected by a logical link if two routers are connected by a lightpath. Due to the limitations on the length of lightpath, the length of a lightpath connecting two routers in a logical topology should be taken into account in the routing path decision. In the graph transformation of phase 2, the length of lightpath increases at least by one whenever a new link is added to the intermediate graph. If we regard the length of lightpath as a cost, then each link in the final graph obtained in phase 2 associates with color and this cost. Even though we consider the limitation of the lightpath length, our algorithm can construct a shortest path accounting for subnets with a slight modification.

References

1. S. Acharya, B. Gupta, P. Risbood, and A. Srivastava, "IP-Subnet Aware Routing in WDM Mesh Networks," Proc. The IEEE Conference on Computer Communications (IEEE Infocom), 2003.
2. D. Awduche, L. Berger, D. Gan, T. Li, G. Swallow, and V. Srinivasan, "RSVP-TE: Extensions to RSVP for LSP tunnels," RFC-3209, Dec. 2001.
3. E. Dijkstra, "A note on two problems in connection with graphs," Numerische Mathematik, 1959.
4. M. Kodialam and T. V. Lakshman, "Integrated Dynamic IP and Wavelength Routing in IP over Networks," Proc. The IEEE Conference on Computer Communications (IEEE Infocom), pp. 358-366, 2001.
5. M. R. Garey and D. S. Johnson, Computers and Intractability - A Guide to the Theory of NP-Completeness, Freeman, California, USA, 1979.
6. R. Guerin, D. Williams, and A. Orda, "QoS Routing Mechanisms and OSPF Extensions," Proc. The IEEE Global Telecommunications Conference (IEEE Globecom), 1997.
7. J. T. Moy, OSPF: Anatomy of an Internet Routing Protocol, Addison-Wesley, 1998.
8. R. Perlman, Interconnections: Bridges, Routers, Switches, and Internetworking Protocols, Addison-Wesley, 1999.
9. T. E. Stern and K. Bala, Multiwavelength Optical Networks: A Layered Approach, Wesley Longman, 1999.
10. M. Schwartz and T. E. Stern, "Routing Techniques used in Computer Communication Network," IEEE Trans. on Comm., vol. COM-28, 1980.

Logical Communication Model and Real-Time Data Transmission Protocols for Embedded Systems with Controller Area Network

Kenya Sato[1] and Hiroyuki Inoue[2]

[1] Department of Computer Systems Design, Doshisha University,
Kyotanabe, Kyoto, 610-0321, Japan
ksato@mail.doshisha.ac.jp

[2] Ubiquitous Laboratories Department, Internet Research Institute, Inc.,
Sinjuku-ku, Tokyo 163-0511 Japan
inoue@iri.co.jp

Abstract. Embedded real-time systems are deployed in a wide range of application domains such as transportation systems, automated manufacturing, home appliances and telecommunications. Originally developed for use in automotive applications, CAN (Controller Area Network) has been adopted by industry as the standard network technology to transmit data for sensors and actuators, due to its fast response and high reliability for applications. We design and implement a logical communication model and a real-time data transmission protocol for embedded systems with CAN. This logical communication model with prioritized data transmission is effective for developing application programs in an embedded system. Owing to the small size of the CAN frame format, there are many restrictions on the implementation of a network protocol on CAN. In an attempt to solve this problem, we describe a concrete implementation of the network protocols in detail to realize the logical communication model.

1 Introduction

Embedded real-time systems are deployed in a wide range of application domains including transportation systems, automated manufacturing, process control, defense, and aerospace. In real-time processing, the urgency of messages to be exchanged over the network can differ [1]. A rapidly changing dimension has to be transmitted more frequently and therefore with fewer delays than other dimensions. Originally developed for use in automotive applications, CAN (Controller Area Network) [2] [3] has been adopted by industry as the standard network technology to transmit data for sensors and actuators. This is due to its fast response and high reliability for applications as demanding as control of anti-lock brakes and airbags. CAN is a serial communications protocol that efficiently supports distributed real-time control with a high level of security.

Since current embedded real-time systems tend to be complicated, a general communication model is required that allows the development of real-time application itself on a system without having to consider the underlying network

C. Kim (Ed.): ICOIN 2005, LNCS 3391, pp. 551–560, 2005.

communication mechanism [4]. Therefore, we define a new logical communication model that includes a priority communication function for real-time embedded systems, and specify data transmission protocol on CAN to realize the model. In this communication model, a logical unit is defined as abstraction of an application, and each application can communicate with other applications directly, and communication messages between applications are prioritized within its dimensions. In addition, the logical communication model is capable of implementing a gateway to connect to different kinds of network protocols. Owing to the small size of the CAN frame format, there are any restrictions on implementing a network protocol onto CAN. Consequently, it is also important to find how to implement a network protocol to realize large functions onto a CAN 8-byte data frame with a 29-bit identifier. In this work, we design and implement this logical communication model and the real-time data transmission protocols on a system to evaluate the model and the protocols.

2 Related Work

2.1 IDB-C

There are several distributed embedded systems based on CAN. One of the systems is the Intelligent Transport System (ITS) Data Bus [5] is a network for telematics and other vehicle devices. The IDB network architecture is shown in Fig. 1, in which there are two types of bus domains; one is the automakerfs proprietary side, and the other is IDB side. The specifications of IDB-C (J2366) suffers several shortcomings: long response time, low effective data transfer rate, high CPU power required, difficulty in realizing a gateway without a logical communication concept, and the requirement for a large message buffer.

CAN includes the original media access control mechanism CSMA/CD in its hardware, which results in small latency, high throughput, and low CPU power. The link layer of IDB-C features a token-passing mechanism implemented on CSMA/CD that is redundant and removes the advantages of each media access

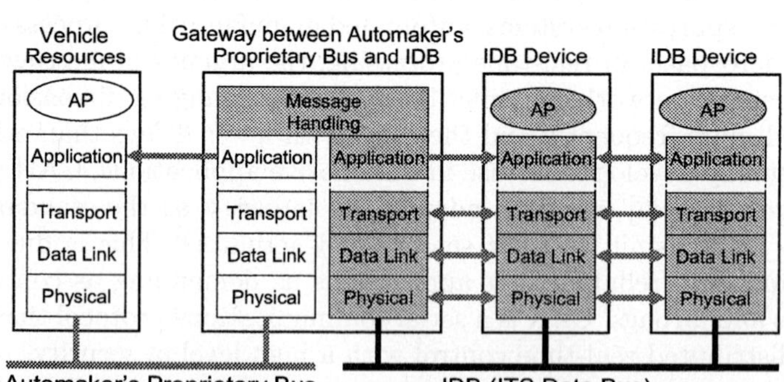

Fig. 1. Network Architecture for IDB (ITS Data Bus)

control mechanism. This is why IDB-C has a long response time (maximum 100 ms), even with CSMA/CD, and a low effective data transfer rate (1/3 of the original CAN throughput). In addition, the network software needs to receive many interruptions to check needless CAN frames, which results in the requirement for high CPU power.

In IDB-C's data-link layer, there is no mechanism to support logical communication. This means that it is difficult to develop application software because each application needs to recognize message-type fields to identify which application is a receiver of the message. Moreover, without logical communication, it becomes difficult for the network software to realize a gateway, because the gateway needs to keep a sender's application information until a receiver application sends back a response. Since many fragmented messages occur in this situation, the network software requires a large message buffer.

2.2 IP over CAN

IP over CAN was a former Internet-Draft [6] in IETF to transfer IP datagrams over CAN. The specification specifies how to address devices on CAN, distribution of Internet addresses among CAN nodes, fragmentation of IP datagrams, and timing requirements. The draft document was expired in September 2001.

We evaluated the efficiency of data transmission with IP over CAN. The length of IP packet header is 20 to 60 bytes. In case that the header length is 24 bytes, it takes four CAN frames to transmit 8-byte data. The effective throughput for IP over CAN is 12.5% of the total throughput for CAN. In addition, the Internet-Draft specification requires two CAN frames for initialization. Since the effective throughput is less than 10%, IP over CAN is not practical for embedded network systems.

3 Logical Communication Model

A logical communication model is capable of directly communicating among application modules in each device with a logical identifier (address). Each application module does not need to consider a physical component in which application modules are located. Instead, by using this mechanism, it is possible to develop a real-time application itself on a system without considering underlying network layers. Because each application does not need to take any physical components into consideration, it is easy to port an application from a certain physical node to a different node in the case that the physical node has enough capability to function and possesses sufficient resources for the application. In addition, by using a logical identifier, it would be possible to contain information about the kind of service (e.g. wheel-speed sensor, G sensor for airbag deployment, door-lock actuator) that each application has, thus an application can send another target application without the need for a special service discovery mechanism.

This logical communication model can solve the IDB-C network problems laid out in the previous section, and results in realization of high throughput with logical communication mechanism for a simple application software implementation.

3.1 Logical Unit

In the logical communication model, a logical unit is defined as an abstraction of an application module, and a physical unit is a physical component. Each logical unit has an identifier that is a logical address, which is different from an identifier of a physical unit, which is a physical address. Fig. 2 shows an example of a logical unit structure.

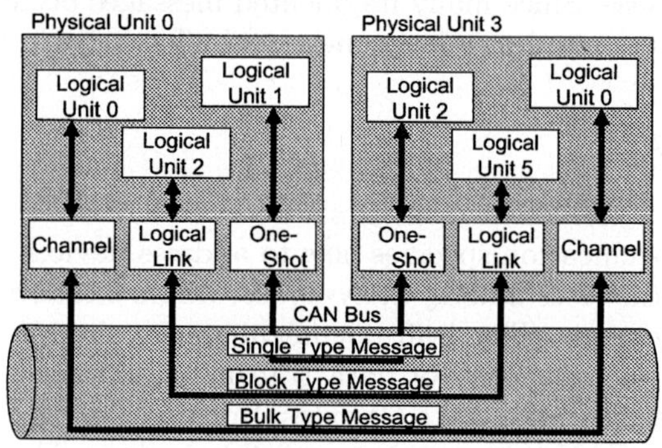

Fig. 2. Structure of Logical Communication Model

In this example, there are three logical units in a single physical unit, and each physical unit is physically connected to the CAN bus. Logical units are present in a Physical Unit. A Logical Unit in a Physical Unit is virtually connected with another Logical Unit in a different Physical Unit, and they can communicate with each other without confirming each other's destination address. This means there is a connection between each logical unit, and this concept is similar to the Berkeley sockets interface. Therefore, only a logical address is needed; a physical identifier is not needed when a logical unit communicates to another logical unit. By using this logical communication model, none of the application modules need to consider a physical component, and they may be transferred to another physical component if needed. To realize the logical communication model, not only an application layer but also a transport layer and a data-link layer are need recognize and control each logical connection between logical units.

3.2 Priority Communication

In real-time processing, the urgency of messages to be exchanged over the network can differ. A rapidly changing dimension has to be transmitted more frequently, and therefore with fewer delays, than other dimensions. For example, in an automotive situation, messages for anti-lock brakes, and airbags have to be transmitted very quickly in comparison with ones for controlling signals to open/close windows, turn on headlights, etc. Moreover, for example, log data to show where a vehicle drives and accumulated diagnostics data are typically large and do not need to be transported quickly, and in this case, efficiency is more important than speed when transport such large data. Therefore, transmission of these types of data should be separated on a communication mechanism.

As shown in Fig. 2, there are three message types transferred through three types of connections. gSingle-typeh messages are transferred through a one-shot connection, while gblock-typeh messages transferred through a logical link, and gbulk-typeh messages transferred through a channel. A single-type message has the highest priority, which means a message arrives at a destination unit with minimum transmission delay. A block-type message has a medium priority with a typical transmission delay, while a bulk message has the lowest priority, with maximum transmission delay; however, is transported with high efficiency, and it is necessary to append additional levels of priority to each channel.

4 Data Link Layer

In this section, we specify a data-link layer protocol on CAN to realize priority communications in this logical communication model described in the previous section. It is important to effectively assign each function onto the CAN frame format. Since the maximum data length of CAN is 8 bytes with 29-bit identifiers, a large header field of a data link protocol drastically reduces communication performance. A data-link layer consists of two sub-layers, a media access control sub-layer and a logical link sub-layer, to control communication, flow, and to check for errors among units. The media access control mechanism of this protocol is simply CSMA/CD on CAN, and this communication model includes an additional logical link layer function in the network.

4.1 Message Type

In this logical communication model, there are four message types defined in the data-link layer: (1) MAC type, (2) single type, (3) block type, and (4) bulk type.

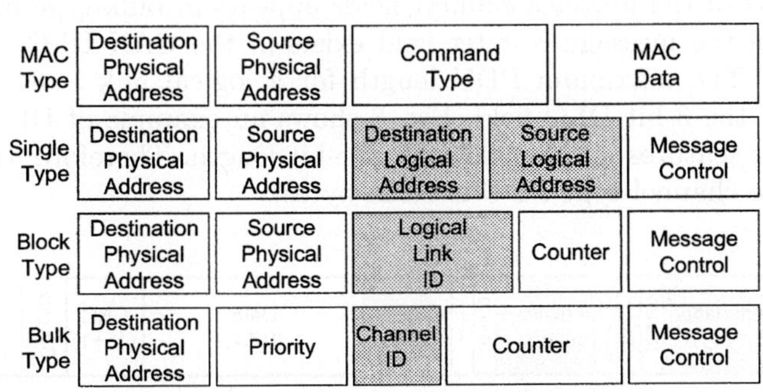

Fig. 3. Frame Format for Data Link Layer

Fig. 3 shows the message fields of each message type in a CAN message frame. The MAC type includes a destination physical address, a source physical address, and command-type and MAC data fields. The single type has a destination physical address, a source physical address, a destination logical address, a source logical address, and message-control fields. The block type features a destination physical address, a source physical address, a logical link ID, and counter and message control fields. The bulk type has a destination physical address, priority, channel ID, and counter and message-control fields.

4.2 Priority Setting

The three types of messages, single-type, block-type, and bulk-type, are transmitted over one-shot connection, logical link, and channel, respectively. A single-type message does not define a logical link or a channel connection. In fact, a logical link ID is declared with a pair of physical nodes and a pair of logical nodes. Channel is a one-way connection between two logical nodes, and a channel ID is allocated for each destination physical node. A MAC-type frame is defined for bus reset, device reset, physical-node join, physical-node leave, status request, status response and so on. A single-type message, with a length of less than or equal to eight bytes, is transferred within a single CAN frame, and this type of message can be utilized for ACK and NACK frames in the transport layer. While each logical unit has a single one-shot connection, logical link or channel, it is possible for a logical unit to have multiple logical connections; that is, a one-shot connection, a logical link and/or a channel to another logical unit.

4.3 Data Link Frame Format

Fig. 4 shows the CAN 2.0B extended frame format as an informative reference. We specify data included in the fields: an 11-bit identifier, an 18-bit identifier extension, and a 4-bit DLC (Data Length Code). The frame format for the data-link layer mapped on CAN 2.0B frame format is shown in Appendix.

The maximum number of logical nodes is 32, shown by five bits. The Logical Link ID field has 10 bits, which means there are 1,024 connections established between two logical nodes. The Channel ID has six bits, taking 64 channels. A larger part of the message counter fields appears in bulk-type messages. The remainder of the message counter field exists in the CAN DLC (Data Length Code) field. The maximum PDU length for a logical link is 64 bytes, which is shown by the 3-bit DLC field. Fig. 8 shows an example of DLC usage. The channel type features another field of four-bit length. Therefore, the maximum length of the channel-type PDU is 1,024 bytes.

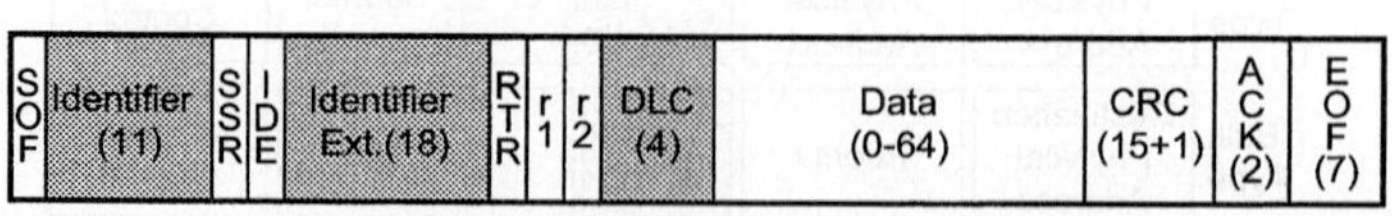

Fig. 4. Frame Format for CAN 2.0B

5 Transport Layer

We also specify a transport layer protocol which is another key function for realizing logical communication in this communication model. A transport layer provides reliable end-to-end data transport. Error checking and other reliability features are handled by the protocols in the transport layer if the underlying networks do not handle them. The transport layer includes buffers for each application as a logical unit. Appendix shows the transport frame format mapping to the CAN frame.

There are two types of transport layer header format: one is the command type and the other is the data type. The command-type format is for setup, ACK for single, ACK for block/bulk and NACK for block/bulk. These types of command messages are sent as single-type LLC messages, which include 1-byte data. There are four types of ACK attributes: no ACK, transport block ACK, Application frame ACK, and complete ACK only. The packet ID number shows a random number as a supplement to determine the difference among each message for sending ACK.

Before sending block/bulk messages, a sender node creates a connection (logical link or channel) to a receiver node. The link descriptor contains the link ID and ACK attribute, and a sender can transmit a message by using the link descriptor. The TPL packet counter describes a sequence number of a fragmented application message. It is 4 bits long; therefore, the maximum transport PDU is 1 k (64x16) bytes for a logical link-type message and 16 k (1,024x16) bytes for a channel-type message.

6 Other Layers

In this research, there are no modifications to the CAN physical layer's specifications. In terms of application layer, there is no defined format for the specific application message layer. Instead, the length of an application message is specified in the transport layer header. However, the header should be simple when considering that the base layer is composed of CAN frames. In our implementation, the maximum length of a block-type message is 1,024 bytes and for a bulk-type message, it is 16 kB.

7 Implementation and Evaluation

7.1 Implementation of Protocols

We have implemented our logical communication model and real-time data transmission protocols described above on the CAN 2.0B format. The microprocessor in this implementation is a Mitsubishi Electric (Renesus Technology) M16C (16 MHz), which contains a CAN interface. The speed of CAN is 250 kbps, which is the same as the IDB-C specification. Fig. 5 presents an outline of the protocol's implementation. We run a 10 μs software timer to measure the message-sending interval.

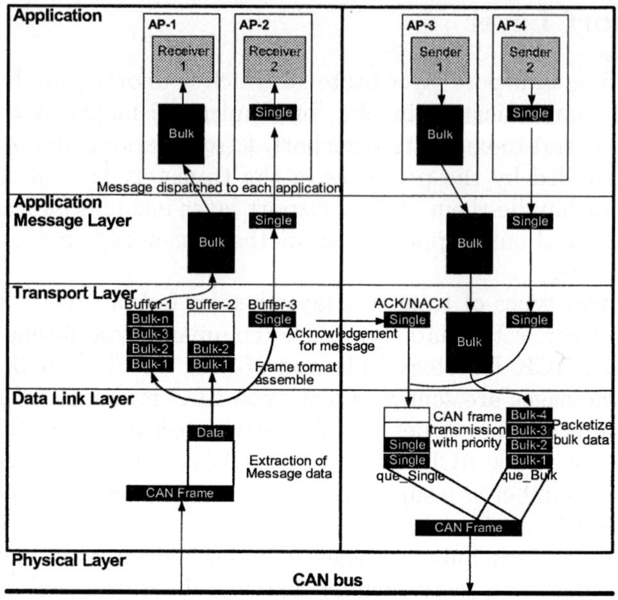

Fig. 5. Software Architecture for Protocol Implementation

7.2 Evaluation of Performance

We evaluated the following times of the network protocol that we implemented, transmission time of a single-type message, transmission time of a bulk-type message, transmission time of a single-type message while a bulk-type message is being sent. In this evaluation, the length of a single-type message is 8 bytes, while that of a bulk-type message is 1,024 bytes. Timing begins when "the application program calls for a send function to send a message," to the point when "the interrupt function for sending the message is completed."

- Transmission time of a single-type message. The average time to transmit a single-type message is 1.59 ms. The interval when a message is transported from the application to the data-link layer is about 1 ms., and it takes about 670 μs. to send out a single CAN frame completely to the bus.
- Transmission time of a bulk-type message. The average time to transmit a bulk-type message is 106.1 ms. the interval when a message is transported from the application to the data-link layer is about 20 ms.
- Transmission time of a single-type message while a bulk-type message is being sent. The average time to transmit a single-type message is 1.71 ms. and the average time to transmit a bulk-type message is 107.4 ms. It takes 512 μs. (128 bits/250 kbps) to transport a single CAN frame. When a single-type message is sent while a bulk-type message is also being sent, the delay in transmitting a single message must be shorter than this 512 μs interval. Because the evaluated delay time for transmitting a single-type message is

about 120 μs, we can confirm that the priority message mechanism of the network protocol works.

- In case of IDB-C type transmission, transmission time of a single-type message while a bulk-type message depends on the timing when a single-type message is sent; however, a single-type message is sent after a bulk-type message is completely sent, and the maximum transmission time of a single-type message is 106.1 ms.

Table 1. Message Transmission Time (Average)

	Single Type	Balk Type
Individual Transmission	1.59 ms	106.1 ms
Complex Transmission	1.71 ms	107.4 ms

The additional delay time for transmitting a bulk-type message when a single-type message is sent while a bulk-type message is also being sent is approximately 1.3 ms. Sending a single-type message that includes an ACK frame occupies the CAN bus for about 1.2 ms., and it takes about 300 μs. to process the ACK frame for the single-type message. Therefore, we also confirm that there is no problem in sending a single-type message because the delay time of 1.3 ms. is a reasonable value. In terms of throughput, while the token-passing IDB-C is 1/3 of the original CAN throughput, our model makes the most of the original CAN throughput. Because the media access control mechanism is simply CSMA/CD on CAN, and the header of the transport frame can be contained in the CAN header, there is no additional overhead to carry our message frame.

8 Conclusion

We have designed and implemented a logical communication model and a real-time data transmission protocol for embedded systems with CAN. Because of the CAN frame format's small size, there are many restrictions on implementing the network protocol on CAN; therefore, we described our implementation of the network protocols to realize our logical communication model onto the CAN 8-byte data frame with a 29-bit identifier. This logical communication model is capable of communication between application modules without considering the underlying network mechanism, and transmission of real-time data with priorities between applications. We implemented the model on an M16C microcontroller, and evaluated it by transmitting real data frames. Results indicated that our logical communication model and real-time data transmission protocol works and solves the current IDB-C problems of long response time, low effective data transfer rate, high CPU power required, difficulty in realizing a gateway without a logical communication concept, and the large message buffer required.

References

1. Nossal, R.: Meeting the Challenges in a Collaborative OEM-Supplier Development of Distributed Embedded Systems, *Distributed Embedded Systems Engineering SP-1885 SAE International*, (2004), 21–27 .
2. International Organization for Standardization: ISO 11898-1, Road Vehicles - Controller Area Network (CAN) - Part 1: Data Link Layer and Physical Signalling, (2003).
3. International Organization for Standardization: ISO 11898-2, Road Vehicles - Controller Area network (CAN) - Part 2: High-Speed Medium Access Unit (2003).
4. Neilsen, M.L.: A Flexible Real-time Transport Protocol for Controller Area Networks, *Proc. International Conference on Parallel and Distributed Processing Techniques and Applications*, (2001), 25–28.
5. Society of Automotive Engineering: J2355, ITS Data Bus Architecture Reference Model Information Report, (1997).
6. Cache, P., and Fiedler, P.: IP over CAN, Internet-Draft of the Internet Engineering Task Force, draft-cafi-can-ip-00.txt, (2001).

Appendix: Mapping to CAN Frame

The frame format for the data link layer and the transport layer mapped on CAN 2.0B frame is shown in the following tables.

Data Link Frame Format

	CAN ID (11)				
Type	28	27	26	25 - 22	21- 18
MAC	0	0	Bcast/ P2PP	Src Phy Addr	Dest Phy Addr
Single	0	1			
Block	1	0	1		
Bulk	1	1	1	Priority	
Reserve (Stream)	0	0	1	Src Phy Addr	

	CAN ID (18)				
Type	17 - 14	13	12	11 - 8	7 - 0
MAC	Cmd Type	Reserve			MAC Data
Single	Src Logical Addr		Dest Logical Addr		Data
Block	Logical Link ID				
Bulk	Channel ID			Msg High Count	
Reserve (Stream)	1111	Stream ID		Msg High Count	

	CAN DLC (4)	
Type	3	2-0
MAC	0	0
Single	1 / 0	DLC
Block	1	Msg Count
	1 / 0	DLC
Bulk	1	Msg Count
	1 / 0	DLC
Reserve (Stream)	1	Msg Low Count
	1 / 0	DLC

CAN 2.0B extended frame format includes an 11-bit identifier, an 18-bit identifier extension, a 4-bit DLC (Data Length Code), and a 0-64 bit data field. In this table, "0" means dominant and "1" means recessive on the CAN arbitration bit.

Transport Frame Format

	CAN ID (8)							
Frame Type	7	6	5	4	3	2	1	0
Setup	0	1	1	res.	Link Setup			
Ack for Single	0	0	0	res.	Packet ID Number			
Ack for Block/Bulk	0	1	0	res.	TPL Packet Count			
Nack for Block/Bulk	0	0	1	res.	TPL Packet Count			
Single Data	1	0	0	Ack Request	Packet ID Number (count)			
Block/Bulk Data (Start)	1	1	1	Re-sync.	TPL Packet Count			
Block/Bulk Data (Cont)	1	1	0	res.	Same as "Start"			
Block/Bulk Data (End)	1	0	1	res.	Same as "Start"			

	CAN Data (64)		
Frame Type	0 - 1	2 - 7	8 - 63
Setup	Ack Mode - No Ack - Transport Block Ack - Application Frame Ack - Complete Ack only	Link ID - Logical Link ID (6bits) - Channel ID (4bits)	N. A.
Ack for Single	N. A.		N. A.
Ack for Block/Bulk	Ack Attribute - Transport Block Ack - Application Frame Ack - Complete Ack	Link ID	N. A.
Nack for Block/Bulk	Ack Attribute - Transport Block Ack - Application Frame Ack - Complete Ack	Link ID	N. A.
Single Data	Data		
Block/Bulk Data (Start)	Data		
Block/Bulk Data (Cont)	Data		
Block/Bulk Data (End)	Data		

DINPeer: Optimized P2P Communication Network

Huaqun Guo[1], Lek Heng Ngoh[2], Wai Choong Wong[1,2], and Ligang Dong[3]

[1] Dept. of Electrical & Computer Engineering, National University of Singapore, 2 Engineering Drive 4, Singapore 117584
{guohq, lwong}@i2r.a-star.edu.sg

[2] Institute for Infocomm Research, A*STAR, 21 Heng Mui Keng Terrace, Singapore 119613
lhn@i2r.a-star.edu.sg

[3] College of Information and Electronic Engineering, Zhejiang Gongshang University
donglg@mail.hzic.edu.cn

Abstract. In this paper, we propose DINPeer middleware to overcome limitations in current peer-to-peer (P2P) overlay systems. DINPeer exploits a spiral-ring method to discover an inner ring with relative largest bandwidth to form a DINloop (Data-In-Network loop). DINPeer further integrates DINloop with P2P overlay network via node state and routing algorithm. The key features of DINPeer include using DINloop to replace a multicast rendezvous point and turning DINloop into a cache to achieve data persistency. Simulations show that DINPeer is able to optimize P2P communication in a number of ways, such as when the size of the DINloop is capped within a limit, it can achieve a better performance than native IP multicast and P2P overlay multicast systems.

1 Introduction

Recent works on P2P overlay network offer scalability and robustness for the advertisement and discovery of services. Pastry [1], Chord [2], CAN [3] and Tapestry [4] represent typical P2P routing and location schemes. Furthermore, there has been a number of works reported on adding multicast schemes and applications on P2P object platform, e.g., Scribe [5] and CAN-Multicast [6]. Compared to IP multicast, P2P overlay multicast has a number of advantages. First, most proposals do not require any special support from routers and can therefore be deployed universally. Second, the deployment of application-level multicast is easier than IP multicast by avoiding issues related to inter-domain multicast. Third, the P2P overlay network is fully decentralized.

However, P2P overlay multicast also suffers some disadvantages. Due to the fact that the underlying physical topology is hidden, using application-level multicast increases the delay to deliver messages compared to IP multicast. A node's neighbors on the overlay network need not be topologically nearby on the underlying IP network and this can lead to inefficient routing.

C. Kim (Ed.): ICOIN 2005, LNCS 3391, pp. 561–570, 2005.

Recently, there are some works that acknowledged the above limitation of P2P overlay network and inferred network proximity information for topology-aware overlay construction [7]. [8] described and compared three approaches in structured overlay networks: proximity routing, topology-based nodeId assignment, and proximity neighbor selection. In [9], proximity neighbor selection was identified as the most promising technique. In our scheme, we adopt proximity neighbor selection, but use a different method in Section 3.

In addition, the routing in the P2P overlay network does not consider the load on the network. It treats every peer as having the same power and the same bandwidth. Further, in P2P tree-based multicast, the multicast trees are built at application level and the rendezvous point (RP) is the root of the multicast tree. The RP can potentially be subjected to overloading and single-point of failure.

In this paper, we propose DINPeer to overcome the above limitations and to facilitate P2P communication. The remainder of this paper is organized as follows: we describe the overview of our solution in Section 2 and present the details of DINPeer in Section 3. Section 4 presents our evaluation metrics and simulation results. Finally we conclude research results in Section 5.

2 Solution Overview

DINPeer is a massively scalable middleware that combines the strengths of multicast and P2P systems. DINPeer exploits a spiral-ring method to discover an inner ring with relative largest bandwidth to form a logical DINloop. DINPeer further integrates the DINloop with the P2P overlay network. DINPeer has two key features. First, DINPeer uses the DINloop instead of a rendezvous point as multicast sources. Second, DINPeer uses the DINloop as a cache in the form of continuously circulating data in the DINloop to achieve data persistency.

To elaborate, a DINloop is shown in Fig. 1 (thick arrow line). *Nodes A*, *B*, *C*, *D* and *E* in the DINloop are referred as DIN Nodes. A sender sends data to a DIN Node and the DIN Node then circulates data in the DINloop. DINPeer can control the lifetime of the message in the DINloop so that during the lifetime of the message, any node can retrieve the data from the DINloop. This also reduces the delay to receive the message and reduces the traffic from the sender to the root. For example, *Host a* injects the data into the DINloop. Then, *Hosts b*, *c*, *d*, *e* and *f* can retrieve the data from the DINloop.

We use a spiral-ring method (described in Section 3) to find the inner ring with relative largest bandwidth. The multi-ring in [10] has some similarity as they chose power nodes as inner ring nodes. However, there are some differences. First, we use the spiral-ring method to find the inner ring. Second, multiple rings in [10] may cause a node to receive multiple copies of a same message. In our scheme, the inner ring is the center that is similar to the root of a tree, while other hosts, which are not in the inner ring, will receive only one copy of a same message. Finally, we only use the inner-most ring and the other hosts are organized into tree topologies for group communication.

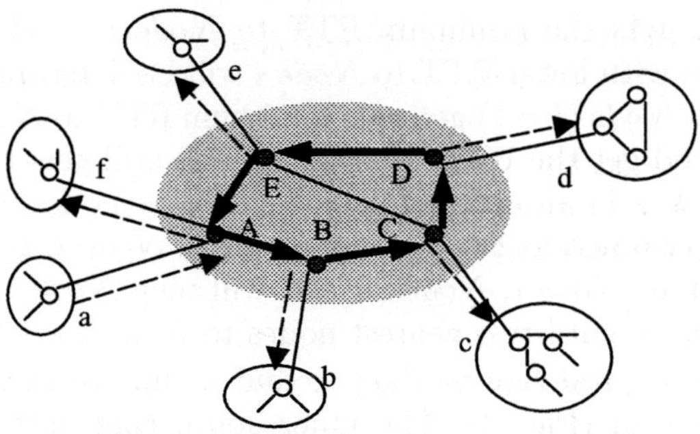

Fig. 1. Illustration of DINloop

3 Details of DINPeer

This section focuses on exploiting an inner ring with relative largest bandwidth to form a general DINloop, integrating DINloop with P2P overlay network via node state and routing algorithm, and optimizing P2P application-level multicast.

3.1 Generalizing DINloop

We exploit a spiral-ring method to find an inner ring to form a DINloop. First, a big outer ring with all nodes is formed. A new node is inserted between two nearest connected nodes. In the beginning, *Node* 0 and *Node* 1 form a ring (Fig. 2a). Then *Node* 2 is added to the ring (Fig. 2b). From *Node* 3 onwards, the two nearest nodes will break the link and the new node is added (Fig. 2c and 2d).

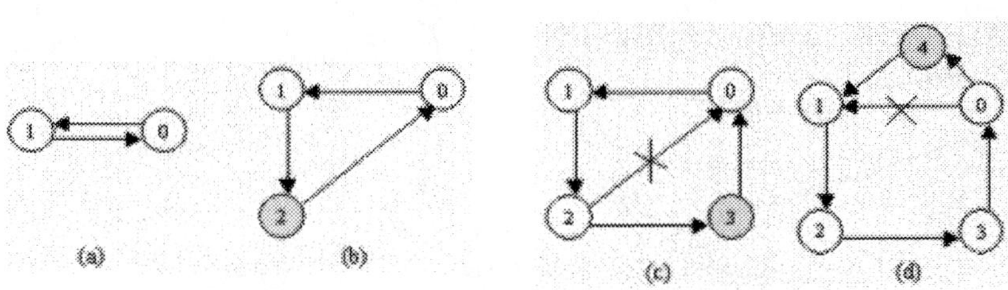

Fig. 2. Automatic formation of ring

The algorithm to obtain two nearest nodes is described here. Assume *Node* i plans to join the ring and knows of a local nearby *Node* k in the ring. *Node* i sends a join message to *Node* k. Then *Node* k and its two adjacent *Node* $(k-1)$ and *Node* $(k+1)$ in the ring will ping *Node* i and get the Round Trip Time

(RTT). If *Node k* gets the minimum RTT to *Node i*, *Node k* and one of its two adjacent nodes with lower RTT to *Node i* will be determined as two nearest nodes to *Node i*. If *Node* $(k+1)$ gets the minimum RTT to *Node i*, *Node* $(k+2)$ will ping *Node i* and get the RTT. If *Node* $(k+1)$ still gets the minimum RTT to *Node i*, *Node* $(k+1)$ and one of its two adjacent nodes with lower RTT to *Node i* will be determined as two nearest nodes to *Node i*. If *Node* $(k+2)$ gets the minimum RTT to *Node i*, *Node* $(k+3)$ will ping *Node i* and get the RTT. The process continues until two nearest nodes to *Node i* are found.

Second, we use a spiral-ring method to find an inner-most ring with relative largest ring bandwidth (Fig. 3). The inner spiral ring must provide a higher bandwidth than the outer ring. The formation of the inner ring is not limited, as links with lower bandwidth are dropped if enough nodes are available and the ring bandwidth is increased.

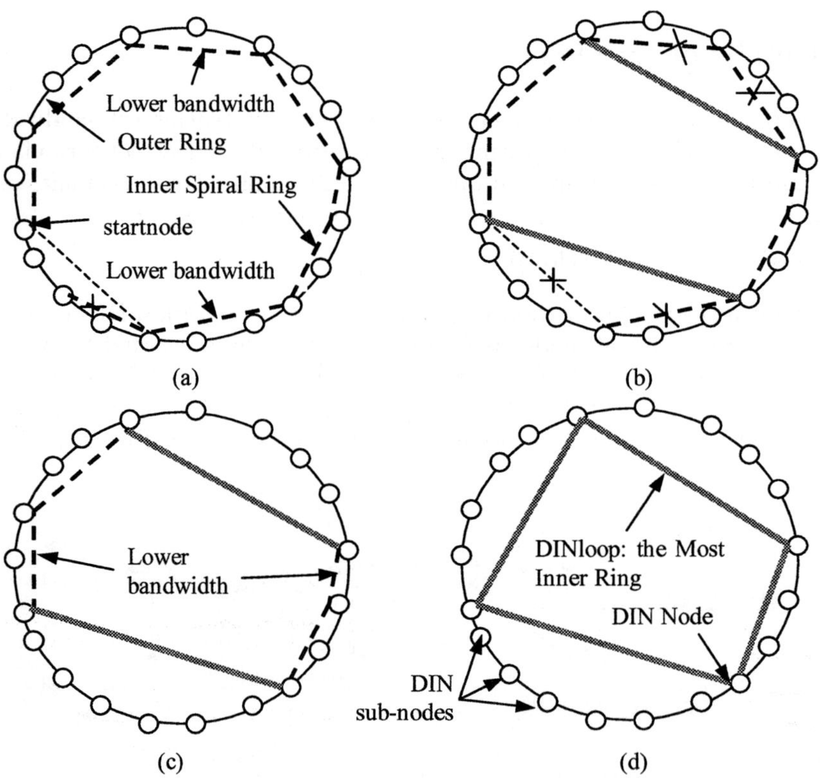

Fig. 3. Spiral-Ring Method

We use f_b as the bandwidth-increasing rate and N as the desirable number of DIN Nodes. Let β be the current inner ring bandwidth. Each recurring process drops inner links where their link bandwidths are less than $\beta \times (1 + f_b)$ (Fig. 3b) by replacing them with further inner links with higher bandwidth. The process is repeated (Fig. 3c) until the desired number of nodes in the inner ring is achieved or the ring bandwidth can no longer be increased (Fig. 3d).

Now we show how fast the recurrence can converge. We assume the outer ring bandwidth is B_0. We assume B_1 is the ring bandwidth of the inner ring at the $1st$ iteration of dropping inner links with the lower bandwidth. We assume B_2 is the ring bandwidth of the inner ring at the $2nd$ iteration of dropping inner links with the lower bandwidth. We assume that after k iterations, B_k is the maximum ring bandwidth of the inner ring. So,

$$B_1 \geq B_0 \times (1 + f_b). \tag{1}$$

$$B_2 \geq B_1 \times (1 + f_b) \geq B_0 \times (1 + f_b)^2. \tag{2}$$

$$B_k \geq B_{k-1} \times (1 + f_b) \geq B_0 \times (1 + f_b)^k. \tag{3}$$

$$k \leq log_{(1+f_b)}(B_k \div B_0). \tag{4}$$

Thus, the iteration step k has a small upper bound and therefore the process converges quickly.

In this way, we find the inner ring with largest bandwidth. This inner ring is used as a DINloop and nodes in this inner ring are used as DIN Nodes. The DINloop can be used as a cache called a DIN cache. The other nodes in DINPeer are called DIN sub-nodes (Fig. 3d). Each DIN sub-node finds a nearest DIN Node and becomes a member of a group associated to this DIN Node (Fig. 4).

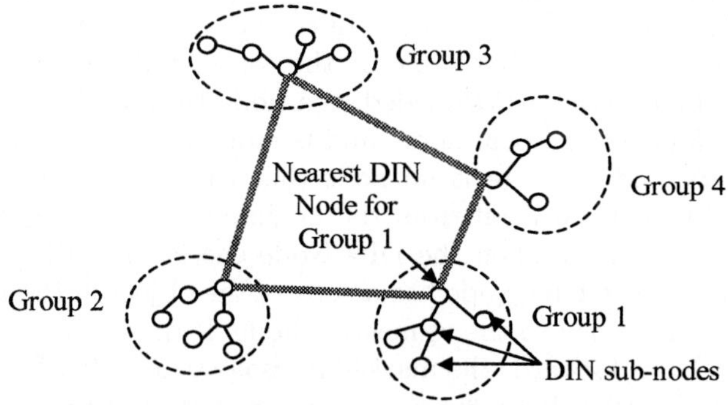

Fig. 4. DIN sub-nodes with associated DIN Nodes

3.2 Integrating with P2P Overlay Network

After the DINloop is formed, the DINloop is integrated with the P2P overlay network [1] via node state and routing algorithm. We elaborate each one next.

Node State. We use the hash function to generate nodeIds similar to Pastry [1]. As shown in Fig. 5, every node maintains a small node state.

The entries of node state are briefly explained as follows. In a DIN subnode, the Nearest-DIN-Node entry is its associated DIN Node, and the entries of Predecessor-DIN-Node and Successor-DIN-Node are empty. In a DIN Node, the Parent-Node entry is empty, and the entries of Predecessor-DIN-Node and Successor-DIN-Node are its two adjacent nodes in the DINloop. The Predecessor-Node and Successor-Node refer to its two adjacent nodes in the outer ring. The entries of Parent-Node and Child-Node are the nodeIds of nodes that are part of a multicast tree (described in Section 3.3). The Bandwidth entry is the end-to-end bandwidth from itself to the neighbor node in the inner ring, if the present node belongs to the inner ring as well. Each node maintains a routing table and a leaf set that are constructed using nodeIds in the same group. The entries at row n of the routing table each refer to a node whose nodeId shares the present node's nodeId (i.e., 20133103 in Fig. 5) in the first n digits, but whose $(n+1)^{th}$ digit is different with the $(n+1)^{th}$ digit in the present node's nodeId. The half leaf set is the set of nodes whose nodeIds are closest smaller than the present node's nodeId, and another half leaf set are the set of nodes whose nodeIds are closest larger than the present node's nodeId in the same group.

Routing Algorithm. The routing algorithm in DINPeer is similar to the routing algorithm in Pastry [1], but it integrates with DIN Nodes and DIN cache. Given a message, the node first checks to see if the key falls within the range of nodeIds covered by its leaf set. If so, the message is forwarded directly to the destination node.

If the key is not covered by the leaf set, the node uses the routing table and forwards the message to a node in its own group that shares a common prefix with the key by at least one more digit. In certain cases, it is possible that there is no such node in the routing table or the associated node is not reachable, in which case the message is forwarded to a node that shares a prefix with the key at least as long as the local node, and is numerically closer to the key than the present node's id. If there is no node that can be routed to, the message is marked to differentiate itself from the beginning message and forwarded to its associated DIN Node. Then the DIN Node checks the DIN cache with this marked message. If the DIN Node has a copy of the object in the DIN cache, the copy of object is returned to the requesting node. If there is no copy in the DIN cache, the DIN Node forwards the marked message to all other DIN Nodes. Then each DIN Node forwards it to the node in its own group respectively. Finally, a copy of the object is returned via the DINloop and the DIN cache is updated. DIN cache is updated using the RLU (Recent Least Use) algorithm.

3.3 Optimizing P2P Application-Level Multicast

In IP multicast, a sender sends data to the rendezvous point and the rendezvous point forwards the data along the multicast tree to all members. In DINPeer, the DINloop with multiple DIN Nodes is used to replace a single rendezvous point. Each DIN Node is the root of its associated multicast tree. Subsequently, every member in its associated group, who wants to join the multicast group, sends a

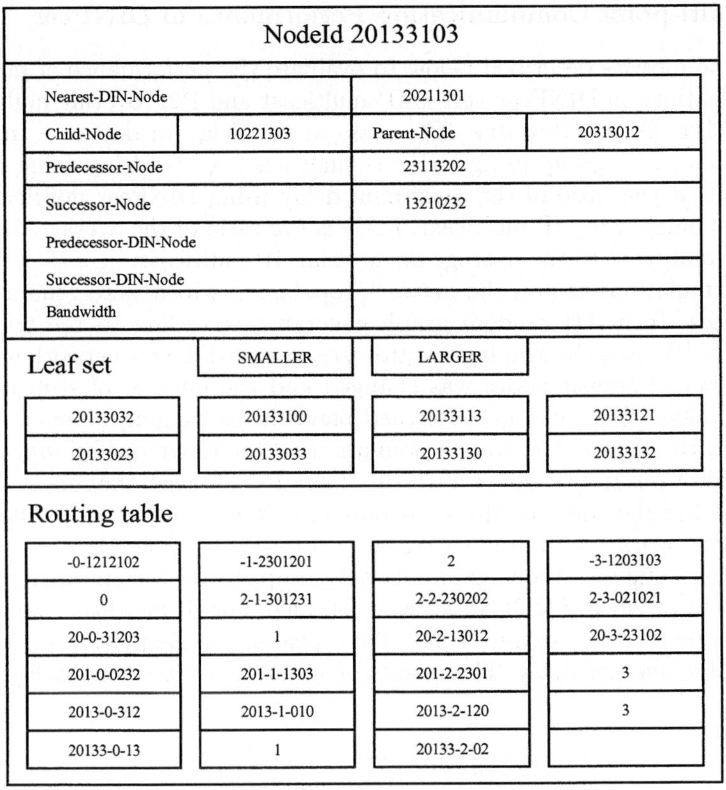

Fig. 5. An example of node state

message to the DIN Node, and the registration is recorded at each node along the path in the P2P overlay network. In this way, the multicast tree is formed. When multicasting, a node sends a multicast message to the nearest DIN Node. The nearest DIN Node then forwards the message to its child-nodes, and the neighbor DIN Node along the DINloop. The neighbor DIN Node forwards the message to its associated child-nodes, and its neighbor DIN Node along the DINloop. The process repeats itself until all DIN Nodes receive the message or when the lifetime of the message expires.

4 Evaluation Results

This section comprises two parts. First, we describe the simulations to demonstrate that DINPeer reduces the delay of multicast. Second, we investigate the impact of data persistency on the performance of multi-point communication.

4.1 Multi-point Communication Performance of DINPeer

We use the metrics described below to evaluate the performance of multi-point communications in DINPeer versus IP multicast and P2P overlay multicast.

Relative Delay Penalty: The ratio of the delay to deliver a message to each member of a group using DINPeer multicast to the delay using IP multicast. RMD is the ratio of the maximum delay using DINPeer multicast to the maximum delay using IP multicast. RAD is the ratio of the average delay using DINPeer multicast to the average delay using IP multicast.

The simulations ran on the network topologies, which were generated using the Georgia Tech [11] random graph generator according to the transit-stub model [12]. We used the graph generator to generate different network topologies. The number of transit nodes was changed and the number of stub nodes was fixed, i.e., 6000. We randomly assigned bandwidths ranging between 1Gbps to 10Gbps to the links in the transit domain, used the range of 100Mbps to 1Gbps for the links from transit domain to stub domains, and used the range of 500kbps to 1Mbps for the links in the stub domains. The topology is similar to one backbone network and multiple access networks. We assumed that all the overlay nodes were members of a single multicast group. Using the spiral-ring method, the different number of DIN Nodes was obtained. For IP multicast, we randomly chose a node as a rendezvous point. We repeated the simulation six times and obtained the average delay. The results of simulation are shown in Fig. 6.

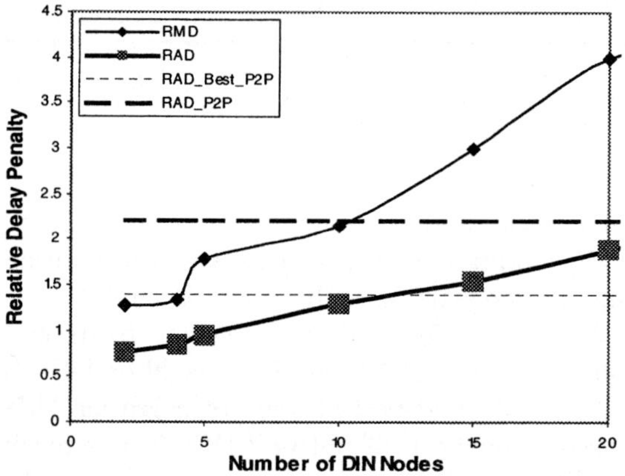

Fig. 6. Delay ratio of message retrieval using DINPeer over IP multicast

From Fig.6, when the number of DIN Nodes is small enough, the average latency in DINPeer overlay network is even lower than that of IP multicast, as RAD is less than 1. On another hand, other topology-aware routing techniques are currently able to achieve an average delay stretch (delay penalty) of 1.4 to 2.2, depending on the Internet topology model [8]. The value of 1.4 reported in [8] is shown in the thin dash-line RAD_Best_P2P in Fig. 6. The value of 2.2 reported in [8] is shown in the thick dash-line RAD_P2P in Fig. 6. Since the simulation conditions between our scheme and other P2P systems are different, the RAD difference in quantity cannot be calculated. We only show the general difference. DINPeer has the opportunity to achieve better performance than IP multicast while other P2P overlay multicast systems are worse than IP multicast. In Fig. 6, DINPeer, as shown below the thin dash-line, is the best among any other P2P overlay multicast systems. Thus, DINPeer provides the opportunity to achieve better performance than other P2P overlay multicast systems when the number of DIN Nodes is within the certain range.

4.2 Performance of Multi-point Communication with Data Persistency

In this sub-section, simulation was conducted to investigate the impact of data persistency on the performance of multi-point communication. We measured the delay to deliver a message to each member of a group using the DINloop and IP multicast respectively. The same network topologies as in the previous sub-section were used here. The results are shown in Fig. 7. RMD_2 is the ratio between the maximum delay from the DINloop to receivers and the maximum delay using IP multicast, and RAD_2 is the ratio between the average delay from the DINloop to receivers and the average delay using IP Multicast. When Fig. 7 was compared with Fig. 6, we found that RMD_2 and RAD_2 in Fig. 7 are lower than RMD and RAD in Fig. 6 respectively. The range of DINPeer achieving better performance than other P2P multicast systems is wider. It is also clear that the delay of retrieving messages directly from the DINloop is less than the delay of getting messages from the sender.

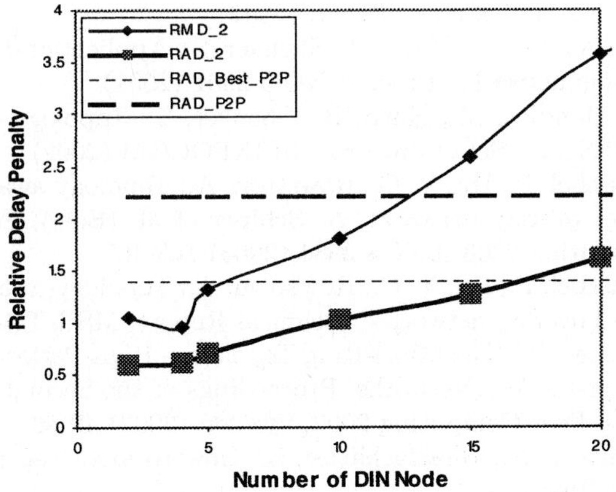

Fig. 7. Delay ratio of message retrieval from DINloop over IP multicast

5 Conclusions

We propose a DINPeer middleware to overcome the existing limitations in current P2P overlay systems. DINPeer exploits the spiral-ring method to discover an inner ring with relative largest bandwidth to form the DINloop. DINPeer further integrates the DINloop with a P2P overlay network. DINPeer has two key features. First, DINPeer uses the DINloop instead of the rendezvous point as multicast sources. Second, DINPeer uses the DINloop as a cache in the form of continuously circulating data in the DINloop to achieve data persistency. Our principle findings are (1) When the number of DIN Nodes is capped within a limit, DINPeer even achieves better performance than native IP multicast and is the best among any other P2P overlay multicast systems reported in the literature. (2) Data persistency further improves the performance of multi-point communication in DINPeer.

References

1. Rowstron, A., Druschel, P.: Pastry: Scalable, distributed object location and routing for large-scale peer-to-peer systems. Proc. of the 18th IFIP/ACM International Conference on Distributed Systems Platforms (Middleware), Germany (2001) 329-350
2. Stoica, I., Morris, R., Karger, D., Kaashoek, M. F., Balakrishnan, H.: Chord: A scalable peer-to-peer lookup service for Internet applications. ACM SIGCOMM 2001, San Deigo, CA (2001) 149-160
3. Ratnasamy, S., Francis, P., Handley, M., Karp, R., Shenker, S.: A scalable content-addressable network. Proceedings of ACM SIGCOMM 2001 (2001)
4. Zhao, B. Y., Kubiatowicz, J., Joseph, A. D.: Tapestry: an infrastructure for fault-tolerant wide-area location and routing. UCB Technical Report (2001)
5. Castro, M., Druschel, P., Kermarrec, A.-M., Rowstron, A.: SCRIBE: a large-scale and decentralized application-level multicast infrastructure. IEEE Journal on Selected Areas in Communications (JSAC) (2002)
6. Ratnasamy, S., Handley, M., Karp, R., Shenker, S.: Application-level multicast using content-addressable networks. Proc. of NGC 2001 (2001)
7. Ratnasamy, S., Handley, M., Karp, R., Shenker, S.: Topologically-Aware Overlay Construction and Server Selection. Proc. of INFOCOM (2002)
8. Castro, M., Druschel, P., Hu, Y. C., Rowstron, A.: Topology-aware routing in structured peer-to-peer overlay networks. A. Schiper et al. (Eds.), Future Directions in Distributed Computing 2003. LNCS 2584 (2003) 103-107
9. Castro, M., Druschel, P., Hu, Y. C., Rowstron, A.: Topology-aware routing in structured peer-to-peer overlay networks. Technical Report, MSR-TR-2002-82 (2002)
10. Junginger, M., Lee, Y.: The Multi-Ring Topology - High-Performance Group Communication in Peer-to-Peer Networks. Proceedings of the Second International Conference on Peer-to-Peer Computing 2002, Sweden (2002) 49-56
11. Zegura, E., Calvert, K., Bhattacharjee, S.: How to model an internetwork. Proc. of IEEE Infocom (1996)
12. Modeling Topology of Large Internetworks. http://www.cc.gatech.edu/projects/gtitm/

The Algorithm for Constructing an Efficient Data Delivery Tree in Host-Based Multicast Scheme

Jin-Han Jeon, Keyong-Hoon Kim, and Jiseung Nam

Dept. of Computer Engineering, Chonnam National University,
Yongbong-dong 300, Puk-gu, Kwangju, Korea
jhjeon23@naver.com
pluit@mdclab.chonnam.ac.kr
jsnam@chonnam.ac.kr

Abstract. To minimize the network resource and to meet the need of related application programs such as Internet Broadcasting or Video conferencing are the main concerns of Host-based Multicast tree construction algorithm. Existing works reduced performance-degrading factors like the duplicate data transmission on the same network link and overhead incurred at host or end-systems. However, they had relatively high RDP(Relative Delay Penalty), which made it difficult to adapt for application programs. In this work, we proposed a DDTA(Data Delivery Tree Adjust) algorithm which can reduce RDP by minimizing the tree depth, but it also could incur performance debasement. Proposed algorithm adapted techniques such as the detection of hosts existing in the same LAN to decrease the RDP and node switching to compensate performance loss. In addition, it suggests schemes for rapid recovery of a data delivery tree and shows that its RDP is lower than existing work.

1 Introduction

Conventional IP multicast has been introduced to support in the network layer the application programs like a VOD system or a Video Conferencing system. In IP multicast, routers play key roles in building and managing multicast tree by performing operations such as group creation, member joining and member leaving using multicast protocols. Protocols operated by routers are DVMRP (Distance Vector Multicast Routing Protocol), PIM-SM/DM (Protocol Independent Multicast Sparse Mode/Dense Mode) and MOSPF (Multicast Open Shortest Path First).

However, IP Multicast has not widely deployed for following reasons. To carry out operations enumerated above, routers are required to maintain per group state, which increases the complexity of them in the implementation and maintenance. To solve these problems and promote the propagation of Multicast, an alternative method known as Host-based Multicast (Overlay Multicast, Application Layer Multicast) [1,3] has been introduced and the relative research is now under way.

C. Kim (Ed.): ICOIN 2005, LNCS 3391, pp. 571–581, 2005.

In Host-based Multicast, hosts not only perform data transmission and reception but also take part in multicast tree building process. Therefore Host-based Multicast, contrary to IP multicast, does not call for changes at the infrastructural level. Besides, it could more easily provide the higher level features such as reliability, congestion control and flow control [1].

However, Host-based Multicast has inevitable performance-degrading factors like the duplicate data transmission to the same link and the overhead caused by hosts that forward data to others because each host cannot grasp the physical network topology. To measure the efficiency of the multicast tree, we would use some values such as RDP(Relative Delay Penalty) [1,5], Tree Cost [3,6] and Link Stress [1,2,3,6] which are regarded as the barometer for evaluating these performance-degrading factors. Accordingly the performance of the Host-based Multicast depends on the data delivery tree building algorithm which minimizes the performance-degrading factors presented above and copes with appropriately at the dynamic state changes of the physical network.

We develop a new data delivery tree-constructing algorithm named DDTA Algorithm and demonstrate the efficiency of DDTA Algorithm through comparison with other existing algorithms. The rest of the paper is structured as follows. We review the related works on Host-based Multicast in Section 2. We provide detailed explanation about DDTA Algorithm in Section 3 and present the performance result from simulations of algorithms in Section 4. In Section 5, we give a brief description about future works and conclusion.

2 Related Work

In Host-based Multicast, the process of building a multicast tree for data transmission is the process of determining the most appropriate parent node of the host which calls for member joining or member rejoining operation to the multicast group. Most mechanisms ever proposed for building a Host-based Multicast tree have two basic steps. The first is a distance measuring step and the next is an overlay date delivery tree construction step. In a distance measuring step, most mechanisms measured the distance among nodes and collected the distance information. The key metrics used for distance measuring among nodes were RTT (Round Trip Time), routing path and available bandwidth between nodes. In an overlay data deliver tree construction step, most mechanisms constructed the overlay data delivery tree or overlay graph based on collected distance information. If the overlay graph was built, existing mechanisms offered the way that converts the overlay graph into the corresponding overlay data delivery tree. The characteristics of existing schemes are as follows.

Narada [1] exploits two-step approach to construct the data delivery tree. In the first step, the new member which requests to be added to the multicast group joins to the multicast group by connecting to the active member of existing group. Next, it builds a new graph called mesh by exchanging the refresh messages between newly joined member and existing neighbors. In the second step, Narada constructs date delivery trees rooted at each corresponding source by converting mesh into trees. After construction of data delivery trees, it measures RTT through the periodical information exchange among the existing members in the mesh and applies that information to the Utility Function of Narada. The performance of the control mesh is improved corresponding to the change of the network state by adding the new links to the mesh or removing the existing links from the mesh. On the other hand, the radical traffic compared with other mechanisms makes it difficult to apply to the large scale multicast group.

In TAG [2], when a new member requests to be added to the multicast group, the root node of the existing data delivery tree obtains the shortest path from the root to the new member. Next it chooses one of its descendant nodes whose path from the root is the longest prefix of the path from the root to the new member as the parent of new member. Compared with other schemes, TAG can construct the most efficient data delivery tree, but sometimes the shortest path between the root and the new member can not be obtained. Besides the hosts of the data delivery tree should hold more information than other methods do in order to compare their paths with other nodes.

As for HMTP [3], the data delivery tree building mechanism is a combination of Host-based Multicast and IP Multicast. The existing IP multicast group is defined as the IP Multicast Island and one member in an island is elected as the representative member called Designated Member (DM) that participates in the data delivery tree building process with the hosts outside of the IP Multicast Island. In this case, the members that belong to the data delivery tree forward data using UDP, while inside the IP Multicast group the DM uses IP Multicast to deliver data. When HMTP constructs the data delivery tree, the Depth First Search is run from the root and then one node in the existing tree that is the nearest to the new member is chosen as the new members parent. Therefore HMTP could group the adjacent members, while it could also produce a skewed tree. At the same time, the depth and RDP of the data delivery tree could be increased compared with other mechanisms.

In Overcast [4], the bandwidth among member is considered as a key metric to determine the appropriate parent of the new member. When a new member requests joining to the multicast group, it compares the bandwidth from the new member to the root with the bandwidth from the new member to each of roots children respectively. Then it chooses the node which brings the greatest bandwidth as its potential parent. Next, the new member begins a series of rounds until it finds its most suitable parent. In each round the new member considers its bandwidth to its potential parent as well as the bandwidth to each of the potential parents children. Therefore the data delivery tree constructed by Overcast could utilize the bandwidth among members more efficiently. On the other hand, it is not easy to measure the exact bandwidth among members and to respond to the state change of the network properly.

HostCast [5] consists of two steps. In the first step, it constructs a data delivery tree and the corresponding control mesh. The data delivery tree is used to deliver the multicast traffic and the control mesh is used to transmit control messages and overlay path measurement packets. In control mesh, each node has two parents. The parent node is the child nodes primary parent in the mesh. The grandparent node is the nodes secondary parent in the mesh. In the second step, Hostcast provides the mechanisms to improve the performance of data delivery tree and its scenarios are as follows. If a member realizes that a secondary parent can potentially provide better QoS than its current primary path and the secondary parent can accept one or more child, Hostcast can improve the performance of data delivery tree by switching its primary parent and the secondary one. At the same time, Hostcast provides the mechanism for avoiding the tree partition.

In Switch Tree [6], when a new member requests joining to the multicast group, Switch Tree chooses the root as its parent and then exploits switch schemes such as switch-sibling, switch-one-hop, switch-two-hop and switch-any to improve the efficiency of the data delivery tree. Such schemes are performed to cope with the network state changes based on RTT among each node. Besides, it also presents a method to minimize the inevitable protocol overhead and the temporary suspension of the data exchange incurred during node switch.

TBCP [7] and ALMI [8] also use RTT as the metric of distance among each node. TBCP attempts to improve the efficiency of the data delivery tree by setting up the nodes that are in the same domain to the same sub tree. In ALMI, session controller constructs the MST(Minimum Spanning Tree) based on round trip delay measured by application program and then MST uses as a data delivery tree.

3 Construction of the Data Delivery Tree

3.1 Architecture

In DDTA algorithm, we uses distributed mechanism to construct and maintain data delivery tree. Each host which participates in host based multicast tree building process could have two types of agents. One is named as MCA (Multicast Control Agent) and the other is named as MRA (Multicast Routing Agent). The MCA is run on the designated host and it only contains brief multicast session information like root node IP address, contents type, contents name and necessary tree control data.

On the other hand, the MRAs are run on the all host of the same multicast group and their roles are (1) to request the join and group creation operations to the MCA and to receive the multicast group root information from MCA, (2) to request leave operation to its neighbor MRAs, (3) to build multicast data delivery tree through the distance measurement with designated MRA and (4) to exchange multicast data with their neighbor nodes. The brief architecture of DDTA is depicted in Fig. 2(a).

3.2 DDTA Algorithm

DDTA Algorithm exploits SBT (Source Based Tree) to build multicast tree so the selection of a root of multicast tree would be required. We assumed that the root is the host that requests the creation of new multicast group to the MCA and at the same time attempts to transmit the data to the same multicast group. After the creation of new multicast group and initialization of its root, newly joined host would determine its parent node to build a data delivery tree.

In DDTA Algorithm, we regard the root as the potential parent node of a new member and set the root an initial search node because the parent node selection process of newly joined or rejoined host starts at the root. The search node would be changed while the DDTA Algorithm is processing. We also name newly joined or rejoined host a request node because that node sends the join request to the MCA. Besides, we select the RTT as the metric of distance between nodes that comprise the multicast group. DDTA Algorithm consists of four steps and the complete DDTA Algorithm is presented in Fig. 1.

```
Proc TreeAdjust (Snode, Rnode)
Begin
If Rnode and one of Snode's child node are in same LAN {
          Set Rnode to the peer node of Snode's child node that is in same LAN with Rnode;
          Exit;
}         // End of step I
If Snode's out-degree permits it to accept another child node {
          Set Rnode to the child node of Snode;
          Exit;
}         //End of step II
Set the longest distance between Snode and its children to Max-dist;
Set Snode's child node that has the longest distance from Snode to designated-Switch-node;
Set the distance between Snode and Rnode to Request-dist;
If Max-dist > Request-dist {
          Set Snode to the parent node of Rnode;
          Call Proc TreeAdjust (Root, designated-Switch-node);
}         //End of step III
Set Snode's child node which satisfies following condition to new Snode;
(( Minimum (Distance between Root node and Snode's children nodes +
           Distance between Snode's children nodes and Rnode) ))
Call Proc TreeAdjust(new-Snode, Rnode);                    //End of step IV
End
Variables
Snode : Search node
Rnode : Request node
```

Fig. 1. DDTA Algorithm

In step I, DDTA Algorithm checks whether the request node and one of search nodes child node are in same LAN or not. IP address and netmask are used for determining if two nodes are in same LAN. If so, it sets the request node to peer node of the child of search node and exits the algorithm (Fig. 2(b)). Otherwise, DDTA Algorithm will perform next step. All nodes selected as peer node do not participate in data relay, but just receive the data sent from their peer. In addition, they do not have any children. Thus, DDTA Algorithm can reduce tree depth and RDP at the same time.

In step II, DDTA Algorithm checks if the available out degree of search node is greater than search nodes current out degree. If so, it sets the request node to the child of the search node and exits the algorithm (Fig. 2(c)). Otherwise, DDTA algorithm will perform next step.

In step III, DDTA Algorithm checks the condition of the node-switching and deals with it if required (Fig. 2(d)). The detailed procedures of step III are as below:

i. Select the search nodes child that has the longest distance from its parent, set selected node a designated switch node and set the distance a Max-dist. A designated switch node is the node selected as a node-switching candidate node.
ii. Measure the distance between search node and request node and set the distance as a request-dist.
iii. If Max-dist is greater than request-dist then DDTA Algorithm, then the node-switching between the designated switch node and request node will occur. Otherwise, DDTA algorithm will perform next step.

As seen in Fig. 2(e), the detailed node-switching procedures are as follow:

i. Set the request node the child of the search node.
ii. Remove the connection between the search node and the designated switch node.
iii. The designated switch node performs the rejoin operation.

In step IV, DDTA Algorithm selects a new search node among the children of a current search node. The new search node is the one of the search nodes children minimizing the sum of the distance between root node and search nodes children and the distance between search nodes children and request node(Fig. 2(f)).

3.3 Tree Management

To maintain the multicast tree, the MCA contains brief information of multicast tree. On the other hand, MRAs contain comprehensive information of their neighbors. In DDTA Algorithm, the MCA and MRAs perform operations like member-join, member-leave, failure detection and recovery using the information that they have. The detailed operations are presented below.

1) Member-join operation

When a host attempts to join the multicast group, the MRA run in that host requests the information of the root to the MCA. In response to the request, the MCA sends the information of the root to the MRA. Next, MRAs build multicast tree using information exchanged between them and DDTA Algorithm.

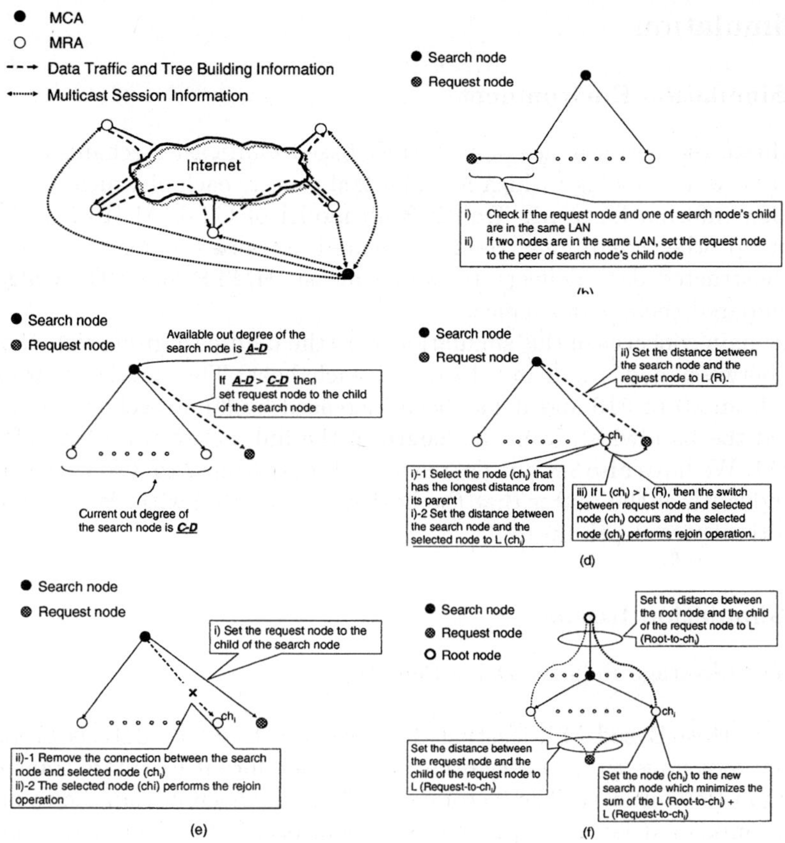

Fig. 2. The Architecture and process of Data Delivery Tree Construction

2) Member-leave operation

When an existing member wants to leave from the multicast group, the MRA of an existing member sends a member-leave request to the MCA and neighbor MRAs. Upon receiving the request, the MCA modifies multicast session information and parent neighbor MRA deletes information of leaving node. If there be child node, then each child MRA requests member-join operation to the MCA.

3) Failure detection and recovery operation

The MRA of each node regards the failure of neighbor MRAs when there is no data transmission during the time set by MCA session information. Upon detecting failure, one of neighbor MRA reports the fact to the MCA and MCA changes multicast session information. The rest processing is the same as the case of member-leave operation.

4 Simulation

4.1 Simulation Environment

To evaluate the performance of each Host based multicast mechanism, we have created network topology which includes 200 routers each of which has 5 LANs and other information using Transit-Stub model of GT-ITM [10,11]. We have also set up other conditions based on created network topology. And then, we have constructed data delivery trees of Unicast, HMTP and DDTA algorithm and compared their performance.

The main concerns of this simulation were the multicast group size, the bandwidth between nodes and out-degree of each host. The multicast group size ranged from 30 to 510 and it has been increased by 30 at each simulation. We obtained the bandwidth from the length of the link between routers offered by GT-ITM. We have also set out-degree of each host from 2 to 8 so that it reflects the bandwidth of the router that the host was connected with. We have studied the depth of tree depth up to 7.

4.2 Simulation Result

1) ARDP (Average Relative Delay Penalty)

In the Host-Based Multicasting data delivery tree, the data is propagated from the root node to the every other node. Therefore it has more transmission delay than IP Multicast. The factor called RDP has been introduced to measure the transmission delay.The equation for measuring RDP is given in the following, where D(Root,i) means the network transmission delay between the root and a certain node indexed i using Unicast or Multicast, while D(Root,i) means the transmission delay between the root and a certain node indexed i using Host-Based Multicast tree.

$$RDP = \frac{D'(Root, i)}{D(Root, i)} \tag{1}$$

The ARDP [5,9] means the Average value of RDP. The equation for measuring ARDP is given in the following.

$$ARDP = \frac{1}{N-1} \sum_{i=0, i \neq Root}^{N} \frac{D'(Root, i)}{D(Root, i)} \tag{2}$$

The ARDP is 1 when the data is transmitted using Unicast. As for both the DDTA Algorithm and HMTP, ARDP grows to the fixed point where the value converges as the number of the host gradually grows (Fig. 3(a)). From the view point of ARDP, we can see that DDTA algorithm gains improvement of 50% in ARDP compared with HMTP.

2) Total Tree Cost

The total tree cost is the sum of transmission delays occurred at the all links when the root node transmits the data to the every other node in the same multicast group.

In Fig. 3(b), the DDTA Algorithm and HMTP improve the total tree cost by 200% compared with Unicast as the number of host gradually increases. When the multicast group size is small, HMTP surpasses DDTA Algorithm by about 5 10% in total tree cost but the difference between them tends to diminish with the hosts increasing in number.

3) Link Stress

The link stress is defined as the number of the times that all links have been used when the root node transmits the data to the every other node in the same multicast group.

In Fig. 3(c), the DDTA Algorithm and HMTP improve the performance of link stress by more than 200% compared with Unicast as the number of host gradually increases. At the same time, HMTP has lower link stress than the DDTA Algorithm by about 15~20%.

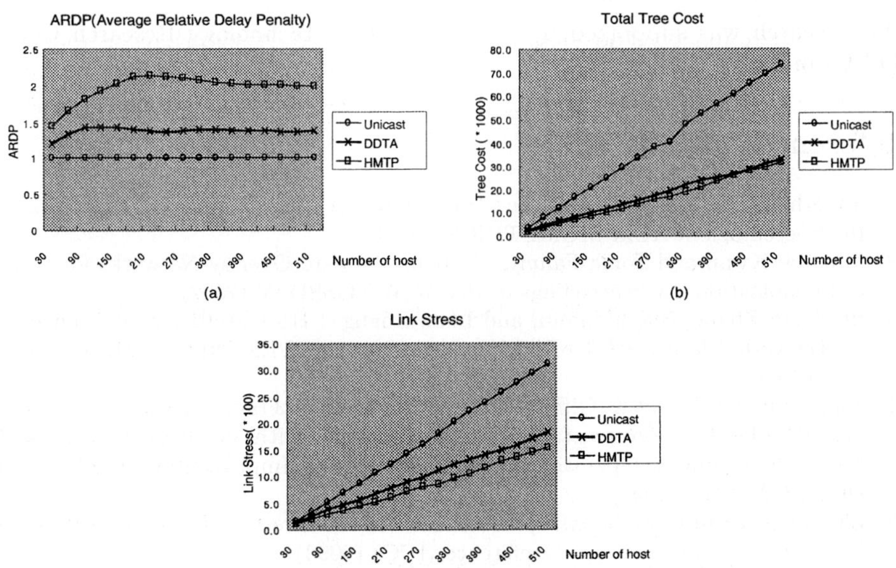

Fig. 3. Simulation Result

According to the analysis of the simulation results, it is clear that DDTA Algorithm improves the performance of ARDP more than HMTP does. From the view point of total tree cost and link stress, HMTP takes a little advantage compared with DDTA Algorithm. In the case of HMTP, grouping happens among nodes existing at the lower level. It leads to excellent performance for link stress. On the other hand, grouping also increases the depth of the tree at the same time. It causes a drop in the performance of ARDP and becomes a drawback compared with DDTA Algorithm.

5 Conclusion and Future Work

In this paper, we have presented DDTA algorithm designed for data delivery using Host-Based Multicast: the simulation result shows that the ARDP of DDTA algorithm is much lower than existing work. To reduce the ARDP, proposed algorithm exploited node-switching and minimized tree depth using same LAN checking and out-degree control.

However, node-switching could produce a temporary suspension of data transmission incurred by the discordance of synchronization between nodes. To solve a problem of node-switching, we are considering the ways that could minimize the number of the node-switching times and the overhead occurred during the node-switching.

Acknowledgement

This research was supported by the Information Technology Research Center (ITRC) of Korea.

References

1. Yang-hua Chu, Sanjay G. Rao, and Hui Zhang, : A Case for End System Multicast. In Proceedings of ACM SIGMETRICS (2000)
2. Minseok Kwon and Sonia Fahmy, : Topology-Aware Overlay Networks for Group Communication. In Proceedings of the ACM NOSSDAV (2002)
3. Beichuan Zhang, Sugih Jamin, and Lixia Zhang, : Host Multicast: A Framework for Delivering Multicast To End Users. In Proc. of IEEE INFOCOM, New York, NY, (2002)
4. John Jannotti, David K. Gifford, Kirk L. Johnson, M. Frans Kaashoek, and James W. O'Toole, Jr., : Overcast: Reliable Multicasting with an Overlay Network. In 4 th Symposium on Operating Systems Design and Implementation(OSDI), San Diego, CA, USA (2000)
5. Zhi Li and Prasant Mohapatra, : HostCast: A New Overlay Multicasting Protocol. IEEE Int. Communications Conference (ICC) (2003)
6. David A. Helder and Sugih Jamin, : End-host multicast communication using switch-trees protocols. In Proc. GP2PC, Berlin, Germany, (2002)
7. Laurent Mathy, Roberto Canonico, and David Hutchison, : An Overlay Tree Building Control Protocol. Proc. of NGC (2001)

8. Dimitrios Pendarakis,Sherlia Shi,Dinesh Verma, and Marcel Waldvogel, : ALMI : An Application Level Multicast Infrastructure. Proc. of the 3rd USNIX Symposium on Internet Technologies and Systems, (2001)
9. W. Wang, D. Helder, S. Jamin and L. Zhang, : Overlay Optimizations for End-host Multicast. In Proc. NGC, (2002)
10. K. L. Calvert, M. B. Doar, and E. W. Zegura, : Modeling Internet Topology. IEEE Communications Magazine 35, 6 (1997)
11. E. Zegura, K. Calvert, and S. Bhattacharjee, : How to model an internetwork. In Proceedings of IEEE INFOCOM (1996)

Phase Synchronization and Seamless Peer-Reconnection on Peer-to-Peer Streaming Systems

Chun-Chao Yeh

Department of Computer Science, National Taiwan Ocean University, Taiwan
ccyeh@mail.ntou.edu.tw

Abstract. To reduce perceived impact, we argue that, in P2P streaming networks, selection of new parent peer for a peer should consider not only network quality (e.g. delay and bandwidth) but also the frame-buffer status between the parent-child peers. When there is large mismatch on the frame-buffer, the child peer is in danger of suffering frame-buffer underflow while playback. Meanwhile, another issue discussed in this paper is phase-skew among peers. Degrees of phase-skew between two peers mean the degrees of presentation-time difference between the two peers. Without proper control, phase-skew could be serious between peers. In this paper we propose an effective buffer management and coordination scheme to avoid these two problems. A prototype P2P streaming system basing on open standards was built to verify our design approaches.

1 Introduction

In P2P streaming system, all participant peers form one or multiple broadcast trees to forward the broadcasting program from source peer to all participant peers. Subject to dynamics of network traffic conditions and peer behavior (join and leave (or failure) randomly), smoothly delivering program streams to each participant peer becomes a challenge, especially when the peer member becomes large. Various design strategies have been proposed to construct and maintain the multicast tree effectively [1,2,3,4,5,6,7]. Also, some research efforts were contributed to reduce the perceived impact of an ancestor change or data packet loss [8,9,10,11,12]. Despite recent research results on the P2P streaming systems, some design issues have not been well addresses, especially when real deployment is considered. In this paper we investigate buffer management issues to provide better streaming quality for peers. To reduce perceived impact, we argue that selection of new parent peer for a peer should consider not only network quality (e.g. delay and bandwidth) but also frame buffer status between both of parent and child peers. Two issues are discussed. One is frame-buffer mismatch; the other is phase-skew. The former would cause viewers perceiving interruptions of program playback when the mismatch is large; the latter would cause viewers perceiving lags of program presentation comparing with others. To our best knowledge, none of previous works has addressed these two issues well. The two

C. Kim (Ed.): ICOIN 2005, LNCS 3391, pp. 582–589, 2005.

problems were either over-simplified or totally ignored. However, in reality they are very likely to happen.

The rest of the paper is organized as follows. In the following section, we give a more detailed discussion on the problems we deal with. In Section 3, we provide our solution schemes. Section 4 presents the prototype system we made and the experiment results. Conclusions are made in Section 5.

2 Problem Descriptions

Maintaining frame buffers in application level is a common approach to smooth out playback interruptions due to transient network faults. Limited frame buffers are allocated to media clients to preload program frames from its upstream. Through out the time of playback, the frames valid in the frame-buffer, referred as *frame window* in the rest of the context, are updated to keep those frames received while not being playback yet.

2.1 Phase-Skew Among Peers

It is quite natural for viewers/listeners to think what the frame presented to she (or he) is the same (or near the same) frame what others would see/heard at the same time when they are watching/listening a broadcasting program. It is more likely true for conventional TV/Radio broadcasting systems. However, people might perceive some degrees of presentation lags comparing with others when they view/listen the program over internet. To evaluate this effect, we refer *phase* of a peer, in the rest of the context, as the time lag between the time a frame is playback on the peer and the time the same frame is playback on source peer (the root node on the broadcast tree). Fig. 1 shows an example. Since in P2P networks, instead of receiving frames from same server, each peer receives frames from its parent peer which in turn receives the frames from other peer, the skews could be larger and more diverse.

2.2 Frame-Buffer Mismatch Between Peers

In P2P networks, peers can join and leave freely. To recover from peer leave, an existing peer should be selected to take the jobs of the leaving peer. Most of previous works considered network quality as a main criterion in the processing of choosing the new parent peer, while we argue status of frame-window should be another important factor to be taken into account. Fig. 2 shows an example. Assume at time t, peer *b* leaves (or fails). It seems feasible to have peer *d* to replace peer *b* as shown in Fig. 2(a). However, when the status of frame-window is taken into account, we might find that peer *d* is not a good candidate. Assume at the moment peer *b* leaves (or fails), the frame-windows of the four peers are as those shown in Fig. 2(b). If we replace peer *b* with peer *d*, we would find that peer *d* are will miss some frames since the next frame peer *d* needed (frame for time 10 in this example) is not available in its new parent peer (peer *a*), as shown

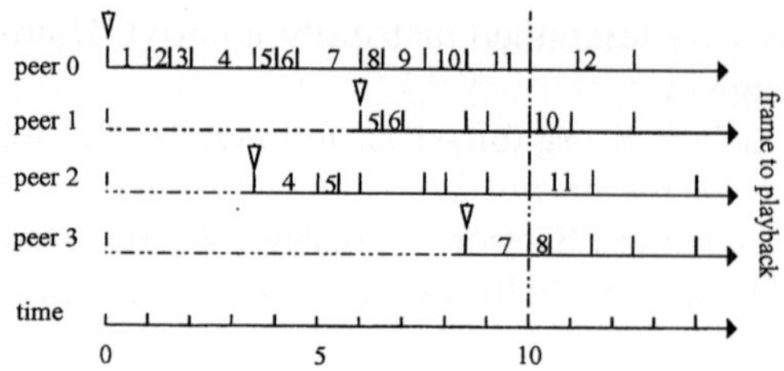

Fig. 1. Phase-skew between peers. Three peers (*peer*1, *peer*2, *peer*3) join a multicasting tree rooted by *peer*0 at different time. Each peer starts the playback with different starting frame (frames 5, 4, and 7, for the three peers respectively). Since the three peers are not in phase, they view same frames at different time. For example, at time 10, the frames presented to the three peers are 10, 11, and 8 respectively

in Fig. 2(c). Under such a circumstance, the downstream peer (peer d), would suffer program interruption due to frame loss. The situation could become more serious over time when more peer-reconnection operations involve.

In general, possible relations of two frame-windows are those shown in Fig. 3. Assume at some moment, peer a is selected as the new parent peer of peer b. And assume peer a receives streaming data from its parent peer at same speed as it consumes the frames in the frame-buffer for playback. For the cases (a) and (b) in Fig. 3, since the immediate next frame (frame $j' + 1$) peer b should have from peer a would not be available by the time it needs for playback, peer b would suffer missing a sequence of frames during playback; for cases (c) and (e), the immediate next frame peer b needed could be available but with some waiting time. During the waiting time, peer b continues to consume frames in the frame-buffer. Consequently the size of frame-window would reduce, and so as for all its descendent peers. This puts all these peers in danger of buffer underflow. Finally, for the cases of (d) and (f) in Fig. 3, the new parent peer can forward the immediate next frame (required by the child peer) and all the following frame smoothly.

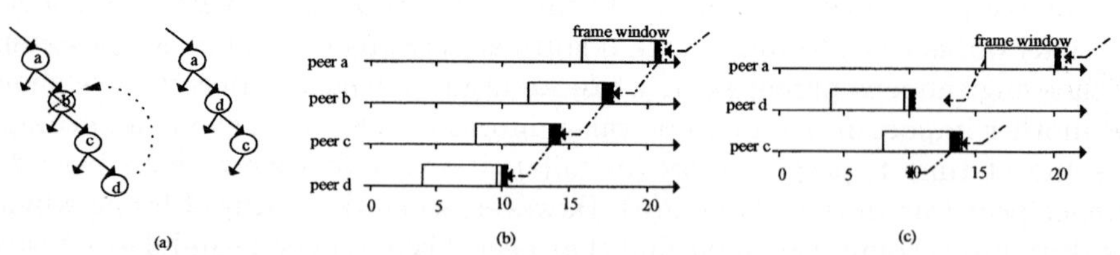

Fig. 2. An example of peer reconnection (a) the multicast tree before and after the peer b fails. (b) frame-window statues of the four peers before peer b fails (c) frame-window statues of the nodes after peer b fails

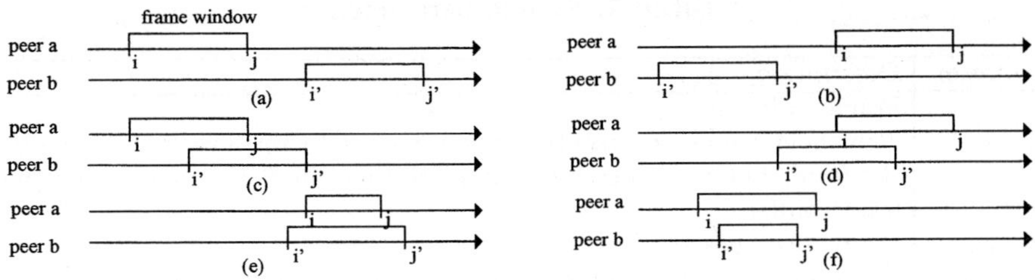

Fig. 3. Possible relations of two frame-windows. (a)-(b) disjoin; (c)-(d) partial overlay; (e)-(f) cover

3 Our Approaches

Our approaches to deal with the problems are basing on effective frame-buffer management. Our approaches make no assumption on the underlying P2P networks or on the mechanisms to maintain the multicast trees. Existing P2P streaming systems can integrate our approaches to their systems easily.

3.1 System Models for Frame-Buffer Management

In general, several factors determine the contents of frame-buffer in each participant peer: (1) the frame-buffer size, (2) the starting frame the peer received from its upstream peer when it joins the multicast tree, (3) the frame rate the peer received from its upstream peer, (4) the frame rate the peer consumes the frame for playback, (5) the individual frame rate the peer forwards the frames to each of its downstream peers, (6) the network delay between the peer and its upstream peer, and (7) the frame-retransmission mechanisms. Among these, we assume each peer is required to reserve a fixed size of memory space as frame-buffer, which is organized as a circular queue. Meanwhile, since frame-retransmission mechanisms are mainly for recovery of transient faults but not particularly designed for the problems we discussed, we ignore the effects in the following discussion. Other factors are taken into accounted and parameterized with notations summarized in Table 1. A new join peer should conduct a join-process before it starts to receive the broadcast streaming. When the new join peer successfully find a peer as its parent peer, the parent peer starts to forward the streaming data to the new join peer at a rate of $R_{catchUp}$.

The system model applies for all peers except for the source peer which by definition is the root node of the multicast tree. We assume the source peer always keeps its frame buffer full. A possible strategy to commit the assumption is to let the source peer fills up its frame-buffer before it start to playback and/or accept peer join requests. Also, according to this assumption, it is not hard to find that the synchronized frame rate, r_{sync}, is equal to $R_{playback}$, $R_{catchUp}$ or zero. The case for $r_{sync} = 0$ is due to the peer or one of its ancestor leave (or fail).

Table 1. System parameters

Notation	Description
B	Frame buffer size.
T_{init}	Threshold value, representing the time lag to retrieve first frame in the frame-buffer for playback after the time the first frame enters the frame-buffer.
$R_{playback}$	Frame playback rate.
$R_{catchUp}$	Catch-up frame rate, representing data forwarding speed of a parent peer to its child peer when a child peer connects to the parent peer. The parent peer will maintain the forwarding speed till the time the next frame for the child peer reaches the end of the parent peer's frame buffer. Then, the parent peer changes its forwarding speed to synchronized rate, r_{sync}.
r_{sync}	Synchronized frame rate, representing data forwarding speed of a parent peer to its child peer at the same pace as the speed the parent peer received the data streaming from its immediate upstream.
$d_{i,j}$	Network delay between peer i and peer j.

3.2 Phase Synchronization

Our approach to achieve phase synchronization is quite straightforward. In stead of using a complicate synchronization mechanism to coordinate among peers, we impose the synchronization mechanism, each time, on two peers only: a new join peer and its parent peer. To synchronize the phase of a new join peer with its parent peer, we should guarantee the time the first frame presents to the viewers on the new join peer is equal to (or nearly equal to) the time the same frame presents to the viewers on the parent peer. Since the frame buffer is organized as circular queue, the first frame into the frame-buffer of the new peer is the frame same as or later than the first frame the parent peer sends to the new peer, referred as $startingframe$. Consequently, proper choice of the $startingframe$ for a new join peer on the parent peer is the key to achieve phase synchronization between the two peers. Assume a new join peer i successfully finds a peer j as its parent peer and the parent peer j gets ready to send the first frame (the starting frame) to peer i. The time t_{phase}, defined as the elapsed time between the time the starting frame was sent out from peer j and the time the frame would be retrieved from the frame-buffer of peer i, can be obtained by

$$t_{phase} = d_{i,j} + T_{init} \ . \tag{1}$$

And, the starting frame s for the new peer i can be obtained by

$$s = f(t_0 + t_{phase}) \ , \tag{2}$$

where the function $f(x)$ indicates the frame with designated presentation time of x, and t_0 is the designated presentation time of the first frame in the frame-buffer of the parent peer.

Equations (1) and (2) provide a simple formula for a peer to find the starting frame on its frame-buffer for a new join peer. If the starting frame is available on the parent peer, then the parent peer proceeds to forward the consequence of streaming data starting from the starting frame. If it is not, it waits if the starting frame can arrive in a short time. It is worthy of noting that the calculation of the starting frame needs to be conducted only during peer-join process. For the case a peer switches to a new parent peer (for example, due to leave or failure of its parent peer), the new parent peer needs not recalculate the starting frame for the peer. Information about the starting frame the peer should have is provided by the re-join peer along with the peer's reconnection requests.

3.3 Seamless Peer-Reconnection

To provide seamless peer-reconnection we should prevent frame-buffer mismatch between two peers in peer-reconnection processes so that with high probability frames can be playback smoothly. Two things should be done. First, we should validate the frame-buffer status between the new parent peer and the child peer in peer-reconnection process. The second is to keep the frame-windows aligned among peers most of the time. For frame-buffer checking, we develop a check rule as Lemma 1.

Lemma 1. *Peer i can be selected as new parent peer of peer j, if*

(1) $e_j > b_i$ *and* $b_j < e_i$, *and*
(2) $e_i > e_j$ *or* $(e_j - b_j)/R_{playback} > (e_j - e_i)/r_i + T_{threshold}$,

where (b_i, e_i) and (b_j, e_j) are first and last frame number at the frame buffers of the peer i and peer j respectively, $R_{playback}$ is the playback rate defined before, r_i is the receiving frame rate of peer i from its parent peer, $T_{threshold}$ is a constant value. The first rule of Lemma 1 is to prevent the cases of selecting a parent peer with its frame-window disjoins with the frame-window of peer i (as the cases shown in Fig. 3(a) or Fig. 3(b)). The second rule is to make sure the new parent peer be able to provide the streaming data following the last frame available in the frame-buffer of peer i in time.

Lemma 1 is a check rule for general cases. On the P2P streaming systems utilizing the buffer management mechanism we proposed, the check rule can be more likely to meet, since the phase synchronization mechanism discussed before would make the frame-windows nearly align in the long run. Two frame windows, (b_i, e_i) and (b_j, e_j) are said to be nearly aligned, if $|b_i - b_j| < \delta_1$ and $|e_i - e_j| < \delta_2$ where δ_1 and δ_2 are a small value comparing with frame window size.

Lemma 2. *In the proposed P2P system model, the frame-window of a peer i will nearly align with that of its parent peer j within a constant time $T_{align} = d_{i,j} + B/(R_{catchUp} - R_{playback})$, if no network or system faults happen to both of the peers during the time.*

Lemma 3. *In the proposed P2P system model, the frame-window of a peer i will nearly align with all those of its ancestor peers within a constant time* $T_{align} = n * d + B/(R_{catchUp} - R_{playback})$*, if no network (or system) fault happens to all these peers during the time, where n is the level of the peer in the multicast tree and d is the maximum network delay between peers.*

4 Implementation and Experimental Results

A prototype system to provide P2P streaming services is built to verify our design approaches. The prototype system is basing on SpreadIt [1] as a building block. We made some enhancements to the original source code to handle peer join/leave/reconnection smoothly, and integrated the proposed buffer management scheme to it. The prototype system tightly works with open standards, such as RTP/RTSP for media streaming transport protocols, and Darwin/QTP (Apple Quick Time Player) for media server and client player. When a peer joins or reconnects to a new parent peer, the frame-window of the peer is sent along with its join/reconnection requests carried in RTSP packages.

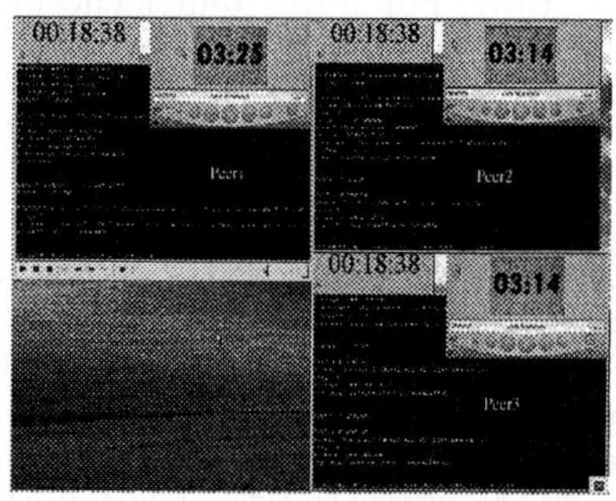

Fig. 4. A snapshot for all the three peers after they successfully join the program stream. During experiments, we recorded the screen of each peer. The video clips taken from the three peers are put together for comparison (*peer*1 is in up-left, *peer*2 is in up-right, and *peer*3 is in down-right). The broadcasted program is a digit timer. Results show that the playbacks on all the three peers (except the source peer, *peer*1, which is 9 to 10 seconds ahead others for frame preloading) are synchronized. Both media players on the *peer*1 and *peer*2 were showing the frame of 03:14 at the time 18:38

Some experiments with limited system size of three were done for real tests. The P2P middleware we implemented (NtouPCast) and a Quick Time media-player (QTP) were included in each peer. Initially only *peer*1(as source peer) are available in the system. Then, we had *peer*2 and *peer*3 join at different time. Fig. 4 shows a snapshot of the results, which shows all peers (except the source peer) are synchronized in phase. The phase difference between the source peer and other peers is the buffer size (in the unit of playback time). We also did some experiments on the cases of peer leave and failure, and the results showed the prototype system can work well.

5 Conclusions

While many P2P streaming systems have been proposed, most of them focused on structure construction and maintenance, error resilience, and security. Issues on frame buffer management to reduce degrees of frame-window mismatch between peers and phase-skew among peers were ignored or over simplified. We point out the problems and investigate some design principles. An in-phase scheme by carefully selecting the starting frame for a new join peer was proposed to reduce phase-skew. Also, under the in-phase scheme and the proposed frame forwarding control mechanism, we make all peers' frame-windows nearly align and consequently reduce the buffer mismatch during peer reconnection. Our research results complement current available research results contributed by other researchers. The proposed approaches make no assumption on the underlying P2P networks or on the mechanisms to maintain the multicast trees. Existing P2P streaming systems can integrate our approaches to their systems easily.

References

1. Deshpande, H., Bawa, M., Garcia-Molina, H.: Streaming live media over peers. Tech. Rep. 2002-21, Stanford University (2002)
2. Bawa, M., Deshpande H., Garcia-Molina, H.: Transience of peers and streaming media. ACM SIGCOMM Computer communications Review, Vol. 33 , No. 1 (2003) 107–112
3. Chu, Y.-H., Rao, S.G., Zhang, H.: A case for end system multicast. In Proc. of ACM 2000 SIGMETRICS Conf. (2000) 1–12
4. Chu, Y.-H., Rao, S.G., Seshan, S., Zhang, H.: Enabling conferencing applications on the internet using an overlay multicast architecture. In Proc. of ACM 2001 SIGCOMM Conf. (2001) 55–67
5. Banerjee, S., Bhattacharjee, B., Kommareddy, C.: Scalable application layer multicast. In Proc. of ACM 2002 SIGCOMM Conf. (2002) 205–217
6. Tran, D.A., Hua, K.A., Do, T.T.: Zigzag: An effcient peer-to-peer scheme for media streaming. In Proc. of IEEE 2003 INFOCOM Conf. (2003)
7. Guo, Y., Suh, K., Kurose, J., Towsley, D.: P2Cast: peer-to-peer patching scheme for VoD service. In Proc. Of ACM 2003 WWW Conf. (2003) 301–309
8. Padmanabhan, V.N., Wang, H.J., Chou, P.A., Sripanidkulchai, K.: Distributing streaming media content using cooperative networking. In Proc. of ACM 2002 NOSSDAV Conf. (2002) 177–186
9. Rejaie, R., Ortega, A.: PALS: peer-to-peer adaptive layered streaming. In Proc. Of ACM 2003 NOSSDAV Conf. (2003) 153–161
10. Cui, Y., Nahrstedt, K.: Layered peer-to-peer streaming. In Proc. Of ACM 2003 NOSSDAV Conf. (2003) 162–171
11. Xu, D., Hefeeda, M., Hambrusch, S., Bhargava, B.: On peer-to-peer media streaming. In Proc. of IEEE 2002 ICDCS Conf. (2002) 363–371
12. Zhang, R., Hu, Y.C.: Borg: a hybrid protocol for scalable application-level multicast in peer-to-peer networks. In Proc. Of ACM 2003 NOSSDAV Conf. (2003) 172–179

3Sons: Semi-structured Substrate Support for Overlay Network Services

Hui-shan Liu, Ke Xu, Ming-wei Xu, and Yong Cui

Department of Computer Science and Technology, Tsinghua University, Beijing, 100084, China.
{liuhs, xuke, xmw, cy}@csnet1.cs.tsinghua.edu.cn

Abstract. 3sons is a distributed overlay network services framework based on hierarchal semi-structured topology. Its excellent characters include good scalability,high utilization rate of network resources and strong robustness. Every node of 3Sons maintains two static neighbors and $\lfloor logN \rfloor$ dynamic neighbors, which can guarantee network connectivity and improve the success rate of fuzzy lookup. 3Sons can effectively reduce the average lookup length and improve transmission performance by adjusting dynamic neighbors and optimizing forward routes according to current traffic of network. The simulation results show that 3Sons can get higher lookup success rate with shorter average lookup length, and lower fuzzy lookup workload of network, so that it can better support large-scale overlay service.

1 Introduction

Through engineering the software at Internet end hosts, overlay service can be quickly constructed and easily upgraded as well as bringing large-scale wide-area distributed services to the masses. Most of the current researches implied that every overlay service is independent, and hadn't considered the traffic jitter and the decline of network performance due to independent resources management and congestion control. If we can design a general substrate framework to support multiple services, we will get fundamental solution to these problems.

3Sons (Semi-structured Substrate Support for Overlay Network Services) is an overlay network service scheme based on hierarchical semi-structured distributed topology, which can support multiple overlay services. Because of adoption unified substrate routing, 3Sons has excellent characters as good scalability, high utilization rate of network resource and strong robustness. Besides substrate routing, we can also construct specific routing policies based on point-to-point data transmission. 3Sons-based services can lead to high lookup success rate with short average lookup length, and low fuzzy lookup workload of the whole network.

C. Kim (Ed.): ICOIN 2005, LNCS 3391, pp. 590–599, 2005.

2 Related Works

To realize special functions, many overlay services had been designed(e.g. P2P, application layer multicast, Service-oriented fast route recovery[1] and QoS guarantee[2], etc.). Those services always adopt different topology structures.

Fully connected topology structure The RON(Resilient Overlay Network) nodes monitor the functioning and quality of the Internet paths among themselves, and use this information to decide whether to route packets directly over the Internet or by way of other RON nodes. Every RON node takes part in distributed route protocol to exchange routing metrics. In the network of N nodes, Every node need to maintain $N-1$ neighbors' information, which leads RON's poor scalability.

Maintaining finite neighbors To improve the scalability, it is important to limit the number of neighbors of each node. The normal policies are described in this section:

Unstructured overlays organized nodes in a random graph, e.g. Gnutella[11] and Napster[12]. Each node is equal and maintains several neighbors selected at random. Unstructured overlays always use flooding or random walks[13] algorithms to find where the destination is. It's simple enough to implement but more query levels are needed to increase lookup success rate. In some cases, the query may fail even if the goal node has existed. In addition, the unstructured topology structure can't support those services efficiently because the accurate keyword lookup is unavailable.

Structured overlays assigned keyword to responsible node and maintained keyword-correlated neighbors, e.g. CAN[14] and Chord[15], etc. In the structured overlays, we can adopt cache techniques to improve lookup efficiency. However, it brings relatively heavy control workload for keeping neighbors' specific position under the dynamic environment.

3 Topology Organization

To maintain topology of network, the node must exchange information with neighbors periodically, which brings control workload. The workload keeps direct relations with the number of static neighbors[16].

Three features that distinguish Chord[15] from many other peer-to-peer lookup protocols are its simplicity, provable correctness, and provable performance. Though each Chord node maintains a successor to guarantee the system scalability, the average lookup length is relatively long. Optimized Chord is simple, routing a key through a sequence of $O(logN)$ other nodes toward the destination. A Chord node requires information about $O(logN)$ other nodes for efficient routing, but performance degrades gracefully when that information is out of date. This is important in practice because nodes will join and leave arbitrarily, and static $O(logN)$ state may be hard to maintain.

So, We have chosen Chord as basic structure to design our 3Sons. with the structured and unstructured topological advantages in combination. We call the structure as semi-structured topology and show it in figure 1.

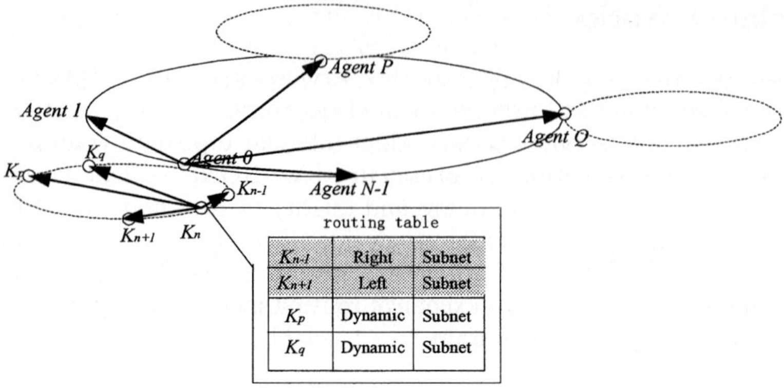

Fig. 1. Hierarchical topology of S-Chord

The 3Sons node has maintained two static neighbors to construct foundational annular topology. Forwarding to static neighbor clockwise or counter-clockwise, we can reach all nodes of subnet and guarantee the connectivity of the network. We have also used hierarchical structure to reduce the number of nodes of subnet. Even if we use flooding or random walks to realize fuzzy lookup, we can still obtain higher lookup success rate than unstructured overlay.

We maintain $\lfloor logN \rfloor$ dynamic neighbors besides two static neighbors in 3Sons. The dynamic neighbors choose the routing with smaller hop-count as new dynamic neighbor according to the present traffic in the network. So, the average lookup length is reduced obviously among the whole network. To decrease the influence of dynamic change, we choose dynamic neighbors at random and don't need to keep the special relation with the local node ID. The result of comparison among 3Sons, Chord and optimized Chord is shown in table 1. In dynamic environment, every node of 3Sons only maintains two static neighbors validity to keep topological integrality, and the control workload is obviously low. Otherwise, the invalid dynamic neighbors will be updated by dynamic neighbor adjusting algorithm.

Table 1. Comparisons among 3Sons,Chord and optimized Chord

character	Chord	optimized Chord	3Sons
static neighbor	1	logN	2
dynamic neighbor	0	0	logN
average lookup length	long	short	short
flexible	strong	weak	strong
control workload	low	high	low

In 3Sons, the physically closed nodes construct subnet through distributed self-organizing method. Each subnet elects one node with good performance as agent, and all subnet agents form the upper network of the hierarchical structure. Both subnet and the upper sub network use semi-structured topology structure. The identification of node is made up of two parts in 3Sons, the subnet ID and the local ID, which are produced by applying IP address to SHA-1 [4].

4 System Structure and Component Design

In this section, we will describe the topology maintenance, substrate support routing and dynamic neighbor adjusting in detail.

4.1 Topology Maintenance

This module keeps topological integrality in distributed self-organizing method.

- **New node joining** The applicant broadcasts the probe packet carrying node ID and IP address and use TTL to restraint the probe range. According to apperception, the applicant confirms its static neighbors and chooses $\lfloor logN \rfloor$ dynamic neighbors at random . If the applicant has not collected enough information to create connection with left and right neighbors, it will neglect this application and repeat again after some time. During this process, lots of the nodes in different subnets might send apperceptions to new applicant nodes which may choose one of the subnets randomly to achieve the joining operation.
- **Agent bootstrap** We adopt the procedure similar to Yallcast and YOID[6]. Suppose that 3Sons system has a related DNS domain name, and the domain name can be parsed into one or several bootstrap node IP address. New agent sends the joining application to its left and right agent neighbors, and creates the connection.
- **Node quit** The adjoint nodes use keep-alive packets to confirm forwarding path validity. The keep-alive packet carries information of static neighbors. So the node can keep the topological integrality when it breaks down or quit the system.

4.2 Substrate Support Routing

Three kinds of substrate topological route methods are offered to support different overlay services.

- **precise routing**: Each node chooses the next hop node according to local routing table, whose ID is the closest to destination's. If the destination is not in same subnet, the node will regard the local agent as the middle node. If there are no closer nodes, the local host will consider that the destination can not be reached. Every 3Sons node needs to maintain $2 + \lfloor logN \rfloor$ routing table items, so, space and lookup complexity are all $o(2 + \lfloor logN \rfloor)$. Because we have maintained $\lfloor logN \rfloor$ dynamic neighbors, with the stability of the traffic of network, the average lookup length is close to optimized Chord.

- **flooding routing algorithm**: 3Sons node duplicates and forwards the query to all neighbors of routing table. We suppose that a node sends the query to x new node each time. After flooding k times, the query will be transmitted to $\sum_{i=1}^{k} x^i$ nodes. The flooding level should not be less than $log_x(N - \frac{N+1}{x})$ to ensure the query reach most nodes of the subnet. In practice, we fetch the upper limit $log_x(N)$. Every 3Sons node maintains two static neighbors and $\lfloor logN \rfloor$ dynamic neighbors. To simplify our design, we only consider the impact on flooding levels of the dynamic neighbor's count. It is L_F that we define the flooding routing levels to ensure the higher lookup success rate.

$$L_F = \lceil \frac{log_2 N}{log \lfloor log_2 N \rfloor} \rceil \tag{1}$$

- **random walks routing algorithm**: 3Sons node randomly duplicates and forwards the query to one or two neighbors of routing table. So, we can think approximately that node sends the query to 1.5 new node each time. According to the similar analysis of the flooding routing algorithm, random walks routing is defined as L_R to ensure the higher lookup success rate.

$$L_R = \lceil \frac{log_2 N}{log 1.5} \rceil \tag{2}$$

4.3 Dynamic Neighbor Adjusting

Every 3Sons node checks the traffic statistics regularly, if it finds that the flow sent out from node A via itself to node B exceeds certain threshold, it will send the notice to node A and asks node A to create a direct route to node B. Node A selects the dynamic routing table item whose traffic statistic value is minimum, then replace next hop of the item with node C. A consultation is needed between node A and C to ensure the connect be correctly built. Through dynamic adjustment of the neighbors, we can reduce the average lookup length effectively.

5 Simulation and Result Analysis

This section presents a detailed evaluation of the 3Sons using simulations. We implemented a simple discrete event-based simulator which assigns each application level hop a unit delay. To reduce overhead and enable the simulation of large networks, the simulator does not model any queuing delays or packet loss on links. The simplified simulation environment was chosen for two reasons: first, it allows the simulations to scale to a large (up to 20K) number of nodes, and secondly, this evaluation is not focused on proximity routing depended on link status. Since our basic design is similar in spirit to Chord, we believe that heuristics for performing proximity-based routing can be adapted easily to 3Sons.

Experiment 1 In the subnet with 2000 nodes, we measure the influence of different dynamic neighbor's count to average lookup length on precise routing, and the result is shown in figure 2.

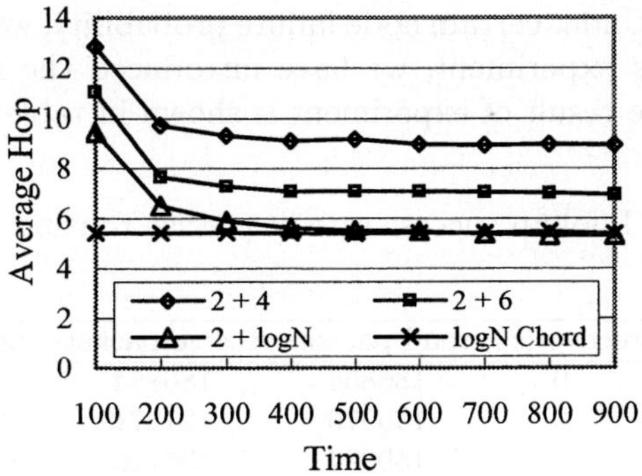

Fig. 2. Average lookup length in different dynamic neighbor's count

Because of the use of the dynamic neighbor adjusting algorithms, the forwarding path of larger traffic has been optimized as time goes. The average lookup length of the whole network decreases obviously. 3Sons adopts the policy of maintaining $\lfloor logN \rfloor$ dynamic neighbors, and its average lookup length is close to or slightly superior to optimized Chord.

Experiment 2 In the subnet with 20000 nodes, we measure the influence of different subnet scale to average lookup length on precise routing, and the result is shown in figure 3.

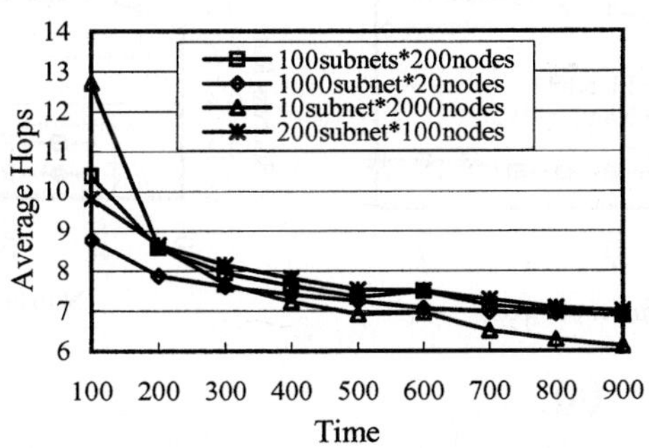

Fig. 3. Average lookup length in different subnet scale

As the number of dynamic neighbors of 3Sons node is the logarithm of subnet scale, it increases slowly with the increase of the subnet scale. In addition, since we have used the dynamic neighbor adjusting algorithms, the average lookup length is not sensitive to subnet scale. In the following experiments, we set the node number as 200 in each subnet, and there are 100 subnets in the 20000-node network.

Experiment 3 Under certain node failure probability, we measure the lookup success rate. In this experiment, we have introduced the recovery probability with fault node. The result of experiment is shown in table 2.

Table 2. Lookup success rate in precise routing algorithm

fault(%)	recover(%)	send packets	receive packets	success rate
0	0	186854	186854	1
5	0	149970	121271	0.808635
	10	159094	136710	0.859303
	20	164549	147184	0.894469
10	0	123260	85105	0.690451
	10	137704	104571	0.75939
	20	147217	119216	0.809798

With the increase of node fault probability, the lookup success rate to single copy will drop by a large margin. If the node can be recovered fast, the lookup success rate will obviously be improved.

Experiment 4 To compute flooding levels and random walks levels in different subnet scale. The result is shown in figure 4.

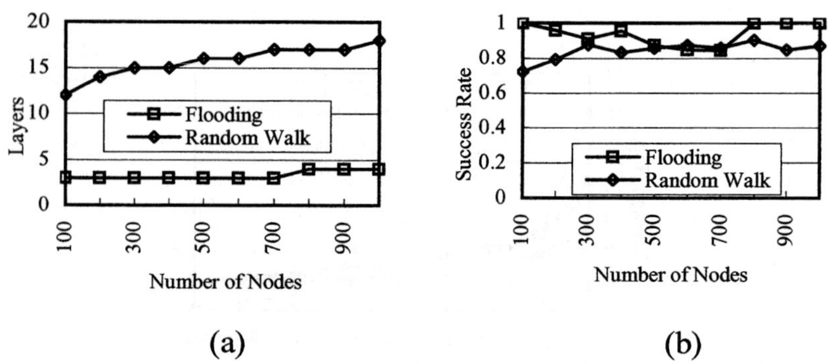

Fig. 4. (a)flooding levels; (b)random walks levels

With the enlargement of the subnet scale, the number of dynamic neighbor will rise slowly, leading to slow growth of the flooding and Random walks algorithm level at the same time. A higher lookup success rate of single copies is ensured.

Because composition is limited on space, we can only illustrate 3Sons system briefly. Please consult the technique report(http://netlab.cs.tsinghua.edu.cn/) to find the detail.

6 How Does 3Sons Support the Services?

Overlay services can directly use substrate supporting routing algorithms for communication. They may also construct specific routing policies based on point-to-point data transmission. In this section, we take keyword accurate lookup, fuzzy lookup and multicast as examples to explain how to support different services in 3Sons.

- **keyword accurate lookup based on 3Sons**: We use function H (SHA-1[4]) to transform the keyword *key* to keyword ID, and transmit the query whose destination subnet ID is the keyword ID to destination agent. If the destination subnet ID and the local host subnet ID are the same, the local host forwards the query to the next hop neighbor whose ID is the closest to the destination ID. Otherwise, the local host forwards the query to local agent. The agent transmits the query to next hop agent whose ID is the closest to destination subnet ID, until the query reaches agent whose subnet ID is closest to the destination subnet ID. Then, 3Sons node forwards the query whose destination ID is the keyword ID from agent to destination. We store the keyword in final node. The procedure of keyword lookup is similar to keyword storage.
- **keyword fuzzy lookup based on 3Sons**
  - **keyword storage** The keyword storage of fuzzy lookup is similar to accurate lookup's.
  - **keyword indexing** Agent keeps the keyword ID table whose index is from 0 to 255 and indexes all keywords stored in the subnet[5]. The basic indexing scheme is to split each string to be indexed into "n-grams": distinct n-length substrings. For example, a keyword "Semi-structured" could be split into thirteen trigrams:Sem, emi,mi -,i-s, -st,str,tru, ruc,uct,ctu, utr,ure, red. To reduce memory workload of agent, we use hash function H_2 to map these trigrams to index of keyword ID table, and register the keyword ID in the table item. We use $Maj(a, b, c)$ in SHA-1 to balance keyword ID count in every table item. Supposing the substring is "abc", the equation of function H_2 is shown as follows:

$$H_2 = (a \wedge b) \oplus (b \wedge c) \oplus (a \wedge c) \quad (3)$$

  - **keyword lookup** (1)The source node forwards query to local agent; (2)Carrying on the flooding in the upper subnet area, we determine the flooding levers as $L_F + 1$ because it will cause all keywords in the subnet ignored if the query can not be transmitted to relevant agent. (3)The agent match function returned TRUE creates new query to local subnet with flooding levels L_F. Keyword match function is used to judge whether the keyword is stored in local subnet. Lower flooding level is with lower query success rate, however, higher flooding level with higher query success rate brings heavy workload of network and greater average lookup length as well. In addition, it is very difficult to confirm the levels of flooding in Gnutella. However, in 3Sons system, agent can confirm the flooding levels of upper subnet area and subnet by equation (1), and win higher query success rate with smaller workload of network.

- **keyword transfer**: When new subnets or nodes join and depart from the network, it is necessary to check keywords of relevant node neighbors. The procedure is similar to Chord.

– **multicast based on 3Sons**: The service of multicast group maintains the multicast tree, and uses precise routing offered by 3Sons to realize point-to-point data transmission. We use one keyword to identify the multicast group, and a node to store the keyword is the root of n-tree.

- **confirm the root in different layer**: The multicast group is divided into three layers. 1)The service provider is the root of the first layer(P_{root}); 2)The agent in the same subnet is the root of the second layer(A_{root}) and connected with P_root to decrease the depth of the whole tree; 3)The agents of other subnets are the roots of the third layer(S_{root}).
- **join procedure**: When the node applies to join the multicast group, it sends the application to P_{root} node. Each node of the forwarding path checks whether it is the root itself before forwarding. If it is, it balances the tree and inserts the new node into the group. If the root has not become a member, it sends a new application to P_{root}. Otherwise, it continues to transmit the application to the root of tree until finishes the join procedure.
- **balance the tree**:When the root receives an application, the balance procedure starts. The application is delivered from root to leaf. The node of forwarding path inserts the applicant into children set when the number of its children is less than n, otherwise, it forwards the application to the child-tree which has smallest scale. If we define N_A and N_S as the members in upper subnet and subnet, then the upper limitation of the depth of multicast tree is D_M.

$$D_M = \lceil log_n N_S * (n-1) + 1 \rceil + \lceil log_n N_A * (n-1) + 1 \rceil + 1 \quad (4)$$

- **member quit normally**: All members of multicast group use keep-alive packet to assure the father and children nodes available. If a node is going to quit the group normally, it selects a leaf from child-tree randomly to replace its position.
- **member break down**: If a node breaks down and leave the group without informing its children, its children will rejoin the tree by repeating the joining procedure.

7 Conclusion

We proposed hierarchical semi-structured overlay topology, and set up 3Sons which is distributed overlay service support scheme. Its excellent characters are good scalability,high utilization rate of network resource and strong robustness. 3Sons node maintains a few dynamic neighbors besides two static neighbors. The dynamic neighbors can choose the route with smaller hop-count as new route according to the present traffic in the network. So, the average lookup length is reduced and the query success rate is increased. We described how 3Sons supports large-scale services and gets good performance. In this paper, we proposed how to confirm the number of dynamic neighbors, flooding level and random level to ensure the higher lookup success rate be achieved.

8 Acknowledgement

This work was supported by the National Nature Science Foundation of China (No 60473082, No 60403035) and National Key Fundamental Research Plan (973) of China (No 2003CB314801).

References

1. David, A., Hari, B., Frans, K., Robert,M.: Resilient Overlay Networks.In Proceedings of ACM Symposium on Operating Systems Principles (SOSP).2001
2. Lakshminarayanan, S., Ion S., Hari B., Randy H.:OverQoS: Offering Internet QoS Using Overlays. ACM SIGCOMM Computer Communications Review.2003
3. Akihiro, N., Larry,P., Andy, B.: A Routing Underlay for Overlay Networks.ACM SIGCOMM Computer Communications Review.2003
4. FIPS 180-1. Secure Hash Standard. U.S. Department of Commerce/NIST, National Technical Information Service, SpringLeld, VA, Apr.1995
5. Witten, I. H., Moffat, A., and Bell, T. C.:Managing Gigabytes: Compressing and Indexing Documents and Images. second ed. Morgan Kaufmann. 1999
6. FRANCIS, P.:Yoid: Extending the internet multicast architecture. http://www.icir.org/yoid/docs/yoidArch.ps. 2000
7. Savage, S., Collins, A., Hoffman, E., Snell, J., Anderson, T. :The End-to-End Effects of Internet Path Selection. Proceedings of the ACM SIGCOMM. August. 1999. 289-299
8. Banerjee, S., Bhattacharjee, B., Kommareddy, C.:Scalable Application Layer Multicast. Proceedings of the ACM SIGCOMM. August. 2002. 205-217
9. Chu, Y.-H., Rao, S. G., Zhang, H.: A Case for End System Multicast. Proceedings the ACM SIGMETRICS. June. 2000. 1-12
10. Jannotti, J., Gifford, D. K., Johnson, K. L., Kaashoek, M. F., O'Toole Jr, J. W.: Overcast: Reliable Multicasting with an Overlay Network. Proceeding of the USENIX OSDI. October. 2000
11. GNUTELLA: http://gnutella.wego.com
12. NAPSTER: http://www.napster.com
13. Lv, Q., Cao, P., Cohen, E., Li, K., Shenker, S.: Search and replication in unstructured peer-to-peer networks. Proceeding of the 16th international conference on Supercomputing. June. 2002
14. Sylvia, R., Paul, F., Mark, H., Richard, K., Scott, S.: A Scalable Content-Addressable Network. Proceeding of the ACM SIGCOMM. 2001
15. Ion, S., Robert, M., David, K., M.Frans, K., Hari, B.: Chord A Scalable Peer-to-peer Lookup Service for Internet Applications. Proceeding of the ACM SIGCOMM. 2001
16. Liu, H.-sh., Xu, M.W., Xu, k., Cui Y.: How to evaluate scalability of packet switching network?. Proceeding of the IEEE TENCON. 2004

Catalog Search for XML Data Sources in Peer-to-Peer Systems

Ying Yang[1,2] and Jia-jin Le[1]

[1] College of Information , University of DongHua
Shanghai, 200051, P.R.China
yingy2004@126.com
[2] College of Computer & Electron Information, University of Guangxi
Guangxi, 530004, P.R.China

Abstract. A core challenge in peer-to-peer systems is efficient location of large numbers of nodes or data sources. This paper proposes a novel catalog search for the large distributed XML data sources based on consistent hashing, and the improved algorithm is given to speed the key location. Utilizing this model with data summary can quickly process queries for XML repositories. Results from performance simulation show that our approach has significant improvement.

1 Introduction

Web search requires a large number of nodes to provide desired results. Therefore, the core problem is efficient location of data sources in Peer-to-Peer systems, which have no central point of failure and no central repository necessary to maintain. This paper presents a novel catalog search for a large distributed XML data sources. In addition, it can be used to perform other tasks such as query optimization.

The rest of the paper is organized as follows. In the next section we briefly discuss related work on Web search or lookup and the contributions of this paper. In section 3 preliminary definitions are given in detail. Section 4 describes our system model for XML data sources in P2P systems. Section 5 presents the performance evaluation. The conclusion and the future work are given in final section.

2 Related Work

There are many representatives of distributed catalog search such as Napster, Gnutella and Chord etc. Napster and Gnutella [5] provide a search based on the keyword. Napster uses a central index, which results in a single point of failure. Gnutella floods each query over the whole system that leads to high processing costs in large system. Chord [3], which built on Distributed Hash Tables (or DHTs), forwards message based on numerical difference with the destination address, however, it makes no explicit effort to achieve good network locality and allows only simple key for query.

C. Kim (Ed.): ICOIN 2005, LNCS 3391, pp. 600–608, 2005.

Our engine makes use of data items that often appear in queries, such as metadata and word characteristic of a specific node. Based on consistent hashing approach, these data items can be mapped to the corresponding nodes and these identifiers can be used to direct queries. When nodes join or leave the system, they exchange information through node routing table. Therefore, the identifier hashing ring drives selection of promising data sources. The contributions of this paper are:

- The consistent hashing approach is utilized to construct a catalog search in P2P systems
- The catalog search for querying large XML repositories is proposed to determine which nodes or data sources should receive queries based on query content.
- The model can provide good scalability when new nodes join or leave, and good load balance fairly across the participating nodes.

3 Preliminary Definition

In this paper, node data are defined as data summary (or catalog information) in order to be known by other nodes in a P2P network. The data summary is defined as follow:

Definition 3.1 : $N_i(1 \leq i \leq n)$ denote the n nodes, $D_i(1 \leq i \leq m)$ denote the m data items of a data source for the N_i node, the data summary set $C_i = \{(K_j, S_{ij}) | S_{ij}$ is a summary of K_j on node $N_i\}(1 \leq j \leq m)\}$.

The key items K_j are present in the data items D_i. Each S_{ij} is the data summary corresponding to K_j and depends on the data of node N_i. The catalog service determines which nodes a query Q should execute on using the key items K_j and *map*(). The key items K_j are extracted in D_i from a query.

Definition 3.2: the function $map() = \{Q\} \times \{\{C_i | 1 \leq i \leq n\}\} \longrightarrow \{N_i | 1 \leq i \leq n\}$, uses K_j in catalog information C_i to examine the relevant sets of data summaries in order to determine the nodes storing data relevant to Q.

Theorem3.1: for a given query Q, map $(Q, \{C_i\}) = \{N | P1 \bigvee P2\}$.

Proof: parameter P1 is a non-empty results set generated by Executing Q on N, on the other hand, parameter P2 is the final results set generated by Executing part of Q on N. It covers the case in which Q requires a join or an intersection of data across different nodes. Therefore, for a given query Q, the final result consists of $\{N | P1 \bigvee P2\}$.

In this paper, the design of model are used in the large network of distributed XML data repositories, where D_i is a set of XML documents, and K_j is a set of the most frequent element tags and attribute names, as well as their parent or children (or ancestor and descendant) in D_i .Each S_{ij} is a data summary corresponding to K_j and depends on the data of node N_i. For example, a data summary for the element "price" on node N_i might contain all the unique paths lead to "price". As follow, table 3.1 give a data sample using a simple Xpath [11] to illustrate how the above definitions is used in XML repositories.

	Path in XML data
N_1	Provider/product/title, producer/product/price
N_2	Country/catalog/product/title Country/superstor/product/title
...	...
N_8	Provider/product/title, producer/product/price
N_9	Provider/product/title, producer/product/price

N_i	K_i	element	Data summary
N_1	K_3	price	$\{(S_{1,price,}),(S_{8,rice}),(S_{9,price})\}$
N_2	K_2 K_4	catalog title	$\{(S_{2,cata})\}$ $\{(S_{1,titl}),(S_{2,titl}),(S_{8,titl}),(S_{9,titl})\}$
...		...	...
N_8	K_1	product	$\{(S_{1,pro}),(S_{2,pro}),(S_{8,pro}),(S_{9,pro})\}$
N_9	K_5	Superst-ore	$\{(S_{2,store})\}$

Table 3.1: Path in XML data of node Table 3 .2: The data summary of K_j

Elements such as "product", " price", " title", as well as their parent element "catalog", "superstore", are designed as keys K_j according to element frequent occurrences. Each summary S_{ij} contains a set of all possible paths in the data table that lead to K_j . For example, data summaries of nodes containing the element "title" are described as $S_{1,title} = \{product/title\}$,$S_{2,title} = \{catalog/product/title, superstore/product/title\}$ etc. Similarly the data summaries table of XML data can be obtained in table 3.2.

4 System Model

4.1 Consistent Hashing Ring

Definition 4.1: A consistent hash function f is : $2^B \times I \longrightarrow B.f_v(i)$ is the bucket to the data item i assigned in view v , $f_v(i) \subseteq v$. A consistent hash family F is a set of consistent hash functions and a random consistent hash function f is a function drawn at random from a particular consistent hash family.

i is a data item and I is the set of data items, B is the set of buckets, I_n is the number of data items ($I_n =| I |$), v is a view in any subset of the buckets B, The consistent hash family F have the properties of smooth and good load balance [6]. They can fairly spread data items to the relevant bucket.

In our model, an m-bit identifier is assigned to each node and key. Node identifier obtains through hashing the node's IP address and port and key identifier obtains through hashing the key. Identifiers are ordered in an identifier circle module 2^m. A identifier key k is mapped to the closest node whose identifier is equal to or follows k, denoted by successor (k_j). If identifiers are represented as a circle of numbers from 0 to $2^m - 1$,then successor (k_j) is the first node clockwise from k_j.

For the network of large distributed XML repositories above (section 2). An identifier ring with m=6 (mod 2^6) bits is chosen. By hashing XML documents, identifiers for the set of nodes $\{N_1, N_2, N_3, N_4, N_5, N_6, N_7, N_8, N_9\}$ are generated, denoted as $\{n6, n13, n18, n25, n33, n41, n46, n50, n59\}$, while by hashing elements of XML document, identifiers for the set of keys $\{K_1, K_2, K_3, K_4, K_5\}$ are obtained, denoted as $\{k47, k10, k5, k11, k57\}$. Consistent hashing maps keys to nodes in the identifier ring, each key is assigned to its successor node, which is the nearest node travelling the ring clockwise. For example, key 5 would be located at node 6, since the successor of identifier 5 is node 6. Similarly, keys 10 and 11 at node 13 etc, as above figure 4.1.

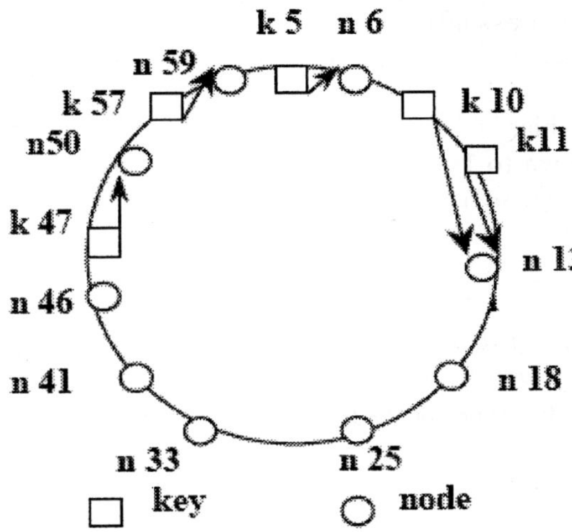

Figure 4.1: The identified hashing ring

4.2 Key Location

Lookup can be performed in an identified hashing ring. Queries for identifier keys can be passed around the circle via these successor pointers until they encounter the nodes obtaining the desired identifier. This approach is simple but very low efficient. Take the lookup for key 57 through node 6 for example, Node 6 will lookup its successor node 13, if key 57 is not found, then node 13 will lookup node 18, analogically, until the node 59 holding key 57 is returned eventually. The query visits every node on the circle among node 6 to 59. An improved algorithm as follow is adopted to speed the key location. For m-bit key/node identifier, let each node n maintain an additional routing table with m entries. This routing table enables each node to store more information about near nodes succeeding it on the identifier circle than other nodes farther away.

Let *point*[*k*] is the first node on circle that succeeds $(n + 2^{k-1})), (1 \leq k \leq m, \text{mod } 2^m)$, successor is the next node of node n in the identifier circle, predecessor is the previous node of node n on the identifier circle. The *ith* entry at node n in the routing table contains the identity of the first node point $[i] = \mathit{successor}(n + 2^{i-1}), (1 \leq i \leq m, \text{mod } 2^m)$, node point[i] is called the *ith* point of node n. Note that the first point of n is the immediate successor of n on the circle. Table 4.1 shows the routing table of node 6(m=6,mod 26) in the identifier ring (Figure4.1). The first point of node 6 points to node 13, since node 13 is the first node that succeeds (6+20)=7. Similarly, the second point of node 6 points to node 13, since node 13 is the first node that succeeds (6+21)=8. The last point of node 6 points to node 41, because node 41 is the first node that succeeds (6 + 25) = 38.

Algorithm key-location
Input: queried key id and node n on which lookup starts
Output:the node n' holding the key
1. If $(id \in (n, successor))$
2. $n' = successor$;
3. else
4. n.routing- table;
5. for $i = m$ down to 1
6. if $(point[i] \in (n, id))$
7. $n' = point[i]$;
8. go to 14 ;
9. else
10. for $i = m$ down to 1
11. if the largest $point[i]id$
12. then $n =$ the largest $point[i]$;
13. go to 1 ;
14. Output the result n'

Entry	Entry→point[i]
n $6+2^0$	n 7 →n 13
n $6+2^1$	n 8 →n 13
n $6+2^2$	n 10 →n 13
n $6+2^3$	n 14 →n 18
n $6+2^4$	n 22 →n 25
n $6+2^5$	n 38 →n 41

Table 4.1: The routing table of n 6

The improved algorithm can be performed in an iterative style. In this style, a node asks a series of nodes for information from their routing tables, each time moving closer to the desired successor on the identifier ring. For example, suppose node 6 wants to find the node holding key 57. Since the largest point of node 6 that precedes 57 is node 41, node 6 will ask node 41 to resolve the query. In turn, node 41 will determine the largest point in its routing table that precedes 57, supposing node 50. Finally, node 50 will discover that its own successor, node 59, succeeds key 57, and thus will return node 59 to node 6.

On average, only $\lceil \log_{2^b} N \rceil$ rows are populated in the routing table. The choice of b involves a trade-off between the size of the routing table (approximately $\lceil \log_{2^b} N \rceil \times (2^b - 1)$ entries) .With a value of b=4 and 10^6 nodes, a routing table contains on average 75 entries and the expected number of routing hops is 5. While with 10^9 nodes, the routing table contains on average 105 entries and the number of routing hops is 7.

4.3 Catalog Query for XML Repositories

A complicated Xpath query Q can be decomposed to the set of simple queries $\{q1, q2, ... qn\}$[9][11]. The system can handle regular Xpath queries such as the form $Q = /q1[e_1]/q2[e_2]/.../qn[e_n]$ op value. Given a Xpath query, the catalog service engine will determine which nodes in the system should receive the query. The algorithm of catalog query for XML repository is given as follow:

Algorithm: catalog query
Input: A query Q $\{q_1, q_2, ... q_n\}$ and $E\{e_1, e_2, ...e_n\}$ is the set of keys extracted elements or attributes from Q
Output: The data summaries S_{ij} satisfying Q
1. Let$N = \Phi, K\{k1, k2, ...kn\}$ is the identifier of E
2. Call algorithm key-location , then return $N_C\{(k_1, n_1), (k_2, n_2), ...(k_n, n_n), (k_i, n_i)\}$, the set of the identifier pair on identifier ring
3. Pick the next q_i, e_i, visit the node N_C and table3.2, then retrieve the set of N_i
4. Let $N = N \bigcap N_i$, or $N = N_i$, if $N = \Phi, Q = Q - \{q_i, e_i\}$
5. If $Q \neq \Phi$ go to 3
6. Execute the N satisfying Q
7. Output the data summaries S_{ij} satisfying Q

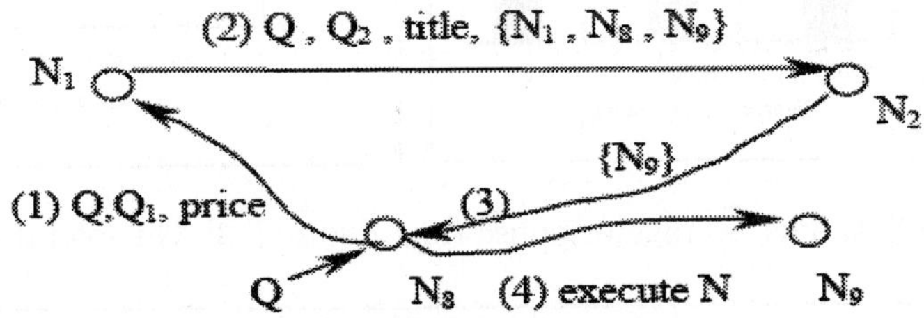

Figure 4.2

For example, Q is a query of "$//product/title = "SONY - TV"/price$" , which retrieve the price of product by title "SONY-TV", It can be divided into two branches: Q_1 is the query of "$//product/price$" and Q_2 is the other query of "$//product/title = "SONY - TV""$" . Both branches must be satisfied and the result of query will be sent only to those nodes that have both paths in their repositories. Suppose query Q starts from nodeN_8 . Through the above algorithm step1 and step 2, we can find the key of element price is K_3 and is located on node N_1 . Similarly, the key of title is K_4 and is located on node N_2

the key of product is K_1 and is located on node N_8 . Figure 4.2 describes the produce of query from step 3 to the final execution.

Firstly, Q is sent to N_1 based on price and Q_1, then the set $N_{11} = \{N_1, N_8, N_9\}$ is produced. Secondly, based on Q_2 . Q and N_{11} are forwarded to N_2 , which is relevant to title. Q_2 contains a value predicate "SONY-TV" on title. Therefore, both structure and value summaries are utilized to select relevant nodes. Suppose only N_9 satisfied and produce the set $N_{21} = \{N_9\}$. Finally, according to the theorem 3.1 in section 3, N_8 sends Q to N_9 for execution because it is the only node appearing in $N_{11} \bigcap N_{21}$.

5 Performance Evaluation

Our simulation experiments are to evaluate the validity of our model and ability of processing catalog queries. Three different distributed networks obtaining respectively 1000, 5000,10000 nodes were set to verify the performances of the average response times of queries, load distribution and the effect of node join or leave.

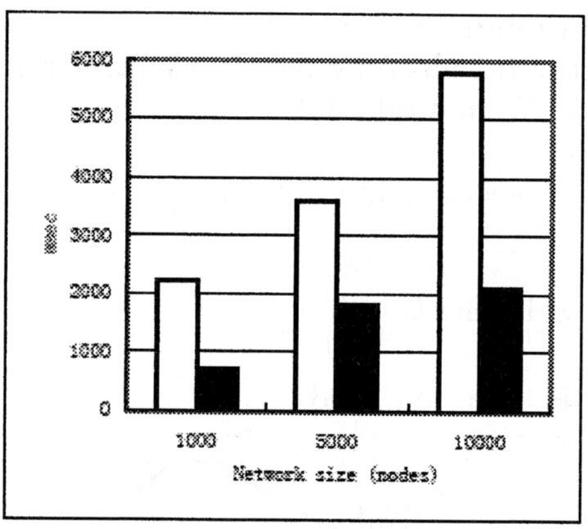

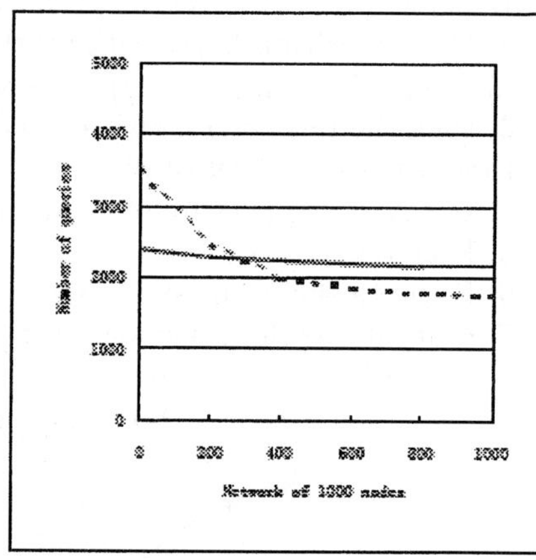

Figure 5.1: The average response time Figure 5.2: The load distribution

Query rate (per second)	Average number of Query failure	Average number of timeout
1000	0	0
2000	0	0
3000	0	1
5000	1	1
10000	1	2

Table 5.1: The effect of join or leave of nodes

Figure5.1 shows the average response times of our model using the improved algorithm (black histogram) and DHT alone, (blank histogram) without any provision for adapting to query workload. Obviously, the response times using our model are greatly reduced.

Figure5.2 shows the load distributions of queries across the network of 5000 nodes. The real line represents for the load distribution of our model and the broken line for DHT. The different nodes take charge almost the same number of queries in real line. This indicates our model can spread more fairly queries to different nodes than DHT and has better load balance.

In table5.1, we evaluate the scalability of the model. When nodes join or leave, queries at different rates across network of 10000 nodes and the number of query failure can be tested. From the table we note that there are few failures, timeouts may still occur during the query operation.

6 Conclusion and the Future Work

In this paper, we present the catalog search for XML data sources in Peer-to-Peer network. In essence, we adopt consistent hashing approach to map the keys of catalog information to the closest node. The research model is based on query processing for the large distributed XML repositories. Results from simulation experiment evaluate the good performance of the model. We need to develop the future research for join algorithms, as well as various query processing techniques.

References

1. L. Galanis, Y. Wang, S. R. Jeffery, D. J. DeWitt. Processing Queries in a Large Peer-to-Peer System, CAiSE 2003, Klagenfurt, Austria, June 2003, pp.273-288.
2. A. Crespo, H. Garcia-Molina. Routing Indices for distributed Systems. ICDCS 2002. Vienna, Austria. IEEE Computer Society p.23, July 2-5, 2002.
3. Gnutella, resources, http://gnutella.wego.com . Napster, http://www.napster.com
4. M. Harren, J. M. Hellerstein, R. Huebsch, B. T. Loo, S. Shenker, I. Stoica. Complex Queries in DHT-based Peer-to-Peer Networks. IPTPS '02, Cambridge, MA, USA, March 2002.
5. I. Stoica, R. Morris, D. Karger, M.F. Kaashoek, H. Balakrishnan. Chord: A Scalable Peer-to-Peer Lookup Service for Internet Applications. In Proc. SIGCOMM 2001, USA, August 2001, pp. 149-160.
6. Karger, D. Lehman, E. Leighton, F. Levine, M. Lewin, D. and Panigrahy, R. Consistent hashing and random trees: Distributed caching protocols for relieving hot spots on the World Wide Web. In Proceedings of the 29th Annual ACM Symposium on Theory of Computing, TX, USA, May 1997, pp. 654-663.
7. B.Yang, H.Garcia-Molina. Designing a Super-Peer Network, In Proc. ICDE 2003.
8. A. Rowstron, P. Druschel, Pastry. Scalable, distributed object location and routing for large-scale peer-to-peer systems. IFIP/ACM Intl. Conference on Distributed Systems Platforms. Germany, November 2001, PP. 329-350.

9. Quan Zhong Li, BongKi Moon. Indexing and Querying XML Data for Regular Path Expressions. Proceedings of the 27th VLDB Conference, Roma, Italy, September 2001, pp. 361-370
10. V. Papadimos, D. Maier, K. Tufte. Distributed Query Processing and Catalogs for Peer-to-Peer Systems . CIDR 2003, CA, USA, January , 2003.
11. XML path language (Xpath) 2.0 $http://www.w3.org/TR/Xpath20/$

Modeling and Analysis of Impatient Packets with Hard Delay Bound in Contention Based Multi-access Environments for Real Time Communication

Il-Hwan Kim[1], Kyung-Ho Sohn[2], Young Yong Kim[2], and Keum-Chan Whang[2]

[1] LG R&D Complex 533, Hogye-1dong, Dongan-gu, Anyang-shi, Kyongki-do 431-749, Korea
ilhkim@lge.com

[2] Department of Electrical and Electronic Engineering, Yonsei University, Shinchon-Dong, Seodaemoon-Ku, Seoul 120-749, Korea
{heroson7, y2k, kcwhang}@yonsei.ac.kr

Abstract. In this paper, we study problem of impatient packets in multimedia multi-access communication channel. Impatient users generally mean the users who leave the system if their service is not started before their respective deadlines. Their characteristic is perfect match to that of the packets with hard delay bound in real-time communication. In real-time communication, one of the most critical performance measure is the percentage of packets that are transmitted within hard delay bound. We assume contention based reservation ALOHA type MAC protocol, which is basic framework in next generation multimedia wireless LAN, and develop analytical model for the performance evaluation of defection probability of impatient packets. Our results show that proposed model is strong tool to evaluate packet loss probability due to expiration of deadline in contention based MAC protocols, which matches well with the simulation results.

1 Introduction

We consider a problem of impatient packets (customers) in multi-access communication channel. Impatient users mean the users who leave the system if their service is not started before their respective deadlines. The performance of queuing system is studied and evaluated in terms of defection probability of them, where defection probability is defined as portions of packets who leave the system before their service begin due to impatience [1][2]. One application of this problem is the transmission of packets with hard delay bound over shared wireless channel in real-time communication system. In the real-time communication system, timing constraints are one of the most important characteristics. Real-time packet, which is not transmitted within the specific deadline, is useless for both sender and receiver. Its characteristic is perfect match to the behavior of impatient customers.

C. Kim (Ed.): ICOIN 2005, LNCS 3391, pp. 609–618, 2005.

For the last few decades, numerous multi-access protocols have been proposed and developed for distributed users to efficiently share the single multi-access communication channel. Initially, many research efforts were mainly concentrated on data service in LAN's or satellite communication network. They may be classified into several categories according to their rule and function. [3][4] (i.e. fixed assignment, demand assignment, random access, and reservation). Primary performance objectives of these protocols have been high channel throughput, low average access delay. However, in recent integrated service environment, multi-access protocols that can support real-time applications have been studied and their performances have been evaluated in terms of tail distribution of delay instead of throughput or average delay[5][6].

In real time communication environment, the distribution of access delay rather than average access delay is what is important. For this reason, previously proposed multi-access protocols without respect to this characteristic may not be particularly suited for real-time communication, but have been partially used for transmission of real-time messages so far. As an example, we generally transmit real-time messages under IEEE 802.11 DCF mode despite IEEE 802.11 PCF mode is suitable for real-time service, which is not implemented in practice. Therefore, the performance measuring tool of real time communication in reservation protocols should be strongly required in current situation.

In this paper, we analyze the performance of reservation Aloha protocol, which is foundation for many contention based WLAN or cellular multi-access protocol, with impatient customer model. Our main focus here is to study the impatient packets' behavior induced by delay from contentions in a reservation sub-frame. In other words, the performance of contention scheme in the reservation protocol for real-time system is analyzed. This paper is organized as follow. Section II contains the system model. In section III, the system model is modified for simplicity of analysis and then is analyzed using impatient customer model. Section IV validates the accuracy of analysis by comparing the results of analysis and simulation. Conclusions are presented in section V.

2 System Model

2.1 Reservation Based Aloha Protocol

In the Aloha-reservation channel, we assume a synchronized system structure. Time is divided into fixed-length slots. Duration of a slot is identical with the transmission time of a packet. Users will thus start transmissions of messages only at times coinciding with starting times of the synchronized time slots. Moreover, a fixed-frame structure is considered. As shown in Fig. 1, the slots are organized into frames with F slots in Aloha-reservation channel. Each frame consists of a reservation sub-frame with K slots and a data sub-frame with L slots. A reservation slot is again divided into V small slots. In a reservation period, users contend on the KV mini-slots in a slotted ALOHA mode. Users who succeed in making reservation can transmit exclusively packets for allocated slots of a data sub-frame. In this case, we assume one common queue for all users that the queue discipline is FCFS according to the order in which reservation requests are received. Then a user whose reservation packet is successfully transmitted enters instantly a common queue. The user transmits the packet for allocated slots of a data sub-frame by the queue discipline of system.

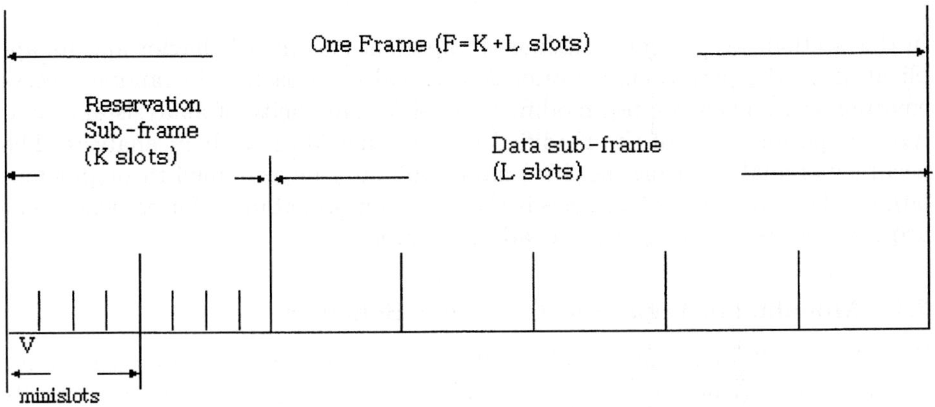

Fig. 1. Fame structure of Aloha-reservation channel

2.2 Model Formulation

We consider infinite population. Users arrive at a system according to a Poisson process with rate λ. Each user has single packet with limited deadline γ. If the transmission of a packet is not started before his deadline runs out, it leaves the system. In this paper, we study for the case that the deadline of all users is absolutely identical.

Packets that arrive in the system are required to make reservations for the transmission in a reservation period. They will send a reservation packet containing information about its identity. This packet is shorter than a regular data packet.

For the reservation packet transmission procedure, the following contention-based reservation protocol, namely slotted ALOHA, is employed. Users can transmit a reservation packet at random times within certain reservation periods. In other words, they transmit a reservation packet in a randomly selected one of KV mini-slots in the reservation sub-frame. Each reservation packet occupies a mini-slot. If two or more reservation packets are transmitted at the same mini slot, the reservation packets are collided with each other. Then, user who finds the failure of transmission of his reservation packet decides whether it retransmits or not with probability β in the next reservation period.

The channel is assumed to be error-free except for collisions. The round-trip propagation delay is not considered. However, we assume that users arriving within a reservation period must wait until the next reservation period to transmit their reservation packets.

3 Modeling of Packet Loss Using Impatient Customer Model

In this section, we present modeling of impatient customer behavior and its application to the performance evaluation of real time packets in random access environments. Then, we use modified model for simplicity of analysis and analyze the performance of the modified model using Markov chain analysis. The accuracy of analysis using modified model will later be examined through simulation. The main focus of analysis is the defection probability of users who don't acquire reservation before their deadline expires.

3.1 Modeling of Impatient Customer Behavior

From Boxma[2], some exact result for the $M/G/m{+}G$ queue has been derived for the case of exponential services. It should be noted that hardly any exact results are known even for $M/G/m$ queue, therefore it is not easy to get exact solution for general $M/G/m{+}G$ queue. Therefore we started with $M/M/m{+}G$ queue. Let's consider the Markov process $\{(N(t), \eta(t)), t \geq 0\}$ for the M/M/m+G queue, here

$N(t) = n$ when the number of customers at time t equals n and $0 \leq n \leq m{-}1$:

$N(t) = L$ when the number of customers at time t exceeds m-1:

$\eta(t)$ is the time that a customer with infinite patience would have to wait for service. It is strictly positive when $N(t) = L$, and it equals zero otherwise. Define in the steady-state situation, which exists iff $\lambda\overline{F}(\infty) < m/\beta$

$$P_j := \lim_{t->\infty} \Pr\{N(t) = j,\, \eta(t) = 0\}, \quad j = 0, ..., m-1$$

$$v(x) := \lim_{t\to\infty} \lim_{dx\to 0} \Pr\{N(t) = L,\, x < \eta(t) \leq x + dx\}/dx$$

From the Chapman-Kolmogorov equations for $P_j,\ j = 0, ..., m-1$, it follows, with offered traffic load $\rho := \lambda\beta$

$$P_j := \frac{\rho^j}{j!} P_0, \quad j = 0, ..., m-1$$

$$v(0) = \lambda P_{m-1}$$

$$v(x) = v(0) \exp[\lambda \int_{u=0}^{\infty} \overline{F}(u) du - mx/\beta], \quad x > 0$$

The normalizing condition $\sum_{j=0}^{m-1} P_j + \int_0^{\infty} v(x) dx = 1$ yields,

$$P_0 := [1 + \rho + \frac{\rho^2}{2!} + ... + \frac{\rho^{m-1}}{(m-1)!}(1 + \lambda J)]^{-1}$$

where

$$J := \int_0^\infty \exp[\lambda \int_0^\infty \overline{F}(u)du - mx/\beta]dx$$

The overflow probability π is given by

$$\pi = \int_0^\infty F(x)v(x)dx$$

hence

$$\pi = (1 - \frac{m}{\rho})(1 - \sum_0^{m-1} P_j) + P_{m-1}$$

Therefore if one can specify the delay distribution of some system, defection probability can be readily available with the defection probability.

3.2 Modification of Model

In the previous section, we assumed impatient users with limited deadline in multi-access communication system using a reservation protocol. Because the distribution of access delay affects mainly the performance of the system, it is of great importance for real time application. To find the distribution of access delay, we use the discrete Markov model with a finite state space. In this case, we have some problem in using the discrete Markov model owing to state vector that has many states. However, it is hard for us to use the analysis method which is shown in [7] as well as the simple Markov model with a single state space due to users who exist in the system until their deadline expire and contention of users in the reservation process.

To solve the problem and simplify the analysis, we adopt the following assumptions. First, the system model with batch arrivals is considered. Users arrive simultaneously at only the beginning of a frame in the system and can transmit their reservation packets during the reservation period of the frame. The number of users who arrive at a time is an i.i.d Poisson random variable with rate λ. Let d denote the number of reservation periods in which a user participates before their deadline expires. If not batch arrivals, new arrivals are uniformly distributed over any time frame and user who newly arrives will wait for the half frame on average until the first reservation period. Therefore, $d = (\gamma - T_{frame}/2)/T_{frame}$, where T_{frame} is the frame length. Second, the limited distribution of a Poisson random variable is considered and user's deadline is restricted to several times of frame length. These constraints shall later be applied to obtain numerical results.

3.3 State Probability

We defined that d denote the number of reservation periods in which a user participates before their deadline expires. Let n_k^i be a random variable representing the number of backlogged packets, which fail in making reservation during i reservation periods at the beginning of k th frame. Then, n_k^d, n_k^0denote the number of defections and arrivals. Now the state vector is defined as $\mathbf{N}_k(n_k^d, \cdots, n_k^0)$ which denotes the state of users at the beginning of k th frame. The state space consists of an infinite set of $(x_d, \cdots, x_0),\ \ 0 \leq x_i \leq \infty$.

Let us define $\mathbf{\Pi}$ be the steady-state probability vector where $\pi_\mathbf{x}$ is the steady-state probability of finding the system in state $\mathbf{N}_k(n_k^d, \cdots, n_k^0)$, defined as

$$\pi_\mathbf{x} = \lim_{k \to \infty} \Pr[\mathbf{N}_k(n_k^d = x_d, \cdots, n_k^0 = x_0)], \quad 0 \leq \forall x_i \leq \infty$$

Let $\mathbf{P}$ be the vector that represents the transition probability matrix of the state vector. An entry $p_{\mathbf{x},\mathbf{z}}$ of $\mathbf{P}$ is the one-step transition probability that there is state $(z_d, \cdots, z_0)$ at the beginning of the frame, given there was state $(x_d, \cdots, x_0)$ in the system at the beginning of the previous frame, derived as

$$p_{\mathbf{x},\mathbf{z}} = \Pr\{\mathbf{N}_{k+1}(n_{k+1}^d = z_d, \cdots, n_k^0 = z_0) | \mathbf{N}_k(n_k^d = x_d, \cdots, n_k^0 = x_0)\}$$

$$0 \leq z_d \leq x_{d-1},\ \ldots,\ 0 \leq z_1 \leq x_0,\ 0 \leq z_0 \leq \infty$$

Then, $\mathbf{P}$ is obtained by solving the set of equations.

$$\mathbf{\Pi} = \mathbf{\Pi} \cdot \mathbf{P}$$

$$\sum_{\mathbf{x}} \pi_\mathbf{x} = 1 \tag{1}$$

To calculate the one-step state transition probability, we shall first compute the distribution of the number of reservation success and the distribution of the number of backlogged users who determine to retransmit their reservation packet. The number of reservation success depends on the total number of users who participate in reservation process. Then, its distribution is the following.

P[probability that r out of n users have their reservations which don′t conflict with others in V minislots]

$$P\left[r|n, V\right] = \frac{(-1)^r V! n!}{r! V^n} \sum_{k=r}^{\min(V,n)} (-1)^k \frac{(V-k)^{n-k}}{(k-r)!(V-k)!(n-k)!}$$

$$P(t|n) = \binom{n}{t} \beta^t (1-\beta)^{n-t}$$

$$P(a) = \frac{(\lambda T_{frame})^a}{a!} e^{-\lambda T_{frame}} \tag{2}$$

Where $P(t|n)$ is the probability that out of backlogged users transmit their reservation packets and $P(a)$ is the probability that a users newly arrive at the system at the beginning of frame.

Let us define variables, which are used to compute the one-step transition probability

t_i The number of users who transmit the reservation packet among x_i
r_i The number of users who succeed in making reservation among x_i
t The total number of users who transmit their reservation packet
r The total number of users who succeed in making reservation
Then, it can be represented as follows.

$$r_i = x_i - z_{i+1}, \quad i = 0, \cdots, d-1$$

$$t = \sum_{i=0}^{d-1} t_i, \quad r = \sum_{i=0}^{d-1} r_i$$

Using the equation (2) and the above variable, we can derive the one-step transition probability $p_{\mathbf{x},\mathbf{z}}$. This probability is derived as

$$P_{\mathbf{x},\mathbf{z}} = \begin{cases} P(z_0) \times \sum\limits_{t_{d-1}=r_{d-1}}^{x_{d-1}} \cdots \sum\limits_{t_1=r_1}^{x_1} \prod\limits_{j=1}^{d-1} p[t_j|x_j] \times t_j\mathrm{C}r_j \times x_0\mathrm{C}r_0 \times P[r|t,V]/t\mathrm{C}r & \\ & \forall r_i > 0, 0 \le r \le V \\ 0 & otherwise \end{cases} \tag{3}$$

where $n\mathrm{Ck} = \binom{\mathrm{n}}{\mathrm{k}}$ is the number of cases that we select k out of n.

3.4 Defection Probability

Combining results of equation (1) and (3), we can obtain the steady state probability. Now we calculate the defection probability by making use of the steady state probability. The defection probability P_d is the probability that a user who newly arrives in the system fails in making the reservation until his deadline is time out. Let us define A_k be the number of new arrivals at k th frame and D_k be the number of users who don't acquire the reservation among A_k. The defection probability is derived the following.

$$P_d = \lim_{k \to \infty} \frac{\sum\limits_{i=0}^{k} D_i}{\sum\limits_{i=0}^{k} A_i} = \frac{E[D_k]}{E[A_k]} \tag{4}$$

$$E[D_k] = \sum_{x_d=0}^{\infty} \sum_{x_{d-1}=0}^{\infty} \cdots \sum_{x_1=0}^{\infty} \sum_{x_0=0}^{\infty} x_d \cdot \pi_{\mathbf{x}}$$

$$E[A_k] = \sum_{a=0}^{\infty} a \cdot p(a) = \lambda T_{frame}$$

where $p(a)$ is defined in (2).

4 Numerical Result

In this section we compare derived analytic results with simulation results. The system with $F = 5$, $K = 1$, $V = 4$, $L = 4$, $T_{frame} = 20$ msec is considered. To obtain numerical result, the limited distribution of an i.i.d Poisson random variable with rate λ is supposed and user's deadline is restricted to several times of frame length to reduce the size of state vector. The simulations were carried out in the model that defined in section 2.

Fig. 2, 3 show the defection probability vs. arrival rate and retransmission probability for analysis and simulation. $\beta = 0.5$ and $\lambda = 30.0$ users/sec is assumed in each Figure and $\gamma = 70$ msec in both of them. In Table.1, the defection probability according to deadline γ is presented, where $\beta = 0.5$ and $\lambda = 30.0$ users/sec. From Fig. 2, we can observe that as the arrival rate increases, the performance decreases because of collisions in fixed reservation mini-slots. Fig. 3 and Table.1 are shown that the performance is improved with the increase of retransmission probability β and deadline γ because the number of reservation periods in which user participates gradually increases. However, this improvement of performance will decrease along the increase of arrival rate because the probability of reservation success is lower as the number of users is increases more than a certain number. The comparison between analytical and simulated results is shown that the assumption and analysis is fairly reasonable. With wide range of parameters, the analytical results agree well with simulated results.

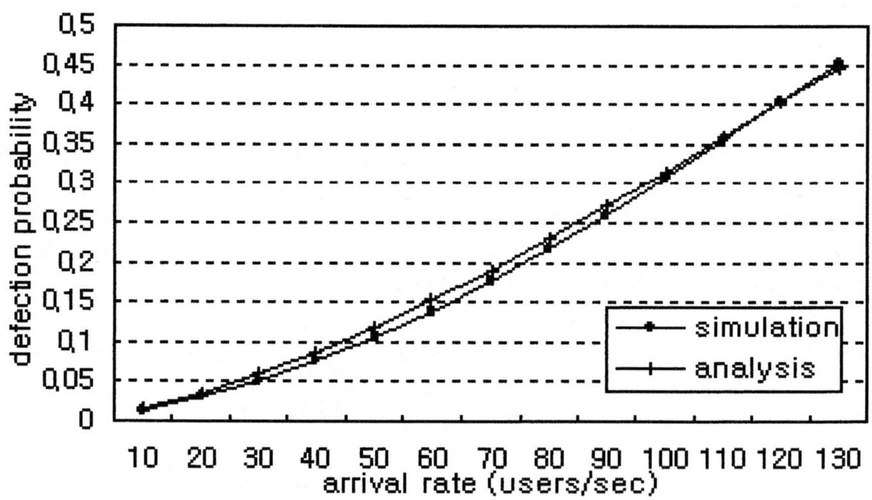

Fig. 2. Defection Probability versus Arrival Rate ($\beta = 0.5, \gamma = 70$ msec)

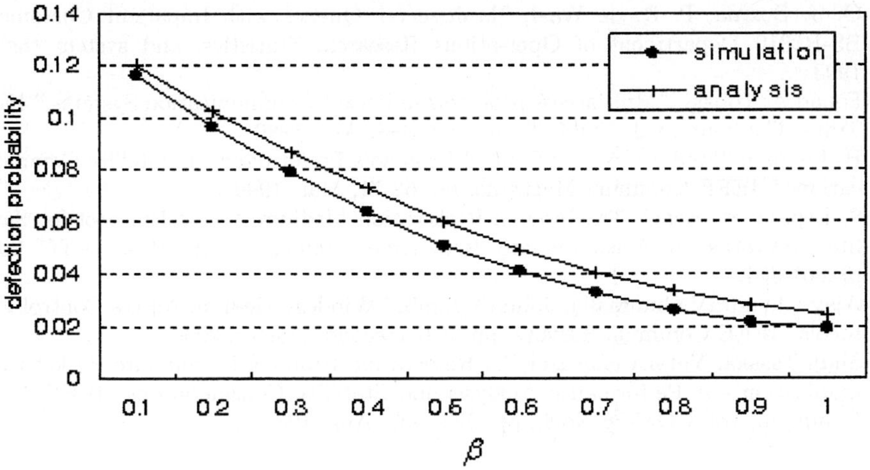

Fig. 3. Defection Probability versus β ($\lambda = 30.0$ users/sec, $\gamma = 70$ msec)

Table 1. Defection Probability versus Deadline γ ($\lambda = 30.0$ users/sec, $\beta = 0.5$)

Deadline γ(msec)	Defection Probability(%)	
	simulation	analysis
50	0.0788	0.0929
70	0.0513	0.0604
90	0.0329	0.0388
110	0.0209	0.0244
130	0.0133	0.0152

5 Conclusions

We studied a problem of impatient packets with hard delay bound in contention based multi-access communication channel. The model of Aloha-reservation protocol in real-time system was modified and analyzed in this paper. Simulations have been performed and compared with analytical results to verify the assumptions and approximations. Simulated results indicated that our approximation and analysis is fairly accurate in wide range of system parameter values. Future works may include extension of the model to multi-class traffic types.

References

1. F. Baccelli, G. Hebuterne, "On Queues with Impatient Customers", PERFORMANCE'81 , North-Holland Publishing Company, pp. 159- 179, 1981.

2. O. J. Boxma, P. R. de Waal, "Multiserver Queues with Impatient Customers" BS-R9319 Department of Oper-ations Research, Statistics, and system theory, 1993
3. Fouad A. Tobagi, "Multiaccess protocols in Packet Communication Systems," IEEE Trans. Commun., vol. COM-28, pp. 468- 488, Apr. 1980.
4. H. Peyravi, "Medium Access Control Protocols Performance in Satellite Communications," IEEE Commun. Magazine, pp. 62-71, Mar. 1999
5. S. Lepaja, K. Bengi, "A Random-Reservation Medium Access Protocol for Satellite Networks to Accommodate Real-Time Traffic," IEEE Proc. VTC 2001. pp.861-865.
6. Ajay ChandraV.Gummalla, John O. Limb, "Wireless Medium Access Control Protocols" IEEE Commun. Surveys, pp. 2-15, second quarter. 2000.
7. Shuji Tasaka, Yutaka Ishivashi, "A Reservation Protocol for Satellite Packet Communication - A Performance Analysis and Stability Considerations" IEEE Trans. Commun., vol. COM-36 No.8., pp. 920- 927, Aug. 1988.

Bidirectional FSL3/4 on NEDIA (Flow Separation by Layer 3/4 on Network Environment Using Dual IP Addresses)*

Kwang-Hee Lee and Hoon Choi

Department of Computer Engineering, Chungnam National University
220 Gung-dong, Daejeon 305-764, Korea
{khlee, hchoi}@ce.cnu.ac.kr

Abstract. A home network may have various equipment such as home appliances, PCs and other small electronic devices. Current trend is to make these devices have Internet connectivity. NAT (Network Address Translation) is a short-term solution of IP depletion problem and is widely used to construct home networks. Most of the existing NAT techniques support only unidirectional communication, i.e., connections from a local network to the public network. Though AVES (Address Virtualization Enabling Service) technique provides a bidirectional communication service to the NAT network, it is restrictive in a sense that it requires application level gateways which not only degrades performance but also is vulnerable in security. In this paper, we propose FSL3/4 aware DNS server for supporting bidirectional communication between NEDIA and external network. We also analyze the proposed method by comparing it with other methods which support bidirectional communication.

1 Introduction

NAT (Network Address Translation) method [1][2] is a short-term solution for the IP depletion problem. It is widely used for home networks or SOHO (Small Office Home Office) networks. Home or SOHO networks constructed by NAT technology share a public IP address for connecting Internet, and it supports initiation of unidirectional connections from local network to Internet. Recent deployment of new applications such as remote home control, virtual home, peer-to-peer services requires bidirectional connections between local network side and Internet side. SOHO network also requires bidirectional connections because many application servers such as WWW, FTP, SMTP are located inside of its network. However, home network or SOHO network configured by the NAT technology cannot support this requirement unless having a special ALG (Application Level Gateway) [3]. A well known NAT technology, AVES

* This research was supported by the program for training of Graduate Students in Regional Innovation which was conducted by the Ministry of Commerce, Industry and Energy of the Korean Government.

C. Kim (Ed.): ICOIN 2005, LNCS 3391, pp. 619–626, 2005.

(Address Virtualization Enabling Service) [4] supports a bidirectional communication service to the NAT network by using DNS_ALG (Domain Name System Application Level Gateway) [3]. However, AVES requires additional operation such as MTU (Maximum Transmission Unit) discovery and IP-in-IP encapsulation which causes frequent packet fragmentation.

In this paper, we expand FSL3/4 (Flow Separation by Layer 3/4) on NEDIA (Network Environment using Dual IP Addresses) as our previous study [5] to provide bidirectional communication capability between a private network and external public IP network, i.e. communication can be initiated by both sides. We describe the mechanism in detail and show the performance of the proposed method by comparing it with other methods that support bidirectional communication. Bidirectional FSL3/4 on NEDIA has many advantages than basic NAT with DNS_ALG, NAPT (Network Address Port Translation) with port forwarding or AVES.

2 Related Works

2.1 Bidirectional NAT

Basic NAT supports bidirectional communication between local network and global network by using FQDN (Fully Qualified Domain Name) and DNS_ALG [3]. However, it shows poor IP address reusability because one public IP address is dedicated to one private IP address while connecting a local host to a global host. Therefore, the number of incoming communication session is limited to the number of public IP addresses assigned to the NAT router. **NAPT(Network Address Port Translation)** cannot support bidirectional communication by using FQDN because it uses TCP/UDP port as a demultiplexing key and DNS Query packet does not contain TCP/UDP port. NAPT uses Port Forwarding Table for supporting incoming data flow which is initiated from an external host. Port Forwarding Table is a preconfigured table for incoming connection which is destined to a local Internet server such as WWW and FTP servers. It consists of a port number and a private IP address. However, if there are more than one Internet server of the same service in the local network, NAPT cannot support bidirectional communication because the same port number will be used for those local Internet servers.

2.2 AVES

AVES (Address Virtualization Enabling Service) supports a bidirectional communication service to NAT network [4]. It is composed of an AVES-aware DNS, waypoints and AVES-aware NAT daemons. The key idea behind AVES is to virtualize non-IP hosts such as private IP hosts, IPv6 hosts by a set of IP addresses assigned to the waypoints. The waypoint act as a relay to connect IP hosts to non-IP hosts. To relay packets from an external IP host to a local non-IP host, AVES performs translation of packet's destination address and encapsulates the packet for passing through IP tunnel which is required for routing between AVES network and NAT network.

2.3 FSL3/4 on NEDIA

FSL3/4 (Flow Separation by Layer 3/4) on NEDIA (Network Environment using Dual IP Addresses) is the new public IP address sharing technique presented by the authors [5]. It distinguishes data flow by only referring to packet's L3/L4 information such as protocol ID, source MAC (Media Access Control) address, destination address, source port, and destination port without any modification of these information. It also does not need ALG processing, thus it leaves application layer payload transparent. FSL3/4 on NEDIA also performs L2 forwarding for transmitting incoming packet to NEDIA host. L2 forwarding is to transmit packet using the MAC address of the destination host without L3 routing process.

3 Bidirectional FSL3/4 on NEDIA

3.1 Method

To support bidirectional communication in FSL3/4 on NEDIA, we propose a FSL3/4 aware DNS server module. FSL3/4 aware DNS server is built in FSL3/4 router. It may be a plain DNS server such as BIND [6] with a new set of API added to the DNS module for the communication with FSL3/4 module. FSL3/4 aware DNS server performs general DNS server's functionalities except the zone transfer with master DNS server. It acts as a local DNS server for NEDIA hosts. This DNS server communicates with FSL3/4 router by calling API only when DNS query arrives from external DNS server to get global IP address which is assigned to FSL3/4 router's external interface to make a DNS response packet.

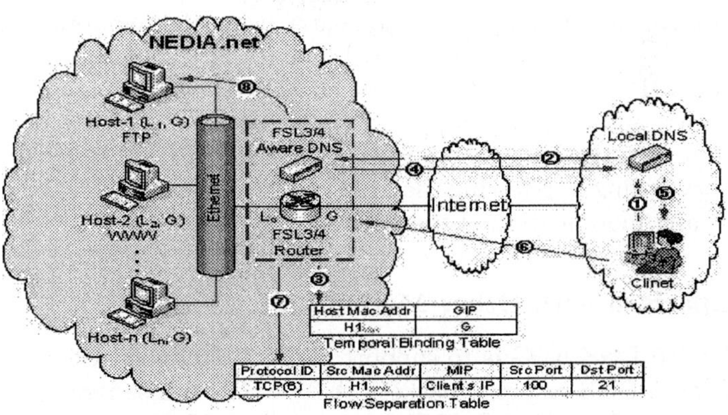

Fig. 1. Bidirectional FSL3/4 on NEDIA

Figure 1 shows the procedure. Suppose that a client wants to ftp with a server in NEDIA and the host's FQDN (Fully Qualified Domain Name) is **ftp.nedia.net**.

1. An external client requests DNS query for resolving FQDN "ftp.nedia.net" to its local DNS server.
2. Local DNS server performs general FQDN resolving process and gets IP address of FSL3/4 aware DNS server. Local DNS server sends DNS query for "ftp.nedia.net" to FSL3/4 aware DNS server.
3. To response DNS query from the external DNS server, FSL3/4 aware DNS server searches its database such as a zone file and obtains the private IP address "L1" of FQDN "ftp.nedia.net". FSL3/4 aware DNS server requests the global IP address to FSL3/4 data flow separation module for making a response RR (Resource Record) in DNS. FSL3/4 data flow separation module performs ARP (Address Resolution Protocol) process for the finding host MAC address of the private IP address "L1" and creates a temporal binding entry with timeout 2 seconds. The temporal binding entry consists of host MAC address and GIP (Global IP address which is assigned to FSL3/4 router's external interface). FSL3/4 data flow separation module responses GIP "G" to FSL3/4 aware DNS server.
4. FSL3/4 aware DNS server creates a response RR in which RDATA field is filled with "G" and TTL (Time To Live) field with "0". It transmits the response RR to the external DNS server which had sent DNS query. To prevent caching of the response RR in external DNS server, TTL field in RR must be set to zero.
5. When the local DNS server receives the response RR from the FSL3/4 aware DNS server, it processes the response RR and transmits IP address of FQDN "ftp.nedia.com" to the client application.
6. The client's ftp application tries to connect the FTP server in NEDIA of which IP address is known as "G" by DNS query.
7. Connection request from the external client arrives at FSL3/4 router. FSL3/4 router records a data flow separation entry using the temporal binding entry and packet's header from the client. The data flow separation entry consists of protocol ID, source MAC address, client's IP address as multiplexing IP, source port, and destination port.
8. FSL3/4 router transmits the packet to the destined FTP server in NEDIA by L2 forwarding in referring the MAC address of the FTP server in NEDIA without any modification of the packet.

3.2 Comparison with Other Methods

Basic NAT with DNS_ALG is to support bidirectional connection using FQDN and it modifies a DNS response packet's A type RR. It also requires DNS server in the local network and DNS_ALG module in the NAT router. DNS server in the local network is a plain DNS server and performs general DNS functionality. DNS_ALG module running on NAT router replaces a private IP address with a public IP address of A type RR's RDATA field in DNS. However, this technique has poor IP address reusability because one public IP address is dedicated to one private IP address. Therefore, the number of incoming communication sessions is limited by the number of public IP addresses assigned to NAT router.

Because DNS_ALG modifies the IP address in the DNS packet, Basic NAT with DNS_ALG cannot support DNSSEC (DNS Security Extensions) between a local DNS server and a master DNS server.

NAPT with port forwarding uses preconfigured port forwarding table, which is configured by network administrator, for routing externally initiated connections. This table consists of private IP address and TCP/UDP service port. If there are multiple application servers using the same well-known service port, NAPT with port forwarding cannot distinguish incoming connections with the same well-known service port. Therefore, NAPT with port forwarding is not flexible.

AVES uses FQDN for identifying non-IP hosts and supports a bidirectional communication service to the NAT network which cannot support bidirectional communication without DNS_ALG. To support incoming connection to the NAT network, AVES must have information about NAT network such as local host's private IP address, NAT router's public IP address, etc. This requirement reduces privacy and autonomy of NAT network. AVES performs translation of packet's destination address and packet encapsulation for passing through IP tunnel between AVES network and NAT network. Therefore, AVES requires additional operation such as MTU discovery and IP encapsulation which causes frequent packet fragmentation. Because AVES acts as a relay for only incoming connection which is initiated from an external IP host to NAT network, triangle routing problem occurs between a local non-IP host and an external IP host.

On the other hand, Bidirectional FSL3/4 on NEDIA supports N:N connection using just one public IP address and does not modify the DNS packet. These characteristics enable DNSSEC session between FSL3/4 aware DNS server and master DNS server if it is needed. It preserves privacy and autonomy of private network and supports transport-mode IPSec [7] without any ALG in both outgoing and incoming directions.

4 Implementation and Performance Evaluation

To implement a bidirectional FSL3/4 router, we simply installed FSL3/4 aware DNS server into the FSL3/4 router and added two APIs; request_GIP() and response_GIP().

To compare the performance of bidirectional FSL3/4 on NEDIA with AVES, NAPT with port forwarding, and normal routing, we used an experimental environment shown in Figure 2. The experiment measured the RTT (Round Trip Time) using ping program, the average transfer bandwidth and file transfer time using ftp program. The measurement of RTT is to show how long it takes per packet by each method supporting bidirectional communication. File transfer is to show the packet forwarding performance of each method.

Figure 3 shows the result of measuring average RTT by AVES, bidirectional FSL3/4 on NEDIA and normal routing method. This experimentation shows average end-to-end delay which is the total elapsed time after sending a packet till receiving for the packet. It also includes the delay traversed the methods for

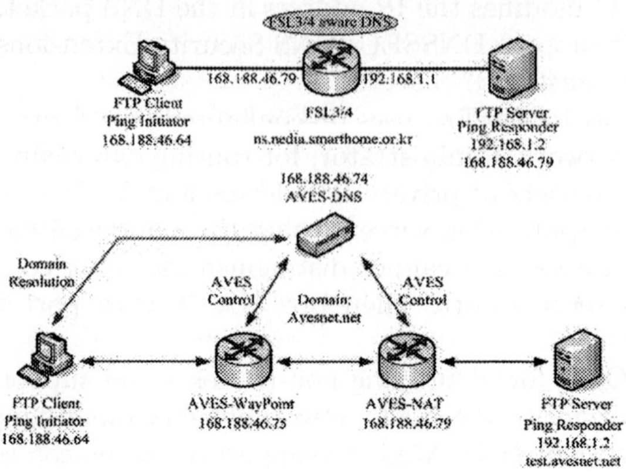

Fig. 2. Experimental Environment for Evaluating the Performance

supporting bidirectional communication. We did not carry out the experiment for NAPT with port forwarding because NAPT with port forwarding does not accept externally initiated ICMP echo messages [8].

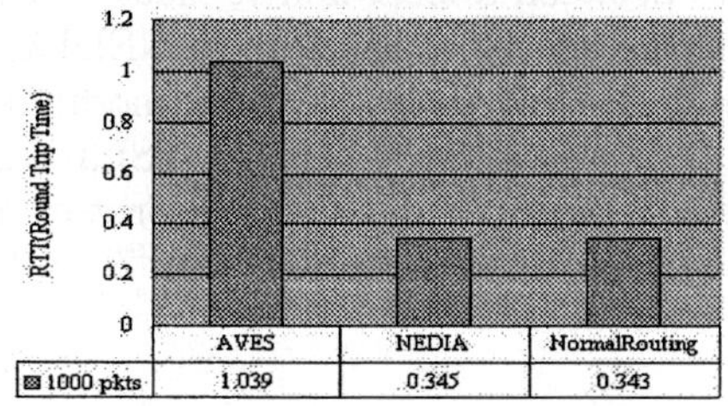

Fig. 3. Round Trip Time

In Figure 3, AVES gives the longest average RTT than other methods because a packet destined to a local host must be modified at AVES-Waypoint for setting proper destination IP address (private IP address of destination local host) in ICMP message and encapsulating for routing between AVES-Waypoint and AVES-NAT router. The modification of packet requires additional processing delay such as checksum operation. The encapsulation of packet also requires additional packet processing delay. The average RTT of AVES consists of the delay of resolving FQDN, the delay of traversing AVES network, the delay of creating a response ICMP echo message at a local host, and the delay of going through NAT router. The delay of resolving FQDN is composed of AVES session setup time for routing a packet destined to NAT network and ordinary domain name resolution time. On the other hand, bidirectional FSL3/4 on NEDIA is simple compared with AVES. The average RTT of the proposed method consists of the delay of resolving FQDN, the delay of going through FSL3/4 router both incoming and outgoing directions and the delay of creating a response ICMP echo message at a local host. The delay of resolving FQDN in this case is ordinary domain name resolution time. Bidirectional FSL3/4 on NEDIA performs L2 forwarding rather than L3 routing process when the FSL3/4 router forwards incoming packets to a local host. This characteristic enables the bidirectional FSL3/4 on NEDIA to have almost the same routing performance as the normal routing case. By normal routing, we mean the routing in the public IP address domain which does not require NAT processing. From the result of measurement of RTT, we confirmed that our method has only one third RTT compared with AVES and almost the same RTT compared with normal routing. We also could find the negative effect of both the modification of packet and the encapsulation of packet that appears in AVES method.

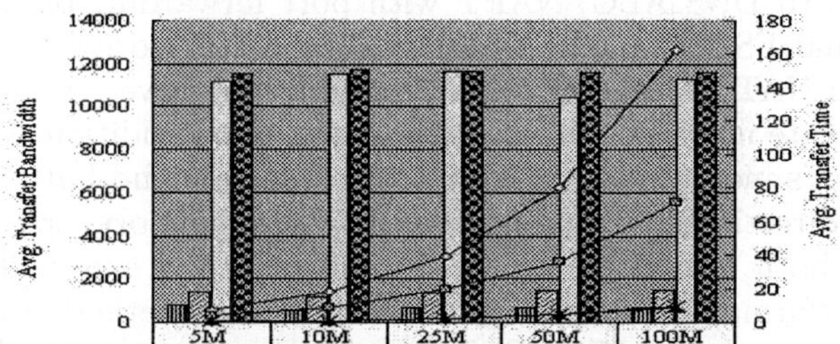

	5M	10M	25M	50M	100M
AVES	769.65	573.59	671.63	652.18	643.18
NAPT	1427.8	1187.65	1336.79	1437.63	1458.89
NEDIA	11202.74	11573.69	11650.84	10452.31	11296.88
Normal Routing	11573.69	11768.53	11650.84	11609.57	11610.85
—◇— AVES	6.81	18.28	39.03	80.39	163.03
—□— NAPT	3.67	8.83	19.61	36.47	71.88
—✱— NEDIA	0.47	0.91	2.27	5.02	9.28
—▲— Normal Routing	0.45	0.89	2.25	4.52	9.03

Fig. 4. File Transfer Time and Bandwidth

Figure 4 shows the packet forwarding performance of each method for transferring a data file to the local FTP server with respect to various file size from 5Mbyte to 100Mbyte. In this experimentation, AVES has the longest transfer time and the lowest transfer bandwidth compared with other methods. This is because AVES performs the modification of all packets destined to NAT network and packet encapsulation using IP in IP, for instance. As increasing file size, AVES shows an exponential increase of the average transfer time compared with bidirectional FSL3/4 on NEDIA. NAPT has smaller average transfer time and higher average transfer bandwidth compared with AVES, but it also has an exponential increase of the average transfer time compared with Bidirectional FSL3/4 on NEDIA because NAPT performs packet modification to route packets. Bidirectional FSL3/4 on NEDIA does not need any packet modification to route packets, and it performs L2 forwarding to route incoming packets destined to a local host which takes shorter time than L3 routing. These important characteristics enable it to have powerful forwarding performance and to support transport mode IPSEC session in both directions. Based on the above experiments, we could conclude that our method is simpler and has superior routing performance than AVES and NAPT with port forwarding to support bidirectional communication for small private networks.

5 Conclusions and Remarks

In this paper, we proposed an IP address sharing mechanism based on the FSL3/4 aware DNS server to support bidirectional communication. The mechanism of Bidirectional FSL3/4 on NEDIA can be simply implemented by installing FSL3/4 aware DNS server into the FSL3/4 router. FSL3/4 aware DNS server can be implemented by just adding two APIs to a free DNS server source such as BIND. Bidirectional FSL3/4 on NEDIA has many advantages than basic NAT with DNS_ALG, NAPT with port forwarding or AVES. First of all, Bidirectional FSL3/4 on NEDIA has a great IP address reusability compared with Basic NAT with DNS_ALG. Second, it preserves the privacy and autonomy of private network. Third, it does not require additional processing such as DNS packet's modification in NAT or IP tunneling and Path MTU Discovery to provide bidirectional communication. Forth, it supports transport mode IPSec session in both directions, i.e. locally initiated and externally initiated. Fifth, routing performance is superior to other popular methods such as NAPT with port forwarding and AVES. One shortcoming of our method may be scalability. While one AVES system can support many NAT networks simultaneously, our system is dedicated for one private network. This restriction enables a private network to preserve privacy and autonomy of the private network on the other hand. Overall, Bidirectional FSL3/4 on NEDIA is believed more efficient technology than NAT with DNS_ALG, NAPT with Port Forwarding or AVES.

References

1. P. Srisuresh and M. Holredge, "IP Network Translator (NAT) Terminology and Consid-erations," RFC2663, IETF, August 1999.
2. P. Srisuresh and K. Egevang, "Traditional IP Network Address Translation (Traditional NAT)," RFC3022, IETF, January 2001.
3. P P. Srisuresh, G. Tsirtsis, P. Akkiraju and A. Heffernan, "DNS extensions to Network Address Translators (DNS_ALG)," RFC2694, IETF, September 1999.
4. T. S. Eugene Ng, Ion Stoica, Hui Zhang, "A Waypoint Service Approach to Connect Heterogeneous Internet Address Spaces," USENIX Annual Technical Conference 2001, Boston, MA, June 2001.
5. Kwang-Hee Lee, Hoon Choi, "FSL3/4 on NEDIA (Flow Separation by Layer 3/4 on Network Environment using Dual IP Addresses," LNCS 3090, pp. 1015-1024, 2004.
6. Internet Systems Consortium, "BIND (Berkeley Internet Name Domain)," http://www.isc.org.
7. S. Kent and R. Atkinson, "Security Architecture for the Internet Protocol," RFC2401, IETF, November 1998.
8. J. Postel, "Internet Control Message Protocol," RFC 792, September 1981.

A Packet-Loss Recovery Scheme Based on the Gap Statistics*

Hyungkeun Lee and Hyukjoon Lee

Kwangwoon University, Seoul, Korea
{hklee, hlee}@daisy.kw.ac.kr

Abstract. Packet losses are bursty in nature since the dominant reason is temporary overload situations in the shared resources over networks. To increase the effectiveness of forward error correction (FEC) schemes, adaptive FEC schemes have been suggested, where the amount of redundancy is controlled according to current network status. In this paper, we propose a gap- based adaptive FEC scheme, which is motivated by the Markov gap model for packet losses, and show that the gap-based adaptive FEC scheme performs better than the adaptive FEC scheme based on the Gilbert model.

1 Introduction

In packet-switching networks such as the Internet, most packet losses occur during temporary overload situations. Packet losses are bursty in nature due to the limited resources at the intermediate nodes in such networks [1]. They introduce significant degradation in QoS of various services such as multimedia applications that have time constraints.

To recover from packet losses, there are two basic mechanisms available: automatic repeat request (ARQ) in which lost packets are retransmitted, and packet-level forward error correction (FEC) in which redundant packets along with the original data packets are transmitted [2]. Packet-level FEC is more appropriate for multimedia applications than ARQ, and the design of efficient schemes for packet loss recovery using packet-level FEC has been investigated recently [3][4]. The capacity of FEC to recover from packet losses highly depends on the packet-loss behavior, and FEC schemes are more effective when packet losses are not bursty. Transmission of additional redundant packets increases the probability of recovering lost packets, but it also increases the bandwidth requirements. Furthermore, the appropriate amount of redundancy of FEC is hard to decide because of packet-loss burstiness.

To increase the effectiveness of FEC schemes, adaptive FEC schemes have been investigated [4][5][6], where the current network status is monitored to control the FEC mechanism. It minimizes transmission overhead in the case of low packet losses, and increases the possibility to recover lost packets in the

* This work was supported by Grant No. R01-2001-00349 from the Korea Science & Engineering Foundation and the Research Grant of Kwangwoon University in 2004.

C. Kim (Ed.): ICOIN 2005, LNCS 3391, pp. 627–634, 2005.

presence of high network congestion. This approach is based on the assumption that packet losses are bursty and the bursts last for a long enough period. The packet-loss traces in [1] show that the above assumption is usually valid in most networks. Understanding the packet loss behavior is crucial for the proper design of adaptive FEC to recover packet losses since a strong correlation between consecutive packet losses causes the degradation in performance of FEC schemes. A more accurate model for packet losses is required, which will allow the design of efficient communication framework and better quality of service in different applications.

In this paper, we propose a gap-based adaptive FEC scheme, which is motivated by the Markov gap model for packet losses. The Markov gap model has been shown to be more accurate compared to the Markov chain models. Performance evaluation is carried out by the traces experimentally obtained over actual networks. The results show that the gap-based adaptive FEC scheme performs better than the static FEC scheme and the Gilbert-based adaptive FEC in terms of packet loss rate and overhead. The remainder of this paper is organized as follows. Packet-loss models are briefly described in Section 2. Section 3 discusses the Gilbert-based adaptive FEC algorithm and proposes the gap-based adaptive FEC algorithm. The performance evaluation is presented in Section 4. Finally, concluding remarks are made in Section 5.

2 Packet Loss Modeling

The objective of packet loss modeling is to characterize the probabilistic behavior of packet losses. The packet loss process is represented as a binary sequence $\{L_i\}$ where $L_i = 1$ if the i-th packet is lost, and $L_i = 0$ otherwise, as shown in Figure 1. The packet-loss sequences obtained from the records of actual packet transmission over networks are called traces. We can define two terms, gaps and clusters in the sequence $\{L_i\}$, where a gap is defined as the loss-free run between two losses and a cluster is defined as the loss run in a similar manner. The integer sequences $\{G_k\}$ and $\{C_k\}$ describes the consecutive gaps and clusters, respectively, in the sequence $\{L_i\}$, as shown in Figure 1, where numbers indicate the lengths of gaps or clusters.

The key task of modeling the packet loss process is to find statistical characteristics of these two alternating sequences, $\{G_k\}$ and $\{C_k\}$. In modeling the packet loss process, the correlation between neighboring gap lengths, or between neighboring cluster lengths in $\{L_i\}$ is significant, since the non-renewal property affects the packet loss behavior. In this context, this non-renewal property is important for the design and accurate characterization of FEC schemes.

2.1 Markov Chain Models

In order to build an accurate model for packet losses, a k-th order Markov chain model is widely used. The Bernoulli model and the Gilbert model are special cases of the k-th order Markov chain model with $k = 0$ and $k = 1$,

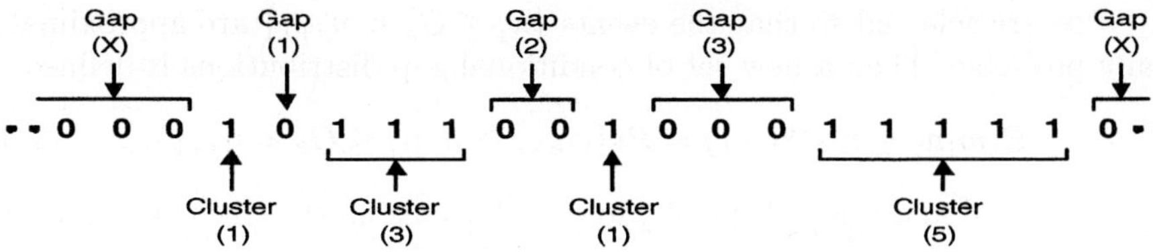

Fig. 1. An Example of Packet Loss Process

respectively. Since the complexity of the model increases significantly with the order, k, the first-order Markov chain model with two states such as the Gilbert model [7] is widely used. This model consists of one good state and one bad state with corresponding loss probabilities for the two states and P_G, P_B respectively, and transition probabilities p and q between them. While in the good state, transmissions are received incorrectly with probability P_G, and while in the bad state, transmissions are received incorrectly with probability P_B. For the Gilbert model for packet losses, it is common to assume that $P_G = 0$, $P_B = 1$ [6] and we also make the same assumption in this paper.

The average loss rate is the ratio of the number of lost packets to the total number of transmitted packets. The average burst length b is defined as the mean cluster length, $b = 1/q$. The steady-state average packet loss rate π is a function of the probabilities p and q,

$$\pi = \frac{p}{p+q}. \tag{1}$$

The Gilbert model considers a trace as being composed of alternating two states whose periods are geometrically distributed and independent of each other, i.e., a renewal process.

2.2 Markov Gap Model

The Markov gap model was proposed in [10] to represent the stochastic behavior of the packet losses. The unconditional gap distribution $P(m)$ is defined as the anticumulative first- order probability distribution of the sequence $\{G_k\}$,

$$P(m) = Pr\{G_k \geq m\}, \qquad m \geq 0 \tag{2}$$

with $P(0) = 1$. The conditional gap distribution $P(m|n)$ is then defined as the anticumulative conditional probability distribution of G_{k+1} given $G_k = n$, i.e.,

$$P(m|n) = Pr\{G_{k+1} \geq m|G_k = m\}, \quad m \geq 0 \text{ and } n \geq 0 \tag{3}$$

In order to reduce the number of such distributions, the gaps are grouped into r sets of gap lengths in the intervals $[n_j, n_{j+1})$ where the integers $\{n_j\}$ with $n_1 = 0$,

$n_{r+1} = \infty$ are selected so that the events $\{n_j \leq G_k < n_{j+1}\}$ are approximately equally probable. Then a new set of conditional gap distributions is defined as

$$P(m|n_j \leq n < n_{j+1}) = Pr\{G_{k+1} \geq m|n_j \leq G_k < n_{j+1}\}. \quad (4)$$

The conditional and unconditional gap distributions must satisfy the relation,

$$P(m) = \sum_{j=1}^{r} P(m|n_j \leq n < n_{j+1})[P(n_j) - P(n_{j+1})] \quad (5)$$

The Markov gap model assumes that the gap length sequence $\{G_k\}$ is a discrete- time, integer-valued Markov process of the first order with conditional probability distributions.

2.3 Model Construction and Evaluation

In [10], we construct the Markov gap model, the Bernoulli model and the Gilbert model based on actual collected trace sets and compare their accuracy, and it is shown that the renewal assumption is not valid for the packet loss process because the conditional gap distributions are different from unconditional gap distributions. It is also clearly seen that $P(m, n)$, the probability of m or more packet losses in a block of n consecutive packets, obtained from the Markov gap model is closer to that obtained directly from the data set than that obtained from the Gilbert model or the Bernoulli model.

3 Adaptive Error Correction Strategies

Packet-level FEC techniques are generally based on the use of error correcting codes such as erasure codes where additional redundant packets are transmitted for packet-loss recovery. In static FEC schemes, the sender adds a fixed number of redundant packets to original data packets to recover packet losses at the receiver. However, the optimum number of redundant packets is hard to find and most packet networks operate under dynamic conditions. Therefore, static FEC schemes might waste the bandwidth of networks under low loss conditions due to overprotection as well as perform worse during the congestion period due to a fixed capability of loss recovery. Adaptive FEC schemes generally increase the efficiency in terms of bandwidth by adapting their degree of redundancy according to network conditions.

Several adaptive FEC techniques have been proposed in [4][5] and [6]. It has been established that adaptive FEC schemes are able to perform better than static FEC schemes when the policy for redundancy control is properly selected.

3.1 Gilbert-Based Adaptive FEC

The redundancy control schemes based on the Gilbert model have been investigated under the unicast or multicast environment [5][6]. In these schemes, it is assumed that network behavior conforms to the Gilbert model and the current status can be predicted by the results of previous transmissions. The probabilities of receiving correctly at least m packets out of a block of n packets $D(m,k)$ can be obtained from [6]. Based on these probabilities of the constructed Gilbert model, an integer k where $D(m,k)$ is greater than a given threshold K_p is chosen as the size of a transmission block. The preset threshold K_p is the probability with which at least m packets are expected to arrive successfully.

This approach performs better with a relatively large window size. However, estimation with a large window lacks prompt responsiveness to dynamic changes in the network status, and the computational complexity increases as the size of the block increases since the computation of cumulative block loss probabilities has the complexity of the order of the block size n^2.

3.2 Gap-Based Adaptive FEC

Gap lengths in packet loss traces show the correlation between neighboring gap lengths as described in the previous section, which implies that the previous and current gap lengths might form good bases for the prediction of the next gap lengths. Therefore, an adaptive FEC scheme based on gap lengths can be an alternative approach for adaptive FEC. This approach is based on heuristic methods using the correlation of gap lengths and is expected to be more responsive to network load changes.

Gap-length information is fed back as a function of time. It generates a set $\mathbf{L} = \{l_0, \cdots, l_t\}$, where l_0 and l_i represent the current ongoing gap length and i-th past gap length, respectively. The integer t is the maximum number of gap lengths that are used in decision-making. Our goal is to estimate the current network condition and the next expected gap length, and control the degree of redundancy using them. We can compute the local mean μ_l and local standard deviation σ_l of gap lengths over $\mathbf{L}$ as well as the global mean μ_g and global standard deviation σ_g for the entire time period or for a very long window. The mean ratio and deviation ratio are defined as

$$R_m = \frac{\mu_l}{\mu_g}, \quad \text{and} \quad R_d = \frac{\sigma_l}{\sigma_g}. \tag{6}$$

Based on these statistics, it is possible to perform a hypothesis test to detect congestion periods during the transmission of packets. In this case, two thresholds K_m and K_d for the mean ratio and the deviation ratio, respectively, can be used in the hypothesis test as follows: Decide Congestion state, if $R_m \leq K_m$ and $R_d \leq K_d$, where K_m and K_d are the threshold values. This hypothesis test is used to decide whether or not the system is in the congestion state, while the decision of reverting back to the non-congestion state is based on the current gap length l_0 and the global mean μ_g. When the current gap length becomes larger than the global mean of gap lengths, the network is declared to be in the non-congestion state. Congestion detection algorithm is described in Figure 2.

Two different types of FEC schemes using gap lengths are utilized in this adaptive scheme, one is used in the congestion state and the other is used in the

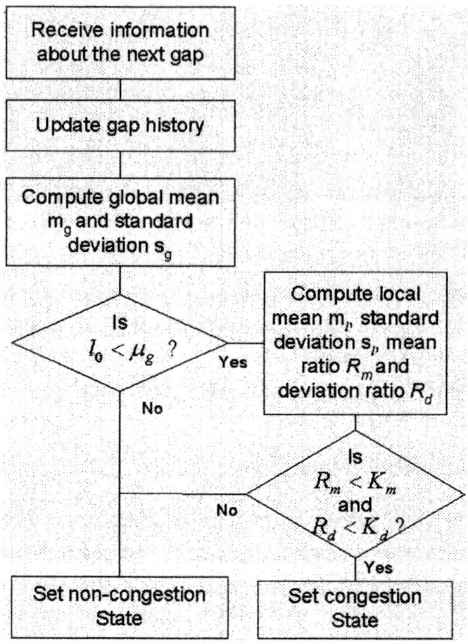

Fig. 2. Flow Chart of the Congestion Detection Algorithm

non-congestion state in networks. When congestion is detected, a set of weights is defined as $\mathbf{W} = \{w_0, \cdots, w_t\}$, where w_0 and w_i are the weights for the current gap length and i-th past gap length, respectively. The vector $\mathbf{W}$ is determined through the investigation of gap sequences where weights are determined to better predict the next gap lengths by employing a number of gap-length sequences during congestion periods. Then, the expected gap length is predicted as the following weighted average, $d = \mathbf{L}\mathbf{W}^T$, where d is the expected gap length. The size of the transmission block n is the number of packets which are encoded as a block, where $n \in \{k+1, \cdots, k+r\}$ and r is the maximum degree of redundancy. The value of n is determined as the smallest integer in $\{k+1, \cdots, k+r\}$ that satisfies

$$\left\lceil \frac{n}{d+1} \right\rceil \leq n - k. \tag{7}$$

This will enable us to recover up to $n-k$ packet losses in the next block consisting of n packets. This heuristic procedure is based on our prediction that at most $\left\lceil \frac{n}{d+1} \right\rceil$ packets will be lost in the next transmission block of length n.

When non-congestion state is detected, gap lengths tend to be large. Therefore, important gap lengths are limited to a few recent ones, since they provide enough information about current network conditions. If the current gap length l_0 is larger than or equal to the value of $k+r$, redundancy is decreased by one. Otherwise, the next gap length l_1 is also examined, and if l_0 and l_1 are smaller than the value of $k+r$, redundancy is increased by one.

4 Performance Evaluation

The task of evaluating adaptive error control is very challenging due to the fact that the environment is hard to analyze and the adaptive behavior of the system [11]. We evaluate the performance of adaptive FEC schemes using a trace-based evaluation methodology. Due to the use of the trace-base method for performance evaluation, it is not possible to fix one of the performance metrics for all the algorithm and then examine the other performance metric. For performance evaluation we used sets of traces collected over Mbone [1], and Figure 3 shows one of the results about the performance of adaptive FEC schemes. In order

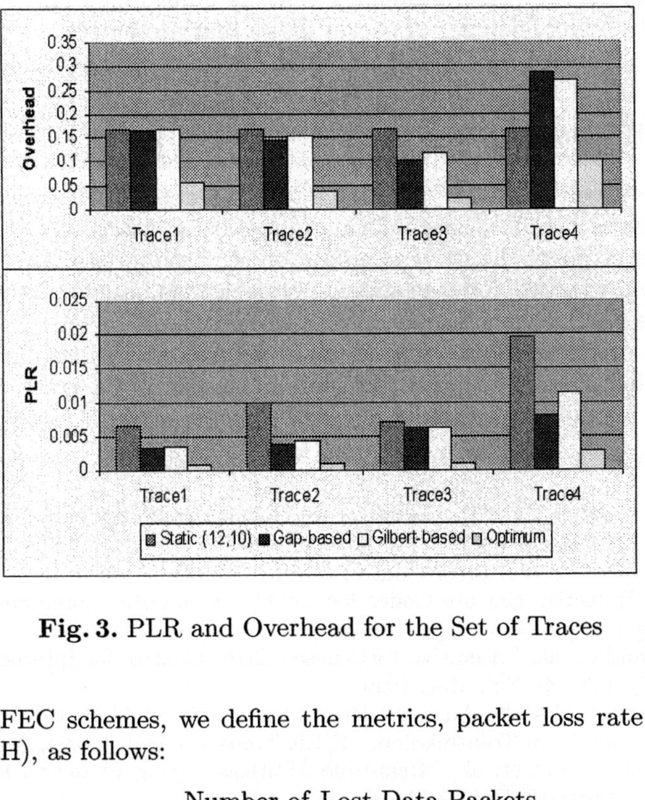

Fig. 3. PLR and Overhead for the Set of Traces

to evaluate FEC schemes, we define the metrics, packet loss rate (PLR) and overhead (OH), as follows:

$$\text{PLR} = \frac{\text{Number of Lost Data Packets}}{\text{Number of Lost Transmitted Packets}}, \tag{8}$$

and

$$\text{OH} = \frac{\text{Number of Tramsmitted Parity Packets}}{\text{Number of Transmitted Data and Parity Packets}}. \tag{9}$$

In this section, for convenience we will denote the static FEC scheme as SFEC, the Gap-based adaptive FEC scheme as Gap-AFEC and the Gilbert-based adaptive FEC scheme as Gil-AFEC. For the first trace in Figure 3, the PLR of Gap-AFEC is 52.03% less than the PLR of SFEC and 10.64% less than the PLR of Gil-AFEC. The overhead of Gap-AFEC is 0.6% and 1.78% less than the overheads of SFEC and Gil-AFEC, respectively. The second trace also shows that Gap-AFEC performs better than SFEC and Gil-AFEC in terms of both PLR and overhead. The results of experiments on the above two sets of traces indicate that Gap-AFEC recovers more lost packets using fewer redundant packets than SFEC and Gil-AFEC. Improvement for these two traces can be attributed to the fact that packet losses are highly correlated and, therefore, gap lengths are also highly correlated. For the fourth trace, the PLR of Gap-AFEC is 58.28% less than the PLR of SFEC at the expense of 71.93% more overhead, and 28.16% less than the PLR of Gil-AFEC at the expense of 6.15% more overhead. When we examine the results of the optimum scheme for this trace, we observe that the overhead for the optimum scheme is quite high. This indicates that this trace has a large number of packet losses that are bursty. Therefore, higher overhead for both Gil-AFEC and Gap-AFEC schemes is inevitable.

5 Conclusions

We have proposed a scheme to recover packet losses with the use of adaptive FEC techniques based on gap information. Gap statistics provide suitable information to estimate the network condition that changes temporally, and then the redundancy control for transmission of packets is based on the gap process. An adaptive FEC scheme we have proposed here, the Gap-based adaptive FEC with two modes of operation depending on network conditions, performs better than the Gilbert-based adaptive FEC scheme as well as the static FEC scheme.

References

1. M. Yajnik, and et. al., "Measurement and Modeling of the Temporal Dependence in Packet Loss," IEEE INFOCOM 99, Mar. 1999.
2. S. Lin and D. Costello, Error Control Coding, Prentice Hall, 1983.
3. L. Rizzo, "Effective Erasure Codes for Reliable Computer Communication Protocols," Comp. Comm. Rev., Apr. 1997.
4. J. Bolot, and et. al., "Adaptive FEC-based Error Control for Internet Telephony," IEEE INFOCOM 99, NY, Mar. 1999.
5. S. Yuk, and et. al., "An Adaptive Redundancy Control Method for Erasure-code-based Real-time Data Transmission," IEEE Trans. on Multimedia, Sep. 2001.
6. D. Rubenstein, and et. al., "Real-time Multicast Using Proactive Forward Error Correction," TR98-19, Univ. of Mass., Mar. 1998.
7. E. N. Gilbert, "Capacity of a Burst-Noise Channel," Bell Sys. Tech. J., Sep. 1960.
8. F. Swarts and H. C. Ferreira, "Markov Characterization of Digital Fading Mobile VHF Channels," IEEE Trans. of Vehicular Tech., Nov. 1994.
9. L. N. Kanal and A. R. K. Sastry, "Models for Channels with Memory and their Applications to Error Control," Proc. of IEEE, Jul. 1978.
10. H. Lee and P. K. Varshney, "Gap-based Modeling of Packet Losses over the Internet," 10th IEEE on MASCOTS, Oct. 2002.
11. D. Eckhardt and P. Steenkiste, "A Trace-based Evaluation of Adaptive Error Correction for a Wireless Local Area Network," Mob. Net. & Apps., 1999.

Flow Classification for IP Differentiated Service in Optical Hybrid Switching Network

Gyu Myoung Lee and Jun Kyun Choi

Information and Communications University (ICU)
103-6, Munji-dong, Youseong-ku, Daejeon, Korea
{gmlee, jkchoi}@icu.ac.kr

Abstract. In a new optical hybrid switching environment which combined Optical Burst Switching (OBS) and Optical Circuit Switching (OCS), we propose flow-level service classification scheme for IP differentiated service. In particular, this classification scheme classifies incoming IP traffic flows into short-lived and long-lived flows for Quality of Service (QoS) provisioning according to traffic characteristics such as flow bandwidth, loss and delay. In this hybrid switching, the short-lived flows including delay sensitive traffic use OBS and the long-lived flows including loss-sensitive traffic use OCS. Therefore, optical hybrid switching network can take advantages of both switching technologies using the proposed flow classification scheme. The aim of proposed technique is to maximize network utilization while satisfying user's QoS requirements

1 Introduction

There are two kinds of optical switching technologies in IP over optical network that combine the optical and the electronic worlds. From the optical switching technology point of view, it is known that the Optical Circuit Switching (OCS) networks achieve low bandwidth utilization with burst traffic such as Internet traffic. So, sophisticated traffic grooming mechanism is needed to support statistical multiplexing of data from different users. On the other hand, Optical Burst Switching (OBS) technology has been emerging to utilize resources and transport data more efficiently than the existing circuit switching [1]-[4]. OBS is accepted as an alternative switching technology due to the limitation of optical devices that do not support buffering.

The OBS and OCS have the advantages and disadvantages in performance point of view. So we can consider the so-called hybrid switching. The optical hybrid switching [5]-[6] is a new switching technique which combines OCS and OBS to take advantages of both switching technologies and to improve their performance degradation. The OCS module of optical hybrid switching can avoid the several overheads for long-lived flows and reuse the current OCS network technology. On the other hand, the OBS can improve the resource utilization for short-lived flows such as bursty IP traffic. In this paper, we consider a combined OCS and OBS system in a hierarchical Quality of Service (QoS) mapping architecture.

C. Kim (Ed.): ICOIN 2005, LNCS 3391, pp. 635–642, 2005.

One of the today's most pressing challenges in designing IP networks is the provisioning of QoS. Therefore, we propose the flow-level service classification scheme for IP differentiated service. This scheme classifies incoming IP traffic flows into short-lived and long-lived flows for QoS provisioning according to traffic characteristics in an optical hybrid switching environment. The incoming IP traffic flows divided into premium service, assured service and best-effort service for IP differentiated service as described in [7]. Short-lived flows are composed of a few packets and better suited for OBS which has a short-delay characteristic than OCS. Long-lived traffic typically indicate loss-sensitive or real-time video streams that are better suited for circuit (or wavelength) switching which has an advantage of loss-less through connection establishment. Therefore, the optical hybrid switching technique using the proposed flow classification scheme takes advantages of both OBS and OCS. The aim is to maximize network utilization while satisfying user's QoS requirements.

The remainder of the paper is organized as follows. In Section 2, we explain the optical hybrid switching system. In Section 3, we propose the new flow-level service classification scheme in optical hybrid switching networks and propose QoS provisioning algorithm for IP differentiated service in this network. Then, in Section 4, we give numerical results for the proposed network.

2 Optical Hybrid Switching System

The optical hybrid switching is a new switching technique which combines OCS and OBS to take advantages of both switching technologies and to eliminate their disadvantages. OCS has advantages of supporting QoS and Traffic Engineering, on the other hand, OBS has advantages of improving utilization of network for bursty IP traffic.

The objective of optical hybrid switching scheme is to effectively transport the long-lived and short-lived flows at optical edge switching node. The OCS module of optical hybrid switching system can avoid control packet processing, frequent switch fabric reconfiguration, and burst assembly/de-assembly for long-lived flows. The OCS module can reuse the current OCS network control and hardware. The OBS module of optical hybrid switching system can improve the resource utilization for short-lived flows.

Fig. 1 shows the example of optical hybrid switching system in optical hybrid switching network which consists of OBS switch and OCS switch.

3 Flow Classification for IP Differentiated Service

3.1 QoS Classification for IP Differentiated Service

We can classify the incoming traffic types using the value of flow bandwidth threshold (Bth) in the relationship of flow bandwidth and the number of packets. Short-lived flows (e.g., flow bandwidth $<$ Bth) are composed of a few packets such as e-mail, light-loaded FTPs and so on. These flows are better suited for

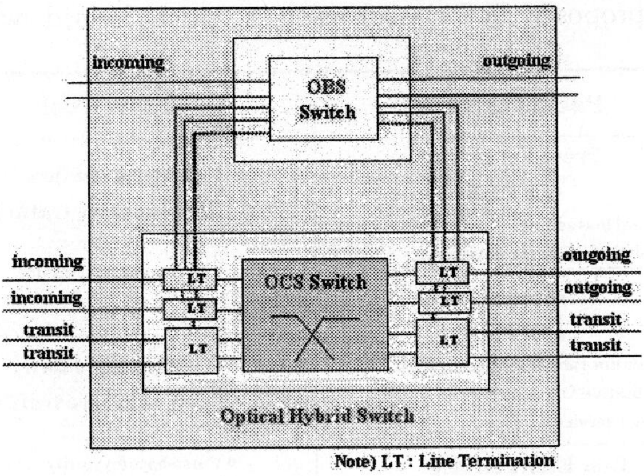

Fig. 1. The architecture for optical hybrid switching module of optical router

OBS. Long-lived flows (e.g., flow bandwidth > Bth) contain a large number of packets, that is, stream media. These flows are better suited for OCS. We can consider other type of traffic. For example, big burst such as very high-load FTPs and images require very high bandwidth for a short period of time and require special reservation. This case is better suited for wavelength routed OBS (WR-OBS) [8]. The reservation of this switching scheme is made for the entire burst before it is transmitted.

Table 1 shows the proposed QoS classification in optical hybrid switching network with hierarchical QoS mapping architecture. The packet level QoS is divided into three services for IP differentiated service [7],[9]. We propose the flow level QoS which classifies incoming IP differentiated service into long-lived flows and short-lived flows.

Services of flow level are divided into long-lived flow and short-lived flow. The long-lived flow including loss-sensitive traffic use OCS for guaranteed service and the short-lived flow including delay-sensitive traffic use OBS for class-based priority service. The details of the proposed flow-level classification will be explained in the next section.

In flow-level QoS classification, we identify the following functions to be implemented at the optical edge router. The functions for OCS are the aggregation of incoming IP differentiated service flows into fewer flows at rates corresponding to lightpath traffic carrying capacity and the mapping of aggregated incoming IP differentiated service flows onto lightpath that correspond to the QoS of the aggregated flows. Similarly the functions for OBS are the creation of data bursts using burst assembling process and the mapping of three priority classes for class-based priority service.

Table 1. The proposed QoS classification in optical hybrid switching network

Packet level	Flow level
Premium service (EF PHB) • Virtual leased line • Bandwidth pipe for data service	**Long-lived flow (loss sensitive traffic)** • Guaranteed service
Assured service (AF PHB) • Minimum rate guarantee service • Qualitative Olympic service • Funnel service	**Short-lived flow (delay sensitive traffic)** • Class-based priority service
Best Effort service (Default PHB)	

Note) EF: Expedited Forwarding, AF: Assured Forwarding

3.2 QoS Provisioning for IP Differentiated Service Using Flow Classification

Here, we propose a QoS provisioning algorithm for IP differentiated service using flow-level service classification scheme. Fig. 2 shows the flow classification algorithm for optical hybrid switching. The long-lived flows are composed of the premium service and loss-sensitive service. The short-lived flows are composed of the delay sensitive service and best effort service. Otherwise, we check the flow bandwidth and compare this with the threshold value (Bth). The details of the QoS provisioning algorithm using flow classification as shown in Fig 2 are shown in Fig. 3.

In the case of short-lived traffic flows, data burst is created in burst assembler module which has a separate buffer per class and generates Burst Control Protocol (BCP) packet. And then the scheduling and the class-based resource reservation function are simultaneously performed. The scheduler performs the class-based priority queuing. In resource reservation, higher priority bursts are assigned longer offsets than lower priority bursts using BCP [10]. Finally, after electro-optical conversion, data burst cut through intermediate nodes without being buffered. This increases the utilization of network through OBS for short-lived flows.

In the case of long-lived traffic flows, we assume that these flows have the highest priority and a great influence on network performance. These flows are aggregated in flow aggregator and then the QoS and resource constraints of aggregated flows which are related to traffic parameters and available wavelengths are checked. The admitted traffic flows are allocated the requested resource through static Routing and Wavelength Assignment (RWA) [11] which is executed off-line with average traffic demands and predetermined shortcut route.

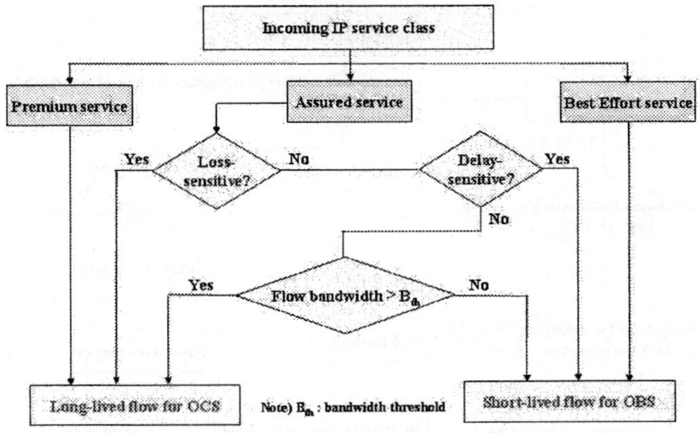

Fig. 2. Flow classification algorithm for optical hybrid switching in optical edge router

Finally, a lightpath LSP is established and after electro-optical conversion, these flows are transmitted.

In the above proposed algorithm, the aim is to maximize network utilization while satisfying user's QoS requirements in a hybrid switching environment taking advantage of both OBS and circuit switching technologies.

4 Numerical Results

We model optical hybrid switching as a queue in a Markovian environment. The burst generation process is assumed to follow a two-state Markov Modulated Poisson Process (MMPP) [12], and The admission and completion of the long-lived traffic stream OCS connections is modeled as an M/M/k/k process. As shown in Fig. 4, the bandwidth available to OBS bursts is dependent upon the number of OCS sessions active on the hybrid switching system, and it fluctuates in accordance with OCS traffic loading.

The analysis was performed for an optical hybrid switching system of particular parameters. There are 120 available capacity units of link. Out of the 120 available, 10 capacity units are reserved exclusively for OBS bursts only. Each link is 10Gbps. The mean duration of an OCS session is 3 minutes. The OCS load is chosen so that the capacity available to the OCS connection has a utilization of 10%, 30% and 50%.

In Fig. 5, we present the result for the mean delay versus utilization for different OCS load. When OCS load is 30%, the mean delay is rapidly increased for high utilization (over 0.7). Thus, this result indicates that in order to operate an optical hybrid system with reasonable burst delays the utilization must be kept below 70%. On the other hand, When OCS load is 50%, the mean delay is

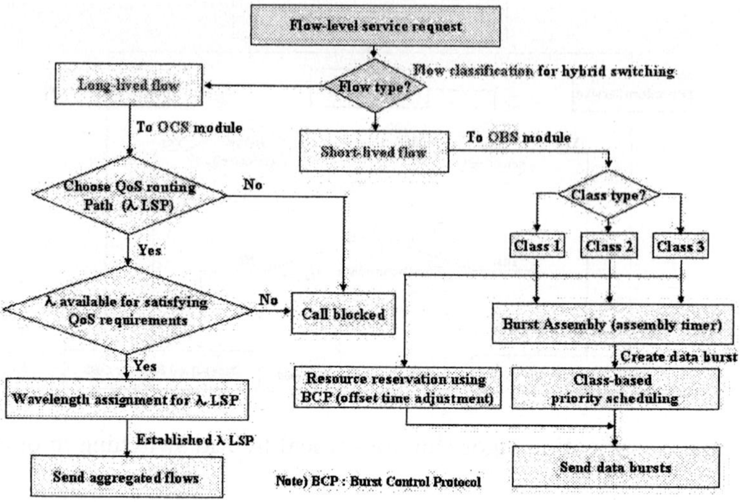

Fig. 3. QoS provisioning algorithm for IP differentiated service using flow classification

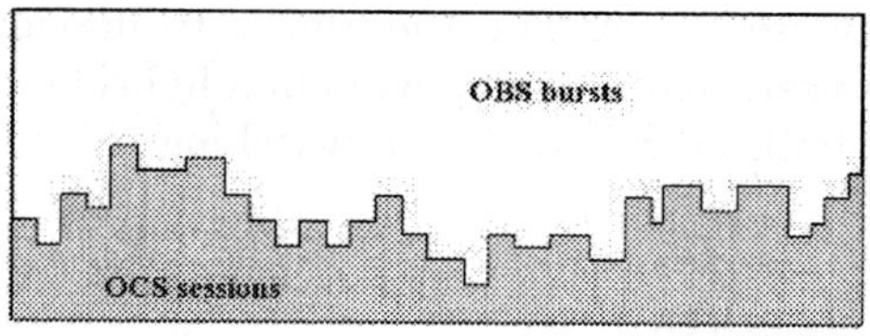

Fig. 4. Resource sharing model for OBS bursts and OCS sessions

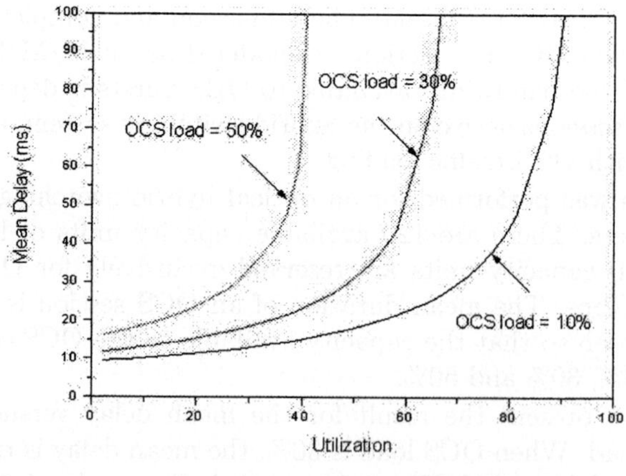

Fig. 5. Mean delay versus utilization for OBS bursts

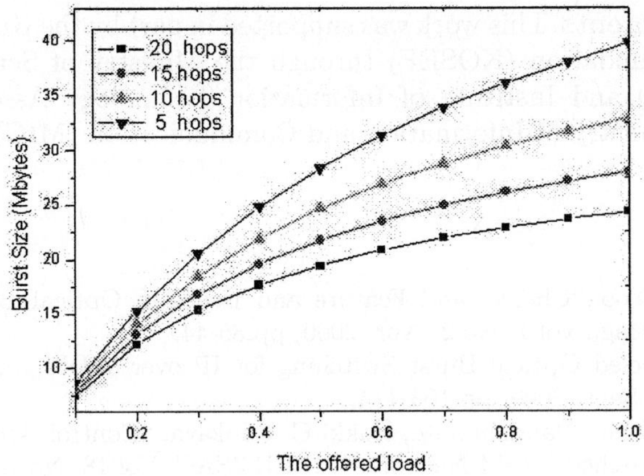

Fig. 6. Burst size vs. the offered load for different hop count (end-to-end delay=100ms)

rapidly increased at the low utilization (over 0.4). Thus, these results for different OCS load indicate that the mean delay depends on OCS load.

Next, we present result of end-to-end performance for OBS. In particular we would like to show you the relation of burst size concerning end-to-end delay constraints [13]. Fig. 6 shows the burst size versus the offered load when the end-to-end delay is 100ms. Here, we want to show the effect of hop count change. When the hop distance is 20 hops then the burst size should be less than about 20Mbps for the worst case to satisfy the end-to-end constraint (100ms). On the other hand, when the hop distance is just 5 hops then the constraint of burst size is reduced.

5 Conclusions

In this paper, we have proposed QoS provisioning algorithm using flow-level service classification in a new optical hybrid switching system which combines OBS and OCS. To support IP differentiated service in optical hybrid switching network, the proposed flow classification scheme classifies the incoming IP differentiated service flow into long-lived and short-lived flows. The aim is to maximize network utilization while satisfying user's QoS requirements. In particular, we have considered flow classification scheme that should be implemented cost-effectively and easily in optical hybrid switching system. We also have shown the performance results for delay characteristic of optical hybrid switching. For the further study, we would like to show the comparison result of OCS, OBS and optical hybrid switching.

Acknowledgements This work was supported in part by the Korea Science and Engineering Foundation (KOSEF) through the Ministry of Science and Technology (MOST) and Institute of Information Technology Assessment (IITA) through the Ministry of Information and Communication (MIC), Korea.

References

1. C. Qiao, M. Yoo.:"Choice, and Feature and Issues in Optical Burst Switching", Optical Net. Mag., vol.1, No.2, Apr. 2000, pp.36-44.
2. C. Qiao.:"Labeled Optical Burst Switching for IP over WDM Integration", IEEE Comm. Mag., Sept. 2000, pp.104 114.
3. Yijun Xiong, Marc Vandenhoute, Hakki C. Cankaya.:"Control Architecture in Optical Burst-Switched WDM Networks", IEEE JSAC, Vol.18, No.10, Oct. 2000.
4. Ilia Baldine, George N. Rouskas, Harry G. Perros, Dan Stevension.:"JumpStart: A Just-in-time Signaling Architecture for WDM Burst-Switching Networks", IEEE Comm. Mag., Feb. 2002.
5. Gyu Myoung Lee, Bartek Wydrowski, Moshe Zukerman, Jun Kyun Choi, Chuan Heng Foh.:"Performance evaluation of optical hybrid switching system", Proceedings of Globecom'2003, vol.5, pp.2508-2512, December 2003
6. Chunsheng Xin, Chunmming Qiao, Yinghua Ye, Sudhir Dixit.:"A hybrid optical switching approach", Proceedings of Globecom'2003, vol.7. pp.3808-3812, December 2003.
7. Panos Trimintzios, et al.:"A management and control architecture for providing IP differen-tiated services in MPLS-based networks", IEEE Comm. Mag., pp.80-88, May 2001.
8. M. Dueser, I. de Miguel, P. Bayvel and D. Wischik.:"Timescale analysis for wavelength-routed optical burst switched (WR-OBS) networks", Proceedings of OFC 2002. March 2002.
9. S. Blake, D. Black, M. Carlson, E. Davies, Z. Wang, W. Weiss.:"An Architecture for Dif-ferentiated Services", RFC 2475, Dec 1998.
10. H. L. Vu and M. Zukerman.:"Blocking Probability for Priority Classes in Optical Burst Switching Networks", IEEE Communications Letters, vol. 6, no. 5, May 2002, pp. 214-216.
11. Admela Jukan, et.al.:"Service-specific resource allocation in WDM networks with quality constraints", IEEE JSAC, pp. 2051-2061, Oct. 2000.
12. H. Heffes and D.M. Lucantoni.:"A Markov modulated characterization of packetized voice and data traffic and related statistical multiplexer performance", IEEE JSAC, vol. SAC-4, no-6, Sep. 1986.
13. ITU-T Recommendation Y.1541.:"Network performance objectives for IP-based services", May 2002.

Supporting Differentiated Service in Mobile Ad Hoc Networks Through Congestion Control

Jin-Nyun Kim, Kyung-Jun Kim, and Ki-Jun Han*

Department of Computer Engineering, Kyungpook National University, 1370 Sankyuk-dong, Book-gu, Daegu, 702-701, Korea
{duritz, kjkim}@netopia.knu.ac.kr
kjhan@bh.knu.ac.kr

Abstract. Differentiated services (DiffServ) has been widely accepted as the service model to adopt for providing quality-of-service (QoS) over the next-generation IP networks. There is a growing need to support QoS in mobile ad hoc networks. Supporting DiffServ in mobile ad hoc networks, however, is very difficult because of the dynamic nature of mobile ad hoc networks, which causes network congestion. The network congestion induces long transfer packet delay and low throughput which make it very difficult to support QoS in mobile ad hoc networks. We propose DiffServ module to support differentiated service in mobile ad hoc networks through congestion control. Our DiffServ module uses the periodical rate control for real time traffic and also uses the best effort bandwidth concession when network congestion occurs. Network congestion is detected by measuring the packet transfer delay or bandwidth threshold of real time traffic. We evaluate our mechanism via a simulation study. Simulation results show our mechanism may offer a low and stable delay and a stable throughput for real time traffic in mobile ad hoc networks.

1 Introduction

Differentiated services (DiffServ) [1] has been widely accepted as the service model to adopt for providing quality-of-service (QoS) over the next-generation IP networks. DiffServ uses the concept of Per Hop Behaviors (PHBs), which provide different levels of QoS to aggregated flows. This is done by classifying individual traffic flows into various service levels desired before entering the DiffServ network domain. Within the DiffServ domain, flows of the same class are aggregated and treated as one flow. Each aggregated flow is given a different treatment, in terms of network resources assigned, as described by the PHB for that class.

There are three PHBs such as Expedited Forwarding (EF) [2], [3] service, Assured Forwarding (AF) service and Best Effort service. Service level agreements (SLAs) contain delay and throughput requirements among others like reliability

* Correspondent author

C. Kim (Ed.): ICOIN 2005, LNCS 3391, pp. 643–651, 2005.

requirements [4]. The EF PHBs provide low loss, low delay, and low jitter services for real time traffic that represents traffic like video or voice. We will use EF traffic as the same term with real time traffic within this paper.

A mobile ad hoc network is formed by a group of wireless stations without infrastructure. There is a growing need to support real time traffic in mobile ad hoc networks. This, However, is very challenging because mobile ad hoc networks represent dynamic nature, which causes unexpected network congestion as illustrated in Fig. 1 [6]. In this figure, we can see that network congestion is induced when a mobile station moves, which may consequently cause high delay and low throughput. Consequently, the QoS guarantee of real time flows is violated.

In this paper, we propose a DiffServ module to support differentiated service in mobile ad hoc networks through congestion control. Our DiffServ module uses the periodical rate control for real time traffic and the best effort bandwidth concession when network congestion occurs.

The organization of this paper is as follows. In Section 2, we describe our DiffServ module and congestion detection and congestion control mechanism. In Section 3, we represent simulation model, simulation parameters and some results. Finally, conclusions are offered in Section 4.

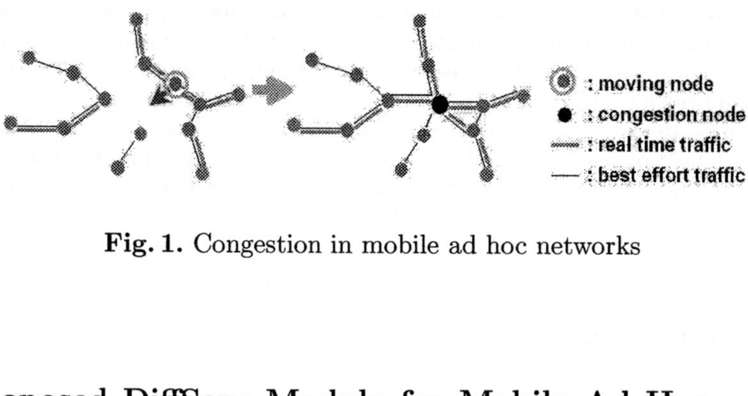

Fig. 1. Congestion in mobile ad hoc networks

2 Proposed DiffServ Module for Mobile Ad Hoc Networks

The most dominant factor of packet transfer delay in networks is queuing delay. So, delay and jitter are minimized when queuing delays are minimized. The intent of the EF PHB is to provide a PHB in which EF marked packets usually encounter short or empty queues. EF Service should provide minimum delay and jitter [2].

According to RFC 3246 [2] which discusses the departure time of EF traffic, a node that supports EF on an interface I at some configured rate R must satisfy the following condition for the j-th packet:

$$d(j) \leq F(j) + E \tag{1}$$

where d(j) is an actual departure time, F(j) is a target departure time, and E is a tolerance that depends on the particular node characteristics. E provides an upper bound on (d(j)-F(j)) for all j.

F(j) is defined iteratively by

$$F(0) = 0, \qquad d(0) = 0 \qquad \text{for all } j > 0: \tag{2}$$

$$F(j) = max[a(j), min(d(j-1), F(j-1) + \frac{L(j)}{R}. \tag{3}$$

where a(j), L(j), and R denote an arrival time, the packet length, and the EF configured rate, respectively.

The rate at which EF traffic is served at a given output interface should be at least the arrival rate, independent of the offered load of non-EF traffic to that interface [2]. The relationships between EF traffic's input rate and output rate in each node are represented in the following three cases:

1. input rate > output rate
2. input rate < output rate
3. input rate = output rate

In case 1, it is difficult to support EF service because of the higher queuing delay. In case 2, the queuing delay is minimal so that high quality is provided to EF traffic. Non-EF traffic, however, is starved. Also, the delay and throughput of EF traffic can fluctuate because the output rate of EF traffic is disturbed. Finally, case 3 is considered as an ideal case for EF traffic. We assert that the input and output rate of EF traffic should be the same.

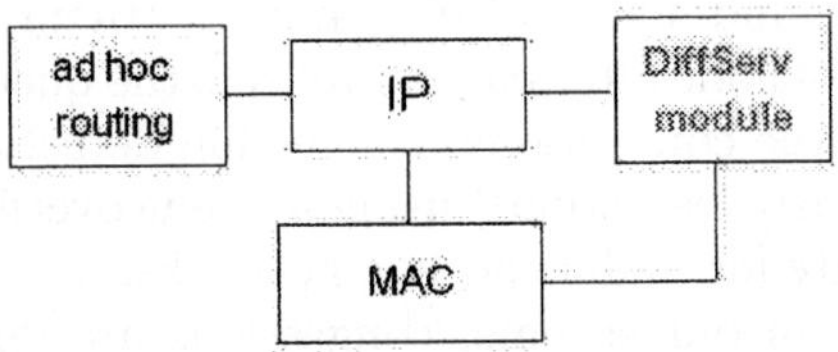

Fig. 2. DiffServ module

A path of real time traffic is established by the QoS extensions [7] of routing protocols such as AODV (Ad hoc On-demand Distance Vector) [10] and OLSR (Optimized Link State Routing protocol) [11]. QoS routing protocols find an optimal path in meeting the delay and bandwidth requirements of real time traffic.

The proposed DiffServ module exists in the IP layer together with routing protocol as illustrated in Fig. 2. There is an interface between the DiffServ module and MAC for their interoperation. The DiffServ module controls the output rate of traffic using a MAC delay or bandwidth usage provided through the interface. The objective of our DiffServ module is guaranteeing the QoS requirement of already established real time traffic. The DiffServ module has two main roles:

1. It periodically regulates the output rate of real time traffic to meet bandwidth requirements. In other words, the output rate is maintained the same as the input rate. This rate maintenance provides not only the ideal EF service as previously described but also a stable throughput and delay of real time traffic. The rate adjustment can be implemented by using the token (leaky) bucket [9].
2. When congestion occurs, the bandwidth of best effort traffic is conceded to real time traffic in order to prevent the QoS requirement penalty. Fig. 3 illustrates the conceding of best effort bandwidth to real time traffic.

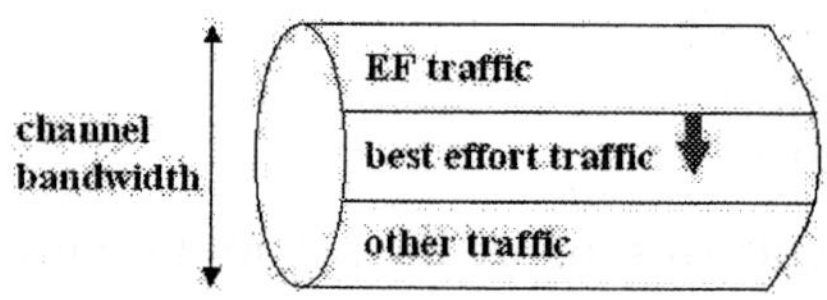

Fig. 3. The concession of best effort bandwidth

If queues remain short and empty relative to the buffer space available, packet loss and queuing delay is kept to a minimum. Since EF traffic usually encounters short or empty queues, and node mobility induces obscurity of queue utilization (i.e., the queue length of node after moving reflects the queue length of at position right before moving), the conventional congestion detection method (e.g., drop tail, RED (Random Early Detection)) using a queue overflow or queue threshold value is not appropriate for mobile ad hoc networks.

For these reasons, in our scheme, congestions are detected by monitoring when the delay and bandwidth utilization of real time packets exceed a given threshold. Packet delay and bandwidth utilization can be simply measured at the congestion node by using the timestamp in a packet and counting amount of packet received per second, respectively. Also, the recent research, BLUE [8] shows that RED congestion avoidance algorithm using a queue length estimate to detect congestion has inherent weakness. Queue lengths do not directly relate to the true level of congestion in the real packet networks. BLUE use the packet loss and link utilization history for congestion detection.

When a node detects congestion, it sends out congestion notification messages in the direction of the source node of the real time flow as illustrated in Fig. 4. The notification messages are broadcast because of wireless medium characteristics. First, one-hop upstream nodes receiving the notification messages concede the bandwidth allocated for their best effort flows to their real time flows to relieve a congested situation. At this time, if the congestion is resolved, all congestion control processes end and congestion notification messages are no longer relayed in the direction of source nodes. Usually, the congestion can be solved at one-hop upstream nodes of congestion node because many one-hop upstream nodes (three nodes in Fig. 4) simultaneously reduce the rate of their best effort traffic.

If the congestion is not relieved, however, the congestion notification messages are continuously relayed in the direction of source nodes. So, two-hop, three-hop, ..., upstream nodes receiving the notification messages reduce their output rates of their best effort flows. Through this process, if the congestion is solved all processes successfully end, and if the congestion is not solved the notification messages are continuously relayed in direction of source nodes until congestion is solved. So, light congestion is simply relieved at one-hop upstream nodes, but the heavy congestion is relieved after many upstream nodes reduce their best effort output rates.

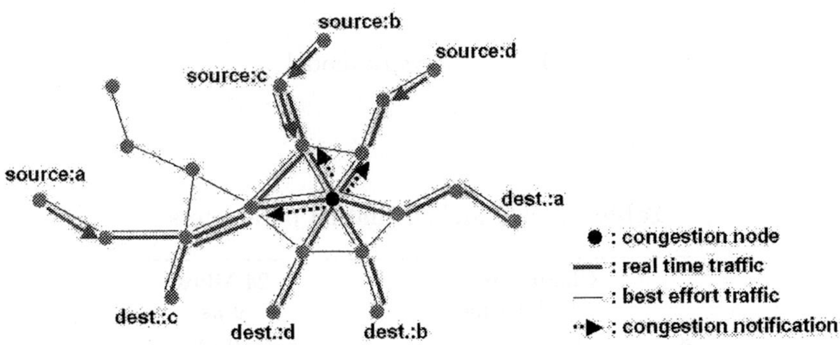

Fig. 4. Congestion control in mobile ad hoc networks

3 Simulation

We evaluated our mechanism via simulation. Fig. 5 illustrates the network model used in the simulation. We have modeled only one congestion node and it's three upstream nodes from network of Fig. 4 because only three one-hop upstream nodes of congestion node can completely relieve the congestion in our simulation. Also, the rate reduction amount of the best effort bandwidth at each node can determine more relay of congestion notification message. This is network design choice. The simulation in mobility situation is future work.

In Fig. 5, three nodes simultaneously access a channel in order to communicate with a congested node. It is assumed that a contention-based service by the IEEE 802.11 Distributed Coordination Function (DCF) [5] channel access mode, based on the carrier sense multiple access with collision avoidance (CAMA/CA), is used to contend for the medium for each packet transmission. When a packet arrives at the MAC layer, the MAC listens to the channel and defers access to the channel according to CSMA/CA algorithm. When the MAC acquires access to the channel, then packets are exchanged.

We have simulated the 802.11 DCF as time slot based. The length of data packet is assumed as 80 bytes which is equivalent to 26 μs at the channel bit rate of 24 Mbps. The DCF and simulation parameters are reported in Table. 1. Each node is modeled as a perfect output buffered device, that is, one which delivers packets immediately to the appropriate output queue as illustrated in Fig. 6.

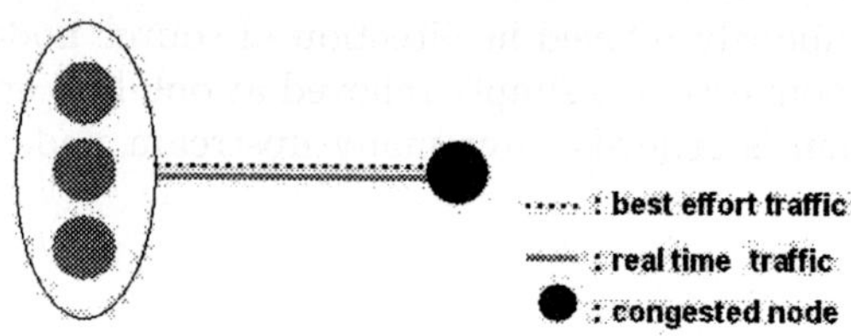

Fig. 5. network model

Table 1. DCF and simulation parameters

Channel bit rate	24 Mbps
Slot time	9 μs
SIFS	16 μs
DIFS	34 μs
Length (size) of contention window	0∼63 μs (8)
ACK transmission time	5 μs
Data packet transmission time	27 μs (80bytes)

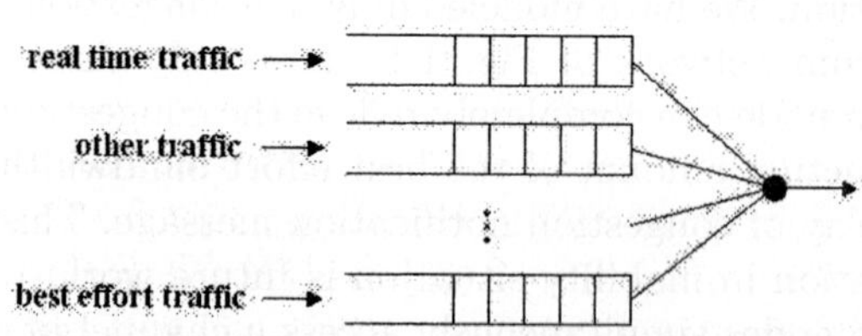

Fig. 6. Model of node with prioritized queues

The focus of the experiments is whether our DiffServ module guarantees the QoS requirements of real time traffics as the best effort traffic load is increased.

Fig. 7 shows throughput and delay when the output rate of real time traffic is maintained the same as the input rate. The input rate of real time traffic is 1Mbps that represents a bandwidth requirement. In the experiment, the best effort traffic load continuously increased while the rate of real time traffic is maintained at 1 Mbps. As a result of experiment, the throughput of the best effort traffic increases up to some point and after that point, is saturated to almost 5 Mbps. Furthermore, the delay of the best effort traffic suddenly increases after the saturation point. We can see that the throughput of real time traffic is maintained successfully meeting the bandwidth requirement. The delay performance of real time traffic also shows a relatively stable pattern. Assuming that the bandwidth requirement is 1 Mbps and the node-to-node delay requirement is 10 ms, then the network is not congested. A congestion control mechanism for real time traffic is not needed.

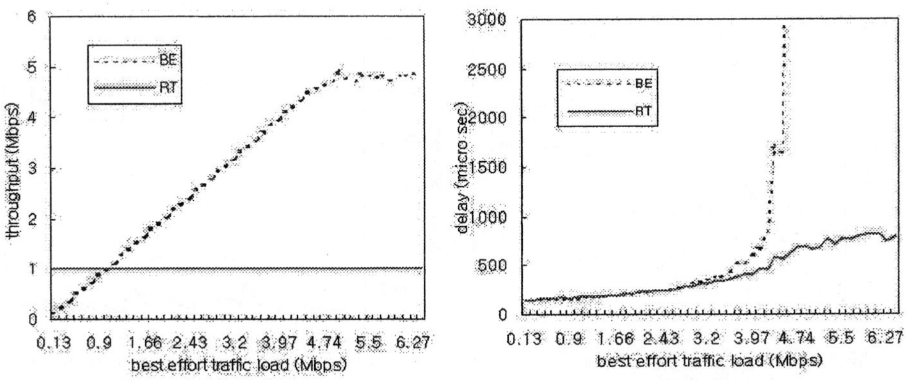

(a)Throughput (b)Delay

Fig. 7. The throughput and delay when the real time bandwidth requirement is 1 Mbps

If the bandwidth requirement of real time traffic is changed to 3.027Mbps, however, we can observe congestion as shown in Fig. 8. In this case, the bandwidth requirement of real time is satisfied when the offered loads of best effort are relatively low. As the best effort traffic load increases, however, the throughput of real time traffic decreases and the delay also terribly increases, which induces the QoS violation of real time traffic.

Fig. 9 shows the throughput and delay performances with our congestion control mechanism. When the bandwidth requirement of real time traffic is 3.027Mbps, the throughput of real time traffic is stably maintained at 3.027Mbps and the delay is also maintained at stably low values as a result of the best effort bandwidth concession. In Fig. 9, the crooked points represent that congestion

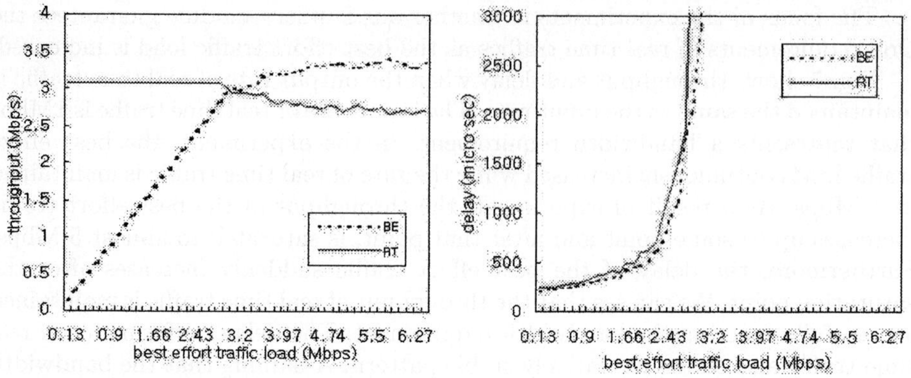

(a)Throughput (b)Delay

Fig. 8. The throughput and delay with no congestion control when the real time bandwidth requirement is 3.027 Mbps

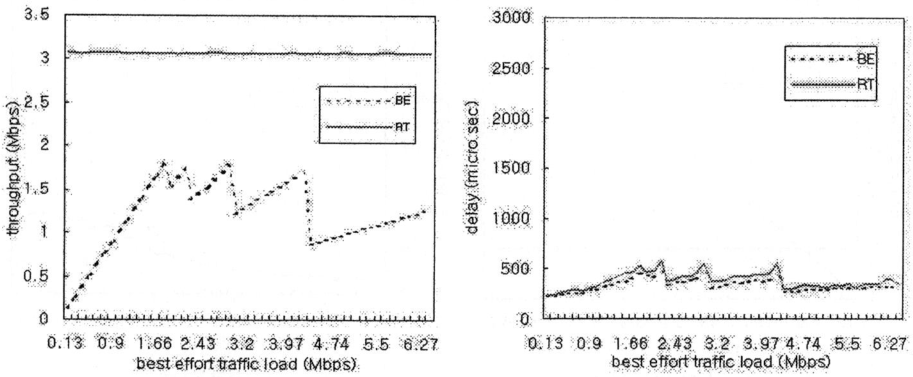

(a)Throughput (b)Delay

Fig. 9. The throughput and delay with congestion control when the real time bandwidth requirement is 3.027 Mbps

control is performed. Whenever congestion is detected, the bandwidth of best effort traffic is conceded to real time traffic through its rate reduction. As previously described, congestion is detected by a delay threshold value of real time packet.

4 Conclusion

In this paper, we proposed DiffServ module, supporting service differentiation in mobile ad hoc networks through rate regulation and congestion control. In our scheme, for real time traffic, we regulated its output rate the same as the input rate. This regulation produced stable throughput and delays. The congestion was detected by measuring the delay or bandwidth utilization of real time traffic and comparing it with some threshold values. The congestion was controlled by conceding the best effort bandwidth to real time traffic. We verified our DiffServ mechanism through simulation. The experiment results showed that our mechanism could offer stable throughput and stably low delays for real time traffic.

Acknowledgement
Academic Research Program supported by Ministry of Information and Communication in Republic of Korea

References

1. S. Blake, D. Black, M. Carlson, E. Davies, Z. Wang, and W. Weiss.: An architecture for differentiated services. IETF, RFC 2475, Dec (1998)
2. B. Davie, A. Charny, J.C.R. Bennett, K. Benson, J.Y. Le Boudec, W. Courtney, S. Davari, V. Firoiu, and D. Stiliadis.: An expedited forwarding PHB. IETF, RFC 3246, March (2002)
3. A. Charny, J.D.R. Bennett, K. Benson, J.Y. Le Boudec, A. Chiu, W. Courtney, S. Davari, V. Firoiu, C. Kalmanek, and K.K. Ramakrishnan.: Supplemental information for the new definition of the EF PHB. IETF, RFC 3247, March (2002)
4. T. C-K. Hui and C.-K. Tham.: Adaptive provisioning of differentiated services networks based on reinforcement learning. IEEE Trans. Syst. Man. Cybern. C, vol. 33, no. 4, pp. 492-501, Nov. (2003)
5. IEEE 802.11 Working Group.: Wireless LAN medium access control (MAC) and physical layer (PHY) specifications. IEEE standard 802.11, June (1999)
6. Gahng-Seop Ahn, Andrew T. Campbell, Andras Veres, and Li-Hsiang Sun.: Supporting service differentiation for real-time and best-effort traffic in stateless wireless ad hoc networks. IEEE Trans. Mobile computing, vol. 1, July-Sept. (2002)
7. S. Chen and K. Nahrstedt.: Distributed quality-of-service routing in ad hoc networks. IEEE JSAC, vol. 17, no. 8, Aug. (1999)
8. Wu-chang Feng, Kang G. Shin, Dilip D. Kandlur and Debanjan Saha.: The BLUE Active Queue Management Algorithms. IEEE/ACM Trans. Networking, Aug. (2002)
9. Mohamed A. El-Gency, Abhijit Bose, Kang G. Shin.: Evolution of the internet QoS and support for soft real-time applications. Proc. IEEE, vol. 91, no. 7, Jury (2003)
10. C. Perkins, E. Belding-Royer and S. Das.: Ad hoc On-Demand Distance Vector (AODV) Routing. IETF, RFC 3561, July (2003)
11. T. Clausen and P. Jacquet.: Optimized Link State Routing Protocol (OLSR). IETF, RFC 3626, Oct. (2003)
12. Satyabrata Chakrabarti and Amitabh Mishra.: QoS issues in ad hoc wireless networks. IEEE communications magazine, Feb. (2001)
13. Y. Dong, D. Makrakis, T. Sullivan.: Network congestion control in ad hoc IEEE 802.11 wireless LAN. CCECE 2003-CCGEI 2003, May (2003)
14. Prasant Mohapatra, Jian Li, and Chao Gui.: QoS in mobile ad hoc networks. IEEE Wireless Communication, June (2003)

HackSim: An Automation of Penetration Testing for Remote Buffer Overflow Vulnerabilities

O-Hoon Kwon[1], Seung Min Lee[1], Heejo Lee[2], Jong Kim[1],
Sang Cheon Kim[3], Gun Woo Nam[3], and Joong Gil Park[3]

[1] Dept. of Computer Science & Engineering,
Pohang University of Science and Technology
{dolphin, puhaha, jkim}@postech.ac.kr
[2] Dept. of Computer Science & Engineering, Korea University
heejo@korea.ac.kr
[3] National Security Research Institute
{wsound, daemon99, jgpark}@etri.re.kr

Abstract. We propose an extensible exploit framework for automation of penetration testing (or pen-testing) without loss of safety and describe possible methods for sanitizing unreliable code in each part of the framework. The proposed framework plays a key role in implementing HackSim a pen-testing tool that remotely exploits known buffer-overflow vulnerabilities. Implementing our enhanced version of HackSim for Solaris and Windows systems, we show the advantages of our sanitized pen-testing tool in terms of safety compared with existing pen-testing tools and exploit frameworks. This work is stepping toward a systematic approach for substituting difficult parts of the labor-intensive pen-testing process.

1 Introduction

Vulnerability scanning is deployed to check known vulnerabilities on a single system or a series of systems in a network. There are a number of scanning tools which are available publicly or commercially [1]. Penetration testing (or pen-testing) is a goal-oriented method similar to "catch-the-flag" that attempts to gain privileged access to a system using pre-conditional means that a potential attacker could manipulate. A tester, sometimes known as an ethical hacker, generally uses the same methods and tools used by attackers to undermine network security. Afterward, penetration testers report on the exploitable vulnerabilities they found and suggest strengthening steps needed to make their client's systems more secure [2,10,11]. Most security consulting firms provide pen-testing services by red teams or ethical hackers [3,4], and the market volume for these services is expected to grow substantially [5,19].

Vulnerability scanners provide automated scanning with user-friendly interfaces and extensible structures for updating new vulnerabilities. In addition, scanning is conducted using safe methods that do not produce unexpected impact on target systems, at the expense of false-positive results. Pen-testing is performed manually using the same methods a real attacker employs. Such a

C. Kim (Ed.): ICOIN 2005, LNCS 3391, pp. 652–661, 2005.

time consuming task as pen-testing provides visible and useful results from a deep investigation of a target system. However, pen-tesing may leave behind security holes or cause unintended damage to the system [6]. Thus, this safety problem is an obstacle in automating pen-testing procedures. Because recent pen-testing tools or exploit frameworks for pen-testing do not provide a sanitization method to deal with unreliable exploit code, safety of their pen-testing cannot be guaranteed [10,11,12,13].

Therefore, in this paper, we propose an extensible exploit framework as a foundation for automating pen-testing with safeguards and describe considerations for sanitizing unreliable code in each part of the framework. Also, we implement HackSim, an automated pen-testing tool, as the prototype system of the proposed framework. Current implementations of HackSim are able to exploit four well-known vulnerabilities in Solaris and three in Windows. Nevertheless, it is easy to add new vulnerability tests to HackSim. Also, we show two examples of sanitized exploit code that do not negate the benefits of pen-testing.

The remainder of the paper is organized as follows: We describe related works and the differences underlying our work in Section 2. Overall system architecture, the extensible exploit framework for pen-testing and the design consideration for sanitizing each part of exploit framework is presented in Section 3. We examine implementation issues and implementation results in Section 4. Finally, we summarize this paper and give concluding remarks in Section 5.

2 Related Works

Testing methodologies can be classified into two categories: blackbox testing and whitebox testing. Blackbox testing is used when the tester has no prior knowledge of a system. On the other hand, whitebox testing is used when the tester knows everything about the system – like a glass house in which everything is visible.

Blackbox testing is a very useful method for finding unpublished vulnerabilities and it can be performed quickly using automated tools such as SPIKE [9]. Using it, we can collect information that is necessary for testing vulnerabilities and we also can obtain exploit codes for the vulnerabilities that we want to check. However, such a tool terminates the system or service because it carries out random attacks in order to know whether the testing has succeeded or not. Thus, blackbox testing is inappropriate for finding potential vulnerabilities when the purpose of pen-testing is not to terminate system or service but to safely find vulnerabilities.

Several commercial pen-testing tools and open-sourced exploit frameworks using whitebox testing have been proposed. Canvas and Core Impact are commercial pen-testing tools that include a network scanner and exploit framework [10,11]. Also, open source projects such as Metasploit and LibExploit provide exploit frameworks for pen-testing [13,12]. These frameworks include libraries of common routines and tools to generate shellcode [4].

[4] In case of exploit codes for remote targets, shellcode is defined as a set of instructions injected into an exploited program and then executed on remote targets.

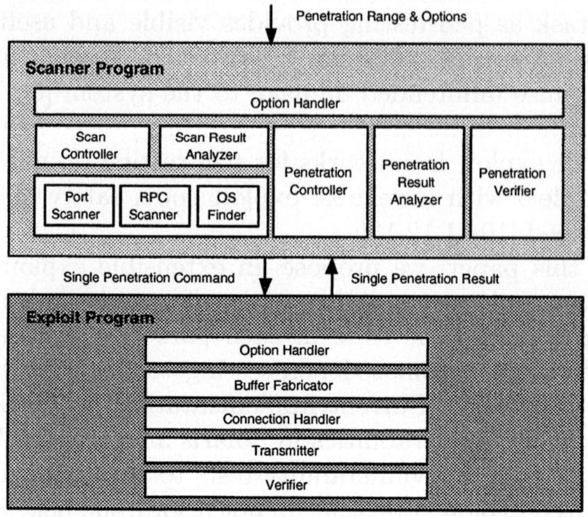

Fig. 1. HackSim Architecture Overview

Existing pen-testing tools and exploit frameworks have pros and cons. Those released as open source have not been fully verified in terms of safety. They produce lots of unreliable exploit code which causes pen-testing to fail and may lead to operating system and application crashes. Therefore, the safety of exploit code is highly required, but existing tools and frameworks are not concerned about the safety of exploit codes. They do not try to verify the safety of their pen-testing procedures.

In order to achieve appealing results for administrators and reliable pen-testing, we designed and implemented an automated pen-testing tool supporting the usability and safety of vulnerability scanners as well as the correctness of manual pen-testing.

3 HackSim Design

3.1 Architecture Overview

A top-down approach is used for describing our proposed system. First, we present the overall system architecture for automated pen-testing, which consists of two parts: "scanner program" and "exploit program". The scanner program is for processing user inputs and preparing corresponding parameters being passed to the exploit program. The exploit program is for launching the exploit codes in an extensible and safe way, which is described in the following subsections.

Components in the overall architecture are shown in Fig. 1. The lower part is for the exploit program, and the upper part is for the scanner program. When the scanner program generates penetration commands after getting user inputs such as the target range and other option values, the scanner program invokes the exploit program with commands generated by the scanner program.

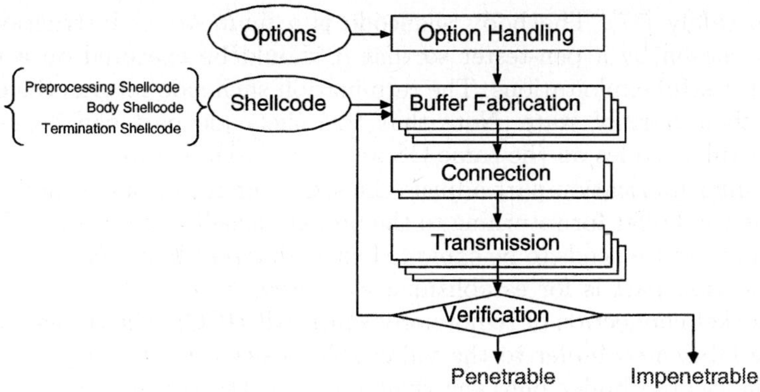

Fig. 2. Structure of the extensible exploit framework

The scanner program includes concise scanning functionalities, as shown in the dotted box in Fig. 1. The simple scanner performs OS fingerprinting and port scanning for checking the availability of targets and investigating information about them before penetration.

3.2 Extensible Framework for Exploit Codes

The exploit program of our pen-testing tool is designed as an extensible exploit framework for representing various exploit codes. Exploit codes are organized quite differently by the class of vulnerability. Among many classes of vulnerabilities, we confine our efforts to the remote buffer overflow vulnerability. The reason for this is that buffer overflows accounted for more than 50 percent of CERT advisories from 1996 to 2001 and 40 percent of the 20 most critical internet security vulnerabilities identified by the SANS Institute and the FBI [14,15].

From the analysis of public exploit codes for remote buffer overflow vulnerabilities, we can build an exploit framework that consists of five functions and two input data as illustrated in Fig. 2. This framework plays a key role as an engine for pen-testing and provides extensibility as a common platform for adding new exploit codes.

The option handling part is in charge of handling options for individual exploit codes in the framework. Different options in each exploit code are integrated into common interfaces in the option handling part.

The shellcode part is a set of instructions to be injected into an exploited program and executed on a target when the exploit works. It consists of the preprocessing shellcode, the body shellcode and the termination shellcode. The preprocessing shellcode is the preparation code to be executed before executing the body shellcode. For example, if the body shellcode contains a null character, the preprocessing shellcode should use a technique to avoid it. In Windows, if the body shellcode makes use of dynamic libraries, it also supports techniques such as PEB, SEH or TOPSTACK to retrieve function APIs(Application Programming

Interface) safely [17]. The body shellcode is a main set of instructions to be explicitly chosen by a pen-tester so that it should be executed on a target in case of successful exploitations. The termination shellcode returns the exploited hosts to their normal state. Note that this shellcode part can be reused for different exploit codes on the same OS and CPU architecture.

The buffer fabrication part adjusts the size of an input buffer, and writes an address on the buffer for returning to the prepared shellcode position. Therefore, every target service needs to be exploited using different buffer fabrication codes. The connection part is for establishing a connection to a remote service using either a socket connection or remote procedure call(RPC). The transmission part sends the fabricated buffer to the vulnerable position in the target service. To put the fabricated buffer onto the right position, the transmission part should talk to the target service with its own protocol until the overflow is caused.

The verification part returns the result after executing the shellcode. The result informs whether the penetration succeeds or not. If the penetration succeeds, the framework reports that the corresponding service is 'penetrable'. Otherwise, the penetration starts over from the buffer fabrication, as indicated by the outer arrow in Fig. 2. In each trial, the return address stored on the run-time stack is written again incrementally in order to fit the exploitable position. If the number of trials is not set by a tester, the exploit program repeats until it succeeds. However, if the penetration fails in the system that this brute force attack is not allowed, the framework reports that the corresponding service is 'impenetrable' at a trial.

3.3 Design Consideration for Sanitizing Exploit Codes

Pen-testing is performed using the same methods employed by an attacker. Consequently, it executes coded commands after exploiting vulnerable hosts. Thus, we have to make sure that this code is trustworthy and safe. In this section, we consider the sanitization of exploit codes to guarantee these needs.

Pen-testing should provide the assurance of safety. Safety can be handled by two parts: system part and service part. System safety means the safety of whole system and includes the safety of all services in the system. Service safety should meet the availability and the reliability of the service. Availability means that the service is in operation and reliability means that the service operates correctly. That is, we should verify that pen-testing against a service does not affect the safety of other services and the system by creating back doors, propagating worms, etc. In order to accomplish safe pen-testing, some considerations for sanitizing each module in Fig. 2 are described as follows.

The buffer fabrication part is very important in some cases, especially when a multi-threaded service in Windows is terminated after one of its threads generates an unhandled exception. Mis-prediction of a return address causes an application to crash and Windows does not allow brute force attacks. Some methods support relatively safe jumps to the shell code area by using the register. This method also works well in a multi-threaded environment [16].

The function of the connection and transmission part is to deliver a shellcode to the service of a remote system. In these parts, pen-testing will fail if there are problems caused by a tester or by other factors such as the network environment, service availability, etc. We take it for granted that there is no problem connecting to the service of a remote system and just concentrate on user faults to sanitize these parts. It is helpful to support libraries or modules that are divided by protocol and to contain functions making communication requests easily.

As mentioned before, shellcode can be divided into three parts: preprocessing, body and termination. Among them, the body shellcode is the one to let the tester execute arbitrary commands on the target machine. Therefore, this part should be carefully checked so that the body shellcode does not affect the integrity of the system and successfully sends results. Also, because the system call number or the address of a kernel service in system library is different in various operating systems, the body shellcode using kernel services should be checked carefully.

The termination shellcode returns the exploited service to its normal state. Most public exploit codes do not consider the importance of the termination shellcode for the safety of pen-testing. An infected thread or process rarely goes back to its normal state and polluted data makes it impossible to return a system to its normal state. The best choice is always to terminate the thread or process safely. To achieve this goal, we have to handle important tasks like releasing resources used by the thread and restoring data in shared memory, etc. It is not always a necessary task but the system might be impaired if we close our eyes to this matter. After that, we should terminate the thread by calling the corresponding function to exit it.

In some cases, not all modules of the extensible exploit framework need to be inspected, but it is highly recommended that the shellcodes are sanitized carefully, especially the body shellcode and the termination shellcode.

4 Implementation

This section describes the implementation issues of HackSim, based on previous system design. HackSim was implemented on Linux operating systems using C and Java for the exploit framework and the scanner program, respectively. It can do pen-testing against Windows and Solaris operating systems.

4.1 Modularizing Exploit Codes

To implement a prototype of the proposed exploit framework, we collected publicly available remote exploit codes for well-known vulnerabilities in Solaris and Windows. We analyzed them and selected 7 exploit codes for rapid prototyping of the proposed framework [14]. The characteristics of the seven vulnerabilities and their exploits are listed on Table 1. The number of exploit codes is small, but they cover the main exploits of Solaris and Windows. Their connection

Table 1. Exploit codes used for implementing the exploit framework

CVE Index	Target	OS	Connection	Vulnerability	Shellcode
CVE-2001-0236	snmpXdmid	Solaris	RPC	Stack Overflow	findsocket
CVE-2001-0797	telnetd	Solaris	Socket	Stack Overflow	bindsocket
CVE-2001-0803	dtspcd	Solaris	Socket	Stack Overflow	cmdshell
CVE-2002-0033	cachefsd	Solaris	RPC	Heap Overflow	findsocket
CAN-2003-0352	RPC-DCOM	Windows	RPC	Stack Overflow	bindsocket
CAN-2003-0533	LSASS	Windows	RPC	Stack Overflow	bindsocket
CAN-2003-0719	IIS-PCT	Windows	Socket	Stack Overflow	bindsocket

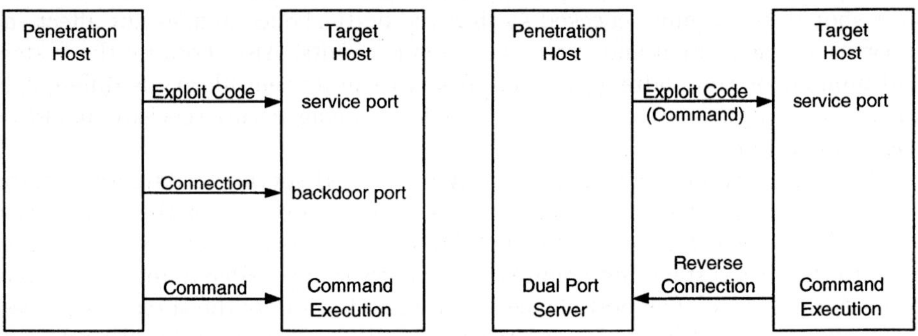

Fig. 3. Two ways for executing specified commands on a remote target host

methods included both Socket and RPC, and their vulnerabilities included both stack and heap overflow. In addition, their shellcodes manipulate commonly-used codes such as findsocket, cmdshell, or bindsocket [5].

To integrate diverse exploit codes into a single framework, the most important part is the shellcode. The proposed system uses the same shellcode for all the different exploit codes. Hence, the verification of executing the shellcode is integrated into the framework using the same code for each exploit code.

There are two ways for executing specified commands on a remote host using shellcodes as shown in Fig. 3. One is to execute commands on a root shell acquired by connecting to a backdoor port that is opened on a target host by shellcodes such as bindsocket. The other is to execute commands directly within shellcodes invoking a root shell. We selected the latter to avoid leaving any backdoors. We manually replaced the shellcodes of four exploit codes for Solaris with a common shellcode "cmdshell" executing reverse telnet commands on a target host [8]. Also, we replaced shellcodes of three exploit codes for Windows with a common shellcode "connectback", which provides the same result as executing reverse telnet commands [13].

[5] Assembly codes of findsocket, cmdshell, bindsocket are described in [7].

4.2 Sanitizing Exploit Codes

Sanitizing the Body Shellcode

If public exploit codes are used to perform penetration testing as they are, unexpected security holes may be left on target systems. For example, in Table 1, exploit codes for 'telnetd' and 'dtspcd' services may leave backdoors on target systems. The following code comes out of the exploit code for 'dtspcd' services, and this shows how the exploit code creates a backdoor.

```
cmd[] = "echo \"ingreslock stream tcp nowait root /bin/sh sh -i
\"$>$/tmp/.x; /usr/sbin/inetd -s /tmp/.x;/bin/rm -f /tmp/.x";
execl("/bin/sh","/bin/sh","-c", cmd, 0);
```

If the above code is executed on a target system, a backdoor port is opened on the system like the left figure of Fig. 3. Unless the backdoor port is closed explicitly, it will remain a severe security hole.

In order to execute specified commands without leaving any backdoor on a target system, we can use the following reverse telnet code [8].

```
cmd[] = "telnet target 1234 | /bin/sh | telnet target 5678";
execl("/bin/sh","/bin/sh","-c", cmd, 0);
```

Reverse telnet also allows the execution of commands on a compromised system even when the system is protected by a firewall. This is due to the fact that the security policy for incoming traffic is usually stricter than that of outgoing traffic. For the purpose of satisfying the requirements of safety and utilizing the benefits of reverse telnet, we used this technique in order to sanitize every exploit codes for Solaris.

But, if the above command is used in shellcode, the exploit code does not work because the second telnet session fails to write 'stdout' messages on the target hosts. Accordingly, in order to redirect 'stdout' messages of the second telnet session without printing out any messages, we added a '| sleep 1' command to *cmd*[].

Sanitizing the Termination Shellcode

If public exploit codes or the exploit framework of Metasploit is used in order to perform pen-testing on the LSASS [18] vulnerability, unintended damage on target systems occurs.

That is, they succeed in penetrating the systems, but if the command window is closed, the target host is rebooted in one minute. The reason is that a multi-threaded process in Windows is terminated when one of its threads generates an unhandled exception. This problem can be solved by adding the 'ExitThread' function to the end of the termination shellcode instead of 'ExitProcess'.

Also, we analyzed the RPCRT4 thread model because the exploit codes for the LSASS vulnerability affects the RPCRT4 thread. The analysis is focused on

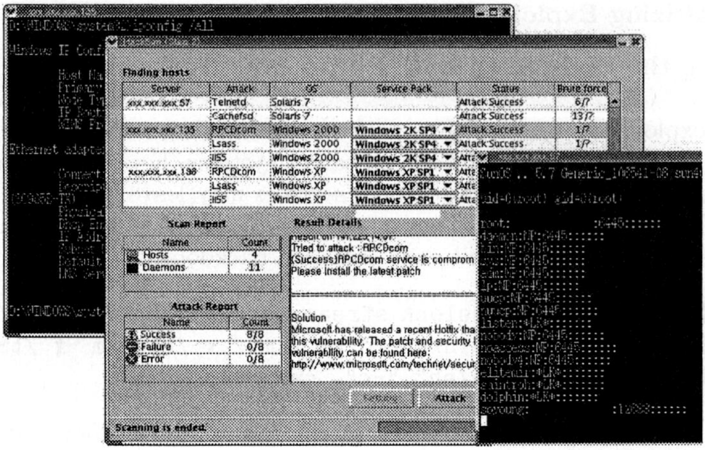

Fig. 4. A result of automated penetration testing using HackSim

whether the LSASS service works correctly after one thread is terminated using the 'ExitThread' function call. From this analysis, we got the positive result that the service is recovered and works well.

However, one problem remains. The exploit succeeds in Windows 2000 but not always in Windows XP. The reason is that one value in the data part of LSASRV.dll is changed to a wrong value during the exploit. The vulnerable function changes this value to zero when the last byte of shellcode is not a new line character, 0x0A. This zero value causes the following exploit or request to fail. This problem can be solved easily by assigning a non zero value to the data area or by adding a new line character to the last position of the shellcode.

4.3 Implementation Results

HackSim provides a high level of automation for labor-intensive pen-testing. Selectable options allow testing a wider range of targets and provide professional pen-testers with a flexible testing environment. Also, the result of penetration appears in the status window and provides collective evidence with higher accuracy than existing scanners. When the exploit works, the tool provides a privileged access on the newly created window in order to provide an evidence about exploited targets. By terminating the window, all connections are simply cleaned up without leaving behind any security hole. Fig. 4 shows pen-testing results using HackSim.

In addition, HackSim provides the extensibility for newly found vulnerabilities. This tool supports remote buffer overflow vulnerabilities that are used in the recent most worms. Also, it includes a sanitized shellcode that can be used commonly for all exploit codes. Therefore, HackSim can be easily extended to support exploit codes for newly found remote buffer overflow vulnerabilities.

5 Conclusion

In this paper, we have proposed an extensible exploit framework for an automation of pen-testing without loss of safety and described considerations for sanitizing unreliable codes in each part of the framework. Furthermore, a penetration testing tool, HackSim, is implemented on the basis of this framework. The enhanced HackSim can perform automated penetration testing without loss of safety and collect probed evidence if it succeeded in penetrating a system. Experiments against Solaris and Windows systems have shown how safely we can retrieve the most important information using automated pen-testing.

From the fact that penetrating a network is often done by exploiting a well-known weakness, this study is one step toward confirming the usefulness of automated pen-testing. The extension of HackSim to enhance the extensibility for newly found vulnerability and support the automation of the sanitization process remains for future work.

References

1. Joel Snyder, "How Vulnerable?," Information Security Magazine, Mar. 2003.
2. Pete Herzog, Open-Source Security Testing Methodology Manual(OSSTMM) 2.1, Institute for Security and Open Methodologies, 2003.
3. B. J. Wood and Ruth A. Duggan, "Red Teaming of Advanced Information Assurance Concepts," *DARPA Information Survivability Conference and Exposition (DISCEX)*, pp.112-118, 2000.
4. Chanrles C. Palmer, "Ethical Hacking," *IBM Systems Journal 3*, pp.769-780, 2001.
5. N. Wingfield, "It Takes a Hacker," *Wall Street Journal*, Mar. 11, 2002.
6. B. Skaggs, B. Blackburn, G. Manes, and S. Shenoi, "Network Vulnerability Analysis," *IEEE Midwest Symposium on Circuits and Systems (MWSCAS-2002)*, 2002.
7. UNIX Assembly Codes Development for Vulnerabilities Illustration Purposes, The Last Stage of Delirium Research Group, 2001, http://lsd-pl.net.
8. J. Scambray, S. McClure, and G. Kurtz *Hacking Exposed. 2nd Ed.* pp. 319-321, Osborne: McGraw Hill, 2001.
9. Dave Aitel, "The Advantages of Block-Based Protocol Analysis for Security Testing", 2002, http://www.immunitysec.com/resources-papers.shtml
10. CANVAS Homepage, http://www.immunitysec.com/products-canvas.shtml.
11. CORE IMPACT Homepage, http://www.coresecurity.com.
12. LibExploit Homepage, http://www.packetfactory.net/Projects/libexploit.
13. Metasploit Homepage, http://www.metasploit.com.
14. Common Vulnerabilities and Exposures Homepage, http://www.cve.mitre.org.
15. The 20 Most Critical Internet Security Vulnerabilities, Version 4.0, October 8, 2003, http://www.sans.org/top20.
16. Jack Koziol, Dave Aitel, David Litchfield, Cris Anley, Sinan Eren, Neel Mehta, and Riley Hassell, *The Shellcoder's Handbook Discovering and Exploiting Security Holes*, pp. 49-53, Wiley Publishing, Inc., 1997.
17. Win32 Assembly Components, The Last Stage of Delirium Research Group, 2002, http://lsd-pl.net.
18. LASSS Vulnerability, http://www.microsoft.com/technet/security/bulletin/MS04-011.mspx.
19. TruSecure Homepage, http://www.trusecure.com.

Cocyclic Jacket Matrices and Its Application to Cryptography Systems*

Jia Hou and Moon Ho Lee

Chonbuk National University, Institute of Information&Communication, chonju, 561-756, Korea
{jiahou, moonho}@chonbuk.ac.kr

Abstract. We describe the extended Hadamard matrices named Jacket in terms of combinatorial designs, and show that the center weighted Jacket matrices belong to a class of cocyclic matrices. Additionally, a simple public key exchange system on cocylcic Jacket matrices is proposed.

1 Introduction

Recently, Lee derived Jacket transform from the Walsh-Hadamard transform (WHT) and discrete Fourier transform (DFT) [5,6]. A generalized Jacket transform including both the WHT and a variant of the DFT for signals was described in [7]. The purpose of this paper is to show that the center weighted Jacket matrices belong to the classes of cocyclic matrices and generalized Butson Hadamard [1-4]. Different from the conventional matrices, the inverse form of Jacket matrices is only from the entrywise inverse and transpose.

Center Weighted Jacket Matrices: An $2^n \times 2^n$ matrix $[J]_{2^n}$ is of the form [5,6]

$$[J]_{2^n} = [J]_{2^{n-1}} \otimes [H]_2, \quad n \geq 3, \tag{1}$$

where $[J]_{2^2} = \begin{bmatrix} 1 & 1 & 1 & 1 \\ 1 & -w & w & -1 \\ 1 & w & -w & -1 \\ 1 & -1 & -1 & 1 \end{bmatrix}$, $w \neq 0$, and $[H]_2 = \begin{bmatrix} 1 & 1 \\ 1 & -1 \end{bmatrix}$.

Theorem 1: Assuming that G is a finite group of order v. A *cocycle* is a set of map which has [3, 4]

$$\varphi(g,h)\varphi(gh,k) = \varphi(g,hk)\varphi(h,k), \tag{2}$$

where $g, h, k \in G$ and $\varphi(1,1) = 1$. Then the cocycle φ over G is naturally displayed as a *cocyclic matrix* M_φ. This is a $v \times v$ matrix whose rows and columns are indexed by the elements of G, such that the entry in row g and column h is $\varphi(g,h)$.

* This work was supported by University IT Research Center Project, Ministry of Information & Communications, Korea.

C. Kim (Ed.): ICOIN 2005, LNCS 3391, pp. 662–668, 2005.

In previous work, generalized cocyclic Hadamard and different sets were reported and analyzed [3,4]. In this paper, we investigate a simple polynomial index representation on $GF(2^n)$ for constructing the cocylcic Jacket matrices.
A polynomial index on $GF(2^n)$: A set of index is defined by a recursive extension by using

$$G_{2^n} = G_{2^{n-1}} \otimes G_{2^1} \tag{3}$$

For given $G_2 = \{1, a\}$ and $\{1, b\}$, we can obtain

$$G_{2^n} = G_{2^{n-1}} \otimes G_{2^1} = \{1, b\} \otimes \{1, a\} = \{1, a, b, ab\}, \tag{4}$$

where the symbols $a, b, c, ...$ have $a^2 = b^2 = c^2 = ... = 1$. Further, the generalized extension way is illustrated in Fig.1. And this group G_{2^n} can be one to one mapped into polynomial Glaois field $GF(2^n)$, as shown in Table 1.

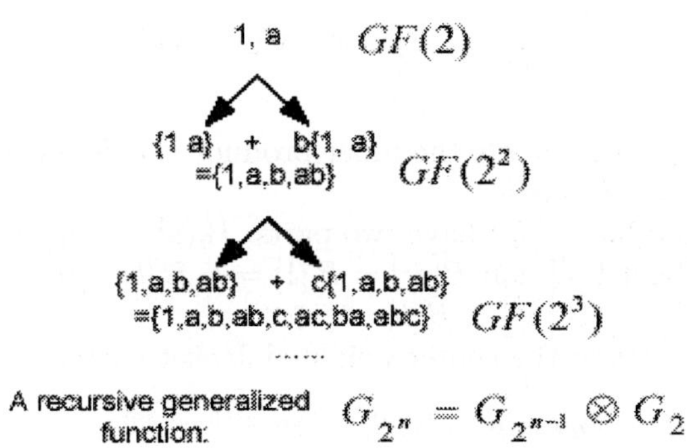

Fig. 1. Polynomial index extension.

Table 1. Representation G_{2^3} to $GF(2^3)$

Symbol	Binary	Exponential	Polynomial
1	000	0	0
a	001	α^0	1
b	010	α^1	x
c	100	α^2	x^2
ab	011	α^3	$x+1$
bc	110	α^4	x^2+x
abc	111	α^5	x^2+x+1
ac	101	α^6	x^2+1

2 Cocyclic Jacket Matrices

The center weighted Jacket matrices could be easily mapped by using a simple binary index representation [7,8]

$$sign = (-1)^{<g,h>} \tag{5}$$

where $< g, h >$ is the binary inner product. Such $g = (g_{n-1}g_{n-2}...g_0)$ and $h = (h_{n-1}h_{n-2}...h_0)$, and $< g, h >= g_0h_0+g_1h_1+...+g_{n-1}h_{n-1}$, where $g_t, h_t \in \{0, 1\}$. In the proposed polynomial index, we can use a special computation to present the binary inner product $< g, h >$,

$$< g, h > \triangleq (B\left[P_0(gh)\right] \oplus B\left[P_1(gh)\right] ... \oplus B\left[P_t(gh)\right]) \tag{6}$$

where $P_t(gh)$ denotes the tth part of the product of gh, $\oplus$ is mod 2 addition and the function $B\left[x\right]$ is defined by

$$B(x) = \begin{cases} 0 & x \in G_{2^n} - \{1\} \\ 1 & x = 1 \end{cases} \tag{7}$$

Example 1: If $g = a$, $h = ab$, the inner product $< g, h >$ can be calculated as
Step 1: $gh = a \cdot ab = a^2b$
Step2: The product of gh have two parts, $P_0(a^2b) = a^2$and $P_1(a^2b) = b$
Step 3: The bent function $B[a^2] = B\left[1\right] = 1$, $B\left[b\right] = 0$
Step 4: $< g, h >= B\left[a^2\right] \oplus B\left[b\right] = 1 \oplus 0 = 1.$
The weighted factors in the center weighted Jacket pattern can be presented by

$$weight = (i)^{(g_{n-1} \oplus g_{n-2})(h_{n-1} \oplus h_{n-2})} \tag{8}$$

where $i = \sqrt{-1}$. By directly using the polynomial index we can define the *weight* function equals to [7,8]

$$weight = (i)^{f(g)f(h)} \tag{9}$$

and

$$f(x) = \begin{cases} 1 & if (x_{n-1}x_{n-2}) \in \{a, b\} \\ 0 & others \end{cases} \tag{10}$$

where $(x_{n-1}x_{n-2}) \in GF(2^2)$, and $a, b \in \{1, a, b, ab\} \in GF(2^2)$.
Example 2: let $g = c$, $h = ac$, in $GF(2^3)$, we can calculate the $f(g)$ and $f(h)$ by using
Step 1: $g = c = (g_{n-1}g_{n-2}g_0) = (g_2g_1g_0) = (100)$

$$h = ac = (h_{n-1}h_{n-2}h_0) = (h_2h_1h_0) = (101)$$

Step 2: $(g_{n-1}g_{n-2}) = (10) = b \in GF(2^2)$

$$(h_{n-1}h_{n-2}) = (10) = b \in GF(2^2)$$

Step 3: $f(g) = 1$, $f(h) = 1$ from (10)
Step 4: $weight = (i)^{f(g)f(h)} = i$

Thus a center weighted Jacket pattern can be denoted by

$$[J]_{(g,h)} = sign \cdot weight = (-1)^{<g,h>}(i)^{f(g)f(h)} \tag{11}$$

Example 3: A four by four Jacket matrix with polynomial index can be mapped as below.

$$\begin{array}{c|c} g,h & 1 \ a \ b \ ab \\ \hline \begin{matrix} 1 \\ a \\ b \\ ab \end{matrix} & \begin{bmatrix} 1 & 1 & 1 & 1 \\ 1 & -w & w & -1 \\ 1 & w & -w & -1 \\ 1 & -1 & -1 & 1 \end{bmatrix} \end{array}$$

According to the pattern of (1), it is clearly that $\varphi(1,1) = 1$, and

$$\varphi(g,h) = (-1)^{<g,h>}(i)^{f(g)f(h)} \tag{12}$$

further we have

$$\begin{aligned} \varphi(g,h)\varphi(gh,k) &= (-1)^{<g,h>}(i)^{f(g)f(h)}\left((-1)^{<gh,k>}(i)^{f(gh)f(k)}\right) \\ &= (-1)^{<g,h>\oplus<gh,k>}(i)^{f(g)f(h)\oplus f(gh)f(k)} \end{aligned} \tag{13}$$

In the polynomial index mapping, the binary presentation of the product of two indexes equals to the addition of the binary presentation of each index, such as

$$Binary(gh) = ((g_{n-1} \oplus h_{n-1}), (g_{n-2} \oplus h_{n-2}), ... (g_0 \oplus h_0)) \tag{14}$$

Example 4: $g = a$, $h = b$ in $GF(2^2)$, $gh = ab => (1,1)$, it equals to

$$((g_1 \oplus h_1), (g_0 \oplus h_0)) = ((0 \oplus 1), (1 \oplus 0)) = (1,1)$$

where $(g_1g_0) = (0,1) = a$,$(h_1h_0) = (1,0) = b$. Based on (14), we now prove

$$< g,h > \oplus < gh,k >=< g,hk > \oplus < h.k > \tag{15}$$

Proof: $< g,h > \oplus < gh,k >= (g_{n-1}h_{n-1} \oplus g_{n-2}h_{n-2} \oplus ... \oplus g_0h_0) \oplus$

$$\begin{aligned} &((gh)_{n-1}k_{n-1} \oplus (gh)_{n-2}k_{n-2} \oplus ... \oplus (gh)_0k_0) = ((g_{n-1}h_{n-1} \oplus g_{n-2}h_{n-2} \oplus ... \oplus g_0h_0) \\ &\oplus ((g_{n-1} \oplus h_{n-1})k_{n-1} \oplus (g_{n-2} \oplus h_{n-2})k_{n-2} \oplus .. \oplus (g_0 \oplus h_0)k_0)) \\ &= (g_{n-1}(h_{n-1} \oplus k_{n-1}) \oplus g_{n-2}(h_{n-2} \oplus k_{n-2}) \oplus ... \oplus g_0(h_0 \oplus k_0)) \\ &\oplus (h_{n-1}k_{n-1} \oplus h_{n-2}k_{n-2} \oplus ... \oplus h_0k_0) =< g,hk > \oplus < h,k > \end{aligned} \tag{16}$$

and we obtain

$$(-1)^{<g,h>\oplus<gh,k>} = (-1)^{<g,hk>\oplus<h.k>} \tag{17}$$

Similarly, we can prove that

$$f(g)f(hk) \oplus f(h)f(k) = f(g)f(h) \oplus f(gh)f(k) \tag{18}$$

Proof: $f(g)f(hk) \oplus f(h)f(k) = (g_{n-1} \oplus g_{n-2})((h_{n-1} \oplus k_{n-1}) \oplus (h_{n-2} \oplus k_{n-2}))$
$\oplus (h_{n-1} \oplus h_{n-2})(k_{n-1} \oplus k_{n-2}) = (g_{n-1} \oplus g_{n-2})(k_{n-1} \oplus k_{n-2}) \oplus (h_{n-1} \oplus h_{n-2})$

$$= (g_{n-1} \oplus g_{n-2})(h_{n-1} \oplus h_{n-2}) \oplus ((g_{n-1} \oplus g_{n-2}) \oplus (h_{n-1} \oplus h_{n-2})) (k_{n-1} \oplus k_{n-2})) = f(g)f(h) \oplus f(gh)f(k) \quad (19)$$

And the function has

$$(i)^{f(g)f(hk) \oplus f(h)f(k)} = (i)^{f(g)f(h) \oplus f(gh)f(k)} \quad (20)$$

As a result, we can prove that any Jacket pattern from

$$\varphi(g,h) = (-1)^{<g,h>}(i)^{f(g)f(h)} \quad (21)$$

has

$$\begin{aligned} \varphi(g,h)\varphi(gh,k) &= (-1)^{<g,h>}(i)^{f(g)f(h)}\left((-1)^{<gh,k>}(i)^{f(gh)f(k)}\right) \\ &= (-1)^{<g,h> \oplus <gh,k>}(i)^{f(g)f(h) \oplus f(gh)f(k)} \\ &= \varphi(g,hk)\varphi(h,k) = (-1)^{<g,hk> \oplus <h.k>}(i)^{f(g)f(hk) \oplus f(h)f(k)} \end{aligned} \quad (22)$$

In the Example 3, we can take $g = a$, $h = b$, $gh = ab$, $k = 1$ and $\varphi(g,h) = \varphi(a,b) = -i$, $\varphi(gh,k) = \varphi(ab,1) = 1$, $\varphi(g,hk) = \varphi(a,b) = -i$, $\varphi(h,k) = \varphi(b,1) = 1$, thus we have

$$\varphi(g,h)\varphi(gh,k) = \varphi(g,hk)\varphi(h,k) = (-i) \times 1 = -i \quad (23)$$

3 A Simple Application to Cryptography System

The cocyclic Jacket matrices will be very useful in combination theory and designs. Especially, the results will be applied for cryptography. Public key cryptography provides a radical departure from all that has gone before. More important, public key cryptography is asymmetric, involving the use of two separate keys, in contrast to symmetric encryption, which uses only one key. The use of two keys has profound consequences in the areas of confidentiality, key distribution, and authentication, as we shall see [9]. Stronger security for public-key distribution can be achieved by providing tighter control over the distribution of public keys from the directory. In order to protect the transmitted information, we propose a simple public key exchange system by using cocyclic Jacket matrices. The encryption algorithm is denoted as follows.
Public Key Exchange Cryptography Algorithm:

Step 1: The authentication (network center) generates two random number $g, h \in GF(2^n)$, and a authentication key $k \in GF(2^n)$.

Step 2: Generated information

$\{g, h, hk\}$for user A; $\{h, k, gh\}$for user B.

The transmitted information for every user will hide at least one authentication information. For example, for user A, we transmit $\{g, h, hk\}$, the information $\{k\}$ is hidden in $\{hk\}$, it will be used to protect the information from the partial wiretapping.

Step 3: Jacket matrices generation
Different weight factors should be considered to protect the information as a private confusion key.

Step 4: User A Key Generation
Select Private: $n_A = \varphi(g, h)$; Calculate public: $P_A = \varphi(g, hk)$.
$\varphi()$:Cocyclic function on Jacket matrices.

Step 5: User B Key Generation
Select Private: $n_B = \varphi(h, k)$; Calculate public: $P_B = \varphi(gh, k)$.

Step 6: The public key exchange
Send P_A to user B; Send P_B to User A.

Step 7: Generation of Secret Key by User A and User B
User A: $K_A = n_A \times P_B$; User B: $K_B = n_B \times P_A$;
If $K = K_A = K_B$, the Key exchange is success.
Since the $\varphi()$ is cocyclic function on Jacket matrices, we can easily prove that

$$\begin{aligned} K_A &= n_A \times P_B = \varphi(g,h)\varphi(gh,k) = \varphi(h,k)\varphi(g,hk) \\ &= K_B = n_B \times P_A = K \end{aligned} \quad . \tag{24}$$

The cocyclic function public key exchange system is shown in Fig.2. For example, according to (23), we have $n_A = \varphi(a, b)$, $P_A = \varphi(a, b)$, $n_B = \varphi(b, 1)$, and $P_B = \varphi(ab, 1)$, thus we obtain

$$n_A \times P_B = \varphi(g,h)\varphi(gh,k) = (-w) \times 1 = n_B \times P_A = K = -w \tag{25}$$

where we use a weight factor w in previous introduction. Note that the weight factor w can be any inverseable nonzero values. To protect the transmitted information, the generated secret key K can be a polyphase sequence from the cocylcic Jacket matrices.

4 Conclusion

We generalize the cocyclic Jacket matrices by using the combinatory theory in this paper. As a result, the Cocyclic Hadamard can also be proved, if the weighted factor w changed to 1. Based on a novel index group and binary mapping for Jacket matrices we show that it can be efficiently applied to cryptography or other areas, similarly to the generalized Hadamard and generalized Butson matrices. Additionally, a simple public key cryptography system according to the cocyclic Jacket matrices is proposed, it shows that the proposed algorithm will efficiently give the protection of the consumers.

References

1. A.T.Butson: Nonlinear oscillations and boundary-value problems for Hamiltonian systems. Proc. Amer. Math. Soc. **13** (1963) 894–898
2. C.J. Colbourn and J,H, Dinitz: The CRC Handbook of Combinatorial Designs. CRC Press, Boca Raton.(1996)

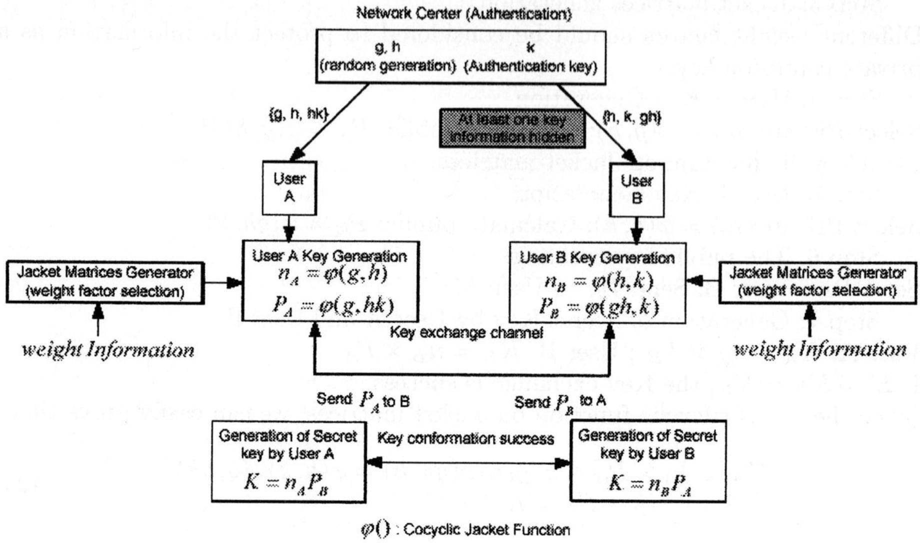

Fig. 2. The public key exchange system by using cocyclic function on Jacket matrices.

3. D.A. Drake: Partial λ-geometries and generalized Hadamard matrices over groups. Canad.J. Math. **31** (1979) 617–627
4. K.J. Horadam and P. Udaya: Cocyclic Hadamard codes. IEEE Trans. Inform. Theory.**46** no.4, (2000) 1545-1550
5. Moon Ho Lee: The center weighted Hadamard transform. IEEE Trans. Circuits Syst.. **36** no.2, (1980) 1247–1249
6. Moon Ho Lee:A new reverse jacket transform and its fast algorithm. IEEE Trans. Circuits Syst. II. **47** no.1, (2000) 39–47
7. Moon Ho Lee, B. Sunder Rajan and J. Y. Park:A generalized reverse jacket transform. IEEE Trans. Circuits Syst. II. **48** no.7, (2001) 684–690
8. Moon Ho Lee, J. Y. Park and S. Y. Hong: Simple Binary Index Generation for Reverse Jacket Sequence. In Proceedings of International Symposium on Information Theory and Applications (ISITA 2000). **1** Hawaii, USA, Nov. 5-8,(2000) 329–433
9. William Stallings:A generalized reverse jacket transform. Cryptography and Network Security. Pearson Education, Inc. U.S. (2003)

Design and Implementation of SIP Security

Chia-Chen Chang[1], Yung-Feng Lu[1], Ai-Chun Pang[1,2], and Tei-Wei Kuo[1,2]

[1] Department of Computer Science and Information Engineering
{r91086, d93023}@csie.ntu.edu.tw
[2] Graduate Institute of Networking and Multimedia
National Taiwan University, Taipei, Taiwan 106, ROC
{acpang, ktw}@csie.ntu.edu.tw

Abstract. Session Initiation Protocol (SIP) is the Internet Engineering Task Force (IETF) standard for IP telephony. It is the most promising candidate as a signaling protocol for the future IP telephony services. Comparing with the reliability provided by the traditional telephone systems, there is an obvious need to provide a certain level of quality and security for IP telephony. The problem of security is tightly related to the signaling mechanisms and the service provisioning model. Therefore, there is a very hot topic in the SIP and IP telephony standardization for security support. In this paper, we propose a security mechanism for SIP-based voice over IP (VoIP) systems. Specifically, we focus on the design and implementation of SIP authentication for VoIP users. The performance issues for our secure SIP-based VoIP systems are also investigated. By means of a real testbed implementation, we provide an experimental performance analysis of our SIP security mechanisms, and compare the performance with the methods described in standardization.

1 Introduction

The Session Initiation Protocol (SIP) [1] proposed by Internet Engineering Task Force (IETF) is a signaling protocol for establishing real-time multimedia calls and sessions. SIP defines four basic classes of network entities: (1) registrar servers, (2) redirect servers, (3) proxy servers, and (4) user agents (UA). A registrar server is responsible for keeping the registration information of the user. A redirect server is a server that accepts SIP requests, maps the destination address to zero or more new addresses, and returns the translated address to the originator of the request. A proxy server either handles those requests itself or forwards them to other servers. The user agent represents an application that contains both the user agent client (i.e., calling party) and user agent server (i.e., called party).

It is widely guaranteed that the traditional public switched telephone network (PSTN) has a good level of quality of service (QoS) and security. If new IP telephony architectures such as that defined in SIP would like to replace the PSTN, they shall provide the same basic telephony service with a comparable level of QoS and network security. SIP defines some security mechanisms such as Digest [2] and S/MIME [3] (which will be elaborated in the following section).

C. Kim (Ed.): ICOIN 2005, LNCS 3391, pp. 669–678, 2005.

SIP messages carry MIME bodies where the MIME standard includes mechanisms for securing MIME contents to ensure both integrity and confidentiality.

In Section 2, we describe some security considerations on the VoIP network, and elaborate on the existing solutions for SIP security. In Section 3, we introduce our design and implementation for SIP security, and describe our proposed system architecture and the components used in the architecture. Then, we perform the experiments based on our implementation testbed, and evaluate the performance of our security mechanism and compare with that of the existing solutions. Finally, we conclude this paper and future works.

2 Related Works

2.1 Security Considerations

The considerations examine a set of classic threat models that broadly identify the security needs of SIP. It is anticipated that SIP will be used on the public Internet. Attackers on the network are able to modify the packet and wish to steal services, eavesdrop on communications, or disrupt sessions. We list the threats to SIP and the corresponding security requirements as follows.

Eavesdropping is the most common attack to a network system. If the traffic is not encrypted, the content of the traffic is exposed to the public. *Confidentiality* is a security service against eavesdropping. Tampering is the attacker actively modifying the traffic to cheat the other party and obtain some benefit. *Integrity* is security requirement against this attack. The implementation against this kind of attack can be achieved by adding a messages authentication code (MAC) to the packet. Replaying attack is the attack that an attacker replays some previous eavesdropped packets to other parties. To prevent these attacks, the timestamp is typically used to guarantee the Freshness of the messages.

2.2 The Existing Security Mechanisms and Limitations

SIP signaling between multiple users consists of requests and responses. SIP messages contain some important information that a user or a server wants to keep private. The existing security mechanisms in standardization are listed as follows.

- *HTTP Basic* [2]: authentication is an industry-standard method that is used to gather user name and password information. Due to this weak security, the usage of "Basic" authentication has been deprecated. Digest Authentication. As for HTTP Digest authentication, although it solves the problem in Basic authentication, it is susceptible to chosen plaintext attack. The attacker can gather many responses from different nonces, and then tries out the password with brute force.
- Regarding the use of the *S/MIME* (Secured Multipurpose Internet Mail Extension), SIP message carries S/MIME bodies to ensure both integrity and confidentiality. However, it is susceptible to man-in-middle (MIMD) attack

when the first exchange of keys. Another disadvantage of S/MIME used by SIP is that a very large size of S/MIME messages will be generated when the SIP tunneling mechanism is used (please see section 23.4 of RFC 3261) [1].

- *IPSec* (IP Security) [4] is standardized by IETF. The purpose of IPSec is to protect IP packets, which can be accomplished by ESP and AH. IPSec provides confidentiality and integrity of communication in the IP layer and authenticates each other in communicating parties. Although IPsec has been deployed widely, IPsec only authenticates machines, not users.
- *TLS* (Transport Layer Security) [5] is the transport layer protocol. TLS is used to provide encryption and integrity services to higher-level network applications. However, it may be arduous for a local proxy server to maintain many simultaneous long-lived TLS connections with numerous UAs. Furthermore, the TLS is a kind of public-key based cryptosystems and is also susceptible to man-in-the-middle attack.

3 System Implementation

3.1 Network Architecture

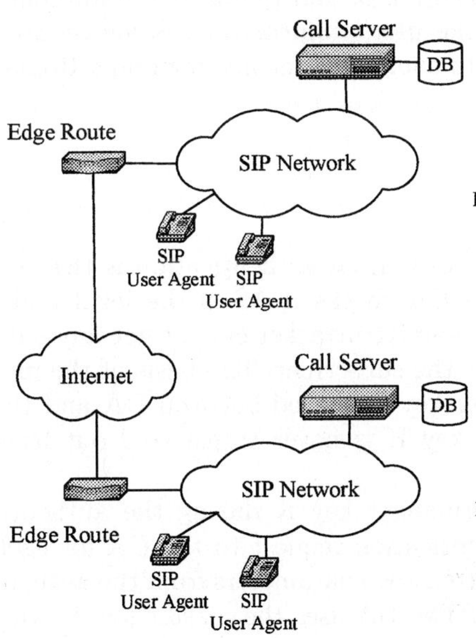

Fig. 1. The NTP VoIP Network Architecture

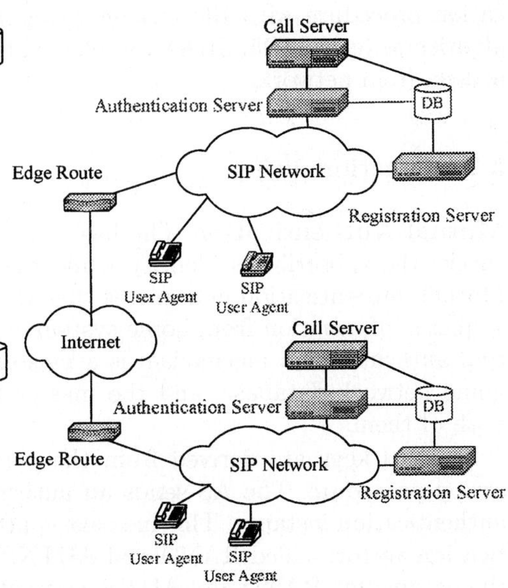

Fig. 2. The Proposed Architecture

Figure 1 shows the network architecture of a NTP (National Telecommunications Development Program) VoIP platform which conducts R&D in advanced telecommunication technologies, for the promotion and development of telecommunication industry. The functionalities of the network entities in this architecture are described as follows.

Call Server (CS) is located in a service provider's network and provides call control functions. The CS is an integration of SIP proxy and registrar server, and includes call originating, terminating or forwarding functions. An open-source CS from IPTel (http://www.Iptel.org), named SER (SIP Express Router) is used in this architecture. *Database* (DB) is located in a service provider's network, and co-located with a CS. *User Agent*(UA) is an application program residing in the user's device. The functions of a UA have been elaborated in Section 1, and we do not re-iterate them. The source of our UAs comes from CCL (Computer & Communication Research Laboratories)/ITRI (Industrial Technology Research Institute).

Based on Figure 1, we introduce and implement two new network entities, Authentication Server (AS) and Registration Server (RS), to provide a secure authentication mechanism (see Figure 2). *Authentication Server* (AS) is responsible for authenticating VoIP users. The mutual authentication is provided by our AS to prevent a fake server from impersonating a legal server. *Register Server* (RS) is a web-based interface for users to apply for a new account. Also, the functionalities of UA and DB are extended to support our proposed mechanism. AS retrieves the user information by querying DB, and performs the authentication procedure with the UA by using the user's information. Some security algorithms (e.g., MD5, SHA) are selected to avoid attacks, like Spoofing, Replay attack, from network.

3.2 Functionalities

Mutual Authentication The basic idea of mutual authentication is that AS checks the subscriber's identity while the UA checks if AS is the legal node. Mutual authentication guarantees that the active attacker cannot get any subscriber's information from home system or the subscriber. The basis of the mutual authentication mechanism is a master key K shared between UA and the home network database, and the master key K is never transferred out from both of them.

Several keys are derived from the permanent key K during the authentication procedure. The AS sends an authentication request to the UA for each authentication instance. This message contains two parameters from the authentication vector, called RAND and AUTN. The UA uses the master key K with the parameters RAND and AUTN as inputs, and carries out a computation like the generation of authentication vectors in AS. This process for authentication vector (AV) generation contains executions of several algorithms, which will be described in the following sections. As the result of the computation, the UA is able to verify whether the parameter AUTN was indeed generated in a legal AS. At the same time, the computed parameter RES is sent back to the AS in the authentication response. By using the authentication response, the AS is able to compare the user response RES with the expected response XRES from the authentication vector.

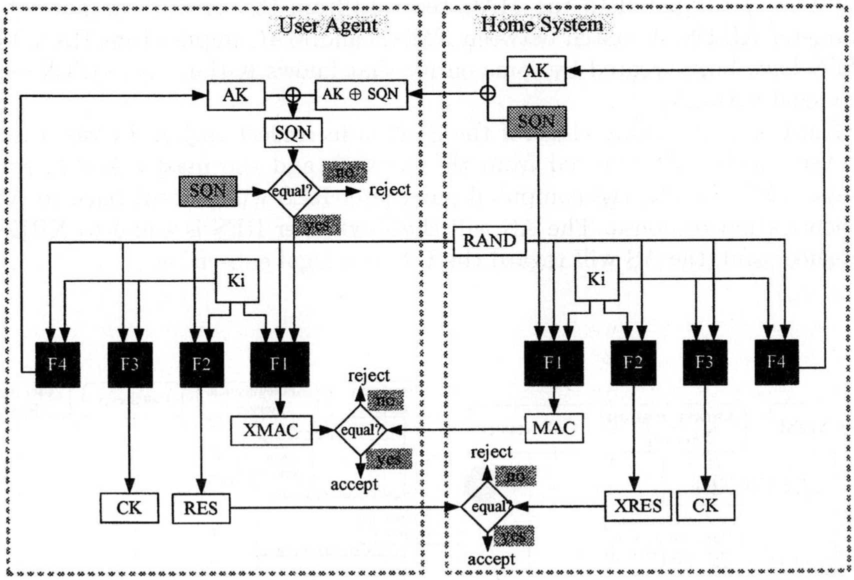

Fig. 3. Key Generation Function

Authentication Vector Generation in HS The process begins by selecting a sequence number SQN (note that SQN is in an increasing order). The purpose of the sequence number is to guarantee that the generated AV is fresh. In parallel, a random string RAND is generated.

The computation for generating the authentication vector relies on one-way functions. In our system, four one-way functions, SHA256, MD5, SHA128 and SHA1, are used to compute the authentication vector. For convenience, those functions are denoted as F1, F2, F3 and F4. The function F1 differs from the other three in the number of input parameters. It takes three input parameters: master key K, random number RAND and sequence number SQN. The other functions F2, F3 and F4 take only K and RAND as inputs. The output of F1 is Message Authentication Code (MAC), and the outputs of F2, F3, and F4 are, respectively, XRES, CK, AK. The authentication vector consists of the parameters RAND, XRES, CK and AUTN. The last one is obtained by concatenating two different parameters: MAC and the result of bit-by-bit exclusion or operation on SQN and AK.

Authentication Handling in UA The same functions are involved on this side, but in a slightly different order. The function F4 has to be computed before the function F1 because F4 is used to conceal SQN. This concealment is needed in order to prevent eavesdroppers from getting information about the user identity through SQN. The output of the function F1 is marked as XMAC on the user side. This is compared to the MAC received from the network as part of the parameter AUTN. A match between XMAC and MAC implies that RAND and AUTN have been created by some entity that knows K (i.e., the AUTN comes from legal network).

The UA shall simply check if the SQN is in correct region. In yes, the UA will use the RAND received from the network and the master key K as the inputs of F2. Finally, the computed parameter RES will be sent back to AS in authentication response. The AS will check whether RES is equal to XRES. If it is successful, the AS will regard the UA as a legal subscriber.

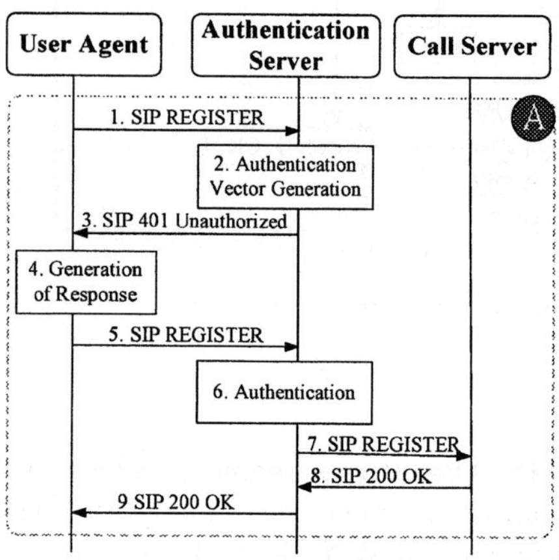

Fig. 4. The Registration Procedure

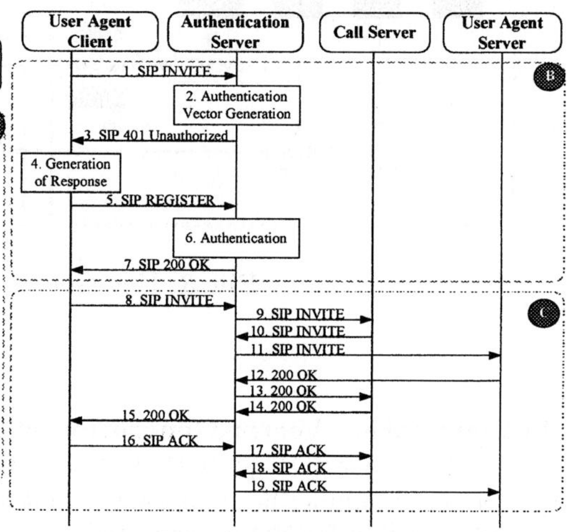

Fig. 5. The Call Setup Procedure

Message Flow *Scenario 1: Registration procedure (user not registered yet)*
Figure 4 shows the registration signaling flow when the user is not registered yet. The subscriber will be authenticated by the AS. If the authentication procedure is successful, the AS will notify the Call Server about the location of the UA. For convenience, we denote this part as A.

Scenario 2: Call setup procedure
Figure 5 shows the message flow of connection establishment between UAC and UAS. In the first part (Steps 1∼7), AS asks UAC to start the authentication procedure as mentioned in Part A of the registration procedure. The second part (Steps 8∼19) is the normal SIP connection establishment procedure, and the details are omitted.

3.3 Security Analysis

Since it is susceptible to MITM attack in public-key based systems in the first key exchange, the pre-share key (i.e., the master key K) based cryptosystem is adopted in our security mechanism to avoid MITM attack. Furthermore, Table 1 provides a mapping between the potential security threats of an SIP-based application system and the countermeasures for our security mechanism.

Table 1. Relation of Against Threats, Keys, and Countermeasures for Our Security Mechanism

Against Threats	Keys	Countermeasures
Replay Attack	Sequence Number (SQN)	- The client is able to detect that the request is old by checking SQN. - We create a key, AK to conceal SQN from eavesdropping.
Chosen Plaintext Attack	Message Authentication Code (MAC)	- By checking the MAC, the fake authentication procedure will be refused. - It provides mutual authentication between the server and the UA.
Eavesdropping Attack	Cipher Key (CK)	- The sensitive data will be encrypted, and the CK will be changed in each authentication procedure to assure that the CK is fresh.

4 Performance and Security Feature Analysis

In order to experiment with advanced security features in SIP, we have implemented the authentication functions embedded in the CCL SIP UA. A high-performance, configurable and free IPTel SIP Express Router (SER), is adopted for the SIP server. The testbed consists of PCs acting as SIP UA, SIP AS, and SIP call server. The UAs and AS run in separate PCs (Pentium IV 2.4GHz/256 MB RAM) equipped with Windows XP professional operating system. The call server runs in a PC (Intel XEON 2.4GHz /512MB RAM) equipped with GMU/Linux (kernel 2.4.22) operating system. Those PCs are connected to a Fast Ethernet switch (100MB).

4.1 Performance Evaluation

The goal of this experiment is to saturate the processing capability of the AS and measure its maximum throughput.

In this experiment, the SIP UA is a multithreaded application that generates SIP REGISTER procedure, and each thread in the SIP UA generates a series of SIP REGISTER procedure. That is, a new REGISTER procedure is generated

Table 2. The Processing Capability of an AS

Threads	1	2	3	4	5	6
Average RTT (s)	0.112	0.231	0.378	0.537	0.711	0.917
Total throughput (s^{-1})	8.929	8.658	7.937	7.449	7.032	6.543

immediately when the previous procedure is completed (by receiving a 200 OK message). Then the round-trip time (RTT) for each transaction is measured. Table 2 shows the average RTT of one thread is 0.112 second. It is apparent that our system can serve 8.9 threads per second. Furthermore, when six threads simultaneously send requests to our AS, the average RTT for one thread is 0.917 second. In other words, our system still can serve 6.54 threads per second.

Table 3. Experiment Result under IPSec mechanism

Phase	Time (ms)
Security tunnel establishment	1022.27
Call establishment	31.44
Location update	2.29
IKE negotiation	702.12

The goal of the following experiments is to measure the RTTs of a single REGISTER procedure and INVITE procedure for our security mechanism and compare those with pure SIP procedures (i.e., without incorporating the authentication operation). In REGISTER part, the definition of RTT is the time interval between sending a REGISTER message and finally receiving an OK message. In INVITE part, the definition of RTT is the time interval between sending an INVITE message and finally receiving an OK message.

If the UA makes a location update to Call Server directly and makes a connection request through Call Server without activating the authentication procedure. The average RTT is 26.611 milliseconds. On the other hand, if our authentication mechanism is incorporated into this procedure (see Figure 5), the average RTT for Part B in Figure 5 is 112.102 milliseconds, and the average RTT in Part C in Figure 5 is 31.421 milliseconds. Furthermore, we compare our purposed architecture with an engineering solution that an original SIP using IPSEC to protect authentication packets. Table 3 presents the measured time interval for different phases (i.e., security tunnel establishment, call establishment, location update and IKE negotiation) of IPSec.

Based on the data shown in Table 3, Figures 6 and 7 compare the performance of our security mechanism with that of IPSec based on the accumulative signaling overhead. We assume that the subscriber uses the SIP phone in the continued 10 hours and he makes a phone call every hour. Figure 6 shows that the cost

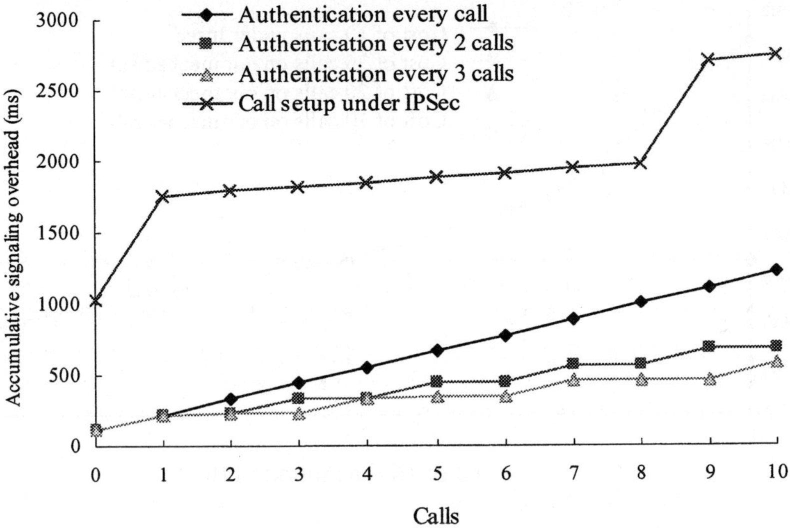

Fig. 6. Accumulative Signaling Overhead: Our Security Mechanism vs. IPSec
†Note: the default IKE negotiation performs every eight hours.

(i.e., accumulative signaling overhead) of IPSec increases substantially when the IKE negotiation is performed. We found that the total time of performing the authentication process for every call is lower than IPSec. Furthermore, we can flexibly change the frequency of authentication operation in our system. Therefore, as the frequency of the activated authentication procedure decreases, the system cost is decreased. Figure 7 illustrates the cost for different frequencies of IKE negotiations under IPSec. When IKE negotiation is performed every four or less than 4 hours, the cost of IPSec is larger than that of our security mechanism with 30 calls. However, when the frequency of IKE negotiation for IPSec is decreased, the cost of IPSec significantly decreases.

5 Conclusions

Session Initiation Protocol (SIP) is the most promising candidate as a signaling protocol for the future IP telephony services. Comparing with the reliability provided by the traditional telephone systems, there is an obvious need to provide a certain level of quality and security for voice over IP (VoIP) networks. In this paper, the main security aspects related to SIP-based VoIP services were discussed. We have designed and implemented a security mechanism that provides mutual authentication, confidentiality, and encryption for a SIP-based VoIP system. We have also experimented different security procedures for our developed security mechanism. The experimental results have showed that our system not only provides good level security, but also takes less time to achieve authentication procedure and call establishment.

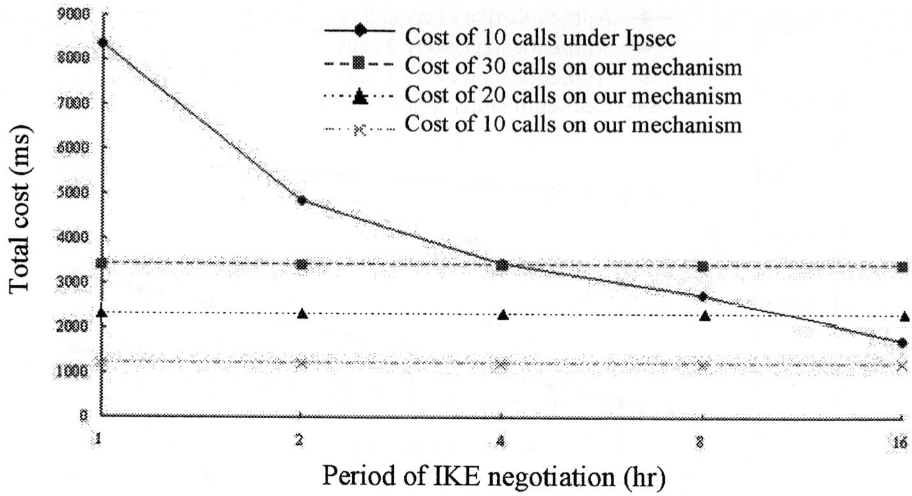

Fig. 7. The Cost (i.e., Accumulative Signaling Overhead) under Different Frequencies of IKE Negotiations for IPSec

Acknowledgement

This work was supported in part by National Science Council under contract NSC 93-2213-E-002-093 and NSC 93-2213-E-002-025, Intel, Microsoft and CCL/ITRI.

References

1. J. Rosenberg, et. al., "SIP: Session Initiation Protocol", IETF RFC 3261, June 2002.
2. J. Franks, et. al., "HTTP Authentication: Basic and Digest Access Authentication", IETF RFC 2617, June 1999.
3. S. Dusse, et. al. "S/MIME Version 3 Message Specification" IETF RFC 2633, June 1999.
4. Kent, S., and R. Atkinson, "Security Architecture for the Internet Protocol", RFC 2401, November 1998.
5. T. Dierks et. al. "The TLS Protocol Version 1.0", IETF RFC 2246, January 1999.
6. 3GPP. 3rd Generation Partnership Project; Technical Specification Group Services and System Aspects; 3G Security; Security architecture (Release 6). 3GPP TS 33.102 V6.1.0 (2004-06).
7. 3GPP. 3rd Generation Partnership Project; Technical Specification Group Services and System Aspects; 3G Security; Access security for IP-based services (Release 6). 3GPP TS 33.203 V6.3.0 (2004-06).
8. Stefano Salsano, Luca Veltri, and Donald Papalilo, "SIP Security Issues: The SIP authentication procedure and its processing load", IEEE Network, November/December 2002.
9. H. Kaaranen, A. Ahtiainen, L.Laitinen, S.Naghian, V. Niemi, "UMTS Networks Architecture, Mobility and Services," John Wiley & Sons, 2001.

Algorithm for DNSSEC Trusted Key Rollover

Gilles Guette, Bernard Cousin, and David Fort

IRISA, Campus de Beaulieu, 35042 Rennes CEDEX, FRANCE
{gilles.guette, bernard.cousin, david.fort}@irisa.fr

Abstract. The Domain Name System Security Extensions (DNSSEC) architecture is based on public-key cryptography. A secure DNS zone has one or more keys and signs its resource records with these keys in order to provide two security services: data integrity and authentication. These services allow to protect DNS transactions and permit the detection of attempted attacks on DNS.
The DNSSEC validation process is based on the establishment of a chain of trust between zones. This chain needs a secure entry point: a DNS zone whose at least one key is trusted. In this paper we study a critical problem associated to the key rollover in DNSSEC: the trusted keys rollover problem. We propose an algorithm that allows a resolver to update its trusted keys automatically and in a secure way without any delay or any break of the DNS service.

1 Introduction

During the last decade, the Internet Engineering Task Force (IETF) has developed the DNS security extensions (DNSSEC). The first standard document designing DNSSEC is the RFC 2535 [1]. Many RFCs and drafts have updated the RFC 2535 and according to the experience given by implementations, this protocol is today enhanced [2] [3] [4].

The DNSSEC architecture uses public key cryptography to provide integrity and authentication of the DNS data. Each node of the DNS tree, called a *zone*, owns at least a key pair used to secure the zone records with digital signatures.

In order to validate DNSSEC records, a resolver builds a chain of trust [5] by walking through the DNS tree from a secure entry point [6] (typically a top level zone) to the zone queried. Each secure entry point (SEP), is statically configured in a resolver: the resolver knows at least one key of the zone which is taken as SEP, this key is called a *trusted key*. A resolver is able to build a chain of trust if it owns a secure entry point for this query and if there are only secure delegations from the secure entry point to the zone queried.

The lifespan of keys used to secure DNS zone is not infinite, because old keys become weak. Consequently, the zone keys must be renewed periodically, that is to say a new zone key is added to the zone file and an old one is deleted from the zone file. This process is called key rollover [7].

Key rollover only updates keys on the name server, resolvers that have configured the old key as trusted are not notified that this key has been deleted.

C. Kim (Ed.): ICOIN 2005, LNCS 3391, pp. 679–688, 2005.

Consequently, the static key configuration in a resolver raises some problems of consistency between keys deployed in a zone and trusted keys configured in a resolver for this zone.

In section 2 we present the notations use in this paper and the DNSSEC validation process. Then, in section 3 we describe the trusted key rollover problem and finally we present our solution to this problem in section 4.

2 Validation Process in DNSSEC

2.1 Definitions

In this subsection are explained the notations used in the document.

- A DNS domain X is the entire subtree beginning at the node X.
- A DNS zone is a node of the DNS tree. A zone name is the concatenation of the nodes labels from this node to the root of the DNS tree. A zone contains all the not delegated DNS names ended by the zone name. For example, the zone `example.com.` contains all the not delegated names `X.example.com.` where X can be composed by several labels.
- A zone can delegate the responsibility of a part of its names. The zone `example.com.` can delegate all the names ended by `test.example.com.` to a new zone. This zone is named the `test.example.com.` zone (cf. Fig 1).

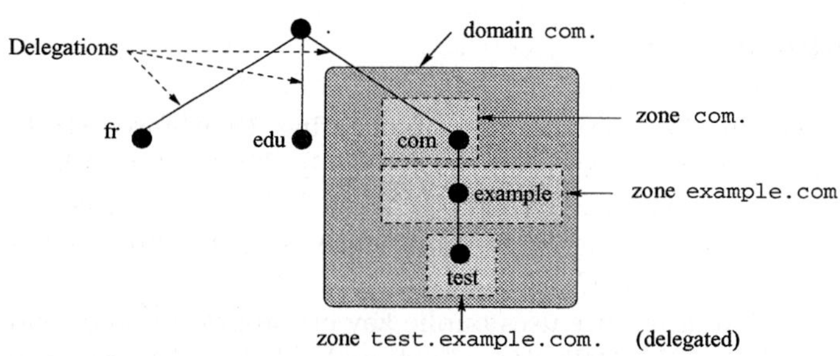

Fig. 1. DNS domains and DNS zones.

- RR means Resource Record, the basic data unit in the Domain Name System. Each RR is associated to a DNS name. Every RR is stored in a zone file and belongs to a zone.
- Resource records with same name, class and type fields form a RRset. For example the DNSKEY RRs of a zone form a DNSKEY RRset.
- DNSKEY($key1$) is the RR which describes the key named *key1*.
- RRSIG$(X)_y$ is the RR which is the signature of the RR X generated with the private part of key y.
- A trusted key is the public part of a zone key which is configured in a resolver.
- A Secure Entry Point is a zone for which the resolver trusts at least key.

2.2 DNS Entities

Three entities with distinct roles are present in the DNS architecture: the name servers, the resolvers and the cache servers (see [8,9]).

The name server. The name server is authoritative on a DNS zone. It stores resource records in its DNS zone file. Every resource record is associated to a DNS name. The name server receives DNS queries on a DNS name and answers with the resource records contained in its zone file.

The resolver. The resolver is the local entity that receives a request from an application and sends DNS queries to the appropriate name server. After having performed the name resolution, the resolver sends back the answer to the application.

The cache server. The cache server is not authoritative on any zone. The scalability of DNS is based on the use of cache servers. These cache servers only forward queries and cache the valid responses until they timeout.

2.3 DNSSEC Chain of Trust

DNS security extensions define new resource records in order to store keys and signatures needed to provide integrity and authentication.

Each secured zone owns at least one zone key. The public part of this key is stored in a DNSKEY resource record. The private part of this key is kept secret and should be stored in a secure location. The private part of the key is used to generate a digital signature of each resource record in the zone file. These signatures are stored in a RRSIG resource record. A resource record is considered valid when the verification of *at least one* of its associated RRSIG RR is complete. Figure 2 shows the signature verification process.

In order to verify the signature of a resource record, the resolver cyphers the RRSIG RR with the public key of the zone contained in the DNSKEY RR present in the DNS answer message. If the result of this operation, called `Resource Record'` in figure 2 is equal to `Resource Record` present in the DNS response message, the signature is verified. Thus, the resource record is valid.

During the signature verification process, the zone key is needed and must be verified too. This allows to avoid the use of a fake key sent in a message forged by a malicious person. To trust a zone key, DNSSEC uses the DNS-tree model to establish a chain of trust [5] beginning from a secure entry point [6] to the queried zone. To create this chain, a verifiable relation between child zone and parent zone must exist: this is the role of the Delegation Signer resource record (DS RR) [10]. This record, stored in the parent zone, contains information allowing the authentication of one child zone key. Figure 3 shows the validation of a delegation.

Once the signatures (the signature of the DS RR provided by the parent zone and the signature of the DNSKEY provided by the child zone) are verified, the resolver checks that information contained in one DS RR identifies one key in the child zone. If one DS RR identifies one DNSKEY RR in the child zone, a link of the chain of trust is built and the name resolution progress to secure the next

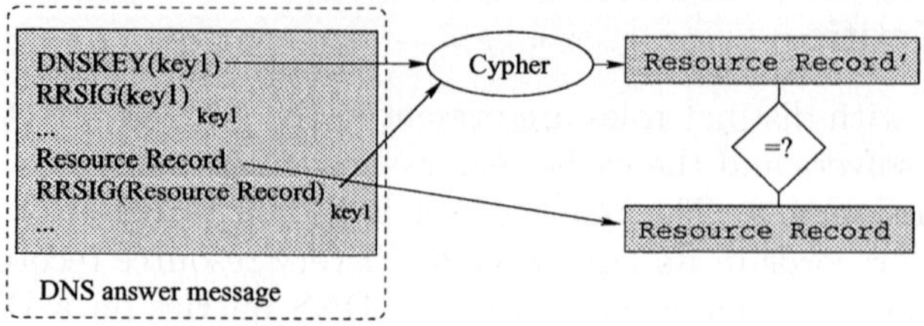

Fig. 2. The signature verification process.

link in the DNS tree. If there is no valid DS RR that identifies a valid DNSKEY RR in the child zone, the chain of trust is broken and the name resolution is unsecure.

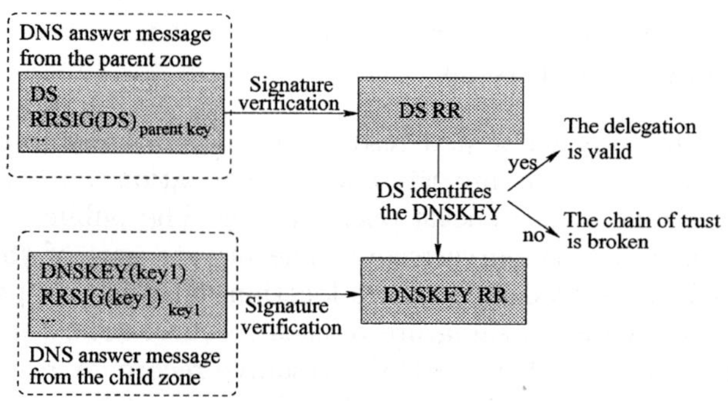

Fig. 3. The delegation verification process.

The Delegation Signer model introduces a distinction between two types of keys: the Zone Signing Keys (ZSK) and the Key Signing Keys (KSK). A ZSK signs all types of record in the zone file and a KSK signs only the DNSKEY RRs present in the zone file. This distinction minimizes the burden produced when a key is changed in the child zone, because only KSKs have an associated DS RR in the parent zone file. Hence, a DNSKEY RR can be considered valid only by verifying of a signature generated by a KSK.

3 The Trusted Key Rollover Problem

The validation process of resource records based on the DNSSEC chain of trust needs a secure entry point to start. The RFC 2535 [1] specifies that a DNSSEC resolver must be configured with at least a public key which authenticates one zone as a secure entry point. If DNSSEC is totally deployed on the whole DNS tree, only one secure entry point is sufficient, *i.e.* the root zone. But due to deployement constraints, the current model that seems to emerge is an *island of security* model with DNS zones and DNSSEC zones at the same time in the DNS tree. An island of security is a subtree of the DNS tree totally secured with DNSSEC (each zone of this subtree has a signed zone file). Consequently, a resolver needs at least one secure entry point for the apex of each island of security in order to perform secure name resolution for any zone.

Moreover, to maintain a good level of security, zone keys have to be renewed at regular intervals, to prevent against key disclosure. And consequently, trusted keys must be updated in resolvers to keep consistency between the key in the zone file of a name server and the trusted key in the resolvers.

Currently, the rollover of a trusted key in the resolver's configuration file is done manually by the administrator, this implies risks of misconfiguration and an interruption of the service between the moment the zone rolls its keys and the moment the administrator changes the resolver's configuration file. At the present time, there is no automated trusted key rollover. When a zone decides to renew one of its zone keys, there is no mechanism to notify the resolvers that this key is going to be removed from its zone file. When a key is removed from its zone file, all resolvers that have configured this key as trusted fails all DNSSEC validation using this key.

Without an automated procedure to notify resolvers that a trusted key is going to be removed, name resolution can fail at any time even if a chain of trust exists from the root to the resource record queried.

In the next section, we propose a modification of the DNSKEY resource record and an algorithm to automatically update the trusted keys present in the resolver configuration file.

4 A Mechanism for Trusted Key Rollover

4.1 The DNSKEY Resource Record Wire Format

Figure 4 shows the DNSKEY resource record format.

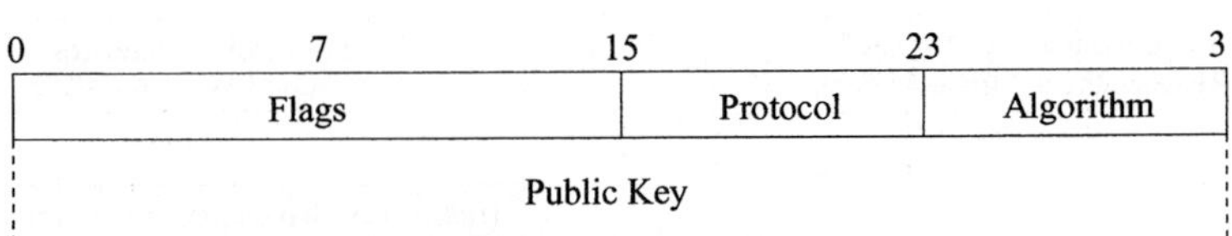

Fig. 4. The DNSKEY resource record wire format.

Only two bits of the Flags field are used. These bits are the seventh bit and the fifteenth bit. Bit 7 of the Flags field is the Zone Key flag. If bit 7 is set, then the DNSKEY record holds a DNS zone key. If bit 7 has value 0, then the DNSKEY record holds some other type of DNS public key (such as a public key used by TKEY [11]). Bit 15 of the Flags field is the Secure Entry Point flag, described in [6]. If bit 15 has value 1, then the DNSKEY record holds a key intended for use as a secure entry point. The other bits of the field, bits 0-6 and 8-14 are reserved: these bits must have value 0.

The Protocol field specifies the algorithm allowed to use the DNSKEY RR. Since the publication of the RFC 3445 [12] this field must have value 3.

The Algorithm field identifies the public key's cryptographic algorithm and determines the format of the Public Key field that holds the public key material. The format depends on the algorithm of the key being stored.

We propose to call the *under changes* bit, the first of the reserved bits of the Flag field and to give this bit the following meaning: when this bit is set the key contained in the DNSKEY RR is under changes and is going to be removed from the DNS zone file.

4.2 The Automated Trusted Key Rollover Algorithm

When a zone renewed one of its keys, it sets the *under changes* bit in the DNSKEY RR containing this key. Piece of advice for the duration of the rollover period must be found in [13].

During the rollover period, when a resolver asks for the DNSKEY RR, it retrieves the resource record with its *under changes* bit set. Three distinct periods could be defined for each key: before the rollover of the key, during the rollover of the key and after the rollover of the key. Before and after the rollover, the key is not under changes so the bit is not set. During the rollover the *under changes* bit is set to one.

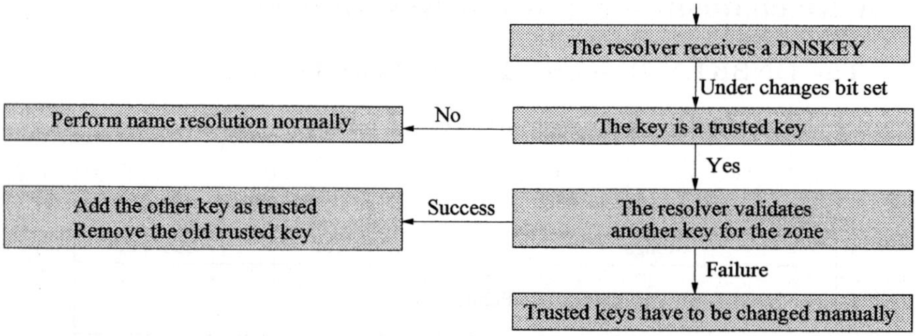

Fig. 5. The *best case* resolver behavior.

Figure 5 shows the best case resolver behavior. When a resolver receives a DNSKEY RR with the *under changes* bit set, it checks if this key is configured as trusted. If this is not the case, this key rollover will have no impact on the resolver behavior and the resolver must continue normally the current name resolution. If the received key is a trusted key for the resolver, it tries to validate another key of the zone (see section 4.3). If it succeeds, the old key is removed and the new key is added in the resolver configuration file. If the new key validation fails, the resolver raises a warning and the change in the resolver's configuration file must be made manually by the administrator.

On some resolvers, requests on a given zone could be largely spaced out, more than one month for example. Consequently, a resolver can perform no name resolution on a zone during the rollover period of a trusted key. This implies that the trusted key configured in such resolver may have been renewed and removed from the zone file during this period of *resolver inactivity.*

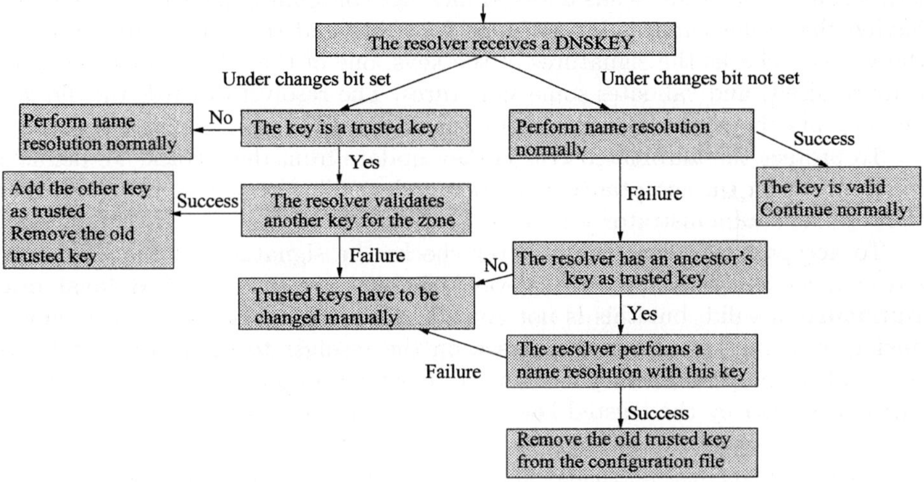

Fig. 6. The complete resolver behavior.

Figure 6 shows the two cases that can arise: the resolver has another trusted key belonging to an ancestor's zone of the zone that has changed its zone key, or the resolver has not such a trusted key. The left part of the Figure 6 is the *best case* resolver behavior presented in Figure 5, the right part takes into account the reception of a DNSKEY RR with the *under changes* bit not set.

If the *under changes* bit is not set the resolver performs name resolution normally. If the name resolution fails the resolver must try to update its trusted keys using another key for the name resolution. If the resolver cannot performed a DNSSEC resolution either because it has no valid trusted key or because the resolution is not secure (some data are retrieved from unsecured DNS zone) the resolver can not update its configuration file and the trusted key update must be made manually.

Algorithm presented on Figure 6 shows that in some cases a manual administrative action is needed and a totally automated solution does not exist. This is due to possible misconfiguration in any DNS zone or due to the possibility for a resolver to not have sent query to a zone during its rollover period, and hence, to not have received the notification of key changes.

4.3 Security Considerations and Validation of a New Trusted Key

Automated update of trusted keys in a resolver is a critical mechanism for DNSSEC and must be resistant to compromised key. An attacker that owns a compromised key can try to place new fake keys as trusted in a target resolver using the automated trusted key rollover mechanism, as describe now.

Firstly, the target resolver has a trusted key set including the compromised key. Then, the attacker sends a forged message containing the compromised key having the *under changes* bit set, new fake keys and their signatures. Finally, the resolver checks the signatures of the keys, one of these keys is trusted (but compromised) and validates some signatures. The resolver discards the old key and accepts the new fake keys sent by the attacker.

To protect the automated trusted key update from this attack, we define a requirement for the acceptance of a new key depending on the security level that the resolver's administrator wants.

To accept a new key, the resolver checks the signatures of the DNSKEY RRset it has received. The normal validation process succeeds if **at least one signature** is valid, but this is not enough to ensure security because compromising only one key allows the attack on the resolver to be fruitful. So to be resistant to compromised key the resolver must accept a new key only if k signatures generated by the trusted key owns by the resolver for this zone are valid. If one of this signature is not correct the resolver rejects the new key.

If the resolver has n trusted keys for a given zone, our automated trusted key rollover resists to $n - k$ compromised keys. Because, an attacker that owns $n - k$ compromised key for this zone can only generate $n - k$ valid signatures and the resolver will accept new keys only if there are n valid signatures.

Some local policy parameter may be defined by administrators to ensure a sufficient level of security. This parameter defines the number of trusted keys an administrator wants to configure to a given zone. If an administrator wants a low security level, he sets this parameter to one and configures only one trusted key for a zone. But, if he wants a higher security level he sets this parameter to any number included between 2 and the number of keys present in this zone.

Moreover, a resolver removes or accepts keys as soon as the DNSKEY RRset is received, our method does not implies extra delays or time constraints for resolvers or servers during the rollover period.

5 Related Work

5.1 Description

Recently, Mike St Johns has published an IETF draft [14] describing a method for automated trusted key rollover. This draft defines a revocation bit in the

Flags field of the DNSKEY resource record. When this bit is set, it means that the key must be removed from resolver's trusted key set. This revocation is immediate. Figure 7 shows the principles of this method.

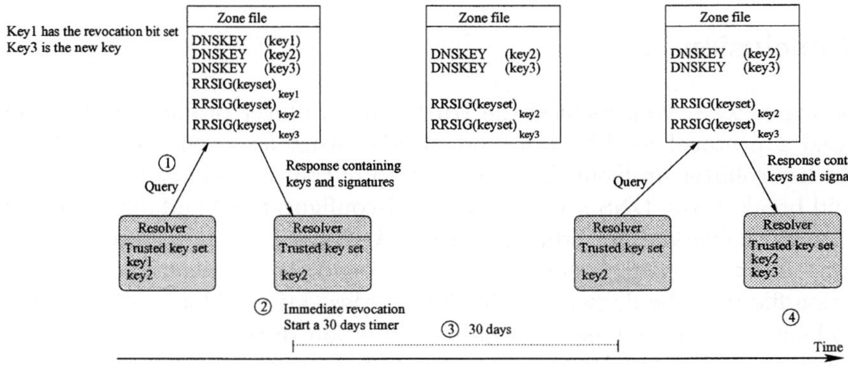

Fig. 7. Acceptation method with 30 days timer.

Firstly, the resolver sends a query about a DNS name ① and receives DNSKEY RR with the revocation bit set. The revocation is immediate and the resolver updates its trusted keys set ②. To avoid attack when a key is compromised, the resolver starts a timer of 30 days and keeps a trace of all the keys that signed the DNSKEY RRset ③. If the resolver received a response containing a DNSKEY RRset without the new keys but validly signed before the expiration of the timer, the resolver considers this information as a proof that something goes wrong (attack or misconfiguration). Consequently, it stops the acceptance of the new keys and resets the timer.

In the same way, if all the keys that have signed the DNSKEY RRset is revoked before the timer expiration, the resolver stops the acceptance of the new keys. Once the timer expires, the next time the resolver receives a valid DNSKEY RRset containing the new key it accepts this key as trusted ④.

5.2 Comparison

The major problem with the method described in the draft [14] is that it introduces extra delay (30 days) and requires a lot of management precaution for the zone administrator. Indeed this method implies a lot of constraints about the minimal number of keys in a zone, the number of trusted keys and the rollover frequency due to the 30 days timer.

Moreover, with this method, when a server creates a new key, this key is pending and can not be used during 30 days after its creation. This implies that if a zone owns n keys and rolls a keys every $\frac{30}{n-1}$ days, all resolvers will be unable to update its trusted key set for this zone.

Our method does not suffer from this problem because acceptance and revocation of keys are made as soon as the DNSKEY RRset is received. Our method does not imply changes or constraints for the DNS authoritative server management.

6 Conclusion

In this paper we have presented the trusted key rollover problem and we have proposed a protocol modification and an algorithm for resolvers to solve the trusted key rollover problem. The automation of the trusted key rollover allows to avoid break in the DNS service due to misconfigurations and allows to minimize the work overload for administrators. We have exposed that in some case the trusted key rollover cannot be automated and the modification of the configuration file must be done manually. These cases arise seldom and are limited to machines that do not perform name resolution during the rollover period. Moreover, the algorithm described in this paper is resistant to compromised key and places the choice of security level in the resolver's administrator side, which is a natural way to protect a resolver.

References

1. Eastlake, D.: Domain Name System Security Extensions. RFC 2535 (1999)
2. Arends, R., Larson, M., Massey, D., Rose, S.: DNS Security Introduction and Requirements. Draft IETF, work in progress (2004)
3. Arends, R., Austein, R., Larson, M., Massey, D., Rose, S.: Protocol Modi.cations for the DNS Security Extensions. Draft IETF, work in progress (2004)
4. Arends, R., Austein, R., Larson, M., Massey, D., Rose, S.: Resource Records for the DNS Security Extensions. Draft IETF, work in progress (2004)
5. Gieben, R.: Chain of Trust. Master's Thesis, NLnet Labs (2001)
6. Kolkman, O., Schlyter, J., Lewis, E.: Domain Name System KEY (DNSKEY) Resource Record (RR) Secure Entry Point (SEP) Flag. RFC 3757 (2004)
7. Guette, G., Courtay, O.: KRO: A Key RollOver Algorithm for DNSSEC. In: International Conference on Information and Communication (ICICT'03). (2003)
8. Mockapetris, P.: Domain Names - Concept and Facilities. RFC 1034 (1987)
9. Albitz, P., Liu, C.: DNS and BIND. fourth edn. O'Reilly & Associates, Inc., Sebastopol, CA. (2002)
10. Gundmundsson, O.: Delegation Signer Resource Record. RFC 3658 (2003)
11. Eastlake, D.: Secret Key Establishment for DNS (TKEY RR). RFC 2930 (2000)
12. Massey, D., Rose, S.: Limiting the Scope of the KEY Resource Record (RR). RFC 3445 (2002)
13. Kolkman, O., Gieben, R.: DNSSEC operational practices. Draft IETF, work in progress (2004)
14. StJohns, M.: Automated Updates of DNSSEC Trust Anchors. Draft IETF, work in progress (2004)

A Self-organized Authentication Architecture in Mobile Ad-Hoc Networks

Seongil Hahm[1], Yongjae Jung[1], Seunghee Yi[1], Yukyoung Song[1], Ilyoung Chong[2], and Kyungshik Lim[3]

[1] School of Electrical Engineering and Computer Science
Seoul National University, Seoul, 151-742, Korea
{siham, yjjung, shyi, songyk}@popeye.snu.ac.kr
[2] School of Computer Science and Engineering
Hankuk University of Foreign Studies, Seoul, 130-791, Korea
iychong@hufs.ac.kr
[3] School of Electrical Engineering and Computer Science
Kyungpook National University, Kyungpook, 702-701, Korea
kslim@kyungpook.ac.kr

Abstract. Network security is considered one of the obstacles that prevent the wide deployment of ad-hoc wireless networks. Among many security problems, authentication is the most fundamental problem that has to be solved before any others. Some previous solutions requiring centralized entities were not scalable and others requiring physical encounter took a long time. We propose a new architecture, called the Secure Overlay Network (SON), for fully distributed authentication. The SON has scalability because it is constructed in a self-organizing manner. The SON also has NPC-reachability which makes the SON robust against Sybil attacks, and guarantees authentication service between any two nodes in the SON. Both NPC-reachability and simulation results confirm the effectiveness of our architecture.

1 Introduction

Network security consists of several services such as authentication, confidentiality, integrity, non-repudiation, and access control. Among these services, authentication, which ensures the true identities of peer nodes, is the most fundamental service because other services depend on the sure authentication of communication entities. However, authentication in ad-hoc networks is considered to be more difficult than that in wired networks due to the limited physical protection of broadcast medium, the frequent route changes caused by mobility, etc.

Conventional wired networks mainly use a globally trusted Certificate Authority (CA). Each node is assumed to know the predetermined CA and a node registers its public key to the CA. The CA signs and issues a certificate of the subscribed node. This signed certificate ensures that the node uses the correct public key. The architecture requiring a CA can offer powerful security solutions to several network areas, but is not quite a suitable solution for ad-hoc wireless

C. Kim (Ed.): ICOIN 2005, LNCS 3391, pp. 689–696, 2005.

networks, which have no support of any infrastructure. Nevertheless, a CA can be applied to large ad-hoc wireless networks in a distributed manner [1, 2, 9] while there are still issues about service availability, robustness against attacks, and scalability.

We propose a new architecture for authentication in ad-hoc networks called the Secure Overlay Network (SON). Based on Threshold Cryptography [10], it offers better fault tolerance than other mechanisms based on non-threshold cryptography since no single entity is entrusted to perform the whole security task. The SON has three beneficial properties. First, it has scalability because it is designed in a fully distributed manner; there are no any central points such as a CA server even during the bootstrapping phase of ad-hoc networks. Second, it has availability because it is robust against up to $m-1$ Sybil attackers [4]. Finally, compared with authentication solutions [3, 8] that do not guarantee authentication service between any two nodes even in the same secure domain, our architecture assures any two nodes of their authentication service if they belong to the same secure domain, the SON. Our SON-construction procedures and mobility help to shorten the time necessary for all nodes to join the unified SON.

The rest of this paper is organized as follows. Section 2 presents related works. We will describe how to construct a SON and to provide authentication service over the SON in Section 3. Then we will show you the simulation results in Section 4 and finalize this paper with conclusions in section 5.

2 Related Works

Several authentication mechanisms for ad-hoc wireless networks have already been proposed. Zhou and Hass [1] identified the vulnerability of using a centralized CA for authentication in ad-hoc networks and proposed a method with multiple CAs based on Threshold Cryptography [10]. These multiple CAs have secret shares of a Certificate Authority Signing Key (CASK) while no CAs individually know the whole complete CASK, which can be known only when CAs of more than m collaborate. Therefore, this method can support the network security against up to $m-1$ collaborative compromised nodes. While Zhou and Hass's method improved the robustness of the authentication system, it depended on the offline authority which elects n CAs ($n \geq m$) during the bootstrapping phase. Furthermore, it has poor availability because if $n-m+1$ CAs have been compromised, uncompromised $m-1$ CAs that are left can't provide authentication services anymore.

Kong and et. al. [2, 9] proposed another authentication method based on threshold secret sharing [6]. After the bootstrapping phase, a new node can join the network at any time through self-initialization: it can obtain its own secret share of CASK with the help of m local neighbor nodes. Even though this approach enhances scalability and availability, it still depends on an offline authority during the bootstrapping phase. Furthermore, Narasimha and et. al. [7] pointed out the defect of [2, 9], that is, secret sharing based on RSA-signature does not provide an important property known as verifiability. They proposed the method for group admission control in peer-to-peer systems which are given a trustable CA. It is based on DSA-signature [11] which has verifiability.

There is another approach on the literature. Habaux and et. al. proposed the scheme based on a chain of public-key certificates [8], which is scalable and self-organized. However, first of several drawbacks, Habaux's approach does not guarantee authentication service between any two nodes even though they are in the same secure domain, but provides only probabilistic guarantee. There is also a storage problem because each node has to store relatively many other nodes' certificates.

Recently, Capkun and et. al [3] proposed an authentication method and asserted that mobility helps the security. The key idea is that if two nodes are in the vicinity of each other, they can establish a security association (SA) by exchanging appropriate cryptographic material through a secure channel with the short transmission range. However, this direct solution takes a long time because it requires a node to encounter every node that it wants to communicate with.

3 System Model

3.1 Assumptions and Basics of Secret Sharing

We use threshold secret sharing [6, 2] based on RSA signature [5] in our architecture. We assume each node has the same functionality and responsibility, which is suitable for ad-hoc networks. A node i has an RSA private and public key pair (d_i, e_i) with modulus n_i. Also, node i randomly selects a Lagrange interpolating polynomial of degree $m-1$, $f_i(x) = a_{i(m-1)}x^{m-1} + \cdots + a_{i1}x + d_i$, where a_{ij} is the coefficient of the jth power-monomial of $f_i(x)$, so that $f_i(0)$ is its private key d_i. Node i also issues a raw (or unsigned) Certificate X_i in the format of $(id, e_i, T_{issue}, T_{expire})$, which explains that this certificate was issued at T_{issue} and will expire at T_{expire}.

Our architecture is based on the concept of Neighbors for Authentication (NA); if node i and j establish an NA relationship, they partially attest each other's identity. Establishment of an NA relationship is based on their physical encounter. To support this mechanism, we assume that each device is equipped with some local detection mechanism to identify intruders among its one-hop neighborhood [12]. Suppose node i and j are in the vicinity of each other. If they judge each other to be trustworthy with help of the intrusion detection mechanism, they then exchange their secret shares $(P_{ij} = f_i(j) \bmod n_i, P_{ji} = f_j(i) \bmod n_j)$ and NA lists each other through a short range channel such as infrared interfaces. They also establish a symmetry key by means of Diffie-Helmen key exchange. After the exchange of the secret shares and the symmetry key, two nodes become an NA for each other. Nodes which have at least m NAs can organize a SON and an NA relationship represents connectivity over the SON. A node can communicate with its NAs securely regardless of physical topologies because all packets are encrypted by the shared symmetry keys.

The coalition of m NAs of node i can collaboratively restore node i's private key d_i like the following equation, but no one alone knows complete d_i: $d_i = \sum_{j=1}^{m}(P_{iv_j} \cdot l_{iv_j}(0) \bmod n_i) \pmod{n_i} \equiv \sum_{j=1}^{m} SS_{iv_j} \pmod{n_i}$ where $l_{iv_j}(0)$ are the Lagrange coefficients which are publicly known [2]. Each NA v_j can compute SS_{iv_j} by means of Lagrange interpolation. SS_{iv_j} will be used as a signing key to partially sign node i's certificate.

3.2 Intrusion Model and Objective

Giving infinite power to an intruder simply makes any security design meaningless. Kong and et. al. [2, 9] considered a realistic intrusion model in the system: time is divided into intervals of length T and during any time interval T, intruders cannot compromise or control m or more nodes. Once a node is compromised, all the information including its own private key is exposed to malicious intruders. The problem is that gradual break-ins into m nodes over several consecutive time intervals may be possible and therefore long-lived RSA-key pairs are not sufficient. To resist such gradual break-ins, node i has to periodically change its RSA-key pair and reissue its raw certificate X_i. If the RSA key pair of a compromised node is changed by a periodic key-changing rule, the node can escape from the control of the intruders. Otherwise, no security mechanisms are robust against gradual break-ins. Compared with proactive secret sharing of [2, 9, 13] which changes secret shares of all nodes, each node over a SON just changes secret shares held by its NAs into secret shares of its new private key. Therefore, this difference makes the SON more scalable than others [1, 2, 7, 9, 13].

The main objective of this paper is to prevent intruders from joining a SON. Because the intruders can break or control up to $m-1$ nodes during any time interval T, the SON has to be robust against the coalitions of up to $m-1$ compromised nodes and the intruders.

3.3 Primitive-SON and NPC-reachability

A primitive-SON is the smallest SON and is self-organized without the help of an offline authority; this property gives the SON scalability and makes deployment of the SON easier. If a network administrator wants to protect the network against attacks of up to $m-1$ collaborative compromised nodes, a primitive-SON has to be composed of at least $m+1$ nodes. In Figure 1, node A, B, C, D form a primitive-SON where every pair among A, B, C, D has an NA relationship when m=3.

A node informs its existing NAs of the triple (j, e_j, NA_list$_j$) via a NA-addition unicast message whenever it establishes a new NA relationship with any node j. Suppose the establishment of the NA relationship between node B and D is the last step to construct the primitive-SON. At this time, both node B and D simultaneously notice that they and their NAs can organize a primitive-SON and then notify their NAs of the creation of the primitive-SON via a SON-creation unicast message including SON-member list. Node A and C trust this creation if they receive the SON-creation message twice. This redundant notice

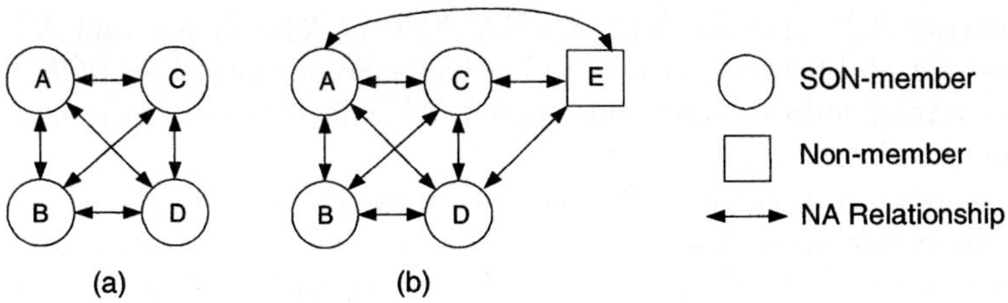

Fig. 1. (a)Primitive-SON which consists of node A, B, C, and D when m=3 and (b)its procedure of node-joining

can not only prevent node B and D from deceiving each other, but also improve reliability of informing all member nodes of the creation.

Authentication over the SON is based on m Node-disjoint Partial certificate Chains (NPC), while the scheme of [8] uses a single certificate chain. For example, NPCs between node A and D in Figure 1 are $A-C-D,\ A-D,\ and\ A-B-D$. Existence of m NPCs between two nodes over the SON makes them trustworthy to each other; their identities are assured by at least m SON-members which are also trusted by other m SON-members. Otherwise, existence of m sponsors for a node's identity can not be guaranteed because of Sybil attacks.

We can easily observe that there are always m NPCs between any two nodes over the primitive-SON. We call this property of the SON NPC-reachability. That is, the primitive-SON is robust against attacks of up to $m-1$ compromised nodes. Therefore, we designed the node-joining and SON-combining procedures in order to keep NPC-reachability of the primitive-SON.

3.4 Procedure of Node-Joining

A node can become a member of a SON when it has m NAs with any m members of the SON. Suppose non-member node E has already had an NA relationship with node C and D as shown in Figure 1. Lastly, node E establishes an NA relationship with node A. Then node E solicits node A for a permission to join the SON by asserting its NAs to be the SON-members. In order to confirm this assertation, node A separately requests node C and D to send partial certificates signed by secret shares of node E ($X_E^{SS_{EC}}$, $X_E^{SS_{ED}}$) to itself. Now node A has to examine whether node E has m NAs with m SON-members. Otherwise, the SON is no more NPC-reachable. This examination is very simple because all nodes over the SON trust each other. Sponsor C of node E appends the message authentication code (MAC) of its ID and nonce to $X_E^{SS_{EC}}$ using its private key for message integrity. After passing in the integrity check, node A then computes $X_E^{SS_{EC}+SS_{ED}+SS_{EA}}$ including its holding $X_E^{SS_{EA}}$. To obtain $X_E^{d_E}$ from $X_E^{SS_{EC}+SS_{ED}+SS_{EA}}$, we apply the K-bounded Coalition Offsetting[4] [2]. Node A

[4] The computation of m^N (mod N) in K-bounded Coalition Offsetting has extremely computation overhead [7], therefore, we can adopt DSA signature [11] instead of RSA signature.

then decrypts $X_E^{d_E}$ into X'_E with e_E $((X_E^{d_E})^{e_E} = X'_E)$. If X_E and X'_E are the same, node A can believe node E's identity because at least m SON members have guaranteed node E to be trustworthy. We call this entire process the NPC-reachability test.

Finally, node A broadcasts the SON-join message of (E, e_E, NA_list$_E$), which informs all members of node E's new join. We use broadcasting rather than unicasting to improve efficiency in terms of bandwidth when there are many SON-members. However, broadcasting is not a reliable method for transmitting control messages; we can adopt Lou's method [14] to improve reliability while maintaining efficiency.

3.5 Procedure of SON-combining

Two SONs can be merged into one large SON if there are m distinct NA relationships between the two SONs. In Figure 2, when node A encounters node E, they exchange SON-member list. Both of them can notice that they belong to the different SON by comparing the SON-member list. If they establish an NA relationship, node A and E broadcast the NA-addition messages of (E, e_E, NA_list$_E$, SON_list$_E$) and (A, e_A, NA_list$_A$, SON_list$_A$), respectively. Suppose node B also establishes an NA relationship with node E after receiving the NA-addition message from node A. However, they do not broadcast any NA-addition messages because this NA relationship does not make an additional NPC between two primitive-SONs. Suppose the establishment of the NA relationship between node C and G is the last step to construct the combined SON. At this time, both node C and G confirm the existence of m distinct NAs between their SONs to keep NPC-reachability of the combined SON. They then broadcast the SON-combined message of (m distinct NA pairs, combined SON-member list). The minimum number of broadcasting for SON-combining is just m regardless of the size of SONs: this makes the SON keep scalability. The procedure of SON-combining can accelerate the construction of the unified SON which contains all nodes.

Once the unified SON including all nodes has completely been organized, the SON assures any two nodes of their authentication service because both procedures of node-joining and SON-combining keep the NPC-reachability of SONs and then finally the ultimate unified SON has NPC-reachability.

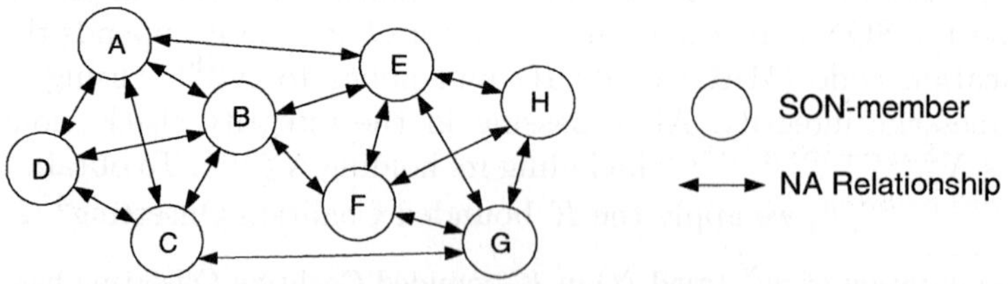

Fig. 2. Procedure of SON-combining between two primitive-SONs where m=3

4 Simulations

The dominant time-consuming job is to construct the unified SON. Therefore, the interesting criterion is the time necessary to construct the unified SON. We run a number of simulations using ns-2 simulator with 802.11 MAC and AODV routing protocols. An NA relationship can be established when two nodes are located in close proximity. We use the random walk mobility model with various pause time and maximum speed [3].

First, our architecture is evaluated with 250 x 250, 500 x 500, and 750 x 750 m^2 square area of ad-hoc network. In each case, 50 nodes move around with 0.5 second pause time and 20m/s maximum speeds. The transmission range of the secure channel is 5 meters while that of the data channel is fixed to 250 meters. Figure 3 shows that in case of 250 x 250 m^2 network, three primitive-SONs are created and are merged at around 200 seconds. Because the node density is high, the unified SON is self-organized within about 210 seconds. In case of 500 x 500 m^2 network, four primitive-SONs are initialized at around 220 seconds and the construction of the unified SON takes about 890 seconds. In case of 750 x 750 m^2 network, the first primitive-SON is created at around 830 seconds, and the growth speed is the slowest because of the smallest node density. However, we can observe that the SON growth speeds up in an instant after two small SONs merge at 1170 seconds. Based on these results, we can conclude that the node density is the key factor that determines the period of the SON construction.

To analyze the effect of node mobility into the SON construction time, we perform more simulations with different mobility pattern of 1.5 second pause time and 10m/s maximum speeds. From Figure 4, high mobility helps to shorten the SON construction time.

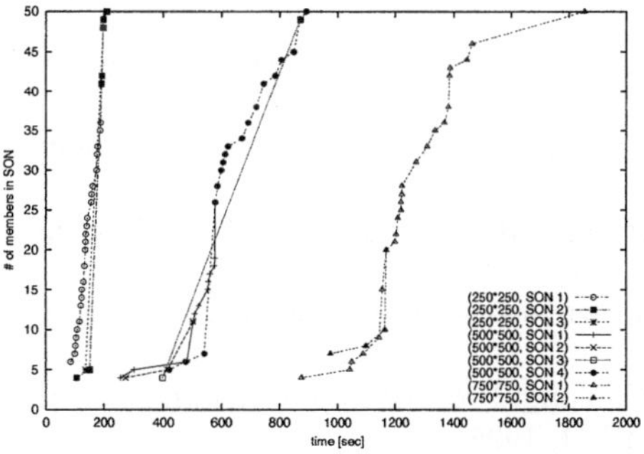

Fig. 3. The growth patterns of the SON as a function of time when network size varies. A separate line presents a distinct SON. SON-combining is represented by the merging of lines

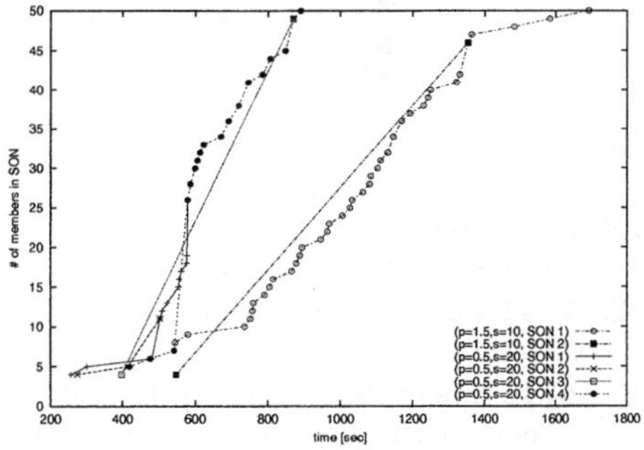

Fig. 4. The growth patterns of the SON as a function of time when the mobility pattern varies

5 Conclusions

In this paper, we propose a new authentication architecture called the SON for ad-hoc wireless networks. The proposed method can provide authentication services in a fully distributed manner. Because the SON can operate without infrastructures or offline authorities, it is more deployable than the methods proposed in [1, 2, 7, 9]. The SON is also robust against up to $m-1$ collaborative compromised nodes and assures any two members of their authentication service if they belong to the same SON. These beneficial properties come from NPC-reachability of the SON. We will further devise the compromised node detection mechanism that operates in a fully distributed manner and the procedure of SON-splitting when some nodes leave.

References

1. L. Zhou, Z. J. Hass: Securing Ad Hoc Networks. IEEE Network, 12(6):24-30, 1999
2. J. Kong, P. Zerfos, H. Luo, S. Lu, L. Zhang: Providing robust and ubiquitous security support for mobile ad hoc networks. In IEEE ICNP, 2001
3. S. Capkun, J. P. Hubaux, L. Buttyan: Mobility Helps Security in Ad Hoc Networks. In ACM MobiHoc, 2003
4. J. Douceur: The Sybil attack. In IPTPS, 2002
5. W. Stallings: Cryptography and Network Security. Prentice Hall, 3rd edition
6. A. Shamir: How to Share a Secret. Communications of the ACM, 22(11):612-613, 1999
7. M. Narasimha, G. Tsudik, J. H. Yi: On the Utility of Distributed Cryptography in P2P and MANETs: the Case of Membership Control. In IEEE ICNP, 2003
8. J. Hubaux, L. Buttyan, S. Capkun: The Quest for Security in Mobile Ad Hoc Networks. In ACM MobiHoc 2001
9. H. Luo, P. Zerfos, J. Kong, S. Lu, L. Zhang: Self-securing Ad Hoc Wireless Networks. In ISCC 2002
10. Y. Desmedt: Threshold cryptography. European Transactions on Telecommunications, 5(4):449-457, 1994.
11. R. Gennaro, S.Jarecki, H.Krawczyk, T.Rabin: Robust threshold DSS signatures. In EUROCRYPT, 1996.
12. Y. Zhang, W. Lee: Intrusion detection in wireless ad hoc networks. In ACM MobiCom, 2000
13. A. Herzberg, S. Jarecki, H. Krawczyk, M. Yung: Proactive secret sharing, or: How to cope with perpetual leakage. In CRYPTO, 1995.
14. W. Lou, J. Wu: Double-Covered Broadcast (DCB): A Simple Reliable Broadcast Algorithm in MANETs. In IEEE INFOCOM 2004

Throughput Enhancement Scheme in an OFCDM System over Slowly-Varying Frequency-Selective Channels

Kapseok Chang and Youngnam Han

Wireless Information Transmission System Laboratory,
Information and Communications University,
103-6, Munji-dong, Yuseong-gu, Daejeon, 305-714, Korea
{changks, ynhan}@icu.ac.kr

Abstract. We propose multilevel transmit power allocation (MTPA) scheme and outer-loop control scheme to increase bandwidth efficiency and to guarantee a required bit error rate (BER). The system considered is orthogonal frequency-code division multiplexing (OFCDM) under slowly-varying frequency-selective Rayleigh fading channels. In the proposed schemes, modulation level and transmit power for each subcarrier is simultaneously controlled corresponding to channel condition to maximize transmitted bits per symbol (BPS) while keeping both total transmit power and BER constant. Computer simulations elucidate that our proposed schemes are effective in terms of throughput and complexity.

1 Introduction

Orthogonal frequency division multiplexing (OFDM) realizing multi-carrier (MC) modulation has been adopted to achieve high data rate with the robustness against frequency-selective fading [1],[2]. MC modulation in combination with the spread-spectrum technique offers promising multiple access systems for future mobile radio communications, known as MC code division multiple access (MC-CDMA) and OFDM-CDMA. In this paper, the system of OFCDM, that is, OFDM code division multiplexing (OFDM-CDM), is applied to compensate for the poor performance of uncoded OFDM systems [3]. OFCDM is a multiplexing scheme. The aforementioned MC-CDMA is a special case of OFCDM, where each data symbol is spread over multiple subcarriers, exploiting additional frequency and time diversity. In this system, to avoid a decrease in bandwidth efficiency due to spreading, encoded data symbols are superimposed. As a result, the spreading does not change the bit rate of system.

It has been widely studied in [4]-[6] to maximize throughput while guaranteeing a given required BER under additive white Gaussian noise (AWGN) or time-invariant fading for discrete multitone (DMT) systems. These MC adaptive modulation techniques are efficient to increase throughput by changing appropriate modulation scheme corresponding to the estimated channel state information for each subcarrier. However, [4] and [5] with finite granularity provide suboptimal solutions by using water-filling method, Hughes-Hartogs algorithm, and

C. Kim (Ed.): ICOIN 2005, LNCS 3391, pp. 697–706, 2005.

flat-energy algorithm. [6] exploiting a Lagrange-multiplier bisection search converges faster to the optimal solution, but still computationally infeasible. Also, to apply these techniques, we should limit total transmit power (i.e., peak power). Uncontrolled peak power may saturate the transmitter amplifier if not carefully designed. In addition, unpredictable intercarrier interference (ICI) might be incurred. Thus, the peak power should be limited to reduce ICI in [7].

Moreover, when we employ an adaptive modulation technique, we should decide the minimum required signal to interference plus noise ratio (SINR) value (i.e., switching level) for each modulation scheme to maintain a given required BER. The switching level should be adaptive to the time-varying fading channel because of the reliability of estimated SINR. So, the optimization for the switching level for each modulation scheme is a challenge.

In this paper, an efficient, computationally simple MTPA scheme controlling both modulation level and power allocated to each subcarrier is proposed to increase bandwidth efficiency. Also, an outer-loop control scheme which may sustain a given BER is proposed under slow channel variation.

The rest of the paper is organized as follows. In Sect. 2, we provide a wireless channel model and signal representation for the OFCDM system under study. The MTPA scheme is proposed in Sect. 3. Outer-loop control scheme is next depicted in Sect. 4. In Sect. 5, simulation environments and results are presented. Finally, concluding remarks are given in Sect. 6.

2 Channel and System Model

2.1 Channel Model

The wide sense stationary uncorrelated scattering channel (WSS-USC) is considered in this paper [8]. The time impulse response $T(n;\tau)$ of the channel is

$$T(n;\tau) = \sum_{p=0}^{L-1} h(n,p)\delta(\tau - \tau_p), \tag{1}$$

where L is the number of multipaths, $h(n,p)$ is the path gain of the pth multipath components at sample time n, which is a complex Gaussian random process, and τ_p is the corresponding path delay.

2.2 System Model

A block diagram of OFCDM system under consideration is shown in Fig. 1. At the transmitter, information bits are mapped onto complex modulated symbols. The mapped symbols are fed into different subcarriers, applying code spreading and inverse fast Fourier transform (IFFT) to generate OFCDM signals. The post-IFFT signals undergo parallel-to-serial conversion and cyclic prefix (CP) addition, followed by RF-modulation on the carrier frequency f_c before being transmitted through each fading channel with AWGN. For OFDM, vector **B**

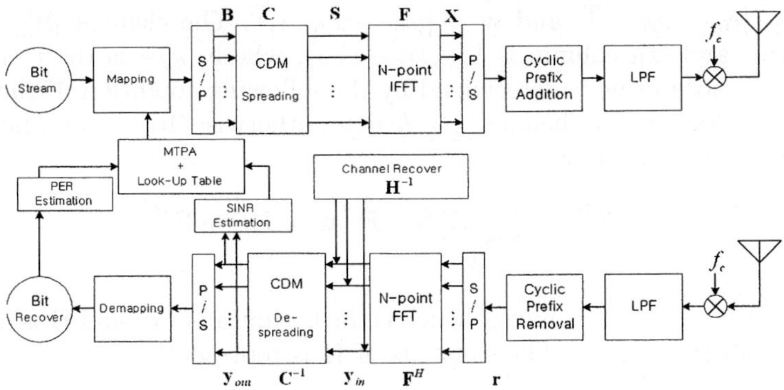

Fig. 1. The Block Diagram of OFCDM system considered.

of length N carries each of subcarrier symbols, with $\mathbf{B} = [b_0 \cdots b_{N-1}]^{\mathrm{T}}$, where the element b_k is the complex modulated symbol for subcarrier k. Assume that $E[b_k b_m^*] = \delta(k,m)P_k$, which means symbols on different subcarriers are independent of each other, with the average power P_k for subcarrier k. And $E[b_k] = 0$. In OFCDM, $\mathbf{S} = \mathbf{CB}$, where $\mathbf{C}$ is an $N \times N$ Walsh-Hadamard matrix and $\mathbf{S} = [s_0 \cdots s_{N-1}]^{\mathrm{T}}$, whose element represents a spread symbol [3]. At the receiver, the received OFCDM signals are filtered by its local oscillator frequency f_c. After that, the cyclic prefix is first removed and then each subcarrier corresponding to the received signal is coherently detected with fast Fourier transform (FFT), followed by channel recover, code despreading, and SINR estimation. After that, parallel to serial is performed, and then de-mapped for bit recovering, followed by average packet error rate (PER) measure. In the MTPA block, we allocate the modulation scheme and transmit power for each subcarrier based on estimated PER and individual SINR. In addition, the switching level for each modulation scheme, the look-up table as shown in Tab. 1, is updated by the outer-loop control presented in Sect. 4.

The lowpass equivalent transmitted signal vector is, using IFFT,

$$\mathbf{X} = \mathbf{FS}, \tag{2}$$

where $\mathbf{X} = [x_0 \cdots x_{N-1}]^{\mathrm{T}}$, and the elements of $\mathbf{F}$, N-point IFFT unitary matrix, are defined as $F_{n,k} = (1/\sqrt{N})e^{j(2\pi nk/N)}$ $(n = 0, ..., N-1,\ k = 0, ..., N-1)$. The sampled received signal in Fig. 1, after passing through CP removed where the CP length is no less than the maximum channel delay τ_{L-1}, can be represented as

$$r_n = \sum_{p=0}^{L-1} h(n,p)x_{n-\tau_p} + w_n, \tag{3}$$

where $h(n,p)$ is defined at Eq. 1, and w_n is a complex AWGN with zero mean and variance $N_0(=\sigma_w^2)$. The matrix form of Eq. 3 is

$$\mathbf{r} = \mathbf{hX} + \mathbf{w}, \tag{4}$$

where $\mathbf{r} = [r_0 \cdots r_{N-1}]^{\mathrm{T}}$, and $\mathbf{w} = [w_0 \cdots w_{N-1}]^{\mathrm{T}}$. The element $[\mathbf{h}]_{n,c}$ of $\mathbf{h}$ in the nth row and cth column is $h(n, (n-c)_N)$, where $(k)_N$ is the residue of k modulo N, whose value is determined by the following condition: If there exists $\tau_p = (n-c)_N$ for some p, then $[\mathbf{h}]_{n,c} = h(n,p)$; otherwise $[\mathbf{h}]_{n,c} = 0$. The received signal vector, after FFT, is

$$\mathbf{y} = \mathbf{F}^H\mathbf{r} = \underbrace{\mathbf{F}^H\mathbf{h}\mathbf{F}}_{\mathbf{H}}\mathbf{S} + \mathbf{F}^H\mathbf{w} = \mathbf{HS} + \mathbf{F}^H\mathbf{w}, \tag{5}$$

where $\mathbf{F}^H$ is N-point FFT matrix which is the hermitian of $\mathbf{F}$, and $\mathbf{H}$ is frequency response matrix of fading. The elements of $\mathbf{H}$ is represented as

$$H_{k,m} = \frac{1}{N}\sum_{n=0}^{N-1}\sum_{p=0}^{L-1} h(n,p)e^{-j\frac{2\pi\tau_p m}{N}}e^{j\frac{2\pi(m-k)}{N}}, \tag{6}$$

where $k = 0, ..., N-1$, and $m = 0, ..., N-1$. Note that $\mathbf{H}$ becomes diagonal matrix as fading goes to time-invariant. The output signal vector after channel recover in Fig. 1 is

$$\mathbf{y}_i = \mathbf{H}^{-1}\mathbf{y} = \mathbf{S} + \mathbf{H}^{-1}\mathbf{F}^H\mathbf{w}, \tag{7}$$

where $\mathbf{y}_{in} = [y_{in,0} \cdots y_{in,N-1}]^{\mathrm{T}}$. And the final output signal vector after code despreading is

$$\mathbf{y}_{out} = \mathbf{C}^{-1}\mathbf{y}_{in} = \mathbf{B} + \mathbf{C}^{-1}\mathbf{H}^{-1}\mathbf{F}^H\mathbf{w}, \tag{8}$$

where $\mathbf{y}_{out} = [y_{out,0} \cdots y_{out,N-1}]^{\mathrm{T}}$. From $\mathbf{F}^H = \mathbf{F}^{-1}$, $\mathbf{C}^{-1}\mathbf{H}^{-1}\mathbf{F}^H$ in Eq. 8 is $(\mathbf{FHC})^{-1}$. Let $\mathbf{G} = (\mathbf{FHC})^{-1}$. The elements of $\mathbf{G}$ are defined as $g_{k,m}$ ($k = 0, ..., N-1, m = 0, ..., N-1$). The received resultant SINR value γ_k for subcarrier k can be obtained as

$$\gamma_k = \frac{\zeta_k \cdot P_k}{N_0}, \tag{9}$$

where $\zeta_k = 1/\sum_{m=0}^{N-1}|g_{k,m}|^2$.

3 Proposed MTPA Scheme

Unlike conventional adaptive modulation (CAM) scheme allocating the same power for each subcarrier, our MTPA scheme can enhance much more throughput by allocating different powers under a given total transmit power. There are cases where the allocated power is excessive over the required SINR for some subcarriers or fails to achieve any valid transmission. The use of those excessive or ineffective power can be made for throughput enhancement. We propose a computationally efficient multilevel transmit power reallocation scheme, which consists of the following two steps.

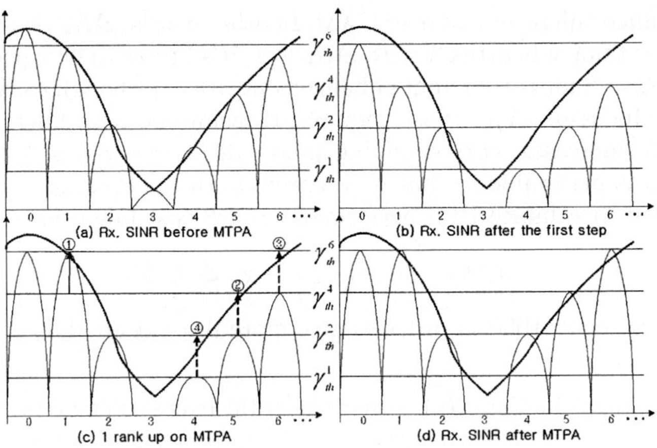

Fig. 2. MTPA process.

Step 1 Suppose the received SINR for each subcarrier is represented as in Fig. 2(a). We compare it with the predefined minimum SINR $\gamma_{th}^{R_k}$ of maximum modulation scheme R_k ($= 0, 1, 2, 4, 6$ meaning None, BPSK, QPSK, 16QAM, and 64QAM, respectively) to sustain the required BER for subcarrier k. Next, as shown in Fig. 2(b), the received SINR is fit down to the SINR switching level for each modulation as you can see the look-up table in Tab. 1. The total power that can be reallocated, ΔP_T^{re}, can be represented as

$$\Delta P_T^{re} = P_T - \sum_{k=0}^{N-1} (P_k - P_{th}^{R_k}) \text{ [dB]}, \tag{10}$$

where P_T is the total transmit power, P_k is current transmit power defined in Eq. 9, and $P_{th}^{R_k}$ denotes the minimum transmit power required for R_k, which can be expressed as $\gamma_{th}^{R_k}/\beta_k$ with $\beta_k = N_0/\zeta_k$. After ΔP_T^{re} is calculated, to reallocate the final transmit power and modulation scheme for subcarrier k, we introduce the cost function $J(k)$. The priority for reallocation will be given to the subcarrier with the smallest $J(k)$. When either $\gamma_k < \gamma_{th}^1$ or $\gamma_{th}^6 \leq \gamma_k$ for all k, no further allocation can be made. Two types of cost functions are considered as follows.

- MTPA1 The cost function $J_1(k)$ is expressed as

$$J_1(k) = \Delta\gamma_k \text{ [dB]}, \tag{11}$$

  where $\Delta\gamma_k = \gamma_{th}^{R'_k} - \gamma_k$ with R'_k increased by one rank. For example, for $R_k = 1$ and 2, R'_k equals 2 and 4, respectively. This scheme is very simple because the difference between $\gamma_{th}^{R'_k}$ and γ_k is just considered.
- MTPA2 Since MTPA1 cannot consider the information of the channel gain ζ_k and the difference ΔR_k between R'_k and R_k, there may be not much BPS

performance enhancement over CAM. In other words, ΔR_k belongs to $\{1, 2\}$. So, rather than subcarrier k with $\Delta R_k = 1$, it's better to give priority to that with $\Delta R_k = 2$ for throughput enhancement. Also, although some subcarriers provide the same $\Delta\gamma_k$, those required transmit powers differ according to ζ_k: As ζ_k increases, corresponding power decreases. So, subcarrier k with higher ζ_k is preferable to that with lower ζ_k. In MTPA2, ζ_k and ΔR_k being considered to enhance BPS performance, each cost function is defined as

$$J_2(k) = J_1(k) - \zeta_k - \chi \cdot \Delta R_k \text{ [dB]}, \tag{12}$$

where χ is set to 1000 in order to give more priority to ΔR_k than ζ_k.

Step 2 As shown in Fig. 2(c), using the total excess/deficit power and each cost function, we can reallocate power to the subcarrier with the smallest cost, and increase modulation level by one, and update the total excess power in Eq. 10. The process will be repeated until either the excess power gets close to zero or all subcarrier's modulation schemes has reached to 64QAM. We can see the final result of MTPA as shown in Fig. 2(d).

4 Outer-Loop Control

If we assume a non-adaptive threshold scheme, the mobility should be estimated before deciding the optimal switching level for each modulation scheme. It becomes more complicated if the multipath channel is involved. However, our outer-loop control does not require such a priori channel information, but it is able to adapt to new environment automatically.

The SINR switching level for each modulation scheme in Tab. 1 should be adapted to maintain a target PER $\wp_{th}$ corresponding to the required BER. $\wp_{th}$ is updated for each PER measure period when both the selection number n_{R_k} $(R_k = 1, 2, 4, 6)$ and the total selection number K for all modulation schemes are also updated. The target PER $\wp_{th}$ is determined by

$$\wp_{th} = \left(\frac{n_1}{K} \cdot \wp_{th}^1 + \frac{n_2}{K} \cdot \wp_{th}^2 + \frac{n_4}{K} \cdot \wp_{th}^4 + \frac{n_6}{K} \cdot \wp_{th}^6 \right), \tag{13}$$

where $\wp_{th}^{R_k}$ $(R_k = 1, 2, 4, 6)$ indicates the target PER for each modulation scheme, each of which corresponds to the required BER and may be invariant even over slow time-variation of fading. With $\wp_{th}$, we can control $\gamma_{th}^{R_k}$ by the following two schemes. In scheme 1, when the estimated PER $\hat{\wp}$ is less than $\wp_{th}$, the SINR switching level for each modulation scheme can be controlled by

$$\gamma_{th}^{R_k} = \gamma_{th}^{R_k} - \delta_d, \tag{14}$$

where δ_d represents adjustment value for the decrease of current SINR level. For $\hat{\wp} \geq \wp_{th}$, each switching level increases by

$$\gamma_{th}^{R_k} = \gamma_{th}^{R_k} + \delta_u, \tag{15}$$

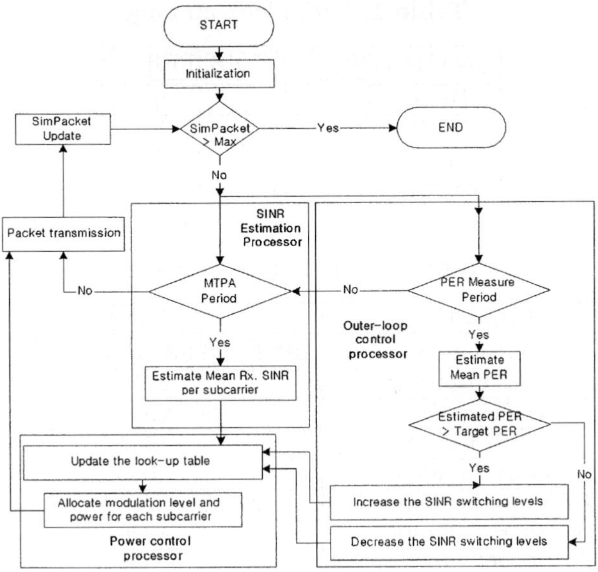

Fig. 3. Simulation model.

where δ_u represents adjustment value for the increase of current SINR level. Scheme 2 is the same as scheme 1 except that δ_d and δ_u are multiplied by n_{R_k}/K, respectively, to improve the reliability of decision for dominant modulation scheme having higher n_{R_k}/K than any other scheme. We employ the selecting suitable values of δ_d and δ_u presented as $\delta_d = \delta_u \wp_{th}/(1 - \wp_{th})$ in [9].

5 Simulation

5.1 Simulation Model and Parameters

Fig. 3 represents the flow chart of the proposed MTPA and outer-loop control. It consists of three processors. In SINR estimation processor, for each MTPA period, γ_k for each subcarrier is estimated. In outer-loop control processor, for each outer-loop period, n_{R_k}, K, $\hat{\wp}$, and $\wp_{th}$ are updated. Based on these, we update the look-up table as shown in Tab. 1 which is an initial loop-up table based on AWGN. Then, in MTPA processor, to support given BER ($=10^{-3}$), Using the updated look-up table, we decide the modulation scheme and power reallocation for each subcarrier. The detailed simulation parameters are illustrated in Tab. 2.

5.2 Simulation Results

Fig. 4 shows the BER performances of several schemes, where SNR on X-axis means the SNR value for each subcarrier in case of fixed 16QAM and CAM. The fixed 16QAM has the worst performance among all five schemes because each

Table 1. Initial look-up table

Modulation	SINR $\wp_{th}^{R_k}$ [dB]	PER $\wp_{th}^{R_k}$
BPSK	7	0.01
QPSK	10	0.02
16QAM	17	0.04
64QAM	23	0.07

Table 2. Simulation parameters

Parameter	Value
Modulation	BPSK, QPSK, 16QAM, 64QAM
Carrier frequency	2 GHz
Traffic model	Real time service
FFT/IFFT size	512
Packet duration	$1ms$
PER measure period	$20ms$ (=MTPA period)
one symbol duration	$100\mu s$
Cyclic prefix interval	$18.08\mu s$
Maximum Doppler freq., f_d	$1 \sim 10$ Hz
Power delay profile	Indoor Channel B in ITU-R
Fading	Jake's Rayleigh fader

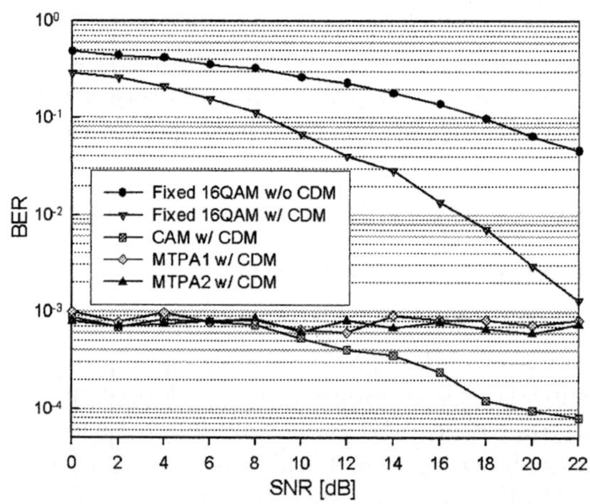

Fig. 4. BER vesus SNR: $f_d = 1.85$ Hz.

subcarrier's channel state was not considered, which leads to occur burst errors. However, for both CAM and MTPAs, the burst error generation probability gets decreased due to adaptive allocation of modulation scheme according to channel state. It leads to satisfy the required BER on the whole SNR range. The reason why two MTPA schemes have higher BER performance over CAM is that only

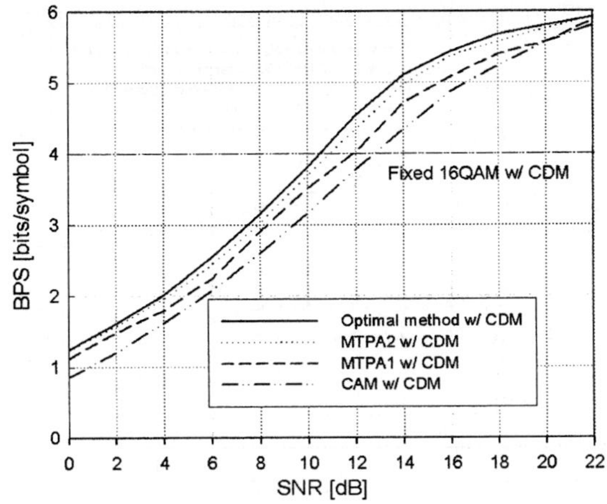

Fig. 5. BPS vesus SNR: $f_d = 1.85$ Hz.

the minimum transmit power needed for each subcarrier is reallocated to maximize throughput. As shown in Fig. 5, compared to fixed modulation scheme, as input SNR increases, the average BPS performances of CAM, MTPA1, MTPA2, and optimal method get increased. This is because the selection probability in high-order modulation scheme gets larger in particular subcarrier. Especially, the average BPS performance in MTPA2 is 15% $\sim$ 20% higher than that of CAM. It means that the surplus powers from unallocated subcarriers have been effectively distributed to increase overall system throughput. Also, it can be shown that MTPA2 provides the BPS performance, which is very close to the optimal value [6]. Also, in Fig. 6, it is shown that our proposed outer-loop scheme 2 maintains the required BER, unlike other schemes over varying maximum Doppler frequency.

5.3 Computation Time

Assuming β_k is precomputed, the optimal scheme [6] requires $2N \cdot \log_2(N \log_2 N)$ additions and $(2N+1) \cdot \log_2(N \log_2 N)$ multiplies (including divisions). The resulting computation time depends on the number of iterations ($\approx \log_2(N \log_2 N)$). On the other hand, for MTPA2, $3.5N$ additions (including subtractions) and $2.25N$ multiplies are needed. It indicates that the computation time of our scheme is shorter by about $0.7 \log_2(N \log_2 N)$ times than that of the optimal scheme, where the computation time for the optimal method becomes huge when the number of subcarriers, N, gets larger.

6 Conclusion

In the OFCDM system, we have proposed the low-complexity MTPA to simultaneously allocate both modulation level and transmit power for each subcar-

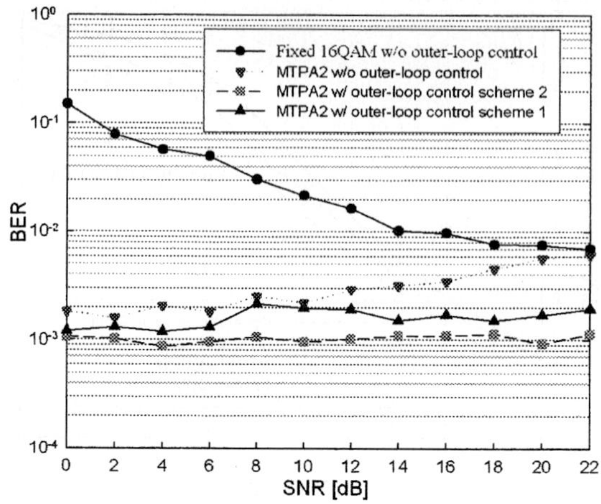

Fig. 6. BER vesus SNR: $f_d = 1 \sim 10$ Hz.

rier with total transmit power constraint. Simulation reveals that our proposed scheme achieved 15 % to 20 % higher BPS over CAM. Also, we have suggested the outer-loop control method which may sustain a given required BER by updating the look-up table under slowly-varying channels.

References

1. 3GPP: Feasibility study for OFDM for uTRAN enhancement (Release 6). 3G TR25.892 V0.2.0 (2003)
2. Cimini Jr., L.J.: Analysis and Simulation of a Digital Mobile Channel using Orthogonal Frequency Division Multiplexing. IEEE Trans. Commun., Vol. 33, No. 7. (1985) 665–675
3. Kaiser, Stefan: OFDM Code-Division Multiplexing in Fading Channels. IEEE Trans. Commun., Vol. 50, No. 8. (2002) 1266–1273
4. Chow, P.S., Cioffi, J.M., Bingham, J.A.C.: A Practical Discrete Multitone Transceiver Loading Algorithm for Data Transmission over Spectrally Shaped Channels. IEEE Trans. Commun., Vol. 43. (1995) 773–775
5. Leke, A., Cioffi, J.M.: A Maximum Rate Loading Algorithm for Discrete Multitone Modulation Systems. in Proc. IEEE GLOBECOM'97, (1997) 1514–1518
6. Krongold, B.S., Ramchandran, Kannan, Jones, D.L.: Computationally Efficient Optimal Power Allocation Algorithms for Multicarrier Communication Systems. IEEE Trans. Commun., Vol. 48, No. 1. (2000) 23–27
7. Choi, W.J., Cheong, K.W., Cioffi, J.M.: Adaptive Modulation with Limited Peak Power for Fading Channels. IEEE 51st VTC2000 Spring, Vol. 3. Tokyo (2000) 2568–2572
8. Stüber, G.L.: Principles of Mobile Communications. Reading, MA: Kluwer Academic (1996)
9. Lee, J.S., Arnott, R., Hamabe, K., Takano, N.: Adaptive Modulation Switching Level Control in High Speed Downlink Packet Access Transmission. IEE 3G Mobile Communication Technologies (2002) 8–10

Soft QoS-based Vertical Handover Between cdma2000 and WLAN Using Transient Fluid Flow Model

Yeong M. Jang

School of Electrical Engineering
Kookmin University
861-1, Jeongneung-dong, Songbuk-gu, Seoul 136-702, Korea
yjang@kookmin.ac.kr

Abstract. This paper proposes a transient(predictive) connection admission control(CAC) scheme using the transient quality of service(QoS) measure for vertical handover between cdma2000 and WLAN. We derive the transient outage probability as the QoS measure using the fluid flow model. We need an approximate approach using fluid flow model for real-time CAC applications. Based on the outage measure, we compare soft QoS-based transient outage performance against traditional hard QoS-based transient outage performance. Numerical results show that the predictive CAC is a promising approach for vertical handover between cdma2000 and WLAN.

1 Introduction

A future wireless service provision will be characterized by global mobile access at anywhere and anytime. These mobile communication systems will consist of different types of wireless networks, each providing varying access bandwidth and coverage level. Two of the most largely deployed packet-switched technologies are wireless local area network(WLAN) based on IEEE 802.11b standards and third-generation wireless wide area networks such as cdma2000 and UMTS(WCDMA). The two technologies offer characteristics that complement each other perfectly. So if users could seamlessly roam across the two networks, the performance and flexibility of wireless data services would be dramatically improved.
Several approaches have been proposed for interworking between WLAN and cellular networks. The 3GPP and 3GPP2 have been standardizing specifications on the interworking between 3G and WLAN[1]. The IEEE 802.21 Working Group has been establishing the requirements for media independent handover services under the heterogeneous systems[2]. For an effective interworking, we need to study a various techniques such as authentication protocol including AAAs, intelligent selection algorithm, QoS vertical handover algorithm, and so on[3][4][5]. But, in this paper, we especially focus at QoS aware vertical handover algorithm between cdma2000 and WLAN.

C. Kim (Ed.): ICOIN 2005, LNCS 3391, pp. 707–716, 2005.

In the proposed interworking system, vertical handover connections must be assigned higher priority over other connections, because vertical handover connection users are more sensitive to outage probability than new connection. Thus, vertical handover connections and horizontal handover connections are treated differently in the point of radio resource management. Therefore, CAC scheme is necessary to support QoS requirements of vertical handover connection. So we propose a soft QoS-based vertical handover scheme for Mobile IPv6-based interworking system.
As for the queuing models, there are essentially three solution methods: the matrix analytic method, probability generating function(pgf), and fluid flow approximation. The most attractive approach, because of its mathematical simplicity and tractability, approximates the actual arrival process to the buffer by a continuous fluid flow approximation[6]. Generally, however, queuing analysis such as bufferless fluid flow models, provide only steady state results due to the complexity of modeling transient behavior. So, for traffic modeled as a superposition of On-Off sources, we approach a transient fluid flow model for the proposed CAC scheme.
This paper is organized as follows. In section 2, we describe an interworking architecture and traffic model which is assumed in our algorithm. We propose the soft QoS-based scheme for CAC under interworking environment, and analyze the outage probability in section 3. In section 4, we discuss the numerical results. Finally, we conclude the paper in section 5.

2 Interworking Architecture for Vertical Handover

2.1 Interworking Architecture

Fig. 1 shows the cdma2000-WLAN interworking architecture. There are two handover procedures between cdma2000 and WLAN: horizontal handover and vertical handover. A user initially connects to AP(Access Point) of WLAN and request vertical handover(λ_V) into ANTS(Access Network Transceiver Subsystem) of cdma2000. When a cdma2000 user moves to adjacent cdma2000 system, a user requests handover(λ_H). Also, MN(Mobile Node) has dual mode interfaces[7][8]. In loose coupling interworking approach, Mobile IP[9] mechanisms must be implemented in the MNs and be installed on the network devices(for example, ANC) of WLAN and cdma2000. This approach provides IP mobility for the roaming between cdma2000 and WLAN.

2.2 Traffic Model

Each source is modeled as an On-Off source[10]. We assume that a series of fixed packets arrive in the form of a continuous stream of bits and use a fluid model. We also assume that the 'OFF' and 'ON' periods for sources are both exponentially distributed with parameters λ_H, λ_V, and μ, respectively. The transition rate from 'ON' to 'OFF' is μ, and from 'OFF' to 'ON' is λ_H or λ_V. Hence the average length of the 'ON' and 'OFF' periods is $\frac{1}{\lambda_H}$ or $\frac{1}{\lambda_V}$ and $\frac{1}{\mu}$, respectively. In this traffic model, when a source is 'ON', it generates fixed packets with a constant interarrival time, $\frac{1}{R_p}$ seconds/bit. When the source is 'OFF', it does not generate any packets.

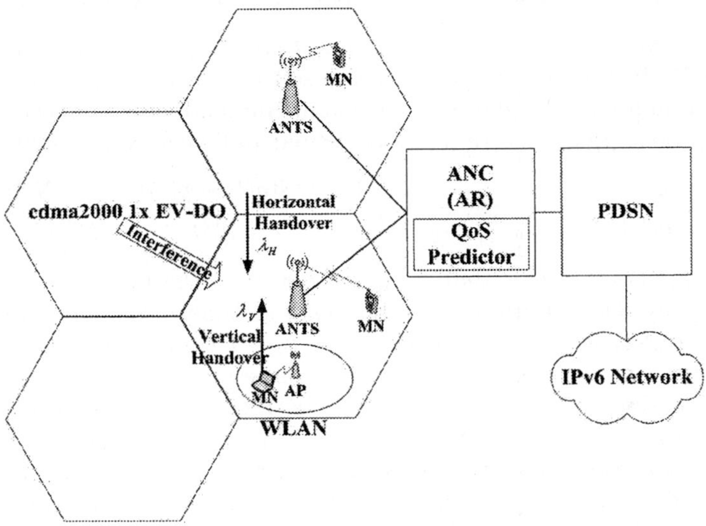

Fig. 1. Interworking architecture between cdma2000 and WLAN.

3 Proposed Soft QoS-based Vertical Handover

3.1 Soft QoS-based CAC

The objective of this paper is that vertical handover connections are allowed to satisfy their performance requirements in interworking architecture. As it were, traditional resource allocation service(e.g. hard QoS) provide only static resource contract. But the vertical handover needs also to coordinate service negotiations that adjust the service requirements dynamically to ensure QoS requirement. The soft QoS is satisfied with dynamic traffic requirement. That is, if the QoS required by the handover connection cannot be supported, the handover connection scales its performance within the given range specified by the requirement of the handover connection, called the critical bandwidth ratio, ξ[11][12]. ξ is a value that results in minimum acceptable satisfaction. In general, ξ for video on demand (VoD) is $0.6 \sim 0.8$ and ξ for background data is $0.2 \sim 0.6$. Namely when the handover caused to congestion in ANC, the soft QoS control improves the satisfaction of undersatisfied connections while maintaining the overall satisfaction of active connections as highly as possible.

We next provide an outline of the soft QoS-based CAC algorithm. The connection is admitted into the radio access network if the QoS requirements can be met for both the new connection and the existing connections. A soft QoS-based CAC algorithm is implemented in the ANC and executed for each radio cell. A flow chart of the soft QoS-based CAC scheme is depicted in Fig. 2.
The proposed CAC scheme based on soft QoS is processed as follows.

1. *Calculate the maximum number of connections*, N_{max}: Including the transient voice activity factor, the number of active sources, and taking into account the inference from adjacent radio cells, the maximum number of connections, N_{max}, which may be supported within any particular radio cell can be found[13].

$$N_{max} = \frac{W}{p(t)(f+1)R_p(\frac{E_b}{I_o})_{req}} + \frac{1}{f+1} - \frac{i(q(t)-p(t))}{p(t)} - \frac{N_o W}{p(t)(f+1)S}, \quad (1)$$

where N_{max} should larger than i, we assume that the power of the reference MN is initially 'ON' at time 0 and still 'ON' at time t. We compute the reference cell interference power by considering $N-1$ connections in the reference cell. We also compute the other cells interference power by assuming that there are N connections in each other cells. Let W(Hz) be the spreading bandwidth, $(E_b/I_o)_{req}$ the required target value, R_p be the peak transmission rate, and f represents the other cells interference as a fraction of the interference from the reference cell. Recall that N denotes the total number of connections in the reference cell. S denotes the total power received from each MNs at the ANC. A perfect power control at each ANC ensures that the total power received from each MNs within that cell is limited to S. N_o stands for the background thermal noise spectral density.

2. *Update the* N: For a new (N+1)-st connection request, we update $N \leftarrow N+1$ for the fluid flow model. The number N represents the number of the existing connections.

3. *Compute (predict) hard QoS-based transient outage probability*, $P_{out}^{H}(t)$: Based on the traffic parameters, λ_H, λ_V and μ, the number of existing connections in the reference radio cell, N, and the number of active connections at the current time $t = 0$, i, predict the outage probability at time t, $P_{out}^{H}(t)$. If $P_{out}^{H}(t) > QoS_{out}$ and new connection is horizontal handover, new connection is rejected.

4. *Compute(predict) soft QoS-based transient outage probability*, $P_{out}^{S}(t)$: Based on soft QoS scheme, peak rate decrease from R_p to R_p^*, which is ξR_p, and the total number of connection increase from N to N^* . After that, $P_{out}^{S}(t)$ is calculated, and if $P_{out}^{S}(t) > QoS_{out}$, new connection is rejected.
Details are provided in next subsection.

3.2 Transient Outage

We will use a statistical bufferless fluid flow model to predict the probability that outage occurs at time t(round trip delay between MN and ANTS) based

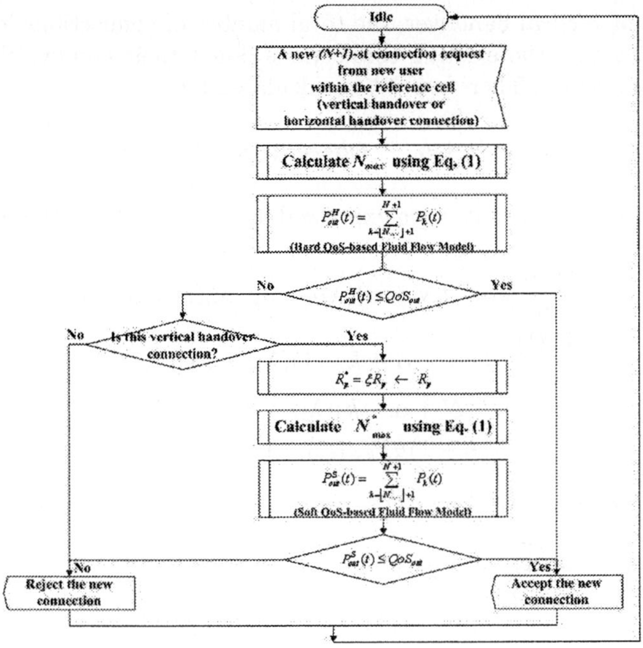

Fig. 2. The proposed soft QoS-based CAC algorithm.

on the traffic statistical behavior at time 0. Equivalently, we can express the capacity of the cdma2000 uplink, as the maximum number of simultaneously active connections, $\lfloor N_{max} \rfloor$, that can be supported on the uplink. The floor function $\lfloor x \rfloor$, also called the greatest integer function, gives the largest integer less than or equal to x. We assume that N On-Off sources in each radio cell share the capacity, $N_{max}R_p$, of a uplink of cdma2000 system. Let $\Lambda(t)$ denote the aggregate arrival rate from $Y(t)$ active connections. According to a fluid model, outages occurs at time t when $\Lambda(t)$ exceeds the link capacity $N_{max}R_p$. Since we are interested in transient outage performance, a formula involving the backward Kolmogorov equations of the process is used(see Fig. 3).

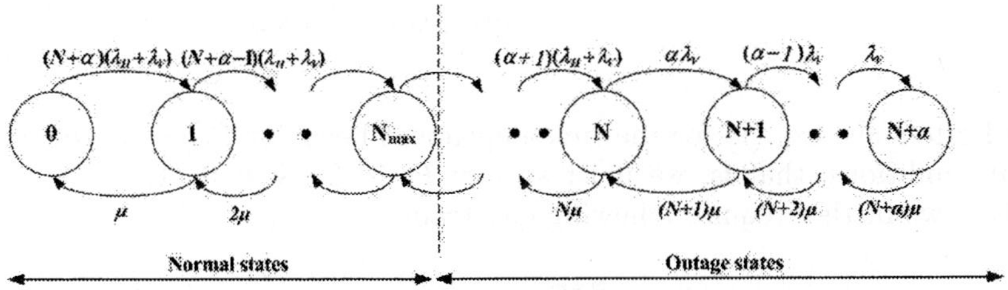

Fig. 3. State transition diagram.

And in case of vertical handover, the total number of connections increase from N to N^* , because the existing connections donor their bandwidth to vertical handover connection. Therefore N^* is calculated by

$$N^* = N + \alpha, \quad where \ \ \alpha = \lfloor (1 - \xi) N \rfloor . \tag{2}$$

The transitions among states are expressed as a set of differential equations.

$$\frac{dP_o(t)}{dt} = -(N + \alpha)(\lambda_H + \lambda_V)P_0(t) + \mu P_1(t), \tag{3}$$

$$\begin{aligned} \frac{dP_k(t)}{dt} &= (N + \alpha - k + 1)(\lambda_H + \lambda_V)P_{k-1}(t) \\ &-[(N + \alpha - k)(\lambda_H + \lambda_V) + k\mu]P_k(t) \\ &+(k+1)\mu P_{k+1}(t), \ \ 1 \le k \le N - 1 \end{aligned} \tag{4}$$

$$\frac{dP_N(t)}{dt} = \alpha(\lambda_H + \lambda_V)P_{N-1}(t) - (\lambda_V + N\mu)P_N(t) + (N+1)\mu P_{N+1}(t), \tag{5}$$

$$\begin{aligned} \frac{dP_{N+j}(t)}{dt} &= (\alpha - j + 1)\lambda_V P_{N+j-1}(t) \\ &-[(\alpha - j)\lambda_V + (N + j)\mu]P_{N+j}(t) \\ &+(N + j + 1)\mu P_{N+j+1}(t), \ 1 \le j \le \alpha \end{aligned} \tag{6}$$

$$\frac{dP_{N+\alpha}(t)}{dt} = \lambda_V P_{N+\alpha-1}(t) - (N + \alpha)\mu P_{N+\alpha}(t). \tag{7}$$

We recognize the above equations (3)~(7) as the backward Chapman-Kolmogorov equations. In matrix form, they can be written as

$$\frac{dP(t)}{dt} = \mathbf{A}P(t) \tag{8}$$

where *P(t)* is the column vector $(P_0(t),\ P_1(t), \cdots,\ P_k(t), \cdots,\ P_{N+\alpha}(t))^T$.$P_k(t)$ represents the probability of having k active sources in the reference cell at time t. $\mathbf{A}$ is a $(N + \alpha + 1) \times (N + \alpha + 1)$ matrix:

$$\mathbf{A} = \tag{9}$$

$$\begin{bmatrix} -(N+\alpha)(\lambda_H+\lambda_V) & \mu & 0 & 0 & 0 & \cdots & 0 \\ (N+\alpha)(\lambda_H+\lambda_V) & -(N+\alpha-1)(\lambda_H+\lambda_V)-\mu & 2\mu & 0 & 0 & \cdots & 0 \\ 0 & \cdots & \cdots & \cdots & \cdots & \cdots & 0 \\ \cdots & \cdots & \alpha(\lambda_H+\lambda_V) & -(\lambda_V+N\mu) & (N+1)\mu & \cdots & 0 \\ 0 & \cdots & \cdots & \cdots & \cdots & \cdots & \cdots \\ 0 & 0 & \cdots & \cdots & \cdots & \lambda_V & -(N+\alpha)\mu \end{bmatrix}$$

In order to solve Eq. (8) for the time-dependent behavior $P_k(t)$, we require our initial conditions; that is, we must specify $P_k(0)$ for $k = 0,\ 1, \cdots, N + \alpha$. In addition we further require following constraint:

$$\sum_{k=0}^{N+\alpha} P_k(t) = 1 \tag{10}$$

Thus we can find the predictive conditional state probability, $P(t)$, by using the eigenvalues of matrix **A**:

$$P(t) = \mathbf{V} \begin{bmatrix} e^{-s_1 t} & 0 & 0 & \cdots & 0 \\ 0 & e^{-s_2 t} & 0 & \cdots & 0 \\ \cdots & \cdots & \cdots & \cdots & \cdots \\ 0 & 0 & \cdots & e^{-s_{N+\alpha} t} & 0 \\ 0 & 0 & \cdots & 0 & e^{-s_{N+\alpha+1} t} \end{bmatrix} \mathbf{V}^{-1} P(0) \tag{11}$$

where s_1, $s_2, \cdots$, $s_{N+\alpha+1}$ are the eigenvalues of matrix **A** and $s_1 = 1$. **V** stands for the right eigenvectors of matrix **A**. $P(0)$ is the column vector $(P_0(0),\ P_1(0),\ \cdots, P_i(0),\ ,\cdots, P_{N+\alpha}(0))^T$ with $P_i(0) = 1$ because the number of active connection at time 0 are $Y(0) = i$. The conditional transient outage probability is then given by

$$\begin{aligned} P_{out}(t) &= P(BER \geq 10^{-3} | Y(0)) = P(\Lambda(t) \geq N_{max} R_p | Y(0)) \\ &= \sum_{k=\lfloor N_{max} \rfloor + 1}^{N+\alpha} P_k(t) \leq QoS_{out} \end{aligned} \tag{12}$$

where $\Lambda(t) = R_p k$ denotes the aggregate arrival rate of the k active connections at time t. QoS_{out} is the QoS requirement for outage probability.

3.3 Application to CAC

Hard QoS-based CAC: Using this admission rule Eq. (12), a new $(N+1)$-st connection is established. We update $N \leftarrow N+1$. We then admit the new connection if this connection is vertical handover, and only if, the condition in Eq. (13) is met. If new connection is horizontal handover, this connection is rejected.

$$N_{Hard} = max\{N | P_{out}^{H}(t) = \sum_{k=\lfloor N_{max} \rfloor + 1}^{N+1} P_k(t) \leq QoS_{out}\}. \tag{13}$$

Soft QoS-based CAC: Let N^*_{Soft} be the maximum number of connections that can be supported in a cell such that the probability of outage i.e., the number of simultaneously bursting connections, exceeds $\lfloor N^*_{max} \rfloor$ with probability less than QoS_{out}. Based on the QoS requirements, N^*_{Soft} will be computed as follows:

$$N^*_{Soft} = max\{N | P_{out}^{S}(t) = \sum_{k=\lfloor N^*_{max} \rfloor + 1}^{N^*+1} P_k(t) \leq QoS_{out}\}. \tag{14}$$

To satisfy the QoS, the radio resource management function at ANC would accept any new connections as long as $N^*_{Soft} > N$, the number of current connections.

4 Numerical Results

Some numerical results have been generated. As an example, consider the following system: $R_p = 96kbps$, $(E_b/I_o)_{req} = 7dB$, $f = 0.33$, $W = 3.84Mbps$(for cdma2000), $S = 60dBm$, $N_o = -166dBm/Hz$. For soft QoS scheme, we assume $\xi = 0.6$ which is a typical value for video traffic.
In Fig. 4, we describe the predicted outage probability in a radio cell as a function of the prediction time(in second) for various values of the initial conditions. We assume $\lambda_H = 0.4, \lambda_V = 0.1$, and $\mu = 0.833$. According to prediction time, it is shown clearly that soft QoS scheme is better performance than traditional hard QoS scheme. Also, we observe that *Y(0)* is larger, outage probability is also larger.
In Fig. 5, we simulate the outage probability according to arrival rate of vertical handover connection, λ_V. We assume $t = \infty$(steady state) and $Y(0) = 2$. We obtain that the performance of soft QoS scheme is better than hard QoS scheme.

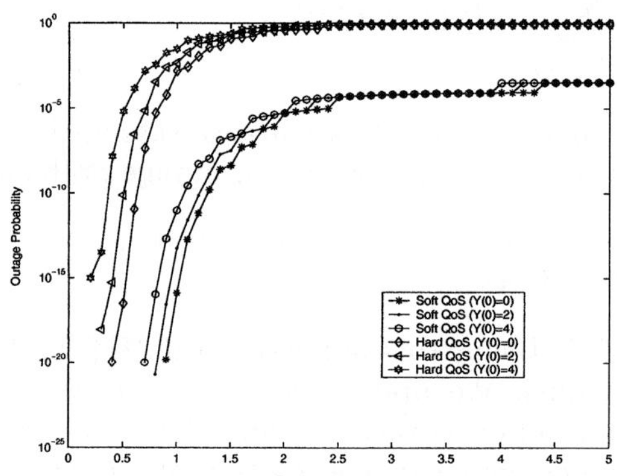

Fig. 4. Outage probability according to prediction time.

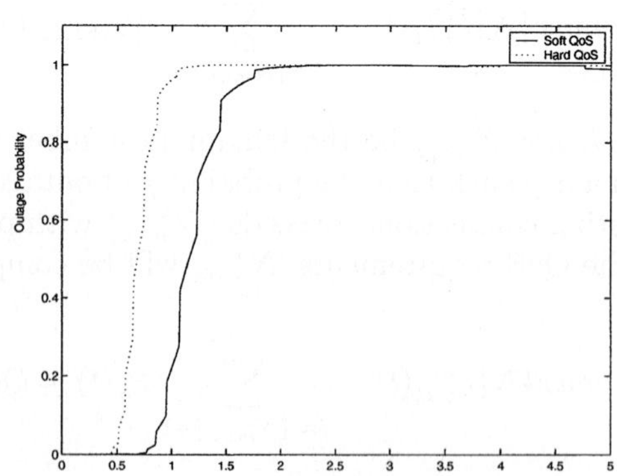

Fig. 5. The outage probability versus the arrival rates.

When the arrival rate of vertical handover connection is 1 calls/sec, outage probability of soft QoS scheme is 0.21 and outage probability of hard QoS scheme is 0.96, respectively. So, soft QoS based CAC scheme has better performance.
Now, we simulate the number of connections according to arrival rate of vertical handover connection. We assume $\lambda_V = 0.4, \mu = 0.833, Y(0) = 0$ and $QoS_{out} = 10^{-3}$. Fig. 6 illustrates that as arrival rate of vertical handover connection increases, the maximum number of connections in soft QoS scheme is larger than hard QoS scheme. At time $t = 1$, soft QoS scheme admit connections around 90, but hard QoS scheme admit 52 connections. Soft QoS scheme can admit more connections around 73%.

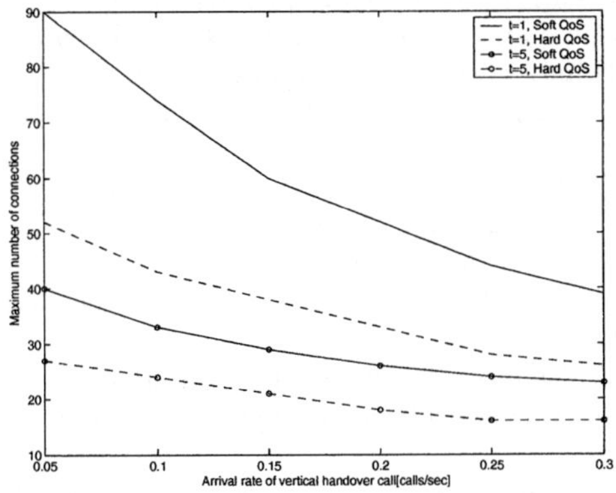

Fig. 6. The maximum number of connections versus the arrival rates.

From our results, we clearly see that the outage probability based on soft QoS is smaller than the outage probability based on hard QoS. Also the proposed soft QoS scheme always accommodate more connections. For the smaller value of ξ, the quality of the individual connection during the vertical handover procedure is slightly deteriorated. Therefore, we conclude that soft QoS-based scheme is more efficient than hard QoS-based scheme.

5 Conclusions

The WLAN and cdma2000 technologies provided complementary environments for mobile packet data users. To integrate two heterogeneous networks, Mobile IPv6-based loose coupling interworking approach was considered. We proposed CAC scheme based on soft QoS concept that is allowed to satisfy their performance requirements in interworking architecture. We focused on the transient uplink performance of cdma2000 system with burst On-Off sources, and developed a fluid flow approximation method for computing transient outage probability for such a system. We could reduce the outage probability using soft QoS-based scheme for vertical handover connections. Therefore, the proposed soft QoS scheme was more efficient than traditional hard QoS scheme under the heterogeneous interworking environments.

Acknowledgement

This work was supported by the KOSEF through the grant No. R08-2003-000-10922-0.

References

1. 3GPP TS 23.234 V6.1.0, 3rd Generation Partnership Project; Technical Specification Group Services and System Aspects; 3GPP system to Wireless Local Area Network (WLAN) interworking; System description (Rel. 6), June 2004
2. 21-04-0087-07-0000, IEEE P802.21; Media Independent Handover Service Draft Technical Requirements, Aug. 2004
3. G.M. Koien and T. Haslestad, 'Security Aspects of 3G-WLAN Interworking,' *IEEE Communications Magazine*, Nov. 2003
4. Wen-Tsuen Chen et al., 'An Adaptive Scheme for Vertical Handoff in Wireless Overlay Networks,' *International Conference on Parallel and Distributed Systems (ICPADS)*, July 2004
5. R. Inayat et al., 'A Seamless Handoff for Dual-interfaced Mobile Devices in Hybrid Wireless Access Networks,' *Advanced Information Networking and Applications*, 2004
6. D. Anick, D. Mitra, and M. M. Sondhi,'Stochastic Theory of a Data-handling System with Multiple Sources,' *Bell System Technical Journals*, 61(8), pp. 1871-1894, Oct. 1982
7. M. Buddhikot et al., 'Design and Implementation of a WLAN/CDMA2000 Interworking Architecture,' *IEEE Communications Magazine*, Nov. 2003
8. Young J. Lee et al., 'The Strategy for Interworking between WLAN and cdma2000,' *Contribution Document for IEEE Plenary Meeting*, Nov. 2003
9. D. Johnson, C. Perkins and J. Arkko, 'Mobility Support for IPv6,' RFC 3775, June 2004
10. P. T. Brady, 'A Statistical Analysis of On-Off Patterns in 16 conversations,' *Bell System Technical Journals*, Vol. 47, pp. 73-91, Jan. 1968
11. D. Reininger and R. Izmailov, 'Soft quality-of-service Control for Multimedia Traffic on ATM Networks,' *The Proceeding of IEEE ATM Workshop*, May 1998
12. Sung H. Kim and Yeong M. Jang, 'Soft QoS-Based Vertical Handover Scheme for WLAN and WCDMA Networks Using Dynamic Programming Approach,' *CIC2002, LNCS 2524*, Nov. 2002
13. Yeong M. Jang and J. Ahn, 'A Connection Admission Control using Transient Outage Probability in CDMA Systems,' *IEEE VTC-Fall* , Vol. 3, pp. 24-28, Sept. 2000

Distributed Mobility Prediction-Based Weighted Clustering Algorithm for MANETs

Vincent Bricard-Vieu and Noufissa Mikou

LIRSA, Faculté des Sciences Mirande
9, av A. Savary
BP 47870, 21078 Dijon Cedex
{vincent.bricard-vieu, nmikou}@u-bourgogne.fr

Abstract. In this paper, we propose a new distributed Mobility Prediction based Weighted Clustering Algorithm based on an on-demand distributed clustering algorithm for multi-hop packet radio networks. These types of networks, also known as *mobile ad hoc* networks (MANETs) are dynamic in nature due to the mobility of the nodes. The association and dissociation of nodes to and from *clusters* perturb the stability of the network topology, and hence reconfiguration of the system is often unavoidable. However, it is vital to keep the topology stable as long as possible. The nodes called *cluster-heads* form a *dominant set* and determine the topology and its stability. Simulation experiments are conducted to evaluate performances of our algorithm and compare them to those of the weighted clustering algorithm (WCA), which does not consider prediction. Results show that our algorithm performs better than WCA, in terms of *updates* of the dominant set, *handovers* of a node between two clusters and *average number of clusters* in a dominant set.

1 Introduction

The rapid advancement in mobile computing platforms and wireless communication technology lead us to the development of protocols for easily deployable wireless networks typically termed wireless ad hoc networks. These networks are used where fixed infrastructures are non-existent or have been destroyed. They permit the interconnectivity between workgroups moving in urban or rural area. They can also help in collaborative operations, for example, distributed scientific research and rescue.

A multi-cluster, multi-hop wireless network should be able to dynamically adapt itself with the changing networks configurations. Some nodes, known as *cluster-heads*, are responsible for the formation of *clusters* each consisting of a number of nodes (analogous to *cells* in a cellular network) and maintenance of the topology of the network. The set of cluster-heads is also called *Dominant set.* A cluster-head is responsible of resource allocation to all nodes belonging to its cluster. Due to the dynamic nature of the mobile nodes, their association and dissociation to and from clusters perturb the *stability* of the network and thus reconfiguration of cluster-heads is unavoidable.

C. Kim (Ed.): ICOIN 2005, LNCS 3391, pp. 717–724, 2005.

The paper is organized as follows. Section 2 describes previous clustering algorithms. In section 3 we propose a new distributed Mobility Prediction-based Weighted Clustering Algorithm and compare in section 4, using simulations, its performances to those of the Weighted Clustering Algorithm (WCA)[1]. Section 5 concludes our study.

For our simulations, we use GloMoSim[2]. GloMoSim is a discrete event parallel environment based on PARSEC (PARallel Simulation Environment for Complex systems)[3].

2 Related Works

Current algorithms for the construction of clusters contained in many routing protocols, as well as clustering heuristics, such as the lowest-identifier[4], the *highest-degree*[5][6] and the *Linked-Cluster Algorithm* (LCA)[7][8], have proactive strategies. By *proactive*, we mean that they require a constant refresh rate of cluster dependent information. That introduces a significant background control overhead even if there is no data to send. The major difficulty comes from node mobility, which has an impact on the position of the nodes and on the neighborhood information, which is essential for clustering. To ensure the correct collection of neighborhood information, existing clustering solutions rely on periodic broadcast of the neighbor list. Mobility causes adjacency relations to change. As well as in Lowest Distance Value (LDV) and the Highest In-Cluster Traffic (ICT)[9], depending on nodes movement and traffic characteristics, the criterion values used in the election process can keep on varying for each node, and hence result in instability.

Proposed by Chatterjee et al[1], the Weighted Clustering Algorithm (WCA) works differently of the algorithms described above since it is only invoked on demand by isolated nodes. Moreover, to determine the cluster-head nodes, that algorithm considers the ideal number of nodes that a cluster can handle, the mobility (speed of nodes), the distance between a node and its neighbors and the battery power. WCA assigns weights to these different parameters.

The cluster-head election procedure is invoked at the time of systhem activation, and also when the current *dominant set* is unable to cover all the nodes. Every invocation of the election algorithm does not necessarily mean that all the cluster-heads in the previous *dominant set* are replaced by new ones. If a node detaches itself from its current head-cluster and attaches itself to another cluster-head, then involved cluster-heads update their member list instead of invoking the election algorithm. See detailed description of WCA in appendix A and in [1]. After the election, all the nodes are in clusters with a cluster-head in each cluster and each node has a list constituted by its neighbors and the set of all the cluster-heads.

All nodes continuously monitor the signal strength of a *Hello* messages received from the cluster-head. When the distance between the node and its cluster-head increases, the signal strength decreases. In that case, the mobile has to notify its current cluster-head that it is no longer able to attach itself to it. The node tries to handover to a neighboring cluster which cluster-head is the first found in its list. If the node goes into a region not covered by any cluster-head, then the WCA election procedure is invoked and a new dominant set is obtained.

Unfortunately, these periodic *hello* messages induce a high communication overhead.

3 Our Algorithm: MPWCA

As mentionned above, the overhead induced by WCA is very high, since it uses a large part of bandwidth for building and maintaining the dominant set (discovery of neighbors, election process, signal strength monitoring), which cannot be used for useful data transmissions.

To avoid that overhead, we propose to increase the duration between two *hello messages*. Due to nodes mobility, the topology is always changing. Increasing the duration between two *hello* messages will lead to link breaking since a node can go out of its cluster without knowing it. So, we propose a distributed mobility prediction-based mechanism using the past movements of the cluster-heads to replace the missing informations given by frequent *hello* messages.

3.1 Description

Our estimation algorithm starts after the election of the cluster-heads, when the ordinary nodes are monitoring the signal strength of packets from their cluster-head. It works as follow :

1. The cluster-head periodically sends informations about its position and its speed in *Hello* messages. When an ordinary node receives these *Hello* messages, it stores the informations about its cluster-head into a list named *past information list.* The stored informations are :
   - the position of the cluster-head in cartesian coordinates (x, y, z).
   - the speed of the cluster-head.
2. If an ordinary node has less than two past informations about its cluster-head stored in its list, during the time between two *Hello* messages, it waits for the next *hello* message (step 1). Otherwise, it uses the past information list to estimate the current position and speed of its cluster-head and store it in a list named *prediction list.* Since the time interval between two *hello* messages can be very large, the ordinary node could make more than one estimation which will be stored in its prediction list. In this case, the prediction list containing previous estimations is appended to the past information list, to make other estimations. The computing of an estimated position is detailed in subsection 3.2.
3. Using its estimation, the ordinary node decides if it should stay in its current cluster or not. To this goal, the ordinary node computes the distance to the estimated position of its cluster-head. It compares this distance to the

transmission range, which is the same for all nodes. If the distance is less than transmission range, the ordinary stays in its cluster. Otherwise, it tries to handover to another neighboring cluster-head. Even if it cannot find another cluster-head in its neighborhood, it stays in its current cluster, waiting for the next estimation or the next *Hello* message, thus to avoid updates of the dominant set which are not needed, due to false estimations.

The ordinary nodes make estimations until they receive a new *hello* message from their cluster-head. After the ordinary nodes have received a new *hello* message (step 1), their prediction list is cleared.

Since an ordinary node estimates position using past information, it needs to take "fresh information". So, the past information list has a finite size (in our experiments, we choose to keep at most 10 positions). When the list is full, the next inserted information drives the oldest information out of the list.

When either a handover or an update of the dominant set and then a new election occurs, both the past information list and the prediction list are cleared.

3.2 Estimation Computation

All ordinary nodes need to estimate the position of their cluster-head. To explain the estimation computation, we suppose that a given ordinary node has already estimated positions stored in its prediction list. To estimate the next position of its cluster-head, the ordinary node appends its prediction list to its past information list to form a bigger list L. Let us note $P = \{p_i = (x_i, y_i, z_i)\}$ the list of the N last positions of its cluster-head (in cartesian coordinates) stored in L. In our experiments we choose N such as $2 \leq N \leq 10$

We compute the $N-1$ vectors

$$\boldsymbol{p_i p_{i+1}} = (x_{i+1} - x_i, y_{i+1} - y_i, z_{i+1} - z_i) \; for \; i = 1..(N-1)$$

Then, we compute the average moving vector

$$\boldsymbol{D} = \frac{1}{N-1} \sum_{i=1}^{N-1} \boldsymbol{p_i p_{i+1}}$$

Finally, the next estimated position p_{N+1} is computed by translating the last position p_N (either monitored or previously estimated) by the vector $\boldsymbol{D}$ ($\boldsymbol{p_N p_{N+1}} = \boldsymbol{D}$). The position p_{N+1} is then stored in the prediction list.

4 Performance Evaluation

Using simulations, we show that our algorithm (MPWCA) performs better than WCA in terms of number of *updates* of a dominant set, number of successful *handovers* of a node in a cluster and *average number of clusters*.

4.1 Simulation Study

We simulate two systems of 50 and 100 nodes respectively on a 1000 $m \times 1000$ m area. The nodes have a transmission range of 100 m and 200 m. The nodes can randomly move in all possible directions with speed varying uniformly between 0 and one parameter representing the maximum value of the speed.

The cluster-head election takes place at the start of the simulation and when a node can no longer be covered by the dominant set. See details of the election procedure in annexe A and in [1]. For this election, we assume that each cluster-head can handle $\delta = 3$ nodes (ideal degree) in its cluster in terms of resource allocation. In this election, parameters w_1 and w_2 (see appendix A) are higher than w_3 and w_4 because we want properties of connectivity and promiscuity with neighbors to be more important for a *good* cluster-head than low mobility and high battery energy. In our experiments, the values used are $w_1 = 0.7$, $w_2 = 0.2$, $w_3 = 0.05$ and $w_4 = 0.05$.

When all cluster-heads are chosen, they start sending *hello* messages with a period of $2s$. Then, ordinary nodes start estimations (step 2 to step 3 described in section 3). In our experiments, they make two estimations before receiving the next *hello* message. After that, the prediction list is cleared and the algorithm is then in step 1.

To measure performances of our system, we consider three metrics :

- the number of *updates* of the dominant set.
- the number of successful *handovers* between two clusters.
- the *average number of clusters* in the dominant set, which characterizes the load of clusters.

These three parameters are studied as a function of the maximum speed of the nodes.

4.2 Simulation Results

In our simulation experiments, we choose values 1 m/s, 5 m/s, 10 m/s, 20 m/s, 30 m/s and 40 m/s for the maximum speed of nodes. The nodes move randomly and uniformly in all possible directions. On figure 1 we can see that the number of succesful handovers increases while speed increases. Due to the mobility, nodes do not always stay in the same cluster. But, there are less changes when nodes have a high transmission range or when they move slowly. We can also observe that our algorithm allows a higher number of successful handovers than WCA.

As well as for handovers, figure 2 shows that the number of updates of the dominant set increases while speed increases, due to mobility. We can see that our algorithm gives better results for this metric, since WCA involves more updates of the dominant set than our algorithm, and the cost of these udpates is higher in terms of resources allocation such as CPU and bandwidth.

The higher the number of nodes in a cluster is, the more the dominant set is stable. Nevertheless, each cluster should not be overloaded, because nodes need to load the cluster-head to comunicate. The usually used number of nodes

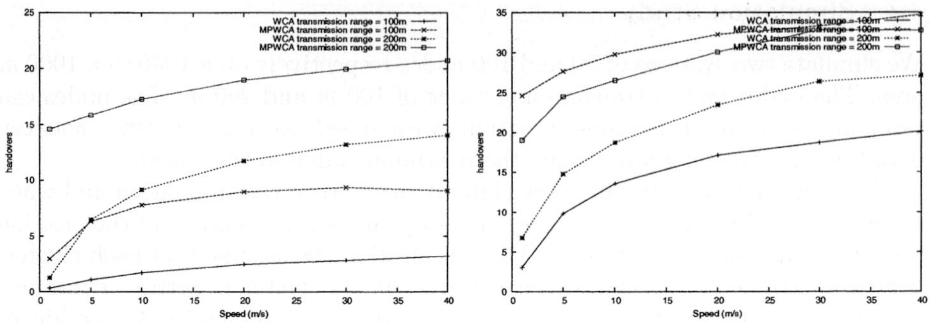

(a) 50 nodes (b) 100 nodes

Fig. 1. number of successful handovers vs maximum speed

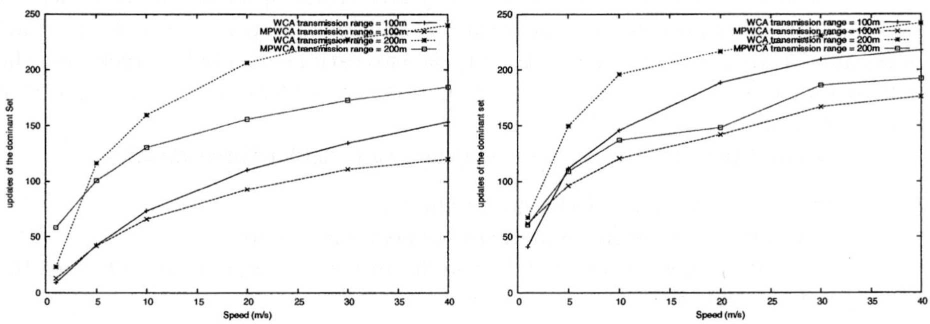

(a) 50 nodes (b) 100 nodes

Fig. 2. number of updates of the dominant set vs maximum speed

that a cluster-head can handle ideally is three nodes (in addition to the cluster-head). Figure 3 shows that the number of clusters increases when speed increases because, due to fast mobility, the dominant set tends to be unstable, but we can notice that with our algorithm, the number of clusters increases slower than with WCA . We can note that our algorithm (with an average value of 2.7 nodes per cluster) is closer to this ideal number than WCA(with an average value 2.5 nodes per cluster).We can also see that when the transmission range is high (200 m instead of 100 m), the number of clusters is low, due to the fact that cluster-head can cover much more ordinary nodes with a value of 200 m for transmission range than with a value of 100m.

5 Conclusion and Further Works

In this paper we propose a new distributed Mobility Prediction-based Weighted Clustering Algorithm (MPWCA). To limit the overhead induced by control mes-

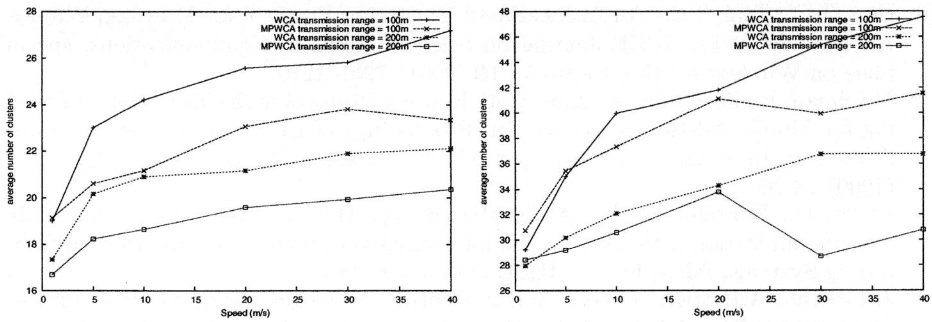

(a) 50 nodes (b) 100 nodes

Fig. 3. average number of clusters vs maximum speed

sages such as *Hello* messages, we increase the interval between two messages. During this long time, nodes try to estimate the movement of their cluster-head and then anticipate handovers, to avoid link breaks.

Using GloMoSim [2], we simulate two networks of 50 and 100 nodes respectively, uniformly distributed on a $1000m \times 1000m$ area. To compare the performances of our new distributed mobility prediction-based weighted clustering algorithm with the Weighted Clustering Algorithm (WCA), we consider three metrics characterizing the stability of the dominant set : the number of updates of the dominant set, the number of handovers of a node to another cluster and the number of clusters. We show that our algorithm ensures a better stability of the dominant set than WCA.

In future work, we will investigate the performances of our estimation mechanism on a locally-centralized system. By *locally-centralized* we mean that we will always use a cluster-based architecture but the estimation will work on the cluster-head itself instead of ordinary nodes.

References

1. Chatterjee, M., Das, S., Turgut, D.: WCA: A weighted clustering algorithm for mobile ad hoc networks. Journal of Cluster Computing (Special Issue on Mobile Ad hoc Networks) **5** (2002) 193–204
2. Zeng, X., Bagrodia, R., Gerla, M.: GloMoSim: A Library for Parallel Simulation of Large-Scale Wireless Networks. In: Workshop on Parallel and Distributed Simulation. (1998) 154–161
3. Bagrodia, R., Meyer, R.: PARSEC: A Parallel Simulation Environment for Complex System. Computer Magazine (1998)
4. M.Jiang, J.Li, Y.C.Tay: Cluster Based Routing Protocol (cbrp) Function Specifications, IETF Draft (1999)
5. Gerla, M., Tsai, J.: Multicluster, mobile, multimedia radio network. ACM/Baltzer Journal of Wireless Networks **1** (1995) 255–265

6. Hou, T.C., Tsai, T.J.: An Access-Based Clustering Protocol for Multihop Wireless Ad Hoc Networks. IEEE Journal on Selected Areas in Communications, special issue on Wireless Ad Hoc Networks **19** (2001) 1201–1210
7. Mitelman, B., Zaslavsky, A.: Link State Routing Protocol with Cluster Based Flooding for Mobile Ad-Hoc Networks. In: Proceedings of the Workshop on Computer Science an Information Technologies (CSIT), Moscow, Russia, MEPhI Publishing (1999) 28–35
8. Baker, D., Ephremides, A.: A distributed algorithm for organizing mobile radio telecommunication networks. In: 2nd international Conference on Distributed Computing Systems, Paris, France, IEEE (1981) 476–483
9. J.Habetha, A.Hettich, J.Peetz, Y.Du: Central Controller Handover Procedure for ETSI_BRAN HiperLAN2 Ad Hoc Networks and Clustering with Quality of Service Guarantees. In: Mobile and Ad Hoc Networking and Computing (MobiHOC). (2000)

A Cluster-Head Election in WCA

The algorithm for the cluster-head election in WCA is the following :

1. Find the set $N(v)$ of neighbors of each node v (ie. nodes v' such that the distance between v and v' is less than the transmission range of v). Set d_v, the *degree* of v, the cardinality of $N(v)$.
2. Compute the *degree-difference* $\Delta_v = \mid d_v - \delta \mid$ for each node v, where δ is the number of nodes (pre-defined threshold) that a cluster-head can handle ideally.
3. For every node, compute the sum of the distances D_v with all its neighbors
$$D_v = \sum_{v' \in N(v)} dist(v, v')$$
4. Compute the running average of the speed for every node until current time T. This gives a measure of mobility M_v
$$M_v = \frac{1}{T} \sum_{t=1}^{T} \sqrt{(X_t - X_{t-1})^2 + (Y_t - Y_{t-1})^2}$$
where (X_t, Y_t) defines the position of the node v at instant t.
5. Compute the cumulative time P_v during which a node v acts as cluster-head. P_v indicates how much battery power has been consumed, which is assumed more for a cluster-head than an ordinary node.
6. Calculate the combined Weight (W_v) for each node v where
$$W_v = w_1 \times \Delta_v + w_2 \times D_v + w_3 \times M_v + w_4 \times P_v$$
7. Choose that node with the smallest W_v as cluster-head. All neighbors of the chosen cluster-head are no more allowed to participate in the election procedure.
8. Repeat steps 2 to 7 for the remaining nodes which are not yet selected as a cluster-head or assigned to a cluster.

An Efficient Subcarrier and Power Allocation Algorithm for Dual-Service Provisioning in OFDMA Based WiBro Systems

Mohmmad Anas[1], Kanghee Kim[1], Jee Hwan Ahn[2], and Kiseon Kim[1]

[1] Department of Information and Communications,
Gwangju Institute of Science and Technology (GIST),
1 Oryong-dong, Buk-Gu, Gwangju, 500-712, Republic of Korea
Tel:+82-62-970-2264 Fax:+82-62-970-2274
{anas, khkim, kskim}@gist.ac.kr
[2] Electronics and Telecommunications Research Institute (ETRI),
161 Gajeong-dong, Yuseong-Gu, Daejeon, 305-350, Republic of Korea
jhahn@etri.re.kr

Abstract. This paper investigates the problem of resource allocation for quality of service (QoS) support in Orthogonal Frequency Division Multiple Access (OFDMA) based WiBro systems. We identify the key QoS parameters as data rate and bit error rate (BER), which are used to determine the individual traffic demands. We propose a resource allocation algorithm to provide dual-service, Guaranteed Performance (GP) and Best Effort (BE) differentiated on the basis of required QoS. Subcarrier assignment and power allocation are carried out sequentially to reduce the complexity, and GP users are given priority over BE users in assigning subcarrier and allocating power. We present the simulation results of the proposed algorithms applied to frequency selective Rayleigh fading channel with additive white Gaussian noise (AWGN) and OFDMA.

1 Introduction

WiBro (Wireless Broadband) [1], also known as High-speed Portable internet (HPi) is a Korean technology for next generation (NextG) communication systems, to provide high-rate data communication to users with diverse quality of service (QoS) requirements over a wireless channel. In this paper, we consider the resource allocation problem in Orthogonal Frequency Division Multiple Access (OFDMA), which is a modulation and multiple access method for WiBro systems based on IEEE 802.16a [2], to provide service to heterogeneous users.

Different broadband services require different amount of rates and different priorities [4]. For example, it requires more bandwidth to provide video service than one for data service, and in general voice service is given higher priority than either a data or a video service. In response to these diverse requirements network designer may choose to support a variety of services with guaranteed QoS and high bandwidth utilization while servicing maximum number of users.

C. Kim (Ed.): ICOIN 2005, LNCS 3391, pp. 725–734, 2005.

So far several papers [5]-[9] have dealt with the problem of adaptive resource allocation in multiuser Orthogonal Frequency Division Multiplexing (OFDM) system under various constraints. When the requirements for each user's data rate and bit error rate (BER) are given, the subcarrier assignment and transmit power allocation problem become more complex to be analytically solved as compared to when there is no constraint on each user's data rate and BER [5]. The problem in this case should be solved by a nonlinear programming technique [6], which requires high complexity to be implemented in practical. So far several suboptimal algorithms have been proposed to solve the problem such as iterative method in [7] and heuristic methods in [6][8]. In [9], an optimal power allocation is proposed for a determined subcarrier assignment scheme to satisfy each user's data rate proportionally.

In this paper, we consider two types of users, Guaranteed Performance (GP) and Best Effort (BE), differentiated on the basis of required data rate and BER criteria. Applications that require guaranteed QoS, such as bounded BER, and a guarantee on the throughput, are called GP services. On the other hand, applications which are less sensitive to instantaneous variations in available bandwidth and do not require guarantees on the throughput, are called BE services [13]. In this context, we propose an efficient subcarrier and power allocation algorithm to provide dual-service (GP and BE) differentiated on the basis of required rate and BER in an OFDMA system.

Ideally, subcarriers and power should be allocated jointly to achieve the optimal solution. However this poses an extreme computational burden on the Base Station (BS) in order to reach the optimal allocation. Separating the subcarrier and power allocation is a way to reduce the complexity since the number of variables in the objective function is almost reduced by half [9]. Here, to make our problem tractable we separate the subcarrier and power allocation. For subcarrier assignment we modify the suboptimal subcarrier allocation algorithm proposed in [6] to provide services to GP and BE users where, GP users are given priority in assigning subcarriers to that of BE users. In assigning subcarrier we assume that total available power at BS is equally distributed among the subcarriers. For power allocation we propose an algorithm to allocate power to GP users so as to satisfy the data rate requirements of GP users and then allocate the rest of the power equally among the subcarriers assigned to BE users.

The remainder of this paper is organized as follows. In Section 2 we present system model and formulate the subcarrier and power allocation problem. In Section 3 subcarrier assignment and power allocation algorithm to provide service to GP and BE user's is developed. In Section 4, we give simulation results of the proposed algorithms. Section 5 contains the concluding remarks.

2 System Model and Problem Formulation

A schematic diagram of an OFDMA based WiBro system used in this paper is shown in Fig. 1. [1][3]. In the figure, K denotes the total number of users

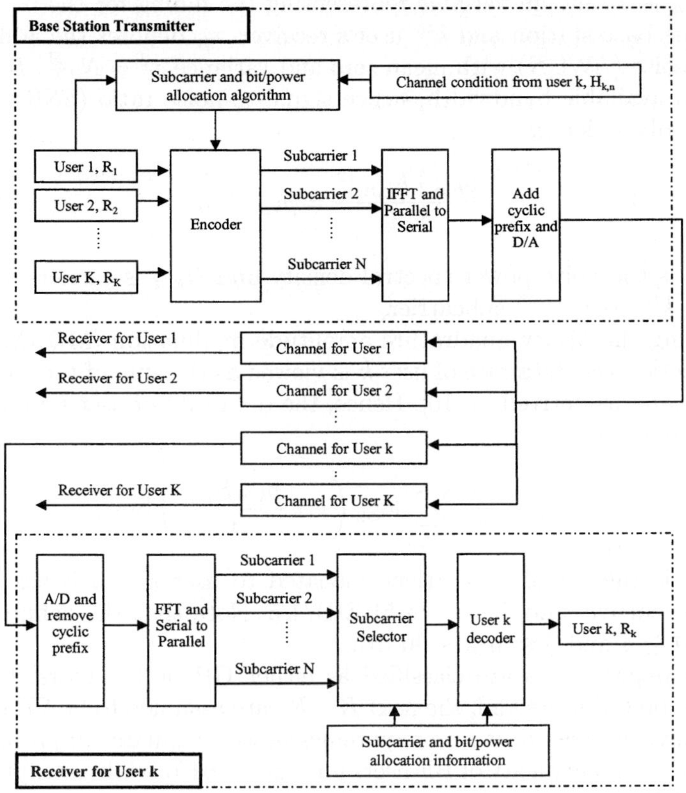

Fig. 1. System Model of a downlink OFDMA System

and N denotes the total number of subcarriers. At the transmitter, the serial data stream from the K users are fed into the encoder block. Using the channel information from all K users, the subcarrier and bit/power allocation algorithm is applied to assign different subcarriers to different users. Here, we assume that a subcarrier at a particular time is not being shared among users. The number of bits and power allocated to each subcarrier is also determined in the process. This information is used to configure the encoder and the input data is encoded and transmitted accordingly. At the receiver, the subcarrier and bit/power allocation information is used to configure the subcarrier selector and decoder to extract the data from the subcarriers assigned to the kth user.

Let us assume that $b_{k,n}$ denotes a set of data symbols for the kth user's n^{th} subcarrier and $p_{k,n}$ is the power allocated to the kth user's n^{th} subcarrier. Under the assumptions above, the transmitted signal from the base station is detected by the kth user's receiver and the decision statistic $z_{k,n}$ for the kth user's n^{th} subcarrier data symbol may be written as,

$$z_{k,n} = b_{k,n}\sqrt{p_{k,n}}h_{k,n} + \eta_n \tag{1}$$

where, $h_{k,n}$ is a random variable representing the fading for the nth subcarrier between the base station and k^{th} user's receiver. η_n denotes the additive white Gaussian noise (AWGN) with mean zero and variance $\sigma^2 = N_0\frac{B}{N}$. B is assumed to be total available bandwidth, hence signal-to-noise ratio (SNR) for the kth user's n^{th} subcarrier is,

$$\frac{p_{k,n} \mid h_{k,n} \mid^2}{N_0\frac{B}{N}} = p_{k,n}H_{k,n} \tag{2}$$

where, N_0 is the noise power spectral density and $H_{k,n}$ is carrier-to-noise ratio (CNR) for k^{th} user's n^{th} subcarrier.

Assuming the M-ary quadrature amplitude modulation (MQAM) and ideal phase detection the data rate of user k is viewed as the sum of the user's subcarriers data rate, as derived in [10]. Hence, the data rate of user k in an OFDMA system is represented by,

$$R_k = \frac{B}{N}\sum_{n\in\Omega_k}\log_2\left(1+\frac{p_{k,n}H_{k,n}}{\Gamma}\right)\text{ bps} \tag{3}$$

where, Ω_k is the set of subcarriers allocated to user k and is assumed to be mutually exclusive, and $\Gamma = -\ln(5\text{BER})/1.5$. Note that the definition of Γ is valid for $M \geq 4$ and $0 \leq \gamma_{k,n} \leq 30$ dB.

In this paper, users are classified as either GP or BE users, first K_1 are assumed to be GP users and, the next $K-K_1$ are assumed to be BE users. Since BE users have no strict data rate requirements, we formulate our problem so as to maximize the sum-capacity of BE users for a given BER while satisfying the data rate and BER requirements of all the GP user's under the total power constraint [11]. Hence, the general optimization problem of interest can be expressed as,

$$\max_{p_{k,n},\Omega_k}\sum_{k=K_1+1}^{K}\sum_{n\in\Omega_k}\frac{B}{N}\log_2\left(1+\frac{p_{k,n}H_{k,n}}{\Gamma_2}\right) \tag{4}$$

$$\text{subject to: }\sum_{n\in\Omega_k}\frac{B}{N}\log_2\left(1+\frac{p_{k,n}H_{k,n}}{\Gamma_1}\right) = R_k$$

$$\sum_{k=1}^{K}\sum_{n\in\Omega_k}p_{k,n} \leq P_{total}$$

$$p_{k,n} \geq 0 \text{ for all } k,n$$

$$R_1 : R_2 : \ldots : R_{K_1} = \gamma_1 : \gamma_2 : \ldots : \gamma_{K_1}$$

$$\Omega_1 \cup \Omega_2 \cup \ldots \cup \Omega_K \subseteq \{1,2,\ldots,N\}$$

where, P_{total} is the total available power; $\Gamma_1 = -\ln(5\text{BER}_1)/1.5$ and $\Gamma_2 = -\ln(5\text{BER}_2)/1.5$ are the SNR gap for GP and BE users respectively; $\{\gamma_i\}_{i=1}^{K_1}$is a set of values proportional to the GP users rate. In this problem, we need to find $p_{k,n}$ and Ω_k to maximize the sum capacity of BE users under the data rate constraints of GP users and total power constraint. As discrete subcarrier assignment is involved in the above problem, it turns to be a hard problem to solve. However if subcarrier assignment Ω_k is known, the dual-service provisioning problem can be converted to a convex optimization problem, similar to the transformation used in [11], for Discrete MultiTone (DMT) Systems.

Hence, for the known subcarrier assignment the power allocation problem to service GP and BE users can be formulated as,

$$\max_{p_{k,n}} \sum_{k=K_1+1}^{K} \sum_{n\in\Omega_k} \frac{B}{N} \log_2\left(1+\frac{p_{k,n}H_{k,n}}{\Gamma_2}\right) \tag{5}$$

$$\text{subject to: } \sum_{n\in\Omega_k} \frac{B}{N} \log_2\left(1+\frac{p_{k,n}H_{k,n}}{\Gamma_1}\right) = R_k$$

$$\sum_{k=1}^{K} \sum_{n\in\Omega_k} p_{k,n} \leq P_{total}$$

$$p_{k,n} \geq 0 \text{ for all } k, n$$

$$R_1 : R_2 : \ldots : R_{K_1} = \gamma_1 : \gamma_2 : \ldots : \gamma_{K_1}$$

$$\Omega_1 \cup \Omega_2 \cup \ldots \cup \Omega_K \subseteq \{1, 2, \ldots, N\}$$

3 Resource Allocation for Dual-Service Provisioning

Transmit power allocation algorithm is carried out sequentially after subcarrier assignment. The optimization problem in (5) is a convex function of power and can be solved using Lagrangian multiplier techniques [13]. The optimal power allocation solution comes out to be well known waterfilling solution in frequency domain [12]. The price for the optimal solution is obviously the computational time and complexity, which makes them impractical for real-time systems. In the following subsections we present a subcarrier assignment algorithm assuming equal power allocation, and an efficient power allocation algorithm for known subcarrier assignment, for dual-service provisioning in OFDMA based WiBro systems [1][2].

3.1 Subcarrier Assignment with Equal Power Allocation

To support the dual class (GP and BE) users, we here modify the suboptimal subcarrier assignment algorithm proposed in [6]. In the proposed subcarrier assignment algorithm we give priority to GP users in assigning subcarriers to that of BE users. In assigning subcarrier we assume that total available power at BS is equally distributed among the subcarriers, as is assumed in [6]. Since power is equally distributed among the subcarriers, we shall refer to this method of subcarrier assignment as proposed-EQ. The proposed subcarrier assignment algorithm to provide service to combined GP and BE users is represented as follows:

Proposed-EQ Algorithm:

1. Initialization (enforce zero initial conditions)
   (a) set $R_k = 0$, $\Omega_k = \Phi$ for all $k = \{1, 2, .., K_1, K_1+1, .., K\}$ and $A = \{1, 2, \ldots, N\}$
   (b) $p = P_{total}/N$: Equal power allocation
2. for $k = 1$ to K_1 (allocate best subcarrier to each GP user) {
   (a) find n satisfying $| H_{k,n} | \geq | H_{k,j} |$ for all $j \in A$
   (b) let $\Omega_k = \Omega_k \cup \{n\}$, $A = A - \{n\}$
   (c) $R_k = R_k + \frac{B}{N} \log_2 \left(1 + \frac{pH_{k,n}}{\Gamma_1}\right)$ }
   (d) while $A \neq \Phi$, repeat step 2. until the rate requirements of GP users are fulfilled
3. for $k = K_1 + 1$ to K (allocate left subcarriers to BE users) {
   (a) find n satisfying $| H_{k,n} | \geq | H_{k,j} |$ for all $j \in A$
   (b) let $\Omega_k = \Omega_k \cup \{n\}$, $A = A - \{n\}$
   (c) $R_k = R_k + \frac{B}{N} \log_2 \left(1 + \frac{pH_{k,n}}{\Gamma_2}\right)$ }
4. while $A \neq \Phi$ (iteratively give lowest rate BE user first choice) {
   (a) find k satisfying $R_k \leq R_i$ for all i, $K_1 + 1 \leq i \leq K$
   (b) for the found k, find n satisfying $| H_{k,n} | \geq | H_{k,j} |$ for all $j \in A$
   (c) for the found k and n, let $\Omega_k = \Omega_k \cup \{n\}$, $A = A - \{n\}$
   (d) $R_k = R_k + \frac{B}{N} \log_2 \left(1 + \frac{pH_{k,n}}{\Gamma_2}\right)$ }

Proposed-EQ is a more general algorithm than proposed in [6]. Omission of step 2.(d) and step 3 in the aforementioned proposed-EQ algorithm reduces it to the method in [6].

3.2 Transmit Power Allocation for Known Subcarrier Assignment

Here, we propose a power allocation algorithm to deal with the high computational complexity issue for dual-service provisioning. To users demanding strict QoS (*e.g.*, GP users), resources (power) are allocated according to the optimal approach, while for the users with loose QoS requirements (*e.g.*, BE users) we can save the computations by using lower complexity algorithm like equal power allocation scheme.

In the proposed algorithm we subdivide the power allocation procedure for GP and BE users. To quantify the amount of combined resources (subcarrier and power), we assume based on the reasonable assumption made in [8] that the amount of power assigned to the users should be proportional to the number of subcarriers allocated. We use optimal waterfilling solution to allocate power to GP users [9][13], and an equal power allocation scheme for BE users. The equal power distribution among subcarriers is shown to be near optimal in [5] for the sum capacity maximization problem under total power constraints. Fig. 2. summarizes the proposed power allocation algorithm. It assumes that subcarrier assignment is known and is determined by aforementioned proposed-EQ algorithm. We shall refer to this method as proposed-RA, where RA stands for resource allocation. Details of the proposed power allocation scheme are described as follows:

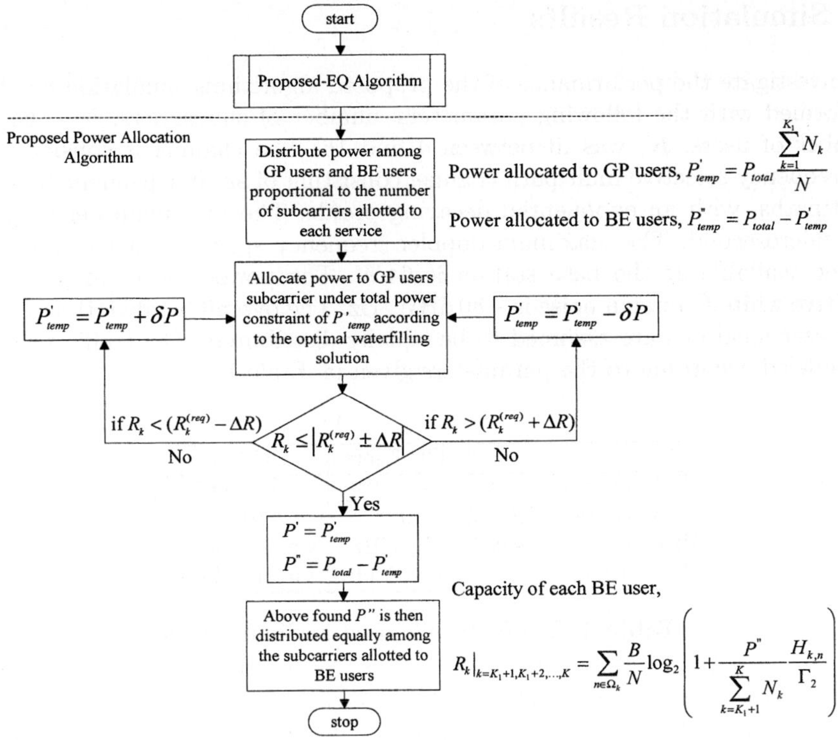

Fig. 2. Proposed-RA Algorithm

Proposed Power Allocation Algorithm:

1. Initialization (quantify the amount of combined resources *i.e.*, subcarrier and power)

   (a) estimate power allocated to GP users, $P'_{temp} = P_{total} \frac{\sum_{k=1}^{K_1} N_k}{N}$
   (b) estimate power allocated to BE users, $P''_{temp} = P_{total} - P'_{temp}$
2. for $k = 1$ to K_1 (allocate power to individual GP users using waterfilling solution){
   (a) allocate power to GP users under total power constraint P'_{temp}
   (b) check if $R_k \leq | R_k^{(req)} \pm \Delta R \mid \}$
   (c) if not then increase or decrease the P'_{temp} by δP (*i.e.*, $P'_{temp} = P'_{temp} \pm \delta P$) and repeat step 2.
   (d) $P' = P'_{temp}$, and $P'' = P_{total} - P'_{temp}$
3. for $k = K_1 + 1$ to K (allocate power to BE users using equal power allocation)

$$R_k = \sum_{n \in \Omega_k} \frac{B}{N} \log_2 \left(1 + \frac{P''}{\sum_{k=K_1+1}^{K} N_k} \frac{H_{k,n}}{\Gamma_2} \right)$$

4 Simulation Results

To investigate the performance of the proposed algorithms simulation has been performed with the following parameters: number of subcarriers, $N = 64$; the number of users, K, was in between 4 and 16. The channel is considered to be frequency selective multipath channel consisting of six independent Rayleigh multipaths, with an exponential decaying profile. The maximum delay spread is 5 microsecond. The maximum doppler frequency spread is 30 Hz. The total power available at the base station is 64 W. The power spectrum density of additive white Gaussian noise is -80 dBW/Hz. The overall bandwidth is 1 MHz. The user locations are assumed to be equally distributed. The traffic behavior is modeled according to the parameters given in Table 1.

	GP Users	BE Users
Number of Users	First 50%	Last 50%
Required BER	10^{-5}	10^{-3}
Required Capacity	1 bps/Hz	Not Applicable
Example	Voice, Video	Internet Data

Table 1. Traffic Profile used in Simulations

Fig. 3. shows an example of capacity comparison between proposed-EQ and proposed-RA algorithms vs. user index. We notice that GP users adapt to their data rate requirements (*i.e.,* 1 bps/Hz) after proposed-RA algorithm, while for the BE users rate distribution is found to be almost same as that of after proposed-EQ. This is because we use equal power allocation method for BE users in both proposed-EQ and proposed-RA algorithms.

Fig. 4. compares min-user's capacity of GP and BE users vs. number of users. Here, proposed-RA-GP and proposed-EQ-GP represent the GP user's performance using proposed-RA and proposed-EQ algorithms respectively, and similar representation stands true for BE users. We notice that the min-user's capacity of GP users remains constant while those of BE users decreases as the number of users increases. This trend can easily be understood from Fig. 3., as the number of users (K) increases the BE user's capacity decreases and hence the min-user's capacity of BE users. BE user's capacity performance is also compared with that of the min-user's capacity performance of fixed time division multiple access (TDMA) *i.e.,* a fixed time slot is allotted to each user in TDMA. We notice that adaptive resource allocation performs better for smaller number of users than for higher number of users. This is because, as the number of users increases, more resources (subcarrier and power) are needed to fulfill the rate requirements of GP users while BE users are left with lesser resources, and hence we notice the decrease in min-user's capacity gain over TDMA with the increase in number of users.

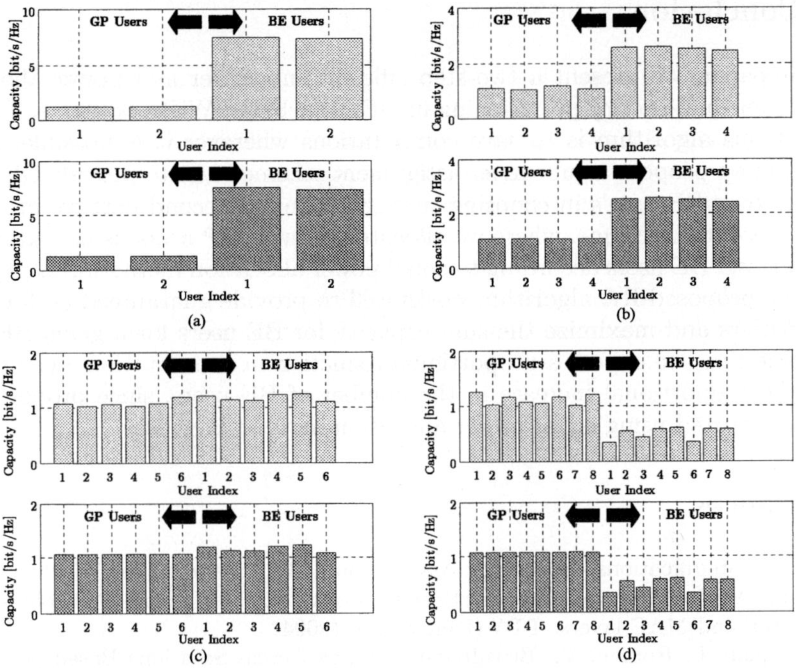

Fig. 3. Example capacity performance comparison between proposed-EQ (figure above) and proposed-RA (figure below) vs. user index. (a) $K = 4$, (b) $K = 8$, (c) .

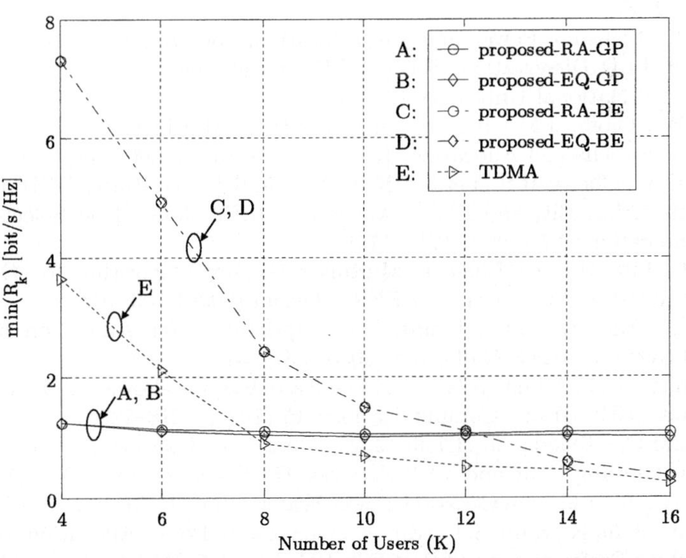

Fig. 4. min-user's capacity of GP and BE users vs. number of users

5 Conclusion

In this paper, we present a two-step efficient subcarrier and power allocation scheme for dual-service provisioning in OFDMA based WiBro systems. The key idea of this algorithm is to save computations wherever it is possible. In the first part we propose a subcarrier assignment scheme so as to provide GP user's priority over BE user's in choosing subcarrier. In the second part we propose a power allocation scheme, where we allocate power to GP users using waterfilling solution and BE users according to equal power allocation method. Result shows that the proposed-RA algorithm works well to provide guaranteed performance to GP users and maximize the sum capacity for BE users for a given BER. At the same time we expect a proportional reduction in computational complexity with the proportional increase in the number of BE users, since proportion of users supported using equal power allocation method increases.

References

1. Telecommunications Technology Association (TTA) Standard for Wireless Broadband (WiBro) Portable Internet: Specifications for 2.3 GHz band Portable Internet - PHY and MAC layers. TTA (Korea) Std. (2004)
2. Koffman, I., Roman, V.: Broadband Wireless Access Solutions Based on OFDM Access in IEEE 802.16. IEEE Communications Magazine **40** (2002) 96–103
3. Nee, R.V., Prasad, R.: OFDM for Wireless Multimedia Communications. Artech House, Boston (2000)
4. Bahl, P., Chlamtac, I., Farago, A.: Resource Assignment for Integrated Services in Wireless ATM Network. International Journal of Communication Systems **11** (2000) 29–41
5. Jang, J.: Transmit Power and Bit Allocations for OFDM Systems in a Fading Channel. PhD Dissertation, School of Electrical Engineering and Computer Science, Seoul National University (2003)
6. Rhee, W., Cioffi, J.M.: Increase in Capacity of Multiuser OFDM System Using Dynamic Subchannel Allocation. IEEE VTC-Spring (2000) 1085–1089
7. Wong, C.Y., Cheng, R.S., Letaief, K.B., Murch, R.D.: Multiuser OFDM with Adaptive Subcarrier, Bit, and Power Allocation. IEEE Journal on Selected Areas in Communications **17** (1999) 1747–1758
8. Yin, H., Liu, H.: An Efficient Multiuser Loading Algorithm for OFDM-based Broadband Wireless Systems. IEEE Globecom (2000) 103–107
9. Shen, Z., Andrews, J.G., Evans, B.L.: Optimal Power Allocation in Multiuser OFDM Systems. IEEE Globecom (2003) 337–341
10. Goldsmith, A.J., Chua, S.G.: Variable-Rate Variable-Power MQAM for Fading Channels. IEEE Trans. Communications **45** (1997) 1218–1230
11. Hoo, L.M.C., Tellado, J., Cioffi, J.M.: Dual QoS Loading Algorithm for DMT Systems Offering CBR and VBR Services. IEEE Globecom (1998) 25–30
12. Yu, W., Cioffi, J.M.: On Constant Power Water-filling. IEEE ICC (2001) 1665–1669
13. Anas, M., Kim, K., Shin, S.J., Kim, K.: QoS Aware Power Allocation for Combined Guaranteed Performance and Best Effort Users in OFDMA Systems. IEEE ISPACS (2004) 477–481

P-MAC: Parallel Transmissions in IEEE 802.11 Based Ad Hoc Networks with Interference Ranges

Dongkyun Kim and Eun-sook Shim

Department of Computer Engineering,
Kyungpook National University, Daegu, Korea
dongkyun@knu.ac.kr, esshim@monet.knu.ac.kr

Abstract. IEEE 802.11 prohibits an exposed node from transmitting any packet until the end of its NAV (Network Allocation Vector). Some trials have been proposed to enable an exposed node, called secondary sender, to transmit its packets in parallel with a primary sender which reserved a wireless channel in advance through RTS/CTS exchange. However, they did not cope with the existence of interference ranges, while they considered only that of transmission ranges. We therefore propose our P-MAC (Parallel MAC) protocol to enable an exposed node to determine whether or not it can succeed in transmitting its data without any collision. Simulation study proves that P-MAC is superior to other schemes in terms of performance metrics.

1 Introduction

Recently, the interest in MANET (Mobile Ad Hoc Networks) has increased because of the proliferation of small, inexpensive, portable, and mobile personal computing devices. MANET is a wireless network where all nomadic nodes are able to communicate each other through packet relaying service of intermediate nodes. In particular, the MANET working group in IETF [1] has been trying to standardize its routing protocols. In addition to the routing protocols, a medium access control protocol at link layer is needed to enable the data transmission over a common radio channel which has the collision problem of access to a shared medium among contending nodes. In general, the carrier sense multiple access (CSMA) protocols have been used in the packet radio network. However, since carrier sensing is sensitive to location of nodes in MANET, the well-known hidden and exposed terminal problems can occur. For the purpose of resolving the hidden terminal problem, various approaches such as MACA (Multiple Access with Collision Avoidance) [2] have been developed by introducing the exchange of RTS (Request-To-Send) and CTS (Clear-To-Send) messages before actual data transmission. Furthermore, because of the absence of mechanism to enable a reliable data transmission in MACA, DFWMAC (Distributed Foundation Wireless MAC) protocol used in IEEE 802.11 [3] adds the transmission of ACK packet to this basic MACA protocol, that is, four-way exchange, RTS-CTS-DATA-ACK. However, although the RTS/CTS exchange partially addresses the

C. Kim (Ed.): ICOIN 2005, LNCS 3391, pp. 735–744, 2005.

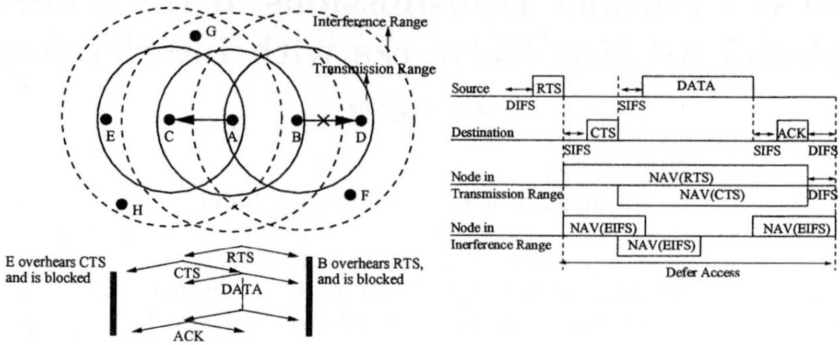

Fig. 1. The RTS-CTS-DATA-ACK handshake in IEEE 802.11.

hidden terminal problem, the exposed terminal problem should still be resolved. As shown in Figure 1, a node needs to sense whether the channel is idle or not for a DIFS (Distributed Inter-Frame Space) interval before attempting an RTS transmission and an SIFS (Short Inter-Frame Space) interval before sending an ACK packet and a CTS packet, respectively. After the idle time of DIFS, the sender (e.g., node A) transmits an RTS packet and waits for a corresponding CTS packet from a receiver (e.g., node C), which requires the sender's neighboring nodes (e.g., node B) to defer their transmission until a DATA packet transmission is completed [1]. When a receiver (e.g., node C) receives the RTS packet successfully, it sends a CTS packet to the sender after an SIFS interval, which requires the receiver's neighboring nodes to defer their transmission through their NAVs. As node E is *hidden* from node A, this CTS packet is capable of avoiding a collision at node C. Receiving the CTS message allows the sender to transmit its DATA packet and awaits an ACK packet for the transmitted DATA packet. Receiving the ACK packet means the completion of a successful transmission. According to IEEE 802.11 standard, a large frame is transmitted using the exchange of RTS-CTS-DATA-ACK. Furthermore, it enables a small frame to be transmitted with the two-way handshake of DATA-ACK, instead of the four-way handshake.

Although we depend on the basic handshaking, the existence of interference range which is distinguished from transmission range makes the MAC protocol more complex. The nodes located within an interference range of a sender cannot decode the data packet successfully because it is simply undecodable. IEEE 802.11 therefore allows the nodes located within an interference range of an RTS-sending or a CTS-sending node to simply defer their transmission trials with their own NAVs set to EIFS(Extended Inter-Frame Space) values. As shown in Figure 1, suppose that node B needs to send a packet to node D. As node

[1] The duration for which neighboring nodes should be silent is set through NAV (Network Allocation Vector) value.

B is *exposed* to node A, which forces node B to defer its transmission to node D, node B waits for a completion of node A's transmission. However, if ACK transmissions between node C to node A and node D to node B are synchronized, two transmissions between node A to node C and node B to node D will be enabled in parallel. Therefore, the increase of concurrent and simultaneous transmissions can improve the MAC performance.

However, current IEEE 802.11 does not permit any possible parallel transmissions among neighbors in MANET. Several research works therefore attempted to mitigate the exposed terminal problem by enabling parallel transmissions [4] [5] [6]. However, there exists no trial to address the problem in the situation where the interference and transmission ranges coexist. In this paper, we therefore propose an efficient P-MAC (Parallel MAC) protocol to allow neighbors to transmit their packets in parallel in face of the existence of interference range.

The rest of this paper is organized as follows. A brief description of related works is presented in Section 2. Section 3 describes our proposed P-MAC followed by performance evaluation in Section 4. Finally, some concluding remarks with future plans are given in Section 5.

2 Related Works

In MACA-P(Medium Access via Collision Avoidance with Enhanced Parallelism) [4], a control gap (T_{DATA}) is introduced between the RTS/CTS exchange and the subsequent DATA/ACK exchange of a primary connection, that is, the first pair of transmissions. During the T_{DATA}, a secondary connection performs the exchange of its own RTS/CTS, schedules a parallel DATA transmission and finally synchronizes ACK transmission with the primary connection. However, the length of this control gap can affect the performance of the protocol as well as the number of parallel transmissions.

In EMAC(Enhanced MAC) [5], the fragmentation technique for a large MAC frame is exploited for parallel transmissions. During the exchange of DATA/ACK packets for the first transmitted fragment of a primary connection, the secondary pair of transmissions finishes the exchange of its own RTS/CTS and schedules the DATA/ACK transmission for further fragments synchronized with those of the first pair. However, the size of a fragment can affect the performance with the assumption that a MAC frame should be large enough to be fragmented.

MENP(Mitigating the Exposed Node Problem) [6] utilizes the traffic statistics showing that approximately 50 % of all packets on the Internet are small packets below 100 bytes in size. In particular, as mentioned before, IEEE 802.11 allows the transmission of a small packet to rely on the exchange of DATA/ACK packets without a prior exchange of RTS/CTS packets. Therefore, during the transmission of a large DATA, a sender exposed to the first pair of transmissions can start a parallel transmission for a small packet after carrier sensing of the first pair's actual DATA transmission if the ACK returning to the exposed sender is synchronized with the ACK transmitted to the sender of the first pair (see Figure 2).

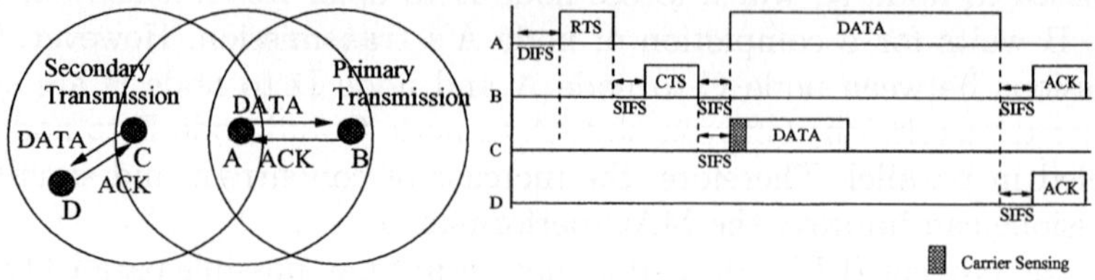

Fig. 2. The synchronization of ACK packets.

3 P-MAC: Parallel MAC

MACA-P requires several additions to the RTS/CTS frame formats. Prior to a parallel transmission, a new control frame such as RTS' is needed to update T_{DATA} value. EMAC relies on the fragmentation capability with additional control packets like PRTS and PCTS in order to synchronize the parallel transmissions. For the purpose of minimizing the modification to the basic IEEE 802.11 protocol, the concept of the MENP protocol is therefore exploited in our work. As mentioned earlier, MENP makes good use of the observation that traffic statistics reveals that many small packets below 100 bytes in size are transmitted on the Internet. These small packets can potentially be sent in parallel by the exposed nodes during transmitting a large DATA of a primary connection.

3.1 Recognition of Exposedness

In MENP, for the purpose of performing a parallel transmission, a node should be able to determine whether the node is a real exposed node to the first pair of transmissions. According to the IEEE 802.11 standard, node A_{over} overhearing an RTS becomes aware of the time when the actual DATA transmission between a sender and a receiver begins (denoted by RESET_NAV), which is a duration of CTS_TIME + 2 * SIFS_TIME + 2 * SLOT_TIME where the CTS_TIME is calculated from the length of the CTS and the rate at which the previous frame was received [6]. If the actual DATA transmission is sensed after RESET_NAV, node A_{over} concludes that it is a real exposed node and schedules its parallel transmission. Otherwise, node A_{over} decides that the trial of sender's transmission failed and attempts to take an action accordingly, which is beyond the scope of our work and refer to [7] for the issue. We also exploit this approach as a mechanism to detect whether or not a node is a real exposed node and we furthermore require an additional sensing period for parallel transmissions, which is described in Section 3.3 in detail.

3.2 Problem Caused by Existence of Interference Range

It is known that when a node transmits a packet with its transmission power, two ranges are formed, namely transmission range and interference range. Other nodes within a transmission range of a sender are able to receive the packet and process it. Otherwise, the nodes within an interference range out of a transmission range of a sender cannot decode the packet successfully and the packet is recognized as an undecodable signal. Therefore, the existence of an interference range can fail the MENP protocol's mechanism as shown in Figure 4 (a). Without the interference range, during the transmission of a large DATA from node A to node B, a small DATA is allowed to be transmitted from node C to node D in parallel with two synchronized ACKs. However, the interference range produces a serious collision at node B because node B is located within an interference range of node C. Therefore, the DATA from node A to node B will be interfered with the undecodable signal from node C. The transmission of node A can also interfere the DATA received by node D because node D is located within an interference range of node A.

3.3 Description of Proposed P-MAC Protocol

In this section, we describe our P-MAC to enable parallel transmission in spite of the existence of interference ranges. We assume that the transmission ranges of all nodes are the same and furthermore the interference ranges of all nodes are the same. It is furthermore assumed that the interference range is 1.6 times of a transmission range [2]. A primary sender and a primary receiver mean a sender and a receiver of the first pair of transmissions, respectively. If a primary receiver is located within an interference range of a node exposed to a primary sender, the exposed node should not transmit its frame although it needs to send a small DATA, because it can collide with the DATA received by a primary receiver.

The key feature of our P-MAC is how to detect whether or not the exposed node can make a collision at a primary receiver. Due to the symmetric characteristic of an interference range [3], it is enough that the exposed node determines whether or not it is located within an interference range of the primary receiver. Overhearing an RTS, the exposed node expects that the primary receiver will send its CTS packet SIFS_TIME right after receiving the RTS. If an exposed node is located within a transmission range of a primary receiver, it will succeed in receiving the CTS. Otherwise, if it is located within an interference range of a primary receiver (that is, out of the transmission range), it will receive an undecodable signal. Therefore, the undecodable signal can be utilized to prohibit the exposed node from performing a parallel transmission even though it needs to send a small DATA. In addition, since it is possible that an exposed node receives an undecodable signal caused by some trials of transmission from other nodes in its neighborhood, the exposed node should not try to transmit its small DATA in parallel because the trial can disturb other transmissions due to the symmetric characteristic of transmission range or interference range. A brief diagram for state transition of the operation of a secondary sender is described

[2] If the transmission range is 380 meters, its interference range is set to 608 meters.
[3] If a node exposed to a primary receiver is within an interference range of the primary receiver, the primary receiver is also located within that of the exposed node.

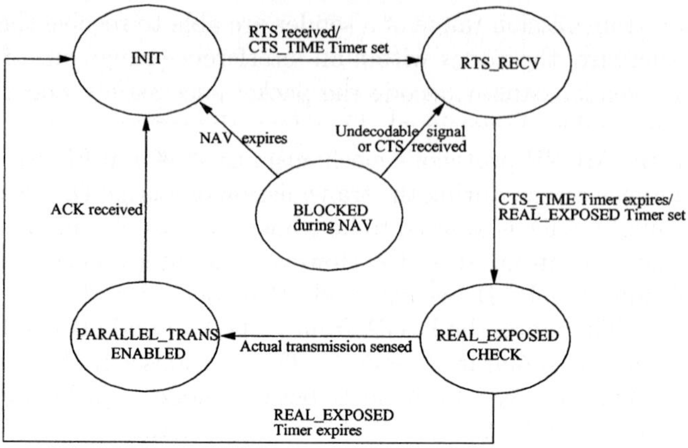

Fig. 3. A brief diagram for state transition of the operation of a secondary sender.

in Figure 3. Only when an exposed node is not located within an interference range of a primary receiver and senses the silence of a wireless channel during a CTS_TIME right after receiving the RTS, it can become a secondary sender with a chance to transmit its small DATA safely without disturbing the transmission of a primary connection. Therefore, by utilizing an additional period of carrier sensing as shown in Figure 3, that is, CTS_TIME, the silence of the channel during CTS_TIME allows the exposed node to determine that it can transmit its small DATA in parallel.

In addition, as mentioned in Section 3.1, before transmitting a small DATA, a secondary sender checks if the actual DATA transmission from the primary sender begins (REAL_EXPOSED timer is used in Figure 3.). If the transmission begins, it means that it is a real exposed node and schedules its parallel transmission for small packets. Otherwise, the secondary sender decides that the trial of sender's transmission failed and attempts to take an action accordingly, which is beyond the scope of our work and refer to [7] for the issue.

Besides, a primary sender can be located within an interference range of a secondary receiver when the secondary receiver returns its ACK to the secondary sender. Therefore, although two ACKs of a primary receiver to a primary sender and a secondary receiver to a secondary sender are synchronized, a collision at the primary sender can occur. For the purpose of avoiding the collision at a primary sender for the returned ACKs, our P-MAC requires that the returned ACK from a secondary receiver to secondary sender should be transmitted before the end of the DATA transmission for the primary connection. Particularly, since there exist many nodes exposed to a primary sender which want to perform their parallel transmissions for their small DATA packets and the collisions at the secondary senders can occur due to their ACKs, the times when the ACK frames from secondary receivers are transmitted are randomly selected before the end of the

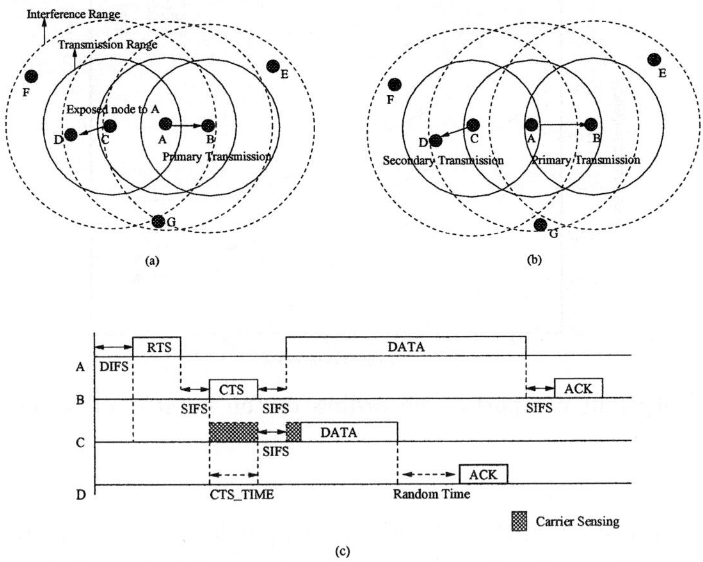

Fig. 4. The Illustration of P-MAC protocol.

DATA transmission of a primary sender in order to minimize such collisions as shown in Figure 4 (c).

Figure 4 (a) shows a case where node C exposed to a primary sender, node A, can not become a secondary sender to transmit its small DATA in parallel because it can produce a collision at a primary receiver, node B. However, when Figure 4 (a) is applied to the MENP protocol, node C will transmit its small packets indiscriminately, which results in performance degradation. In Figure 4 (b), our P-MAC allows node C to perform its parallel transmission because node B is not within an interference range of node C. Figure 4 (c) shows an example for which the exposed node, node C, performs its carrier-sensing activity.

4 Performance Evaluation

We evaluated our P-MAC protocol using our event-driven simulator, which operates IEEE 802.11 DCF (Distributed Coordination Function) [3]. The MAC parameters such as Inter Frame Spaces and the length of RTS and CTS were set according to IEEE 802.11 standard. We compared our P-MAC with the basic IEEE 802.11 and MENP in terms of performance metrics. The transmission range and interference range were set to 380 and 608 meters, respectively. Each simulation ran 200 seconds and we plotted graphs using average values of 10 runs.

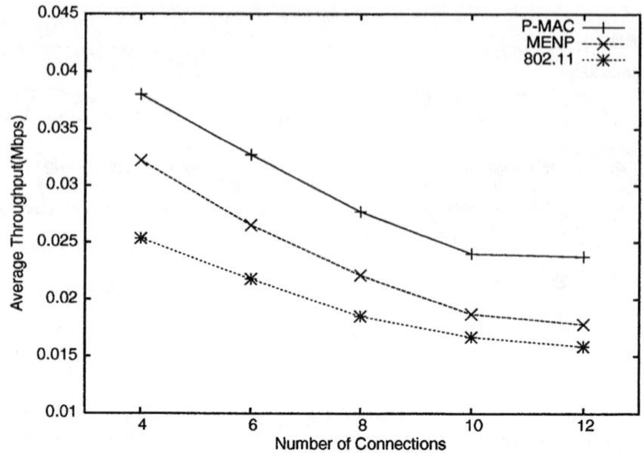

Fig. 5. Throughput comparison according to the number of nodes for a ring topology.

First, we investigated the average throughput for a dual ring topology consisting of outer and inner rings. We assumed that the nodes on an inner-ring send their DATA packets to the nodes on an outer-ring. Some nodes on an inner ring were selected as primary senders which transmit their packets of 1024 bytes in size and we selected the others as the nodes which want to transmit their packets of 512 bytes in size. We observed the average throughput by varying the number of nodes on the rings. For all schemes, a large number of nodes in the rings increase the collision due to high contention to a shared wireless channel , which results in the degradation of throughput as shown in Figure 5. In IEEE 802.11, any parallel transmission is not enabled due to its well-known, NAV-based exposed terminal problem. Although MENP allows some secondary senders to transmit their packets in parallel with transmissions of primary senders, it does not cope with the existence of interference ranges. Therefore, an indiscriminate initiation of a transmission of an exposed node disturbs the primary transmission if the primary receiver is located within an interference range of the secondary sender. However, our P-MAC protocol permits the secondary sender to perform its parallel transmission only if its transmission is independent of the primary transmission, that is, only when the secondary node recognizes that it is a really exposed node and out of interference range of the primary receiver through the additional sensing technique mentioned in Section 3.3, which results in better performance than the others.

Second, we measured the performance using a random topology with 70 nodes which are randomly located in the area of 1000 m x 1000 m. We used the same simulation parameters as in the ring topology. In this simulation, we performed throughput comparison of three protocols according to the number of connections among nodes. A connection is defined as a pair of sender and receiver in one-hop wireless link which is randomly selected for each packet transmission. A large number of connections among nodes cause high contention and many collisions to the wireless media among the connections, which produces

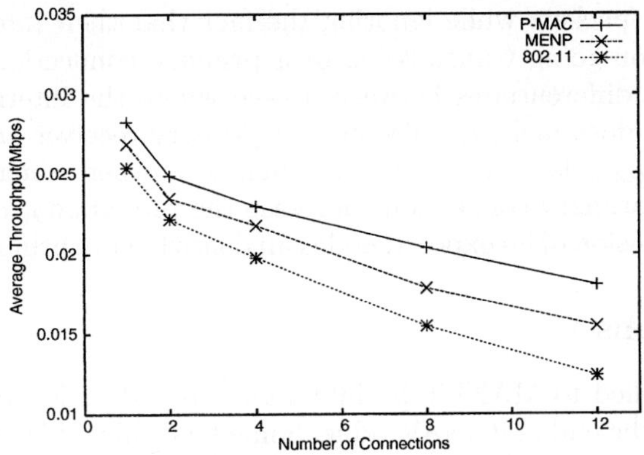

Fig. 6. Throughput comparison according to the number of connections for a random topology.

lower throughput than a small number of connections (see Figure 6). Due to the same reason mentioned in the previous simulation, P-MAC shows the best performance of throughput.

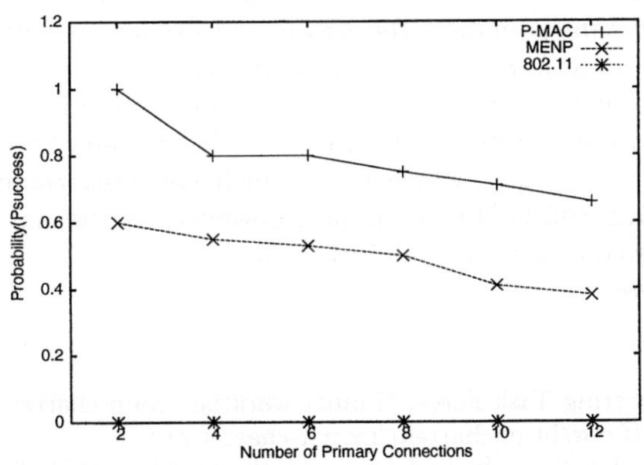

Fig. 7. $P_{Success}$ comparison according to the number of primary connections for a random topology.

Finally, we investigated the probability, $P_{Success}$ according to the number of primary connections (see Figure 7). $P_{Success}$ is defined as a probability that a node exposed to a primary sender can succeed in transmitting its packet to its receiver in parallel without producing any collision at a primary receiver. In IEEE 802.11, all nodes exposed to a primary sender cannot be assigned any chance to transmit its data until the end of a primary connection. Therefore, in all cases, the $P_{Success}$ are all zeros. In MENP, all exposed nodes are trying to transmit their packets while ignoring the fact that their interference ranges can disturb the on-going transmission of a primary connection. However, our P-MAC protocol differentiates between a case where the interference range of an exposed node does make a collision at a primary receiver and a case where it does not. $P_{Success}$ is therefore higher than any other scheme. Note that a large number of primary connections decrease the probability, $P_{Success}$ because a parallel transmission of an exposed node can disturb more primary connections.

5 Conclusions

IEEE 802.11 applied to MANET prohibits an exposed node from transmitting any packet until the end of its NAV value. Some trials like MACA-P, EMAC and MENP have been proposed to enable an exposed node, called secondary sender, to transmit its packets in parallel with a primary sender which reserved a wireless channel in advance through RTS/CTS exchange. All those protocols however did not cope with the existence of interference ranges which could disturb primary connections in the network. In particular, we exploited a concept of MENP to minimize the modification to IEEE 802.11 standard while providing a mechanism to perform the parallel transmissions of primary and secondary senders. MENP allows an indiscriminate initiation of a transmission of a secondary connection without considering that it can disturb a primary transmission when the primary receiver is located within an interference range of the secondary sender. Our P-MAC protocol therefore enables the secondary sender to perform its parallel transmission after determining that its transmission is independent of a primary transmission, that is, only when the secondary node recognizes that it is a really exposed node and out of interference range of the primary receiver through an additional sensing technique, which results in better performance than MENP and the basic IEEE 802.11. In this paper, however, we did not consider nodes' mobility, which is considered as our future work.

References

1. Internet Engineering Task Force, "Manet working group charter," http://www.ietf.org/html.charters/manet-charter.html.
2. P. Karn, "MACA-a New Channel Access Method for Packet Radio", ARRL/CRRL Amateur Radio 9th Computer Networking Conference, pp. 134-140, ARRL, 1990.
3. IEEE Computer Society LAN MAN Standards Committee. Wireless LAN MAC and PHY Specification, IEEE Std 802.11-1997.
4. A. Acharya, A. Mishra, and S. Bandal, "MACA-P: A MAC for Concurrent Transmission in Multi-hop Wireless Networks", In Proc. of IEEE PerCom, 2003.
5. A. Velayutham, H. Wang, "Solution to the Exposed Node Problem of 802.11 in Wireless Ad-Hoc Networks," http://www.cs.iastate.edu/vel/research/E-MAC.pdf
6. Deepanshu Shukla, Leena Chandran-Wadia and Sridhar Iyer, "Mitigating the exposed node problem in IEEE 802.11 ad hoc networks", In Proc. of IEEE ICCCN, 2003.
7. Saikat Ray, Jeffrey B. Carruthers and David Starobinski,"RTS/CTS-Induced Congestion in Ad Hoc Wireless LANs", In Proc. of IEEE WCNC, 2003.

A Pattern-Based Predictive Indexing Method for Distributed Trajectory Databases

Keisuke Katsuda[1*], Yutaka Yanagisawa[2], and Tetsuji Satoh[1,2]

[1] Graduate School of Information Science and Technology, Osaka University, Japan
k-katuda@ist.osaka-u.ac.jp
[2] NTT Communication Science Laboratories, NTT Corporation, Japan

Abstract. Recently, it has become possible to collect large amounts of trajectory data of moving objects by using sensor networks. To manage such trajectory data, we have developed a distributed trajectory database composed of a server and many sensor nodes deployed over wide areas. The server manages the trajectory data of each moving object by using indices. However, since each sensor node cannot send trajectory data to the server all the time, the server does not always manage indices for the current trajectory data. In other words, the server is delayed in answering queries for current data because it has to forward each query to the sensor nodes to answer them. This is defined as a *delay problem.* To avoid this problem, we propose a pattern-based predictive indexing method for the database to answer queries in real time. This method uses past motion patterns of moving objects to predict the future locations of moving objects. In this paper, we describe the method and evaluate it with practical trajectory data. We conclude that the technique can predict the future locations of moving objects well enough in real time and show optimal parameters for prediction.

1 Introduction

In recent years, various types of applications using the trajectory data of moving objects have been developed [1] [2] and have attracted attention because they allow us to obtain high accurate trajectories using positioning devices on sensor networks. Applications include forecasting traffic congestion, management of taxis and trucks, automatic switching of point-of-purchase advertisements, and so on. These systems must deal efficiently with a large amount of trajectory data (see Fig. 1). However, since the amount of trajectory data has been growing rapidly year by year and such data are managed over wide areas, it is difficult to manage all it in a single database [3].

Therefore, we have developed a distributed trajectory database (DTDB) that stores trajectory data in distributed environments as sensor networks. DTDB consists of a server and many sensor nodes connected to that server. Each sensor node has a positioning device and a database that stores the obtained position

---

* Corresponding author

C. Kim (Ed.): ICOIN 2005, LNCS 3391, pp. 745–754, 2005.

data. Nodes do not send all of the trajectory data to the server but only the data necessary for the server to generate indices. Using the indices, the server can answer a *window query* to find the objects intersecting a query window during a past time interval, even though the server does not store all of the obtained trajectory data.

However, there is a problem associated with distributing data. Since each sensor node does not send the trajectory data to the server in real time, the server may have to wait for the data from the sensors to generate indices, which are used to answer window queries. Therefore, the server may answer the queries late. We call this the *delay problem.* To avoid this problem, the server must predict future trajectory data and generate predictive indices corresponding to future trajectory data. Using predictive indices, the server can answer a *future window query* to find objects intersecting a query window not only during past time but future time.

We propose a pattern-based predictive indexing method for the future position of moving objects. In this paper, we describe a method that uses the past motion patterns of moving objects extracted from past trajectories. Moreover, we develop a DTDB prototype to evaluate our proposed method with practical trajectory data on rickshaws in Nara, Japan. In this evaluation, we investigate the effects of variations in the length of trajectory data for prediction, the data granularity, and the transmission interval of sensor nodes on the prediction.

The rest of the paper is organized as follows. In Section 2, we describe DTDB and the delay problem in detail. Section 3 describes our proposed method. In Section 4, we evaluate our method with comprehensive experiments using trajectory data from rickshaws. Section 5 introduces related work and explains differences from our method. Finally, Section 6 concludes the paper with a discussion of future work.

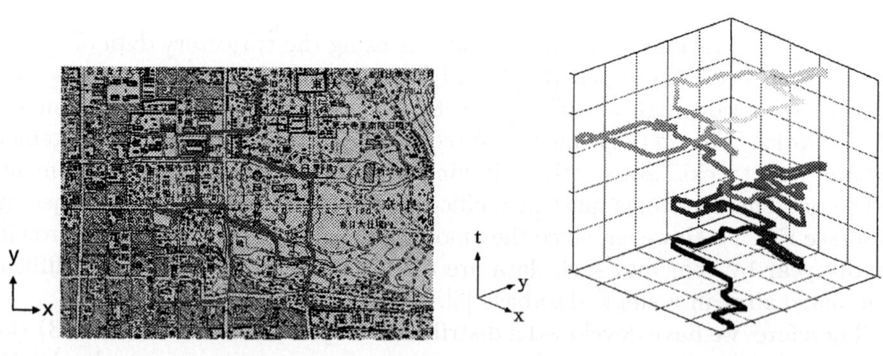

(a) x-y plane (b) x-y-t space

Fig. 1. Sample trajectory data of a rickshaw

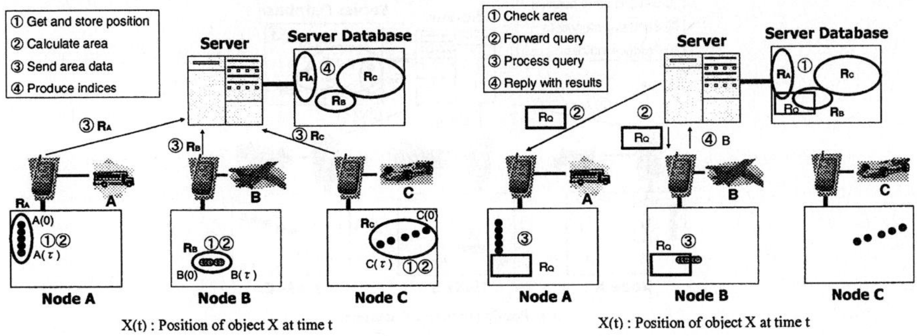

(a) Generating indices (b) Processing a query

Fig. 2. Distributed trajectory database

2 Distributed Trajectory Database

2.1 Overview

We define trajectory data as the sequence of both the position and the time of a moving object. Fig. 1(a) shows the trajectory of a rickshaw moving around Nara. In Fig. 1(b), the same trajectory is projected in x-y-t space.

We consider a trajectory database that comprises both a server database and many positioning devices embedded within a moving object [4]. When each positioning device obtains the location of an object, it sends the data to the server. In other words, the server collects all the trajectory data of all objects to answer a *window query* [5]. Thus, the database is a system that is effective enough to answer queries for moving objects in real time.

However, since sensors are becoming cheaper and smaller and sensor networks are growing, in the future it will become more difficult to manage all trajectory data at a single location. Therefore, we have developed a distributed trajectory database, a distributed version of a trajectory database.

DTDB comprises a server and many sensor nodes connected to the server. Each sensor node has both a positioning device and a database. The former is embedded within a moving object and stores obtained trajectory data in its embedded database. Fig. 2(a) illustrates the process by which the server database generates indices to the data stored at the embedded databases. In the example, there are three sensor nodes: A, B, and C. Each node obtains the position of a moving object and stores in its database at each time interval and calculates the maximum area within which an object moves at regular interval τ. Moreover, each sensor node sends area $\Re$ to the server at interval τ. On the other hand, the server database generates indices from received area $\Re$. Each index at $t = \tau$ indicates the area to which the object moved within $0 \leq t \leq \tau$. In Fig. 2(a), R_A,

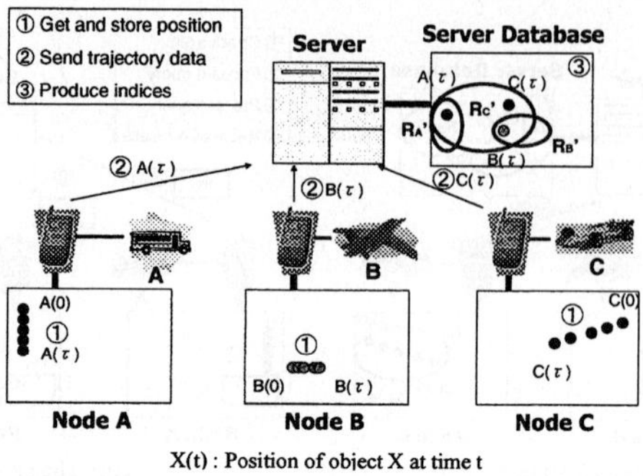

Fig. 3. Producing predictive indices

R_B, and R_C are the areas within which objects A, B, and C respectively move $0 \leq t \leq \tau$.

Next, we illustrate the process by which the server database retrieves the data indicated by a given query window. A query is also given as area R_Q in Fig. 2(b). When the server receives a query, it verifies whether R_Q overlaps the areas stored in the server database. If R_A and R_B overlap with R_Q, the server forwards R_Q to sensor nodes A and B. When the sensor nodes receive a query, they process it in their embedded databases. Finally, each sensor node receiving a query replies with the results of the query to the server. In this case, the result of R_Q is B. In this process, the server can efficiently retrieve any object by using indices.

2.2 Delay Problem and Approach

In this section, we describe the *delay problem* that occurs in DTDB. Before describing it, we state two assumptions.

- The server receives data and generates indices to the data ($(n-1)\tau \leq t \leq n\tau$) at $t = n\tau$ ($n \in \mathbb{N}$), where $\mathbb{N}$ is the set of all natural numbers.
- The server manages the indices to the data ($0 \leq t \leq n\tau$) at $t = n\tau$.

Therefore, at $t = n\tau + j$, where $j \in \mathbb{N}$ and $j < \tau$, the server can search for the data at $t \leq n\tau$. If the server searches for the data at $t = n\tau + j$, it has to wait until $t = (n+1)\tau$, when indices corresponding to the data at $t = n\tau + j$ are generated. As a result, a delay of $\tau - j$ occurs. We call this the *delay problem* in DTDB.

To avoid this problem, we introduce a method that predicts data from $n\tau + 1$ to $(n+1)\tau - 1$, using already received data. By applying this prediction technique, the server produces predictive indices to answer queries for data that have not been received yet.

We show the process by which the server produces predictive indices in Fig. 3. In this example, sensor nodes A, B, and C obtain their position at every time

Table 1. Definition of symbols

X_t	position of X at t ($X_t = (x, y)$)
S_X	CID sequence of X in ascending time
$S_X(n)$	n th CID of S_X
$\lvert S_X \rvert$	element number of S_X
L	element number of CID sequence for prediction

interval, and each node stores trajectory data in their embedded databases. They also send trajectory data to the server at every regular interval τ. At $t = n\tau$, the server generates indices corresponding to the data ($n\tau < t \leq (n+1)\tau$) using the past trajectory data at the time when the server received the trajectory data. Suppose that now $t = \tau$; the three circles R'_A, R'_B, and R'_C are the areas within which each object will move $\tau < t \leq 2\tau$. The server database uses these predictive indices to answer a query for $\tau < t \leq 2\tau$.

To produce a predictive index, the server must predict the positions at which an object will be in the future. We describe the proposed prediction technique to calculate the future positions of moving objects in Section 3 and evaluate the technique in Section 4.

3 Pattern-Based Predictive Indexing Method

In this section, we describe our pattern-based predictive indexing method, which assumes that an object tends to move along the trajectories of other moving objects. Based on this assumption, the positions of objects can be predicted by using motion patterns extracted from the trajectories of other moving objects.

In the following we explain the process for extracting motion patterns from trajectories. First, the server divides the entire area into a grid with several small cells; each cell has a cell identification label (CID). After receiving the trajectory data from sensors, the server records the CID at the point where each object enters. The server manages the CID sequences as the motion patterns of moving objects. We define the notation to describe how our method predicts future locations of moving objects in Table 1.

For predicting the future positions of moving object X, the server compares the last several CID sequences of X with all stored past CID sequences of all objects. The server obtains the CID subsequence most similar to the sequence of X by comparing CID sequences of every object with S_X. As a result, the server uses the next CID of the obtained CID subsequence as a cell to which X will move in the future. In Fig. 4, we show an algorithm that predicts the most probable cell to which an object will move in the future.

Fig. 5 shows an example of the prediction technique. In Fig. 5(a), object X moves around a grid divided into 9 cells. The small circles show the positions of object X at $t = cn$, $c, n \in \mathbb{N}$, and c const. The numbers from 00 to 22 indicate CIDs. The server manages the CID sequence of X as $S_X = \langle 00, 10, 20, 21, 22, 12, 02, 01 \rangle$. In Fig. 5(b), there are two positions of X: at $t =$

```
INPUT : CID sequence SET S = {S1,...,Sn} /* CIDs of all objects */
        CID sequence Sq /* CIDs to be predicted */
        CID SET Θ = {θ1,...,θm}
        int L /* |Sq| */
OUTPUT : CID c /* the most probable cell where object q will move */
CID Prediction(S,Sq,Θ){
 int Ps[m] (for each s in S), i, j, k
   for each s in S{
     for(i = |s|; i ≥ L; i−−){
       if(Sq(L) == s(i))then {
         Ps[s(i+1)] = i;
         j = i−1;
         for(k = L−1; k ≥ L; k−−){
           if(Sq(k) == s(j))then
             Ps[s(i+1)] = Ps[s(i+1)] + j;
           j = j−1}}}}
 if(all Ps == 0)then /* there is no similar trajectory */
   return Sq(L); /* the same cell as predecessor */
 else
   return c where maximum Ps[θs]  (for each s in S, θ in Θ);}
```

Fig. 4. Algorithm obtaining most probable area

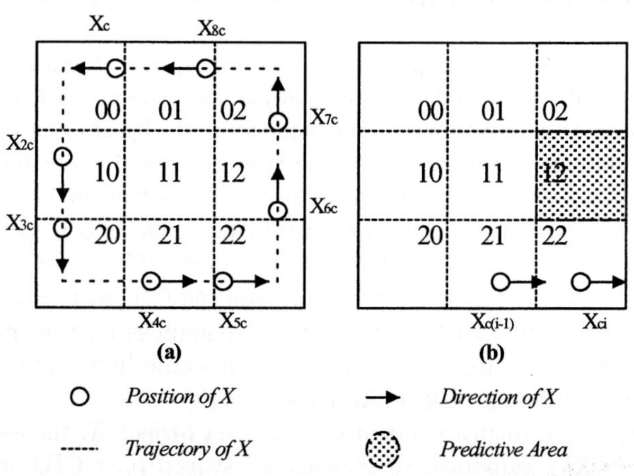

Fig. 5. Example of indexing moving objects

$c(i-1)$ and $t = ci$ $(i \in \mathbb{N})$. Suppose that $t = ci$, then the CID sequence to be compared is $S_q = \langle 21, 22 \rangle$ if $L = 2$. The server calculates that X will move to the area of CID($S_X(6) = 12$) at $t = c(i+1)$ by using that algorithm (Fig. 4).

Table 2. Defenitions of evaluation symbols

P_{inter}	transmission interval (second)
P_{len}	length of a CID sequence for prediction
P_{grid}	number of cells in the grid
Ω	amount of trajectory data (byte)
ω	amount of predictable trajectory data (byte)
P_{acc}	predictive accuracy ($= \omega/\Omega$)

4 Performance Study

In this section, we describe experiments conducted to evaluate the performance of our proposed method for a DTDB, using a DTDB prototype system.

4.1 Settings

The system comprises of many embedded databases equipped with GPS and a server database. Each sensor obtains its position at any time and manages the data in its own embedded database. Every embedded database sends data to the server database at regular intervals. The server database generates indices corresponding to the data stored at embedded databases using the received data. It can also deal with *future window queries* from users using the indices. First, the server database identifies the embedded database managing the data that will be used for the query results. Second, the server forwards the query to all identified sensors. After receiving the query, the sensors process it in their database and send the results to the server. Consequently, query results can be answered.

In this evaluation we use the trajectory data from 10 rickshaws during 4 days in the city of Nara, Japan. Each trajectory data has 20,000 position data values at 20,000 seconds. The notation used for the evaluation is shown in Table 2. The system targeted in this paper is assumed to be an application operating with sensor networks composed of many battery-powered sensors. Therefore, sensors must reduce the number of transmissions and the amount of transmitted data. It is important to predict as much data as possible in the most effective manner. Therefore, it is desirable to obtain high predictive accuracy at large transmission intervals, using only a little information for prediction and grids that are divided into as many cells as possible. Consequently, we compare P_{acc} by varying P_{inter}, P_{len}, and P_{grid}.

4.2 Results

We show simulation results in Fig. 6 by plotting average values taken from four days of simulations in each scenario. Fig. 6(a) shows P_{acc} for P_{inter} under three different P_{len} (P_{grid} is kept constant at 30×30). The figure indicates that P_{acc} generally tends to increase while P_{inter} decreases. Moreover, P_{acc} tends to

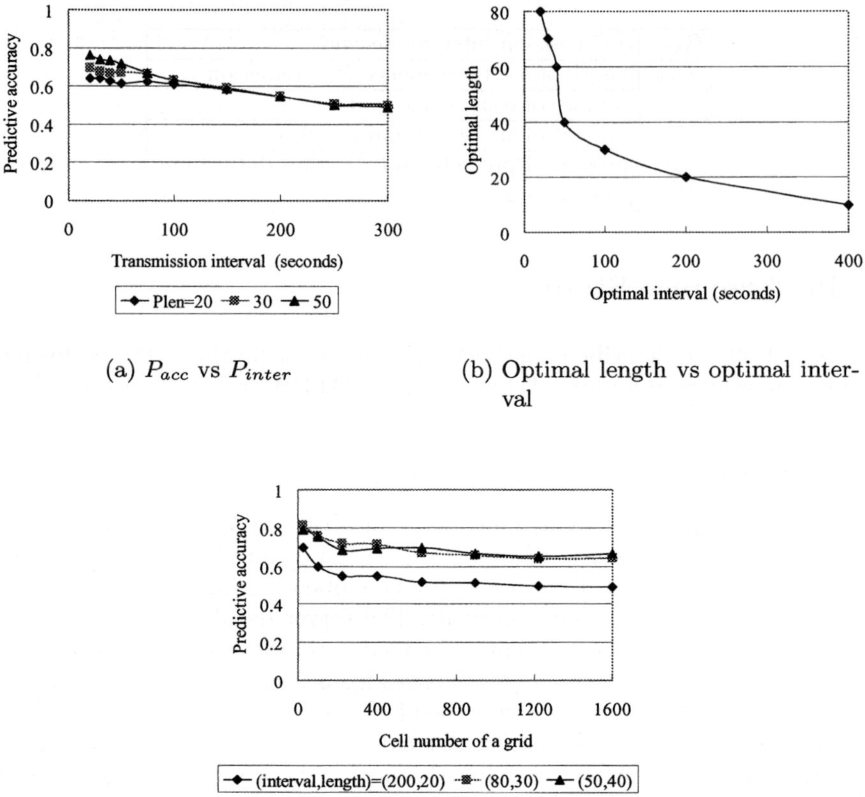

(a) P_{acc} vs P_{inter}

(b) Optimal length vs optimal interval

(c) P_{acc} vs P_{grid}

Fig. 6. Experiment results

increase with P_{len} for $P_{inter} < 200$. On the contrary, P_{acc} does not increase even if P_{len} increases where $P_{inter} > 200$. These results suggest that there is a limit at which P_{acc} does not increase any more, regardless of increases of P_{len} in each P_{inter}. For example, when $P_{inter} = 200$, the limit is $P_{acc} = 0.55$, and the minimum length of CID sequences for prediction is $P_{len} = 20$. We call P_{inter}, P_{len} at such a limit point *optimal interval* and *optimal length*, as shown in Fig. 6(b), the optimal length decreases exponentially with increasing optimal intervals. Thus, changes of the optimal length are large for small optimal interval (< 50) and small for large optimal intervals (> 200).

Moreover, we experimentally evaluate the effects of P_{grid} on P_{acc}. Fig. 6(c) shows P_{acc} for P_{grid} under three different groups of optimal intervals and lengths: (200,20), (80,30), (50,40). If the server divides the grid into many cells, the areas of each cell are small, confining the predictive area to a small area. Fig. 6(c) indicates that P_{acc} decreases at most by 20% if P_{grid} increases to $P_{grid} = 1,600$. Therefore, we can increase the number of cells to a large number in the system without requiring high predictive accuracy.

Consequently, we can obtain an effective system by using these results. For example, where $P_{inter} = 50$, $P_{len} = 40$, and $P_{grid} = 1,600$, our proposed method can obtain about 70% predictive accuracy, and the predictive area is confined to $1/1,600$ of the entire area.

5 Related Work

In this section, we give an overview of related work and show the advantages of our method by comparing it with other approaches. Many publications related to our proposed method have attempted to retrieve moving objects from a database system.

Modern database applications dealing with moving objects are usually managed by a *spatial-temporal database management system* (STDBMS). Recently, STDBMS research has attracted a great deal of attention [5] [6] [7] [8] [9] [10] [11] [12] [13]. In STDBMS, the location of a moving object is represented as a function of time, and the database stores such function parameters as velocity and location. The system is updated only when an object changes any of its moving parameters. To manage the locations of moving objects, many indexing methods have been proposed [8] [10] [11] [12] [13]. However, in STDBMS the sensor nodes must send parameters to the server whenever they change. Since such moving objects as people and cars rarely go straight for a long while, sensor nodes have to send parameters to the server frequently. On the contrary, in our database (DTDB), sensor nodes must send trajectory data to the server at constant intervals, however they need not send so frequently.

Also, several papers describe predictive indexing methods [10] [11] [12] [13] that process future queries in moving object or trajectory databases. These indexing methods predict the future locations of moving objects using only positions or velocities of objects. Therefore, these methods cannot predict the locations of moving objects that continually turn. However, since our indexing method predicts the future locations of moving objects using the past trajectory patterns of moving objects, our method can also predict the locations of moving objects that continually turn.

6 Conclusion

In this paper, we proposed a pattern-based predictive indexing method for DTDB and evaluated it by using a prototype DTDB system. As a result of the experiments, we obtained optimal values of both transmission intervals of sensor nodes and the length of trajectory data for prediction.

We have every confidence that our proposed method will locate moving objects well in real time. Currently, we are planning to incorporate trajectory data from such additional objects as cars and pedestrians. We are also investigating other predictive indexing methods using destination and purpose of moving.

References

1. Laube, P., Imfeld, S.: Analyzing relative motion within groups of trackable moving point objects. In: Proceedings of GIScience 2002 Conference, Boulder, CO, USA, Springer-Verlag Heidelberg (2002) 132–144
2. Vazirgiannis, M., Wolfson, O.: A spatio temporal model and language for moving objects on road networks. In Jensen, C.S., Schneider, M., Seeger, B., Tsotras, V.J., eds.: Proceedings of SSTD 2001. Volume 2121 of Lecture Notes in Computer Science., Springer-Verlag (2001) 20–35
3. Papadias, D., Zhang, J., Mamoulis, N., Tao, Y.: Query processing in spatial network databases. In: Proceedings of the 29th VLDB Conference, Berlin, German (2003) 12–23
4. Yanagisawa, Y., Akahani, J., Satoh, T.: Shape-based similarity query for trajectory of mobile objects. In: Proceedings of the 4th International Conference on Mobile Data Management, Melbourne, Australia (2003) 63–77
5. Saltenis, S., S.Jensen, C., T.Leutenegger, S., Lopez, M.A.: Indexing the positions of continuously moving objects. In: Proceedings of SIGMOD Conference. (2000) 331–342
6. Guttman, O.: R-trees: a dynamic index structure for spatial searching. In: Proceedings of SIGMOD'84 Conference. (1984) 47–57
7. Zhang, Q., Lin, X.: Clustering moving objects for spatio-temporal selectivity estimation. In: Proceedings of the fifteenth conference on Australasian database. Volume 27. (2004) 123–130
8. Kollios, G., Tsotras, V.J., Gunopulos, D., Delis, A., Hadjieleftheriou, M.: Indexing animated objects using spatiotemporal access methods. Knowledge and Data Engineering **13** (2001) 758–777
9. Agarwal, P.K., Arge, L., Erickson, J.: Indexing moving points. In: Proceedings of Symposium on Principles of Database Systems. (2000) 175–186
10. Tao, Y., Sun, J., Papadias, D.: Selectivity estimation for predictive spatio-temporal queries. In: Proceedings of International Conference on Data Engineering. (2003) 417–428
11. Choi, Y.J., Chung, C.W.: Selectivity estimation for spatio-temporal queries to moving objects. In: Proceedings of the 2002 ACM SIGMOD international conference on Management of data, Madison, Wisconsin, USA, ACM SIGMOD international conference on Management of data table of contents, ACM Press (2002) 440–451
12. Hadjieleftheriou, M., Kollios, G., Tsotras, V., Gunopulos, D.: On-line discovery of dense areas in spatio-temporal databases. In: Proceedings of the 8th SSTD Conference, Santorini, Greece (2003) 306–324
13. Hadjieleftheriou, M., Kollios, G., Tsotras, V.J.: Performance evaluation of spatio-temporal selectivity estimation techniques. In: Proceedings of 15th International Conference on Scientific and Statistical Database Management, Cambridge, Massachusetts, USA, IEEE Computer Society (2003) 202–211

Implementing an JAIN Based SIP System for Supporting Advanced Mobility*

Jong-Eon Lee[1], Byung-Hee Kim[2], Dae-Young Kim[1], Si-Ho Cha[3], and Kuk-Hyun Cho[1]

[1] Dept. of Computer Science, Kwangwoon University, Korea
{jelee, dykim, khcho}@cs.kw.ac.kr
[2] Elite Multimedia Lab., LG Electronics Inc, Korea
apolo@lge.com
[3] NMS Lab., Network Business Group, WarePlus Inc., Korea
sihoc@wareplus.com

Abstract. Mobile IP(MIP) and SIP have been proposed to support mobility in the wireless internet working at different layers of the protocol stack. However MIP has some problems such as triangle routing, the need of each host's home address, the overhead of tunneling and the lack of wide deployment. Thus we proposed a scheme for supporting mobility based on SIP in this research. A novel SIP system to provide a hierarchical registration has been designed according to this scheme. Our SIP system has been established by implementing JAIN technologies which follow next generation network standards to support the mobility of wireless terminal. This system successfully satisfied ITU-T recommandation.

1 Introduction

In general, mobility management in the wireless environments may involve terminal, personal, session, and service mobility. MIP is basically a network layer solution that provides continuous media support when users move around, dealing with the terminal mobility problems. However, MIP by itself dose not provide device independent personal, session and service mobility. What is worse, MIP suffers from several limitations such as triangle routing, the need of each host's home address, and the overhead of tunneling. So MIP is not suitable for delay sensitive real-time applications. To solve these limitations, MIP derivatives such as MIP-RO(MIP with Route Optimization) and MIPv6 have been suggested but these solutions still have problems such as overhead of tunneling and additional option field, and the lack of wide deployment [1] [2].

SIP is an application layer signaling protocol which is used for establishing and tearing down multimedia sessions. It has been standardized by IETF for internet telephone calls. Components in SIP are user agent(UA), proxy server, and redirect server. Because these SIP components are similar to MIP components

* The present Research has been conducted by the Research Grant of Kwangwoon University in 2004.

C. Kim (Ed.): ICOIN 2005, LNCS 3391, pp. 755–765, 2005.

such as Home Agent(HA) and Foreign Agent(FA), SIP can support terminal mobility. In addition to terminal mobility, SIP also supports other mobility concepts(personal, session, service) and could compensate for the current lack of wide deployment of MIP. What is more, SIP is widely accepted as the protocol which can support multimedia service and call setup for next generation network [3] [4].

In this paper, we classified and defined mobility in wireless environments and suggested the detail idea for each mobility case. To support effective terminal mobility, we introduced a hierarchical mobility management scheme. We designed JAIN SIP mobility management system(JSMAN) that supports this suggested idea for each mobility case and the hierarchical mobility management scheme. The JSMAN is composed of UAs and servers and adopts JAIN technologies which follow the standards of next generation network to support the mobility of wireless terminal.

This paper is structured as follows. Section 2 describes categories of mobility, the proposed scheme for supporting mobility, and JAIN SIP. Section 3 discusses the solution for supporting mobility and designs components of the proposed JSMAN. Section 4 presents the implementation of JSMAN and the experimental results. Finally in section 5 we conclude the paper.

2 Related Work

2.1 Mobility

- Personal Mobility : Personal mobility allows same logical address to a user located at different terminals. The user can use terminals either at the same time or alternate between them. In MIP case, it is difficult to support personal mobility because MIP uses home address for user identification. However SIP can easily support personal mobility because it uses sip-url such as an e-mail address to identify each user.
- Session Mobility : Session mobility allows a user to maintain a media session even while changing his terminals. In SIP case, there are two ways to support session mobility, one way is third-party call control(3PCC) and the other is the REFER mechanism [5] [6].
- Service Mobility : Service mobility allows a user to maintain access to their services even while moving or changing devices and network service providers.
- Terminal Mobility : Terminal mobility refers to an end user's ability to use users own terminal regardless location and the ability of the network to maintain the user's ongoing communication as the user moves across subnets. This mobility can be defined either in the same administrative(micro mobility) or in different administrative(macro mobility) domains.

In micro mobility, it is important to reduce the high handoff latency by handling mobility within micro mobility regions with low latency local signaling. To deal with this problem, Hierarchical mobility management schemes have been proposed. Schemes are as followings.

- MIP-RR(MIP Regional Registration) : MIP-RR involves the fewest modifications to MIP. In a foreign network, the two-level mobility hierarchy contains the upper layer GFA(Gateway Foreign Agent) and several lower layer RFAs(Regional Foreign Agent) [7].
- TeleMIP : TeleMIP uses MIP as macro mobility and uses IDMP(Inter Domain Mobility Management Protocol) as micro mobility. IDMP offers intra domain mobility by multiple COAs that are taken care of subnet agents and the mobility agent at the subnet and domain level respectively [8].
- HMIPv6(Hierarchical Mobile IPv6 mobility management) : HMIPv6 uses MAP(Mobility Anchor Point) as a subnet agent in IPv6 networks [9].
- HMSIP(Hierarchical Mobile SIP) : HMSIP uses SIP based hierarchical registration and mobility agent for micro mobility supports [10].

2.2 JAIN SIP

JAIN is a standard technology for next generation networks. JAIN APIs enable the rapid development of next generation communication products and services [11]. JAIN SIP is a set of JAIN API and it was produced by JCP(Java Community ProcessSM). JAIN SIP's objective is to provide standard SIP interfaces. Fig. 1 shows the JAIN SIP architecture and the execution process.

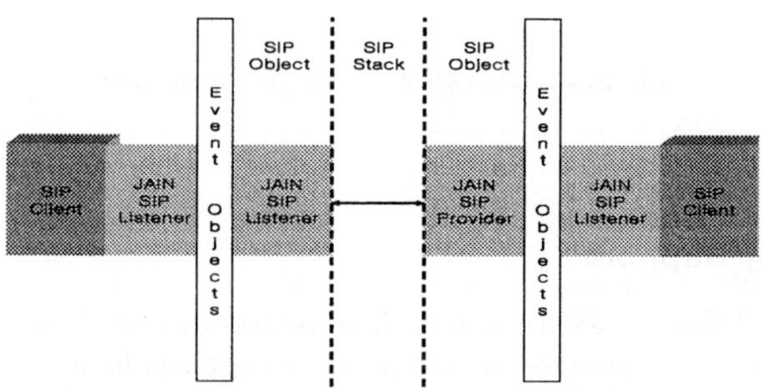

Fig. 1. JAIN SIP Architecture

- JAIN SIP Provider : JAIN SIP Provider is defined as the entity that provides an application access to the services of the SIP stack.
- JAIN SIP Events : SIP messages are encapsulated as message objects that are passed between the JAIN SIP Provider and the JAIN SIP Listener.
- JAIN SIP Listener : Within the API, the JAIN SIP Listener is defined as the entity that uses the services provided by the JAIN SIP Provider.

3 Design JSMAN

Fig. 2 shows the hierarchical architecture of JSMAN network. The network is composed of domain sets and each domain has sub-domain networks. JSMAN servers are deployed in this network and execute a mobility management procedure. JSMAN UA on an end terminal is attached to JSMAN server by wired link or by wireless link via an AP(access point). JSMAN uses the hierarchical mobile management scheme for terminal mobility. Therefore, JSMAN could reduce the handoff latency [2] [10].

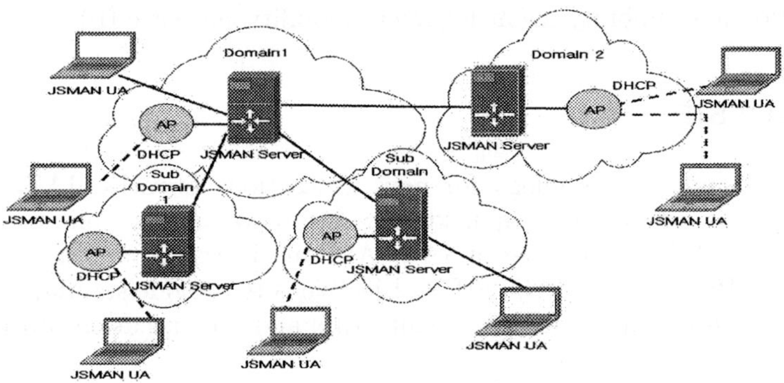

Fig. 2. Overall System Architecture

3.1 Mobility Support

Terminal Mobility : JSMAN UA does periodically OS(Operating System) polling. If IP address changes, JSMAN UA recognizes handoff occurrence. JSMAN UA sends Re-INVITE message to CH(Correspondent Host) and then sends REGISTER message to JSMAN Server. Fig. 3 shows a UML(Unified Modeling Language) sequence diagram for terminal mobility.

Service Mobility User's services information should be stored in XML documents to support appropriate user's service lists. So the Contact header field of message should be adjusted to include user's service information and this information should be included in the Contact header field of REGISTER message.

Personal Mobility : To support personal mobility, JSMAN Server should know information of user's devices. To accomplish a personal mobility, SIP INVITE message is sent to user's all the devices using a same SIP logical address(sip-url such as an e-mail address to identify each user) and user can receive a call

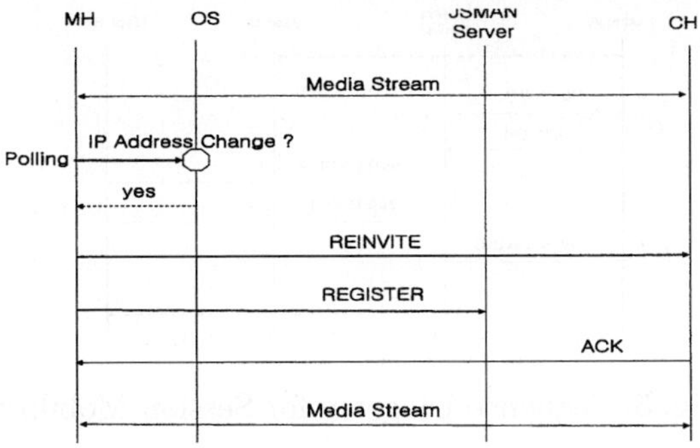

Fig. 3. Sequence Diagram for Terminal Mobility

regardless of his location and using devices. Thus a user can select his device which he wants. Fig. 4 shows a UML sequence diagram for personal mobility.

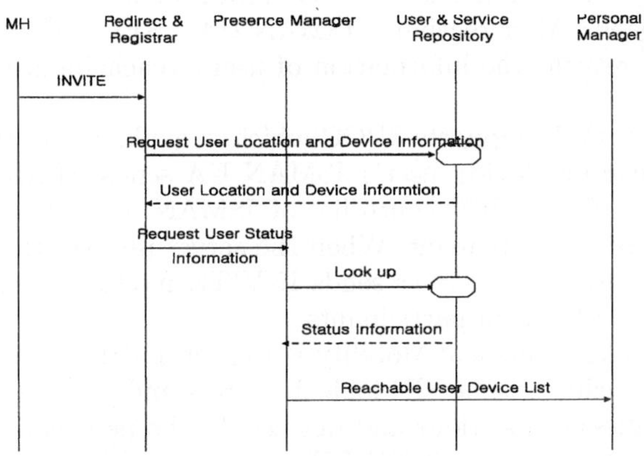

Fig. 4. Sequence Diagram for Personal Mobility

Session Mobility : We designed JSMAN to support session mobility by using 3PCC method. 3PCC can create and manipulate calls between different participants. Fig. 5 shows a UML sequence diagram for session mobility.

3.2 JSMAN Server

JSMAN Server operates on JVM and is designed and implemented with JAIN SIP API. JSMAN Server can support user's location, registration and presence information, etc. And it has an ability to support mobility regardless of locations and devices. Fig. 6 shows overall architecture of JSMAN Server. Below, we describe the operations of major components.

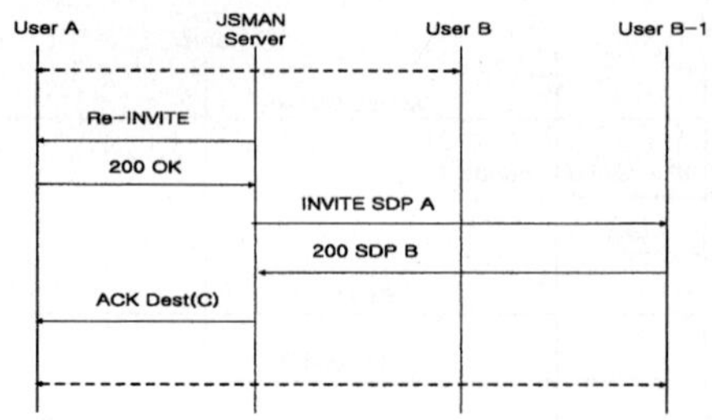

Fig. 5. Sequence Diagram for Session Mobility

- Redirect & Registrar : Redirect module extracts user's information from User & Service Repository, provides user's current location and operates as a redirect server. Registrar module, which registers user to JSMAN server, consists of RegistrarAccess class and RegistrarTable class. It extracts user's information from registered user lists in XML document form and then composes registered user table based on the information.
- Service Manager : After receiving JSMAN UA's REGISTER message, Service Manager returns the information of user's reachable lists from service repository.
- 3PCC Controller : It supports 3PCC which is explained in the 3.1. If user wants to change his device, user's JSMAN UA sends ALARM message to JSMAN server. Then 3PCC controller of JSMAN server sends Re-INVITE message to another participant. When the server receives the 200 OK message, 3PCC controller of server sends INVITE message to CH and a new session is created between participants.
- Personal Manager : Personal Mobility is to send INVITE message to user's all the devices using a same SIP logical address and so the user can receive the call regardless of locations and devices. In the personal mobility, if JSMAN UA on a terminal sends INVITE message to JSMAN server, location module of JSMAN server extracts user's information such as device lists and presence status from user & service repository. Then personal manager sends INVITE messages to all the reachable users according to this extracted presence information.
- User & Service Repository : User & Service Repository is a repository to store user information and their device information.

3.3 JSMAN User Agent(UA)

JSMAN user agent which is designed to support instant messaging and voice call is implemented using JAIN SIP API. The architecture of user agent designed in this paper is shown in Fig. 7.

- Listening Point : Listening point receives and sends SIP Messages from/to lower network stacks.

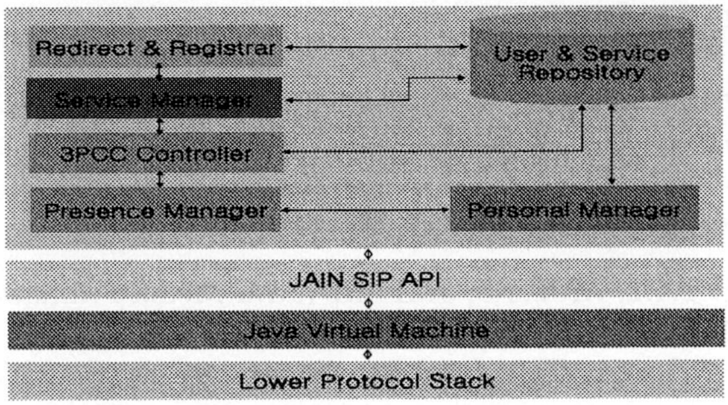

Fig. 6. Architecture of JSMAN Server

– Handoff Polling : Handoff Polling module polls operating system to know changes of UA's IP address. In the SIP, polling mechanism is used to support terminal mobility. If IP address is changed, JSMAN UA knows that user has moved into new region, this module sends Re-INVITE message to CH and sends REGISTER message to JSMAN server. As soon as CH receives this Re-INVITE message, it sends ACK message to MH and so the sessions between users are kept without losing connection.
– Session Mobility : Session Mobility module sends ALARM messages about session mobility to the 3PCC controller of JSMAN server. And 3PCC controller which receives these messages controls session movement.
– Session Manager : JSMAN UA maintains media stream session between users using Session Manager. This module provides creating and deleting operations of session between users and it adds or removes the sessions to/from SessionList. And if user's session is terminated, Session Manager deletes the user's session list.
– Session : Session which includes instant messaging or voice call is created by user's request and these created sessions are maintained by Session Manager.

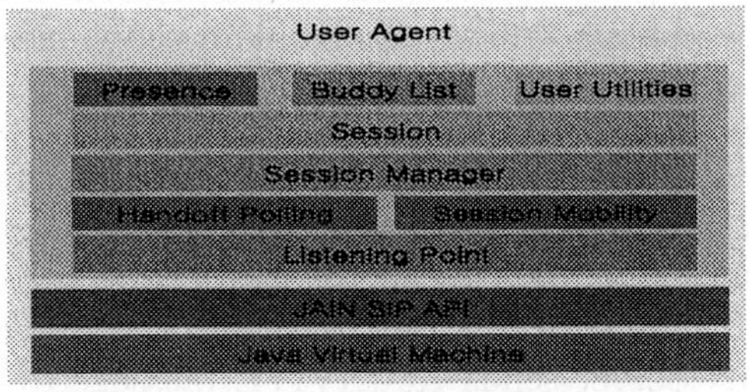

Fig. 7. Architecture of JSMAN User Agent

4 Implementation

The JSMAN system in this paper is implemented on Windows 2000 server and the information needed for sever and client is stored in XML documents. JAIN SIP API is used to implement JSMAN system and other implementation environments are as followings.

- J2SE 2 SDK 1.4.1, ANT 1.5, JMF 2.1 (Java Media Framework)
- JAIN SIP API 1.2
- Window 2000 Server, CPU Intel Pentium-4 2.0GHz, RAM 512MB

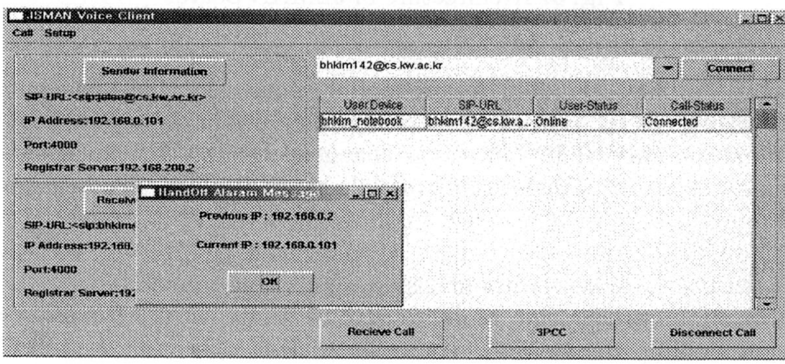

Fig. 8. Handoff(Terminal Mobility)

4.1 Mobility Execution Results

Terminal mobility is demonstrated in Fig. 8. While JSMAN UA polls operating system whether IP of UA changes, if IP address is changed, JSMAN UA knows handoff occurrence and shows handoff alarm messages. As we can see in the fig.9(a), a user using sip:bhkim@cs.kw.ac.kr registered three devices, those are bhkim_notebook, bhkim_desktop and bhkim_cellphone and current status of each device is "online", "busy" and "offline", so it is not necessary to send INVITE message to bhkim_cellphone of which status is offline. JSMAN UA on bhkim_notebook accepted the INVITE message from jelee@cs.kw.ac.kr and then bhkim_notebook is connected to jelee@cs.kw.ac.kr. Accordingly, jelee_desktop stops ringing. If user (bhkim@cs.kw.ac.kr) wants to change device during connections, JSMAN UA on bhkim_notebook sends ALARM message to JSMAN Server. And JSMAN Server sends Re-INVITE message to jelee_notebook. JSMAN Server received 200 OK message from jelee_notebook then this server sends INVITE message to bhkim_desktop. Thus a new session between bhkim_desktop and jelee_notebook is connected. Fig. 10 shows 3PCC controller execution.

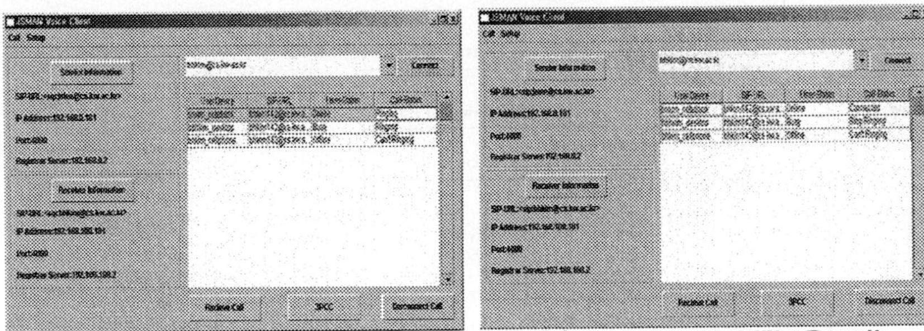

(a) Personal Mobility Ringing (b) Personal Mobility Result

Fig. 9. Handoff(Personal Mobility)

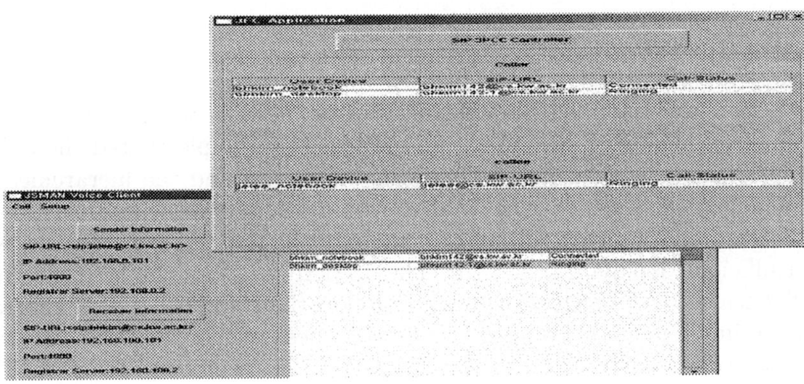

Fig. 10. 3PCC Execution (Session Mobility)

4.2 Performance Evaluation

The diagram in the fig. 11 shows various delay time during call setup. Connection delay time means the delay from INVITE message of user A to 180 Ringing message of user B. Response signal delay time means the delay in which the receiver(user B) holds up the phone and sends 200 OK message to UA and receives ACK message. Call termination delay means the delay in which user A send BYE message to user B and receive 200 OK message from user B.

In this paper, the performance test for JSMAN is done on delay time. We assumed that users walked with normal speed and the number of users who are connected to JSMAN server is not more than 1000. And performance test was conducted on two situations(no-handoff and handoff). The performance result in the fig. 12 shows the delays of JSMAN system for voice call. This result is satisfied with E.721 recommendation of ITU-T [12] [13].

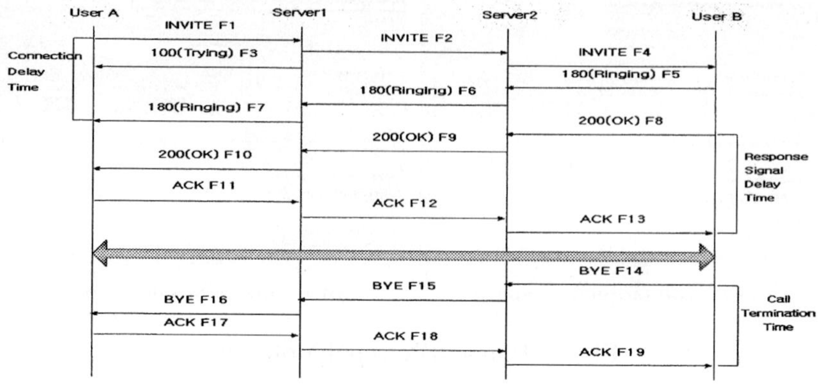

Fig. 11. Call Setup and Termination Delay Time

5 Conclusion

In this paper, we described the categories of mobility in wireless environments and compared MIP with SIP. Also, we proposed and implemented the JSMAN to supports the suggested idea for each mobility case and the hierarchical mobility management scheme. The JSMAN could take advantage of existing SIP infrastructure and support all kinds of mobility in the current networks which have insufficient deployment of MIP. The JSMAN satisfies the next generation network because JAIN technologies which follow the standard of next generation network is implemented. To show the effectiveness of our JSMAN, the system was tested to demonstrate the suitability for E.721 recommendation of ITU-T. To support TCP traffics and more effective micro mobility, future works will be to re-design and implement our JSMAN which will harmonically operates in FMIPv6 and HMIPv6 environments.

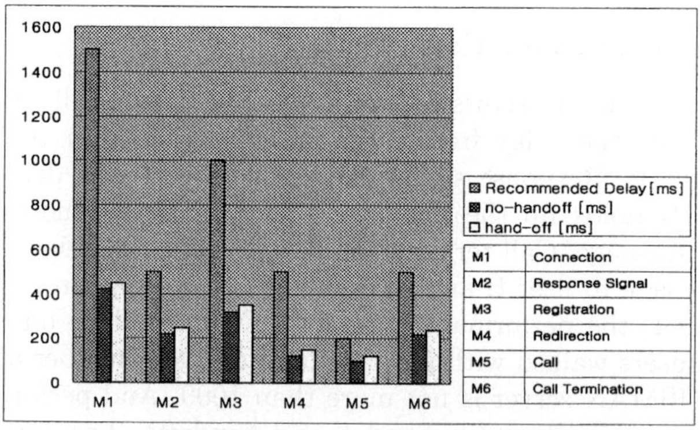

Fig. 12. Performance Result

References

1. C. E. Perkins, D. B. Johnson, "Route Optimization in Mobile IP", IETF work in progress draft-ietf-mobileip-optim-11.txt, Sep. 2001.
2. H. Schulzrinne et al, "Performance of IP Micro-Mobility Management Scehemes using Host Based Routing.", WPMC 2001, Sep. 2001.
3. H. Schulzrine and J. Rosenberg, "The session initiation protocol: Internet-centric signaling", IEEE Communications Magazine, Oct. 2000.
4. M. Handley, H. Shulzrinne, E .Schooler and J. Rosenberg, "SIP:Session Initiation Protocol", RFC 3261, IETF, Nov. 2000.
5. H. Schulzrinne and E. Wedlund, "Application-layer mobility using SIP", Service Portability and Virtual Customer Environments, 2000 IEEE , Dec. 2000.
6. J. Rosenberg, J. Peterson, H. Schulzrine and G.Gamarillo, "Third Party Call Control in SIP", Internet Draft IETF, Nov. 2001.
7. E.Gustafsson, A.Jonsson and C.Perkins "Mobile IP Regional Registration", Internet Draft(work in progress), IETF, March 2001.
8. S. Das, A. Misra, and P. Agrawal, "TeleMIP: telecommunications-enhanced mobile IP architecture for fast intradomain mobility", IEEE Personal Comms, Aug. 2000.
9. I. Vivaldi et al, "Fast handover algorithm for hierarchical mobile IPv6 macro-mobility management", Communications, 2003. APCC 2003. Sept. 2003
10. D. VaIi, S. Paskalis, A. Kaloxylos and L. Merakos, "An efficient micro-mobility solution for SIP networks", GLOBECOM 03. IEEE , Volume: 6 , Dec. 2003.
11. The JAIN/SIP 1.3 Specification(JSR 32), SUN Microsystems.
12. ITU-T, "Network grade of service parameters and target values for circuit-switched services in the evolving ISDN", ITU-T E.721, Feb. 1999.
13. I. D. Curio and M. Lundan, "SIP Call Setup Delay in 3G Networks", ISCC02, July 2002.
14. Si-Ho Cha, Jong-Eon Lee, Jae-Oh Lee, WoongChul Choi, Kuk-Hyun Cho, "Policy-based Differentiated QoS Provisioning for DiffServ Enabled IP Networks", Springer-Verlag's Lecture Notes in Computer Science (LNCS) Vol. 3090, June 2004.

The Content-Aware Caching for Cooperative Transcoding Proxies

Byoung-Jip Kim, Kyungbaek Kim, and Daeyeon Park

Department of Electrical Engineering & Computer Science,
Division of Electrical Engineering,
Korea Advanced Institute of Science and Technology (KAIST),
373-1 Kusong-dong Yusong-gu, Taejon, 305-701, Korea
{bjkim, kbkim}@sslab.kaist.ac.kr
daeyeon@ee.kaist.ac.kr

Abstract. The Web is rapidly increasing its reach beyond the desktop to various devices and the transcoding proxy is appeared to support web services efficiently. Recently, the cooperative transcoding proxy architecture is proposed to improve the system performance to cope with the scalability problem of a stand-alone transcoding proxy. However, because of the multiple versions, the communication protocol of the cooperative caches is very complex and causes additional delay to find best version for a requested object.

In this paper, we propose efficient cooperative transcoding proxy architecture which uses the content-aware caching. The main purpose of the proposed system is simplifying the communication protocol of cooperative caches. We associates a home proxy for each URL and the home proxy is responsible for transcoding and maintaining multiple version of an URL. This mechanism reduces the amount of messages exchanged and communication latency involved. To prevent the hot-spot problem, each proxy cache has the private cache which stores the recently requested objects. We examine the performance of the proposed system by using trace based simulation with Simjava and show the effective enhancement of the cooerative transcoding proxy system.

1 Introduction

In recent years, the technologies of the network and the computer have developed enormously and the diverse devices such as PDAs, mobile phones, TVs and etc which are connected to the network with various ways such as wired or wireless interfaces. These diverse devices have been able to use the web contents, but some clients can not use the web contents directly because their capabilities differ from those of the web content provider's expectation. For these clients, the content adaptation, called the *transcoding*, is needed. This transcoding transforms the size, quality, presentation style, and etc of the web resources to meet the capabilities of the clients. The main features of the transcoding can be summarized with two. First is that multiple versions exist for the same web content

C. Kim (Ed.): ICOIN 2005, LNCS 3391, pp. 766–775, 2005.

due to the diverse client demand. Second is that the transcoding is a very computational task. These two features of the transcoding bring many issues to design a transcoding system.

The existing approaches of the transcoding system can be classified into three categories broadly, depending on the entity that performs the transcoding process: *client-based*, *server-based*, and *intermediary-based* approaches. In the client-based approaches, the transcoding is performed in client devices and the transcoder has direct access to the capabilities of the various devices. However, these approaches are extremely expensive due to the limited connection bandwidth and computing power of clients. Conversely, in the server-based approaches, the content server transforms objects into multiple versions on online or offline. These approaches preserve the original semantic of the content and reduce the transcoding latency during the time between the client request and the server response. However, keeping the multiple versions of an object wastes too much storage and the content providers actually can not provide all kind of versions of contents for the diverse clients. In the intermediary-based approaches, edge servers or proxy servers can transform the requested object into a proper version for the capability of the client before it sends the object to the client. These approaches need additional infrastructures in the network and the additional information (e.g., client capability information, semantic information of contents). Although these additional needs exist, this intermediary-based approaches address the problems of the client-based and server-based approaches and many researches have been emerged.

Although the intermediary-based approaches are considered most appropriate due to their flexibility and customizability, they have some system issues to be addressed. Because the costly transcoding has to be performed on demand in proxy servers, the scalability problem arises. To address the scalability problem and improve the system performance, researchers have proposed caching the transcoding results. Because of the cached results, we can reduce repeated transcoding tasks and the system performance can be improved. Recently, the cooperative transcoding proxy architecture is proposed to improve the system performance[3,4]. However, applying the traditional cooperative caching directly to the transcoding proxy architecture is not efficient due to inherent problems of the content transcoding such as multiple versions of contents[1,5]. Because of the multiple versions, the communication protocol of cooperative caches is more complex than existing protocols, such as ICP and causes additional delay which is incurred by finding more similar version of an object during the time for discover the object in cooperative caches. Additionally, each cooperative caches consumes too much storage to store redundant multiple versions for the same object. These hurdles decrease the system performance and utilization.

In this paper, we propose the efficient cooperative transcoding proxy architecture which uses the content-aware caching. The main purpose of the proposed system is simplifying the communication protocol of cooperative caches. To cope with the problem which is caused by the multiple version, we propose that every version for an object is stored at one designated proxy together. Each transcoding proxy is mapped with the hashed value of the URL and a proxy stores its transcoded result at the designated proxy which is mapped with the URL of the requested object. By using this concept, we gather the whole of the version for an object in its designated proxy, so called a *home proxy*, and find the best version for an object deterministically. According to this behavior, every version of an object resides at one designated proxy and the discovery process becomes simple.

While the proxies store objects deterministically, they should fear for the hot spot problem; a small fraction of objects will be hot which could lead to excessive load at nodes which are their homes. To prevent this overload, we divide a cache storage into two; *public storage* and *private storage*. The general content-aware caching mechanism uses the public storage which is used to find the best version of the requested object. The private storage contains the hot objects of the local clients. If the requested object is found in the private storage of a local proxy, there is no need to check the public storage of the home proxy. That is, we reduce the load of the home proxies of hot objects.

Moreover, we refine the cooperation mechanism for the proposed system to perform more efficiently. There are three main processes: the discovery process, the transcoding process and the delivery process. We exploit the characteristics of the content-aware caching to make the discovery process simpler than the previous process and design the transcoding process to increase the performance of the proxies. By using the redirection in the delivery process, we reduce the network traffic which is needed to manage the system.

We evaluate the performance of the proposed system by using trace based simulation. We use the Simjava to simulate the cooperative transcoding proxy system. We compare the system response time, the cache hit ratio and the communication cost between the previous cooperative caching and the proposed content-aware caching and show that the performance increases when the content-aware caching is used.

The rest of this paper is organized as follow. Section 2 briefly represent the related works and the problem of them. Section 3 presents our proposed architecture for cooperative transcoding proxy. The performance evaluation is on section 4. Finally, we concludes this paper on section 5.

2 Background

In recent years, some proposals have exploited both of transcoding and caching to reduce the resource usage at the proxy server, especially for transcoding time. The main idea of these approaches is that caching the transcoding results improves the system performance by reducing the repeated transcoding operation. Moreover, some studies extend a stand-alone transcoding proxy to cooperate each other to increase the size of the community of clients. This cooperative caching increases the hit ratio of the cache system by cooperating with each other caches for discovery, transcoding and delivery. As result of the increased hit ratio of system, it reduces not only the repeated transcoding operations, but also the user perceived latency for an object.

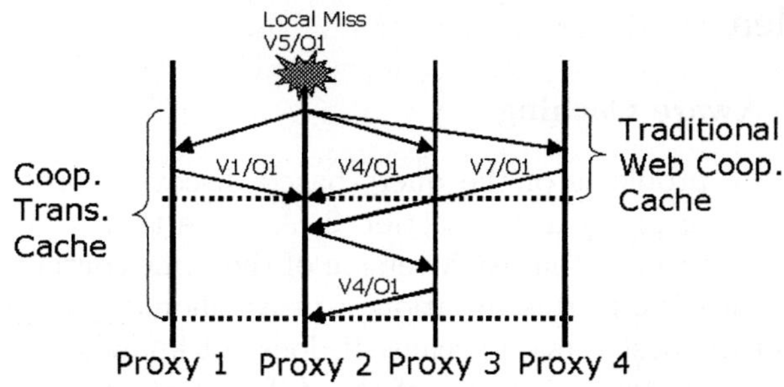

Fig. 1. The discovery process of the cooperative transcoding proxies

The transcoding proxy should manage the multiple version of an object, because the transcoding results depend on the various capability of clients. According to the multiple versions, there are two types of hit; *exact hit* and *useful hit*. The exact hit means the proxy cache contains the exact version required by the client, and the useful hit means the proxy cache does not have the exact version of the requested object but contains a more detailed and transcodable version of the requested object that can be transformed to obtain a less detailed version that meets the client request.

These two types of hits make the communication protocol of cooperative caches, especially the discovery protocol, more complex than existing protocols, such as ICP. Figure 1 shows the difference of the discovery process between the cooperative transcoding proxies and the traditional web proxies. The proxy 2 gets a request for an object, O1, whose version is the version 5, V5, and misses the object, then the proxy 2 sends queries to other proxies to find the object. If we use the traditional web proxies, the discovery process is over after getting any object from any proxy. In this figure, the proxy 1 or 3 returns the object O1 to the proxy 2 and the process is over. However, if we use the transcoding proxies, we should consider not only the object but also the version, then we have to wait for the best version that minimize the transcoding operation. In this figure, though the proxy 1 and 3 return the objects with version V1 and V4, the proxy 2 does not know that the proxy 4 has the exact version and has to wait for the responses from proxy 4. After the proxy 2 gets all responses form all proxies, it chooses the proxy which has the best version, in this figure the proxy 3, and sends a query to get the object itself. This behavior takes for long time to determine the best version that minimize the transcoding operation because a local transcoding proxy have to wait for potentially better version. Also, it generates enormous query messages to discover an object. That is, each transcoding proxy has to process redundant query messages for every discovery request.

3 Main Idea

3.1 Content-Aware Caching

We mentioned the problems of the query-based discovery protocol of the cooperative transcoding proxy in the section 2. A cause that a proxy waits for a potentially better version is that each version of the same content resides irregularly at different proxies. In this situation, a proxy should send queries to every proxy to discover the best version because it does not know which proxy has the best version of the content. However, if the different versions of a content are cached together at the designated proxy, a proxy can determine the best version of a content with only one query message.

Each proxy has to store transcoded versions of an object at designated proxy being aware of the object. We would refer this caching scheme as *content-aware caching*. In content-aware caching, a transcoding proxy stores its transcoded results at a designated proxy according to the URL of objects. Then, every version of the same URL is stored at a designated proxy. We would refer this designated proxy as a *home proxy* of an objects. Each URL is mapped into its home proxy by using URL hashing. The 128bit ID space is generated by the hash function which balances the ID with high probability such as SHA-1 and each proxy which is the participant of the system manages the partial ID space which is determined by the proxy node ID that is computed by hashing the unique value of node such as an ip address. Each object has the object ID which is obtained by hashing the URL of the object and is stored at the home proxy that manages the object ID.

This content-aware caching has several advantages. First, a proxy can discover the best version of an object deterministically. A proxy can find the best version at a home proxy with only one query and does not wait for potentially better version after it receives an exact or useful hit message. Second, the redundant query processing is reduced significantly. In the previous system, a proxy sends a query to every peer proxy, and each proxy which receives a query performs the query processing to find a proper version of an object. However, in the content-aware caching system, only home proxy performs query processing. Third, the network traffic is reduced because the number of query messages is reduced significantly.

3.2 Prevention of Hot Spot

When we use the content-aware caching, the request for an object is always forwarded to the home proxy. If the hot spot for an object occurs, the home proxy which has the responsibility for the object has to deal with every request and the home proxy is overloaded and out of order. To prevent this case, we divide a cache storage into two : *public storage* and *private storage*. The public storage is used to store the every version of an object for the content-aware caching and the private storage is used to store the hot objects of the local clients. That is, a proxy stores the clusters of version for objects whose object IDs are managed

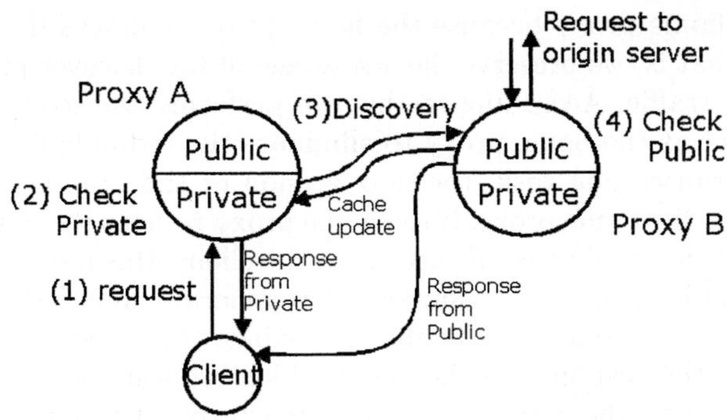

Fig. 2. Overall of the Content-aware caching system

by itself in the public storage and caches the frequently requested objects from the local clients in the private storage.

Figure 2 shows the overall of the cache architecture and the cooperation mechanism. When a proxy receives a request (1), it first checks its private storage (2). If a exact hit occurs, the proxy returns the object to the client. However, if either a local useful hit or a local miss occurs, the proxy tries to discover a matched transcoded object at the home proxy of the requested object (3). Then, the home proxy checks its public storage for the object (4). If there is the object which is exact or useful, the home proxy transforms the object into the exact version and returns the object to the client, and updates the private storage of the local proxy if the home proxy decides that the object is frequently requested. Otherwise, if a miss occurs, this proxy gets the new object from the origin server. According to this behavior, the hot objects reside at the local private storage with high probability and we can reduce the excessive load of the home proxies of the hot objects.

3.3 Cooperation Mechanism

There are mainly three cooperation processes: the discovery process, the transcoding process, and the delivery process. The first, the discovery process takes advantages of the content-aware caching. In our proposed system, different versions of the same content are cached together at the public storage of the home proxy. If a proxy gets a request and a miss occurs at the private storage, it need not send queries to every peer proxies but send only one query to a home proxy of the requested URL. Therefore, in the discovery process, we can reduce not only the query messages but also the waiting time for finding a potentially better version. Moreover, because the home proxy could have almost versions of an object, the exact hit ratio increases and the system performance would increase.

When we find the useful object in the public storage of the home proxy, we should decide the location of the transcoding. If the local proxy which gets the request from the client performs the transcoding, it has to update the public storage of the home proxy because the home proxy manages the whole version of an object. That is, we preserve the advantage of the discover process by using this redundant traffic. According to this, we performs transcoding task for the requested object at the home proxy to eliminate the redundant transmission.

After the transcoding task, the new version of the object is stored at the public storage of the home proxy. If the home proxy returns the new object to the local proxy and the local proxy returns it to the client, this indirect transmission causes the redundant object transmission that generally makes the response time long. To prevent this redundant traffic and reduce the response time, the home proxy redirects the response to the client which request the object. When the local proxy forward the request to the home proxy, the forwarding message includes the redirection information. However, this redirection mechanism can cause the hot spot problem at the home proxy. To cope with this problem, the home proxy has to update the private storage of the local proxy. If the exact hit occurs at the public storage, the home proxy checks how frequently the object is requested. If the object is decided as a hot object, the home proxy sends this object to the local proxy which requests it and the local proxy stores it at the private storage. This cache update policy compensates the effect of the hot objects with the local private storage.

4 Evaluation

4.1 Simulation Setup

We simulate the cooperative transcoding proxy architecture to evaluate its performance. We use Simjava to simulate the architecture. Simjava is a toolkit for building working models of complex systems. It is based around a discrete event simulation kernel [6].

We try to reflect the real environment in our simulation as accurate as possible. We examine the previous papers on the cooperative caching to extract the simulation parameters [7]. The size of a cache storage is 300MB and 30% of the total storage is assigned to the public storage. We use 4 caches which are cooperated with each other and use the LRU replacement policy. The establishing HTTP connection takes 3 msec and the cache lookup needs 1.5 msec. The processing the ICP query need 0.3 msec and the hashing calculation for the content-aware caching takes 0.6 msec. The transmission time to content server takes 300 msec as average and the transmission time in local backbone network takes 40 msec as average. The simulation parameters about the transcoding operation are extracted from the previous paper [5]. The transcoding takes 150 msec as the mean value and 330 msec as the 90th percentile.

We use a trace file of IRCache [2]. The trace date is October 22, 2003 and the total duration is 1 day. The total number of requests is 416,015 and the mean request rate is 4.8 requests per second. We assume that the 100 clients use one proxy cache and consider a classification of the client devices on the basis of their capabilities of displaying different objects and connecting to the assigned proxy

Device type	PC	Laptop	TV Browser	PDA	Mobile phone
Percentage	40%	15%	15%	15%	15%

Table 1. Client device types and population

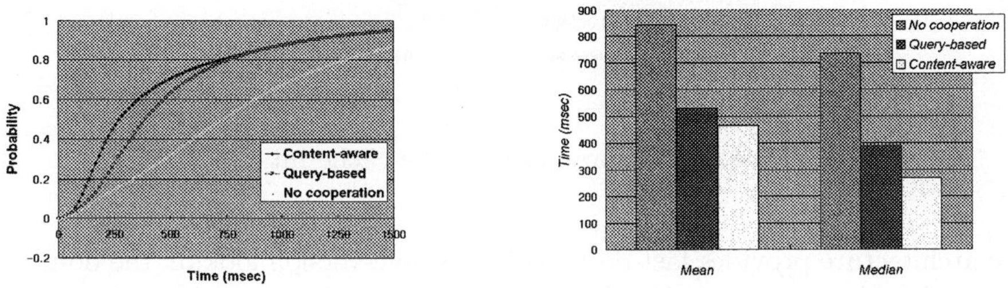

(a) Cumulative distribution (b) Mean and Median

Fig. 3. The comparison on the system response time

server. The classes of devices range from high-end workstations/PCs which can consume every object in its original form, to mobile phones with very limited bandwidth and display capabilities. We introduce five classes of clients. Table 1 shows the client types and their population.

4.2 System Response Time

The system response time is the time between sending requests of clients and receiving of responses of clients and it is generally used as a criterion of system performance. Figure 3 shows the comparison on the system response time. It shows clearly that the cooperative architecture provides better performance than the stand-alone architecture. Also, it shows that the content-aware cooperation architecture provides better performance than the query-based cooperation architecture. The 90th percentile of the response time is similar between the content-aware architecture and the query-based architecture. However, the median of the response time is much better in the content-aware architecture.

The main reason of the different response times is the cooperative discovery protocol. The query-based discovery protocol has a problem of a long decision time due to the two-phase lookup. To address this problem, we proposed content-aware caching mechanism and this significantly reduces the decision time in the multiple-version lookup. Therefore, the performance of the system increases.

However, the 90th percentile is similar because the global cache miss causes the long round-trip time to the original server. This long round-trip time is the system bottleneck for both architectures. Although the content-aware coopera-

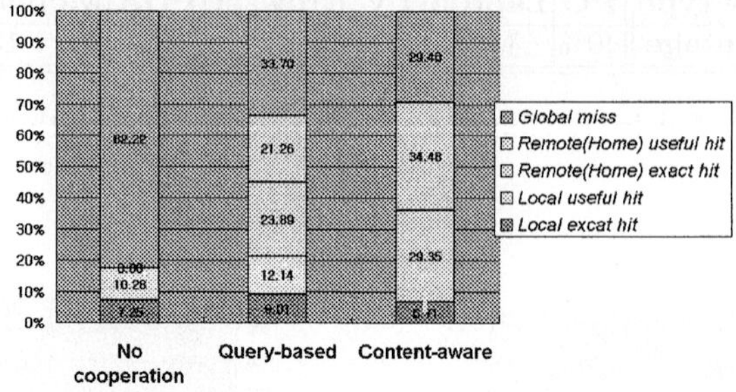

Fig. 4. The comparison on the cache hit ratio

tive architecture provides fast decision in multiple-version lookup, the dominant factor of the long transmission time from content server to a proxy server causes long user response time. Therefore, high hit ratio of a proxy cache is important.

4.3 Cache Hit Ratio

Cache hit ratio is the important factor that affects the system performance of the transcoding proxy. A cache in the multiple-version environment has three different event: an exact hit, a useful hit, and a miss. The high exact hit ratio improves the system performance by eliminating the transcoding overhead that generally involves a long processing time. The high useful hit ratio improves the system performance by reducing the redundant transcoding process. The high cache miss ratio degrades the system performance since this case needs the content transmission from a content sever to a transcoding proxy and the complete transcoding task.

Figure 4 shows the cache hit ratio of each scheme. The cooperation schemes provide much higher ratio of both an exact hit and a useful hit. The hit ratio of the content-aware scheme is slightly higher than the query-based scheme. In the query-based scheme, the exact hit ratio of the local cache is 9.01% and the exact hit ratio of the remote cache is 23.89%. In the content-aware scheme, the exact hit ratio in the private storage is only 6.71% which is smaller than the query-based but the exact hit ratio of the public storage is 29.35% which is much bigger than the query-based. Even if the local exact hit ratio of the content-aware scheme is smaller, the main factor of high exact hit ratio is the remote exact hit ratio for both schemes. In this case, the global lookup process of the query-based scheme causes the long decision time due to two-phase lookup in the multiple-version environment mentioned in the section 2. However, the content-aware scheme finds the exact object with the simple discovery process which takes only one query to the home proxy. Therefore, the content-aware scheme can provide better performance than the query-based scheme.

We can see that the useful hit ratio is increased in case of the content-aware cooperation architecture. The reason is that each useful version is clustered to be discovered directly in the content-aware cooperation architecture and they use the cache storage more efficiently without the redundant copies. Additionally, in the content-aware scheme, the local useful hit ratio is 0 because the private storage is used to find the exact objects only.

5 Conclusions

In this paper, we propose the efficient cooperative transcoding proxy architecture which uses the content-aware caching. We cope with the problem which is caused by the multiple version environment by using the content-aware caching, which means that every version for an object is stored at one designated proxy together. The proposed architecture makes the communication protocol between each proxies simpler, especially the discovery process and reduces the number of messages which are used to maintain the cache system such as ICP queries and object responses. Moreover, because of gathering all versions of an object at one proxy, the exact hit ratio increases and the performance of the system increases too. This architecture has an improvement of 20 percentage points of response time, an increase of 10 percentage points of cache hit ratio, and improved scalability on bandwidth consumption. Though the many advantages exist, the hot spot problem can be appeared. We prevent this problem by using the private cache and the cache update policy. The detail of the cache update policy is our ongoing work.

References

1. V. Cardellini, M. Colajanni, R. Lancellotti, and P. S. Yu. *A distributed architecture of edge proxy servers for cooperative transcoding* In Proc. of 3rd IEEE Workshop on Internet Applications, June 2003.
2. IRCache project, 2003. http://www.irchache.net
3. A. Maheshwari, A. Sharma, K. Ramamritham, and P. Shenoy. *TransSquid: Transcoding and caching proxy for heterogeneous e-commerce environments* In Proc. of 12th IEEE Int'l Workshop on Research Issues in Data Engineering, pages 50-59, Feb. 2002.
4. A. Singh, A. Trivedi, K. Ramamritham, and P. Shenoy. *PTC: Proxies that transcode and cache in heterogeneous Web client environments* In Proc. of 3rd Int'l Conf. on Web Information Systems Engineering, Dec. 2002.
5. C. Canali, V. Cardellini, M. Colajanni, R. Lancellotti, P. S. Yu. *Cooperative Architectures and Algorithms for Discovery and Transcoding of Multi-version Content* In Proc. of 8th Int'l Workshop on Web Content Caching and Distribution, Sep. 2003.
6. F. Howell, R. McNab. *SimJava: a discrete event simulation package for Java with applications in computer systems modeling* In Proc. 1st Int'l Conference on Web-based Modelling and Simulation, San Diego CA, Society for Computer Simulation, Jan. 1998.
7. C. Lindemann, O. P. Waldhorst. *Evaluating cooperative web caching protocols for Emerging Network Technologies* In Proc. of Int'l Workshop on Caching, Coherence and Consistency, 2001.

A JXTA-based Architecture for Efficient and Adaptive Healthcare Services*

Byongin Lim, Keehyun Choi, and Dongryeol Shin

School of Information and Communication Engineering
Sungkyunkwan University
440-746, Suwon, Korea, +82-31-290-7125
{lbi77, gyunee, drshin}@ece.skku.ac.kr

Abstract. JXTA is a Peer-to-Peer application development infrastructure that enables developers to easily create service oriented software. This paper presents a low-cost, patient-friendly JXTA-based healthcare system, which is comprised of medical sensor modules in conjunction with wireless communication technology. In particular, the proposed system supports a wide range of services including mobile telemedicine, patient monitoring, emergency management and information sharing between patients and doctors or among the healthcare workers involved. For this purpose, the proposed system provides not only a systematic computing environment between the BAN (Body Area Network) and the subsystem within the hospital, but also participates in a JXTA network with the BAN's PDAs through peers called JXTA relays. Information within the hospital is shared under a P2P environment by means of JXTA grouping technology. The proposed system is shown to be adequate for a ubiquitous environment which offers effective service management as well as continuous, remote monitoring of the patient's status.

1 Introduction

Increased economic growth has brought about an increase of "lifestyle-related" diseases (e.g., diabetes and high blood pressure), to the extent that this has now become a serious problem. Such diseases which need long-term treatment require considerable cost, time and effort. However, information and communication technology can play an important role in achieving cost reduction and efficiency improvement in healthcare delivery systems. In particular, the introduction of wireless technology has paved the way for the implementation of mobile healthcare systems which enable ambulatory patients to lead a normal daily life. For example, the mobile healthcare system proposed in [1, 2] allows the mobile and remote monitoring of the maternal and fetal vital signs, while pregnant women proceed with their daily life activities, thereby obviating the need for them to be hospitalized for monitoring. It also enables health professionals to provide their patients with health services, irrespective of their location.

* This work was supported by CUCN Grant, Korea.

C. Kim (Ed.): ICOIN 2005, LNCS 3391, pp. 776–785, 2005.

However, conventional healthcare systems, including the Mobihealth [1] project, utilize a central server for the look up of information. Furthermore, the possibility of offering seamless service is limited by the network connectivity of wireless devices or resource-constraints. In addition, traditional systems only focus on organic communication and service utilization between patients and hospital, whereas they ignore the systematic communication and sharing of information between healthcare workers in the hospital. For these reasons, these systems cannot cope with acute situations dynamically and promptly. To resolve these problems, we propose a JXTA-based healthcare system, which operates in a peer-to-peer (P2P) environment so as to offer seamless service by distributing the healthcare services among the healthcare workers. This sharing of information about the medical treatment that patients receive between the healthcare workers in the JXTA environment improves their ability to cope with dynamic situations, which in turn makes it possible to offer more efficient medical services. Also, the system architecture enables the continuous monitoring of the patient's health condition 'anytime and anywhere', thereby reducing long-term costs and improving quality of service. In this paper, we focus on the system architecture design which consists of the BAN and subsystem within the regional hospital, along with JXTA relays to link the BAN to the subsystem within the regional hospital under the JXTA service platform. The remainder of the paper is organized as follows. Section 2 discusses related research. Section 3 describes the design issues, while the system architecture is described in section 4 and the system implementation is presented in section 5. Finally, this paper is concluded in section 6.

2 Related Research

Traditionally related projects include the @HOME (Remote home monitoring of patients) project, which targets next generation health services using UMTS, Bluetooth and ubiquitous medical sensors for patient monitoring in the recovery phase and for chronically ill patients, while also addressing the problems of patient compliance. However, their basic platform concepts are the classical architecture of the personal computer. The software that makes the peripheral devices useful runs on the CPU of the wearable computer, not the peripheral devices themselves. The Spot Computer from MIThril from MIT Media Lab [3], or LART (Linux Advanced Radio Terminal) developed at Delft University of Technology [4] all have modular concepts, use the Intel StrongARM microprocessor and are capable of running on Linux operating systems. These modular concepts and powerful platforms provide good bases for the development of wearable computers. These wireless-enabled devices are seen as a potentially powerful means of improving the quality of health monitoring and self-diagnosis. However, most telemedicine units allow the transmission of vital signs such as the ECG, SPO2, temperature and blood pressure, but not the glucose level. Consequently, we are currently in the process of developing a unit dealing with the blood glucose level in the context of the wireless environment. The approach that we adopted is also based on the modular concept, as in the case of the above methods. We focus on the use of a modular hardware design, and are attempting to build a healthcare system for diabetes patients which operates over the CDMA public wireless network and JXTA service platform. For this purpose, we implement software for the purpose of monitoring the patient's status in the BAN, as well as a JXTA application used for managing the patient's information, medical status and personal information.

3 Design Issues

In this section, we will look at the technical requirements of our system. The system approach taken aims at providing a wide range of services, which include mobile telemedicine, patient monitoring, emergency management and information sharing between patients and doctors or among the healthcare workers involved [5]. Also, we describe the technical issues which motivate to the design of a low-cost, patient-friendly healthcare system and then, in the next section, we describe the system that is needed to satisfy these requirements.

3.1 Mobile Telemedicine and Emergency Management

Many medical errors result from a lack of correct and complete data at the time of service, which can produce wrong diagnoses and drug interaction problems. It is important for healthcare workers to know have access to critical information about their patients before an emergency situation occurs. Wireless communication technology allows patient to send real-time data about the patient's condition to be sent to the hospital in real-time. For example, ambulance personnel can send real-time information in advance to a hospital while en route for the hospital, thus gives allowing the hospital staff to pre-evaluate the patient's condition and prepare for his or her treatment. In the same way, a patient can alert the hospital of an emergency situation using wireless communication, such as via a cellular network. Thus, intelligent emergency management and mobile telemedicine rely on wireless network technology (e.g. CDMA), as well as the systematic sharing of information between the BAN and the subsystem within the hospital.

3.2 Patient Monitoring

Wireless technology and personal area networks make it possible to continuously monitor a patient's health condition 'anytime and anywhere'. This is achieved with by the integration of the PDA and sensors, as well as related software, to a wireless BAN. The information provided by these sensors can also be integrated into a PDA that also contains the user's medical history. In this way, the patients themselves can monitor their health condition for self-care their states and immediately notify the healthcare workers at the nearest hospital, of an emergency service situation arising from a critical change in their status. For example, a blood glucose monitoring system can help everyone with would be of great benefit both to diabetes patients and to all those involved in their treatment. The information transferred via the sensors attached to these patients and wireless communication would make it possible to judge provide daily control of their condition, and by allowing changes to be made in their meals, physical activity, or medications, in order to improve their blood glucose levels. Thus, the efficient patient monitoring which could be achieved by means of medical sensor technology in conjunction with wireless communication technology. However, the resource constraints associated with small devices and network connectivity problems must first need to be solved.

3.3 Information Sharing Between Healthcare Workers in the Hospital

Among of the most time consuming activities associated with the public health sector are documentation and data exchange. Furthermore, not only the management of the patient's records in physician's private practices or hospitals, but also the documentation and data processing involved in emergency cases of an emergency is a of critical situation importance [6]. In a hospital, the healthcare workers form a peer group within the institution. Together they can create a secure peer group with various access rights, so as to allow the sharing of the patient's data (e.g. his or her patient UUID, patient sensor data, etc) and other kind of information (e.g. a first-aid, drugs to prescribe, etc). For example, medical information can be exchanged between patients, doctors or healthcare workers who are willing to share their data under the P2P environment. By making certain aspects of the sharable patient's data available to them, other doctors can examine the situation of patients with conditions similar to those of their own patients, thus helping them to make better decisions on the medical treatment to be administered to their own patients. Thus, the ideal healthcare system should enable the efficient processing of information and be able to cope with dynamic situations systematically, by allowing the sharing of information between groups by the JXTA grouping mechanism. Furthermore, the JXTA grouping service makes it possible to share data in a secure manner under the P2P environment. This is the main reason for which the healthcare system is based on the JXTA platform.

4 Proposed System Architecture

4.1 System Architecture

Our work is most closely related to the previous study described in [1]. The information between the BAN and the hospital is transferred by means of cellular CDMA communication. The wireless healthcare system consists of two regions, viz. the Healthcare region and the Hospital region, as shown in Fig. 1.

The healthcare BAN consists of sensors, a Wireless Interface, PDA communication, and various facilities. Depending on which type of patient data need

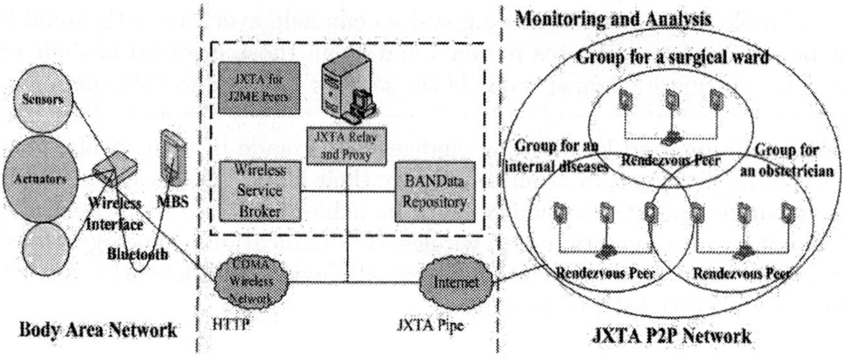

Fig. 1. System Architecture.

to be collected, his or her basic medical information can be integrated into the BAN. Communication between the different entities takes place within a BAN adopted Bluetooth network. The gateway that facilitates extra-BAN communication utilizes CDMA wireless networks. Sensors, each of which consists of multiple devices, are extended to have wireless intra-BAN communication capabilities. These sensors send their signals to the PDA, and actuators receive the control signals from the PDA. The information between the BAN and the JXTA relay peers is transferred by means of cellular CDMA communication. The Hospital region is composed of the JXTA P2P network that supports the doctor's mobility and dynamic service management modules. A P2P network distributes the information among the member nodes, instead of concentrating it at a single server. If any one peer fails, the service is still available from another peer member. The question of how to allow peers to safely communicate with each other is an important issue in this system. By creating peer groups, authorized groups of peers can be allowed to share a specific patient's data and other kinds of information. Also, peer group boundaries can be used to define the extent of this collaboration, when searching for a healthcare service within the group's content. A JXTA relay peer is defined among the two regions, in order to provide a systematic computing environment between the Healthcare region and the Hospital region. JXTA relays act as proxies for the individual patients' PDAs, as well as taking care of all the heavier tasks on their behalf.

4.2 Body Area Network

This section explains the system specification and operation in a body area network. We take the blood glucose data which consist of the basic information (6 bytes) and supplementary information (12 bytes). An additional 12 bytes, consisting of the Patient's Name, Machine ID and Patient's ID, are required to complete the service protocol. Fig. 2 shows the format of the data which is transferred via the Wireless Interface to the PDA.

The service platform supports patient-friendly applications. The application is targeted at direct interaction with our system users and can range from simple

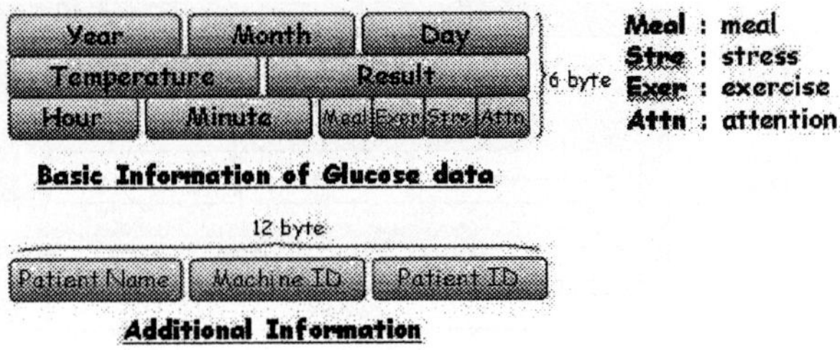

Fig. 2. Blood Glucose Data Format.

viewer applications that provide a graphical display of the BAN data to complicated applications that analyze the data. The PDA operates in several modes, as in the case of the system described in [1]. The first is a "store mode", in which the data is stored for a certain period of time, before being sent to the hospital at regular intervals. The second is the "process mode", in which the data is processed and the patient is provided with some first level information. The third is the "transmission mode", in which a connection is established to the hospital in order to send the data. Because these applications can be deployed on their own PDA, the patients can not only monitor their health condition, but can also ask the hospital for help in an emergency, and thus be treated in time. All of the data from the sensors are recorded on the PDA and transferred to the JXTA relay peer in the remote medical center, using the CDMA wireless network. In this way, the system can support the mobility of the patients and provide them with appropriate services everywhere. The patient sends a query to the JXTA relay peer to receive hospital services. The JXTA relay peer receives this query, which is propagated to the JXTA network in the hospital. The JXTA relay peer then accesses the related service group in the hospital. Also, all service responses are collected by the JXTA relay peer, which optimizes the transmission, in terms of trim and XML code parsing. Because the heavier tasks are performed by specific members of the peer group called JXTA relay peers, the patients can obtain medical service, without the JXTA porting of their own PDA. For this reason their own PDAs do not have to deal with polling, link status sensing, or neighbor detection messages. This is very useful for mobile devices, since it saves energy and bandwidth.

4.3 Service Grouping Mechanism in Hospital

In this section, we explain how service groups operating under the JXTA P2P network environment are formed and managed in a hospital region. Before introducing the service grouping mechanism of the hospital region, however, the JXTA architecture is presented, as follows. The JXTA architecture [7] consists of three layers: the platform layer, the service layer and the application layer, as shown in Fig. 3.

The platform layer is also known as the JXTA core layer, and provides those elements which are absolutely essential to every P2P solution. The service layer

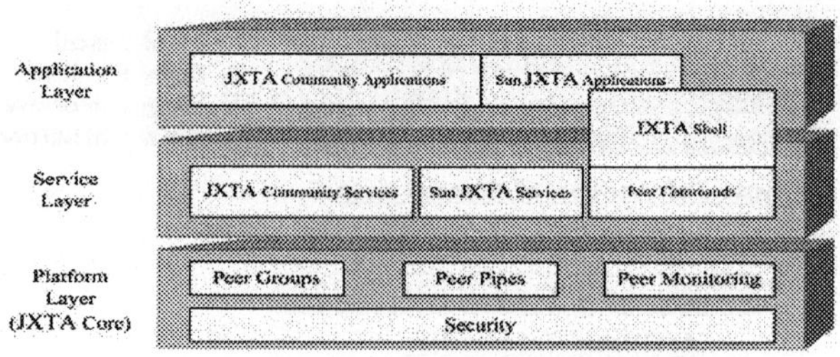

Fig. 3. The JXTA 3-layer architecture.

incorporates the network services, including both those which are essential and those which are desirable but may not be absolutely necessary for the P2P network to operate. These include communicating with a peer, searching for resource and peers, sharing documents between peers, peer authentication, and Public Key Infrastructure services. The application layer provides common P2P applications, such as instant messaging, by building on the capabilities of the service layer. JXTA Services are organized in peer groups. Collaborative service and functionality of the peer form peer groups. All of these groups share their resources and perform their own services, while cooperating with the other groups. The working mechanism of a peer group is as follows [8, 9]. Searching and sharing are done on the peer group level, i.e. shared content is only available to the peer group. When creating a peer group, the healthcare workers involved initialize the local secure environment, and then publish a peer group advertisement, which includes the peer group ID and peer group medical service. In order for a service consumer to join in the medical service group, he or she must first locate the peer group (e.g. surgical group) advertisement. Once the correct group advertisement has been located, the corresponding peer group is created using the parent group. In JXTA, the membership service is used to apply for peer group membership, joining a peer group, and exiting from a peer group. The membership service allows a peer to establish an identity within a peer group. Once an identity has been established, a credential is available, which allows the peer to prove that he or she has obtained that identify rightfully. Identities are used by services to determine the capabilities which should be offered to peers. In short, JXTA implements a peer group environment. A peer group is a collection of peers that have agreed upon a common set of services. Peers organize themselves into peer groups and all communication is constrained to the group members who are identified by their peer group ID. Each peer group can establish its own membership policy, thus allowing highly secure and protected operations to be supported. Also, the JXTA grouping mechanism supports the sharing of medical information and collaborative medical work. Thus, the system improves the ability of the healthcare workers to cope with dynamic situation, which in turn makes it possible to offer more efficient medical services.

5 Implementation

In this section, we describe the design and implementation of the hardware system based on the BAN, as shown in Fig. 4.

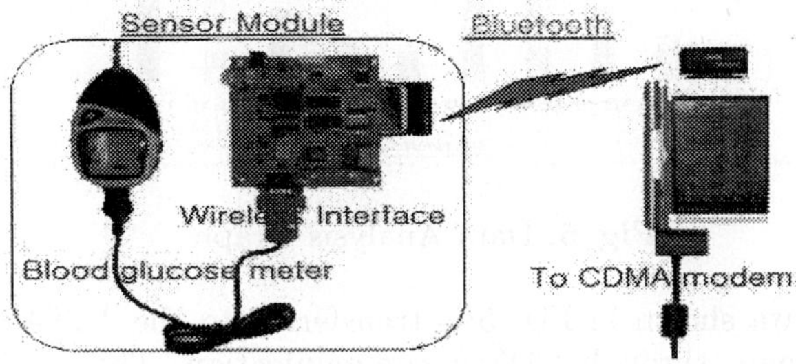

Fig. 4. Body Area Network.

The measurement device used to determine the glucose value is a blood glucose meter, which is embedded and portable, meaning that the patient is able to use it outdoors. It consists of mechanical parts and a chemical sensor, and is made by Allmedicus Ltd. The Wireless Interface device is divided into two parts. The first part is the operation processing part composed of a 32bit X-scale CPU. The second part is composed of a wireless network interface of the compact flash (CF) type, which supports the Bluetooth connection. The information retrieved from the blood glucose meter is then transferred to the Wireless Interface device via serial communication. In this way, the wireless Interface acquires the blood glucose information from the blood glucose meter and then establishes communication between the user and the medical center through the CDMA wireless network. The PDA recognizes the user's glucose level and informs the user of his or her condition, as well as transferring blood glucose contexts to the JXTA relay peer in the remote medical center or hospital. The service platform supports two types of applications. The first type is targeted at direct interaction with the system users and can range from simple viewer applications that provide a graphical display of the BAN data to complicated applications that analyze the data. These applications can be deployed either on the PDA or on the subsystem within a hospital under the JXTA environment. The second type of applications adds various functionalities to the core service, such as the management of the patient's information, on demand real-time data streaming and the JXTA discovery services used for searching for a doctor or group. In our system, all of the data from the blood glucose meter are recorded on the PDA, before being transferred to the JXTA relay peer at the remote medical center. The PDA receives the data concerning the blood glucose level from the sensor and interprets the user's health conditions and behaviors based on this information. The graph below sequentially shows ten time series plots based on the user's glucose value recorded in the PDA. In this graph, a blue bar denotes normal status, a yellow bar implies a glucose level slightly over the standard value and a red bar indicates critical status.

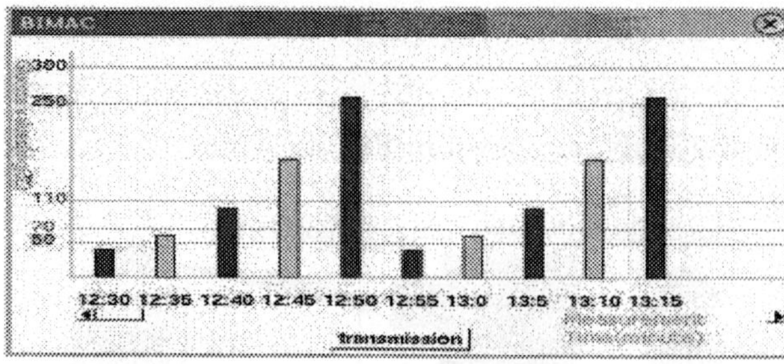

Fig. 5. Data Analysis Graph.

The blood data shown in Fig. 5 is transferred to the JXTA application in the hospital domain through CDMA communication. The JXTA application interface shown in Fig. 6 makes it possible for the doctors in the hospital to examine the list of patients, as well as the supplementary information available for them described in Fig. 2.

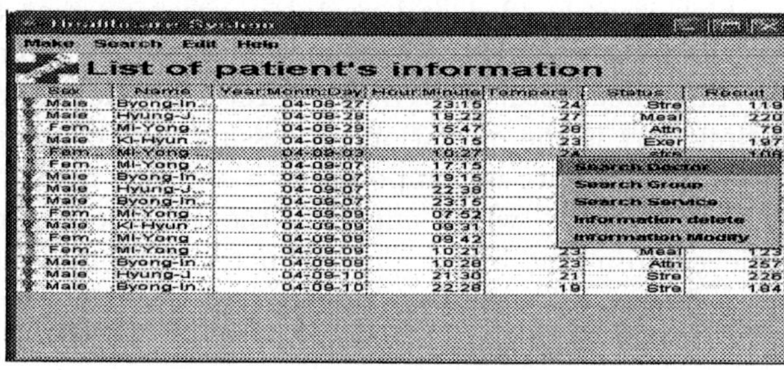

Fig. 6. Data Management JXTA Application.

As shown in Fig. 6, double-clicking on the appropriate item causes a graph of the patient's status to be displayed, whereas right clicking the same item causes a pop-up menu to be displayed, which consists of a submenu allowing the medical information and medical services to be selected for this patient. Selecting the "search for doctor" menu and typing the appropriate patient's ID leads to a request for data being sent to either the attending physician or other doctor assigned to this patient. In this way, the required information can be found by the discovery mechanism of JXTA. The Search for Group and Search for Service menu provide the means of locating a particular group or group service, respectively. Selecting a group by means of the Search Group menu provides a means of sharing the patient's data with all of the members of this group, while the concept of group service allows more flexible medical service to be offered. The two remaining menus provide the doctor with the possibility to delete or modify the patient's information. In this study, we implement software for monitoring the patient's status, in the BAN as well as in the JXTA application used for managing the patient's information; medical status and personal information. In particular, we utilize the JXTA Platform to develop a medical-service management tool in a hospital domain. The use of the JXTA Platform with its discovery and grouping mechanisms enables us to offer efficient and adaptive medical services.

6 Conclusion

As we have shown in this paper, the proposed healthcare system resolves a number of the technical challenges which arise when attempting to provide patients with efficient and adaptive healthcare services. In the proposed system, a P2P network distributes service and information among the member nodes, instead of concentrating it on a central server. This paradigm offers significant advantages in service and information sharing. The key software design method that we employed involved the offering of medical information and service to certified service customers using the JXTA grouping mechanism. This sharing of information concerning the medical treatment and patients in the JXTA environment improves the ability of the healthcare workers to cope with dynamic situations, thus improving the quality of the medical service that they can provide. The proposed system can greatly facilitate the delivery of a wider range of medical services and improve the productivity of the healthcare practitioners involved, as well as increasing their ability to offer low-cost, patient-friendly medical service. We are currently developing a more sophisticated version of the JXTA grouping mechanism, while simultaneously performing simulations and evaluating the performance of the existing system. Finally, we will expand our healthcare system to a multi-center situation, suitable for a ubiquitous environment, as well as improving it so as to offer high-quality medical service anytime and anywhere.

References

1. Nikolay Dokovsky, Aart van Halteren, Ing Widya, "BANip: enabling remote healthcare monitoring with Body Area Networks", International Workshop on scientific engineering of Distributed Java applications, November 27-28, 2003.
2. N. Maglaveras, et al, "Home care delivery through the mobile telecommunications platform: the Citizen Health System (CHS) perspective", International Journal of Medical Informatics 68 (2002), pp.99-111.
3. MIThril, Project Home Page. Massachusetts Institute of Technology, http://www.media.mit.edu/wearables/mithril.
4. Lart, Project Home Page. TU Delft, http://www.lart.tudelft.nl, 2001.
5. Upkar Varshney, "Pervasive Healthcare", IEEE Computer, December 2003(Vol. 36, No. 12). pp. 138-140 (2003)
6. Nico M, and Thomas M, "JXTA: A Technology Facilitating Mobile Peer-To-Peer Networks", MobiWac2002, 12 Oct.2002, pp. 7-13
7. JXTA v2.0 Protocols Specification. Sun Microsystems, Inc, March 2003.
8. Project JXTA v2.0: Java Programmer's Guide, Sun Microsystems, May 2003, pp 102-118
9. Zupeng Li, Yuguo Dong, Lei Zhuang, Jianhua Huang, "Implementation of Secure Peer Group in Peer-to-Peer Network", Proceedings of ICCT2003, pp 192-195.

An Architecture for Interoperability of Service Discovery Protocols Using Dynamic Service Proxies

Sae Hoon Kang[1], Seungbok Ryu[1], Namhoon Kim[1], Younghee Lee[1], Dongman Lee[1], and Keyong-Deok Moon[2]

[1] School of Engineering, Information and Communications University, Daejeon, Korea
{kang, sbryu, nhkim, yhlee, dlee}@icu.ac.kr
[2] Electronics and Telecommunications Research Institute, Daejeon, Korea
{kdmoon}@etri.re.kr

Abstract. Although all existing service discovery middlewares provide similar functionality, they are incompatible with one another due to differences in approach and architecture. We believe that co-existence of various service discovery middlewares will be indispensable since their target services and network environments are quite different to each other. Considering the future pervasive computing environment, the interoperability between them must be essential toward minimum user distraction under the heterogeneous systems. As the requirements of this interoperability system under these environment, complete translation, accommodation of changes of legacy middlewares, and simplicity of managing available services must be very important aspects. We propose a novel architecture which satisfies these requirements, using dynamic service proxy concept. While this concept provides various advantages, it has a notable drawback such that it needs to prepare many proxy codes per each service. To alleviate the load, we design and implement a Service Code Development Toolkit. We implemented a sample application and proposed architecture, and the results show that this architecture fully satisfies targeted design objectives.

1 Introduction

Since Mark Weiser introduced his vision of ubiquitous computing [1] in 1991, our computing environments have been changing rapidly. Since the proliferation of mobile devices such as PDAs, laptops, and cellular phones is growing, and in order to provide effective services to the user in the face of restricted resources and limited functions, these devices should establish a connection to nearby networks and cooperate with other services provided by other devices in the network. Unlike enterprise networks, in a ubiquitous computing environment, we can not depend on the existence of a system administrator for configuration. A service discovery middleware not only frees users from redundant administrative and configuration work, it also allows for the configuration and discovery of

C. Kim (Ed.): ICOIN 2005, LNCS 3391, pp. 786–795, 2005.

all devices in the network, and can be used by other devices automatically. Accordingly, service discovery researches are essential to the success of ubiquitous computing [2].

To the best of our knowledge, several service discovery middlewares have been proposed including Jini [3], UPnP [4], SLP [5], Salutation [6], HAVi [7], and Bluetooth SDP [8]. All of these middlewares provide similar functionality, i.e., automated discovery of required services. However, they are quite different in their target service types and suppose different network environments. For example, HAVi was developed to control digital AV streams in IEEE1394, while UPnP was introduced to control devices in TCP/IP. None of them has fully matured enough to dominate the market; consequently, it is unlikely that these middlewares will converge into one standard. As a result, we need an architecture which allows a client to discover and use its desired service in a network, regardless of middleware.

The interoperability among heterogeneous service discovery middlewares increases the user's satisfaction through improving service availability, that is, the possibility of discovering a desired service, as long as a device exists in the network. Benefits from the interoperable system also expedite the emergence of new compound services, boasting better functionality by extending the range of service selection.

To support interoperability among different service discovery middlewares, several approaches [9] [10] [11] [12] have been proposed. These approaches, however, have some limitations to the requirements such as translation loss, maintenance costs, the management cost of service information, ease of development, consciousness of other middleware, the modification of existing service, and accommodation of current evolving middleware.

In this paper, we propose a new architecture that provides interoperability among various service discovery middlewares which satisfy above requirements. By introducing a dynamic service proxy, we provide interoperability on the service level, not on the protocol level. We also achieve a minimal loss during translating protocols, and accommodate the change of legacy service discovery middleware without any modification of the existing system thereby supporting interoperability. In our architecture, no information for service discovery and invocation is managed, with the exception of the current service lists in a network. Once a dynamic service proxy is created on a middleware, all responsibilities to that service for discovery and invocation are given to that middleware. For example, if a dynamic service proxy is created on Jini, Jini Lookup Service is responsible for discovery of that service which is invoked using Jini RMI.

Despites of many advantages, our approach may require a little bit troublesome task because each service needs its own dynamic service proxies for each client of different middleware, respectively, which means that many dynamic service proxy codes should be implemented. In fact, much redundant work and codes are needed to develop a dynamic service proxy. To reduce the developer's work load, we design and implement a development toolkit. In order to verify the proposed architecture, we implement several dynamic service proxies for interoperability among Jini, UPnP and X10 services. Through them, we show that interoperability among many service discovery protocols can be applied effectively, and that the dynamic service proxy can be developed easily.

The rest of this paper is organized as follows. Section 2 presents a survey of related work, while Section 3 presents design issues of our approach and the proposed architecture. Section 4 presents Proxy Code Development Toolkit for developing proxy code. Section 5 describes evaluation of our proposed architecture comparing with other approaches, and some concluding remarks are provided in Section 6.

2 Related Work

There are several approaches to support interoperability among heterogeneous middlewares which are classified into four categories: 1) one-to-one protocol bridge; 2) one-to-many protocol conversion; 3) common integrated middleware; and 4) service level proxy. Examples of the one-to-one protocol bridge include Jini-HAVi Bridge by Philips, Sony and Sun, and SLP-Jini Bridge [9] by Sun. As these researches deal with the interoperability between only two middlewares, when more than two different middlewares exist in a network, we need more bridges. In order to overcome problems of the one-to-one protocol bridge, the one-to-many protocol conversion approach is proposed in [10]. This approach requires two protocol conversion steps because the service request of a client-side protocol is translated into a common protocol, and then retranslated into a server-side protocol. Also introduced is an intermediate protocol called Virtual Service Gateway (VSG) protocol, which connects middleware to other middleware, and the Protocol Conversion Manager (PCM), that converts the protocol of a local middleware component into that of VSG and vice versa.

Protocol conversion approaches have limitations in terms of development costs and translation completeness. Developing a bridge which can translate whole protocol specifications is very expensive and difficult. As well, since none of these middlewares is a superset of the others, a complete translation is impossible. For example, HAVi focuses on multimedia streams while UPnP is developed to control devices such as home appliances. As such, a gap exists between their functions.

In addition, current middleware are still being developed and may require modifications or enhancements in order to accommodate new types of services or support new types of killer applications which were not considered in the early stage of development. In that case, legacy bridges must also reflect changes in current middleware, making the bridge maintenance cost extremely high. For translating protocols into other protocols information of heterogeneous services, such as service locations and contexts, are required. Information of services should be managed by a directory-like component with an integrated method.

Another approach for interoperability proposed in [11] suggests building a common integrated middleware which is able to accommodate current legacy middlewares. The strong point of this approach is that a new middleware can be participated in the framework by simply adding a module for the protocol conversion. On the other hand, developing an integrated middleware requires additional cost and more time than developing protocol conversion bridges. Another problem is that current service discovery middlewares are in the development stages and are therefore not fixed. As such it is a substantial burden to version-up the protocol conversion modules whenever the legacy middlewares are modified.

The proxy based approach is proposed in [9] and [12]. The Jini-SLP bridge proposed in [9] was developed to support thin servers that lack the resources to host JVM. The client over Jini middleware downloads from the thin server jar file, which contains manifests such as the name of the driver class, and instantiates the driver object. The client then uses a service offered by the thin server using a driver object. For this approach, the client is specially designed to use the thin server; that is, a client application developer should be aware of this approach when advancing and rewriting client codes. Another proxy based approach, the Jini/UPnP framework, is described in [12]. The Jini/UPnP framework allows Jini clients to use UPnP services, and UPnP clients to use Jini services, without any modification, through the introduction of a virtual service that interacts with a real service in performing a service function. Even though this approach shows similarities to our approach, one difference is that it does not consider extending system to accommodate new service discovery middleware. Since this approach is designed to support interoperability between Jini and UPnP only, service advertisement and discovery module does not separated from the system. This may cause the degradation of system load when lots of proxies are running at same time.

3 Proposed Architecture

Here we discuss design issues in the development of an interoperability support system and our proposed architecture.

3.1 Design Issues

In designing our architecture, we consider the following five points.

The first is to maximize the completeness of interoperability. As described above, providing complete interoperability among heterogeneous service discovery middlewares is almost impossible since none of the service discovery middlewares is a superset or subset of the others, and each of them has their own focus. Nevertheless, we must try to reduce the gap among bridged middlewares. In this paper, we suggest an interoperability support mechanism based on the service level, not the protocol level, which allows for the possibility of more customized support to each specific service.

The second is to accommodate modifications of existing service discovery middlewares. Since current middleware are still in the development stages, they have not been fixed. As such, modifications or enhancements may be required to accommodate new types of services or support new types of killer applications which were not considered in the early stage of development. Current protocol-based approaches have a drawback in this point: if a legacy middleware changes, the protocol bridges for that middleware should also be changed, which may present serious problems for both bridge developers and users. While bridge developers must try to cover the change of legacy middleware, users of a bridge are not able to use new services which have been developed over the new version of middleware. To avoid this problem, the interoperability system should not be affected by the change of the legacy middleware.

The third is to accommodate new emerging service discovery middlewares. At present, many service discovery middlewares are being developed and proposed. We can easily expect that some of them will be widely used such as Jini and UPnP. Therefore, we should consider the support of interoperability with future service discovery middlewares.

The forth is to minimize the load for the interoperability system to manage information services spread over various middlewares. In the protocol level approaches, integrated information management such as virtual service repository in [10] is needed for protocol conversion. This kind of job may result in placing a heavy burden on the system.

The fifth is to avoid the requirement of existing client or service programs to be conscious of whether a desired service has been accessed through the interoperability system or not. Clients or services expect that their corresponding entities are running, as they are developed using the same method defined by their protocols. However, if a client must know that a desired service is being accessed only through the interoperability system, the client program must be rewritten. It is not easy to modify already existing programs to adapt to the system. If a client should be modified or rewritten to use a service running in different middleware, the version-up cost of legacy entities will increase in proportion to their volume.

3.2 Architecture

Our approach for interoperability support is based on service proxy. Considering design issues described in Section 3.1, we reach the conclusion that interoperability is provided based on a service-to-service level, i.e., service proxy. This approach boasts a lot of advantages: by providing interoperability at the service-to-service level, we can translate a service more perfectly in terms of customization. In addition, we do not need to modify the interoperability system when the legacy middleware changes. Even if a middleware changes, backward compatibility is guaranteed so that the service proxy remains unaffected, and this brings another advantage, the minimization of maintenance costs. Once a proxy is generated the responsibility for service advertisement, discovery, and invocation (with the exception of removing the proxy) remains to the original middleware of the proxy because the proxy is treated as the same as any other service in the middleware. For example, once a service proxy is created on Jini, it register itself with available Lookup Server and other Jini clients also discover

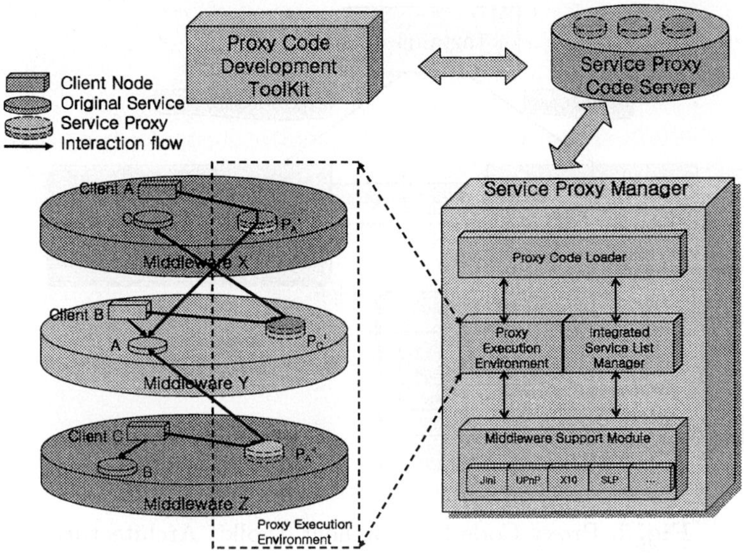

Fig. 1. Service proxy-based interoperability support architecture

and invoke that proxy service using Lookup Server and RMI. This means that interoperability system does not need to participate to those processes.

Our framework consists of three components: Service Proxy Manager (SPM), Service Proxy (SP), Service Proxy Code Server (SPCS), and Proxy Code Development Toolkit (PCDK). Figure 1 shows our proposed architecture for interoperability support. The SPM is responsible for creating and removing service proxy. It is usually located on the server-like host (for example, home server or home gateway) in the network; the SP connects the original service and the client which is running on different middleware; the SPCS manages service proxy codes made by the proxy code developer and transfers the proxy codes to the SPM when the SPM requests the corresponding proxy codes to a specific service; and the PCDK is used to help the proxy code developer reduce redundant work and save time, enabling greater convenience.

When a new service appears in one of the middleware environments, the SPM detects it with the help of the Integrated Service List Manger (ISLM). The ISLM manages a list of services which are running on each middleware by polling each middleware services periodically. When it detects a service appears or disappears on one of the middlewares it give a notification to the SPM so that corresponding service proxies created or removed on each middleware.

The Proxy Code Loader in the SPM requests the proxy code for a discovered new service and downloads it if it is available in the SPCS. The SPM then generates a service proxy using a downloaded code and executes it in the Proxy Execution Environment (PEE). The PEE provides execution environment where SPs execute and it provides multiple middleware environments using the Mid-

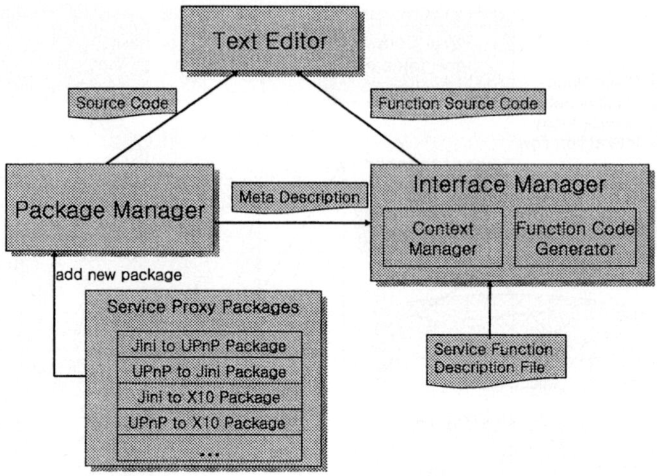

Fig. 2. Proxy Code Development Toolkit Architecture

dleware Support Module. The shaded part of the Figure 1 depicts the PEE. The generated proxy can then communicate among different middlewares. The Middleware Support Module can be easily added when a new middleware is developed.

In Figure 1, when a new *Service A* appears in *Middleware Y*, the ISLM detects it and makes the Proxy Code Loader download the corresponding proxy code for *Service A* and execute it on the PEE. In this case, proxy codes of *Service A* which is running in the *Middleware X* and *Z* (that is *PA'* and *PA"* respectively) are available at the SPCS, and service proxy *PA'* and *PA"* are generated. *Client B* has no problem in using *Service A* because they are on the same middleware. Meanwhile, *Client A* or *C* can access *Service A* throughout service proxy *PA'* and *PA"*. On the other hand, there is only one available code for *Service C*. In this case, only one service proxy is generated on the *Middleware Y*.

When an original service leaves the network, the ISLM detects it and asks the Proxy Code Loader to remove generated service proxies from each middleware. This prevents too many proxies from running in the PEE simultaneously and the load of our interoperability support system can be reduced.

4 Proxy Code Development Toolkit

In this section, we review the Proxy Code Development Toolkit in order to develop a Service Proxy Code more easily. Even though our approach has a lot of advantages compared to others, it requires as many as M*(N-1) service proxy codes, where M is the number of services and N is the number of middlewares, since our approach is based on service proxy. Requiring many service proxy codes seems to be a burden for developers and to pose a serious problem in terms of the extensibility of our system. But please note that proxy codes are made by many people such as device developers, service developers and application programmers. Once they develop a proxy code for a service, they register it with well-known Service Proxy Code Server and many applications share it. To help proxy code developers, we suggest a development toolkit for service proxy code. Much redundant work is done during the development of codes which we intend to reduce through the use of the Proxy Code Development Toolkit.

In this paper, we describe the Proxy Code Development Toolkit very simply (more detailed information can be found in [14]). The most important factor we should consider for designing and developing a toolkit is that the service proxy developer be able to implement service proxies easily and quickly. In addition, if a new middleware appears, it must be able to easily add a module for it into our system. As depicted in Fig.2, to meet the above requirements, our toolkit consists of four components: Text Editor, Interface Manager, Package Manager, and Service Proxy Packages: The Text Editor provides the interface upon which the developer can program the code of a service proxy. The Interface Manager manages common properties of a middleware for implementing a service proxy easily; it also manages the interface for generating the function code of service proxies. The Package Manager manages the Service Proxy Package, delivers the source code of the framework to the Editor, and gives other information about the framework to the Interface Manager. In Service Proxy Packages, there are basic modules and descriptions for bridging two middlewares, for example, UPnP to Jini bridging or Jini to UPnP bridging. Using this toolkit, we can develop a proxy code with programming 10% of total lines.

5 Evaluation

Here we compare existing service interoperability approaches and our proposed architecture. For evaluating these approaches, we consider the following 8 factors. (Please refer back to Section 3.1 for the reasons we chose these factors.)

1. Completeness of interoperability
2. Accommodation of changes of legacy middlewares
3. Accommodation of new emerging middleware
4. The load of interoperability system
5. No Consciousness of desired service's middleware
6. Interoperability support between more than 3 kinds of middlewares
7. Architectural Complexity
8. Development Cost
   (a) Initial
   (b) New service type
   (c) Change of legacy middleware
   (d) New middlewares

Protocol conversion-based schemes [9][10][11] have congenital limitations. Translating whole protocol specification into other protocols is very complex

Table 1. Comparison of various interoperability support approaches

Factors \ Approaches		Protocol-to-protocol conversion	Integrated middleware	Proxy-based	Dynamic service proxy
Factor 1		Low	Medium	High	High
Factor 2		No	No	Yes	Yes
Factor 3		No	No	No	Yes
Factor 4		High	High	Medium	Low
Factor 5		No	No	Yes[9], No[12]	No
Factor 6		Yes	Yes	No	Yes
Factor 7		Complex	Complex	Simple	Simple
Factor 8	(a)	Very High	Very High	Low	Low
	(b)	Low	Low	High $N*(M-1)$	High $N*(M-1)$
	(c)	High	High	Very low	Very Low
	(d)	Very High	Very High	High	Low

and requires huge costs in the early states. Whenever a change of legacy middlewares occurs, the whole system should also be modified, which may present serious problems due to small errors during modification. The protocol-based approach also requires common descriptions for each service in order to mediate two middlewares. The problem remains of who will provide descriptions.

On the other hand, the proxy-based system has a lot of advantages as shown in Factor 1, 2, 4 and 5 in Table 1. However, it requires as many as M*(N-1) service proxy. In [9] and [12], they do not describe these issues. To overcome this drawback, we design and implement a Proxy Code Development Toolkit.

Especially, comparing with other proxy-based approaches, our approach has advantages in terms of number of middlewares that can be supported by the interoperability system. Our proposed system supports interoperability between more than three different types of middlewares. Also separating service advertisement and discovery part from the interoperability support system in architecture level, we achieve that our system can be easily expanded to accommodate new emerging service discovery middlewares. The only thing for the accommodation is to add new middleware support module.

We implement a sample application for testing our scheme. Participant entities for this application include the AXIX UPnP camera, X10 light, and a door bell service written over Jini. The scenario is that if a visitor pushes a door bell, then the Jini door bell service gives notification to a person in the room, and he/she identifies the person and decides whether to allow him/her to open the door using the UPnP camera service and X10 light service located on the side of the door. For the development of proxy codes, we found that we do not need to spend much time (less than 3 hours), and the duration gets shorter and shorter in line with the development of more proxy codes. The number of lines which are really programmed by developers are less than 10% of total lines of proxy codes.

6 Conclusion

We have presented a new architecture to provide a more complete and simple interoperability among various service discovery middlewares without any modification of existing services and clients. Our approach provides more complete interoperability and easy accommodation to the change of legacy middlewares, separating the service advertisement and discovery part from the interoperability support system -. We also reduced the load of integrated information management of services by giving all the responsibility for advertising, discovering, leasing, and service invoking.We implemented a prototype with sample application which consists of UPnP, X10, and Jini services. The experiment result shows that our approach is especially suitable to accommodate many different kinds of middlewares for interoperability. And the Proxy Code Development Toolkit greatly helps to develop proxy codes. It turns out that it can sufficiently alleviate the drawback of our approach that is the load of preparation of proxy codes per each service In the future, we will test our approach on more kinds of middleware and enhance the convenience of our Proxy Code Development Toolkit.

References

1. Mark Weiser: The Computer for the Twenty-First Century. Scientific American, September (1991) 94-10
2. T. Kindbrg, A. Fox: System Software for Ubiquitous Computing. IEEE Pervasive Computing, January-March, (2002) 70-81
3. Ken Arnold et al.: The Jini Specification, V1.0, Addison-Wesley (1999)
4. Universal Plug and Play specification v1.0. available online at http://www.upnp.org/
5. Service Location Protocol Version 2. Internet Engineering Task Force (IETF), RFC 2608, June (1999)
6. Salutation Architecture Specification. available online at http://www.salutation.org/specordr.htm
7. Specification of the Home Audio/Video Interoperability (HAVi) Architecture, V1.1, HAVi, Inc., May 15, (2001)
8. Specification of the Bluetooth System. available at http://www.Bluetooth.com/developer/specification/specification.asp
9. Erik Guttman, James Kempf: Automatic Discovery of Thin Servers: SLP, Jini and the SLP-Jini Bridge. IECON, San Jose, (1999)
10. Hiroo Ishikawa et al: A Framework for Connecting Home Computing Middleware. Proc. of IWSAWC2002
11. Kyeong-Deok Moon, Younghee Lee, and Yong-Sung Son, Chae-Kyu Kim: Universal Home Network Middleware Guaranteeing Seamless Interoperability among the Heterogeneous Home Network Middleware. IEEE Transactions on Consumer Electronics, Vol. 49, No. 3, AUGUST (2003) 546-553
12. J. Allard, V. Chinta, S. Gundala, G.G Richard: Jini Meets UPnP: An Architecture for Jini/UPnP Interoperability. 2003 Symposium on Applications and the Internet, Orlando, (2003)
13. Seungbok Ryu: Home Network Interoperability System Using Dynamic Service Proxy, Master's thesis, School of Engineering, Information and Communications University, August (2004)

A Quality of Relay-Based Incentive Pricing Scheme for Relaying Services in Multi-hop Cellular Networks

Ming-Hua Lin and Chi-Chun Lo

Institute of Information Management, National Chiao-Tung University,
1001 Ta-Hseuh Road, Hsinchu, Taiwan, R.O.C.
mhlin@iim.nctu.edu.tw, cclo@faculty.nctu.edu.tw

Abstract. Cooperation among nodes is a critical prerequisite for the success of the relaying ad-hoc networks. Providing incentives for mobile nodes to forward data packets for others has received increasing attention. In this paper, we propose a Quality of Relay (QoR)-based pricing scheme to determine the price of the feedback incentives for intermediate nodes based on the individual importance of each mobile node contributing to successful hop-by-hop connections. Simulation results indicate that the QoR-based pricing scheme results in higher service availability than the fixed-rate pricing scheme under different relationships between price of feedback and willingness of forwarding packets. Moreover, the proposed pricing scheme shifts incentives from the nodes of low importance to the nodes of high importance in the networks so that it enhances service availability with only a slight increase in relaying costs.

1 Introduction

Multi-hop cellular networks that integrate the characteristics of both cellular and mobile ad hoc networks have received increasing attention. Several benefits have been investigated from this new family of networks [2, 4, 12, 14]: (i) reducing the number of the fixed antennas; (ii) conserving the energy consumption of the mobile device; (iii) reducing the interference with other mobile nodes; (iv) enhancing the service area of the network; (v) increasing the capacity of the cell.

In the hybrid networks, the communication between the mobile node and the base station is relayed by a number of other mobile nodes. Therefore, cooperation among nodes is a critical prerequisite for the success of the relaying ad-hoc networks. Since forwarding data for others incurs the consumption of battery energy and the delay of its own data, the assumption of spontaneous willingness to relay data is unrealistic for autonomous mobile nodes [16]. Some research [7–16] has described how to stimulate intermediate nodes to forward data packets in multi-hop networks. Most works focus on its protocol and security aspects or just employ fixed-rate pricing on number of packets or volume of traffic forwarded. The major advantage of the fixed-rate pricing is that billing and accounting processes are simple. However, the price of the feedback incentives is independent

C. Kim (Ed.): ICOIN 2005, LNCS 3391, pp. 796–805, 2005.

of the degree of each mobile node supporting relaying connections. Such approach cannot react effectively to the individual impact of each mobile node on service availability of the multi-hop cellular networks.

Service availability and operational costs are two major concerns of a network provider for adopting multi-hop cellular networks. Monetary incentives not only influence the motivation of the intermediate nodes supporting relaying services but represent the costs of providing connection services in multi-hop cellular networks. Therefore, the network provider should give appropriate feedback incentives to the intermediate nodes. In this paper, we propose a Quality of Relay (QoR)-based incentive pricing scheme to encourage collaboration based on individual degree of each mobile node contributing to successful hop-by-hop connections. Simulation results indicate that the proposed QoR-based incentive pricing scheme results in higher service availability than the fixed-rate pricing scheme under different relationships between price of feedback and willingness of forwarding packets. Moreover, the proposed scheme shifts incentives from the nodes of low importance to the nodes of high importance so that it enhances service availability with only a slight increase in relaying costs.

The rest of this paper is organized as follows. In section 2, we review existing multi-hop cellular network models and incentive schemes. Section 3 describes the detail of the proposed QoR-based pricing scheme. Section 4 presents the simulation results and discussions. Finally, conclusions are made in section 5.

2 Literature Review

2.1 Multi-hop Cellular Network Model

Although many approaches in the literature have been proposed to improve the performance of cellular networks and multi-hop networks in isolation, more and more research focuses on integrating the cellular and multi-hop network models to leverage the advantages of each other.

Opportunity Driven Multiple Access (ODMA) is an ad hoc multi-hop protocol that the transmissions from mobile hosts to the base station are broken into multiple wireless hops, thereby reducing transmission power [1, 2]. Aggélou et al. describe an Ad Hoc GSM (A-GSM) system that presents a network layer platform to accommodate relaying capability in GSM cellular networks [3]. The authors extend the standard GSM radio interface with sufficiently flexible capabilities to support relaying. Qiao et al. present a network model called iCAR that integrates the cellular infrastructure and ad-hoc relaying technologies [4]. The proposed architecture places a number of Ad-hoc Relaying Stations (ARS) at strategic locations to relay data from one cell to another cell. Load balancing among different cells in the iCAR system not only increases system capacity, but also reduces transmission power for mobile terminals. Wu et al. propose a scheme called Mobile-Assisted Data Forwarding (MADF) to add an ad-hoc overlay to the fixed cellular infrastructure and special channels are assigned to connect users in a hot cell to its neighboring cold cells [5]. The authors find that under a certain delay requirement, the throughput can be greatly improved. Luo et al. propose a Unified Cellular and Ad-Hoc Network (UCAN) architecture to enhance the cell throughput. Each mobile device in the UCAN model has both 3G cellular link and IEEE 802.11-based peer-to-peer links. The 3G base station forwards packets for destination clients with poor channel quality to proxy clients with better channel quality [6].

2.2 Incentive Scheme

Much research has discussed the incentive schemes in pure ad hoc or hybrid ad hoc networks. The approaches can be classified into detection-based and motivation-based.

The detection-based approach finds out misbehaving nodes and mitigates their impact in the networks. Marti et al. describe two techniques to improve network throughput by detecting misbehaving nodes and mitigating their impact in ad hoc networks [7]. They use a watchdog to identify misbehaving nodes and a pathrater to avoid routing packets through these nodes. Although the proposed solution fosters cooperation in ad hoc networks, it does not castigate malicious nodes but rather mitigates the burden of forwarding for others. Michiardi et al. suggest a mechanism called CORE based on reputation to enforce cooperation among nodes and prevent denial of service attacks due to selfishness [8]. The request from the entity with negative reputation will not be executed. Buchegger et al. propose a protocol called CONFIDANT to detect and isolate misbehaving nodes, thus making it unattractive to deny cooperation [9]. Both two methods discourage misbehavior by identifying and punishing misbehavior nodes. However, they do not involve using positive cooperation incentives in their methods.

The motivation-based approach provides incentives to foster positive cooperation in ad hoc networks. Buttyán et al. use a virtual currency called nuglets as incentives given to cooperative nodes in every transmission [10]. The proposed models do not discuss the number of nuglets should be feedback to the intermediate nodes. Buttyán et al. also propose a mechanism based on credit counter to stimulate packet forwarding [11]. The number of feedback nuglets depends on the number of forwarding packets in this method. In [12], Jakobsson et al. present a micro-payment scheme that fosters collaboration and discourages dishonest behavior in multi-hop cellular networks. Packet originators associate subjective reward levels with packets according to the importance of the packet. Lamparter et al. propose a charging scheme in hybrid cellular and multi-hop networks, which would be beneficial for Internet Service Provider (ISP) and the ad hoc nodes and thus motivates cooperation among mobile nodes [13]. The charging scheme is based on volume-based pricing models. A fixed price per unit is rewarded for forwarding traffic irrespective of the network conditions. In [14], the authors propose an incentive mechanism based on a charging/rewarding scheme in multi-hop cellular networks. Both the charge of sending data and the reward of forwarding data depend on the packet size in the proposed method.

In our previous work [15], we have proposed a dynamic incentive pricing scheme to maximize the revenue of the network provider based on the actual network conditions. The proposed scheme adjusts the price of the feedback incentives according to the total number of the mobile nodes and do not consider the individual contribution of each mobile node.

3 QoR-based Incentive Pricing Scheme

Most of the motivation-based approaches in the literature just employ fixed-rate pricing for relaying services. However, because each mobile node has different effects on supporting hop-by-hop connections, the base station should give more incentives to the nodes of high importance so that it can make more mobile nodes connect to the base station successfully.

3.1 Supply Function for Providing Relaying Services

Pricing is an inducer for suppliers to provide services. The price of the incentives can affect the motivation of mobile nodes providing relaying services and is usually characterized by a supply function that represents the reaction of mobile nodes to the change of the price [17]. The general supply function describes that the producers are willing to produce more goods as the price goes up. Here we consider three forms for the supply function as follows [18]:

$$S_1 : S(p_v) = \frac{p_v}{p_{\max}} \quad 0 \le p_v \le p_{\max}, \tag{1}$$

$$S_2 : S(p_v) = \begin{cases} e^{-\left(\frac{p_{\max}}{p_v} - 1\right)^2} & \text{when } 0 < p_v \le p_{\max} \\ 0 & \text{when } p_v = 0, \end{cases} \tag{2}$$

$$S_3 : S(p_v) = \begin{cases} \frac{1}{(\frac{p_{\max}}{p_v} - 1)^4 + 1} & \text{when } 0 < p_v \le p_{\max} \\ 0 & \text{when } p_v = 0, \end{cases} \tag{3}$$

where p_{max} is the maximum price that the network provider can feedback, p_v is the price of the feedback incentives for node v per unit of relay data. In our scheme, p_v is adjusted based on the degree of node v contributing to service availability in the multi-hop cellular networks. The proposed pricing scheme enhances service availability by increasing the price of the feedback incentives for the mobile nodes that affect more relaying connections. $S(p_v)$ denotes the possibility of node v accepting the price to forward data packets. Note that $S(0) = 0$, which means that node v will not relay traffic for others if no feedback is provided for relaying services. The willingness of forwarding packets increases as the price of feedback increases. For $p_v = p_{max}$, we have $S(p_{max}) = 1$, which means that the maximum price is acceptable to all mobile nodes to provide relaying services. Figure 1 illustrates the difference between the three supply functions with various supply flexibility. S_1 represents a linear relationship between price of feedback and willingness of forwarding packets. S_2 and S_3 begin low for small p_v, then increase rapidly as p_v gets into a mid-range. When prices are low, S_1 is more sensitive to price changes. When prices are in the middle range, S_3 is much more sensitive than the others to small price changes.

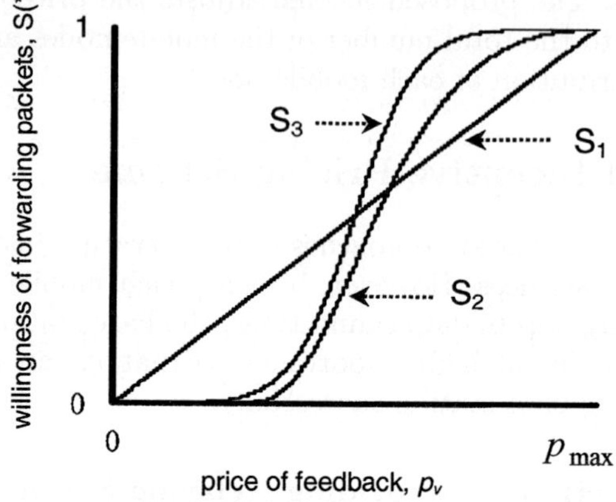

Fig. 1. The supply functions of price of feedback and willingness of forwarding packets

3.2 Proposed QoR-based Incentive Pricing Scheme

In this paper, we focus only on a single base-station cell as indicated in Fig. 2. The base station can enhance the service area by adopting relaying connections supported by the mobile nodes.

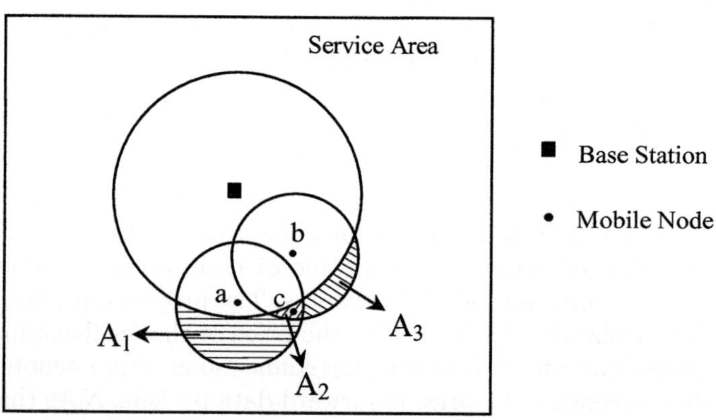

Fig. 2. An example of multi-hop cellular networks with a single base-station

In multi-hop cellular networks, data packets must be relayed hop by hop from a given mobile node to a base station, thus the path availability from a mobile node to the base station depends on the individual willingness of each mobile node to forward packets on the routing path. Let M_x be the set of intermediate nodes on the path from node x to the base station, then the path availability between node x and the base station, PA_x, is defined as follows:

$$PA_x = \prod_{v \in M_x} S(p_v). \tag{4}$$

Since networking services provided by the base station are available for the mobile nodes outside the coverage of the base station when the mobile nodes can set up a hop-by-hop connection to the base station successfully, we define the service availability of the relaying networks as the probability that a mobile node outside the coverage of the base station but inside the service area can connect to the base station.

In order to evaluate the degree of a mobile node contributing to the service availability of the multi-hop cellular networks, we introduce a new metric called Quality of Relay (QoR) as follows:

$$QoR_v = \sum_{i \in C_v} \frac{1}{RI_i}, \tag{5}$$

C_v is the set of positions inside the coverage of the node v where the mobile node requires hop-by-hop connections to reach the base station, RI_i is the relay index (RI) of position i that is defined to be the number of mobile nodes capable of relaying traffic for a mobile node staying in position i. As the example indicated in Fig. 2, $RI_{i \in A_1}$ is 1 because only node a can relay data for the mobile nodes reside in area A_1 ; $RI_{i \in A_2}$ is 2 because both node a and node b can relay data for the mobile nodes reside in area A_2 . The degrees of node a and node b contributing to the service availability of networks are evaluated by their QoR values as follows:

$$\begin{aligned} QoR_a &= \sum_{i \in C_a} \frac{1}{RI_i} = \sum_{i \in (A_1 \cup A_2)} \frac{1}{RI_i} = (A_1 * \frac{1}{RI_{i \in A_1}}) + (A_2 * \frac{1}{RI_{i \in A_2}}) \\ &= A_1 * \frac{1}{1} + A_2 * \frac{1}{2} \end{aligned} \tag{6}$$

$$\begin{aligned} QoR_b &= \sum_{i \in C_b} \frac{1}{RI_i} = \sum_{i \in (A_2 \cup A_3)} \frac{1}{RI_i} = (A_2 * \frac{1}{RI_{i \in A_2}}) + (A_3 * \frac{1}{RI_{i \in A_3}}) \\ &= A_2 * \frac{1}{2} + A_3 * \frac{1}{1} \end{aligned} \tag{7}$$

From above equations, QoR_a is greater than QoR_b because the coverage of area A_1 is larger than that of area A_3. There are two conditions that node v has a higher QoR value:

- The node v has larger C_v, which means it can support larger coverage where mobile nodes necessitate hop-by-hop connections to reach the base station.
- The position inside C_v has lower RI value, which means the mobile nodes inside C_v can be supported by fewer nodes. That is, the node v can provide relaying services to the mobile nodes that others cannot support.

Consequently, a node with a higher QoR value represents that it has more contributions to the relaying capability of the networks, so that its high willingness of forwarding data packets can enhance the service availability of the networks. Since the higher QoR value represents that the resource of the mobile node is more valuable, the base station should apply the QoR value of a mobile node as a reference to give incentives for increasing the willingness of providing relaying services.

Let N be the set of intermediate nodes capable of forwarding data for mobile nodes to reach the base station, $AQoR$ be the average QoR value of all nodes in N, that is,

$$AQoR = (\sum_{v \in N} QoR_v)/(\sum_{v \in N} 1). \tag{8}$$

Then, the proposed QoR-based incentive pricing scheme assigns the price of the feedback incentives for node v, p_v, as follows:

$$p_v = p_0 + (QoR_v - AQoR) * \frac{R_p}{R_{QoR}}, \tag{9}$$

where $R_p = \min\{p_0, p_{\max} - p_0\}$
$R_{QoR} = \max\{\max_{v \in N}\{QoR_v\} - AQoR, AQoR - \min_{v \in N}\{QoR_v\}\}$

p_0 is the price adopted in the fixed-rate pricing method. The proposed scheme employs p_0 as a basic price and derives p_v according to the difference between QoR_v and $AQoR$. The parameter $\frac{R_p}{R_{QoR}}$ aims to adjust p_v in the interval $[0, p_{max}]$.

4 Simulation Results and Discussions

We evaluate the performance of the proposed QoR-based incentive pricing scheme in terms of service availability with different supply functions described in section 3. The simulation environment is a rectangular region of size 400 units by 400 units with a single base station located in the central point. The radius of the base station is 150 units and the radius of each mobile node is 100 units. For different number of mobile nodes randomly distributed in the rectangular region, the simulator computes the service availability of the networks, that is, the probability that a mobile node outside the coverage of the base station can connect to the base station successfully. Since no routing topology is pre-constructed, herein we assume the mobile nodes randomly select one of the neighboring nodes that have relaying paths to the base station.

We compare the proposed incentive pricing scheme with the fixed-rate pricing scheme. We adopt p_0 ($p_0 = S^{-1}(0.5)$, $p_0 = S^{-1}(0.4)$) as the fixed price in the fixed-rate pricing scheme and the basic price in the proposed pricing scheme. In Figs. 3 through 5, we observe that the QoR-based incentive pricing scheme results in higher service availability than the fixed-rate pricing scheme under various number of mobile nodes for different supply functions. According to Figs.

3 through 5, we summarize the percentage of improvement in service availability from the fixed-rate pricing scheme to the proposed QoR-based pricing scheme for different supply functions in Table 1. Since costs is one of major concerns that the network provider adopts multi-hop cellular networking model, we also list the percentage of increase in relaying costs per connection that feedback to the intermediate nodes in Table 1. By examining Table 1, we notice:

- The increase in service availability obtained in p_O ($p_O = S^{-1}(0.4)$) is more significant for S_2 and S_3. Because mobile nodes are more sensible to the change of price of feedback in $p_O = S^{-1}(0.4)$ than that in $p_O = S^{-1}(0.5)$ for both S_2 and S_3. The increase obtained by different basic prices are not obviously distinct for the linear function S_1 with constant supply flexibility.
- The QoR-based pricing scheme results in higher relaying costs per connection than the fixed-rate pricing scheme. Because the QoR-based pricing scheme gives more incentives to the nodes that affect more relaying connections to enhance service availability. However, the proposed scheme also decreases incentives for the node of low impact on relaying connections. Consequently, the increase in relaying costs is much lower than that in service availability.

5 Conclusions

Service availability and operational costs are two major concerns of a network provider adopting multi-hop cellular networking technology. In this paper, we present a QoR-based incentive pricing scheme to enhance service availability by adjusting the price of feedback incentives based on the degree of the mobile nodes contributing to relaying services. The proposed method increases incentives for nodes of high importance and decreases the incentives for node of low importance so that it enhances service availability with only a slight increase in relaying costs. Simulation results indicate that the QoR-based pricing scheme results in higher service availability than the fixed-rate pricing scheme under different forms for supply function of price of feedback and willingness of forwarding packets.

Table 1. Percentage of increase in service availability and relaying costs per connection from fixed-rate pricing to QoR-based pricing for different supply functions

	S_1		S_2		S_3	
p_o	$S_1^{-1}(0.4)$	$S_1^{-1}(0.5)$	$S_2^{-1}(0.4)$	$S_2^{-1}(0.5)$	$S_3^{-1}(0.4)$	$S_3^{-1}(0.5)$
Increase in service availability	10.05%	11.22%	23.64%	10.07%	34.76%	15.82%
Increase in relaying costs	4.12%	4.39%	4.41%	4.05%	4.44%	4.62%

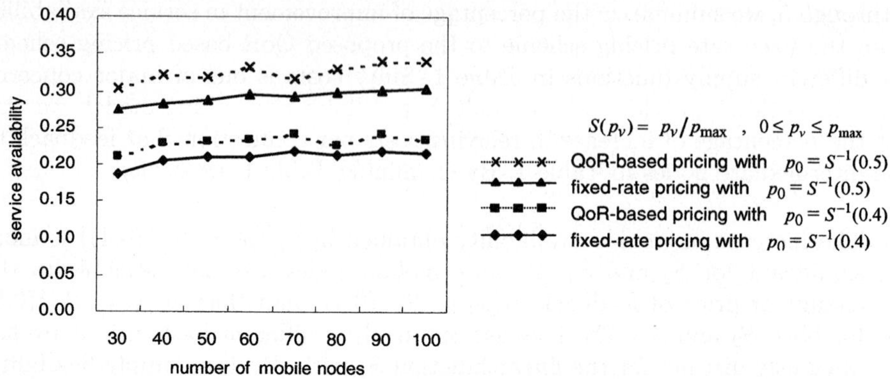

Fig. 3. Comparison of service availability by fixed-rate pricing and QoR-based pricing under different number of mobile nodes with supply function S_1

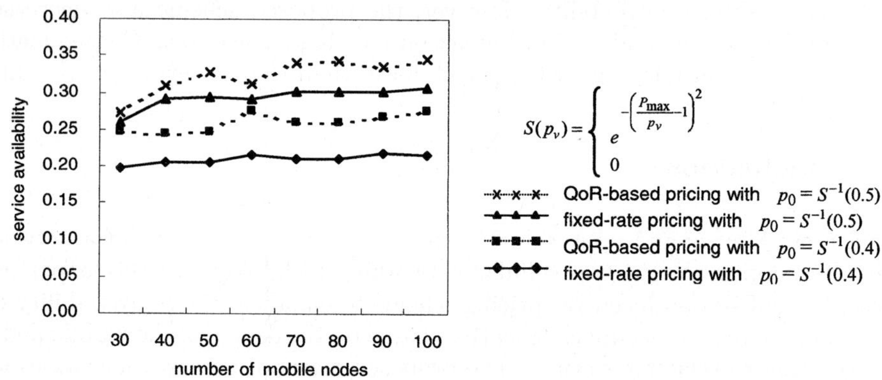

Fig. 4. Comparison of service availability by fixed-rate pricing and QoR-based pricing under different number of mobile nodes with supply function S_2

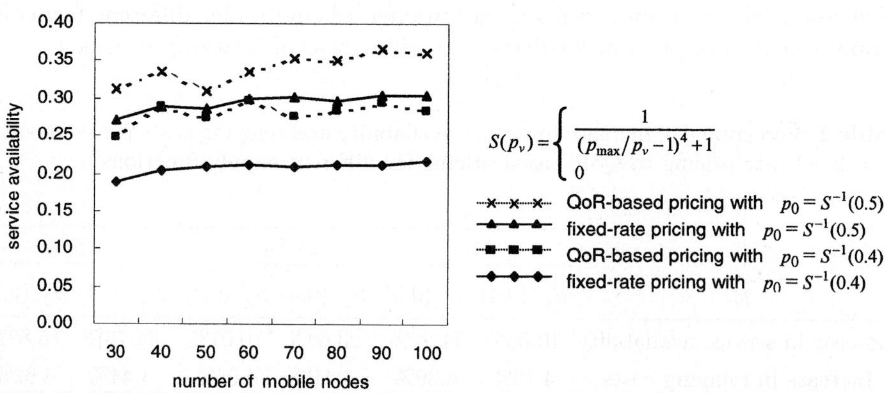

Fig. 5. Comparison of service availability by fixed-rate pricing and QoR-based pricing under different number of mobile nodes with supply function S_3

References

1. 3G TR 25.924 V 1.0.0. 3GPP TSG-RAN; Opportunity Driven Multiple Access, Dec. 1999.
2. Rouse, T., Band, I., McLaughlin, S.:Capacity and Power Investigation of Opportunity Driven Multiple Access (ODMA) Networks in TDD-CDMA Based Systems, Proc. of IEEE ICC, pp.3202-3206, Apr. 2002.
3. Aggélou, G. N., Tafazolli, R.: On the Relaying Capacity of Next-Generation GSM Cellular Networks, IEEE Personal Communications, pp.40-47, Feb. 2001.
4. Qiao, C., Wu, H.: iCAR: an Intelligent Cellular and Ad-hoc Relay System, Proc. of IEEE IC3N, pp.154-161, Oct. 2000.
5. Wu, X., Chan, S.H., Mukherjee, B.: MADF: A novel approach to add an ad-hoc overlay on a fixed cellular infrastructure, Proc. of IEEE WCNC, Sep. 2000.
6. Luo, H., Ramjee, R., Sinha, P., Li, L., Lu, S.: UCAN: A Unified Cellular and Ad-hoc Network Architecture, ACM MOBIHOC 2003, Jun. 2003.
7. Marti, S., Giuli, T. J., Lai, K., Baker, M.: Mitigating routing misbehavior in mobile ad hoc networks, Proc. of ACM MOBICOM, pp.255-265, Aug. 2000.
8. Michiardi, P., Molva, R.: Core: A COllaborative REputation mechanism to enforce node cooperation in Mobile Ad Hoc Networks, Proc. of the sixth IFIP Communications and Multimedia Security Conference, Sep. 2002.
9. Buchegger, S., Boudec, J.Y.L: Performance Analysis of the CONFIDANT Protocol: Cooperation Of Nodes - Fairness In Dynamic Ad-hoc Networks, ACM MOBIHOC 2002, Jun. 2002.
10. Buttyán, L., Hubaux, J.P.: Enforcing Service Availability in Mobile Ad Hoc WANs, Proc. of ACM MOBIHOC, Aug. 2000.
11. Buttyán, L., Hubaux, J.P.: Stimulating cooperation in self-organizing mobile ad hoc networks, ACM/Kluwer MONET, Vol. 8, No. 5, Oct. 2003.
12. Jakobsson, M., Hubaux, J.P., Buttyán, L.: A micropayment scheme encouraging collaboration in multi-hop cellular networks, Proc. of Financial Crypto 2003.
13. Lamparter, B., Paul, K., Westhoff, D.: Charging Support for Ad Hoc Stub Networks, Journal of Computer Communication, Elsevier Science, Vol. 26, Issue 13, Aug. 2003.
14. Ben Salem, N., Buttyán, L., Hubaux, J.P., Jakobsson, M.: A Charging and Rewarding Scheme for Packet Forwarding in Multi-Hop Cellular Networks, ACM MOBIHOC 2003, Jun. 2003.
15. Lin, M.H., Lo, C.C.: A Dynamic Incentive Pricing Scheme for Relaying Services in Multi-hop Cellular Networks, Lecture Notes in Computer Science, Vol. 3090, pp.211-220, 2004.
16. Ileri, O., Mau, S.C., Mandayam, N.B.: Pricing for Enabling Forwarding in Self-Configuring Ad Hoc Networks, Proc. of IEEE WCNC, pp.1034-1039, Mar. 2004.
17. Hou, J., Yang, J., Papavassiliou, S.: Integration of Pricing with Call Admission Control to meet QoS Requirements in Cellular Networks, IEEE Transactions on Parallel and Distributed Systems, vol. 13, no. 9, pp.898-910, Sep. 2002.
18. Fishburn, P.C., Odlyzko, A.M.: Dynamic Behavior of Differential Pricing and Quality of Service Options for the Internet, ICE'98, pp. 128-139.

A Dynamic Path Identification Mechanism to Defend Against DDoS Attacks

GangShin Lee[1], Heeran Lim[2], Manpyo Hong[2], and Dong Hoon Lee[1]

[1] Center for Information Security Technologies(CIST), Korea University, Seoul, Korea
kslee@kisa.or.kr
donghlee@korea.ac.kr

[2] Internet Immune System Laboratory, Ajou University, Suwon, Korea
lhr5456@empal.com
mphong@ajou.ac.kr

Abstract. Many Researchers have tried to design mechanisms to resist Distributed Denial of Service(DDoS) attacks. Unfortunately, any of them has not been satisfactory. Recently, Yaar et al.[1] suggested Pi (short for Path Identifier) marking scheme as one of solutions to thwart DDoS attacks, which is fast and effective in dropping the false positive and negative packets from users and attackers. They make use of the IP Identification field of which length is 16 bits as marking section. Every router en-route to the victim marks 1-bit or 2-bits by wrapping method sequentially. The victim drops the false positive and negative packets according to the attack markings list. The performance of Pi is measured for marking bit size of 1 or 2 bits. This paper suggests the method to decide the marking bit size dynamically in accordance with the number of hop counts. The performance is quite improved, compared with the existing one.

1 Introduction

Many Researchers have tried to design mechanisms to resist Distributed Denial of Service(DDoS) attacks[1,3,4,5,6,7,8]. In the 25th of January, 2003, the MS-SQL Slammer worm impacts on Korea backbone networks heavily because of the high speed networks, the absence of DNS root servers, and the appliance of the inappropriate packet filtering techniques, etc. According to the CAIDA report, 90% of the vulnerable MS-SQL servers was plagued in 10 minutes worldwide[9]. So, this type of attack must be blocked quickly. There are several methods to defend against DDoS attracks. For example, IP traceback[7,10,11,12,13,14] , pushback[8,15], etc. But these methods have a shortcoming not to drop attack packets immediately because these need much time to collect the enough packets and to reconstruct the path. A. Yaar et al. suggested the new method to drop the attack packets on a per packet basis immediately using the static marking scheme[1]. Better performance can be expected if the Pi marking scheme suitable for every packet is applied to the routers en-route to the victim respectively. This

C. Kim (Ed.): ICOIN 2005, LNCS 3391, pp. 806–813, 2005.

paper suggests the dynamic bit marking scheme. The remainder of the paper is organized as follows : in Section 2 we review and analysis previous researches and propose the new idea. In Section 3 we design the dynamic Pi marking scheme, each router's marking algorithms, and the filtering scheme in a given network topologies. In Section 4 we compare the results from the A. Yaar et al.'s and the dynamic marking bit scheme. In Section 5 there are conclusion and future works.

2 Previous Researches

Three papers about the Pi marking and filtering scheme are published by now[1,6, 16]. In the first paper[1], the method is based on a per packet, not a per flow, and not a per network basis. Every router marks 1 bit or 2 bits on the 16 bit IP Identification field of packets en-route using TTL. By using the deterministic characteristics - all packets traversing the same path carry the same marking value, the victim can drop the attack packets from the upstream router immediately. The paper concludes that the 2 bit marking scheme is better than the 1 bit one in the performance of filtering packets in the average 15 hop count network topology. But this scheme is surmised less effective because of the static properties. In the second paper [6], the author tries to find the most proper marking size n (=1, 2, and so on) for the given Internet data set from CAIDA's Skitter Map because the marking size n is the most important parameter. It is concluded that, for less than 13 hop counts, 2-bit marking scheme's false ratio is lower than 1-bit. This scheme is fixed also even if the appropriate number of hop counts is tried to be found to decide the scheme. In the third paper [16], the Pi marking scheme not using TTL is experimented. In the case of using TTL, there are garbage basically on the marked position of the Pi Identification field of the packets because legacy routers don't mark the ones en-route. It turns out to raise false positive. In this paper, only the routers of this scheme mark the Pi value of the router's IP address on the right bit of the Identification field of the packets after shifting Pi value to the left direction. This paper results in better performance. But the static scheme is applied also. If we can decide the marking scheme n=1 or 2 according to the distance from the source to the destination, we can expect better Pi packet filtering performance. In this paper, the dynamic Pi marking scheme is applied for the better performance, and the differences between this scheme and the existing scheme is analyzed.

3 Design Schemes

3.1 Assumptions

This paper assumes the followings. (1) Every router has the same marking scheme or not in order to experiment to the legacy ratio. (2) The initial value of TTL is affordable to 255 because the TTL field length is 8 bits. (3) There are no changes in the network topologies for experiment because the appropriate hop count x must be fixed after x is acquired dynamically.

3.2 Pi Marking Scheme

The Basic Scheme

The left 1 bit of the 16 bit IP Identification field is used to register the decided marking scheme - "1" for the 1 bit marking scheme and "00" for the 2 bit marking scheme because of the usefulness to modulate TTL with even number. We use the wrapping method[3] as a marking one basically. Therefore the limited marking space causes routers close to the victim to overwrite the markings of routers farther away from the victim. In the case of n=1, the maximum number of markings is 15 if no rewriting occurs. It is reason that the marking starts at the 2nd bit from the left. The position to be marked is decided to the value of TTL modulo 15. In this case, the marking space is $2^{15} = 32768$ because the possible marking bit size is 15. This marking scheme can be more [double] exploitable to the saturation attack than in 16-bit marking space. The position to be marked is TTL modulo 15 In the case of n=2, the right 14-bits is used for marking except the left 2-bits because of multiple by 2. The maximum number of markings is 7 if no rewriting occurs. It is reason that the marking starts at the 3rd bit from the left. The position to be marked is (TTL modulo (14/2)) * 2 In this case, the marking space is $2^{14} = 16384$, which is 1/4 of 2^{16}. So, this marking scheme can be more exploitable to the saturation attack than in 16-bit marking space. Less the marking space, higher the possibility of rewritings.

Algorithm to Find the Appropriate Hop Count Deciding the Pi Marking Scheme

Let's discuss the dynamic bit marking scheme. It is necessary to find the appropriate hop count x deciding the Pi scheme in a given network topology. Because network topology is not changed frequently, x can be used for a long time as long as the network topology is unchanged. Precisely, x can be managed as a global parameter. To find x, we must calculate the false rate when n=1 and n=2 as A. Yaar et al. experimented. Maybe, the false rate for n=1 is higher than for n=2 under x. On the contrary, the false rate for n=2 is higher than for n=1. By applying the curve fitting method to the false rate, we can acquire the appropriate polynomials for n=1 and 2. Also, we can find the real number x such that $a_1x^k + a_2x^{k-1} + \cdots + a_k \ \ (if\ n = 1) \ = \ b_1x^j + b_2x^{j-1} + \cdots + b_j(if\ n = 2)$(See Fig.1). It is not good that x is approximated to the nearest natural number because it is not easy to decide which marking scheme is more appropriate when the approximated natural number is the same to the later traceroute results. For example, if x is the one of 9.7 or 10.3, then x will be 10. To send any packet, we do traceroute and have the hop count 10 as the distance. In this case, we can not decide the marking scheme. To find the initial value of x, it is necessary to create some data set to be analyzed because there are no data set at first. Therefore any meaningless value is assigned to x initially. Later, we find the appropriate x using Fig.1 after analyzing the data set collected for some period. So, the source host writes the value of the Pi marking scheme on the left 1-bit of the Identification field of packets which will be sent.

---

[3] According to the TTL modulation, the marking position is moved to the right, at the end, moved to the first position. The marking position is circulated.

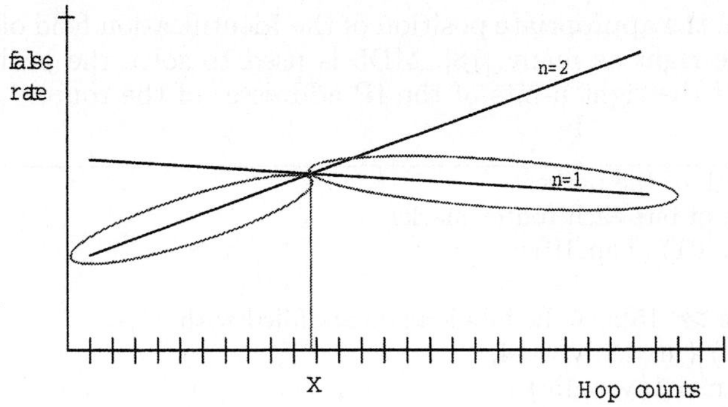

Fig. 1. The appropriate hop count deciding Pi marking scheme n=1 or 2

```
/* Algorithm to Initialize Pi and to decide the value of the Pi marking scheme */
Pts= Pi mark of the packets
InitializeOfPiMark(Pts, hop_cnt) /* x : the appropriate hop count */
{
    extern x;
    Pts = 0; /* initialize */
    if ((real)hop_cnt > x) Pts = (1 ≪ (16 - 1)); /* in the case of n=1 */
};
```

To find x, 10,000 packets are made from CAIDA's Skitter Map. Then, false ratio is found for every hop count for n=1, 2 respectively. We know that x is about 10.3.

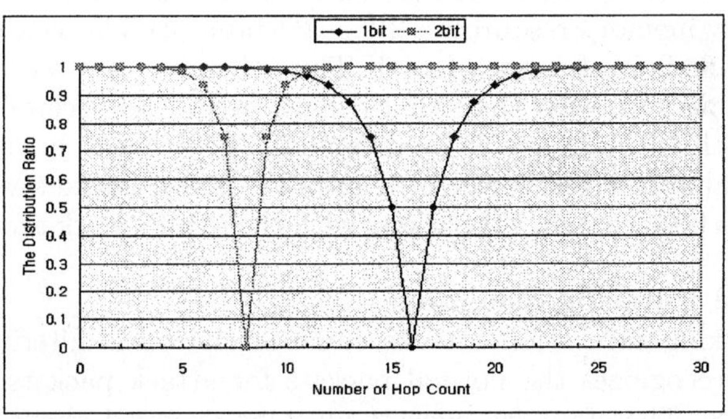

Fig. 2. False ratio of 1-bit and 2-bit for every hop count

Router's Pi Marking Algorithm

The routers en-route confirm the value of the Pi marking scheme and mark its marking value on the Identification field of incoming packets. The position to be marked is decided according to TTL modulo. The marking router hashes its IP address with MD5, accepts the right n-bits, and marks the value of the right n-bits on the appropriate position of the Identification field of the incoming packets to the right as many [18]. MD5 is used to solve the problem that the distribution of the right n-bits of the IP addresses of the routers can be highly skewed[18].

```
Pts= Pi mark of the packets
n = number of bits each router marks
Pimark(Pts, TTL, Cur_IP)
{
    z = (Pts ⋙ 15); /* the bits leftside are filled with '0' */
    if (z = 0) { n = 2; y = 14;
    } else { n = 1; y = 15;}
    m = 2n - 1;
    b = markingbits(Curr_IP) & m; /* markingbits(Curr_IP) = MD5(Curr_IP) to
    normalize the distribution */
    bitpos = (TTL mod [y/n]) n;
    b ≪ bitpos; m ≪ bitpos;
    return((Pts & ~m) | b);
}
```

Filtering Scheme

This section describes how the victim can make use of the Pi marks to filter incoming packets during DDoS attacks. Here, two methods are applied.

- **Basic Filtering Scheme**
  This scheme is to record the markings of identified attack packets and to drop subsequent incoming packets matching any of those markings. It has some characteristics as follows [1]. (1) This filter provides little flexibility to the victim. (2) This filter has very fast attack reaction time. (3) This filter requires few memory resources : $2^{15} + 2^{14}$ bits. (4) The victim has two vectors. The one is in the case of n=1. The i-th value of a bit-vector of length 2^{15} is 0 if packets with the i-th Pi mark are to be accepted, 1 if packets with the i-th Pi mark are to be dropped. The other is in the case of n=2. The i-th value of a bit vector of length 2^{14} is 0 if packets with the i-th Pi mark are to be accepted, 1 if packets with the i-th Pi mark are to be dropped.

- **Threshold Filtering Scheme**
  If the marking saturation attacks come in the basic filtering scheme, the victim misrecognizes the normal packets for attack packets and drops the normal ones. Therefore the packets must be dropped above the some level of attack packet ratio. This is only the threshold(T_i). The threshold is as follows: $T_i = a_i/(a_i + u_i)$ where a_i is the number of attack packets and u_i is the number of user packets for $0 < i < 2^{15}$ if n=1 and $0 < i < 2^{14}$ if n=2. The packet filtering on the dynamic Pi scheme environment can be deployed not only on the ISP's side of the last hop link, but also at end-host in the ISP[4].

[4] In the Kim et al.'s paper, it is suggested that it is appropriate to apply 2 bit marking scheme on the ISP's side because the distance(hop counts en-route) is short [16].

4 Simulation and Results

4.1 Experiment

The followings are experimented for n=1, 2, and the proposed scheme respectively. (1) We choose 5,000 paths at random from one of our Internet data sets of CAIDA's Skitter Map. (2) Each end-host at a path sends three packets to the victim in learning phase [1]. (3) The victim makes the attack markings list. (4) Each end-host at a path sends one packet to the victim in attack phase [1].

4.2 Simulation

For the marking scheme n=1, 2, and the dynamic Pi marking scheme, we represent user acceptance ratio and attacker acceptance ratio in graphs. We can see acceptance ratio gap for each scheme in graphs. The acceptance ratio gap should be found for the threshold 0, 0.25. We can choose the appropriate Pi marking scheme according to the threshold values. The properties and performance of the marking schemes will be compared each other.

4.3 Results

Fig.3 describes the user(normal) and attacker(attack) acceptance ratio for each scheme. We see that the ratio gap of the proposed scheme is bigger than the n=1, n=2. We know that the drop ratio is high for abnormal packets and low for normal packets (user packets). It means that the filtering performance of the proposed scheme is high. In Fig.4 the acceptance ratio gap is presented when the routers are legacy rate 0.25.

5 Conclusion

In A. Yaar et al's paper, the static marking scheme is proposed. It has demerits which can not reflect the variation of the distance from the source host to the victim. On the other hand, this paper shows that if the appropriate marking scheme is adopted dynamically according to the network properties and the distance distribution we can expect better results. In this paper, there are some problems to be solved. First, the router's traceroute service is prohibited occasionally against hacker's attacks. As a result, we cannot count the distance. The acceptance ratio will be studied in the network topology including the routers which don't provide traceroute service. Second, it is necessary to study the method using the left 2-bits for registering the Pi marking scheme. For example, "00","01","10","11" can be registered on the left 2-bits of the Identification field of packets. Assume that only "01" and "10" are used for Pi marking. Then it has a merit that "00" and "11" packets[5] are dropped absolutely but demerit that the marking space is

[5] The bits for registering the Pi marking scheme can be filled with "00" for legacy router and "11" for the routers which don't provide traceroute service. In this case, the value of the Pi marking scheme registering bits can not be used to decide whether the packets are normal or not any more.

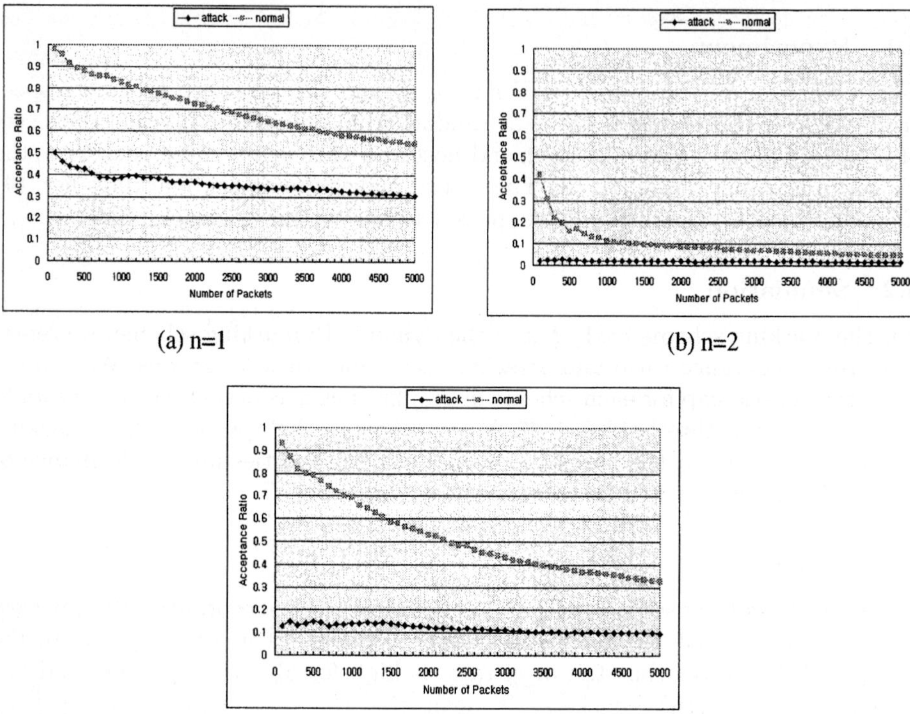

(a) n=1 (b) n=2

(c) proposed

Fig. 3. Pi Filtering with 0% threshold

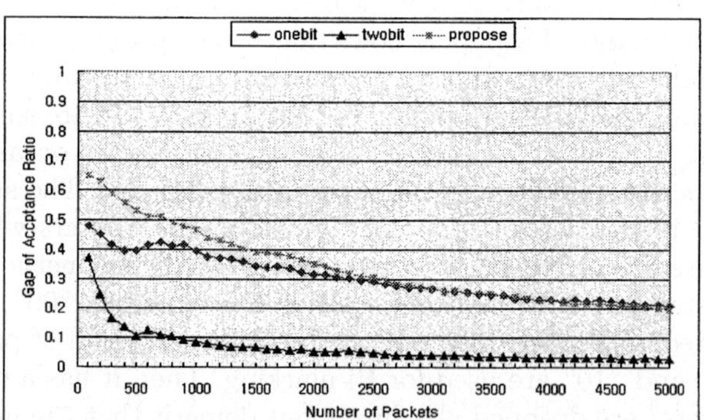

Fig. 4. the Pi filtering performances of Pi marking schemes for legacy ratio

small. Last, it is necessary to study deeply the relationship between the marking space and the performance. The fact that the marking space is small means that the possibility of the same valued marking packets is high. The drop rate against attack packets is smaller as long as the acceptance ratio gap is smaller.

References

1. Yaar, A., Perrig, A., Song, D.: Pi: A Path Identification Mechanism to Defend against DDoS Attacks. *Proceeding of Symposium on Security and Privacy 2003.* (2003) 93-107.
2. H. Burch and B. Cheswick. Internet watch: Mapping the Internet. Computer, 32(4):97-98, Apr. 1999.
3. Denial of Service Attacks, CERT (1997).
4. XiaoFeg Wang, Michael K. Reiter. Defending Against Denial-of-Service Attacks with Puzzle Auctions, *In Proceedings of the 2003 Security and Privacy Symposium*, May 2003.
5. Ryan naraine, Massive. DDoS Attack Hit DNS Root Servers, eSecurityPlanet.com (Oct 2002) http://www.esecurityplanet.com/trends/article.php/10751_1486981
6. Heeran Lim, Manpyo Hong, Effective Packet Marking Approach to Defend Against DDoS Attack, In *Proceeding of ICCSA 2004.* 2004.
7. Chen, Z., Lee, M.: An IP traceback technique against denial-of-service attacks. In *Proceeding of 19th Annual Computer Security Applications Conference*, 96-104, 2003.
8. J. Ioannidis and S. M. Bellovin. Implementing Pushback:Router-based defense against DDoS attacks, In *Proceeding of the Symposium on Network and Distributed Systems Security (NDSS 2002)*, Feb. 2002.
9. Caida, Inside the Slammer Worm, http://www.caida.org/outreach/papers/2003/sapphire2/
10. M. Adler. Tradeoffs in probabilistic packet marking for IP traceback. In *Proceeding of 34th ACM Symposium on Theory of Computing (STOC)*, 2002.
11. D. Dean, M. Franklin, and A. Stubblefield. An algebraic approach to IP traceback. *ACM Transactions on Information and System Security*, May 2002.
12. A. C. Snoeren, C. Partridge, L. A. Sanchez, C. E. Jones, F. Tchakountio, S. T. Kent, and W. T. Strayer. Single-packet IP traceback. *IEEE/ACM Transactions on Networking (ToN)*, 10(6), Dec. 2002.
13. A. C. Snoeren, C. Partridge, L. A. Sanchez, C. E. Jones, F. Tchakountio, S. T. Kent, and W. T. Strayer. Hash-based IP traceback. In *Proceeding of the ACM SIGCOMM 2001 Conference*, pages 3-14, Aug. 2001.
14. D. X. Song and A. Perrig. Advanced and authenticated marking schemes for IP traceback. In *Proceedings of IEEE INFOCOMM 2001*, April 2001.
15. R. Mahajan, S. M. Bellovin, S.Floyd, J. Ioannidis, V. Paxon, S. Shenker. Controlling high bandwidth aggregates in the network. *CCR*, 32(3):62-73, July 2002.
16. Soon-Dong Kim, Man-Pyo Hong, Dong-Kyoo Kim, A Study on Marking Bit Size for Path Identification Method: Developing the Pi Filter at the End Host, In *Proceeding of ICCSA 2004.* 2004.
17. Caida. Skitter. http://www.caida.org/tools/measurement/skitter/, 2004.
18. R. L. Rivest. The MD5 message digest algorithm. RFC 1321, Internet Activities Board, Internet Privacy Task Force, Apr. 1992.

A Secure Mobile Agent Protocol for AMR Systems in Home Network Environments

Seung-Hyun Seo, Tae-Nam Cho, and Sang-Ho Lee

Department of Computer Science and Engineering,
Ewha Womans University,
11-1 Daehyun-dong, Seodaemun-ku, Seoul 120-750, Korea
{happyday, tncho, shlee}@ewha.ac.kr

Abstract. Home network environments provide telemetering services through AMR (Automatic Meter Reading) systems. For households' convienience, the AMR system automatically inspects gas, water, and electricity meters at the remote site. Since this telemetering information is associated with billing, it must be secure. So, it is necessary to design a security protocol for AMR systems. In this paper, we adapt discrete logarithm based multi-signcryption to an elliptic curve based multi-signcryption protocol, for the purpose of efficiency. And, we propose a secure mobile agent protocol for AMR systems using the elliptic curve based multi-signcryption. Our protocol efficiently provides user authentication, integrity and confidentiality of telemetering information.

Keywords. mobile agent, multi-signcryption, home network, AMR system, security.

1 Introduction

Recently, interest of home networks has rapidly increased. A home network is the configuration of two or more home devices enabling the mutual transfer and sharing of communications and data. It can provide information service, entertainment service, control service, education service, and medical service to home users. Among these home network services, the control service includes functions of telemetering, home appliance control and remote control[8,9].

An AMR system is a telemetering system that automatically inspects various kinds of meters such as electricity meters, water meters and gas meters at a remote site through a specific communications medium, and without a need for human meter readers. So, using an AMR system, we can prevent mistakes in billing due to human error. With an AMR system in place criminals will also be unable to disguise themselves as meter readers[1,2].

Since the charges for gas, water and power are associated with usage, data collected through telemetering must not be forged or modified. So, the AMR system must provide user authentication and integrity of the telemetering data. Moreover, the AMR system must provide confidentiality of the telemetering data

C. Kim (Ed.): ICOIN 2005, LNCS 3391, pp. 814–823, 2005.

to prevent criminals, such as burglar, from monitoring usage as a way of predicting when householders will be absent. However, most existing AMR systems don't provide security services such as confidentiality, integrity and user authentication. Even if a few AMR systems provide confidentiality, they don't guarantee user authentication and integrity[1,2]. Therefore, it is necessary to develop a security protocol that provides user authentication, integrity and confidentiality of the telemetering data in home network environments. In this paper, we propose a secure mobile agent protocol using an elliptic curve based multi-signcryption for AMR systems in home network environments.

Multi-signcryption protocol is an extension of signcryption protocol[10] for multi-users. Since it fulfills both the functions of encryption and digital multi-signature for multi-users, it efficiently provides user authentication, message integrity, and confidentiality[4,6,7]. So, we choose a multi-signcryption protocol to provide security services for AMR systems. But, in a low resource environment, due to the low computational cost and storage cost of EC (Elliptic Curve) based protocols, the natural choice for cryptographic protocols would be an EC implementation[3]. So, for efficiency, we have adapted a DL (Discrete Logarithms) based multi-signcryption protocol to the EC based multi-signcryption protocol. We call it EC Multi-Signcryption protocol, and we have used it to design a secure mobile agent protocol. A mobile agent can migrate from host to host and act autonomously, so, it can collect a household's telemetering data without continuous network connection between the AMR server and home gateways[9]. We expect that the mobile agent will be useful in improving efficiency for home network environments. Moreover, using EC Multi-Signcryption, our mobile agent protocol provides user authentication, integrity and confidentiality of the telemetering data.

The rest of this paper is organized as follows. In section 2, we describe background concepts and related works. In section 3, we present a basic solution for secure AMR systems. In section 4, we propose a secure mobile agent protocol using EC Multi-Signcryption for AMR systems in home network environments. In section 5, we discuss the security of our mobile agent protocol. And then we analyze performance of the basic solution and our mobile agent protocol. Finally, we draw our conclusions.

2 Background Concepts and Related Works

In this section, we define the notations for this paper and describe the telemetering services in home network environments. And then, we briefly explain the concepts of multi-signcryption protocols and mobile agents.

2.1 Notations

- $Home_i$: the i-th home gateway which belongs to the i-th home
- $Cent$: the management center of an apartment complex
- SP : the service provider that provides gas, electricity, and water

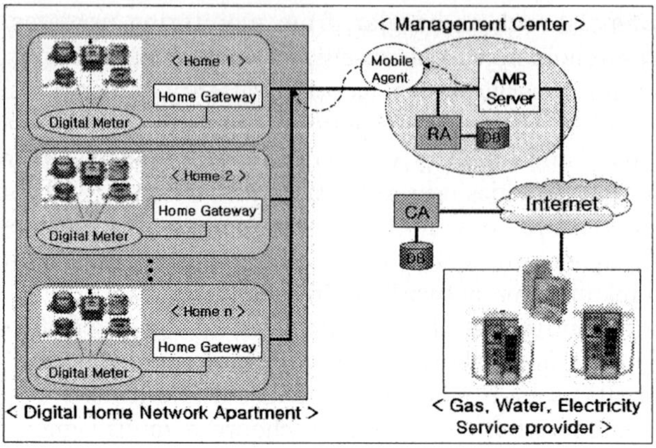

Fig. 1. Home network environments for telemetering service

- $E_{a,b}$: an elliptic curve over a finite field $GF(p^m)$, either with $p \geq 2^{150}$, $m = 1$ or $p = 2$, $m \geq 150$ ($E_{a,b}$: $y^2 = x^3 + ax + b (p > 3)$, $E_{a,b}$: $y^2 + xy = x^3 + ax^2 + b (p = 2)$, $4a^3 + 27b^2 \neq 0 \pmod{p}$)
- q : a large prime number whose size is approximately of $|p^m|$
- G : a point with order q which is chosen randomly from the points on $E_{a,b}$
- $ENC_K(\cdot), DEC_K(\cdot)$: the encryption and decryption algorithms of a private key cipher system with the key K
- $H(\cdot), hash(\cdot)$: a one-way hash function
- x_i : the secret key of the i-th householder who uses the $Home_i$, $x_i \in_R [1, ..., q-1]$
- Y_i : the public key of the i-th householder who uses the $Home_i$, $Y_i = x_i G$

2.2 Telemetering Service in Home Network Environments

We propose a mobile agent protocol that provides secure telemetering services for the households of a cyber apartment complex. The cyber apartment is a relatively new concept in housing, characterized by built-in broadband services and home network services. In the cyber apartment, many homes and offices of the apartment complex are connected with each other through a LAN. A cyber apartment is connected with other home networks through the Internet, and households can access Internet services through their own home gateways[9]. The home gateway controls the digital meters for gas, electricity, water and so on, and it manages this usage data. The AMR server of a management center collects and manages the telemetering data of all households in the complex. We assume that an agent platform such as Java is embedded in the AMR server and the home gateways. After the SP transmits the total usage and rates to the $Cent$, the $Cent$ collects each household's telemetering data by using a mobile agent, or MA. The $Cent$ allocates the gas, power, water charges according to each household's usage. And then, the $Cent$ transmits the information on the each household's usage and charges to the SP. The home network environments, where the secure telemetering services are provided, are shown in Figure 1. We assume that a secure channel such as SSL is established between the $Cent$ and the SP, and they can communicate securely. Besides, the $Cent$, the households, and the SP register their public keys at the CA (Certification Authority), and receive their certificates from the CA.

2.3 A Multi-signcryption Protocol

The multi-signcryption protocol is a cryptographic method that fulfills both the functions of secure encryption and digital multi-signature for multi-users, at a cost smaller than that required by multi-signature-then-encryption[4,6,7].

Recently, Mitomi and Miyaji first proposed a multi-signcryption protocol which combined a multi-signature with the encryption function[4]. However, since their protocol can not provide message confidentiality, it cannot prevent a malicious attacker from obtaining the information in the messages. Pang, Catania and Tan proposed a modified multi-signcryption protocol to achieve message confidentiality[6]. However, since their protocol fixes the order of multi-signers beforehand, it does not satisfy the need for order flexibility. Moreover, it cannot provide non-repudiation. Seo and Lee analyzed the weaknesses of these previous multi-signcryption protocols and proposed a new multi-signcryption protocol[7]. We call it the Seo-Lee protocol. Their protocol provides not only message confidentiality, non-repudiation and order flexibility but also other requirements for secure and flexible multi-signcryption. Moreover, It is more efficient than any other protocols. Therefore, in this paper, we adapt a DL based Seo-Lee protocol to the EC based multi-signcryption protocol, and use it to design our secure mobile agent protocol.

2.4 A Mobile Agent

A mobile agent is a program or an object that consists of code, data, and its current execution state. It can migrate autonomously from host to host during its execution, and perform computations on behalf of the user. So, using a mobile agent, we can reduce network traffic, overcome network latency, and allow for increased asynchrony between clients and servers[9]. In this paper, we use a mobile agent to collect the telemetering data efficiently. By using the mobile agent, which can migrate autonomously among home gateways and the AMR server, it is unnecessary to continuously maintain a network connection between the AMR server and the home gateways. So, remote interactions and network traffic can be reduced.

3 A Basic Solution

In this section, we present a basic solution for secure telemetering services by applying an EC based signature protocol to existing AMR systems. Since existing

AMR systems provide only confidentiality of data, we append the EC-DSS (Elliptic Curve based Digital Standard Signature) scheme[5] to the existing AMR system for user authentication and integrity of data.

We assume that the existing AMR system already establishes a common secret key K_i between $Home_i$ and the AMR server of the $Cent$, and provides confidentiality through a private key cipher algorithm with K_i. Our basic solution is as follows.

[EC-DSS Generation and Encryption phase]

1. $Home_i$ generates a signature on the telemetering data M_i as follows:
   (a) $Home_i$ chooses random $k_i \in_R [1, ..., q-1]$, and computes $r_i = k_i G \pmod q$
   (b) $Home_i$ computes $s_i = (H(M_i) + r_i x_i) \cdot k_i^{-1} \pmod q$
2. $Home_i$ encrypts M_i with K_i, i.e, it generates $C_i = ENC_{K_i}(M_i)$.
3. $Home_i$ sends (r_i, s_i, C_i, ID_i) to the $Cent$.

[EC-DSS Verification and Decryption phase]

1. After the $Cent$ receives $(r_1, s_1, C_1, ID_1), (r_2, s_2, C_2, ID_2), ..., (r_n, s_n, C_n, ID_n)$ from home gateways, it decrypts the C_i and obtains the telemetering data M_i of $Home_i$.
2. $Cent$ verifies the signature (r_i, s_i) of $Home_i$ as follows:
   (a) $Cent$ computes $r_i^{'} = (H(M_i)G + r_i Y_i) \cdot s_i^{-1} \pmod q$.
   (b) $Cent$ checks $r_i = r_i^{'}$.

4 A Secure Mobile Agent Protocol Using EC Multi-signcryption

In this section, we modify DL based multi-signcryption into EC based multi-signcryption, and we propose a secure mobile agent protocol for telemetering services in home network environments. Our protocol consists of four procedures such as registration procedure, mobile agent creation procedure, mobile agent execution procedure, and mobile agent arrival procedure. It provides confidentiality and integrity for the telemetering data, and user authentication using EC Multi-Signcryption. An overview of the proposed protocol is shown in Figure 2.

4.1 Registration Procedure

In this procedure, each householder $U_i (1 \leq i \leq n)$ registers his own public key and address at the management center, $Cent$.

1. U_i gives his public key certificate and address information to the $Cent$.
2. After the $Cent$ checks U_i's identity and address, it stores U_i's identity ID_i, public key Y_i, address, and $Home_i$ information in the database of the RA (Registration Authority).

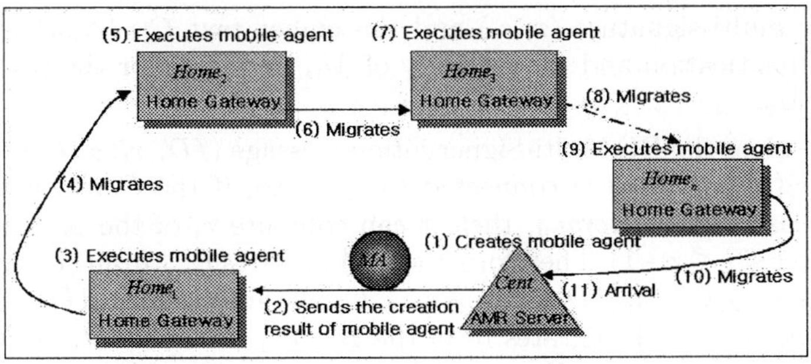

Fig. 2. Overview of the proposed protocol

4.2 Mobile Agent Creation Procedure

In this procedure, the *Cent* calls a mobile agent MA and determines the migration path of MA, $MA_{route} = Home_1||Home_2||...||Home_n$. Then it creates a telemetering request message *req*, and generates a signature on *req* as follows:

1. *Cent* chooses $k_C \in_R [1, ..., q-1]$ and computes $R_C = k_C G$.
2. *Cent* computes $r_C = H(req||ID_C||R_C) \pmod q$ and $s_C = (x_C + r_C) \cdot k_C^{-1} \pmod q$.

Cent gives *req*, MA_{route}, and signature, (ID_C, r_C, s_C) to the MA, and the MA migrates to the first household's home gateway, $Home_1$ with them.

4.3 Mobile Agent Execution Procedure

1. After the MA has migrated to $Home_i (1 \le i \le n)$, $Home_i$ checks the *req* and MA_{route}.
2. $Home_i$ verifies the *Cent*'s signature and generates the EC Multi-Signcryption on its telemetering data, M_i as follows:

   [**Verification phase of the *Cent*'s signature**]
   (a) $Home_i$ computes $R'_C = {s_C}^{-1} \cdot (Y_C + r_C G) = {s_C}^{-1} \cdot (x_C + r_C)G = k_C G$.

   (b) $Home_i$ checks whether $H(req||ID_C||R'_C) \pmod q = r_C$, or not. If the equation holds, then it performs the following EC Multi-Signcryption phase. Otherwise, it reports the failure to the *Cent*.

   [**EC Multi-Signcryption phase**]
   (a) $Home_i$ chooses $k_i \in_R [1, ..., q-1]$, and computes a session key $K_i = hash(k_i \cdot Y_C) = hash(k_i \cdot x_C G)$ by using the *Cent*'s public key and k_i.
   (b) $Home_i$ computes the signature $r_i = H(M_i||ID_i||K_i) + r_{i-1} \pmod q$ and $s_i = (x_i + r_i) \cdot {k_i}^{-1} \pmod q$ by using received $r_{i-1} (1 \le i \le n, r_0 = r_C)$ from MA. And, it generates $C_i = ENC_{K_i}(ID_i||M_i)$ by encrypting (ID_i, M_i) with K_i. The EC Multi-Signcryption message is composed of

the multi-signature (r_i,s_i) and the cipher text C_i. (r_i,s_i) are for user authentication and the integrity of M_i, and C_i is for the confidentiality of M_i.

3. $Home_i$ gives the EC Multi-Signcryption message (ID_i, r_i, s_i, C_i) to the MA. Here, $r_i (1 \leq i \leq n)$ is connected to r_{i-1}. So, if the $Cent$ knows only r_n of the last signer, $Home_n$, then it can compute r_i of the previous signers, $Home_i (1 \leq i \leq n-1)$. Therefore, the MA removes r_{i-1} from $(ID_1, s_1, C_1), ...,$ $(ID_{i-2}, s_{i-2}, C_{i-2}), (ID_{i-1}, r_{i-1}, s_{i-1}, C_{i-1})$, and it stores (ID_i, r_i, s_i, C_i).
4. If $i = n$, then MA migrates from the $Home_n$ to the $Cent$. Otherwise, the MA migrates from the $Home_i$ to $Home_{i+1}$.

4.4 Mobile Agent Arrival Procedure

After the MA finishes the travels of the migration path MA_{route}, it arrives at the $Cent$.

1. MA gives $(ID_1, s_1, C_1), ..., (ID_{n-1}, s_{n-1}, C_{n-1})$, and (ID_n, r_n, s_n, C_n) to the $Cent$.
2. $Cent$ performs the following EC Multi-UnSigncryption to verify and decrypt the EC Multi-Signcryption message.

   [**EC Multi-UnSigncryption phase**]

   (a) For $i = n, ..., 3, 2, 1$, $Cent$ computes the session key K_i' using its private key x_C, $Home_i$'s public key Y_i, and (r_i, s_i).
      i. $Cent$ computes $u_i = x_C \cdot {s_i}^{-1} \pmod q$ and $K_i' = hash(u_i \cdot r_i G + u_i Y_i) = hash((r_i + x_i) \cdot u_i G) = hash(x_C k_i G)$.
      If $K_i' = K_i$, then the $Cent$ can decrypt C_i. And it can obtain the telemetering data M_i and ID_i of the $Home_i$.
      ii. $Cent$ computes $r_{i-1} = r_i - H(M_i || ID_i || K_i') \pmod q$. If the signature, r_{i-1}, is recovered then the $Cent$ lets $i = i - 1$ and performs steps i and ii again.

   (b) If the verification is finished correctly then the $Cent$ can confirm its own signature, $r_C (= r_0)$.
3. If the EC Multi-UnSigncryption phase is performed successfully and all telemetering data $M_1, ..., M_n$ of $Home_1, .., Home_n$ are decrypted, then the $Cent$ stores $M_1, ..., M_n$.
4. $Cent$ terminates the MA's execution.

5 Analysis of the Proposed Protocol

In this section, we analyze the security of our mobile agent protocol according to the security requirements of message confidentiality, message integrity, user authentication, non-repudiation, and robustness. Then we analyze the efficiency of our protocol in comparison with the basic solution.

5.1 Security Analysis

1. **Message confidentiality:** Message confidentiality means that it is computationally infeasible for a malicious attacker to gain any partial information on the content of the EC Multi-Signcryption message. In our protocol, if an attacker intercepts the mobile agent, MA, and searches the data in MA, then he can obtain the EC Multi-Signcryption messages $(ID_1, s_1, C_1), (ID_2, s_2, C_2), ..., (ID_n, s_n, r_n, C_n)$ of the telemetering data $M_1, M_2, ..., M_n$. And the attacker can compute $s_i{}^{-1} \cdot (r_i \cdot G + Y_i) = k_i G (1 \le i \le n)$ from the EC Multi-Signcryption messages. But, since the attacker cannot know $Cent$'s private key, x_C, he cannot compute session keys due to the difficulty of the elliptic curve discrete logarithm problem[5]. Therefore, it is computationally infeasible for the attacker to gain any information of the telemetering data, $M_1, M_2, ..., M_n$. Our protocol provides confidentiality for the telemetering data.
2. **Message Integrity:** Message integrity means that the communicated EC Multi-Signcryption messages cannot be manipulated by unauthorized attackers without being detected. Assume that a malicious attacker modifies $Home_i$'s telemetering data and tries to forge $Home_i$'s $(1 \le i \le n)$ EC Multi-Signcryption message, (ID_i, r_i, s_i, C_i). The attacker can create the forged telemetering data M_i' by modifying M_i of $Home_i$. And then, he chooses $k_i' \in_R [1, ..., q-1]$ and can compute the session key $K_i' = hash(k_i' \cdot Y_C) = hash(k_i' \cdot x_C G)$ by using the $Cent$'s public key and k_i'. Moreover, the attacker can use the r_{i-1} by eavesdropping on the MA, and he can generate signature $r_i' = H(M_i' || ID_i || K_i') + r_{i-1} \pmod q$. But, since the attacker cannot know the U_i's private key x_i, he cannot compute $s_i' = (x_i + r_i') \cdot k_i'^{-1} \pmod q$. Even if he chooses a random x_i' and computes $s_i'' = (x_i' + r_i') \cdot k_i'^{-1} \pmod q$, the $Cent$ can verify that s_i'' is forged signature in the EC Multi-UnSigncryption phase. Therefore, the attacker cannot modify the telemetering data and cannot forge the EC Multi-Signcryption message. So, our protocol provides integrity for the telemetering data.
3. **User authentication:** User authentication means the process whereby one party is assured of the identity of the second party involved in a protocol, and of whether the second party has actually participated. In our protocol, the $Cent$ can confirm the identity of householder, U_i, through the ID_i included in the EC Multi-Signcryption message. In the EC Multi-UnSigncryption phase, the $Cent$ can assure that U_i actually participated. So, our protocol provides user authentication.
4. **Non-repudiation:** Non-repudiation means that neither householders nor the $Cent$ can falsely deny later the fact that he generated a EC Multi-Signcrytion message. In our protocol, non-repudiation is provided as follows. Since each EC Multi-Signcryption message includes the householder U_i's $(1 \le i \le n)$ private key, x_i, anyone who does not know x_i cannot generate an EC Multi-Signcryption message instead of U_i. Therefore, if $Home_i$ of U_i generates the EC Multi-Signcryption, he cannot falsely deny later the fact that he generated it.

5. **Robustness:** Robustness means that if the signature verification on a message fails, then it prevents such unauthentic messages from damaging a receiver. In our protocol, After the *Cent* receives the EC Multi-Signcryption message from the *MA*, if the verification of $K_i' = hash(x_C \cdot {s_i}^{-1} \cdot r_i G + x_C \cdot {s_i}^{-1} \cdot Y_i) = hash(x_C k_i G)$ fails, then the *Cent* cannot compute the session key, K_i. So, since it cannot decrypt the cipher text C_i, it can prevent damage by an unauthentic message or malicious code in the *MA*. Therefore, our protocol provides robustness.

5.2 Efficiency Analysis

We evaluate our protocol from a point of view of computational cost and communication overhead, and compare our protocol with the basic solution. We use the number of point multiple and modular multiplication to measure the computational cost, and the communicated message size to measure the communication overhead.

For convenience, we assume the following conditions: (1) we denote the number of home gateways by n and the message size by $|M|$ bits; (2) the size of q is set to 160 bits; (3) the output size of the cryptographic hash functions is 160 bits.

In the basic solution, since all $Home_i$s transmit EC Multi-Signcryption messages $(ID_i, r_i, s_i, C_i)(1 \leq i \leq n)$ to the AMR server of the *Cent* at the some time, a network bottleneck can be happened. The total communication overhead of the basic solution is $n \cdot |M| + n \cdot |q| + n \cdot |H(.)| = n \cdot (|M| + 320)$. But, in our protocol, the total EC Multi-Signcryption messages from $Home_1$ to $Home_n$ are $(ID_1, s_1, C_1), ..., (ID_{n-1}, s_{n-1}, C_{n-1}), (ID_n, r_n, s_n, C_n)$, and the communication overhead is $n \cdot |M| + (n+1) \cdot |q| = n \cdot (|M| + 160) + 160$. So, when compared with the basic solution, our protocol reduces the communication overhead to, at most, 50%. The amount of EC Multi-Signcryption messages to be stored in the AMR server can also be reduced to, at most, 50%. Moreover, since the *MA* migrates autonomously and transfers EC Multi-Signcryption messages either between $Home_i$ and $Home_{i+1}$ or between $Home_i$ and the AMR server, the total remote interaction and network traffic can be reduced between them.

In the computational cost of our protocol and the basic solution, the point multiple is 1 for $Home_i(1 \leq i \leq n)$ and $2n$ for the AMR server. In the case of 160-bit modular multiplication, our protocol is 1 for $Home_i(1 \leq i \leq n)$ and $2n$ for the AMR server, but the basic solution is 2 for $Home_i(1 \leq i \leq n)$ and n for the AMR server.

We have, so far, assumed that the same secret key K_i established previously between the $Home_i(1 \leq i \leq n)$ and the AMR server in the basic solution, and evaluated the efficiency of the basic solution without computational and communication costs for key establishment. However, key establishment is complex, it results in heavy computational cost and communication overhead. If the secret key is fixed in the basic solution, "key freshness" cannot be provided. If the basic solution simply refreshes the secret key periodically, then it can provide "key freshness." But it has another security problem, i.e. it cannot provide "forward secrecy" or "backward secrecy", and it is not secure against "known-key attack"[5]. Therefore, if we add a key establishment phase to the basic solution for overcoming these security problems, then the computational cost and communication overhead of the basic solution increase, and the efficiency decreases.

Unlike the basic solution, our protocol does not need a key establishment phase. So, our protocol is more efficient than the basic solution.

6 Conclusions

AMR systems efficiently provide telemetering services in home network environments. Since telemetering data such as gas, water, and electricity usage should not be forged or modified, a security protocol that guarantees confidentiality, integrity, and user authentication is necessary.

In this paper, we proposed a new secure mobile agent protocol for AMR systems in home network environments. To provide efficient security services, we adapted DL based multi-signcryption to EC based multisigncryption. And then we used this scheme to propose a secure mobile agent protocol. Our protocol can efficiently provide user authentication, non-repudiation, robustness, integrity and confidentiality of telemetering data. We expect that our protocol will be adopted to provide security services for AMR systems.

References

1. AMR system, http://www.nuritelecom.com/.
2. AMR system, http://www.meters.co.kr/english/index-01.htm/.
3. C. Boyd, P. Montague, and K. Nguyen, "Elliptic Curve Based Password Authenticated Key Exchange Protocols," *In Proceedings of ACISP 2001*, *Lecture Notes in Computer Science*, pages 487-501, Springer-Verlag, 2001.
4. S. Mitomi and A. Miyaji, "A General Model of Multisignature Schemes with Message Flexibility, Order Flexibility, and Order Verifiability," *IEICE Transaction on Fundamentals*,Vol. E84-A, No. 10, pages 2488-2499, 2001.
5. A. J. Menezes, P. C. Oorschot and S. A. Vanstone, *Handbook of Applied Cryptography*, CRC, 1997.
6. X.Pang, B.Catania, and K-L, Tan, "Securing Your Data in Agent-Based P2P Systems," *In Proceedings of Eight International Conference on Database Systems for Advanced Applications(DASFAA '03)*, 2003.
7. S. Seo and S. Lee, "Secure and Flexible Multi-signcryption Scheme," *In Proceedings of ICCSA 2004 ,Lecture Notes in Computer Science 3046*, pages 689-697, Springer-Verlag, 2004.
8. S. Tak, S. Dixit, and E. K. Park, "An End-to-End Home Network Security Framework," *Computer Communications*, Vol. 27, pages 412-422, Elsevier, 2004.
9. J. Yoo and D. Lee, "Scalable Home Network Interation Model Based on Mobile Agents," *In Proceedings of PerCom 2003*, 2003.
10. Y. Zheng, "Digital Signcryption or How to Achieve Cost (Signature & Encryption) $\ll$ Cost (Signature) + Cost (Encryption)," *Advances in Cryptology - Crypto'97*, *Lecture Notes in Computer Science*, Vol. 1294, pages 165-179, Springer-Verlag, 1997.

MDS: Multiplexed Digital Signature for Real-Time Streaming over Multi-sessions

Namhi Kang and Christoph Ruland

University of Siegen, Institute for Data Communications Systems
Hoelderlin-str. 3, 57076 Siegen, Germany
{kang, ruland}@nue.et-inf.uni-siegen.de

Abstract. We propose an efficient scheme called *MDS (Multiplexed Digital Signature)* to digitally sign on real-time stream of which application especially requires multiple sessions. A typical scenario is that a source multicast multimedia contents over the Internet using several RTP/RTCP sessions. With a system using a previously proposed stream authentication scheme directly, both the computation and the transmission overhead are linearly increased in proportional to the number of sessions to be opened. This is mainly because existing schemes have only taken a single session into account. MDS is well suited for supporting data origin authentication efficiently in such a scenario.

1 Introduction

Many Internet applications have already taken advantage of the multimedia streaming technology. It is out of doubt that real-time streaming will be more and more employed by virtue of the capability to timely transmit multimedia contents. Examples of applications include multimedia contents distribution service, on-line education, news feeds, and others. In such applications, multicast enables the source to transmit multimedia streams to a group of receivers efficiently. However, multicast is more complicated to deploy than unicast and there exist several problems that still remain to be solved [1]. Among of those challenging problems, we focus on security issues, especially, on *'Data Origin Authentication'* (DOA) problem and its effects on the performance of an application which needs multiple sessions to deliver multimedia contents.

In multicast, stream authentication is one of the most difficult concerns. Message authentication code (MAC) is a typical primitive to support DOA in unicast [2]. However, it is difficult to apply MAC directly to multicast since any member of group who shares the key can impersonate the source. Applying digital signatures instead of MAC can solve this problem since only the source is able to bind its identity to the signature. The trade-off is nevertheless that there exist critical performance problems when an asymmetric cryptography primitive is employed to real-time stream. To overcome this problem, several schemes have been proposed over the years (see section 2.2).

However, there are still several limitations when such schemes are used for the case where the source streams multimedia contents over *'multiple sessions'* (see

C. Kim (Ed.): ICOIN 2005, LNCS 3391, pp. 824–834, 2005.

section 3.2). All of existing schemes do not consider multiple sessions but treat a single session (it is referred to as the *'session based treatment'* in this paper). Most multimedia applications are necessary to establish multiple sessions to transmit several different types of media such as video and audio simultaneously. This is due to the fact that each media content has its own characteristics and requirements, for example, different transmission rate, packet size, and coding algorithm [3].

In this paper, consequently, we propose a new approach to overcome limitations of existing schemes and address a couple of gains of the proposed scheme in such a scenario. In section 2, we discuss multimedia stream and existing stream authentication solutions. Our approach is described in section 3 including several limitations of session-based treatment in existing schemes. In section 4, we analyze the proposed scheme from the viewpoint of performance and compare with other schemes. Finally, we conclude this paper in section 5.

2 Preliminaries

2.1 Multimedia Stream

The term *'multimedia'* may be defined in a various aspects. In this paper, *'real-time multimedia'* is referred to as the integrated set of time-sensitive media contents such as video and audio. We suppose that media contents are transmitted over separate session allocated for each one (this is usual scenario in Internet world [3]). In order to define multimedia stream formally, we assume that there exist k different media contents consisting of a series of packets, P^s_m, of which allocated session is denoted $\mathcal{S}^s$, where s indicates a session index and m (also l and n below) denotes a sequence number of the packet.

Definition 1. *Let s^{th} session be $\mathcal{S}^s = \{P^s_m, P^s_{m+1}, P^s_{m+2}, \cdots\}$, where $0 < s \leq k$ and $s, k, l, m, n \in \mathbb{N}$. Multimedia stream $\mathcal{M}$ is defined as*

$$\mathcal{M} = \{S^1, \cdots, S^s, \cdots, S^k\} = \{\{P^1_l, \cdots\}, \cdots, \{P^s_m, P^s_{m+1}, \cdots\}, \cdots, \{P^k_n, \cdots\}\}.$$

2.2 Previous Work

In Multicast, DOA is intended for all receivers to ensure that the received data is coming from the claimed source rather than from any member of the group that is called group authentication (GA) [5]. Existing streaming authentication solutions that we consider in this paper are divided into two categories: MAC based approaches [4,5] and digital signature based approaches [4,6,7,8,9,10,11].

TESLA [4] and the Multiple-MACs scheme [5] use MAC as an underlying primitive. In TESLA, the source attaches a MAC computed using a key known only to itself to each packet. The key is disclosed after every receiver gets the MAC corresponding to the key. In the Multiple-MACs scheme, the source computes l MACs using l different keys that are a whole set of keys and then attaches those MACs to all packets. Each receiver verifies a part of the arriving MACs using a subset of l keys that the source holds (not all MACs).

Most schemes based on digital signature employ a means of amortizing a signature over a set of packets (called a block). There are two conspicuous differences between them: one is the method of setting a block of packets for amortizing, and the other is the way to achieve the loss tolerant property.

Wong and Lam proposed the star and tree scheme (referred to as the WL's scheme in this paper) [6] based on Merkle's hash tree technique [12]. The basic idea of the WL's scheme is to sign a block hash which is the root of the hash tree representing all packet's hash in a block so that a signature (called a block signature) amortizes all packets in a block.

Hash chaining technique is an alternative way to apply amortizing signature over a set of packets. Hash chaining was introduced in [7], where only the first packet including hash of the next packet is digitally signed. Thereafter, the source continuously sends each packet which contains the hash value of its next packet. This scheme is efficient but it is not loss tolerant and the source must know the entire stream in advance. In order to overcome these shortcomings, a couple of solutions (e.g. [4] and [8]) have been proposed recently.

SAIDA [9], PM scheme [10], and PRABS [11] are streaming authentication schemes using erasure code to achieve space efficiency. An erasure code is used to divide authentication information such as hash values and a signature into several small chunks. The source attaches a chunk to a packet. If pre-specified number of chunks are arrived at the receiver side, the authentication information can be re-constructed. The drawback of these schemes is higher computational cost caused by applying an erasure code than others (see Fig. 3).

3 Our Approach

3.1 Overview of MDS

We propose MDS (Multiplexed Digital Signature) scheme, where the source multiplexes all sessions into a *temporary* session to amortize a single signature over a set of packets (a block) that are dispersed throughout several sessions (refer to Fig. 2). As a result, a time expensive signature is generated once a block regardless of the number of sessions. We do not mean nevertheless that all sessions are merged into a single session to transmit. Only authentication procedures are performed in a single module. In other words, a media packet is transmitted over its originally allocated session after authentication procedures. In case of using RTP, a session manager and an authentication module are inserted between RTP and the transport layer (normally UDP) so that a single authentication module covers all of RTP sessions in cooperation with a session manager.

In particular, the most important concern of MDS is that a packet should be verifiable individually regardless of packet losses. This is due to the fact that a receiver can select a subset of sessions (not all). In MDS, all packets of a session(s) that the receiver does not select are regarded as the packet loss. To support individual packet authentication, we employ WL's Tree scheme [6]. not select are regarded as the packet loss. Therefore, the most important concern of MDS is that a packet should be verifiable individually regardless of packet losses. To support individual packet authentication, we employ WL's Tree scheme [6].

3.2 Limitations of Session Based Treatment

There are several limitations when the source applies an existing schemes to multiple sessions for supporting DOA. Firstly, in case of MAC based approaches (i.e. TESLA and Multiple-MACs), there is no means to multiplex several sessions since they handle each packet individually. Therefore, the computation and the transmission overhead are increased as much as the number of sessions to be opened. In TESLA, moreover, all receivers must make synchronization with a source and the source must generate and send signed information about the session whenever a receiver adds a session(s) according to its network condition dynamically. It can be vulnerable to denial of service (DoS) attacks against the source because an attacker easily sends lots of malicious requesting for session establishment. In addition, TESLA uses multiple key chains to cope with different delay in heterogeneous multicast: several MACs are calculated over a packet with different keys and each key is disclosed with different time lag [4]. Therefore, the increase of the cost depends on the number of delay sets.

Secondly, in schemes using an amortizing signature over a block (i.e. WL scheme, PM scheme, SAIDA), the source can reduce the computational cost. However, such a way brings about delayed and bursty transmission as illustrated in Fig. 1, where the results are from our simulation. To simulate, we implemented prototypes of WL's scheme in the NSv2 simulator [13] and employed Crypto++ library [14]. Fig. 1, (a) shows the effect of block size (namely, the number of packets in a block to be amortized) on the bursty sending. More block size leads to higher burstiness. The total delay time at the source is illustrated in (b) of Fig. 1. These drawbacks become higher in a setting of multiple sessions.

Finally, the common drawback of session based treatment in multiple sessions is highlighted at the control session (i.e. RTCP session). In a setting of RTP, the source sends RTCP SR (Sender Report) to offer a means to synchronize between different contents (e.g. lip-synchronization between audio and video). RTCP packets are sent over different session from the data session periodically (no more than 5% of the data channel capacity [3]). Therefore, the verification delay becomes higher than those of the data session if the source applies the same block size of the data session to the control session.

3.3 Notations and Definition of MDS

Suppose that a *source* opens k different sessions and receivers know each session information containing security association such as cryptographic algorithm and its corresponding key and the size of a block denoted BS. We assume that the key was certified by proper means. We use the following notations and a definition.

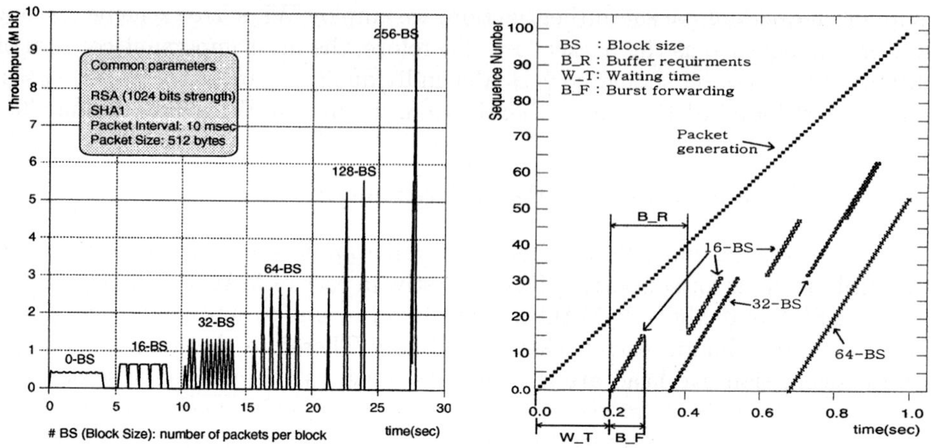

(a) Burstiness vs. block size

(b) Outbound waiting delay

Fig. 1. Limitations of session based WL's Tree scheme

- $< sk, pk >$: certified key pair (the private and the public key respectively).
- $GR_g = \{U_u | (0 \leq u < v)\}$: group containing v users, where U_u and g denotes a group user and a group index respectively.
- P_m^s: m^{th} packet in s^{th} session, where $0 \leq s < k$ and $m \geq 0$.
- b, p: block index (BI) and packet position index (PI) in a block.
- MP_p^s: p^{th} multiplexed packet, which belongs to s^{th} session originally.
- σ_b: block signature for the bth block.
- rH^b : root of the authentication hash tree of the bth block.
- $\{sH_p^b\}$: sibling hash values on the path from p^{th} packet to the root of hash tree of the bth block.

Definition 2. *MDS consists of five operations: MDS.GenKey, MDS.SeMux, MDS.Sig, MDS.SeDemux, and MDS.Ver operation, where*

- *MDS.GenKey*, the key pair generation operation, outputs sk and pk to be used for signature generation and signature verification respectively;
- *MDS.SeMux*, the session multiplexing operation, takes P_m^s as an input based on the packet generation time regardless of the session, and then attaches b and p to P_m^s in order to form MP_p^s;
- *MDS.Sig*, the signing operation, takes a data to be signed and sk of the signer as input, then outputs the signature;
- *MDS.SeDemux*, the session demultiplexing operation, takes MP_p^s as input, then outputs P_m^s by use of b and p contained within MP_p^s;
- *MDS.Ver*, the verification operation, takes a candidate signature, a data to be verified, and pk as input, then returns one of r=\{*true, false*\}.

3.4 Outbound Processing

Fig. 2, (a) illustrates an example where a source distributes three RTP sessions. A session manager handles three RTP sessions so that all packets in a block are able to access common authentication information of the block regardless of their session.

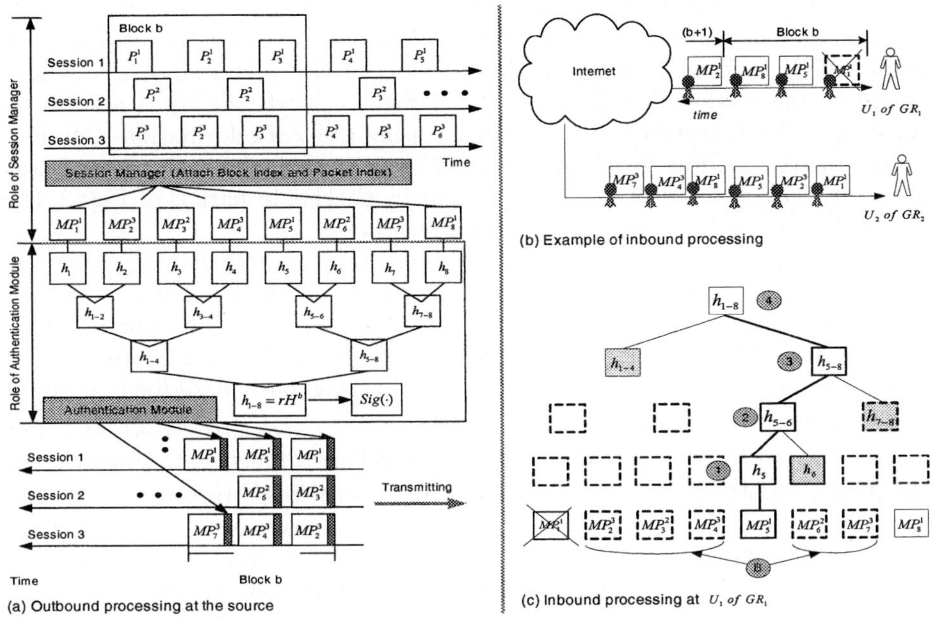

Fig. 2. MDS processing

The source first constructs a hash tree to calculate the block hash value, rH^b, for all packets in the bth block then signs over rH^b. In order to support individual packet authentication, the source transmits a packet containing an amortizing signature along with sibling hash values of each node in the packet's path to the root of the hash tree. Outbound procedures are as follows. The *source* repeats these procedures for the next block till end of all sessions.

1. Session Manager of the *source* calls $MDS.SeMux(P^s_m)$.
2. $MDS.SeMux(P^s_m)$ counts the number of P^s_m to construct a block and then attaches b and p to form MP^s_p.
3. If the number of counter meets BS, then the authentication module of the *source* constructs an authentication hash tree.
4. Generates signature, namely $\sigma_b = MDS.Sig(rH^b, sk)$.
5. The *source* forms all packets in a block to be transmitted, $\forall^{BS}_{p=0}\{[MP^s_p||\{sH^b_p\}||\sigma_b] = [P^s_m||b||p||\{sH^b_p\}||\sigma_b]\}$, then resets variables of $MDS.SeMux$ such as count value, b, and p, where $[a||b]$ denotes the concatenation of a and b.

3.5 Inbound Processing

In our approach, the source must allow a receiver to verify a packet individually even though each receiver selects different set of sessions. Fig. 2, (b) illustrates an example of such a scenario, where there are two users belonging to different group: U_1 of GR_1 and U_2 of GR_2. Suppose that U_1 decides to receive a session $\mathcal{S}^1$ and U_2 selects two sessions, $\mathcal{S}^1$ and $\mathcal{S}^3$. In addition, the first packet, MP_1^1, of $\mathcal{S}^1$ for U_1 was lost in transit, so that MP_5^1 is the first arrived packet at U_1.

The U_1 first calls $MDS.SeDemux(MP_5^1)$ operation to demultiplex. The packet MP_5^1 is formed as $[P_2^1||b||5||\{sH_5^b\}||\sigma_b]$, where p is 5 since it was fifth packet in the session manager (see (a) of Fig. 2). Next, the U_1 is able to reconstruct a hash tree to verify the packet using a set of information contained within the packet. Fig. 2, (c) shows how U_1 reconstructs the hash tree to verify a packet MP_5^1 without other packets (namely, packets in part B of Fig. 2, (c)) that belong to other sessions that U_1 did not choose.

The packet MP_5^1 contains h_6, $h_{7\sim 8}$, and $h_{1\sim 4}$ as sH_5^b (sibling hash values as illustrated by the gray colored box in (c) of Fig. 2) and includes the packet position ($p = 5$) of the bth block. Therefore, U_1 can compute $h_{1\sim 8}'$ which is the root hash value over which the source applied $MDS.Sign(\cdot)$. The procedures are followed along with the numbering path described in (c) of Fig. 2, namely $h_5'=H(MP_5^{1'})$, $h_{5\sim 6}'=H(h_5'||h_6)$, $h_{5\sim 8}'=h_{5\sim 6}'||H(h_{7\sim 8})$, and then $h_{1\sim 8}'=H(h_{1\sim 4}||h_{5\sim 8}')$. Thereafter U_1 calls $MDS.Ver(\cdot)$ function to check the authenticity, $r = MDS.Ver(h_{1\sim 8}', \sigma_b, pk)$. If and only if the return value r is true, the packet is authentic. Otherwise, the packet is discarded. Now, U_1 of G_1 knows a couple of verified hash values on the tree (h_5, h_6, $h_{5\sim 6}$, $h_{7\sim 8}$, $h_{1\sim 4}$, and $h_{5\sim 8}$). Therefore, U_1 only examines whether the calculated hash value $h_{7\sim 8}'$ and the cashed value after the arrival of the MP_8^1 since $h_{7\sim 8}$ has verified already.

3.6 Space Overhead Consideration

In this subsection, we address two efficient ways to reduce the space overhead of the basic MDS. Firstly, we apply a *TCR* (Target Collision Resistance) hash function [15] instead of *ACR* (Any Collision Resistance) to MDS under the assumption that there exist UOWHFs (Universal One-Way Hash Functions) as did [11] and [16]. The major advantage of TCR hash function compared with ACR hash function is that the birthday attack, which is the best attack on ACR hash function families, does not directly apply to TCR hash function since a message is specified before the hash function is given an attacker. Unlike TCR, an attacker who wishes to find a collision pair of an ACR can freely select two different messages in advance: M and M' that map to the same hash, namely $h(M) = h(M')$. As a result, TCR hash function of size $L/2$ can satisfy with the same secure level as those of ACR hash function of size L.

Secondly, we propose to use a separate session (denoted as an authentication session) of which purpose is transmitting a block signature σ_b and a *key*. A *key*, which is used for specifying a TCR hash function for a block, should be transmitted to meet the secure condition of a TCR hash function [15]. This way is efficient (see (b) of Fig. 3) especially in the scenario where the loss rate is low or there exists a mechanism capable of setting high priority on a packet containing a signature. In this case, all receivers must select an authentication session and the source sends a signature packet contained $[b||\sigma_b||key]$ multiple times or set with high priority in order that the signature must be arrived at the receiver. This is the basic assumption of stream authentication schemes using hash chaining such as [4] or [8]. A packet of a data session contains hash value, $H(\sigma_b||key)$, instead of σ_b to verify the integrity of the corresponding signature packet. The value $H(\sigma_b||key)$ is also used to protect against DoS on receiver in the case where an attacker can insert a malicious signature packet(s) into an authentication session. To mount a DoS attack, an attacker should replace $H(\sigma_b||key)$ of all packets in a block with $H(\acute{\sigma_b}||key)$ corresponding to modified signature $\acute{\sigma_b}$. Such efforts are equivalent to those of modifying all packets in a setting of the sign-each packet approach.

4 Performance Analysis

4.1 Multiplexing Gains of MDS

We denote the packet generation rate as λ_s (packets per second) and the total delay at the *source* as ω_s, where s indicates the session index. ω_s consists of T_{Block} and T_{Auth} denoted as the buffering time to set a block and the total amount of time it takes to build a hash tree and calculate authentication information respectively. In WL's Tree, for example, T_{Auth} is equivalent to $(2BS-1)\cdot T_h+T_{sig}$, where T_h and T_{sig} are referred to as the time it takes to hash and sign respectively. We note that other delay factors such as coding delay and network delay are not considered in this paper to concentrate on the context of security.

Delay gain: In a setting of amortizing signature, the *source* needs $\omega_s = T_{Block}+T_{Auth}$ per session before sending. In MDS, the *source* needs equivalent delay to WL's scheme in terms of T_{Auth}, but the *source* is able to reduce the term BS/λ_s of T_{Block} to $BS/\sum_{s=1}^{k}\lambda_s$. For example, MDS requires only $\omega = 116.5\ ms$ in a setting of $BS = 64$ and four sessions with $\lambda_1 = 50(G.711\ audio)$, $\lambda_2 = 33(G.723.1\ audio)$, $\lambda_3 = 100(ASF\ video)$, $\lambda_4 = 366(MPEG\ video)$. (Compare to $\omega_1 = 1280\ ms$, $\omega_2 = 1920\ ms$, $\omega_3 = 640\ ms$, and $\omega_4 = 173\ ms$ in a setting of session-based WL's Tree scheme.)

Buffer space gain: To generate an amortizing signature over all packets in a block, the *source* should buffer BS packets to set a block and a couple of packets that are arrived during the time T_{Auth}. In MDS, the *source* prepares buffer space for at most $[BS + (\sum_{s=1}^{k} \lambda_s) \cdot T_{Auth}] \cdot m_{ax}\{PL\}$, where $m_{ax}\{PL\}$ denotes the maximum size of a packet in a block. On the other hand, in the setting of session based approach, the *source* needs approximately k times more space since each session requires its own buffer space. That is, the source needs $\sum_{s=1}^{k}[(BS+\lambda_s\cdot T_{Auth})\cdot PL_s]$, where PL_s denotes the packet size in a session s.

Burstiness shaping gain: In a block based scheme, all packets in a block can be transmitted simultaneously after getting the authentication information. In the worst case of the session based treatment, the *source* transmits $k * BS$ bursty packets in an instant. Moreover, the recipient also suffers from bursty arriving of all sessions he selects. In MDS, however, at most BS packets are transmitted at once. In addition, from a recipient point of view, only a few packets that belong to the selected sessions are arrived at once since BS packets are dispersed throughout all established sessions.

4.2 Comparison with Other Schemes

To evaluate MDS scheme, we used Crypto++ library [14] on 2.4GHz Linux computer, RSA (1024-bits strength) and SHA1 as the underlying primitives, and 1024-bytes as the packet length. In RSA, the verification cost is lower than that of signing (signing/verification cost is 6.29/0.32msec in our experiments), therefore it is not a critical problem that MDS requires more verifying processing than WL's Tree (we do not illustrate it because of the length limit).

In MDS and WL scheme as well, the packet loss rate is not important since each packet is verified individually. However, in other schemes such as SAIDA and EMSS, the packet loss rate effects on the performance. In our experiments, we assumed up to 50% losses per block for SAIDA and set 6 edges for EMSS (namely, 6 hashes per packet). If the packet loss rate is increased, then it leads to lower performance in such schemes.

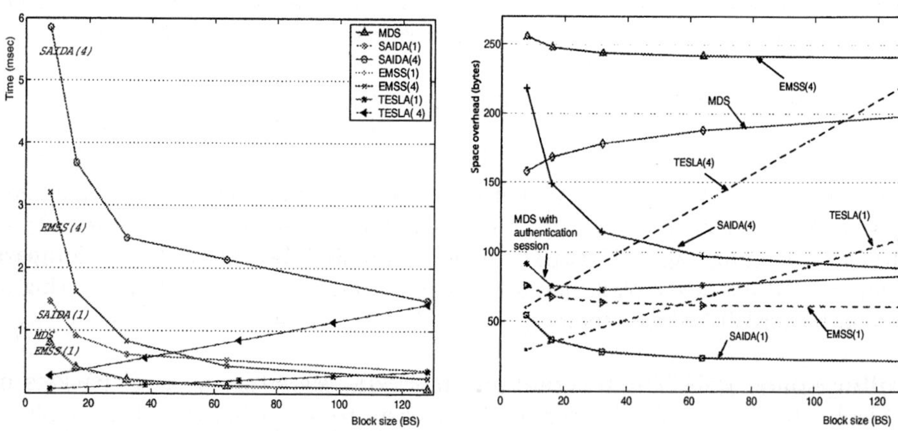

(a) Computational cost (b) Space overhead

Fig. 3. Performance comparison

Fig. 3 describes the results of comparison of MDS with three other schemes (SAIDA, EMSS, and TESLA). The values in the bracket of the figure denote the number of sessions (1 or 4 sessions). In the case of TESLA, the x-axis indicates the number of different delay groups ranging from 2 to 10 (not BS). Fig. 3, (a) shows that MDS is very efficient with respect to computational cost and (b) illustrates comparisons of space overhead. It particularly shows that MDS with a separate authentication session reduces space overhead by a factor of 2 in comparison with the basic MDS scheme after 32 of BS. As a results, the figure shows that more sessions result in higher overhead in other schemes.

5 Conclusion

We proposed efficient MDS scheme to solve data origin authentication problem that is regarded as a hard challenge owing to the requirements of time sensitive delivery, unreliability, and heterogeneity in a multimedia multicast scenario. In this paper, we have shown that the proposed MDS scheme is efficient for multimedia streaming over multiple sessions. MDS can be a reasonable solution to reduce effects of digital signature on multimedia applications from the viewpoint of the performance.

References

1. C. Diot, B. Levine, B. Lyles, H. Kassem and D. Balensiefen, Deployment Issues for the IP Multicast Service and Architecture, IEEE Network, Jan./Feb. 2000.
2. S. Kent and R. Atkinson, Security Architecture for the Internet Protocol, IETF RFC2401, Nov. 1998.
3. Schulzrinne, H., Casner, S., Frederick, R. and V. Jacobson, RTP: A Transport Protocol for Real-time Applications, IETF RFC 3550, July 2003.
4. A. Perrig, R. Canetti, J. D. Tygar and D. Song, Efficient Authentication and Signing of Multicast Streams over Lossy Channels, IEEE Security and Privacy Symposium, May. 2000.
5. R. Canetti, J. Garay, G. Itkis, D. Micciancio, M. Naor and B. Pinkas. Multicast Security: A Taxonomy and Some Efficient Constructions. Infocom'99, 1999.
6. C. K. Wong and S. S. Lam, Digital Signatures for Flows and Multicasts, IEEE/ACM Trans. Networking, vol.7(4), pp. 502-513, Aug. 1999.
7. R. Gennaro and P. Rohatgi. How to Sign Digital Streams. Lecture Notes in Computer Science, vol. 1294, pages 180-197, 1997.
8. P. Golle and N. Modadugu, Authenticating streamed data in the presence of random packet loss, NDSS'01, pp. 13-22, Feb. 2001.
9. J. M. Park and E. K. P. Chong, Efficient multicast stream authentication using erasure codes, ACM Trans. Inf. Syst. Secur. vol 6(2), pp. 258-285, 2003.
10. A. Pannetrat and R. Molva. Efficient multicast packet authentication. In Proceedings of the Symposium on NDSS 2003.
11. C. Karlof, N. Sastry, Y. Li, A. Perrig, and J.D. Tygar. Distillation Codes and Applications to DoS Resistant Multicast Authentication. In Pro. of NDSS 2004.
12. R. Merkle, Protocols for public key cryptosystems, In Proc. IEEE Symposium on Research in Security and Privacy, pp. 122-134, Apr. 1980.

13. The Network Simulator(ns-2). http://www.isi.edu/nsnam/ns/.
14. Crypto++ class library, http://www.eskimo.com/ weidai/cryptlib.html.
15. M. Bellare, P. Rogaway, Collision-Resistant Hashing:Towards Making UOWHFs Practicla, LNCS 1294, Springer-Verlag, 1997, pp 470-484.
16. P. Rohatgi, A Compact and Fast Hybrid Signature Scheme for Multicast Packet Authentication, in Proc. of 6th ACM Conference on CCS, Nov. 1999.

The Improved Risk Analysis Mechanism in the Practical Risk Analysis System

SangCheol Hwang[1], NamHoon Lee[2], Kouichi Sakurai[3], GungGil Park[2], and JaeCheol Ryou[1]

[1] The Department of Computer Science, Chungnam National University, Daejeon, KOREA
{schwang, jcryou}@home.cnu.ac.kr

[2] National Security Research Institute 62-1 Hwa-am-dong, Yu-seong-gu Daejeon, 305-718, Republic of Korea
{nhlee, jgpark}@etri.re.kr

[3] Faculty of Information Science and Electrical Engineering, Kyushu University, 6-10.1, Hakozaki, Higashi-ku, Fukuoka, 812-8581, JAPAN
sakurai@csce.kyushu-u.ac.jp

Abstract. The risk analysis system has a mechanism to evaluate and analysis the potential risk level in an organization IT system. To evaluate the Risk Level, it must be calculated the essential vulnerability that appear in various assets of organization, threats for these assets. These elements, vulnerabilities, threats and assets are the important factor to evaluate the risk level in an organization In this paper, we describe about design and implementation of a system using the practical risk analysis process that we propose. Furthermore we suggest the security countermeasure choice algorithm against the risk we found in an organization. Especially, The Security Countermeasure choice algorithm is implemented by using the Genetic-Algorithm restricted by some important factor. In this paper, we describe the design and implementation idea of the suggested genetic-algorithm module. Finally, We propose the main idea of the practical risk analysis process and the system using the risk analysis process that we propose in this paper.

Keywords: risk analysis, risk management, praha

1 Introduction

The Risk analysis process is the process that evaluates the threat, various vulnerabilities and risk in an organization's management system, personnel and information system. And the Risk measurement is that selects the suitable security counter-measure after processing risk analysis. And recently, as increase the importance of managemental and technical information system, the importance of risk analysis and risk measurement process is increasing more and more. Especially, The Republic of Korea is recognizing necessity about risk analysis in various organization by enforcement of information and communication base protection law July, 2001, and operate main information and communication base infra structure is achieving risk analysis regularly. As well as the purpose

C. Kim (Ed.): ICOIN 2005, LNCS 3391, pp. 835–842, 2005.

of risk analysis process research that inter-est about risk analysis is scientific increasing so, risk analysis methodology or development of an automatic risk analysis system that is active practical use really in actuality business is consisting. However, "What" about process is decided in re-search subject of universal risk analysis methodology, and is state that the justice is definite mathematically but contents definition about "How" is vague. The problem is appearing Britain BSI's "BS7799" that is known well methodology worldwide or United States of America Carnegie melon university's "OCTAVE" on this methodology. Therefore, it is very difficult that apply existent risk analysis methodology to the existing information and communication organization to get substantial risk analysis. In this paper, we suggest that the design of practical risk analysis process and de-scribe the implementation of our risk analysis automation tool. And we suggest main design of risk decrease algorithm and different view extension that is used in a risk analysis automation system that take advantage of our practical risk analysis methodology.

2 The Research of the Existing Risk Analysis Methods

Information protection administration area did and is to ISO/IEC-13335 (GMIT), BS_7799 (ISO/IEC-17799), Carnegie melon University's SSE-CMM etc. that is international standard to confirm detail about ancient temple existent risk analysis and risk management process of existing risk analysis methodology, and risk management area is IRM index hand that the country develops and IRM index hand that organ develops, and did CRMM that is general common use product and risk management methodology that use in BDSS to investigation target. The existent methods to have these processes have following some problem. First, the pre-process phase and administration support function that existent most methods manages whole project of risk analysis process are lacked. Estimation target organization's inside staff, target organization's outside person and many risk analysis appraisers of risk analysis are participated. Also, many debates and discussion and opinion harmony are gone, and transfers of many documents and estimation period of overlong time are needed. For this, we need administration phase and function of risk analysis project on estimation process interior. Second, the existent methods are not executing Risk analysis of TOP-DOWN form. TOP-DOWN risk analysis is method to divide and enforces process by high level and low level risk analysis process usually. In high level risk analysis step, we analyze observance availability of state standard such as ISO/IEC 13335 or BS7799. In Low level risk analysis process, we analyze risk with each asset threats, vulnerability. General high level risk analysis pumice is come to black-box analysis method, and low level risk analysis can say as White-box analysis method. ISO/IEC-13335 is risk analysis step of high level, but can not say that step about details level follows TOP-DOWN risk analysis process because it is not. Third, the existent risk analysis methodology is poor relatively process about choice and selection of security countermeasure about grasped risk. Main purpose of risk analysis actualizes effort to analyze risk that exist to establishment, and reduce this. Therefore, choice of security countermeasure and analysis of residual risk about grasped risk are important problem. However, a concrete methodology and technological process presentation are lacked in the existent risk analysis methodology and problem is revealed because being depending many parts on question of risk estimation target organization interior.

3 The Improved Risk Analysis Process

It is the risk analysis methodology that can be applied in various information and communication system environment that PRAHA process to develop in our research institute and this methodology. It is an improved risk analysis process is independent general methodology has no concerned with various user's special environment. This PRAHA risk analysis methodology has been developed as the practical risk analysis process is including the information and communication system environment of Republic of Korea, security policy, security guide etc. Also, this PRAHA risk analysis methodology does get together the advantage of risk analysis methodology possessing in various area and developed methodology for developing risk analysis framework and process to have an efficiency. The advantage of the PRAHA methodology is the following.
- Advantage containment the existing risk analysis method
- The Risk analysis methodology selection possibility by organization's security policy and risk level
- The realistic achievement of general risk analysis formality is easy
- The possibility of evaluation asset value according to the attribute of risk analysis target asset
- Threat, vulnerability analysis is possible according to the second asset classification

- The improved Objectivity of PRAHA risk analysis by presenting the concrete method and procedure
- Most sutiable countermeasure selection through cost effective research security management level evaluation

That is, Praha risk analysis process uses TOP-DOWN risk analysis model and Two-step risk analysis model, Low level risk analysis and High level risk analysis. Also, Praha process is different from another risk analysis methodology in method-ology about system and asset evaluation, methodology threat analysis, methodology and Security Countermeasure choice methodology. Especially, to minimize opinion deviation among risk analyzers, it uses "Opinion collection, control method (Cyber Delphi techniques)".

4 The Architecture of the PRAHA Method

The Praha Risk analysis process is the same as following. The Praha risk analysis process is consist of two steps. One is the high level risk analysis process and the other is low level risk analysis process.

4.1 The High Level Risk Analysis Process

The high level risk analysis is that manages estimation project and use security control item that appear in security management standard and evaluate the organization's " Risk management level as quantitative. Also, execute interview to the estimated organization's each person include process to grasp assets. According to analysis result of high level risk analysis for the estimated, we decide to do more detailed low level risk analysis process or not. The question investigation activity estimates estimation target organization's risk management level according to BS-7799 standard. The high level risk analysis is consist of question using web system. This result is used to continue the low level risk analysis. In this system, Using various risk evaluation question and security control item of BS-7799, and evaluate the result through presented method. Also, to do improve usage, it does to do question in summary question, general question, and details question and that it improves flexibility of high position analysis. Also, PRAHA system offers distribution of question, administrative and analysis function. If question investigation's result and present security management level is good, that organization executes basic security control. The following picture shows the high level risk analysis process.

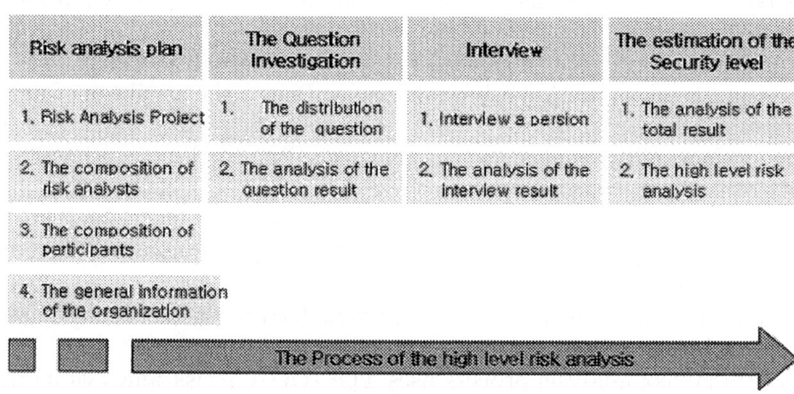

Fig. 1. The high level risk analysis process in the PRAHA methodology

4.2 The Low Level Risk Analysis Process

The low level risk analysis process is used an estimation method per system unit in this PRAHA risk analysis automation system. It is a process to produce risk level evaluating estimated organization's asset a system unit as quantitative. Because it is not easy to say the effect that evaluating by individual asset unit in business, it has more difficult problem than estimation of system unit relatively.

Threat analysis has some difference according to establishment of threat domain, but than individuation asset unit when establish threat connected with business, system unit estimation is easy relatively. While technological vulnerability is easy individuation asset unit estimation in vulnerability analysis, but management and physical vulnerability analysis estimation of system unit relatively easy. And the security countermeasure analysis does together vulnerability analysis. Now, risk analysis method to select in the PRAHA system is the method to evaluate estimation target unit asset of system unit as quantitative and produce risk level. The following process shows the low level risk analysis in the Praha system. This methodology's main idea is to evaluate risk level per system unit

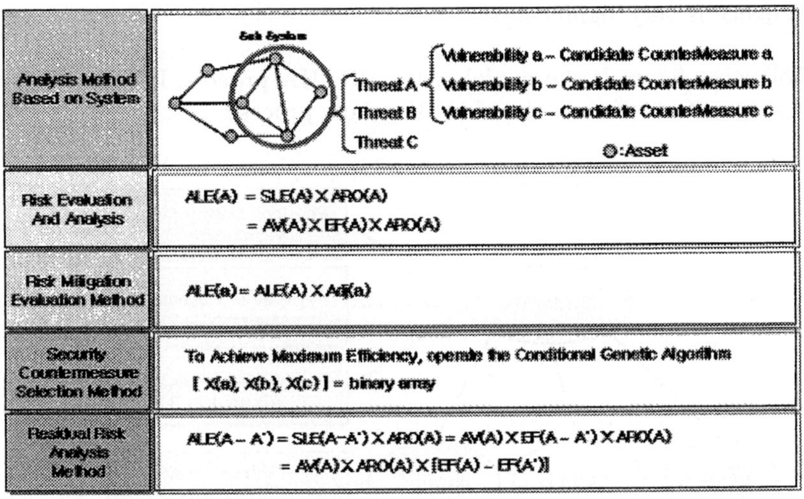

Fig. 2. The low level risk analysis process in the PRAHA methodology

including various assets in an organization. So the following fomula is produced per system.

$$\text{ALE} = \text{SLE X ARO}$$

We use the factor, EF (EF: Asset's Exposure Factor) to calculate SLE. If value of system is AV (Asset Value) and degree that this system is exposed in threat is EF, SLE is consisted by times of this two factor.

$$\text{SLE} = \text{AV X EF}$$

To produce ALE, it must be included ARO, AV, EF etc. in risk analysis achievement process (risk analysis process). Usually, we must predict the amount of loss cost by various threats. And then, we must calculate EF by various vulnerabilities. This complex calculation process is possible by the Praha low level risk analysis process.

4.3 Exposure Factor(EF) Calculation

The Risk analysis system(PRAHA Tool) has the module and algorithm to calculate the exposure factor composed with various vulnerabilities in many systems. However, it is very hard to estimate among various vulnerabilities related with threats and assets because the number of whole vulnerabilities related with system or individual asset is variable. So, first we make following assumptions and propose a reasonable exposure factor calculation method. The Assumptions in calculating EF are as following.
- The Number of whole vulnerabilities that exist to system can not define
- The vulnerability level affects greater risk level tan vulnerability number
- Even if level of vulnerability is low but many same vulnerabilities are found, the influence is very significance

The Following figure is the model of the PRAHA process. In PRAHA process, the expo-sure factor is calculated in the whole process.

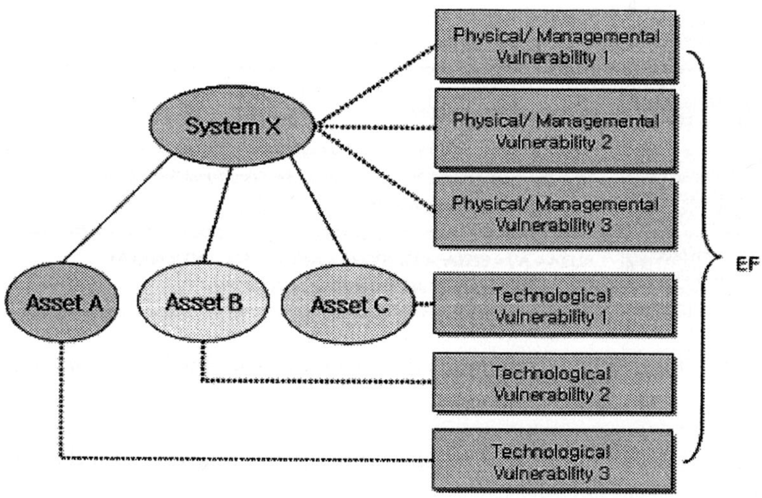

Fig. 3. The EF calculation

4.4 The Choice of Security Countermeasures and Calculation of the Residual Risk

Usually, the optimization solving process is used widely looking for the best result of a given problem. In PRAHA system, A choice of suitable security countermeasure about an organization's system - asset - Threats-vulnerability is the same case. Therefore, this solving process is the problem of an association optimization, and the problem of this optimization becomes NP hard-problem.

Finally, it is the same problem that choose the most suitable case. As many candidate security countermeasures exist, the complexity of this problem increases by complexity of exponential function. Therefore, it is more efficient that remove evil of most suitable in the determined time using limited Genetic-Algorithm, than simple method that list solution of problem one by one. Choice problem of security countermeasure can be solved with Knapsack problem. Because it is the problem that choice of the most suitable security countermeasure gathering of the security countermeasures with limited condition. The condition and assumptions applied in this methodology are the followings. First, if one candidate countermeasure is chosen, there is case that necessarily should choose another candidate security countermeasure exists. Second, at the same time elect candidate security countermeasures do not calculate duplicating expense be-cause is same candidate countermeasure over. Third, because Security countermeasure exposition parturition at risk analysis process in Praha risk analysis methodology established and suppose that necessary datum is applied in countermeasure choice beforehand. Security countermeasure choice algorithm that such three restricted conditions are applied can be composed as the next following equation.

$$P' : \max imize \sum_{i=0}^{n'} pi' xi'$$

$$subject\ to\ \sum_{i=1}^{n'} wi' xi' \leq C$$

$$xi' = \{0,1\}, i = 1..n'$$

n' = the index of the candidate group, when we bind same candidate countermeasure by the same one.

$Pi' = pi \times i$'s number of the candidate countermeasure

xi' : true or false, the choice of the P

wi : the cost of the countermeasure in case of selecting P

C : the maximum cost of the countermeasure group

Fig. 4. The suitable countermasure choice

At present, many risk analysis methodologies have simple risk level that depend on specialist's opinion. But this new methodology to use genetic algorithm was com-posed to determine suitable countermeasure and cyber Delphi mechanism.

PRAHA v1.0 using genetic algorithm creates security countermeasure, but that these created contents are suitable countermeasure in present situation. These countermeasures doesn't exact to that organization's situation. The genetic algorithm does look for most suitable countermeasure but it is difficult to confirm the exact answers. Therefore, there is the purpose to find the most suitable countermeasure to the organization's situation. And it is necessary that this new methodology is applied to many real world organization's system. Till now, in our simulation, the Risk analyst system selects suitable countermeasure on the basis of risk analysis result in PRAHA v1.0 in our opinion.

5 Conclusion and Future Work

Generally, there are two methods, quantitative method and qualitative method, to calculate risk level in risk analysis. Quantitative method is specific comparatively being done numerical value, but have to do value and risk for organization's assets exactly numerical value that is difficult. While, Qualitative method has the advantage that it is easy to evaluate relative assets and risk but data is abstract being not numerical value. This PRAHA risk analysis system includes Delphi method to get organization's assets threats and vulnerabilities as quantitative by specialist group. To use a practical risk analysis tool, it must make more suitable result of security countermeasure. And, it is needed the complementation to restriction condition of algorithm for suitable security countermeasure choice to each organization with applications of more field.

References

1. British Stands Institution(BSI), BS-7799, 1999
2. CSE, " A Guide to Security Risk Management for IT systems", Government of Canada, Communications Security Establishment(CSE), 1996
3. NIST, "CC ToolBox Reference Manual" Version 6.0
4. GStonebumer, et al. "Risk Management Guide for Information Technology System" NIST SP-800.30, NIST, 2002.1
5. M. Timms, "A Practical Approach to Risk Assessment", Compsec Computer Security Conference'90, 1990.10
6. Z. Ruthber, et al., "Guide to Auditing for Controls and Security: A System Development LifeCycle Approach", NBS Special Publication 500-153, 1998.4
7. NIST IR-4387, "Simplified Risk Analysis Guideline", NIST, 1990
8. GAO, "Informatin Security Risk Assessment - Practices of Leading Oranizations", - Case Study 1, GAO/AIMD-00-03, 1999.11
9. GAO. "Information Security Risk Assessment - Practices of Leading Organizations", - Case Study 3, GAO/AIMD-00-03 1999.11

A Fast Defense Mechanism Against IP Spoofing Traffic in a NEMO Environment*

Mihui Kim and Kijoon Chae

Dept. of Computer Science and Engineering, Ewha Womans University, Korea
mihui@ewhain.net, kjchae@ewha.ac.kr

Abstract. The boundary of a distributed denial of service attack, one of the most threatening attacks in a wired network, now extends to wireless mobile networks, following the appearance of a DDoS attack tool targeted at mobile phones. Many protocols and architectures for mobile networks were designed without regard to the possibility of a DDoS attack. Moreover, the existing defense mechanisms against such attacks in a wired network are not effective in a wireless mobile network, because of differences in their characteristics. In this paper, we propose a fast defense mechanism against IP spoofing traffic for mobile networks. IP spoofing is one of the features of a DDoS attack against which it is most difficult to defend. Among the various mobile networks, we focus on the Network Mobility standard that is being established by the NEMO Working Group in the IETF. Our defense consists of the following five processes: speedy detection, filtering of attack packets, identification of attack agents, isolation of attack agents, and notification of neighboring routers. We simulated and analyzed the effects on normal traffic of moving attack agents, and the results of applying our defense to a mobile network. Our experimental results show that our mechanism provides a robust defense.

1 Introduction

Currently, distributed denial of service (DDoS) attacks are considered one of the most threatening of attacks against wired networks. This type of attack is a relatively simple, yet very powerful, technique for attacking Internet and system resources. In a DDoS attack, distributed multiple agents consume critical resources at the target within a short time. As a side effect, such attacks cause network congestion en route from the source to the target, thus disrupting normal Internet operations and causing many users to lose their connection.

The damage from DDoS attacks is no longer limited to wired networks. Although the functionality of most mobile devices is extremely limited and largely non-programmable, the first virus targeted at mobile phones has already appeared. In addition, the SMS flooder has emerged as the first DDoS attack tool against mobile phones. This tool commands all infected Microsoft Outlook software to send short messages (SMS-messages) to the specified victim's mobile

* This research was supported by University IT Research Center Project.

C. Kim (Ed.): ICOIN 2005, LNCS 3391, pp. 843–852, 2005.

phone in order to inundate it. The potential hazard is not only the blocking of communications but also the high financial cost when pricing is usage-based [1]. Damage from these attacks is expected to become increasingly serious as mobile devices become high-performance machines. Moreover, wireless networks under design or recently introduced, such as sensor networks, will be more susceptible to these attacks.

Long an issue of interest, network mobility technology is now being realized with the foundation of the NEMO (Network Mobility) Working Group (WG) in the IETF. This WG is concerned with managing the mobility of an entire network that changes, as a unit, its point of attachment to the Internet. The WG defines NEMO as "NEtwork MObility" or "a NEtwork that is MObile". There are three types of node in a NEMO: local fixed nodes (LFNs), local mobile nodes (LMNs), and visiting mobile nodes (VMNs). The MRs access the Internet from access routers (ARs) on visited links. Most importantly, a NEMO can be nested in another NEMO and can be multihomed. This structure provides one or multi-hop wireless links, as well as a tree-like hierarchy.

The characteristics of a NEMO, the multi-hop wireless links and mobility, pose many security challenges, such as in the case of a mobile ad hoc network (MANET). First, wireless channels are intrinsically more susceptible to attacks such as passive eavesdropping, active signal interference, and jamming similar to a DDoS attack. Second, continuous and unpredictable mobility clouds the distinction between normalcy and anomaly, thus making it difficult to detect malicious behaviors. Also, owing to mobility, intrusion detection systems must broadly deploy for accurate detection and must cooperate effectively [2]. Third, existing centralized approaches to security on a wired network are inefficient on a NEMO, such as a MANET, because of the possible problems posed by high mobility and scalability. In addition, multi-hop communication over an error-prone wireless channel exposes the data transmission to high loss rates [3]. Fourth, cell phones and small routers can become mobile routers (MRs) that provide the features of a NEMO, and most mobile devices have limited processing power. Therefore, the security mechanisms on a NEMO should be light, for low power consumption, but also robust. Finally, compromise of a MR that performs basic NEMO operations can also compromise all the nodes under its charge in the NEMO; thus, a compromised MR in a NEMO can cause far more widespread damage than a compromised node in a MANET.

Of the many possible features of DDoS attacks, source IP address spoofing is among the most common. IP spoofing creates particular difficulties for network managers, because it increases the number of flows by varying the source address and concealing the identity of the attack agent. DDoS attacks using IP spoofing also pose a threat in a NEMO environment, where detection of and defense against such attacks is far more difficult because the attacker can move.

In this paper, we will simulate and analyze the effects of an IP spoofing attack on a mobile network. Through the simulated results and an analysis of the characteristics of a DDoS attack on a NEMO, we will adapt and extend previously proposed detection and identification defense mechanisms against spoofing packets on a wired network [4], to the NEMO environment. A previously proposed mechanism was implemented, tested using strong DDoS attack tools on a real network, and confirmed to be an effective design. Initially, we set the following design goals for defense against spoofing attacks on a wireless network, particularly on a NEMO.

- **Speedy detection&filtering** of the source-side network as soon as possible
- **Identification&isolation of attack agents** for prompt follow-up measures
- **Notification of neighboring MRs** to proactively isolate the attacking traffic

This paper is divided into four sections. Section 2 introduces the proposed defense mechanism. In Section 3, we evaluate our mechanism and explain our analysis of the simulation results. Finally, we present a brief conclusion.

2 Fast Defense Mechanism

Our defense against spoofing traffic on a mobile network consists of five parts: speedy detection, filtering of attack packets, identification of attack agents or NEMOs including attack agents, isolation of attack agents' traffic in the lower layer, and notification of neighboring routers. The last two steps are especially important on a mobile network in order to decrease overall damage to the entire network, owing to the wireless environment and the mobility of nodes, respectively. Although there can be spoofing in several layers, we assume that the spoofing traffic is spoofing the source IP address, as in the case of DDoS attacks in wired networks. This defense could be adapted to handle spoofing of other layers in the same manner. Also, we assume that the defense mechanism is applied to the source network. Although it would be possible to use the defense mechanism in the target network, using the defense at each source network is more efficient and can decrease the damage before the target is shut down. We will explain each part of our defense against spoofing traffic in the following sub-sections.

2.1 Detection and Filtering

As it is very difficult to prevent spoofing attacks, the first priority is rapid detection. In a NEMO environment, we cannot assume that there is at least one detection agent per router, because a NEMO can be very small, as with a PAN. Therefore, we consider two cases: in one case, all MRs perform detection and filtering; in the other, one agent per top level mobile router (TLMR)/AR performs these roles.

In the first case, MRs can choose one of two different detection mechanisms, according to the number of served nodes. One method uses the configurable network addresses, and the other uses the rate of change of IP addresses for a single MAC address. Each attack detection condition is shown in Table 1 and 2.

Table 1. Detection Condition using Configurable Address Information

$(sIP \notin CAi \quad AND \quad dIP \notin CAi) \quad OR$	$(Condition1)$
$(sIP \in RAi \quad AND \quad dIP \notin CAi) \quad OR$	$(Condition2)$
$(sIP \in DAi)$	$(Condition3)$

- *sIP*: Source IP address of a packet.
- *dIP*: Destination IP address of a packet.
- *CAi*={*Ca1*, *Ca2*,...,*Cak*}: Configurable IP addresses in the shared media.
- *DAi*={*Da1*, *Da2*,...,*Dal*}: Deniable IP addresses in the shared media.
- *RAi*={*Ra1*, *Ra2*,...,*Ram*}: Directly connected router IP addresses.

The first method makes use of the characteristic that most spoofing attack tools rotate a specific range of IP addresses as the source address. Thus, these addresses include one or more non-configurable address, such as a router address, deniable address, or a different subnet address. This method needs to discover the configurable address, router address, and deniable address. In a NEMO environment, when a NEMO attempts to become nested, its network addresses can be conveyed to its parent MR or AR as a step in the handoff. The first method may not be as fast as the second, but the required storage is smaller and the middle MRs in the nested structure can easily perform the detection feature using the same mechanism.

The second method uses the characteristic that although spoofing attack tools rapidly change the source address, the source MAC address doesn't change during an IP spoofing attack. Test results indicate that the rate of change is about 0.074 ms for the TFN2k DDoS attack tool. Therefore, the admissible *Tcng_rate* can be set to an interval such as 2 sec, considering a case that a person changes the IP address of a machine. This method provides faster detection, but it is limited to monitoring general server hosts, because routers forward packets with various source IP addresses for the same MAC address. If a middle MR in a nested NEMO uses this method as the detection feature, it should determine whether the served node is an MR or a general host. It should also determine this when an MR that wants to be served at the middle MR performs the handoff process. For these two methods, proper and fast detection has been established using well-known, powerful spoofing tools on a real network [4].

In the case explained above, all MRs perform detection and filtering. This assumption is ideal, but is not always the reality. The possible detection/ filtering/identification processes of the second case are explained in the next section.

2.2 Identification and Isolation of Attack Nodes

One reason that DDoS attacks are difficult to defend against is that most DDoS attacks use IP spoofing to conceal distributed attack agents. Although this makes it difficult to take basic measures against such attacks, it is important to quickly and accurately identify attack agents in order to minimize possible damage. To make the identification, the defending agent should have an IP2MAC table that

Table 2. Detection using the Rate of Change of the IP address for a MAC address

$\mid t[sIP_1, sMAC_1]Pi \; - \; t[sIP_2, sMAC_1]Pj \mid \; \leq \; Tcng\_rate$
• $sMAC$: Source MAC address of a packet. • $t[sIP, sMAC]Pj$: Time of discovering the packet Pj having sIP and $sMAC$. • $Tcng\_rate$: Upper threshold for change interval for spoofed packets. This is the change interval (seconds) of source IP addresses for a source MAC address.

includes the mapping of an IP address to a MAC address. If MRs/ARs were to perform this task, they could use an ARP table to find the real IP address of an attack agent.

As explained above, when all MRs perform the defense feature, a MR can quickly determine the attack agents by using an ARP table. However, if only one detection agent per TLMR or per AR performs the detection feature, the detection agent needs to determine the NEMO/MR that includes the attack agent by means of its ARP or IP2MAC table. In that case, it notifies the NEMO/MR of the detected attack in order to transfer the identification job, and it filters the traffic from the NEMO/MR that includes the attack agent until it finds the exact attack agents. This detection and notification work is iterated over the downward path of nested NEMOs until it finds the attack agents. When the MR finds the attack agents, it notifies the parent-MRs regarding the identification, with a request to forward the normal traffic of the NEMO. If this notification is conveyed to the first detection agent, only the attack traffic is filtered.

The result of isolating attack agents' traffic in the lower layer is that the MAC layer denies grant of the channel to the attack agents; for example, in the case of 802.11, it refuses to send a clear to send (CTS) message in response to a request to send (RTS) message. This is important in a wireless environment that uses a CSMA/CA like 802.11, because granting a channel to attack agents considerably affects the transmission rate of other nodes, even while filtering of attack traffic occurs at the IP layer.

2.3 Notification of Neighboring Routers

This part of defense against attack is specific to a NEMO, and is necessary because NEMOs that include attack agents, or the attack agents themselves, can move to other MRs/AR. To decrease the time for detection and identification at the handed-off parent MRs/AR, the first agent that detects and identifies the attack agent provides its neighboring MRs/AR with information about the attack agents, such as the real IP/MAC address. If all MRs are performing the defense feature, if the attack agents are mobile, the MR that is handling the attack agent provides other MRs/AR that may subsequently be affected with the attack information about the attack agent. If there is only one detection agent per TLMR/AR, it provides neighboring detection agents with the attack information.

3 Evaluation

We have suggested an architecture for defense against spoofing attacks that consists of five processing steps. In this section, we will evaluate the performance of each step from various points of view.

3.1 Detection and Identification Speed

First, we consider how quickly the spoofing traffic can be detected. Our previous evaluation of detection performance in a wired network resulted in detection and identification times of less than 50 milliseconds for all three suggested schemes [4]. Although the environment considered here is wireless rather than wired, the detection and identification mechanism for a MR is almost the same as the monitoring agent in [4]. This result is dependent on the number of received packets, the capacity of the monitoring agent, whether the IP addresses of attack packets are entered in the IP2MAP table in the case of scheme2/3 [4], the spoofed address, the speed of attack, and so on. In order to explain this relationship, we define the symbols that we use in the equations, as shown in Table 3.

Table 3. Equations related with detection/identification speed

$$\begin{aligned} Tdt &\leq 2 \cdot uTcmp \cdot |CAi| + uTcmp \cdot |RAi| + uTcmp \cdot |DAi| + Tsp \\ &= uTcmp \cdot (2 \cdot |CAi| + |RAi| + |DAi|) + Tsp \qquad (1) \\ Tident &\leq uTcmp \cdot |IP2MAC| \cdot isThere(macAdr) \qquad (2) \\ Tdt &\leq uTcmp \cdot |IP2MAC| \cdot 2 \qquad (3) \\ Tdt &\leq uTcmp \cdot \log(|IP2MAC|) \cdot 2 \qquad (4) \end{aligned}$$

- *Tdt*: Time for detecting spoofing attack.
- *Tident*: Time for identifying the real IP address of spoofing attack agent.
- *uTcmp*: Unit time for comparing IP/MAC address with an entity of address set.
- *isThere(macAdr)*: a function that returns 1 if there is an entry for a MAC address in the *IP2MAC* table, otherwise 0. If this function outputs 0, the identification process fails.
- *Tsp*: Spent time that the first spoofing packet with impossible configurable address appears after the attack is mounted.
- *IP2MAC*: IP & MAC address mapping table that the detection/identification agent manages.

There are two ways to detect a spoofing attack such as the one described in Section 3. One way is to use network configuration information. In this case, we can represent the relationship between the detection/identification time and influential elements as shown in equations (1) and (2) in Table 3. Another way is to check the rate of change of source IP addresses for a given source MAC address. We can represent the relation between detection time and influential elements as shown in Equation (3) in Table 3. In this case, the identification time does not need to include additional time, because the agent can immediately determine the IP address of the DDoS attack agent in the detection process. Of course, these upper bound times can decrease if the sets (e.g. CAi) or the *IP2MAC*table are managed as a tree data structure. For example, Equation (3) can be changed into Equation (4).

3.2 Simulation Results

To analyze the influence on a NEMO of spoofing attack traffic, we configured a simulation network and scenario as shown in Figure 1, and used Glomosim. We assumed route optimization on the NEMO, so that traffic from node 9 to node 7 would be routed in the sequence MR3, AR, MR2, MR1, node 7. Using this scenario, we observed changes in Forwarding Rate(FR:bytes per second) at the MRs/AR, and the ratio of normal/abnormal traffic in the throughput of the MRs/AR. This resulted in Figures 2 and 3 in a 2-Mbps link.

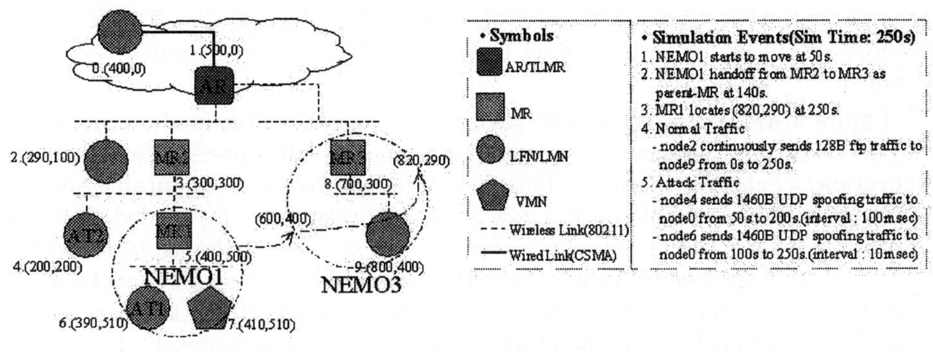

Fig. 1. Simulated Network Configuration and Scenario

We can analyze the result of Figure 2 as follows. AR and MR3 forward FTP traffic from node 2 to node 9, from the start of the simulation until 50 s. At 50 s, the DDoS attack traffic on node 4 is added, and MR2 forwards the attack traffic to node 0. The FR of MR3 is decreased because of the FTP traffic from node 2 and the attack traffic from MR2. This FR decrease in normal traffic indicates the damage caused by the absence of a proper attack defense mechanism. To worsen the scenario, DDoS attack traffic on node 6 is added from 100 seconds, thus accelerating the decrease in the FR of normal traffic at MR3. The decrease ratio increases because the attack on node 6 is more severe than the attack on node 4. In addition, the attack traffic from node 6 increases the FR of MR1/MR2 on the route to the destination (node 0). This situation changes when the NEMO1 that forwards attack traffic from node 6 hands off from MR2 to MR3 at 140 seconds. For this reason, the FR of MR3 increases owing to the addition of the attack traffic. After the movement of the NEMO1 that contains node 6 to generate attack traffic, the FR of AR increases slightly. The reason for this will be explained later using Figure 3. At 200 seconds, the generation of attack traffic on node 4 finishes. As a result, the competition for AR decreases and the FR of MR1/MR3 increases slightly, but the FR of AR decreases slightly.

To experiment in an environment with sufficient bandwidth in relation to the application traffic, we simulated the same events as in Figure 2 using the link bandwidth (11 Mbps). In this case, because the competition for the channel is small, the FR variation is small and more attack traffic is transmitted. Therefore, in the absence of a proper DDoS attack defense mechanism, attack traffic occupies much of the throughput of the MRs/AR that forward the attack traffic, because heavy traffic and flows are generated within a short time. This causes a decrease in the amount transmitted for normal applications, and the effect is more severe when link bandwidth is small and thus the competition for the channel is intense.

Figure 3 depicts the traffic variations of normal/abnormal traffic at the AR, for the case shown in Figure 2. The results indicate that the more broadly distributed the sources of the DDoS attack and the more severe the attack traffic, the greater the FR of parent MRs/AR is occupied by attack traffic. This situation is more severe in the case of small link bandwidth or severe competition for a channel.

Figure 4(a) depicts the traffic variations for each router in the case where the detection and filtering feature are performed at all routers, using the configuration/scenario of Figure 1. This detection feature uses configurable address information. As a result, only a few spoofed packets are forwarded to other nodes twice (50 s,100 s). The first occasion (50 s) is when the attack starts at node 4, so the MR2 detects the attack after forwarding two spoofed packets and then filters the later attack traffic. The second occasion (100 s) is when the attack starts at node 6, so the MR1 also detects the attack. However, the handoff (140 s) of NEMO1 does not affect detection as expected, because the parent MR(MR1) of the attack node (node 6) moves with it. Another prominent result is that the single L3 filtering feature at the parent MR of the attack agent is insufficient to isolate the influence of the attack, as shown. The forwarding rate for normal traffic at AR/MR3 is influenced by the start/end of the attack and the attack intensity. Thus, the forwarding rate for normal traffic at AR/MR3 decreases a little at the start (50 s) of the attack at node 4, and decreases a great deal because of the addition (100 s) of a severe attack at node 6. It also increases a little because of the end (200 s) of the node 4 attack. These effects of the attack are more severe because all of the nodes under an AR use the same radio frequency. If each NEMO used a different radio frequency and its radio resources were carefully managed to prevent overlap with the radio resources of its neighbors, the influence of the attack on normal traffic transmission could be decreased. However, as a fundamental measure, an L2 filtering feature should ensure that channel access authority is not given to attack nodes at the MAC layer, for example, by cutting off CTS messages for the attack nodes in the case of an 802.11 MAC.

Figure 4(b) depicts the results of adding an L2 isolation feature to the defense mechanisms of Figure 4(a). If the MR detects the attack, it filters the attack traffic and does not give the L2 access authority. As a result, even though a double attack is mounted, the effects of the attack traffic are considerably decreased.

Table 4 summarizes the packet statistics at each router for attack detection deployment. The Spoofed Packets is the count of packets that the MR decides are spoofed packets and filters, and the Unicast Packets is the count of packets that the MR decides are unicast packets and forwards. In the second detection deployment, the detection is performed only at the AR, which notifies the downstream routers with the detection information so that they can identify the attack agents. Consequently, all forwarded packets at MR1/2 are spoofed packets, and in the tally of the Unicast Packets for MR1/2, packets are misjudged. However, two deployments both showed fast detection results, with the attack being detected after only a few spoofed packets were forwarded. In the first case, the attack was detected more quickly because the MR that included the attack agent promptly detected the attack.

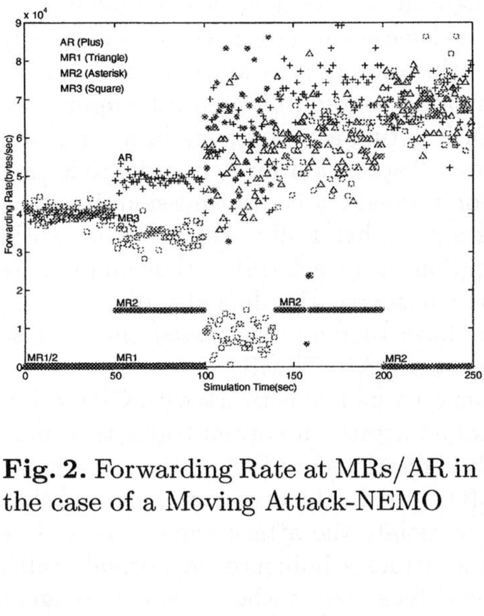

Fig. 2. Forwarding Rate at MRs/AR in the case of a Moving Attack-NEMO

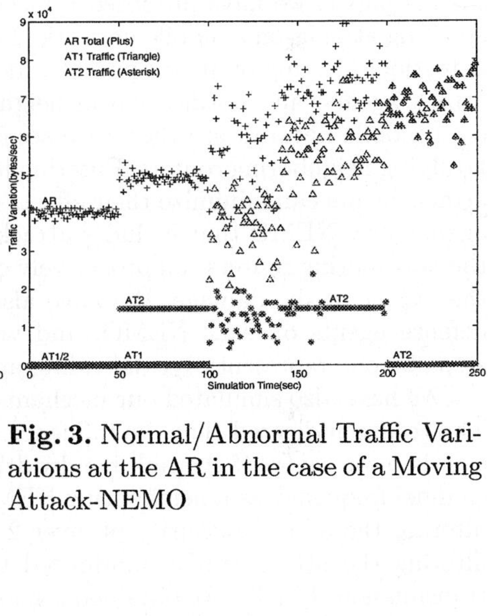

Fig. 3. Normal/Abnormal Traffic Variations at the AR in the case of a Moving Attack-NEMO

Table 4. Packet statistics for attack detection deployment (simulation time: 140 s)

Node Id	Detection at all Routers		Detection at the AR Only	
	Unicast Packet	*Spoof Packet*	*Unicast Packet*	*Spoof Packet*
1 (AR)	15470	0	15525	3
3 (MR2)	2	909	6	913
5 (MR1)	1	2783	9	2701
8 (MR3)	15458	0	15522	0

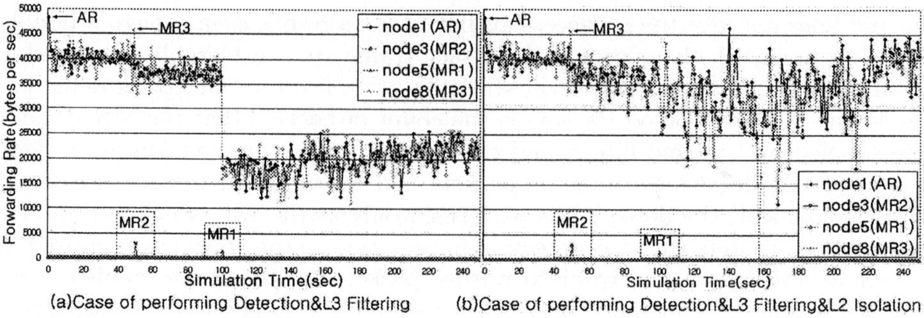

Fig. 4. Forwarding Rate of Each Router

4 Conclusion

In this paper, we have proposed a mechanism that rapidly defends against IP spoofing attacks on a mobile network. The defense mechanism consists of speedy detection, filtering of attack packets, identification of attack agents, isolation of attack agents, and notification of neighboring routers. Of greatest importance on a mobile network are the processes of isolating attack agents in layer 2 and notifying neighboring routers. Filtering layer 2 access authority in order to isolate attack agents can minimize their effects on normal traffic transmission. If attack agents, or a NEMO that includes attack agents, hand off to their own network, the neighboring routers can proactively cut off the attack traffic, thus minimizing the extent of the damage. We have also considered the difficulty of deploying defense agents on each NEMO, and we have therefore suggested the defense steps by one representative defending agent per AR/TLMR.

We have also simulated our mechanism on a mobile network with Glomosim. The results indicated that the influence of an attack on normal traffic transmission is more severe if the link bandwidth is restricted and when the same radio channel frequency as a neighboring NEMO is used. The results also showed that filtering the access authority of layer 2 to isolate the attack agents, as well as filtering the attack traffic, minimized the attack's influence on normal traffic transmission. Finally, we obtained a good defense result when a detection agent was implemented at an AR.

References

1. Xianjun Geng, et al.: Defending Wireless Infrastructure Against the Challenge of DDoS Attacks. Mobile Networks and Applications, vol. 7 (2002) 213-223
2. Wenjing Lou, Wei Liu and Yuguang Fang: SPREAD: Enhancing Data Confidentiality in Mobile Ad Hoc Networks. IEEE INFOCOM (2004)
3. Jiejun Kong, et al.: Providing Robust and Ubiquitous Security Support for Mobile Ad-Hoc Networks. IEEE ICNP (2001)
4. Mihui Kim and Kijoon Chae: Detection and Identification Mechanism against Spoofed Traffic Using Distributed Agents. ICCSA, LNCS 3043 (2004) 673-682

A Novel Traffic Control Architecture Against Global-Scale Network Attacks in Highspeed Internet Backbone Networks

Byeong-hee Roh[1], Wonjoon Choi[1], Myungchul Yoon[2], and Seung W. Yoo[1]

[1] Graduate School of Information and Communication, Ajou University,
San 5 Wonchon-dong, Youngtong-gu, Suwon, 443-749, Korea
{bhroh, mecgebi, swyoo}@ajou.ac.kr
[2] Mobile Extend Inc. 1107 Greenvill, Seocho-dong, Seocho-gu, Seoul, Korea
mc_yoon@naver.com

Abstract. In this paper, we propose a global traffic control architecture to isolate network attacks from normal traffic in the backbone networks designed to serve normal traffic only. Unlike existing methods based on individual packets or flows, the proposed traffic control methods are operated on the aggregate traffic level, so the computational complexity can be significantly reduced, and they are applicable to develop a global defense architecture against attacks to network infra-structure. Experimental results show that the proposed scheme can detect the network attack symptoms very exactly and quickly, and protect the network resources as well as the normal traffic flows very efficiently.

1 Introduction

Recently, we have experienced several troubles of the Internet services due to various network attacks. Attacks on the Internet infrastructure can lead to enormous destructions, since different infrastructure components of the Internet have implicit trust relationship with each other.

Though there have been much of works on network attacks and corresponding solutions such as [1], [2], [3], and [4], they have still the following problems. First, much of those methodologies have been focused on detecting and reacting to network attacks at individual end-networks for their own safety. However, since global-scale network attacks are far more defined and visible in the backbone before it spreads out towards targets, it might be more efficient that to detect and react the symptoms of those global-scale network attacks are done in backbone domain. Second, detection and control by those methods are done based on individual packet or flow levels so that they require much higher computational complexities. Accordingly, they can not be directly applied to the high-speed Internet backbone networks. Finally, since attack mechanisms and tools continue to improve and evolve, there is a limitation to develop a solution against each attack type. As one promising direction to react those evolving various attack patterns, Chang suggested to develop a global defense infrastructure[4]. However, it is only suggestion, but does not provide any actual mechanism or architecture.

C. Kim (Ed.): ICOIN 2005, LNCS 3391, pp. 853–863, 2005.

In this paper, we propose a new global traffic control architecture against network attacks in high-speed Internet backbone networks. In the proposed architecture, attack symptoms detection is carried out at aggregate traffic level, not at individual packet of flow level. Then, by the proposed traffic control, the attack traffic is entirely cut of from the normal traffic in the backbone network.

The rest of the paper is organized as follows. In section 2, our network attack symptom detection method at aggregate traffic level is illustrated. In Section 3, the proposed global traffic control architecture against the network attacks is presented. Then, experimental results are shown in Section 4, and finally we conclude the paper in Section 5.

2 Detection of Attack Symptoms Based on Traffic Measurement

According to Houle and Weaver[1]'s survey on the trends in deployment, use, and impact of variety of denial of service (DoS) attacks, most of attack tools alter packets' major attributes such as source/destination IP addresses and port numbers for different purposes of attacks. By regarding the degree of alteration of those attributes, we investigated the network attack traffic pattern from the captured traffic in an Internet backbone network administrated by National Computerization Agency, Korea[5]. From the investigation, two measures have been derived to detect network attack symptoms at aggregate traffic level[6]. The two measures are the packet count-to-the traffic volume ratio (CVR) and the self-similarity represented by the Hurst parameter. Some important attack traffic characteristics that are used for deriving those two measures are illustrated in the Appendix.

The Hurst parameter can be estimated by using periodogram method[7]. For the discrete-time sequence with l samples X_0, X_1,...,X_{l-1}, the periodogram is defined as

$$S(\omega) = \frac{1}{2\pi N}\left|\Sigma_{k=0}^{l-1} X_k e^{jk\omega}\right|^2 \tag{1}$$

The relationships between S(ω) and ω can be written as

$$\log_{\omega \to 0} S(\omega) = a_0 \log|\omega| + a_1 \tag{2}$$

That is, when S(ω)is plotted against ω on a log-log plot, it can be approximated by a straight line with a slope a_0. Then, the Hurst parameter is given by $H = (1 - a_0)/2$. It is noted that Eq.(1) is based on the discrete-time Fourier transform (DFT). Li et al[8] showed that the queuing performances in networks can be dominated by the average power spectrum obtained by DFT of the input traffic. From the researches related to [8] and the results shown in Appendix, we use the average power spectrum(APS) as one of the measures for detecting the network attack symptoms. The average power spectrum is defined as follows. Let us assume that the time is divided into a constant period of Δ, then Δ is a basic unit for traffic measurement. Let c_n and v_n (n=0,1,2,...) be the packet count and the traffic volume of the aggregate traffic measured during n-th Δ duration, respectively. Let $L\Delta$ be the detection period, which is the time duration that the detection algorithm is applied to. That is, each detection period consists of unoverlapped L consecutive Δs. And, we define the packet counts and traffic volumes measured during m-th detection period as the vectors $\boldsymbol{c}_m = [c_{mL}, c_{mL+1}, ..., c_{(m+1)L-1}]$ and $\boldsymbol{v}_m = [v_{mL}, v_{mL+1}, ..., v_{(m+1)L-1}]$ (m=0,1,2,...), respectively. Then, we have the average power spectrum for the vector $\boldsymbol{c}_m$ as following

$$\overline{P}(m) = \Sigma_{k=0}^{L-1} \phi_{mk} \tag{3}$$

where $\Psi_m = [\phi_{m0}, \phi_{m1}, ..., \phi_{m(L-1)}]$ is obtained through DFT of $\boldsymbol{c}_m$. That is, $\Psi_m = L^{-2}|DFT(\boldsymbol{c}_m)|^2$.

It is noted that the average power is a measure for reflecting the effect of the self-similarity due to the network attacks. Besides the self-similarity, in [6], it is shown that the ratio of the packet count to the traffic volume is significantly changed when the network attacks are added. The packet count-to-the traffic volume ratio (CVR) is given by

$$\overline{R}(m) = \frac{\boldsymbol{c}_m \cdot \boldsymbol{e}}{\boldsymbol{c}_m \cdot \boldsymbol{e}} \tag{4}$$

where $\boldsymbol{e} = [1, 1, ..., 1]^T$, and $[\bullet]^T$ indicates a transpose matrix.

The method to detect the network attack symptoms considering two measures given in Eq.(3) and Eq.(4) is as follows. Let $x_p(m)$ and $x_r(m)$ be the weighted averages of the APS and the CVR measured at m-th detection period, respectively, and given by

$$x_p(m+1) = \alpha_p x_p(m) + (1 - \alpha_p)\overline{P}(m) \qquad , m = 0, 1, 2, ... \tag{5}$$

$$x_r(m+1) = \alpha_r x_r(m) + (1 - \alpha_r)\overline{R}(m) \qquad , m = 0, 1, 2, ... \tag{6}$$

where α_p and α_r are the constant values between 0 and 1.

In general, it is known that the normal traffic flows in a stationary state vary within a forecastable range. Let m_p and δ_p be the average and the tolerance for the weighted average of the APS of the normal traffic within a certain stationary state, respectively. The m_p is calculated by the conventional way to obtain an arithmetic mean, and the δ_p is determined by an administrative policy of the network operators. Similarly, we define m_r and δ_r as the average and the tolerance for the weighted average of the CVR of the normal traffic within a certain stationary state, respectively. It is assumed that the tolerances δ_p and δ_r are determined based on the normal traffic flows prior to the actual measurement for the detection. The case when the measured $x_p(m)$ and/or $x_r(m)$ exceed the tolerances δ_p and δ_r, we can assume that it may be a symptom of the network attacks. In order to determine the situation that the network infrastructure is currently being attacked, we define the three states such as NORMAL, ALERT, and ATTACK. The NORMAL state is the state that there is no attack. The

ALERT state is the one that the attack symptom is in question but the decision of the attack symptom is not completed. And, in the ATTACK state, it is inferred that the network resources are being attacked. The transition between those states and the main algorithm to detect the attack symptom considering those states are illustrated as follows.

<variables>

attack_count : counter for representing the degree of attack
Alert_Threshold : threshold value to change between ALERT and ATTACK states
Attack_Threshold : maximum value of *attack_count*
state : current state of the algorithm

<main algorithm>

at the end of every detection period, update $x_p(\cdot)$ and $x_r(\cdot)$ by using (5) and (6) considering $x_p(\cdot)$ and $x_r(\cdot)$, the state at the detect period is determined by the following sequence.

```
if ( state == NORMAL )
   if ( ( xp(·) > δp AND xr(·) ≤ δr ) OR ( xp(·) ≤ δp AND xr(·) > δr) )
      state = ALERT;
      attack_count += 1;
   elseif ( xp(·) > δp AND xr(·) > δr )
      state = ALERT;
      attack_count += 2;
   endif
elseif ( state == ALERT )
   if ( ( xp(·) > δp AND xr(·) ≤ δr ) OR ( xp(·) ≤ δp AND xr(·) > δr ) )
      attack_count += 1;
   elseif ( xp(·) > δp AND xr(·) > δr) )
      attack_count += 2;
   elseif ( xp(·) ≤ δp AND xr(·) ≤ δr )
      attack_count -= 2;
   else
      attack_count -= 1;
   endif
   if ( attack_count > Alert_Threshold )
      state = ATTACK;
   elseif ( attack_count ≤ 0 )
      state = NORMAL;
      attack_count = 0;
   endif
elseif ( state == ATTACK )
   if ( xp(·) > δp AND xr(·) > δr) )
      attack_count = MIN ( attack_count+1, Attack_Threshold);
   elseif ( xp(·) ≤ δp AND xr(·) ≤ δr )
      attack_count -= 2;
   else
      attack_count -= 1;
   endif
   if ( attack_count ≤ Alert_Threshold )
      state = ALERT ;
   endif
endif
```

---

3 Proposed Global Traffic Control Architecture

Fig.1(a) shows the proposed global traffic control architecture for protecting the resources of network infrastructure by isolating network attacks using the detection method described in Section 2. At the boundary of the Internet backbone

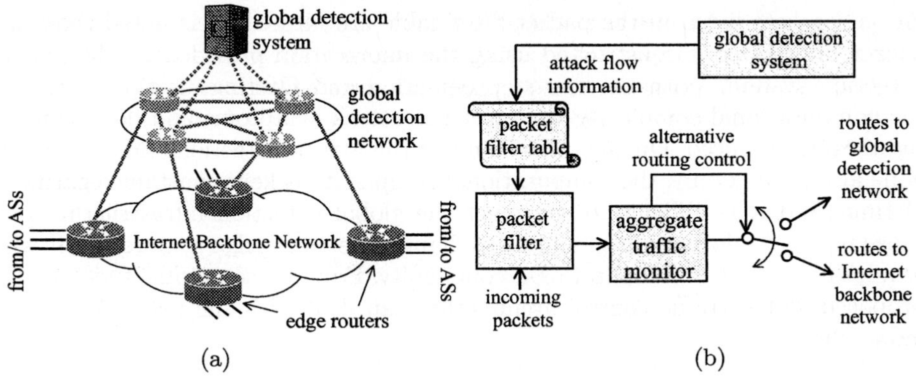

Fig. 1. Global traffic control architecture against network attacks: (a) architecture (b) traffic control mechanism at edge routers

network, there are edge routers with the control mechanism as shown in Fig.1(b). Each link at an edge router has a packet filter unit and an aggregate traffic monitor unit. The packet filter unit filters attack packets flowed into the backbone network according to the packet filter table. In the packet filter table, there exist the list of attack flows to be filtered, which is provided by the global detection system in the global detection network. With the filtered aggregate traffic, the traffic monitor unit detects whether the further attack symptoms exist by using the detection method described in Section 2. It is noted that at initial state, there is no list in the packet filter table, so the filtered aggregate traffic is just same as the incoming traffic into the packet filter unit. If any attack symptom is detected, all packets are alternatively routed to the global detection network. If no attack is detected, packets are forwarded through the backbone network according to a normal routing policy.

In the global detection network shown in Fig.1(a), the packet forwarding process to deliver packets to their destination networks is also carried out. However, egress edge routers with links from both the backbone and the global detection networks treat packets from the backbone network with higher priority than those from the global detection network. That is, packets from the global detection network can be forwarded only if there exist an available bandwidth after the packets from the backbone network are forwarded first. The global detection system classifies attack flows from the traffic flowed into the global detection network by using the detection method such as [2] or others. The detected attack flow information is reflected in the packet filter table at each ingress edge router, then the packet filter unit at the edge routers can filter the attack packets by referring to the packet filter table.

In the global traffic control architecture, firstly the attack symptoms are detected for the aggregate traffic at each ingress edge router on each inbound link to the backbone network, and then the flow level attack detection is carried out for those detected aggregate traffic only in the global detection network. In addition, packets listed in the packet filter table are filtered. It is noted that the packet filter table is constructed using the information provided by the global detection system. With the two step detection and filtering, as time goes by, the computational complexity to detect the attack symptoms and flows can be significantly reduced. On the other hand, if the attack detection and filtering is done at each link only, the computational complexity is kept constant regardless of time, and it is difficult to construct the global detection infrastructure because much of information should be exchanged between edge routers and it will require much higher computational complexity. From these points of views, the proposed global traffic control architecture can deal with the network attacks more efficiently.

4 Experimental Results

For the experiment, we artificially generated normal and attack traffic according to the known statistics as in [7] and [5], respectively. That is, for the parameters of the normal traffic, we used 0.9 for the Hurst parameter and 9 Kpackets per second for the average packet count. For those of the attack traffic, we used 0.99 for the Hurst parameter and 5.3 Kpackets per second for the average packet count. We had the packet size of each generated packet followed the distribution shown in [5]. For the self-similar traffic generation, we used the method proposed in [9].

Efficiency of Aggregate Traffic Level Attack Detection First, we carried out an experiment to show the efficiency of the attack detection method at aggregate traffic level. For the experiment, normal traffic was generated during about one hour. Then, attack traffic was additionally generated for 30 minutes from 10 minutes after the normal traffic generation had started. For testing the effectiveness of the proposed method, we define the following performance measures

$$Exactness = (T_{detected}/T_{actual}) \times 100(\%) \tag{7}$$

$$ErrorRatio = (T_{error}/T_{actual}) \times 100(\%) \tag{8}$$

$$DetectDelay = T_{detected} - T_{actual} \tag{9}$$

where T_{actual} and $T_{detected}$ are the time duration that the attack traffic is appeared in the test sequence used for the experiment and the time duration that the attack is detected by the proposed algorithm, respectively. T_{error} is the time duration that the detection is not correct, which consists of the duration classified into ATTACK state though it is not the actual attack duration and the duration that is not classified into ATTACK state though it is within the actual attack duration. And, T_{actual} and $T_{detected}$ are the start time of the attack in the test sequence and the time that the attack is firstly detected by the algorithm, respectively.

Table 1. Efficiency of the attack detection algorithm

$L \setminus \Delta$	10 msec	100 msec	1 sec
10	99.98%/0.001/300msec	99.77%/0.016/3sec	97.71%/0.275/30sec
100	99.77%/0.016/3sec	97.71%/0.274/30sec	n/a
1000	97.71%/0.274/30sec	n/a	n/a

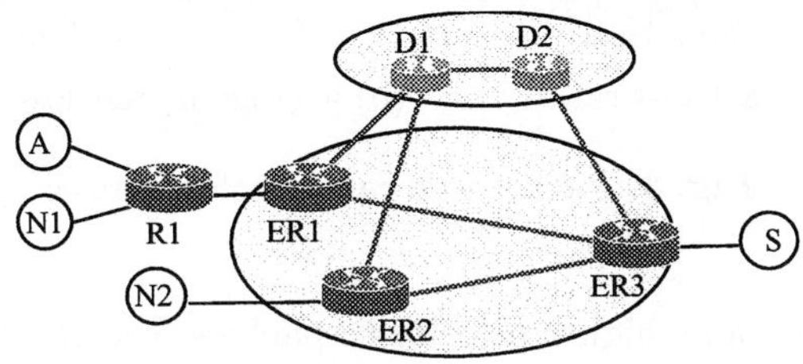

Fig. 2. Network topology for the experiment

Table.1 shows the experimental results for the detection algorithm varying the unit time Δ and the detection period L. The format of A/B/C in each cell in Table.1 represents the performance measures such as Exactness/ Error-Ratio/ DetectDelay. The parameters for obtaining Table.1 are as follows: α_p= α_r=0.9, *Alert_Threshold*=5, *Attack_Threshold*=10. For δ_p and δ_r, the maximum values of weighted averages obtained from the normal traffic test sequence by using Eq(5) and Eq(6), respectively, are used. As we can imagine intuitively, the smaller the values of Δ is, the better the performance measures such as Exactness and ErrorRatio are. While the bigger the value of L is, the larger the DetectDelay is. However, it is noted that as Δ and L are decreased, the computational complexity will be increased, and it will give the more load to the system. For any treatable parameters, the exactness of the detection is greater than 97%.

Performance of the Global Traffic Control Architecture. The performances of the proposed global traffic control architecture are evaluated by using ns-2 simulator[10]. The simulation topology is shown in Fig.2. The backbone network consists of three edge routers such as ER1, ER2 and ER3, and there are two routers D1 and D2 in the global detection network. R1 is the router outside the backbone network. We assign each link's bandwidth as 100 Mbps, the queue length between ER3 and S as 200 packets, and other queues as infinite. Nodes N1 and N2 generate normal packets destined for node S, and node A does attack packets for node S also. To show the efficiency of the proposed architecture, we set the following three scenarios. In all scenarios, all packets generated from N2 are delivered to S through ER2 and ER3. As queueing disciplines, CBQ (class-based queueing) is used at ER3, and FCFS is used at other routers. For the

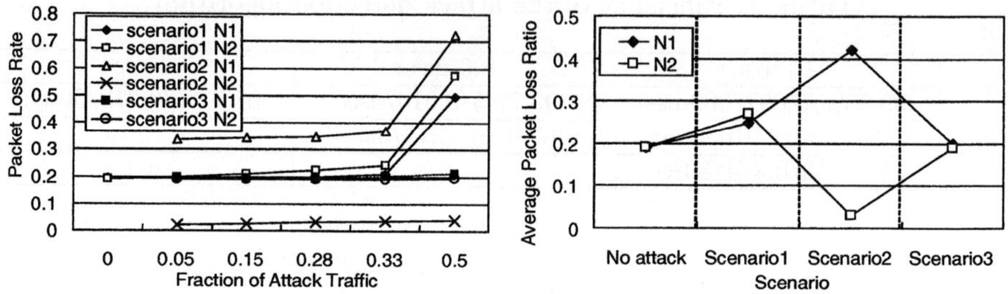

(a) packet loss ratio (b) average packet loss ratio

Fig. 3. Packet loss ratio performances

CBQ, we assign the same higher priority to packets directly delivered from ER1 and ER2, and the lower priority to packets from D2.

- Scenario 1 : All packets from N1 and A are delivered to S from ER1 to ER3 directly.
- Scenario 2: All packets from A and N1 are alternatively routed to D1 at ER1, and then delivered to ER3 through D2.
- Scenario 3 : All packets from A are filtered at ER1, and those from N1 are delivered to ER3 through ER1 directly.

It is noted that the Scenario 1 corresponds to the case that no attack detection and control mechanisms are not applied to. Scenario 2 is the case that the attack from a link is detected at the aggregate traffic level, and the packets from the link are alternatively routed to the global detection network for protecting resources in the backbone network, but global attack flows are not classified yet. Scenario 3 illustrates that packets from the attack flows classified by the global detection system are filtered. For those three scenarios, we evaluate the packet loss ratio and the end-to-end delay performances, which are shown in Fig.3 and Fig.4. For obtaining the results of Fig.3 and Fig.4, the traffic rates from node N1 and N2 are fixed, while the attack traffic rate from A is varied. So, the horizontal axes of Fig.3 and Fig.4 are the fraction of the attack traffic to the total normal traffic in packet counts.

As we can see from Fig.3(a), as the attack traffic increases, the loss ratio of packets from N1 and N2 also increases in Scenario 1. Especially, the loss ratio of N2 shows slightly larger than that of N1. The reason of this phenomenon is as follows: Since priorities of packets from all nodes N1, N2 and A are same in Scenario 1, ER3 treats all packets according to FCFS discipline for two links, so the loss probability from N2 may be increased. In Scenario 2, while the loss ratio of N2 packets are kept at a constant level lower than that of Scenario 1, the loss ratio of N1 increases significantly. This is because the N1 packets are alternatively routed to D1 with attack packets and they are treated at ER3 with lower priority than N2 packets. In Scenario 3, the loss ratios of N1 and N2 packets are very similar as those in Scenario 1 when there is no attack

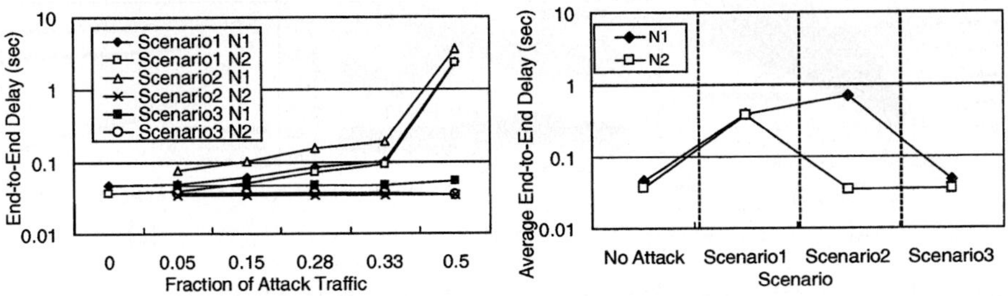

(a) end-to-end delay (b) average delay

Fig. 4. End-to-end delay performances

traffic. Due to the insertion of attack packets in the link between R1 and ER1, the loss ratios of N1 packets show slightly larger values compared with those of Scenario 1. The average statistics for Fig.3(a) is shown in Fig.3(b). From Fig.3(b), we can see that the quality of service(QoS) on the links without attack traffic degraded during Scenario 1 period, but the QoS is recovered through scenario 2 and 3 period. It is noted that since the attack detection is done at aggregate traffic level, the duration of Scenario 1 can end in very short period. In addition, though the node that shares the link with any attack traffic experiences the QoS degradation in Scenario 1 and Scenario 2 periods, its QoS is recovered after Scenario 3 period starts. Fig.4 shows the end-to-end delay performances for the proposed architecture. As we can see from Fig.4, the trend of the end-to-end delay is very similar to that of the packet loss ratio.

From the experimental results, we can see that the proposed global traffic control architecture can protect the normal traffic in the backbone networks very efficiently. That is, when attack traffic is inserted to a link, though there are some performance degradation of the normal traffic on the same link during some short period before the global detection system classify those attacks, the performance of the normal traffic can be recovered in soon. It is noted that while the performance degradation is experienced in the normal traffic shared same link with the attacks, the other normal traffic with different links without attacks are not affected by those attacks. Likewise, the proposed global traffic control architecture can not only isolate the attacks, but also protect the normal traffic from those attacks.

5 Conclusion

In this paper, we proposed a global traffic control architecture against network attacks to isolate the attack traffic from the normal traffic. The architecture is based on the detection of network attack symptoms at the aggregate traffic level, so the architecture can be operated with very lower computational complexities. We showed the effectiveness and the applicability of the proposed methods by experiments.

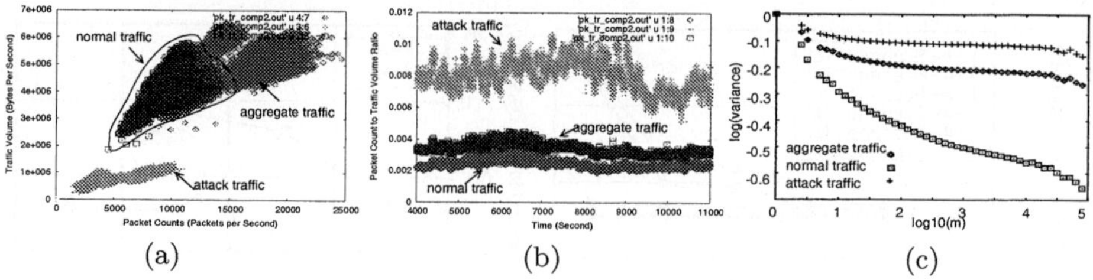

Fig. 5. Attack traffic characteristics:(a) relative statistics between packet counts and traffic volumes (b) CVR characteristics (c) VTP characteristics

Backbone networks have their own network management systems to monitor and maintain the networks. By cooperating the detection methodology based on the traffic measurement with those network management systems, it is possible to develop a global defense infrastructure. The proposed global traffic control architecture can be used for constructing the global defense infrastructure. The more practical control mechanism to cooperate with network management systems by extending the proposed control architecture requires further study. In addition, we derived the measures for detecting attack symptoms by using captured packet data empirically. So, to investigate the traffic model for reflecting the variety of network attacks and to develop more practical control method to detect and deal with those attacks require further study.

Appendix: Attack Traffic Characteristics

In this Appendix, it is shown some important attack traffic characteristics that are used for deriving two measures illustrated in Section 2. Traffic data for the investigation were captured on two trans-pacific T3 links connecting the U.S. and a Korean Internet Exchange with the help of National Computerization Agency, Korea[5]. Then, we classified network attack packets from the captured data by using the method proposed in [2] into three classes such as the attack traffic as the sequence of packets classified for network attacks among captured packets, the normal traffic as the sequence with remaining packets, and the aggregate traffic as the sequence of whole packets. Among the traffic characteristics for those three traffic classes, the relationships between the traffic volume and the packet counts are depicted in Fig.5(a), in which all values are measured in 1 second period. From Fig.5(a), we can see that when the attacks are added, the degree of packet count variation in the aggregate traffic becomes much higher than that of traffic volume variation. Fig.5(b) shows the CVR in a certain time period, in which more attacks are detected than in other time period. As we can expect, the CVR of the attack traffic is much higher than that of the normal traffic. Accordingly, the CVR of the aggregate traffic is getting increase by adding the attack traffic as shown in Fig.5(b).

It has been well known that the Ethernet traffic is statistically self-similar[7]. The self-similar nature of a traffic flow can affect the development of network congestion control schemes as well as the source traffic characterization. It is noted that the higher the self-similarity of the traffic is, the more the network performances such as link utilization, throughput, and so on, are affected. In order to find out how the self-similar nature of the normal traffic is affected by the attack traffic, in Fig.5(c), it is shown the variance-time-plot (VTP) [7] for estimating Hurst parameters for each traffic type. We can see that the attack traffic shows higher self-similarity than the normal traffic, and that the self-similarity of the aggregate traffic is increased by adding the attack traffic. This indicates that since the addition of the attack traffic increases not only the traffic volume in networks, but also the self-similarity of the aggregate traffic, it can significantly affect the network performances more than in the situation that the same amount of the normal traffic is added.

Acknowledgements. This work was supported by grant (No. R05-2004-000-10824-0) from Ministry of Science & Technology, Korea.

References

1. Houle K., Weaver J.: Trends in Denial of Service Attack Technology, CERT Coordination Center (2001)
2. Kim H., et al.: Fast Classification, Calibration, and Visualization of Network Attacks on Backbone Links, Tech. Report, http://net.korea.ac.kr (2003)
3. Chakrabarti A., Manimaran G.: Internet Infrastructure Security: A Taxonomy, IEEE Networks, vol.16, no.6 (2002)
4. Chang R.: Defending Against Flooding-Based Distributed Denial-of-Service Attacks: A tutorial, IEEE Communications Magazine, Oct. (2002) 42-51
5. Jeon Y., Roh B., Kim J.: Traffic Characterization For Network Attack Flows On the Internet Backbone Links," Internet Computing 2004, Las Vegas, June (2004)
6. Choi W.,Roh B., Yoo S.W.: Measures For Detecting Network Attacks At The Aggregate Traffic Level On High-Speed Internet Backbone Links, SAM'04, Las Vegas, June. (2004)
7. Leland W.E., Taqqu M.S., Willinger W., Wilson D.V.: On the Self-Similar Nature of Ethernet Traffic (extended version), IEEE/ACM Tr. Networking, Feb. (1994)
8. Li S., Hwang C.L.: Queue Response to Input Correlation Functions: Discrete Spectral Analysis, IEEE/ACM Tr. Networking, Oct. (1993) 522-533
9. Paxon V.: Fast, Approximate Synthesis of Fractional Gaussian Noise for Generating Self-Similar Network Traffic, ACM SIGCOMM CCR, vol.27, no.5, (1997)
10. The network simulator version 2, ns-2, available at http://www.isi.edu/nsnam/ns/

An Enhancement of Transport Layer Approach to Mobility Support*

Moonjeong Chang[1], Meejeong Lee[1], Hyunjeong Lee[2],
Younggeun Hong[3], and Jungsoo Park[3]

[1] Dept. of Computer Engineering, Ewha Womans University, Seoul 121-791, Korea
{mjchang, lmj}@ewha.ac.kr
[2] Dept. of Electrical Engineering, University of Nevada, Reno, NV 89557, USA
hyunlee@unr.edu
[3] Protocol Engineering Center, ETRI, Daejeon 305-350, Korea
{yghong, pjs}@etri.re.kr

Abstract. In this paper, we propose an approach to transport layer mobility support leveraging the SCTP extension dubbed dynamic address reconfiguration. Timing issues related to the end-to-end address management, and a novel error recovery mechanism associated with a handover are discussed. The error recovery time of proposed mechanism is analyzed and compared to that of the plain SCTP for handover cases. Finally, through a series of simulations, the performance of the proposed SCTP enhancements over plain IPv6 is compared with the MIPv6 and its variants (i.e., HMIPv6 and FMIPv6) with TCP Reno on top of them. The simulation results present the performance gains of the pro-posed error recovery mechanism, which is possible within the context of transport layer mobility management.

1 Introduction

Realizing seamless mobility is required for the next generation IPv6 [1] Internet users. Currently, starting from Mobile IPv6 (MIPv6) [2], several approaches [3, 4] to the seamless mobility support have been proposed at the network layer. Network layer approaches may be beneficial as mobility support can be done transparently to transport layers or upper layers. However, these approaches counter to a few undesirable characteristics: network architecture complexity due to additional special entities, overhead due to triangular routing and tunneling, complicated security issues, etc. Furthermore, the transport layer may not be able to optimally control the transmissions because it is transparent to the handover as well as the new network situations [2, 3, 4].

On the other hand, an end-to-end approach based on transport layer might alleviate these problems, while such an approach requires a mechanism to map a new end-point address to the existing connection. Stream Control Transmission Protocol (SCTP) [5] is among standard transport layer protocols and provides

* This work was supported by grant No. R04-2004-000-10073-0 from the Basic Research Program of the Korea Science and Engineering Foundation

C. Kim (Ed.): ICOIN 2005, LNCS 3391, pp. 864–873, 2005.

a good potential for the dynamic address mapping problem. Inherently, SCTP supports multi-homed end hosts with more than one IP address allocated to it [5]. Standardization within IETF is in progress to extend this protocol to provide a method to dynamically reconfigure IP address information on an existing connection [6]. Therefore, in this paper, we propose a novel transport layer approach to mobility support based on SCTP. Leveraging the SCTP dynamic address reconfiguration feature, we first design a mechanism to change a data packet destination address as well as means to map an endpoint address to an existing association. Then, an efficient error recovery mechanism is proposed to optimize the network performance upon handover. A thorough simulation study will be followed to evaluate and validate the proposed approach compared to other network layer approaches.

The rest of the paper is organized as follows. Section 2 presents our transport layer approach to mobility support. A comprehensive simulation study of various schemes including MIPv6, Hierarchical MIPv6 (HMIPv6)[3], Fast Handover for MIPv6 (FMIPv6)[4] and the proposed approach is given in Section 3. Finally, we conclude our work in Section 4.

2 A Transport Layer Mobility Support Scheme

This section describes our approach to transport layer mobility support leveraging the SCTP extension dubbed dynamic address reconfiguration [6]. This extension is used to dynamically add or remove addresses from an ongoing transport layer association. In the proposed scheme, we are particularly interested in mobile terminals (MTs) equipped with a single wireless network interface that are currently the most common feature.

An SCTP endpoint is considered multi-homed if there are more than one transport addresses that can be used as a destination address to reach that endpoint. With these multi-homed endpoints, one endpoint shall select one of the multiple destination addresses of a multi-homed peer endpoint as the primary path to transmit data packets. The rest of paths can be used as backup in the case of failure of the primary address, or for retransmissions of lost packets [5]. In fact, the proposed approach allows the correspondent terminal (CT) to transmit data packets only through a primary path at a given time even though the CT maintains several IP addresses for the MT with which it is communicating because we focus on MTs equipped with a single network inter-face. In the proposed scheme, the end-to-end mobility support can be achieved by the aid of Address Configuration Change (ASCONF) and Address Configuration Acknowledgement (ASCONF-ACK) control chunks, which may contain different re-quest parameters for the peer [6]. These parameters signal:

- AddIP: the address specified is to be added to the existing SCTP association,
- DELETEIP: the address specified is to be removed from the existing SCTP assciation, or
- Set Primary: the specified address is marked as the primary address to send data

In the following sections 2.1 and 2.2, we first discuss timing issues for the MT to send its CTs ASCONF chunks with these three different types of parameters in a handover situation. In addition, we present an efficient error recovery mechanism associated with a handover leveraging the fact that the transport layer perceives the MT handovers.

2.1 Timing Issues for End-to-End Address Management

As illustrated in Fig. 1, the MT first performs the layer-2 handover and then obtains an IP address as it moves towards the new sub-network. For the MT equipped with a single network interface, once the layer-2 handover begins, the MT cannot receive any more data packet from the previous Access Point (AP). Thus, it needs to update the CTs on its new IP address as soon as possible. Consequently, upon detect-ing a new IP address, the associated ADDIP and Set Primary embedded in the ASCONF chunk need to be sent immediately. On the other hand, the related DELETEIP might be triggered at one of the following options:

1. After the MT completely moves into the new location (i.e., the previous AP is unreachable any longer)
2. As soon as it acquires a new IP address (i.e., at the same time the MT triggers ADDIP and Set Primary)
3. When the MT successfully receives the first data packet through the new primary path

We note that the time instant that the MT informs the CT of the DLETEIP is not critical to the handover latency. The handover performance is determined by the time instants that data transmissions through a new path become possible and data transmissions through the previous path become unavailable, respectively. These events are independent on when to trigger the DELETEIP. In fact, the former depends on when the Set Primary is performed and the latter when the layer-2 handover starts. It is however worth noting that protocol processing overhead is directly related to when the DELETEIP is triggered.

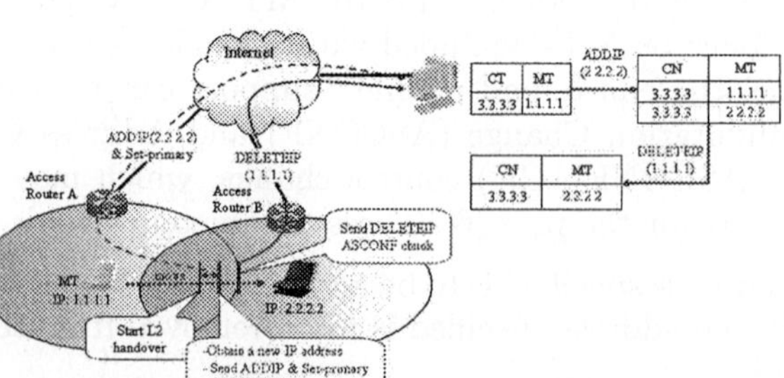

Fig. 1. Operation of proposed scheme

As a normal movement, if the MT keeps moving from a pervious AP to a new AP, the strategy (2) is desirable because three signaling messages (i.e., ADDIP, Set Primary, and DELETEIP) are aggregated into a single ASCONF chunk. Thus, this strategy is beneficial in terms of the number of ASCONF/ASCONF-ACK control chunks per handover and also the interrupt overhead within the communication protocol stacks resulting from the control chunk reception by the CT. However, the strategy (2) bundles up three operations and thus imposes unnecessary processing over-heads on the MT and the CT with unnecessary ADDIP and DELETEIP if the ping-pong patterned movement repeats. The strategy (1) is suitable to the ping-pong movement case due to its ability to separate the Set Primary triggering. The strategy (3) has the same amount of overhead as the strategy (1) if the ping-pong movement incurs another handover before MT receives the first data packet through the current path. Otherwise, the strategy (3) triggers both ADDIP and Set Primary because the DELETEIP has been already performed and thus incurs the same amount of processing overhead as the strategy (2) does. In the proposed scheme, we use the strategy (1) by which the ping-pong phenomenon can be gracefully dealt.

2.2 An Efficient Error Recovery Mechanism for Handover

In this section, as an extension of SCTP, we present a novel approach to achieve efficient error recovery mechanism for handover. The proposed approach and the current SCTP standards will be analyzed with regards to the handover latency as well as the time duration to recover lost packets. The handover latency is defined as the elapsed time between the beginning of layer-2 handover and the arrival of the first packet via new route. In our analysis, we assume that packet loss is only due to hand-over and all of the acknowledgements (ACKs) sent by the MT for packets that have arrived at it before the handover are successfully delivered to the CT. SCTP employs error and flow/congestion control algorithms similar to those used in TCP. While supporting multi-homing, SCTP maintains a receive window size per association, and a congestion window size and an outstanding data size per path. In mobile environments, this implies that the congestion window size of the new path is not affected by the ACK sent over the old path before handover. Following the slow start phase, SCTP reduces the congestion window size to one in the case of retransmission time-out (as TCP does) and sets it to two for the new path, respectively.

As illustrated in Fig. 2(a), CT may not receive any ACK for a while, after receiving the last ACK sent over to the old path, because of the losses caused by the handover. Then, the CT may exhaust its available window, set its retransmission timer, and thus temporarily stop transmitting. Later, it retransmits the first lost packet as the timer expires. If this happens after the CT receives the ASCONF chunk, the packet would be successfully sent over the new path. But if not, the packet would be sent over to the old path, and lost again, resulting in second timeout, and so on. That is, the number of retransmissions for the packet can be more than once depending on the time that the CT receives the ASCONF chunk. For each timeout, the timer doubles its timeout threshold value [5]. As shown in Fig. 2(a), if the CT suffers from n^{th} retransmission timeout before the corresponding ASCONF chunk, the handover latency $H_{timeout}$ is

$$H_{timeout} = PAT + D^{new}_{ASCONF} + \alpha + D^{new}_{data} \approx \sum_{i=0}^{n} 2^i \times TO, \tag{1}$$

, where $0 \leq \alpha \leq 2^n \times TO$.
In (1), PAT (Path Acquisition Time) is the elapsed time for the MT between the be-ginning of the layer-2 handover and the acquisition of a new IP address. D^x_y denotes the delay to deliver y through the path x, TO denotes the initial retransmission timeout value, and α equals the length of time interval from the time instant that the CT receives the ASCONF chunk till the instance that the current retransmission timer expires. Following the congestion control mechanism of the current SCTP, the new path cannot be used even after the arrival of the ASCONF chunk unless the current retransmission timer expires. This side effects result in exponentially increasing hand-over latency with the number of retransmission timeout. Furthermore, every retransmission timeout reduces the congestion window to one and this value increases following the slow start phase. Therefore, given that $l\,(>0)$ packets are lost and i retransmission timeouts occur during handover before the CT receives the corresponding ASCONF chunk, the total time delay ($L_{timeout}$) at the CT for the lost packet recovery after the arrival of the ASCONF chunk is

$$L_{timeout} = \alpha + (1 + \lfloor \log_2 l \rfloor) \times RTT. \tag{2}$$

, where and RTT denotes the round trip time.

On the other hand, as illustrated in Fig. 2(b), the CT may proceed its transmission over the new path without experiencing retransmission timeout. In this case, the handover latency is minimum with the original SCTP. We formulate this handover latency $H_{no-timeout}$ as in the following:

$$H_{no-timeout} = PAT + D^{new}_{ASCONF} + D^{new}_{data}. \tag{3}$$

In this case, the lost packets are recovered by Fast Retransmission mechanism because ACKs are arriving continuously. Since SCTP Fast Retransmission mechanism requires 4 duplicate ACKs, which takes at least 2RTTs after receiving the ACONF, the recovery of losses caused by handover may start only after 2RTT passes after receiving the ASCONF. Detailed proof of this assertion is omitted due to page limitation. After receiving 4 duplicate ACKs, SCTP applies Fast Recovery congestion control. In this kind of handover case, the congestion window is supposed to be less than 2 packets when the Fast Recovery starts, and Fast Recovery phase increases the size of congestion window by one packet for each RTT.

Based on this observation, now we formulate the relationship between the number of lost packets and its recovery time normalized by RTT in Equation (4). The sequence of minimum number of lost packets that requires $k \cdot RTT$ for error recovery, for $k \geq 2$, is a sequence of progression of differences with the initial term being 2 and the progression difference being $(k + 2)$. Hence, the minimum number of lost packets, which is denoted by S_k, that requires $k \cdot RTT$ recovery time is

$$S_k = 2 + \sum_{i=1}^{k-1} (i + 2) = \frac{k^2 + 3k}{2}, \quad for \quad k \geq 2. \tag{4}$$

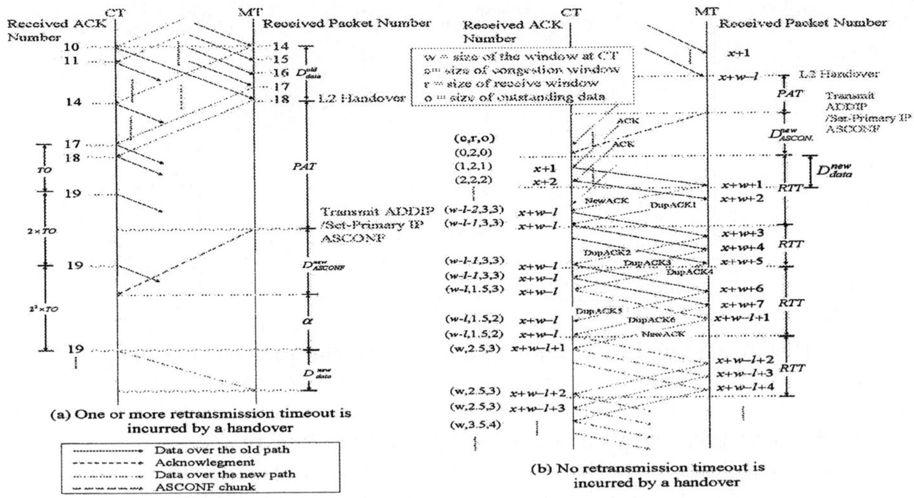

(a) One or more retransmission timeout is incurred by a handover

(b) No retransmission timeout is incurred by a handover

Fig. 2. Error recovery for handovers by the original SCTP

We replace k with *(k-1)* in (5) in order to include the case with $k = 1$, and obtain

$$S_{k-1} = \frac{(k-1)^2 + 3(k-1)}{2} = \frac{k^2 + k - 2}{2}, \quad for \quad k - 1 \geq 1. \tag{5}$$

Then, the number of lost packets becomes

$$l = \lfloor \frac{k^2 + k - 2}{2} \rfloor. \tag{6}$$

Solving the equation (7) for k, k is obtained as follows:

$$k = \lfloor \frac{-1 + \sqrt{1 - 4(-2 - 2l)}}{2} \rfloor = \lfloor \frac{\sqrt{8l + 9} - 1}{2} \rfloor, \quad for \quad l \geq 1. \tag{7}$$

As a result, without retransmission timeout, the total time delay $L_{no-timeout}$ at the CT for the lost packet recovery after the arrival of the ADDIP/Set-Primary ASCONF chunk is

$$L_{no-timeout} = (2 + \lfloor \frac{\sqrt{8l - 9} - 1}{2} \rfloor) \times RTT, \quad for \quad l \geq 1. \tag{8}$$

As formulated in Equation (8), the total time delay to recover all of the lost packets is considerable even with no retransmission timeout.

In view of this, we propose a simple but very effective means to minimize the lost packet recovery time as well as the handover latency by capitalizing on the transport layer awareness of the mobility. The proposed approach requires changes of SCTP module only at the CT part. Upon receipt of this ASCONF chunk, the CT immediately stops its timer for the last packet transmitted over the old path and sends the packet with the lowest sequence number among the packets which are never sent to the MT yet. This packet is referred to as a probe packet in this paper. SCTP standards allow sending one packet even with zero receiver window if the outstanding data size is zero. Since the outstanding data

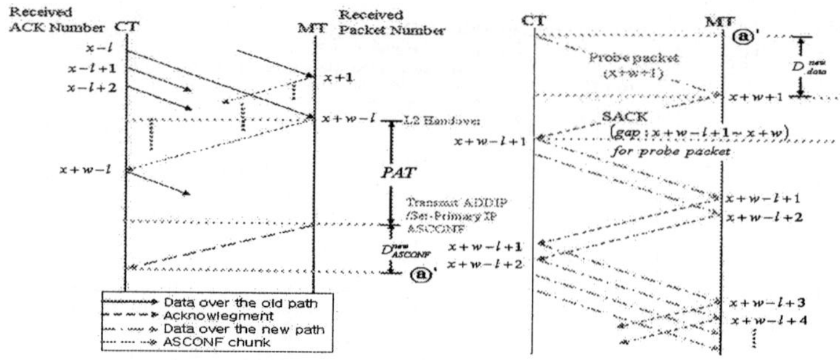

Fig. 3. Error recovery for handovers by the proposed scheme

size of the new path is always zero, transmitting this probe packet is always legitimate with respect to the existing SCTP flow control regardless of the receiver window size. Since SCTP deploys Selective ACK (SACK), the ACK for the probe packet from the MT contains the information about the last packet received by the MT through the old path. Receiving the ACK for the probing packet, the CT starts transmitting the packets that are lost during handover with the slow start congestion control. Note that slow start congestion control assures that the transmission on the new path is not only as friendly as a newly starting SCTP flow but also it grasps the available bandwidth on the new path exponentially fast. Fig. 3 illustrates the error recovery procedure of the proposed mechanism when handover happens.

The handover latency of the proposed approach, H_{probe} is

$$H_{probe} = PAT + D^{new}_{ASCONF} + D^{new}_{data}. \tag{9}$$

Obviously, the proposed approach reduces the handover latency by α ($0 \leq \alpha \leq 2^i \times TO$) compared to the original SCTP when retransmission timeout occurs.

The total time delay L_{probe} to recover lost packets after receiving the ASCONF chunk is

$$L_{probe} = \{1 + \lfloor \log_2(l+1) \rfloor\} \times RTT. \tag{10}$$

Therefore, the proposed approach indeed reduces the packet loss recovery time by E:

$$E = \begin{cases} L_{timeout} - L_{probe} = \alpha, \quad where 0 \leq \alpha \leq 2^i \times TO, \\ \quad \text{if (number of rtx timeout >0)} \\ L_{no-timeout} - L_{probe} = [1 + \lfloor \frac{\sqrt{8l+9}-1}{2} \rfloor - \lfloor \log_2(l+1) \rfloor] \times RTT, \\ \quad \text{if (number of rtx timeout = 0)} \end{cases} \tag{11}$$

3 Performance Evaluation

Series of simulations were conducted for the performance comparisons between the proposed SCTP enhancements over plain IPv6, and MIPv6 and its variants

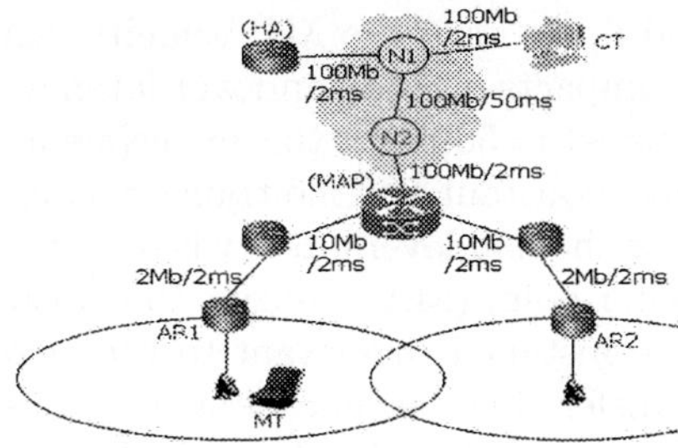

Fig. 4. A network model used in the simulation

(i.e., HMIPv6, and FMIPv6) with TCP Reno, the most widely used connection oriented transport layer protocol, on top of them. The various previously proposed TCP enhancement mechanisms for mobile environments are not taken into account in our simulation since all of those mechanisms mainly focus on avoiding spurious congestion window reduction caused by handover whereas the proposed transport performance enhancement mechanism mainly focuses on the error recovery aspect. Therefore, the simulation results will present the performance gains of the proposed error recovery mechanism, which is possible in the context of transport layer mobility management, compared to the performance of plain TCP Reno whose session mobility is supported by MIPv6 and its variants. We construct a simulation model using Network Simulator 2.27 [8].

The network model used in the simulation study is illustrated in Fig. 4. The entities noted in the parenthesis in Fig. 4 are required to implement the network layer approaches. The transmission range of AR1 and AR2 has been set to 250m and the distance to pass through the overlapping area of these two adjacent ARs' coverage to 50m. The MT randomly moves around the coverage areas of AR1 and AR2 following Waypoint Mobility Model. In various scenarios, we measure the handover latency and the time duration to transfer the file of 20MB, referred to as a file transfer time in the sequel. In each simulation, we varied the velocity of the MT, PAT, and the internet delay. The internet delay will be varied by changing delay in the path between N1 and N2.

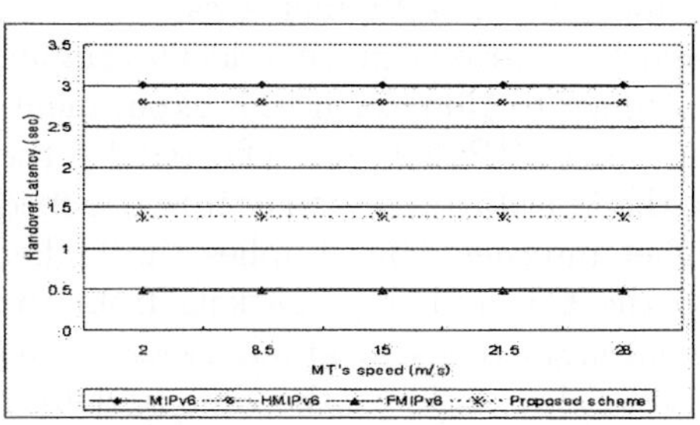

Fig. 5. Handover latency for different moving speed of MT

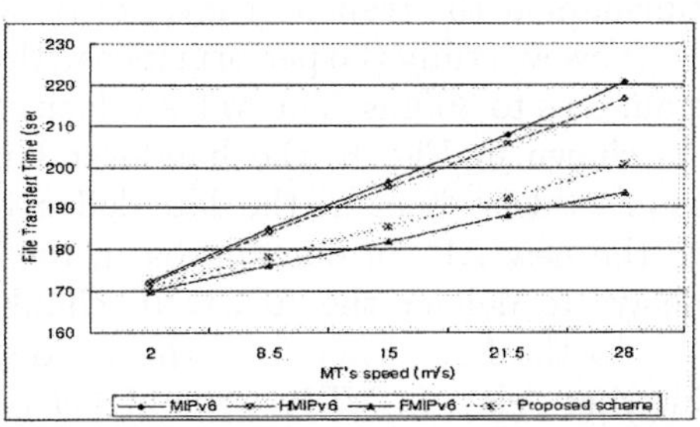

Fig. 6. File transfer time for different moving speed of MT

As shown in Fig. 5 and 6, we vary the MT velocity range from 2 m/sec to 28 m/sec and examine the impacts on the handover latency and the file transfer time. The internet delay was set to 50ms for this experiment. For all schemes, the handover latency was almost constant but the transfer time increases as the MT moves faster. This is because the handover latency is determined by the time that the CT or the mobility anchor point (MAP) begins to transmit data packets over the new path and this time instant is irrelevant to the velocity of the MT. On the other hand, the file transfer time becomes longer because the handover rate of the MT increases as it moves faster. As shown in Fig. 5, FMIPv6 yields the lowest handover latency. It is expected because the MT can receive data packets through the tunneling between previous access router and new access router as soon as it performs the layer-2 handover [4]. Thus, its handover latency is about the layer-2 handover latency. The handover latency of the proposed approach based on SCTP presents the second lowest one. In our scheme, data packets are immediately transmitted to the new IP address as the CT receives ADDIP and Set Primary parameters embedded in the ASCONF chunk. In both MIPv6 and HMIPv6, if timeout occurs due to a lost packet resulting from the handover, even after receiving Binding Updates (BU), the CT may not be able to transmit data to the new IP address and has to postpone it until the current retransmission timer expires. By the way, in HMIPv6, BU can be processed at the local MAP, which is located much closer to the MT than the CT is [3]. Thus, BU can be completed faster in HMIPv6 than in MIPv6. As a result, HMIPv6 yields lower latency than MIPv6 does. We note that the handover latency directly influences the file transfer time and thus the latter becomes longer if the former is higher as shown in Fig. 6.

In order to precisely investigate the reason for the different handover latency of each scheme, we then examine the performance while varying PAT from 0 sec to 5 sec. The MT velocity and the internet delay were set to 15 m/sec and 50ms, respectively. As shown in Fig. 7, in our approach, the handover latency gradually increases proportional to PAT and those values are almost same as PAT values. This result confirms the results displayed in Fig. 5 in which the handover latency is also almost the same as PAT. Unlike our scheme, the handover latency in MIPv6 or HMIPv6 has the step pattern jumping with the amount of each jump being twice of its previous jump. This is because the number of retransmission timeouts at the transport layer of the CT increases as PAT increases.

Now we study the performance with regards to various internet delays ranging from 5ms to 300ms. The MT's velocity is still set to 15m/sec in this experiment. As shown in Fig. 8, the handover latency of FMIPv6 is not affected by the internet delay because the data delivery to the MT starts as soon as it is attached to the new AP. In our approach, the higher internet delay implies the higher delays to deliver the ASCONF chunk to the CT and data packets from the CT to the MT. Therefore, the handover latency comes to about twice of the internet delay. In MIPv6, the higher internet delay prolongs the time to deliver BU, data packets and ACKs, is also related to the TCP retransmission timeout mechanism indicated in the previous experiment, and results in overall the worst

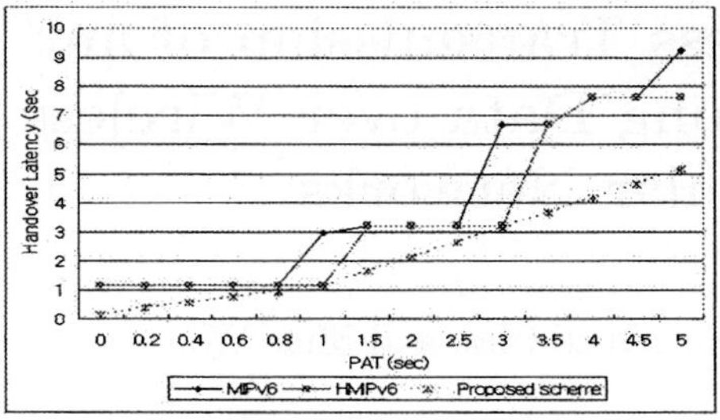

Fig. 7. Handover latency for different PAT

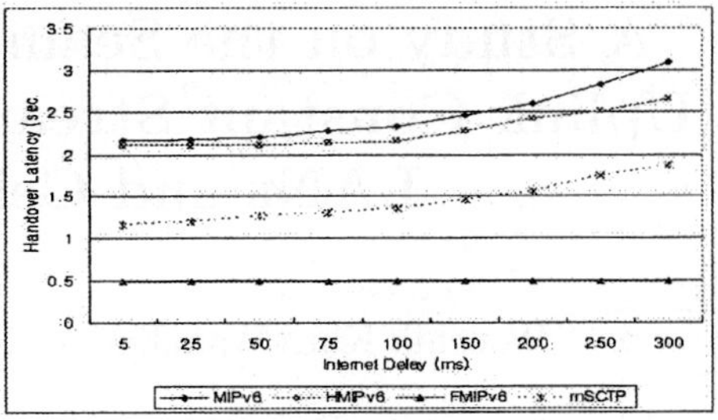

Fig. 8. Handover latency for different Internet delay

performance. HMIPv6 is also affected, but not as seriously as MIPv6 does, by the relationship between the internet delay and retransmission timeout due to the delay to deliver ACKs.

4 Conclusion

In this paper, we propose an approach to transport layer mobility support leveraging the SCTP extension dubbed dynamic address reconfiguration. Timing issues related to the end-to-end address management, and an error recovery mechanism associated with a handover are discussed. The error recovery time of proposed mechanism is analyzed and compared to that of the plain SCTP. Finally, through a series of simulations, the performance of the proposed SCTP enhancements over plain IPv6 is compared with the MIPv6 and its variants with TCP Reno on top of them. The simulation results present the performance gains of the proposed error recovery mechanism, which is possible within the context of transport layer mobility management, compared to the plain TCP Reno whose session mobility is supported by MIPv6 and its variants.

References

1. S. Deering, R. Hinden: Internet Protocol, Version 6 Specification, RFC 2460 (1998)
2. C.Perkins: IP Mobility Support for IPv6, RFC3775 (2004)
3. H. Soliman, C. Catelluccia: Hierarchical Mobile IPv6 mobility management (HMIPv6), drafti-ietf-mipshop-hmipv6-02.txt (2004)
4. R. Koodli: Fast Handovers for Mobile IPv6, draft-ietf-mipshop-fast-mipv6-02.txt (2004)
5. R. Stewart, et al.,:Stream Control Transmission Protocol, RFC 2960 (2000)
6. R. Stewart: Stream Control Transmission Protocol Dynamic Address Reconfiguration, draft-ietf-tsvwg-addip-sctp-09.txt (2004)
7. C.Perkins: IP Mobility Support for IPv4, RFC3344 (2002)
8. http://pel.cis.udel.edu/download

A Study on the Seamless Transmission of an Uplink Constant Streaming Data over Wireless LANs and Cellular Networks

Wooshik Kim, Wan Jin Ko, HyangDuck Cho, and Miae Woo

Sejong University, Department of Information and Communication Engineering
98 Kunja-dong, Kwangjin-ku, Seoul, Korea
wskim@sejong.ac.kr

Abstract. Recently, with the advent of the wireless Internet, IPv6, and P2P communications, the seamless transmission of a constant streaming data through wireless Internet has been widely accepted as a new research area. In this paper, we address a seamless transmission of constantly generated streaming data from a moving object to a remote site. We divide the whole wireless communication environments into an indoor environment, where the main communication method is the WLAN and an outdoor, where the main communication method is CDMA cellular Network and assume that the mobile server is able to move anywhere. We develop algorithms, implement them on a PDA-based hardware platform, and show some of the results. We think that, with some improvement of the performances, this work can be applied to many services such as P2P communication, file sharing services such as Napster or Soribada, and constant monitoring of the mobile objects.

1 Introduction

Recently, the file sharing services such as Napster or Soribada are spreading widely. In most cases of P2P communications, we assume that servers containing information such as multimedia files, are fixed in a wired network. However, with the development of the technologies in the portable devices such as the notebook,PDA, etc, the mobile cellular networks, and the WLAN, is is possible to operate mobile servers. In running mobile servers, one of the most crucial problem is the mobility of the servers. If the mobile servers are not equipped with handover, then the mobile servers are not able to function as serves.

In this paper, we consider the mobility of the mobile servers, i.e., the problem of handover of mobile servers. Many researchers think that the IP handover is one of the solutions. To do this, however, in addition to the techniques in IP and higher layers, something has to be done in the lower layer such as physical layer on a mobile terminal. For example, the terminal should check the signal strength from the BTS always and decide the time when the best time for the handover is. This is basically the Mobile Controlled Handover.

We can assume that the mobile servers can be in any environment. Among them, the two most important networks are the cellular networks and WLANs.

C. Kim (Ed.): ICOIN 2005, LNCS 3391, pp. 874–883, 2005.

The WLANs are rapidly replacing with the existing wired LANs. It is expected that, sooner or later, the WLANs become the most popular method for accessing Internet and voice communication [1,2]. WLANs provide 11, 54, or 100 Mbps of speed and the reliability in accessing. Also, once the infra is established, then we do not need to pay virtually extra costs. However, since the coverage is not wide, it is not easy to build a wide area network such as the cellular. On the other hand, the wireless cellular phone is the most popular communication method. Most cellular networks have a nationwide coverage. This, however, needs to pay expenses. Another weak point is that the data rate is not high. Thus, the two methods have their strong points and weak points as well. Exploiting the strong points of these methods, we can build more reliable and cheap communication methods. If the handover between these environments are developed, then we can realize the mobile servers in P2P communications.

In this paper, we present some basic results for the implementation of the seamless transmission of constantly generated low speed data through both Wireless LAN and CDMA public cellular network. We consider also the data transmission over each of the environment as well as the handover between the two environments. The organization of this paper is as follow. After an introduction of the problem that we address in this paper, we present the overview of a WLAN environment and a CDMA cellular network. Then we consider the transmission of data at the two environments and the handover between these two environments. Finally, we show some results.

2 System Organization

2.1 Overall Scenario

The wireless communication environments to be considered in this paper are given in Figure 1. As we can see in this figure, a mobile server has a data source that generates low speed constant data. These data are sent to a remote client over any possible network. In this figure, a human has a moving server in a PDA and is assumed to be moving around the two different environments: the indoor environment and the outdoor environment. In the indoor environment, the main communication method is assumed to be the Wireless LAN. In the outdoor environment, the main communication method is the CDMA public cellular service. In the following sections, we review the characteristics of each of the two networks briefly.

2.2 Overview of the WLAN Network

The WLAN is a modified Wired LAN network for the existing LAN users to provide the mobility. In WLAN, there are two different modes; Ad Hoc mode and Infrastructure mode [3,4,5]. In this paper, we consider only the 'Infrastructure mode' where the data server and the clients are not in the same network so that the data are sent through at least one public network [3]-[6]. In other words, the

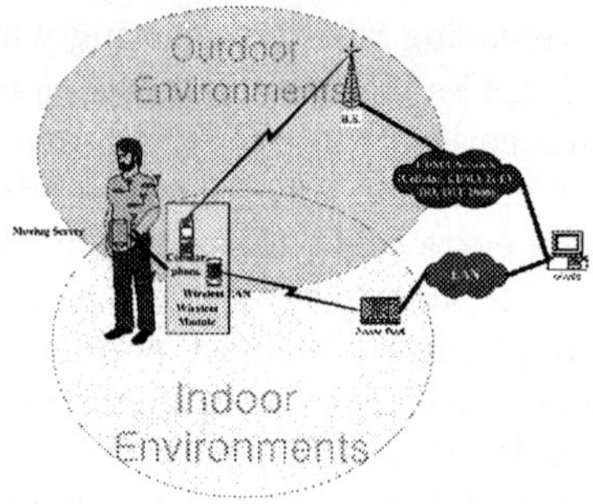

Fig. 1. Overall Scenario of the Communication Environments

data in the server are sent to a remote client through at least one AP (Access point). The WLAN works to the standard IEEE 802.11b, and uses CSMA-CD [3]-[7]. The WLAN is as fast as 10 Mbps (or 100 Mbps) and as reliable as the wired LAN.

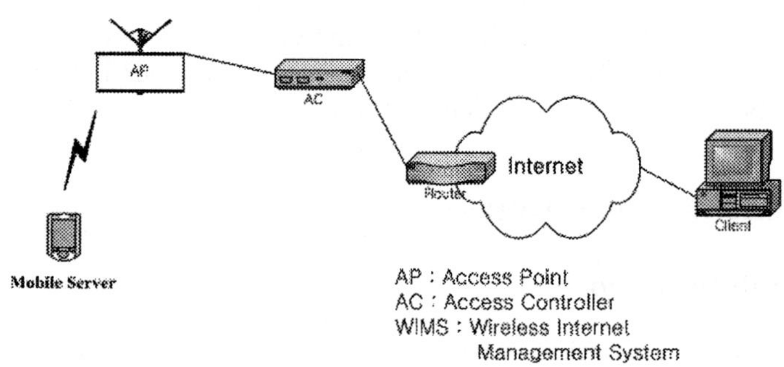

Fig. 2. Overall View of a Wireless LAN Network

2.3 Overview of a Cellular CDMA Network

There are several cellular services such as GSM, DCS-1800, IS-95, CDMA 2000 1X, etc. In this paper, we assume that the main communication method is the CDMA public cellular network. The cellular networks have strong points as well as weak points. The strong point is that they are accessible from anywhere. In other words, the cellular networks has a nationwide coverage and can be reached from everywhere, i.e., outdoor as well as indoor. It can even be reached from the abroad through roaming. This wide coverage has a weak point that the link quality varies from place to place and so it can be bad in some places. This can cause errors during the transmission. Another one is that, if we send data for a while, then the cost becomes huge. Also, the speed of the transmission is not high and limited, and is about 10 to 14.4 kbps or 144 kbps which is much lower than 11Mbps or 100 Mbps of WLAN. These limitations, however, are expected to be resolved in a near future since the speed will increase rapidly as new cellular services such as CDMA 2000 1X EVDO, IMT 2000, etc., are deployed.

Figure 3 is an overall diagram of the data transmission through CDMA Network [8]. As we can see in this figure, the data at the mobile server are sent to the nearest BTS (Base Transceiver System) through a CDMA modem. The data that come into the CDMA network are sent to the RAS (Remote Access Server) through the PPP (Point to Point Protocol), which is is more efficient than the SLIP in terms of stability and flexibility, for TCP/IP communication [9]. In RAS, a user certification is performed. After finishing the certification process, we can send the data through TCP/IP.

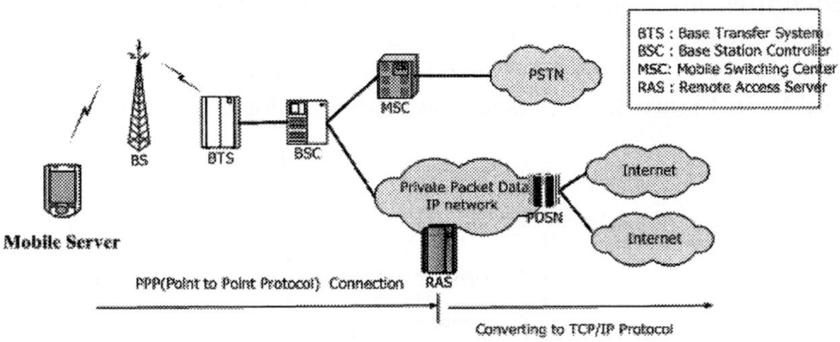

Fig. 3. Overall structure of the CDMA network

3 Seamless Transmission over WLAN and CDMA

3.1 Registration

To begin communication, the first thing to do is the registration. We assume that the WLAN has higher priority to CDMA because the cellular network is accessible from anywhere and the WLAN has better performances. In sensing which service area the mobile server is in, the most important criterion is the signal strength (SS) from the APs. The mobile server is constantly monitoring the signal strength, analyzes the current state, and takes action accordingly. In this section, we address a procedure of sensing the service area of WLAN.

A. Power on: When the power is on, then the mobile server automatically enters into an initialization process. During the initialization process, both the server and the remote client open two independent ports: one for the WLAN (Port A) and the other for the Cellular Network (Port B). As soon as the connection between the AP and the mobile server has been established, the mobile server constantly checks the strength of the signal from the AP. If SS is 0, there are two possible situations. The one is that the server is moving from the WLAN service area out into a no-service area. The other situation is that it is already outside of the WLAN service area. In the latter case, since the connection to RAS has been already established, the server keeps sending data through CDMA. In the former case, the server initiates the registration process to establish the connection and then sends data through CDMA network.

If the SS value is greater than 0, then there are two possible occasions. The one is when the user is about to cross the boundary of the WLAN service area and moves out of the WLAN service area. The other is that the mobile server resides in the WLAN service area already.

B. Procedure for Establishing a Connection over WLAN: If the measured SS is greater than 0 and if the mobile server does not have a connection over WLAN, then it needs to establish a connection to the remote client over the WLAN. In Figure 4, we present the procedure for establishing a connection over WLAN.

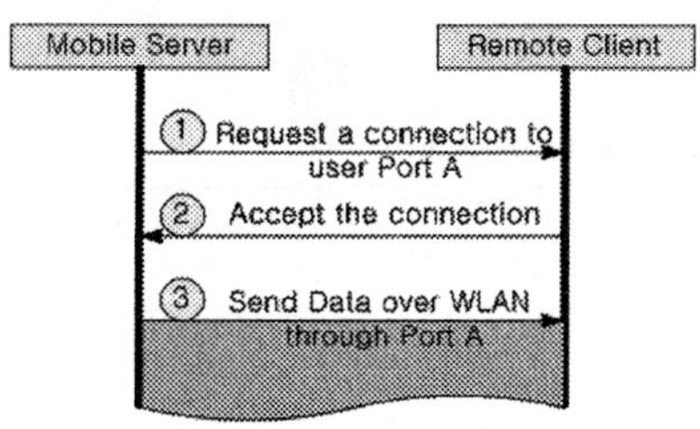

Fig. 4. Procedure for establishing a connection over WLAN

The explanation of each procedure is as follows.

1. The mobile server requests a connection to the client using Port A.
2. The client realizes that the server is now in the service area of WLAN and sends an acceptance notice for the request.
3. The mobile server begins transmission of the data to the remote client through WLAN.

C. Procedure for Closing the Connection over WLAN and Establishing a Connection over the Cellular Network: While transmitting the data in the WLAN area, if the SS becomes 0, then it can be said that the client moves out from the WLAN area and is no longer possible to main-tain the connection. In Figure 5, the procedure is presented.

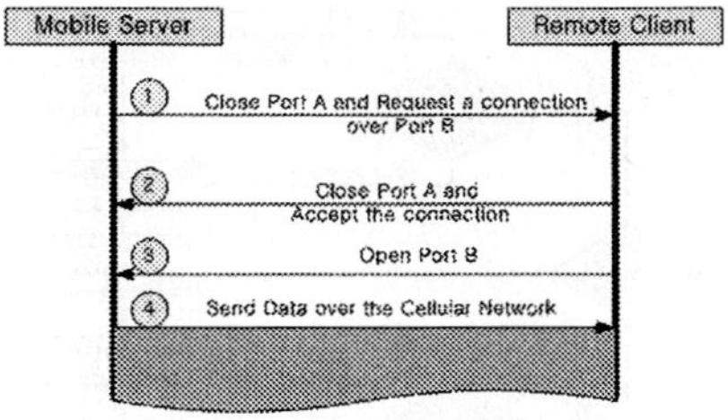

Fig. 5. Procedure for Establishing a Connection over the Cellular Network

1. The mobile server closes the connection of Port A and sends a request for connection to the remote client through Port B.
2. The client realizes that the mobile server moved out of the WLAN service area, closes Port A immediately, and accepts the mobile server's request through Port B.
3. The mobile server retransmits data over the cellular network through Port B.

3.2 Handover

If the mobile server moves out of the service area of WLAN and crosses over the boundary of the service area, then the flow of the data through WLAN may be stopped and lost its track. To prevent this, if the mobile server leaves the service area, we must switch to the cellular network. Figure 6 shows flow chart for the handover between the two communication environments. A detailed explanation of the procedure is as follows.

1. If the SS checked is larger than 0, the mobile server is in the WLAN service area. In this case, the next step is to check whether the mobile server has a connection to the cellular network.
2. The mobile server disconnects to the RAS immediately, closes port B, and be-gins transmission of data through Port A over WLAN.
3. If the mobile server is not connected to CDMA network, then it means that the server is transmitting data over WLAN and goes to the next step.
4. If the SS is 0, then the server is out of the WLAN area and checks whether it has a connection with the cellular network.
5. Same as before.
6. If the mobile server is connected to CDMA network, then the mobile server transmits data as maintain the connection to CDMA network.
7. If not, then the mobile server moves out of the WLAN area. As the RAS connection program is executed automatically, transmission of data is achieved through CDMA network and port B.

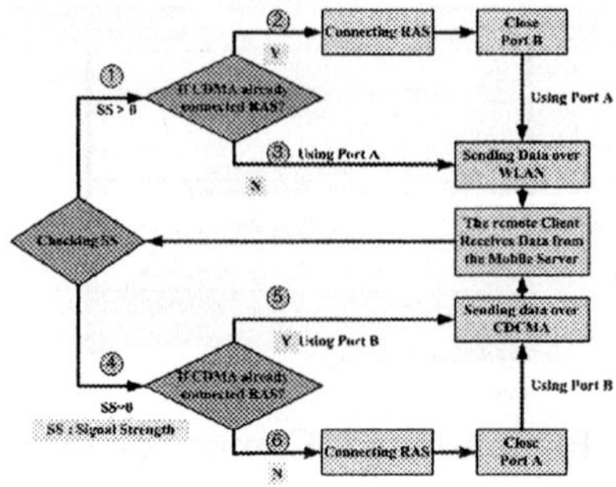

Fig. 6. Procedures for the handover WLAN

4 Implementation and Results

4.1 Hardware Platform

The hardware platform implemented in this paper is as follows. We use a Compaq iPAQ 5450 PDA as the mobile server. This PDA has two modems; the one is a built-in WLAN modem and the other is a CDMA modem in a cradle. The OS of the PDA is PocketPC 2002. To develop PDA related software, we used the embedded Visual C++3.0 and Platform Builder. For the client side, Visual C++6.0 is used for the development of the client related programs. The program for monitoring the strength of signal is developed through NDIS (Network Driver Interface Specification) and SDK (Software Development Kit).

4.2 Comparison of the Characteristics of the Two Communication Environments

In the WLAN environment, the speed and quality of the communication is very good and so there is hardly a problem. However, in the CDMA communication environment, transmission delays become big occasionally. (Since the data are transmitted through TCP/IP, they are transmitted without an error.) These delays are not fatal in general, since eventually the data will be transmitted without any loss, but may cause inconveniences.

Figure 7 is the measured throughput of each network. In this figure, the dotted (red) lines are a moving average of the throughput and are calculated at every 10 seconds for WLAN and 20 seconds for CDMA. In Figure 7 (a), the MA (Moving Average) is nearly constant for the WLAN and the throughput does not vary much. This is due to stability and good link quality. On the other hand, in Figure 7 (b), the throughput in CDMA varies much because the expired packets cause retransmissions which in turn cause burst data. The average throughput in WLAN and CDMA network are 3.195 (Kbps) and 3.787 (Kbps) and are nearly the same. The throughput of WLAN varies 512 bps to 4.3 kbps. On the other hand, in CDMA network, the throughput varies from 512 bps to 31.2 kbps. This comes from the fact that the link quality of the CDMA network varies from

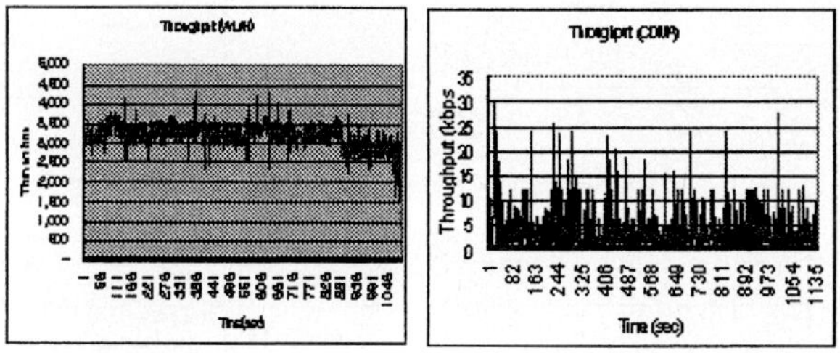

(a) WLAN (b) Cellular (CDMA)

Fig. 7. Throughput

place to place. In Figure 8 and Figure 9, we present the comparison between the WLAN and the CDMA network. Figure 8 shows the histogram of the number of pack-ets per transmission in WLAN (a) and in CDMA network (b). As we can see in this figure, the most of the data in the WLAN are transmitted in 1-5 packets. On the other hand in CDMA, the number of packets vary from 1 to 128.

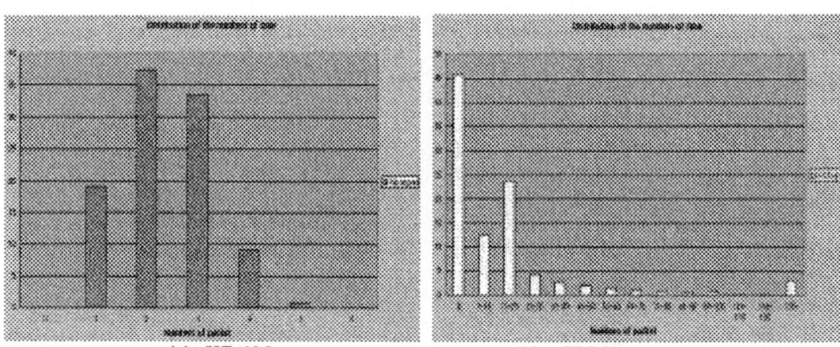

(a) WLAN (b) CDMA

Fig. 8. Histogram of the number of packets per transmission

Figure 9 shows the distribution of the arrival time of packets in WLAN (a) and in CDMA network (b). As we can see in this figure, the most of the packets in WLAN arrives within 1.4 seconds. On the other hand, the packets in the CDMA network take more than 30 seconds.

4.3 Handover

The implementation of handover on the mobile server (PDA) is done by switching operation between the RAS and the WLAN network through switching the port A and B. Figure 10 and 11 shows the sequence number of the packets and traffic coming into the remote client. As we can see in these figures, basically the handover operation works fine. However, after the data are coming through

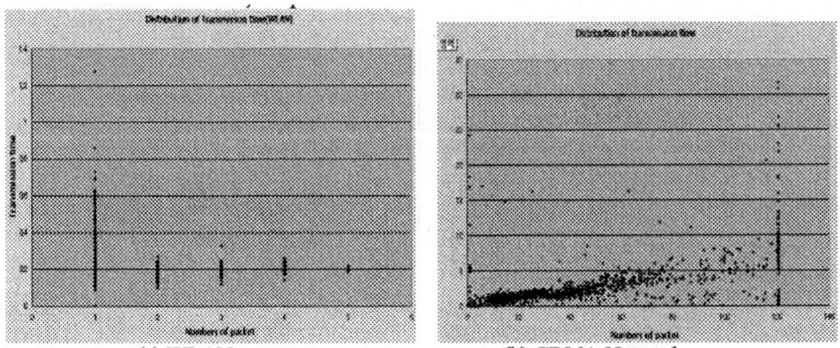

(a) WLAN (b) CDMA Network

Fig. 9. Distribution of the arrival time of the packets in WLAN and CDMA network

the WLAN, the data having new sequence number comes into the remote client through the CDMA network. Be-tween these two interval, there is an about 10 seconds of break, which is the transition time during the handover process. This period of time is basically the same as the connection time to CDMA network via RAS and is a typical phenomenon in hard handoff. During this, the mobile server does not transmit data to a remote client and there is no packet loss.

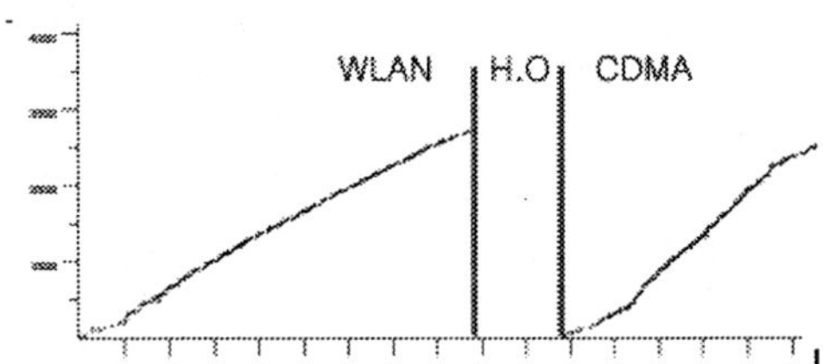

Fig. 10. Sequence Number of the Packets coming into the Remote Client

5 Conclusion

In this paper we developed and implemented data transmission through WLAN in the indoor environment, CDMA public network in the outdoor environment, as well as the handover between the two environments. We developed programs and implemented on a real hardware platform for transmitting data and switching between the modems for handover to transmit the same data to a remote client. We found that basically the handover operation works well although there are some glitches during the transition. This problem comes from the fact that this handover is basically the MCHO (Mobile Controlled Handover) and a hard handover. This problem, however, is expected to be solved if we use a buffer or if we adopt a soft handover.

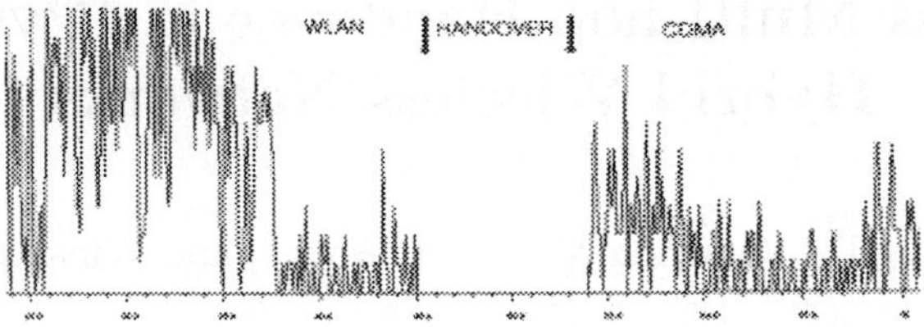

Fig. 11. Traffic Coming into the Remote Client

We expect that, with some improvement of performance, the result presented here can be used in the many services including P2P (Peer-to-peer) Communication, File sharing services such as Napster or Soribada, and Monitoring of data coming from mobile sources. We also think that this will be of great helpful for the IP handover, because to perform the handover in the Network Layer or IP layer, something has to be done in the lower level such as handset terminal level, or physical layer, and is basically the same as the Mobile Controlled Handover given in this paper.

Acknowledgement This work is supported by the ABRC (Advanced Biometric Research Center) sponsored by KOSEF and CUCN (National Center of Excellence in Ubiquitous Computing and Networking).

References

1. J.Kallioukulju, P. Meche, et. al., "Radio Access Selection for Multistandard Terminals", IEEE Communications Magazine, pp116-124, Oct. 2000.
2. Atlsd Research Group, 'WLAN MARKET FORECAST, 2002 4.8
3. "Requirements and Architectures for Interworking between HIPERLAN/2 and 3rd Generation Cellular Systems", ETSI 2001.8.
4. IEEE 802.11f-D3.1, "Draft-Recommended Practice for Multi-Vendor Access Point Protocol Across Distribution Systems Supporting IEEE802.11 Operation", April.2002.
5. IEEE802.11a, "Wireless LAN Medium Access Control (MAC) and Physical Layer (PHY) specifications: High-speed Physical Layer in the 5GHz Band", 1999 RFC 2865, "Remote Authentication Dial In User Service (RADIUS)", June. 2000.
6. J. Ala-Laurila, J. Mikkonen, J. Rinnemaa, "Wireless LAN access Network Architecture for Mobile Operators", IEEE Communications Magazine, pp82-89, Nov.2001.
7. ANSI/IEEE std.802.1D "IEEE Standard for Information technology Telecommunications and Information exchange between systems Local and metropolitan area networks-Common Specification-Media access (MAC) bridge", 1998
8. H. Kim, et al., "Principles of IMT-2000 Mobile Communications", Jin Han Publishing Com-pany. 9. W. Richard Stevens, "TCP/IP Illustrated, Volume 1", Addison Wesley, 1994.
9. W. Richard Stevens, "TCP/IP Illustrated, Volume 1", Addison Wesley, 1994.

Seamless Multi-hop Handover in IPv6 Based Hybrid Wireless Networks

Tonghong Li[1], Qunying Xie[2], Jing Wang[2], and Winston Seah[1]

[1] Institute for Infocomm Research, 21 Heng Mui Keng Terrace, Singapore 119613
{lith, winston}@i2r.a-star.edu.sg
[2] Department of ECE, National University of Singapore, Singapore 117576
{g0202596, jing.wang}@nus.edu.sg

Abstract. When a mobile device is accessing Internet, it is important to have undisrupted communication with the wired network in the event of handover. This paper studies multi-hop handover in IPv6 based hybrid wireless networks. A generic handover scheme with notification mechanism is proposed. Simulation results under different scenarios demonstrate that this scheme can reduce the packet loss and handover delay without incurring too much signaling overhead during the multi-hop handover.

1 Introduction

Mobile ad hoc networks (MANETs) are becoming popular due to the abundance of mobile devices, the speed and the convenience of deployment, and the independence of network infrastructure. It is desired that MANETs are interconnected to fixed IP networks so that the Internet services can be offered to MANET nodes. In such scenarios, commonly known as hybrid ad hoc networks, mobile nodes (MNs) are viewed as an easily deployable extension to the existing infrastructure. Gateways (GWs) are installed, which can be used by MNs to seamlessly communicate with nodes in the fixed network.

Recently, much work [1] [2] has been done on providing Internet connectivity for mobile nodes in hybrid ad hoc networks. Although different aspects including the global address configuration, gateway discovery, and communication in different scenarios are addressed, the multi-hop handover is still an open issue. Considering multiple GWs in hybrid networks, many proposals extended mobile IP to achieve seamless handover for MNs one hop away from GWs; however, work on handover for MNs multi-hop away from GWs is very limited.

An integrated protocol for IPv6-based hybrid wireless multi-hop networks based on Cellular IP and AODV [3] has been proposed [4]. This protocol extends micro-mobility management into ad hoc network and interconnects the ad hoc network efficiently into the existing network infrastructure. Two different handover schemes, proxy-based and proxy-disabled are included in the protocol to allow the protocol to adapt to different networking requirements. However, its handover schemes are designed for intra GW handover, where MN still uses

C. Kim (Ed.): ICOIN 2005, LNCS 3391, pp. 884–893, 2005.

the same GW when changing its base station due to its movement. The network architecture used is different from the scenario studied here, as there are base stations between MANETs and GWs.

In [5], different gateway discovery approaches for IPv4-based hybrid ad hoc network are compared in various scenarios by means of simulation. Multi-hop handover is performed if MN changes its GW while communicating with a corresponding node (CN) in the Internet, which is categorized into forced handover and route optimization handover. Forced handover occurs whenever the path between the MN and the GW is disrupted during data transmission. The following GW discovery process may result in the detection of a new GW, which will consequently result in a handover. On the other hand, if the MN detects that a shorter path to the Internet becomes available while communicating with a CN, the active path will be optimized. In case the shorter path is via a different IGW, a route optimization handover occurs. Though the performance of multi-hop handover under different gateway discovery approaches is evaluated, no scheme is proposed to provide smooth multi-hop handover without incurring too much signaling overhead.

This paper studies multi-hop handover in IPv6-based hybrid ad hoc networks. Our hybrid ad hoc network architecture is constructed by integrating Mobile IPv6 (MIPv6) with MANET, which is very similar to the architecture used in [5]. We assume the routing protocol for MANET is AODV, though our handover scheme can be applied to any other on-demand MANET protocol. The main contribution of this paper is a seamless multi-hop handover scheme to reduce the packet loss and handover delay without incurring too much signaling overhead during the handover. For the purpose of studying the performance, we also developed extensions to NS2 for simulating hybrid networks based on Hierarchical MIPv6 (HMIPv6) [6] and AODV. To our knowledge, this is the first simulation tool that integrates HMIPv6 and AODV in hybrid ad hoc network.

The remainder of the paper is organized as follows: section 2 introduces our hybrid network architecture, the gateway discovery and registration procedures used in our architecture are also discussed. In section 3 we describe our proposed seamless multi-hop handover scheme. The results of simulations are shown in section 4. Finally, section 5 gives some conclusions and draws some future directions.

2 IPv6 Based Hybrid Ad Hoc Network

Figure 1 shows our IPv6-based hybrid network architecture. MNs in the multi-hop wireless networks use AODV to communicate with each other. GWs are installed, which connect MANETs and wired networks. HMIPv6 is used in access networks due to its smooth local mobility management feature. We assume each MN has a unique IP address to be identified in wireless and wired networks; this address is called home address conforming to Mobile IP. MNs and GWs understand HMIPv6 and AODV, and they identify each other by their home addresses.

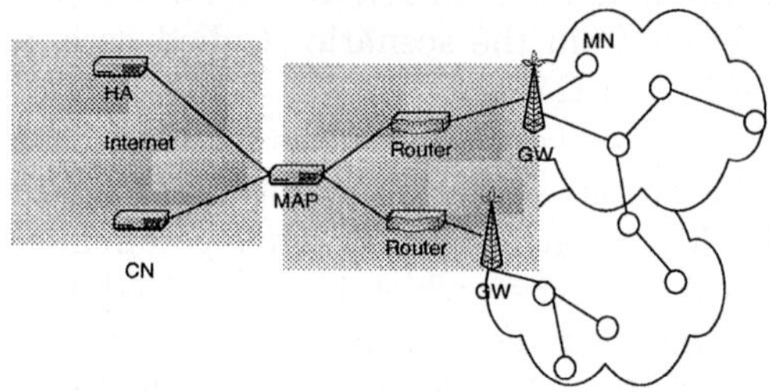

Fig. 1. Hybrid Multi-hop Wireless Network

2.1 Hybrid Gateway Discovery

To connect with the Internet, a MN must find a GW. In our hybrid network, we use a hybrid GW discovery scheme to realize GW discovery. Each GW broadcasts the Router Advertisement (RA) messages within N hops in a fixed advertisement interval. N is called the flooding range, which can be adjusted by setting the TTL field in the IP header of the RA message. The larger N is, the larger the overhead of RA broadcasting is.

For each MN in the RA flooding range, it records the address of Mobility Anchor Point (MAP), the address of the GW, the advertisement sequence number and advertisement lifetime in its GW table upon receiving RA messages. It also sets up a route to the GW. This route allows MN to update its routes to the GW if the RA message arrives along a shorter path and to refresh the route entries if the route is already known. A unique broadcast ID is used to prevent the broadcast of RA messages that a MN has already seen before.

For each MN beyond the RA flooding range, it will broadcast a solicitation for GW discovery if Internet access is required, as it cannot receive RA messages. Upon receiving the solicitation, the GW will unicast a RA message to the MN. Solicitation also sets up a reverse route to the MN, ensuring that unicast RA message sent out in response do not generate unnecessary RREQ messages. When MN receives this RA message, the procedure will be the same as the MN in the RA flooding range.

2.2 Registration

After MN receives a RA message, MN will add an entry in its GW table and choose one of ⟨MAP, GW⟩ to register with. The choice can be based on criteria such as distance, cost or other information contained in RA messages. In our scheme, we select the GW with the shortest hop count. After that, the MN auto-configures a unique RCoA (Regional Care of Address) and LCoA (Local Care of Address) which will be contained in Binding Update (BU) messages.

When GW receives BU from MN, it will record the MN at its MN table, which is used to keep track of registered MNs for making routing decisions.

When MAP receives BU message, it will create an entry in its binding cache to bind the MN's RCoA to LCoA, and set a registration time for this entry. Similarly, the HA will create an entry to bind the MN's home address to RCoA, and set a registration time for this entry as well. MNs must periodically register with the HA and the MAP to refresh the entries. GWs forward BU acknowledgements they receive from the MAP back to the MN. Once the MN receives the acknowledgement, it will set the lifetime field in the entry. To maintain the registration, the mobile node must re-register before the lifetime expires. Through periodical registrations, the bi-direction path to the registered GW from the MN can be refreshed periodically.

For each MN beyond the RA flooding range, it registers with the current GW until there is no path to this GW. In this case, its GW table is checked in order to find an alternative GW for registration. If it fails, the GW discovery is initiated to discover a new GW for registration.

3 Seamless Multi-hop Handover Scheme

In traditional wireless access network, MNs have link-layer connections with access points, and handover is normally defined as the service disruption period between the disconnection with previous access point and establishing connection with a new access point. For MNs multi-hop away from GWs, as in our hybrid wireless network, the handover issue is much more complicated. We define the multi-hop handover as a route change from MN to the registered GW, which may occur when a MN itself or any of the intermediate MNs moves and breaks the active route to the registered GW during the MN's communication with a CN in the wired network. In normal MIP, the handover process includes Layer 2 handover and Layer 3 handover. In our hybrid wireless network, MNs that are multi-hop away from a GW do not have link-layer connections with the GW and thus handover is only performed at Layer 3.

3.1 Multi-hop Handover Type

Considering different situations, multi-hop handover can be as follows:

• Intra-GW handover & Inter-GW handover

Intra-GW handover occurs when MN still registers with the same GW although the route to the current GW is broken; while Inter-GW handover occurs when the MN registers with a new GW. When Intra-GW handover happens, no mobile IP operation is involved. However, for Inter-GW handover, mobile IP operations like BU operations are required.

• Compulsory handover & Optimized handover

A compulsory handover occurs when MN detects a route break to its current GW. The optimized handover occurs when MN's route to its current GW is still active, but a better (more optimized) route is available and selected.

Combining the above two categories of handovers, we can classify the handover into four types:

1. Compulsory Intra-GW handover: when the route to the current GW breaks, but there is an alternative path to the current GW. This type of handover is handled by the recovery procedure of AODV.

2. Compulsory Inter-GW handover: when there is no path to the current GW and the subsequent GW discovery process results in the detection of a new GW.

3. Optimized Intra-GW handover: no route break, but a MN uses a shorter path to the current GW when a new RA message from the current GW arrives from a shorter path. This type of handover is completely handled by the RA message processing in GW discovery procedure.

4. Optimized Inter-GW handover: no route break, a MN uses a shorter path to another GW when it receives a new RA message with less hop count from another GW.

Type 2 handover has a bad impact on the performance of communication because the delay and overhead of discovering a new GW and a route to it are very high. As a result, we should reduce the rate of its occurrence. On the other hand, a timely type 4 handover is also very important. In multi-hop scenarios, using shorter paths can reduce the packet end-to-end delay; moreover, a shorter path to a new GW may indicate a potential future compulsory Inter-GW handover, thus using optimized Inter-GW handover can prevent a future route break.

3.2 Our Enhanced Multi-hop Handover Scheme

The objective of our handover scheme is to achieve low handover latency and low packet loss without incurring excessive overhead. For this purpose, we propose the following approaches:

(1) Proactively maintain a bi-direction route between MN and its registered GW.

In AODV, when the path to the current GW is broken, the source node does not re-initiate the route discovery unless there is traffic to the current GW, which leads to the long delay for packet transmission. To solve this problem, we proactively maintain a bi-direction route between the MN and its registered GW, which can be realized as follows: the bi-direction route can be refreshed each registration through BU and BU acknowledgement messages. Between two consecutive registrations, the route discovery is initiated immediately upon the receiving Route Error message for the bi-direction route irrespective of the presence of traffic between the GW and MN.

(2) Dependant Notifying Mechanism

For MN at the Nth hop from its current GW, it will immediately inform its dependants with ⟨previous GW, new GW⟩ information by broadcasting a handover notification message when performing an optimized Inter-GW handover. To be MN's dependant must satisfy two conditions: (1) it is registered in the same GW as MN; (2) it uses this MN as the next hop in the route to its registered GW. When its dependant receives the message, it constructs a route to the new GW and registers with it; a new notification message is also broadcasted.

In this way, future compulsory Inter-GW handovers can be avoided, which can reduce the handover delay and overhead significantly.

For a MN beyond N hops of its GW, it will immediately notify its dependants by using a new type of Route Error message called "GW_ Error" when performing a compulsory Inter-GW handover. The GW_ Error also includes the ⟨previous GW, new GW⟩ information. When a dependant receives this error message, it will construct a route to the new GW and register with it if no alternative path to the previous GW can be discovered by AODV's repair process. Similarly, a new "GW_ Error" message is broadcasted again if a dependant decides to switch to the new GW. In this way, the dependants can perform compulsory Inter-GW handover without the start of GW discovery, which results in the reduction of handover delay and overhead.

(3) Intermediate nodes reply to solicitation request for the GW if they have active routes to the GW.

When a MN outside the RA flooding range wants to connect to the Internet, it will send out a solicitation with its destination set to a GW multicast address. When a GW receives the solicitation, it will unicast a RA message to the MN. Previously, the intermediate node just forwards the solicitation. To let intermediate nodes reply to solicitation can reduce the GW discovery latency and overhead.

(4) Intermediate nodes hear the RA message unicasted by GW.

When the intermediate node forwards a unicasted RA message, it can update its GW table's entry with respect to this GW.

(5) Optimized Inter-GW handover stability management.

This is to prevent high frequent oscillations and decrease the probability of a MN registering with a GW that is only temporarily better. MN will not start an optimized Inter-GW handover immediately even if the new GW is nearer than the original one. Instead, it will only perform this type of handover after receiving at least two consecutive RA messages about the new GW.

We name the handover scheme with the above approaches as "enhanced HMIPAODV (E-HMIPAODV)", and the scheme without these approaches as "plain HMIPAODV (P-HMIPAODV)". We will compare two schemes through simulation in section 4.

4 Simulation

MobiWan [7] is a simulation tool based on NS (version ns-2.1b6) meant to simulate Mobile IPv6 under large Wide-Area Networks (both local-area mobility and global-area mobility). In order to enable it to support AODV and HMIPv6, the Network and MIPv6 agents are modified. The Network agent is replaced by an AODV agent, which integrates the handover schemes described in section 3. The MIPv6 agent is responsible for tasks like maintaining binding cache and sending Binding Update.

The purpose of the simulation is to study the performance of our enhanced multi-hop handover scheme in hybrid network. The protocol metrics used are: (1)

handover latency; (2) packet loss ratio; and (3) control overhead. These metrics have been examined under different network scale (small & large), mobility level (changing pause time), and other related network parameters (e.g. RA interval, RA flooding range, etc).

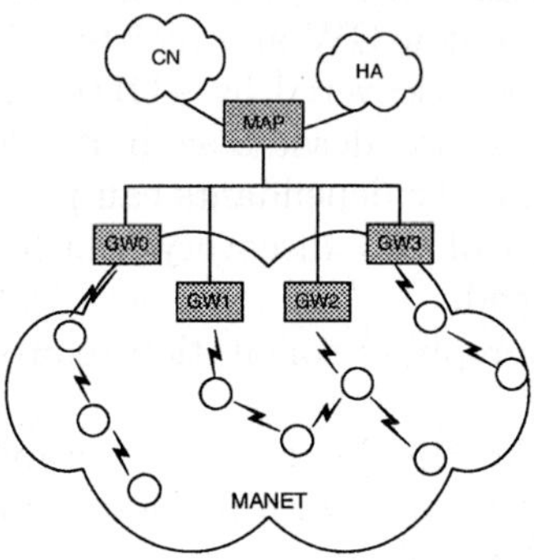

Fig. 2. Simulation scenario

Fig 2 shows our simulation scenario, where the wired network consists of a cloud of five CNs (CN0 to CN4), HA for all MNs, one MAP, and four GWs. In the wireless network, we study two topologies, a small network with 30 MNs over 600x600m area and a large network with 50 MNs over 1000x1000m area. To simplify the simulation, we make the whole wireless network belong to one MAP, thus there will not be handover between MAPs. Five of the MNs are CBR sources, and the five CNs are CBR sinks. Each source node sends constant bit rate (CBR) traffic with sending rate 10 packet/s (packet size is 50bytes). MN's movement complies with the random waypoint mobility model. All the simulations are done with the maximum speed set to 10m/s, the pause time is changed to simulate different level of mobility.

4.1 Impact of Mobility on Performance

Each Mobile node moves randomly with speed uniformly distributed in the range (0, 10)m/s, the pause time is set to [5, 10, 20, 100, 200, 300, 400]s in each simulation respectively. RA flooding range is set to 1 and RA interval is set to 10s for all simulations in this set.

For small network with 30 MNs, Fig. 3(a) shows that E-HMIPAODV has less average handover latency than P-HMIPAODV. For both schemes, the handover latency decreases with increasing pause time. As for traffic performance, Fig. 3(b) shows that E-HMIPAODV has less packet loss than P-HMIPAODV under different mobility levels. We also noticed that its improvement is more significant when pause time is shorter. Figs. 3(c) and 3(d) show both schemes' overhead, which is measured as the total number of transmissions of control packet during

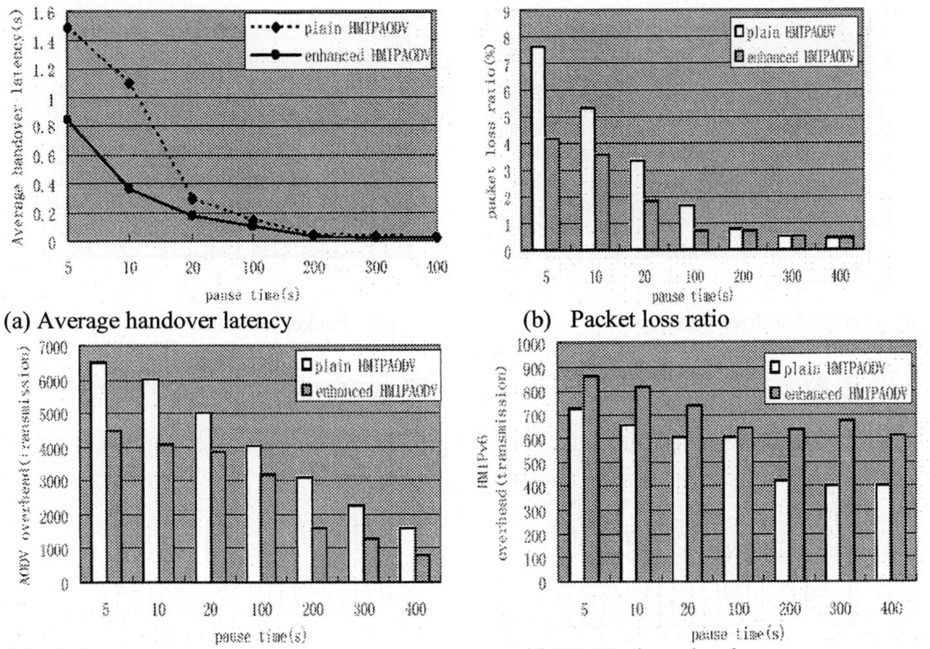

(a) Average handover latency (b) Packet loss ratio

(c) AODV overhead (d) HMIPv6 overhead

Fig. 3. Performance of both schemes in a small network with 30 MNs

simulation times. E-HMIPAODV reduces AODV control overhead because of its notification mechanism and allowing intermediate nodes reply with GW route information. However, it introduces more HMIPv6 control messages because E-HMIPAODV will perform more Inter-GW handover than P-HMIPADOV.

Figs. 4(a)∼4(d) show the results in large network with 50 MNs, which are similar to the results in the small network with 30 MNs. However, we note that performance improvement of E-HMIPAODV is greater in larger networks and under higher mobility.

4.2 Impact of Frequency of RA

To examine both schemes' performance against the frequency of RA messages, we use the large network topology with the RA flooding range set to 1 and MN's pause time set to 10s.

We find that the handover performance of both schemes have little difference when the sending rate of RA is very high or node's mobility is very low. We also observe that the RA sending rate has impact on the performance of both schemes. The higher the RA sending rate is, the lower the handover delay is. However, a high RA sending rate leads to high protocol overhead. There is an optimal value of RA interval considering both handover delay and protocol

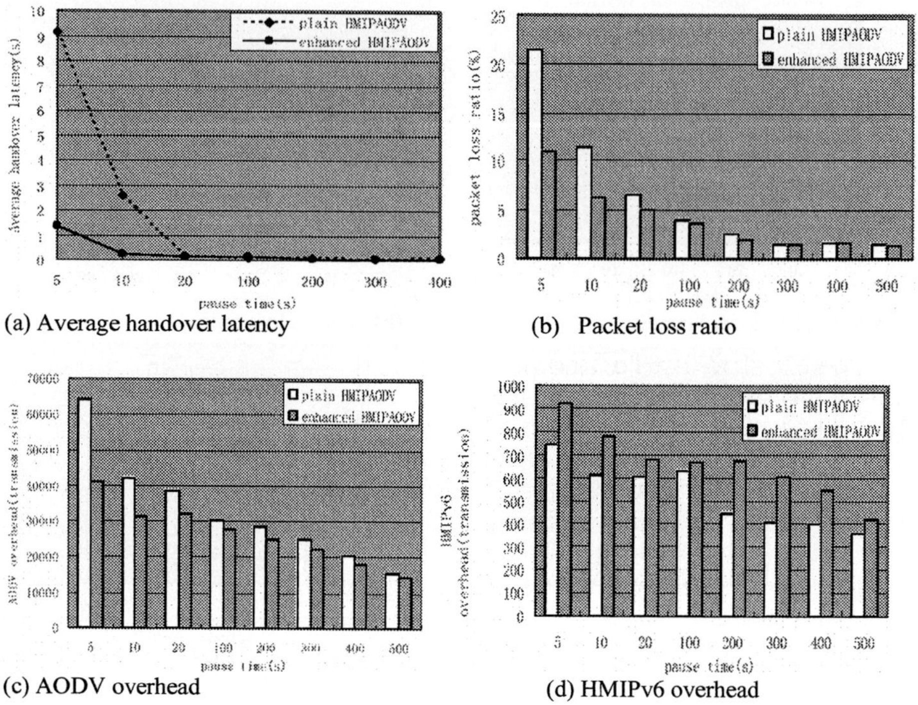

(a) Average handover latency (b) Packet loss ratio

(c) AODV overhead (d) HMIPv6 overhead

Fig. 4. Performance of both schemes in a big network with 50 MNs

overhead, which depends on the node mobility and node's traffic pattern. In our scenario, the optimal value of RA interval is 10s.

4.3 Impact of RA Flooding Range

We also study the impact of RA flooding range on both schemes' performance. We use the large network topology with RA interval set to 10s and MN's pause time set to 10s. We observe that E-HMIPAODV outperforms P-HMIPAODV at lower N. However when N is large, both schemes have the similar results. That is because the entire area is likely to be covered by the flooding area of RA messages when N is large. While under pure proactive GW discovery approach, both schemes have no difference.

On the other hand, packet loss and handover delay can decrease dramatically as N increases in both schemes. This can be explained as follows: when N is small, there is only a small area in which the MN can receive GW information. The other MNs use the on-demand approach to discover GWs. Hence when handover occurs, these MNs need longer time to reestablish the GW route compared with MNs in the flooding range. By increasing N, more MNs are covered by the flooding range. As a result, handover due to route failures drops, while handover due to route optimization, which does not incur packet loss, becomes more frequent.

However, when N is large, the overhead of RA broadcasting is high, which leads to heavy protocol overhead. Therefore, RA flooding range should be adjusted according to different network scenarios in a similar approach as [8] to keep the overheads at a reasonable level while maintaining low delay and packet loss. In this study, we set the RA flooding range as 3. Note that when data traffic is high, excessive RA messages can compete for resources and adversely affect the network's performance.

5 Conclusion and Future Work

This paper studies multi-hop handover in IPv6-based hybrid ad hoc network. We define multi-hop handover in hybrid network and propose approaches to reduce multi-hop handover latency while minimizing protocol overhead. The key to provide a smooth handover for MNs in hybrid network is to reduce the occurrence rate of compulsory Inter-GW handover as well as its handover latency. We propose to use notification mechanism to solve this problem. Through simulation, we show our handover scheme can reduce the handover latency and packet loss without incurring too much overhead. For the future work, downlink handover need to be studied, and one way to improve it is the Next-GW prediction mechanisms. Detailed algorithms can be designed and implemented as an extension of the existing handover scheme to achieve seamless handover performance.

References

1. Y. Sun, et al, "Internet connectivity for ad hoc mobile networks", International Journal of Wireless Information Networks special issues on mobile ad hoc networks, Vol.9, No.2, April 2002, pp.75-88.
2. Y. C. Tseng, et al, "Mobile IP and ad hoc networks: an integration and implementation experience", IEEE Computer, vol.36, No.5, May 2003, pp. 48-55.
3. C.Perkins, et al, "Ad hoc On-Demand Distance Vector (AODV) Routing", RFC 3561 in IETF, July 2003.
4. V. Typpo, "Micro-Mobility within Wireless Ad hoc netowrks: Towards Hybrid Wireless Multihop Networks", Diploma thesis, Department of Electrical Engineering, University of Oulu, Oulu, Finland, 2001.
5. M. Ghassemian, et al, "Performance Analysis of Internet Gateway Discovery Protocols in Ad Hoc Networks", WCNC 2004, At-lanta, USA, April 2004.
6. H. Soliman, et al, "Hierarchical MIPv6 mobility Management (HMIPv6)", Internet Draft, work in progress, June 2003.
7. MOTOROLA Labs Paris & INRIA PLANETE, "MobiWan: NS-2 extensions to study mobility in Wide-Area IPv6 Networks", http://www.inrialpes.fr/planete/pub/mobiwan, May 2002
8. Hwee-Xian Tan,et al, "Dynamically Adapting Mobile Ad Hoc Routing Protocols to Improve Scalability", the IASTED International Conference on Communication Systems and Networks (CSN2004), Marbella, Spain, Sep 2004.

Route Optimization in Nested Mobile Network Using Direct Tunneling Method*

Jungwook Song[1], Sunyoung Han[1**], Bokgyu Joo[2], and Jinpyo Hong[3]

[1] Dept. of Computer Science and Engineering, Konkuk University
1 Hwayang, Gwangjin, Seoul 143-701, Korea
{swoogi, syhan}@cclab.konkuk.ac.kr

[2] Dept. of Computer Information and Communications, Hongik University
300 Shinan-Ri, Jochiwon, Chungnam 339-701, Korea
bkjoo@hongik.ac.kr

[3] Dept. of Information and Communications Engineering,
Hankuk University of Foreign Studies
San 89, Wangsan, Mohyun, Yongin, Kyonggi 449-791, Korea
jphong@hufs.ac.kr

Abstract. We are noticing the emergence of various internet-ready electronic devices as well as popular laptops and personal digital assistances, and most of us want to access the Internet while moving our locations. Recently, new internet protocol IPv6 is being extended to support not only host mobility but also network mobility. Real situation will be a nested mobile network, where mobile networks would be nested as they change locations. Most serious problem in a nested mobile network is the complexity of routing path of packets, and the complexity grows as the nesting level increases. In this paper, we propose 'direct tunneling method', which delivers packets through optimized path even when mobile networks are nested. We show the effectiveness of our method by simulation results.

1 Introduction

As rapid progress of information and communication technology, there are increasing sorts of electronics devices that can access wireless network while they move their locations. When IPv6 (Internet Protocol version 6)[1] protocol that guarantees enough IP addresses is widely deployed, not only desktop computers but also notebook computers, personal digital assistances (PDAs), mobile phones, and even home appliances will be connected to the Internet. If IPv6 protocol prevails and wireless network infrastructure is established, it would be a common situation that tens or hundreds of mobile nodes move their locations at the same time. Because existing Mobile IP and Mobile IPv6 have been designed

* This research was supported by the MIC(Ministry of Information and Communication), Korea, under the ITRC(Information Technology Research Center) support program supervised by the IITA(Institute of Information Technology Assessment)

** Corresponding author

C. Kim (Ed.): ICOIN 2005, LNCS 3391, pp. 894–903, 2005.

to support host mobility only[2], they do not smoothly support the concurrent movement of many hosts, i.e. network mobility. To complement this weakness, IETF NEMO (network mobility) Working Group is studying an extension of Mobile IPv6 focusing on this issue[3,4,5].

In a simple case, single (mobile) network moves to another location. In general situation, however, more than one networks and hosts change their locations simultaneously. Direct application of IETF's NEMO basic support protocol to this situation causes complicated routing path from a mobile network node to the external communicating node, called the correspondent node[5]. Therefore, it is important to optimize the routing path of packets in nested mobile network.

In this paper, we try to solve the routing optimization problem in nested mobile network using direct tunneling method. Instead of by opening bidirectional tunnel between a mobile router and its home agent, the packet routing path can be optimized by opening direct tunnel from the mobile router to the correspondent node that is communicating with a mobile network node below the mobile router. The correspondent node catches the path of packets that were passed through nested mobile routers, and sends packets back in the reverse path. Thus, we can optimize the routing path in both directions.

The rest of this paper is organized as follows. Section 2 presents an overview of network mobility, several key elements needed to understand this paper, and description of nested mobile network. Section 3 describes how direct tunneling method operates. Section 4 presents simulation models and the results. Finally, section 5 gives our concluding remarks.

2 Overview of NEMO and Components

The purpose of Mobile IP is to allow a mobile node to continue communication, without interruption, by keeping connection states of transport layer and upper layer, even if the mobile node moves to another location. The home agent and the correspondent node maintains up-to-date information on the location of the mobile node, so that the mobile mode and the correspondent node could communicate through optimized path. In NEMO architecture, a mobile router and its whole subnet is the unit of movement, and the subnet might consist of fixed nodes, mobile nodes and another mobile networks.

2.1 Network Mobility Support

In Mobile IPv6, the home address and the care-of-address of a mobile node are associated in the binding caches of the home agent and the correspondent node. When Mobile IPv6 is applied to a mobile network, it can forward packets from the correspondent node to the mobile router, but it does not know the path from the mobile router to the mobile node. So, in order to support network mobility, IETF NEMO Working Group extended the Binding Cache and the Binding Update of Mobile IPv6. A mobile router sends a Binding Update that consists of care-of-address and mobile network prefix instead of home address.

The home agent maintains this information in its Binding Cache. Using mobile network prefix, the packet that is forwarding to a mobile network is tunneled to the care-of-address of the mobile router by the home agent, and the packet that is forwarding to outside of mobile network is tunneled to the home agent by the mobile router. This is the "NEMO Basic Support Protocol" proposed by IETF NEMO Working Group[5]. Several keywords and key elements are as follows[6].

2.2 Keywords and Key Elements

Binding Update (BU) A message indicating a mobile node's current mobility binding, and in particular its care-of-address.

Care-of-Address (CoA) An IP address associated with a mobile node while visiting a foreign link; the subnet prefix of this IP address is a foreign subnet prefix. A packet addressed to the mobile node which arrives at the mobile node's home network when the mobile node is away from home and has registered a Care-of-Address will be forwarded to that address by the Home Agent in the home network.

Correspondent Node (CN) An IPv6 node that communicates with a mobile network node.

Home Agent (HA) A router on a mobile node's home link with which the mobile node has registered its current care-of-address. While the mobile node is away from home, the home agent intercepts packets on the home link destined to the mobile node's home address, encapsulates them, and tunnels them to the mobile node's registered care-of-address.

Mobile Network Prefix A bit string that consists of some number of initial bits of an IP address which identifies the entire mobile network within the Internet topology.

Mobile Node (MN) An IP node capable of changing its point of attachment to the network. A Mobile Node may either be a Mobile Host (no forwarding functionality) or a Mobile Router (forwarding functionality).

Mobile Network An entire network, moving as a unit, which dynamically changes its point of attachment to the Internet and thus its reachability in the topology. The mobile network is composed of one or more IP-subnets and is connected to the global Internet via one or more Mobile Routers (MR).

Mobile Network Node (MNN) Any node (host or router) located within a mobile network, either permanently or temporarily.

Mobile Router (MR) A router capable of changing its point of attachment to the network, moving from one link to another link. The MR is capable of forwarding packets between two or more interfaces, and possibly running a dynamic routing protocol modifying the state by which it does packet forwarding.

2.3 Nested Mobile Network

It is called a nested mobile network when mobile network moves to a subnet of another mobile network. Note that we could apply NEMO basic support protocol to the nested mobile network.

Fig. 1 shows a nested mobile network formed as follows: (1) a mobile network with mobile router MR1 moves new location and connects to the Internet through AR (Access Router). (2) Then, another mobile network with mobile router MR2 moves to MR1's subnet. (3) Finally, a mobile node VMN moves to MR2's subnet. Each time the movement occurs, mobile routers and mobile node send Binding Update to their home agent to notify their new location.

2.4 Pinball Routing in the Nested Mobile Network

In NEMO basic support protocol, nodes that belong to a mobile network exchange packets through bidirectional tunnel between mobile router and home agent. The scheme enables them to exchange packets without any problem, even if the mobile network node and the correspondent node do not aware of the existence of network mobility. However, as the nesting level of a mobile network increases, so is the number of bidirectional tunnels between mobile routers and home agents that participate in packet exchange. If there are many tunnels, routing path of packet becomes more complex.

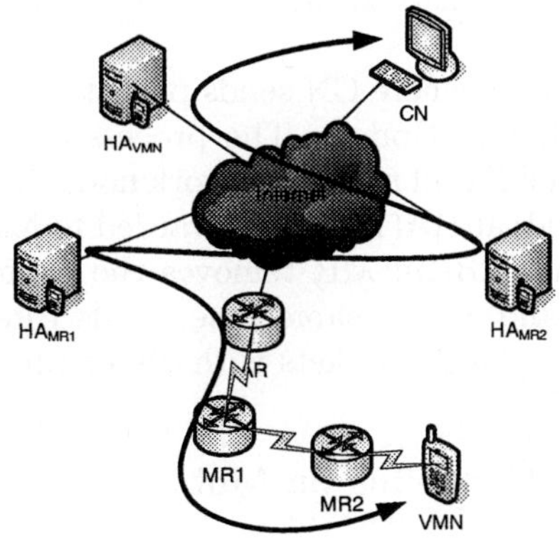

Fig. 1. Nested Mobile Network and Pinball Routing

Fig.1 shows complicated routing path of packet in nested mobile network. Here, mobile node VMN updates its location to CN and HA_{VMN}, so that CN and VMN can exchange packets bypassing HA_{VMN}. However, the packets exchanged between CN and VMN still have to visit HA_{MR1} and HA_{MR2} which are home agents of MR1 and MR2, respectively. It is the Pinball Routing that packets visit all home agents of mobile routers in nested mobile network. As the nesting level of mobile network increases, probability of packet loss or error and propagation delay will be increased.

3 Direct Tunneling Method

Direct tunneling method we propose in this paper differs from the NEMO basic support protocol proposed by IETF NEMO Working Group. Our method transmits all packets through unidirectional tunnel between the mobile router and the correspondent node. The packet from mobile network node to the correspondent node is tunneled on the mobile router, and the correspondent node sends a packet to the mobile network node with a routing header that is one of IPv6 extension headers. To achieve this, we modified Binding Update process and tunneling on mobile router. We also added storage Binding Cache for mobile network, and inclusion of a routing header on the correspondent node.

3.1 Modification of Binding Update Process

When a mobile router detects its movement, it gets allocation of a care-of-address and sends its new location to the home agent. Then, the home agent can properly forward packets that were sent to a mobile network node. In the same way, the mobile router sends Binding Update to the correspondent node when a mobile network node sends packets to a correspondent node or a correspondent node sends packets to a mobile network node. The correspondent node that received Binding Update can send packets to the mobile network node through optimized path.

Fig. 2 shows the process where CN sends packet to LH in a mobile network, and then MR sends Binding Update. The process achieves optimized routing path of packets between CN and mobile network node. In the figure, a packet the CN has sent to LH arrives at HA(1); it is forwarded to MR through bidirectional tunnel between HA and MR(2); MR removes the tunnel and it forwards the original packet to LH, and, at the same time, sends Binding Update to CN(3); now, MR and CN can exchange packets with direct tunnel between them(4).

3.2 Modification of Tunneling on Mobile Router

In the direct tunneling method, a mobile router sends packet through direct tunnel to a correspondent node instead of sending through a tunnel to its home agent. A mobile router maintains a 'CN table' to keep the information if Binding

Update was sent to a CN or not. The mobile router must initialize the CN table when it moves to another location.

All mobile routers in nested mobile network, which are in the path of packet passing by, should open direct tunnel to correspondent nodes. In this scheme, the mobile routers do not care about the types of mobile network node, whether it is plain host or not: it applies the same direct tunneling method on all packets passing by.

Fig. 3 (a) shows multistage-tunneled packet created when VMN sends Binding Update to its home agent HA_{VMN} in the architecture shown in Fig. 1.

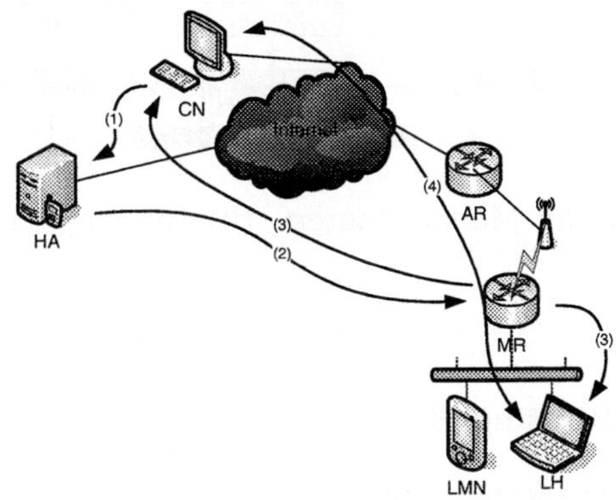

Fig. 2. Modified Binding Update on Mobile Router

3.3 Addition of Binding Cache for Mobile Network

In NEMO basic support protocol, Binding Cache is modified so that the home agent can hold mobile network prefix. In our scheme, we introduced 'NEMO Binding Cache' that holds a chain of mobile routers' care-of-addresses, instead of modifying the existing Binding Cache in MIPv6. This data structure holds all information coming from mobile routers. By adding NEMO Binding Cache both to home agents and to all correspondent nodes communicating with mobile network nodes, they all know optimized path. To maintain the information on nested mobile network, a list or a linked list structure is suitable for NEMO Binding Cache.

NEMO Binding Cache constructed according to Binding Updates from mobile routers holds addresses of all intermediate mobile routers as well as the final destination network address. Using this information, a correspondent node can directly send packets bypassing mobile routers' home agents even if mobile networks are nested. Mobile router's home agent can also send packets directly bypassing the next mobile routers' home agents. Fig. 3 shows Binding Cache in NEMO basic support(b), and NEMO Binding Cache proposed in this paper(c).

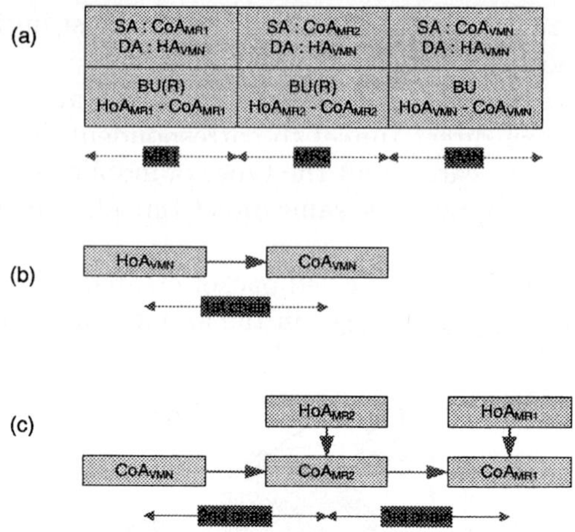

Fig. 3. (a) Tunneling on Mobile Routers, (b) Binding Cache on HA_{VMN}, (c) NEMO Binding Cache

3.4 Inclusion of Routing Header on CN

A correspondent node creates and maintains NEMO Binding Cache based on Binding Updates from mobile routers. Using information in Binding Cache and NEMO Binding Cache, the correspondent node organizes a routing header which contains optimized path, which allows packets to be sent directly to the communicating mobile network node.

Fig. 4 shows IPv6 packet header from HA_{VMN} to VMN using information in Fig. 3 (b) and (c). SA (Source Address) of packet is HA_{VMN} and DA (Destination Address) is CoA_{MR1}. The care-of-addresses of the next two nodes MR2 and VMN are designated in the routing header. The last header is Mobile IPv6 destination option header for VMN.

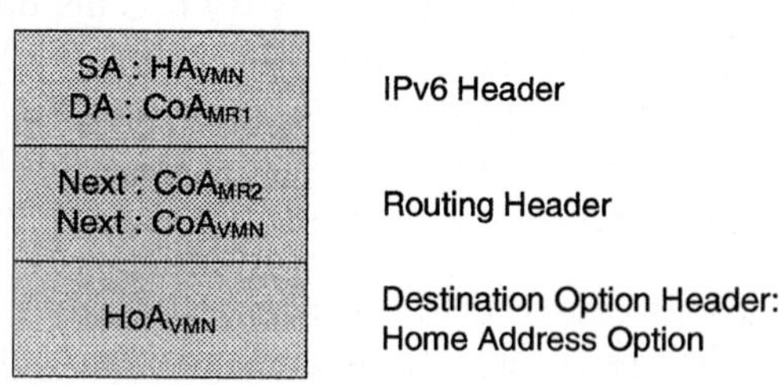

Fig. 4. Example of Routing Header on CN

4 Simulation and the Result

In order to evaluate the performance of direct tunneling method, we performed simulation. The nested mobile network architecture shown in Fig. 5, which is similar to Fig. 1 we introduced at the beginning, is one of two simulation models we used.

In the figure, HA_{MR1} and HA_{MR2} are home agents of mobile router MR1 and MR2, respectively; LH is a mobile network node of MR2's network; and CN is the correspondent node of LH. The path (a) is the packet routing path when NEMO basic support protocol is applied, and path (b) is the packet routing path of direct tunneling method proposed in this paper.

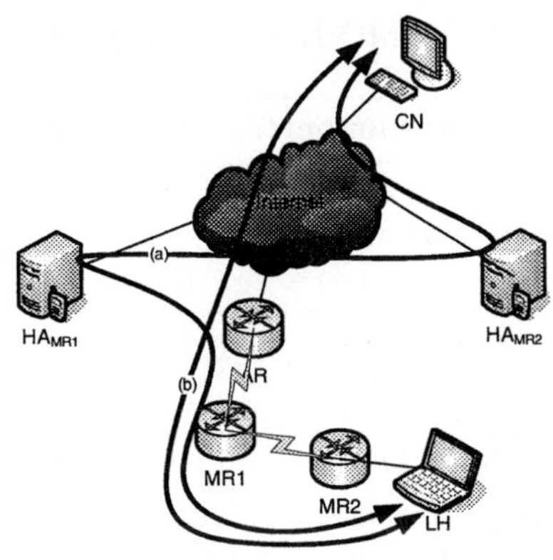

Fig. 5. Simulation Model

4.1 Simulation Models

For comparison, we performed simulation for both models with OMNeT++, which is open architecture simulation environment for communication network[8]. Simulation models are path (a) and path (b) of packet routing in Fig. 5. We measure the RTT (Round Trip Time) of both paths. Delay time of each link between nodes is $5ms$; processing time in each node, such as tunneling process, de-tunneling process, and handling messages, is $10\mu s$. Internet cloud in Fig. 5 represents IPv6 network, and its delay time (of a packet passing the cloud) is selected random values ranging between $45ms$ and $55ms$. Because selected random values follow normal distribution, the average is $50ms$. Thus random values do not have any effect to the result.

We have done simulations on totally six sub-models: three cases for NEMO basic support protocol and three for direct tunneling method. General model is the same as shown in Fig. 5. The first case is for the configuration with just one mobile network, the second case is for the configuration with two nested mobile networks, and the last case is for with three nested mobile networks.

4.2 Result

Statistical results of simulation are shown in Table 1 and Table 2. When there is single mobile network, the average RTT of direct tunneling method is about 1.9 times less than NEMO basic support protocol; when there are two nested mobile networks, it is 2.7 times less; it is 3.4 times less when there are three nested mobile networks. We could carefully predict that this gap between NEMO basic support protocol and direct tunneling method will be getting bigger as the nesting level increases.

Table 1. NEMO Basic Support Protocol

$Factor$	$MinimumRTT$	$MaximumRTT$	$AverageRTT$
One Mobile Network	$230ms$	$270ms$	$250ms$
Two Mobile Networks	$352ms$	$405ms$	$380ms$
Three Mobile Networks	$477ms$	$540ms$	$510ms$

Table 2. Direct Tunneling Method

$Factor$	$MinimumRTT$	$MaximumRTT$	$AverageRTT$
One Mobile Network	$120ms$	$140ms$	$130ms$
Two Mobile Networks	$130ms$	$150ms$	$140ms$
Three Mobile Networks	$140ms$	$160ms$	$150ms$

Fig. 6 shows the trend of RTT as the nesting level increases in mobile networks. We fit average RTTs to linear equation of nesting level for each case. Graph (a) is result for NEMO basic support protocol, and graph (b) is result for direct tunneling method. Graph (b) shows that RTT increases slowly as increments in nesting level. But, graph (a) shows that RTT increases sharply. Fitting equation is as follows.

$$y = ax + b$$

5 Conclusions

In this paper, we proposed the direct tunneling method which optimizes the routing path in nested mobile network. From the result of simulation, we argue that our method performs better than NEMO basic support protocol proposed by IETF NEMO Working Group.

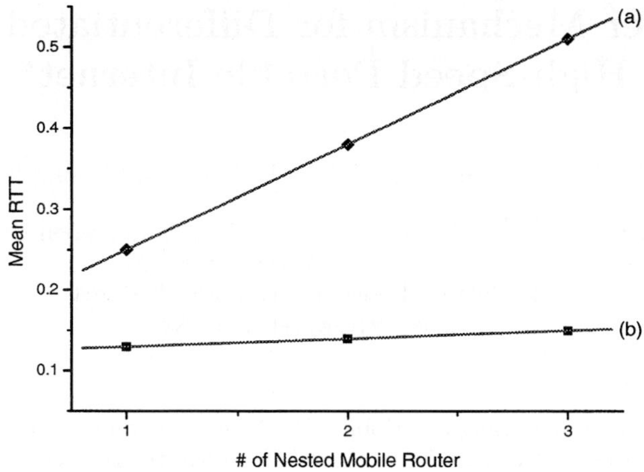

Fig. 6. The trends of RTT as the nesting level increases in mobile networks.

Direct tunneling method opens the same number of tunnels as NEMO basic support protocol, but it establishes direct path bypassing mobile routers' home agents. To establish the optimized path, we modified Binding Update process on mobile router and added the NEMO Binding Cache to correspondent node and home agent, while only the Binding Cache of home agent is extended for NEMO basic support protocol. To achieve better performance, we also made some extension on the procedures of network elements.

We expect our method will have larger impacts on overall performance with the simple extension. Meanwhile, more investigation is needed to find ways to reduce the number of tunnels to lighten the load on mobile routers.

References

1. Deering, S., Hinden, R.: Internet Protocol, Version 6 (IPv6) Specification: RFC2460, IETF, 1998
2. Johnson, D., Perkins, C., Arkko, J.: Mobility Support in IPv6: RFC3775, IETF, 2004
3. Ernst, T.: Network Mobility Support Goals and Requirements: Internet Draft, IETF, 2003.
4. Ernst, T., Lach, H-Y.: Network Mobility Support Terminology: Internet Draft, IETF, 2003
5. Vijay Devarapalli, Ryuji Wakikawa, Alexandru Petrescu, Pascal Thubert: Network Mobility (NEMO) Basic Support Protocol: Internet Draft, IETF, 2004
6. Manner, J., Kojo, M.: Mobility Related Terminology: RFC3753, IETF, 2004
7. Conta, A., Deering, S.: Generic Packet Tunneling in IPv6 Specification: RFC 2473, IETF, 1998
8. András Varga: OMNeT++: http://www.omnetpp.org

Handover Mechanism for Differentiated QoS in High-Speed Portable Internet*

Ho-jin Park[1], Hwa-sung Kim[1], Sang-ho Lee[2], and Young-jin Kim[2]

[1] Department of Electronic Communications Eng., Kwangwoon Univ., Korea
{sanzini, hwkim}@daisy.kw.ac.kr
[2] IP Mobility Research Team, ETRI, Korea
{shlee, yjkim}@etri.re.kr

Abstract. When using the Mobile IP, which is the representative technology to secure the mobility in general IP networks, the packet loss during the handover is inescapable. To remedy the packet loss problem, the smooth handover was introduced. However, this solution is also flawed. In particular, the smooth handover causes the packet sequence to be disrupted during the packet forwarding procedure. In turn, it may result in the degradation of the network performance. The same problem also occurs in the HPi (High-speed Portable Internet) system; the next generation portable IP service system. The HPi system, which provides the high speed data service just like xDSL and leased line in wired internet, aims to guarantee the portability, mobility, and the differentiated service based on IEEE 802.16. In this paper, we will propose a handover mechanism and a packet sequence control algorithm that prevent the packet loss and the out-of-sequence packet problem for differentiated service in the HPi system.

1 Introduction

Demands for Internet access in wireless and mobile environments are increasing parallel to the diffusion of the personal portable terminals. As a result of these demands, the HPi system that secures the portability and high data rate service is being developed by ETRI, Samsung, SK telecom and other companies in Korea [1]. As shown in Fig. 1, the HPi system consists of HPi-AT(HPi-Access Terminal), HPi-AP(HPi-Access Point) and PAR(Packet Access Router) connected to the IP networks. More specifically, the HPi system offers three types of service class: RtPS(Real time Polling Service), NrtPS(Non-real Time Polling Service), and BE(Best Effort service)[1]. Thus, the HPi-AT using the IP service of specific class should guarantee the proper quality of service during the handover. However, the packet loss and the out-of-sequence packet problem during the handover are inescapable in the HPi system.

The smooth handover was proposed to minimize the packet loss during the handover in general IP networks [2][3]. While the smooth handover minimizes

* This work was supported by University IT Research Center Project and the Research Grant from Kwangwoon University in 2004

C. Kim (Ed.): ICOIN 2005, LNCS 3391, pp. 904–911, 2005.

the packet loss, another problem of out-of-sequence packet is introduced. More specifically, the TCP sends duplicated ACK to the sending host. The sender, in turn, regards it as the network congestion if a MN (Mobile Node) receives out-of-sequence packets. Therefore, to solve this problem, another handover mechanisms based on the buffering at the cross-over node or the new FA (Foreign Agent) was proposed [4][5].

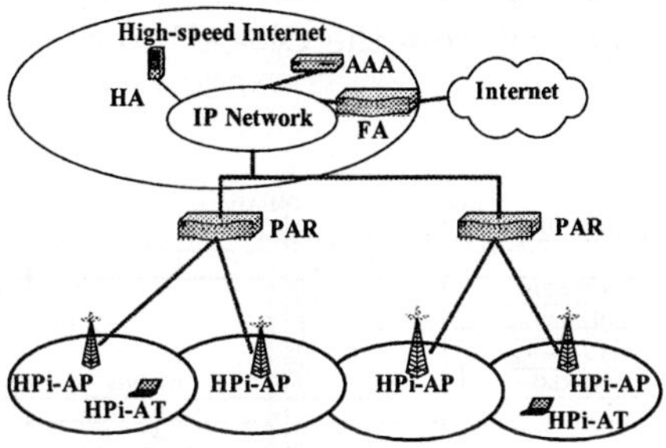

Fig. 1. HPi System architecture

The basic HPi handover mechanism assumes that the packet buffering is performed only at Old-AP where the HPi-AT was originally attached during the handover [1]. However, it neither solves the out-of-sequence packet problem nor considers the service classes. Therefore, we propose an enhanced Old-AP buffering handover mechanism for the BE and NrtPS service classes. In addition, we also propose the PAR buffering handover mechanism based on the cross-over node buffering and the New-AP handover mechanism based on new FA buffering for the RtPS service class.

This paper is organized as follows. Section 2 will explain the overview of basic HPi handover mechanism. Section 3 will describe the proposed handover mechanism and packet sequence control algorithm. Section 4 will present the simulation results. Finally, section 5 will present the conclusion.

2 HPi Handover Mechanism and Current Problem

When an HPi-AT is connected to an HPi-AP, it measures the wireless signal strength from the currently connected HPi-AP continuously with NBR-ADV (Neighbor Advertisement) message, SCN-REQ (Scanning interval allocation Request) and DL-MAP (Downlink MAP) message.

When the HPi-AT triggers the handover to the HPi-AP, it sends HO-REQ (Handover Request) message containing the list of HPi-APs that have the signal strength stronger than the constant level. The Old-AP then negotiates with the New-AP whether the HPi-AT is acceptable in the New-AP by using HOreq (HO requset), HOind (Handover indication) and HOcnf (Handover confirmation) messages. The PAR that is informed by the New-AP delivers HOrsp (Handover

Response) message, including the result of the handover request and the cell address of the New-AP, to the Old-AP. Consequently, the Old-AP starts packet buffering for downlink channel. The HPi-AT sends HO-IND (Handover Indication) message that indicates the handover initiation to the Old-AP. Then the context information of the HPi-AT is passed from the Old-AP to the New-AP through ACIind (AP Context Information indication) and ACIcnf (AP Context Information confirmation) messages. After sending ACIcnf message to the Old-AP, the PAR routes all the packets toward the AT, to the New-AP, and the buffered packets are forwarded from the Old-AP to the New-AP [1].

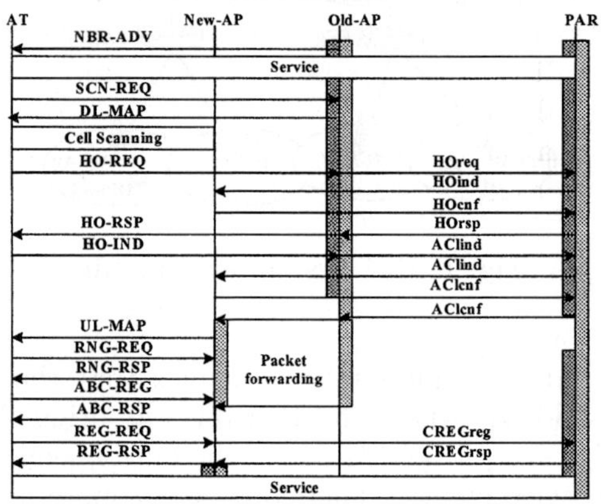

Fig. 2. Inter-AP handover in HPi system

The New-AP can receive the two kinds of packet streams during the handover. The packet stream, which is denoted as the Old-Stream, is sent from the buffering node. Another that is denoted as the New-Stream is sent from the outside network after sending ACIcnf message to the PAR. However, both the Old-Stream and the New-Stream may be delivered at the same time to the New-AP through the same interface of the PAR. As a result, the Old and New-Stream may be mixed up in the New-AP and create the out-of-sequence packet problem. These out-of-sequence packets disturb UDP and TCP applications. Especially, in TCP, the out-of-sequence packet problem degrades the performance of TCP; as a result, the service quality of HPi-AT becomes low during and after the handover.

Next, the procedure for connecting the wireless link between the HPi-AT and the New-AP is performed with RNG-REQ (Ranging Request), RNG-RSP (Ranging Response), ABC-REQ (AT Basic Capability Request), ABC-RSP (AT Basic Capability Response), REG-REQ (Registration Request) and CREGeg (Cell Registration Request) messages. Finally, after the creation of the new traffic session is achieved with REQ-RSP (Registration Response) message, the service for the HPi-AT is started. However, the AT may still receive out-of-ordered packets. Fig. 2 shows the HPi basic handover mechanism.

3 Proposed Handover Mechanism

3.1 Old-AP Buffering Handover and Sequence Control Mechanism

In order to solve the out-of-sequence packet problem, the New-AP must be able to differentiate the Old-Stream from the New-Stream. For this reason, we add a "START" message that should be sent with HOrsp message by the PAR at the beginning of buffering for the Old-Stream. Furthermore, we also add a "LAST" message that should be sent with ACIcnf message when the PAR stops buffering. The New-AP can not distinguish the mixed stream packets until LAST message arrives even though it already received START message. Therefore, an additional sequence control mechanism in the New-AP is proposed. We use the packet ID (IP packet identification field in IP header) to solve the out-of-sequence packets problem.

As shown in Fig. 3, the New-AP decides whether the newly received packet belongs to the Old-Stream or the New-Stream by comparing the input packet ID with the New-First pointer in the buffer of the New-AP. The New-First and the New-Last pointers are initially set to the bigger IDs among the first and the second arrived packets. Then the New-Last is replaced with the packet ID of the newly received New-Stream packet. Fig. 4 shows the proposed sequence control mechanism. Until the START message is received in the New-AP, all packets for the HPi-AT must be the New-Stream. And after receiving the LAST message, all packets for the HPi-AT belong to the New-Stream. If the two packets received first belong to the Old-Stream, and the newly received Old-Stream packet has the packet ID value between the New-First and the New-Last, it means that the initial New-First value is wrong. The New-First value is then modified as the smallest packet ID among the New-Stream in buffer. However, due to the overhead of sequence control mechanism, the Old-AP buffering handover mechanism is not suitable for the RtPS.

3.2 PAR Buffering Handover Mechanism

PAR is the cross-over node of the Old and the New-AP. Moreover, it can be the other candidate for the buffering to solve the packet loss and the out-of-sequence problems during the handover. In the case of PAR buffering, the packet loss and the out-of-sequence problem do not occur because all the packets of the mixed streams are buffered sequentially in the PAR buffer. As compared to the Old-AP buffering handover mechanism, the PAR buffering minimizes the packet forwarding delay. With PAR, there is no additional delay required for sequence control at the New-AP and the packet forwarding path is shorter than the Old-AP buffering handover mechanism. It is for this reason why the PAR buffering handover mechanism is suitable for the RtPS. While currently, in the real HPi

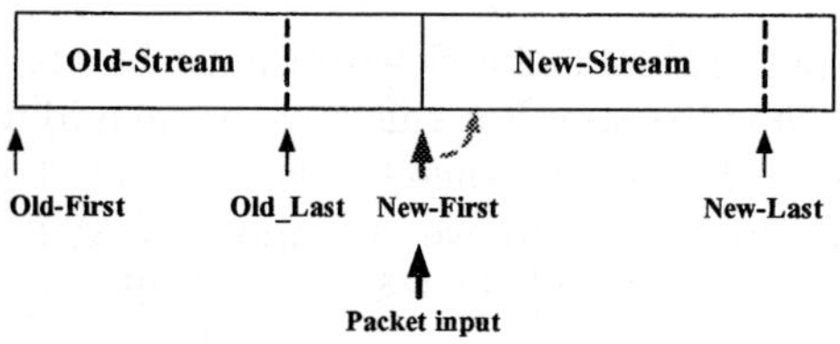

Fig. 3. Basic concept of proposed mechanism

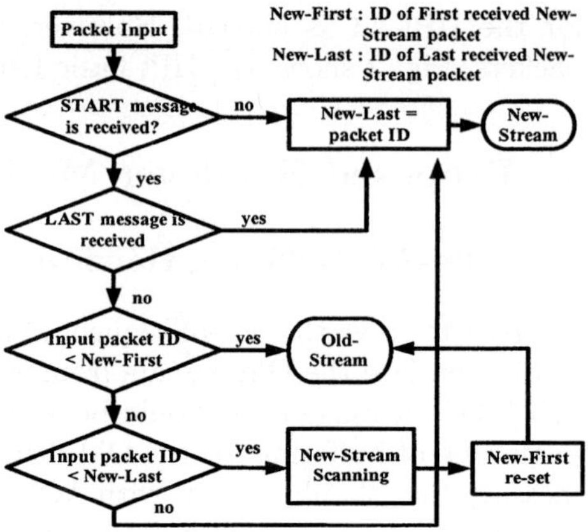

Fig. 4. Packet sequence control mechanism in New-AP

system, there is no buffer in the PAR because of the bottleneck problem [1], we can simulate and compare the PAR buffering handover mechanism with other two buffering handover mechanisms.

3.3 New-AP Buffering Handover

The HPi-AT can anticipate the New-AP by using the signal strength of the neighboring HPi-APs. Based on this fact, we propose a New-AP buffering mechanism. In order to adopt this mechanism, however, all context information exchanged between the HPi-AT and the HPi-AP should be done prior to initiating the handover. Therefore, several new messages such as Horeq-Ext (Handover request Extension), Hoind-Ext (Handover indication Extension) and Horeq-Ext (Handover response Extension) must be defined in order to deliver the information about the handover request and the context information of HPi-AT.

As shown in Fig. 5, the PAR sends the Horsp-Ext message to the Old-AP and forwards all packets from outside of the PAR to the New-AP. As a result, the New-AP that receives the packets of the HPi-AT, starts packet buffering and transmits the packets as soon as it creates a new traffic session. Thus, all packets to the HPi-AT are buffered sequentially in the New-AP during the handover. The New-AP buffering mechanism minimizes the packet forwarding delay and solves the out-of-sequence packet problem. For this reason, the New-AP buffering handover mechanism is suitable for the RtPS.

4 Analysis

The simulation was performed using NS-2 simulator to compare the performance of the proposed three handover mechanisms of Old-AP, PAR and New-AP buffering handover in terms of UDP and TCP. Fig. 6 shows the network topology used

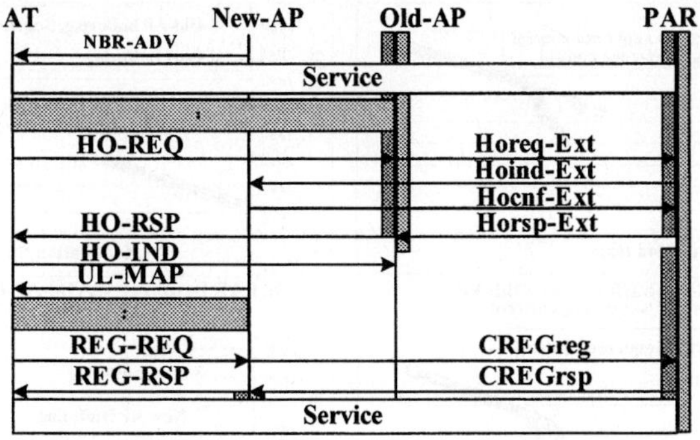

Fig. 5. New-AP buffering handover mechanism

for the simulation. For simplicity, we assume the HPi-AT moves from HPi-AP1 and reaches HPi-AP3 through HPi-AP2. Then it goes back.

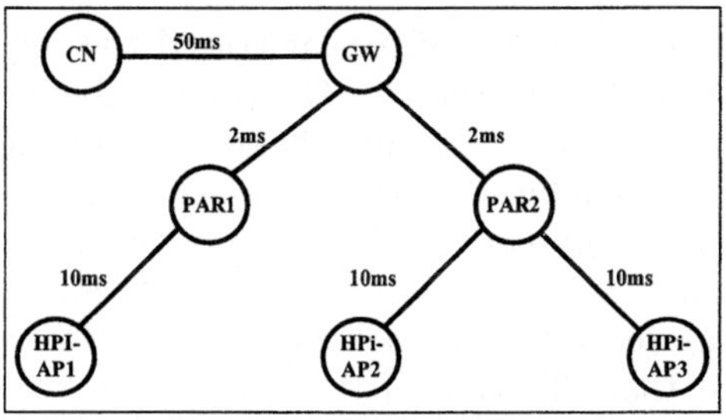

Fig. 6. Simulation Topology

Fig. 7 shows the simulation results that traced the packet ID of the UDP traffic. UDP is an unreliable protocol, so it does not react to the packet loss and the out-of-sequence packets. Therefore, the degradation of service quality may be happening in higher level applications due to the lost packets and out-of-sequence packets. Fig. 7.a shows the trace of packets IDs that HPi-AT has received in each case: no buffering; Old-AP buffering; and Old-AP buffering equipped with New-AP sequence control. In the case of no buffering, there is considerable packet loss during the handover. In the case of the Old-AP buffering, the HPi-AT also receives the out-of sequence packets. Conversely, when Old-AP buffering is used with New-AP sequence control, the out-of-sequence packet problem is almost eliminated. As shown in Fig. 7.b, Fig. 7.c and Fig. 7.d, when PAR buffering or New-AP buffering is used, both the packet loss and the out-of-sequence problems did not occur. Furthermore, the HPi-AT could receive the buffered packet faster than the Old-AP buffering.

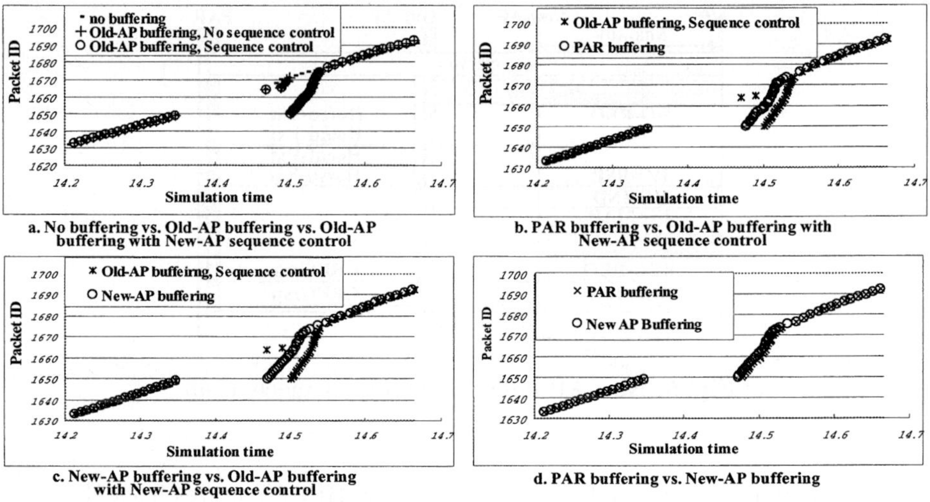

Fig. 7. UDP: Packet ID numbers received by MN, Data rate=0.2Mbps

Fig. 8 shows the simulation results that traced the sequence number of the TCP traffic and the TCP throughput. TCP is a reliable protocol and performs the error recovery for the packet loss and the out-of-sequence delivery. As shown in Fig. 8.a and Fig. 8.d, among the three proposed handover mechanisms, HPi-AT receives the smallest number of TCP packets during the same time interval when no buffering handover is used. Moreover, Fig. 8.b, Fig. 8.c and Fig. 8.d show that HPi-AT can receive more packets when the PAR buffering or the New-AP buffering is used than the case of the Old-AP buffering with the New-AP sequence control. This results from the fact that the Old-AP buffering mechanism with the New-AP sequence control mechanism, requires more packet forwarding and sequence control delays than the two other mechanisms.

5 Conclusion

The HPi system tries to provide the differentiated quality of service according to the three classes of BE, NrtPS, and RtPS. Therefore, the proper handover mechanism is needed to solve the packet loss and the out-of-sequence problems according to the service classes. In this paper, we have proposed the handover mechanisms that include the packet buffering procedure in each case of the Old-AP buffering, the PAR buffering and the New-AP buffering. Furthermore, we also proposed a new packet sequence control mechanism that should be performed in the New-AP, because the Old-AP buffering causes the out-of-sequence packet problem.

In order to compare the performance of three proposed handover mechanisms, a simulation was performed. The simulation results show that all three proposed mechanisms solve both of the packet loss and the out-of-sequence problems.

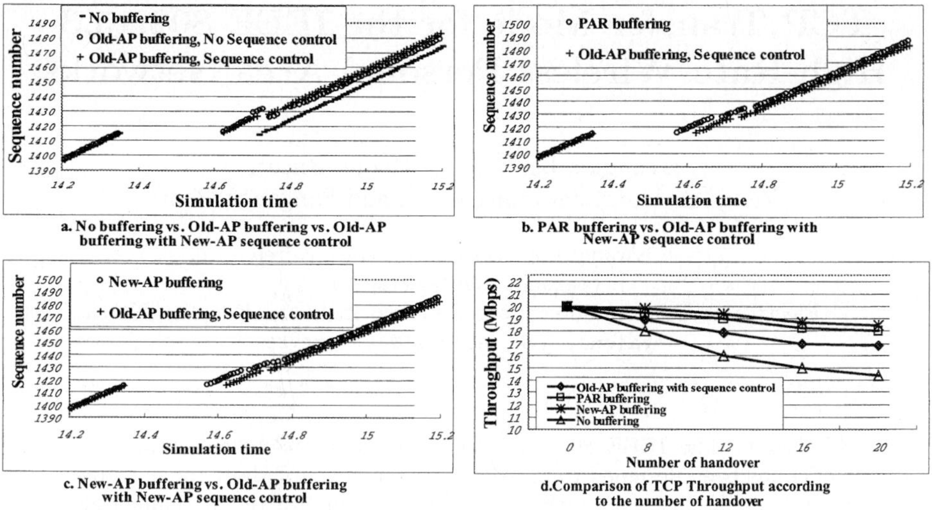

Fig. 8. TCP: TCP sequence numbers received by MN and TCP throughput, Data rate=0.2Mbps

However, the three buffering mechanisms show different delays to deliver the in-order packets to the HPi-AT. The Old-AP buffering, which shows the longest delay can be used for the BE or the NrtPS because they are less sensitive to delay. Moreover, the PAR buffering or the New-AP buffering, which shows the least delay, is suitable for the RtPS that is sensitive to delay. Therefore, in TCP, we confirm that more packets can be received by adopting packet sequence control mechanism in New-AP than only buffering in Old-AP. Moreover, we also confirm the same result in the case of the PAR buffering handover mechanism and the New-AP buffering mechanism.

References

1. "The HPi Handover Specification", ETRI (Electronics and Telecommunications Research Institute) 2003.
2. C. Perkins, "IP mobility support, Internet RFC 2002, Oct. 1996.
3. C. Perkins and K.Y. Wang, "Optimized Smooth Handoffs in Mobile IP", Computers and Communications, July 1999.
4. D. Tandjaoui, N Badache, A.Bouabdallah , H.Bettahar, H Seba, INC-2-2002 "Towards a smooth Handoff for TCP and real time applications in wireless network", INC2002 proceedings (International Network Conference), UK, July 2002.
5. D. Tandjaoui, N Badache, H.Bettahar, A.Bouabdallah, H Seba "Performance enhancement of smooth Handoff in mobile IP by reducing packets disorder" ISCC2003 proceedings (8th IEEE Symposium on Computers and Communications)Kemer-Turkey ISCC2003.

TCP Transfer Mode for the IEEE 802.15.3 High-Rate Wireless Personal Area Networks

Byungjoo Lee[1], Seung Hyong Rhee[1],
Yung-Ae Jeon[2], Jaeyoung Kim[2], and Sangsung Choi[2]

[1] Kwangwoon University, Seoul, Korea
{parang, shr}@kw.ac.kr
[2] Electronics Telecommunications Research Institute, Daejoen, Korea
{yajeon, jyk, sschoi}@etri.re.kr

Abstract. The IEEE 802.15.3 WPAN (Wireless Personal Area Network) has been designed to provide a very high-speed short-range transmission capability with QoS provisions. The unidirectional channel allocations for the guaranteed time slots, however, often result in poor throughput when a higher layer protocol such as TCP requires a full-duplex transmission. In this paper we propose a mechanism, called TCP transfer mode, that provides the bidirectional transmission capability between TCP sender and receiver for the channel time allocations (CTAs) of the high-rate WPAN. As our scheme does not require additional control messages nor additional CTAs, the throughput of a TCP connection on the high-rate WPAN can be greatly improved. Our simulation results show that the proposed scheme outperforms any methods of TCP transmission according to the current standard of the WPAN.

1 Introduction

The emerging high-rate wireless personal area network (WPAN) technology, which has been standardized[1] and being further enhanced by the 15.3 task group in IEEE 802 committee, will provide a very high-speed short-range transmission capability with quality of service (QoS) provisions. The QoS capability is provided by the channel time allocations using TDMA; if a DEV (device) needs channel time on a regular basis, it makes a request for isochronous channel time. Asynchronous or non-realtime data is sup-posed to use CAP (Contention Access Period) which adopts CSMA/CA for the medium access.

Among the high-rate applications expected to be prevalent in the near future, the high quantity file transfer using TCP will also occupy a large portion of the traffic transmitted in the WPAN environment. The unidirectional channel allocations for the guaranteed time slots, however, often result in poor throughput because a TCP connection requires a full-duplex transmission channel. In order to transmit the TCP traffic according to the current standard of the high-rate WPAN, one of the following three methods can be adopted. First, it can be transmitted during CAP. However, as the duration of the CAP is determined by the piconet coordinator (PNC) and communicated to the DEVs via the beacon,

C. Kim (Ed.): ICOIN 2005, LNCS 3391, pp. 912–922, 2005.

it is very hard for the devices to estimate the available bandwidth for the TCP connection. Second, the TCP connection may request a guaranteed time slot and use it for the bidirectional TCP data and acknowledgment packets. Clearly it will cause frequent collisions at the MAC layer between TCP sender and receiver, and thus significantly degrade the transmission performance. Finally, they may request two CTAs, one for the TCP data and another for the TCP acknowledgment. Due to the dynamic nature of the TCP flow control, it is very hard to anticipate or dynamically allocate the size of the CTAs.

Recently, a lot of work has been done by many researchers in the area of the high-rate WPAN. However, few attempts have been made at the problem of non-realtime TCP transmission so far. In this paper, a mechanism, called TCP transfer mode, that provides the bidirectional transmission capability between TCP sender and receiver on the guaranteed time slots of the high-rate WPAN. If a CTA is declared to be in the TCP transfer mode, the source DEV alternates between transmit mode and receive mode so that the destination DEV is able to send data (TCP ACK) in the reverse direction. Our mechanism is transparent to the MAC entity: the source DEV regularly makes transitions between transmit and receive mode, and the destination DEV sends data only when the CTA is in the TCP mode. In addition, as our scheme does not require additional control messages nor additional CTAs, the throughput of a TCP connection on the high-rate WPAN can be greatly improved.

The remaining part of this paper is organized as follows. After introducing related works and the high-rate WPAN protocol in chapter 2, we describe the three methods of TCP transmission under the current standard in chapter 3. In chapter 4, we propose a new transmission mode that allows bidirectional TCP transfer on the guaranteed time slots. Simulation results are provided and discussed in chapter 5, and finally chapter 6 concludes the paper.

2 Preliminaries

2.1 IEEE 802.15.3 High-Rate WPAN

The IEEE 802.15.3 WPAN has been designed to provide a very high-speed short-range transmission capability with QoS provisions[7]. Besides a high data rate, the standard will provide low power and low cost solutions addressing the needs of portable consumer digital images and multimedia applications. Fig. 1 shows several components of an IEEE 802.15.3 piconet. The piconet is a wireless ad hoc network that is distinguished from other types of networks by its short range and centralized operation. The WPAN is based on a centralized and connection-oriented networking topology. At initialization, one device (DEV) will be required to assume the role of the coordinator or scheduler of the piconet. It is called PNC (piconet coordinator). Its duty includes allocating network resources, admission control, synchronization in the piconet, providing quality of services, and managing the power save mode.

The superframe of the piconet consists of periods as follows. In the first period, the PNC transmits a beacon frame which contains all the necessary

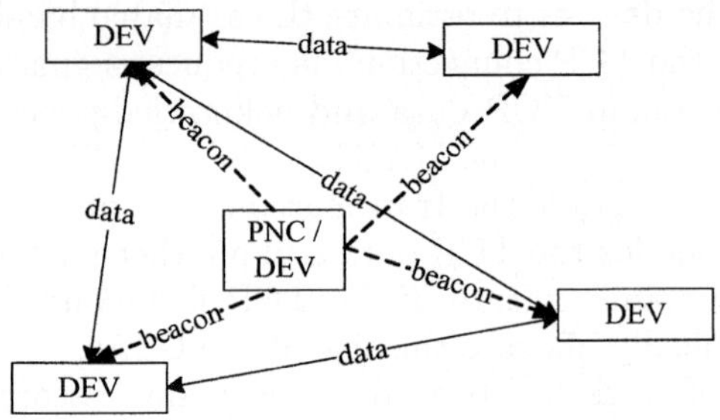

Fig. 1. IEEE 802.15.3 piconet components

information to maintain the piconet. All the DEVs in the piconet receive the beacon frame and synchronize their timer with the PNC. The beacon frame is used to carry control information to the entire piconet and the allocation of channel time. In the second period, optional CAP (Contention Access Period) is located for the purposes of association request/response, channel time request/response and possible exchange of asynchronous traffic using CSMA method. CTA (Channel Time Allocation) period is in the third period and is the most part of the superframe. This period is used for isochronous streams and asynchronous data transfer. The CTAP adopts a TDMA method and allocates guaranteed time slots for each DEV. All transmission opportunities during the CTAP begin at predefined times, which is relative to the actual beacon transmission time and extends for predefined maximum durations. Those allocation information is communicated in advance from the PNC to the respective devices using the traffic mapping information element conveyed by the beacon. During its scheduled CTA, a DEV may send a number of arbitrary data frames with the restriction that aggregate duration of these transmissions which are not exceed the scheduled duration limit.

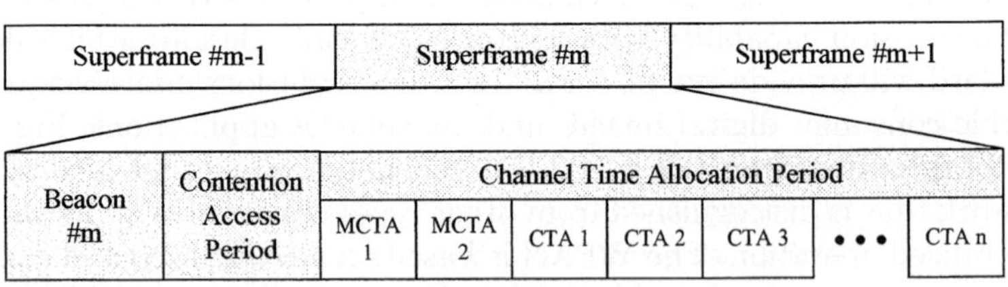

Fig. 2. IEEE 802.15.3 superframe

2.2 Related Works

Recently, a lot of work has been done by many researchers in the area of the high-rate WPAN. However, few attempts have been made at the problem of non-real time TCP transmission so far. In the [2], authors proposed a MAC protocol that enhances the TCP transmission in TDMA-based satellite networks. This is an approach of TCP throughput enhancement using the modified MAC protocol. Similarly, [3] proposed a mechanism for enhancement of TCP transfer via satellite environment. In the [4], authors propose a MAC layer buffering method to improve handoff performance in the Bluetooth WPAN system in order to improve a TCP/IP performance. It can minimize the negative effects of TCP exponential backoff algorithm during handoff and no duplicate packets occur due to MAC layer buffering method. In [5], although authors are not concerned with TCP transmission, they proposed an *Application-Aware MAC* mechanism which considers the status of higher layer. To the best of our knowledge, however, there have been no research that MAC layer supports efficient TCP transfer in the high-rate WPAN.

3 TCP Transmissions in the WPAN

In this chapter, we describe three possible methods of TCP transmission with the MAC protocol of the current WPAN standard which contains no mention on higher layer protocols. The performance of TCP transmission using each possible method will be discussed and compared. Except the three methods discussed in this chapter, TCP traffic can be transmitted during CAP. However, as the duration of the CAP is determined by the PNC and communicated to the DEVs via the beacon, it is very hard for the devices to estimate the available bandwidth for the TCP connection. We will consider only the methods of using CTAP in this paper. In order to transmit the TCP traffic using CTAP according to the current standard of the high-rate WPAN, one of the three methods in this chapter can be adopted.

Fig. 3 shows a TCP transmission process using immediate ACK policy in high-rate WPAN. First, TCP data packet comes from the higher layer and is processed at the MAC layer. The sender DEV sends the MAC frame to the receiver DEV via wireless interface. At the MAC layer of TCP receiver, it receives the MAC frame and sends a MAC ACK to the sender. The TCP sink that accepted TCP data frame sends a TCP ACK packet to the TCP sender. The TCP sender that received this TCP ACK packet sends MAC ACK frame for the TCP ACK. TCP sender transmits next TCP data packet to the receiver when it received TCP ACK. All TCP transmissions are achieved by this way in the high-rate WPAN.

3.1 TCP Transmission Via on CTA

According to the current MAC protocol, one may use a single (unidirectional) CTA for a TCP connection. This will result in poor throughput because a TCP

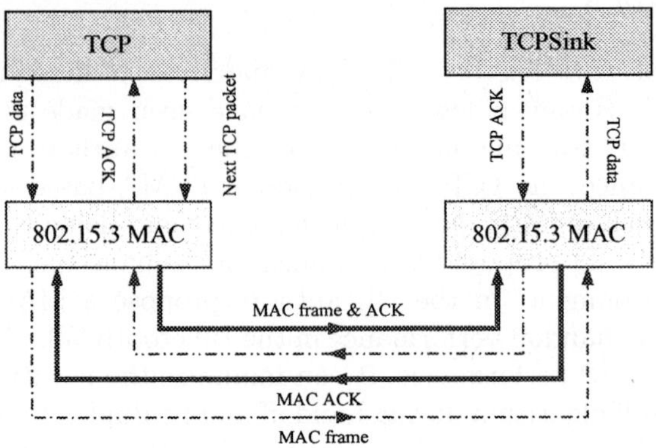

Fig. 3. TCP transmission in IEEE 802.15.3 high-rate WPAN

connection requires a full-duplex transmission channel. TCP traffic can not be transmitted between the sender and receiver using this method, because the CTA is defined as unidirectional and the TCP receiver has no way of sending TCP ACKs to the sender. Thus transmission of transport layer ACK is impossible and the connection can not be maintained. Fig. 4 depicts the case where a single CTA is allocated to the TCP sender and the TCP receiver has no way of sending back the ACKs.

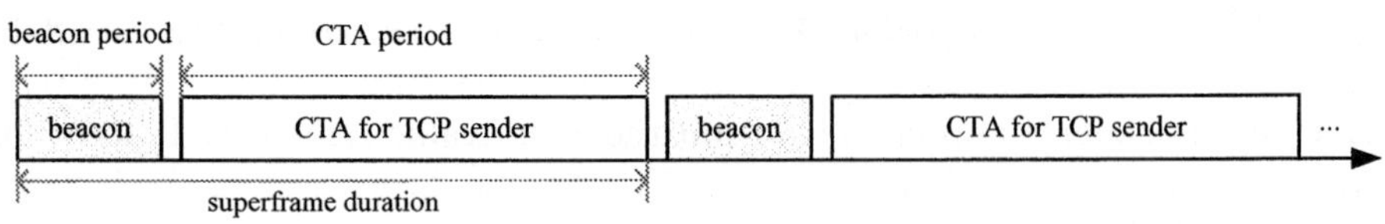

Fig. 4. Single CTA in the high-rate WPAN

3.2 Allocating Two CTAs for TCP Data/ACK

The PNC may allocate extra CTA for the TCP receiver, so that the receiver is able to send back the necessary ACKs. In this method, the TCP sender transmits data during its own CTA, and the receiver transmits ACKs also during its allocated CTA. The problem here is that, due to the dynamic nature of the TCP flow control, it is very hard to anticipate or dynamically allocate the size of those CTAs. In addition, TCP sender waits an ACK packet after sending data up to the window size, and the receiver can not send an ACK before its CTA comes. Therefore, this method may waste the two CTAs because exact channel time allocation is almost impossible due to the dynamic property of the TCP connection. Fig. 5 explains this method: TCP sender is assumed to transmit data packets during CTA1, and the receiver sends ACK packets during CTA2. In this

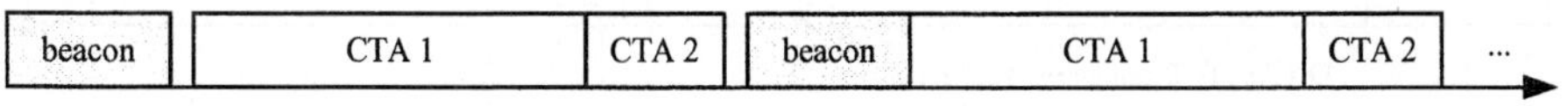

Fig. 5. Two CTAs for a TCP connection

method, the throughput can be very different according to the ratio of the durations of the two CTAs. If the ratio of the allocated CTAs does not consider the current status of the TCP connection, those CTAs can be seriously wasted. We performed a simple simulation to see this problem of using two separate CTAs. The sum of two CTAs is fixed to 4000 and the ratio is varied. In the simulation result of Fig. 6, we can see that the throughput drops significantly because of the CTA waste according to the CTA ratio. Again, the problem of using two separate CTAs for a TCP connection is that it is extremely hard to adjust the ration of the two CTAs according to the dynamics of a TCP connection.

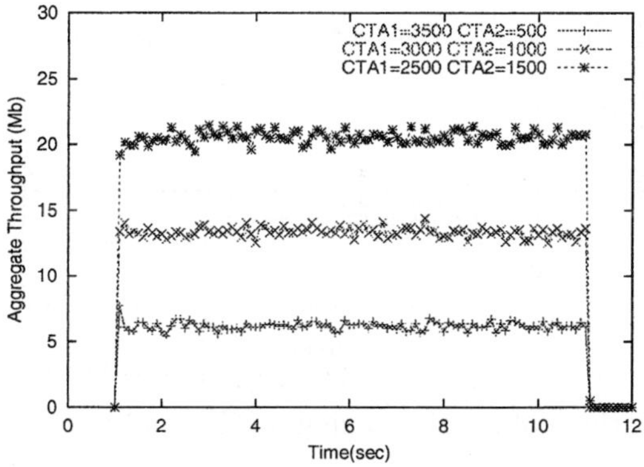

Fig. 6. TCP throughputs for the different ratio of two CTAs

3.3 Sharing a Single CTA

TCP connection may request a single guaranteed time slot and use it for the bidirectional TCP data and acknowledgment packets. A single CTA is shared between TCP sender and receiver. That is, the sender send TCP data to the receiver and the receiver send TCP ACK to the sender during a single CTA.

Clearly it will cause frequent collisions at the MAC layer between TCP sender and receiver, and significantly degrade the transmission performance.

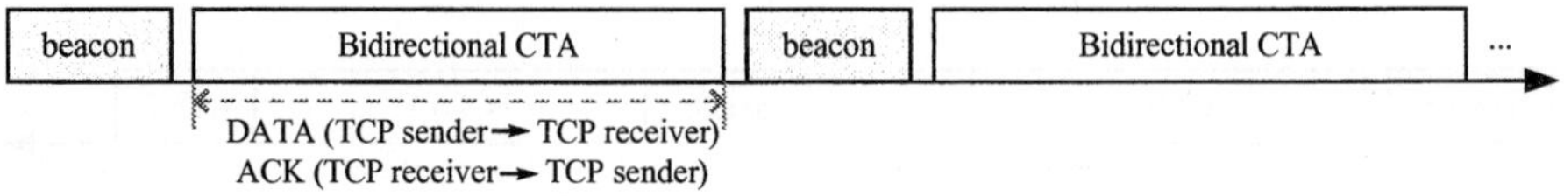

Fig. 7. Sharing a single CTA between two device

Fig. 8 depicts a comparison between sharing a single CTA and allocating two CTAs. The result shows that more throughput can be achieved by sharing a single CTA method. Again, however, the performance degradation is inevitable because of the frequent collision between the two devices. Thus we need an elaborated way of scheduling the single CTA between TCP sender and receiver.

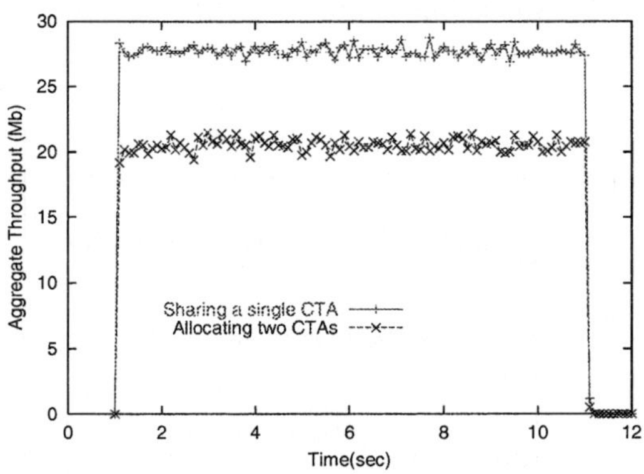

Fig. 8. Comparison of sharing a single CTA and allocating two CTAs

4 TCP Transfer Mode

In this chapter, we propose *TCP transfer mode* which can maintain the throughput of a TCP connection, avoiding the collisions due to a single CTA that is shared between TCP sender and receiver. The sender device informs PNC that it will send TCP data when it makes a channel time request. For this purpose, we have defined *T*CP Enable bit using the reserved bits in 15.3 MAC header. The PNC responds to the request, and then broadcast the beacon frame with the information on the newly allocated CTA for the TCP connection. The CTA information contains the stream index field which tells that the CTA is allocated to a TCP connection and the CTA will be used according to the TCP Transfer mode between the transmitter and the receiver. The TCP stream index and CTA block are depicted in Fig. 9.

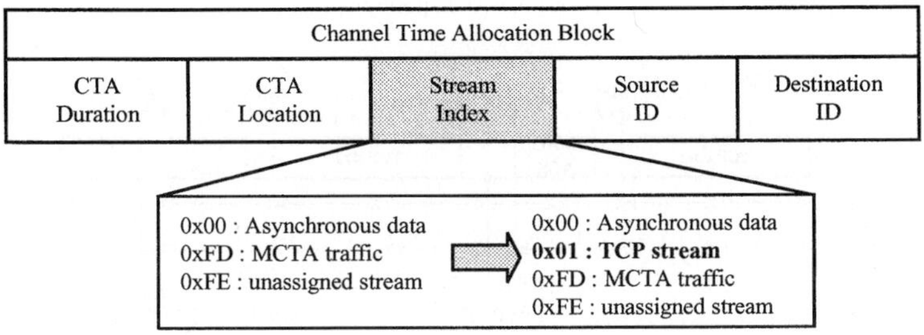

Fig. 9. Stream Index field and value in CTA block

Three ACK policies are available in the current standard of the IEEE 802.15.3 high-rate WPAN: Immediate-ACK, No-ACK and Delayed-ACK policy. We consider only Immediate-ACK and No-ACK policy in this paper. We describe the TCP transfer mode with the No-ACK case first. The (TCP) sender changes its radio interface from TX mode to RX mode immediately after it sends a (TCP) data. Then if it senses a frame in the reverse direction during the SIFS (short inter frame space), it is possibly TCP ACK from the TCP receiver. Otherwise if the channel is idle during the SIFS, TCP sender returns back to the TX mode and will be transmitting TCP data continually. If the sender receives a TCP ACK from TCP receiver, it maintains the radio interface status as RX. After receiving the frame (possibly TCP ACK), it returns back to the TX mode if there is no more frame during SIFS from the receiver. Two time diagrams in Fig. 10 shows the operations of proposed TCP transfer mode for the no-ACK plocy and immediate-ACK policy, respectively.

5 Performance Evaluation

5.1 Simulation Environment

We have implemented the TCP transfer mode using the CMU wireless extension[9] of the ns-2 network simulator. The parameters used for the simulation are summarized in Table 1. We have assumed that the DEVs are fixed during the simulation and associated to the piconet before the simulation begins. We have chosen the channel bit rate as 100Mbps in order to provide an enough TCP window size.

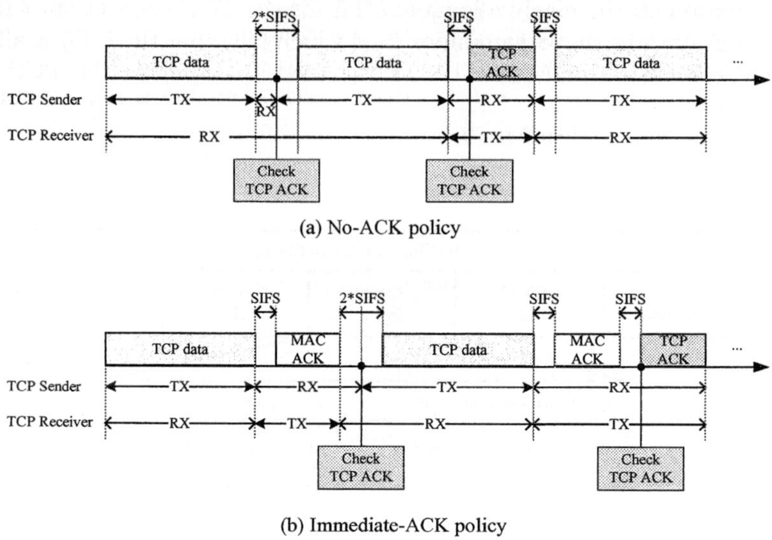

Fig. 10. Difference of ACK policies: (a) No-ACK policy (b) Imnediate-ACK policy

5.2 Results

In this section we evaluate the performance of the proposed TCP transfer mode. According to the current standard of the HR-WPAN, using a single CTA yields the best performance for a TCP connection. Fig. 11 shows that, although the single CTA method outperforms the multiple CTA method, the aggregate throughput is saturated with about 28Mbps out of 100Mbps. As explained in the previous chapter, this low throughput is due to the collisions of TCP data and TCP ACKs between the transmitter and the receiver; TCP data and ACK packets are framed to MAC frames in MAC layer, and as the MAC entity could not distinguish between TCP data and ACK, it transmits MAC frames in own backlogged

Table 1. Simulation parameters

Attribute	Value
Bandwidth	100Mbps
Number of flows	1 flow
Request CTA duration	4000 μsec
MAC ACK policy	Immediate-ACK policy
TCP packet size	1000 bytes
TCP window size	20

queue whenever the wireless medium is idle. It causes severe collisions with peer DEVs and degrades the performance.

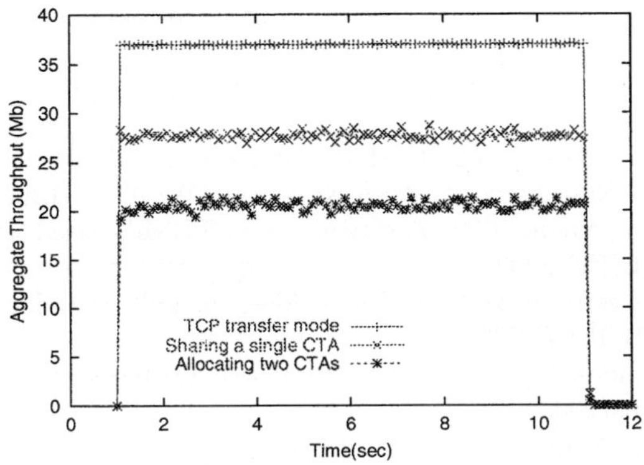

Fig. 11. Aggregate throughputs of various TCP transmission mechanisms

Fig. 11 depicts that the TCP transfer mode proposed in this paper achieves 38Mbps of throughput in the same simulation environment. This is because our method avoids the collision between the transmitter and receiver using the inter-frame spaces where the transmitter checks whether the receiver sends back the TCP ACK frames. As a result, the TDMA-based time slots allocated to the TCP connection provides a bidirectional transmission capability, avoiding packet losses and retransmissions.

As the transmitter checks the possible frame sent back by the receiver after every frame it sends, the TCP ACK arrives at the transmitter with a negligible delay and it has almost no effect on the TCP performance. As mentioned in the previous chapter, the TCP transfer mode also outperforms the case where the CAP period is adopted for the TCP transmission; the PNC can not estimate the exact size of CAP period required for the TCP transmission in a superframe.

6 Conclusion

We have proposes the TCP transfer mode for the IEEE 802.15.3 HR-WPAN in order to maintain the throughput of a TCP connection which suffer from severe collisions in the unidirectional time slots. We have designed the bidirectional transmission mechanism on a single CTA. Extensive simulation results show that the proposed scheme achieves significantly higher aggregate throughput than the possible methods which are available under the current standard. Also, our scheme requires only a small modification to the current MAC standard, and provides the perfect backward compatibility. Our future plan includes the TCP transfer mode with channel errors and the delayed ACK policy.

Acknowledgments

The authors would like to thank Mr. Wangjong Lee for helping us in the simulation of the IEEE 802.15.3 piconet using ns-2.

References

1. Wireless Medium Access Control (MAC) and Physical Layer (PHY) Specifications for High Rate Wireless Personal Area Networks (WPAN), IEEE Std, Sep. 2003
2. J. Zhu, S. Roy: Improving TCP Performance in TDMA-based Satellite Access Networks: ICC2003 IEEE(2003)
3. J. Neale, A. Mohsen: Impact of CF-DAMA on TCP via Satellite Performance: GLOBECOM2001 IEEE(2001)
4. W. Lee et al.: Handoff Provisioning in Bluetooth Wireless Personal Area Networks: IEEE Transactions(2003)
5. S. Rhee et al.: An Application-Aware MAC Scheme for A High-Rate Wireless Personal Area Network: IEEE WCNC2004 (2004)
6. H. Balakrishnan, S. Seshan, E. Amir, R.Kantz: Improving TCP/IP Performance Over Wireless Networks: ACM MOBICOM 95(1995)
7. P. Gandolfo, J. Allen: 802.15.3 Overview/Update: The WiMedia Alliance(2002)
8. R. Managhanram, M. Demirhan: Performance and simulation analysis of 802.15.3 QoS: IEEE 802.15-02/293(2002)
9. The CMU Monarch Project: Wireless and mobile extension to ns Snapshot Release 1.1.1: Carnegie Mellon University(1999)

How to Determine MAP Domain Size Using Node Mobility Pattern in HMIPv6*

Jin Lee[1], Yujin Lim[2], and Jongwon Choe[3]

[1] Standardiztion & System Research Group (SSRG), Mobile Communication Technology Research Lab., CTO, LG Electronics Inc.,
LG R&D Complex, 533, Hogyel-Dong, Dongan-Gu, Anyang-City, Kyongki-Do, 431-749, Korea
jins978@lge.com

[2] Department of Information Media, University of Suwon, Suwon, Korea
yujin@suwon.ac.kr

[3] Department of Computer Science, Sookmyung Women's University, Seoul, Korea
choejn@sookmyung.ac.kr

Abstract. In this paper we present an adaptive determination method of the MAP domain size that can reduce the communication cost between a mobile node and correspondent nodes by assigning a different size of the MAP domain by each node in hierarchical mobile IPv6. Every node is categorized by the residence time in one subnet within the previous visited MAP domain. According to the node's mobility pattern, a slow-moving node can have a small size of MAP domain and a fast-moving node can have a large size of MAP domain as well. Analysis experiments are presented in this paper to evaluate the performance of the proposed method and compare it with the existing scheme.

1 Introduction

Recently, as the number of wireless devices is increasing extremely, it is getting important to reduce signaling cost and to support seamless mobility while the devices move among subnets. Mobile IPv6 [1] enables a mobile node (MN) not only to constitute its own address by itself but also to solve the lack of the addresses in Mobile IPv4. However, it causes to inform Home Agent (HA) of a nodes location whenever a node moves among subnets. In fact, Mobile IPv6 handles global area mobility and local area mobility identically. According to [2], a hierarchical concept that separates micro mobility from macro mobility is preferred because 69% of a users mobility is local.

Hierarchical Mobile IPv6 (HMIPv6) is proposed to support local mobility. HMIPv6 differentiates global (inter-site) from local (intra-site) management by introducing a Mobility Anchor Point (MAP) which functions as the local Home Agent [3]. When a mobile node is changing its position within a MAP domain,

* This Research was supported by the Sookmyung Women's University Research Grants 2004.

C. Kim (Ed.): ICOIN 2005, LNCS 3391, pp. 923–931, 2005.

it is only required to notify the MAP of its new location. For a MAP selection, it is recommended to choose the farthest MAP from a node in HMIPv6. It means that a large size of MAP domain helps to decrease signaling handoff cost.

In this paper, we propose a determination method of MAP domain size that is based on node's mobility pattern. The proposed method is called a Speed based HMIPv6, namely S-HMIPv6. Unlike the existing method, our proposal provides various size of MAP domain for a variety of different nodes mobility pattern. In [4], we can anticipate how fast the node is. In S-HMIPv6, we assign a minimum number of subnets for a MAP domain to a slow-moving node. We record the time whenever a node sends a binding update message to get an interval time, which means "a residence time" in a subnet. Our method uses Access Routers (ARs) as a new MAP in the visited domain to share load and each node comes to have different size of MAP domain.

The performance of the proposed method is evaluated analytically by calculation of the communication cost. Since we only focus on efficiency of an adaptive MAP domain size, we do not care of a packet arrival rate in evaluation.

This paper first presents the mobile IPv6 and Hierarchical Mobile IPv6. Section 3 describes our purposed method. In section 4 we present the analytic evaluation of expression in both Hierarchical Mobile IPv6 and the proposed method. In Section 5 we show the comparison result by applying numerical examples in the expression. Finally, we conclude in section 6.

2 Related Works

In this section, we will see mobile IPv6, focused in mobility [2] and HMIPv6(Hierarchical Mobile IPv6) proposed to provide seamless mobility, which is a method for local mobility management to reduce the registration time and signaling messages [6].

2.1 Mobile IPv6

Mobile IPv6 is proposed for mobility management in IPv6 networks. In Mobile IPv6, a MN(Mobile Node) configures a new *Care of Address* (CoA) whenever it moves to other subnets and registers the CoA to HA(Home Agent) by sending a *Binding Update* (BU) message. This binding update message includes MN's Home Address and the CoA. If Binding procedure succeeds, the HA transmits packets through a tunnel which is established between a HA and a MN. Therefore, Mobile IPv6 can enable MNs to communicate with each *Correspondent Node* (CN) if CNs know a MN address. The MN may send a BU to its CNs so that CNs can send packets directly without HA. However, this binding procedure in Mobile IPv6 causes overload that is to send BU messages to remote HA and CNs whenever a MN moves even to a near place. Also, if a MN is far from CNs, hand off can be delayed.

2.2 Hierarchical Mobile IPv6

HMIPv6 handles the hierarchical mobility management architecture to improve handoff performance and to reduce the mobility management of signaling load of Mobile IPv6. MAP is introduced to provide different local mobility (within a site) from global mobility. The MAP can be located in any level of a hierarchical network of routers.

A MN moving into a MAP domain configures two kinds of CoAs; Regional CoA (RCoA) and Link Local CoA (LCoA). The former one is the address on the MAPs subnet and the latter one is on link address. A MN binds its LCoA with RCoA by sending a BU to the MAP. Packets are sent to the MAP, then the MAP tunnels packets to the MNs LCoA. If the MN moves to other subnets within the current MAP domain, it only needs to register its LCoA with the MAP. Since the RCoA does not change as long as the MN moves within one MAP domain, the CNs can communicate with the MN without the MN's LCoA.

The boundary of a MAP domain is defined by the AR's advertising the MAP information to the attached MNs. If the MN changes its current address within a local MAP domain, it only needs to register the new LCoA with the MAP. Hence, the RCoA is registered as the CNs and the HA does not change.

3 S-HMIPv6

In HMIPv6, we can reduce the signaling cost by using the MAP. When a MN selects the MAP located in any level of a hierarchical network, the MN considers the distance between the MN and the MAP. It is recommended to choose the farthest MAP to avoid frequent re-registrations. It would reduce the probability of changing the MAP and informing all CNs and the HA of its location change. This is particularly significant for fast-moving MNs that will perform frequent handoffs. However, this procedure to choose the farthest MAP is to overwork. At the same time, handoff delay can be occurred during the MAP selection [5]. In addition, slow-moving MNs do not need to organize a large size of MAP domain while fast-moving MNs need it to reduce handoff signaling cost. Therefore, we propose "A Determination Method Of MAP Domain Size Based On The Node Mobility Pattern".

3.1 MAP Selction

When a MN moves out of the current subnet and moves into a new subnet, the MN chooses a default AR of the subnet as its MAP. Therefore, a MNs RCoA and LCoA are the same at this time and also we assume that the MAP domain is composed of one subnet. This uses distributed environment concept in [6]. Any AR can be acted as a MAP for MNs. Overload in HMIPv6 can be concentrated on some of the MAPs while S-HMIPv6 can distribute local management overheads on all MAPs over the networks.

The MN uses its moving history. By measuring the BU message interval, the MN's residence time in one subnet can be calculated approximately [4]. Table 1

```
Count =0;
T_previous= 0;
I = 0;

While (The MN's RCoA is unchanged) {
  If (MN crosses subnets within the current MAP domain) {
     MN sends BU (including its LCoA) message to MAP
     Count = Count + 1;
     I = I +1;
     Store the time, T_now
     Get Interval time T_I between T_now and T_previous
  }
  T_previous=T_now;
}
```

Fig. 1. Calculation process of the MN's resident time.

shows how to get the average residence time of the MN. $E(T_K)$ means the average residence time when the MN visits subnets at I times and 'Count' indicates the visited times to subnets. Also, T_I means the residence time of MN at the visiting subnet and given by

$$E(T_K) = \frac{1}{count} \sum_{I=1}^{count} T_I. \quad (1)$$

3.2 Determination of the MAP Domain Size

Table 2 describes the procedure to decide the MAP domain size according to the MN's residence time. The $E(T_{MNSlow})$ and $E(T_{MNFast})$ are average node residence time computed by the speed of `MNFast` and `MNSlow` used in [7]. If $E(T_K)$ of a MN is more than 200sec, we classify it into `MNSlow` and assign K_{min}. Also if $E(T_K)$ is less than 14sec, the MN is considered as `MNFast`, then set Kmax for the MAP domain size. For others, we allocate the result number of the computation that is to increase K with the rate of increase relatively. Consequently, the MN increases the number of subnets by K.

$$K_{min} + \left(\frac{E(T_{MNSlow}) - E(T_K)}{E(T_{MNSlow})} \right) \cdot \Delta X \quad (2)$$

4 Performance Analysis

To evaluate the performance of S-HMIPv6 and to compare it with HMIPv6, we analytically derived the communication cost between CNs and the MN. First, we call the existing HMIPv6 as D-HMIPv6. We suppose limited network of 100 subnets and also the fixed size of one subnet. In addition, we assume that there is the only one AR in a subnet. We introduce some parameters and assumptions for analysis as follows:

```
// E(T_MNSlow) : 200sec, E(T_MNFast) : 14sec
// K_max : the minimum number of subnets for the MAP domain.
// K_min : the maximum number of subnets for the MAP domain.

  ΔX =K_max - K_min;

  If  E(T_K) >= E(T_MNSlow)
then MN is MNSlow with K set as K_min
  Else if  E(T_K) <= E(T_MNFast)
    then MN is MNFast with K set as K_max
  Else MN is MNOrdinary with K set as equation (2)
```

Fig. 2. Determination process of K.

- The number of whole subnets: $N(1 \leq N \leq 100)$
- The number of subnet for one MAP domain: $K(1 \leq K \leq 100)$
- The standard distance of one subnet: 200meter
- **D**istance Field **O**rdering **C**ost: DOC
- **M**AP Domain Size **D**ecision **C**ost based on MN Speed: MDC
- The average resident time within the MAP domain having K subnets: T_K
- The probability of global and local location update in D-HMIPv6: P_G^D, P_L^D
- The probability of global and local location update in S-HMIPv6: P_G^S, P_L^S
- The cost of global and local location update: C_G, C_L

In D-HMIPv6, the probability to move out of other $(N-1)$ subnets can be achieved as follows [6], i.e., assume that a MN performs a global location update at movement m. Then, we have

$$P_G^m = \frac{N-K}{N-1} \cdot \left(\frac{K-1}{N-1}\right)^{m-2} \tag{3}$$

where $2 \leq m \leq \infty$.

Then, it can be shown that the expectation of M as follows:

$$E(M) = \sum_{m=2}^{\infty} m P_G^m = 1 + \frac{N-1}{N-K}. \tag{4}$$

Throughout $E(M)$, P_G^D and P_L^D can be computed as follows:

$$P_G^D = \frac{1}{E(M)}, \; P_L^D = \frac{E(M)-1}{E(M)}. \tag{5}$$

On the other hand, we assume that all AR can be the MAP for the MN in S-HMIPv6. A MN does not update its global location until the MN visits each number of K subnets while a MN in D-HMIPv6 can move out of the current MAP domain at less than K times. When a MN moves from the $(K-1)$th to the Kth subnet, the expectation of movement, m, is given as follows:

$$
\begin{aligned}
E(M)_{K-1\to K} &= \sum_{n=1}^{\infty} n \cdot \left(\frac{K-2}{N-1}\right)^{n-1} \cdot \frac{N-K+1}{N-1} \\
&= \frac{N-1}{N-K+1} \\
E(M) &= (E(M)_{1\to 2} + E(M)_{2\to 3} \cdots + E(M)_{K-1\to K}) + E(M)_{D-HMIPv6} \\
&= 1 + (N-1)\sum_{i=1}^{K} \frac{1}{N-i}.
\end{aligned} \tag{6}
$$

Therefore, P_G^S and P_L^S are given as follows:

$$
P_G^S = \frac{1}{E(M)}, \; P_L^S = \frac{E(M)-1}{E(M)}. \tag{7}
$$

The following equations are the numerical expression of the communication cost in S-HMIPv6 and D-HMIPv6.

$$
\begin{aligned}
C_{Total} &= C_{LocationUpdate} + C_{PacketDelivery} \\
C_{LocationUpdate}^{D-HMIPv6} &= P_G^D \cdot (C_G + \mathbf{DOC}) + P_L^D \cdot C_L \\
C_{LocationUpdate}^{S-HMIPv6} &= P_G^S \cdot (C_G + \mathbf{MDC}) + P_L^S \cdot C_L \\
C_{PacketDelivery}^{D-HMIPv6,S-HMIPv6} &= \delta \cdot K \cdot (\alpha N(MN) + \beta Log(N(AR))) \\
&\quad + \lambda \left(D_{CN\to MAP} + D_{MAP\to AR}\right).
\end{aligned} \tag{8}
$$

The cost can be described as the sum of the location update cost and packet delivery cost. We also suggest two costs: DOC is the cost to choose a MAP in D-HMIPv6 and MDC is the cost to classify a MN's mobility pattern in S-HMIPv6. And we consider MDC is less than the global update cost.

Packet delivery cost depends on the number of ARs and MNs. If the number of ARs is increased, the complexity of the IP lookup and routing table lookup is high. Also the complexity of IP address lookup is proportional to the logarithm of the length of the routing table [8]. The δ and β are weight factors. We exclude the first packet transmission to HA.

Table 1. Parameters for analysis

Weight			Location Update Cost		Distance Cost		Pkt. delivery rate
δ	α	β	C_G	C_L	$D_{CN\to MAP}$	$D_{MAP\to AR}$	λ
0.1	0.3	0.7	75	25	12	8	0.5

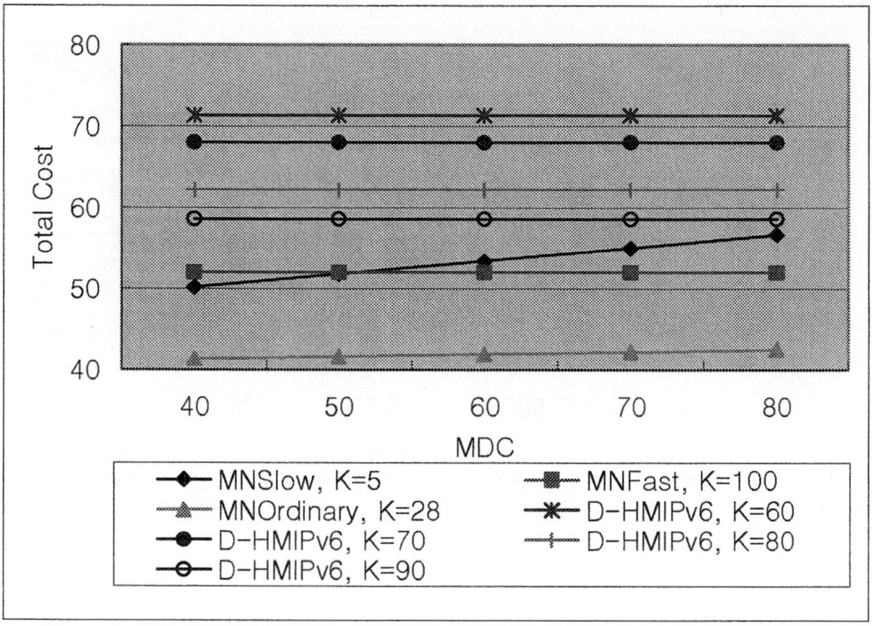

Fig. 3. Total cost depending on the cost of **MDC** on S-HMIPv6.

5 Numerical Examples

In this section, we analyze the result from Equ. (8) to put into numerical examples. We investigate the minimum number of subnets for slow moving MNs. The cost of MDC is diversified from 10 to 80. As we have mentioned before, we assume the cost of MDC is the less than that of the global update cost. Table 1 shows the attributes used on the packet delivery cost.

The proposed S-HMIPv6 differentiates the MAP domain size depending on MN's mobility pattern. For Slow MNs, we find out the minimum number of subnets in Fig. 3. As K is more than 5, a numerical difference of total cost is not enough. So we set 5 as the minimum number of subnets.

On the other hand, we determine 100 for the MAP domain size of the fast MN since the MN does not need to register its location globally. Although too large scale of the MAP domain might overload the packet delivery cost, we assign the whole number of subnets for a fast MN since a fast-moving node could change its location extremely often.

The result of the Fig. 4 shows the total cost comparison of each `MNSlow`, `MNOrdinary`, and `MNFast` on the S-HMIPv6 and D-HMIPv6. In D-HMIPv6, we assume the number of MAP option messages is regular with 30. So, the sorting cost of the distance field in the MAP option message is regarded as 30log30 that is the average sorting cost. Also, for D-HMIPv6 we measure each cost as the MAP domain size is from 60 to 90. We analyze the total cost when MDC

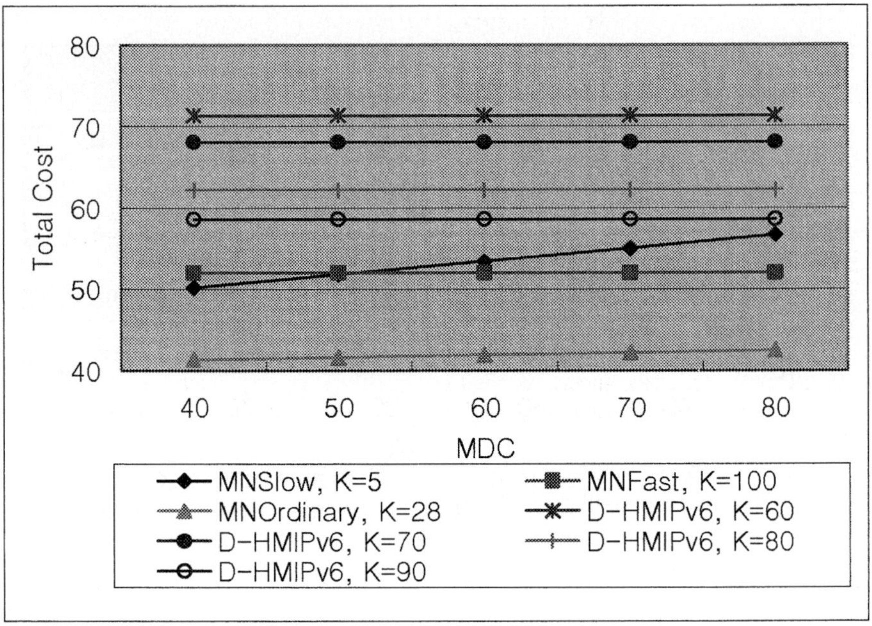

Fig. 4. Total cost of the Each `MLSlow`, `MNOrdinary`, and `MNFast` on D-HMIPv6, S-HMIPv6.

is increasing till 80 while DOC is fixed as 30log30. In S-HMIPv6, we assign 5, 100 and 28 as the MAP domain size for the `MNSlow`, `MNFast`, and `MNOrdinary`, respectively.

The `MNOrdinary` whose the average residence time is 150s has 28 subnets for the MAP domain comparatively. On the other hand, D-HMIPv6 only concerns a large MAP domain regardless of MN mobility pattern. As seen in Fig. 4, S-HMIPv6 is beneficial in reducing the total communication cost.

6 Conclusion and Future Works

In this paper, we have proposed a determination method of the MAP domain size based on the MN mobility pattern. We have performed analytical approach and compared with the previous approach. In order to evaluate the performance of the proposed method, we present to assign different size of the MAP domain to each MN. Analytical results demonstrated that the proposed method reduced the total communication cost although it needed some works additionally to discern the mobility pattern of each MN.

In fact, it is not easy to find minimum cost that is the sum of the location update cost and packet delivery cost because two costs are with the tradeoff relation. We still need to make efforts to find the most potent value to decrease the total cost.

For the future work, we will study a reliable measuring of the MN's residence time. Predicting the MN's speed needs to be very careful. Aside from the MN's speed, the cost of MDC in S-HMIPv6 has to be the minimum. MOBOPTS (IP Mobility Optimization) Group is going to study the scheme to discover the MAP in HMIPv6.

References

1. D. Johnson, C. Perkins, and J. Arkko: Mobility Support in IPv6. IETF Internet Draft, draft-ietf-mobileip-21.txt, (2003)
2. G. Kirby: Locating and the User. Communication International, (1995)
3. H. Soliman, C. Castelluccia. K. El-Malki, and L. Bellier: Hierarchical MIPv6 mobility management (HMIPv6). IETF Internet Draft, draft-ietf-mipshop-hmipv6-00.txt, (2003).
4. K. Kawano, K. Kinoshita, and K. Murakami: A multilevel hierarchical distributed IP mobility management scheme for wide area networks. in Proc. of International Conference on Computer Communications and Networks (ICCCN'2002), (2002) 480-484.
5. Yi Xu, Henry C. J. Lee, and Vrizlynn L. L. Thing: A Local Mobility Agent Selection Algorithm for Mobile Networks. IEEE International Conference on Communications (ICC'2003), (2003) 1074-1079.
6. J. Xie, Ian F. Akyildiz: A Nobel distributed dynamic location management scheme for minimizing signaling costs in mobile ip. IEEE Transaction on mobile computing, 1(3), 2002.
7. M. Bandai and I. Sasase: A Load balancing Mobility Management for Multilevel Hierarchical Mobile IPv6 Networks: in Proc. IEEE PIMRC'2003, (2003).
8. H. Y Tzeng and T. Przygienda: On fast address-lookup algorithms. IEEE Jrnl. on Selected Areas in Comm., 17(6), (1999), 1067-1082.

Author Index

Lecture Notes in Computer Science

For information about Vols. 1–3271

please contact your bookseller or Springer

Vol. 3391: C. Kim (Ed.), Information Networking. XVII, 936 pages. 2005.

Vol. 3387: J. Cardoso, A. Sheth (Eds.), Semantic Web Services and Web Process Composition. VIII, 148 pages. 2005.

Vol. 3386: S. Vaudenay (Ed.), Public Key Cryptography - PKC 2005. IX, 436 pages. 2004.

Vol. 3385: R. Cousot (Ed.), Verification, Model Checking, and Abstract Interpretation. XII, 483 pages. 2005.

Vol. 3382: J. Odell, P. Giorgini, J.P. Müller (Eds.), Agent-Oriented Software Engineering V. X, 239 pages. 2004.

Vol. 3381: P. Vojtáš, M. Bieliková, B. Charron-Bost, O. Sýkora (Eds.), SOFSEM 2005: Theory and Practice of Computer Science. XV, 428 pages. 2004.

Vol. 3376: A. Menezes (Ed.), Topics in Cryptology – CT-RSA 2005. X, 385 pages. 2004.

Vol. 3368: L. Paletta, J.K. Tsotsos, E. Rome, G. Humphreys (Eds.), Attention and Performance in Computational Vision. VIII, 231 pages. 2005.

Vol. 3363: T. Eiter, L. Libkin (Eds.), Database Theory - ICDT 2005. XI, 413 pages. 2004.

Vol. 3362: G. Barthe, L. Burdy, M. Huisman, J.-L. Lanet, T. Muntean (Eds.), Construction and Analysis of Safe, Secure, and Interoperable Smart Devices. IX, 257 pages. 2004.

Vol. 3360: S. Spaccapietra, E. Bertino, S. Jajodia, R. King, D. McLeod, M.E. Orlowska, L. Strous (Eds.), Journal on Data Semantics II. XI, 223 pages. 2004.

Vol. 3358: J. Cao, L.T. Yang, M. Guo, F. Lau (Eds.), Parallel and Distributed Processing and Applications. XXIV, 1058 pages. 2004.

Vol. 3357: H. Handschuh, M.A. Hasan (Eds.), Selected Areas in Cryptography. XI, 354 pages. 2004.

Vol. 3356: G. Das, V.P. Gulati (Eds.), Intelligent Information Technology. XII, 428 pages. 2004.

Vol. 3353: J. Hromkovič, M. Nagl, B. Westfechtel (Eds.), Graph-Theoretic Concepts in Computer Science. XI, 404 pages. 2004.

Vol. 3352: C. Blundo, S. Cimato (Eds.), Security in Communication Networks. XI, 381 pages. 2004.

Vol. 3350: M. Hermenegildo, D. Cabeza (Eds.), Practical Aspects of Declarative Languages. VIII, 269 pages. 2005.

Vol. 3348: A. Canteaut, K. Viswanathan (Eds.), Progress in Cryptology - INDOCRYPT 2004. XIV, 431 pages. 2004.

Vol. 3347: R.K. Ghosh, H. Mohanty (Eds.), Distributed Computing and Internet Technology. XX, 472 pages. 2004.

Vol. 3345: Y. Cai (Ed.), Ambient Intelligence for Scientific Discovery. XII, 311 pages. 2005. (Subseries LNAI).

Vol. 3344: J. Malenfant, B.M. Østvold (Eds.), Object-Oriented Technology. ECOOP 2004 Workshop Reader. VIII, 215 pages. 2004.

Vol. 3342: E. Şahin, W.M. Spears (Eds.), Swarm Robotics. X, 175 pages. 2004.

Vol. 3341: R. Fleischer, G. Trippen (Eds.), Algorithms and Computation. XVII, 935 pages. 2004.

Vol. 3340: C.S. Calude, E. Calude, M.J. Dinneen (Eds.), Developments in Language Theory. XI, 431 pages. 2004.

Vol. 3339: G.I. Webb, X. Yu (Eds.), AI 2004: Advances in Artificial Intelligence. XXII, 1272 pages. 2004. (Subseries LNAI).

Vol. 3338: S.Z. Li, J. Lai, T. Tan, G. Feng, Y. Wang (Eds.), Advances in Biometric Person Authentication. XVIII, 699 pages. 2004.

Vol. 3337: J.M. Barreiro, F. Martin-Sanchez, V. Maojo, F. Sanz (Eds.), Biological and Medical Data Analysis. XI, 508 pages. 2004.

Vol. 3336: D. Karagiannis, U. Reimer (Eds.), Practical Aspects of Knowledge Management. X, 523 pages. 2004. (Subseries LNAI).

Vol. 3335: M. Malek, M. Reitenspieß, J. Kaiser (Eds.), Service Availability. X, 213 pages. 2005.

Vol. 3334: Z. Chen, H. Chen, Q. Miao, Y. Fu, E. Fox, E.-p. Lim (Eds.), Digital Libraries: International Collaboration and Cross-Fertilization. XX, 690 pages. 2004.

Vol. 3333: K. Aizawa, Y. Nakamura, S. Satoh (Eds.), Advances in Multimedia Information Processing - PCM 2004, Part III. XXXV, 785 pages. 2004.

Vol. 3332: K. Aizawa, Y. Nakamura, S. Satoh (Eds.), Advances in Multimedia Information Processing - PCM 2004, Part II. XXXVI, 1051 pages. 2004.

Vol. 3331: K. Aizawa, Y. Nakamura, S. Satoh (Eds.), Advances in Multimedia Information Processing - PCM 2004, Part I. XXXVI, 667 pages. 2004.

Vol. 3330: J. Akiyama, E.T. Baskoro, M. Kano (Eds.), Combinatorial Geometry and Graph Theory. VIII, 227 pages. 2005.

Vol. 3329: P.J. Lee (Ed.), Advances in Cryptology - ASIACRYPT 2004. XVI, 546 pages. 2004.

Vol. 3328: K. Lodaya, M. Mahajan (Eds.), FSTTCS 2004: Foundations of Software Technology and Theoretical Computer Science. XVI, 532 pages. 2004.

Vol. 3327: Y. Shi, W. Xu, Z. Chen (Eds.), Data Mining and Knowledge Management. XIII, 263 pages. 2004. (Subseries LNAI).

Vol. 3326: A. Sen, N. Das, S.K. Das, B.P. Sinha (Eds.), Distributed Computing - IWDC 2004. XIX, 546 pages. 2004.

Vol. 3323: G. Antoniou, H. Boley (Eds.), Rules and Rule Markup Languages for the Semantic Web. X, 215 pages. 2004.

Vol. 3322: R. Klette, J. Žunić (Eds.), Combinatorial Image Analysis. XII, 760 pages. 2004.

Vol. 3321: M.J. Maher (Ed.), Advances in Computer Science - ASIAN 2004. XII, 510 pages. 2004.

Vol. 3320: K.-M. Liew, H. Shen, S. See, W. Cai (Eds.), Parallel and Distributed Computing: Applications and Technologies. XXIV, 891 pages. 2004.

Vol. 3317: M. Domaratzki, A. Okhotin, K. Salomaa, S. Yu (Eds.), Implementation and Application of Automata. XII, 336 pages. 2004.

Vol. 3316: N.R. Pal, N.K. Kasabov, R.K. Mudi, S. Pal, S.K. Parui (Eds.), Neural Information Processing. XXX, 1368 pages. 2004.

Vol. 3315: C. Lemaître, C.A. Reyes, J.A. González (Eds.), Advances in Artificial Intelligence – IBERAMIA 2004. XX, 987 pages. 2004. (Subseries LNAI).

Vol. 3314: J. Zhang, J.-H. He, Y. Fu (Eds.), Computational and Information Science. XXIV, 1259 pages. 2004.

Vol. 3313: C. Castelluccia, H. Hartenstein, C. Paar, D. Westhoff (Eds.), Security in Ad-hoc and Sensor Networks. VIII, 231 pages. 2004.

Vol. 3312: A.J. Hu, A.K. Martin (Eds.), Formal Methods in Computer-Aided Design. XI, 445 pages. 2004.

Vol. 3311: V. Roca, F. Rousseau (Eds.), Interactive Multimedia and Next Generation Networks. XIII, 287 pages. 2004.

Vol. 3309: C.-H. Chi, K.-Y. Lam (Eds.), Content Computing. XII, 510 pages. 2004.

Vol. 3308: J. Davies, W. Schulte, M. Barnett (Eds.), Formal Methods and Software Engineering. XIII, 500 pages. 2004.

Vol. 3307: C. Bussler, S.-k. Hong, W. Jun, R. Kaschek, D.. Kinshuk, S. Krishnaswamy, S.W. Loke, D. Oberle, D. Richards, A. Sharma, Y. Sure, B. Thalheim (Eds.), Web Information Systems – WISE 2004 Workshops. XV, 277 pages. 2004.

Vol. 3306: X. Zhou, S. Su, M.P. Papazoglou, M.E. Orlowska, K.G. Jeffery (Eds.), Web Information Systems – WISE 2004. XVII, 745 pages. 2004.

Vol. 3305: P.M.A. Sloot, B. Chopard, A.G. Hoekstra (Eds.), Cellular Automata. XV, 883 pages. 2004.

Vol. 3303: J.A. López, E. Benfenati, W. Dubitzky (Eds.), Knowledge Exploration in Life Science Informatics. X, 249 pages. 2004. (Subseries LNAI).

Vol. 3302: W.-N. Chin (Ed.), Programming Languages and Systems. XIII, 453 pages. 2004.

Vol. 3300: L. Bertossi, A. Hunter, T. Schaub (Eds.), Inconsistency Tolerance. VII, 295 pages. 2005.

Vol. 3299: F. Wang (Ed.), Automated Technology for Verification and Analysis. XII, 506 pages. 2004.

Vol. 3298: S.A. McIlraith, D. Plexousakis, F. van Harmelen (Eds.), The Semantic Web – ISWC 2004. XXI, 841 pages. 2004.

Vol. 3296: L. Bougé, V.K. Prasanna (Eds.), High Performance Computing - HiPC 2004. XXV, 530 pages. 2004.

Vol. 3295: P. Markopoulos, B. Eggen, E. Aarts, J.L. Crowley (Eds.), Ambient Intelligence. XIII, 388 pages. 2004.

Vol. 3294: C.N. Dean, R.T. Boute (Eds.), Teaching Formal Methods. X, 249 pages. 2004.

Vol. 3293: C.-H. Chi, M. van Steen, C. Wills (Eds.), Web Content Caching and Distribution. IX, 283 pages. 2004.

Vol. 3292: R. Meersman, Z. Tari, A. Corsaro (Eds.), On the Move to Meaningful Internet Systems 2004: OTM 2004 Workshops. XXIII, 885 pages. 2004.

Vol. 3291: R. Meersman, Z. Tari (Eds.), On the Move to Meaningful Internet Systems 2004: CoopIS, DOA, and ODBASE, Part II. XXV, 824 pages. 2004.

Vol. 3290: R. Meersman, Z. Tari (Eds.), On the Move to Meaningful Internet Systems 2004: CoopIS, DOA, and ODBASE, Part I. XXV, 823 pages. 2004.

Vol. 3289: S. Wang, K. Tanaka, S. Zhou, T.W. Ling, J. Guan, D. Yang, F. Grandi, E. Mangina, I.-Y. Song, H.C. Mayr (Eds.), Conceptual Modeling for Advanced Application Domains. XXII, 692 pages. 2004.

Vol. 3288: P. Atzeni, W. Chu, H. Lu, S. Zhou, T.W. Ling (Eds.), Conceptual Modeling – ER 2004. XXI, 869 pages. 2004.

Vol. 3287: A. Sanfeliu, J.F. Martínez Trinidad, J.A. Carrasco Ochoa (Eds.), Progress in Pattern Recognition, Image Analysis and Applications. XVII, 703 pages. 2004.

Vol. 3286: G. Karsai, E. Visser (Eds.), Generative Programming and Component Engineering. XIII, 491 pages. 2004.

Vol. 3285: S. Manandhar, J. Austin, U.B. Desai, Y. Oyanagi, A. Talukder (Eds.), Applied Computing. XII, 334 pages. 2004.

Vol. 3284: A. Karmouch, L. Korba, E.R.M. Madeira (Eds.), Mobility Aware Technologies and Applications. XII, 382 pages. 2004.

Vol. 3283: F.A. Aagesen, C. Anutariya, V. Wuwongse (Eds.), Intelligence in Communication Systems. XIII, 327 pages. 2004.

Vol. 3282: V. Guruswami, List Decoding of Error-Correcting Codes. XIX, 350 pages. 2004.

Vol. 3281: T. Dingsøyr (Ed.), Software Process Improvement. X, 207 pages. 2004.

Vol. 3280: C. Aykanat, T. Dayar, İ. Körpeoğlu (Eds.), Computer and Information Sciences - ISCIS 2004. XVIII, 1009 pages. 2004.

Vol. 3279: G.M. Voelker, S. Shenker (Eds.), Peer-to-Peer Systems III. XI, 300 pages. 2004.

Vol. 3278: A. Sahai, F. Wu (Eds.), Utility Computing. XI, 272 pages. 2004.

Vol. 3275: P. Perner (Ed.), Advances in Data Mining. VIII, 173 pages. 2004. (Subseries LNAI).

Vol. 3274: R. Guerraoui (Ed.), Distributed Computing. XIII, 465 pages. 2004.

Vol. 3273: T. Baar, A. Strohmeier, A. Moreira, S.J. Mellor (Eds.), <<UML>> 2004 - The Unified Modelling Language. XIII, 454 pages. 2004.

Vol. 3272: L. Baresi, S. Dustdar, H. Gall, M. Matera (Eds.), Ubiquitous Mobile Information and Collaboration Systems. VIII, 197 pages. 2004.